HANDBOOK
OF
BATTERIES

HANDBOOK OF BATTERIES

David Linden   Editor

Thomas B. Reddy   Editor

Third Edition

McGraw-Hill

New York   Chicago   San Francisco   Lisbon   London
Madrid   Mexico City   Milan   New Delhi   San Juan   Seoul
Singapore   Sydney   Toronto

Library of Congress Cataloging-in-Publication Data

Handbook of batteries / David Linden, Thomas B. Reddy.—3d ed.
     p.   cm.
    Rev. ed. of: Handbook of batteries / David Linden, editor in chief. 2nd c1995.
    Includes bibliographical references and index.
    ISBN 0-07-135978-8
       1. Electric batteries—Handbooks, manuals, etc. I. Title: Handbook of
batteries. II.
    Linden, David, III. Reddy, Thomas B.

    TK2901.H36 2001
    621 31′242—dc21

                                            2001030790

4 5 6 7 8 9 0 DOC/DOC 1 0 9

ISBN 0-07-135978-8

The sponsoring editor for this book was Steve Chapman and the production supervisor was Sherri Souffrance. It was set in Times Roman by Pro-Image Corporation.

Printed and bound by R. R. Donnelley & Sons Company.

This book is printed on acid-free paper.

McGraw-Hill books are available at special quantity discounts to use as premiums and sales promotions, or for use in corporate training programs. For more information, please write to the Director of Special Sales, Professional Publishing, McGraw-Hill, Two Penn Plaza, New York, NY 10121-2298. Or contact your local bookstore.

CONTENTS

PART 5   Advanced Batteries for Electric Vehicles and Emerging Applications

PART 7 Appendices

Index follows Appendices

CONTRIBUTORS

Vaidevutis Alminauskas *U.S. Naval Surface Warfare Center, Crane Division*

Austin Attewell *International Power Sources Symposium, Ltd.*

Terrill B. Atwater *Power Sources Division, U.S. Army CECOM*

William L. Auxer *Pennsylvania Crusher Corp.*

Christopher A. Baker *Acme Electric Corp., Aerospace Division*

Gary A. Bayles *Consultant (formerly with Northrup-Grumann Corp.)*

Stephen F. Bender *Rosemount, Inc.*

Asaf A. Benderly *Harry Diamond Laboratories, U.S. Army (retired)*

Jeffrey W. Braithwaite *Sandia National Laboratories*

John Broadhead *U.S. Nanocorp and U.S. Microbattery*

Ralph Brodd *Broddarp of Nevada, Inc.*

Jack Brill *Eagle-Picher Technologies, LLC*

Curtis Brown *Eagle-Picher Technologies, LLC*

Paul C. Butler *Sandia National Laboratories*

Anthony G. Cannone *Rutgers University and University of Medicine and Dentistry of New Jersey*

Joseph A. Carcone *Sanyo Energy Corp.*

Arthur J. Catotti *General Electric Co. (retired)*

Allen Charkey *Evercel Corp.*

David L. Chua *Maxpower, Inc.*

Frank Ciliberti *Duracell, Inc. (retired)*

Dwayne Coates *Boeing Satellite Systems*

John W. Cretzmeyer *Medtronic, Inc. (retired)*

Jeffrey R. Dahn *Dalhousie University, Canada*

Josef David-Ivad *Technische Universitat, Graz, Austria*

James M. Dines *Eagle-Picher Industries, Inc. (retired)*

James D. Dunlop *Comsat Laboratories (retired)*

Phillip A. Eidler *Eaton Corp.*

Grant M. Ehrlich *International Fuel Cells*

Ron J. Ekern *Rayovac Corp. (retired))*

William J. Eppley *Maxpower, Inc.*

Rex Erisman *Eagle-Picher Technologies, LLC*

John M. Evjen   *Consultant (formerly with General Electric Co.)*

John Fehling   *Bren-Tronics, Inc.*

Michael Fetcenko   *Ovonic Battery Co.*

H. Frank Gibbard   *H Power Corp.*

Allan B. Goldberg   *U.S. Army Research Laboratory*

Patrick G. Grimes   *Grimes Associates*

Robert P. Hamlen   *Power Sources Division, U.S. Army CECOM*

Ronald O. Hammel   *Consultant (formerly with Hawker Energy Products, Inc.)*

Robert J. Horning   *Valence Technology, Inc.*

Gary L. Henriksen   *Argonne National Laboratory*

Sohrab Hossain   *LiTech, LLC*

James C. Hunter   *Eveready Battery Co., Inc. (deceased)*

John F. Jackovitz   *University of Pittsburgh*

Andrew N. Jansen   *Argonne National Laboratory*

Alexander P. Karpinski   *Yardney Technical Products, Inc.*

Peter A. Karpinski   *PAK Enterprises*

Arthur Kaufman   *H Power Corp.*

Sandra E. Klassen   *Sandia National Laboratories*

Visvaldis Klasons   *Consultant (formerly with Catalyst Research Corp.)*

Ralph F. Koontz   *Magnavox Co. (retired)*

Karl Kordesch   *Technische Universitat, Graz, Austria*

Han C. Kuo   *NEXCell Battery Co., Taiwan*

Charles M. Lamb   *Eagle-Picher Technologies, LLC*

Duane M. Larsen   *Rayovac Corp. (retired)*

Peter Lex   *ZBB Technologies, Inc.*

David Linden   *Consultant (formerly with U.S. Army Electronics Command)*

R. David Lucero   *Eagle-Picher Technologies, LLC*

Dennis W. McComsey   *Eveready Battery Co., Inc.*

Doug Magnusen   *GP Batteries, USA*

Sid Megahed   *Rechargeable Battery Corp. (deceased)*

Ronald C. Miles   *Johnson Controls, Inc.*

Elliot M. Morse   *Eagle-Picher Industries, Inc. (retired)*

Denis Naylor   *Duracell, Inc. (deceased)*

Arne O. Nilsson   *Consultant (formerly with SAFT NIFE and Acme Electric)*

James E. Oxley   *Oxley Research, Inc.*

Boone B. Owens   *Corrosion Research Center, University of Minnesota*

Joseph L. Passaniti   *Rayovac Corp.*

Stefano Passerini   *Dipartimento Energia, Divisione Technologie Energetiche Avanzate, Italy*

David F. Pickett *Eagle-Picher Technologies, LLC*

Thomas B. Reddy *Consultant, Rutgers University and University of Medicine and Dentistry of New Jersey*

Terrence F. Reise *Duracell, Inc.*

Alvin J. Salkind *Rutgers University and University of Medicine and Dentistry of New Jersey*

Robert F. Scarr *Eveready Battery Co., Inc. (retired)*

Stephen F. Schiffer *Lockheed Martin Corp.*

Paul M. Skarstad *Medtronic, Inc.*

Phillip J. Slezak *Eveready Battery Co., Inc.*

John Springstead *Rayovac Corp.*

Patrick J. Spellman *Rayovac Corp. (retired)*

Philip C. Symons *Electrochemical Engineering Consultants, Inc.*

Russell H. Toye *Eveready Battery Co., Inc.*

Forrest A. Trumbore *University of Medicine and Dentistry of New Jersey*

Darrel F. Untereker *Medtronic, Inc.*

Steven P. Wicelinski *Duracell, Inc.*

PREFACE

Since the publication of the second edition of the *Handbook of Batteries* in 1995, the battery industry has grown remarkedly. This growth has been due to the broad increase in the use of battery-operated portable electronics and the renewed interest in low- or zero-emission vehicles and other emerging applications with requirements that can best be met with batteries. Annual worldwide battery sales currently are about $50 billion, more than double the sales of a decade ago.

This growth and the demand for batteries meeting increasingly stringent performance requirements have been a challenge to the battery industry. The theoretical and practical limits of battery technology can be a barrier to meeting some performance requirements. Batteries are also cited as the limiting factor for achieving the application's desired service life. Nevertheless, substantial advances have been made both with improvement of the performance of the conventional battery systems and the development of new battery systems. These advances have been covered by significant revisions and updating of each of the appropriate chapters in this third edition of the *Handbook*.

Recent emphasis on the performance of the primary alkaline manganese dioxide battery has been directed toward improving its high-rate performance to meet the requirements of the new digital cameras and other portable electronics. The new high-rate (Ultra or Premium) battery was first sold in 2000 and already commands about 25% of the market.

The lithium primary battery continues its steady growth, dominating the camera market and applications requiring high power and performance over long periods of time. It now accounts for over $1 billion in annual sales.

Development has been most active in the area of portable rechargeable batteries to meet the needs of the rapidly growing portable electronics market. The portable nickel-metal hydride battery, which was becoming the dominant rechargeable battery replacing the nickel-cadmium battery, is itself being replaced by the newer lithium-ion battery. Recognizing the significance of this new technology, a new chapter, Chapter 35 "Lithium-ion Batteries," has been added to the third edition of the *Handbook*.

The revived interest in electric vehicles, hybrid electric vehicles, and energy storage systems for utilities has accelerated the development of larger-sized rechargeable batteries. Because of the low specific energy of lead-acid batteries and the still unresolved problems with the high temperature batteries, the nickel-metal hydride battery is currently the battery system of choice for hybrid electric vehicles. This subject is covered in another new chapter, Chapter 30 "Propulsion and Industrial Nickel-Metal Hydride Batteries."

The inherent high energy conversion efficiency and the renewed interest in fuel cell technology for electric vehicles has encouraged the development of small subkilowatt fuel cell power units and portable fuel cells as potential replacements for batteries. Because of this new interest, Part 6 "Portable Fuel Cells" has been added that includes two new chapters, Chapters 42 and 43, covering portable fuel cells and small subkilowatt fuel cells, respectively. Large fuel cells are beyond the scope of this third edition of the *Handbook*. Much information has been published about this subject; see references listed in Appendix F.

Several editorial changes have been instituted in the preparation of this edition of the *Handbook*. The term "specific energy" is now used in place of gravimetric energy density (e.g., Wh/kg). The term "energy density" now refers to volumetric energy density (Wh/L). Similarly, "specific power" (W/kg) and power density (W/L) refer to power per unit weight and volume, respectively.

Another point that has been defined more clearly in this edition is the distinction between a "cell" and a "battery." Manufacturers most commonly identify the product they offer for sale as a "battery" regardless of whether it is a single-cell battery or a multicell one. Accordingly, we have defined the cell as "the basic electrochemical unit providing a source of electrical energy by direct conversion of chemical energy. The cell consists of an assembly of electrodes, separators, electrolyte, container, and terminals." The battery is defined as "the complete product and consists of one or more electrochemical cells, electrically connected in an appropriate series/parallel arrangement to provide the required operating voltage and current levels, including, if any, monitors, controls and other ancillary components (e.g., fuses, diodes, case, terminals, and markings)."

In this edition, the term "cell" has been used, almost universally by all of the authors, when describing the cell components of the battery and its chemistry. Constructional features have been described as either cells, batteries, or configurations depending on the particular choice of the author. This has not been uniformly edited, as it does not appear to cause any confusion. The term "battery" has been generally used when presenting the performance characteristics of the product. Usually the data is presented on the basis of a single-cell battery, recognizing that the performance of a multicell battery could be different, depending on its design. In some instances, in order not to mislead the reader relative to the performance of the final battery product, some data (particularly in Chapter 35 on lithium-ion batteries and in the chapters in Part 5 on advanced batteries) is presented on a "cell" basis as hardware, thermal controls, safety devices, etc., that may ultimately be added to the battery (and have not been included in the cell) would have a significant impact on performance.

This third edition of the *Handbook of Batteries* has now grown to over 1400 pages, recognizing the broad scope of battery technology and the wide range of battery applications. This work would not have been possible without the interest and contributions of more than eighty battery scientists and engineers who participated in its preparation. Their cooperation is gratefully acknowledged, as well as the companies and agencies who supported these contributing authors.

We also acknowledge the efforts of Stephen S. Chapman, Executive Editor, Professional Book Group, McGraw-Hill Companies for initiating this project and the McGraw-Hill staff, and Lois Kisch, Tom Reddy's secretary, for their assistance toward its completion. We further wish to express out thanks to our wives, Rose Linden and Mary Ellen Scarborough, for their encouragement and support and to Mary Ellen for the editorial assistance she provided.

David Linden
Thomas B. Reddy

P · A · R · T · 1

PRINCIPLES OF OPERATION

CHAPTER 1
BASIC CONCEPTS

David Linden

1.1 COMPONENTS OF CELLS AND BATTERIES

A battery is a device that converts the chemical energy contained in its active materials directly into electric energy by means of an electrochemical oxidation-reduction (redox) reaction. In the case of a rechargeable system, the battery is recharged by a reversal of the process. This type of reaction involves the transfer of electrons from one material to another through an electric circuit. In a nonelectrochemical redox reaction, such as rusting or burning, the transfer of electrons occurs directly and only heat is involved. As the battery electrochemically converts chemical energy into electric energy, it is not subject, as are combustion or heat engines, to the limitations of the Carnot cycle dictated by the second law of thermodynamics. Batteries, therefore, are capable of having higher energy conversion efficiencies.

While the term "battery" is often used, the basic electrochemical unit being referred to is the "cell." A battery consists of one or more of these cells, connected in series or parallel, or both, depending on the desired output voltage and capacity.*

The cell consists of three major components:

1. The anode or negative electrode—the reducing or fuel electrode—which gives up electrons to the external circuit and is oxidized during the electrochemical reaction.

2. The cathode or positive electrode—the oxidizing electrode—which accepts electrons from the external circuit and is reduced during the electrochemical reaction.

---

*Cell vs. Battery: A *cell* is the basic electrochemical unit providing a source of electrical energy by direct conversion of chemical energy. The cell consists of an assembly of electrodes, separators, electrolyte, container and terminals. A *battery* consists of one or more electrochemical cells, electrically connected in an appropriate series/parallel arrangement to provide the required operating voltage and current levels, including, if any, monitors, controls and other ancillary components (e.g. fuses, diodes), case, terminals and markings. (Although much less popular, in some publications, the term "battery" is considered to contain two or more cells.)

Popular usage considers the "battery" and not the "cell" to be the product that is sold or provided to the "user." In this 3rd Edition, the term "cell" will be used when describing the cell component of the battery and its chemistry. The term "battery" will be used when presenting performance characteristics, etc. of the product. Most often, the electrical data is presented on the basis of a single-cell battery. The performance of a multicell battery will usually be different than the performance of the individual cells or a single-cell battery (see Section 3.2.13).

3. The electrolyte—the ionic conductor—which provides the medium for transfer of charge, as ions, inside the cell between the anode and cathode. The electrolyte is typically a liquid, such as water or other solvents, with dissolved salts, acids, or alkalis to impart ionic conductivity. Some batteries use solid electrolytes, which are ionic conductors at the operating temperature of the cell.

The most advantageous combinations of anode and cathode materials are those that will be lightest and give a high cell voltage and capacity (see Sec. 1.4). Such combinations may not always be practical, however, due to reactivity with other cell components, polarization, difficulty in handling, high cost, and other deficiencies.

In a practical system, the anode is selected with the following properties in mind: efficiency as a reducing agent, high coulombic output (Ah/g), good conductivity, stability, ease of fabrication, and low cost. Hydrogen is attractive as an anode material, but obviously, must be contained by some means, which effectively reduces its electrochemical equivalence. Practically, metals are mainly used as the anode material. Zinc has been a predominant anode because it has these favorable properties. Lithium, the lightest metal, with a high value of electrochemical equivalence, has become a very attractive anode as suitable and compatible electrolytes and cell designs have been developed to control its activity.

The cathode must be an efficient oxidizing agent, be stable when in contact with the electrolyte, and have a useful working voltage. Oxygen can be used directly from ambient air being drawn into the cell, as in the zinc/air battery. However, most of the common cathode materials are metallic oxides. Other cathode materials, such as the halogens and the oxyhalides, sulfur and its oxides, are used for special battery systems.

The electrolyte must have good ionic conductivity but not be electronically conductive, as this would cause internal short-circuiting. Other important characteristics are nonreactivity with the electrode materials, little change in properties with change in temperature, safety in handling, and low cost. Most electrolytes are aqueous solutions, but there are important exceptions as, for example, in thermal and lithium anode batteries, where molten salt and other nonaqueous electrolytes are used to avoid the reaction of the anode with the electrolyte.

Physically the anode and cathode electrodes are electronically isolated in the cell to prevent internal short-circuiting, but are surrounded by the electrolyte. In practical cell designs a separator material is used to separate the anode and cathode electrodes mechanically. The separator, however, is permeable to the electrolyte in order to maintain the desired ionic conductivity. In some cases the electrolyte is immobilized for a nonspill design. Electrically conducting grid structures or materials may also be added to the electrodes to reduce internal resistance.

The cell itself can be built in many shapes and configurations—cylindrical, button, flat, and prismatic—and the cell components are designed to accommodate the particular cell shape. The cells are sealed in a variety of ways to prevent leakage and dry-out. Some cells are provided with venting devices or other means to allow accumulated gases to escape. Suitable cases or containers, means for terminal connection and labeling are added to complete the fabrication of the cell and battery.

1.2 CLASSIFICATION OF CELLS AND BATTERIES

Electrochemical cells and batteries are identified as primary (nonrechargeable) or secondary (rechargeable), depending on their capability of being electrically recharged. Within this classification, other classifications are used to identify particular structures or designs. The classification used in this handbook for the different types of electrochemical cells and batteries is described in this section.

1.2.1 Primary Cells or Batteries

These batteries are not capable of being easily or effectively recharged electrically and, hence, are discharged once and discarded. Many primary cells in which the electrolyte is contained by an absorbent or separator material (there is no free or liquid electrolyte) are termed "dry cells."

The primary battery is a convenient, usually inexpensive, lightweight source of packaged power for portable electronic and electric devices, lighting, photographic equipment, toys, memory backup, and a host of other applications, giving freedom from utility power. The general advantages of primary batteries are good shelf life, high energy density at low to moderate discharge rates, little, if any, maintenance, and ease of use. Although large high-capacity primary batteries are used in military applications, signaling, standby power, and so on, the vast majority of primary batteries are the familiar single cell cylindrical and flat button batteries or multicell batteries using these component cells.

1.2.2 Secondary or Rechargeable Cells or Batteries

These batteries can be recharged electrically, after discharge, to their original condition by passing current through them in the opposite direction to that of the discharge current. They are storage devices for electric energy and are known also as "storage batteries" or "accumulators."

The applications of secondary batteries fall into two main categories:

1. Those applications in which the secondary battery is used as an energy-storage device, generally being electrically connected to and charged by a prime energy source and delivering its energy to the load on demand. Examples are automotive and aircraft systems, emergency no-fail and standby (UPS) power sources, hybrid electric vehicles and stationary energy storage (SES) systems for electric utility load leveling.

2. Those applications in which the secondary battery is used or discharged essentially as a primary battery, but recharged after use rather than being discarded. Secondary batteries are used in this manner as, for example, in portable consumer electronics, power tools, electric vehicles, etc., for cost savings (as they can be recharged rather than replaced), and in applications requiring power drains beyond the capability of primary batteries.

Secondary batteries are characterized (in addition to their ability to be recharged) by high power density, high discharge rate, flat discharge curves, and good low-temperature performance. Their energy densities are generally lower than those of primary batteries. Their charge retention also is poorer than that of most primary batteries, although the capacity of the secondary battery that is lost on standing can be restored by recharging.

Some batteries, known as "mechanically rechargeable types," are "recharged" by replacement of the discharged or depleted electrode, usually the metal anode, with a fresh one. Some of the metal/air batteries (Chap. 38) are representative of this type of battery.

1.2.3 Reserve Batteries

In these primary types, a key component is separated from the rest of the battery prior to activation. In this condition, chemical deterioration or self-discharge is essentially eliminated, and the battery is capable of long-term storage. Usually the electrolyte is the component that is isolated. In other systems, such as the thermal battery, the battery is inactive until it is heated, melting a solid electrolyte, which then becomes conductive.

The reserve battery design is used to meet extremely long or environmentally severe storage requirements that cannot be met with an "active" battery designed for the same performance characteristics. These batteries are used, for example, to deliver high power for relatively short periods of time, in missiles, torpedoes, and other weapon systems.

1.2.4 Fuel Cells

Fuel cells, like batteries, are electrochemical galvanic cells that convert chemical energy directly into electrical energy and are not subject to the Carnot cycle limitations of heat engines. Fuel cells are similar to batteries except that the active materials are not an integral part of the device (as in a battery), but are fed into the fuel cell from an external source when power is desired. The fuel cell differs from a battery in that it has the capability of producing electrical energy as long as the active materials are fed to the electrodes (assuming the electrodes do not fail). The battery will cease to produce electrical energy when the limiting reactant stored within the battery is consumed.

The electrode materials of the fuel cell are inert in that they are not consumed during the cell reaction, but have catalytic properties which enhance the electroreduction or electro-oxidation of the reactants (the active materials).

The anode active materials used in fuel cells are generally gaseous or liquid (compared with the metal anodes generally used in most batteries) and are fed into the anode side of the fuel cell. As these materials are more like the conventional fuels used in heat engines, the term "fuel cell" has become popular to describe these devices. Oxygen or air is the predominant oxidant and is fed into the cathode side of the fuel cell.

Fuel cells have been of interest for over 150 years as a potentially more efficient and less polluting means for converting hydrogen and carbonaceous or fossil fuels to electricity compared to conventional engines. A well known application of the fuel cell has been the use of the hydrogen/oxygen fuel cell, using cryogenic fuels, in space vehicles for over 40 years. Use of the fuel cell in terrestrial applications has been developing slowly, but recent advances has revitalized interest in air-breathing systems for a variety of applications, including utility power, load leveling, dispersed or on-site electric generators and electric vehicles.

Fuel cell technology can be classified into two categories

1. Direct systems where fuels, such as hydrogen, methanol and hydrazine, can react directly in the fuel cell
2. Indirect systems in which the fuel, such as natural gas or other fossil fuel, is first converted by reforming to a hydrogen-rich gas which is then fed into the fuel cell

Fuel cell systems can take a number of configurations depending on the combinations of fuel and oxidant, the type of electrolyte, the temperature of operation, and the application, etc.

More recently, fuel cell technology has moved towards portable applications, historically the domain of batteries, with power levels from less than 1 to about 100 watts, blurring the distinction between batteries and fuel cells. Metal/air batteries (see Chap. 38), particularly those in which the metal is periodically replaced, can be considered a "fuel cell" with the metal being the fuel. Similarly, small fuel cells, now under development, which are "refueled" by replacing an ampule of fuel can be considered a "battery."

Fuel cells were not included in the 2nd Edition of this Handbook as the technical scope and applications at that time differed from that of the battery. Now that small to medium size fuel cells may become competitive with batteries for portable electronic and other applications, these portable devices are covered in Chap. 42. Information on the larger fuel cells for electric vehicles, utility power, etc can be obtained from the references listed in Appendix F "Bibliography."

1.3 OPERATION OF A CELL

1.3.1 Discharge

The operation of a cell during discharge is also shown schematically in Fig. 1.1. When the cell is connected to an external load, electrons flow from the anode, which is oxidized, through the external load to the cathode, where the electrons are accepted and the cathode material is reduced. The electric circuit is completed in the electrolyte by the flow of anions (negative ions) and cations (positive ions) to the anode and cathode, respectively.

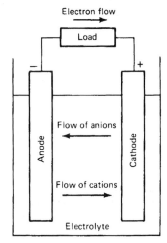

FIGURE 1.1   Electrochemical operation of a cell (discharge).

The discharge reaction can be written, assuming a metal as the anode material and a cathode material such as chlorine (Cl_2), as follows:

Negative electrode: anodic reaction (oxidation, loss of electrons)

$$Zn \rightarrow Zn^{2+} + 2e$$

Positive electrode: cathodic reaction (reduction, gain of electrons)

$$Cl_2 + 2e \rightarrow 2Cl^-$$

Overall reaction (discharge):

$$Zn + Cl_2 \rightarrow Zn^{2+} + 2Cl^- (ZnCl_2)$$

1.3.2 Charge

During the recharge of a rechargeable or storage cell, the current flow is reversed and oxidation takes place at the positive electrode and reduction at the negative electrode, as shown in Fig. 1.2. As the anode is, by definition, the electrode at which oxidation occurs and the cathode the one where reduction takes place, the positive electrode is now the anode and the negative the cathode.

In the example of the Zn/Cl_2 cell, the reaction on charge can be written as follows:

Negative electrode: cathodic reaction (reduction, gain of electrons)

$$Zn^{2+} + 2e \rightarrow Zn$$

Positive electrode: anodic reaction (oxidation, loss of electrons)

$$2Cl^- \rightarrow Cl_2 + 2e$$

Overall reaction (charge):

$$Zn^{2+} + 2Cl^- \rightarrow Zn + Cl_2$$

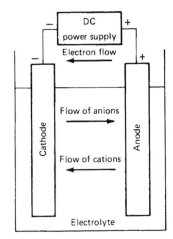

FIGURE 1.2   Electrochemical operation of a cell (charge).

1.3.3 Specific Example: Nickel-Cadmium Cell

The processes that produce electricity in a cell are chemical reactions which either release or consume electrons as the electrode reaction proceeds to completion. This can be illustrated with the specific example of the reactions of the nickel-cadmium cell. At the anode (negative electrode), the discharge reaction is the oxidation of cadmium metal to cadmium hydroxide with the release of two electrons,

$$Cd + 2OH^- \rightarrow Cd(OH)_2 + 2e$$

At the cathode, nickel oxide (or more accurately nickel oxyhydroxide) is reduced to nickel hydroxide with the acceptance of an electron,

$$NiOOH + H_2O + e \rightarrow OH^- + Ni(OH)_2$$

When these two "half-cell" reactions occur (by connection of the electrodes to an external discharge circuit), the overall cell reaction converts cadmium to cadmium hydroxide at the anode and nickel oxyhydroxide to nickel hydroxide at the cathode,

$$Cd + 2NiOOH + 2H_2O \rightarrow Cd(OH)_2 + 2Ni(OH)_2$$

This is the discharge process. If this were a primary non-rechargeable cell, at the end of discharge, it would be exhausted and discarded. The nickel-cadmium battery system is, however, a secondary (rechargeable) system, and on recharge the reactions are reversed. At the negative electrode the reaction is:

$$Cd(OH)_2 + 2e \rightarrow Cd + 2OH^-$$

At the positive electrode the reaction is:

$$Ni(OH)_2 + OH^- \rightarrow NiOOH + H_2O + e$$

After recharge, the secondary battery reverts to its original chemical state and is ready for further discharge. These are the fundamental principles involved in the charge–discharge mechanisms of a typical secondary battery.

1.3.4 Fuel Cell

A typical fuel cell reaction is illustrated by the hydrogen/oxygen fuel cell. In this device, hydrogen is oxidized at the anode, electrocatalyzed by platinum or platinum alloys, while at the cathode oxygen is reduced, again with platinum or platinum alloys as electrocatalysts. The simplified anodic reaction is

$$2H_2 \rightarrow 4H^+ + 4e$$

while the cathodic reaction is

$$O_2 + 4H^+ + 4e \rightarrow 2H_2O$$

The overall reaction is the oxidation of hydrogen by oxygen, with water as the reaction product.

$$2H_2 + O_2 \rightarrow 2H_2O$$

1.4 THEORETICAL CELL VOLTAGE, CAPACITY, AND ENERGY

The theoretical voltage and capacity of a cell are a function of the anode and cathode materials. (See Chap. 2 for detailed electrochemical theory.)

1.4.1 Free Energy

Whenever a reaction occurs, there is a decrease in the free energy of the system, which is expressed as

$$\Delta G^0 = -nFE^0$$

where F = constant known as Faraday ($\approx$96,500 C or 26.8 Ah)
n = number of electrons involved in stoichiometric reaction
E^0 = standard potential, V

1.4.2 Theoretical Voltage

The standard potential of the cell is determined by the type of active materials contained in the cell. It can be calculated from free-energy data or obtained experimentally. A listing of electrode potentials (reduction potentials) under standard conditions is given in Table 1.1. A more complete list is presented in Appendix B.

The standard potential of a cell can be calculated from the standard electrode potentials as follows (the oxidation potential is the negative value of the reduction potential):

Anode (oxidation potential) + cathode (reduction potential) = standard cell potential.

For example, in the reaction $Zn + Cl_2 \rightarrow ZnCl_2$, the standard cell potential is:

$$
\begin{aligned}
Zn &\rightarrow Zn^{2+} + 2e &\quad -(-0.76 \text{ V}) \\
Cl_2 &\rightarrow 2Cl^- - 2e &\quad \underline{1.36 \text{ V}} \\
&\quad E° = &\quad 2.12 \text{ V}
\end{aligned}
$$

The cell voltage is also dependent on other factors, including concentration and temperature, as expressed by the Nernst equation (covered in detail in Chap. 2).

1.4.3 Theoretical Capacity (Coulombic)

The theoretical capacity of a cell is determined by the amount of active materials in the cell. It is expressed as the total quantity of electricity involved in the electrochemical reaction and is defined in terms of coulombs or ampere-hours. The "ampere-hour capacity" of a battery is directly associated with the quantity of electricity obtained from the active materials. Theoretically 1 gram-equivalent weight of material will deliver 96,487 C or 26.8 Ah. (A gram-equivalent weight is the atomic or molecular weight of the active material in grams divided by the number of electrons involved in the reaction.)

The electrochemical equivalence of typical materials is listed in Table 1.1 and Appendix C.

The theoretical capacity of an electrochemical cell, based only on the active materials participating in the electrochemical reaction, is calculated from the equivalent weight of the reactants. Hence, the theoretical capacity of the Zn/Cl_2 cell is 0.394 Ah/g, that is,

$$
\begin{aligned}
Zn \quad &+ Cl_2 \quad \longrightarrow \quad ZnCl_2 \\
(0.82 \text{ Ah/g}) \quad &(0.76 \text{ Ah/g}) \\
1.22 \text{ g/Ah} \quad &+ 1.32 \text{ g/Ah} = 2.54 \text{ g/Ah or } 0.394 \text{ Ah/g}
\end{aligned}
$$

Similarly, the ampere-hour capacity on a volume basis can be calculated using the appropriate data for ampere-hours per cubic centimeter as listed in Table 1.1.

The theoretical voltages and capacities of a number of the major electrochemical systems are given in Table 1.2. These theoretical values are based on the active anode and cathode materials only. Water, electrolyte, or any other materials that may be involved in the cell reaction are not included in the calculation.

TABLE 1.1 Characteristics of Electrode Materials*

Material	Atomic or molecular weight, g	Standard reduction potential at 25°C, V	Valence change	Melting point, °C	Density, g/cm³	Electrochemical equivalents Ah/g	g/Ah	Ah/cm³‡
Anode materials								
H_2	2.01	0 / −0.83†	2	—	—	26.59	0.037	
Li	6.94	−3.01	1	180	0.54	3.86	0.259	2.06
Na	23.0	−2.71	1	98	0.97	1.16	0.858	1.14
Mg	24.3	−2.38 / −2.69†	2	650	1.74	2.20	0.454	3.8
Al	26.9	−1.66	3	659	2.69	2.98	0.335	8.1
Ca	40.1	−2.84 / −2.35†	2	851	1.54	1.34	0.748	2.06
Fe	55.8	−0.44 / −0.88†	2	1528	7.85	0.96	1.04	7.5
Zn	65.4	−0.76 / −1.25†	2	419	7.14	0.82	1.22	5.8
Cd	112.4	−0.40 / −0.81†	2	321	8.65	0.48	2.10	4.1
Pb	207.2	−0.13	2	327	11.34	0.26	3.87	2.9
$(Li)C_6^{(1)}$	72.06	∼−2.8	1	—	2.25	0.37	2.68	0.84
$MH^{(2)}$	116.2	−0.83†	2	—	—	0.45	2.21	—
CH_3OH	32.04	—	6	—	—	5.02	0.20	—
Cathode materials								
O_2	32.0	1.23 / 0.40†	4	—	—	3.35	0.30	
Cl_2	71.0	1.36	2	—	—	0.756	1.32	
SO_2	64.0	—	1	—	—	0.419	2.38	
MnO_2	86.9	1.28‡	1	—	5.0	0.308	3.24	1.54
NiOOH	91.7	0.49†	1	—	7.4	0.292	3.42	2.16
CuCl	99.0	0.14	1	—	3.5	0.270	3.69	0.95
FeS_2	119.9	—	4	—	—	0.89	1.12	4.35
AgO	123.8	0.57†	2	—	7.4	0.432	2.31	3.20
Br_2	159.8	1.07	2	—	—	0.335	2.98	
HgO	216.6	0.10†	2	—	11.1	0.247	4.05	2.74
Ag_2O	231.7	0.35†	2	—	7.1	0.231	4.33	1.64
PbO_2	239.2	1.69	2	—	9.4	0.224	4.45	2.11
$Li_xCoO_2^{(3)}$	98	∼2.7	0.5	—	—	0.137	7.29	—
I_2	253.8	0.54	2	—	4.94	0.211	4.73	1.04

*See also Appendixes B and C.
†Basic electrolyte: all others, aqueous acid electrolyte.
‡Based on density values shown.
(1) Calculations based only on weight of carbon.
(2) Based on 1.7% H_2 storage by weight.
(3) Based on x = 0.5; higher valves may be obtained in practice.

TABLE 1.2 Voltage, Capacity and Specific Energy of Major Battery Systems—Theoretical and Practical Values

Battery type	Anode	Cathode	Reaction mechanism	Theoretical values[†]				Practical battery[‡]		
				V	g/Ah	Ah/kg	Specific energy Wh/kg	Nominal voltage V	Specific energy Wh/kg	Energy density Wh/L
Primary batteries										
Leclanché	Zn	MnO_2	$Zn + 2MnO_2 \rightarrow ZnO \cdot Mn_2O_3$	1.6	4.46	224	358	1.5	85[4]	165[4]
Magnesium	Mg	MnO_2	$Mg + 2MnO_2 + H_2O \rightarrow Mn_2O_3 + Mg(OH)_2$	2.8	3.69	271	759	1.7	100[4]	195[4]
Alkaline MnO_2	Zn	MnO_2	$Zn + 2MnO_2 \rightarrow ZnO + Mn_2O_3$	1.5	4.46	224	358	1.5	145[4]	400[4]
Mercury	Zn	HgO	$Zn + HgO \rightarrow ZnO + Hg$	1.34	5.27	190	255	1.35	100[6]	470[6]
Mercad	Cd	HgO	$Cd + HgO + H_2O \rightarrow Cd(OH)_2 + Hg$	0.91	6.15	163	148	0.9	55[6]	230[6]
Silver oxide	Zn	Ag_2O	$Zn + Ag_2O + H_2O \rightarrow Zn(OH)_2 + 2Ag$	1.6	5.55	180	288	1.6	135[6]	525[6]
Zinc/O_2	Zn	O_2	$Zn + \tfrac{1}{2}O_2 \rightarrow ZnO$	1.65	1.52	658	1085	—	—	—
Zinc/air	Zn	Ambient air	$Zn + (\tfrac{1}{2}O_2) \rightarrow ZnO$	1.65	1.22	820	1353	1.5	370[6]	1300[6]
Li/$SOCl_2$	Li	$SOCl_2$	$4Li + 2SOCl_2 \rightarrow 4LiCl + S + SO_2$	3.65	3.25	403	1471	3.6	590[4]	1100[4]
Li/SO_2	Li	SO_2	$2Li + 2SO_2 \rightarrow Li_2S_2O_4$	3.1	2.64	379	1175	3.0	260[5]	415[5]
$LiMnO_2$	Li	MnO_2	$Li + Mn^{IV}O_2 \rightarrow Mn^{IV}O_2(Li^+)$	3.5	3.50	286	1001	3.0	230[5]	535[5]
Li/FeS_2	Li	FeS_2	$4Li + FeS_2 \rightarrow 2Li_2S + Fe$	1.8	1.38	726	1307	1.5	260[5]	500[5]
Li/$(CF)_n$	Li	$(CF)_n$	$nLi + (CF)_n \rightarrow nLiF + nC$	3.1	1.42	706	2189	3.0	250[5]	635[5]
Li/I_2[(3)]	Li	I_2(P2VP)	$Li + \tfrac{1}{2}I_2 \rightarrow LiI$	2.8	4.99	200	560	2.8	245	900
Reserve batteries										
Cuprous chloride	Mg	CuCl	$Mg + Cu_2Cl_2 \rightarrow MgCl_2 + 2Cu$	1.6	4.14	241	386	1.3	60[7]	80[7]
Zinc/silver oxide	Zn	AgO	$Zn + AgO + H_2O \rightarrow Zn(OH)_2 + Hg$	1.81	3.53	283	512	1.5	30[8]	75[8]
Thermal	Li	FeS_2	See Section 21.3.1	2.1–1.6	1.38	726	1307	2.1–1.6	40[9]	100[9]

Secondary batteries

System	Anode	Cathode	Reaction mechanism							
Lead-acid	Pb	PbO_2	$Pb + PbO_2 + 2H_2SO_4 \rightarrow 2PbSO_4 + 2H_2O$	2.1	8.32	120	252	2.0	35	70[10]
Edison	Fe	Ni oxide	$Fe + 2NiOOH + 2H_2O \rightarrow 2Ni(OH)_2 + Fe(OH)_2$	1.4	4.46	224	314	1.2	30	55[10]
Nickel-cadmium	Cd	Ni oxide	$Cd + 2NiOOH + 2H_2O \rightarrow 2Ni(OH)_2 + Cd(OH)_2$	1.35	5.52	181	244	1.2	35	100[5]
Nickel-zinc	Zn	Ni oxide	$Zn + 2NiOOH + 2H_2O \rightarrow 2Ni(OH)_2 + Zn(OH)_2$	1.73	4.64	215	372	1.6	60	120
Nickel-hydrogen	H_2	Ni oxide	$H_2 + 2NiOOH \rightarrow 2Ni(OH)_2$	1.5	3.46	289	434	1.2	55	60
Nickel-metal hydride	MH[1]	Ni oxide	$MH + NiOOH \rightarrow M + Ni(OH)_2$	1.35	5.63	178	240	1.2	75	240[5]
Silver-zinc	Zn	AgO	$Zn + AgO + H_2O \rightarrow Zn(OH)_2 + Ag$	1.85	3.53	283	524	1.5	105	180[10]
Silver-cadmium	Cd	AgO	$Cd + AgO + H_2O \rightarrow Cd(OH)_2 + Ag$	1.4	4.41	227	318	1.1	70	120[10]
Zinc/chlorine	Zn	Cl_2	$Zn + Cl_2 \rightarrow ZnCl_2$	2.12	2.54	394	835	—	70	—
Zinc/bromine	Zn	Br_2	$Zn + Br_2 \rightarrow ZnBr_2$	1.85	4.17	309	572	1.6	70	60
Lithium-ion	Li_xC_6	$Li_{(i-x)}CoO_2$	$Li_xC_6 + Li_{(i-x)}CoO_2 \rightarrow LiCoO_2 + C_6$	4.1	9.98	100	410	4.1	150	400[5]
Lithium/manganese dioxide	Li	MnO_2	$Li + Mn^{IV}O_2 \rightarrow Mn^{IV}O_2(Li^+)$	3.5	3.50	286	1001	3.0	120	265
Lithium/iron disulfide[2]	Li(Al)	FeS_2	$2Li(Al) + FeS_2 \rightarrow Li_2FeS_2 + 2Al$	1.73	3.50	285	493	1.7	180[11]	350[11]
Lithium/iron monosulfide[2]	Li(Al)	FeS	$2Li(Al) + FeS \rightarrow Li_2S + Fe + 2Al$	1.33	2.90	345	459	1.3	130[11]	220[11]
Sodium/sulfur[2]	Na	S	$2Na + 3S \rightarrow Na_2S_3$	2.1	2.65	377	792	2.0	170[11]	345[11]
Sodium/nickel chloride[2]	Na	$NiCl_2$	$2Na + NiCl_2 \rightarrow 2NaCl + Ni$	2.58	3.28	305	787	2.6	115[11]	190[11]

Fuel cells

System	Anode	Cathode	Reaction mechanism							
H_2/O_2	H_2	O_2	$H_2 + \frac{1}{2}O_2 \rightarrow H_2O$	1.23	0.336	2975	3660	—	—	—
H_2/air	H_2	Ambient air	$H_2 + (\frac{1}{2}O_2) \rightarrow H_2O$	1.23	0.037	26587	32702	—	—	—
Methanol/O_2	CH_3OH	O_2	$CH_3OH + \frac{3}{2}O_2 \rightarrow CO_2 + 2H_2O$	1.24	0.50	2000	2480	—	—	—
Methanol/air	CH_3OH	Ambient air	$CH_3OH + (\frac{3}{2}O_2) \rightarrow CO_2 + 2H_2O$	1.24	0.20	5020	6225	—	—	—

† Based on active anode and cathode materials only, including O_2 but not air (electrolyte not included).

* These values are for single cell batteries based on identified design and at discharge rates optimized for energy density, using midpoint voltage. More specific values are given in chapters on each battery system.

(1) MH = metal hydride, data based on 1.7% hydrogen storage (by weight).
(2) High temperature batteries.
(3) Solid electrolyte battery (LiI_2 (P2VP)).
(4) Cylindrical bobbin-type batteries.
(5) Cylindrical spiral-wound batteries.
(6) Button type batteries.
(7) Water-activated.
(8) Automatically activated 2- to 10-min rate.
(9) With lithium anodes.
(10) Prismatic batteries.
(11) Value based on cell performance, see appropriate chapter for details.

1.4.4 Theoretical Energy*

The capacity of a cell can also considered on an energy (watthour) basis by taking both the voltage and the quantity of electricity into consideration. This theoretical energy value is the maximum value that can be delivered by a specific electrochemical system:

$$\text{Watthour (Wh)} = \text{voltage (V)} \times \text{ampere-hour (Ah)}$$

In the Zn/Cl_2 cell example, if the standard potential is taken as 2.12 V, the theoretical watthour capacity per gram of active material (theoretical gravimetric specific energy or theoretical gravimetric energy density) is:

$$\text{Specific Energy (Watthours/gram)} = 2.12 \text{ V} \times 0.394 \text{ Ah/g} = 0.835 \text{ Wh/g or } 835 \text{ Wh/kg}$$

Table 1.2 also lists the theoretical specific energy of the various electrochemical systems.

1.5 SPECIFIC ENERGY AND ENERGY DENSITY OF PRACTICAL BATTERIES

The theoretical electrical properties of cells and batteries are discussed in Sec. 1.4. In summary, the maximum energy that can be delivered by an electrochemical system is based on the types of active materials that are used (this determines the voltage) and on the amount of the active materials that are used (this determines ampere-hour capacity). In practice, only a fraction of the theoretical energy of the battery is realized. This is due to the need for electrolyte and nonreactive components (containers, separators, electrodes) that add to the weight and volume of the battery, as illustrated in Fig. 1.3. Another contributing factor is that the battery does not discharge at the theoretical voltage (thus lowering the average

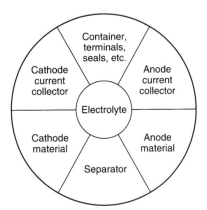

FIGURE 1.3 Components of a cell.

---

*The energy output of a cell or battery is often expressed as a ratio of its weight or size. The preferred terminology for this ratio on a weight basis, e.g. Watthours/kilogram (Wh/kg), is "specific energy"; on a volume basis, e.g. Watthours/liter (Wh/L), it is "energy density." Quite commonly, however, the term "energy density" is used to refer to either ratio.

voltage), nor is it discharged completely to zero volts (thus reducing the delivered ampere-hours) (also see Sec. 3.2.1). Further, the active materials in a practical battery are usually not stoichiometrically balanced. This reduces the specific energy because an excess amount of one of the active materials is used.

In Fig. 1.4, the following values for some major batteries are plotted:

1. The theoretical specific energy (based on the active anode and cathode materials only)
2. The theoretical specific energy of a practical battery (accounting for the electrolyte and non-reactive components)
3. The actual specific energy of these batteries when discharged at 20°C under optimal discharge conditions

These data show:

• That the weight of the materials of construction reduces the theoretical energy density or of the battery by almost 50 percent, and

• That the actual energy delivered by a practical battery, even when discharged under conditions close to optimum, may only be 50 to 75 percent of that lowered value

Thus, the actual energy that is available from a battery under practical, but close to optimum, discharge conditions is only about 25 to 35 percent of the theoretical energy of the active materials. Chapter 3 covers the performance of batteries when used under more stringent conditions.

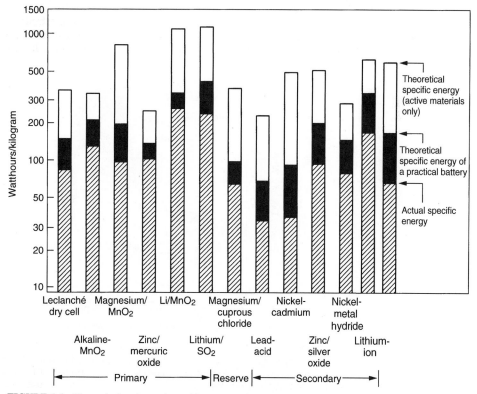

FIGURE 1.4   Theoretical and actual specific energy of battery systems.

These data are shown again in Table 1.2 which, in addition to the theoretical values, lists the characteristics of each of these batteries based on the actual performance of a practical battery. Again, these values are based on discharge conditions close to optimum for that battery.

The specific energy (Wh/kg) and energy density (Wh/L) delivered by the major battery systems are also plotted in Fig. 1.5(a) for primary batteries and 1.5(b) for rechargeable batteries. In these figures, the energy storage capability is shown as a field, rather than as a

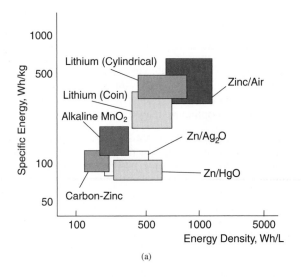

(a)

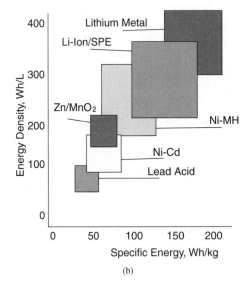

(b)

FIGURE 1.5   Comparison of the energy storage capability of various battery systems (a) Primary batteries; (b) Rechargeable batteries. (*From Ref 1*)

single optimum value, to illustrate the spread in performance of that battery system under different conditions of use.

In practice, as discussed in detail in Chap. 3, the electrical output of a battery may be reduced even further when it is used under more stringent conditions.

1.6   UPPER LIMITS OF SPECIFIC ENERGY AND ENERGY DENSITY

Many advances have been made in battery technology in recent years as illustrated in Fig. 1.6, both through continued improvement of a specific electrochemical system and through the development and introduction of new battery chemistries. But batteries are not keeping pace with developments in electronics technology, where performance doubles every 18 months, a phenomenon known as Moore's Law. Batteries, unlike electronic devices, consume materials when delivering electrical energy and, as discussed in Secs. 1.4 and 1.5, there are theoretical limits to the amount of electrical energy that can be delivered electrochemically by the available materials. The upper limit is now being reached as most of the materials that are practical for use as active materials in batteries have already been investigated and the list of unexplored materials is being depleted.

As shown in Table 1.2, and the other such tables in the Handbook, except for some of the ambient air-breathing systems and the hydrogen/oxygen fuel cell, where the weight of the cathode active material is not included in the calculation, the values for the theoretical energy density do not exceed 1500 Wh/kg. Most of the values are, in fact, lower. Even the values for the hydrogen/air and the liquid fuel cells have to be lowered to include, at least, the weight and volume of suitable containers for these fuels.

The data in Table 1.2 also show that the specific energy delivered by these batteries, based on the actual performance when discharged under optimum conditions, does not exceed 450 Wh/kg, even including the air-breathing systems. Similarly, the energy density values do not exceed 1000 Wh/L. It is also noteworthy that the values for the rechargeable systems are lower than those of the primary batteries due, in part, to a more limited selection of materials that can be recharged practically and the need for designs to facilitate recharging and cycle life.

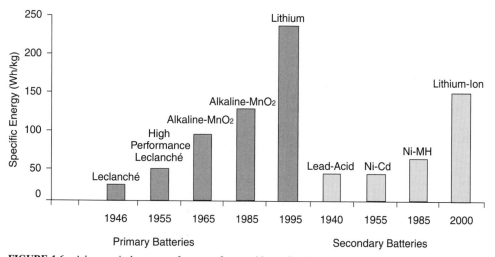

FIGURE 1.6   Advances in battery performance for portable applications.

Recognizing these limitations, while new battery systems will be explored, it will be more difficult to develop a new battery system which will have a significantly higher energy output and still meet the requirements of a successful commercial product, including availability of materials, acceptable cost, safety and environmental acceptability.

Battery research and development will focus on reducing the ratio of inactive to active components to improve energy density, increasing conversion efficiency and rechargability, maximizing performance under the more stringent operating and enhancing safety and environment. The fuel cell is offering opportunities for powering electric vehicles, as a replacement for combustion engines, for use in utility power and possibly for the larger portable applications (see Chap. 42). However, the development of a fuel cell for a small portable applications that will be competitive with batteries presents a formidable challenge.

REFERENCES

1. Ralph J. Broad, "Recent Developments in Batteries for Portable Consumer Electronics Applications," *Interface* **8**:3, Fall 1999, Electrochemical Society, Pennington, NJ.

CHAPTER 2
ELECTROCHEMICAL PRINCIPLES AND REACTIONS

John Broadhead and Han C. Kuo

2.1 INTRODUCTION

Batteries and fuel cells are electrochemical devices which convert chemical energy into electrical energy by electrochemical oxidation and reduction reactions, which occur at the electrodes. A cell consists of an anode where oxidation takes place during discharge, a cathode where reduction takes place, and an electrolyte which conducts the electrons (*via* ions) within the cell.

The maximum electric energy that can be delivered by the chemicals that are stored within or supplied to the electrodes in the cell depends on the change in free energy ΔG of the electrochemical couple, as shown in Eq. (2.5) and discussed in Sec. 2.2.

It would be desirable if during the discharge all of this energy could be converted to useful electric energy. However, losses due to polarization occur when a load current i passes through the electrodes, accompanying the electrochemical reactions. These losses include: (1) activation polarization, which drives the electrochemical reaction at the electrode surface, and (2) concentration polarization, which arises from the concentration differences of the reactants and products at the electrode surface and in the bulk as a result of mass transfer.

These polarization effects consume part of the energy, which is given off as waste heat, and thus not all of the theoretically available energy stored in electrodes is fully converted into useful electrical energy.

In principle, activation polarization and concentration polarization can be calculated from several theoretical equations, as described in later sections of this chapter, if some electrochemical parameters and the mass-transfer condition are available. However, in practice it is difficult to determine the values for both because of the complicated physical structure of the electrodes. As covered in Sec. 2.5, most battery and fuel cells electrodes are composite bodies made of active material, binder, performance enhancing additives and conductive filler. They usually have a porous structure of finite thickness. It requires complex mathematical modeling with computer calculations to estimate the polarization components.

There is another important factor that strongly affects the performance or rate capability of a cell, the internal impedance of the cell. It causes a voltage drop during operation, which also consumes part of the useful energy as waste heat. The voltage drop due to internal impedance is usually referred to as "ohmic polarization" or *IR* drop and is proportional to the current drawn from the system. The total internal impedance of a cell is the sum of the ionic resistance of the electrolyte (within the separator and the porous electrodes), the electronic resistances of the active mass, the current collectors and electrical tabs of both electrodes, and the contact resistance between the active mass and the current collector. These resistances are ohmic in nature, and follow Ohm's law, with a linear relationship between current and voltage drop.

When connected to an external load R, the cell voltage E can be expressed as

$$E = E_0 - [(\eta_{ct})_a + (\eta_c)_a] - [(\eta_{ct})_c + (\eta_c)_c] - iR_i = iR \qquad (2.1)$$

where
E_0 = electromotive force or open-circuit voltage of cell
$(\eta_{ct})_a, (\eta_{ct})_c$ = activation polarization or charge-transfer overvoltage at anode and cathode
$(\eta_c)_a, (\eta_c)_c$ = concentration polarization at anode and cathode
i = operating current of cell on load
R_i = internal resistance of cell

As shown in Eq. (2.1), the useful voltage delivered by the cell is reduced by polarization and the internal IR drop. It is only at very low operating currents, where polarization and the IR drop are small, that the cell may operate close to the open-circuit voltage and deliver most of the theoretically available energy. Figure 2.1 shows the relation between cell polarization and discharge current.

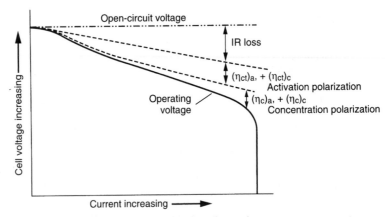

FIGURE 2.1 Cell polarization as a function of operating current.

Although the available energy of a battery or fuel cell depends on the basic electrochemical reactions at both electrodes, there are many factors which affect the magnitude of the charge-transfer reaction, diffusion rates, and magnitude of the energy loss. These factors include electrode formulation and design, electrolyte conductivity, and nature of the separators, among others. There exist some essential rules, based on the electrochemical principles, which are important in the design of batteries and fuel cells to achieve a high operating efficiency with minimal loss of energy.

1. The conductivity of the electrolyte should be high enough that the IR polarization is not excessively large for practical operation. Table 2.1 shows the typical ranges of specific conductivities for various electrolyte systems used in batteries. Batteries are usually designed for specific drain rate applications, ranging from microamperes to several hundred amperes. For a given electrolyte, a cell may be designed to have improved rate capability, with a higher electrode interfacial area and thin separator, to reduce the IR drop due to electrolyte resistance. Cells with a spirally wound electrode design are typical examples.

2. Electrolyte salt and solvents should have chemical stability to avoid direct chemical reaction with the anode or cathode materials.

TABLE 2.1 Conductivity Ranges of Various
Electrolytes at Ambient Temperature

Electrolyte system	Specific conductivity, $\Omega^{-1}\ cm^{-1}$
Aqueous electrolytes	1–5×10^{-1}
Molten salt	$\sim 10^{-1}$
Inorganic electrolytes	2×10^{-2}–10^{-1}
Organic electrolytes	10^{-3}–10^{-2}
Polymer electrolytes	10^{-7}–10^{-3}
Inorganic solid electrolytes	10^{-8}–10^{-5}

3. The rate of electrode reaction at both the anode and the cathode should be sufficiently fast so that the activation or charge-transfer polarization is not too high to make the cell inoperable. A common method of minimizing the charge-transfer polarization is to use a porous electrode design. The porous electrode structure provides a high electrode surface area within a given geometric dimension of the electrode and reduces the local current density for a given total operating current.

4. In most battery and fuel cell systems, part or all of the reactants are supplied from the electrode phase and part or all of the reaction products must diffuse or be transported away from the electrode surface. The cell should have adequate electrolyte transport to facilitate the mass transfer to avoid building up excessive concentration polarization. Proper porosity and pore size of the electrode, adequate thickness and structure of the separator, and sufficient concentration of the reactants in the electrolyte are very important to ensure functionality of the cell. Mass-transfer limitations should be avoided for normal operation of the cell.

5. The material of the current collector or substrate should be compatible with the electrode material and the electrolyte without causing corrosion problems. The design of the current collector should provide a uniform current distribution and low contact resistance to minimize electrode polarization during operation.

6. For rechargeable cells it is preferable to have the reaction products remain at the electrode surface to facilitate the reversible reactions during charge and discharge. The reaction products should be stable mechanically as well as chemically with the electrolyte.

In general, the principles and various electrochemical techniques described in this chapter can be used to study all the important electrochemical aspects of a battery or fuel cell. These include the rate of electrode reaction, the existence of intermediate reaction steps, the stability of the electrolyte, the current collector, the electrode materials, the mass-transfer conditions, the value of the limiting current, the formation of resistive films on the electrode surface, the impedance characteristics of the electrode or cell, and the existence of the rate-limiting species.

2.2 THERMODYNAMIC BACKGROUND

In a cell, reactions essentially take place at two areas or sites in the device. These reaction sites are the electrode interfaces. In generalized terms, the reaction at one electrode (reduction in forward direction) can be represented by

$$aA + ne \rightleftharpoons cC \qquad (2.2)$$

where a molecules of A take up n electrons e to form c molecules of C. At the other electrode, the reaction (oxidation in forward direction) can be represented by

$$bB - ne \rightleftharpoons dD \qquad (2.3)$$

The overall reaction in the cell is given by addition of these two half-cell reactions

$$aA + bB \rightleftharpoons cC + dD \qquad (2.4)$$

The change in the standard free energy ΔG^0 of this reaction is expressed as

$$\Delta G^0 = -nFE^0 \qquad (2.5)$$

where F = constant known as the Faraday (96,487 coulombs)
E^0 = standard electromotive force

TABLE 2.2 Standard Potentials of Electrode Reactions at 25°C

Electrode reaction	E^0, V	Electrode reaction	E^0, V
$Li^+ + e \rightleftharpoons Li$	−3.01	$Tl^+ + e \rightleftharpoons Tl$	−0.34
$Rb^+ + e \rightleftharpoons Rb$	−2.98	$Co^{2+} + 2e \rightleftharpoons Co$	−0.27
$Cs^+ + e \rightleftharpoons Cs$	−2.92	$Ni^{2+} + 2e \rightleftharpoons Ni$	−0.23
$K^+ + e \rightleftharpoons K$	−2.92	$Sn^{2+} + 2e \rightleftharpoons Sn$	−0.14
$Ba^{2+} + 2e \rightleftharpoons Ba$	−2.92	$Pb^{2+} + 2e \rightleftharpoons Pb$	−0.13
$Sr^{2+} + 2e \rightleftharpoons Sr$	−2.89	$D^+ + e \rightleftharpoons \frac{1}{2}D_2$	−0.003
$Ca^{2+} + 2e \rightleftharpoons Ca$	−2.84	$H^+ + e \rightleftharpoons \frac{1}{2}H_2$	0.000
$Na^+ + e \rightleftharpoons Na$	−2.71	$Cu^{2+} + 2e \rightleftharpoons Cu$	0.34
$Mg^{2+} + 2e \rightleftharpoons Mg$	−2.38	$\frac{1}{2}O_2 + H_2O + 2e \rightleftharpoons 2OH^-$	0.40
$Ti^+ + 2e \rightleftharpoons Ti$	−1.75	$Cu^+ + e \rightleftharpoons Cu$	0.52
$Be^{2+} + 2e \rightleftharpoons Be$	−1.70	$Hg^{2+} + 2e \rightleftharpoons 2Hg$	0.80
$Al^{3+} + 3e \rightleftharpoons Al$	−1.66	$Ag^+ + e \rightleftharpoons Ag$	0.80
$Mn^{2+} + 2e \rightleftharpoons Mn$	−1.05	$Pd^{2+} + 2e \rightleftharpoons Pd$	0.83
$Zn^{2+} + 2e \rightleftharpoons Zn$	−0.76	$Ir^{3+} + 3e \rightleftharpoons Ir$	1.00
$Ga^{3+} + 3e \rightleftharpoons Ga$	−0.52	$Br_2 + 2e \rightleftharpoons 2Br^-$	1.07
$Fe^{2+} + 2e \rightleftharpoons Fe$	−0.44	$O_2 + 4H^+ + 4e \rightleftharpoons 2H_2O$	1.23
$Cd^{2+} + 2e \rightleftharpoons Cd$	−0.40	$Cl_2 + 2e \rightleftharpoons 2Cl^-$	1.36
$In^{3+} + 3e \rightleftharpoons In$	−0.34	$F_2 + 2e \rightleftharpoons 2F^-$	2.87

When conditions are other than in the standard state, the voltage E of a cell is given by the Nernst equation,

$$E = E^0 - \frac{RT}{nF} \ln \frac{a_C^c a_D^d}{a_A^a a_B^b} \tag{2.6}$$

where a_i = activity of relevant species
R = gas constant
T = absolute temperature

The change in the standard free energy ΔG^0 of a cell reaction is the driving force which enables a battery to deliver electrical energy to an external circuit. The measurement of the electromotive force, incidentally, also makes available data on changes in free energy, entropies and enthalpies together with activity coefficients, equilibrium constants, and solubility products.

Direct measurement of single (absolute) electrode potentials is considered practically impossible.[1] To establish a scale of half-cell or standard potentials, a reference potential "zero" must be established against which single electrode potentials can be measured. By convention, the standard potential of the H_2/H^+(aq) reaction is taken as zero and all standard potentials are referred to this potential. Table 2.2 and Appendix B list the standard potentials of a number of anode and cathode materials.

2.3 ELECTRODE PROCESSES

Reactions at an electrode are characterized by both chemical and electrical changes and are heterogeneous in type. Electrode reactions may be as simple as the reduction of a metal ion and incorporation of the resultant atom onto or into the electrode structure. Despite the apparent simplicity of the reaction, the mechanism of the overall process may be relatively complex and often involves several steps. Electroactive species must be transported to the electrode surface by migration or diffusion prior to the electron transfer step. Adsorption of electroactive material may be involved both prior to and after the electron transfer step. Chemical reactions may also be involved in the overall electrode reaction. As in any reaction, the overall rate of the electrochemical process is determined by the rate of the slowest step in the whole sequence of reactions.

The thermodynamic treatment of electrochemical processes presented in Sec. 2.2 describes the equilibrium condition of a system but does not present information on nonequilibrium conditions such as current flow resulting from electrode polarization (overvoltage) imposed to effect electrochemical reactions. Experimental determination of the current-voltage characteristics of many electrochemical systems has shown that there is an exponential relation between current and applied voltage. The generalized expression describing this relationship is called the Tafel equation,

$$\eta = a \pm b \log i \tag{2.7}$$

where η = overvoltage
i = current
a, b = constants

Typically, the constant b is referred to as the Tafel slope.

The Tafel relationship holds for a large number of electrochemical systems over a wide range of overpotentials. At low values of overvoltage, however, the relationship breaks down and results in curvature in plots of η versus $\log i$. Figure 2.2 is a schematic presentation of a Tafel plot, showing curvature at low values of overvoltage.

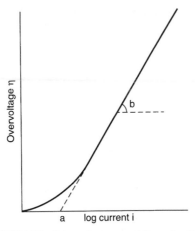

FIGURE 2.2 Schematic representation of a Tafel plot showing curvature at low overvoltage and indicating significance of parameters a and b.

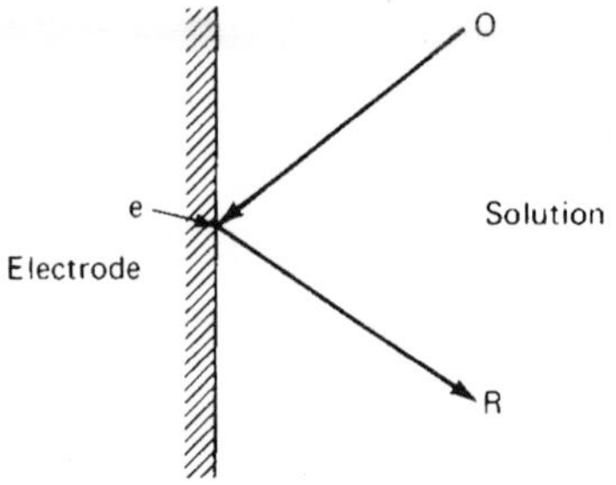

FIGURE 2.3 Simplified representation of electroreduction process at an electrode.

Success of the Tafel equation's fit to many experimental systems encouraged the quest for a kinetic theory of electrode processes. Since the range of validity of the Tafel relationship applies to high overvoltages, it is reasonable to assume that the expression does not apply to equilibrium situations but represents the current-voltage relationship of a unidirectional process. In an oxidation process, this means that there is a negligible contribution from reduction processes. Rearranging Eq. (2.7) into exponential form, we have

$$i = \exp\left(\pm \frac{a}{b}\right) \exp \frac{\eta}{b} \tag{2.8}$$

To consider a general theory, one must consider both forward and backward reactions of the electroreduction process, shown in simplified form in Fig. 2.3. The reaction is represented by the equation

$$O + ne \rightleftharpoons R \tag{2.9}$$

where O = oxidized species
  R = reduced species
  n = number of electrons involved in electrode process

The forward and backward reactions can be described by heterogeneous rate constants k_f and k_b, respectively. The rates of the forward and backward reactions are then given by the products of these rate constants and the relevant concentrations which typically are those at the electrode surface. As will be shown later, the concentrations of electroactive species at the electrode surface often are dissimilar from the bulk concentration in solution. The rate of the forward reaction is $k_f C_O$ and that for the backward reaction is $k_b C_R$. For convenience, these rates are usually expressed in terms of currents i_f and i_b, for the forward and backward reactions, respectively,

$$i_f = nFAk_f C_O \tag{2.10}$$

$$i_b = nFAk_b C_R \tag{2.11}$$

where A is the area of the electrode and F the Faraday.

Establishing these expressions is merely the result of applying the law of mass action to the forward and backward electrode processes. The role of electrons in the process is established by assuming that the magnitudes of the rate constants depend on the electrode potential. The dependence is usually described by assuming that a fraction αE of the electrode potential is involved in driving the reduction process, while the fraction $(1 - \alpha)E$ is effective in making the reoxidation process more difficult. Mathematically these potential-dependent rate constants are expressed as

$$k_f = k_f^0 \exp \frac{-\alpha n FE}{RT} \tag{2.12}$$

$$k_b = k_b^0 \exp \frac{(1 - \alpha)n FE}{RT} \tag{2.13}$$

where α is the transfer coefficient and E the electrode potential relative to a suitable reference potential.

A little more explanation regarding what the transfer coefficient α (or the symmetry factor β, as it is referred to in some texts) means in mechanistic terms is appropriate since this term is not implicit in the kinetic derivation.[2] The transfer coefficient determines what fraction of the electric energy resulting from the displacement of the potential from the equilibrium value affects the rate of electrochemical transformation. To understand the function of the transfer coefficient α, it is necessary to describe an energy diagram for the reduction-oxidation process. Figure 2.4 shows an approximate potential energy curve (Morse curve) for an oxidized species approaching an electrode surface together with the potential energy curve for the resultant reduced species. For convenience, consider the hydrogen ion reduction at a solid electrode as the model for a typical electroreduction. According to Horiuti and Polanyi,[3] the potential energy diagram for reduction of the hydrogen ion can be represented by Fig. 2.5 where the oxidized species O is the hydrated hydrogen ion and the reduced species R is a hydrogen atom bonded to the metal (electrode) surface. The effect of changing the electrode potential by a value of E is to raise the potential energy of the Morse curve of the hydrogen ion. The intersection of the two Morse curves forms an energy barrier, the height of which is αE. If the slope of the two Morse curves is approximately constant at the point of intersection, then α is defined by the ratio of the slope of the Morse curves at the point of intersection

$$\alpha = \frac{m_1}{m_1 + m_2} \tag{2.14}$$

where m_1 and m_2 are the slopes of the potential curves of the hydrated hydrogen ion and the hydrogen atom, respectively.

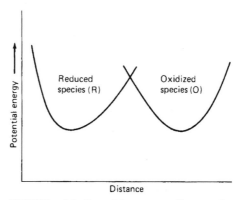

FIGURE 2.4 Potential energy diagram for reduction-oxidation process taking place at an electrode.

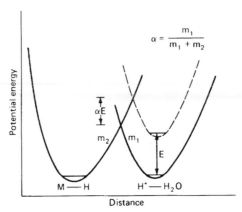

FIGURE 2.5 Potential energy diagram for reduction of a hydrated hydrogen ion at an electrode.

There are inadequacies in the theory of transfer coefficients. It assumes that α is constant and independent of E. At present there are no data to prove or disprove this assumption. The other main weakness is that the concept is used to describe processes involving a variety of different species such as (1) redox changes at an inert electrode (Fe^{2+}/Fe^{3+} at Hg); (2) reactant and product soluble in different phases [$Cd^{2+}/Cd(Hg)$]; and (3) electrodeposition (Cu^{2+}/Cu). Despite these inadequacies, the concept and application of the theory are appropriate in many cases and represent the best understanding and description of electrode processes at the present time. Examples of a few values of α are given in Table 2.3.[4]

TABLE 2.3   Values of Transfer Coefficient α at 25°C [4]

Metal	System	α
Platinum	$Fe^{3+} + e \rightarrow Fe^{2+}$	0.58
Platinum	$Ce^{4+} + e \rightarrow Ce^{2+}$	0.75
Mercury	$Ti^{4+} + e \rightarrow Ti^{3+}$	0.42
Mercury	$2H^+ + 2e \rightarrow H_2$	0.50
Nickel	$2H^+ + 2e \rightarrow H_2$	0.58
Silver	$Ag^+ + e \rightarrow Ag$	0.55

From Eqs. (2.12) and (2.13) we can derive parameters useful for evaluating and describing an electrochemical system. Equations (2.12) and (2.13) are compatible both with the Nernst equation [Eq. (2.6)] for equilibrium conditions and with the Tafel relationships [Eq. (2.7)] for unidirectional processes.

Under equilibrium conditions, no net current flows and

$$i_f = i_b = i_0 \tag{2.15}$$

where i_0 is the exchange current. From Eqs. (2.10)–(2.13), together with Eq. (2.15), the following relationship is established:

$$C_O k_f^0 \exp \frac{-\alpha n F E_e}{RT} = C_R k_b^0 \exp \frac{(1 - \alpha)nFE_e}{RT} \tag{2.16}$$

where E_e is the equilibrium potential.

Rearranging,

$$E_e = \frac{RT}{nF} \ln \frac{k_f^0}{k_b^0} + \frac{RT}{nF} \ln \frac{C_O}{C_R} \tag{2.17}$$

From this equation we can establish the definition of formal standard potential E_C^0, where concentrations are used rather than activities,

$$E_C^0 = \frac{RT}{nF} \ln \frac{k_f^0}{k_b^0} \tag{2.18}$$

For convenience, the formal standard potential is often taken as the reference point of the potential scale in reversible systems.

Combining Eqs. (2.17) and (2.18), we can show consistency with the Nernst equation,

$$E_e = E_C^0 + \frac{RT}{nF} \ln \frac{C_O}{C_R} \tag{2.19}$$

except that this expression is written in terms of concentrations rather than activities.

From Eqs. (2.10) and (2.12), at equilibrium conditions,

$$i_0 = i_f = nFAC_O k_f^0 \exp \frac{-\alpha nFE_e}{RT} \tag{2.20}$$

The exchange current as defined in Eq. (2.15) is a parameter of interest to researchers in the battery field. This parameter may be conveniently expressed in terms of the rate constant k by combining Eqs. (2.10), (2.12), (2.17), and (2.20),

$$i_0 = nFAkC_O^{(1-\alpha)}C_R^{\alpha} \tag{2.21}$$

The exchange current i_0 is a measure of the rate of exchange of charge between oxidized and reduced species at any equilibrium potential without net overall change. The rate constant k, however, has been defined for a particular potential, the formal standard potential of the system. It is not in itself sufficient to characterize the system unless the transfer coefficient is also known. However, Eq. (2.21) can be used in the elucidation of the electrode reaction mechanism. The value of the transfer coefficient can be determined by measuring the exchange current density as a function of the concentration of the reduction or oxidation species at a constant concentration of the oxidation of reduction species, respectively. A schematic representation of the forward and backward currents as a function of overvoltage, $\eta = E - E_e$, is shown in Fig. 2.6, where the net current is the sum of the two components.

For situations where the net current is not zero, that is, where the potential is sufficiently different from the equilibrium potential, the net current approaches the net forward current (or, for anodic overvoltages, the backward current). One can then write

$$i = nFAkC_O \exp \frac{-\alpha nF\eta}{RT} \tag{2.22}$$

Now when $\eta = 0$, $i = i_0$, then

$$i = i_0 \exp \frac{-\alpha nF\eta}{RT} \tag{2.23}$$

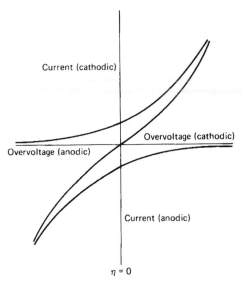

FIGURE 2.6 Schematic representation of relationship between overvoltage and current.

and

$$\eta = \frac{RT}{\alpha n F} \ln i_0 - \frac{RT}{\alpha n F} \ln i \qquad (2.24)$$

which is the Tafel equation introduced earlier in a generalized form as Eq. (2.7).

It can now be seen that the kinetic treatment here is self-consistent with both the Nernst equation (for equilibrium conditions) and the Tafel relationship (for unidirectional processes).

To present the kinetic treatment in its most useful form, a transformation into a net current flow form is appropriate. Using

$$i = i_f - i_b \qquad (2.25)$$

substitute Eqs. (2.10), (2.13), and (2.18),

$$i = nFAk \left[C_O \exp \frac{-\alpha n F E_C^0}{RT} - C_R \exp \frac{(1 - \alpha) n F E_C^0}{RT} \right] \qquad (2.26)$$

When this equation is applied in practice, it is very important to remember that C_O and C_R are concentrations at the surface of the electrode, or are the effective concentrations. These are not necessarily the same as the bulk concentrations. Concentrations at the interface are often (almost always) modified by differences in electric potential between the surface and the bulk solution. The effects of potential differences that are manifest at the electrode-electrolyte interface are given in the following section.

2.4 ELECTRICAL DOUBLE-LAYER CAPACITY AND IONIC ADSORPTION

When an electrode (metal surface) is immersed in an electrolyte, the electronic charge on the metal attracts ions of opposite charge and orients the solvent dipoles. There exist a layer of charge in the metal and a layer of charge in the electrolyte. This charge separation establishes what is commonly known as the "electrical double layer."[5]

Experimentally, the electrical double-layer affect is manifest in the phenomenon named "electrocapillarity." The phenomenon has been studied for many years, and there exist thermodynamic relationships that relate interfacial surface tension between electrode and electrolyte solution to the structure of the double layer. Typically the metal used for these measurements is mercury since it is the only conveniently available metal that is liquid at room temperature (although some work has been carried out with gallium, Wood's metal, and lead at elevated temperature).

Determinations of the interfacial surface tension between mercury and electrolyte solution can be made with a relatively simple apparatus. All that are needed are (1) a mercury-solution interface which is polarizable, (2) a nonpolarizable interface as reference potential, (3) an external source of variable potential, and (4) an arrangement to measure the surface tension of the mercury-electrolyte interface. An experimental system which will fulfill these requirements is shown in Fig. 2.7. The interfacial surface tension is measured by applying pressure to the mercury-electrolyte interface by raising the mercury "head." At the interface, the forces are balanced, as shown in Fig. 2.8. If the angle of contact at the capillary wall is zero (typically the case for clean surfaces and clean electrolyte), then it is a relatively simple arithmetic exercise to show that the interfacial surface tension is given by

$$\gamma = \frac{h\rho gr}{2} \tag{2.27}$$

where γ = interfacial surface tension
ρ = density of mercury
g = force of gravity
r = radius of capillary
h = height of mercury column in capillary

The characteristic electrocapillary curve that one would obtain from a typical electrolyte solution is shown in Fig. 2.9. From such measurements and, more accurately, by AC impedance bridge measurements, the structure of the electrical double layer has been determined.[5]

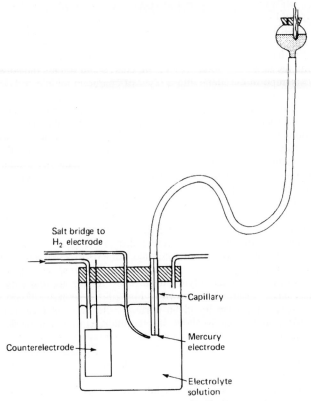

FIGURE 2.7 Experimental arrangement to measure interfacial surface tension at mercury-electrolyte interface.

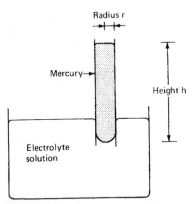

FIGURE 2.8 Close-up of mercury-electrolyte interface in a capillary immersed in an electrolyte solution

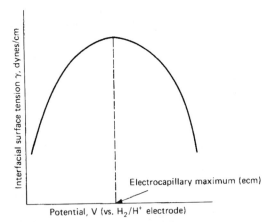

FIGURE 2.9 Generalized representation of an electro-capillary curve.

Consider a negatively charged electrode in an aqueous solution of electrolyte. Assume that at this potential no electrochemical charge transfer takes place. For simplicity and clarity, the different features of the electrical double layer will be described individually.

Orientation of solvent molecules, water for the sake of this discussion, is shown in Fig. 2.10. The water dipoles are oriented, as shown in the figure, so that the majority of the dipoles are oriented with their positive ends (arrow heads) toward the surface of the electrode. This represents a "snapshot" of the structure of the layer of water molecules since the electrical double layer is a dynamic system which is in equilibrium with water in the bulk solution. Since the representation is statistical, not all dipoles are oriented the same way. Some dipoles are more influenced by dipole-dipole interactions than by dipole-electrode interactions.

Next, consider the approach of a cation to the vicinity of the electrical double layer. The majority of cations are strongly solvated by water dipoles and maintain a sheath of water dipoles around them despite the orienting effect of the double layer. With a few exceptions, cations do not approach right up to the electrode surface but remain outside the primary layer of solvent molecules and usually retain their solvation sheaths. Figure 2.11 shows a typical example of a cation in the electrical double layer. The establishment that this is the most likely approach of a typical cation comes partly from experimental AC impedance measurements of mixed electrolytes and mainly from calculations of the free energy of approach of an ion to the electrode surface. In considering water-electrode, ion-electrode, and ion-water interactions, the free energy of approach of a cation to an electrode surface is strongly influenced by the hydration of the cation. The general result is that cations of very large radius (and thus of low hydration) such as Cs^+ can contact/adsorb on the electrode surface, but for the majority of cations the change in free energy on contact absorption is positive and thus is against the mechanism of contact adsorption.[6] Figure 2.12 gives an example of the ion Cs^+ contact-adsorbed on the surface of an electrode.

It would be expected that because anions have a negative charge, contact adsorption of anions would not occur. In analyzing the free-energy balance of the anion system, it is found that anion-electrode contact is favored because the net free-energy balance is negative. Both from these calculations and from experimental measurements, anion contact adsorption is found to be relatively common. Figure 2.13 shows the generalized case of anion adsorption

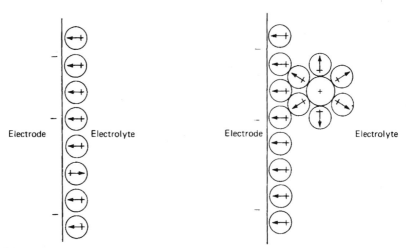

FIGURE 2.10 Orientation of water molecules in electrical double layer at a negatively charged electrode.

FIGURE 2.11 Typical cation situated in electrical double layer.

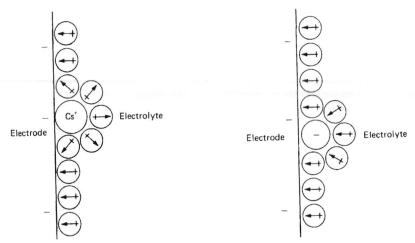

FIGURE 2.12 Contact adsorption of Cs^+ on an electrode surface.

FIGURE 2.13 Contact adsorption of anion on an electrode surface.

on an electrode. There are exceptions to this type of adsorption. Calculation of the free energy of contact adsorption of the fluoride ion is positive and unlikely to occur. This is supported by experimental measurement. This property is utilized, as NaF, as a supporting electrolyte* to evaluate adsorption properties of surface-active species devoid of the influence of adsorbed supporting electrolyte.

Extending out into solution from the electrical double layer (or the compact double layer, as it is sometimes known) is a continuous repetition of the layering effect, but with diminishing magnitude. This "extension" of the compact double layer toward the bulk solution is known as the Gouy-Chapman diffuse double layer.[5] Its effect on electrode kinetics and the concentration of electroactive species at the electrode surface is manifest when supporting electrolyte concentrations are low or zero.

The end result of the establishment of the electrical double-layer effect and the various types of ion contact adsorption, is directly to influence the real (actual) concentration of electroactive species at an electrode surface and indirectly to modify the potential gradient at the site of electron transfer. In this respect it is important to understand the influence of the electrical double layer and allow for it where and when appropriate.

The potential distribution near an electrode is shown schematically in Fig. 2.14. The inner Helmholtz plane corresponds to the plane which contains the contact-adsorbed ions and the innermost layer of water molecules. Its potential is defined as ϕ^i with the zero potential being taken as the potential of the bulk solution. The outer Helmholtz plane is the plane of closest approach of those ions which do not contact-adsorb but approach the electrode with a sheath of solvated water molecules surrounding them. The potential at the outer Hemholtz plane is defined as ϕ^o and is again referred to the potential of the bulk solution. In some texts ϕ^i is defined as ϕ^1 and ϕ^o as ϕ^2.

---

*A supporting electrolyte is a salt used in large excess to minimize internal resistance in an electrode chemical cells, but which does not enter into electrode reactions.

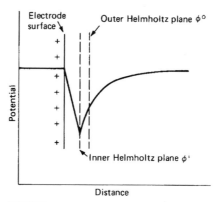

FIGURE 2.14 Potential distribution of positively charged electrode.

As mentioned previously, the bulk concentration of an electroactive species is often not the value to be used in kinetic equations. Species which are in the electrical double layer are in a different energy state from those in bulk solution. At equilibrium, the concentration C^e of an ion or species that is about to take part in the charge-transfer process at the electrode is related to the bulk concentration by

$$C^e = C^B \exp \frac{-zF\phi^e}{RT} \tag{2.28}$$

where z is the charge on the ion and ϕ^e the potential of *closest approach* of the species to the electrode. It will be remembered that the plane of closest approach of many species is the outer Hemholtz plane, and so the value of ϕ^e can often be equated to ϕ^o. However, as noted in a few special cases, the plane of closest approach can be the inner Helmholtz plane, and so the value of ϕ^e in these cases would be the same as ϕ^i. A judgment has to be made as to what value of ϕ^e should be used.

The potential which is effective in driving the electrode reaction is that between the species at its closest approach and the potential of the electrode. If E is the potential of the electrode, then the driving force is $E - \phi^e$. Using this relationship together with Eqs. (2.26) and (2.28), we have

$$\frac{i}{nFAk} = C_O \exp \frac{-z_O E\phi^e}{RT} \exp \frac{-\alpha nF(E - \phi^e)}{RT}$$

$$- C_R \exp \frac{-z_R F\phi^e}{RT} \exp \frac{(1 - \alpha)nF(E - \phi^e)}{RT} \tag{2.29}$$

where z_O and z_R are the charges (with sign) of the oxidized and reduced species, respectively. Rearranging Eq. (2.28) and using

$$z_O - n = z_R \tag{2.30}$$

yields

$$\frac{i}{nFAk} = \exp \frac{(\alpha n - z_O)F\phi^e}{RT}\left[C_O \exp \frac{-\alpha n FE}{RT} - C_R \exp \frac{(1 - \alpha)nFE}{RT}\right] \quad (2.31)$$

In experimental determination, the use of Eq. (2.26) will provide an apparent rate constant k_{app}, which does not take into account the effects of the electrical double layer. Taking into account the effects appropriate to the approach of a species to the plane of nearest approach,

$$k_{app} = k \exp \frac{(\alpha n - z_O)F\phi^e}{RT} \quad (2.32)$$

For the exchange current the same applies,

$$(i_0)_{app} = i_0 \exp \frac{(\alpha n - z_O)F\phi^e}{RT} \quad (2.33)$$

Corrections to the rate constant and the exchange current are not insignificant. Several calculated examples are given in Bauer.[7] The differences between apparent and true rate constants can be as great as two orders of magnitude. The magnitude of the correction also is related to the magnitude of the difference in potential between the electrocapillary maximum for the species and the potential at which the electrode reaction occurs; the greater the potential difference, the greater the correction to the exchange current or rate constant.

2.5 MASS TRANSPORT TO THE ELECTRODE SURFACE

We have considered the thermodynamics of electrochemical processes, studied the kinetics of electrode processes, and investigated the effects of the electrical double layer on kinetic parameters. An understanding of these relationships is an important ingredient in the repertoire of the researcher of battery technology. Another very important area of study which has major impact on battery research is the evaluation of mass transport processes to and from electrode surfaces.

Mass transport to or from an electrode can occur by three processes: (1) convection and stirring, (2) electrical migration in an electric potential gradient, and (3) diffusion in a concentration gradient. The first of these processes can be handled relatively easily both mathematically and experimentally. If stirring is required, flow systems can be established, while if complete stagnation is an experimental necessity, this can also be imposed by careful design. In most cases, if stirring and convection are present or imposed, they can be handled mathematically.

The migration component of mass transport can also be handled experimentally (reduced to close to zero or occasionally increased in special cases) and described mathematically, provided certain parameters such as transport number or migration current are known. Migration of electroactive species in an electric potential gradient can be reduced to near zero by addition of an excess of inert "supporting electrolyte," which effectively reduces the potential gradient to zero and thus eliminates the electric field which produces migration. Enhancement of migration is more difficult. This requires that the electric field be increased so that movement of charged species is increased. Electrode geometry design can increase migration slightly by altering electrode curvature. The fields at convex surfaces are greater than those at flat or concave surfaces, and thus migration is enhanced at convex curved surfaces.

The third process, diffusion in a concentration gradient, is the most important of the three processes and is the one which typically is dominant in mass transport in batteries. The analysis of diffusion uses the basic equation due to Fick[8] which defines the flux of material crossing a plane at distance x and time t. The flux is proportional to the concentration gradient and is represented by the expression:

$$q = D\frac{\delta C}{\delta x} \tag{2.34}$$

where q = flux
D = diffusion coefficient
C = concentration

The rate of change of concentration with time is defined by

$$\frac{\delta C}{\delta t} = D\frac{\delta^2 C}{\delta x^2} \tag{2.35}$$

This expression is referred to as Fick's second law of diffusion. Solution of Eqs. (2.34) and (2.35) requires that boundary conditions be imposed. These are chosen according to the electrode's expected "discharge" regime dictated by battery performance or boundary conditions imposed by relevant electroanalytical technique.[9] Several of the electroanalytical techniques are discussed in Sec. 2.6.

For application directly to battery technology, the three modes of mass transport have meaningful significance. Convective and stirring processes can be employed to provide a flow of electroactive species to reaction sites. Examples of the utilization of stirring and flow processes in batteries are the circulating zinc/air system, the vibrating zinc electrode, and the zinc-chlorine hydrate battery. In some types of advanced lead-acid batteries, circulation of acid is provided to improve utilization of the active materials in the battery plates.

Migration effects are in some cases detrimental to battery performance, in particular those caused by enhanced electric fields (potential gradients) around sites of convex curvature. Increased migration at these sites tends to produce dendrite formations which eventually lead to a short-circuit and battery failure.

2.5.1 Concentration Polarization

Diffusion processes are typically the mass-transfer processes operative in the majority of battery systems where the transport of species to and from reaction sites is required for maintenance of current flow. Enhancement and improvement of diffusion processes are an appropriate direction of research to follow to improve battery performance parameters. Equation (2.34) may be written in an approximate, yet more practical, form, remembering that $i = nFq$, where q is the flux through a plane of unit area. Thus,

$$i = nF\frac{DA(C_B - C_E)}{\delta} \tag{2.36}$$

where symbols are defined as before, and C_B = bulk concentration of electroactive species, C_E = concentration at electrode, A = electrode area, δ = boundary-layer thickness, that is, the layer at the electrode surface in which the majority of the concentration gradient is concentrated (see Fig. 2.15).

When $C_E = 0$, this expression defines the maximum diffusion current, i_L, that can be sustained in solution under a given set of conditions,

$$i_L = nF\frac{DAC_B}{\delta_L} \tag{2.37}$$

where δ_L is the boundary-layer thickness at the limiting condition. It tells us that to increase i_L, one needs to increase the bulk concentration, the electrode area, or the diffusion coefficient. In the design of a battery, an understanding of the implication of this expression is important. Specific cases can be analyzed quickly by applying Eq. (2.36), and parameters such as discharge rate and likely power densities of new systems may be estimated.

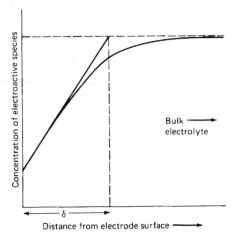

FIGURE 2.15 Boundary-layer thickness at an electrode surface.

Assume that the thickness of the diffusion boundary layer does not change much with concentration. Then $\delta_L = \delta$ and Eq. (2.36) may be rewritten as

$$i = \left(1 - \frac{C_E}{C_B}\right) i_L \tag{2.38}$$

The difference in concentration existing between the electrode surface and the bulk of the electrolyte results in a concentration polarization. According to the Nernst equation, the concentration polarization or overpotential η_c, produced from the change of concentration across the diffusion layer, may be written as

$$\eta_c = \frac{RT}{nF} \ln \frac{C_B}{C_E} \tag{2.39}$$

From Eq. (2.38) we have

$$\eta_c = \frac{RT}{nF} \ln \left(\frac{i_L}{i_L - i}\right) \tag{2.40}$$

This gives the relation of concentration polarization and current for mass transfer by diffusion. Equation (2.40) indicates that as i approaches the limiting current i_L, theoretically the overpotential should increase to infinity. However, in a real process the potential will increase only to a point where another electrochemical reaction will occur, as illustrated in Fig. 2.16. Figure 2.17 shows the magnitude of the concentration over-potential as a function of i/i_L with $n = 2$ at 25°C, based on Eq. (2.40).

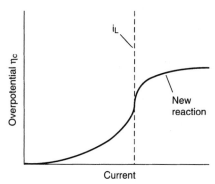

FIGURE 2.16   Plot of overpotential η_c vs. current i.

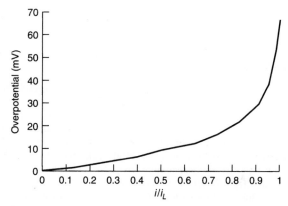

FIGURE 2.17   Magnitude of concentration overpotential as a function of i/i_L, with $n = 2$ at 25°C, based on Eq. (2.40).

2.5.2 Porous Electrodes

Electrochemical reactions are heterogeneous reactions which occur on the electrolyte-electrolyte interface. In fuel cell systems, the reactants are supplied from the electrolyte phase to the catalytic electrode surface. In battery systems, the electrodes are usually composites made of active reactants, binder and conductive filler. In order to minimize the energy loss due to both activation and concentration polarizations at the electrode surface and to increase the electrode efficiency or utilization, it is preferred to have a large electrode surface area. This is accomplished with the use of a porous electrode design. A porous electrode can provide an interfacial area per unit volume several decades higher than that of a planar electrode (such as 10^4 cm^{-1}).

A porous electrode consists of porous matrices of solids and void spaces. The electrolyte penetrates the void spaces of the porous matrix. In such an active porous mass, the mass-transfer condition in conjunction with the electrochemical reaction occurring at the interface is very complicated. In a given time during cell operation, the rate of reaction within the pores may vary significantly depending on the location. The distribution of current density within the porous electrode depends on the physical structure (such as tortuosity, pore sizes), the conductivity of the solid matrix and the electrolyte, and the electrochemical kinetic parameters of the electrochemical processes. A detailed treatment of such complex porous electrode systems can be found in Newman.[10]

2.6 ELECTROANALYTICAL TECHNIQUES

Many steady-state and impulse electroanalytical techniques are available to the experimentalist to determine electrochemical parameters and assist in both improving existing battery systems and evaluating couples as candidates for new batteries.[11] A few of these techniques are described in this section.

2.6.1 Cyclic Voltammetry

Of the electroanalytical techniques, cyclic voltammetry (or linear sweep voltammetry as it is sometimes known) is probably one of the more versatile techniques available to the electrochemist. The derivation of the various forms of cyclic voltammetry can be traced to the initial studies of Matheson and Nicols[12] and Randles.[13] Essentially the technique applies a linearly changing voltage (ramp voltage) to an electrode. The scan of voltage might be ± 2 V from an appropriate rest potential such that most electrode reactions would be encompassed. Commercially available instrumentation provides voltage scans as wide as ± 5 V.

To describe the principles behind cyclic voltammetry, for convenience let us restate Eq. (2.9), which describes the reversible reduction of an oxidized species O,

$$O + ne \rightleftharpoons R \tag{2.9}$$

In cyclic voltammetry, the initial potential sweep is represented by

$$E = E_i - vt \tag{2.41}$$

where E_i = initial potential
t = time
v = rate of potential change or sweep rate (V/s)

The reverse sweep of the cycle is defined by

$$E = E_i + v't \tag{2.42}$$

where v' is often the same value as v. By combining Eq. (2.42) with the appropriate form of the Nernst equation [Eq. (2.6)] and with Fick's laws of diffusion [Eqs. (2.34) and (2.35)], an expression can be derived which describes the flux of species to the electrode surface. This expression is a complex differential equation and can be solved by the summation of an integral in small successive increments.[14–16]

As the applied voltage approaches that of the reversible potential for the electrode process, a small current flows, the magnitude of which increases rapidly but later becomes limited at a potential slightly beyond the standard potential by the subsequent depletion of reactants. This depletion of reactants establishes concentration profiles which spread out into the solution, as shown in Fig. 2.18. As the concentation profiles extend into solution, the rate of diffusive transport at the electrode surface decreases and with it the observed current. The current is thus seen to pass through a well-defined maximum, as illustrated in Fig. 2.19. The peak current of the reversible reduction [Eq. (2.9)] is defined by

$$i_p = \frac{0.447 \, F^{3/2} \, An^{3/2} \, D^{1/2} \, C_O \, v^{1/2}}{R^{1/2} \, T^{1/2}} \qquad (2.43)$$

The symbols have the same identity as before while i_p is the peak current and A the electrode area. It may be noted that the value of the constant varies slightly from one text or publication to another. This is because, as mentioned previously, the derivation of peak current height is performed numerically.

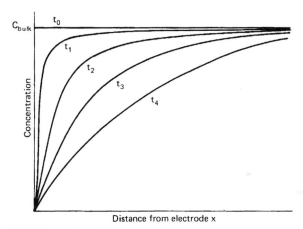

FIGURE 2.18   Concentration profiles for reduction of a species in cyclic voltammetry, $t_4 > t_0$.

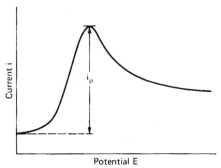

FIGURE 2.19   Cyclic voltammetry peak current for reversible reduction of an electroactive species.

A word of caution is due regarding the interpretation of the value of the peak current. It will be remembered from the discussion of the effects of the electrical double layer on electrode kinetics that there is a capacitance effect at an electrode-electrolyte interface. Consequently the "true" electrode potential is modified by the capacitance effect as it is also by the ohmic resistance of the solution. Equation (2.41) should really be written in a form which described these two components. Equation (2.44) shows such a modification,

$$E = E_i - vt + r(i_f + i_c) \qquad (2.44)$$

where r = cell resistance
i_f = faradic current
i_c = capacity current

At small values of voltage sweep rate, typically below 1 mV/s, the capacity effects are small and in most cases can be ignored. At greater values of sweep rate, a correction needs to be applied to interpretations of i_p, as described by Nicholson and Shain.[17] With regard to the correction for ohmic drop in solution, typically this can be handled adequately by careful cell design and positive feedback compensation circuitry in the electronic instrumentation.

Cyclic voltammetry provides both qualitative and quantitative information on electrode processes. A reversible, diffusion-controlled reaction such as presented by Eq. (2.9) exhibits an approximately symmetrical pair of current peaks, as shown in Fig. 2.20. The voltage separation ΔE of these peaks is

$$\Delta E = \frac{2.3\, RT}{nF} \qquad (2.45)$$

and the value is independent of the voltage sweep rate. In the case of the electrodeposition of an insoluble film, which can be, subsequently, reversibly reoxidized and which is not governed by diffusion to and from the electrode surface, the value of ΔE is considerably less than that given by Eq. (2.45), as shown in Fig. 2.21. In the ideal case, the value of ΔE for this system is close to zero. For quasi-reversible processes, the current peaks are more separated, and the shape of the peak is less sharp at its summit and is generally more rounded, as shown in Fig. 2.22. The voltage of the current peak is dependent on the voltage sweep rate, and the voltage separation is much greater than that given by Eq. (2.45). A completely

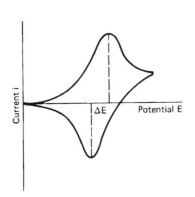

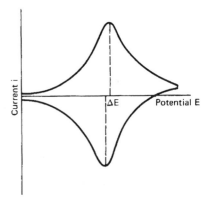

FIGURE 2.20 Cyclic voltammogram of a reversible diffusion-controlled process.

FIGURE 2.21 Cyclic voltammogram of electroreduction and reoxidation of a deposited, insoluble film.

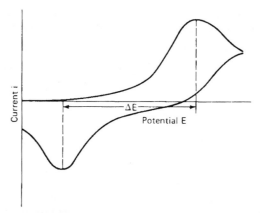

FIGURE 2.22 Cyclic voltammogram of a quasi-reversible process.

irreversible electrode process produces a single peak, as shown in Fig. 2.23. Again the voltage of the peak current is sweep-rate dependent, and, in the case of an irreversible charge-transfer process for which the back reaction is negligible, the rate constant and transfer coefficient can be determined. With negligible back reaction, the expression for peak current as a function of peak potential is[17]

$$i_p = 0.22nFC_O\, k_{app} \exp\left[-\alpha\, \frac{nF}{RT}(E_m - E^o)\right] \tag{2.46}$$

where the symbols are as before and E_m is the potential of the current peak. A plot of E_m versus $\ln i_p$, for different values of concentration, gives a slope which yields the transfer coefficient α and an intercept which yields the apparent rate constant k_{app}. Though both α and k_{app} can be obtained by analyzing E_m as a function of voltage sweep rate v by a reiterative calculation, analysis by Eq. (2.46) (which is independent of v) is much more convenient.

For more complex electrode processes, cyclic voltammetric traces become more complicated to analyze. An example of one such case is the electroreduction of a species controlled by a preceding chemical reaction. The shape of the trace for this process is shown in Fig. 2.24. The species is formed at a constant rate at the electrode surface and, provided the diffusion of the inactive component is more rapid than its transformation to the active form, it cannot be depleted from the electrode surface. The "peak" current is thus independent of potential and resembles a plateau.

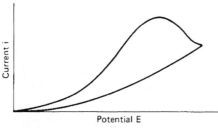

FIGURE 2.23 Cyclic voltammogram of an irreversible process.

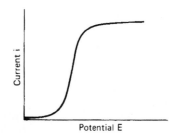

FIGURE 2.24 Cyclic voltammogram of electroreduction of a species controlled by a preceding chemical reaction.

Cyclic voltammograms of electrochemical systems can often be much more complicated than the traces presented here. It often takes some ingenuity and persistence to determine which peaks belong to which species or processes. Despite these minor drawbacks, the cyclic voltammetric technique is a versatile, and relatively sensitive, electroanalytical method appropriate to the analysis of systems of interest to battery development. The technique will identify reversible couples (desirable for secondary batteries), it provides a method for measuring the rate constant and transfer coefficient of an electrode process (a fast rate constant indicates a process of possible interest for battery development), and it can provide a tool to help unravel complex electrochemical systems.

2.6.2 Chronopotentiometry

Chronopotentiometry involves the study of voltage transients at an electrode upon which is imposed a constant current. It is sometimes alternately known as galvanostatic voltammetry. In this technique, a constant current is applied to an electrode, and its voltage response indicates the changes in electrode processes occurring at its interface. Consider, for example, the reduction of a species O as expressed by Eq. (2.9). As the constant current is passed through the system, the concentration of O in the vicinity of the electrode surface begins to decrease. As a result of this depletion, O diffuses from the bulk solution into the depleted layer, and a concentration gradient grows out from the electrode surface into the solution. As the electrode process continues, the concentration profile extends further into the bulk solution as shown in Fig. 2.25. When the surface concentration of O falls to zero (at time t_6 in Fig. 2.25), the electrode process can no longer be supported by electroreduction of O. An additional cathodic reaction must be brought into play and an abrupt change in potential occurs. The period of time between the commencement of electoreduction and the sudden change in potential is called the transition time τ. The transition time for electroreduction of a species in the presence of excess supporting electrolyte was first quantified by Sand,[18] who showed that the transition time τ was related into the diffusion coefficient of the electroactive species,

$$\tau^{1/2} = \frac{\pi^{1/2} n F C_O D^{1/2}}{2i} \tag{2.47}$$

where D is the diffusion coefficient of species O and the other symbols have their usual meanings.

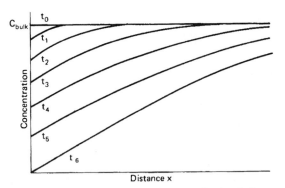

FIGURE 2.25   Concentration profiles extending into bulk solution during constant-current depletion of species at an electrode surface, $t_6 > t_0$.

Unlike cyclic voltammetry, the solution of Fick's diffusion equations [Eqs. (2.34) and (2.35)] for chronopotentiometry can be obtained as an exact expression by applying appropriate boundary conditions. For a reversible reduction of an electroactive species [Eq. (2.9)], the potential-time relationship has been derived by Delahay[19] for the case where O and R are free to diffuse to and from the electrode surface, including the case where R diffuses into a mercury electrode,

$$E = E_{\tau/4} + \frac{RT}{nF} \ln \frac{\tau^{1/2} - t^{1/2}}{t^{1/2}} \tag{2.48}$$

In this equation $E_{\tau/4}$ is the potential at the one-quarter transition time (the same as the polarographic half-wave potential in the case of a mercury electrode) and t is any time from zero to the transition time. The trace represented by this expression is shown in Fig. 2.26.

The corresponding expression for an irreversible process[20] with one rate-determining step is

$$E = \frac{RT}{\alpha n_a F} \ln \frac{nFC_O k_{app}}{i} + \frac{RT}{\alpha n_a F} \ln \left[1 - \left(\frac{t}{\tau} \right)^{1/2} \right] \tag{2.49}$$

where k_{app} is the apparent rate constant, n_a is the number of electrons involved in the rate-determining step (often the same as n, the overall number of electrons involved in the total reaction), and the other symbols have their usual meanings. A plot of the logarithmic term versus potential yields both the transfer coefficient and the apparent rate constant.

In a practical system, the chronopotentiogram is often less than ideal in the shape of the potential trace. To accommodate variations in chronopotentiometric traces, measurement of the transition time can be assisted by use of a construction technique, as shown in Fig. 2.27. The transition time is measured at the potential of $E_{\tau/4}$.

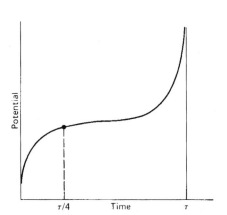

FIGURE 2.26 Potential curve at constant current for reversible reduction of an electroactive species.

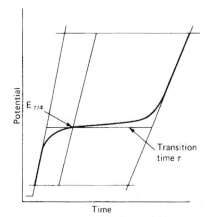

FIGURE 2.27 Construction of transition time τ for a chronopotentiogram.

To analyze two or more independent reactions separated by a potential sufficient to define individual transition times, the situation is slightly more complicated than with cyclic voltammetry. Analysis of the transition time of the reduction of the nth species has been derived elsewhere[21,22] and is

$$(\tau_1 + \tau_2 + \cdots + \tau_n)^{1/2} - (\tau_1 + \tau_2 + \cdots + \tau_{n-1})^{1/2} = \frac{\pi^{1/2}nFD_n^{1/2}C_n}{2i} \qquad (2.50)$$

As can be seen, this expression is somewhat cumbersome.

An advantage of the technique is that it can be used conveniently to evaluate systems with high resistance. The trace conveniently displays segments due to the IR component, the charging of the double layer, and the onset of the faradaic process. Figure 2.28 shows these different features of the chronopotentiogram of solutions with significant resistance. If the solution is also one which does not contain an excess of supporting electrolyte to suppress the migration current, it is possible to describe the transition time of an electroreduction process in terms of the transport number of the electroactive species[23,24]

$$\tau^{1/2} = \frac{\pi^{1/2}nFC_OD_s^{1/2}}{2i(1 - t_O)} \qquad (2.51)$$

where D_s is the diffusion coefficient of the salt (not the ion) and t_O the transport number of the electroactive species. This expression can be useful in battery research since many battery systems do not have supporting electrolyte.

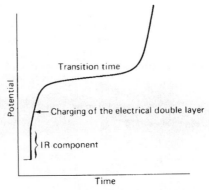

FIGURE 2.28   Chronopotentiogram of a system with significant resistance.

2.6.3 Electrochemical Impedance Spectroscopy (EIS) Methods

The two preceding electroanalytical techniques, one in which the measured value was the current during imposition of a potential scan and the other a potential response under an imposed constant current, owe their electrical response to the change in impedance at the electrode-electrolyte interface. A more direct technique for studying electrode processes is to measure the change in the electrical impedance of an electrode by electrochemical impedance spectroscopy (EIS). To relate the impedance of the electrode-electrolyte interface to electrochemical parameters, it is necessary to establish an equivalent circuit to represent the dynamic characteristics of the interface.

Establishment of equivalent circuits attempting to describe electrode processes dates to the turn of the century with the derivation by Warburg[25] of an equation for the faradaic impedance of diffusion-controlled processes at a planar electrode. In fact, the impedance of the diffusion process is often referred to as the "Warburg impedance." In 1903, Krüger[26] realized that the double-layer capacity had influence on the impedance of the electrode interface and derived an expression for its effect. Much later the technique was developed and adapted to the study of electrode kinetics,[27–31] with emphasis on the charge-transfer process. The technique has been found to be extremely useful in evaluating several different electrode processes at a planar electrode and has been analyzed by several authors. The charge-transfer step has been considered and analyzed by Randles,[28] Grahame,[32] Delahay,[33] Baticle and Perdu[34,35] and Sluyters-Rehbach and Sluyters,[36] while Gerischer[37,38] and Barker[39] considered coupled homogeneous and heterogeneous chemical reactions. Adsorption processes can be studied by this technique, and the description of the method used to measure adsorption has been given by Laitinen and Randles,[40] Llopis et al.,[41] Senda and Delahay,[42] Sluyters-Rehbach et al.,[43] Timmer et al.,[44] Barker,[39] and Holub et al.[45] An excellent EIS publication has been written by J. R. MacDonald[46] that analyses the solid/solution interface by equivalent circuitry to calculate the interfacial resistance, capacitance, and inductance and relates these to reaction mechanisms.

The expressions defining the various electrochemical parameters are relatively straightforward to derive but are complex in format, in particular when capacity effects are considered in the presence of strongly adsorbed electroactive species. Derivation of expressions relevant to the previously mentioned processes will not be given here. Only one of the more applicable analyses, and the one which is the most straightforward to handle, will be presented. The problem is that the majority of the processes require a transmission-line analysis to give a closed-system solution for a nonplanar electrode. We shall consider the system without adsorption and without complications of homogeneous series reactions where the impedance can be represented by a circuit diagram as shown in Fig. 2.29. In this analysis, due to Sluyters and coworkers, the electrode process is evaluated by the analytical technique of complex plane analysis. Here the capacitive component $1/\omega C$ is plotted versus the resistive component of the cell. Figure 2.30 shows a typical plot, which displays kinetic control only. The interdependence of the capacitive and resistive components yields a semicircle, with the top yielding the charge-transfer resistance r_{ct},

$$\omega_m = \frac{1}{r_{ct}C_{nf}} \tag{2.52}$$

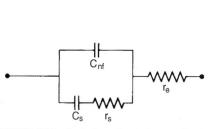

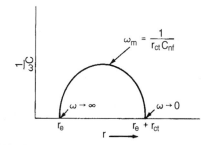

FIGURE 2.29 Equivalent circuit for a cell where the cell impedance is kinetically controlled and is localized at the working electrode by using a large, unpolarized countrelectrode. C_{nf}—nonfaradaic capacitance; C_s, r_s—faradaic components of impedance; r_e—electrolyte resistance.

FIGURE 2.30 Complex plane analysis of cell impedance for a charge-transfer process with kinetic control at a planar electrode.

where C_{nf} is the nonfaradaic capacitance and $r_{ct} = RT/nF(i_O)_{app}$. The intercept of the semi-circle and the abscissa gives the electrolyte resistance r_e and again r_{ct}. If the electrode process is governed by both kinetic and diffusion control, a somewhat different plot is observed, as shown in Fig. 2.31. In this plot, the linear portion of the curve corresponds to the process where diffusion control is predominant. From this plot, in addition to the previously mentioned measurement, the extrapolated linear portion gives a somewhat complex expression involving the diffusion coefficients of the oxidized and reduced species,

$$\text{Intercept} = r_e + 2S^2 C_{nf} \tag{2.53}$$

where

$$S = \frac{RTL}{n^2 \, F^2 \sqrt{2}} \tag{2.54}$$

and

$$L = \frac{1}{C_O \sqrt{D_O}} + \frac{1}{C_R \sqrt{D_R}} \tag{2.55}$$

D_O being the diffusion coefficient of the oxidized species and D_R that of the reduced species. Treatment of this system assumes that we can write the equivalent circuit of kinetic and diffusion control as shown in Fig. 2.32, where the diffusion component of the impedance is given by the Warburg impedance W. It should also be noted that the derivation applies to a planar electrode only. Electrodes with more complex geometries such as porous electrodes require a transmission-line analysis.

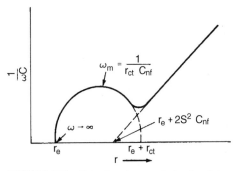

FIGURE 2.31 Complex plane analysis of cell impedance for a charge-transfer process with both kinetic and diffusion control at a planar electrode.

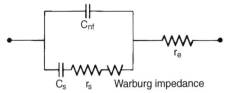

FIGURE 2.32 Equivalent circuit for an electrode process limited by both charge-transfer kinetics and diffusion processes. The diffusion portion of the impedance is represented by the Warburg impedance, the other circuit components are the same as in Fig. 2.29.

The AC impedance technique coupled to the complex plane method of analysis is a powerful tool to determine a variety of electrochemical parameters. To make the measurements, instrumentation is somewhat more complex than with other techniques. It requires a Wheatstone bridge arrangement with series capacitance and resistance in the comparison arm, a tuned amplifier/detector, and an oscillator with an isolation transformer. A Wagner* ground is required to maintain bridge sensitivity, and a suitably large inductance should be incorporated in the electrode polarization circuit to prevent interference from the low impedance of this ancillary circuitry. Sophisticated measurement instruments or frequency response analyzers with frequency sweep and computer interface are currently available such as the Solartron frequency response analyzers. Data obtained can be analyzed or fitted into proper equivalent circuit using appropriate software.

This technique has found considerable acceptance in measuring fast rate constants, and it has been extended to include faradaic rectification and second-harmonic generation and detection. These advanced techniques extend the range of AC impedance measurements to the evaluation of charge-transfer rate constants with values of 10 cm/s or greater. Unfortunately the description of these techniques is beyond the scope of this chapter, but can be found in advanced texts on electrochemical methods.[11] In addition to applications in electrode kinetics, the technique has been used to study the corrosion and passivation of metals,[47–49] to characterize battery electrode reactions, as well as for a simple nondestructive check of the state of charge of batteries.[50–53] This method is therefore finding increasing utility in the study of battery systems. EIS is a powerful tool to evaluate battery electrode materials and has recently been shown to describe metal hydride anodes[54] and lithium intercalation anodes[65] and provide rate constants for the respective electrode processes.

2.6.4 Polarography

The technique of polarographic analysis is one of the most widely employed and the electroanalytical technique with the longest history of use. Polarography[56–59] utilizes a dropping mercury electrode (DME) as the sensing electrode to which a slowly changing potential is applied, usually increasing in the negative direction, such that for each mercury drop, the potential remains essentially constant. The output current from the DME is typically displayed on a chart recorder, and the current magnitude is a measure of the concentration of electroactive species in solution. Because of the periodic nature of the DME, the current-measuring circuit includes a damping capacitor with a time constant such that the oscillations in current due to the charging current of the DME are minimized. Figure 2.33 shows a typical polarogram of an electroactive species, where i_m is the polarographic mean diffusion current. The expression relating the concentration of electroactive species to the mean diffusion current is given by the Ilkovic equation.[60,61]

$$I_m = 607nD^{1/2}Cm^{2/3}t_d^{1/6} \tag{2.56}$$

where i_m = mean diffusion current, μA
$\qquad n$ = number of electrons involved in overall electrode process
$\qquad D$ = diffusion coefficient of electroactive species
$\qquad C$ = concentration of electroactive species, mmol/L
$\qquad m$ = mercury flow rate, mg/s
$\qquad t_d$ = mercury drop time, s

---

*A Wagner ground maintains a corner of a bridge at ground potential without actually connecting it directly to ground. This helps to eliminate stray capacitances to ground and maintains bridge sensitivity over a wide range of frequencies.

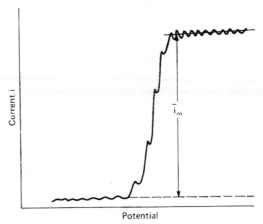

FIGURE 2.33 Polarogram of an electroactive species. i_m—polarographic mean diffusion current.

Analysis of the solutions usually is performed by calibrations with known standards rather than by using Eq. (2.56). Occasionally the polarogram of a solution may show a current spike at the beginning of the current plateau. This effect is due to a streaming phenomenon around the DME and can be suppressed by the addition of a *small* quantity of a surface-active compound such as gelatin or Triton X-100.

Solutions containing several electroactive species can be conveniently analyzed from one polarogram, provided the potential separation, the half-wave potential,* is sufficient to distinguish the current plateaus. Figure 2.34 shows a polarogram of an aqueous solution containing several species in a potassium chloride supporting electrolyte. The polarographic method is often found as a standard instrumental technique in electrochemical laboratories.

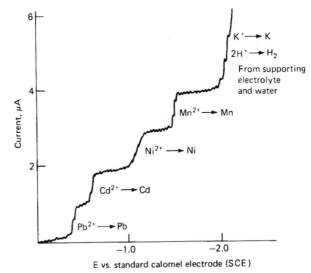

FIGURE 2.34 Polarogram of reduction of several electroactive species in a solution of $0.1M$ potassium chloride as supporting electrolyte.

---

*The half-wave potential ($E_{1/2}$) is the electrode potential of the DME at which the value of i is one-half of its limiting value (i_m).

2.6.5   Electrodes

Several electroanalytical techniques have been discussed in the preceding sections, but little was mentioned about electrodes or electrode geometries used in the various measurements. This section deals with electrodes and electrode systems.

A typical electroanalytical cell has an indicator electrode (sometimes called a working electrode), a counter electrode, and, in most cases, a reference electrode. The counter electrode is a large, inert, relatively unpolarized electrode situated a suitable distance from the indicator electrode and sometimes separated from it by a sintered glass disk or some other porous medium which will allow ionic conduction but prevent gross mixing of solutions surrounding the counter and working electrodes. The reference electrode provides an unpolarized reference potential against which the potentials of the indicator electrode can be measured. Reference electrodes are usually constructed in a separate vessel and are connected to the cell via a salt bridge composed of an electrolyte with a common anion or cation with the supporting electrolyte solution in the cell. The concentration of the salt bridge electrolyte is also chosen to be approximately the same as the supporting electrolyte. This minimizes liquid junction potentials which are established between solutions of different composition or concentration. Contamination of the cell solution with material from the salt bridge is minimized by restricting the orifice of the salt bridge by a Luggin capillary. Typical reference electrode systems are Ag/AgCl, Hg/Hg$_2$Cl$_2$, and Hg/HgO. For a comprehensive treatise on the subject of reference electrodes, the reader is referred to an excellent text by Ives and Janz.[62]

Indicator electrodes have been designed and fabricated in many different geometrical shapes to provide various performance characteristics and to accommodate the special requirements of some of the electroanalytical techniques. The simplest of the indicator electrode types is the planar electrode. This can be simply a "flag" electrode, or it can be shielded to provide for linear diffusion during long electrolysis times. Figure 2.35 shows a selection of planar-type indicator electrodes. Included in this selection is a thin-film cell (Fig. 2.35e) in which two planar electrodes confine a small quantity of solution within a small gap between the working and counter electrodes. The counter electrode is adjusted by a micrometer drive, and the reference salt bridge is located, as shown, as a radial extension of the cell. This cell has the advantage that only a small quantity of solution is required, and in many cases it can be used as a coulometer to determine n, the overall number of electrons involved in an electrode reaction.

Indicator electrodes not requiring a planar surface can be fabricated in many different geometries. For simplicity and convenience of construction, a wire electrode can be sealed into an appropriate glass support and used with little more than a thorough cleaning. For more symmetrical geometry, a spherical bead electrode can often be formed by fusing a thin wire of metal and shriveling it into a bead. In some cases, is possible to form a single crystal of the metal. Figure 2.36 shows various electrodes with nonplanar electrode-electrolyte interfaces.

Several electroanalytical methods require special electrodes or electrode systems. A classic example is the DME, first used extensively in polarography (Sec. 2.6.4). The DME is established by flowing ultrapure mercury through a precision capillary from a mercury reservoir connected to the capillary by a flexible tube. Mercury flow rates and drop times are adjusted by altering the height of the mercury head so that mercury drops detach every 2 to 7 s, with typical polarographic values being in the range 3 to 5 s. Figure 2.37 shows a schematic representation of a DME system. In addition to use of the DME in polarography, the electrode finds application when a fresh, clean mercury surface is necessary for reproducible electroanalytical determinations. In this respect, the DME can be used as a "stationary" spherical mercury electrode suitably "frozen" in time by a timing circuit initiated by the voltage pulse at the birth of the drop and triggered at a time short of the natural drop time of the DME. With a suitably short transient impulse (voltage, current, and so on) an output response can be obtained from what is essentially a hanging mercury drop electrode.

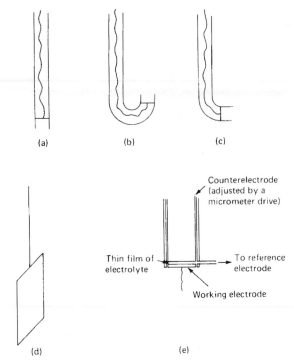

FIGURE 2.35 Various planar working electrodes, some with shielding to maintain laminar diffusion of electroactive species to surface.

Other special types of electrodes are the rotating disk electrode and the rotating ring-disk electrode. Figure 2.38 shows both rotating disk and ring-disk electrodes. In both cases, rotation of the electrode establishes a flow pattern which maintains a relatively constant diffusion layer thickness during reduction or oxidation of an electroactive species and provides a means of hydrodynamically varying the rate at which electroactive species are brought to the outer surface of the diffusion layer. In the ring-disk electrode, the ring can be used either as a "guard ring" to ensure laminar diffusion to the disk or as an independent working electrode to monitor species generated from the disk electrode or the generation of transient intermediate species. Details of the derivation of current-voltage curves as a function of rotation speed have been given by Riddiford[63] for the disk electrode, while more detail is given by Yeager and Kuta[64] and Pleskov and Filinovski[65] on the ring-disk electrode, with emphasis being given by the latter authors to the detection of transient species. Details of the current-voltage characteristics are not given here. It is sufficient to say that limiting currents for a fixed potential can be obtained by varying the rotation speed of the electrode from which potential-dependent rate constants can be determined. Ring currents may be determined either at a fixed potential such as to reverse the process leading to the formation of the species of interest or scanning through a range of potentials to detect a variety of species electroactive over a range of potentials. Quantitative measurements with the ring of the ring-disk electrode are difficult to interpret because the capture fraction N of the ring is difficult to evaluate.[66]

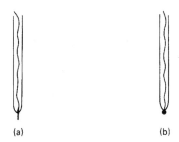

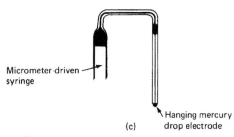

FIGURE 2.36   Working electrodes with nonplanar surfaces.

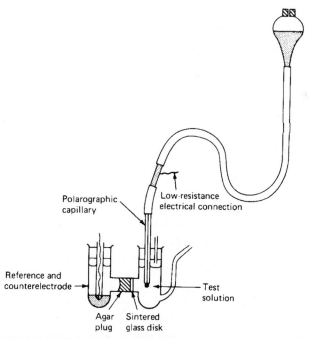

FIGURE 2.37   Dropping mercury electrode complete with polarographic H cell.

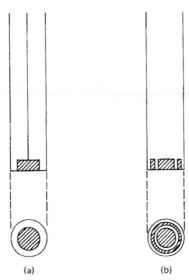

FIGURE 2.38 Rotating electrodes. (*a*) Disk electrode. (*b*) Ring-disk electrode.

REFERENCES

1. J. O'M. Bockris and A. K. N. Reddy, *Modern Electrochemistry,* vol. 2, Plenum, New York, 1970, p. 644.
2. H. H. Bauer, *J. Electroanal. Chem.* **16:**419 (1968).
3. J. Horiuti and M. Polanyi, *Acta Physicochim. U.S.S.R.* **2:**505 (1935).
4. J. O'M. Bockris and A. K. N. Reddy, op. cit., p. 918.
5. P. Delahay, *Double Layer and Electrode Kinetics,* Interscience, New York, 1965.
6. J. O'M. Bockris and A. K. N. Reddy, op. cit., p. 742.
7. H. H. Bauer, *Electrodics,* Wiley, New York, 1972, p. 54, table 3.2.
8. A. Fick, *Ann. Phys.* **94:**59 (1855).
9. P. Delahay, *New Instrumental Methods in Electrochemistry,* Interscience, New York, 1954.
10. J. S. Newman, *Electrochemical Systems,* 2d ed., Prentice-Hall, Englewood Cliffs, N.J., 1991.
11. E. B. Yeager and J. Kuta, "Techniques for the Study of Electrode Processes," in *Physical Chemistry,* vol. IXA: *Electrochemistry,* Academic, New York, 1970, p. 346.
12. L. A. Matheson and N. Nichols, *J. Electrochem. Soc.* **73:**193 (1938).
13. J. E. B. Randles, *Trans. Faraday Soc.* **44:**327 (1948).
14. A. Sevcik, *Coll. Czech, Chem. Comm.* **13:**349 (1948).
15. T. Berzins and P. Delahay, *J. Am. Chem. Soc.* **75:**555 (1953).
16. P. Delahay, *J. Am. Chem. Soc.* **75:**1190 (1953).
17. R. S. Nicholson and I. Shain, *Anal. Chem.* **36:**706 (1964).
18. H. J. S. Sand, *Phil. Mag.* **1:**45 (1901).
19. P. Delahay, *New Instrumental Methods in Electrochemistry,* op. cit., p. 180.
20. P. Delahay and T. Berzins, *J. Am. Chem. Soc.* **75:**2486 (1953).
21. C. N. Reilley, G. W. Everett, and R. H. Johns, *Anal. Chem.* **27:**483 (1955).

22. T. Kambara and I. Tachi, *J. Phys. Chem.* **61:**405 (1957).

23. M. D. Morris and J. J. Lingane, *J. Electroanal. Chem.* **6:**300 (1963).

24. J. Broadhead and G. J. Hills, *J. Electroanal. Chem.* **13:**354 (1967).

25. E. Warburg, *Ann. Phys.* **67:**493 (1899).

26. F. Krüger, *Z. Phys. Chem.* **45:**1 (1903).

27. P. Dolin and B. V. Ershler, *Acta Physicochim. U.S.S.R.* **13:**747 (1940).

28. J. E. B. Randles, *Disc. Faraday Soc.* **1:**11 (1947).

29. B. V. Ershler, *Disc. Faraday Soc.* **1:**269 (1947).

30. B. V. Ershler, *Zh. Fiz. Khim.* **22:**683 (1948).

31. K. Rozental and B. V. Ershler, *Zh. Fiz. Khim.* **22:**1344 (1948).

32. D. C. Grahame, *J. Electrochem. Soc.* **99:**370C.

33. P. Delahay, *New Instrumental Methods in Electrochemistry,* op. cit., p. 146.

34. A. M. Baticle and F. Perdu, *J. Electroanal. Chem.* **12:**15 (1966).

35. A. M. Baticle and F. Perdu, *J. Electroanal. Chem.* **13:**364 (1967).

36. M. Sluyters-Rehbach and J. H. Sluyters, *Rec. Trav. Chim.* **82:**525, 535 (1963).

37. H. Gerischer, *Z. Phys. Chem.* **198:**286 (1951).

38. H. Gerischer, *Z. Phys. Chem.* **201:**55 (1952).

39. G. C. Barker, in E. B. Yeager (ed.), *Trans. Symp. Electrode Processes* (Philadelphia, 1959), Wiley, New York, p. 325.

40. H. A. Laitinen and J. E. B. Randles, *Trans. Faraday Soc.* **51:**54 (1955).

41. J. Llopis, J. Fernandez-Biarge, and M. Perez-Fernandez, *Electrochim. Acta* **1:**130 (1959).

42. M. Senda and P. Delahay, *J. Phys. Chem.* **65:**1580 (1961).

43. M. Sluyters-Rehbach, B. Timmer, and J. H. Sluyters, *J. Electroanal. Chem.* **15:**151 (1967).

44. B. Timmer, M. Sluyters-Rehbach, and J. H. Sluyters, *J. Electroanal. chem.* **15:**343 (1967).

45. K. Holub, G. Tesari, and P. Delahay, *J. Phys. Chem.* **71:**2612 (1967).

46. J. R. MacDonald, *Impedance Spectroscopy, Emphasizing Solid Materials and Systems,* Wiley, New York, 1987, pp. 154–155.

47. F. Mansfeld, *Advances in Corrosion Science and Technology,* vol. 6, Plenum, New York, p. 163.

48. I. Epelboin, C. Gabrielli, M. Keddam, and H. Takenouti, in F. Mansfeld and U. Bertocci (eds.), *Proc. ASTM Symp. on Progress in Electrochemical Corrosion Testing* (May 1979), p. 150.

49. S. Haruyama and T. Tsuru, in R. P. Frankenthal and J. Kruger (eds.), *Passivity of Metals,* Electrochemical Soc., Princeton, New Jersey, p. 564.

50. N. A. Hampson, S. A. G. R. Karunathlaka, and R. Leek, *J. Appl. Electrochem.* **10:**3 (1980).

51. S. Sathyanarayana, S. Venugopalan, and M. L. Gopikanth, *J. Appl. Electrochem.* **9:**129 (1979).

52. S. A. G. R. Karunathlaka, N. A. Hampson, R. Leek, and T. J. Sinclair, *J. Appl. Electrochem.* **10:** 357 (1980).

53. M. L. Gopikanth and S. Sathyanarayana, *J. Appl. Electrochem.* **9:**369 (1979).

54. C. Wang, "Kinetic Behavior of Metal Hydride Electrode by Means of AC Impedance," *J. Electrochem. Soc.* **145:**1801 (1998).

55. M. D. Levi, K. Gamolsky, D. Aurbach, U. Heider, and R. Oesten, "On Electrochemical Impedance Measurements of $Li_xCo_{0.2}Ni_{0.8}O_2$ and Li_xNiO_2 Intercalation Electrodes," *Electrochem. Acta* **45:**1781 (2000).

56. I. M. Kolthoff and J. J. Lingane, *Polarography,* 2d ed., Interscience, New York, 1952.

57. J. Heyrovsky and J. Kuta, *Principles of Polarography,* Czechoslovak Academy of Sciences, Prague, 1965.

58. L. Meites, *Polarographic Techniques,* 2d ed., Interscience, New York, 1965.

59. G. W. C. Milner, *The Principles and Applications of Polarography and Other Electroanalytical Processes,* Longmans, London, 1957.

60. D. Ilkovic, *Coll. Czech. Chem. Comm.* **6:**498 (1934).

61. D. Ilkovic, *J. Chim. Phys.* **35:**129 (1938).

62. D. J. G. Ives and G. J. Janz, *Reference Electrodes, Theory and Practice.* Academic, New York, 1961.

63. A. C. Riddiford, "The Rotating Disc System," in P. Delahay and C. W. Tobias (eds.), *Advances in Electrochemistry and Electrochemical Engineering,* Interscience, New York, 1966, p. 47.

64. E. B. Yeager and J. Kuta, op. cit., p. 367.

65. Y. V. Pleskov and V. Y. Filinovskii, in H. S. Wroblowa and B. E. Conway (eds.), *The Rotating Disc Electrode,* Consultants Bureau, New York, 1976, chap. 8.

66. W. J. Albery and S. Bruckenstein, *Trans. Faraday Soc.* **62:**1920, 1946 (1966).

BIBLIOGRAPHY

General

Bauer, H. H.: *Electrodics,* Wiley, New York, 1972.

Bockris, J. O'M., and A. K. N. Reddy: *Modern Electrochemistry,* vols. 1 and 2, Plenum, New York, 1970.

Conway, B. E.: *Theory and Principles of Electrode Processes,* Ronald Press, New York, 1965.

Gileadi, E., E. Kirowa-Eisner, and J. Penciner: *Interfacial Electrochemistry,* Addison-Wesley, Reading, Mass., 1975.

Newman, J. S.: *Electrochemical Systems,* 2d ed., Prentice-Hall, Englewood Cliffs, N.J., 1991.

Sawyer, D. T., and J. L. Roberts: *Experimental Electrochemistry for Chemists,* Wiley, New York, 1974.

Transfer Coefficient

Bauer, H. H.: *J. Electroanal. Chem.* **16:**419 (1968).

Electrical Double Layer

Delahay, P.: *Double Layer and Electrode Kinetics,* Interscience, New York, 1965.

Electroanalytical Techniques

Delahay, P.: *New Instrumental Methods in Electrochemistry,* Interscience, New York, 1954.

Yeager, E. B., and J. Kuta: "Techniques for the Study of Electrode Processes," in *Physical Chemistry,* vol. IXA: *Electrochemistry,* Academic, New York, 1970.

Polarography

Heyrovsky, J., and J. Kuta: *Principles of Polarography,* Czechoslovak Academy of Sciences, Prague, 1965.

Kolthoff, I. M., and J. J. Lingane: *Polarography,* 2d ed., Interscience, New York, 1952.

Meites, L.: *Polarographic Techniques,* 2d ed., Interscience, New York, 1965.

Milner, G. W. C.: *The Principles and Applications of Polarography and Other Electroanalytical Processes,* Longmans, London, 1957.

Reference Electrodes

Ives, D. J. G., and G. J. Janz: *Reference Electrodes, Theory and Practice,* Academic, New York, 1961.

Electrochemistry of the Elements

Bard, A. J. (ed.): *Encyclopedia of Electrochemistry of the Elements,* vols. I–XIII, Dekker, New York, 1979.

Organic Electrode Reactions

Meites, L., and P. Zuman: *Electrochemical Data,* Wiley, New York, 1974.

AC Impedance Techniques

Gabrielli, G.: *Identification of Electrochemical Processes by Frequency Response Analysis,* Tech. Rep. 004/83, Solartron Instruments, Billerica, Mass., 1984.

Macdonald, J. R. (ed.): *Impedance Spectroscopy,* Wiley, New York, 1987.

CHAPTER 3
FACTORS AFFECTING BATTERY PERFORMANCE

David Linden

3.1 GENERAL CHARACTERISTICS

The specific energy of a number of battery systems was listed in Table 1.2 These values are based on optimal designs and discharge conditions. While these values can be helpful to characterize the energy output of each battery system, the performance of the battery may be significantly different under actual conditions of use, particularly if the battery is discharged under more stringent conditions than those under which it was characterized. The performance of the battery under the specific conditions of use should be obtained before any final comparisons or judgments are made.

3.2 FACTORS AFFECTING BATTERY PERFORMANCE

Many factors influence the operational characteristics, capacity, energy output and performance of a battery. The effect of these factors on battery performance is discussed in this section. It should be noted that because of the many possible interactions, these effects can be presented only as generalizations and that the influence of each factor is usually greater under the more stringent operating conditions. For example, the effect of storage is more pronounced not only with high storage temperatures and long storage periods, but also under more severe conditions of discharge following storage. After a given storage period, the observed loss of capacity (compared with a fresh battery) will usually be greater under heavy discharge loads than under light discharge loads. Similarly, the observed loss of capacity at low temperatures (compared with normal temperature discharges) will be greater at heavy than at light or moderate discharge loads. Specifications and standards for batteries usually list the specific test or operational conditions on which the standards are based because of the influence of these conditions on battery performance.

Furthermore it should be noted that even within a given cell or battery design, there will be performance differences from manufacturer to manufacturer and between different versions of the same battery (such as standard, heavy-duty, or premium). There are also performance variables within a production lot, and from production lot to production lot, that are inherent in any manufacturing process. The extent of the variability depends on the process controls as well as on the application and use of the battery. Manufacturers' data should be consulted to obtain specific performance characteristics.

3.2.1 Voltage Level

Different references are made to the voltage of a cell or battery:

1. The *theoretical voltage* is a function of the anode and cathode materials, the composition of the electrolyte and the temperature (usually stated at 25°C).
2. The *open-circuit voltage* is the voltage under a no-load condition and is usually a close approximation of the theoretical voltage.
3. The *closed-circuit voltage* is the voltage under a load condition.
4. The *nominal voltage* is one that is generally accepted as typical of the operating voltage of the battery as, for example, 1.5 V for a zinc-manganese dioxide battery.
5. The *working voltage* is more representative of the actual operating voltage of the battery under load and will be lower than the open-circuit voltage.
6. The *average voltage* is the voltage averaged during the discharge.
7. The *midpoint voltage* is the central voltage during the discharge of the cell or battery.
8. The *end* or *cut-off voltage* is designated as the end of the discharge. Usually it is the voltage above which most of the capacity of the cell or battery has been delivered. The end voltage may also be dependent on the application requirements.

Using the lead-acid battery as an example, the theoretical and open-circuit voltages are 2.1 V, the nominal voltage is 2.0 V, the working voltage is between 1.8 and 2.0 V, and the end voltage is typically 1.75 V on moderate and low-drain discharges and 1.5 V for engine-cranking loads. On charge, the voltage may range from 2.3 to 2.8 V.

When a cell or battery is discharged its voltage is lower than the theoretical voltage. The difference is caused by IR losses due to cell (and battery) resistance and polarization of the active materials during discharge. This is illustrated in Fig. 3.1. In the idealized case, the discharge of the battery proceeds at the theoretical voltage until the active materials are consumed and the capacity is fully utilized. The voltage then drops to zero. Under actual conditions, the discharge curve is similar to the other curves in Fig. 3.1. The initial voltage of the cell under a discharge load is lower than the theoretical value due to the internal cell resistance and the resultant IR drop as well as polarization effects at both electrodes. The voltage also drops during discharge as the cell resistance increases due to the accumulation of discharge products, activation and concentration, polarization, and related factors. Curve 2 is similar to curve 1, but represents a cell with a higher internal resistance or a higher discharge rate, or both, compared to the cell represented by curve 1. As the cell resistance or the discharge current is increased, the discharge voltage decreases and the discharge shows a more sloping profile.

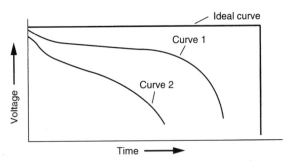

FIGURE 3.1 Characteristic discharge curves.

The specific energy that is delivered by a battery in practice is, therefore, lower than the theoretical specific energy of its active materials, due to:

1. The average voltage during the discharge is lower than the theoretical voltage.

2. The battery is not discharged to zero volts and all of the available ampere-hour capacity is not utilized.

As specific energy equals

$$\text{Watthours/gram} = \text{Voltage} \times \text{Ampere hours/gram}$$

the delivered specific energy is lower than the theoretical energy as both of the components of the equation are lower.

The shape of the discharge curve can vary depending on the electrochemical system, constructional features, and other discharge conditions. Typical discharge curves are shown in Fig. 3.2. The flat discharge (curve 1) is representative of a discharge where the effect of change in reactants and reaction products is minimal until the active materials are nearly exhausted. The plateau profile (curve 2) is representative of two-step discharge indicating a change in the reaction mechanism and potential of the active material(s). The sloping discharge (curve 3) is typical when the composition of the active materials, reactants, internal resistance, and so on, change during the discharge, affecting the shape of the discharge curve similarly.

Specific examples of these curves and many others are presented in the individual chapters covering each battery system.

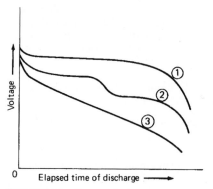

FIGURE 3.2 Battery discharge characteristics—voltage profiles.

3.2.2 Current Drain of Discharge

As the current drain of the battery is increased, the *IR* losses and polarization effects increase, the discharge is at a lower voltage, and the service life of the battery is reduced. Figure 3.3*a* shows typical discharge curves as the current drain is changed. At extremely low current drains (curve 2) the discharge can approach the theoretical voltage and theoretical capacity. [However, with very long discharge periods chemical deterioration during the discharge can become a factor and cause a reduction in capacity (Sec. 3.2.12).] With increasing current drain (curves 3–5) the discharge voltage decreases, the slope of the discharge curve becomes more pronounced, and the service life, as well as the delivered ampere-hour or coulombic capacity, are reduced.

If a battery that has reached a particular voltage (such as the cutoff voltage) under a given discharge current is used at a lower discharge rate, its voltage will rise and additional capacity or service life can be obtained until the cutoff voltage is reached at the lighter load. Thus, for example, a battery that has been used to its end-of-life in a flash camera (a high drain

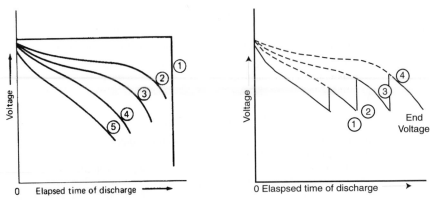

FIGURE 3.3 (*a*) Battery discharge characteristics—voltage levels. (*b*) Discharge characteristics of a battery discharged sequentially from high to lower discharge rates.

application) can subsequently be used successfully in a quartz clock application which operates at a much lower discharge rate. This procedure can also be used for determining the life of a battery under different discharge loads using a single test battery. As shown in Fig. 3.3*b*, the discharge is first run at the highest discharge rate to the specified end voltage. The discharge rate is then reduced to the next lower rate. The voltage increases and the discharge is continued again to the specified end voltage, and so on. The service life can be determined for each discharge rate, but the complete discharge curve for the lower discharge rates, as shown by the dashed portion of the each curve, obviously is lost. In some instances a time interval is allowed between each discharge for the battery to equilibrate prior to discharge at the progressively lower rates.

"C" Rate.*   A common method for indicating the discharge, as well as the charge current of a battery, is the C rate, expressed as

$$I = M \times C_n$$

---

*Traditionally, the manufacturers and users of secondary alkaline cells and batteries have expressed the value of the current used to charge and discharge cells and batteries as a multiple of the capacity. For example, a current of 200 mA used to charge a cell with a rated capacity of 1000 mAh would be expressed as $C/5$ or $0.2\ C$ (or, in the European convention as $C/5$ A or $0, 2$ CA). This method for designation of current has been criticized as being dimensionally incorrect in that a multiple of the capacity (e.g. ampere-hours) will be in ampere-hours and not, as required for current, in amperes. As a result of these comments, the International Electrotechnical commission (IEC) Sub-committee SC-21A has published a "Guide to the Designation of Current in Alkaline Secondary Cell and Battery Standards (IEC 61434)" which described a new method for so designating this current. In brief, the method states that the current (I) shall be expressed as

$$I_t(A) = C_n(\text{Ah})/1(h)$$

where I = is expressed in amperes
$\quad\ C_n$ = is the rated capacity declared by the manufacturer in ampere-hours, and
$\quad\ n$ = is the time base in hours for which the rated capacity is declared.

For example, a battery rated at 5 Ah at the 5 hour discharge rate (C_5 (Ah)) and discharged at $0.1 I_t$(A) will be discharged at 0.5 A or 500 mA.
For this Handbook, the method discussed in the text, and not in this footnote, will be used.

where I = discharge current, A

C = numerical value of rated capacity of the battery, in ampere-hours (Ah)

n = time, in hours, for which rated capacity is declared

M = multiple or fraction of C

For example, the $0.1C$ or $C/10$ discharge rate for a battery rated at 5 Ah is 0.5 A. Conversely, a 250-mAh battery, discharged at 50 mA, is discharged at the $0.2C$ or $C/5$ rate, which is calculated as follows:

$$M = \frac{I}{C_n} = \frac{0.050}{0.250} = 0.2$$

To further clarify this nomenclature system the designation for a $C/10$ discharge rate for a battery rated at 5 Ah at the 5 hour rate is:

$$0.1\ C_5$$

In this example, the $C/10$ rate is equal to 0.5 A, or 500 mA.

It is to be noted that the capacity of a battery generally decreases with increasing discharge current. Thus the battery rated at 5 Ah at the $C/5$ rate (or 1 A) will operate for 5 h when discharged at 1 A. If the battery is discharged at a lower rate, for example the $C/10$ rate (or 0.5 A), it will run for more than 10 h and deliver more than 5 Ah of capacity. Conversely, when discharged at its C rate (or 5 A), the battery will run for less than 1 h and deliver less than 5 Ah of capacity.

Hourly Rate. Another method for specifying the current is the *hourly rate.* This is the current at which the battery will discharge for a specified number of hours.

"E" Rate. The constant power discharge mode is becoming more popular for battery-powered applications. A method, analogous to the *"C* rate," can be used to express the discharge or charge rate in terms of power:

$$P = M \times E_n$$

where P = power (W),

E = numerical value of the rated energy of the battery in watthours (Wh),

n = time, in hours, at which the battery was rated

M = multiple or fraction of E.

For example, the power level at the $0.5E_5$ or $E_5/2$ rate for a battery rated at 1200 mWh, at the $0.2E$ or $E/5$ rate, is 600 mW.

3.2.3 Mode of Discharge (Constant Current, Constant Load, Constant Power)

The mode of discharge of a battery, among other factors, can have a significant effect on the performance of the battery. For this reason, it is advisable that the mode of discharge used in a test or evaluation program be the same as the one used in the application for which it is being tested.

A battery, when discharged to a specific point (same closed-circuit voltage, at the same discharge current, at the same temperature, etc.) will have delivered the same ampere-hours to a load regardless of the mode of discharge. However, as during the discharge, the discharge current will be different depending on the mode of discharge, the service time or "hours of discharge" delivered to that point (which is the usual measure of battery performance) will, likewise, be different.

Three of the basic modes under which the battery may be discharged are:

1. *Constant Resistance:* The resistance of the load remains constant throughout the discharge (The current decreases during the discharge proportional to the decrease in the battery voltage)

2. *Constant Current:* The current remains constant during the discharge.

3. *Constant Power:* The current increases during the discharge as the battery voltage decreases, thus discharging the battery at constant power level (power = current × voltage).

The effect of the mode of discharge on the performance of the battery is illustrated under three different conditions in Figs. 3.4, 3.5 and 3.6.

Case 1: Discharge loads are the same for each mode of discharge at the start of discharge
In Fig. 3.4, the discharge loads are selected so that at the start of the discharge the discharge current and, hence, the power are the same for all three modes. Figure 3.4*b* is a plot of the voltage during discharge. As the cell voltage drops during the discharge, the current in the case of the constant resistance discharge, reflects the drop in the cell voltage according to Ohm's law:

$$I = V/R$$

This is shown in Fig. 3.4*a*.
In the case of a constant current discharge, the current remains the same throughout the discharge. However, the discharge time or service life is lower than for the constant resistance case because the average current is higher. Finally, in the constant power mode, the current increases with decreasing voltage according to the relationship:

$$I = P/V$$

The average current is now even higher and the discharge time still lower.
Figure 3.4*c* is a plot of the power level for each mode of discharge.

Case 2: "Hours of discharge" is the same for each mode of discharge
Figure 3.5 shows the same relationships but with the respective discharge loads selected so that the discharge time or "hours of service" (to a given end voltage) is the same for all three modes of discharge. As expected, the discharge curves vary depending on the mode of discharge.

Case 3: Power level is the same for each mode of discharge at the end of the discharge
From an application point of view the most realistic case is the assumption that the power under all three modes of operation is the same at the end of the discharge (Fig. 3.6). Electric and electronic devices require a minimum input power to operate at a specified performance level. In each case, the discharge loads are selected so that at the end of the discharge (when the cell reaches the cutoff voltage) the power output is the same for all of the discharge modes and at the level required for acceptable equipment performance. During the discharge, depending on the mode of discharge, the power output equals or exceeds the power required by the equipment until the battery reaches the cutoff voltage.
In the constant-resistance discharge mode, the current during the discharge (Fig. 3.6*b*) follows the drop in the battery voltage (Fig. 3.6*a*). The power, $I \times V$ or V^2/R, drops even more rapidly, following the square of the battery voltage (Fig. 3.6*c*). Under this mode of discharge, to assure that the required power is available at the cutoff voltage, the levels of current and power during the earlier part of the discharge are in excess of the minimum required. The battery discharges at a higher current than needed, draining its capacity rapidly, which will result in a shorter service life.

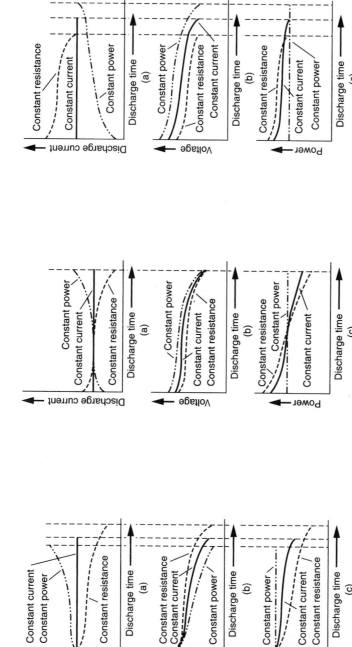

FIGURE 3.4 Discharge profiles under different discharge modes; same current and power at start of discharge. (*a*) Current profile during discharge. (*b*) Voltage profile during discharge. (*c*) Power profile during discharge.

FIGURE 3.5 Discharge profiles under different discharge modes; same discharge time (*a*) Current profile during discharge. (*b*) Voltage profile during discharge. (*c*) Power profile during discharge.

FIGURE 3.6 Discharge profiles under different discharge modes: same power at end of discharge. (*a*) current profile during discharge. (*b*) Voltage profile during discharge. (*c*) Power profile during discharge.

In the constant-current mode, the current is maintained at a level such that the power output at the cutoff voltage is equal to the level required for acceptable equipment performance. Thus both current and power throughout the discharge are lower than for the constant-resistance mode. The average current drain on the battery is lower and the discharge time or service life to the end of the battery life is longer.

In the constant-power mode, the current is lowest at the beginning of the discharge and increases as the battery voltage drops in order to maintain a constant-power output at the level required by the equipment. The average current is lowest under this mode of discharge, and hence, the longest service time is obtained.

It should be noted that the extent of the advantage of the constant-power discharge mode over the other modes of discharge is dependent on the discharge characteristics of the battery. The advantage is higher with battery systems that require a wide voltage range to deliver their full capacity.

3.2.4 Example of Evaluation of Battery Performance Under Different Modes of Discharge

Thus, in evaluating or comparing the performance of batteries, because of the potential difference in performance (service hours) due to the mode of discharge, the mode of discharge used in the evaluation or test should be the same as that in the application. This is illustrated further in Fig. 3.7.

Figure 3.7a shows the discharge characteristics of a typical "AA"-size primary battery with the values for the discharge loads for the three modes of discharge selected so that the hours of discharge to a given end voltage (in this case, 1.0 volt) are the same. This is the same condition shown in Fig. 3.5b. (This example illustrates the condition when a resistive load, equivalent to the average current, is used, albeit incorrectly, as a "simpler" less-costly test to evaluate a constant current or constant power application.) Although the hours of service to the given end voltage is obviously the same because the loads were pre-selected, the discharge current vs. discharge time and power vs. discharge time curves (see Figs. 3.5a and c respectively), show significantly different characteristics for the different modes of discharge.

Figure 3.7b shows the same three types of discharge as Fig. 3.7a, but on a battery that has about the same ampere-hour capacity (to a 1.0 volt end voltage) as the battery illustrated in Fig. 3.7a. The battery illustrated in Fig. 3.7b, however, has a lower internal resistance and a higher operating voltage. Note, by comparing the voltage vs. discharge time curves in Fig. 3.7b, that, although the voltage level is higher, the "hours of discharge" obtained on the constant resistance discharge to the 1.0 volt cutoff in Fig. 3.7b is about the same as obtained on Fig. 3.7a. However, the hours of service obtained on the constant current discharge and, particularly, the constant power discharge are significantly higher.

In Fig. 3.7a, the discharge loads were deliberately selected to give the same "hours of service" to 1.0 volt at the three modes of discharge. Using these same discharge loads, but with a battery having different characteristics, Fig. 3.7b shows that different "hours of service" and performance are obtained with the different modes of discharge. To the specified 1.0 volt end voltage, the longest "hours of service" are obtained with the constant power discharge mode. The shortest service time is obtained with the constant resistance discharge mode and the constant current mode is in the middle position. This clearly illustrates that, on application tests, where performance is measured in "hours of service," erroneous results will be obtained if the mode of discharge used in the test is different from that used in the application.

It is recognized that the performance differences obtained when comparing batteries are directly dependent on the differences in battery design and performance characteristics. With batteries that are significantly different in design and characteristics, the performance obtained on test will be quite different as shown in the comparisons between Figs. 3.7a and

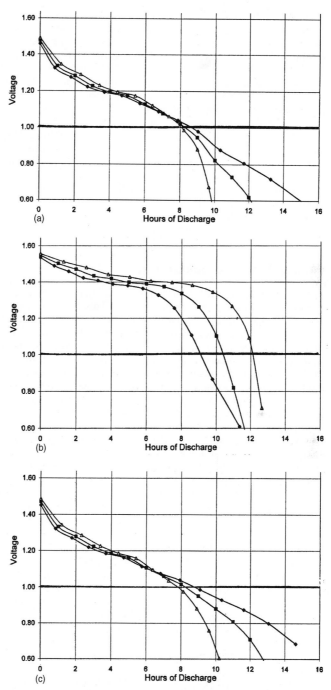

FIGURE 3.7 Characteristics of a "AA" size primary battery discharged under constant resistance, constant current and constant power conditions at 5.9 ohms, ——•——; 200 milliamperes, ——■——; and 235 milliwatts, ——▲——. See Sec. 3.2.4 for detailed description.

3.7*b*. When the batteries are similar, the performance differences obtained on any of the modes of discharge may not be large and may not appear to be significantly different. However, just because the difference in this case are small, it should not lead to the false assumption that testing under a discharge mode different from the application would give accurate results.

This is illustrated in Fig. 3.7*c*, which shows the discharge characteristics of another battery which has a slightly higher capacity and higher internal resistance than the one shown in Fig. 3.7*a*. Although the differences are small, a careful comparison of the Fig. 3.7*a* with Fig. 3.7*c* at the different modes of discharge does show a different behavior in the "hours of discharge" obtained to the specified 1.0 volt end voltage. Under the constant power mode, the "hours of discharge" show a slight decrease comparing Fig. 3.7*c* with Fig. 3.7*a*. While there is a slight increase under the constant current and constant resistance discharge modes.

(Note: The influence of end voltage should be noted. As 1.0 volt was used as the end-voltage in determining the load values for these examples, this end voltage should be used in making comparisons. If discharged to lower end voltages, the service life for the constant resistance mode increases compared to the other modes because of the lower current and power levels. However, these lower values may be inadequate for the specified application).

3.2.5 Temperature of Battery During Discharge

The temperature at which the battery is discharged has a pronounced effect on its service life (capacity) and voltage characteristics. This is due to the reduction in chemical activity and the increase in the internal resistance of the battery at lower temperatures. This is illustrated in Fig. 3.8, which shows discharges at the same current drain but at progressively increasing temperatures of the battery (T_1 to T_4), with T_4 representing a discharge at normal room temperature. Lowering of the discharge temperature will result in a reduction of capacity as well as an increase in the slope of the discharge curve. Both the specific characteristics and the discharge profile vary for each battery system, design, and discharge rate, but generally best performance is obtained between 20 and 40°C. At higher temperatures, the internal resistance decreases, the discharge voltage increases and, as a result, the ampere-hour capacity and energy output usually increase as well. On the other hand, chemical activity also increases at the higher temperatures and may be rapid enough during the discharge (a phenomenon known as *self-discharge*) to cause a net loss of capacity. Again, the extent is dependent on the battery system, design and temperature.

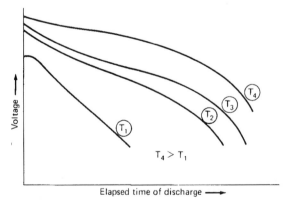

FIGURE 3.8   Effect of temperature on battery capacity. T_1 to T_4—increasing temperatures.

Figures 3.9 and 3.10 summarize the effects of temperature and discharge rate on the battery's discharge voltage and capacity. As the discharge rate is increased, the battery voltage (for example, the midpoint voltage) decreases; the rate of decrease is usually more rapid at the lower temperatures. Similarly, the battery's capacity falls off most rapidly with increasing discharge load and decreasing temperature. Again, as noted previously, the more stringent the discharge conditions, the greater the loss of capacity. However, discharging at high rates could cause apparent anomalous effects as the battery may heat up to temperatures much above ambient, showing the effects of the higher temperatures. Curve T_6 in Fig. 3.10 shows the loss of capacity at high temperatures at low discharge rates or long discharge times due to self-discharge or chemical deterioration. It also shows the higher capacity that may be obtained as a result of the battery heating at the high rate discharge.

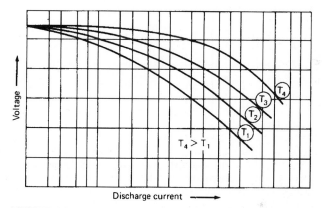

FIGURE 3.9   Effect of discharge load on battery midpoint voltage at various temperatures, T_1 to T_4—increasing temperatures; T_4—normal room temperature

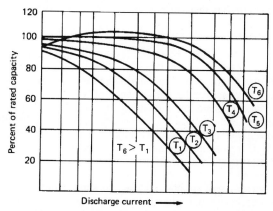

FIGURE 3.10   Effect of discharge load on battery capacity at various temperatures. T_1 to T_6—increasing temperatures; T_4—normal room temperature

3.2.6 Service Life

A useful graph employed in this Handbook summarizing the performance of each battery system, presents the service life at various discharge loads and temperatures, normalized for unit weight (amperes per kilogram) and unit volume (amperes per liter). Typical curves are shown in Fig. 3.11. In this type of presentation of data, curves with the sharpest slope represent a better response to increasing discharge loads than those which are flatter or flatten out at the high current drain discharges.

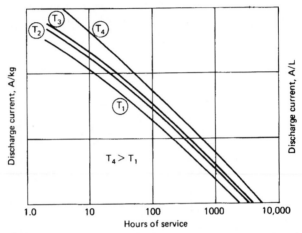

FIGURE 3.11 Battery service life at various discharge loads and temperatures (log–log scale). T_1 to T_4—increasing temperature.

Data of this type can be used to approximate the service life of a given cell or battery under a particular discharge condition or to estimate the weight or size of a battery required to meet a given service requirement. In view of the linearity of these curves on a log-log plot, mathematical relationships have been developed to estimate the performance of batteries under conditions that are not specifically stated. Peukert's equation,

$$I''t = C$$

or
$$n \log I + \log t = \log C$$

where I is the discharge rate and t the corresponding discharge time, has been used in this manner to describe the performance of a battery. The value n is the slope of the straight line. The curves are linear on a log-log plot of discharge load versus discharge time but taper off at both ends because of the battery's inability to handle very high rates and the effect of self-discharge at the lower discharge rates. A more detailed explanation of the use of these graphs in a specific example is presented in Fig. 14.13. Other mathematical relationships have been developed to describe battery performance and account for the non-linearity of the curves.[1]

Other types of graphs are used to show similar data. A Ragone plot, such as the one illustrated in Fig. 6.3, plots the specific energy or energy density of a battery system against the specific power or power density on a log-log scale. This type of graph effectively shows the influence of the discharge load (in this case, power) on the energy that can be delivered by a battery.

3.2.7   Type of Discharge (Continuous, Intermittent, etc.)

When a battery stands idle after a discharge, certain chemical and physical changes take place which can result in a recovery of the battery voltage. Thus the voltage of a battery, which has dropped during a heavy discharge, will rise after a rest period, giving a sawtooth-shaped discharge, as illustrated in Fig. 3.12. This can result in an increase in service life. However, on lengthy discharges, capacity losses may occur due to self-discharge (see Sec. 3.2.12). This improvement, resulting from the intermittent discharge, is generally greater after the higher current drains (as the battery has the opportunity to recover from polarization effects that are more pronounced at the heavier loads). In addition to current drain, the extent of recovery is dependent on many other factors such as the particular battery system and constructional features, discharge temperature, end voltage, and length of recovery period.

The interactive effect on capacity due to the discharge load and the extent of intermittency is shown in Fig. 8.11. It can be seen that the performance of a battery as a function of duty cycle can be significantly different at low and high discharge rates. Similarly, the performance as a function of discharge rate can be different depending on the duty cycle.

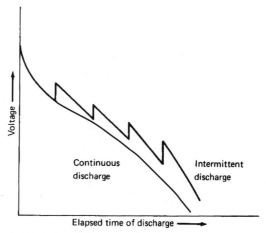

FIGURE 3.12   Effect of intermittent discharge on battery capacity.

3.2.8   Duty Cycles (Intermittent and Pulse Discharges)

Another consideration is the response of the battery voltage when the discharge current is changed during the discharge, such as changing loads from receive to transmit in the operation of a radio transceiver. Figure 3.13 illustrates a typical discharge of a radio-transceiver, discharging at a lower current during the receive mode and at a higher current during the transmit mode. Note that the service life of the battery is determined when the cut-off or end voltage is reached under the higher discharge load. The average current cannot be used to determine the service life. Operating at two or more discharge loads is typical of most electronic equipment because of the different functions they must perform during use.

Another example is a higher rate periodic pulse requirement against a lower background current, such as backlighting for an LCD watch application, the audible trouble signal pulse in the operation of a smoke detector, or a high rate pulse during the use of a cell phone or computer. A typical pulse discharge is plotted in Fig. 3.14. The extent of the drop in voltage

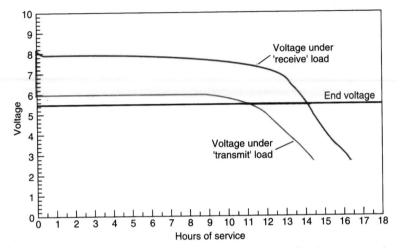

FIGURE 3.13 Typical discharge characteristics of a battery cycling between transmit and receive loads.

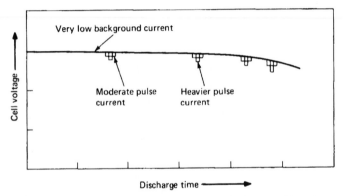

FIGURE 3.14 Typical discharge characteristics of a battery subjected to a periodic high rate pulse.

depends on the battery design. The drop in voltage for a battery with lower internal resistance and better response to changes in load current will be less than one with higher internal resistance. In Fig. 3.14, note that the voltage spread widens as the battery is discharged due to the increase in internal resistance as the battery is discharged.

The shape of the pulse can vary significantly depending on the characteristics of the pulse and the battery. Figure 3.15 shows the characteristics of 9-volt primary batteries subjected to the 100 millisecond audible trouble signal pulse in a smoke detector. Curve A shows the response of a zinc-carbon battery, the voltage dropping sharply initially and then recovering. Curves B and C are typical of the response of a zinc/alkaline/manganese dioxide battery, the voltage initially falling and either maintaining the lower voltage or dropping slowly as the pulse discharge continues.

The type of response shown in Fig. 3.15*a* is also typical of batteries that have developed a protective or passivating film on an electrode, the voltage recovering as the film is broken during the discharge (see Sec. 3.2.12 on Voltage Delay). The specific characteristics, how-

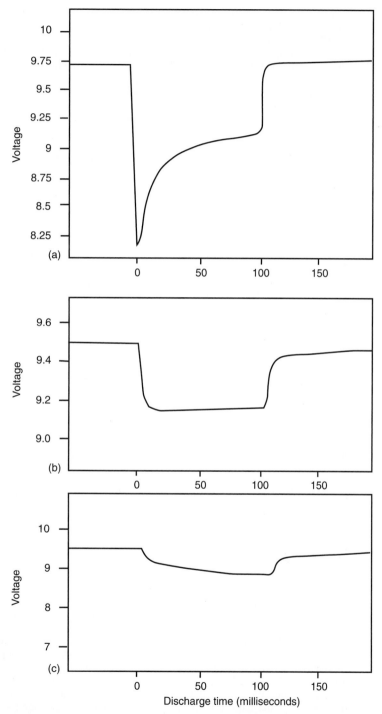

FIGURE 3.15 Discharge characteristics of a 9-volt battery subjected to a 100 ms pulse (smoke detector pulse tests): (*a*) zinc-carbon battery; (*b*) and (*c*) zinc/alkaline/manganese dioxide battery.

ever, are dependent on the battery chemistry, design, state of discharge, and other factors, related to the battery's internal resistance at the time of the pulse and during the pulse (also see Chap. 2 on internal resistance and polarization).

The performance of a battery under pulse conditions can be characterized by plotting the output power of the pulse against the load voltage, measuring the power delivered to the load by the short term pulse over the range of open circuit to short circuit.[2] Peak power is delivered to the load when the resistance of the external is equal to the internal resistance of the battery. Figure 3.16 is a power vs. load voltage plot of the pulse characteristics of an undischarged zinc/alkaline/manganese dioxide battery ("AA"-size) at the end of constant voltage pulses of 0.1 and 1 second. The lower values of power for the longer pulse is indicative of the drop in voltage as the pulse length increases. Figure 10.15 shows a similar plot, with the 1-second pulse taken at different depths of battery discharge.

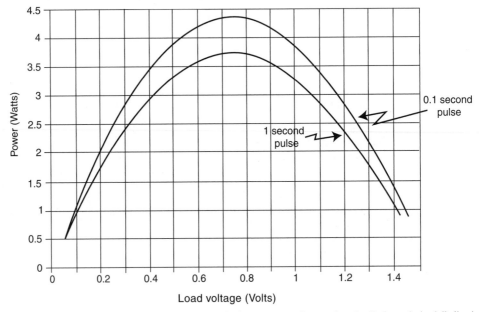

FIGURE 3.16   Power vs. Load voltage at the end of a constant voltage pulses (undischarged zinc/alkaline/manganese dioxide "AA" size battery. (*From Ref 2.*)

3.2.9 Voltage Regulation

The voltage regulation required by the equipment is most important in influencing the capacity or service life obtainable from a battery. As is apparent from the various discharge curves, design of the equipment to operate to the lowest possible end voltage and widest voltage range result in the highest capacity and longest service life. Similarly, the upper voltage limit of the equipment should be established to take full advantage of the battery characteristics.

Figure 3.17 compares two typical battery discharge curves: curve 1 depicts a battery having a flat discharge curve; curve 2 depicts a battery having a sloping discharge curve. In applications where the equipment cannot tolerate the wide voltage spread and is restricted, for example, to the −15% level, the battery with the flat discharge profile gives the longer service. On the other hand, if the batteries can be discharged, to lower cutoff voltages, the service life of the battery with the sloping discharge is extended and could exceed that of the battery with the flat discharge profile.

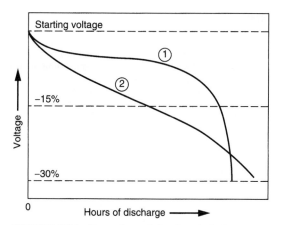

FIGURE 3.17 Comparison of flat ① and sloping ② discharge curves.

Discharging multicell series-connected batteries to too low an end voltage, however, may result in safety problems. It is possible, in this situation, for the poorest cell to be driven into voltage reversal. With some batteries this could result in venting or rupture.

In applications where only a narrow voltage range can be tolerated, the selection of the battery may be limited to those having a flat discharge profile. An alternative is to use a voltage regulator to convert the varying output voltage of the battery into a constant output voltage consistent with the equipment requirements. In this way, the full capacity of the battery can be used with inefficiency of the voltage regulator the only energy penalty. Figure 3.18 illustrates the voltage and current profiles of the battery and regulator outputs. The input from the battery to the regulator is at a constant power of 1 W, with the current increasing as the battery voltage drops. With an 84% conversion efficiency, the output from the regulator is constant at a predetermined 6 V and 140 mA (constant power = 840 mW).

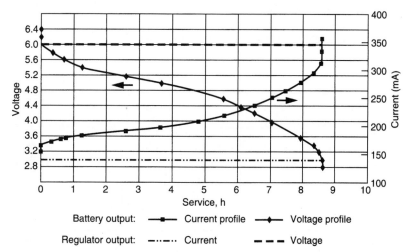

FIGURE 3.18 Characteristics of a voltage regulator. Battery output—1 W; regular output—840 mW.

3.2.10 Charging Voltage

If a rechargeable battery is used (for example, as a standby power source) in conjunction with another energy source which is permanently connected in the operating circuit, allowance must be made for the battery and equipment to tolerate the voltage of the battery on charge. Figure 3.19 shows the charge and discharge characteristics of such a battery. The specific voltage and the voltage profile on charge depend on such factors such as the battery system, charge rate, temperature, and so on.

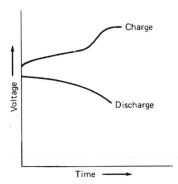

FIGURE 3.19 Typical voltage profile on charge and discharge.

If a primary battery is used in a similar circuit (for example, as memory backup battery), it is usually advisable to protect the primary battery from being charged by including an isolating or protective diode in the circuit, as shown in Fig. 3.20. Two diodes provide redundancy in case one fails. The resistor in Fig. 3.20*b* serves to limit the charging current in case the diode fails.

The charging source must also be designed so that its output current is regulated during the charge to provide the needed charge control for the battery.

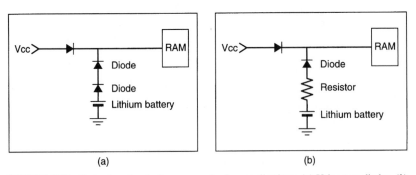

FIGURE 3.20 Protective circuits for memory backup applications. (*a*) Using two diodes. (*b*) Using diode and resistor.

3.2.11 Effect of Cell and Battery Design

The constructional features of the cell and battery strongly influence its performance characteristics.

Electrode Design.   Cells that are designed, for example, for optimum service life or capacity at relatively low or moderate discharge loads contain maximum quantities of active material. On the other extreme, cells capable of high-rate performance are designed with large electrode or reaction surfaces and features to minimize internal resistance and enhance current density (amperes per area of electrode surface), often at the expense of capacity or service life.

For example, two designs are used in cylindrical cells. One design known as the bobbin construction, is typical for zinc-carbon and alkaline-manganese dioxide cells. Here the electrodes are shaped into two concentric cylinders (Fig. 3.21*a*). This design maximizes the amount of active material that can be placed into the cylindrical can, but at the expense of surface area for the electrochemical reaction.

The second design is the "spirally wound" electrode construction, typically used in sealed portable rechargeable batteries and high-rate primary and rechargeable lithium batteries (Fig. 3.21*b*). In this design, the electrodes are prepared as thin strips and then rolled, with a separator in between, into a "jelly roll" and placed into the cylindrical can. This design emphasizes surface area to enhance high-rate performance, but at the expense of active material and capacity.

Another popular electrode design in the flat-plate construction, typically used in the lead-acid SLI and most larger storage batteries (Fig. 3.21*c*). This construction also provides a large surface area for the electrochemical reaction. As with the other designs, the manufacturer can control the relationship between surface area and active material (for example, by controlling the plate thickness) to obtain the desired performance characteristics.

A modification of this design is the bipolar plate illustrated in Fig. 3.21*d*. Here the anode and cathode are fabricated as layers on opposite sides of an electronically conductive but ion-impermeable material which serves as the intercell connector.

Most battery chemistries can be adapted to the different electrode designs, and some in fact, are manufactured in different configurations. Manufacturers choose chemistries and designs to optimize the performance for the particular applications and markets in which they are interested.

In Fig. 3.22 the performance of a battery designed for high-rate performance is compared with one using the same electrochemical system, but optimized for capacity. The high-rate batteries have a lower capacity but deliver a more constant performance as the discharge rate increases.

Hybrid Designs.   "Hybrid" designs, which combine a high energy power source with a high-rate power source, are now becoming popular. These hybrid systems fulfill applications more effectively (e.g. higher total specific energy or energy density), than using a single power source. The high energy power source is the basic source of energy, but also charges a high-rate battery which handles any peak power requirement that cannot be handled efficiently by the main power source. Hybrid designs are being considered in many applications, ranging from combining a high energy, low rate metal/air battery or fuel cell with a high rate rechargeable battery to electric vehicles, using an efficient combustion engine with a rechargeable battery to handle starting, acceleration and other peak power demands (also see Fig. 6.13).

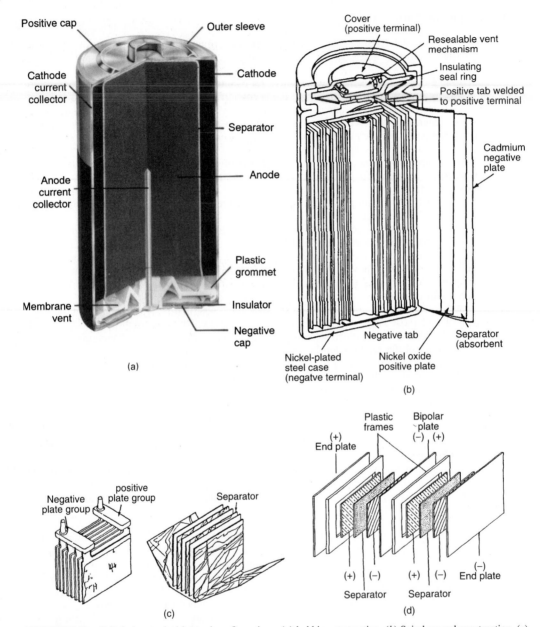

FIGURE 3.21 Cell design; typical internal configurations. (*a*) bobbin construction. (*b*) Spiral wound construction. (*c*) Flat-plate construction. (*d*) Bipolar-plate construction.

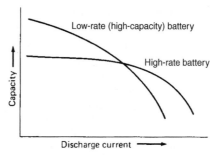

FIGURE 3.22   Comparison of performance of batteries designed for high- and low-rate service.

Shape and Configuration.   The shape or configuration of the cell will also influence the battery capacity as it affects such factors as internal resistance and heat dissipation. For example, a tall, narrow-shaped cylindrical cell in a bobbin design will generally have a lower internal resistance than a wide, squat-shaped one of the same design and may outperform it, in proportion to its volume, particularly at the higher discharge rates. For example, a thin "AA" size bobbin type cell will have proportionally better high rate performance than a wider diameter "D" size cell. Heat dissipation also will be better from cells with a high surface-to-volume ratio or with internal components that can conduct heat to the outside of the battery.

Volumetric Efficiency versus Energy Density.   The size and shape of the cell or battery and the ability to effectively use its internal volume influence the energy output of the cell. The volumetric energy density (watthours per liter) decreases with decreasing battery volume as the percentage of "dead volume" for containers, seals, and so on, increases for the smaller batteries. This relationship is illustrated for several button-type cells in Fig. 3.23. The shape of the cell (such as wide or narrow diameter) may also influence the volumetric efficiency as it relates to the amount of space lost for the seal and other cell construction materials.

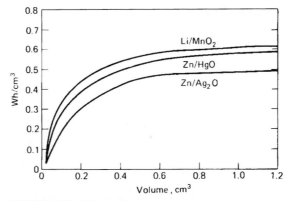

FIGURE 3.23   Energy density, in watthours per cubic centimeter, of button batteries as a function of cell volume. (*From Paul Ruetschi, "Alkaline Electrolyte-Lithium Miniature Primary Batteries," J. Power Sources, vol. 7, 1982.*)

Effect of Size on Capacity. The size of the battery influences the voltage characteristics by its effect on current density. A given current drain may be a severe load on a small battery, giving a discharge profile similar to curve 4 or 5 of Fig. 3.3, but it may be a mild load on a larger battery with a discharge curve similar to curve 2 or 3. Often it is possible to obtain more than a proportional increase in the service life by increasing the size of the battery (for paralleling cells) as the current density is lowered. The absolute value of the discharge current, therefore, is not the key influence, although its relation to the size of the battery, that is, the current density, is significant.

In this connection, the alternative of using a series-connected multicell battery versus a lower voltage battery, with fewer but larger cells and a voltage converter to obtain the required high voltage, should be considered. An important factor is the relative advantage of the potentially more efficient larger battery versus the energy losses of the voltage converter. In addition, the reliability of the system is enhanced by the use of a smaller number of cells. However, all pertinent factors must be considered in this decision because of the influences of cell and battery design, configuration, and so on, as well as the equipment power requirements.

3.2.12 Battery Age and Storage Condition

Batteries are a perishable product and deteriorate as a result of the chemical action that proceeds during storage. The design, electrochemical system, temperature, and length of storage period are factors which affect the shelf life or charge retention of the battery. The type of discharge following the storage period will also influence the shelf life of the battery. Usually the percentage charge retention following storage (comparing performance after and before storage) will be lower the more stringent the discharge condition. The self-discharge characteristics of several battery systems at various temperatures are shown in Fig. 6.7 as well as in the chapters on specific battery chemistries. As self-discharge proceeds at a lower rate at reduced temperatures, refrigerated or low-temperature storage extends the shelf life and is recommended for some battery systems. Refrigerated batteries should be warmed before discharge to obtain maximum capacity.

Self-discharge can also become a factor during discharge, particularly on long-term discharges, and can cause a reduction in capacity. This effect is illustrated in Fig. 3.10 and 3.24. More capacity will be delivered on a discharge at a light load than on a heavy load. However, on an extremely light load over a long discharge period, capacity may be reduced due to self-discharge.

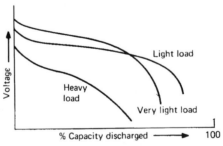

FIGURE 3.24 Effect of self-discharge on battery capacity.

Some battery systems will develop protective or passivating films on one or both electrode surfaces during storage. These films can improve the shelf life of the battery substantially. However, when the battery is placed on discharge after storage, the initial voltage may be low due to the impedance characteristics of the film until it is broken down or depassivated by the electrochemical reaction. This phenomenon is known as "voltage delay" and is illustrated in Fig. 3.25. The extent of the voltage delay is dependent on an increases with increasing storage time and storage temperature. The delay also increases with increasing discharge current and decreasing discharge temperature.

The self-discharge characteristics of a battery that has been or is being discharged can be different from one that has been stored without having been discharged. This is due to a number of factors, such as the discharge rate and temperature, the accumulation of discharge products, the depth of discharge, or the partial destruction or reformation of the protective film. Some batteries, such as the magnesium primary battery (Chapter 9), may lose their good shelf-life qualities after being discharged because of the destruction of the protective film during discharge. Knowledge of the battery's storage and discharge history is needed to predict the battery's performance under these conditions.

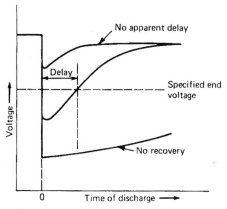

FIGURE 3.25   Voltage delay

3.2.13   Effect of Battery Design

The performance of the cells in a multicell battery will usually be different than the performance of the individual cells. The cells cannot be manufactured identically and, although cells are selected to be "balanced," they each encounter a somewhat different environment in the battery pack.

The specific design of the multicell battery and the hardware that is used (such as packaging techniques, spacing between the cells, container material, insulation, potting compound, fuses and other electronic controls, etc.) will influence the performance as they effect the environment and temperature of the individual cells. Obviously, the battery materials add to its size and weight and the specific energy or energy density if the battery will be lower than that of the component cells. Accordingly, when comparing values such as specific energy, in addition to being aware of the conditions (discharge rate, temperature, etc.) under which these values were determined, it should be ascertained whether the values given are for cells, single cell batteries or multicell batteries. Usually, as is the case in this Handbook, they are on the basis of a single-cell battery unless specified otherwise.

A comparison of the performance of single-cell and multicell batteries can be made by reviewing the data in Table 1.2 on single-cell batteries with the data in Table 6.5 on 24-volt multicell batteries.

Battery designs that retain the heat dissipated by the cells can improve the performance at low temperatures. On the other hand, excessive buildup of heat can be injurious to the battery's performance, life and safety. As much as possible, batteries should be designed thermally to maintain a uniform internal temperature and avoid "hot spots."

In the case of rechargeable batteries, cycling could cause the individual cells in a battery peak to become unbalanced and their voltage, capacity or other characteristics could become significantly different. This could result in poor performance or safety problems, and end-of-charge or discharge control may be necessary to prevent this.

The influence of battery design and recommendations for effective battery design are covered in Chap. 5.

REFERENCES

1. R. Selim and P. Bro, "Performance Domain Analysis of Primary Batteries," *Electrochemical Technology, J. Electrochem. Soc.* **118**(5) 829 (1971).

2. D. I. Pomerantz, "The Characterization of High Rate Batteries," *IEEE Transactions Electronics*" **36**(4) 954 (1990).

CHAPTER 4
BATTERY STANDARDIZATION

Frank Ciliberti and Steven Wicelinski

4.1 GENERAL

The standardization of batteries started in 1912, when a committee of the American Electrochemical Society recommended standard methods of testing dry cells. This eventually led to the first national publication in 1919 issued as an appendix to a circular from the National Bureau of Standards. It further evolved into the present American National Standards Institute (ANSI) Accredited Standards Committee C18 on Portable Cells and Batteries. Since then, other professional societies have developed battery related standards. Many battery standards were also issued by international, national, military, and federal organizations. Manufacturers' associations, trade associations, and individual manufacturers have published standards as well. Related application standards, published by the Underwriters Laboratories, the International Electrotechnical Commission, and other organizations that cover battery-operated equipment may also be of interest.

Table 4.1*a* to *d* lists some of the widely known standards for batteries. Standards covering the safety and regulation of batteries are listed in Table 4.11.

TABLE 4.1a   International Standards (IEC—International Electrotechnical Commission)

Publication	Title	Electrochemical systems
IEC 60086-1, IEC 60086-2	Primary Batteries; Part 1 General, and Part 2, Specification Sheets	Zinc-carbon Zinc/air Alkaline-manganese dioxide Mercuric oxide Silver oxide Lithium/carbon monofluoride Lithium/manganese dioxide Lithium/copper oxide Lithium/chromium oxide Lithium/thionyl chloride
IEC 60086-3	Watch Batteries	
IEC 60095	Lead-Acid Starter Batteries	Lead-acid
IEC 60254	Lead-Acid Traction Batteries	Lead-acid
IEC 60285	Sealed Nickel-Cadmium Cylindrical Rechargeable Single Cells	Nickel cadmium
IEC 60509	Sealed Nickel-Cadmium Button Rechargeable Single Cells	Nickel-cadmium
IEC 60622	Sealed Nickel-Cadmium Prismatic Rechargeable Single Cells	Nickel-cadmium
IEC 60623	Vented Nickel-Cadmium Prismatic Rechargeable Cells	Nickel-cadmium
IEC 60952	Aircraft Batteries	Nickel-cadmium Lead-acid
IEC 60986	Stationary Lead-Acid Batteries	Lead-acid
IEC 61056	Portable Lead-Acid Cells and Batteries	Lead-acid
IEC 61150	Sealed Nickel-Cadmium Rechargeable Monobloc Batteries in Button Cell Design	Nickel-cadmium
IEC 61436	Sealed Nickel-Metal Hydride Rechargeable Cells	Nickel-metal hydride
IEC 61440	Sealed Nickel-Cadmium Small Prismatic Rechargeable Single Cells	Nickel-cadmium

Note: See Table 4.11a for IEC Safety Standards.

TABLE 4.1b   National Standards (ANSI—American National Standards Institute)

Publication	Title	Electrochemical systems
ANSI C18.1M, Part 1	Standard for Portable Primary Cells and Batteries with Aqueous Electrolyte	Zinc-carbon Alkaline-manganese dioxide Silver oxide Zinc/air
ANSI C18.2M, Part 1	Standard for Portable Rechargeable Cells and Batteries	Nickel-cadmium Nickel-metal hydride Lithium-ion
ANSI C18.3M, Part 1	Standard for Portable Lithium Primary Cells and Batteries	Lithium/carbon monofluoride Lithium/manganese dioxide

Note: see Table 4.11a for ANSI Safety Standards.

TABLE 4.1c   U.S. Military Standards (MIL)

Publication	Title	Electrochemical systems
MIL-B-18	Batteries Non-Rechargeable	Zinc-carbon, mercury
MIL-B-8565	Aircraft Batteries	Various
MIL-B-11188	Vehicle Batteries	Lead-acid
MIL-B-49030	Batteries, Dry, Alkaline (Non-Rechargeable)	Alkaline-manganese dioxide
MIL-B-55252	Batteries, Magnesium	Magnesium
MIL-B-49430	Batteries, Non-Rechargeable	Lithium/sulfur dioxide
MIL-B-49436	Batteries, Rechargeable, Sealed Nickel-Cadium	Nickel-cadmium
MIL-B-49450	Vented Aircraft Batteries	Nickel-cadmium
MIL-B-49458	Batteries, Non-Rechargeable	Lithium/manganese dioxide
MIL-B-49461	Batteries, Non-Rechargeable	Lithium/thionyl chloride
MIL-B-55130	Batteries, Rechargeable, Sealed Nickel-Cadmium	Nickel-cadmium
MIL-B-81757	Aircraft Batteries	Nickel-cadmium
MIL-PRF-49471	Batteries, Non-Rechargeable, High Performance	Various

TABLE 4.1d   Manufacturers' and Professional Associations

Publication	Title	Battery type covered
Society of Automotive Engineers		
SAE AS 8033	Aircraft Batteries	Nickel-cadmium
SAE J 537	Storage Batteries	Lead-acid
Battery Council International	Battery Replacement Data Book	Lead-acid

4.2   INTERNATIONAL STANDARDS

International standards are rapidly gaining in importance. This has been further accelerated by the creation of the European Common Market and the 1979 Agreement on Technical Barriers to Trade. The latter requires the use of international standards for world trade purposes.

The International Electrotechnical Commission (IEC) is the designated organization responsible for standardization in the fields of electricity, electronics, and related technologies. Promoting international cooperation on all questions of electrotechnical standardization and related matters is its basic mission. This organization was founded in 1906 and consists of 50 national committees that represent more than 80% of the world's population and 95% of the world's production and consumption of electricity. The International Standards Organization (ISO) is responsible for international standards in fields other than electrical. IEC and ISO are gradually adopting equivalent development and documentation procedures while ever closer ties are being established between these two international organizations.

The American National Standards Institute (ANSI) is the sole U.S. representative of the IEC through the United States National Committee (USNC). This committee coordinates all IEC activities in the United States. It also serves as the U.S. interface with emerging regional standards-developing bodies such as CENELEC, PASC, CANENA, COPANT, ARSO, and other foreign and national groups. ANSI does not itself develop standards; rather it facilitates development by establishing consensus among accredited, qualified groups. These standards are published as U.S. National Standards (see Table 4.1(b)).

To further its overall mission, the objectives of the IEC are to:

1. Efficiently meet the requirements of the global marketplace
2. Ensure maximum use of its standards and conformity assessment schemes
3. Assess and improve the quality of products and services covered by its standards
4. Establish conditions for interchangeability
5. Increase the efficiency of electrotechnical industrial processes
6. Contribute to the improvement of human health and safety
7. Work towards protection of the environment

The objectives of the international battery standards are:

1. To define a standard of quality and provide guidance for its assessment
2. To ensure the electrical and physical interchangeability of products from different manufacturers
3. To limit the number of battery types
4. To provide guidance on matters of safety

The IEC sponsors the development and publication of standard documents. This development is carried out by working groups of experts from participating countries. These experts represent consumer, user, producer, academia, government, and trade and professional interests in the consensus development of these standards. The Groups of Experts in IEC working on battery standards are:

TC 21: Rechargeable Batteries

TC 35: Primary Batteries

The designation for the ANSI Committee on Portable Cells and Batteries is C18.

Table 4.1a lists the IEC standards that pertain to primary and secondary batteries. Many countries utilize these standards either by simply adopting them in toto as their national standards or by harmonizing their national standards to the IEC standards.

Table 4.1b lists the ANSI battery standards. When feasible, the two groups harmonize the requirements in their standards.

4.3 CONCEPTS OF STANDARDIZATION

The objective of battery interchangeability is achieved by specifying the preferred values for the physical aspects of the battery, such as dimensions, polarity, terminals, nomenclature and marketing. In addition, performance characteristics, such as service life or capacity, may be described and specified with test conditions for verification.

It is the inherent nature of batteries, in particular primary batteries, that replacements will at some time be required. A third-party end-user of the equipment typically replaces the battery. It is therefore essential that certain characteristics of the battery be specified by standard values—size, shape, voltage and terminals. Without a reasonable match of at least these parameters there can be no interchangeability. These characteristics are absolute requirements in order to fit the appliance receptacle, to make proper contact, and to provide the proper voltage. In addition to the end-user's need for replacement information, the original equipment manufacturer (OEM) appliance designer must have a reliable source of information about these parameters in order to design a battery compartment and circuits that will accommodate the tolerances on battery products available for purchase by the end-user.

4.4   IEC AND ANSI NOMENCLATURE SYSTEMS

It is unfortunate that the various standards identified in Table 4.1 do not share the same nomenclature system. The independent nomenclature systems of the various battery manufacturers even worsen this situation. Cross-references, however, are generally available from battery manufacturers.

4.4.1   Primary Batteries

The IEC nomenclature system for primary batteries, which became effective in 1992, is based on the electrochemical system and the shape and size of the battery. The letter designations for the electrochemical system and the type of cell remain the same as in the previous IEC system for primary batteries. The new numerical designations are based on a diameter/height number instead of the arbitrary size classification used previously. The first digits specify the diameter of the cell in millimeters and the second the height of the cell (millimeters times 10). An example is shown in Table 4.2a. The codes for the shape and electrochemical system are given in Tables 4.2b and 4.2c, respectively. For reference, the ANSI letter codes for the electrochemical systems are also listed in Table 4.2c. The ANSI nomenclature system does not use a code to designate shape.

Nomenclature for existing batteries was grandfathered. Examples of the nomenclature for some of these primary cells and batteries are shown in Table 4.3a. Examples of the IEC nomenclature system for primary batteries are shown in Table 4.3b.

TABLE 4.2a   IEC Nomenclature System for Primary Batteries, Example

Nomenclature	Number of cells	System letter (Table 4.2c)	Shape (Table 4.2b)	Diameter, mm	Height, mm	Example
CR2025	1	C	R	20	2.5	A unit round battery having dimensions shown and electro-chemical system letter C of Table 4.2c (Li/MnO_2)

TABLE 4.2b   IEC Nomenclature for Shape, Primary Batteries

Letter designation	Shape
R	Round-Cylindrical
P	Non-Round
F	Flat (layer built)
S	Square (or Rectangular)

TABLE 4.2c Letter Codes Denoting Electrochemical System of Primary Batteries

ANSI	IEC	Negative electrode	Electrolyte	Positive electrode	Nominal voltage (V)
(1)	—	Zinc	Ammonium chloride, Zinc chloride	Manganese dioxide	1.5
	A	Zinc	Ammonium chloride, Zinc chloride	Oxygen (air)	1.4
LB	B	Lithium	Organic	Carbon monofluoride	3
LC	C	Lithium	Organic	Manganese dioxide	3
	E	Lithium	Non-aqueous inorganic	Thionyl chloride	3.6
LF	F	Lithium	Organic	Iron sulfide	1.5
	G	Lithium	Organic	Copper dioxide	1.5
A[2]	L	Zinc	Alkali metal hydroxide	Manganese dioxide	1.5
Z[3]	P	Zinc	Alkali metal hydroxide	Oxygen (air)	1.4
SO[4]	S	Zinc	Alkali metal hydroxide	Silver oxide	1.55

Notes:
(1)	No suffix	Carbon-zinc
	C	Carbon-zinc industrial
	CD	Carbon-zinc industrial, heavy duty
	D	Carbon-zinc, heavy duty
	F	Carbon-zinc, general purpose
(2)	A	Alkaline
	AC	Alkaline industrial
	AP	Alkaline photographic
(3)	Z	Zinc/air
	ZD	Zinc/air, heavy duty
(4)	SO	Silver oxide
	SOP	Silver oxide photographic

TABLE 4.3a IEC Nomenclature for Typical Primary Round, Flat, and Square Cells or Batteries*

IEC designation	Nominal battery dimensions, mm					ANSI designation	Common designation
	Diameter	Height	Length	Width	Thickness		
Round batteries							
R03	10.5	44.5				24	AAA
R1	12.0	30.2				—	N
R6	14.5	50.5				15	AA
R14	26.2	50.0				14	C
R20	34.2	61.5				13	D
R25	32.0	91.0				—	F
Flat cells							
F22			24	13.5	6.0		
F80			43	43	6.4		
Square batteries							
S4		125.0	57.0	57.0			
S8		200.0	85.0	85.0			

*Chart is not complete—only a sampling of sizes is shown. Dimensions are used for identification only. Complete dimensions can be found in the relevant specification sheets listed in IEC 60086-2 and in Table 4.9a.

TABLE 4.3*b* Examples of IEC Nomenclature for Primary Batteries

IEC Nomenclature	Number of cells	System letter (Table 4.2*c*)	Shape (Table 4.2*b*)	Cell (Table 4.3*a*)	C, P, S, X, Y	Parallel	Groups in parallel	Example
R20	1	None	R	20	*			A unit round battery using basic R20 type cell and electrochemical system letter (none) of Table 4.2*c*
LR20	1	L	R	20	*			Same as above, except using electrochemical system letter L of Table 4.2*c*
6F22	6	None	F	22	*			A 6-series multicell battery using flat F22 cells and electrochemical system letter (none) of Table 4.2*c*
4LR25-2	4	L	R	25	*		2	A multicell battery consisting of two parallel groups, each group having four cells in series of the R25 type and electrochemical system, letter L, of Table 4.2*c*
CR17345	1	C	R	see Section 4.4.1				A unit round battery, with a diam. of 17 mm and height of 34.5 mm, and electrochemical system, letter C, of Table 4.2*c*

*If required, letters C, P, or S will indicate different performance characteristics and letters *X* and *Y* different terminal arrangements.

4.4.2 Rechargeable Batteries

The documentation for standardization of rechargeable batteries is not as complete as the documentation for primary batteries. Most of the primary batteries are used in a variety of portable applications, using user-replaceable batteries. Hence, the need for primary battery standards to insure interchangeability. Developing such standards have been active projects by both IEC and ANSI for many years.

The early use of rechargeable batteries was mainly with larger batteries, however, usually application specific and multicell. The large majority of rechargeable batteries were lead-acid manufactured for automotive SLI (starting, lighting, ignition) use. Standards for these batteries were developed by the Society for Automotive Engineers (SAE), the Battery Council International (BCI) and the Storage Battery Association of Japan. More recently, rechargeable batteries have been developed for portable applications, in many cases in the same cell and battery sizes as the primary batteries. Starting with the portable-sized nickel-cadmium batteries, IEC and ANSI are developing standards for the nickel-metal hydride and lithium-ion batteries. The currently available standards are listed in Tables 4.1*a* and 4.1*b*.

The IEC has been considering a new nomenclature system, possibly covering both primary and rechargeable batteries but none have yet been published. Table 4.4*a* lists the letter codes that are being considered by IEC and those adopted by ANSI for secondary or rechargeable batteries. The IEC nomenclature system for nickel-cadmium batteries is shown in Table. 4.4*b*. In this system, the first letter designates the electrochemical system, a second letter the shape, the first number of the diameter, and a second number the height. In addition, the letters *L*, *M*, and *H* may be used to classify arbitrarily the rate capability as low, medium, or high. The last part of the designation is reserved for two letters which indicate various tab terminal arrangements, such as CF—none, HH—terminal at positive end and positive sidewall, or HB—terminals at positive and negative ends, as shown in Table 4.4*b*.

TABLE 4.4*a* Letter(s) Denoting Electrochemical System of Secondary Batteries

ANSI	IEC[1]	Negative electrode	Electrolyte	Positive electrode	Nominal voltage (V)
H	H	Hydrogen absorbing alloy	Alkali metal hydroxide	Nickel oxide	1.2
K	K	Cadmium	Alkali metal hydroxide	Nickel oxide	1.2
P	PB	Lead	Sulfuric acid	Lead dioxide	2
I	IC	Carbon	Organic	Lithium cobalt oxide	3.6
I	IN	Carbon	Organic	Lithium nickel oxide	3.6
I	IM	Carbon	Organic	Lithium manganese oxide	3.6

[1] Proposed for portable batteries.

TABLE 4.4*b* IEC Nomenclature System for Rechargeable Nickel-Cadmium Cells and Batteries

Nomenclature[2]	System letter (Table 4.4*a*)	Shape (Table 4.2*b*)	Diameter, mm	Height, mm	Terminals	Example
KR 15/51 (R6)	K	R	14.5	50.5	CF	A unit round battery of the K system having dimensions shown, with no connecting tabs

[2] Nomenclature dimensions are shown rounded off. ( ) indicates interchangeable with a primary battery.
Source: IEC 60285.

4.5 TERMINALS

Terminals are another aspect of the shape characteristics for cells and batteries. It is obvious that without standardization of terminals, and the other shape variables, a battery may not be available to match the receptacle facilities provided in the appliance. Some of the variety of terminal arrangements for batteries are listed in Table 4.5.

When applicable, the terminal arrangement is specified in the standard within the same nomenclature designators used for shape and size. The designators thus determine all interchangeable physical aspects of the cells and batteries in addition to the voltage.

TABLE 4.5 Terminal Arrangements for Batteries

Cap and base	Terminals that have the cylindrical side of the battery insulated from the terminal ends
Cap and case	Terminals in which the cylindrical side forms part of the positive end terminal
Screw types	Terminals that have a threaded rod and accept either an insulated or a metal nut
Flat contacts	Flat metal surfaces used for electrical contact
Springs	Terminals that are flat metal strips or spirally wound wire
Plug-in sockets	Terminals consisting of a stud (nonresilient) and a socket (resilient)
Wire	Single or multistranded wire leads
Spring clips	Metal clips that will accept a wire lead
Tabs	Metal flat tabs attached to battery terminals

4.6 ELECTRICAL PERFORMANCE

In terms of the requirement to provide fit and function in the end product the actual appliance does not require specific values of electrical performance. The correct battery voltage, needed to protect the appliance from overvoltage, is assured by the battery designation. Batteries of the same voltage but having differences in capacity can be used interchangeably, but will operate for different service times. The minimum electrical performance of the battery is therefore cited and specified in the standards either by application or by capacity testing.

1. *Application tests:*   This is the preferred method of testing the performance specified for primary batteries. Application tests are intended to simulate the actual use of a battery in a specific application. Table 4.6a illustrates typical application tests.

TABLE 4.6a Example of Application Tests for R20 Type Batteries

Nomenclature				R20C	R20P	R20S	LR20
Electrochemical system				Zinc-carbon (high capacity)	Zinc-carbon (high power)	Zinc-carbon (standard)	Zinc/manganese dioxide
Nominal voltage				1.5	1.5	1.5	1.5
Application	Load, Ω	Daily period	End point	Minimum average duration†			
Portable lighting (1)	2.2	*	0.9	300 min	320 min	100 min	810 min
Tape recorders	3.9	1 h	0.9	9 h	13 h	4 h	25 h
Radios	10	4 h	0.9	30 h	35 h	18 h	81 h
Toys	2.2	1 h	0.8	4.0 h	6 h	2 h	15 h
Portable lighting (2)	1.5	**	0.9	130 min	137 min	32 min	450 min

*4 min beginning at hourly intervals for 8 h/day; **4 min/15 min, 8 h/day.
†Proposed IEC values.

2. *Capacity (service output) tests:* A capacity test is generally used to determine the quantity of electric charge which a battery can deliver under specified discharge conditions. This method is the one that has been generally used for rechargeable batteries. It is also used for primary batteries when an application test would be too complex to simulate realistically or too lengthy to be practical for routine testing. Table 4.6*b* lists some examples of capacity tests.

Test conditions in the standard must consider and therefore specify the following:

Cell (battery) temperature

Discharge rate (or load resistance)

Discharge termination criteria (typically loaded voltage)

Discharge duty cycle

If rechargeable, charge rate, termination criteria (either time or feedback of cell response) and other conditions of charge

Humidity and other conditions of storage may also be required.

TABLE 4.6*b* Example of Capacity Tests

Nomenclature	(Refer to Table 4.3*b*)	SR54
Electrochemical system	(Refer to Table 4.2*c*)	S
Nominal voltage	(Refer to Table 4.2*c*)	1.55

Application*	Load, kΩ	Daily period	End point	Minimum average duration†
Capacity (rating) test	15	24 h	1.2	580 h

* Application for this battery is watches. As an application test could take up to 2 years to test, a capacity test is specified.
† Value under IEC consideration.

4.7 MARKINGS

Markings on both primary and secondary (rechargeable) batteries may consist of some or all of the printed information given in Table 4.7 in addition to the form and dimension nomenclature discussed.

TABLE 4.7 Marking Information for Batteries

Marking information	Primary batteries	Primary small batteries	Rechargeable round batteries
Nomenclature	×*	×	×
Date of manufacture or code	×	××†	×
Polarity	×	×	×
Nominal voltage	×	××	×
Name of manufacturer/supplier	×	××	×
Sealed, rechargeable— nickel-cadmium			×
Charge rate/time			×
Rated capacity			×

* ×—on battery.
† ××—on battery or package.

4.8   CROSS-REFERENCES OF ANSI IEC BATTERY STANDARDS

Table 4.8 lists some of the more popular ANSI battery standards and cross-references to the international standard publications for primary and secondary batteries.

TABLE 4.8a   ANSI/IEC Cross-Reference for Primary Batteries

ANSI	IEC	ANSI	IEC
13A	LR20	1137SO	SR48
13AC	LR20	1138SO	SR54
!3C	R20S	1139SO	SR42
13CD	R20C	1158SO	SR58
13D	R20C	1160SO	SR55
13F	R20S	1162SO	SR57
14A	LR14	1163SO	SR59
14AC	LR14	1164SO	SR59
14C	R14S	1165SO	SR57
14CD	R14C	1166A	LR44
14D	R14C	1170SO	SR55
14F	R14S	1175SO	SR60
15A	LR6	1179SO	SR41
15AC	LR6	1181SO	SR48
15C	R6S	1184SO	SR44
15CD	R6C	1406SOP	4SR44
15D	R6C	1412AP	4LR61
15F	R6S	1414A	4LR44
24A	LR03	1604	6F22
24AC	LR03	1604A	6LR61
24D	R03	1604AC	6LR61
905	R40	1604C	6F22
908	4R25X	1604CD	6F22
910A	LR1	1604D	6F22
918	4R25-2	5018LC	CR17345
918D	4R25-2	5024LC	CR-P2
1107SOP	SR44	5032LC	2CR5
1131SO	SR44	7000ZD	PR48
1132SO	SR43	7001ZD	PR43
1133SO	SR43	7002ZD	PR41
1134SO	SR41	7003ZD	PR44
1135SO	SR41	7005ZD	PR70
1136SO	SR48		

TABLE 4.8b   ANSI/IEC Select Cross-References for Rechargeable Batteries

ANSI	IEC
1.2K1	KR03
1.2K2	KR6
1.2K3	KR14
1.2K4	KR20

4.9 LISTING OF IEC STANDARD ROUND PRIMARY BATTERIES

The tenth edition of IEC 60086-2 for primary batteries lists over one hundred types with dimensional, polarity, voltage, and electrochemical requirements. The third edition of IEC 60285 for rechargeable nickel-cadmium cells (batteries) lists 18 sizes with diameter and height specified in chart form. Several rechargeable nickel-cadmium and nickel-metal hydride batteries are also packaged to be interchangeable with the popular sizes in the primary replacement market. These have physical shapes and sizes that are identical to primary batteries and have equivalent voltage outputs under load. These batteries carry, in addition to the rechargeable nomenclature, the equivalent primary cell or battery size designations and therefore must comply with the dimensional requirements set forth for primary batteries. Table 4.9a lists the dimensions of round primary batteries and Table 4.9b lists some nickel-cadmium rechargeable batteries that are interchangeable with the primary batteries.

In addition to the many designations in national and international standards, of which there may typically be both old and new versions, there are trade association designations. Cross-reference to many of these may be found in sales literature and point-of-purchase information.

TABLE 4.9a Dimensions of Round Primary Batteries

IEC designation	Diameter, mm		Height, mm	
	Maximum	Minimum	Maximum	Minimum
R03	10.5	9.5	44.5	43.3
R1	12.0	10.9	30.2	29.1
R6	14.5	13.5	50.5	49.2
R14	26.2	24.9	50.0	48.6
R20	34.2	32.2	61.5	59.5
R41	7.9	7.55	3.6	3.3
R42	11.6	11.25	3.6	3.3
R43	11.6	11.25	4.2	3.8
R44	11.6	11.25	5.4	5.0
R48	7.9	7.55	5.4	5.0
R54	11.6	11.25	3.05	2.75
R55	11.6	11.25	2.1	1.85
R56	11.6	11.25	2.6	2.3
R57	9.5	9.15	2.7	2.4
R58	7.9	7.55	2.1	1.85
R59	7.9	7.55	2.6	2.3
R60	6.8	6.5	2.15	1.9
R62	5.8	5.55	1.65	1.45
R63	5.8	5.55	2.15	1.9
R64	5.8	5.55	2.7	2.4
R65	6.8	6.6	1.65	1.45
R66	6.8	6.6	2.6	2.4
R67	7.9	7.65	1.65	1.45
R68	9.5	9.25	1.65	1.45
R69	9.5	9.25	2.1	1.85
R1220	12.5	12.2	2.0	1.8
R1620	16	15.7	2.0	1.8
R2016	20	19.7	1.6	1.4
R2025	20	19.7	2.5	2.2
R2032	20	19.7	3.2	2.9
R2320	23	22.6	2.0	1.8
R2430	24.5	24.2	3.0	2.7
R11108	11.6	11.4	10.8	10.4
R12600	12.0	10.7	60.4	58

TABLE 4.9*b*   Dimensions of Some Popular Nickel-Cadmium Rechargeable Batteries That Are Interchangeable with Primary Batteries*

IEC designation†	Consumer designation	ANSI designation	Diameter, mm		Height, mm	
			Maximum	Minimum	Maximum	Minimum
KR03	AAA	1.2K1	10.5	9.5	44.5	43.3
KR6	AA	1.2K2	14.5	13.5	50.5	49.2
KR14	C	1.2K3	26.2	24.9	50.0	48.5
KR20	D	1.2K4	34.3	32.2	61.5	59.5

*IEC plans to add the following to IEC Standard 61436 on Nickel-Metal Hydride batteries: HR03(AAA), HR6 (AA), HR14(C), HR20(D). These will be interchangeable with the respective primary batteries.

See IEC and ANSI standards referenced in Tables 4.1*a* and 4.1*b* for a more complete listing of rechargeable batteries.

†From IEC Standard 60285.

4.10   *STANDARD SLI AND OTHER LEAD-ACID BATTERIES*

SLI battery sizes have been standardized by both the automotive industry through the Society of Automotive Engineers (SAE), Warrendale, PA and the battery industry through the Battery Council International (BCI), Chicago, IL.[1,2] The BCI nomenclature follows the standards adopted by its predecessor, the American Association of Battery Manufacturers (AABM). The latest standards are published annually by the Battery Council International. Table 4.10 is a listing of the Standard SLI and other lead-acid batteries abstracted from the BCI publication.[3]

TABLE 4.10 Standard SLI and Other Lead-Acid Batteries

	BCI group numbers, dimensional specifications and ratings								
	Maximum overall dimensions							Performance ranges	
	Millimeters			Inches			Assembly figure no.	Cold cranking performance amps. @ 0°F (−18°C)	Reserve capacity min @ 80°F (27°C)
BIC group number	L	W	H	L	W	H			
Passenger car and light commercial batteries 12-volt (6 cells)									
21	208	173	222	$8^3/_{16}$	$6^{13}/_{16}$	$8^3/_4$	10	310–400	50–70
21R	208	173	222	$8^3/_{16}$	$6^{13}/_{16}$	$8^3/_4$	11	310–500	50–70
22F	241	175	211	$9^1/_2$	$6^7/_8$	$8^5/_{16}$	11F	220–425	45–90
22HF	241	175	229	$9^1/_2$	$6^7/_8$	9	11F	400	69
22NF	240	140	227	$9^7/_{16}$	$5^1/_2$	$8^{15}/_{16}$	11F	210–325	50–60
22R	229	175	211	9	$6^7/_8$	$8^5/_{16}$	11	290–350	45–90
24	260	173	225	$10^1/_4$	$6^{13}/_{16}$	$8^7/_8$	10	165–625	50–95
24F	273	173	229	$10^3/_4$	$6^{13}/_{16}$	9	11F	250–585	50–95
24H	260	173	238	$10^1/_4$	$6^{13}/_{16}$	$9^3/_8$	10	305–365	70–95
24R	260	173	229	$10^1/_4$	$6^{13}/_{16}$	9	11	440–475	70–95
24T	260	173	248	$10^1/_4$	$6^{13}/_{16}$	$9^3/_4$	10	370–385	110
25	230	175	225	$9^1/_{16}$	$6^7/_8$	$8^7/_8$	10	310–490	50–90
26	208	173	197	$6^3/_{16}$	$6^{13}/_{16}$	$7^3/_4$	10	310–440	50–80
26R	208	173	197	$6^3/_{16}$	$6^{13}/_{16}$	$7^3/_4$	11	405–525	60–80
27	306	173	225	$12^1/_{16}$	$6^{13}/_{16}$	$8^7/_8$	10	270–700	102–140
27F	318	173	227	$12^1/_2$	$6^{13}/_{16}$	$8^{15}/_{16}$	11F	360–660	95–140
27H	298	173	235	$11^3/_4$	$6^{13}/_{16}$	$9^1/_4$	10	440	125
29NF	330	140	227	13	$5^1/_2$	$8^{16}/_{16}$	11F	330–350	95
27R	306	173	225	$12^1/_{16}$	$6^{13}/_{16}$	$8^7/_8$	11	270–700	102–140
33	338	173	238	$13^5/_{16}$	$6^{13}/_{16}$	$9^3/_8$	11F	1050	165
34	260	173	200	$10^1/_4$	$6^{13}/_{16}$	$7^7/_8$	10	375–650	100–110
34R	260	173	200	$10^1/_4$	$6^{13}/_{16}$	$7^7/_8$	11	675	110
35	230	175	225	$9^1/_{16}$	$6^7/_8$	$8^7/_8$	11	310–500	80–110
36R	263	183	206	$10^3/_8$	$7^1/_4$	$8^1/_8$	19	650	130
40R	278	175	175	$10^{15}/_{16}$	$6^7/_8$	$6^7/_8$	15	600–650	110–120
41	293	175	175	$11^9/_{16}$	$6^7/_8$	$6^7/_8$	15	235–650	65–95
42	242	175	175	$9^1/_2$	$6^{13}/_{16}$	$6^{13}/_{16}$	15	260–495	65–95
43	334	175	205	$13^1/_8$	$6^7/_8$	$8^1/_{16}$	15	375	115
45	240	140	227	$9^7/_{16}$	$5^1/_2$	$8^{15}/_{16}$	10F	250–470	60–80
46	273	173	229	$10^3/_4$	$6^{13}/_{16}$	9	10F	350–450	75–95
47	242	175	190	$9^1/_2$	$6^7/_8$	$7^1/_2$	24(A.F)♦	370–550	75–85
48	278	175	190	$12^1/_{16}$	$6^7/_8$	$7^9/_{16}$	24	450–895	85–95
49	353	175	190	$13^7/_8$	$6^7/_8$	$7^9/_{16}$	24	600–810	140–150
50	343	127	254	$13^1/_2$	5	10	10	400–600	85–100
51	238	129	223	$9^3/_8$	$5^1/_{16}$	$8^{13}/_{16}$	10	405–435	70
51R	238	129	223	$9^3/_8$	$5^1/_{16}$	$8^{13}/_{16}$	11	405–435	70
52	186	147	210	$7^5/_{16}$	$5^{13}/_{16}$	$8^1/_4$	10	405	70
53	330	119	210	13	$4^{11}/_{16}$	$8^1/_4$	14	280	40
54	186	154	212	$7^5/_{16}$	$6^1/_{16}$	$8^3/_8$	19	305–330	60
55	216	154	212	$8^5/_8$	$6^1/_{16}$	$8^3/_8$	19	370–450	75
56	254	154	212	10	$6^1/_{16}$	$8^3/_8$	19	450–505	90

TABLE 4.10 Standard SLI and Other Lead-Acid Batteries (*Continued*)

BCI group numbers, dimensional specifications and ratings

BIC group number	Maximum overall dimensions						Assembly figure no.	Performance ranges	
	Millimeters			Inches				Cold cranking performance amps. @ 0°F (−18°C)	Reserve capacity min @ 80°F (27°C)
	L	W	H	L	W	H			
Passenger car and light commercial batteries 12-volt (6 cells) (*Continued*)									
57	205	183	177	$8\frac{1}{16}$	$7\frac{3}{16}$	$6\frac{15}{16}$	22	310	60
58	255	183	177	$10\frac{1}{16}$	$7\frac{3}{16}$	$6\ \frac{15}{16}$	26	380–540	75
58R	255	183	177	$10\frac{1}{16}$	$7\frac{3}{16}$	$6\frac{15}{16}$	19	540–580	75
59	255	193	196	$10\frac{1}{16}$	$7\frac{5}{8}$	$7\frac{3}{4}$	21	540	100
60	332	160	225	$13\frac{1}{16}$	$6\frac{5}{16}$	$8\frac{7}{8}$	12	305–385	65–115
61	192	162	225	$7\frac{9}{16}$	$6\frac{3}{8}$	$8\frac{7}{8}$	20	310	60
62	225	162	225	$8\frac{7}{8}$	$6\frac{3}{8}$	$8\frac{7}{8}$	20	380	75
63	258	162	225	$10\frac{3}{16}$	$6\frac{3}{8}$	$8\frac{7}{8}$	20	450	90
64	296	162	225	$11\frac{11}{16}$	$6\frac{3}{8}$	$8\frac{7}{8}$	20	475–535	105–120
65	306	192	192	$12\frac{1}{16}$	$7\frac{1}{2}$	$7\frac{9}{16}$	21	650–850	130–165
66	306	192	194	$12\frac{1}{16}$	$7\frac{9}{16}$	$7\frac{5}{8}$	13	650–750	130–140
70	208	179	196④	$8\frac{3}{16}$	$7\frac{1}{16}$	$7\frac{11}{16}$④	17	260–525	60–80
71	208	179	216	$8\frac{3}{16}$	$7\frac{1}{16}$	$8\frac{1}{2}$	17	275–430	75–90
72	230	179	210	$9\frac{1}{16}$	$7\frac{1}{16}$	$8\frac{1}{4}$	17	275–350	60–90
73	230	179	216	$9\frac{1}{16}$	$7\frac{1}{16}$	$8\frac{1}{2}$	17	430–475	80–115
74	260	184	222	$10\frac{1}{4}$	$7\frac{1}{4}$	$8\frac{3}{4}$	17	350–550	75–140
75	230	179	196④	$9\frac{1}{16}$	$7\frac{1}{16}$	$7\frac{11}{16}$④	17	430–690	90
76	334	179	216④	$13\frac{1}{8}$	$7\frac{1}{16}$	$8\frac{1}{2}$④	17	750–1075	150–175
78	260	179	196④	$10\frac{1}{4}$	$7\frac{1}{16}$	$7\frac{11}{16}$④	17	515–770	105–115
79	307	179	188	$12\frac{1}{16}$	$7\frac{1}{16}$	$7\frac{3}{8}$	35	880	140
85	230	173	203	$9\frac{1}{16}$	$6\frac{13}{16}$	8	11	550–630	90
86	230	173	203	$9\frac{1}{16}$	$6\frac{13}{16}$	8	10	430–525	90
90	242	175	175	$9\frac{1}{2}$	$6\frac{7}{8}$	$6\frac{7}{8}$	24	520	80
91	278	175	175	11	$6\frac{7}{8}$	$6\frac{7}{8}$	24	600	100
92	315	175	175	$12\frac{1}{2}$	$6\frac{7}{8}$	$6\frac{7}{8}$	24	650	130
93	353	175	175	$13\frac{7}{8}$	$6\frac{7}{8}$	$6\frac{7}{8}$	24	800	150
94R	315	175	190	$12\frac{3}{8}$	$6\frac{7}{8}$	$7\frac{1}{2}$	24	640	135
95R	394	175	190	$15\frac{9}{16}$	$6\frac{7}{8}$	$7\frac{1}{2}$	24	900	190
96R	242	173	175	$9\frac{9}{16}$	$6\frac{13}{16}$	$6\frac{7}{8}$	15	590	95
97R	252	175	190	$9\frac{15}{16}$	$6\frac{7}{8}$	$7\frac{1}{2}$	15	550	90
98R	283	175	190	$11\frac{3}{16}$	$6\frac{7}{8}$	$7\frac{1}{2}$	15	620	120
99	207	175	175	$8\frac{3}{16}$	$6\frac{7}{8}$	$6\frac{7}{8}$	34	360	50
100	260	179	188	$10\frac{1}{4}$	7	$7\frac{5}{16}$	35	770	115
101	260	179	170	$10\frac{1}{4}$	7	$6\frac{11}{16}$	17	690	115

TABLE 4.10 Standard SLI and Other Lead-Acid Batteries (*Continued*)

								Performance ranges	
	BCI group numbers, dimensional specifications and ratings								
BIC group number	Maximum overall dimensions						Assembly figure no.	Cold cranking performance amps. @ 0°F (−18°C)	Reserve capacity min @ 80°F (27°C)
	Millimeters			Inches					
	L	W	H	L	W	H			
Passenger car and light commercial batteries 6-volt (3 cells)									
1	232	181	238	9⅛	7⅛	9⅜	2	475–650	135–230
2	264	181	238	10⅜	7⅛	9⅜	2	475–650	136–230
2E	492	105	232	19⁷⁄₁₆	4⅛	9⅛	5	485	140
2N	254	141	227	10	5⁹⁄₁₆	8¹⁵⁄₁₆	1	450	135
17HF-*①	187	175	229	7⅜	6⅞	9	2B	—	—
Heavy-duty commercial batteries 12-volt (6 cells)									
4D	527	222	250	20¾	8¾	9⅞	8	490–950	225–325
6D	527	254	260	20¾	10	10¼	8	750	310
8D	527	283	250	20¾	11⅛	9⅞	8	850–1250	235–465
28	261	173	240	10⁵⁄₁₆	6¹³⁄₁₆	9⁷⁄₁₆	18	400–535	80–135
29H	334	171	232	13⅛	6¾	9⅛	10	525–840	145
30H	343	173	235	13½	6¹³⁄₁₆	9¼	10	380–685	120–150
31	330	173	240	13	6¹³⁄₁₆	9⁷⁄₁₆	18(A,T)♦	455–950	100–200
Heavy-duty commercial batteries 6-volt (3 cells)									
3	298	181	238	11¾	7⅛	9⅜	2	525–660	210–230
4	334	181	238	13⅛	7⅛	9⅜	2	550–975	240–420
5D	349	181	238	13¾	7⅛	9⅜	2	720–820	310–380
7D	413	181	238	16¼	7⅛	9⅜	2	680–875	370–426
Special tractor batteries 6-volt (3 cells)									
3EH	491	111	249	19⁵⁄₁₆	4⅜	9¹³⁄₁₆	5	740–850	220–340
4EH	491	127	249	19⁵⁄₁₆	5	9¹³⁄₁₆	5	850	340–420
Special tractor batteries 12-volt (6 cells)									
3EE	491	111	225	19⁵⁄₁₆	4⅜	8⅞	9	260–360	80–105
3ET	491	111	249	19⁵⁄₁₆	4⅜	9³⁄₁₆	9	355–425	130–135
4DLT	508	208	202	20	8³⁄₁₆	7¹⁵⁄₁₆	16L	650–820	200–290
12T	177	177	202	7¹⁄₁₆	6¹⁵⁄₁₆	7¹⁵⁄₁₆	10	460	160
16TF	421	181	283	16⁹⁄₁₆	7⅛	11⅛	10F	600	240
17TF	433	177	202	17¹⁄₁₆	6¹⁵⁄₁₆	7¹⁵⁄₁₆	11L	510	145

TABLE 4.10 Standard SLI and Other Lead-Acid Batteries (*Continued*)

BCI group numbers, dimensional specifications and ratings

BIC group number	Maximum overall dimensions: Millimeters L	Millimeters W	Millimeters H	Inches L	Inches W	Inches H	Assembly figure no.	Performance ranges: Cold cranking performance amps. @ 0°F (−18°C)	Reserve capacity min @ 80°F (27°C)
General-utility batteries 12-volt (6 cells)									
U1	197	132	186	$7\frac{3}{4}$	$5\frac{3}{16}$	$7\frac{5}{16}$	10(X)♦	120–295	23–40
U1R	197	132	186	$7\frac{3}{4}$	$5\frac{3}{16}$	$7\frac{5}{16}$	11(X)♦	220–235	25–37
U2	160	132	181	$6\frac{5}{16}$	$5\frac{3}{16}$	$7\frac{1}{8}$	10(X)♦	120	17
Electric golf car/utility batteries 6-volt (3 cells)									
GC2	264	183	290	$10\frac{3}{8}$	$7\frac{3}{16}$	$11\frac{7}{16}$	2	③	③
GC2H②	264	183	295	$10\frac{3}{8}$	$7\frac{3}{16}$	$11\frac{5}{8}$	2	③	③
Electric golf car/utility batteries 8-volt (4 cells)									
GC8	264	183	290	$10\frac{3}{8}$	$7\frac{3}{16}$	$11\frac{7}{16}$	31	—	—
Electric vehicle batteries 8-volt (4 cells)									
202	388	116	175	$15\frac{1}{4}$	$4\frac{9}{16}$	$6\frac{7}{8}$	30	—	—
Electric vehicle batteries 12-volt (6 cells)									
201	388	116	175	$15\frac{1}{4}$	$4\frac{9}{16}$	$6\frac{7}{8}$	29	—	—
Electric vehicle batteries 12-volt (10 cells)									
203	388	116	175	$15\frac{1}{4}$	$4\frac{9}{16}$	$6\frac{7}{8}$	32	—	—
203R	388	116	175	$15\frac{1}{4}$	$4\frac{9}{16}$	$6\frac{7}{8}$	33	—	—
Ordnance batteries 12-volt (6 cells)									
2H	260	135	230	$10\frac{1}{4}$	$5\frac{9}{16}$	$\frac{9}{16}$	28	—	75
6T	286	267	230	$11\frac{1}{4}$	$10\frac{1}{2}$	$\frac{9}{16}$	27	600	200

♦ Letter in parentheses indicates terminal type.

* Rod end types—Extend top ledge with holes for holddown bolts.

† Ratings for batteries recommended for motor coach and bus service are for double insulation. When double insulation is used in other types, deduct 15% from the rating values for cold cranking performance.

① Not in application section but still manufactured.

② Special-use battery not shown in application section.

③ Capacity test in minutes at 75 amperes to 5.25 volts at 80°F (27°C); cold cranking performance test not normally required for this battery.

④ Maximum height dimension shown includes batteries with raised-quarter cover design. Flat-top design model height (minus quarter covers) reduced by approximately $\frac{3}{8}$ inch (10 mm).

TABLE 4.10 Standard SLI and Other Lead-Acid Batteries (*Continued*)

BCI ASSEMBLY NUMBERS, CELL LAYOUTS, HOLDDOWNS AND POLARITY

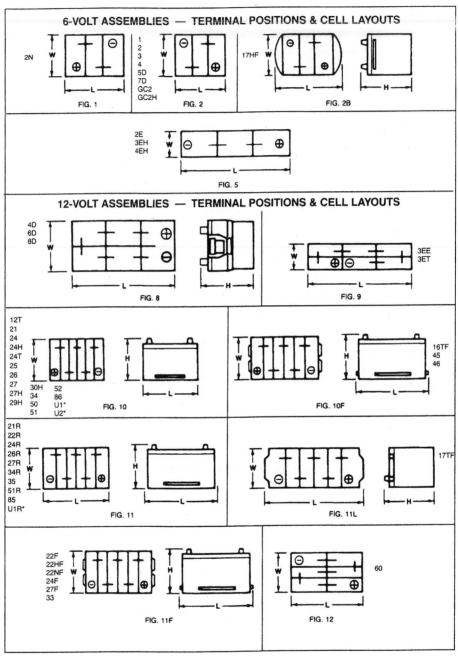

TABLE 4.10 Standard SLI and Other Lead-Acid Batteries (*Continued*)

BCI ASSEMBLY NUMBERS, CELL LAYOUTS, HOLDDOWNS AND POLARITY

12-VOLT ASSEMBLIES — TERMINAL POSITIONS & CELL LAYOUTS

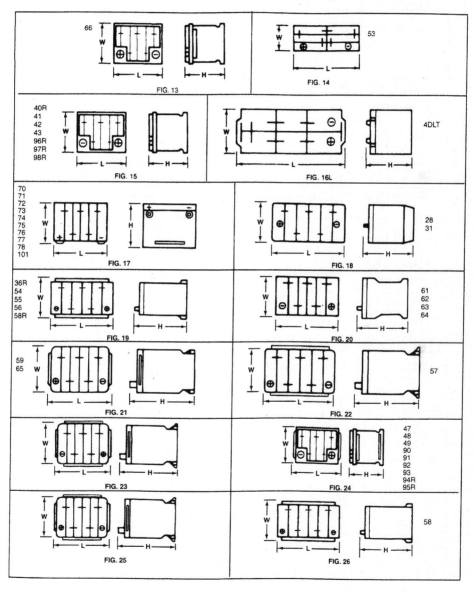

TABLE 4.10   Standard SLI and Other Lead-Acid Batteries (*Continued*)

BCI ASSEMBLY NUMBERS, CELL LAYOUTS, HOLDDOWNS AND TERMINALS

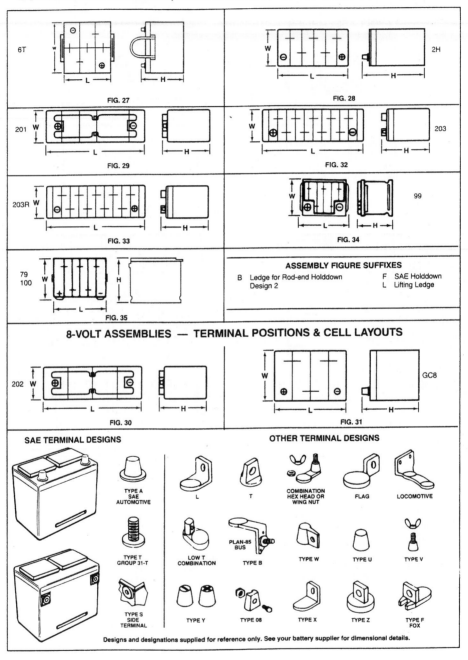

FIG. 27

FIG. 28

FIG. 29

FIG. 32

FIG. 33

FIG. 34

FIG. 35

ASSEMBLY FIGURE SUFFIXES

B   Ledge for Rod-end Holddown Design 2

F   SAE Holddown

L   Lifting Ledge

8-VOLT ASSEMBLIES — TERMINAL POSITIONS & CELL LAYOUTS

FIG. 30

FIG. 31

SAE TERMINAL DESIGNS

OTHER TERMINAL DESIGNS

TYPE A SAE AUTOMOTIVE

L

T

COMBINATION HEX HEAD OR WING NUT

FLAG

LOCOMOTIVE

TYPE T GROUP 31-T

LOW T COMBINATION

PLAN-85 BUS

TYPE B

TYPE W

TYPE U

TYPE V

TYPE S SIDE TERMINAL

TYPE Y

TYPE 08

TYPE X

TYPE Z

TYPE F FOX

Designs and designations supplied for reference only. See your battery supplier for dimensional details.

4.11 REGULATORY AND SAFETY STANDARDS

With the increasing complexity and energy of batteries and the concern about safety, greater attention is being given to developing regulations and standards with the goal to promote the safe operation in use and transport. Stand-alone safety documents on primary and rechargeable batteries have been published by IEC and ANSI. In addition, the Underwriters Laboratories (UL) has published several battery safety standards aimed at the safe operation of UL-approved equipment.[4]

Table 4.11*a* is a list of organizations working on safety standards and the safety standards they prepared that cover various primary and secondary battery systems.

While the various groups involved in developing safety standards are dedicated to the principle of harmonization, there are still differences in the procedures, tests, and criteria between the various standards. It is recommended that users of these standards follow them on a judicious basis, and place their battery or application in the proper context.

Table 4.11*b* is a list of organizations that have focused on the safe transport of various goods and the regulations they have published. These regulations include procedures for the transport of batteries, including lithium batteries.

TABLE 4.11*a* Safety Standards

Publication	Title
American Standards Institute	
ANSI C18.1M, Part 2	American National Standard for Portable Primary Cells and Batteries with Aqueous Electrolyte—Safety Standard
ANSI C18.2M, Part 2	American National Standard for Portable Rechargeable Cells and Batteries—Safety Standard
ANSI C18.3M, Part 2	American National Standard for Portable Lithium Primary Cells and Batteries—Safety Standard
International Electrotechnical Commission	
IEC 60086-4	Primary Batteries—Part 4: Safety for Lithium Batteries
IEC 60086-5	Primary Batteries—Part 5: Safety of Batteries with Aqueous Electrolyte
IEC 61809	Safety for Portable Sealed Alkaline Secondary Cells and Batteries
Underwriters Laboratories	
UL1642	Standard for Lithium Batteries
UL2054	Standard for Household and Commercial Batteries

TABLE 4.11*b* Transportation Recommendations and Regulations

Organization	Title
Department of Transportation (DOT)	Code of Federal Regulations—Title 49 Transportation
Federal Aviation Administration (FAA)	TSO C042, Lithium Batteries (referencing RTCA Document DO-227 "Minimum Operational Performance Standards for Lithium Batteries")
International Air Transport Association (IATA)	Dangerous Goods Regulations
International Civil Aviation Association (ICAO)	Technical Instructions for the Safe Transport of Dangerous Goods
United Nations (UN)	Recommendations on the Transportation of Dangerous Goods Manual of Tests and Criteria

In the United States, this responsibility for regulating the transport of goods rests with the Department of Transportation (DOT) through its Research and Special Programs Administration (RSPA).[5] These regulations are published in the Code of Federal Regulations (CFR49), which include the requirements for the shipment and transport of batteries under all modes of transportation. Under the DOT, the Federal Aviation Administration (FAA) is responsible for the safe operation of aircraft and has also issued regulations covering the use of batteries in aircraft.[6,7] Similar organizations are part of the governments of most countries throughout the world.

Internationally, transport is regulated by such organizations as the International Civil Aviation Association (ICAO),[8] the International Air Transport Association (IATA)[9] and the International Maritime Organization. Their regulations are guided by the United Nations (UN) through their Committee of Experts on the Transport of Dangerous Goods, which has developed recommendations for the transportation of dangerous goods. These recommendations, which also include tests and criteria,[10,11] are addressed to governments and international organizations concerned with regulating the transport of various products. Currently, the UN Committee of Experts is developing new guidelines covering the transport of lithium primary and secondary batteries. The quantity of lithium or lithium equivalent content in each cell and battery will determine which specific rules and regulations are applied concerning the packaging, mode of shipment, marking, and other special provisions.

As these standards, regulations, and guidelines can be changed on an annual or periodic basis, the current edition of each document should be used.

NOTE

It is imperative that only the latest version of each standard be used. Due to the periodic revision of these standards, only the latest version can be relied upon to provide reliable enforceable specifications of battery dimensions, terminals, marking, general design features, conditions of electrical testing for performance verification, mechanical tests, test sequences, safety, shipment, storage, use, and disposal.

REFERENCES

1. Society of Automotive Engineers, 400 Commonwealth Drive, Warrendale, PA 15096. www.sae.org

2. Battery Council International, 401 North Michigan Ave., Chicago, IL 60611. www.batterycouncil.org

3. Battery Council International, *Battery Replacement Data Book,* 2000.

4. Underwriters Laboratories, Inc., 333 Pfingsten Road, Northbrook, IL 60062.

5. Department of Transportation, Office of Hazardous Materials Safety, Research and Special Programs Administration, 400 Seventh St., SW, Washington, DC 20590.

6. Department of Transportation, Federal Aviation Administration, 800 Independence Ave., SW, Washington, DC 20591.

7. RTCA, 1140 Connecticut Ave., NE, Washington, DC 20036.

8. International Civil Aviation Organization, 1000 Sherbrooke St., W., Montreal, Quebec, Canada.

9. International Air Transport Association, 2000 Peel St., Monteal, Quebec, Canada.

10. United Nations, *Recommendation on the Transport of Dangerous Goods,* New York, NY and Geneva, Switzerland.

11. United Nations, *Manual of Tests and Criteria,* New York, NY and Geneva, Switzerland.

CHAPTER 5
BATTERY DESIGN

John Fehling

5.1 GENERAL

Proper design of the battery or the battery compartment is important to assure optimum, reliable, and safe operation. Many problems attributed to the battery may have been prevented had proper precautions been taken with both the design of the battery itself and how it is designed into the battery-operated equipment.

It is important to note that the performance of a cell in a battery can be significantly different from that of an individual cell depending on the particular environment of that cell in the battery. Specifications and data sheets provided by the manufacturers should only be used as a guide as it is not always possible to extrapolate the data to determine the performance of multicell batteries. Such factors as the cell uniformity, number of cells, series or parallel connections, battery case material and design, conditions of discharge and charge, temperature, to name a few, influence the performance of the battery. The problem is usually exacerbated under the more stringent conditions of use, such as high-rate charging and discharging, operation, and extreme temperatures and other conditions which tend to increase the variability of the cells within the battery.

Further, specific energy and energy density data based on cell or single-cell battery performance have to be derated when the weight and volume of the battery case, battery assembly materials, and any ancillary equipment in the battery have to be considered in the calculation.

Another factor that must be considered, particularly with newly developing battery technologies, is the difficulty of scaling up laboratory data based on smaller individual batteries to multicell batteries using larger cells manufactured on a production line.

This chapter will address the issues that the product designer should consider. Cell and battery manufacturers should also be consulted to obtain specific details on their recommendations for the batteries they market.

5.2 DESIGNING TO ELIMINATE POTENTIAL SAFETY PROBLEMS

Batteries are sources of energy and when used properly will deliver their energy in a safe manner. There are instances, however, when a battery may vent, rupture, or even explode if it is abused. The design of the battery should include protective devices and other features which can prevent, or at least minimize, the problem.

Some of the most common causes for battery failure are:

1. Short-circuiting of battery terminals
2. Excessive high rate discharge or charge
3. Voltage reversal, that is, discharging the cells of the battery below 0 V
4. Charging of primary batteries
5. Improper charge control when charging secondary batteries

These conditions may cause an internal pressure increase within the cells, resulting in an activation of the vent device or a rupture or explosion of the battery. There are a number of means to minimize the possibilities of these occurrences.

5.2.1 Charging Primary Batteries

All major manufacturers of primary batteries warn that the batteries may leak or explode if they are charged. As discussed in Sec. 7.4, some primary batteries can be recharged if done under controlled conditions. Nevertheless, charging primary batteries is not usually recommended because of the potential hazards.

Protection from External Charge. The simplest means of preventing a battery from being charged from an external power source is to incorporate a blocking diode in the battery pack, as shown in Fig. 5.1. The diode chosen must have a current rating in excess of the operating current of the device. It should be rated, at a minimum, at twice the operating current. The forward voltage drop of the diode should be as low as possible. Schottky diodes are commonly used because of their typical 0·5-V drop in the forward direction. Another consideration in selecting the diode is the reverse voltage rating. The peak inverse voltage (PIV) rating should be at least twice the voltage of the battery.

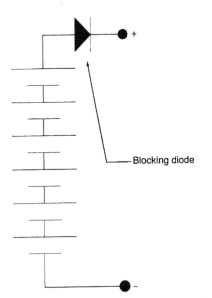

FIGURE 5.1 Battery circuit incorporating a blocking diode to prevent charge.

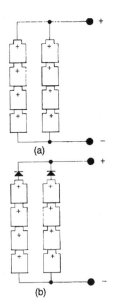

FIGURE 5.2 Series/parallel battery. (*a*) Without diode protection. (*b*) With diode protection.

Protection from Charging within Battery.   When multiple series stacks are paralleled within a battery pack, charging may occur when a detective or a low-capacity cell is present in one of the stacks (Fig. 5.2*a*). The remaining stacks of cells will charge the stack with the defective cell. At best this situation will discharge the good stack, but it could result in rupture of the cells in the weak stack. To avoid this, diodes should be placed in each series string to block charging currents from stack to stack (Fig. 5.2*b*).

When diode protection is used in each series stack, the diode will prevent the stack containing the defective cell from being charged. The diode should have the following characteristics:

1. Forward voltage drop should be as low as possible, Schottky type preferable.

2. Peak inverse voltage should be rated at twice the voltage of the individual series stack.

3. Forward current rating of the diodes should be a minimum of

$$I_{\min} = \frac{I_{op}}{N} \times 2$$

where I_{op} = device operating current
N = number of parallel stacks

5.2.2 Preventing Battery Short-Circuit Conditions

When a battery is short-circuited through the external terminals, the chemical energy is converted into heat within the battery. In order to prevent short-circuiting, the positive and negative terminals of the battery should be physically isolated. Effective battery design will incorporate the following:

1. The battery terminals should be recessed within the external case (Fig. 5.3*a*).

2. If connectors are used, the battery should incorporate the female connection. The connector should also be polarized to only permit for correct insertion (Fig. 5.3*b*).

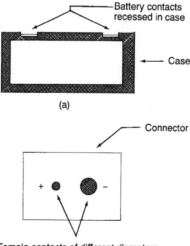

FIGURE 5.3   Design of battery terminals.

Additional Short-Circuit Protection.   In addition it may be also necessary to include some means of circuit interruption. There are a number of devices which can perform this function, including:

1. Fuses or circuit breakers
2. Thermostats designed to open the battery circuit when the temperature or current reaches a predetermined upper limit
3. Positive-temperature-coefficient (PTC) devices that, at normal currents and temperatures, have a very low value of resistance. When excessive currents pass through these devices or the battery temperature increases, the resistance increases orders of magnitude, limiting the current. These devices are incorporated internally in some cells by the cell manufacturer. When using cells with internal protection, it is advisable to use an external PTC selected to accommodate both the current and the voltage levels of the battery application (see Sec. 5.5.1).

5.2.3 Voltage Reversal

Due to variability in manufacturing, capacities will vary from battery to battery. When discharged in a series configuration, the capacity of the weakest cell in the series string of a multicell battery will be depleted before the others. If the discharge is continued, the voltage of the low-capacity cell will reach 0 V and then reverse. The heat generated may eventually cause pressure buildup in the cell and subsequent venting or rupture. This process is sometimes referred to as "forced discharge."

A common test to determine the ability of cells to withstand voltage reversal is the forced-discharge test. The cells are deliberately discharged, at specified currents, to below 0 V by other cells in a series string or by an external power supply to determine whether a venting, rupture, or other undesired safety problem arises.

Some cells are designed to withstand a forced discharge to specified discharge currents. The cells may also be designed with internal protection, such as fuses or thermal cutoff devices, to interrupt the discharge if an unsafe condition develops.

This condition of cell unbalance could be exacerbated with rechargeable cells as the individual cell capacities could change during cycling. To minimize this effect, rechargeable batteries should at least be constructed with "matched" cells, that is, cells having nearly identical capacities. Cells are sorted, within grades, by at least one cycle of charge and discharge. Typically cells are considered matched when the capacity range is within 3%. Recent advances in manufacturing control have reduced the number of cell grades. Some manufacturers have reached the optimal goal of one grade, which negates the need of matching. This information is readily available from the battery companies.

Battery Design to Prevent Voltage Reversal.   Even though matched cells are used, other battery designs or applications can cause an imbalance in cell capacity. One example is the use of voltage taps on cells of a multicell battery in a series string. In this design, the cells are not discharged equally. Figure 5.4 illustrates a battery incorporating voltage taps that could result in voltage reversal.

For illustration assume $I_2 = 3$ A, $I_1 = 2$ A, and the cell capacity is 15 Ah. Cells 3 and 4 are being discharged at the 5-h rate, while cells 1 and 2 are being discharged at the combination of $I_1 + I_2$, or 5 A, a 3-h rate. After 3 h, cells 1 and 2 will be almost depleted of capacity and will go into voltage reversal if the discharge is continued. Many early battery designs using Leclanché-type cells incorporated the use of voltage taps. Batteries with as many as 30 cells in series (45 V) were common with taps typically at 3, 9, 13.5 V, and so on. When the cells with the lower voltage taps were discharged, they could leak. This leakage could cause corrosion, but usually these cells would not be prone to rupture. With the advent

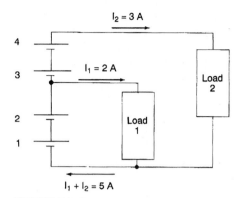

FIGURE 5.4 Battery circuit using voltage taps.

of the high-energy, tightly sealed cells, this is no longer the case. Cells driven into voltage reversal may rupture or explode. In order to avoid problems, the battery should be designed with electrically independent sections for each voltage output. If possible, the device should be designed to be powered by a single input voltage source. DC to DC converters can be used to safely provide for multiple voltage outputs. Converters are now available with efficiencies greater than 90%.

Parallel Diodes to Prevent Voltage Reversal.   Some battery designers, particularly for multicell lithium primary batteries, add diodes in parallel to each cell to limit voltage reversal. As the cell voltage drops below zero volts and into reversal, the diode becomes conducting and diverts most of the current from flowing through the cell. This limits the extent of the voltage reversal to that of the characteristic of the diode. This use of diodes is shown in Fig. 5.6.

5.2.4 Protection of Cells and Batteries from External Charge

Many battery-powered devices are also operated from rectified alternating-current (AC) sources. These could include devices which offer both AC and battery operation or devices which use the battery for backup when the AC power supply fails or is not available.

In the case where the battery is a backup for the main power supply as, for example, in memory backup, the primary battery must be protected from being charged by the main power supply. Typical circuits are shown in Fig. 5.5. In Fig. 5.5*a* two blocking diodes are used redundantly to provide protection in case of the failure of one. A resistor is used in Fig. 5.5*b* to limit the charge current if the diode fails in a closed position. This blocking diode should have the features of a low voltage drop in the forward direction to minimize the loss of battery backup voltage, and a low leakage current in the reverse direction to minimize the charging current.

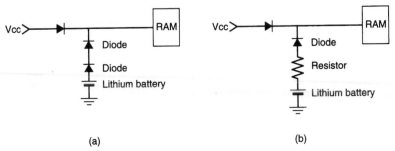

FIGURE 5.5   Protective circuitry for memory backup batteries. (*a*) Using two diodes. (*b*) Using diode and resistor, V_{cc} = power supply voltage.

5.2.5   Special Considerations When Designing Lithium Primary Batteries

Lithium primary batteries contain an anode of elemental lithium (see Chap. 14) and, because of the activity of this metal, special precautions may be required in the design and use of the batteries, particularly when multiple cells are used in the battery pack. Some of the special precautions that should be taken in the design of these batteries, include the following:

1. When multiple cells are required, due to voltage and/or the capacity requirement of the application, they should be welded into battery packs, thus preventing the user mixing cells of different chemistries or capacities if replaceable cells were used.

2. A thermal disconnect device should be included to prevent the build-up of excessive heat. Many of the batteries now manufactured include a PTC or a mechanical disconnect, or both, within the cell. Additional protective thermal devices should be included, external to the cells, in the design of a multicell battery pack.

3. The following protective devices should be included:
   a. Series diode protection to prevent charging must be included
   b. Cell bypass diode protection to prevent excessive voltage reversal of individual cells in a multicell series and/or series parallel configuration
   c. Short circuit protection by means of a PTC, permanent fuse or electronic means, or a combination of all three

4. In order to make the used battery safe for disposal, for some lithium batteries the remaining lithium within the battery must be depleted. This is accomplished by placing a resistive load across the cell pack to completely discharge the battery after use. The resistive load should be chosen to ensure a low current discharge, typically at a five (5) day rate of the original capacity of the battery. This feature has been used mainly in military primary lithium batteries.

Figure 5.6 illustrates a typical schematic showing the use of the safety features discussed.

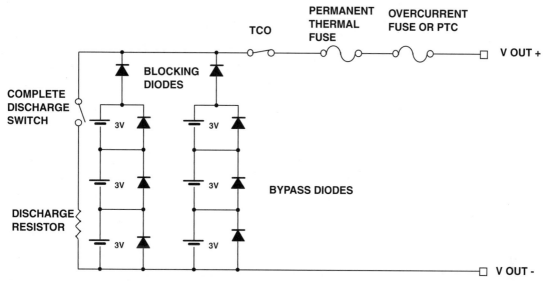

FIGURE 5.6 Lithium primary battery schematic with series and bypass protection.

5.3   BATTERY SAFEGUARDS WHEN USING DISCRETE BATTERIES

5.3.1   Design to Prevent Improper Insertion of Batteries

When designing products using individual single-cell batteries, special care must be taken in the layout of the battery compartment. If provisions are not made to ensure the proper placement of the batteries, a situation may result in which some of the batteries that are improperly inserted could be exposed to being charged. This could lead to leakage, venting, rupture, or even explosion. Figure 5.7 illustrates simple battery-holder concepts for cylindrical and button batteries, which will prevent the batteries from being inserted incorrectly. Figure 5.8 shows several other design options for preventing improper installation.

Two commonly used battery circuits that are potentially dangerous without proper battery orientation are:

1. Series/parallel with one battery reversed (Fig. 5.9). In this circuit, battery 3 has been reversed. As a result, batteries 1–3 are now in series and are charging battery 4. This condition can be avoided, if possible, by using a single series string of larger batteries. Further, as discussed in Sec. 5.2.1, the use of diodes in each series section will at least prevent one parallel stack from charging the other.

2. Multicell series stack with one battery reversed in position (Fig. 5.10). The fourth battery is reversed and will be charged when the circuit is closed to operate the device. Depending on the magnitude of the current, the battery may vent or rupture. The magnitude of the current is dependent on the device load, the battery voltage, the condition of the reversed battery, and other conditions of the discharge.

To minimize the possibility of physically reversing a battery, the proper battery orientation should be clearly marked on the device, with simple and clear instructions. Blind battery compartments, where the individual batteries are not visible, should be avoided. The best practice is to use oriented or polarized battery holders, as discussed previously.

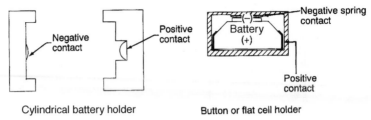

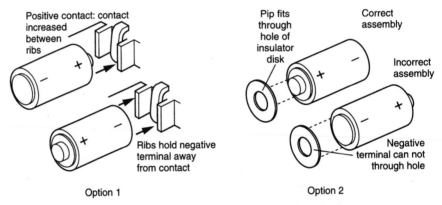

Cylindrical battery holder   Button or flat cell holder

FIGURE 5.7   Battery holders. (Left) Cylindrical. (Right) Button or flat.

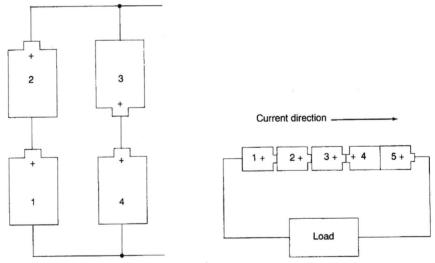

Option 1   Option 2

FIGURE 5.8   Battery contact designs that prevent reverse installation of cells.

FIGURE 5.9   Series/parallel circuit; cell 4 being charged.

FIGURE 5.10   One cell reversed in a series stack; cell 4 being charged.

5.3.2 Battery Dimensions

At times equipment manufacturers may design the battery cavity of their device around the battery of a single manufacturer. Unfortunately the batteries made by the various manufacturers are not exactly the same size. While the differences may not be great, this could result in a cavity design that will not accept batteries of all manufacturers.

Along with variations in size, the battery cavity design must also be able to accommodate unusual battery configurations that fall within IEC standards. For example, several battery manufacturers offer batteries with negative recessed terminals that are designed to prevent contact when they are installed backward. Unfortunately negative recessed terminals will mate only with contacts whose width is less than the diameter of the battery's terminal. Figure 5.11*a* illustrates the dimensional differences between cells with standard and recessed terminals.

The battery cavity should not be designed around the battery of a single manufacturer whose battery may be a unique size or configuration. Instead, cavity designs should be based on International Electrotechnical Commission (IEC) standards and built to accommodate maximum and minimum sizes. IEC and ANSI standards (see Chap. 4) provide key battery dimensions, including overall height, overall diameter, pip diameter, pip height, and diameter of negative cap. Maximum and minimum values are usually specified, as shown in Fig. 5.11*b*. As these standards are revised periodically, the latest edition should be used.

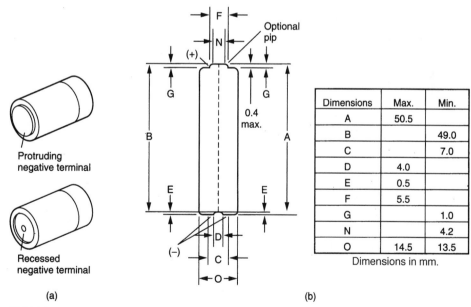

Dimensions	Max.	Min.
A	50.5	
B		49.0
C		7.0
D	4.0	
E	0.5	
F	5.5	
G		1.0
N		4.2
O	14.5	13.5

Dimensions in mm.

FIGURE 5.11   (*a*) Types of battery terminals falling within IEC standards. (*b*) Illustration of typical standard IEC dimensions.

5.4 *BATTERY CONSTRUCTION*

The following constructional features also should be considered in the design and fabrication of batteries:

1. Intercell connections
2. Encapsulation of cells
3. Case configuration and materials
4. Terminals and contact materials

5.4.1 Intercell Connections

Soldering is the method of connection for batteries using Leclanché-type cells. Wires are soldered between the negative zinc can and the adjoining positive cap. This effective method of construction for these cells is still widely used.

Welding of conductive tabs between cells is the preferred method of intercell connection for most of the other battery systems. The tab materials for most applications are either pure nickel or nickel-plated steel. The corrosion resistance of the nickel and its ease of welding result in reliable permanent connections. The resistance of the tab material must be matched to the application to minimize voltage loss. The resistance can be calculated from the resistivity of the material, which is normally expressed in ohm-centimeters,

$$\text{Resistance} = \frac{\text{resistivity} \times \text{length (cm)}}{\text{cross-sectional area (cm}^2)}$$

The resistivity values of nickel and nickel-plated steel are

Nickel	$6.84 \times 10^{-6} \ \Omega \cdot \text{cm}$
Nickel-plated steel	$10 \times 10^{-6} \ \Omega \cdot \text{cm}$

For example, the resistance of a tab with dimensions of 0.635-cm width, 0.0127-cm thickness, and 2.54-cm length is

for nickel,

$$\frac{6.84 \times 10^{-6} \times 2.54}{0.635 \times 0.0127} = 2.15 \times 10^{-3} \ \Omega$$

for nickel-plated steel,

$$\frac{10 \times 10^{-6} \times 2.54}{0.635 \times 0.0127} = 3.15 \times 10^{-3} \ \Omega$$

As is evident, the resistance of the nickel-plated steel material is 50% higher than that of nickel for an equivalent-size tab. Normally this difference is of no significance in the circuit, and nickel-plated steel is chosen due to its lower cost.

Resistance spot welding is the welding method of choice. Care must be taken to ensure a proper weld without burning through the cell container. Excessive welding temperatures could also result in damage to the internal cell components and venting may occur. Typically AC or capacitance discharge welders are used.

Both types of welders incorporate two electrodes, typically made of a copper alloy. A current path is established between the electrodes, melting and fusion of the materials will occur at the interface of the tab and the cell due to resistance heating. Figure 5.12 illustrates the commonly used welding techniques. The method shown in Fig. 5.12a is used in more than 90% of the joints where a tab is welded to a cell surface. Two weld nuggets are formed for each weld action. When welding circular leads to a cell or tab surface, the procedure shown in Fig. 5.12b will result in one weld spot per weld action. The procedure in Fig. 5.12c is commonly used when a tab-to-tab weld or similar joints are needed. This latter method is not recommended for welding to a cell.

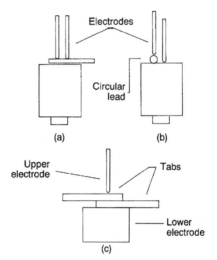

FIGURE 5.12 Various welding configurations used in battery construction.

In all instances the weld should have a clean appearance, with discoloration of the base materials kept to a minimum. At least two weld spots should be made at each connection joint. When the weld is tested by pulling the two pieces apart, the weld must hold while the base metal tears. For tabs the weld diameter, as a rule of thumb, should be three to four times the tab thickness. For example, a 0.125-mm-thick tab should have a tear diameter of 0.375–0.5 mm. Statistical techniques of weld pull strength for process control are helpful, but a visual inspection of the weld diameter must accompany the inspection process.

The least preferred method of battery connection is the use of pressure contacts. Although this technique is used with some inexpensive consumer batteries, it can be the cause of battery failure where high reliability is desired. This type of connection is prone to corrosion at the contact points. In addition, under shock and vibration intermittent loss of contact may result.

5.4.2 Cell Encapsulation

Most applications require that the cells within the battery be rigidly fixed in position. In many instances this involves the encapsulation of the cells with epoxy, foams, tar, or other suitable potting materials.

Care must be taken to prevent the potting material from blocking the vent mechanisms of the cells. A common technique is to orient the cell vents in the same direction and encapsulate the battery to a level below the vent, as shown in Fig. 5.13. If possible the preferred method to keep the cells immobile, within the battery, is through careful case design without the use of potting materials. Although this method may increase initial tooling costs, future labor savings could be realized.

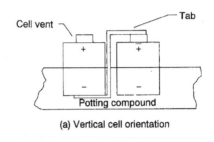

(a) Vertical cell orientation

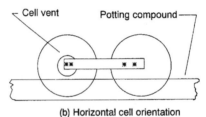

(b) Horizontal cell orientation

FIGURE 5.13 Battery encapsulation techniques. (*a*) Vertical cell orientation. (*b*) Horizontal cell orientation.

5.4.3 Case Design

Careful design of the case should include the following:

1. Materials must be compatible with the cell chemistry chosen. For example, aluminum reacts with alkaline electrolytes and must be protected where cell venting may occur.
2. Flame-retardant materials may be required to comply with end-use requirements. Underwriters Laboratories, the Canadian Standards Association, and other agencies may require testing to ensure safety compliance.
3. Adequate battery venting must allow for the release of vented cell gases. In sealed batteries this requires the use of a pressure relief valve or breather mechanisms.
4. The design must provide for effective dissipation of heat to limit the temperature rise during use and especially during charge. High temperatures should be avoided as they reduce charge efficiency, increase self-discharge, could cause cell venting, and generally are detrimental to battery life. The temperature increase is greater for a battery pack than

for an individual or separated cells as the pack tends to limit the dissipation of heat. The problem is exacerbated when the pack is enclosed in a plastic case. This is illustrated in Fig. 5.14, which compares the temperature rise of groups of cells with and without a battery case.

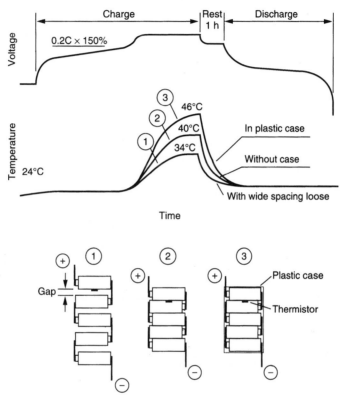

FIGURE 5.14   Temperature increase characteristics during charge of battery pack.

5.4.4   Terminal and Contact Materials

Terminal material selection must be compatible with the environments of the battery contents as well as the surroundings. Noncorrosive materials should be selected. Nickel-cadmium and Nickel–metal hydride batteries typically use solid nickel contacts to minimize corrosion at the terminal contacts.

A number of factors must be considered when specifying contact materials. Several principles apply to the substrate. The normal force provided by the contact must be great enough to hold the battery in place (even when the device is dropped) and to prevent electrical degradation and any resulting instability. Contacts must be able to resist permanent set. This refers to the ability of the contact to resist permanent deformation with a set number of battery insertions. Temperature rise at high current drains due to the resistance of the contact material must be limited. Excessive temperature increase could lead to stress relaxation and loss in contact pressure as well as to the growth of oxide films which raise contact resistance.

A common way to minimize contact resistance is to provide a wiping action of the device contact to the battery contact when the battery is inserted in place. Most camcorder batteries incorporate this feature.

Coatings should be selected to satisfy requirements not met by the substrate material, such as conductivity, wear, and corrosion resistance. Gold is an optimal coating due to its ability to meet most of the requirements. However, other materials may be used. Table 5.1 lists the characteristics of various materials used as contacts.

TABLE 5.1 Contact Materials

Gold plating	Provides the most reliable metal-to-metal contact under all environmental conditions
Nickel (solid)	Provides excellent resistance to environmental corrosion and is second only to gold plating as a contact material; material can be drawn or formed.
Nickel-clad stainless steel	Performs almost as well as solid nickel, with excellent resistance to environmental corrosion
Nickel-plated stainless steel	Very good material; unplated stainless steel is not recommended due to the adverse impact of passive films which develop on stainless steel resulting in poor electrical contact
Inconel alloy	Provides good electrical conductivity and good corrosion resistance; if manufacturers prefer to solder the contact piece in the circuit, soldering may be difficult unless an active flux is used
Nickel-plated cold-rolled steel	The most economical contact material; continuous, nonporous nickel plating of 200 μm is preferred

5.5 DESIGN OF RECHARGEABLE BATTERIES

All of the criteria addressed for the design of primary batteries should be considered for the design of rechargeable batteries.

In addition, multicell rechargeable batteries should be built using cells having matched capacities. In a series-connected multicell battery, the cell with the lowest capacity will determine the duration of the discharge, while the one with the highest capacity will control the capacity returned during the charge. If the cells are not balanced, the battery will not be charged to its designed capacity. To minimize the mismatch, the cells within a multicell battery should be selected from one production lot, and the cells selected for a given battery should have as close to identical capacities as possible. This is especially important with lithium-ion batteries, because due to the need for limiting current during charge, it is not possible to balance the capacity of the individual cells with a top-off or trickle charge.

Furthermore, safeguards must be included to control charging to prevent damage to the battery due to abusive charging. Proper control of the charge process is critical to the ultimate life and safety of the battery. The two (2) major considerations to be addressed include:

1. Voltage and current control to prevent overcharge
2. Temperature sensing and response to maintain the battery temperature within the range specified by the battery manufacturers.

5.5.1   Charge Control

The controls for voltage and current during charge for most batteries are contained in the charger. Nickel–cadmium and nickel–metal hydride batteries may be charged over a fairly broad range of input current, ranging from less than a 0.05C rate to greater than 1.0C. As the charge rate increases, the degree of charger control increases. While a simple, constant current control circuit may be adequate for a battery being charged at a 0.05C rate, it would not suffice at a rate of 0.5C or greater. Protective devices are installed within the battery to stop the charge in the event of an unacceptable temperature rise. The thermal devices that can be used include the following:

1. *Thermistor:* This device is a calibrated resistor whose value varies inversely with temperature. The nominal resistance is its value at 25°C. The nominal value is in the Kohm range with 10K being the most common. By proper placement within the battery pack, a measurement of the temperature of the battery is available and T_{max}, T_{min} and $\Delta T / \Delta t$ or other such parameters can be established for charge control. In addition, the battery temperature can be sensed during discharge to control the discharge, e.g., turn off loads to lower the battery temperature, in the event that excessively high temperatures are reached during the discharge.

2. *Thermostat (Temperature Cutoff, TCO):* This device operates at a fixed temperature and is used to cut off the charge (or discharge) when a preestablished internal battery temperature is reached. TCOs are usually resettable. They are connected in series within the cell stack.

3. *Thermal Fuse:* This device is wired in series with the cell stack and will open the circuit when a predetermined temperature is reached. Thermal fuses are included as a protection against thermal runaway and are normally set to open at approximately 30–50°C above the maximum battery operating temperature. They do not reset.

4. *Positive Temperature Coefficient (PTC) Device:* This is a resettable device, connected in series with the cells, whose resistance increases rapidly when a preestablished temperature is reached, thereby reducing the current in the battery to a low and acceptable current level. The characteristics of the PTC device are shown in Fig. 5.15. It will respond to high circuit current beyond design limits (such as a short circuit) and acts like a resettable fuse. It will also respond to high temperatures surrounding the PTC device, in which case it operates like a temperature cutoff (TCO) device.

Figure 5.16 shows a schematic of a battery circuit, indicating the electrical location of these protective devices. The location of the thermal devices in the battery assembly is critical to ensure that they will respond properly as the temperature may not be uniform throughout the battery pack. Examples of recommended locations in a battery pack are shown in Fig. 5.17. Other arrangements are possible, depending on the particular battery design and application.

Details of the specific procedures for charging and charge control are covered in the various chapters on rechargeable batteries.

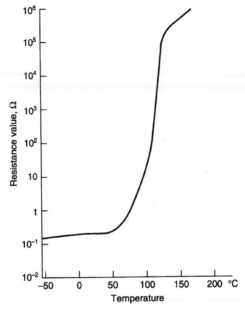

FIGURE 5.15 Characteristics of a typical positive temperature coefficient (PTC) device.

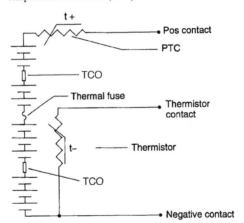

FIGURE 5.16 Protective devices for charge control.

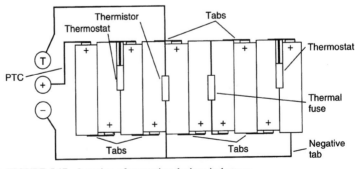

FIGURE 5.17 Location of protective devices in battery.

5.5.2 Example of Discharge/Charge Control

Electronic circuitry can be used to maximize battery service life by cutting off the discharge as close to the specified end or cut-off voltage as possible.[1] Ending the discharge at too high a voltage will result in a loss of a significant amount of battery capacity; ending it at too low an end voltage and, thus, discharging the battery beyond its safe cut-off could cause permanent damage to the battery.[2] Similarly, on charge, accurate control, as discussed above, will enable a maximum charge under safe conditions without damage to the battery.

This is especially important for the lithium-ion battery for which the preferred charge protocol for a high rate charge is to start the charge at a relatively high, usually constant current to a given voltage and then taper charging at a constant voltage to a given current cutoff. Exceeding the maximum voltage is a potential safety hazard and could cause irreversible damage to the battery. Charging to a lower voltage will reduce the capacity of the battery.[3]

Another interesting example is the control of charge for hybrid electric vehicles (HEV). In this application, it is advantageous to obtain close to 100% charge efficiency or charge acceptance rather than maximum battery capacity. The charge acceptance for a battery at the low state-of-charge (SOC) is close to 100%. As the battery is charged, charge acceptance becomes progressively poorer, particularly above 80% SOC.[4] (At full charge, the charge acceptance is zero). In the HEV application using nickel-metal hydride batteries, the charge control keeps the stage-of-charge, under normal driving and regenerative braking conditions, as close to 50% SOC as possible, and preferably within 30 to 70%. At these states-of-charge, the coulombic charge efficiency is very high.[5]

Other useful information also can be included in the chip, such as manufacturing information, chemistry, design data serial number, and cycle count.

5.5.3 Lithium-Ion Batteries

Special controls should be used with Lithium-ion batteries for management of charge and discharge. Typically, the control circuit will address the following items that affect battery life and safety.

Cell Voltage: The voltage of each individual cell in the battery pack is monitored on a continuous basis. Depending on the specific lithium-ion battery chemistry that is used, the upper voltage limit on charge, as specified by the manufacturer, is usually limited between 4.1 to 4.3 volts. On discharge, the cell voltage should not fall below 2.5 to 2.7 volts.

Temperature Control: As with any battery system, high temperature will cause irreversible damage. Internal battery temperature, for most applications, should be kept below 75°C. Temperature cutoff, with a trip of 70°C and reset temperature in the range of 45°C, is routinely used. Temperatures in excess of 100°C could result in permanent cell damage. For this, permanent type fuses are used; typically set for 104°C with a tolerance of +/- 5°C.

Short Circuit Protection: Normally, current limits are incorporated into the protection circuits. As a backup, a PTC device or fuse is placed in series with the battery pack. It is advisable to place the PTC between the pack assembly and the output of the battery. By placing it at this point, the PTC will not interfere with the operation of the upper or lower voltage detection of the electronic control circuit.

5.6 ELECTRONIC ENERGY MANAGEMENT AND DISPLAY— "SMART" BATTERIES

During the past decade, an important development in rechargeable battery technology has been the introduction of the use of electronic microprocessors to optimize the performance of the battery, control charge and discharge, enhance safety and provide the user with information on the condition of the battery. The microprocessor can be incorporated into the battery (the so-called *smart battery*), into the battery charger, or into the host battery-using equipment.

Some of the features include:

1. *Charge control.* The microprocessor can monitor the battery during charge controlling charge rate and charge termination, such as t, V_{max}, $-\Delta V$, ΔT, and $\Delta T/\Delta t$, to cut off the charge or to switch to a lower charge rate or another charge method. Constant current to constant voltage charge can be controlled and options can be incorporated into the chip for pulse charging, "reflex" charging (a brief periodic discharge pulse during charge), or other appropriate control features.

2. *Discharge control.* Discharge control is also provided to control such conditions as discharge rate, end-of-life cutoff voltage (to prevent overdischarge), cell equalization and temperature management. Individual cells, as well as the entire battery pack, can be monitored to maintain cell balance during cycling.

3. *State-of-charge indicator.* These devices, commonly known as "gas gauges," estimate the remaining battery capacity by factoring in such variables as the discharge rate and time, temperature, self-discharge, charge rate and duration. The remaining capacity is normally displayed by a sequence of illuminated LEDs or as a direct output to the device being powered.

4. *Other information.* Data is collected during the life of the battery to update the database and, as needed, make changes to maintain optimum performance. Other information can also be included or collected, such as battery chemistry and characteristics, manufacturing information (battery manufacturer can electronically "stamp" batteries at time of manufacture) and battery history, cycle count and other such data for a complete accounting of the battery's usage.

There are several main elements to consider in the design of smart batteries:[6]

1. *Measurement.* There are several parameters that can be measured directly to provide the basic information for the microprocessor. These include voltage, current, temperature and time. It is important that these measurements be made as accurately as possible to provide the best data for control.

   For example, voltage measurements are critical as the charge control (charge termination) depends on the battery voltage which, in some instances, should be accurate to at least 0.05 volts. An inaccurate measurement could result in under or overcharge which could result in short service life or damage to the battery, respectively. Or, in the case of the lithium-ion battery, overcharge could be a safety hazard. Similarly, on discharge, cutting off the discharge prematurely results in an unnecessarily shortened service life while overdischarge could, again, result in damage to the battery.

   Errors in the measurement of current affect not only the calculation of capacity and the state-of-charge "gas gauge," but influence the termination of charge and discharge as the termination voltages may vary depending on the current. Complicating this measurement is that the current is not a constant value, particularly during the discharge, with multiple power modes and high current pulses as short as milliseconds.

   Likewise, temperature is an important parameter as the performance of batteries is highly temperature dependent and exposure to high temperatures can cause irreversible damage to the cells.

2. *Calculation.* The calculation step covers the procedures for using the measured data as well as the algorithms to estimate battery performance (e.g. capacity at various discharge loads and temperature), charge acceptance, self-discharge, etc. Early "smart battery" electronics used simple linear models for these parameters which severely limited the accuracy in predicting the battery's performance. As noted in the descriptions of battery performance in the various chapters in this Handbook, battery performance, e.g. with respect to current drain and temperature is not linear. Self-discharge, similarly, is a complex relationship influenced at least by temperature, time, state of charge and the discharge load at which it is measured. Further, the performance of even those batteries using the same chemistry, varies with design, size, manufacturer, age, etc. A good algorithm will account for these relationships for control, predicting remaining service life and assuring safe operation.

As an example, Fig. 29.23 is an illustration showing the non-linear relationship of the rate of self discharge with temperature and time.

3. *Communication.* Clear, accurate and secure communication is important between the battery and host charger and the battery-using equipment for each component to obtain data or provide needed data to one of the other components. For example, the battery charger must be informed of the characteristics of the battery it is charging, the battery's state of charge, charge voltage and current requirements, charge-off, etc. The smart battery must also communicate to the user who may require information, such as remaining battery life, power levels, charge time and other characteristics to facilitate the use of the equipment.

4. *Errors.* As discussed under *Measurement,* it is important that parameters be measured accurately as inaccurate ones would not only result in incorrect decisions on the battery's capability but could result in damage to the battery or safety problems. In addition, the Smart Battery provides information on the margin-of-error in the state-of-charge calculation. This function is called "MaxError" and has a range of 0 to 100%. If the "MaxError" displays 20% and the "State-of-Charge" displays 30%, the actual state-of-charge is between 30 and 50%. If this loss of capacity is due to self-discharge while the battery is on stand and the rate of self-discharge does not follow the installed algorithm, the "MaxError" will increase. The user can correct this by, for example, fully charging the battery after a full discharge. This will restore the battery to full capacity and the battery will sense the "Reset" condition and return the MaxError to zero. It is possible to output an alarm when the MaxError has reached a programmed limit, alterting the user that a reset cycle is warranted.

5.6.1 The Smart Battery System

In order to ensure effective communication between the battery and host device, standard methods of communication are required. One method developed by battery manufacturers and microprocessor companies is the Smart Battery System (SBS) based on the System Management Bus. In 1995, the "Smart Battery Data Specification" was released by the SBS Forum. These specifications detail the communications method, protocols and the data interfaces between various devices. These specifications are periodically updated and available on the forum website, *www.sbs-forum.org.*

The goal of the Smart Battery interface is to provide adequate information for power management and charge control regardless of the particular battery's chemistry. Even though the major consumer of the battery information is the user, the system can also take advantage by using power consumption information to better manage its own power use. A charging system will be able to tell the user how long it will take to fully charge the battery.

One possible Smart Battery model is a system consisting of a battery, battery charger and a host (notebook computer, video camera, cellular phone, or other portable electronic equipment) as illustrated in Figure 5.18.

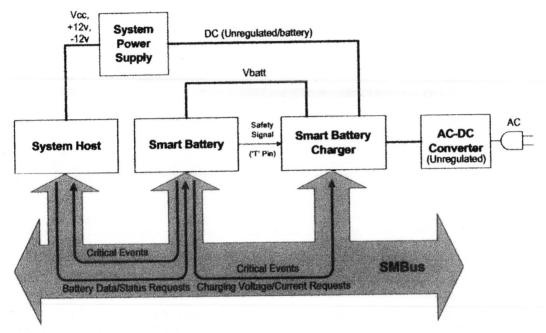

FIGURE 5.18   Outline of a Smart Battery System (*from Reference 7*).

The Smart Battery consists of a collection of cells or single-cell batteries and is equipped with specialized hardware that provides present state, calculated and predicted information to its SMBus Host. These may monitor particular environmental parameters in order to calculate the required data values. The electronics need not be inside the Smart Battery if the battery is not removable from the device.

The Smart Battery communicates with the other devices (such as the SM (System Management) Bus Host and the Smart Battery Charger) via two separate communication interfaces:

The first uses the SMBus CLOCK and DATA lines and is the primary communication channel between the Smart Battery and other SMBus devices. The Smart Battery will provide data when requested, send charging information to the Stuart Battery Charger, and broadcast critical alarm information when parameters (measured or calculated) exceed predetermined limits within the particular Smart Battery.

The other required communication interface is the secondary signaling mechanism or 'Safety Signal' on a Smart Battery pack connector. This is a variable resistance output from the Smart Battery which indicates when charging is permitted. It is meant as an alternate signaling method should the SMBus become inoperable. It is primarily used by the Smart Battery Charger to confirm correct charging.

The Smart Battery Charger is a charging circuit that provides the Smart Battery with charging current and charging voltage to match the Smart Battery's requested requirements. The battery charger periodically communicates with the Smart Battery and alters its charging characteristics in response to information provided by the Smart Battery. This allows the battery to control its own charge cycle. Optionally, the Smart Battery Charger may not allow the Smart Battery to supply power to the rest of the system when the Smart Battery is fully charged and the system is connected to AC power, thus prolonging the life of the battery.

The Smart Battery Charger will also receive critical events from the Smart Battery when it detects a problem. These include alarms for charging conditions or temperature conditions which exceed the limits set within the particular Stuart Battery.

The SM Bus is a specific implementation of a PC-bus that describes data protocols, device addresses and additional electrical requirements that is designed to physically transport commands and information between the components of the Smart Battery system.

The SMBus Host represents a piece of electronic equipment that is powered by a Smart Battery and that can communicate with the Smart Battery. The SMBus Host requests information from the battery and then uses it in the systems power management scheme and/or uses it to provide the user information about the battery's state and capabilities. The SMBus Host will also receive critical events from the Smart Battery when it detects a problem. In addition to the alarms sent to the Smart Battery Charger, it receives alarms for end of discharge, remaining capacity below the user set threshold value and remaining run time below the user set threshold value.

Figure 5.19 is a schematic block diagram of a Smart Battery system, in this case for a three cell Lithium-Ion battery.

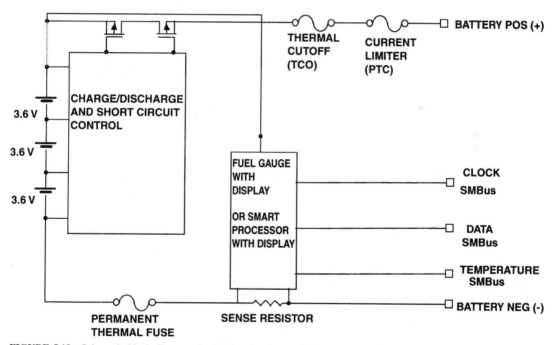

FIGURE 5.19   Schematic block diagram of a Lithium Ion three cell battery with SMBus output or gas gauge display.

The battery incorporates five terminals, battery plus and minus, clock, data and a safety signal, typically temperature. If the battery is hard wired into the host, the electronics need not be inside the battery. When the mechanical construction is complete, it must be programmed with the information such as chemistry, charge current, maximum voltage, etc. The information that is available to the host and charger includes:

1. Charge instructions for voltage, current and temperature
2. Battery design capacity
3. Remaining time to full charge
4. Remaining run time
5. Operating voltage, current, power, and temperature
6. Battery cycle count
7. Manufacturer's name, model, serial number and date of manufacture
8. Other information

5.7 GUIDELINES

In addition to the material covered in the preceding sections, the following should be considered in the design and fabrication of batteries:

1. Allow for the thermal expansion of battery components as well as the change in cell volume which accompanies discharge.
2. Consider the implications of cell leakage on equipment components, intercell weld or solder connections, and other battery components such as potting compounds, wire insulation, and adhesives. Locate the battery compartment in the device, such as to minimize the effects of possible leakage.
3. Wire leads and intercell connectors should be properly isolated and insulated to preclude the development of short circuits during the assembly process as well as during the life of the battery.
4. Intercell and battery connections need to be carefully made to withstand the severity of the equipment environment. Intercell connections shall be welded rather than soldered to avoid heat conduction to the interior of the cell.
5. Handle cells carefully to avoid inadvertent short circuits and discharge.
6. Avoid the use of high-exotherm potting compounds during battery fabrication.
7. Always vent the battery compartment in the device, allowing for release of any buildup of gas pressure. Avoid any confined pockets in the device where these gases may accumulate.

REFERENCES

1. V. Drori, and C. Martinez, "Smart Battery Chips Maximize Usable Battery Capacity and Battery Life," *Battery Power Products and Technology,* April 1999.
2. Section 28.4.2 of this handbook.
3. Section 35.5 of this handbook.
4. Sections 28.5 and 33.5 of this handbook.
5. Section 30.8.7 of this handbook.
6. D. Friel, "How Smart Should a Battery Be," *Battery Power Products and Technology,* March 1999.
7. Smart Battery Data Specification, Rev. 1.1, www.sbs_forum.org.

CHAPTER 6
SELECTION AND APPLICATION OF BATTERIES

David Linden

6.1 GENERAL CHARACTERISTICS

The many and varied requirements for battery power and the different environmental and electrical conditions under which they must operate necessitate the use of a number of different types of batteries and designs, each having the most advantageous performance under specific operational conditions.

Although many advances have been made in battery technology in recent years, as illustrated in Fig. 1.6 and discussed in more detail in Secs. 7.1 and 22.1, both through continued improvement of a specific electrochemical system and the development and introduction of new battery chemistries, there is still no one "ideal" battery that performs optimally under all operating conditions. As a result, over time, many different electrochemical systems and battery types have been and are still being investigated and promoted. However, a relatively small number have achieved wide popularity and significant production and sales volumes. The less conventional systems are typically used in military and industrial applications requiring the specific capabilities offered by these special batteries.

The "ideal" electrochemical cell or battery is obviously one that is inexpensive, has infinite energy, can handle all power levels can operate over the full range of temperature and environmental conditions, has unlimited shelf life, and is completely safe and consumer-proof. In practice, energy limitations do exist as materials are consumed during the discharge of the battery, temperature and discharge rate affect performance and shelf life is limited due to chemical reactions and physical changes that occur, albeit slowly in some cases, during storage. The use of energetic component materials and special designs to achieve high energy and power densities may require precautions during use to avoid electrical and physical abuse and problems related to safety. Further, the influence of the conditions of discharge, charge and the use of the battery, as discussed in Chap. 3, must the considered.

It should be recognized that while the demands of battery-using equipments continually seek smaller and more energetic and powerful batteries, these requirements may not necessarily be met because of the theoretical and practical limits of battery technology.

The selection of the most effective battery and the proper use of this battery are critical in order to achieve optimum performance in an application.

6.2 *MAJOR CONSIDERATIONS IN SELECTING A BATTERY*

A number of factors must be considered in selecting the best battery for a particular application. The characteristics of each available battery must be weighed against the equipment requirements and one selected that best fulfills these needs.

It is important that the selection of the battery be considered at the beginning of equipment development rather than at the end, when the hardware is fixed. In this way, the most effective compromises can be made between battery capabilities and equipment requirements.

The considerations that are important and influence the selection of the battery include:

1. *Type of Battery:* Primary, secondary, or reserve system

2. *Electrochemical System:* Matching of the advantages and disadvantages and of the battery characteristics with major equipment requirements

3. *Voltage:* Nominal or operating voltage, maximum and minimum permissible voltages, voltage regulation, profile of discharge curve, start-up time, voltage delay

4. *Load Current and Profile:* Constant current, constant resistance, or constant power; or others; value of load current or profile, single-valued or variable load, pulsed load

5. *Duty Cycle:* Continuous or intermittent, cycling schedule if intermittent

6. *Temperature Requirements:* Temperature range over which operation is required

7. *Service Life:* Length of time operation is required

8. *Physical Requirements:* Size, shape, weight; terminals

9. *Shelf Life:* Active/reserve battery system; state of charge during storage; storage time a function of temperature, humidity and other conditions

10. *Charge-Discharge Cycle (if Rechargeable):* Float or cycling service; life or cycle requirement; availability and characteristics of charging source; charging efficiency

11. *Environmental Conditions:* Vibration, shock, spin, acceleration, etc.; atmospheric conditions (pressure, humidity, etc.)

12. *Safety and Reliability:* Permissible variability, failure rates; freedom from outgassing or leakage; use of potentially hazardous or toxic components; type of effluent or signature gases or liquids, high temperature, etc.; operation under severe or potentially hazardous conditions; environmentally friendly

13. *Unusual or Stringent Operating Conditions:* Very long-term or extreme-temperature storage, standby, or operation; high reliability for special applications; rapid activation for reserve batteries, no voltage delay; special packaging for batteries (pressure vessels, etc.); unusual mechanical requirements, e.g., high shock or acceleration, nonmagnetic

14. *Maintenance and Resupply:* Ease of battery acquisition, accessible distribution; ease of battery replacement; available charging facilities; special transportation, recovery, or disposal procedures required

15. *Cost:* Initial cost; operating or life-cycle cost; use of critical or exotic (costly) materials

6.3 *BATTERY APPLICATIONS*

Electrochemical batteries are an important power source and are used in a wide variety of consumer, industrial, and military applications. Annual worldwide sales exceed $50 billion.

The use of batteries is increasing at a rapid rate, much of which can be attributed to advancing electronics technology, lower power requirements, and the development of portable devices which can best be powered by batteries. Other contributing factors are the increased demand for battery-operated equipment, the opening of many new areas for battery applications ranging from small portable electronic devices to electric vehicles and utility power load leveling.

Batteries have many advantages over other power sources, as outlined in Table 6.1 They are efficient, convenient, and reliable, need little maintenance, and can be easily configured to user requirements. As a result, batteries are used in an extremely wide range and variety of sizes and applications—from as small as 3 mAh for watches and memory backup to as large as 20,000 Ah for submarine and standby power supplies.

TABLE 6.1 Application of Batteries

Advantages	Limitations
Self-contained power source	High cost (compared with utility power)
Adaptable to user configuration:	Use of critical materials
Small size and weight—portability	Low energy density
Variety of voltages, sizes, and configurations	Limited shelf life
Compatible with user requirements	
Ready availability	
Reliable, low maintenance, safe, minimum, if any, moving parts	
Efficient conversion over a wide range of power demands	
Good power density (with some types)	
Efficient energy-storage device	

6.3.1 Summary of Battery Applications

A generalized summary of battery applications, listing the various battery types and identifying the power level and operational time in which each finds its predominant use, is shown in Fig. 6.1 As with any generalization, there are many instances in which the application of a particular battery will fall outside the limits shown.

Primary batteries are used typically from low to moderately high power drain applications, to a large extent with the familiar flat, button, or cylindrical configurations. They are a convenient, usually relatively inexpensive, lightweight source of packaged power and, as a result, are used in a variety of portable electric and electronic equipment. The so-called "dry-cell" is widely used in lighting devices, toys, radios, cameras, PDA's, and many other such consumer products. Flat or button batteries are popular in watches, calculators, photographic equipment, and as a battery backup for memory preservation. Similarly, primary batteries are used extensively in industrial and military applications to power portable communication, radar, night vision, surveillance, and other such equipment. Larger-sized primary batteries are also produced, mainly for special applications such as navigation aids, standby power, and remote-area uses, where their high capacity and energy density, long shelf life, and freedom from recharging and maintenance are important requisites.

Secondary (rechargeable) batteries are used as energy-storage devices, generally connected to and charged by a prime energy source and delivering their energy to the load on demand. Examples of this type of service are the lead-acid automotive starting, lighting, and ignition (SLI) battery, which is by far the major secondary battery application, hybrid electric vehicles standby electric systems including uninterruptible power systems (UPS) and load leveling. Secondary batteries are also used in applications where they are discharged and recharged subsequently from a separate power source. Examples of this type of service are electric vehicles and many applications, particularly portable devices, such as computers, cellular phones and camcorders, where the secondary battery is preferred in place of primary batteries, either for cost saving as they can be recharged or to handle power levels beyond the capability of conventional primary batteries.

The *special and reserve primary batteries* are used in selected applications usually requiring batteries which are capable of high-rate discharges for short periods of time after long-term storage in an inactivated or "reserve" condition. Many of these are used by the military for munitions and missiles.

The *solid electrolyte batteries* are low-rate batteries, operating in the microampere range, but with extremely long operational and shelf life. They are used in computer memory backup, cardiac pacemakers, and other applications requiring high reliability and extremely long active life.

The *fuel cell* is used in those applications requiring long-term continuous operation. Its major application to date has been as the power source in space flights. Larger sizes are now in development as alternatives to moderate-power engine generators, utility load leveling and electric vehicle propulsion. Most recently, portable-sized air-breathing fuel cells, using fuels, such as hydrogen and methanol stored in containers, are being investigated in the subkilowatt power level alternatives to batteries. (See Part IV.)

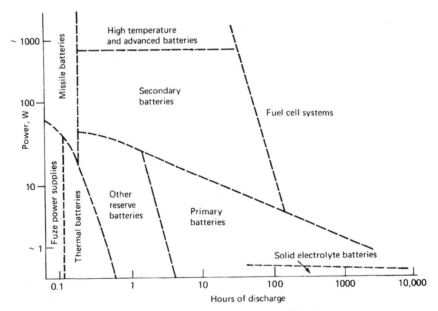

FIGURE 6.1 Predominant application field for various types of batteries.

6.3.2 Portable, Industrial and Electric Vehicle Applications

A listing of a number of the major applications of batteries is given in Table 6.2*a*. These are listed in the following three major categories.

Portable Applications.   This is a rapidly expanding area as many new portable devices are being introduced which are designed to operate only with batteries or, in some instances such as laptop computers, to operate with either batteries or AC line power. Both primary and secondary batteries are used in these portable equipments depending on service life and power requirements, convenience, cost and other factors discussed in Sec. 6.4.

Industrial Applications.   Larger-sized batteries, usually rechargeable, are used in these applications. In many of these applications, the batteries are used as back-up power in the event of an AC power failure. In some instances, such as with munitions, navigation aids and satellites, primary or reserve type batteries are used where an electrical source is not available for recharging. This is another area that is expanding rapidly to meet the demands for uninterrupted power sources (UPS) for computer and other sophisticated systems which require 24/7 operation with extremely high reliability.

Vehicular and Traction Applications.   Batteries have been an important power source for these applications, including starting, lighting and ignition (SLI) application and for the main power source for fork-lift trucks, golf carts and other such vehicles. This also is a growing area for battery application with the goal to replace the internal combustion engine with an environmental-friendly power source or provide a hybrid system which will improve the efficiency of fossil fuel engines and reduce the amount of objectionable and hazardous effluents.

The specific types of batteries in each classification and the major applications and characteristics of the batteries are covered in the appropriate chapters of this Handbook. Table 6.2*a* also lists the Chapters and Sections in this Handbook where some of the applications are discussed in more detail.

The current drains of some of the portable equipments are listed in Table 6.2*b* to illustrate the wide range of the requirements, from microamperes to over an ampere. Although the values are specified in amperes, some of the equipments, such as flashlights, have a resistive load. Many of the newer equipments present a constant current or a constant power load, and in some cases, a constant energy load as the photoflash cameras. In fact, with the more sophisticated equipments, the load is not constant, but will vary, not only in current or power drain, but in the type of load, depending on the particular function of the device that is operating at a particular time.

Note.   As discussed in Chap. 3, when evaluating batteries for a specific application, it is important to use the equipment loads and discharge conditions as inaccurate results will be obtained in these conditions are not correctly simulated. Further, the results could also be significantly different from the rated capacity or generalized performance data and characteristics of a particular battery.

TABLE 6.2a Typical Battery Applications

Portable	Industrial and government	Vehicular
Applicances and household equipment Audio and communication equipment Cameras and photographic equipment Computers and calculators, PDAs Emergency transmitters Hearing aids Implants Lighting Medical appliances Memory back-up Meteorological equipment Meters, test equipment, and instrumentation Signals and alarms Telephones Tools Toys Watches, clocks	Auxillary and emergency (standby power) Electrical energy storage Load leveling Munitions and missiles Navigations aids Oceanographic equipment Railway signaling Satellites and spacecraft Surveillance and detection Uninterruptible power systems	Aircraft batteries Electric vehicles, including golf carts, bicycles Engine starting Industrial and commercial equipment Submarine and underwater propulsion Vehicle (SLI batteries)
References	References	References
Section 6.4.2 Section 7 Section 13 Section 17.6 Section 21 Section 38.3.2, 38.3.3 Section 42.3, 42.4	Section 16.2 Section 23.1, 23.6 Section 24.9.5 Section 32.6 Section 33.8 Chapter 37 Section 38.3.3, 38.4 Section 39.7.2 Section 40.3.2, 40.4.2 Section 41.5.1 Chapter 43	Section 23.1, 23.4, 23.5 Section 24.9.4 Section 25.4, 25.5, 25.6 Section 26.6.4 Section 30.5, 30.8 Section 31.7 Chapter 37 Section 38.5 Section 39.1.1 Section 40.3.2, 40.4.2 Section 41.5.2

TABLE 6.2b   Current Drain in Battery-Operated
Portable Devices

Device	Current drain, mA
Cassette recorders	70–130 (low)
	90–150 (medium)
	100–200 (high)
Disk players	100–350
Calculators: (LCD)	<1
Cameras:	
Photo flash	800–1600
Autowind	200–300
Digital cameras	500–1600
Cellular phones	300–800
Camcorders	700–1000
Computers	
Palm held	400–800
Note book	500–1500
Laptop	800–1000
Fluorescent lamps	500–1000
Flashlights	100–700
Memory	microamperes
Remote controls	10–60
Radios:	
With 9-V battery	8–12 (low volume)
	10–15 (medium)
	15–45 (high)
With cylindrical cells	10–20 (low volume)
	20–30 (medium)
	30–100 (high)
Walkman	200–300
Smoke detector:	
Background	10–15 microamperes
Alarm	10–35
Toys:	
Motorized (radio controlled)	600–1500
Electronic games	20–250
Video games	20–200
TV (portable)	400–700
Travel shaver	300–500
Watches:	
LCD	10–25 (backlight)
LED	10–40

6.4 PORTABLE APPLICATIONS

6.4.1 General Characteristics

The characteristics of the conventional and advanced primary and secondary batteries on a theoretical basis are summarized in Table 1.2. This table also lists the characteristics of each of these batteries, based on the actual performance of a practical battery, under conditions close to optimum for that battery. As discussed in Chaps. 1 and 3:

- The actual capacity available from a battery is significantly less (about 20 to 30%) than the theoretical capacity of the active materials.

- The actual capacity is also less than the theoretical capacity of a practical battery, which includes the weight of the non-energy-producing materials of construction as well as the active materials.

- The capacity of a battery under specific conditions of use could vary significantly from the values listed in Table 1.2. These values are based on designs and discharge conditions optimized for energy density, and while they can be helpful to characterize the energy output of each battery system, the performance of the battery under specific conditions of use should be obtained before any final comparisons or judgments are made.

In general, the capacity of the conventional aqueous secondary batteries is lower than that of the conventional aqueous primary batteries (Table 1.2), but they are capable of performance on discharges at higher current drains and lower temperatures and have flatter discharge profiles (see Figs. 6.2 to 6.6). The conventional primary batteries have and advantage in shelf life or charge retention compared to the secondary batteries (see Fig. 6.7). Most primary batteries can be stored for several years and still retain a substantial portion of their capacity, while most of the conventional secondary batteries lose their charge more rapidly and must be kept charged during storage or charged just before use.

These differences are due to both the electrochemical systems and the cell designs that were selected for the conventional batteries. Typically the more energetic and not readily rechargeable materials, such as zinc, were selected for the primary cells. Also, the bobbin construction was usually used for the primary batteries to obtain higher energy densities while designs providing more surface area lower internal resistance and higher rate capability were used for the rechargeable batteries.

Some of these differences are not longer that distinct as similar chemistries and designers are used for some of the primary and secondary batteries. For example, a primary battery, using the spirally wound electrode design, will have a similar rate capability as the secondary battery using the same design and chemistry. Likewise, the charge retention of that secondary battery will be similar to the charge retention of that primary battery. However, the capacity of the primary battery should always be higher because of the features that have to be added to the secondary battery to achieve effective recharging.

The advantages and general characteristics of primary and secondary batteries for portable applications also are compared in the next section. More detailed characteristics of the different battery types are presented in the appropriate chapter of this handbook. These data, rather than the more generalized data presented in this section, should be used to evaluate the specific performance of each battery.

6.4.2   Characteristics of Batteries for Portable Equipment

Portable, battery-operated electric and electronic equipments once were typically powered by primary batteries. However, the development of small, maintenance-free rechargeable batteries made it possible for secondary batteries to be used in applications which had been almost exclusively the domain of the primary battery. The trade-off, thus, is between a possible lower life-cycle cost of the secondary battery because it can be recharged and reused and the convenience of using a replaceable primary battery.

The development of new portable applications, such as power tools, computer camcorders, PDAs, and cellular phones, with high power requirements accelerated the use of these rechargeable batteries because of their power advantage over conventional types of primary batteries. The newer primary batteries, particularly those using the lithium anode, which have a high specific energy and energy density and good power density, can be used in some of these higher-power applications. Further, advances in electronics technology are gradually reducing the power requirements of a number of these equipments to levels where the primary battery can give acceptable performance. In some applications, therefore, the equipment is being designed to be powered by either a rechargeable or a primary battery, leaving the choice to the user. The user can opt for the convenience, freedom from charging, and longer shelf life of the primary or the potential cost saving with the rechargeable batteries.

Tables 1.2 and 6.3 and the figures in this section compare the performance of the major primary and rechargeable batteries used for portable applications. These comparisons show the following.

1. The primary batteries, particularly the lithium types, depending on the discharge conditions can deliver up to eight times the watthour capacity of the conventional aqueous secondary batteries. Similarly, the new rechargeable batteries using lithium and other high-energy materials will have higher capacities than the conventional secondary batteries. These capacities, however, will most likely be lower than those of the primary batteries using similar chemistries.

2. The aqueous secondary batteries generally have better high-rate performance than the primary batteries. The lithium primary batteries, using spirally wound or other high-rate electrode structures, provide higher output compared to the conventional secondary batteries, even at fairly high rates, because of their better normal temperature performance and generally good performance at high discharge rates. This is illustrated in Fig. 6.2, which compares the performance of the different battery types at different discharge rates. In this figure the service that each of the battery types will deliver at various levels of power density (watts per liter) is plotted. A slope parallel to the idealized line indicates that the capacity of the battery, in watthours, is invariant, regardless of the discharge load. A flatter slope, or one that levels off as the load is increased, indicates a loss of capacity as the discharge rate is increased. This figure shows that the conventional rechargeable or secondary batteries maintain their capacity even at the higher current drains while capacity of the conventional primary batteries begins to drop off at a 20 to 50-h rate. The lithium primary batteries, however, can maintain their advantage over the conventional secondary batteries to fairly high discharge rates because of their significantly higher capacities at the lower discharge rates. The lithium-ion rechargeable battery, which was introduced commercially during the 1990s, has a significantly higher specific energy and energy density compared to the conventional rechargeable batteries, although still below that of the lithium primary batteries. Its high rate and low temperature performance is superior to the nickel-cadmium and nickel-metal hydride batteries because of its higher specific energy, except at the very high discharge rates (see Figs. 6.2 and 6.3). The lithium-ion battery is rapidly becoming the battery of choice in high-end portable equipment and performance advances and cost reduction continue to be achieved as a result of extensive R&D programs (see Chap. 35).

Figure 6.2 is a comparison on a volumetric basis, while Fig. 6.3 is a Ragone plot presenting similar information on a gravimetric basis. This shows a greater advantage for the lithium batteries because of their lower density.

TABLE 6.3 Characteristics of Batteries for Portable Equipment

	Primary batteries			Secondary batteries			
	Zn/alkaline/MnO_2	Li/MnO_2	Li/SO_2	Nickel-cadmium	Lead-acid	Nickel-metal hydride	Lithium-ion
Nominal cell voltage, V	1.5	3.0	3.0	1.2	2.0	1.2	4.1
Specific energy (Wh/kg)	145	230	260	35	35	75	150
Energy density (Wh/L)	400	535	415	100	70	240	400
Charge retention at 20°C (shelf life)	3–5 years	5–10 years	5–10 years	3–6 months	6–9 months	3–6 months	9–12 months
Calendar life, years	—	—	—	4–6	3–8	4–6	5+ yrs
Cycle life, cycles	—	—	—	400–500	200–250	400–500	1000
Operating temperature, °C	−20 to 45	−20 to 70	−40 to 70	−20 to 45	−40 to 60	−20 to 45	−20 to 60
Relative cost per watthour (initial unit cost to consumer)	1	6	5	15	10	25	45

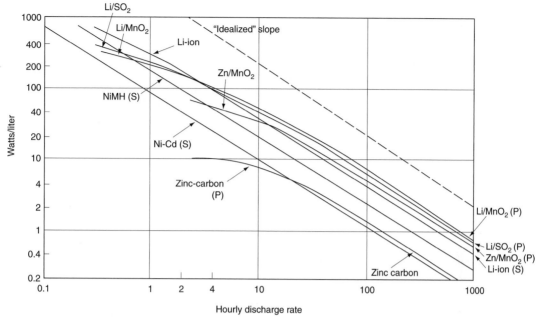

FIGURE 6.2 Performance characteristics of primary (P) and secondary (S) batteries on a volumetric basis, 20°C.

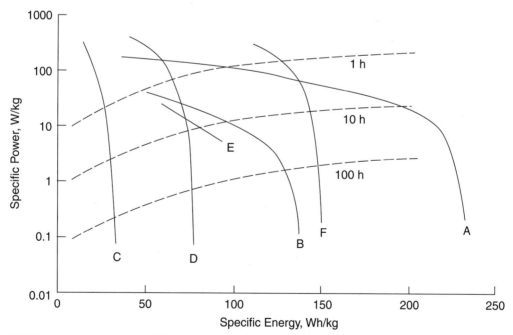

FIGURE 6.3 Performance capabilities of various cylindrical primary (P) and secondary (S) batteries—energy vs. specific power. *A*: Li/MnO$_2$ (P); *B*: Zn/alkaline/MnO$_2$ (P); *C*: Ni-Cd (S); *D*: Ni-MH (S); *E*: Zn/alkaline/MnO$_2$ (S); *F*: Li-ion (S).

3. The conventional aqueous secondary batteries generally have a flatter discharge profile than the primary or lithium batteries. Figure 6.4 compares the discharge curves of several primary and secondary batteries and illustrates this characteristic.

4. The conventional aqueous secondary batteries general have better low-temperature performance than the conventional aqueous primary batteries. The comparative performances on a gravimetric basis (watthours per kilogram) and on a volumetric basis of the various types of batteries at a moderate 20-h rate are shown in Figs. 6.5 and 6.6 respectively. While the specific energy and energy density of the primary batteries is higher at room temperature, their performance drops off more significantly as the temperature is reduced compared to conventional secondary batteries. Again, the lithium batteries have better low-temperature characteristics than the conventional primary batteries, but percentage wise they are poorer compared to the conventional secondary batteries.

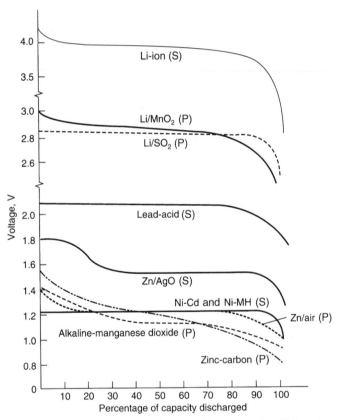

FIGURE 6.4   Discharge profiles of primary (P) and secondary (S) batteries.

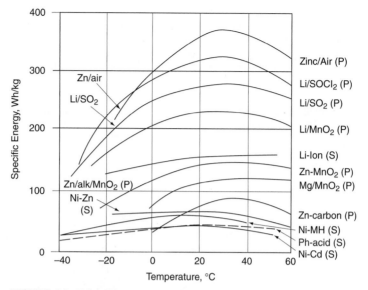

FIGURE 6.5 Effect of temperature on specific energy of primary (P) and secondary (S) batteries.

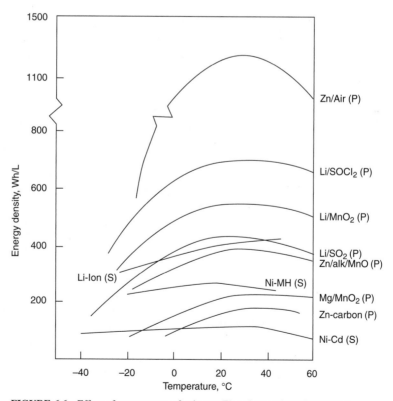

FIGURE 6.6 Effect of temperature of primary (P) and secondary (S) batteries.

5. The charge retention or shelf life of the primary batteries is much superior to that of the conventional aqueous secondary cells. Hence these secondary batteries have to be maintained in a charged condition or recharged periodically to maintain them in a state of readiness. Figure 6.7 compares the charge retention of various types of batteries when stored at different temperatures. The superior charge retention of most of the lithium batteries, both primary and secondary, is illustrated.

6. The secondary batteries are capable of being recharged to their original condition after discharge and reused rather than being discarded. For those applications where recharging facilities are available, convenient, and inexpensive, a lower life-cycle cost can be realized with secondary batteries, even with their higher initial cost, if their full cycle life is utilized.

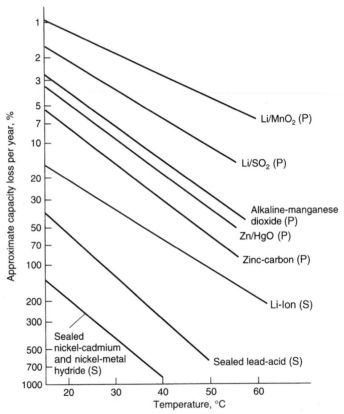

FIGURE 6.7 Shelf life (charge retention) characteristics of various batteries—primary (P), secondary (S).

6.4.3 Cost Effectiveness

Specific cost effectiveness versus life cycle analyses can be used to evaluate the best choice—primary or secondary battery—for a particular application or deployment. Figure 6.8 summarizes such an analysis comparing lithium/sulfur dioxide primary and nickel-cadmium secondary batteries.[1] Figure 6.8*a* depicts a military training situation in which charging facilities are readily available and inexpensive, recharging is convenient, and the batteries are used regularly (thus employing the full cycle life of the secondary battery). In this situation, the initial higher cost of the secondary battery is easily recovered, and the payback or break-even point occurs early. Figure 6.8*b* presents a field situation in which batteries are not used regularly, special charging facilities are required, and recharging is not convenient. In this case, the payback time occurs much later, and the use of secondary batteries may not be cost-effective since the break-even time is close to or beyond the calendar life of the battery.

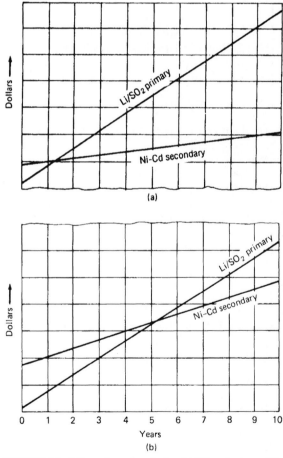

FIGURE 6.8 Life-cycle cost analysis—Li/SO$_2$ primary vs. Ni-Cd secondary cells. (*a*) Military training situation. (*b*) Field situation.

Similar analyses can be made for specific commercial and industrial applications. For example, the cost of the use of zinc/alkaline/manganese dioxide primary batteries versus nickel-cadmium rechargeable batteries in a typical portable electronic device applications are compared in Table 6.4. If the usage rate is low and the drain rate moderate, the primary battery is more cost-effective besides offering the convenience of not having to be periodically recharged.

TABLE 6.4   Cost-Effectiveness of Primary (Zn/alkaline/MnO$_2$) vs. Secondary (Nickel-Cadmium) Batteries

Assumptions
Nominal voltage: 3 V
2 "AA" Zn/MnO$_2$ batteries: $1.60
2 "AA" Ni-Cd batteries: $8.00
Ni-Cd charger: $5.00

a. Comparison at 150 mA

Usage (hrs/day)	Change Ni-Cd (days)	Change Zn/MnO$_2$ (days)	Ni-Cd battery payback (days)
0.5	8	26	210
1.0	4	13	105
2.0	2	6.5	52
4.0	1.0	3.3	36
6.0	0.67	2.2	17
8.0	0.5	1.7	13

b. Comparison in a Walkman Radio

Usage (hrs/week)	Change Ni-Cd (days)	Change alkaline (days)	Ni-Cd battery payback (months)
1	21	49	25
3	7	16	9
5	4	10	5.4
7	3	7	3.7
14	1.5	3.5	1.9
21	1	2.3	1.2
35	.6	1.4	.8

6.4.4 Other Comparisons of Performance

Figure 6.9 provides another illustration comparing the performance of several types of primary and secondary batteries in a typical miliary portable radio transceiver application at several temperatures, based on 1994 data. The primary batteries give the longest service at 20°C, but only the secondary batteries and the Li/SO$_2$ primary battery are capable of performance at the very low temperatures. This equipment is being phased out of use.

A more current example is the performance characteristics of the 24-V military battery type "590" which has been designed in several versions using several different battery chemistries, but all within the same physical envelope. The characteristics of these batteries are listed in Table 6.5. This battery is used in a number of portable military communication and surveillance equipments and the particular type of battery that is used depends on the equipment requirements and deployment. For example, for training and policing, a rechargeable battery would most likely be chosen. In forward, areas and under combat, where lightweight is critical and recharging facilities are not available nor recharging feasible, a primary battery will usually be used. Figure 6.10 compares the performance of the different batteries under an 18-watt discharge load at 25°C. Under heavier discharge conditions (this battery is used ·in equipments requiring as high as a 10-ampere discharge current) and at low temperatures, the zinc/air battery would not perform satisfactorily. The performance of the rechargeable batteries would be comparatively better because of their relatively superior performance under these more stringent operating conditions.

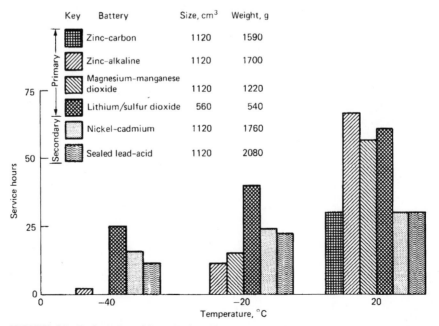

FIGURE 6.9 Performance of batteries in military radio transceiver application. Transmit load, 1 A; receive load, 50 mA; duty cycle, 9:1 receive to transmit (based on 1994 data).

TABLE 6.5   Characteristics of Batteries in "590" Envelope

Voltage: 24 V
Envelope dimensions (cm): 12.7 (H) × 11.2 (L) × 6.2 (W)
Envelope volume (cm³): 885

Battery type	Capacity (Ah)	Weight (kg)	Specific energy (Wh/kg)	Power density (Wh/L)
Primary				
BA-5590 (Li/SO$_2$)	6.8	1.02	160	185
BA-5390 (Li/MnO$_2$)	8.5	1.25	170	240
BA-XX90 (Zn/Air)	10.0	1.00	250	285
Secondary				
BB-690 (Lead-acid)	1.8	1.68	27	50
BB-590 (Ni-Cd)	2.2	1.75	31	62
BB-390 (NiMH)	4.6	1.82	65	135
BB-XX90 (Li-ion, cylindrical cells)	4.5	1.25	105	150
BB-X590 (Li-ion, prismatic cells)	6.0	1.68	100	190

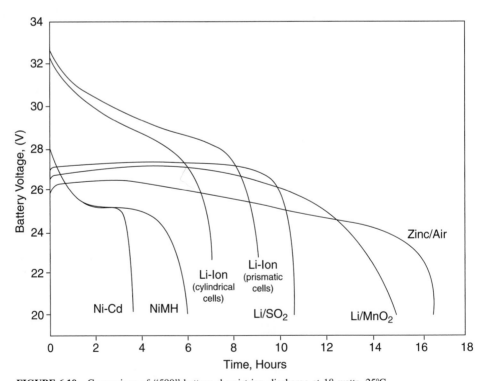

FIGURE 6.10   Comparison of "590" battery chemistries, discharge at 18 watts, 25°C.

Figures 6.11 and 6.12 are two other examples comparing the performance of several "AA-size" primary and rechargeable batteries. In Fig. 6.11, the performance of these batteries is plotted over a range of discharge current. Depending on the battery system, the primary batteries have the advantage at low current drains but lose their advantage as the discharge rate is increased. As shown, low drain applications, such as PDAs and Palm devices, generally use primary batteries, mainly the zinc/alkaline/manganese dioxide battery. Some more sophisticated PDAs, with higher power requirements will use rechargeable batteries. Laptop computers typically use rechargeable batteries.

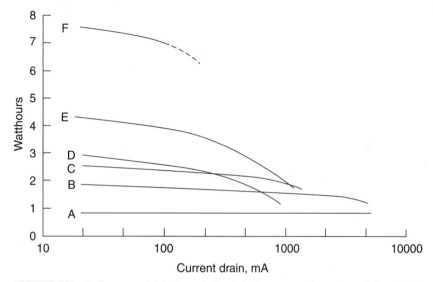

FIGURE 6.11 Performance of AA- (or equivalent) size battery at various current drains at 20°C. *A*: Ni-Cd battery; *B*: Ni-MH battery; *C*: Li-Ion; *D*: zinc-alkaline battery; *E*: Li/MnO$_2$ battery (2/3 size); *F*: Zn/air battery (button configuration). *A–C* are secondary batteries. *D–F* are primary batteries.

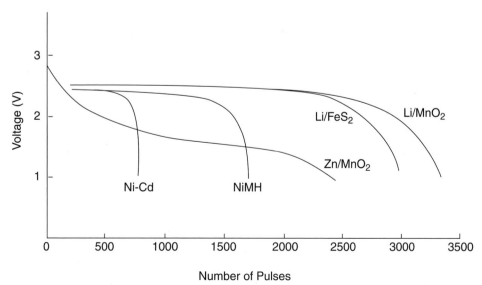

FIGURE 6.12 Photo simulation test; 900 mA, 3 seconds on 27 seconds, OFF, 20°C (Li/MnO$_2$ (2/3 A size), others (AA), 2 cells each).

Figure 6.12 compares the performance of these "AA-size" batteries on a photo simulation test of the flash in an automatic camera. The lithium primary batteries deliver the larger number of cycles. Although the zinc/alkaline/manganese dioxide battery delivers a significant number of pulses, as it operates at a lower voltage, the recycle time is longer than the batteries operating at a higher voltage. (See Figs. 10.1*b* and 14.85.) As the power requirements are increasing with the newer camera models, the battery systems capable of high rate discharge will be favored.

Another comparison of the characteristics of each type of battery is shown in Fig. 6.13 which examines one performance parameter—the total weight of a power sources required to deliver a given power output (in this example, 5 watts) for different lengths of operation. For very short missions, the conventional secondary batteries are lightest because of their high rate capability. At moderate and light loads, the primary and lithium batteries become the lighter ones because of their higher specific energy and energy density under these load conditions.

Figure 6.13 also shows the advantage of a hybrid system: combining a battery with a high specific energy with a rechargeable battery having a high rate capability. The example illustrates a hybrid design using a zinc/air battery (which has a high specific energy at moderate-to-low discharge rates but poor high-rate performance) with a nickel-cadmium battery (which has a low specific energy, but comparatively good performance at high discharge rates). The nickel-cadmium battery is sized to handle the load for the specified time (in this example for 6 minutes) and is then recharged by the zinc/air battery. Curve F shows that at the longer operating times (a low discharge rate for the size of the battery) the zinc/air battery, alone, can handle the load and is the lightest power source. It loses its advantage at the shorter operating times which correspond to a higher discharge rate. For these shorter missions, a rechargeable battery, such as the nickel cadmium battery, is added, resulting in an overall lighter battery (curve G). As the rechargeable battery is sized to handle the load for a specified time, the total weight and size of the hybrid battery is dependent on the time established for the pulse load.

Figure 6.14 shows the distribution of current in another example of a zinc/air, nickel-cadmium hybrid battery, this one designed to handle a transmit load for 2 minutes at 900 mA and a receive load of 50 mA for 18 minutes, similar to the application illustrated in Fig. 6.9. During the "receive" period, the load is carried by the zinc/air battery which, at the same time, charges the nickel-cadmium battery. During the "transmit" period, the load is carried by both batteries.

This hybrid technology is applicable to other batteries, as well as to fuel cells and other power sources, as an efficient and effective way of handling pulse requirements and attaining an optimum system specific energy. It is being considered for use in a wide range of applications, for small portable devices to large systems, such as the hybrid electric vehicle (HEV).

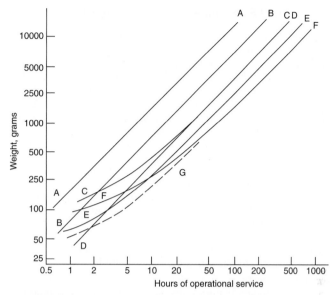

FIGURE 6.13 Comparison of battery systems—weight vs. service life (based on 5-W output and stated energy-weight ratio).

A Nickel-cadmium (S) 35 Wh/Kg
B Nickel-metal hydride (S) 75 Wh/Kg
C Zinc/MnO$_2$ (P) 145 Wh/Kg
D Lithium-ion (S) 150 Wh/Kg

E Lithium/MnO$_2$ (P) 230 Wh/Kg
F Zinc/air (P) 370 Wh/Kg
G Zinc air, Nickel-cadmium hybrid (6 minute pulse)
(P) Primary Battery (S) Secondary Battery

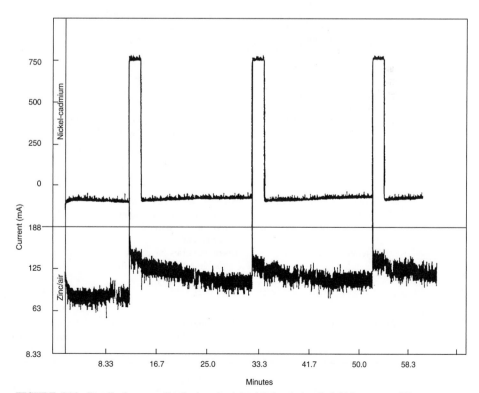

FIGURE 6.14 Detail of current distribution zinc/air nickel-cadmium hybrid battery conditions.

6.4.5 Criteria for Battery Selection—Portable Equipment

A number of different battery systems are available to the equipment designer and user, each offering particular advantages for powering portable electronic equipment. The selection of a battery system, even the first decision of using a primary or secondary battery, may not be clear-cut and is influenced by many factors, such as deployment of the equipment, operating environment, size, weight, required service time, duty cycle, frequency of use, capital investment and cost, and other factors listed in Sec. 6.2. An important consideration, as well, is the end-user's biases and preferences. The selection is further complicated by the overlapping performance characteristics of many of the batteries since the characteristics of a particular battery system or chemistry can be altered by the design features and design changes as the technology develops and matures.

Table 6.6 is a summary, albeit simplified, of criteria that should be considered in making a preliminary determination of the type of battery—primary or secondary—to be used. It is most applicable to comparing the conventional systems and lithium primary batteries. As pointed out in Sec. 6.4, the characteristics of the rechargeable lithium batteries may difference from these generalizations.

1. A primary battery will usually be the battery of choice if the power requirement is low or in applications where the battery will be used infrequently over a long period of time.

2. For applications where the battery will be used frequently (assuming that both primary and secondary batteries are capable of meeting the power requirements of the equipment), the choice should depend on user preference. Will the user opt for the convenience of the primary battery and freedom from the charger, or will the user tolerate the "inconvenience" of the rechargeable battery and opt for its possible lower operating cost?

3. For high discharge rates, high-rate secondary batteries will be the choice as the performance of the primary battery falls off, particularly at lower operating temperatures.

TABLE 6.6 Advantages of Primary and Secondary Batteries

Conditions of use	Secondary batteries	Primary batteries
1) Assuming acceptable load capability Frequent use, repeated cycling	Lower life-cycle cost ($/kWh) if charging is convenient and inexpensive (work force and equipment)	Lighter or smaller—or longer service per "charge" or replacement
		No maintenance or recharging
		Ready availability (for replacement)
Frequent use, low drain capacity or	Aqueous secondary batteries have poor charge retention; have to be charged periodically	Long service per "charge"; cost advantage of secondary disappears
Infrequent use	Li-ion batteries have better charge retention, but still require charging	Infrequent replacement, cost advantage of secondary disappears
		Good charge retention; no need for charging or maintenance
2) Assuming high discharge rates	Best comparative performance at heavy loads	Hybrid battery system may provide longer service, freedom from line power

Table 6.7 illustrates the selection of primary and secondary batteries. As the power requirements of the application increase (going from the bottom of the list to the top) and the size of the battery becomes larger, the trend for the battery-of-choice shifts from the primary to the secondary battery. The primary battery dominates in the lower power applications and where a long service life is required, such as smoke detectors, implants and memory back-up.

TABLE 6.7 Application of Batteries—Primary vs. Secondary

Application	Primary or secondary battery
Utility systems (load-leveling)	SECONDARY
Electric vehicle hybrid (HEV)	
Central telephone system	
Fork lifts	
Electric vehicles (EV)	
Golf carts	
Electric bikes	
Automotive SLI	
Spacecraft power	
Emergency lighting	
Portable tools	
Laptop computers	
Transceivers	
Cordless telephone	
Camcorders	
Portable equipment	
(e.g. Dustbuster, shaver)	
Cellular phone	
PDAs, Palm devices	
Radio	
Pagers	
Flashlight	
Toys	
Cameras	
Instruments	
Hearing aids	
Implants	
Smoke detectors	
Remote controllers	
Memory back-up	PRIMARY

REFERENCE

1. "Cost Effectiveness Comparison of Rechargeable and Throw-away Batteries for the Small Unit Transceiver," U.S. Army Electronics Command, Ft. Monmouth, N.J. Jan. 1977.

P · A · R · T · 2

PRIMARY BATTERIES

CHAPTER 7
PRIMARY BATTERIES—INTRODUCTION

David Linden

7.1 GENERAL CHARACTERISTICS AND APPLICATIONS OF PRIMARY BATTERIES

The primary battery is a convenient source of power for portable electric and electronic devices, lighting, photographic equipment, PDA's (Personal Digital Assistant), communication equipment, hearing aids, watches, toys, memory backup, and a wide variety of other applications, providing freedom from utility power. Major advantages of the primary battery are that it is convenient, simple, and easy to use, requires little, if any, maintenance, and can be sized and shaped to fit the application. Other general advantages are good shelf life, reasonable energy and power density, reliability, and acceptable cost.

Primary batteries have existed for over 100 years, but up to 1940, the zinc-carbon battery was the only one in wide use. During World War II and the postwar period, significant advances were made, not only with the zinc-carbon system, but with new and superior types of batteries. Capacity was improved from less than 50 Wh/kg with the early zinc-carbon batteries to more than 400 Wh/kg now obtained with lithium batteries. The shelf life of batteries at the time of World War II was limited to about 1 year when stored at moderate temperatures; the shelf life of present-day conventional batteries is from 2 to 5 years. The shelf life of the newer lithium batteries is as high as 10 years, with a capability of storage at temperatures as high as 70°C. Low-temperature operation has been extended from 0 to −40C, and the power density has been improved manyfold. Special low-drain batteries using a solid electrolyte have shelf lives in excess of 20 years.

Some of the advances in primary battery performance are shown graphically in Figs. 1.6 and 7.1.

Many of the significant advances were made during the 1970–90 period and were stimulated by the concurrent development of electronic technology, the new demands for portable power sources and the support for the space, military and the environment improvement programs.

During this period, the zinc/alkaline manganese dioxide batery began to replace the zinc-carbon or Leclanche battery as the leading primary battery, capturing the major share of the US market. Environmental concerns led to the elimination of mercury in most batteries without any impairment of performance, but also led to the phasing out of those batteries,

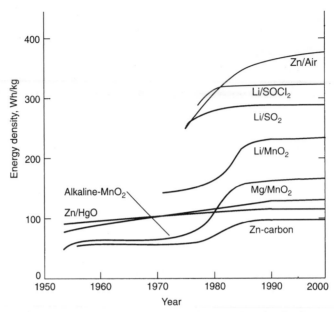

FIGURE 7.1 Advances in development of primary batteries. Continuous discharge at 20°C; 40–60-h rate; AA- or similar size battery.

zinc/mercuric oxide and cadmium/mercuric oxide, that used mercury as the cathodic active material. Fortunately, zinc/air and lithium batteries were developed that could successfully replace these "mercury" batteries in many applications. A major accomplishment during this period was the development and marketing of a number of lithium batteries, using metallic lithium as the anode active material. The high specific energy of these lithium batteries, at least twice that of most conventional aqueous primary batteries, and their superior shelf life opened up a wide range of applications—from small coin and cylindrical batteries for memory backup and cameras to very large batteries which were used for back-up power for missile silos.

Increases in the energy density of primary batteries has tapered off during the past decade as the existing battery systems have matured and the development of new higher energy batteries is limited by the lack of new and/or untried battery materials and chemistries. Nevertheless, advances have been made in other important performance characteristics, such as power density, shelf life and safety. Examples of these recent developments are the high power zinc/alkaline/manganese dioxide batteries for portable consumer electronics, the improvement of the zinc/air battery and the introduction of new lithium batteries.

These improved characteristics have opened up many new opportunities for the use of primary batteries. The higher energy density has resulted in a substantial reduction in battery size and weight. This reduction, taken with the advances in electronics technology, has made many portable radio, communication, and electronic devices practical. The higher power density has made it possible to use these batteries in PDA's, transceivers, communication and surveillance equipment, and other high-power applications, that heretofore had to be powered by secondary batteries or utility power, which do not have the convenience and freedom from maintenance and recharging as do primary batteries. The long shelf life that is now characteristic of many primary batteries has similarly resulted in new uses in medical electronics, memory backup, and other long-term applications as well as in an improvement in the lifetime and reliability of battery-operated equipment.

The worldwide primary battery market has now reached more than $20 billion annually, with a growth rate exceeding 10% annually. The vast majority of these primary batteries are the familiar cylindrical and flat or button types with capacities below 20 Ah. A small number of larger primary batteries, ranging in size up to several thousand ampere-hours, are used in signaling applications, for standby power and in other military and special applications where independence from utility power is mandatory.

7.2 TYPES AND CHARACTERISTICS OF PRIMARY BATTERIES

Although a number of anode-cathode combinations can be used as primary battery systems (see Part 1), only a relatively few have achieved practical success. Zinc has been by far the most popular anode material for primary batteries because of its good electrochemical behavior, high electrochemical equivalence, compatibility with aqueous electrolytes, reasonably good shelf life, low cost, and availability. Aluminum is attractive because of its high electrochemical potential and electrochemical equivalence and availability, but due to passivation and generally limited electrochemical performance, it has not been developed successfully into a practical active primary battery system. It is now being considered in mechanically rechargeable or refuelable aluminum/air batteries and in reserve battery systems. Magnesium also has attractive electrical properties and low cost and has been used successfully in an active primary battery, particularly for military applications, because of its high energy density and good shelf life. Commercial interest has been limited. Magnesium also is popular as the anode in reserve batteries. Now there is an increasing focus on lithium, which has the highest gravimetric energy density and standard potential of all the metals. The lithium anode battery systems, using a number of different nonaqueous electrolytes in which lithium is stable and different cathode materials, offer the opportunity for higher energy density and other advances in the performance characteristics of primary systems.

7.2.1 Characteristics of Primary Batteries

Typical characteristics and applications of the different types of primary batteries are summarized in Table 7.1

Zinc-Carbon Battery. The Leclanché or zinc-carbon dry cell battery has existed for over 100 years and had been the most widely used of all the dry cell batteries because of its low cost, relatively good performance, and ready availability. Cells and batteries of many sizes and characteristics have been manufactured to meet the requirements of a wide variety of applications. Significant improvements in capacity and shelf life were made with this battery system in the period between 1945 and 1965 through the use of new materials (such as beneficiated manganese dioxide and zinc chloride electrolyte) and cell designs (such as the paper-lined cell). The low cost of the Leclanché battery is a major attraction, but it has lost considerable market share, except in the developing countries, because of the newer primary batteries with superior performance characteristics.

TABLE 7.1 Major Characteristics and Applications of Primary Batteries

System	Characteristics	Applications
Zinc-carbon (Leclanché) Zinc/MnO_2	Common, low-cost primary battery; available in a variety of sizes	Flashlight, portable radios, toys, novelties, instruments
Magnesium (Mg/MnO_2)	High-capacity primary battery; long shelf life	Military receiver-transmitters, aircraft emergency transmitters
Mercury (Zn/HgO)	Highest capacity (by volume) of conventional types; flat discharge; good shelf life	Hearing aids, medical devices (pacemakers), photography, detectors, military equipment but in limited use due to environmental hazard of mercury
Mercad (Cd/HgO)	Long shelf life; good low- and high-temperature performance; low energy density	Special applications requiring operation under extreme temperature conditions and long life; in limited use
Alkaline (Zn/alkaline/MnO_2)	Most popular general-purpose premium battery; good low-temperature and high-rate performance; moderate cost	Most popular primary-battery: used in a variety of portable battery operated equipments
Silver/zinc (Zn/Ag_2O)	Highest capacity (by weight) of conventional types; flat discharge; good shelf life, costly	Hearing aids, photography, electric watches, missiles, underwater and space application (larger sizes)
Zinc/air (Zn/O_2)	Highest energy density, low cost; not independent of environmental conditions	Special applications, hearing aids, pagers, medical devices, portable electronics
Lithium/soluble cathode	High energy density; long shelf life; good performance over wide temperature range	Wide range of applications (capacity from 1 to 10,000 Ah) requiring high energy density, long shelf life, e.g., from utility meters to military power applications
Lithium/solid cathode	High energy density; good rate capability and low-temperature performance; long shelf life; competitive cost	Replacement for conventional button and cylindrical cell applications
Lithium/solid electrolyte	Extremely long shelf life; low-power battery	Medical electronics, memory circuits, fusing

Zinc/Alkaline/Manganese Dioxide Battery. In the past decade, an increasing portion of the primary battery market has shifted to the Zn/alkaline/MnO_2 battery. This system has become the battery of choice because of its superior performance at the higher current drains and low temperatures and its better shelf life. While more expensive than the Leclanché battery on a unit basis, it is more cost-effective for those applications requiring the high-rate or low-temperature capability, where the alkaline battery can outperform the Leclanché battery by a factor of 2 to 10 (see Table 7.6). In addition, because of the advantageous shelf life of the alkaline cell, it is often selected for applications in which the battery is used intermittently and exposed to uncontrolled storage conditions (such as consumer flashlights and smoke alarms), but must perform dependably when required. Most recent advances have been the design of batteries providing improved high rate performance for use in cameras and other consumer electronics requiring this high power capability.

Zinc/Mercuric Oxide Battery.   The zinc/mercuric oxide battery was another important zinc anode primary system. This battery was developed during World War II for military communication applications because of its good shelf life and high volumetric energy density. In the postwar period, it was used in small button, flat, or cylindrical configurations as the power source in electronic watches, calculators, hearing aids, photographic equipment, and similar applications requiring a reliable long-life miniature power source. In the past decade, the use of the mercuric oxide battery has about ended due mainly to environmental problems associated with mercury and with its replacement by other battery systems, such as the zinc/air and lithium batteries, which have superior performance for many applications.

Cadmium/Mercuric Oxide Battery.   The substitution of cadmium for the zinc anode (the cadmium/mercuric oxide cell) results in a lower-voltage but very stable system, with a shelf life of up to 10 years as well as performance at high and low temperatures. Because of the lower voltage, the watthour capacity of this battery is about 60% of the zinc/mercuric oxide battery capacity. Again, because of the hazardous characteristics of mercury and cadmium, the use of this battery is limited.

Zinc/Silver Oxide Battery.   The primary zinc/silver oxide battery is similar in design to the small zinc/mercuric oxide button cell, but it has a higher energy density (on a weight basis) and performs better at low temperatures. These characteristics make this battery system desirable for use in hearing aids, photographic applications, and electronic watches. However, because of its high cost and the development of other battery systems, the use of this battery system, as a primary battery, has been limited mainly to small button battery applications where the higher cost is justified.

Zinc/Air Battery.   The zinc/air battery system is noted for its high energy density, but it had been used only in large low-power batteries for signaling and navigational-aid applications. With the development of improved air electrodes, the high-rate capability of the system was improved and small button-type batteries are now used widely in hearing aids, electronics, and similar applications. These batteries have a very high energy density as no active cathode material is needed. Wider use of this system and the development of larger batteries have been slow because of some of their performance limitations (sensitivity to extreme temperatures, humidity and other environmental factors, as well as poor activated shelf life and low power density). Nevertheless, because of their attractive energy density, zinc/air and other metal/air batteries are now being seriously considered for a number of applications from portable consumer electronics and eventually for larger devices such as electric vehicles, possibly in a reserve or mechanically rechargeable configuration (see Chap. 38).

Magnesium Batteries.   While magnesium has attractive electrochemical properties, there has been relatively little commercial interest in magnesium primary batteries because of the generation of hydrogen gas during discharge and the relatively poor storageability of a partially discharged cell. Magnesium dry cell batteries have been used successfully in military communications equipment, taking advantage of the long shelf life of a battery in an undischarged condition, even at high temperatures and its higher energy density. Magnesium is still employed as an anode material for reserve type and metal/air batteries (see Chaps. 17 and 38).

Aluminum Batteries.   Aluminum is another attractive anode material with a high theoretical energy density, but problems such as polarization and parasitic corrosion have inhibited the development of a commercial product. It, too, is being considered for a number of applications, with the best promise as a reserve or mechanically rechargeable battery (see Chaps. 9 and 38).

Lithium Batteries.   The lithium anode batteries are a relatively recent development (since 1970). They have the advantage of the highest energy density, as well as operation over a very wide temperature range and long shelf life, and are gradually replacing the conventional battery systems. However, except for camera, watch, memory backup, military and other niche applications, they have not yet captured the major general purpose markets as was anticipated because of their high cost and concerns with safety.

As with the zinc systems, there are a number of lithium batteries under development, ranging in capacity from less than 5 mAh to 10,000 Ah, using various designs and chemistries, but having, in common, the use of lithium metal as the anode.

The lithium primary batteries can be classified into three categories (see Chap. 14). The smallest are the low-power solid-state batteries (see Chap. 15) with excellent shelf life, and are used in applications such as cardiac pacemakers and battery backup for volatile computer memory, where reliability and long shelf life are paramount requirements. In the second category are the solid-cathode batteries, which are designed in coin or small cylindrical configurations. These batteries have replaced the conventional primary batteries in watches, calculators, memory circuits, photographic equipment, communication devices, and other such applications where its high energy density and long shelf life are critical. The soluble-cathode batteries (using gases or liquid cathode materials) constitute the third category. These batteries are typically constructed in a cylindrical configuration, as flat disks, or in prismatic containers using flat plates. These batteries, up to about 35 Ah in size, are used in military and industrial applications, lighting products, and other devices where small size, low weight, and operation over a wide temperature range are important. The larger batteries are being developed for special military applications or as standby emergency power sources.

Solid Electrolyte Batteries.   The solid-electrolyte batteries are different from other battery systems in that they depend on the ionic conductivity, in the solid state, of an electronically nonconductive compound rather than the ionic conductivity of a liquid electrolyte. Batteries using these solid electrolytes are low-power (microwatt) devices, but have extremely long shelf lives and the capability of operating over a wide temperature range, particularly at high temperatures. These batteries are used in medical electronics, for memory circuits, and for other such applications requiring a long-life, low-power battery. The first solid-electrolyte batteries used a silver anode and silver iodide for the electrolyte, but lithium is now used as the anode for most of these batteries, offering a higher voltage and energy density.

7.3 COMPARISON OF THE PERFORMANCE CHARACTERISTICS OF PRIMARY BATTERY SYSTEMS

7.3.1 General

A qualitative comparison of the various primary battery systems is given in Table 7.2. This listing illustrates the performance advantages of the lithium anode batteries. Nevertheless, the conventional primary batteries, because of their low cost, availability, and generally acceptable performance in many consumer applications, still maintain a major share of the market.

The characteristics of the major primary batteries are summarized in Table 7.3. This table is supplemented by Table 1.2 in Chap. 1, which lists the theoretical and practical electrical characteristics of these primary battery systems. A graphic comparison of the theoretical and practical performance of various battery systems given in Fig. 1.4 shows that only about 25% of the theoretical capacity is attained under practical conditions as a result of design and discharge requirements.

It should be noted, as discussed in detail in Chaps. 1, 3 and 6, that most of these types of data and comparisons are based on the performance characteristics of single-cell batteries and are necessarily approximations, with each system presented under favorable discharge conditions. The specific performance of a battery system is very dependent on the cell and battery design and all of the specific conditions of use and discharge of the battery.

TABLE 7.2 Comparison of Primary Batteries*

System	Voltage	Specific energy (gravimetric)	Power density	Flat discharge profile	Low-temperature operation	High-temperature operation	Shelf life	Cost
Zinc/carbon	5	4	4	4	5	6	8	1
Zinc/alkaline/manganese dioxide	5	3	2	3	4	4	7	2
Magnesium/manganese dioxide	3	3	2	2	4	3	4	3
Zinc/mercuric oxide	5	3	2	2	5	3	5	5
Cadmium/mercuric oxide	6	5	2	2	3	2	3	6
Zinc/silver oxide	4	3	2	2	4	3	6	6
Zinc/air	5	2	3	2	5	5	—	3
Lithium/soluble cathode	1	1	1	1	1	2	2	6
Lithium/solid cathode	1	1	2	2	2	3	2	4
Lithium/solid electrolyte	2	1	5	2	6	1	1	7

*1 to 8—best to poorest.

TABLE 7.3   Characteristics of Primary Batteries

System	Zinc-carbon (Leclanché)	Zinc-carbon (zinc chloride)	Mg/MnO$_2$	Zn/Alk./MnO$_2$	Zn/HgO	Cd/HgO
Chemistry:						
Anode	Zn	Zn	Mg	Zn	Zn	Cd
Cathode	MnO$_2$	MnO$_2$	MnO$_2$	MnO$_2$	HgO	HgO
Electrolyte	NH$_4$Cl and ZnCl$_2$ (aqueous solution)	ZnCl$_2$ (aqueous solution)	MgBr$_2$ or Mg(ClO$_4$) (aqueous solution)	KOH (aqueous solution)	KOH or NaOH (aqueous solution)	KOH (aqueous solution)
Cell voltage, V§:						
Nominal	1.5	1.5	1.6	1.5	1.35	0.9
Open-circuit	1.5–1.75	1.6	1.9–2.0	1.5–1.6	1.35	0.9
Midpoint	1.25–1.1	1.25–1.1	1.8–1.6	1.25–1.15	1.3–1.2	0.85–0.75
End	0.9	0.9	1.2	0.9	0.9	0.6
Operating temperature, °C	−5 to 45	−10 to 50	−20 to 60	−20 to 55	0 to 55	−55 to 80
Energy density at 20°C§:						
Button size:						
Wh/kg				80	100	55
Wh/L				360	470	230
Cylindrical size:						
Wh/kg	65	85	100	145	105	
Wh/L	100	165	195	400	325	
Discharge profile (relative)	Sloping	Sloping	Moderate slope	Moderate slope	Flat	Flat
Power density	Low	Low to moderate	Moderate	Moderate	Moderate	Moderate
Self-discharge rate at 20°C. % loss per year‡	10	7	3	4	4	3
Advantages	Lowest cost; good for noncritical use under moderate conditions; variety of shapes and sizes; availability	Low cost; better performance than regular zinc-carbon	High capacity compared with zinc-carbon; good shelf life (undischarged)	High capacity compared with zinc-carbon; good low-temperature, high-rate performance	High volumetric energy density; flat discharge; stable voltage	Good performance at high and low temperatures; long shelf life
Limitations	Low energy density; poor low-temperature, high-rate performance	High gassing rate; performance lower than premium alkaline batteries	High gassing (H$_2$) on discharge; delayed voltage	Moderate cost but most cost effective at high rates	Expensive moderate gravimetric energy density, poor-low-temperature performance	Expensive, low-energy density
Status	High production, but losing market share	High production, but losing market share	Moderate production, mainly military	High production, most popular primary battery	Being phased out because of toxic mercury	In limited production being phased out because of toxic components except for some special applications
Major types available	Cylindrical single-cell bobbin and multicell batteries (see Tables 8.9 and 8.10)	Cylindrical single-cell bobbin and multicell batteries (see Table 8.9)	Cylindrical single-cell bobbin and multicell batteries (see Table 9.3)	Button cylindrical and multicell batteries (see Tables 10.9 and 10.10)	NLA*	NLA*

*No longer readily available commercially
†See Chap. 14 for other lithium primary batteries.
‡Rate of self-discharge usually decreases with time of storage.
§Data presented are for 20°C, under favorable discharge condition. See details in appropriate chapter.

Zn/Ag₂O*	Zinc/air	Li/SO₂†	Li/SOCl₂†	Li/MnO₂†	Li/FeS₂†	Solid state
Ag_2O or AgO KOH or $NaOH$ (aqueous solution)	Zn O_2 (air) KOH (aqueous solution	Li SO_2 Organic solvent, salt solution	Li $SOC/2$ $SOCl_2$ w/$AlCl_4$	Li MnO_2 Organic solvent, salt solution	Li FeS_2 Organic solvent, salt solution	Li I_2(P2VP) Solid
1.5	1.5	3.0	3.6	3.0	1.5	2.8
1.6	1.45	3.1	3.65	3.3	1.8	2.8
1.6–1.5	1.3–1.1	2.9–2.75	3.6–3.3	3.0–2.7	1.6–1.4	2.8–2.6
1.0	0.9	2.0	3.0	2.0	1.0	2.0
0 to 55	0 to 50	−55 to 70	−60 to 85	−20 to 55	−20 to 60	0 to 200
135	370			230		
530	1300			545		
	Prismatic 300	260	380	230	260	220–280
	Prismatic 800	415	715	535	500	820–1030
Flat	Flat	Very flat	Flat	Flat	Initial drop than flat medium to high	Moderately flat (at low discharge rates)
Moderate	Low	High	Medium (but dependent on specific design)	Moderate	Medium to high	Very low
6	3 (is sealed)	2	1–2	1–2	1–2	<1
High energy density; good high-rate performance	High volumetric energy density; long shelf life (sealed)	High energy density; best low-temperature, high-rate performance; long shelf life	High energy densitty, long shelf life because of protective film	High energy density; good low-temperature, high-rate performance; cost-effective replacement for small conventional type cells	Replacement for Zu/alkaline/MnO₂ batteries for high rate performance	Excellent shelf life (10–20 y); wide operating temperature range (to 200°C)
Expensive, but cost-effective on button battery applications	Not independent of environment—flooding, drying out; limited power output	Pressurized system Potential safety problems, toxic components Shipment regulated	Voltage delay after storage	Available in small sizes; larger sizes being considered Shipment regulated	Higher cost than alkaline batteries	For very low discharge rates; poor low temperature performance
In production	Moderate production, key use in hearing aids	Moderate production, mainly military	Produced in wide range of sizes and designs, mainly for special applications	Increasing consumer production	Produced in "AA" size	In production for special applications
Button batteries (see Table 12.3)	(See Tables 13.2 and 13.3, also Chap. 38)	Cylindrical batteries (see Tables 14.9 and 14.10)	See section 14.6 and Tables 14.11 to 14.13	Button and small cylindrical batteries (see Tables 14.19 to 14.21)	Produced in "AA" size (see Table 14.18)	(See Table 15.6)

7.3.2 Voltage and Discharge Profile

A comparison of the discharge curves of the major primary batteries is presented in Fig. 7.2. The zinc anode batteries generally have a discharge voltage of between about 1.5 and 0.9 V. The lithium anode batteries, depending on the cathode, usually have higher voltages, many on the order of 3 V, with an end or cutoff voltage of about 2.0 V. The cadmium/mercuric oxide battery operates at the lower voltage level of 0.9-0.6 V. The discharge profiles of these batteries also show different characteristics. The conventional zinc-carbon and zinc/alkaline/manganese dioxide batteries have sloping profiles; the magnesium/manganese dioxide and lithium/manganese dioxide batteries have less of a slope (although at lower discharge rates the lithium/manganese dioxide battery shows a flatter profile). Most of the other battery types have a relatively flat discharge profile.

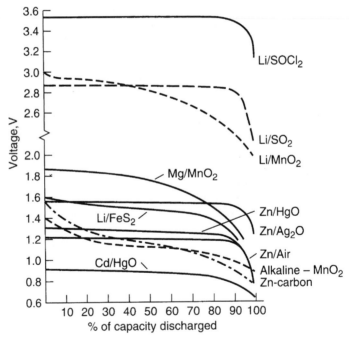

FIGURE 7.2   Discharge profiles of primary battery systems 30–100-h rate.

7.3.3 Specific Energy and Specific Power

Figure 7.3 presents a comparison of the specific energy (also called gravimetric energy density) of the different primary battery systems at various discharge rates at 20°C. This figure shows the hours of service each battery type (normalized to 1-kg battery weight) will deliver at various power (discharge current $\times$ midpoint voltage) levels to an end voltage usually specified for that battery type. The energy density can then be determined by

$$\text{Specific energy} = \text{specific power} \times \text{hours of service}$$

or
$$\text{Wh/kg} = \text{W/kg} \times \text{h} = \frac{\text{A} \times \text{V} \times \text{h}}{\text{kg}}$$

The conventional zinc-carbon battery has the lowest energy density of the primary batteries shown, with the exception, at low discharge rates, of the cadmium/mercuric oxide battery due to the low voltage of the latter electrochemical couple. The zinc-carbon battery performs best at light discharge loads. Intermittent discharges, providing a rest or recovery period at intervals during the discharge, improve the service life significantly compared with a continuous discharge, particularly at high discharge rates.

The ability of each battery system to perform at high current or power levels is shown graphically in Fig. 7.3 by the drop in slope at the higher discharge rates. The 1000-Wh/kg line indicates the slope at which the capacity or energy density of the battery remains constant at all discharge rates. The capacity of most battery systems decreases with increasing discharge rate, and the slope of the linear portion of each of the other lines is less than that of the theoretical 1000-Wh/kg line. Furthermore, as the discharge rate increases, the slope drops off more sharply. This occurs at higher discharge rates for the battery types that have the higher power capabilities.

The performance of the zinc-carbon battery falls off sharply with increasing discharge rate, although the heavy-duty zinc chloride version of the zinc-carbon battery (see Chap. 8) gives better performance under the more stringent discharge conditions. The zinc/alkaline/manganese dioxide battery, the zinc/mercuric oxide battery, the zinc/silver oxide battery, and the magnesium/manganese dioxide battery all have about the same specific energy and performance at 20°C. The zinc/air system has a higher specific energy at the low discharge rates, but falls off sharply at moderately high loads, indicating its low specific power. The lithium batteries are characterized by their high specific energy, due in part to the higher cell voltage. The lithium/sulfur dioxide battery and some of the other lithium batteries are distinguished by their ability to deliver this higher capacity at the higher discharge rates.

Volumetric energy density is, at times, a more useful parameter than gravimetric specific energy, particularly for button and small batteries, where the weight is insignificant. The denser batteries, such as the zinc/mercuric oxide battery, improve their relative position when compared on a volumetric basis, as shown in Table 7.4 and Fig. 7.9. The chapters on the individual battery systems include a family of curves giving the hours of service each battery system will deliver at various discharge rates and temperatures.

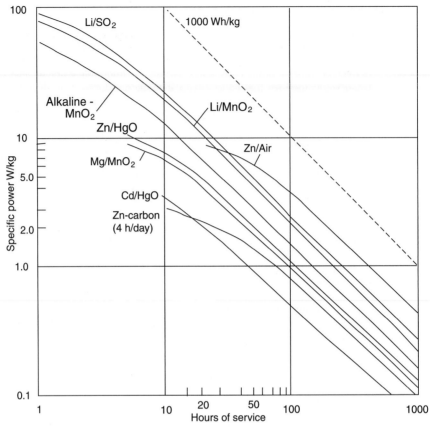

FIGURE 7.3 Comparison of typical performance of primary battery systems—specific power (power density) vs. hours of service.

TABLE 7.4 Comparison of Primary Batteries (Button Configuration)*

System	Voltage, V		Capacity†		Weight, g	Energy density†	
	Nominal	Working	mAh	mWh		mWh/g	Wh/L
Zn/alk/MnO$_2$	1.5	1.25	145	180	2.3	80	360
Zn/HgO	1.35	1.3	180–230	260	2.6	100	470
Zn/Ag$_2$O	1.5	1.55	190	295	2.2	135	575
Zn/AgO	1.5	1.55	245	380	2.2	170	690
Zn/air	1.25	1.25	600	750	1.8	415	1450
Li/FeS$_2$	1.5	1.4	160	220	1.7	130	400
Li/CuO	1.5	1.4	225	315	1.7	135	570
Li/MnO$_2$§	3.0	2.85	160	450	3.3	155	395
Li/Ag$_2$CrO$_4$	3.0	3 to 2.7	130	370	1.7	215	670

 *44 IEC, 1154; 11.6-mm diam.; 5.4-mm high; 0.55-cm^3 volume; these batteries may no longer be available in all chemical systems.
 †At approximately $C/500$ rate, 20°C.
 §$\frac{1}{3}N$ size, equivalent to two 44-size batteries, 11.6-mm diam. by 10.8 mm high.

7.3.4 Comparison of Performance of Representative Primary Batteries

Table 7.4 compares the performance of a number of primary battery systems in a typical button configuration, 1EC size 44, size 44 IEC standard. The data are based on the rated capacity at 20°C at about the $C/500$ rate. The performance of the different systems can be compared, but one should recognize that battery manufacturers may design and fabricate batteries, in the same size and with the same electrochemical system, with differing capacities and other characteristics, depending on the application requirements and the particular market segment the manufacturer is addressing. The discharge curves for these batteries are given in Fig. 7.4.

Table 7.5 summarizes the typical performance obtained with the different primary battery systems for several cylindrical type batteries. The discharge curves for the AA-size batteries are shown in Fig. 7.5, those for the ANSI 1604 9-V batteries in Fig. 7.6.

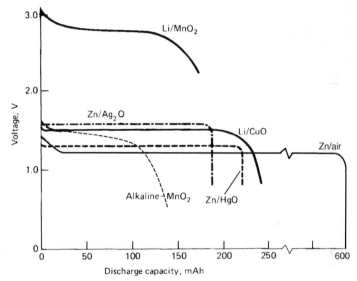

FIGURE 7.4 Typical discharge curves for primary battery systems, 11.6-mm diameter, 5.4 mm high, 20°C. (Li/MnO_2 battery is $\frac{1}{3}$ N size).

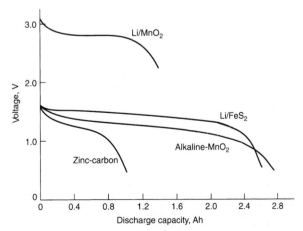

FIGURE 7.5 Typical discharge curves for primary battery systems. AA-size cells, approx. 20-mA discharge rate.* $\frac{2}{3}$A-size battery. (*b*) ANSI 1604 battery 9-V, 250-Ω discharge load.

TABLE 7.5 Comparison of Cylindrical-type Primary Batteries‡

	Zinc-carbon (standard)	Zinc-carbon (heavy-duty ZnCl₂)	Zn/MnO₂ (alkaline)	Zn/HgO	Mg/MnO₂	Li/SO₂	Li/SOCl₂ (bobbin type)	Li/MnO₂	Li/FeS₂
Working voltage, V	1.2	1.2	1.2	1.25	1.75	2.8	3.3	2.8	1.5
D-size cells (54 cm³)									
Ah	4.5	7.0	15	14	7	8	10.2		
Wh	5.4	8.4	18	17.5	12.2	22.4	34		
Weight, g	85	93	138	165	105.	85	100		
Wh/g	65	90	130	105	115	260	340		
Wh/L	100	160	320	325	225	415	675		
N-size cells (3.0 cm³)									
Ah	0.40		0.8	0.8	0.5			1.0*	
Wh	0.48		0.95	1.0	0.87			2.8	
Weight, g	6.3		9.5	12	5.0			13	
Wh/kg	75		100	85	170			215	
Wh/L	145		320	330	290			410	
AA-size cells (7.7 cm³)									
Ah	0.8	1.05	2.85	2.5		1.0	1.6	1.4†	2.6
Wh	0.96	1.25	3.45	3.1		2.8	5.2	3.9	4.35
Weight, g	14.7	15	23.6	30		14	19	17	14.5
Wh/kg	65	84	145	103		200	275	235	300
Wh/L	125	162	400	400		360	670	525	500

*2N size.
†⅔ A size.
‡These batteries may no longer be available in all chemistries.

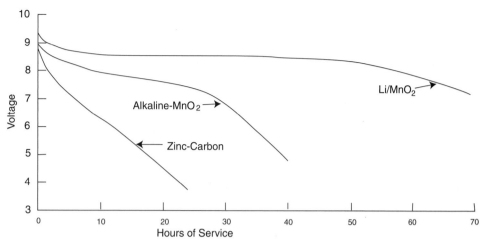

FIGURE 7.6   Typical discharge curves ANSI 1604 battery, 9V, 500 ohm discharge load 20°C.

7.3.5   Effect of Discharge Load and Duty Cycle

The effect of the discharge load on the battery's capacity was shown in Fig. 7.3 and is again illustrated for several primary battery systems in Fig. 7.7. The Leclanché zinc-carbon battery performs best under light discharge loads, but its performance falls off sharply with increasing discharge rates. The zinc/alkaline/manganese dioxide system has a higher energy density at light loads which does not drop off as rapidly with increasing discharge loads. The lithium battery has the highest energy density with reasonable retention of this performance at the higher discharge rates. For low-power applications the service ratio of lithium:zinc (alkaline): zinc-carbon is on the order of 4:3:1. At the heavier loads, however, such as those required for toys, motor-driven applications, and pulse discharges, the ratio can widen to 24:8:1 or greater. At these heavy loads selection of premium batteries is desirable on both a performance and a cost basis.

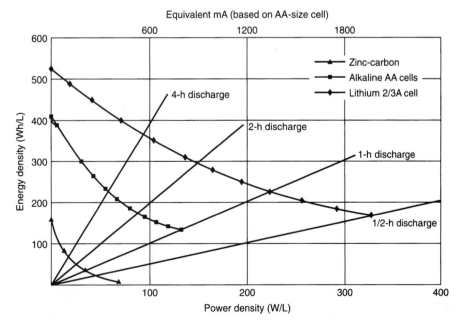

FIGURE 7.7   Comparison of primary battery systems under various continuous discharge loads at 20°C.

7.3.6 Effect of Temperature

The performance of the various primary batteries over a wide temperature range is illustrated in Fig. 7.8 on a gravimetric basis and in Fig. 7.9 on a volumetric basis. The lithium/soluble-cathode systems (Li/SOCl$_2$ and Li/SO$_2$) show the best performance throughout the entire temperature range, with the higher-rate Li/SO$_2$ system having the best capacity retention at the very low temperatures. The zinc/air system has a high energy density at normal temperatures, but only at light discharge loads. The lithium/solid-cathode systems, represented by the Li/MnO$_2$ system, show high performance over a wide temperature range, superior to the conventional zinc anode systems. Figure 7.9 shows an improvement in relative position of the denser, heavier battery systems when compared on a volumetric basis.

(Note: As stated earlier, these data are necessarily generalized and present each battery system under favorable discharge conditions. With the variability in performance due to manufacturer, design, size, discharge conditions, end voltage, and other factors, they may not apply under specific conditions of use. For these details refer to the appropriate chapter for each battery system.)

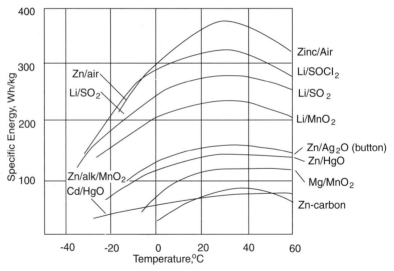

FIGURE 7.8 Specific energy of primary battery systems.

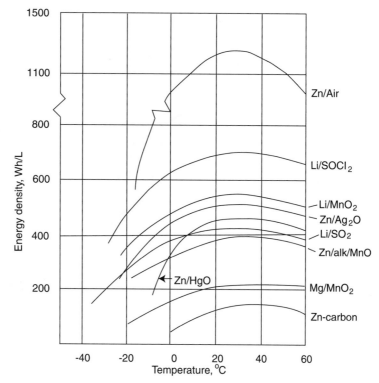

FIGURE 7.9   Volumetric energy density of primary battery systems.

7.3.7 Shelf Life of Primary Batteries

The shelf-life characteristics of the major primary battery systems are plotted in Fig. 7.10 and show the rate of loss (in terms of percentage capacity loss per year) from 20 to 70°C. The relationship is approximately linear when log capacity loss is plotted against log $1/T$ (temperature, °Kelvin). The data assume that the rate of capacity loss remains constant throughout the storage period, which is not necessarily the case with most battery systems. For example, as shown in Chap. 14 for several lithium batteries, the rate of loss tapers off as the storage period is extended. The data are also a generalization of the capability of each battery system under manufacturer-rated conditions because of the many variations in battery design and formulation. The discharge conditions and size also have an influence on charge retention. The capacity loss is usually highest under the more stringent discharge conditions.

The ability to store batteries improves as the storage temperature is lowered. Cold storage of batteries is used to extend their shelf life. Moderately cold temperatures, such as 0°C, was usually used as freezing could be harmful for some battery systems or designs. As the shelf life of most batteries has been improved, manufacturers are no longer recommending cold storage but suggest room temperature storage is adequate provided that excursions to high temperature is avoided.

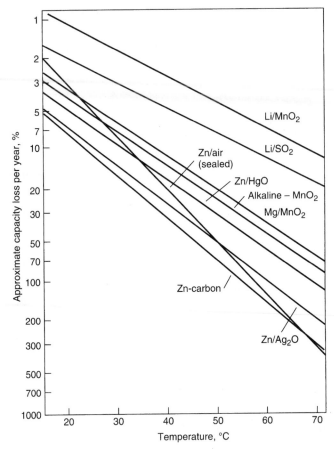

FIGURE 7.10   Shelf-life characteristics of primary battery systems.

7.3.8 Cost

In the selection of the most cost-effective battery for a given application, other factors should also be considered in addition to the initial cost of the battery. These include the battery's performance under the specific conditions of use, operation under other temperature and environmental conditions (if applicable), shelf life, and other parameters that could affect the battery's capabilities. The impact of the discharge rate and duty cycle on the cost of battery operation (cost per hour of service) is shown in Table 7.6, which compares the service life and cost per hour of service of the general-purpose and premium (zinc-chloride) zinc-carbon batteries with the zinc/alkaline-manganese dioxide battery under various regimes. Based on the unit battery cost shown in the table, the general-purpose zinc-carbon battery has the most competitive hourly service cost only on the low-drain intermittent radio application, the conditions most to its liking. However, even with its higher unit cost, the alkaline battery is, by far, the more economical battery to use under stringent high-drain applications such as in toys and electronic games. The use of the zinc-carbon battery is not recommended in the high drawn photoflash or digital camera applications. A similar analysis should be made when evaluating the performance of any candidate battery against the application requirement to determine which battery is the most cost-effective (see Chap. 6).

TABLE 7.6 Comparison of Battery Performance and Cost—Zinc-Carbon vs. Alkaline Manganese Dioxide AA-size Batteries

Type of test	Performance			Cost per hour of service, $		
	General-purpose zinc-carbon	Premium zinc-carbon	Alkaline-manganese dioxide	General-purpose zinc-carbon	Premium zinc-carbon	Alkaline-manganese dioxide
3.9-Ω toy[a]	0.5 h	1.2 h	5 h	0.60	0.33	0.15
43-Ω radio[b]	14 h	27 h	60 h	0.016	0.015	0.013
10-Ω tape	2.5 h	4.7 h	13.5 h	0.12	0.085	0.056
1000 mA photo flash test[d]	NR	NR	210 pulses	NR	NR	0.036
24-Ω remote control[e]		11 h	33 h	—	0.036	0.023
250 mA electronic games[f]		1 h	6 h	—	0.40	0.13
Approximate unit cell cost, $	0.30	0.40	0.75	—	—	—

[a] Toy test: 1 h/day to 0.8 V.
[b] Transistor radio test: 4 h/day to 0.9 V.
[c] Tape player and cassette test: 1 h/day to 0.9 V.
[d] Photo flash test: 10 s/m, 1 h/day to 0.9 V.
[e] Remote control test 15 s/m, 8 h/day to 1.0 V.
[f] Electronic game test 1 h/day to 0.9 V.
NR battery not recommended for this application.
Source: Data based on specification requirements, ANSI C18.1M (2000) "Portable primary cells and batteries with aqueous electrolyte—general and specifications"

7.4 RECHARGING PRIMARY BATTERIES

Recharging primary batteries is a practice that should be avoided because the cells are not designed for that type of use. In most instances it is impractical, and it could be dangerous with cells that are tightly sealed and not provided with an adequate mechanism to permit the release of gases that form during charging. Such gassing could cause a cell to leak, rupture, or explode, resulting in personal injury, damage to equipment, and other hazards. Most primary batteries are labeled with a cautionary notice advising that they should not be recharged.

Technically some primary cells can be recharged for several cycles under carefully controlled charging conditions and usually at low charge rates. However, even if successful, they may not deliver full capacity and may have poor charge retention after recharge. Primary batteries are not designed to be recharged, and charging should not be attempted with any primary battery, unless one is fully aware of the charging conditions, equipment, and risks.

Several of the typical primary battery systems, such as the zinc/alkaline/manganese dioxide system, have been designed in a rechargeable configuration. These batteries are covered in Chap. 36.

CHAPTER 8
ZINC-CARBON BATTERIES
(Leclanché and Zinc Chloride Cell Systems)

Dennis W. McComsey

8.1 GENERAL CHARACTERISTICS

Zinc-carbon batteries have been well known for over a hundred years. The two types of zinc-carbon batteries that are popular now are the Leclanché and zinc chloride systems. Both systems remain among the most widely used of all the primary battery systems worldwide, although their use in the United States and Eurrope is declining. The use of flashlights, portable radios, and other moderate and light drain applications, as well as the absence of a high drain device base, is stimulating the use of zinc-carbon batteries in the emerging third world countries. The battery is characterized as having low cost, ready availability and acceptable performance for a great number of applications.

The zinc-carbon battery industry continues to grow worldwide. The global primary battery market is expected to reach $22 billion in sales by the year 2002. Zinc-carbon battery sales globally are expected to reach $7.2 billion, or a 34% share of the global market. Some details of the zinc-carbon battery market and the global primary battery market are given in Table 8.1.

The current estimate of annual growth for the zinc-carbon global market, through the year 2007, continues to be +5% per year. The expected decline in the zinc-carbon battery market was only realized in the United States with a relatively constant -2% to -5% decline in sales volume per year. This is expected to continue. Asia, emerging third world and Eastern European markets drove the global demand for the inexpensive zinc-carbon battery system. As an example, 80% of all primary batteries presently sold in Eastern and Central Europe are zinc-carbon types. Even in the United States, this system still shows substantial usage with total U.S. sales in 1998 of $370 million dollars.[1-3]

New, heavier drain toy, lighting and communications devices, entering the consumer market continue to stimulate an increased preference for zinc-alkaline cells. This has spawned a segmentation of the zinc-alkaline system resulting in the design of increased power, heavy-duty, zinc-alkaline batteries for those applications. These new applications and continued impact from the use of rechargeable cells will be additional factors impacting zinc-carbon sales in the U.S.

TABLE 8.1 Zinc-Carbon Battery Market

Regional market location	Total primary battery market value by 2002 (billions $US)	Zinc-carbon battery market value by 2002 (billions $US)	Zinc-carbon as a percent of global market (%)
US & Canada	4.4	0.3	6.8
Latin America	1.4	1.0	64.3
Western Europe	3.9	0.9	20.5
Eastern Europe	2.8	1.0	32.1
Asia Pacific	8.6	4.0	45.3
Global Total	21.1	7.2	34.5

Source: Freedonia Group 1999 battery market study.[2]

Historically, the first prototype of the modern dry cell was the Leclanché Wet Cell developed by a telegraphic engineer, Georges-Lionel Leclanché in 1866. The design resulted from the need to provide a more reliable and easily maintained power source for telegraphic offices. The cell was unique in that it was the first practical cell using a single low-corrosive fluid, ammonium chloride, as an electrolyte instead of the strong mineral acids in use at the time. This rendered the cells relatively inactive until the external circuit was connected. The cell was inexpensive, safe, easily maintained and provided excellent shelf (storage) life with adequate performance characteristics.

The cell consisted of an amalgamated zinc bar serving as the negative electrode anode, a solution of ammonium chloride as the electrolyte, and a one-to-one mixture of manganese dioxide and powdered carbon packed around a carbon rod as the positive electrode or cathode. The positive electrode was placed in a porous pot, which was, in turn, placed in a square glass jar along with the electrolyte and zinc bar. By 1876, Leclanché had evolved the design removing the need for the porous pot by adding a resin (gum) binder to the manganese dioxide-carbon mix. In addition he formed this composition into a compressed block by use of hydraulic pressure at a temperature of 100°C. Leclanché's inventiveness brought together the major components of today's zinc-carbon battery and set the stage for conversion from the "wet" cell to the "dry" cell concept.

Dr. Carl Gassner is credited with constructing the first "dry" cell in 1888. It was similar to the Leclanché system except that ferric hydroxide and manganese dioxide were used as the cathode. The "dry" cell concept grew from the desire to make the cell unbreakable and spill-proof. His cell provided an unbreakable container by forming the anode from zinc sheet into a cup, replacing the glass jar. He then immobilized the electrolyte by using a paste containing plaster of Paris and ammonium chloride. The cylindrical block of cathode mix (called a *bobbin*) was wrapped in cloth and was saturated with a zinc chloride-ammonium chloride electrolyte. This reduced local chemical action and improved the shelf life. Gassner, as did others, replaced the plaster of Paris with wheat flour as an electrolyte-gelatinizing agent and demonstrated such a battery as a portable lighting power source at the 1900 World's Fair in Paris. These advances were instrumental in establishing industrial production and commercialization of the "zinc-carbon dry cell" and led to the evolution of "dry-cell" portable power.

From the early 1900s through the 1990s, the portable power industry has been driven to meet the needs of the electric and electronic industries. In the early part of the 20th century, battery-operated telephones, electric doorbells, toys, lighting devices, and countless other applications placed increasing demands on "dry battery" manufacturers. Through the middle of the century the advent of radio broadcasting and World War II military applications further increased that demand significantly. In the latter part of the century, demands for an inexpensive battery to power flashlights, portable transistor radios, electric clocks, cameras, electronic toys, and other convenience applications have maintained the demand.

Zinc-carbon technology has continued to evolve. During much of the 20th century, the system was continually improved. Manganese dioxide, electrolytic and chemical, with higher capacity and substantially higher activity than the natural manganese ores, have been developed. The use of acetylene black carbon as a substitute for graphite has not only provided a more conductive cathode structure, but the higher absorption properties have enhanced the handling characteristics of the cathode powder. Improved manufacturing techniques were implemented that resulted in the production of an improved product at lower costs. A better understanding of the reaction mechanisms, improved separators, and venting seal systems and all have contributed to the present state of the zinc-carbon art.

A significant portion of the technology effort since the 1960s has been directed toward developing the zinc chloride cell system. This design provided substantially improved performance on heavy drain applications over that of the Leclanché cell. During the 1980s to the present time, development effort has been focused on environmental concerns, including the elimination of mercury, cadmium, and other heavy metals from the system. The work the past century, has extended the discharge life and storage life of the zinc-carbon battery over 400% compared to the 1910 version.[3–8]

Most of zinc-carbon cell manufacturing and battery assembly is now done outside of the United States. Manufacturers have opted to consolidate and relocate plants and equipment to achieve cost reductions through the use of economies of scale, low cost labor, and materials. Regional plants are coming of age rather than local country manufacturing facilities. This has occurred because of the improved conditions in global trade, which in many areas has reduced tariffs and duties. As a direct result, cell prices have generally been maintained at steady levels and business opportunities for zinc-carbon batteries have increased globally.

The advantages and disadvantages of zinc-carbon batteries, compared with other primary battery systems, are summarized in Table 8.2. A comparison of the more popular primary cell systems is given in Chap. 7.

TABLE 8.2 Major Advantages and Disadvantages of Leclanché and Zinc-Chloride Batteries

Standard Leclanché battery		
Advantages	Disadvantages	General comments
Low cell cost	Low energy density	Good shelf life if refrigerated
Low cost per watt-hour	Poor low temp service	For best capacity the discharge should
Large variety of	Poor leakage resistance	be intermittent
shapes, sizes,	under abusive conditions	Capacity decreases as the discharge
voltages, and	Low efficiency under high	drain increases
capacities	current drains	Steadily falling voltage is useful if
Various formulations	Comparatively poor shelf life	early warning of end of life is
Wide distribution and	Voltage falls steadily with	important
availability	discharge	
Long tradition of		
reliability		

Standard Zinc-chloride battery		
Advantages	Disadvantages	General comments
Higher energy density	High gassing rate	Steadily falling voltage with discharge
Better low-temperature	Requires excellent sealing	Good shock resistance
service	system due to increased	Low to medium initial cost
Good leak resistance	oxygen sensitivity	
High efficiency under		
heavy discharge		
loads		

8.2 CHEMISTRY

The zinc-carbon cell uses a zinc anode, a manganese dioxide cathode, and an electrolyte of ammonium chloride and/or zinc chloride dissolved in water. Carbon (acetylene black) is mixed with the manganese dioxide to improve conductivity and retain moisture. As the cell is discharged, the zinc is oxidized and the manganese dioxide is reduced. A simplified overall cell reaction is:

$$Zn + 2MnO_2 \rightarrow ZnO \cdot Mn_2O_3$$

In actual practice, the chemical processes which occur in the Leclanché cell are significantly more complicated. Despite the 125 years of its existence, controversy over the details of the electrode reactions continues.[7] A chemical "recuperation" reaction may operate simultaneously with the discharge reactions.[5] This could result in several intermediate states which confuse the reaction mechanisms. Furthermore, the chemistry is complex because MnO_2 is a non-stoichiometric oxide and is more accurately represented as MnO_x, where x typically equals 1.9+. The efficiency of the chemical reaction depends on such things as electrolyte concentration, cell geometry, discharge rate, discharge temperature, depth of discharge, diffusion rates, and type of MnO_2 used. A more comprehensive description of the cell reaction is as follows:[4]

1. For cells with ammonium chloride as the primary electrolyte:

Light discharge: $Zn + 2MnO_2 + 2NH_4Cl \rightarrow 2MnOOH + Zn(NH_3)_2Cl_2$

Heavy discharge: $Zn + 2MnO_2 + NH_4Cl + H_2O \rightarrow 2MnOOH + NH_3 + Zn(OH)Cl$

Prolonged discharge: $Zn + 6MnOOH \rightarrow 2Mn_3O_4 + ZnO + 3H_2O$

2. For cells with zinc chloride as the primary electrolyte:

Light or heavy discharge: $Zn + 2MnO_2 + 2H_2O + ZnCl_2 \rightarrow 2MnOOH + 2Zn(OH)Cl$

or: $4\,Zn + 8MnO_2 + 9H_2O + ZnCl_2 \rightarrow 8MnOOH + ZnCl_2 \cdot 4ZnO \cdot 5H_2O$

Prolonged discharge: $Zn + 6MnOOH + 2Zn(OH)Cl \rightarrow 2Mn_3O_4 + ZnCl_2 \cdot 2ZnO \cdot 4H_2O$

[Note: $2MnOOH$ is sometimes written as $Mn_2O_3 \cdot H_2O$ and Mn_3O_4 as $MnO \cdot Mn_2O_3$. Electrochemical discharge of MnOOH vs zinc (prolonged discharge) does not provide a useful operating voltage for typical applications.]

In the theoretical case, as discussed in Chap. 1, the specific capacity calculates to 224 Ah/kg, based on Zn and MnO_2 and the simplified cell reaction. On a more practical basis, the electrolyte, carbon black, and water are ingredients, which cannot be omitted from the system. If typical quantities of these materials are added to the "theoretical" cell, a specific capacity of 96 Ah/kg is calculated. This is the highest specific capacity a general-purpose cell can have and is, in fact, approached by some of the larger Leclanché cells under certain discharge conditions. The actual specific capacity of a practical cell, considering all the cell components and the efficiency of discharge, can range from 75 Ah/kg on very light loads to 35 Ah/kg on heavy-duty, intermittent discharge conditions.

8.3 TYPES OF CELLS AND BATTERIES

During the last 125 years the development of the zinc-carbon battery has been marked by gradual change in the approach to improve its performance. It now appears that zinc-carbon batteries are entering a transitional phase. While miniaturization in the electrical and electronic industries has reduced power demands, it has been offset by the addition of new features requiring high power, such as motors to drive compact disc players or cassette recorders, halogen bulbs in lighting devices, etc. This has increased the need for a battery that can meet heavy discharge requirements. For this reason, as well as competition from the alkaline battery system for heavy drain applications, many manufacturers are no longer investing capital to improve the Leclanché or zinc-carbon technology. The traditional Leclanché cell construction, which utilizes a starch paste separator, is being gradually phased out and replaced by zinc chloride batteries utilizing paper separators. This results in increased volume available for active materials and increased capacity. In spite of these conversion efforts by manufacturers, a number of third world countries still continue the demand for pasted Leclanché product because of its low cost. The size of that market has prevented a complete conversion. It appears that this situation will continue for the near future.

During this transitional phase, the zinc-carbon batteries can be classified into two types, Leclanché and zinc chloride. These can, in turn, be subdivided into separate general purpose and premium battery grades, in both pasted and paper-lined constructions:

8.3.1 Leclanché Batteries

General Purpose. Application: Intermittent low-rate discharges, low cost. The traditional, regular battery, which is not too different from the one introduced in the late nineteenth century, uses zinc as the anode, ammonium chloride (NH_4Cl) as the main electrolyte component along with zinc chloride, a starch paste separator, and natural manganese dioxide (MnO_2) ore as the cathode. Batteries of this formulation and design are the least expensive and are recommended for general-purpose use and when cost is more important than superior service or performance.

Industrial Heavy Duty. Application: Intermittent medium- to heavy-rate discharges, low to moderate cost. The industrial "heavy-duty" zinc-carbon battery generally has been converted to the zinc chloride system. However, some types continue to include ammonium chloride and zinc chloride ($ZnCl_2$) as the electrolyte and synthetic electrolytic or chemical manganese dioxide (EMD or CMD) alone or in combination with natural ore as the cathode. Its separator may be of starch paste but it is typically a paste-coated paper liner type. This grade is suitable for heavy intermittent service, industrial applications, or medium-rate continuous discharge.

8.3.2 Zinc Chloride Batteries

General Purpose. Application: Low-rate discharges both intermittent and continuous, low cost. This battery has replaced the Leclanché general-purpose battery in all Western countries. It is a true "zinc-chloride" battery and possesses some of the "heavy-duty" characteristics of the premium type. The electrolyte is zinc chloride; however, some manufacturers may add small amounts of ammonium chloride. Natural manganese dioxide ore is used as the cathode. Batteries of this formulation and design are competitive in cost to the Leclanché general-purpose batteries. They are recommended for general-purpose use on both continuous and intermittent discharges and when cost is an important consideration. This battery exhibits a low leakage characteristic.

Industrial Heavy Duty. Application: Low to intermediate-continuous and intermittent heavy-rate discharges; low to moderate cost. This battery has generally replaced the industrial Leclanché heavy-duty battery. It is a true "zinc-chloride" cell and possesses the heavy-duty characteristics of the premium zinc chloride type. The cell electrolyte is zinc chloride; however, some manufacturers may add small amounts of ammonium chloride. Natural manganese dioxide ore is used along with electrolytic manganese dioxide as the cathode. These cells use paper separators coated with cross-linked or modified starches, which enhance their stability in the electrolyte. Batteries of this formulation and design are competitive in cost to the Leclanché heavy-duty industrial batteries. They are recommended for heavy-duty applications where cost is an important consideration. This battery also exhibits a low leakage characteristic.

Extra/Super Heavy Duty. Application: Medium and heavy continuous, and heavy intermittent discharges; higher cost than other zinc-chloride types. The extra/super heavy-duty type of battery is the premium grade of the zinc-chloride line. This cell is composed mainly of an electrolyte of zinc chloride with perhaps a small amount of ammonium chloride, usually not exceeding 1% of the cathode weight. The ore used for the cathode is exclusively electrolytic manganese dioxide (EMD). These cells use paper separators coated with cross-linked or modified starches, which enhance their stability in the electrolyte. Many manufacturers use proprietary separators in almost all their zinc-carbon type batteries. This battery type is recommended when good performance is desired but at higher cost. It also has improved low-temperature characteristics and reduced electrolyte leakage.

In general, the higher the grade or class of zinc-carbon batteries the lower the cost per minute of service. The price difference between classes is about 10 to 25%, but the performance difference can be from 30 to 100% in favor of the higher grades depending upon the application drain.

8.4 CONSTRUCTION

The zinc-carbon battery is made in many sizes and a number of designs but in two basic constructions: cylindrical and flat. Similar chemical ingredients are used in both constructions.

8.4.1 Cylindrical Configuration

In the common Leclanché cylindrical battery (Figs. 8.1 and 8.2), the zinc can serves as the cell container and anode. The manganese dioxide is mixed with acetylene black, wet with electrolyte, and compressed under pressure to form a bobbin. A carbon rod is inserted into the bobbin. The rod serves as the current collector for the positive electrode. It also provides structural strength and is porous enough to permit the escape of gases, which accumulate in the cell, without allowing leakage of electrolyte. The separator, which physically separates the two electrodes and provides the means for ion transfer through the electrolyte, can be a cereal paste wet with electrolyte (Fig. 8.1) or a starch or polymer coated absorbent Kraft paper in the "paper-lined" cell (Fig. 8.2). This provides thinner separator spacing, lower internal resistance and increased active materials volume. Single cells are covered with metal, cardboard, plastic or paper jackets for aesthetic purposes and to minimize the effect of electrolyte leakage through containment.

Construction of the zinc chloride cylindrical battery (Fig. 8.3) differs from that of the Leclanché battery in that it usually possesses a resealable, venting seal. The carbon rod serving as the current collector is sealed with wax to plug any vent paths (necessary for

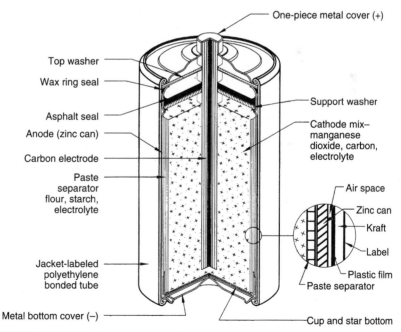

FIGURE 8.1 Typical cutaway view of cylindrical Leclanché battery ("Eveready") paste separator, asphalt seals.

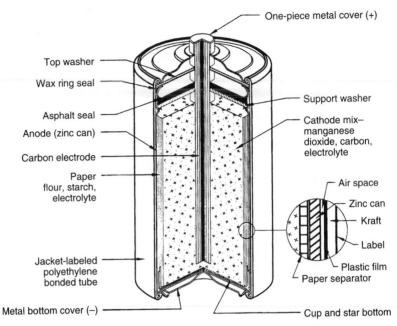

One-piece metal cover (+)

Top washer

Wax ring seal

Asphalt seal

Support washer

Anode (zinc can)

Cathode mix—
manganese
dioxide, carbon,
electrolyte

Carbon electrode

Paper
flour, starch,
electrolyte

Air space

Zinc can

Kraft

Label

Plastic film

Paper separator

Jacket-labeled
polyethylene
bonded tube

Metal bottom cover (−)

Cup and star bottom

FIGURE 8.2 Typical cutaway view of cylindrical Leclanché battery ("Eveready") paper liner separator, asphalt seal.

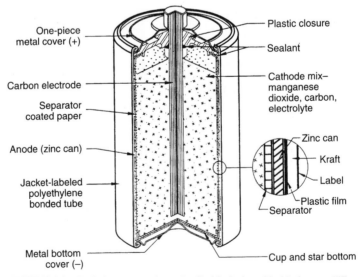

One-piece
metal cover (+)

Plastic closure

Sealant

Carbon electrode

Cathode mix—
manganese
dioxide, carbon,
electrolyte

Separator
coated paper

Zinc can

Kraft

Anode (zinc can)

Label

Plastic film

Jacket-labeled
polyethylene
bonded tube

Separator

Metal bottom
cover (−)

Cup and star bottom

FIGURE 8.3 Typical cutaway view of cylindrical zinc chloride battery ("Eveready") paste separator, plastic seal.

Leclanché types). Venting is then restricted to only the seal path. This prevents the cell from drying out and limits oxygen ingress into the cell during shelf storage. Hydrogen gas evolved from corrosion of the zinc is safely vented as well. In general, the assembly and finishing processes resemble that of the earlier cylindrical batteries.

8.4.2 Inside Out Cylindrical Construction

Another cylindrical cell is the "inside-out" construction shown in Fig. 8.4. This construction does not use the zinc anode as the container. This version resulted in more efficient zinc utilization and improved leakage, but has not been manufactured since the late 1960s. In this cell, an impact-molded impervious inert carbon wall serves as the container of the cell and as the cathode current collector. The zinc anode, in the shape of vanes, is located inside the cell and is surrounded by the cathode mix.

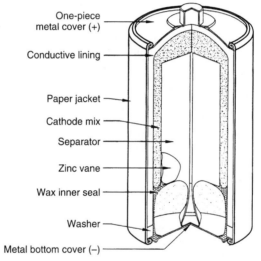

One-piece metal cover (+)

Conductive lining

Paper jacket

Cathode mix

Separator

Zinc vane

Wax inner seal

Washer

Metal bottom cover (−)

FIGURE 8.4 Typical cutaway view of cylindrical Leclanché battery ("Eveready") inside-out construction.

8.4.3 Flat Cell and Battery

The flat cell is illustrated in Fig. 8.5. In this construction, a duplex electrode is formed by coating a zinc plate with either a carbon-filled conductive paint or laminating it to a carbon-filled conductive plastic film. Either coating provides electrical contact to the zinc anode, isolates the zinc from the cathode of the next cell, and performs the function of cathode collector. The collector function is the same as that performed by the carbon rod in cylindrical cells. When the conductive paint method is used, an adhesive must be placed onto the painted side of the zinc prior to assembly to effectively seal the painted surface directly to the vinyl band to encapsulate the cell. No expansion chamber or carbon rod is used as in the cylindrical cell. The use of conductive polyisobutylene film laminated to the zinc instead of the conductive paint and adhesive usually results in improved sealing to the vinyl; however, the film typically occupies more volume than the paint and adhesive design. These methods of construction readily lend themselves to the assembly of multi-cell batteries.

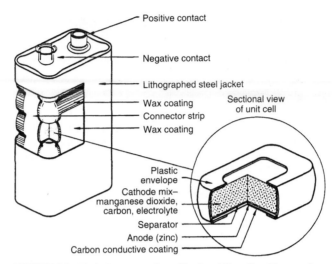

FIGURE 8.5   Typical cutaway view of Leclancé flat cell and battery. (e.g. "Eveready" #216)

Flat cell designs increase the available space for the cathode mix because the package and electrical contacts are minimized, thereby increasing the energy density. In addition, a rectangular construction reduces wasted space in multi-cell assemblies, (which is, in fact, the only application for the flat cell). The volumetric energy density of an assembled battery using flat cells is nearly twice that of cylindrical cell assemblies.

Metal contact strips are used to attach the ends of the assembled battery to the battery terminals; (e.g., 9-V transistor battery). The orientation of the stack subassembly (cathode up or anode up) is only important for each manufacturer's method of assembly. The use of contact strips allows either design mode. The entire assembly is usually encapsulated in wax or plastic. Some manufacturers also sleeve the assembly in shrink film after waxing. This aids the assembly process cleanliness and provides additional insurance against leakage. Cost, ease of assembly, and process efficiencies usually dictate the orientation during the assembly process.

8.4.4 Special Designs

Designs for special applications are currently in use. These designs demonstrate the levels of innovation that can be applied to unusual application and design problems. These are covered in Sec. 8.7.

8.5 CELL COMPONENTS

8.5.1 Zinc

Battery grade zinc is 99.99% pure. Classical zinc can alloys contained 0.3% cadmium and 0.6% lead. Modern lubrication and forming techniques have reduced these amounts. Currently, zinc can alloys with cadmium contain 0.03% to 0.06% cadmium and 0.2% to 0.4% of lead. The content of these metals varies according to the method used in the forming process. Lead, while insoluble in the zinc alloy, contributes to the forming qualities of the can, although too much lead softens the zinc. Lead also acts as a corrosion inhibitor by increasing the hydrogen overvoltage of the zinc in much the same manner as does mercury. Cadmium aids the corrosion-resistance of zinc to ordinary dry cell electrolytes and adds stiffening strength to the alloy. For cans made by the drawing process, less than 0.1% of cadmium is used because more would make the zinc difficult to draw. Zinc cans are commonly made by three different processes:

1. Zinc is rolled into a sheet, formed into a cylinder, and, with the use of a punched-out zinc disk for the bottom, soldered together. This method is obsolete except for the most primitive of assemblies. Last use of this method in the U.S. was during the 1980s in 6 inch cells.

2. Zinc is deep-drawn into a can shape. Rolled zinc sheet is shaped into a can by forming through a number of steps. This method was used primarily in cell manufacturing in the U.S. prior to the relocation and consolidation of U.S. zinc-carbon manufacturing overseas.

3. Zinc is impact extruded from a thick, flat calot. This is now the method of choice. Used globally, this method reshapes the zinc by forcing it to flow under pressure, from the calot shape into the can shape. Calots are either cast from molten zinc alloy or punched from a zinc sheet of the desired alloy.

Metallic impurities such as copper, nickel, iron, and cobalt cause corrosive reactions with the zinc in battery electrolyte and must be avoided particularly in "zero" mercury constructions. In addition, iron in the alloy makes zinc harder and less workable. Tin, arsenic, antimony, magnesium, etc., make the zinc brittle.[4,6]

U.S. federal environmental legislation prohibits the land disposal of items containing cadmium and mercury when these components exceed specified leachable levels. Some states and municipalities have banned land disposal of batteries, require collection programs, and prohibit sale of batteries containing added cadmium or mercury. Some European have similarly prohibited the sale and disposal of batteries containing these materials. For these reasons, levels of both of these heavy metals have been reduced to near zero. This impacts directly upon global zinc can manufacture due to importation of battery products to the U.S. and Europe. Manganese is a satisfactory substitute for cadmium, and has been included in the alloy at levels similar to that of cadmium to provide stiffening. The handling properties of zinc, alloyed with manganese or cadmium are equivalent; however, no corrosion resistance is imparted to the alloy with manganese as is the case with cadmium.

8.5.2 Bobbin

The bobbin is the positive electrode and is also called the black mix, depolarizer, or cathode. It is a wet powder mixture of MnO_2, powdered carbon black, and electrolyte (NH_4Cl and/ or $ZnCl_2$, and water). The powdered carbon serves the dual purpose of adding electrical conductivity to the MnO_2, which itself has high electrical resistance. It also acts as a means of holding the electrolyte. The cathode mixing and forming processes are also important since they determine the homogeneity of the cathode mix and the compaction characteristics associated with the different methods of manufacture. This becomes more critical in the case of the zinc-chloride cell, where the cathode contains proportions of liquid that range between 60% and 75% by volume.

Of the various forming methods available, mix extrusion and compaction-then-insertion are the two used most widely. On the other hand, there is a wide variety of techniques for mixing. The most popular methods are "Cement"-style mixers and rotary mullor mixers. Both techniques offer the ability to manufacture large quantities of mix in relatively short times and minimize the shearing effect upon the carbon black, which reduces its ability to hold solution.

The bobbin usually contains ratios of manganese dioxide to powdered carbon from 3:1 to as much as 11:1 by weight. Also, 1:1 ratios have been used in batteries for photoflash applications where high pulses of current are more important than capacity.

8.5.3 Manganese Dioxide (MnO_2)

The types of manganese dioxide used in dry cells are generally categorized as natural manganese dioxide (NMD), activated manganese dioxide (AMD), chemically synthesized manganese dioxide (CMD), and electrolytic manganese dioxide (EMD). EMD is a more expensive material, which has a gamma-phase crystal structure. CMD has a delta-phase structure and NMDs the alpha and beta phases of MnO_2. EMD, while more expensive, results in a higher cell capacity with improved rate capability and is used in heavy or industrial applications. As shown in Fig. 8.6, polarization is significantly reduced using electrolytic material compared to the chemical or natural ores.[10]

Naturally occurring ores (in Gabon Africa, Greece, and Mexico), high in battery-grade material (70% to 85% MnO_2), and synthetic forms (90% to 95% MnO_2) generally provide electrode potentials and capacities proportional to their manganese dioxide content. Manganese dioxide potentials are also affected by the pH of the electrolyte. Performance characteristics depend upon the crystalline state, the state of hydration, and the activity of the MnO_2. The efficiency of operation under load depends heavily upon the electrolyte, the separator characteristics, the internal resistance and the overall construction of the cell.[4,5]

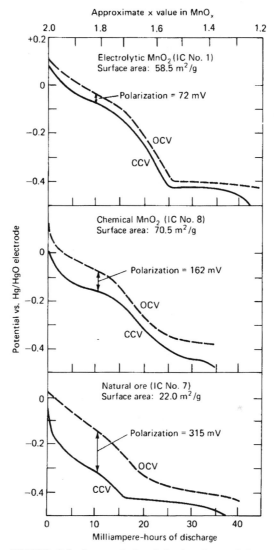

FIGURE 8.6 Open and closed-circuit voltage of three types of manganese dioxide (in 9M KOH). (*From Kozawa and Powers.*)

8.5.4   Carbon Black

Because manganese dioxide is a poor electrical conductor, chemically inert carbon or carbon black is added to the cathode mix to improve its conductivity. This is achieved by coating the manganese dioxide particles with carbon during the mixing process. It provides electrical conductivity to the particle surface and also serves the important functions of holding the electrolyte and providing compressibility and elasticity to the cathode mix during processing.

Graphite was once used as the principal conductive media and is still used to some extent. Acetylene black, by virtue of its properties, has displaced graphite in this role for both Leclanché and zinc chloride cells. One great advantage of acetylene black is its ability to hold more electrolyte in the cathode mix. Caution must be used during the mixing process so as to prevent intense shearing of the black particles as this reduces its ability to hold electrolyte. This is critical for zinc-chloride cells, which contain much higher electrolyte levels than the Leclanché cell. Cells containing acetylene black usually give superior intermittent service, which is the way most zinc-carbon batteries are used. Graphite, on the other hand, serves well for high flash currents or for continuous drains.[4,9]

8.5.5   Electrolyte

The ordinary Leclanché cell uses a mixture of ammonium chloride and zinc chloride, with the former predominating. Zinc-chloride cells typically use only $ZnCl_2$, but can contain a small amount of NH_4Cl to ensure high rate performance. Examples of typical electrolyte formulation for the zinc-carbon battery systems are listed in Table 8.3.

TABLE 8.3   Electrolyte Formulations*

Constituent	Weight %
Electrolyte I	
NH_4Cl	26.0
$ZnCl_2$	8.8
H_2O	65.2
Zinc-corrosion inhibitor	0.25–1.0
Electrolyte II	
$ZnCl_2$	15–40
H_2O	60–85
Zinc-corrosion inhibitor	0.02–1.0

*Electrolyte I based on Kozawa and Powers.[7]
Electrolyte II based on Cahoon.[5]

8.5.6   Corrosion Inhibitor

The classical zinc-corrosion inhibitor has been mercuric or mercurous chloride, which forms an amalgam with the zinc. Cadmium and lead, which reside in the zinc alloy, also provide zinc anode corrosion protection. Other materials like potassium chromate or dichromate, used successfully in the past, form oxide films on the zinc and protect via passivation. Surface-active organic compounds, which coat the zinc, usually from solution, improve the wetting characteristic of the surface unifying the potential. Inhibitors are usually introduced into the cell via the electrolyte or as part of the coating on the paper separator. Zinc cans could be pretreated; however, this is ordinarily not practical.

Environmental concerns have generally eliminated the use of mercury and cadmium in these batteries. These restrictions are posing problems for battery manufacturers in the areas of sealing, shelf storage reliability, and leakage. This is critical for zinc-chloride cells in that the lower pH electrolyte results in the formation of excessive hydrogen gas due to zinc dissolution. Certain classes of materials considered for use to supplant mercury include gallium, indium, lead, tin and bismuth either alloyed into the zinc or added to the electrolyte from their soluble salts. Other organic materials, like glycols and silicates, offer protection alternatives. Additional restrictions on lead use, which are already stringent, may also be imposed in the future.

8.5.7 Carbon Rod

The carbon rod used in round cells is inserted into the bobbin and performs the functions of current collector. It also performs as a seal vent in systems without a positive venting seal. It is typically made of compressed carbon, graphite and binder, formed by extrusion, and cured by baking. It has, by design, a very low electrical resistance. In Leclanché and zinc-chloride cells with asphalt seals, it provides a vent path for hydrogen and carbon dioxide gases, which might build up in and above the cathode during heavy discharge or elevated temperature storage. Raw carbon rods are initially porous, but are treated with enough oils or waxes to prevent water loss (very harmful to cell shelf-life) and electrolyte leakage. A specific level of porosity is maintained to allow passage of the evolved gases. Ideally, the treated carbon should pass internally evolved gases, but not pass oxygen into the cell, which could add to zinc corrosion during storage. Typically this method of venting gases is variable and less reliable then the use of venting seals.[4,6]

Zinc-chloride cells using plastic, resealable, venting seals utilize plugged, non-porous electrodes. Their use restricts the venting of internal gas to only the designed seal path. This prevents the cell from drying out and limits oxygen ingress into the cell during shelf-storage. Hydrogen gas evolved from wasteful corrosion of the zinc is safely vented as well.

8.5.8 Separator

The separator physically separates and electrically insulates the zinc (negative) from the bobbin (positive), but permits electrolytic or ionic conduction to occur via the electrolyte. The two major separator types in use are either the gelled paste or paper coated with cereal or other gelling agents such as methycellulose.

In the paste type, the paste is dispensed into the zinc can. The preformed bobbin (with the carbon rod) is inserted, pushing the paste up the can walls between the zinc and the bobbin by displacement. After a short time, the paste sets or gels. Some paste formulations need to be stored at low temperatures in two parts. The parts are then mixed; they must be used immediately as they can gel at room temperature. Other paste formulations need elevated temperatures (60°C to 96°C) to gel. The gelatinization time and temperature depend upon the concentration of the electrolyte constituents. A typical paste electrolyte uses zinc chloride, ammonium chloride, water, and starch and/or flour as the gelling agent.

The coated-paper type uses a special paper coated with cereal or other gelling agent on one or both sides. The paper, cut to the proper length and width, is shaped into a cylinder and, with the addition of a bottom paper, is inserted into the cell against the zinc wall. The cathode mix is then metered into the can forming the bobbin, or, if the bobbin is preformed in a die, it is pushed into the can. At this time, the carbon rod is inserted into the center of the bobbin and the bobbin is tamped or compressed, pushing against the paper liner and carbon rod. The compression releases some electrolyte from the cathode mix, wetting the paper liner to complete the operation.

By virtue of the fact that a paste separator is relatively thick compared with the paper liner, about 10% or more manganese dioxide can be accommodated in a paper-lined cell, resulting in a proportional increase in capacity.[4,6,10]

8.5.9 Seal

The seal used to enclose the active ingredients can be asphalt pitch, wax and resin, or plastic (polyethylene or polypropylene). The seal is important to prevent the evaporation of moisture and the phenomenon of "air line" corrosion from oxygen ingress.[4,5]

Leclanché cells typically utilize thermoplastic materials for sealing. These methods are inexpensive and easily implemented. A washer is usually inserted into the zinc can and placed above the cathode bobbin. This provides an air space between the seal and the top of the bobbin to allow for expansion. Melted asphalt pitch is then dispensed onto the washer and is heated until it flows and bonds to the zinc can. One drawback to this method of sealing is that it occupies space that could be used for active materials. A second fault is that this type of seal is easily ruptured by excessive generation of evolved gases and is not suitable for elevated temperature applications.

Premium Leclanché and almost all zinc-chloride cells use injection molded plastic seals. This type of seal lends itself to the design of a positive venting seal and is more reliable. Molded seals are mechanically placed onto a swaged zinc can. Many manufacturers have designed locking mechanisms into the seal, void spaces for various sealants and resealable vents. Several wrap the seal and can in shrink wrap or tape to prevent leakage through zinc can perforations. It is very important to prevent moisture loss in the zinc-chloride system, and to vent the evolved gases generated during discharge and storage. The formation of these gases disrupts the separator surface layer significantly and affects cell performance after storage. Use of molded seals in the zinc-chloride cell construction has resulted in the good shelf storage characteristics evidenced by this design.

8.5.10 Jacket

The battery jacket can be made of various components: metal, paper, plastic, polymer films, plain or asphalt-lined cardboard, or foil in combination or alone. The jacket provides strength, protection, leakage prevention, electrical isolation, decoration, and a site for the manufacturer's label. In many manufacturers' designs, the jacket is an integral part of the sealing system. It locks some seals in place, provides a vent path for the escape of gases, or acts as a supporting member to allow seals to flex under internal gas pressures. In the inside-out construction, the jacket was the container in which a carbon-wax collector was impact-molded (Fig. 8.4).

8.5.11 Electrical Contacts

The top and bottom of most batteries are capped with shiny, tin-plated steel (or brass) terminals to aid conductivity, prevent exposure of any zinc and in many designs enhance the appearance of the cell. Some of the bottom covers are swaged onto the zinc can, others are locked into paper jackets or captured under the jacket crimp. Top covers are almost always fitted onto the carbon electrode with interference. All of the designs try to minimize the electrical contact resistance.

8.6 PERFORMANCE CHARACTERISTICS

8.6.1 Voltage

Open-Circuit Voltage.   The open-circuit voltage (OCV) of the zinc-carbon battery is derived from the potentials of the active anode and cathode materials, zinc and manganese dioxide, respectively. As most zinc-carbon batteries use similar anode alloys, the open circuit voltage usually depends upon the type or mixture of manganese dioxide used in the cathode and the composition and pH of the electrolyte system. Manganese dioxides, like EMDs, are of greater purity than the NMDs, which contain a significant quantity of manganite (MnOOH), and thus have lower voltage. Figure 8.7 shows the open circuit voltage for fresh Leclanché and zinc chloride batteries containing various mixtures of natural and electrolytic manganese dioxide.

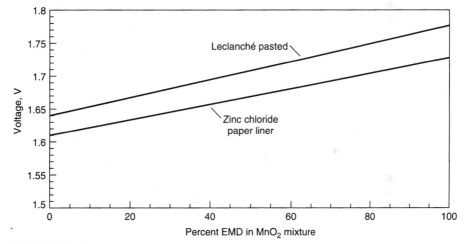

FIGURE 8.7   Comparison of open circuit voltage for batteries using mixtures of natural and electrolytic manganese dioxide.

Closed-Circuit Voltage.   The closed-circuit voltage (CCV), or working voltage, of the zinc-carbon battery is a function of the load or current drain the cell is required to deliver. The heavier the load or the smaller the circuit resistance, the lower the closed-circuit voltage. Table 8.4 illustrates the effect of load resistance on the closed-circuit voltage for D-size batteries in both the Leclanché and zinc-chloride systems.

TABLE 8.4 Initial Closed-Circuit Voltage of a Typical D-size Zinc-Carbon Battery as a Function of Load Resistance at 20°C

Voltage		Load resistance	Initial current, mA	
ZC	LC*	Ω	ZC	LC
1.61	1.56	∞	0	0
1.59	1.52	100	16	15
1.57	1.51	50	31	30
1.54	1.49	25	62	60
1.48	1.47	10	148	147
1.45	1.37	4	362	343
1.43	1.27	2	715	635

*ZC-zinc-chloride battery; LC-Leclanché battery.

The exact value of the CCV is determined mainly by the internal resistance of the battery as compared with the circuit or load resistance. It is, in fact, proportional to $R_1/(R_1 + R_{in})$ where R_1 is the load resistance and R_{in} is the battery's internal resistance. Another factor, important to the battery's ability to sustain the CCV, is the transport characteristic of the cell components, that is, the ability to transport ionic and solid reaction products, and water, to and from the reaction sites. The physical geometry of the cell, its solution volume, electrode porosity, and solute materials are critical characteristics that affect the diffusion coefficient. Transport is enhanced by use of highly mobile ions, high solution volumes, high electrode porosity and high surface area. Transport characteristics are diminished by slow ionic transport, low solution volumes, and barriers of precipitated reaction product which block diffusion paths. (This topic is discussed in greater detail in Chap. 2.) Temperature, age, and depth of discharge greatly affect the internal resistance and transport factors as well.

As zinc-carbon batteries are discharged, the CCV, and to a lesser extent the OCV, drop in magnitude. The drop in OCV is attributable to the decrease in the active material manganese dioxide and the increase in the product of the reaction, manganite. Reduction of the CCV is the result of increased electrical resistance and a decrease in transport characteristic. The discharge curve is a graphic representation of the closed-circuit voltage as a function of time and is neither flat nor linearly decreasing but, as seen in Fig. 8.8, has the character of a single- or double-S curve depending upon the depth of discharge. Figure 8.9 illustrates the shape of typical discharge curves for D-size, general purpose, Leclanché and zinc-chloride batteries.

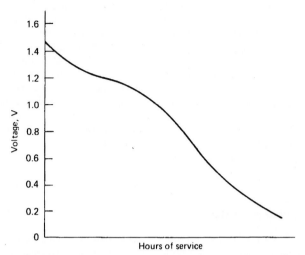

FIGURE 8.8   Typical discharge curve of a Leclanché zinc-carbon battery.

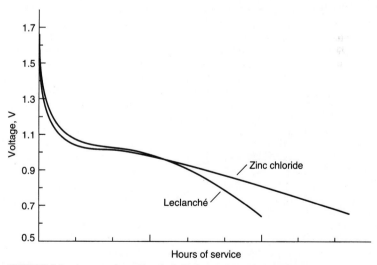

FIGURE 8.9   A comparison of typical discharge curves between Leclanché and zinc-chloride batteries.

End Voltage. The end voltage, or cutoff voltage (COV), is defined as a point along the discharge curve below which no usable energy can be drawn for the specified application.

Typically 0.9 V has been found to be the COV for a 1.5-V cell when used in a flashlight. Some radio applications can utilize the cell down to 0.75 V or lower, while other electronic devices may tolerate a drop to only 1.2 V. Obviously, the lower the end voltage, the greater the amount of energy that can be delivered by the battery. The lower voltage will impact certain applications, like flashlights, resulting in a dimmer light and lower volume and/or range for radios. Devices that can operate only within a narrow voltage range would do better with a battery system noted for a flat discharge curve.

Although a closed-circuit voltage that steadily decreases may present a disadvantage in some applications, it is advantageous where sufficient warning of the end of battery life is required, as in a flashlight.

8.6.2 Discharge Characteristics

Both the Leclanché and zinc-chloride batteries have performance characteristics that show advantages in specific applications, but poor performance in others. A variety of factors influence battery performance (see Chap. 3). It is necessary to evaluate specifics about the application (discharge conditions, cost, weight, etc.) in order to make a proper selection of a battery. Many manufacturers provide data for this purpose.

Typical discharge curves for general-purpose D-size Leclanché and zinc-chloride batteries, of equivalent capacity, discharged 2 h per day at 20°C, are shown in Fig. 8.10. These curves are characterized by a sloping discharge and a substantial reduction in voltage with increasing current. The zinc-chloride construction shows a higher voltage characteristic and more service at the higher current levels. On the 50-mA drain, both constructions provide nearly equivalent performance. This is the result of the depletion of manganese dioxide at the low discharge rates, as most zinc-carbon batteries are cathode limited.

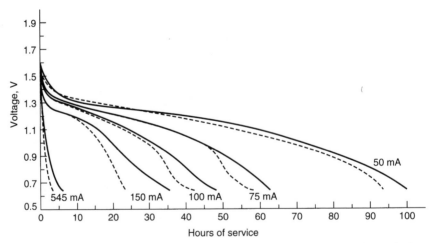

FIGURE 8.10 Typical discharge curves for general-purpose Leclanché zinc-chloride D-size batteries, discharged 2 h/day at 20°C. Solid line—zinc chloride; broken line—Leclanché.

8.6.3 Effect Of Intermittent Discharge

Performance of zinc-carbon batteries varies depending upon the type of discharge. The performance of Leclanché batteries is significantly better when used under intermittent compared to continuous discharge conditions, because (1) a chemical recuperation reaction replaces a small portion of active ingredients during the rest periods; and (2) transport phenomena redistribute reaction products.[5]

Zinc-chloride batteries can support heavier drains and respond to intermittent discharges with longer discharge cycles. This system relies upon its improved transport mechanism to support heavier drains and to redistribute reaction product.

The three-dimensional graph shown in Fig. 8.11 illustrates the general effects of intermittency and discharge rate on the capacity of a general-purpose D-size battery.

On extremely low-current discharges, the benefit of intermittent rest and discharge is minimal for both systems. It is likely that the reaction rate proceeds more slowly than the diffusion rate and results in a balanced condition even during discharge. Under conditions of extremely low rate of discharge, factors such as age will reduce the total delivered capacity. Most applications fall in the moderate- (radio) to high-rate (flashlight) categories, and for these the energy delivered can more than triple when the cell is used intermittently as compared with continuous usage.

The standard flashlight current drains are 300 mA (3.9 Ω per cell) and 500 mA (2.2 Ω per cell), which correspond to two-cell flashlights using PR2 and PR6 lamps, respectively, or three-cell flashlights using PR3 and PR7 lamps, respectively. The beneficial effects of intermittent discharge are clearly shown in Figs. 8.12 and 8.13, which compare Leclanché general-purpose D-size batteries on four different discharge regimens; continuous, light intermittent flashlight, heavy intermittent flashlight, and a 1 h/day cassette simulation test. Table 8.5 lists the ANSI application tests currently being used to evaluate both cell systems.

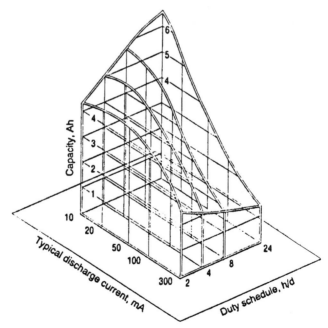

FIGURE 8.11 Battery performance (capacity to 0.9 V) as a function of discharge, load and duty schedule for a general purpose D-size zinc-carbon battery at 20°C. (*From Eveready Battery Energy Data.*)

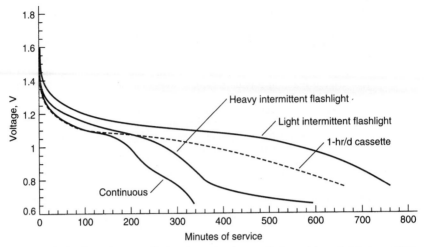

FIGURE 8.12 General purpose D-size zinc-carbon battery discharged through 3.9 ohm at 20°C, under various discharge conditions.

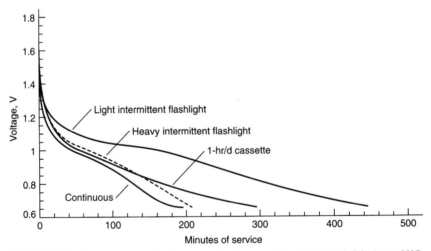

FIGURE 8.13 General purpose D-size zinc-carbon battery discharged through 2.2 ohm at 20°C, under various discharge conditions.

TABLE 8.5 Standard Application Tests Specified in ANSI Battery Specifications

Typical use or test	Discharge schedule
Pulse test (PHOTO)	15 sec ON/min × 24 hrs/day
Portable lighting (GPI)	5 min ON/day
Portable lighting (LIP)	4 min ON/hr × 8 hrs/day
Portable lighting (LANTERN)	0.5 hrs ON/hr × 8 hrs/day
Transistor radios	4 hrs ON/day
Transistor radio (small-9 volt)	2 hrs ON/day
Personal tape recorder, cassette	1 hr ON/day
Toys and motors	1 hr ON/day
Pocket calculator	0.5 hrs ON/day
Hearing aid	12 hrs ON/day
Electronic	24 hrs ON/day

Source: Based on ANSI C18.1M-1999.[11]

8.6.4 Comparative Discharge Curves—Size Effect Upon Heavy Duty Zinc-chloride Batteries

The performance of batteries of different sizes, AAA, AA, C, and D, (see Table 8.9 for a list of cell sizes) are given in Figs. 8.14 to 8.16. Note that the AAA through D-size batteries contain increasing amounts of active materials (zinc and manganese dioxide) with the increase in size. Increasing the size also results in proportionally larger electrode surface areas in the cell and therefore the voltage is maintained at higher levels at the same discharge loads.

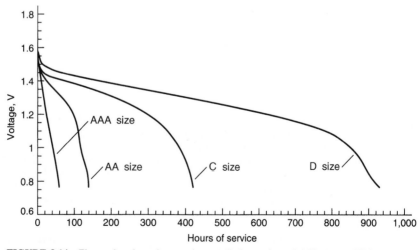

FIGURE 8.14 Zinc-carbon batteries, continuous discharge through 150 ohm at 20°C.

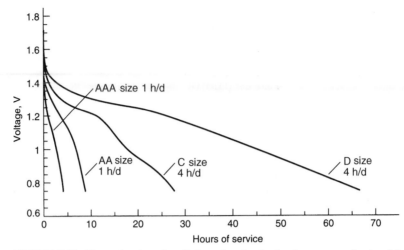

FIGURE 8.15   Zinc-carbon batteries, discharged under simulated cassette application (10 ohm intermittent load) at 20°C.

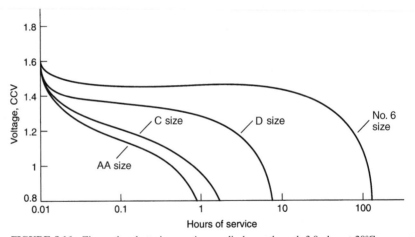

FIGURE 8.16   Zinc-carbon batteries, continuous discharge through 3.9 ohm at 20°C.

Figure 8.14 illustrates a discharge through a relatively high resistance of 150 Ω (about 10 mA) continuously at 20°C. The performance of the D- and C-size batteries on an intermittent discharge would not be too different from that of the continuous discharge, because, for batteries of these sizes, the current drain is low. For the smaller AA and AAA sizes, however, discharge through a 150-Ω load is a heavier load. Under these conditions some benefit could be gained, for Leclanché batteries, using an intermittent discharge since it assists in dissipating reaction product barriers and increases service life. Zinc-chloride batteries would show less evidence of an effect because of their improved transport characteristic.

Figure 8.15 shows the same four battery sizes discharged through a relatively low resistance of 10 Ω (about 150 mA) on a simulated intermittent cassette application. AAA- and AA-size batteries deliver about 30% less service when continuously discharged at this load.

The relative performance of both Leclanché and zinc-chloride AAA, AA, C, and D-size batteries roughly follow a 1:2:8:16 proportion to the 0.9-V cutoff for the low rate and a 1:2:12:24 proportion for the high-rate drain, illustrating the advantage of the lower current density discharge for the larger batteries. The high discharge rate for the general-purpose C- and D- size batteries at 300 mA (3.9 Ω) is shown in Fig. 8.16 compared with the performance of the general purpose larger No. 6 battery for which this discharge rate is low.

8.6.5   Comparative Discharge Curves—Different Battery Grades

Figure 8.17 compares both Leclanché and zinc-chloride general purpose (GP), heavy duty (HD), and the extra/super heavy duty (EHD) D-size batteries (as defined in Sec. 8.3) discharged continuously through a 2.2-Ω load at 20°C. A performance ratio, to the 0.9 V cutoff, of 1.0:1.3 between the Leclanché and zinc-chloride GP batteries was observed. The same ratio for the HD batteries was 1.0:1.5 respectively. Comparison between the Leclanché (LC) and zinc-chloride (ZC) GP, HD, and EHD batteries showed ratios of 1:1.3:2.2:3.4:4.4 for LC,GP:ZC,GP:LC,HD:ZC,HD:ZC,EHD respectively.

Figure 8.18 shows a comparison of the same battery grades discharged intermittently through a 2.2-Ω load on the American National Standards Institute (ANSI) light intermittent flashlight (LIF) test. On this regimen, the performance ratio to the 0.9 V cutoff is 1:1 for the GP batteries, 1:1.3 for the HD, and for the LC,GP:ZC,GP:LC,HD:ZC,HD:ZC,EHD, 1:1:1.7:2.1:2.9, respectively. Testing on the intermittent discharge, which allows for a rest period for recovery, results in increased performance for all batteries and evidences a decreased difference in performance between the grades.

Figure 8.19 illustrates the same battery grades discharged continuously through the lighter 3.9-Ω load. The following ratios were obtained; 1:1.3 GP, 1:1.4 HD, and 1:1.3:2.0:2.8:3.5 between all grades to the 0.9 V cutoff. Less of a difference was observed than that obtained at the heavier 2.2 Ω discharge rate. The slower reaction rate at the lighter drain is evident because of the higher battery voltages. A comparison of an intermittent discharge at 3.9 Ω

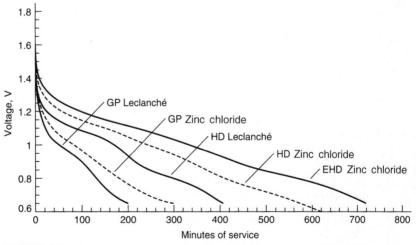

FIGURE 8.17 Comparison of Leclanché and zinc-chloride D-size batteries of various grades, continuously discharged through 2.2 ohm at 20°C. GP-general purpose; HD-heavy duty; EHD-extra heavy duty.

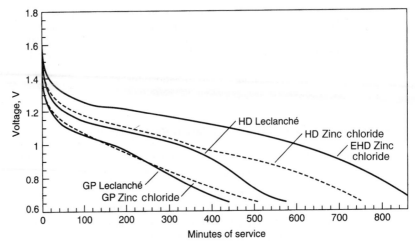

FIGURE 8.18 Comparison of Leclanché and zinc-chloride D-size cells of various grades discharged on the ANSI LIF test (4 min/h, 8 h/day) through 2.2 ohm at 20°C. GP-general purpose; HD-heavy duty; EHD-extra heavy duty.

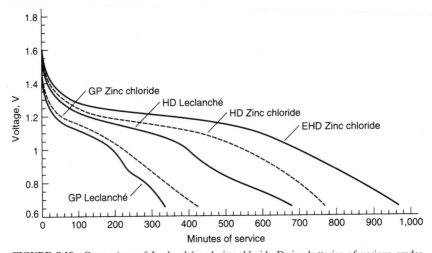

FIGURE 8.19 Comparison of Leclanché and zinc-chloride D-size batteries of various grades continuously discharged through 3.9 ohm at 20°C. GP-general purpose; HD-heavy duty; EHD-extra heavy duty.

ohm 1 hr/day on a simulated cassette test is shown in Fig. 8.20. On this regimen, the performance ratio for the same grouping of battery grades drops to 1:1.1:1.5:2.4:2.7. This reflects the increase in service and tighter grouping for all grades.

These battery grades are compared once again in Fig. 8.21 on a moderate discharge through a 24-Ω resistor for four hours continuously with 20 hours of rest on the ANSI transistor radio and electronic equipment battery tests. At this more moderate discharge load, the performance ratio is even closer, 1:1.4:1.6:1.9:2.0 to 0.9 V cutoff.

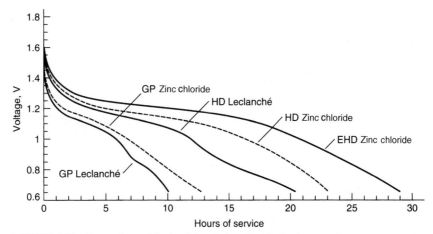

FIGURE 8.20 Comparison of Leclanché and zinc-chloride D-size batteries of various grades discharged on a simulated cassette application, 3.9 ohms, 1hr/day at 20°C. GP-general purpose; HD-heavy duty; EHD-extra heavy duty.

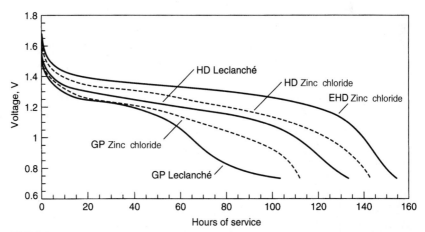

FIGURE 8.21 Comparison of Leclanché and zinc-chloride D-size batteries of various grades discharged through 24 ohm for 4h/day at 20°C. GP-general purpose; HD-heavy duty; EHD-extra heavy duty.

Continuous discharge tends to increase the difference in performance between the different grades of batteries of the same size. The differences between the Leclanché and zinc-chloride systems are evident when tested continuously. Intermittent discharges tend to reduce the differences between systems and grades. Similarly, higher discharge currents tend to increase the performance difference.

Figure 8.22a summarizes the performance of the Leclanché general purpose D-size battery grade discharged continuously to different end voltages. The performance of the zinc-chloride general purpose battery, for the same testing, is shown in Fig. 8.22b. Figures 8.23a and 8.23b present the same performance relationships except for both the Leclanché and zinc-chloride batteries in the heavy duty D-size battery and Fig. 8.24 for the zinc-chloride extra/super heavy duty battery.

Performance differences between the batteries of the same grade but from different manufacturers are shown in Fig. 8.25. There is a difference of about 25% between the best and poorest battery to the 0.9 V cutoff.

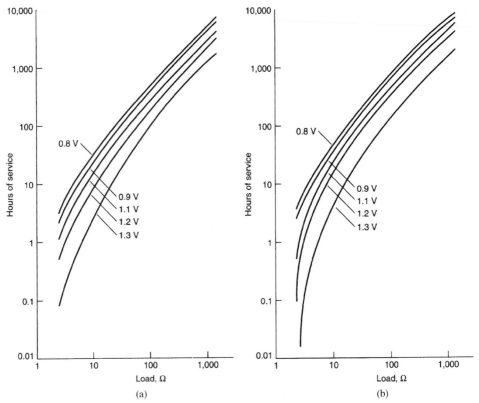

FIGURE 8.22 Discharge load vs. hours of service to different cutoff (end) voltages at 20°C, continuous discharge. (a) General-purpose D-size Leclanché battery construction. (b) General-purpose D-size zinc-chloride battery.

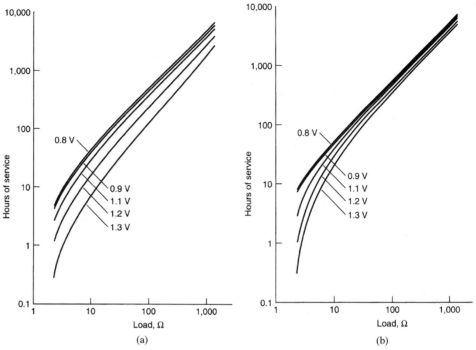

FIGURE 8.23 Discharge load vs. hours of service to different cutoff (end) voltages at 20°C, continuous discharge. (a) Heavy duty D-size Leclanché battery. (b) Heavy duty D-size zinc-chloride battery.

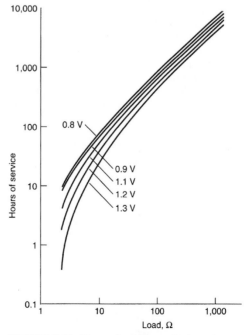

FIGURE 8.24 Hours of service vs. discharge load to different cutoff (end) voltages at 20°C, continuous discharge. Extra heavy duty D-size zinc-chloride battery.

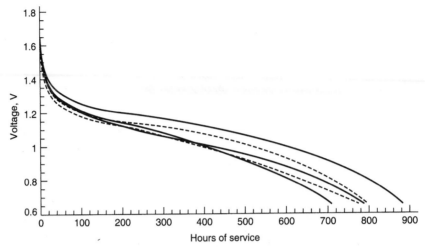

FIGURE 8.25   Comparison of general-purpose D-size zinc-carbon batteries of five different manufacturers, discharged on the ANSI LIF test through 2.2 ohm load at 20°C.

8.6.6   Internal Resistance

Internal resistance R_{in} is defined as the opposition or resistance to the flow of an electric current within a cell or battery, i.e. the sum of the ionic and electronic resistances of the cell components. Electronic resistance includes the resistance of the materials of construction: metal covers, carbon rods, conductive cathode components, and so on. Ionic resistance encompasses factors resulting from the movement of ions within the cell. These include electrolyte conductivity, ionic mobility, electrode porosity, electrode surface area, secondary reactions, etc. These fall into the category of factors that affect the ionic resistance. These factors are encompassed by the term polarization. Other considerations include battery size and construction as well as temperature, age and depth of discharge.

Electronic Resistance.   An approximation of the internal electronic resistance of a battery can be made by determining the OCV and the peak flash current (I) using very low resistance meters. The ammeter resistance should be low enough that the total circuit resistance does not exceed 0.01 ohm and is no more than 10% of the cell's internal resistance. The internal electronic resistance is expressed as:

$$R_{in} = OCV/I$$

where: R_{in} = internal resistance expressed in Ω
OCV = open circuit voltage
I = peak flash current A

A more accurate method of calculation is made using the voltage-drop method. In this method, a small initial load is applied on the battery to stabilize the voltage. A load approximating the application load is then applied. The internal resistance is calculated by:

$$R_{in} = (V_1 - V_2)R_L/V_2$$

where: R_{in} = internal resistance, Ω
V_1 = initial stabilized closed-circuit voltage, V
V_2 = closed circuit voltage reading at the application load, VO
R_L = application load, Ω

The application load time should be kept to a pulse of 5 to 50 ms to minimize effects due to polarization. These methods measure the voltage loss due to the electrical resistance component but do not take into account voltage losses due to polarization.

Ionic Resistance.   The polarization effect is best illustrated by a trace of the Pulse/Time profile as shown in Fig. 8.26. The total resistance (R_T) is expressed using Ohm's Law by

$$R_T = dR = dV/dI$$

which also equals:

$$(V_1 - V_2)/(I_1 - I_2)$$

where: V_1 and I_1 = the voltage and current just prior to pulsing
V_2 and I_2 = the voltage and current just prior to the pulse load removal
dV_3 = total voltage drop shown

The internal resistance of the battery component is expressed as dV_1 and the polarization effect component is the voltage drop dV_2. Since some energy was removed by the pulse, a more correct expression for the battery resistance is the voltage drop expressed by dV_4.

Measurement of the battery voltage drop (dV_4) is very difficult to capture, therefore the pulse duration (dt) is minimized to reduce the polarization effect voltage drop (dV_2). The pulse duration is generally kept in the range of 5 to 50 milliseconds. For accurate and repetitive results, it is recommended that duration times be kept constant by "read and hold" voltage measurements.

Since dV_2 is slightly greater than dV_1, one can see that the resistance due to polarization (R_p) is greater than the internal resistance of the battery (R_{ir}) by the formula

$$R_T = R_{ir} + R_p$$

Partial, light discharge or a light background load prior to the pulse and internal resistance measurements provide equilibration for consistent measurements.

Table 8.6 shows the general relationship of flash current and internal resistance of the more popular cell sizes.

Zinc-carbon batteries perform better on intermittent drains than continuous drains, largely because of their ability to dissipate the effects of polarization. Factors that affect polarization are identified earlier in this section. Resting between discharges allows the zinc surface to "depolarize." One such effect is the dissipation of concentration polarization at the anode surface. This effect is more pronounced as heavier drains and longer duty schedules are

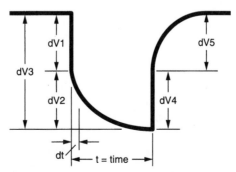

FIGURE 8.26   Voltage pulse/time profile illustration of curve shape and voltage components to calculate internal resistance.

TABLE 8.6   Flash Current and Internal Resistance for Various Battery Sizes

Size	Typical maximum flash current, A		Approximate internal resistance, Ω	
	LC*	ZC*	LC	ZC
N	2.5	...	0.6	...
AAA	3	4	0.4	0.35
AA	4	5	0.30	0.28
C	5	7	0.39	0.23
D	6	9	0.27	0.18
F	9	11	0.17	0.13
9V (battery)	0.6	0.8	5	4.5

*LC-Leclanché, ZC-zinc chloride.
Source: Eveready Battery Engineering Data.[13]

applied. The internal resistance of the zinc-chloride batteries is slightly lower than that of the Leclanché batteries. This results in a smaller voltage drop for a given battery size.

The internal resistance of zinc-carbon batteries increases with the depth of discharge. Some applications use this feature to establish low battery alarms to predict near end of battery life situations (such as in the smoke detector). Fig. 8.27 shows the relative battery internal resistance versus depth of discharge of a 9-V Leclanché battery.

One of the reasons for this increase in internal resistance is the cathode discharge reaction. The porous cathode becomes progressively blocked with reaction products. In the case of the Leclanché system, it is in the form of diammine-zinc chloride crystals; in the case of the zinc-chloride system, it is in the form of zinc oxychloride crystals.

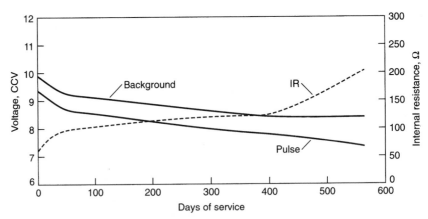

FIGURE 8.27   Comparison of voltage and internal resistance during discharge of a 9-V battery on smoke detector test. Background load = 620,000 ohms continuous; pulse load = 1500 ohms × 10 ms every 40 s.

8.6.7 Effect of Temperature

Zinc-carbon batteries operate best in a temperature range of 20°C to 30°C. The energy output of the battery increases with higher operating temperatures, but prolonged exposure to high temperatures (50°C and higher) will cause rapid deterioration. The capacity of the Leclanché battery falls off rapidly with decreasing temperatures, yielding no more than about 65% capacity at 0°C, and is essentially inoperative below −20°C. Zinc-chloride batteries provide an additional 15% capacity at 0°C or 80% of room temperature capacity. The effects are more pronounced at heavier current drains; a low current drain would tend to result in a higher capacity at lower temperatures than a higher current drain (except for a beneficial heating effect that may occur at the higher current drains).

The effect of temperature on the available capacity of zinc-carbon (Leclanché and zinc-chloride systems) batteries is shown graphically in Fig. 8.28 for both general-purpose (ammonium chloride electrolyte) and heavy-duty (zinc-chloride electrolyte) batteries. At −20°C typical zinc-chloride electrolytes (25% to 30% zinc-chloride by weight) turn to slush. Below −25°C ice formation is likely. Under these conditions, it is not surprising that performance is dramatically reduced. These data represent performance at flashlight-type current drains (300 mA for a D-size cell). A lower current drain would result in a higher capacity than shown. Additional characteristics of this D-size battery at various temperatures are listed in Table 8.7.

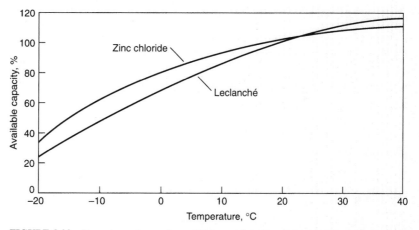

FIGURE 8.28 Percentage of capacity available as a function of temperature, moderate-drain radio-type discharge.

Special low temperature batteries were developed using low freezing-point electrolytes and a design that minimizes internal cell resistance, but they did not achieve popularity due to the superior overall performance of other types of primary batteries. For best operation, at low ambient temperatures, the Leclanché battery should be kept warm by some appropriate means. A vest battery worn under the user's clothing, employing body heat to maintain it at a satisfactory operating temperature was once used by the military to achieve reliable operation at low temperatures.

TABLE 8.7 Temperature Effect on Internal Resistance

Battery size	System*	Resistance, Ω			
		−20°C	0°C	20°C	45°C
Single cell batteries					
AAA	ZC	10	0.7	0.6	0.5
AA	LC	5	0.8	0.5	0.4
AA	ZC	5	0.8	0.5	0.4
C	LC	2	0.8	0.5	0.4
C	ZC	3	0.5	0.4	0.3
D	LC	2	0.6	0.5	0.4
D	ZC	2	0.4	0.3	0.2
Flat cell batteries					
9 V	LC	100	45.0	35.0	30.0
9 V	ZC	100	45.0	35.0	30.0
Lantern batteries					
6 V	LC	10	1.0	0.9	0.7
6 V	ZC	10	1.0	0.8	0.7

*LC = Leclanché, ZC = Zinc chloride.
Source: Eveready Battery Engineering Data.[13]

8.6.8 Service Life

The service life of the Leclanché battery is summarized in Figs. 8.29 and 8.30, which plot the service life at various loads and temperatures normalized for unit weight (amperes per kilogram) and unit volume (amperes per liter). These curves are based on the performance of a general-purpose battery at the average discharge current under several discharge modes. These data can be used to approximate the service life of a given battery under particular discharge conditions or to estimate the size and weight of a battery required to meet a specific service requirement.

Manufacturers' catalogs should be consulted for specific performance data in view of the many cell formulations and discharge conditions. Table 8.8 presents typical data from a manufacturer of two formulations of the AA-size battery.

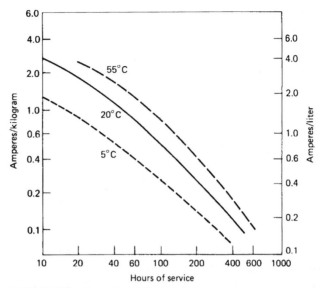

FIGURE 8.29   Service hours for general-purpose zinc-carbon battery, discharged 2 h/day to 0.9 V at 20°C.

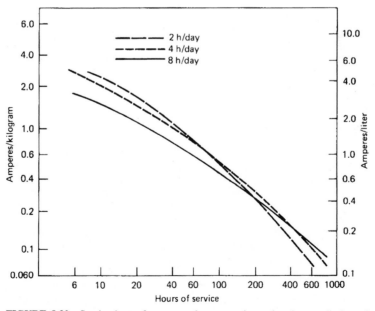

FIGURE 8.30   Service hours for a general-purpose zinc-carbon battery discharged intermittently to 0.9 V at 20°C.

TABLE 8.8 Manufacturer's Data for AA-Size Zinc-Carbon Batteries

Schedule	Drain @1.2 V, mA	Load Ω	Cutoff voltage V					
			1.3	1.2	1.1	1.0	0.9	0.8
Typical service of Eveready no. 1015 general-purpose battery								
			Hours					
4 hr/day	28 mA	43	2	5	12	20	24	27
1 hr/day	120 mA	10	0.1	0.4	1.2	2.6	3.9	4.5
1 hr/day	308 mA	3.9	0.09	0.2	0.4	0.7	0.9	1.0
			Pulses					
15 sec/min/ 24 hr/day (pulse)	667 mA	1.8	6	14	30	51	68	73
Typical service of Eveready no. 1215 superheavy-duty battery								
			Hours					
4 hr/day	28 mA	43	4	10	21	31	36	39
1 hr/day	120 mA	10	0.2	0.4	2.5	5.2	6.4	7.0
1 hr/day	308 mA	3.9	0.1	0.3	0.5	1.2	1.7	1.9
			Pulses					
15 sec/min/ 24 hr/day (pulse)	667 mA	1.8	7	14	30	89	139	160

Source: Eveready Battery Engineering Data.[12]

8.6.9 Shelf-Life

Zinc-carbon batteries gradually lose capacity while idle. This deterioration is greater for partially discharged batteries than for unused batteries and results from parasitic reactions such as wasteful zinc corrosion, chemical side reactions, and moisture loss. The shelf-life or rate of capacity loss is affected by the storage temperature. High temperatures accelerate the loss; low temperatures retard the loss. Refrigerated storage will increase the shelf-life. Figure 8.31 shows the retention of capacity of a zinc-carbon battery after storage at 40, 20, and 0°C. The capacity retention of a zinc-chloride battery is higher than that of the Leclanché type because of the improved separators (coated paper separator-types), sealing systems and other materials used in that design.

Leclanché-type batteries, using the asphalt or pitch-type seals in conjunction with paste-type separators, have the poorest capacity retention. Zinc-chloride batteries, using highly crosslinked starch coated paper separators in conjunction with molded polypropylene or polyethylene seals, provide the best retention.

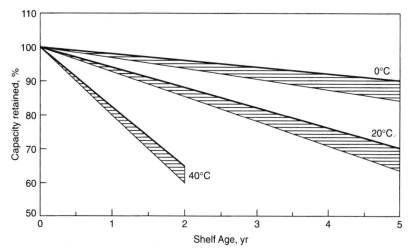

FIGURE 8.31 Capacity retention after storage at 40°C, 20°C and 0°C for paper-lined plastic seal zinc-chloride batteries.

Batteries stored at −20°C are expected to retain approximately 80% to 90% of their initial capacity after 10 years. Since low temperatures retard deterioration, storage at low temperatures is an advantageous method for preserving battery capacity. A storage temperature of 0°C is very effective.

Freezing usually may not be harmful as long as there is no repeated cycling from high to low temperatures. Use of case materials or seals with widely different coefficients of expansion may lead to cracking. When batteries are removed from cold storage, they should be allowed to reach room temperature in order to provide satisfactory performance. Moisture condensation during warm-up should be prevented as this may cause electrical leakage or short-circuiting.

8.7 SPECIAL DESIGNS

The zinc-carbon system is used in special designs to enhance particular performance characteristics or for new or unique applications.

8.7.1 Flat-Pack Zinc/Manganese Dioxide P-80 Battery

In the early 1970s, Polaroid introduced a new instant camera-film system, the SX-70. A major innovation in that system was the inclusion of a battery in the film pack rather than in the camera. The film pack contained a battery designed to provide enough energy for the pictures in the pack. The concept was that the photographer would not have to be concerned about the freshness of the battery as it was changed with each change of film.

Battery Construction. The P-80 battery uses chemistry quite similar to Leclanché round cells, although the shape is unique. Figure 8.32 details one cell (1.5 V). The electrode area is approximately 5.1 cm × 5.1 cm. The zinc anode is coated on a conductive vinyl web.

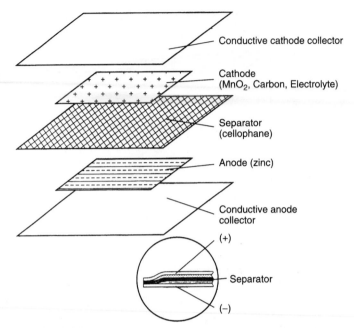

FIGURE 8.32 Exploded view of a single cell from Polaroid P-80 battery pack.

The manganese dioxide is mixed in a slurry which contains the electrolyte salts. The electrolyte is mainly zinc chloride with some ammonium chloride. The anode and the cathode are separated by a thin film of cellophane. The complete 6-volt battery has four cells. The four identical cells are connected by vinyl frames to each other and the aluminum collector plates. A special conductive coating allows the aluminum to bond to the plastic materials.

Battery Parameters.   The key battery parameters of the flat battery are similar to those of the cylindrical one. The flat configuration provides low resistance by virtue of the geometry. The thin layers need to stay in intimate contact to maintain the low resistance and gassing effects have to be minimized.

- *Open-Circuit (No-Load) Voltage:* The open-circuit voltage in this battery is dependent on the manganese dioxide activity and the system pH. The cathode slurry is adjusted to a constant pH to minimize battery to battery voltage variation. For example, the P-80 battery is adjusted so the voltage is 6.40 V at 56 days and 6.30 V after 12 months of shelf storage.
- *Closed-Circuit (On-Load) Voltage:* The closed-circuit voltage is used as an indicator of the battery's capability to deliver energy at high currents. In the case of the P-80 battery, a 1.63-A load is used since that is one of the operating requirements for the camera. The closed-circuit voltage is measured at 55 milliseconds to minimize polarization effects. The normal closed-circuit voltage is 5.58 V at 56 days and 5.35 V after 12 months of shelf storage.
- *Internal Resistance and Voltage Drop (ΔV):* The battery's internal resistance is measured by using the voltage drop or ΔV at a given load for a specified pulse period. A major contributor which effects ΔV is the activity at the zinc surface which is dependent on both

zinc particle size and amount of hydrogen present. The polarization effect occurs when the load is held for some time period, as in the case of charging the camera strobe circuit. Total resistance is then a summation of the two resistances, that is, the internal resistance of the battery and the resistance due to polarization effect, the latter being very time sensitive. To minimize polarization effect resistance, the pulse period for ΔV measurements was minimized.

The 56-day point is of interest, as that is the normal age when the battery is released for assembly into film packs. At that time, every battery is measured for electrical characteristics and defective ones are screened.

The total internal resistance is expressed by the following:

$$R_t = R_i + R_p$$

where: R_i = battery internal resistance, Ω
R_p = polarization resistance effect, Ω

For the P-80 battery, R_i was 0.50 Ω and the R_p was 0.12 Ω.

- *Capacity:* The capacity simulator mimics the energy used to charge the camera strobe. The pulse consists of an open circuit voltage at rest, followed by a pulse at a 2-A load to result in a 50 watt-second (50 Ws) pulse. The 50 Ws cycle test is maintained until the final CCV reaches the 3.7 cut-off voltage, where the number of cycles is determined.

During the time while the 50 Ws load is maintained, the polarization drop occurs. The time to produce the 50 Ws increases with each cycle as the battery is "consumed." A 30-s rest between cycles is used. Initially, the voltage drop is fairly constant; however, near the end of the test, the resistance increases. The test is maintained to 3.7 V to indicate the cut-off point of the camera.

Figure 8.33 illustrates the voltage profile at different discharge loads, and Figure 8.34 shows rate sensitivity of the batteries described above versus capacity.

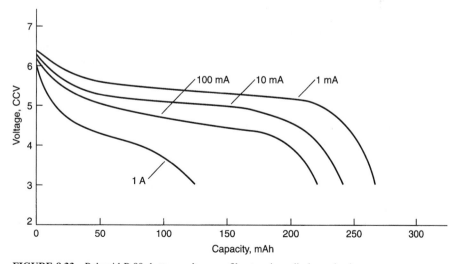

FIGURE 8.33   Polaroid P-80; battery voltage profile at various discharge loads.

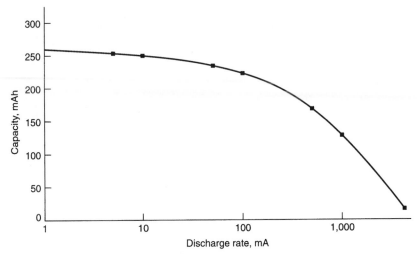

FIGURE 8.34   Polaroid P-80 battery; rate sensitivity vs capacity (to 3.0-V cutoff).

8.8   *TYPES AND SIZES OF AVAILABLE CELLS AND BATTERIES*

Zinc-carbon batteries are made in a number of different sizes with different formulations to meet a variety of applications. The single-cell and multicell batteries are classified by electrochemical system, either Leclanché or zinc chloride, and by grade; general purpose, heavy duty, extra heavy duty, photoflash, and so on. These grades are assigned according to their output performance under specific discharge conditions.

Table 8.9 lists the more popular battery sizes with typical performance at various loads under a two-hour per day intermittent discharge, except for the continuous toy battery test. The performance of these batteries, under several intermittent discharge conditions, is given in Table 8.10.

The AA-size battery is becoming the predominant one and is used in penlights, photoflash and electronic applications. The smaller AAA-size is used in remote control devices and other small electronic applications. The C and D-size batteries are used mainly in flashlight applications and the F-size is usually assembled into multicell batteries for lanterns and other applications requiring these large batteries. Flat cell are used in battery assemblies, in particular, the 9-volt battery used in smoke detectors and electronic applications.

Table 8.11 lists some of the major multicell zinc-carbon batteries that are available commercially. The performance of these batteries can be estimated by using the IEC designation to determine the cell compliment (e.g. NEDA 6, IEC 4R25 battery consists of four F-size cells connected in series). Table 8.12 gives cross-references to the zinc-carbon batteries and manufacturer's designations. The most recent manufacturer's catalogs should be consulted for specific performance data to determine the suitability of their product for a particular application.

TABLE 8.9 Characteristics of Zinc-Carbon Batteries

Size	IEC	ANSI, NEDA	Weight g	Maximum dimensions, mm		Typical service, 2 h/d*			
						Leclanché		Zinc-chloride	
				Diameter	Height	Drain mA	Service h	Drain mA	Service h
N	R1	910	6.2	12	30.2	1	480		
						10	45		
						15	20		
AAA	R03	24	8.5	10.5	44.5	1	—	1	520
						10	—	10	55
						20	—	20	26
AA	R6	15	15	14.5	50.5	1	950	1	1200
						10	80	10	110
						100	4	100	8
						300	0.6	300	1
C	R14	14	41	26.2	50	5	380	5	800
						25	75	20	150
						100	6	100	20
						300	1.7	300	5.5
D	R20	13	90	34.2	61.5	10	400	10	700
						50	70	50	135
						100	25	100	55
						500	3	500	6
F	R25	60	160	34†	92†	25	300	25	400
						100	60	100	85
						500	5.5	500	9
G	R26	—	180	32†	105†	—			
No. 6	R40	905	900	67	170.7	5	8000		
						50	700		
						100	350		
						500	70		

*Typical values of service to 0.9-V cutoff.
†Typical values.

TABLE 8.10 ANSI Standards for Zinc-Carbon and Alkaline-Manganese Dioxide Batteries

Size	Use	Ohms	Schedule	Cutoff voltage	Specification requirements	
					Zinc-carbon batteries	Alkaline-manganese dioxide batteries
					Initial*	Initial*
N					910D	910A
	Portable lighting	5.1	5 min/d	0.9	NA	100 min
	Pager	(10.0	5 sec/hr,	0.9	NA	888 hr
	then	3000.0	3595 sec/h)			
AAA					24D	24A
	Pulse test	3.6	15 sec/min 24 hr/d	0.9	150 pulses	450 pulses
	Portable lighting	5.1	4 min/hr 8 hr/d	0.9	48.0 min	130.0 min
	Recorder	10.0	1 hr/d	0.9	1.5 hr	5.5 hr
	Radio	75.0	4 hr/d	0.9	24.0 hr	48.0 hr
AA					15D	15A
	Pulse test	1.8	15 sec/min 24 hr/d	0.9	100 pulses	450 pulses
	Motor/toy	3.9	1 hr/d	0.8	1.2 hr	5 hr
	Recorder	10.0	1 hr/d	0.9	5.0 hr	13.5 hr
	Radio	43.0	4 hr/d	0.9	27.0 hr	60 hr
C					14D	14A
	Portable lighting	3.9	4 min/hr 8 hr/d	0.9	350 min	830 min
	Toy	3.9	1 hr/d	0.8	5.5 hr	14.5 hr
	Recorder	6.8	1 hr/d	0.9	10.0 hr	24.0 hr
	Radio	20.0	4 hr/d	0.9	30 hr	60.0 hr
D					13D	13A
	Portable lighting	1.5	4 min/15 min 8 hr/d	0.9	150 min	540 min
	Portable lighting	2.2	4 min/hr 8 hr/d	0.9	120 min	950 min
	Motor/toy	2.2	1 hr/d	0.8	5.5 hr	17.5 hr
	Recorder	3.9	1 hr/d	0.9	10 hr	26.0 hr
	Radio	10.0	4 hr/d	0.9	33 hr	90.0 hr
9 Volt					1604D	1604A
	Calculator	180	30 min/d	4.8	380 min	630 min
	Toy	270	1 hr/d	5.4	7 hr	14 hr
	Radio	620	2 hr/d	5.4	23 hr	38 hr
	Electronic	1300	24 hr/d	6.0	NA	NA
	Smoke detector	Currently under consideration.				
6 Volt					908D	908A
	Portable lighting	3.9	4 min/hr 8 hr/d	3.6	5 hr	21 hr
	Portable lighting	3.9	1 hr/d	3.6	50 hr	80 hr
	Barricade	6.8	1 hr/d	3.6	165 hr	300 hr

*Performance after 12 month storage
   zinc-carbon batteries: 80% of initial requirement
   alkaline-manganese dioxide batteries: 90% of initial requirement
Source: ANSI C18.1M-1999.[11]

TABLE 8.11 ANSI/NEDA Dimensions of Zinc-Carbon Batteries*

ANSI	IEC	Diameter, mm		Overall height, mm		Length, mm		Width, mm	
		Max	Min	Max	Min	Max	Min	Max	Min
13C	R20S	34.2	32.3	61.5	59.5				
13CD	R20C	34.2	32.3	61.5	59.5				
13D	R20C	34.2	32.3	61.5	59.5				
13F	R20S	34.2	32.3	61.5	59.5				
14C	R14S	26.2	24.9	50.0	48.5				
14CD	R14C	26.2	24.9	50.0	48.5				
14D	R14C	26.2	24.9	50.0	48.5				
14F	R14S	26.2	24.9	50.0	48.5				
15C	R6S	14.5	13.5	50.5	49.2				
15CD	R6C	14.5	13.5	50.5	49.2				
15D	R6C	14.5	13.5	50.5	49.2				
15F	R6S	14.5	13.5	50.5	49.2				
24D	R03	10.5	9.5	44.5	43.3				
903	—			163.5	158.8	185.7	181.0	103.2	100.0
904	—			163.5	158.8	217.9	214.7	103.2	100.0
908	4R25X			115.0	107.0	68.2	65.0	68.2	65.0
908C	4R25X			115.0	107.0	68.2	65.0	68.2	65.0
908CD	4R25X			115.0	107.0	68.2	65.0	68.2	65.0
908D	4R25X			115.0	107.0	68.2	65.0	68.2	65.0
915	4R25Y			112.0	107.0	68.2	65.0	68.2	65.0
915C	4R25Y			112.0	107.0	68.2	65.0	68.2	65.0
915D	4R25Y			112.0	107.0	68.2	65.0	68.2	65.0
918	4R25-2			127.0	—	136.5	132.5	73.0	69.0
918D	4R25-2			127.0	—	136.5	132.5	73.0	69.0
926	—			125.4	122.2	136.5	132.5	73.0	69.0
1604	6F22			48.5	46.5	26.5	24.5	17.5	15.5
1604C	6F22			48.5	46.5	26.5	24.5	17.5	15.5
1604CD	6F22			48.5	46.5	26.5	24.5	17.5	15.5
1604D	6F22			48.5	46.5	26.5	24.5	17.5	15.5

Source: ANSI C18.1M-1999.[11]

TABLE 8.12 Cross-Reference of Zinc-Carbon Batteries

ANSI	IEC	Duracell	Eveready	Ray-O-Vac	Panasonic	Toshiba	Varta	Military
13C	R20	M13SHD	EV50	GP-D	—	—	—	—
13CD	R20	M13SHD	EV150	HD-D	UM1D	—	—	—
13D	R20	M13SHD	1250	6D	UMIN	R20U	3020	—
13F	R20	—	950	2D	UM1	R20S	2020	BA-30/U
14C	R14	M14SHD	EV35	GP-C	—	—	—	—
14CD	R14	M14SHD	EV135	HD-C	UM2D	—	—	—
14D	R14	—	1235	4C	UM2N	R14U	3014	—
14F	R14	—	935	1C	UM2	R14S	2014	BA-42/U
15C	R6	M15SHD	EV15	GP-AA	—	—	—	—
15CD	R6	M15SHD	EV115	HD-AA	UM3D	—	—	—
15D	R6	M15SHD	1215	5AA	UM3N	R6U	3006	—
15F	R6	—	1015	7AA	UM3	R6S	2006	BA-58/U
24D	R03	—	1212	—	UM4N	—	—	—
24F	R03	—	—	—	—	—	—	—
210	20F20	—	413	—	—	—	—	BA-305/U
215	15F20	—	412	—	15	—	V72PX	BA-261/U
220	10F15	—	504	—	W10E	—	V74PX	BA-332/U
221	15F15	—	505	—	MV15E	—	—	—
900	R25-4	—	735	900	—	—	—	—
903	5R25-4	—	715	903	—	—	—	BA-804/U
904	6R25-4	—	716	904	—	—	—	BA-207/U
905	R40	—	EV6	—	—	—	—	BA-23
906	R40	—	EV6	—	—	—	—	BA-23
907	4R25-4	—	1461	641	—	—	—	BA-429/U
908	4R25	M908	509	941	4F	—	—	BA-200/U
908C	4R25	M908SHD	EV90	GP-6V	—	—	430	—
908CD	4R25	M908SHD	EV90HP	—	—	—	431	—
908D	4R25	M908SHD	1209	944	—	—	430	—
915	4R25	M915	510S	942	—	—	—	BA-803/U
915C	4R25	M915SHD	EV10S	—	—	—	—	—
915D	4R25	M915SHD	—	945	—	—	—	—
918	4R25-2	—	731	918	—	—	—	—
918C	4R25-2	—	EV31	—	—	—	—	—
918D	4R25-2	—	1231	928	—	—	—	—
922	—	—	1463	922	—	—	—	—
926	8R25-2	—	732	926	—	—	—	—
1604	6F22	—	216	1604	006P	—	2022	BA-90/U
1604C	6F22	M9VSHD	EV22	GP-9V	—	—	—	—
1604CD	6F22	M9VSHD	EV122	HD-9V	006PD	—	—	—
1604D	6F22	M9VSHD	1222	D1604	006PN	6F22U	3022	—

Source: Manufacturers' catalogs.

REFERENCES

1. Frost and Sullivan Inc., *U.S. Battery Market,* New York, N.Y. 1997.

2. The Freedonia Group, Inc., Industry Study #1193, *Primary and Secondary Batteries,* Cleveland, Ohio, December 1999.

3. Samuel Rubin, *The Evolution of Electric Batteries in Response to Industrial Needs,* Dorrance, Philadelphia, 1978, chap. 5.

4. George Vinal, *Primary Batteries,* Wiley, New York, 1950.

5. N. C. Cahoon, in N. C. Cahoon and G. W. Heise (eds.), *The Primary Battery,* vol. 2, chap. 1, Wiley, New York, 1976.

6. Richard Huber, in K. V. Kordesh (ed), *Batteries,* vol. 1, chap. 1, Decker, New York, 1974.

7. A. Kozawa and R. A. Powers, "Electrochemical Reactions in Batteries," *J. Chem. Ed.* **49:**587 (1972).

8. R. J. Brodd, A. Kozawa, and K. V. Kordesh "Primary Batteries 1951–1976," *J. Electrochem. Soc.* **125**(7) (1978).

9. M. Bregazzi, *Electrochem. Technol.,* **5:**507 (1967).

10. C. L. Mantell, *Batteries and Energy Systems,* 2d ed., McGraw-Hill, New York, 1983.

11. "American National Standards Specification for Dry Cells and Batteries." ANSI C18.1M-1999, American National Standards Institute, Inc., January 1999.

12. Eveready Battery Engineering Data; Information is available via the Internet at, www.Energizer.com; Technical information website. This data is frequently updated and current.

13. M. Dentch and A. Hillier, Polaroid Corp., *Progress in Batteries and Solar Cells,* vol. 9 (1990).

CHAPTER 9
MAGNESIUM AND ALUMINUM BATTERIES

Patrick J. Spellman, Duane M. Larsen, Ron J. Ekern,
and James E. Oxley

9.1 GENERAL CHARACTERISTICS

Magnesium and aluminum are attractive candidates for use as anode materials in primary batteries.* As shown in Table 1.1, Chap. 1, they have a high standard potential. Their low atomic weight and multivalence change result in a high electrochemical equivalence on both a gravimetric and a volumetric basis. Further, they are both abundant and relatively inexpensive.

Magnesium has been used successfully in a magnesium/manganese dioxide (Mg/MnO_2) battery. This battery has two main advantages over the zinc-carbon battery, namely, twice the service life or capacity of the zinc battery of equivalent size and the ability to retain this capacity, during storage, even at elevated temperatures (Table 9.1). This excellent storability is due to a protective film that forms on the surface of the magnesium anode.

TABLE 9.1 Major Advantages and Disadvantages of Magnesium Batteries

Advantages	Disadvantages
Good capacity retention, even under high-temperature storage	Delayed action (voltage delay)
Twice the capacity of corresponding Leclanché batteries	Evolution of hydrogen during discharge
Higher battery voltage than zinc-carbon batteries	Heat generated during use
Competitive cost	Poor storage after partial discharge

*The use of magnesium and aluminum in reserve and mechanically rechargeable batteries is covered in Chaps. 17 and 38.

Several disadvantages of the magnesium battery are its "voltage delay" and the parasitic corrosion of magnesium that occurs during the discharge once the protective film has been removed, generating hydrogen and heat. The magnesium battery also loses its excellent storability after being partially discharged and, hence, is unsatisfactory for long-term intermittent use. For these reasons, the active (nonreserve) magnesium battery, while used successfully in military applications, such as radio transceivers and emergency or standby equipment, has not found wide commercial acceptance.

Furthermore the use of this magnesium battery is declining significantly, as the present trend is towards the use of lithium primary and lithium-ion rechargeable batteries.

Aluminum has not been used successfully in an active primary battery despite its potential advantages. Like magnesium, a protective film forms on the aluminum, which is detrimental to battery performance, resulting in a battery voltage that is considerably below theoretical and causing a voltage delay that can be significant for partially discharged batteries or those that have been stored. While the protective oxide film can be removed by using suitable electrolytes or by amalgamation, gains by such means are accompanied by accelerated corrosion and poor shelf life. Aluminum, however, has been more successfully used as an anode in aluminum/air batteries. (See chapter 38)

9.2 CHEMISTRY

The magnesium primary battery uses a magnesium alloy for the anode, manganese dioxide as the active cathode material but mixed with acetylene black to provide conductivity, and an aqueous electrolyte consisting of magnesium perchlorate, with barium and lithium chromate as corrosion inhibitors and magnesium hydroxide as a buffering agent to improve storability (pH of about 8.5). The amount of water is critical as water participates in the anode reaction and is consumed during the discharge.[1]

The discharge reactions of the magnesium/manganese dioxide battery are

$$
\begin{array}{lrcl}
\text{Anode} & Mg + 2OH^- & = & Mg(OH)_2 + 2e \\
\text{Cathode} & \underline{2MnO_2 + H_2O + 2e} & = & \underline{Mn_2O_3 + 2OH^-} \\
\text{Overall} & Mg + 2MnO_2 + H_2O & = & Mn_2O_3 + Mg(OH)_2
\end{array}
$$

The theoretical potential of the battery is greater than 2.8 V, but this voltage is not realized in practice. The observed values are decreased by about 1.1 V, giving an open-circuit voltage of 1.9–2.0 V, still higher than for the zinc-carbon battery.

The rest potential of magnesium in neutral and alkaline electrolytes is a mixed potential, determined by the anodic oxidation of magnesium and the cathodic evolution of hydrogen. The kinetics of both of these reactions are strongly modified by the properties of the passive film, its history of formation, prior anodic (and to a limited extent cathodic) reactions, the electrolyte environment, and magnesium alloying additions. The key to a full appreciation of the magnesium electrode lies in an understanding of the predominantly $Mg(OH)_2$ film,[2] the factors which govern its formation and dissolution, as well as the physical and chemical properties of the film.

The corrosion of magnesium under storage conditions is slight. A film of $Mg(OH)_2$ that forms on the magnesium provides good protection, and treatment with chromate inhibitors increases this protection. As a result of the formation of this tightly adherent and passivating oxide or hydroxide film on the electrode surface, magnesium is one of the most electropositive metals to find use in aqueous primary batteries. However, when the protective film is broken or removed during discharge, corrosion occurs with the generation of hydrogen,

$$ Mg + 2H_2O \rightarrow Mg(OH)_2 + H_2 $$

During the anodic oxidation of magnesium, the rate of hydrogen evolution increases with increasing current density due to destruction of the passive film, which exposes more (cathodic) sites on the bared magnesium surface. This phenomenon has often been referred to as the "negative difference effect." An appreciable rate of anodic oxidation of magnesium can only take place on the bare metal surface. Magnesium salts generally exhibit low levels of anion conductivity, and one could theoretically invoke a mechanism wherein OH^- ions migrate through the film to form reaction product $Mg(OH)_2$ at the magnesium-film interface. In practice this does not occur at a sufficiently rapid rate and instead the film becomes disrupted, in all likelihood mechanically, as the result of anodic current flow.[3] A theoretical model for the breakdown of the passive film has been proposed.[4-7] This model involves, successively, metal dissolution at the metal-film interface, film dilatation, and film breakdown. This wasteful reaction is a problem, not only because of the need to vent the hydrogen from the battery and to prevent it from accumulating, but also because it uses water which is critical to the battery operation, produces heat, and reduces the efficiency of the anode.

The efficiency of the magnesium anode is about 60 to 70% during a typical continuous discharge and is influenced by such factors as the composition of the magnesium alloy, battery components, discharge rate, and temperature. On low drains and intermittent service, the anode efficiency can drop to 40 to 50% or less. The anode efficiency also is reduced with decreasing temperature.

Considerable heat is generated during the discharge of a magnesium battery, particularly at high discharge rates, due to the exothermic corrosion reaction (about 82 kcal per gram-mole of magnesium) and the losses resulting from the difference between the theoretical and operating voltage. Proper battery design must allow for the dissipation of this heat to prevent overheating and shortened life. On the other hand, this heat can be used to advantage at low ambient temperatures to maintain the battery at higher and more efficient operating temperatures.

A consequence of the passive film on these metals is the occurrence of a voltage delay— a delay in the battery's ability to deliver full output voltage after it has been placed under load—which occurs while the protective film on the surface of the metal becomes disrupted by the flow of current, exposing bare metal to the electrolyte (see Fig. 9.1). When the current is interrupted, the passive film does indeed reform, but never to the original degree of passivity. Thus both the magnesium and the aluminum batteries are at a significant disadvantage in very low or intermittent service applications.[3] This delay, as shown in Fig. 9.2, is usually less than 1 s, but can be longer (up to a minute or more) for discharges at low temperatures and after prolonged storage at high temperatures.

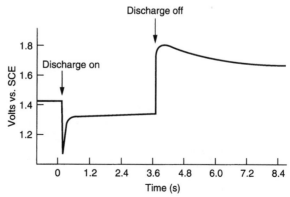

FIGURE 9.1 Voltage profile of magnesium primary battery at 20°C.

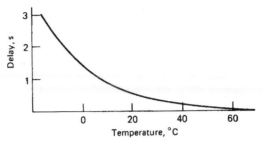

FIGURE 9.2 Voltage delay vs. temperature, Mg/MnO$_2$ battery.

9.2.1 Aluminum

The standard potential for aluminum in the anode reaction,

$$Al \rightarrow Al^{3+} + 3e$$

is reported as -1.7 V. A battery with an aluminum anode should have a potential about 0.9 V higher than the corresponding zinc battery. However, this potential is not attained, and the potential of an Al/MnO$_2$ battery is only about 0.1 to 0.2 V higher than that of a zinc battery. The Al/MnO$_2$ battery never progressed beyond the experimental stage because of the problems with the oxide film, excessive corrosion when the film was broken, voltage delay, and the tendency for aluminum to corrode unevenly. The experimental batteries that were fabricated used a two-layer aluminum anode (to minimize premature failure due to can perforation), an electrolyte of aluminum or chromium chloride, and a manganese dioxide-acetylene black cathode similar to the conventional zinc/manganese dioxide battery. The reaction mechanism is

$$Al + 3MnO_2 + 3H_2O \rightarrow 3MnO \cdot OH + Al(OH)_3$$

9.3 CONSTRUCTION OF Mg/MnO$_2$ BATTERIES

Magnesium/manganese dioxide (nonreserve) primary batteries are generally constructed in a cylindrical configuration.

9.3.1 Standard Construction

The construction of the magnesium battery is similar to the cylindrical zinc-carbon battery. A cross section of a typical battery is shown in Fig. 9.3. A magnesium alloy can, containing small amounts of aluminum and zinc, is used in place of the zinc can. The cathode consists of an extruded mix of manganese dioxide, acetylene black for conductivity and moisture retention, barium chromate (an inhibitor), and magnesium hydroxide (a pH buffer). The electrolyte is an aqueous solution of magnesium perchlorate with lithium chromate. A carbon rod serves as the cathode current collector. The separator is an absorbent kraft paper as in the paper-lined zinc battery structure. Sealing of the magnesium battery is critical, as it must be tight to retain battery moisture during storage but provide a means for the escape of hydrogen gas which forms as the result of the corrosion reaction during the discharge. This is accomplished by a mechanical vent—a small hole in the plastic top seal washer under the retainer ring which is deformed under pressure, releasing the excess gas.[8]

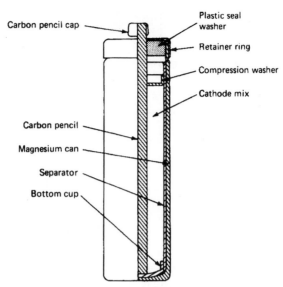

FIGURE 9.3 Cylindrical construction of magnesium primary battery.

9.3.2 Inside-Out Construction

The basis of the inside-out design (Fig. 9.4) is a highly conductive carbon structure which can be molded readily into complex shapes. The carbon structure is formed in the shape of a cup which serves as the battery container; an integral center rod is incorporated to reduce current paths. The cups are structurally strong, homogeneous, and impervious to the passage of liquids and gases and the corrosive effects of the electrolyte. The battery consists of the carbon cup, a cylindrical magnesium anode, a paper separator, and a cathode mix consisting of manganese dioxide, carbon black, and inhibitors with aqueous magnesium bromide or perchlorate as the electrolyte. The cathode mix is packed into the spaces on both sides of the anode and is in intimate contact with the inside and outside surfaces of the anode, the center rod, and the inside surfaces of the cup. This configuration provides larger electrode surface areas. External contacts are made by two metallic end pieces. The positive terminal is bonded during the forming process to the closed end of the carbon cup. The negative terminal, to which the anode is attached, together with a plastic ring forms the insulated closure and seal for the open end of the cup. The entire battery assembly is enclosed in a crimped tin-plated steel jacket.[9–11]

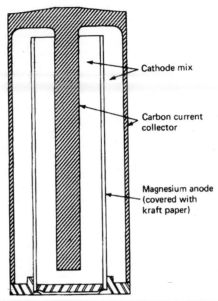

FIGURE 9.4   Inside-out construction of magnesium primary battery. (*Courtesy of ACR Electronics, Inc.*)

9.4   PERFORMANCE CHARACTERISTICS OF Mg/MnO₂ BATTERIES

9.4.1   Discharge Performance

Typical discharge curves for the cylindrical magnesium/manganese dioxide primary battery are shown in Fig. 9.5. The discharge profile is generally flatter than for the zinc-carbon batteries; the magnesium battery also is less sensitive to changes in the discharge rate. The average discharge voltage is on the order of 1.6 to 1.8 V, about 0.4 to 0.5 V above that of the zinc-carbon battery; the typical end voltage is 1.2 V. The performance characteristics of the cylindrical magnesium battery, type 1LM, on continuous and intermittent discharge are summarized in Figs. 9.6 to 9.8 and Table 9.2. The batteries were discharged under a constant-resistance load at 20°C.

Figure 9.6 provides a summary of the battery's performance under continuous load to 1.1-V end voltage.

Figure 9.7 shows the relationship of discharge current to delivered ampere-hour capacity of the battery on continuous constant-current discharge to several end voltages. The intermittent discharge characteristics are illustrated in Table 9.2. The sizable reduction in performance under low-rate or long-term discharge is attributed to the corrosion reaction between the discharging magnesium anode and the cell electrolyte. The reaction, which results in the evolution of hydrogen and the concomitant reduction of water, causes a loss of total cell efficiency. This phenomenon is illustrated in Fig. 9.8, which summarizes the ampere-hour output under continuous discharge of the 1LM cell to an 0.8-V end voltage. This loss of capacity on the low-rate (high-resistance) discharges is evident.

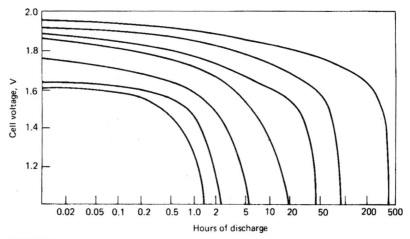

FIGURE 9.5 Typical discharge curves of magnesium/manganese dioxide cylindrical battery at 20°C.

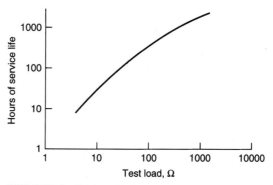

FIGURE 9.6 1LM service life (hours to 1.1 V) vs. test load at room temperature. (*Courtesy of Rayovac Corporation.*)

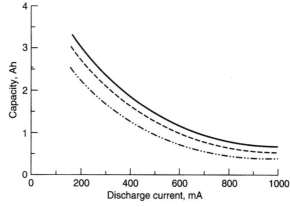

FIGURE 9.7 1LM service life (ampere-hours) vs. constant-current discharge. Dotted line—1.4-V end voltage; dashed line—1.2-V end voltage; solid line—1.0-V end voltage. (*Courtesy of Rayovac Corporation.*)

TABLE 9.2 Performance, in Hours, of 1LM
Batteries on Continuous and Intermittent Discharge

	End voltage	
Type of Discharge	1.1 V	0.8 V
4 Ω, continuous	8.9	9.9
4 Ω, LIFT*	10.7	11.6
4 Ω, HIFT†	11	12
4 Ω, 30 min/h, 8 h/d	9.72	10.60
25-Ω constant resistance		
Continuous	100	104
4 h/d	84.2	88.4
500-Ω constant resistance		
Continuous	1265	1312
4 h/d	752	776

* Light industrial flashlight test, 4 min/h, 8 h/day.
† Heavy industrial flashlight test, 4 min/15 mm, 8 h/ day.
    Source: Rayovac Corporation.

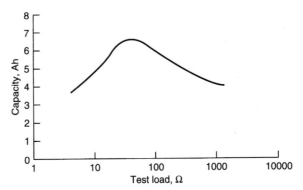

FIGURE 9.8   1LM service life (ampere-hours to 0.8 V) vs. test load. (*Courtesy of Rayovac Corporation.*)

The performance of the magnesium primary battery at low temperatures is also superior to that of the zinc-carbon battery, operating to temperatures of $-20°C$ and below. Figure 9.9 shows the performance of the magnesium battery at different temperatures based on the 20-h discharge rate. The low temperature performance is influenced by the heat generated during discharge and is dependent on the discharge rate, battery size, battery configuration, and other such factors. Actual discharge tests should be performed if precise performance data are needed.

On extended low-rate discharges, the magnesium battery may split open. This rupture is due to the formation of magnesium hydroxide which occupies about one and one-half times the volume of the magnesium. It expands and presses against the cathode mix which has hardened appreciably from the loss of water during the discharge. This opening of the can cause the voltage to rise about 0.1 V, also increasing capacity due to the air that can enter into the reaction.

The service life of the magnesium/manganese dioxide primary battery, normalized to unit weight (kilogram) and volume (liter), at various discharge rates and temperatures is summarized in Fig. 9.10. The data are based on a rated performance of 60 Ah/kg and 120 Ah/L.

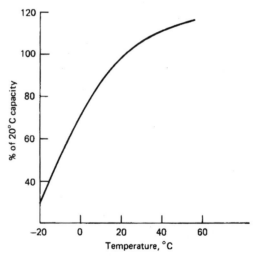

FIGURE   9.9 Performance   vs.   temperature   of magnesium/manganese dioxide cylindrical battery.

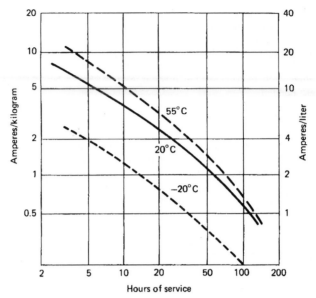

FIGURE 9.10 Service life of magnesium/manganese dioxide primary battery at various discharge rates and temperatures (to 1.2-V/cell end voltage).

9.4.2 Shelf Life

The shelf life of the magnesium/manganese dioxide primary battery at various storage temperatures is compared with the shelf life of the zinc-carbon battery in Fig. 9.11. The magnesium battery is noted for its excellent shelf life. The battery can be stored for periods of 5 years or longer at 20°C with a total capacity loss of 10 to 20% and at temperatures as high as 55°C with losses of about 20%/year.

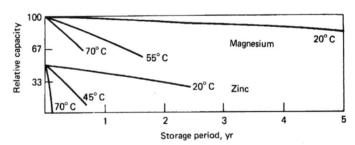

FIGURE 9.11 Comparison of service vs. storage of magnesium/manganese dioxide and zinc-carbon batteries.

9.4.3 Inside-Out Cells

The discharge characteristics of the cylindrical inside-out magnesium primary batteries are shown in Fig. 9.12 for various discharge rates and at 20°C and −20°C. This structure has better high-rate and low temperature performance than the conventional structure. These batteries can be discharged at temperatures as low as −40°C, although at lighter discharge loads at the lower temperatures. Discharge curves are characteristically flat. They also have good and reproducible low-drain, long-term discharge characteristics as they do not split under these discharge conditions. Discharges for a 2½-year duration are realized with a D-size battery at a 270-μA drain at 20°C.

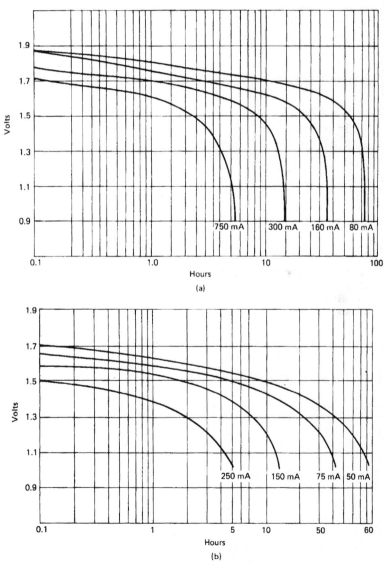

FIGURE 9.12 Typical discharge curves of magnesium inside-out primary battery, D size. (*a*) 20°C. (*b*) −20°C. (*Courtesy of ACR Electronics, Inc.*)

9.4.4 Battery Design

Battery configuration has an important influence on the performance of the magnesium/manganese dioxide battery because of the heat generated during the discharge. As discussed in Sec. 9.2, proper battery design must allow for the dissipation of this heat to prevent overheating, premature dry-out, and shortened performance—or for using this heat to improve performance at low ambient temperatures. In some low-temperature applications it is advantageous to insulate the battery against heat loss. Actual discharge tests will be required to obtain precise performance data under a variety of possible conditions and battery designs.

The battery and equipment design must also consider the hydrogen that is generated during discharge. The hydrogen must be vented and kept from accumulating because hydrogen-air mixtures are flammable above 4.1% hydrogen and explosive above 18% hydrogen.

9.5 SIZES AND TYPES OF Mg/MnO$_2$ BATTERIES

The cylindrical magnesium/manganese dioxide batteries were manufactured in several of the popular standard ANSI sizes, as summarized in Table 9.3. Most of the production of the conventional battery is used for military radio transceiver applications, and mainly in the 1LM size. The batteries are no longer available commercially. Inside-out batteries are no longer manufactured.

TABLE 9.3   Cylindrical Magnesium Prmary Batteries

				Capacity, Ah*	
Battery type	Diameter, mm	Height, mm	Weight, g	Conventional structure†	Inside-out cell‡
N	11.0	31.0	5	0.5	
B	19.2	53.0	26.5	2.0	
C	25.4	49.7	45	—	3.0
1LM§	22.8	84.2	59	4.5	
D	33.6	60.5	105	—	7.0
FD	41.7	49.1	125	—	8.0
No. 6	63.5	159.0	1000	—	65

* 50-h discharge rate.
† Manufacturer: Rayovac Corp.
‡ Manufacturer: ACR Electronics, Inc., Hollywood, Fla. (no longer manufactured).
§ Only size now being manufactured.

9.6 OTHER TYPES OF MAGNESIUM PRIMARY BATTERIES

Magnesium primary batteries have been developed in other structures and with other cathode materials, but these designs have not achieved commercial success. Flat cells, using a plastic-film envelope, were designed but were never produced commercially.

The use of organic depolarizers, such as *meta*-dinitrobenzene (*m*-DNB), in place of manganese dioxide was of interest because of the high capacity that could be realized with the complete reduction of *m*-DNB to *n*-phenylenediamine (2 Ah/g). The discharge of actual batteries, while having a flat voltage profile and a higher ampere-hour capacity than the manganese dioxide battery, had a low operating voltage of 1.1 to 1.2 V per cell. Watthour capacities were not significantly higher than for the magnesium/manganese dioxide batteries. The *m*-DNB battery also was inferior at low temperatures and high current drains. Commercial development of these batteries never materialized.

Magnesium/air batteries were studied, again because of the higher operating voltage than with zinc (see Chap. 38). These batteries, too, were never commercialized. Magnesium, however, is a very useful anode in reserve batteries. Its application in these types of batteries is covered in Chap. 17.

9.7 ALUMINUM PRIMARY BATTERIES

Experimental work on Al/MnO_2 primary or dry batteries was concentrated on the D-size cylindrical battery using a construction similar to the one used for the Mg/MnO_2 battery (Fig. 9.3). The most successful anodes were made of a duplex metal sheet consisting of two different aluminum alloys. The inner, thicker layer was more electrochemically active, leaving the outer layer intact in the event of pitting of the inner layer. The cathode bobbin consisted of manganese dioxide and acetylene black, wetted with the electrolyte. Aqueous solutions of aluminum or chromium chloride, containing a chromate inhibitor, were the most satisfactory electrolytes.

Aluminum active primary batteries were never produced commercially. While the experimental aluminum batteries delivered a higher energy output than conventional zinc batteries, anode corrosion, causing problems on intermittent and long-term discharges and irregularities in shelf life, and the voltage-delay problem restrained commercial acceptance. Aluminum/air batteries are covered in Chap. 38.

REFERENCES

1. J. L. Robinson, "Magnesium Cells," in N. C. Cahoon and G. W. Heise (eds.), *The Primary Battery,* vol. 2, Wiley-Interscience, New York, 1976, chap. 2.

2. G. R. Hoey and M. Cohen, "Corrosion of Anodically and Cathodically Polarized Magnesium in Aqueous Media," *J. Electrochem. Soc.* **105**:245 (1958).

3. J. E. Oxley, R. J. Ekern, K. L. Dittberner, P. J. Spellman, and D. M. Larsen, in *Proc. 35th Power Sources Symp.,* IEEE, New York, 1992, p. 18.

4. B. V. Ratnakumar and S. Sathyanarayana, "The Delayed Action of Magnesium Anodes in Primary Batteries. Part I: Experimental Studies," *J. Power Sources* **10**:219 (1983).

5. S. Sathyanarayana and B. V. Ratnakumar, "The Delayed Action of Magnesium Anodes in Primary Batteries. Part II: Theoretical Studies," *J. Power Sources* **10**:243 (1983).

6. S. R. Narayanan and S. Sathyanarayana, "Electrochemical Determination of the Anode Film Resistance and Double Layer Capacitance in Magnesium-Manganese Dioxide Cells," *J. Power Sources* **15**:27 (1985).

7. B. V. Ratnakumar, "Passive Films on Magnesium Anodes in Primary Batteries," *J. Appl. Electrochem.* **18:**268 (1988).

8. D. B. Wood, "Magnesium Batteries," in K. V. Kordesch (ed.), *Batteries,* vol. 1: *Manganese Dioxide,* Marcel Dekker, New York, 1974, chap. 4.

9. R. R. Balaguer and F. P. Schiro, "New Magnesium Dry Battery Structure," in *Proc. 20th Power Sources Symp.,* Atlantic City, N.J. 1966, p. 90.

10. R. R. Balaguer, "Low Temperature Battery (New Magnesium Anode Structure)," Rep ECOM-03369-F, 1966.

11. R. R. Balaguer, "Method of Forming a Battery Cup," U.S. Patent 3,405,013, 1968.

12. D. M. Larsen, K. L. Dittberner, R. J. Ekern, P. J. Spellman, and J. E. Oxley, "Magnesium Battery Characterization," in *Proc. 35th Power Sources Symp.,* IEEE, New York, 1992, p. 22.

13. L. Jarvis, "Low Cost, Improved Magnesium Battery, in *Proc. 35th Power Sources Symp.,* New York, 1992, p. 26.

CHAPTER 10
ALKALINE-MANGANESE DIOXIDE BATTERIES

Robert F. Scarr, James C. Hunter, and Philip J. Slezak

10.1 GENERAL CHARACTERISTICS

Since its introduction in the early 1960s, the alkaline-manganese dioxide (zinc/KOH/MnO$_2$) battery has become the dominant battery system in the portable battery market. This came about because the alkaline system is recognized as having several advantages over its acidic-electrolyte counterpart, the Leclanché or zinc-carbon battery, the former market leader with which it competes. Table 10.1 summarizes the advantages and disadvantages of alkaline-manganese dioxide batteries compared to zinc-carbon batteries.

The alkaline-manganese dioxide battery is available in two design configurations, (1) as relatively large-size cylindrical batteries and (2) as miniature button batteries. There are also multiple-cell batteries made from various sizes of unit cells. While the alkaline cell is still undergoing change, some developments in the evolution of the present cylindrical cell technology are particularly notable. After the initial concepts of a gelled/amalgamated zinc powder anode in a central compartment and use of vented plastic seals had been established, the first major advance was the butt-seam metal finish which allowed the cell to have greater internal volume. Next came the discovery that organic inhibitors could reduce the rate of gassing caused by contaminants in the zinc anode, resulting in a product with diminished

TABLE 10.1 Major Advantages and Disadvantages of Cylindrical Alkaline-Manganese Dioxide Batteries (Compared to Carbon-Zinc Batteries)

Advantages	Disadvantages
Higher energy density	Higher initial cost
Better service performance: Continuous and intermittent Low and high rate Ambient and low temperature	
Lower internal resistance Longer shelf life Greater resistance to leakage Better dimensional stability	

bulge and leakage. Another major development was the introduction of the plastic label finish and lower profile seal, which permitted a further large increase in the internal volume available for active material and a substantial increase in the capacity of the battery. Perhaps the most significant change to the alkaline cell began in the early 1980s with the gradual reduction of the amount of mercury in the anode and the development of cells containing no added mercury. This trend, which was aided by a substantial improvement in the reliability of cell materials resulting from reduced impurity levels, was driven by worldwide concern over the environmental impact of the materials used in batteries after their disposal.

Developments such as these have enabled the alkaline-MnO_2 battery to gain as much as a 60% increase in specific energy output since its introduction to keep pace with the needs of the consumer. Its leadership position should support further technological improvements, which will ensure continued market dominance.

More recently, development effort has focused on enhancing the performance of the battery at high discharge rates to meet the power demands of new portable electronic equipments, such as digital cameras, camcorders, cellular phones and PDAs. Premium batteries, which have significantly superior performance in these high-rate applications compared with the standard alkaline-manganese dioxide batteries, have recently been introduced to the market. Some examples of the extent of this improvement are illustrated in Fig. 10.1a and a specific example of a simulated photoflash application is shown in Fig. 10.1b. Detailed performance characteristics of these premium batteries are given in Sec. 10.7.

Miniature button-type batteries, using the same zinc/alkaline-manganese dioxide chemistry as cylindrical cells, compete with other miniature battery systems such as mercuric oxide, silver oxide, and zinc/air. Table 10.2 shows the major advantages and disadvantages of miniature alkaline-manganese dioxide batteries in comparison to other miniature batteries.

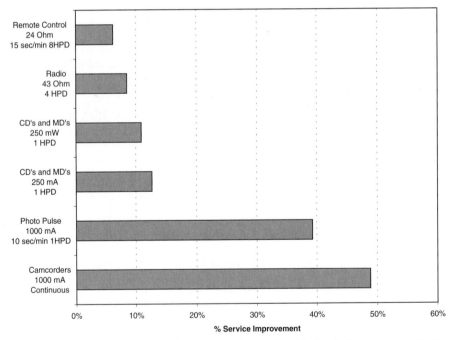

FIGURE 10.1a Typical performance improvements for premium AA-size alkaline-manganese dioxide battery versus standard type AA-size batteries on various simulated device applications. (*Courtesy of Eveready Battery Company.*)

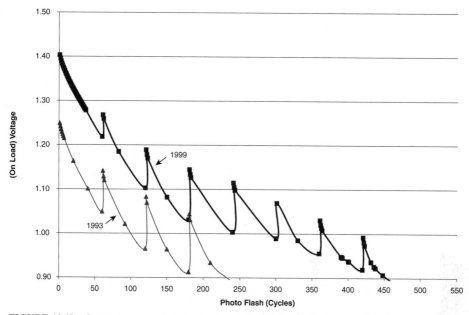

FIGURE 10.1*b*   Comparison of typical simulated photoflash high rate discharge performance characteristics for AA-size alkaline-manganese dioxide batteries produced in 1993 versus 1999. (*Courtesy of Eveready Battery Company.*)

TABLE 10.2   Comparison of Miniature Alkaline-Manganese Dioxide Battery with Other Miniature Systems

Advantages	Disadvantages
Lower cost	Sloping discharge curve prevents its use in some devices
Lower internal resistance	Lower energy density
Better low-temperature performance	
Equivalent leakage resistance	

10.2   CHEMISTRY

The active materials in the alkaline-manganese dioxide cell are electrolytically produced manganese dioxide, an aqueous alkaline electrolyte, and powdered zinc metal. Electrolytic MnO_2 is used instead of either chemical MnO_2 or natural ore because of its higher manganese content, its increased reactivity, and its greater purity. The electrolyte is concentrated caustic, usually KOH in the range of 35 to 52%, which affords greater conductivity and a reduced hydrogen gassing rate compared to the acidic electrolyte of the Leclanché cell. Powdered zinc is used for the anode to provide a large surface area for high-rate capability (that is, to reduce current density) and to distribute solid and liquid phases more homogeneously (to minimize mass-transport polarization of reactant and product).

On discharge, the manganese dioxide cathode undergoes at first a one-electron reduction to the oxyhydroxide,

$$MnO_2 + H_2O + e \rightarrow MnOOH + OH^- \tag{10.1}$$

The MnOOH product forms a solid solution with the reactant, giving rise to the characteristic sloping discharge curve.[1] Of the many structural forms of MnO_2 which exist, only the gamma form functions well as an alkaline cathode because its surface is not prone to become blocked by the reaction product. In forming MnOOH, the cathode expands about 17% in volume. MnOOH can undergo some undesirable chemical reactions as well. In the presence of zincate ion, MnOOH, through its equilibrium with soluble Mn(III), can form the complex compound hetaerolite, $ZnMn_2O_4$. Although electroactive, hetaerolite is not as easily discharged as MnOOH, hence the cell impedance increases. In addition, the $MnOOH/MnO_2$ solid solution can undergo recrystallization into a less active form, resulting in a noticeable loss in cell voltage under certain very slow discharge conditions.[2]

At a lower voltage, MnOOH can then be discharged further, as in the following reaction:

$$3MnOOH + e \rightarrow Mn_3O_4 + OH^- + H_2O \tag{10.2}$$

This reaction produces a flat discharge curve, but it is also slower than the first reduction step and therefore is useful only under low-rate discharge conditions. No additional volume change occurs in the cathode for this reaction. Note that this step provides only one-third of the capacity of the first reaction based on MnO_2. Even further reduction to $Mn(OH)_2$ is possible but not practical.

At its earliest stages, the anode discharge reaction in highly caustic electrolyte produces the soluble zincate ion

$$Zn + 4OH^- = Zn(OH)_4^= + 2e \tag{10.3}$$

At a point depending on the initial composition of the anode and the rate and depth of discharge, however, the electrolyte becomes saturated with zincate, causing the product of the reaction to change to $Zn(OH)_2$. In the water-starved environment of the alkaline anode, zinc hydroxide then slowly dehydrates to ZnO in the following sequence:

$$Zn + 2OH^- = Zn(OH)_2 + 2e \tag{10.4}$$

$$Zn(OH)_2 = ZnO + H_2O \tag{10.5}$$

These forms of oxidized zinc are all equivalent oxidation states which are differentiated little in potential. The transition from one form to another cannot usually be detected in the discharge curve. Under some conditions, where the product of discharge is too densely attached to the surface, passivation of the zinc can occur. Such conditions include high-rate discharge, low temperature, and factors which limit the solubility of ZnO such as low KOH and high zincate concentration. Passivation is not likely to occur with high surface area anodes as these are. While the oxidation of zinc is a facile reaction on its own, contributing little to the impedance of the cell, the addition of metallic mercury enhances the reaction rate even more. The function of mercury in amounts over 0.5% is to promote the electronic adsorption of OH^- ions through facilitation of water adsorption.

Therefore the total reaction of the cell on continuous discharge to the depth of one electron per mole of MnO_2 is

$$2MnO_2 + Zn + 2H_2O = 2MnOOH + Zn(OH)_2 \tag{10.6}$$

Since water is a reactant in this expression, its availability in a water-starved environment is crucial in high-rate discharges. To avoid service limitations, care must be taken to manage the exchange of the water supply from one cell region to another over the short term.

Some battery manufacturers have included additives to the cell (e.g., TiO_2 and $BaSO_4$) to assist in water management for high drain applications. Although the actual mechanisms of these additives are unknown, they are believed to assist in concentration polarization within the cell as evident in the improvement in performance made in these high drain applications.

In contrast, the total cell reaction for light or intermittent drains to 1.33 electrons per mole may be written

$$3MnO_2 + 2Zn = Mn_3O_4 + 2ZnO \tag{10.7}$$

Under these conditions, there is no water management problem.

The initial open-circuit potential of the zinc/alkaline-MnO_2 cell is about 1.5 to 1.65 V, depending on the purity and activity of the cathode material and the ZnO content of the anode. The average voltage during discharge to the functional end voltage of 0.75 V is about 1.2 V.

Besides the intended reactions in the alkaline cell, the anode can undergo undesirable gas-generating reactions as well. Zinc metal is so active that it can reduce water to produce hydrogen gas. Gassing can occur both on long-term storage before the cell is used as well as after partial discharge in proportion to the depth and rate of discharge. In addition to buildup of gas pressure, which causes dimensional distortion and eventual leakage of the cell, this corrosion of zinc causes loss of anode capacity and chemical self-discharge of the cathode by hydrogen gas. The rate of gassing of pure zinc is quite low, but the presence of heavy-metal impurities in trace quantities promotes the gassing rate dramatically by acting as cathodic sites for hydrogen evolution. Gassing may be reduced in several ways: (1) addition of ZnO to the anode to reduce the driving force of the zinc by mass action; (2) addition to the anode of inorganic inhibitors such as certain metal oxides, or use of organic inhibitors, usually end-substituted polyethylene oxide compounds; (3) reduction of impurity levels in cell components; (4) "alloying" of zinc with certain elemental inhibitor metals such as lead or indium; and (5) amalgamation of the zinc with mercury. (This alternative is sharply decreasing in current practice.)

Finally, reaction (10.3) is important to the performance of the anode in another way as well. The expression represents a dynamic equilibrium between zinc and its ions, and it indicates that zinc is continually dissolving and replating throughout the anode at open circuit. The following benefits occur as a result of this action: (1) gas-promoting impurities on the surface of the zinc are coated over, thus diminishing their activity; (2) the particle-to-particle contact of zinc is maintained and improved by the building of zinc metal bridges between them; and (3) the gas-promoting surface of the bare metal collector is coated with zinc, thus reducing its activity. Since many of these functions were once performed by mercury, the importance of the zinc replating reaction is increased as mercury is eliminated.

10.3 CELL COMPONENTS AND MATERIALS

10.3.1 Cathode Components

The composition of a typical alkaline cathode and the purpose of each component are listed in Table 10.3. The cathode is made from a mixture of manganese dioxide and carbon. Other materials may also be added, such as binders (to help hold the cathode together) and water or electrolyte solution (to aid in forming the cathode).

Manganese Dioxide.   Manganese dioxide is the oxidizing component in the cell. To produce an alkaline cell of satisfactory power and long shelf life, the manganese dioxide must be highly active and very pure. The only type of manganese dioxide that is used in commercial alkaline cells is electrolytic manganese dioxide (EMD).

TABLE 10.3 Composition of Typical Cathode

Component	Range, %	Function
Manganese dioxide	79–90	Reactant
Carbon	2–10	Electronic conductor
35–52% aqueous KOH	7–10	Reactant, ionic conductor
Binding agent	0–1	Cathode integrity (optional)

The process for making EMD involves dissolving a manganous compound in acid to produce a solution of manganous ions. If the starting material is a manganese dioxide ore, the ore is first reduced to manganous oxide, then dissolved in sulfuric acid to produce manganous sulfate solution. The solution is treated to remove various harmful impurities, then introduced into a plating cell and electrolyzed. EMD is plated onto an anode, typically made of graphite or titanium, according to the reaction

$$Mn^{2+} + 2H_2O = MnO_2 + 4H^+ + 2e \qquad (10.8)$$

At the same time hydrogen is generated at the cathode, which may be made of copper, graphite, or lead,

$$2e + 2H^+ = H_2 \qquad (10.9)$$

The overall reaction in the EMD plating battery is then

$$Mn^{2+} + 2H_2O = MnO_2 + 2H^+ + H_2 \qquad (10.10)$$

A typical analysis of EMD is shown in Table 10.4. The extremely low level of heavy-metal impurities helps minimize hydrogen gassing at the zinc anode, which might otherwise occur if such impurities were present and were able to migrate to the anode. Other impurities will combine with the manganese sulfate solution during electrolysis, forming undesirable manganese oxide compounds (e.g., cryptomelane) that will reduce the overall effectiveness of the MnO during discharge in the alkaline cell.

TABLE 10.4 Typical Analysis of Electrolytic Manganese Dioxide (EMD)

Component	Typical values*	Component	Typical values*
MnO_2	91.7%	Fe	72 ppm
Mn	60.5%	Ti	<2 ppm
Peroxidation	95.7%	Cr	6 ppm
H_2O, 120°C	1.3%	Ni	2 ppm
H_2O, 120–400°C	3.2%	Co	1 ppm
Real density	4.46 g/cm³	Cu	3 ppm
SO_4^{2-}	0.85%	V	0.5 ppm
C	0.07%	Mo	0.6 ppm
Na	2550 ppm	As	<0.5 ppm
K	235 ppm	Sb	<0.5 ppm

*Based on analysis of 10 samples of alkaline grade cell-grade EMD from five manufacturers.

Carbon.   Since manganese dioxide itself is a poor conductor, carbon is used in the cathode to provide electronic conductivity. The carbon is usually in the form of graphite, although some acetylene black may also be used. The carbon must have low levels of those impurities which might lead to corrosion in the cell. Some natural graphites have been used in alkaline cells. However, with the trend toward making cells with ultralow levels of mercury, there has been increasing use of very pure synthetic graphites. In some recent improvements, thermal and/or chemical treatments of graphite have improved the conductivity of both synthetic and natural graphites leading to higher conductivity of the cathode mixture. This improvement is the result of reducing the number of carbon planes within the individual carbon particles. Conductivity of carbon is lower across carbon planes (C direction) as compared to within carbon planes (A and B directions). The treatments will "peel" away the carbon preferentially along the C direction, thereby decreasing the resistance drop across the carbon particles. Traditional approaches have often improved service by increasing the internal volume of the battery available for active ingredients; however, the treatment of the graphite to improve conductivity has allowed battery manufacturers to make service improvements to the cell within the same cell dimensions. With the increase in graphite conductivity, a decrease in the graphite content can be made which enables an increase in the active manganese dioxide content while maintaining the conductivity of the cathode.

Other Ingredients.   The use of other materials (binders, additives, electrolyte) will depend on the particular manufacturing process used by the battery maker. The ultimate goal is to produce a dense, stable cathode, which has good electronic and ionic conductivity, and discharges efficiently even at high discharge rates.

10.3.2   Anode Components

The composition of a typical alkaline anode and the purpose of each component are listed in Table 10.5. The final three ingredients in the table are optional. Gelling agents are used in nearly all types of alkaline cells, although there have been attempts to utilize pressed powder or binders to form the anode mass as well. Amalgamation levels relative to zinc range from 0 to nearly 6%, but the majority of the cells produced in the "industrialized" countries have no added mercury.

TABLE 10.5   Composition of Typical Alkaline Anode

Component	Range, %	Function
Zinc powder	55–70	Reactant, electronic conductor
35–52% aqueous KOH	25–35	Reactant, ionic conductor
Gelling agent	0.4–2	Electrolyte distribution and immobilization, mix processability
ZnO	0–2	Gassing suppressor, zinc-plating agent
Inhibitor	0–0.05	Gassing suppressor
Mercury	0–4	Gassing suppressor, electronic conductor, discharge accelerator, mix processability

Zinc Powder. Pure zinc is obtained commercially by either a thermal distillation process or electroplating from an aqueous solution. This zinc is converted to battery-grade powder by atomizing a thin stream of the molten metal with jets of compressed air. Particles range in shape from "potatoes" to "dog-bones," and in size from 20 to 820 μm in a log-normal distribution. The median particle diameter ranges from 155 to 255 μm, while the average surface area is about 0.02 m^2/g. Except for intentionally added alloyed metals, the purity of battery-grade zinc is very high. A list of typical impurities is given in Table 10.6. Essentially all battery-grade zinc contains about 500 ppm lead. Other metallic additives which have been alloyed for gassing inhibition or improving mercury distribution are indium, bismuth, and aluminum. Preamalgamated zinc is also available.

TABLE 10.6 Impurity Content of Typical Battery-Grade Zinc Powder

Element	Typical level,* ppm	Maximum level,† ppm	Element	Typical level,* ppm	Maximum level,† ppm
Cd	4.4	21	Sn	0.10	0.44
Fe	4.2	14	Sb	0.090	0.26
Ag	1.6	5.4	Co	0.058	0.18
Cu	1.5	4.3	Mo	0.037	0.13
Ca	0.21	0.85	Mg	0.030	0.12
Si	0.21	0.73	As	0.010	0.044
Ni	0.20	0.58	Hg	0.007	0.025
Al	0.15	0.66	V	0.001	0.0025
Cr	0.12	0.48			

*Based on analysis of 25 samples from seven producers (1990 data).
†Based on 99.7% (3σ) confidence level.

Anode Gel. Starch or cellulosic derivatives, polyacrylates, or ethylene maleic anhydride copolymers are used as gelling agents. The anode cavity is filled with either the complete well-blended anode mixture, or the dry ingredients (using preamalgamated zinc if mercury is needed) to which the electrolyte is added later. As mercury levels are reduced throughout the industry, electrolyte (and water) purity becomes of greater importance. Care must be taken to minimize carbonate, chloride, and iron contamination in particular. Volume fractions of zinc range from 18 to 33%. The lower limit of this range is required to maintain electronic conductivity of the anode, while the upper limit is to avoid the condition where the accumulation of reaction product can block ionic pathways. Densities of anode mixtures are typically in the 2.5 to 3.2-g/cm^3 range, while volume capacities vary from 1.2 to 1.8 Ah/cm^3. The maximum discharge efficiency which can be realized from the zinc ranges from 84 to 94%, depending on cell size and type of operating duty. To avoid hydrogen gassing from the cathode, which would occur if its capacity were exhausted first, cell service is normally designed to be anode limited. As a result of all these factors, anode input capacity is usually established at 96 to 105% of the cathode undergoing a 1.33-electron change.

Elimination of Heavy Metals. Until recent times, the addition of mercury metal to the anode has been widely used to perform several functions in the mix as well as on the collector. These are listed in Table 10.5. However, the industry has reduced or eliminated heavy metals from the battery. The absence of mercury from the anode can lead to reduced service, increased sensitivity to mechanical shock, and increased gassing on initial storage and after partial discharge. It has been necessary to find substitutes for each of the functions of mercury. Such measures have been described in related sections of this chapter. In general, however, the successful elimination of mercury has been aided by the reduction of impurities, particularly iron, in battery-grade materials. [Iron from the can is not normally a problem because it is rendered passive and insoluble by contact with the highly oxidizing cathode EMD (electrolytic manganese dioxide).] In addition, gassing is further controlled by using alloys of zinc containing small amounts of indium, bismuth, aluminum, or calcium. Other measures include modifications to the particle-size distribution and anode mix formulations to reduce anode resistivity and improve zinc discharge reaction kinetics. Even further, some zinc powder and battery manufacturers are developing lead-free alloys as well in order to provide an alternative to the practice of using another heavy-metal additive, lead, for gassing inhibition.

10.3.3 Anode Collectors

The anode collector material in cylindrical alkaline cells is usually cartridge brass in the form of pins or strip. In miniature cells the anode collector is usually a stainless-steel cup whose convex surface is an exterior terminal of the cell. The outer surface of the cup is clad with nickel for good electrical contact while its interior, which encloses the anode, is clad with copper metal. After assembly of either type of cell, the collector surface becomes coated with zinc as a result of the anode plating action described above. Both the electronic conductivity of the anode-to-collector interface and the suppression of gassing in the anode compartment are dependent on this process. In addition to facilitating the zinc-plating action, mercury, if present, would also fulfill this function. Other measures, such as special cleaning methods and/or activator-coating the surface, are taken to promote the natural zinc coating of the collector in mercury-free cells.

10.3.4 Separators

Special properties are required of materials used as separators in alkaline-MnO_2 cells. The material must be ionically conductive but electronically insulating; chemically stable in concentrated alkali under both oxidizing and reducing conditions; strong, flexible, and uniform; impurity-free; and rapidly absorptive. Materials fulfilling these requirements can be cast, woven, or bonded, but most frequently are nonwoven or felted in structure. Accordingly, the most commonly used materials are fibrous forms of regenerated cellulose, vinyl polymers, polyolefins, or combinations thereof. Other types such as gelled, inorganic, and radiation-grafted separators have been tried but have not gained much practical use. Cellulose film such as cellophane is also used, particularly where there is a potential for dendrite growth from the anode.

10.3.5 Containers, Seals, and Finishes

Cylindrical Cell.   The cylindrical alkaline-manganese dioxide cell differs from the Leclanché cell in that the cell container is not an active material in the cell discharge. It is merely an inert container which allows electrical contact to the energy-producing materials inside. The container is generally a can made of a mild steel. It is thick enough to provide adequate strength, without taking up excessive room. It is produced by deep drawing from steel strip stock, and must be of high quality (absence of inclusions or other imperfections).

The inside surface of the steel container makes contact to the cathode. For the cell to discharge well, this must be a very good contact. Depending on the cell construction, the contact to plain steel may or may not be adequate. Sometimes the can inner surface needs to be treated to improve the contact. In some cases the steel is nickel-plated. Alternatively, conductive coatings containing carbon may be placed on the surface. Nickel plating may also be present on the outside surface of the container, either for contact purposes or for appearance.

The seal is typically a plastic material, such as nylon or polypropylene, combined with some metal parts, including the anode collector, to make a seat assembly. It closes off the open end of the cylindrical can, preventing leakage of electrolyte from the cell, and providing electrical insulation between the cathode collector (can) and the anode collector contact.

The cylindrical alkaline cell has some additional parts, collectively referred to as finish. There usually are metal pieces at each end for positive and negative contact. These may be nickel- or tin-plated for appearance and corrosion resistance. There may be a metal jacket around the cell, with a printed label on it. In many recent designs the finish is just a thin plastic jacket or printed label. In the latter type of cell, the use of the thin plastic allows the cell container to be made slightly larger in diameter, which results in a significant increase in cell capacity.

Miniature Cell. The container, seal, and finish materials for the miniature alkaline-manganese dioxide button cell are essentially the same as those for other miniature cells. The can (container and cathode collector) is made of mild steel plated on both sides with nickel. The seat is a thin plastic gasket. The anode cup makes up the rest of the exterior of the cell. The outer surfaces of the can and anode cup are highly finished, with manufacturer identification and cell number inscribed on the can. No additional finish is needed.

10.4 CONSTRUCTION

10.4.1 Cylindrical Configuration

Figure 10.2 shows the construction of typical cylindrical alkaline-manganese dioxide batteries from two manufacturers. Figure 10.3 illustrates the process for assembling the battery. A cylindrical steel can is the container for the cell. It also serves as the cathode current collector. The cathode, a compressed mixture of manganese dioxide, carbon, and possibly other additives, is positioned inside the can in the form of a hollow cylinder in close contact with the can inner surface. The cathode can be formed by directly molding it in the can. Alternatively, rings of cathode material can be formed outside the cell and then pushed into the can. Inside the hollow center of the cathode are placed layers of separator material. Inside of that is the anode, with a metal collector contacting it, and making connection through a plastic seal to the negative terminal of the cell. The cell has top and bottom covers and a metal or plastic jacket applied. The covers serve a dual purpose. Besides providing a decorative and corrosion-resistant finish, they also provide for the proper polarity of the battery. This is necessary because the cylindrical alkaline manganese battery is used as a direct replacement for Leclanché batteries. Leclanché batteries have a flat contact on the negative (zinc can) end, and a button contact on the positive end to accommodate the carbon rod used as current collector. The cylindrical alkaline-manganese dioxide cell is built "inside-out" in relation to the Leclanché cell, with the cell container as the positive current collector and the end of the negative collector protruding from the center of the seal. Therefore to give it an external form similar to the Leclanché battery, the cylindrical alkaline battery must use a flat cover to contact the terminus of the negative collector, and a bottom cover containing the Leclanché positive protrusion in contact with the bottom of the can. (Some manufacturers mold the protrusion into the can itself, and thus do not need the bottom cover.)

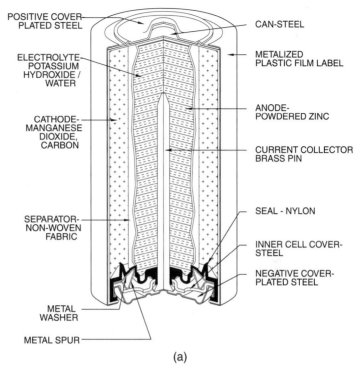

POSITIVE COVER-
PLATED STEEL

CAN-STEEL

ELECTROLYTE-
POTASSIUM
HYDROXIDE /
WATER

METALIZED
PLASTIC FILM LABEL

CATHODE-
MANGANESE
DIOXIDE,
CARBON

ANODE-
POWDERED ZINC

CURRENT COLLECTOR
BRASS PIN

SEAL - NYLON

SEPARATOR-
NON-WOVEN
FABRIC

INNER CELL COVER-
STEEL

NEGATIVE COVER-
PLATED STEEL

METAL
WASHER

METAL SPUR

(a)

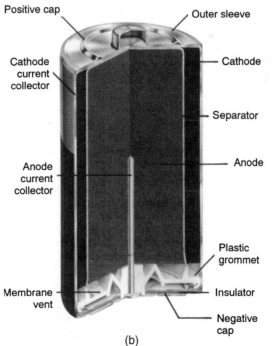

Positive cap

Outer sleeve

Cathode
current
collector

Cathode

Separator

Anode
current
collector

Anode

Plastic
grommet

Membrane
vent

Insulator

Negative
cap

(b)

FIGURE 10.2 Cross section of cylindrical alkaline-manganese dioxide batteries. [(*a*) *From Eveready Battery Engineering Data.*[3] (*b*) *From Duracell Technical Bulletin.*[4]]

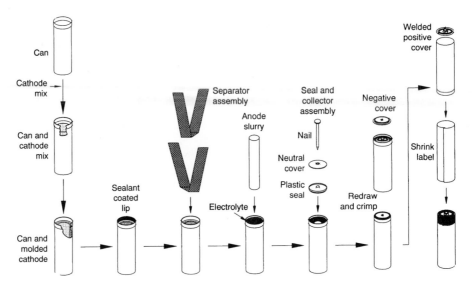

FIGURE 10.3 Assembly process for AA-size cylindrical alkaline-manganese dioxide battery. (*Courtesy of Eveready Battery Company.*)

10.4.2 Button Configuration

The construction of the miniature alkaline-manganese dioxide cell is shown in Fig. 10.4. It is essentially the same as the construction of other miniature alkaline cells. There are a bottom cup with a cathode pellet in it, an anode cup containing the anode mix, one or more round disks of separator material between them, and a plastic seal that is compressed between the bottom cup and the anode cup to prevent leakage.

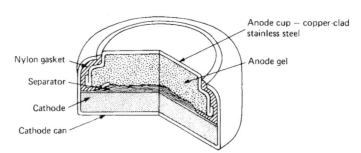

FIGURE 10.4 Cross section of miniature alkaline-manganese dioxide battery. (*From Eveready Battery Engineering Data.[5]*)

10.5 *PERFORMANCE CHARACTERISTICS*

10.5.1 General Characteristics, Comparison with Leclanché Batteries

Alkaline-manganese dioxide batteries have a relatively high theoretical capacity, considerably higher than Leclanché batteries of the same size. There are several reasons for this. The alkaline-manganese dioxide cell uses manganese dioxide of much higher purity and activity than is used in most Leclanché cells. Moreover, the alkaline-manganese dioxide cell can function with a very dense cathode, which contains only a small amount of electrolyte. Furthermore the space taken up by other cell components (separator, current collectors, and so on) is minimized.

In addition to having a high input capacity, these batteries use that capacity very efficiently. The KOH electrolyte has a very high conductivity, and the zinc anode is in the form of a high-area powder (compared to the low-area zinc can used in the Leclanché cell). Consequently the internal resistance of the battery is very low at the beginning of discharge and remains quite low up to the end of its service life (see Sec. 10.5.4).

Alkaline-manganese dioxide batteries perform better than Leclanché batteries under a wide range of conditions. Figure 10.5*a* compares the discharge curve for an alkaline-manganese dioxide battery with that for a Leclanché battery under a single light-drain continuous discharge condition. In this case both types of batteries function efficiently, delivering a significant part of their theoretical capacity. The alkaline-manganese battery outperforms the Leclanché battery simply because of its higher theoretical capacity. Figure 10.5*b* shows a similar comparison of an alkaline-manganese dioxide battery and a Leclanché battery under heavy-drain continuous discharge conditions. Again, the alkaline-manganese dioxide battery outperforms the Leclanché battery, but this time the difference is much larger. The alkaline-manganese dioxide battery still functions quite efficiently due to its superior high-rate capability, but the Leclanché battery is unable to deliver more than a fraction of its theoretical capacity. Other comparisons of these two types of cells are covered in Chap. 7.

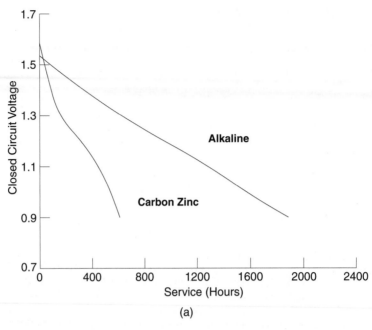

(a)

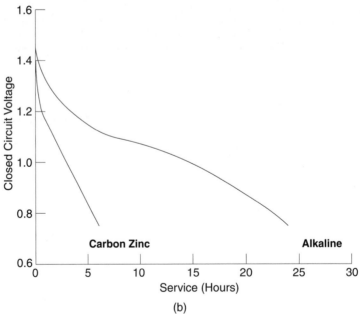

(b)

FIGURE 10.5 Performance comparison of D-size alkaline-manganese dioxide and zinc-carbon batteries. (*a*) Typical light drain test (30 mA continuous test at 20°C). (*b*) Typical heavy drain test (500 mA continuous test at 20°C). (*Courtesy of Eveready Battery Company.*)

10.5.2 Discharge Profile

The voltage profiles for the cylindrical and button alkaline-manganese dioxide batteries are similar. The voltage starts above 1.5 V and decreases gradually during discharge. This occurs because of the nature of the homogeneous-phase discharge of manganese dioxide, discussed in Sec. 10.2. Figure 10.5 illustrates the sloping discharge characteristic of the alkaline-manganese dioxide battery but that it is flatter than that of the zinc-carbon cell. While not necessarily an advantageous characteristic, most devices that use cylindrical alkaline-manganese dioxide batteries at low to moderate drains (radios, flashlights, toys, etc.) are generally tolerant of this discharge characteristic. The sloping discharge can also be advantageous because the gradual decrease in voltage gives the user advance warning of when the battery is nearing the end of its useful life.

Other types of applications and devices that use miniature alkaline batteries are often less tolerant of a sloping discharge profile. Many devices are designed for a particular type of battery, such as zinc/silver oxide. These devices may not function properly when used with miniature alkaline-manganese dioxide batteries which have a sloping discharge profile. But for those devices that will tolerate the voltage profile of the alkaline-manganese dioxide battery, it can provide an economical source of power.

Tables 10.7 and 10.8 present typical service for a particular size, the AA alkaline-manganese dioxide battery, under various resistive loads and discharge schedules. Discharge curves for the AA-size alkaline-manganese dioxide battery under continuous resistive discharge conditions at 20°C are shown in Fig. 10.6. The service to several cutoff voltages as a function of the resistive load is summarized in Fig. 10.7.

Similar curves for constant-current discharges are presented in Fig. 10.8, which shows the discharge curves at rates ranging from the $C/3$ to $C/250$. Figure 10.9 shows the hours of service delivered to various cutoff voltages over a range of discharge currents. Figure 10.10 shows the capacity service hours of the AA-size and D-size batteries as a function of the constant-current load to various cutoff voltages. The midpoint voltages for these discharges are plotted in Fig. 10.11.

TABLE 10.7 Estimated Average Service at 21°C for an AA-Size Alkaline-Manganese Dioxide Battery

Schedule	Typical drains at 1.2 V, mA	Load, Ω	Cutoff voltage, V						
			0.75	0.8	0.9	1.0	1.1	1.2	1.3
			Hours						
24 h/d	0.8	1500	3300	3200	3100	2400	2100	1800	1200
24 h/d	8	150	325	320	310	240	210	180	138
8 h/d	8	150	325	320	310	240	210	180	138
2 h/d	8	150	330	325	310	240	210	180	138
24 h/d	80	15	33	32	28	24	20	13	5
8 h/d	80	15	33	32	28	24	21	14	5
2 h/d	80	15	34	32	28	24	21	15	7
			Minutes						
24 h/d	800	1.5	120	115	86	54	25	12	3

TABLE 10.8   Estimated Average Service at 21°C for Simulated Application Tests for an AA-Size Alkaline-Manganese Dioxide Battery

	Typical drains at 1.2 V, mA	Load, Ω	Cutoff voltage, V			
Schedule			0.75	0.8	0.9	1.0
			Hours			
Radio, 4 h/d	16	75	169	167	155	125
Calculator, 1 h/d	80	15	33	29	26	21
Cassette, 1 h/d	120	10	21	19	17	14
			Minutes			
Camera, 4 min/15 min, 8 h/d, 16 h rest	250 mA constant current					368
			Minutes			
Portable light, 4 min/h, 8 h/d, 16 h rest	308	3.9	404	372	338	292
Toy, continuous	308	3.9	425	399	339	273
Compact disk/TV, 1 h/d	308	3.9	446	404	348	282
			Pulses			
Pulse 15 s/min, 24 h/d	667	1.8	584			

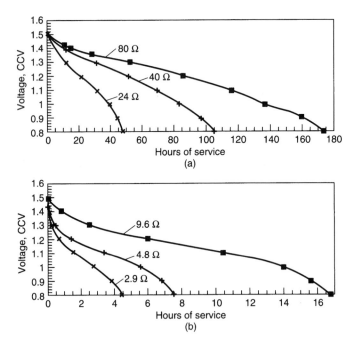

FIGURE 10.6   Typical discharge performance characteristics for AA-size alkaline-manganese dioxide battery. (*a*) continuous moderate-drain discharge at 20°C. (*b*) Continuous heavy-drain discharge at 21°C. (*Courtesy of Eveready Battery Company.*)

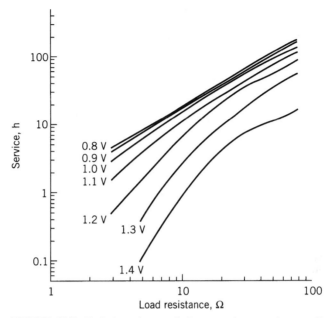

FIGURE 10.7 Typical continuous discharge service to various cutoff voltages at various loads for AA-size alkaline-manganese dioxide battery at 21°C. (*Courtesy of Eveready Battery Company.*)

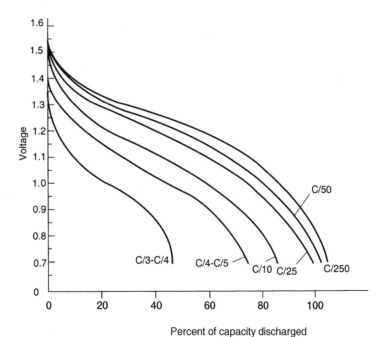

FIGURE 10.8 Typical constant-current discharge curves at 20°C for an AA-size alkaline-manganese dioxide battery; voltage vs. percent of capacity at various discharge rates. (*Courtesy of Duracell, Inc.*)

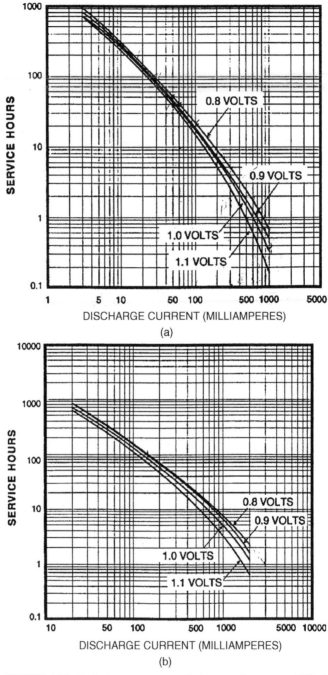

FIGURE 10.9   Typical constant-current discharge performance at 20°C of typical zinc-alkaline batteries; (*a*) AA-size battery. (*b*) D-size battery; service to various cutoff voltages vs. discharge current. (*Courtesy of Duracell, Inc.*)

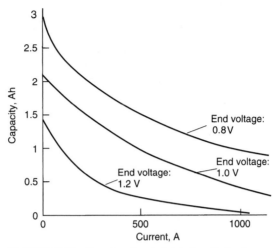

FIGURE 10.10 Capacity of typical AA-size alkaline battery on constant-current discharge at 20°C vs. current drain for various end-point voltages.

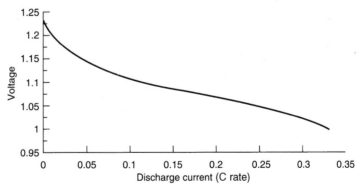

FIGURE 10.11 Typical midpoint voltages during discharge of a typical AA-size alkaline battery at various discharge currents at 20°C, 0.8-V end voltage.

10.5.3 Intermittent Discharge

The capacity of the alkaline-manganese dioxide battery is dependent on the duty cycle during the discharge. As shown in Fig. 10.12, on a light drain the battery delivers almost the same capacity on a continuous discharge as on a highly intermittent one. On heavy drains, however, the battery delivers less capacity on a continuous discharge than on an intermittent discharge as it can recover during the off-periods. In contrast to the Leclanché batteries, where there is a big difference between continuous and intermittent service, the capacity difference for the alkaline batteries is much smaller.

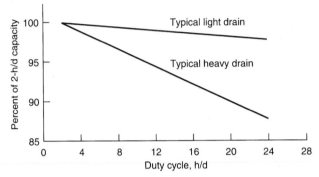

FIGURE 10.12   Comparison of typical output capacities for alkaline-manganese dioxide batteries as a function of duty cycle at light and heavy drains. Highly intermittent schedule of 2 h/d equals 100%. (*From Eveready Battery Engineering Book.*[3])

10.5.4 Internal Resistance

The alkaline-manganese dioxide batteries, because of their construction and highly conductive electrolyte, have a relatively low internal resistance. This low internal resistance is a benefit in applications involving high current pulses. As shown in Fig. 10.13 for two different-size batteries, the internal resistance increases slowly as the voltage decreases during discharge, but more rapidly towards the end of the discharge.

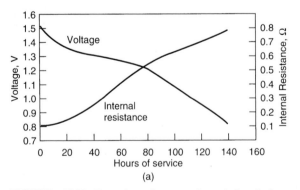

FIGURE 10.13   Internal resistance of typical alkaline-manganese dioxide batteries as a function of service life at 20°C. (*a*) AA-size battery, 62-Ω discharge.

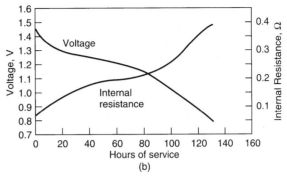

FIGURE 10.13   (*b*) C-size battery, 20-Ω discharge. (*From Duracell Technical Bulletin.*[4]) (*Continued*).

10.5.5   Type of Discharge

Figure 10.14 shows a group of discharge curves illustrating the relative performance obtained from an AA-size alkaline battery under conditions of constant resistance, constant current, and constant power. The values for resistance, current, and power were chosen so that the power provided by the battery at the end-point voltage (0.8 V) was the same in all three cases. As discussed in Sec. 3.2.3, the constant-resistance discharge runs the shortest, because the voltage and current are higher during the earlier stages of discharge. Hence the battery is providing a higher power level earlier in the discharge, with a higher average current and power level during the total discharge. The constant-power discharge runs the longest because the current is lower throughout the discharge. The constant-current discharge is intermediate between the constant-resistance and constant-power curves.

Figure 10.15 shows the power that can be obtained from a typical AA-size alkaline battery as a function of load voltage at various depths of discharge. For any given depth of discharge the maximum power is obtained when the external load resistance equals the battery internal resistance, and this occurs with an output voltage of half the open-circuit voltage. At higher output voltages the available power is less than this maximum power. The height of the curve, hence the power available at a given voltage and depth of discharge, depends largely on the internal resistance of the battery. It also depends on the open-circuit voltage of the battery. As a battery discharges, its open-circuit voltage decreases, and its internal resistance tends to increase. These effects combine to decrease the available power from the battery, as illustrated in Fig. 10.15.

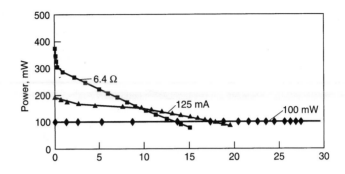

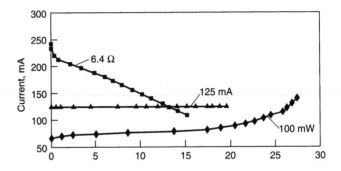

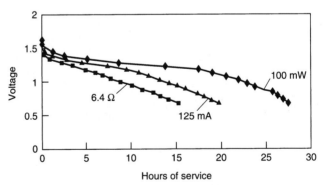

FIGURE 10.14 Comparison of constant-power, constant-current, and constant-resistance discharge curves of typical AA-size alkaline batteries. ■, Constant resistance; ◆, constant power; ▲, constant current.

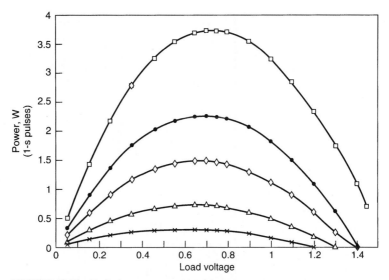

FIGURE 10.15 Typical power removed from AA-size alkaline battery as a function of load voltage, measured at various depths of battery discharge. Total capacity of battery is 2.4 Ah. □, 0 Ah (fresh); ●, 0.54 Ah; ◇, 1.04 Ah; △, 1.54 Ah; ×, 1.82 Ah. (*Courtesy of Duracell, Inc.*)

10.5.6 Effect of Operating Temperature on Discharge Performance

Batteries tend to discharge more efficiently as the operating temperature is increased, at least up to a point. Alkaline-manganese dioxide batteries can be operated at temperatures up to 55°C. As the temperature is decreased, performance decreases accordingly. The lower temperature limit is determined in part by the temperature at which the electrolyte freezes. The alkaline-manganese dioxide battery can operate at temperatures as low as −30°C. The operating range for these batteries is wider than for Leclanché batteries. Moreover there is less variation in output capacity for alkaline-manganese dioxide cells than for Leclanché batteries. Figure 10.16 shows the capacity (service hours) for AA-size and D-size alkaline-manganese dioxide batteries under constant resistance discharges at several temperatures.

A comparison of the performance of the AA-size and D-size batteries in both Figs. 10.9 and 10.16 illustrates the relatively better performance of the AA-size cell under the higher drain and lower temperature discharges. Due to the thinner cross section of the AA-size battery, its internal resistance is relatively lower and, hence, performs proportionally better under these more stringent conditions (also see Sec. 3.2.11). This is shown more clearly in Fig. 10.17 which shows the effect of discharge current on the battery's performance. Comparing the two 20°C curves illustrates the more pronounced drop in capacity for the D-size battery at the higher discharge rates. A similar plot at lower temperatures would confirm the relatively poorer performance of the D-size battery compared to the AA-size cell.

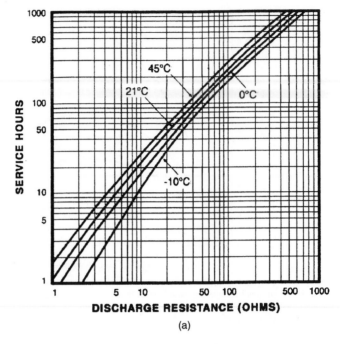

(a)

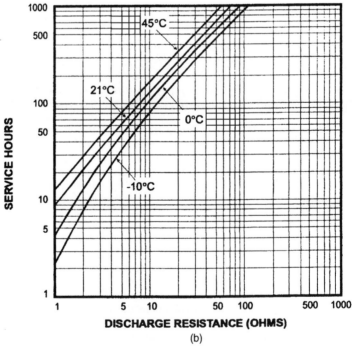

(b)

FIGURE 10.16 Constant resistance discharge performance at typical zinc-alkaline batteries at various temperatures to 0.8 volt cutoff voltage. (*a*) AA-size battery. (*b*) D-size battery (*Courtesy of Duracell, Inc.*)

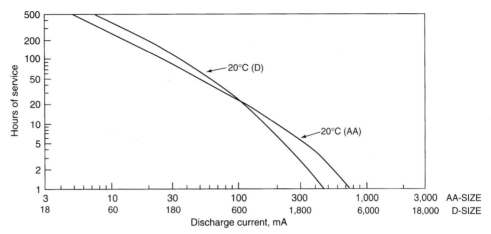

FIGURE 10.17   Effect of load on performance of typical AA-size and D-size alkaline batteries for continuous-current discharge. (End voltage 0.8 V)

10.5.7   Effect of Storage at Various Temperatures on Subsequent Discharge Performance

Undesirable chemical reactions, including self-discharge, corrosion, and degradation of battery materials, can occur during storage of a battery. These reactions will proceed more rapidly if the battery is stored at high temperatures and more slowly at low temperatures. The result of these reactions is decreased battery performance after storage. Alkaline-manganese dioxide batteries tend to maintain their capacity well during storage. In this respect, they are superior to Leclanché batteries. Figure 10.18 shows the projected percent service maintenance for typical alkaline-manganese dioxide batteries on moderate-drain discharge, after being stored for different times at various temperatures. Figure 10.19 shows the storage time needed for a 90% retention of discharge capacity (on moderate-drain discharge) as a function of storage temperature. Over a wide range of temperatures (0–50°C) the plot of the logarithm of storage time vs. storage temperature (as illustrated in Fig. 10.19) is linear. Thus the storage temperature has an exponential effect on the rate of capacity loss of the battery.

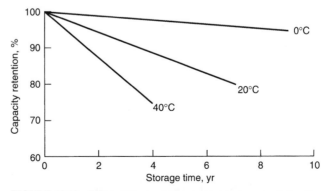

FIGURE 10.18   Effect of time and storage temperature on relative discharge performance (on moderate drain) of alkaline-manganese dioxide batteries. (*From Eveready Battery Engineering Book.*[3])

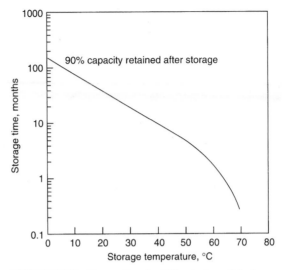

FIGURE 10.19   Storage time for 90% retention of discharge capacity (on moderate drain) vs. storage temperature for typical alkaline-manganese dioxide battery. (*From Eveready Battery Engineering Book.*[3])

While the previous two graphs indicate typical effects of storage on performance, the effects in particular cases will depend, as well, on the specifics of the discharge including the cutoff voltage and the discharge rate. Figure 10.20 illustrates this situation. The percent capacity retention is shown to depend on the discharge rate (capacity retention is higher for lower-rate discharges) as well as on the cutoff voltage (usually the capacity retention is better as the voltage is lower, but not in all cases). In general the more stringent the discharge condition, the greater the loss of capacity.

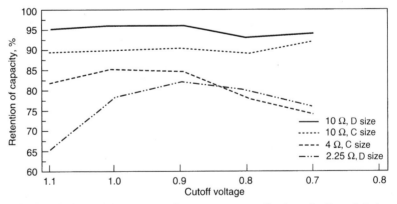

FIGURE 10.20   Capacity retention after storage vs. cutoff voltage for D- and C-size alkaline batteries for different resistive loads.

10.6 BATTERY TYPES AND SIZES

Alkaline-manganese dioxide primary cells and batteries are available in a variety of sizes, in both cylindrical and miniature (button-cell) configurations, as listed in Table 10.9. Some of the unit cell sizes listed are not available as single cells, but are used as components in multiple-cell batteries. Figure 10.21 shows the nominal capacity of various-size batteries as a function of weight and cell volume.

The cylindrical batteries are generally known by the nomenclature of D, C, AA, and so on. In addition, battery manufacturers often have their own identification codes for these batteries. For miniature batteries, manufacturers tend to have their own codes. For both types of batteries there are also nomenclature codes that have been established by different standards agencies, such as the IEC and ANSI.

Multiple-cell batteries (Table 10.10) are made using either miniature or cylindrical cells. In addition, some special multicell batteries are made using flat cells of a type not used in single cell batteries.

TABLE 10.9 Characteristics of 1.5-V Standard Alkaline-Manganese Dioxide Batteries*

Size	IEC	ANSI	Capacity,† Ah	Nominal dimensions			
				Diameter, mm	Height, mm	Weight, g	Volume, cm³
			Cylindrical types				
AAAA	LR61‡	25A	0.56	8	42	65	2.2
N	LR1	910A	0.8	12	29	9	3.3
	LR50		0.56	16	16	11	3.6
AAA	LR03	24A	1.1–1.25	10	44	11	3.8
AA	LR6	15A	2.5–2.85	14	50	23	7.5
C	LR14	14A	7.1–8.4	26	50	66	26
D	LR20	13A	14.3–18	34	61	138	54.4
F	LR25		22	33	91	200	80
			Button types				
	LR41		0.035	7.9	3.6	0.6	0.2
	LR43	1167A	0.100	11.6	4.2	1.4	0.3
	LR44	1166A	0.145	11.6	5.4	2.3	0.5
	LR48		0.060	7.9	5.4	0.9	0.3
	LR53	1129AP	0.160	23.0	5.9	6.8	2.3
	LR54	1168A	0.072	11.6	3.1	1.1	0.3
	LR55	1169A	0.040	11.6	2.1	0.9	0.2

*Values are typical figures based on data from different manufacturers.
†Capacity values based on typical output on rating drain; higher capacity batteries are marketed by some manufacturers.
‡Proposed IEC Nomenclature: LR8D425

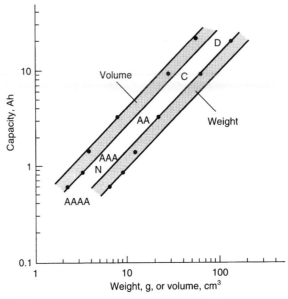

FIGURE 10.21 Capacity vs. weight and volume for variety of sizes of alkaline-manganese dioxide batteries. (*Data from Table 10.9; courtesy of Eveready Battery Company.*)

TABLE 10.10 Characteristics of Multiple-Cell Standard Alkaline-Manganese Dioxide Batteries

Voltage, V	IEC	ANSI	Dimensions, mm	Weight, g	Volume, cm³	Capacity, Ah
3	2LR53	1202AP	24d, 12h	14	5.4	0.16
3	2LR50	1308AP	17d, 42.5h	22	7.4	0.56
4.5		1313AP	41, 17, 11.5	12.2	7.4	0.17
4.5	3LR50	1306AP	17d, 50h	31	11	0.55
4.5	3LR50	1307AP*	17d, 58h	32.5	11.5	0.55
6	4LR44	1414A	13d, 25h	11	3.3	0.15
6	4LR61	1412AP	48, 35.6, 9	32	15	0.57
6	4LR25Y	915A†	110, 67, 67	885	434	22
6	4LR25X	908A‡‖	105, 67, 67	885	434	22
				612	434	13
6		918A‖	137, 125, 73	1900	1123	44
				1270	1123	27
6	4LR20-2	930A	140, 118, 67	1120	883	22
9	6LR61	1604A	49, 26, 17.5	46	21	0.58
12			10.3d, 28.5h	7.4	2.37	0.034

* Similar to battery above, but has snap contacts.
† Screw contacts.
‡ Spring contacts.
‖ Commercially produced batteries using either D- or F-size unit cells. The larger values for capacity and weight are for the batteries containing the larger unit cells.

10.7 PREMIUM ZINC/ALKALINE/MANGANESE DIOXIDE HIGH-RATE BATTERIES

Recently introduced in 1999, these specially designed alkaline-manganese dioxide batteries are capable of better performance at higher discharge rates than the standard batteries. They are marketed to meet the more stringent high power requirements of portable electronic devices (i.e., digital cameras, cell phones, PDAs, etc.) The performance characteristics of the premium AA-size batteries are summarized in Figs. 10.22(*a*) to (*e*) and 10.23(*a*) to (*c*).

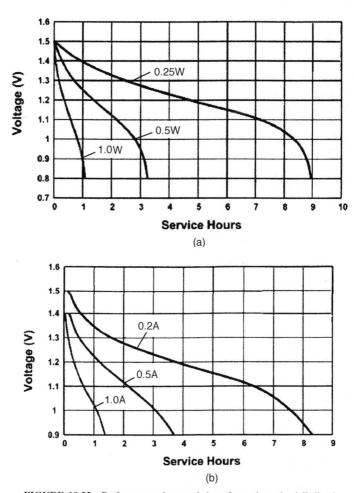

FIGURE 10.22 Performance characteristics of premium zinc/alkaline/manganese dioxide primary batteries AA-size at 20°C. (*a*) Discharge characteristics at various power drains. (*b*) Discharge characteristics at various current drains.

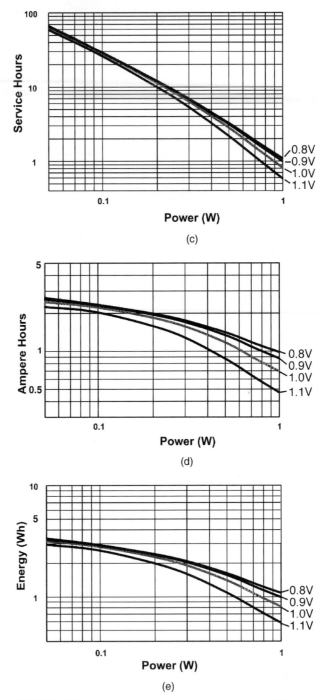

FIGURE 10.22 (*c*) Service hours vs. power drain to specified end voltage. (*d*) Ampere-hours vs. power drain to specified end voltage. (*e*) Energy (Wh) vs. power drain to specified end voltage. (*Courtesy of Duracell, Inc.*) (*Continued*).

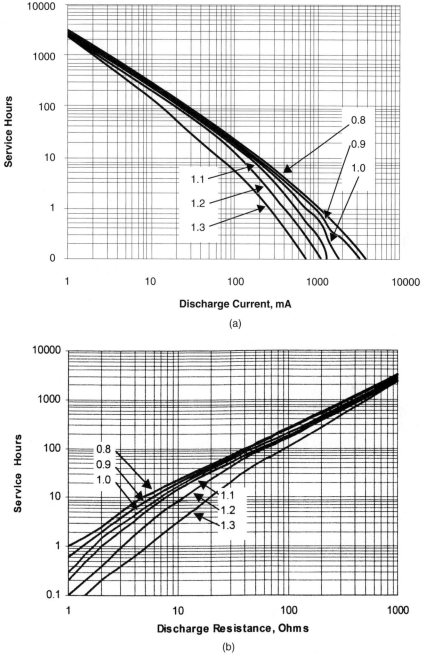

FIGURE 10.23 Performance characteristics of premium zinc/alkaline/manganese dioxide primary batteries AA-size (*a*) Discharge characteristics at various current drains to specified end voltages. (*b*) Discharge characteristics at various resistance drains to specified end voltages.

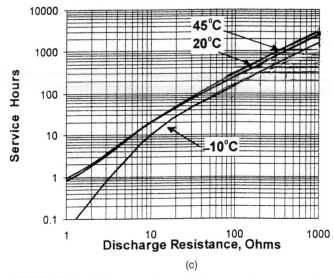

(c)

FIGURE 10.23 (*c*) Constant resistance discharge performance at various temperatures at 0.8 volts. (*Courtesy of Eveready Battery Company.*) (*Continued*).

REFERENCES

1. A. Kozawa and R. A. Powers, *J. Chem. Educ.* **49**:587 (1972).

2. D. M. Holton, W. C. Maskell, and F. L. Tye, in *Proc. 14th International Power Sources Symp.* (Brighton), Pergamon, New York, 1984.

3. *Eveready Battery Engineering Data,* vol. 2a, Eveready Battery Co., St. Louis, Mo.

4. *Duracell Tech. Bull., Alkaline Cells,* Duracell, Inc., Bethel, Conn.

5. *Eveready Battery Engineering Data,* vol. la, Eveready Battery Co., St. Louis, Mo.

CHAPTER 11
MERCURIC OXIDE BATTERIES

Denis Naylor*

11.1 GENERAL CHARACTERISTICS

The alkaline zinc/mercuric oxide battery is noted for its high capacity per unit volume, constant voltage output, and good storage characteristics. The system has been known for over a century, but it was not until World War II that a practical battery was developed by Samuel Ruben in response to a requirement for a battery with a high capacity-to-volume ratio which would withstand storage under tropical conditions.[1,2]

Since that time, the zinc/mercuric oxide battery has been used in many applications where stable voltage, long storage life or high energy-to-volume ratios were required. The characteristics of this battery system were particularly advantageous in applications such as hearing aids, watches, cameras, some early pacemakers and small electronic equipment where it was widely used. The battery has also been used as a voltage reference source and in electrical instruments and electronic equipment, such as sonobuoys, emergency beacons, rescue transceivers, radio and surveillance sets, small scatterable mines and early satellites. These applications, however, did not become widespread, except for military and special uses, because of the relatively higher cost of the mercuric oxide system.

The use of cadmium in place of zinc results in a very stable battery with excellent storage and performance at extreme temperatures due to the low solubility of cadmium in caustic alkali over a wide range of temperatures. However, the cost of the material is high and the cell voltage is low, less than 1.0 V. Hence, the cadmium/mercuric batteries were used, but to a lesser degree, in special applications requiring the particular performance capabilities of the system. These include gas and oil well logging, telemetry from engines and other heat sources, alarm systems, and for operation of remote equipment such as data-monitoring, surveillance buoys, weather stations and emergency equipment.[3]

During the last several years, the market for mercuric oxide batteries has almost completely evaporated, due mainly to environmental problems associated with mercury and cadmium and few are manufactured. They have been removed from the International Electrotechnical Commission (IEC) and the American National Standards Institute (ANSI) standards. In applications, they have been replaced by alkaline-manganese dioxide, zinc/air, silver oxide and lithium batteries.

The major characteristics of these two battery systems are summarized in Table 11.1.

---

*The chapter on Mercuric Oxide Batteries was written by Denis Naylor, now deceased, for the 1st and 2nd Editions. His work was updated and modified for the 3rd Edition by David Linden.

TABLE 11.1   Characteristics of the Zinc/Mercuric Oxide and Cadmium/Mercuric Oxide Batteries

Advantages	Disadvantages
Zinc/mercuric oxide battery	
High energy-to-volume ratio, 450 Wh/L	Batteries were expensive; although widely used in miniature sizes, but only for special applications in the larger sizes
Long shelf life under adverse storage conditions	
Over a wide range of current drains recuperative periods are not necessary to obtain a high capacity from the battery	After long periods of storage, cell electrolyte tends to seep out of seal which is evidenced by white carbonate deposit at seal insulation
High electrochemical efficiency	Moderate energy-to-weight ratio
High resistance to impact, acceleration, and vibration	Poor low-temperature performance
Very stable open-circuit voltage, 1.35 V	Disposal of quantities of spent batteries creates environmental problems
Flat discharge curve over wide range of current drains	
Cadmium/mercuric oxide battery	
Long shelf life under adverse storage conditions	Batteries are more expensive than zinc/mercuric oxide batteries due to high cost of cadmium
Flat discharge curve over wide range of current drains	System has low output voltage (open-circuit voltage = 0.90 V)
Ability to operate efficiently over wide temperature range, even at extreme high and low temperatures	Moderate energy-to-volume ratio
Can be hermetically sealed because of inherently low gas evolution level	Low energy-to-weight ratio
	Disposal of spent batteries creates environmental problem, with both cadmium and mercury being toxic

11.2   CHEMISTRY

It is generally accepted that the basic cell reaction for the zinc/mercuric oxide cell is

$$Zn + HgO \rightarrow ZnO + Hg$$

For the overall reaction, $\Delta G^0 = 259.7$ kJ. This gives a thermodynamic value for E^0 at 25°C of 1.35 V, which is in good agreement with the observed values of 1.34 to 1.36 V for the open-circuit voltage of commercial cells.[4] From the basic reaction equation it can be calculated that 1 g of zinc provides 819 mAh and 1 g of mercuric oxide provides 247 mAh.

Some types of zinc/mercuric oxide cells exhibit open-circuit voltages between 1.40 and 1.55 V. These cells contain a small percentage of manganese dioxide in the cathode and are used where voltage stability is not of major importance for the application.

The basic cell reaction for the cadmium/mercuric oxide cell is

$$Cd + HgO + H_2O \rightarrow Cd(OH)_2 + Hg$$

For the overall reaction, $\Delta G^0 = -174.8$ kJ. This gives a thermodynamic value for E^0 at 25°C of 0.91 V, which is in good agreement with the observed values of 0.89 to 0.93 V. From the basic reaction it can be calculated that 1 g of cadmium should provide 477 mAh.

11.3 CELL COMPONENTS

11.3.1 Electrolyte

Two types of alkaline electrolyte are used in the zinc/mercuric oxide cell, one based on potassium hydroxide and one on sodium hydroxide. Both of these bases are very soluble in water and highly concentrated solutions are used; zinc oxide is also dissolved in varying amounts in the solution to suppress hydrogen generation.

Potassium hydroxide electrolytes generally contain between 30 and 45% w/w KOH and up to 7% w/w zinc oxide. They are more widely used than the sodium hydroxide electrolytes because of their greater operating temperature range and ability to support heavier current drains. For low temperature operation, both the potassium hydroxide and the zinc oxide contents are reduced, and this introduces some instability at higher temperatures with respect to hydrogen generation in the cell.

Sodium hydroxide electrolytes are prepared in similar concentration ranges and are used in cells where low temperature operations or high current drains are not required. These electrolytes are suitable for long-term discharge applications because of the reduced tendency of the electrolyte to seep out of the cell seal after long periods of storage.

Generally only potassium-based alkaline electrolytes are used in the cadmium/mercuric oxide cell. As cadmium is practically insoluble in all concentrations of aqueous potassium hydroxide solutions, the electrolyte can be optimized for low-temperature operation.

The freezing-point curve for caustic potash solutions is shown in Fig. 11.1. It shows that the eutectic with a freezing point below −60°C is 31% w/w KOH, which is the electrolyte most frequently used. Improvements in low-temperature performance have been made in some cases by the addition of a small percentage of cesium hydroxide to the electrolyte.

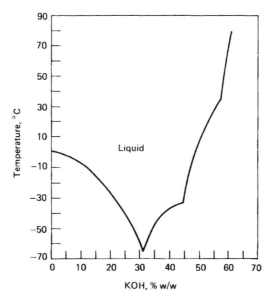

FIGURE 11.1 Freezing-point curve for aqueous caustic potash solutions.

11.3.2 Zinc Anode

Alkaline electrolytes act as ionic carriers in the cell reactions. The reaction at the zinc negative electrode may be written

$$Zn + 4OH^- \rightarrow Zn(OH)_4^{2-} + 2e$$

$$Zn(OH)_4^{2-} \rightarrow ZnO + 2OH^- + H_2O$$

These reactions imply the dissolution of the zinc electrode, with the crystallization of zinc oxide from the electrolyte. The reaction at the anode can be simplified to

$$Zn + 2OH^- \rightarrow ZnO + H_2O + 2e$$

Direct solution of the zinc electrode in the alkaline solution on open circuit is minimized by dissolving zinc oxide in the electrolyte and amalgamating the zinc in the electrode. Mercury levels used in zinc electrodes are usually in the range of 5 to 15% w/w. Great attention is also paid to the impurity levels in the zinc since minor cathodic inclusions in the electrode can drive the hydrogen generation reaction despite the precautions indicated.[5,6]

11.3.3 Cadmium Anode

The reaction at the anode is

$$Cd + 2OH^- \rightarrow Cd(OH)_2 + 2e$$

This implies the removal of water from the electrolyte during discharge, necessitating an adequate quantity of electrolyte in the cell and the desirability of a high percentage of water in the electrolyte. Cadmium has a high hydrogen overvoltage in the electrolyte, and so amalgamation is neither necessary nor desirable, since the electrode potential is some 400 mV less electropositive than zinc.

Cadmium metal powders as produced conventionally are unsuitable for use as electrode materials. Activated cadmium anodes are produced by (1) electroforming the anode, (2) electroforming powder by a special process followed by pelleting, or (3) precipitating by a special process as a low-nickel alloy and pelleting. All of these processes have been used by different manufacturers to give cells with various performance parameters.[7]

11.3.4 Mercuric Oxide Cathode

At the cathode, the overall reaction may be written

$$HgO + H_2O + 2e \rightarrow Hg + 2(OH)^-$$

Mercuric oxide is stable in alkaline electrolytes and has a very low solubility. It is also a nonconductor, and adding graphite is necessary to provide a conductive matrix. As the discharge proceeds, the ohmic resistance of the cathode falls and the graphite assists in the prevention of mass agglomeration of mercury droplets. Other additives which have been used to prevent agglomeration of the mercury are manganese dioxide, which increases the cell voltage to 1.4–1.55 V, lower manganese oxides, and silver powder, which forms a solid-phase amalgam with the cathode product.

Graphite levels usually range from 3 to 10% and manganese dioxide from 2 to 30%. Silver powder is used only in special-purpose cells because of cost considerations, but may be up to 20% of the cathode weight. Again, great care is taken to obtain high-purity materials for use in the cathode. Trace impurities soluble in the electrolyte are liable to migrate to the anode and initiate hydrogen evolution. An excess of mercuric oxide capacity of 5 to 10% is usually maintained in the cathode to "balance" the cell and prevent hydrogen generation in the cathode at the end of discharge.

11.3.5 Materials of Construction

Materials of construction for the zinc/mercuric oxide cells are limited not only by their ability to survive continuous contact with strong caustic alkali, but also by their electrochemical compatibility with the electrode materials. As far as the external contacts are concerned, these are decided by corrosion resistance, compatibility with the equipment interface with respect to galvanic corrosion, and, to some degree, cosmetic appearance. Metal parts may be homogeneous, plated metal, or clad metal. Insulating parts may be injection-, compression-, or transfer-molded polymers or rubbers.

With the exception of the anode contact (where slight modification of the top/anode interface is necessary), materials for the cadmium/mercuric oxide cell are generally the same as for the zinc/mercuric oxide cell. However, because of the wide range of storage and operating conditions of most applications, cellulose and its derivatives are not used, and low-melting-point polymers are also avoided. Nickel is usually used on the anode side of the cell and also, conveniently, at the cathode.

11.4 CONSTRUCTION

The mercuric oxide batteries were manufactured in three basic structures—button, flat, and cylindrical configurations. There are several design variations within each configuration.

11.4.1 Button Configuration

The button configuration of the zinc/mercuric oxide battery is shown in Fig. 11.2. The top is copper or copper alloy on the inner face and nickel or stainless steel on the outer face. This part may also be gold plated, depending on the application. Within the top is a dispersed

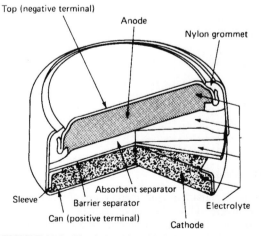

FIGURE 11.2 Zinc/mercuric oxide battery—button configuration. (*Courtesy of Duracell, Inc.*)

mass of amalgamated zinc powder ("gelled anode"), and the top is insulated from the can by a nylon grommet. The whole top-grommet-anode assembly presses down onto an absorbent which contains most of the electrolyte, the remainder being dispersed in the anode and cathode. Below the absorbent is a permeable barrier, which prevents any cathode material from migrating to the anode. The cathode of mercuric oxide and graphite is consolidated into the can, and a sleeve support of nickel-plated steel prevents collapse of the cathode mass as the battery discharges. The can is made of nickel-plated steel, and the whole cell is tightly held together by crimping the top edge of the can as shown.

The cadmium/mercuric oxide button battery uses a similar configuration.

11.4.2 Flat-Pellet Configuration

A form of a larger-sized zinc/mercuric oxide battery is shown in Fig. 11.3. In these cells the zinc powder is amalgamated and pressed into a pellet with sufficient porosity to allow electrolyte impregnation. A double top is used, with an integrally molded polymer grommet, as a safeguard to relieve excessive gas pressures and maintain a leak-resistant structure. The outer top is of nickel-plated steel, and the inner top is nickel-plated steel but tin plated on its inner face. This cell also uses two nickel-plated steel cans with an adaptor tube between the two, the seal being effected by pressing the top-grommet assembly against the inner can and crimping over the outer can. A vent hole is pierced into the outer can so that if gas is generated within the cell, it can escape between the inner and outer cans, any entrained electrolyte being absorbed by the paper adaptor tube.

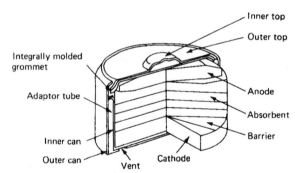

FIGURE 11.3 Zinc/mercuric oxide battery—flat-pellet configuration. (*Courtesy of Duracell, Inc.*)

11.4.3 Cylindrical Configuration

The larger cylindrical zinc/mercuric oxide battery is constructed from annular pressings, as shown in Fig. 11.4. The anode pellets are rigid and pressed against the cell top by the neoprene insulator slug. A number of variations of the cylindrical cell were used with dispersed anodes, where contact with the anode is made either by a nail welded to the inner top or a spring extending from the base insulator to the top.

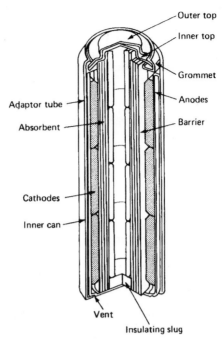

FIGURE 11.4   Zinc/mercuric oxide battery cylindrical configuration. (*Courtesy of Duracell, Inc.*)

11.4.4   Wound-Anode Configuration

Another design of the zinc/mercuric oxide battery which operates better at low temperatures is the wound-anode or jelly-roll structure shown in Fig. 11.5. Structurally the cell is similar to the flat cell shown in Fig. 11.3, but the anode and absorbent have been replaced by a wound anode which consists of a long strip of corrugated zinc interleaved with a strip of absorbent paper. The paper edge protrudes at one side and the zinc strip at the other. This provides a large surface area anode. The roll is held in a plastic sleeve and the zinc is amalgamated in situ. The paper swells in the electrolyte and forms a tight structure, which is compressed in the cell at the assembly stage with the zinc edge in contact with the top.

Electrolyte formulations can be adjusted for low-temperature operation, long storage life at elevated temperature, or a compromise between the two. The performance is optimized by careful adjustment of the anode geometry.

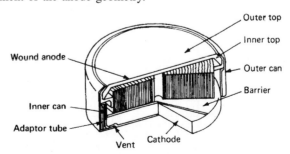

FIGURE 11.5   Zinc/mercuric oxide battery—wound-anode configuration. (*Courtesy of Duracell, Inc.*)

11.4.5 Low-Current-Drain Structures

Batteries designed for operation at low current drain require modification of the structure to prevent internal electrical discharge paths from forming from the conductive materials in both anode and cathode. After partial discharge, metallic mercury globules are particularly troublesome in this respect. The problem can be minimized by the use of silver powder in the cathode.

All available passages through which material could form an electrical track need to be blocked if long-term discharge is to be realized. A typical button battery for watch applications used multiple barrier layers and a polymer insulator washer effectively sealing off the anode from the cathode by compressing these layers against the support ring. These batteries are discharged at the 1 to 2-year rate.[8]

11.5 PERFORMANCE CHARACTERISTICS OF ZINC/MERCURIC OXIDE BATTERIES

11.5.1 Voltage

The open-circuit voltage of the zinc/mercuric oxide battery is 1.35 V. Its voltage stability under open-circuit or no-load conditions is excellent, and these batteries have been widely used for voltage reference purposes. The no-load voltage is nonlinear with respect to both time and temperature. A voltage-time curve is shown in Fig. 11.6. The no-load voltage will remain within 1% of its initial value for several years. A voltage-temperature curve is shown in Fig. 11.7. Temperature stability is even better than age stability. From -20 to $+50°C$ the total no-load voltage range is in the region of 2.5 mV.

Batteries containing manganese dioxide as a cathode additive to the mercuric oxide do not show the no-load voltage stability illustrated in Figs. 11.6 and 11.7.

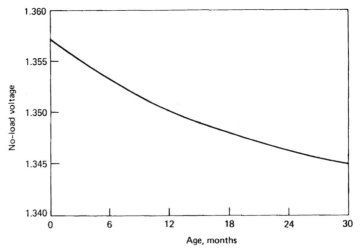

FIGURE 11.6   No-load voltage vs. time, zinc/mercuric oxide battery, 20°C.

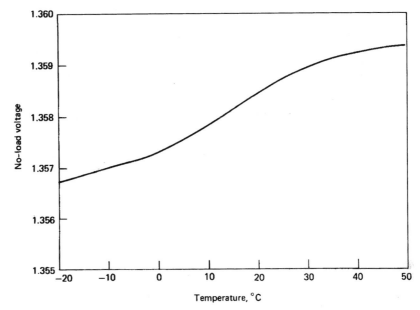

FIGURE 11.7   No-load voltage vs. temperature, zinc/mercuric oxide battery.

11.5.2 Discharge Performance

A flat discharge curve, characteristic of the zinc/mercuric oxide battery, is shown in Fig. 11.8 for a pressed-powder anode battery at 20°C. The end-point voltage is generally considered to be 0.9 V, although at higher current drains the batteries may discharge usefully below this voltage. At low current drains the discharge profile is very flat and the curve is almost "squared off."

The capacity or service of the zinc/mercuric oxide battery is about the same on either continuous or intermittent discharge regimes over the recommended current drain range, irrespective of the duty cycle.

Under overload conditions, however, a considerable shift in available capacity can be realized by the use of "rest" periods, which may increase service life considerably.

Problems are not encountered at low rates of discharge with batteries designed for the purpose unless a high-current-drain pulse is superimposed on a continuous low-drain base current; special designs are necessary to cope with this situation.

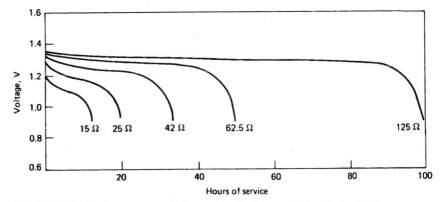

FIGURE 11.8   Discharge curves, zinc/mercuric oxide battery, 1000 mAh size, 20°C.

11.5.3 Effect of Temperature

The zinc/mercuric oxide battery is best suited for use at normal and elevated temperatures from 15 to 45°C. Discharging batteries at temperatures up to 70°C is also possible if the discharge period is relatively short. The zinc/mercuric oxide battery generally does not perform well at low temperatures. Below 0°C, discharge efficiency is poor unless the current drain is low. Figure 11.9 shows the effect of temperature on the performance of two types of zinc/mercuric oxide batteries at nominal discharge drains.

The wound-anode or "dispersed"-powder anode structures are better suited to high rates and low temperatures than the pressed-powder anode.

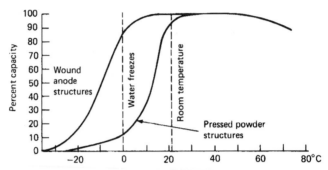

FIGURE 11.9   Effect of temperature on performance of zinc/mercuric oxide batteries.

11.5.4 Impedance

Impedance for the zinc/mercuric oxide button batteries was usually measured at a frequency of 1 kHz because of their use in hearing-aids.[9]

The impedance curve is almost a mirror image of the voltage discharge curve, rising very steeply at the end of the useful discharge life, as illustrated in Fig. 11.10. The value obtained is frequency-dependent to some degree, particularly above 1 MHz, and a fixed frequency has to be specified. A frequency versus impedance curve under no-load conditions is shown in Fig. 11.11.

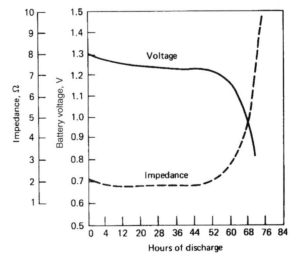

FIGURE 11.10   Internal impedance, zinc/mercuric oxide battery, 350 mAh size, 20°C, 1 kHz, 250-Ω load.

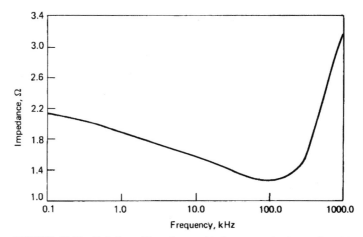

FIGURE 11.11 Variation of impedance with frequency, zinc/mercuric oxide battery, 210 mAh size, 20°C.

11.5.5 Storage

Zinc/mercuric oxide batteries have good storage characteristics. In general they will store for over 2 years at 20°C with a capacity loss of 10 to 20% and 1 year at 45°C with about a 20% loss. Storage at lower temperatures, such as down to −20°C, will, as with other battery systems, increase storage life.

The storability will depend on the discharge load and also on the cell structure. Failure in storage is usually due to the breakdown of cellulosic compounds within the cell which, at first, results in a reduction of the limiting-current density at the anode. Further breakdown produces low-drain internal electrical paths and a real loss of capacity due to self-discharge. Eventually, complete self-discharge can occur, but at 20°C and below these processes take many years.

Long storage lives are within the capabilities of the mercuric oxide system. For example, a wound-anode cell with a noncellulosic barrier has a capacity loss in the region of only 15% over 6 years. With cells designed for long-term storage, dissolution of mercuric oxide from the cathode and its transfer to the anode become a significant factor in capacity loss.

11.5.6 Service Life

The performance of the zinc/mercuric oxide cell at various temperatures and loads is summarized in Figs. 11.12 and 11.13 on a weight and volume basis. These data, based on the performance of a 800 mAh battery with a dispersed anode, can be used to approximate the performance of a zinc/mercuric oxide battery.

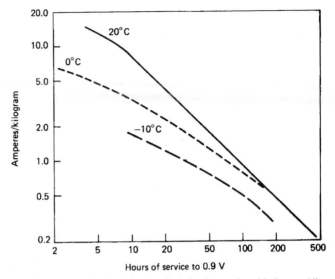

FIGURE 11.12 Service life of typical zinc/mercuric oxide battery (dispersed anode) on a weight basis.

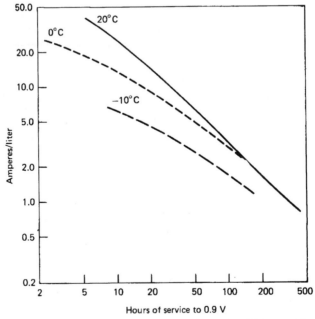

FIGURE 11.13 Service life of typical zinc/mercuric oxide battery (dispersed anode) on a volume basis.

11.6 PERFORMANCE CHARACTERISTICS OF CADMIUM/ MERCURIC OXIDE BATTERIES

11.6.1 Discharge

An outstanding feature of the cadmium/mercuric oxide battery is its ability to operate over a wide temperature range. The usual operating range is from -55 to $+80°C$, but with the low gassing rate and thermal stability of the cell, operating temperatures to 180°C have been achieved with special designs.

Figure 11.14 shows the discharge curves for a typical button battery at various temperatures. Excellent voltage stability and flat discharge curves are characteristic of these but at a low operating voltage (open-circuit voltage is only 0.9 V). Figure 11.15 shows the effect of temperature on the capacity at various discharge loads. A high percentage of the 20°C capacity is available at the lower temperatures. The end-point voltage is usually taken as 0.6 V, although at higher current densities and lower temperatures more useful life can be obtained to lower end voltages.

The performance of the cadmium/mercuric oxide battery is summarized in Figs. 11.16 and 11.17 on a weight and volume basis, respectively. The data were derived from the performance of typical button batteries.

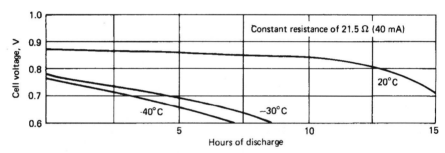

FIGURE 11.14   Discharge curves—cadmium/mercuric oxide button battery (500-mAh size).

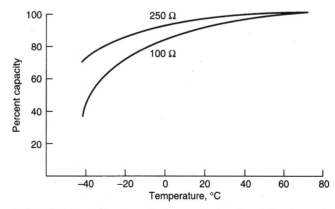

FIGURE 11.15   Effect of temperature on capacity of a cadmium/mercuric oxide button battery (3000-mAh size).

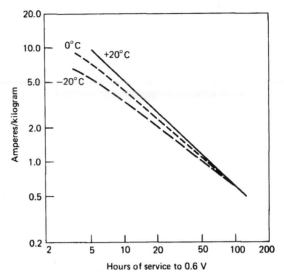

FIGURE 11.16 Service life of typical cadmium/mercuric oxide batteries on a weight basis.

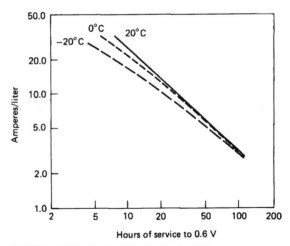

FIGURE 11.17 Service life of typical cadmium/mercuric oxide, batteries on a volume basis.

11.6.2 Storage

Storage life over the temperature range of −55 to +80°C is remarkably good, and if the barrier-absorbent system is designed to withstand elevated-temperature storage, the major self-discharge mechanism is by dissolution of the mercuric oxide and its transfer to the anode. A shelf life of 10 years at ambient temperatures with less than 20% capacity loss is within the capabilities of the system. Elevated-temperature storage is exceptionally good (approximately 15% loss per year at 80°C), and since neither electrode should generate hydrogen, the cells can be hermetically sealed with minimal risk of electrolyte leakage or cell distortion.[8]

REFERENCES

1. C. L. Clarke, U.S. Patent 298,175 (1884).

2. S. Ruben, "Balanced Alkaline Dry Cells," *Proc. Electrochem. Soc. Gen. Meeting,* Boston, Oct. 1947.

3. B. Berguss, "Cadmium-Mercuric Oxide Alkaline Cell," *Proc. Electrochem. Soc. Meeting,* Chicago, Oct. 1965.

4. P. Ruetschi, "The Electrochemical Reactions in Mercuric Oxide-Zinc Cell," in D. H. Collins (ed.), *Power Sources,* vol. 4, Oriel Press, Newcastle-upon-Tyne, England, 1973, p. 381.

5. D. P. Gregory, P. C. Jones, and D. P. Redfearn, "The Corrosion of Zinc Anodes in Aqueous Alkaline Electrolytes," *J. Electrochem. Soc.* **119:**1288 (1972).

6. T. P. Dirkse, "Passivation Studies on the Zinc Electrode," in D. H. Collins (ed.), *Power Sources,* vol. 3, Oriel Press, Newcastle-upon-Tyne, England, 1971, p. 485.

7. D. Weiss and G. Pearlman, "Characteristics of Prismatic and Button Mercuric Oxide-Cadmium Cells," *Proc. Electrochem. Soc. Meeting,* New York, Oct. 1974.

8. P. Ruetschi, "Longest Life Alkaline Primary Cells," in J. Thompson (ed.), *Power Sources,* vol. 7, Academic, London, 1979, p. 533.

9. S. A. G. Karunathilaka, N. A. Hampson, T. P. Haas, R. Leek, and T. J. Sinclair. "The Impedance of the Alkaline Zinc-Mercuric Oxide Cell. I. Behaviour and Interpretation of Impedance Spectra." *J. Appl. Electrochem.* **11** (1981).

CHAPTER 12
SILVER OXIDE BATTERIES

Sid A. Megahed, Joseph Passaniti and John C. Springstead

12.1 GENERAL CHARACTERISTICS

The energy density of the zinc/silver oxide system (zinc/alkaline electrolyte/silver oxide) is among the highest of all battery systems making it ideal for use in small, thin "button" batteries. The monovalent silver oxide battery will discharge at a flat, constant voltage at both high- and low-current rates. The battery has long storage life, retaining more than 95% of its initial capacity after 1 year of room temperature. It also has good low temperature discharge capabilities, delivering about 70% of its nominal capacity at 0°C and 35% at −20°C. These features have enabled the zinc/silver oxide battery to be an important micro-power source for electronic devices and equipment, such as watches, calculators, hearing aids, glucometers, cameras, and other applications that require small, thin, high-capacity, long-service-life batteries that discharge at a constant voltage. The use of this battery system in larger sizes is limited by the high cost of silver.

The primary zinc/silver oxide batteries are manufactured mainly in the button cell configuration in a wide range of sizes from 5 to 250 mAh. Most of these batteries are now prepared from the monovalent silver oxide (Ag_2O).

The divalent silver oxide (AgO) has the advantage of a higher theoretical capacity and the capability of delivering about a 40% higher capacity in the same battery size than the monovalent silver oxide, but it has the disadvantages of dual voltage discharge curves and greater instability in alkaline solutions. The gain of more service hours for the same weight of silver, a cost advantage over monovalent silver oxide, brought the divalent oxide into limited commercial use about a decade or so ago. They were marketed as "Ditronic" or "Plumbate" batteries using heavy metals to stabilize the reactivity of the divalent oxide. However, early in the 1990's, these batteries were withdrawn from the market because of the environmental restrictions that were imposed on the use of these metals. The information on the use of the divalent silver oxide has been included in this chapter, nevertheless, for reference and possible future applications of silver oxide batteries.

The major advantages and disadvantages of the zinc/monovalent silver oxide battery are summarized in Table 12.1.

TABLE 12.1   Major Advantages and Disadvantages of Zinc/Silver Oxide Primary Batteries

Advantages	Disadvantages
High energy density	Use limited to button and miniature cells because of high cost
Good voltage regulation, high rate capability	
Flat discharge curve can be used as a reference voltage	
Comparatively good low-temperature performance	
Leakage and salting negligible	
Good shock and vibration resistance	
Good shelf life	

12.2 BATTERY CHEMISTRY AND COMPONENTS

The zinc/silver oxide cell consists of three active components: a powdered zinc metal anode, a cathode of compressed silver oxide, and an aqueous electrolyte solution of potassium or sodium hydroxide with dissolved zincates. The active components are contained in an anode top, cathode can, separated by a barrier and sealed with a gasket.

The overall electrochemical reaction of the zinc/monovalent silver oxide cell is

$$Zn + Ag_2O \rightarrow 2Ag + ZnO \qquad (1.59 \text{ V})$$

The zinc/divalent silver oxide cell has a two-step electrochemical reaction:

$$Step\ 1: \quad Zn + AgO \rightarrow Ag_2O + ZnO \qquad (1.86 \text{ V})$$

$$Step\ 2: \quad Zn + Ag_2O \rightarrow 2Ag + ZnO \qquad (1.59 \text{ V})$$

12.2.1 Zinc Anode

Zinc is used for the negative electrode in aqueous alkaline batteries because of its high half-cell potential, low polarization, and high limiting current density (up to 40 mA/cm^2 in a cast electrode). Its equivalent weight is fairly low, thus resulting in a high theoretical capacity of 820 mAh/g. The low polarization of zinc allows for a high discharge efficiency 85 to 95% (the ratio of useful capacity to theoretical capacity).

Zinc purity is very important. Zinc is thermodynamically unstable in aqueous alkali, reducing water to hydrogen:

$$Zn + H_2O \rightarrow ZnO + H_2$$

Zinc alloys containing copper, iron, antimony, arsenic, or tin are known to increase the zinc corrosion rate while zinc alloyed with cadmium, aluminum, or lead will reduce corrosion rates.[1,2] If enough gas pressure is generated, the sealed button cell could leak or rupture. In commercial use, the high surface area zinc powder is amalgamated with small amounts of mercury (3 to 6%) to bring the corrosion rate within a tolerable limit.

The oxidation of zinc in the anode is a complicated phenomenon. It is generally accepted that the overall negative reactions are[3,4]

$$Zn + 2OH^- \rightarrow Zn(OH)_2 + 2e \qquad E^0 = +1.249 \text{ V}$$

$$Zn + 4OH^- \rightarrow ZnO_2^{2-} + 2H_2O + 2e \qquad E^0 = +1.215 \text{ V}$$

Electrolyte gelling agents such as polyacrylic acid, potassium or sodium polyacrylate, sodium carboxymethyl cellulose or various gums are generally blended into the zinc powder to improve electrolyte accessibility during discharge.

12.2.2 Silver Oxide Cathode

Silver oxide can be prepared in three oxidation states:[2] monovalent (Ag_2O), divalent (AgO), and trivalent (Ag_2O_3). The trivalent silver oxide is very unstable and is not used for batteries. The divalent form had been used in button cells, generally mixed with other metal oxides. The monovalent silver oxide is the most stable and is the one now used for commercial primary batteries.

The monovalent silver oxide is a very poor conductor of electricity. Without any additives, a monovalent silver oxide cathode would exhibit a very high cell impedance and an unacceptably low closed-circuit voltage (CCV). To improve the initial CCV, the monovalent silver oxide is generally blended with 1 to 5% powdered graphite. However, as the cathode continues to discharge, the silver metal produced by the reaction helps to keep the internal cell resistance low and the CCV high:

$$Ag_2O + H_2O + 2e \rightarrow 2Ag + 2OH^- \qquad E^0 = +0.342 \text{ V}$$

The theoretical capacity of the monovalent silver oxide is 231 mAh/g by weight or 1640 Ah/L by volume. The addition of graphite reduces the cathode capacity due to lower packing density and lower silver oxide content.

Compared to the other silver oxides, the monovalent silver oxide is stable to decomposition in alkaline solutions. Some decomposition to silver metal may occur due to the impurities brought into the cathode material by the graphite. The decomposition rate is dependent upon the source of the graphite, the amount of graphite blended into the cathode, and the cell storage temperature. Greater graphite impurities and higher cell storage temperatures result in greater silver oxide decomposition rates.[5]

To reduce the amount of silver in the cell or to alter the appearance of the discharge curve, other cathode active additives may be blended into the monovalent silver oxide mix. One common additive is manganese dioxide (MnO_2). With increasing amounts of MnO_2 added to the cathode, the voltage curve can be altered from being constant throughout the discharge to where the voltage will gradually decrease as the cathode nears depletion. This gradual drop in voltage can serve as an indicator of silver oxide depletion as the cell is nearing its end of life.

Another additive that can serve a dual function is silver nickel oxide ($AgNiO_2$). Silver nickel oxide is produced by the reaction of nickel oxyhydroxide (NiOOH) with monovalent silver oxide in hot aqueous alkaline solution:[6,7]

$$Ag_2O + 2NiOOH \rightarrow 2AgNiO_2 + H_2O$$

The dual feature of silver nickel oxide is that it is electrically conductive, like graphite, as well as cathode active, like MnO_2. Silver nickel oxide has a coulometric capacity (263 mAh/g), higher than Ag_2O, and will discharge against zinc at 1.5 V (Figs. 12.1 and 12.2). Silver nickel oxide can replace both the graphite and part of the monovalent silver oxide, reducing the cost of the cell.

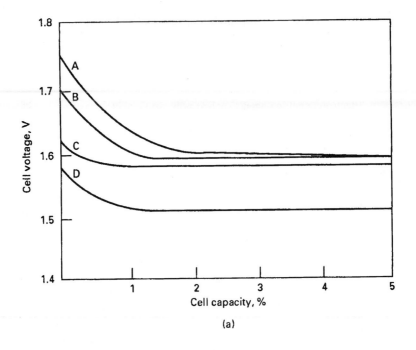

(a)

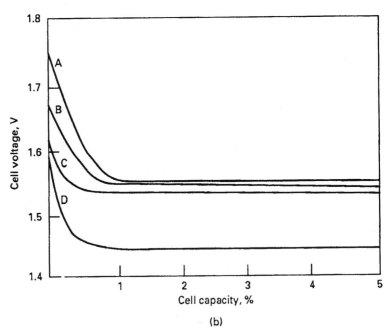

(b)

FIGURE 12.1   Closed-circuit voltage of various zinc/silver oxide chemistries, 7.8 × 3.6 mm button type battery. A: Zn-"double-treatment" AgO; B: Zn-Ag$_2$O; C: Zn-AgO/ silver plumbate; D: Zn-AgNiO$_2$. (*a*) Discharge on 150 Ω, 21°C. (*b*) Discharge on 15 kΩ, 21°C.

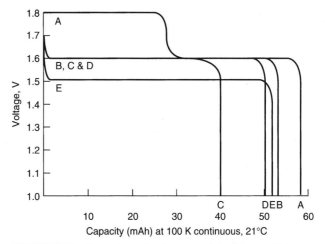

FIGURE 12.2 Comparison of performance of zinc/silver oxide button batteries (various silver oxide cathodes). Size—7.8 × 3.6 mm; A—Zn/AgO; B—double-treatment approach; C—Zn/Ag₂O; D—Zn/plumbate; E—Zn/AgNiO₂.

Although the divalent silver oxide has a higher theoretical capacity (432 mAh/g by weight or 3200 Ah/L by volume) than the monovalent silver oxide, the use of the divalent form in button batteries has been limited and no longer marketed commercially.[2,8] This is due primarily to its instability in alkaline solutions and the fact that it exhibits a two-step discharge (Fig. 12.2).

Divalent silver oxide is unstable in alkaline solutions, decomposing to monovalent silver oxide and oxygen gas:

$$4AgO \rightarrow 2Ag_2O + O_2$$

This instability can be improved by the addition of lead or cadmium compounds[9–12] or by the addition of gold to the divalent silver oxide.[13]

The zinc/divalent silver oxide battery exhibits a two-step discharge curve. The first occurs at 1.8 V, corresponding to the reduction of AgO to Ag₂O:

$$2AgO + H_2O + 2e \rightarrow Ag_2O + 2OH^- \qquad E^0 = +0.607 \text{ V}$$

As the discharge continues, the voltage drops to 1.6 V, corresponding to the reduction of Ag₂O to Ag:

$$Ag_2O + H_2O + 2e \rightarrow 2Ag + 2OH^- \qquad E^0 = +0.342 \text{ V}$$

The overall electrochemical reaction of the Zn/AgO cell is

$$Zn + AgO \rightarrow Ag + ZnO$$

This two-step discharge is not desirable for many electronic applications where tight voltage regulation is required.

The elimination of the two-step discharge has been resolved by several methods.[11,14–16] One previously used commercial approach, shown schematically in Fig. 12.3, was to treat a compressed pellet of AgO with a mild reducing agent such as methanol. The treatment forms a thin outer layer of Ag₂O around a core of AgO. The treated pellet is consolidated into a can and then reacted with a stronger reducing agent such as hydrazine. The hydrazine reduces a thin layer of silver metal across the pellet surface. The cathode produced by this process has only silver metal and Ag_2O in contact with the cathode terminal. The layer of Ag_2O masks the higher potential of the AgO, while the thin, conductive silver layer reduces the cell impedance as the Ag_2O is resistive. In use, only the monovalent silver oxide voltage is observed yet the cell delivers the greater capacity of the divalent silver oxide. Even with the surface treatments, the cells delivered 20 to 40% more hours of service with the same weight of silver than the standard zinc/monovalent silver oxide cells.

Cells produced by this "double-treatment" process are termed Ditronic™ cells. Figure 12.2 shows the benefit of the Ditronic™ design in a button cell. The treated cell has about a 30% capacity advantage over a conventional Ag_2O cathode at the same operating voltage. The AgO cell, without this treatment, produces a two-step discharge.

This "double-treatment" method has the disadvantage that the control of the treatment process is critical to the shelf life of the cell. Reducing the outer surface to either only monovalent silver oxide or to only silver metal does not have the advantages of the dual process (Fig. 12.4). The same is true if either coating is not sufficiently thick (Table 12.2). In storage, the cells will eventually exhibit the phenomenon referred to as "voltage up" and "impedance up." During storage, the silver layer is slowly oxidized by the divalent silver oxide back to resistive monovalent silver oxide:

$$Ag + AgO \rightarrow Ag_2O$$

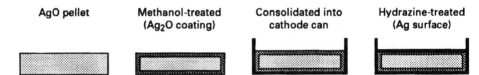

FIGURE 12.3 "Double-treatment" method for divalent silver oxide.

TABLE 12.2 "Double-Treatment" Method for Divalent Silver Oxide—Effect of Thickness of Coating

Ag_2O thickness around each pellet, mm	Ag thickness on the consolidation surface, mm	Final cathode capacity, mAh/g	Voltage level at months		
			1 month	3 months	6 months
0.2	0.12	372	1.73	1.77	1.80
0.6	0.12	360	1.61	1.63	1.71
1.0	0.12	326	1.60	1.59	1.59
0.2	0.24	360	1.60	1.59	1.59
0.6	0.24	348	1.60	1.59	1.59
1.0	0.24	315	1.60	1.59	1.59

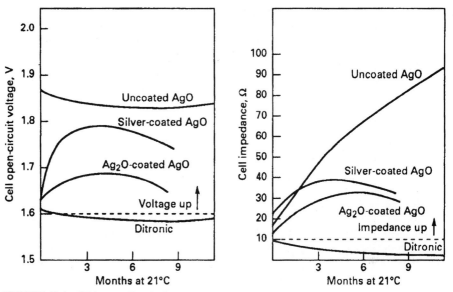

FIGURE 12.4 "Voltage-up" and "impedance-up" of Zn/AgO battery during 1 year storage at 21°C, 7.8 × 3.6 mm button type battery.

As the metallic silver layer is depleted, the cell demonstrates an increase in open-circuit voltage and impedance. The high impedance, due to high internal resistance, results in a low closed-circuit voltage and a nonfunctional cell.

A second approach to eliminate the two-step discharge was through the use of "silver plumbate" as a cathode additive material.[17] Silver plumbate cathode material was prepared by reacting an excess of divalent silver oxide powder with lead sulfide (PbS) in a hot alkaline solution. The product of the reaction is a mixture of remaining divalent silver oxide (AgO), monovalent silver oxide (Ag_2O), and silver plumbate ($Ag_5Pb_2O_6$). The sulfur is oxidized to the sulfate and is washed from the reaction product:

$$2PbS + 19AgO + 4NaOH \rightarrow Ag_5Pb_2O_6 + 7Ag_2O + 2Na_2SO_4 + 2H_2O$$

The AgO particles after the reaction retain a core of AgO but have an outer coating of monovalent silver oxide and silver plumbate. The silver plumbate compound is conductive, stable, and cathode active. The Ag_2O serves to mask the AgO while the conductive $Ag_5Pb_2O_6$ improves the cell impedance. As the silver plumbate is not oxidized by the AgO, as is silver, the cathode impedance remains low throughout the cell life.

Cathode material prepared by the silver plumbate process will discharge through the following reactions (Fig. 12.5):

$$2AgO + H_2O + 2e \rightarrow Ag_2O + 2OH^- \qquad E^0 = +0.607 \text{ V}$$

$$Ag_2O + H_2O + 2e \rightarrow 2Ag + 2OH^- \qquad E^0 = +0.342 \text{ V}$$

$$Ag_5Pb_2O_6 + 4H_2O + 8e \rightarrow 5Ag + 2PbO + 8OH^- \qquad E^0 = +0.2 \text{ V}$$

$$PbO + H_2O + 2e \rightarrow Pb + 2OH^- \qquad E^0 = -0.580 \text{ V}$$

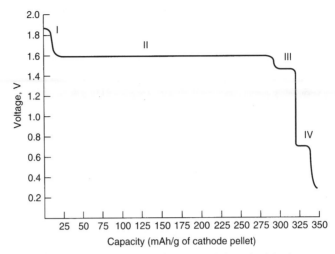

FIGURE 12.5 Discharge curve of cathode-limited zinc/plumbate system. 300 Ω continuous in flooded beaker cell, 21°C. Cathode pellet weight—0.12 g. Reaction I (1.8 V)—$2AgO + H_2O + 2e \rightarrow Ag_2O + 2OH^-$; reaction II (1.6 V)—$Ag_2O + H_2O + 2e \rightarrow 2Ag + 2OH^-$; reaction III (1.4 V)—$Ag_5Pb_2O_6 + 4H_2O + 8e^- \rightarrow 5Ag + 2PbO + 8OH^-$; reaction IV (0.7 V)—$PbO + H_2O + 2e \rightarrow Pb + 2OH^-$.

Not all of these steps are observed when the silver plumbate cathode material is used in cells (Fig. 12.2). The open-circuit voltages (OCV) of these cells are found to be stable at about 1.75 V. However, once placed on discharge, the cell voltage quickly drops to the monovalent silver oxide operating voltage of about 1.6 V, eliminating the AgO plateau (Fig. 12.1). As button cells are anode limited, the cell is depleted in zinc capacity before the $Ag_5Pb_2O_6$ and PbO reduction reactions can occur.

The silver plumbate approach has advantages over the dual-treatment process as the treatment is simpler while still retaining a capacity advantage over monovalent silver oxide. The product from the silver plumbate reaction (8% lead sulfide) has a coulometric capacity of 345 to 360 mAh/g.

The silver plumbate process has the disadvantage that the button cells do contain a small amount of lead, 1 to 4% of the cell weight. An alternate approach is to use bismuth sulfide in place of the lead sulfide in the material preparation reaction.[18] The product of the reaction has the advantages of the silver plumbate material but without the toxicity of lead. Bismuth is not considered toxic and has many medical and cosmetic applications, being used both externally and internally in the body.[19]

The product of the reaction of bismuth sulfide with divalent silver oxide is believed to be silver bismuthate ($AgBiO_3$):

$$Bi_2S_3 + 28AgO + 6NaOH \rightarrow 2AgBiO_3 + 13Ag_2O + 3Na_2SO_4 + 3H_2O$$

Like the silver plumbate compound, the silver bismuth compound is conductive and cathode active. The monovalent silver oxide produced by the reaction coats the divalent silver oxide particles while the conductive silver bismuthate reduces the cell impedance, maintaining a high cell CCV. The silver bismuthate will discharge against zinc in alkaline solutions at about 1.5 V. Therefore, in anode-limited button cells only the monovalent silver oxide voltage is observed.

Unlike monovalent silver oxide systems, additives such as graphite or manganese dioxide cannot be added to the divalent silver oxide. Graphite enhances the decomposition of AgO to Ag_2O and oxygen. Manganese dioxide is readily oxidized by AgO to alkali-soluble manganate compounds.

12.2.3 Electrolyte

The electrolytes used for zinc/silver oxide cells are based upon 20 to 45% aqueous solutions of potassium hydroxide (KOH) or sodium hydroxide (NaOH). Zinc oxide (ZnO) is dissolved in the electrolyte as the zincate to help control zinc gassing. The zinc oxide concentration varies from a few percent to a saturated solution.

The preferred electrolyte for button cells is potassium hydroxide (KOH). Its higher electrical conductivity[20,21] allows cells to discharge over a wide range of current demands (Fig. 12.6). Sodium hydroxide (NaOH) is used mainly for long life cells not requiring a high-rate discharge (Fig 12.7). The sodium hydroxide exhibits less creep and such cells are less apt to leak than the potassium hydroxide cells. Leakage is evidenced as frosting or salting around the seal. However, this characteristic of potassium hydroxide solutions has been corrected in recent years due to improvements in seal technology by most manufacturers.

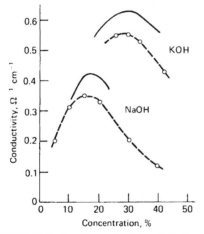

FIGURE 12.6 Specific conductivity of alkaline hydroxide solutions. Solid line—25°C, Ref. 5; dotted line—15°C, Ref. 6.

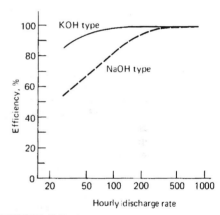

FIGURE 12.7 Dependence of discharge efficiency on discharge rate of zinc/silver oxide button battery at 20°C.

12.2.4 Barriers and Separators

A physical barrier is required to keep the anode and silver cathode apart in the tight volume constraints of a button cell. Failure of the barrier will result in internal cell shorting and cell failure. A silver oxide cell requires a barrier with the following properties:

1. Permeable to water and hydroxyl ions
2. Stable in strong alkaline solutions
3. Not oxidized by the solid silver oxide or dissolved silver ions
4. Retards the migration of dissolved silver ions to the anode

Because of the slight solubility of silver oxides in alkaline electrolyte, little work was done with zinc/silver oxide cells until 1941 when André[22] suggested the use of a cellophane barrier. Cellophane prevents migrating silver ions from reaching the anode[23,24] by reducing them to insoluble silver metal. The cellophane is oxidized and destroyed in the process, making it less effective for long-life cells.

Many types of laminated membranes are presently available. A commonly used alternate barrier material is prepared from a radiation graft of methacrylic acid onto a polyethylene membrane.[23,24] The graft makes the film wettable and permeable to the electrolyte. Studies have shown that lower resistance polyethylene barrier membrane is suitable for high-rate KOH cells while higher resistance polyethylene is suitable for low-rate NaOH cells. Cellophane is used in conjunction with the grafted membrane as a sacrificial barrier. The lamination of cellophane to either side of the polyethylene membrane results in a synergistic action for stopping silver migration.[15]

A separator is commonly used in conjunction with a barrier membrane layer as added protection to the barrier. It is located between the barrier and anode cavity and is multifunctional both during cell manufacture and in performance. Separators in zinc/silver cells are typically fibrous woven or non-woven polymers such as polyvinyl alcohol (PVA). The fibrous nature of the separator gives it stability and strength that protects the more fragile barrier layers from compression failure during cell closure or penetration of zinc particles through the membranes themselves. The separator also acts in controlling dimensional stress in the barrier layers which is often derived during original rollup of cellophane or in lamination with polyethylene membranes. As barrier membranes wet out, this stress is relieved. Differential swelling may then occur which could result in a gap that would allow silver ion migration to the anode. The separator acts to prevent the formation of these gaps.

12.3 CONSTRUCTION

Figure 12.8 is a cross-sectional view of a typical zinc/silver oxide button type battery.

Zinc/silver oxide button cells are designed as anode limited; the cell has 5 to 10% more cathode capacity than anode capacity. If the cell were cathode limited, a zinc-nickel or zinc-iron couple could form between the anode and the cathode can resulting in the generation of hydrogen.

The cathode material for zinc/silver oxide cells is monovalent silver oxide (Ag_2O) mixed with 1 to 5% graphite to improve the electrical conductivity. The Ag_2O cathode material may also contain manganese dioxide (MnO_2) or silver nickel oxyhydroxide ($AgNiO_2$) as cathode extenders. The cathode could also be prepared from a controlled blend of divalent silver oxide and monovalent silver oxide with either silver plumbate ($Ag_5Pb_2O_6$) or silver metal to reduce the AgO cathode voltage and cell impedance, but this is no longer a commercial process. A small amount of polytetrafluoroethylene (Teflon™) may be added to the mix as a binder and to ease pelleting.

The anode is a high surface area, amalgamated, gelled zinc metal powder housed in a top cup which serves as the external negative terminal for the cell. The top cup is pressed from a triclad metal sheet: the outer surface is a protective layer of nickel over a core of steel. The inner surface that is in direct contact with the zinc is high-purity copper or tin. The cathode pellet is consolidated into the positive can, which is formed from nickel-plated steel and serves as the positive terminal for the cell. To keep the anode and cathode separated, a barrier disk of cellophane or a grafted plastic membrane is placed over the consolidated cathode. The entire system is wetted with potassium or sodium hydroxide electrolyte.

The gasket serves to seal the cell against electrolyte loss and to insulate the top and bottom cups from contact. The gasket material is made from an elastic, electrolyte-resistant plastic such as nylon. The seal may be improved by coating the gasket with a sealant such as polyamide or bitumen to prevent electrolyte leakage at the seal surfaces.

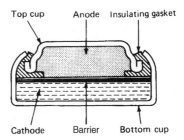

FIGURE 12.8 Cutaway view of typical zinc/silver oxide button type battery.

12.4 *PERFORMANCE CHARACTERISTICS*

12.4.1 Open-Circuit Voltage

The open-circuit voltage (OCV) of the Zn/Ag_2O battery is about 1.60 V but will vary slightly (1.595 to 1.605 V) with electrolyte concentration, concentration of zincate in the electrolyte, and temperature exposure.[25] The reaction of carbon dioxide with the silver oxide during battery manufacturing can raise the OCV to 1.65 V due to the formation of silver carbonate. The increase in voltage, however, is temporary and will drop to the operating voltage level of 1.58 V in a watch within seconds. The depth of discharge has little effect on the OCV of a monovalent silver battery; a partially used battery has the same OCV as a new one.

The OCV of the zinc/divalent silver oxide battery will vary from 1.58 to 1.86 V depending on the ratio of Ag to Ag_2O to AgO in the cathode. The OCV will decrease with greater Ag_2O to AgO ratios and with the presence of silver metal in the cathode. With divalent silver oxide batteries, the depth of discharge does have an effect on the OCV; a partially used battery will have more Ag_2O and silver metal than a new one and will thus have a lower OCV.

12.4.2 Discharge Characteristics

Figure 12.9 exhibits the typical curves for a 11.6 × 3.0 mm sized monovalent silver oxide battery on constant resistance discharge. These are typical voltage curves; the discharge voltage profiles of other size batteries would be similar. The service life will vary depending upon the size of the battery and the applied load.

The discharge characteristics of the various types of silver oxide batteries are also covered in Sec. 12.2.2.

The zinc/silver oxide button battery is capable of operating over a wide temperature range. The battery can deliver more than 70% of its 20°C capacity at 0°C and 35% at −20°C, at moderate loads. At heavier loads, the loss is greater. Higher temperatures tend to accelerate capacity deterioration, but temperatures as high as 60°C can be tolerated for several days with no serious effect.

Figures 12.10 and 12.11 show the initial closed-circuit voltage of representative sizes of zinc/silver oxide batteries at various loads and temperatures.

Figure 12.12 shows the pulse performance of treated divalent silver oxide batteries using sodium hydroxide and potassium hydroxide as electrolytes. The low-rate cell is better suited for analog watches (2000 Ω, 7.8 ms/s) even at −10°C, while the high-rate cell is better suited for LCD watches with backlight (100 Ω, 2 s/h to 8 h/day). Some applications demand a heavy-current pulse of short duration in addition to the low background current, e.g., electronic shutter mechanism of a still camera, LED watch, LCD watch with a backlight, and some analog watches. In these cases, the minimum voltage at the end of the pulse must be met to ensure proper device function.

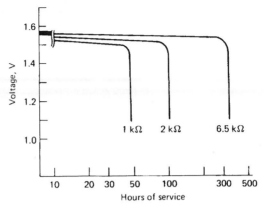

FIGURE 12.9 Typical discharge curves of zinc/silver oxide battery at 20°C, 11.6 × 3.0 mm size.

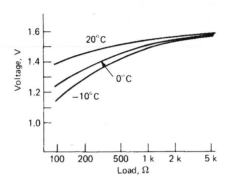

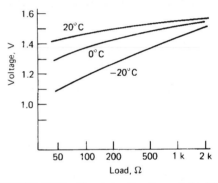

FIGURE 12.10 Closed-circuit voltage of zinc/silver oxide battery, 7.9 × 2.7 mm size, Type 396.

FIGURE 12.11 Closed-circuit voltage of zinc/silver oxide battery, 11.6 × 5.35 mm size.

The manufacturers of these two types of batteries do not distinguish them by service life tests. In fact, similar mAh output is obtained at loads lighter than the 500 hr rate. The industry uses pulse CCV tests to differentiate the higher rate version from the low rate version. Different producers use different CCV loads, durations, and minimum voltages. For example, at Rayovac, the following tests are used

Battery type	1196	1176
Rayovac number	376	377
CCV load (ohms)	100	2000
Minimum CCV @ 150 mS	1.00V	1.50V

The impedance of a zinc/silver oxide battery is influenced primarily by the conductive diluents in the cathode, the barrier resistivity and the electrolyte type and concentration. These factors are balanced by battery manufacturers to obtain the desired values required to meet the applications. As the cell is discharged, the impedance will decline as the resistive silver oxide is reduced to conductive metallic silver (Fig. 12.13).

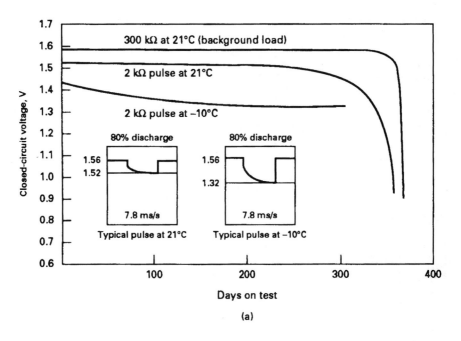

(a)

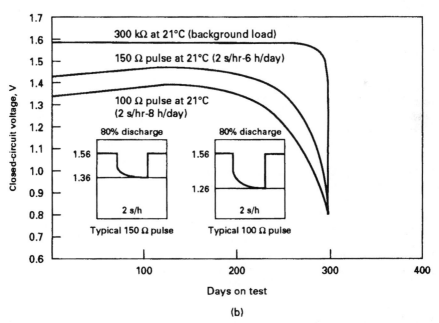

(b)

FIGURE 12.12 Closed-circuit voltage curves for Zn-AgO cells with NaOH and KOH as electrolytes, 7.8×3.6 mm size. (*a*) NaOH cell, analog watch text. (*b*) KOH cell, LCD watch with backlight test.

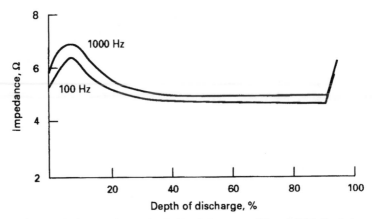

FIGURE 12.13   Impedance of a Zn/Ag_2O battery at 100 and 1000 Hz during discharge, 11.6 × 5.35 mm size.

12.4.3   Shelf Life

Major improvements were needed in seal technology and in cell stability to extend the shelf life of watch batteries to five years. The effect of temperature and humidity on leakage of button batteries was reported by Hull.[26] Leakage was caused by mechanical means (improper seal, fibers in the seal, scratches) or electrochemical means (high oxygen content or high humidity). Batteries are now designed to operate watches for 5 years without leakage.

Stability of these batteries after high temperature storage or prolonged storage at room temperature is influenced by cathode stability and barrier selection. With monovalent silver oxide cathode, gassing in aqueous potassium or sodium hydroxide at 74°C is not a problem. With modified cathodes, however, such as divalent silver oxide, silver plumbate or silver nickel oxide, gas suppression is necessary. CdS, HgS, SnS_2, or WS_2 were found to reduce oxygen evolution while BaS, NiS, MnS, and CuS increased oxygen evolution from AgO.[11]

Failure on shelf of these zinc/silver oxide batteries is closely connected with barrier selection. Cellulosic membranes were used for many years in Zn/Ag_2O cells but their use in Zn/AgO cells was unsuccessful because of massive silver diffusion. While solubility of AgO and Ag_2O was reported[24] to be the same (4.4×10^{-4} mol/L in $10N$ NaOH), AgO decomposition to Ag_2O occurred spontaneously thus resulting in more silver diffusion with Zn/AgO cells than with Zn/Ag_2O cells. The small amount of soluble silver, reaching the zinc caused accelerated corrosion and hydrogen evolution. In addition, silver was plated in the barrier forming electronic shorts which internally discharge the cell. Laminated Permion membranes have been used to stop silver migration to the zinc. Figure 12.14 shows an Arrhenius plot of the storage characteristics of various low- and high-rate zinc/silver oxide systems. The data shows that 10 years of storage at 21°C is possible.

12.4.4   Service Life

Figure 12.15 is a monograph which can be used to calculate the service life of the various sized batteries at various current drains at 20°C.

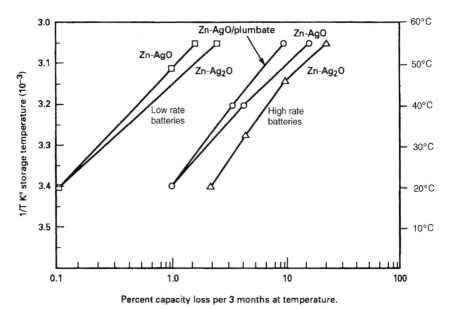

FIGURE 12.14 Arrhenius plot of various zinc/silver oxide chemistries. 11.6 × 5.35 size battery, 6500 Ω continuous discharge to 0.9 V at 21°C. Projected 10% loss: □ >10 years, ○ >5 years, △ >3 years.

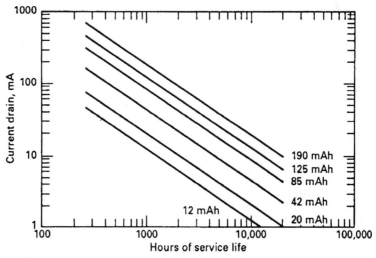

FIGURE 12.15 Service life of zinc/silver oxide batteries at 20°C.

12.5 CELL SIZES AND TYPES

The characteristics of commercially available zinc/monovalent silver oxide button batteries are summarized in Table 12.3.

TABLE 12.3 Characteristics of Commercially Available Zinc/Silver Oxide Batteries

Rayovac model number	ANSI*	IEC†	Drain	Rated load KΩ	Nominal capacity (mAh)	Approximate volume (cm³)	Maximum dimensions Millimeters dia. × ht.	Approximate weight (grams)
376	1196SO	SR626	high	47	26	0.09	6.8 × 2.6	0.40
361	1173SO	SR58	high	30	22	0.10	7.9 × 2.1	0.44
396	1163SO	SR59	high	45	35	0.13	7.9 × 2.7	0.56
392‡	1135SO	SR41	high	15	35	0.17	7.8 × 3.6	0.61
393	1137SO	SR48	high	15	90	0.26	7.8 × 5.4	1.04
370	1188SO	SR69	high	45	35	0.15	9.5 × 2.1	0.60
399	1165SO	SR57	high	20	53	0.19	9.5 × 2.7	0.79
391	1160SO	SR55	high	15	43	0.22	11.6 × 2.1	0.83
389	1138SO	SR54	high	15	85	0.32	11.6 × 3.0	1.21
386	1133SO	SF43	high	6.5	120	0.44	11.6 × 4.2	1.56
357	1131SO	SR44	high	6.5	190	0.57	11.6 × 5.35	2.22
357XP	1184SO	SR44	ultra high	0.62	190	0.57	11.6 × 5.35	2.22
337	NA	SR416	low	100	8.3	0.02	4.8 × 1.65	0.13
335	1193SO	SR512	low	150	6	0.03	5.8 × 1.25	0.13
317	1185SO	NA	low	70	11	0.04	5.8 × 1.65	0.18
379	1191SO	NA	low	70	14	0.06	5.8 × 2.15	0.23
319	1186SO	NA	low	70	16	0.07	5.8 × 2.7	0.26
321	1174SO	SR65	low	70	14	0.06	6.8 × 1.65	0.24
364	1175SO	SR60	low	70	19	0.08	6.8 × 2.15	0.33
377	1176SO	SR66	low	45	26	0.09	6.8 × 2.6	0.40
346	1164SO	SR721	low	100	9.5	0.06	7.9 × 1.25	0.23
341	1192SO	SR714	low	68	13	0.07	7.9 × 1.45	0.30
315	1187SO	SR67	low	70	16	0.08	7.9 × 1.65	0.32
362	1158SO	SR58	low	70	22	0.10	7.9 × 2.1	0.44
397	1164SO	SR59	low	45	35	0.13	7.9 × 2.7	0.56
329	NA	NA	low	20	36	0.15	7.9 × 3.1	0.60
384‡	1134SO	SR41	low	15	35	0.17	7.8 × 3.6	0.61
373	1172SO	SR68	low	45	24	0.12	9.5 × 1.65	0.44
371	1171SO	SR69	low	45	35	0.15	9.5 × 2.1	0.61
395	1162SO	SR57	low	20	53	0.19	9.5 × 2.7	0.81
394	1161SO	SR45	low	15	64	0.26	9.5 × 3.6	0.96
366	1177SO	SR1116	low	30	30	0.17	11.6 × 1.65	0.70
381	1170SO	SR55	low	20	43	0.22	11.6 × 2.1	0.80
390	1159SO	SR54	low	15	85	0.32	11.6 × 3.0	1.21
344	1139SO	SR42	low	15	105	0.38	11.6 × 3.6	1.35
301	1132SO	SR43	low	6.8	110	0.44	11.6 × 4.2	1.68

*ANSI = American National Standards Institute.
†IEC = International Electrotechnical Commission.
‡Available as hybrid formulation Ag_2O/MnO_2 blended cathodes.
Source: Rayovac Corp.

REFERENCES

1. F. Kober and H. West, "The Anodic Oxidation of Zinc in Alkaline Solutions," Extended Abstracts, The Electrochemical Society, Battery Division 12, 66–69 (1967).

2. A. Fleischer and J. Lander (eds.), *Zinc-Silver Oxide Batteries,* Wiley, New York, 1971.

3. W. M. Latimer, *Oxidation Potentials,* Prentice-Hall, Englewood Cliffs, New Jersey, 1952.

4. D. R. Lide (editor-in-chief), *Handbook of Chemistry and Physics,* 73d ed., CRC Press, Boca Raton, Fla., 1992.

5. A. Shimizu and Y. Uetani, "The Institute of Electronics and Communication Engineers of Japan," Tech. Paper CPM79-55, 1979.

6. T. Nagaura and T. Aita, U.S. Patent 4,370,395 (1981).

7. T. Nagaura, *New Material AgNiO$_2$ for Miniature Alkaline Batteries, Progress in Batteries and Solar Cells,* Vol. 4, pp. 105–107 (1982).

8. E. A. Megahed, Small Batteries for Conventional and Specialized Applications, "The Power Electronics Show and Conference," San Jose, Calif., pp. 261–272 (1986).

9. C. C. Cahan, U.S. Patent 3,017,448 (1959).

10. P. Ruetschi, in *Zinc-Silver Oxide Batteries,* edited by A. Fleischer and J. J. Lander, Wiley, New York, p. 117 (1971).

11. E. A. Megahed and C. R. Buelow, U.S. Patent 4,078,127 (1978).

12. A. Tvarusko, *J. Electrochem. Soc.* **116:**1070A (1969).

13. S. M. Davis, U.S. Patent 3,853,623 (1974).

14. E. A. Megahed, C. R. Buelow, and P. J. Spellman, U.S. Patent 4,009,056 (1977).

15. E. A. Megahed and D. C. Davig, "Long Life Divalent Silver Oxide-Zinc Primary Cells for Electronic Applications," *Power Sources,* Vol. 8, Academic, London, 1981.

16. E. A. Megahed and D. C. Davig, "Rayovac's Divalent Silver Oxide-Zinc Batteries," *Progress in Batteries and Solar Cells.* Vol. 4, pp. 83–86 (1982).

17. E. A. Megahed and A. K. Fung, U.S. Patent 4,835,077 (1989).

18. J. L. Passaniti, E. A. Megahed, and N. Zreiba, U.S. Patent 5,389,469 (1994).

19. "Bismuth," chapter from *Minerals, Facts, and Problems,* Bureau of Mines Bulletin 675, U.S. Department of the Interior (1985).

20. E. J. Rubin and R. Babaoian, "A Correlation of the Solution Properties and the Electrochemical Behavior of the Nickel Hydroxide Electrode in Binary Aqueous Alkali Hydroxides," *J. Electrochem. Soc.* **118:**428 (1971).

21. "Kagaku Benran," Maruzen, Tokyo, 1966.

22. H. André, *Bull. Soc. Franc. Elect.* **6:**1, 132 (1941).

23. V. D'Agostino, J. Lee, and G. Orban, "Grafted Membranes," in *Zinc-Silver Oxide Batteries,* edited by A. Fleischer and J. J. Lander, Wiley, New York, 1971, pp. 271–281.

24. R. Thornton, "Diffusion of Soluble Silver-Oxide Species in Membrane Separators," General Electric Final Report, Schenectady, New York (1973).

25. S. Hills, "Thermal Coefficients of EMF of the Silver (I) and the Silver (II) Oxide-Zinc-45% Potassium Hydroxide Systems," *J. Electrochem. Soc.* **108:**810 (1961).

26. M. N. Hull and H. I. James, "Why Alkaline Cells Leak," *J. Electrochem. Soc.* **124:**332–339 (1977).

CHAPTER 13
ZINC/AIR BATTERIES—BUTTON CONFIGURATION

Steven F. Bender, John W. Cretzmeyer, and Terrence F. Reise

13.1 GENERAL CHARACTERISTICS

Zinc/air batteries[a] use oxygen from ambient atmosphere to produce electrochemical energy. Upon opening the battery to air, oxygen diffuses into the cell and is used as the cathode reactant. The air passes through the cathode to the interior cathode active surface in contact with the cell's electrolyte. At the active surface, the air cathode catalytically promotes the reduction of oxygen in the presence of an aqueous alkaline electrolyte. The catalytic air electrode is not consumed or changed in the process. Since one active material lies outside of the cell, the majority of the cell's volume contains the other active component (zinc), thus on a unit volume basis, zinc/air batteries have a very high energy density.

For many applications zinc/air technology offers the highest available energy density of any primary battery system. Other advantages include a flat discharge voltage, long shelf life, safety and ecological benefits, and low energy cost. Since the batteries are open to the ambient atmosphere, a factor limiting universal applications of zinc/air technology is the tradeoff between long service life (high environmental tolerance) and maximum power capability (lower environmental tolerance). The major advantages and disadvantages of this battery type are summarized in Table 13.1.

The effect of atmospheric oxygen as a depolarizing agent in electrochemical systems was first noted early in the nineteenth century. However, it was not until 1878 that a battery was designed in which the manganese dioxide of the famous Leclanché battery was replaced by a porous platinized carbon/air electrode. Limitations in technology prevented the commercialization of zinc/air batteries until the 1930s. In 1932, Heise and Schumacher constructed alkaline electrolyte zinc/air batteries which had porous carbon air cathodes impregnated with wax to prevent flooding. This design is still used almost unchanged for the manufacture of large industrial zinc/air batteries. These batteries are noted for their very high energy densities but low power output capability. They are used as power sources for remote railway signaling and navigation aid systems. Broader application is precluded by low current capability and bulk (see Chap. 38).

---

[a]Rechargeable and larger metal/air batteries are covered in Chap. 38.

TABLE 13.1   Major Advantages and Disadvantaes of Zinc/Air (Button) Batteries

Advantages	Disadvantages
High energy density	Not independent of environmental conditions:
Flat discharge voltage	"Drying out" limits shelf life once opened to air
Long shelf life (sealed)	"Flooding" limits power output
No ecological problems	Limited power output
Low cost	Short activated life
Capacity independent of load and temperature when within operating range	

The thin, efficient air cathode used in today's zinc/air button batteries is the result of fuel-cell research and was made possible through the emergence of fluorocarbons as a polymer class. The unique surface properties of these materials, hydrophobicity combined with gas porosity, made possible the development of thin, high-performance gas electrodes using a Teflon bonded catalyst structure[1] and a hydrophobic cathode composite.[2] A method for continuous fabrication of this composite was developed in the early 1970s.[3]

Early efforts to apply zinc/air battery technology were focussed on portable military applications. After further development, the technology was commercialized for consumer applications, and this resulted in the development of small form factor batteries that are primarily used today. The most successful applications for zinc/air batteries have been in medical and telecommunication applications. Zinc/air batteries are now the leading power source for miniature hearing aids. In hospitals, 9-volt zinc/air batteries power cardiac telemetry monitors used for continuous patient monitoring. Other multi-cell zinc/air batteries are used to power bone growth stimulators for mending broken bones. In the telecommunication area zinc-air batteries are used for communication receivers such as pagers, e-mail devices, and wireless messaging devices. Recently, zinc/air coin-type batteries were employed in wireless telecon headsets that use the Bluetooth, low power digital wireless protocol. Larger size batteries (see Chap. 38) are being developed for cellular phones and laptop computers.

13.2 CHEMISTRY

The more familiar types of primary alkaline systems are the zinc/manganese dioxide, zinc/mercuric oxide, and zinc/silver oxide batteries. These, typically, use potassium or sodium hydroxides, in concentrations from 25 to 40% by weight, as the electrolyte, which functions primarily as an ionic conductor and is not consumed in the discharge process. In simple form, the overall discharge reaction for these metal oxide cells can be stated as

$$MO + Zn \rightarrow M + ZnO$$

During the discharge, the metal oxide (MO) is reduced, either to the metal as shown or to a lower form of an oxide. Zinc is oxidized and, in the alkaline electrolyte, usually reacts to form ZnO. Thus it can be seen that, at 100% efficiency, electrochemically equivalent amounts of metal oxide and zinc must be present. Therefore, an increase in capacity of any cell must be accompanied by an equivalent increase in both cathodic and anodic materials.

In the zinc/oxygen couple, which also uses an alkaline electrolyte, it is necessary to increase only the amount of zinc present to increase cell capacity. The oxygen is supplied from the outside air which diffuses into the cell as it is needed. The air cathode acts only as a reaction site and is not consumed. Theoretically, the air cathode has infinite useful life and its physical size and electrochemical properties remain unchanged during the discharge. The reactions of the air cathode are complex but can be simplified to show the cell reactions as follows:

Cathode	$\frac{1}{2}O_2 + H_2O + 2e \rightarrow 2OH^-$	$E^0 = 0.40$ V
Anode	$Zn \rightarrow Zn^{2+} + 2e$	
	$Zn^{2+} + 2OH^- \rightarrow Zn(OH)_2$	$E^0 = 1.25$ V
	$Zn(OH)_2 \rightarrow ZnO + H_2O$	
Overall reaction	$Zn + \frac{1}{2}O_2 \rightarrow ZnO$	$E^0 = 1.65$ V

The reaction chemistry has a rate-limiting step which affects reaction kinetics and hence the performance. This step relates to the oxygen reduction process, wherein peroxide-free radical (O_2H^-) formation occurs:

Step 1 $\qquad\qquad\qquad\qquad O_2 + H_2O + 2e \rightarrow O_2H + OH$

Step 2 $\qquad\qquad\qquad\qquad O_2H \rightarrow OH + \frac{1}{2}O_2$

The decomposition of the peroxide to hydroxide and oxygen is a key rate-limiting step in the reaction sequence. To accelerate the reduction of the peroxide species and the overall reaction rate, the air cathode is formulated using catalytic compounds which promote the reaction in step 2. These catalysts are typically metal compounds or complexes such as elemental silver, cobalt oxide, noble metals and their compounds, mixed metal compounds including rare earth metals, and transition metal macrocyclics, spinels, manganese dioxide, phtalocyanines or perovskites.[4,5,6]

13.3 CONSTRUCTION

Small form factor zinc/air batteries exist today in a number of types ranging from button to coin sizes. Although small batteries predominate, higher capacity zinc/air batteries for portable applications are currently being developed with shapes ranging from cylindrical round cells (AA and AAA)[7,8] up to small prismatic batteries. These larger batteries are likely to have internal capacities in the range of 4–6 Ah (See Chap. 38). The discussion in this chapter will focus on button designs, with capacities less than 1.2 Ah.

The construction features of zinc/air button cells are quite similar to those of other commercially available zinc anode button cells. The zinc anode material is generally a loose, granulated powder mixed with electrolyte and, in some cases, a gelling agent to immobilize the composite and ensure adequate electrolyte contact with zinc granules. The shape or morphology of the zinc granules plays a role in achieving better inter-particle contact and hence creating a lower internal electrical resistance in the anode pack. High surface area zinc granules are preferred for better performance. The metal can halves housing the cathode and anode active materials also act as the terminals, insulation between the two containers being provided by a plastic (gasket). A cut-away view of a typical zinc/air button battery is shown in Fig. 13.1.

A schematic representation of a typical zinc/air button battery is given in Fig. 13.2. A zinc/metal oxide battery is shown for comparison. The reason for increased energy density in the zinc/air battery is illustrated graphically by comparing the anode compartment volumes. The very thin cathode of the zinc/air battery (about 0.5 mm) permits the use of twice as much zinc in the anode compartment as can be used in the metal oxide equivalent. Since the air cathode theoretically has infinite life, the electrical capacity of the battery is determined only by the anode capacity, resulting in at least a doubling of energy density.

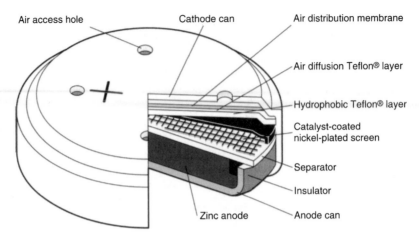

FIGURE 13.1 Typical zinc/air button battery (components not to scale) (*Courtesy of Duracell, Inc.*)

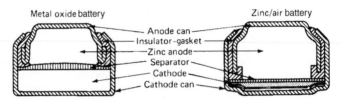

FIGURE 13.2 Cross section of metal oxide and zinc/air button batteries (*Courtesy of Duracell, Inc.*)

A portion of the total volume available internally for the anode must be reserved to accommodate the expansion that occurs when zinc is converted to zinc oxide during the discharge. This space also provides additional tolerance to sustained water gain during operating conditions. Referred to as the anode free volume, it is typically 15 to 25% of the total anode compartment volume.

Figure 13.3 shows a magnified cross-sectional view of the cathode region of the zinc/air battery. The cathode structure includes the separators, catalyst layer, metallic mesh, hydrophobic membrane, diffusion membrane, and air-distribution layer. The catalyst layer contains carbon blended with oxides of manganese to form a conducting medium. It is made hydrophobic by the addition of finely dispersed Teflon particles. The metallic mesh provides structural support and acts as the current collector. The hydrophobic membrane maintains the gas-permeable waterproof boundary between the air and the cell's electrolyte. The diffusion membrane regulates gas diffusion rates (not used when an air hole controls gas diffusion). Finally the air distribution layer distributes oxygen evenly over the cathode surface.

Through advances in the technology, air cathode construction has improved with the introduction of a dual layer approach.[9] The dual layer cathode consists of coating the screen current collector with a blend of low surface-area carbon black and Teflon particles to produce a hydrophobic cathode layer with good electrical contact to the screen. The second layer, which is coated onto the first layer and which contacts the electrolyte in the cell, is produced from a blend of high surface-area, carbon black, Teflon powder and a catalyst. The resulting high surface area of the second layer promotes better access to the electrolyte and facilitates better oxygen catalysis. The first layer promotes good electrical contact to the screen and provides a better hydrophobic barrier to prevent electrolyte penetration and to slow water evaporation loss. Prior to cathode coating, some manufacturers roughen the screen current collector to increase surface area and achieve better screen to cathode mix contact.[10]

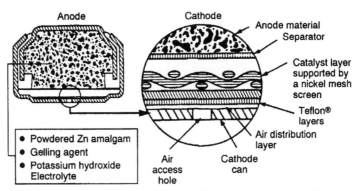

FIGURE 13.3   Key constructional features of zinc/air button batteries. (*Courtesy of Duracell, Inc.*)

An air-excess hole on the positive terminal of a zinc/air battery provides a path for oxygen to enter the cell and diffuse to the cathode catalyst sites. The rate at which oxygen and other gases transfer into or out of the cell is regulated either by the hole area or by the porosity of the diffusion membrane at the surface of the cathode layer. Regulating oxygen diffusion sets a limit to a zinc/air battery's maximum continuous-current capability, because the operating current is directly proportional to oxygen consumption (5.81×10^{-5} cm^3 of oxygen per milliampere-second) used in button cell cathodes.

Limiting current, while being dependent on the air availability and the active electrode surface area, is also a function of the catalytic activity of the cathode noted in Sec. 13.2. The cathode discussed uses a metal oxide catalyst, MnO_2, a common transition metal oxide used in button cell cathodes.[11,12] Studies of catalytic activity have shown that some valence states of the oxides of manganese promote faster peroxide decomposition, leading to faster reaction kinetics and better cell performance.[13] Once achieving maximum catalytic activity, the next step is to optimize cathode porosity to achieve good oxygen transport. Cathode porosity must be balanced between oxygen penetration and the retardation of water vapor loss from the electrolyte. The design of the cathode must also take into consideration the end application for the battery. This will help determine how the cathode should be designed to insure maximum energy output under normal operating conditions.

If only oxygen transfer rates mattered, gas diffusion in zinc/air cells would not be regulated, resulting in higher operating current capability. Regulation is necessary because other gases, most importantly water vapor, can enter or leave the cell. If not properly controlled, undesirable gas transfer can cause a degradation in cell power capability and service life.

Water vapor transfer is generally the dominant form of gas transfer performance degradation. This transfer occurs between the cell's electrolyte and the ambient (Fig. 13.4). The aqueous electrolyte of a zinc/air cell has a characteristic water vapor pressure. A typical electrolyte consisting of 30% potassium hydroxide by weight is in equilibrium with the ambient at room temperature when the relative humidity is approximately 60%. A cell will lose water from its electrolyte on drier days and gain water on more humid days. In the extreme, either water gain or water loss can cause a zinc/air battery to fail before delivering full capacity. A smaller hole or lower diffusion membrane porosity yields greater environmental tolerance because water transfer rates are reduced, resulting in a longer practical service life.

The maximum continuous-current capability of a zinc/air battery, as determined by gas diffusion regulation, is typically specified as the limiting current, denoted by I_L. The relationship between gas transfer regulation, limiting current, and service life is illustrated in Fig. 13.5.

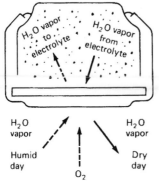

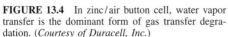

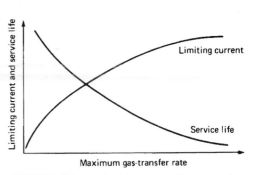

FIGURE 13.4   In zinc/air button cell, water vapor transfer is the dominant form of gas transfer degradation. (*Courtesy of Duracell, Inc.*)

FIGURE 13.5   In zinc/air button cell, gas transfer regulation determines limiting current and useful service life. (*Courtesy of Duracell, Inc.*)

It should be noted that under conditions of continuous discharge, the limiting current would not be sustained indefinitely. It will gradually begin to decline as the voltage falls and internal impedance increases. The limiting current will thus vary depending on the state of charge of the battery. The limiting currents shown in Table 13.2 represents the maximum current that is achievable within the first 3 minutes of fresh cell discharge, and hence is representative of only the early stage of cell discharge.

13.4   *PERFORMANCE CHARACTERISTICS*

13.4.1   Cell Sizes

Zinc/air button and coin batteries are available in a variety of sizes. Capacities range from about 40 to 1100 mAh. Table 13.2 lists the physical and electrical characteristics of some available batteries. The smaller sizes are commonly used as hearing-aid batteries, the medium to larger ones for continuous-drain applications such as pager or telemetry devices.

Zinc/air batteries for hearing aid applications continue to be improved to meet the more stringent needs of new devices and user requirements. High rate zinc/air batteries have been developed for example, the Rayovac Proline High Power batteries, designed for better air access thus improving power output, the potential trade-off being shorter operating life on low rate drain due to higher water vapor loss or gain from the cell. (See Section 13.4.10.)

Batteries have also been designed for greater service life by maximizing the amount of zinc in the cell. The zinc content in these batteries is maximized by creating the largest allowable internal cell volume while not exceeding standard external cell dimensions.[14] Table 13.2*b* lists some characteristics of these batteries. The zinc content is maximized without compromising the internal free volume needed for anode expansion as zinc metal becomes converted to zinc oxide, as this would lead to premature end of life. Designers of zinc/air button cells will experience challenges in the future as digital hearing aids emerge which will require greater power and energy to operate compared to existing models.

The first commercially successful zinc/air coin cell was introduced in 1989. The 2330 coin cell has become the predominant power source for credit-card-style pagers.[15] The 675 zinc/air battery is the most common BTE hearing aid power source, and also serves as battery for the Timex wristwatch Beep pager. Portable Holter telemetry heart monitors the "9-volt" zinc/air batteries to transmit patient data via portable radio transmitters to base station monitors.

Data on two zinc/air multicell batteries are listed in Table 13.3.

TABLE 13.2   Characteristics of Zinc/Air Button and Coin Batteries

a. Standard Single-Cell Batteries, 1.4 Volts

Generic type	IEC No.	ANSI no.	Nominal diameter (mm)	Nominal height, (mm)	Average weight (g)	Rated capacity, (mAh)	Standard drain (mA)	Limiting current (mA)	Typical useful service life (months)
5*	PR63		5.7	2.0	—	42	0.4	—	1–2
10*	PR70	7005ZD	5.7	3.5	0.3	70	0.4	2	1–2
312*	PR41	7002ZD	7.7	3.9	0.6	134	0.8	7	1–2
13*	PR48	7000ZD	7.7	5.2	0.9	260	0.8	12	1–2
675*	PR44	7003ZD	11.4	5.2	1.8	600	2	22	2–3
2330**	PR2330		23.2	3.0	—	960	4	—	1–2
630**	PR1662		16.0	6.2	3.5	1100	4	—	2–3

(*)*Source:*   Duracell, a Gillette Company).
(**)*Source:*   Panasonic Matsushita) Product literature (coin batteries).

b. High Capacity Zinc/Air Single-cell Batteries, 1.4 Volts

Generic type	ANSI no.	Max. ANSI diameter (mm)	Max. ANSI height (mm)	Rated capacity (mAh)
10	7005ZD	5.8	3.6	90
312	7002ZD	7.9	3.6	160
13	7000ZD	7.9	5.4	290
675	7003ZD	11.6	5.4	630

Source: Rayovac Corporation, Ultra Hearing Aid Batteries.

TABLE 13.3   Zinc/Air Multicell Batteries

Generic type	ANSI no.	Battery voltage (V)	Maximum dimensions, (mm)	Average weight (g)	Rated capacity, (mAh)	Standard drain (mA)	Typical useful service life (months)
DA164	—	5.6	16.8 (Dia) × 44.5 (H)	20	950	3	1–2
DA146	7004Z	8.4	26.5 (L), 17.5 (W) 48.5 (H)	36	1500	6	1–2

Source: Duracell, Inc.

13.4.2   Voltage

The nominal open-circuit voltage for a zinc/air battery is 1.4 V. The initial closed-circuit voltage at 20°C ranges from 1.15 to 1.35 V, depending on the discharge load. The discharge is relatively flat, with 0.9 V the typical end voltage.

13.4.3   Energy Density

The zinc/air button batteries have a high volumetric energy density compared with other primary battery systems. They range from approximately 442 Wh/L for a PR63 (size 5 hearing aid battery) to 970 Wh/L for a PR1662 battery.

13.4.4   Discharge Characteristics

Discharge profiles for the DA13 and DA675-size zinc/air button batteries are presented in Fig. 13.6. As the air cathode in the cell is not chemically altered during discharge, the voltage remains quite stable. Data for the zinc/mercuric oxide and zinc/silver oxide batteries are provided for comparison. On continuous discharge at the loads shown, the zinc/air battery will deliver twice the service of the other batteries of the same size. The DA13 battery, which has less than half the volume of the other batteries, outperforms the larger metal oxide batteries.

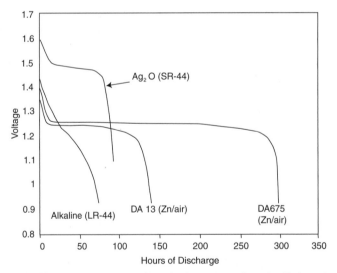

FIGURE 13.6   Discharge profiles of primary button batteries discharged at 620 ohm and 20°C. LR-44, SR-44 and DA-675 (11.6 mm × 5.4 mm) and DA-13 (7.9 mm × 5.4 mm) (*Courtesy of Duracell, Inc. Company*)

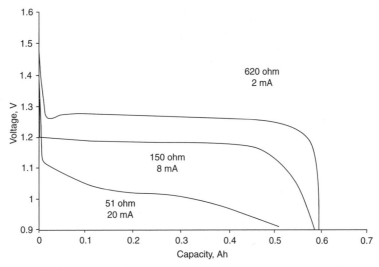

FIGURE 13.7 Discharge curves for a 675 size zinc/air battery discharged at three loads at 20°C. The average discharge current is also shown (*Courtesy of Duracell, Inc.*)

A set of discharge curves, typical of the performance of zinc/air button batteries at 20°C, is presented in Fig. 13.7. The typical discharge curves are relatively flat. The cell delivers about the same capacity over the current range shown when operating below the limiting current. The operating voltage is lower at the higher discharge currents, as is typical with all battery systems. Discharge drains near the limiting current show increasing polarization of the battery voltage and deliver less than the rated capacity.

13.4.5 Voltage-Current Performance

The degree of oxygen access to the cathode and the catalytic activity of the cathode material generally define the voltage-current profile of a zinc/air battery. Oxygen access is defined by the degree of air access to the battery. The higher the access of air to the battery, the greater is the power output. This generally means that increasing the number of air access holes in the battery case will increase the power capability. The consequences of high air access is the increased opportunity for greater water vapor loss due to evaporation from the cell. A higher rate of water vapor loss results in faster dry-out leading to increased electrolyte concentration and higher internal impedance. Thus, more air access shortens the useful operating service life of the battery, once the battery is opened to the air. Zinc/air batteries can be designed for maximum power output and shorter operating life by maximizing air access. A longer operating life can be achieved by minimizing air access, but at the expense of cell power output.

The impact of air access on the limiting current of a zinc/air battery is demonstrated in Fig. 13.8. Limiting current is the maximum current output for a given set of conditions, i.e. temperature, relative humidity and cell design parameters. The limiting current is typically determined by potentiostatically polarizing the battery to a voltage of 1.1 V and measuring the height of the maximum output current within 1 to 5 minutes after the start of the test. In Fig. 13.8, the effect on increasing the number of air access holes in a 675 size zinc/air battery on the limiting current is shown.

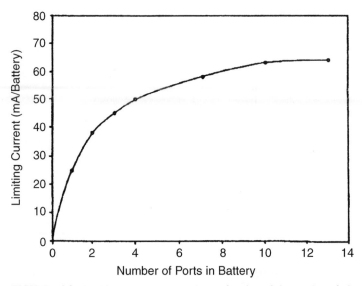

FIGURE 13.8 Limiting current measured as a function of the number of air holes in a 675-size zinc/air cell. Limiting current increases with increased air access. (*Courtesy of Rayovac, Inc.*)

The ability of the cathode to catalyze the oxygen reaching the reactive sites on the active surface of the cathode also determines the degree of maximum power output from a battery. The use of catalytic additives to the carbon mix making up the air cathode is a common practice. The use of catalytic agents such as MnO_x compounds is typical. Figure 13.9 illustrates the discharge characteristics of a carbon cathode in a 675 size cell with and without a manganese oxide catalyst. The manganese-containing cathode produces a cell with a higher average operating voltage.

The voltage-current profiles for various sized button batteries are shown in Fig. 13.10. As can be seen, the voltage falls rapidly when continuous currents above the limiting current are applied. This occurs because the cell has become oxygen-starved. It is consuming oxygen at a faster rate than the rate at which oxygen is entering the cell. Voltage will fall until the load current is reduced to an equilibrium condition.

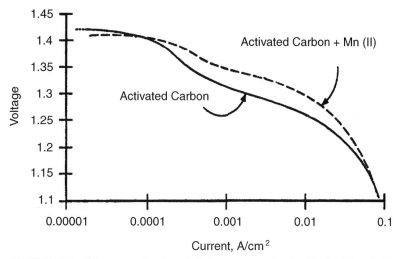

FIGURE 13.9   Voltage as a function of discharge current density. The Mn(II) cathode delivers higher performance due to greater catalytic activity. (*Courtesy Rayovac, Inc.*)

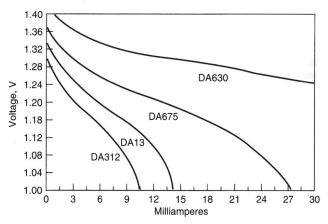

FIGURE 13.10   Voltage-current profiles for various zinc/air button batteries at 20°C (*Courtesy of Duracell, Inc.*)

13.4.6 Cell Internal Impedance

The effects of discharge level and signal frequency on the internal impedance characteristics of the DA13 and DA675 button batteries are presented in Fig. 13.11.

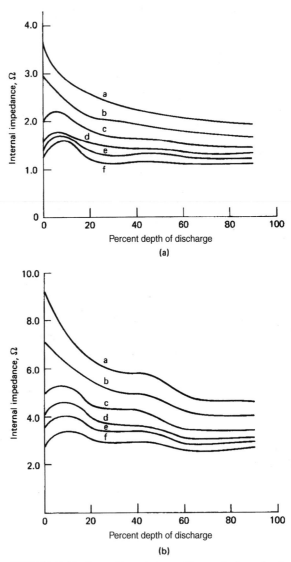

FIGURE 13.11 Internal impedance profiles vs. discharge level. (*a*) 675-size battery. (*b*) 13-size battery. Signal frequency: *a* = 50 Hz; *b* = 100 Hz; *c* = 250 Hz; *d* = 500 Hz; *e* = 1000 Hz; *f* = 3000 Hz. (*Courtesy of Duracell, Inc.*)

13.4.7 Pulse Load Performance

Zinc/air cells can handle pulse currents much higher than the limiting current (I_L), the current level depending on the nature of the pulse. This capability results from a reservoir of oxygen that builds up within the cell when the load is below the limiting current.

Figure 13.12 illustrates general voltage profiles for various pulse waveforms. In Fig. 13.12a a pulse current double the magnitude of I_L is applied. However, the pulses are spaced so that the average current is less than I_L. Therefore the average rate of oxygen ingress is sufficient to sustain the average load current. Furthermore, the pulses are of short duration so that the battery does not become oxygen-starved before the end of a single pulse. The voltage profile shows the ripple effect of the pulse current, but the battery maintains a continuous useful average operating voltage. If the peak pulse current is increased so that the average load current is greater than I_L, the battery will eventually become oxygen-starved and voltage will decline, as shown in Fig. 13.12b.

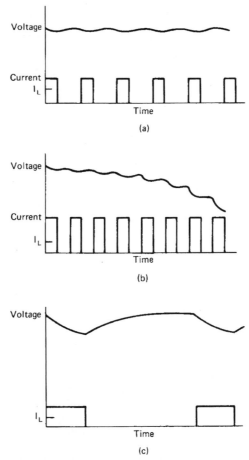

FIGURE 13.12 Pulse load performance of zinc/air button batteries (a) $I_{av} < I_L$, short pulse duration, (b) $I_{av} > I_L$, short pulse duration; (c) $I_{av} < I_L$, long pulse duration. (*Courtesy of Duracell, Inc.*)

In choosing a battery for a particular application, the designer must ensure that the average current for the duration of the pulse load is less than I_L. The designer must also check for the possibility of the battery becoming oxygen-starved during the application of a single pulse. This condition is illustrated in Fig. 13.12c. In this figure the average current is less than I_L, but the battery becomes oxygen-starved and the voltage declines sharply during the time that a pulse is applied because of the long pulse duration. The voltage recovers between pulses, maintaining an average close to the average of Fig. 13.12a. A method for improving pulse performance by incorporating an air reservoir into the zinc/air battery has been proposed. Figure 13.13 illustrates the performance of a DA675 battery subjected to sustained current drains above and below the limiting current.

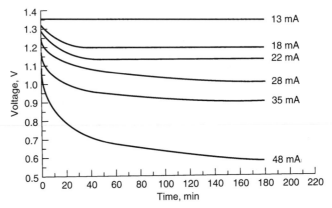

FIGURE 13.13   Voltage-time response of 675-size zinc/air batteries subjected to continuous loads above and below limiting current. (*Courtesy of Duracell, Inc.*)

13.4.8 Effect of Temperature

The effect of temperature on discharge performance at various discharge rates is illustrated in Fig. 13.14. This degradation of performance is caused primarily by a reduction in ionic diffusion capability through the electrolyte. Rather high concentrations of potassium or sodium hydroxide (25 to 40% by weight) are used to obtain good electrical conductivity. This high concentration is subject to changes in viscosity and a general lowering of ionic mobility as the temperature drops.

The operating voltage also varies with temperature. At lower temperatures and a fixed current, the voltage is lower than when the battery is discharged at room temperature or warmer. Although the optimum discharge temperature is between 10 and 40°C, the battery can still function at temperatures below 10°C, but at lower voltages and energy densities. The operating voltage of the zinc/air battery at various discharge loads and temperatures is shown in Fig. 13.15.

It should be noted that at low discharge rates and moderate temperatures, the effect of temperature is minimal. In selecting a zinc/air system that must perform at low temperatures it is important to consider the required current density in order to preclude failure due to diffusion limitations.

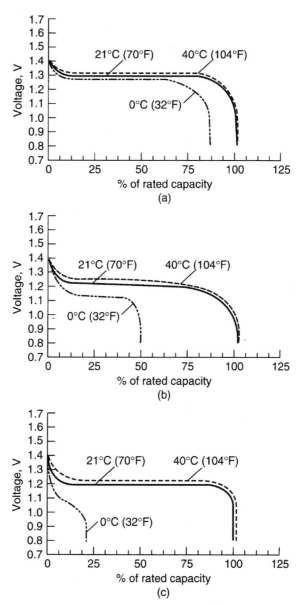

FIGURE 13.14 Comparison of performance of zinc/air batteries at various temperatures. (*a*) Discharged at 300-h rate. (*b*) Discharged at 75-h rate. (*c*) Discharged at 50-h rate. (*Courtesy of Duracell, Inc.*)

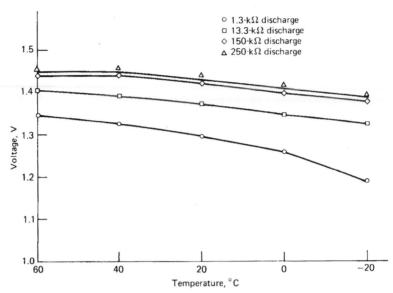

FIGURE 13.15   Effect of temperature on voltage—DA675 zinc/air battery. (*Courtesy of Duracell, Inc.*)

13.4.9 Storage Life

Four principal mechanisms affect the capacity of zinc/air batteries during storage and operating service. One mechanism, self-discharge of the zinc (corrosion), is an internal reaction; the other three are caused by gas transfer. The gas transfer mechanisms are direct oxidation of the zinc anode, carbonation of the electrolyte and electrolyte water gain or loss.

During storage the air access holes of a zinc/air battery can be sealed to prevent gas transfer decay. A typical material for sealing a battery is a polyester tape. Note that, unlike conventional batteries, one of the zinc/air cell's reactants, oxygen, is sealed outside the cell during storage. This characteristic gives zinc/air batteries excellent shelf-life performance.

The primary mechanism affecting the shelf-life of a zinc/air battery is the self-discharge reaction. Zinc is thermodynamically unstable in an alkaline solution (electrolyte) and reacts to form zinc oxide (discharged zinc) and hydrogen gas. (An additional advantage of zinc/air batteries is that hydrogen evolved during the reaction is vented through the sealing tape to prevent the pressure buildup that can cause cell deformation in conventional batteries.) This reaction is controlled by additives to the zinc. Results of shelf-life evaluation of DA675 zinc/air batteries at room ambient conditions over a 5-year storage period are presented in Table 13.4. Capacity retention over this period is 85% of initial capacity, yielding an average capacity loss per year of less than 3%.

Elevated temperatures will increase the rate of the self-discharge reaction dramatically. The capacity loss after 28 days of storage at 54°C, averaged about 3%. Storage life at high temperatures can be optimized through tradeoffs between other performance parameters and choice of cell and battery design components.

TABLE 13.4 Capacity Retention vs. Storage Time at 20°C*—Zinc/Air Battery, 675-Size

Storage, years	Average capacity, mAh	% change from initial capacity	Average % change per annum
0	523	0	0
1.8	497	−5.0	−2.8
5.0	452	−13.5	−2.7

* 1.8-year data are actual test data: 5-year data are estimates based on actual studies conducted on larger cell sizes.

13.4.10 Factors Affecting Service Life

The combination of the effects of self-discharge and gas transfer degradation determines the service life performance of a zinc/air battery. For most applications water transfer is the dominant factor. However, under some conditions electrolyte carbonation and direct oxidation mechanisms can adversely affect performance.

Carbonation of Electrolyte. Carbon dioxide, which is present in the atmosphere at a concentration of approximately 0.04%, reacts with an alkaline solution (electrolyte) to form an alkali metal carbonate and bicarbonate. Zinc/air batteries can be satisfactorily discharged using a carbonated electrolyte, but there are two disadvantages of extreme carbonation: (1) vapor pressure of the electrolyte is increased, aggravating water vapor loss in low-humidity conditions, and (2) crystals of carbonate formed in the cathode structure may impede air access, eventually causing cathode damage with subsequent deterioration of cathode performance. As indicated in Fig. 13.16 carbonation must be extreme to be detrimental to cell performance in most applications.[16]

Direct Oxidation. The zinc anode of a zinc/air battery can be oxidized directly by oxygen which enters the cell and dissolves in and diffuses through the electrolyte. Measurements of the effect of direct oxidation on cell capacity have been predicted experimentally,[17] using advanced microcalorimetry techniques to predict direct oxidation effects on DA675 type cells (cells open, not sealed). This research indicates a capacity loss of less than 1.5% of rated capacity per year at 25°C and less than 5% per year at 37.5°C.

The combined effect of oxidation and self-discharge on fresh batteries, open to the ambient, yields a predicted capacity loss of less than 4% per year. However, the typical useful service life for a zinc/air battery as presented in Table 13.2a is about 3 months.

Effect of Water Vapor Transfer on Service Life. The decay mechanism determining the useful service life is water vapor transfer. It occurs when a partial pressure difference exists between the vapor pressure of the electrolyte and the surrounding environment. As indicated previously, a cell with a typical electrolyte, consisting of 30% concentration of potassium hydroxide, will lose water when the humidity at room temperature is below 60% and will gain water at humidities above 60%. Excessive water loss increases the concentration of the electrolyte and can eventually cause the cell to fail because of inadequate electrolyte to

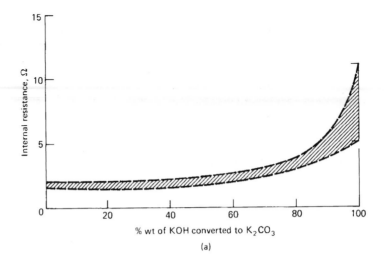

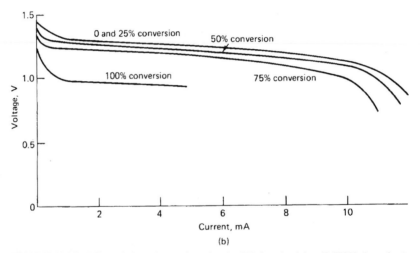

FIGURE 13.16   Effect of electrolyte carbonation in 675-size zinc/air cell (KOH electrolyte). (*a*) On internal impedance at 20°C (4000 Hz). (*b*) On voltage-current profile at 20°C. (*From Ref. 8; courtesy of Duracell, Inc.*)

maintain the discharge reaction. Excessive water gain dilutes the electrolyte, which also reduces conductivity. Furthermore, the catalyst layer of the air cathode will flood under sustained water gain conditions, reducing electrochemical activity and eventually causing cell failure.

The design of a zinc/air battery can be optimized to compensate for water transfer for a specific set of operating conditions. Tradeoffs can be made between the volume and composition of the electrolyte, the amount of zinc, and the degree of gas diffusion regulation to maximize service life. For most applications and cell types, diffusion control is the most important variable.

The balance between performance and service life as related to water transfer rates is addressed in Table 13.5. This table compares the weight loss and impedance changes of four commercially available 312-size zinc/air batteries, typically used in hearing aids. The batteries were exposed to 14 and 28 days of ambient air storage after which time their water weight loss and 1-kHz AC impedance were measured. The variation in this table is primarily related to battery design. Battery A, for example, was designed with a high air diffusion cathode for better power capability. However, the low barrier to gas diffusion allows for more rapid water loss from the cell. Subsequently the A battery is not designed for a low-rate discharge since it will reach end of life earlier due to increased water loss and internal impedance. On the other hand, battery D is designed for lower gas diffusion rates, which allows for more of a medium power capability, but will deliver longer service under low-rate drain conditions as a result of its lower rate of water loss. Batteries B and C in Table 13.5 lie in the midrange of A and D, achieving neither the highest power capability nor the longest operating life. The specific performance characteristics for the 312-size zinc/air batteries are useful in matching power capabilities to the amplification requirements in common hearing aids to achieve the best performance level.

TABLE 13.5 Water Weight and Battery Impedance Changes for Four Commercialy Available 312-Size Zinc/ Air Button Batteries Exposed to Ambient Conditions for 14–28 Days

	Cumulative water weight loss, mg			
	A	B	C	D
Storage time*				
14 days	9.1	5.9	7.3	0.3
28 days	12.2	7.5	9.3	1.0
Impedance, Ω†				
Initial	2.3	4.7	3.8	4.3
Final (28 days)	70.8	18.0	15.1	8.8
Relative change	×30	×4	×4	×2

*Relative Humidity during storage—55 ± 5%.
†AC impedance measured at 1000 Hz and 10 mA.

The effect of diffusion control on water vapor transfer rates for three zinc/air battery sizes is presented in Table 13.6. Actual instantaneous water transfer rates at a given environment will vary throughout the life of the battery, depending on the length of exposure. As a battery gains or loses water, transfer rates will decrease because the electrolyte equilibrium relative humidity moves toward the ambient relative humidity. The figures presented in Table 13.6 represent average rates computed over the indicated exposure period. Actual average rates over the life of the battery would be lower.

The effect of continued moisture loss or gain from the electrolyte is illustrated in Fig. 13.17. This figure shows an intermittent discharge of a zinc/air battery at 30, 60, and 90% relative humidity. The least amount of moisture transfer occurs at the 60% relative humidity condition and the battery delivers the maximum service.

Table 13.7 lists data for zinc/air 675-size battery tested after 14 and 28 days of storage, unsealed and exposed to various relative humidity conditions. The effects of electrolyte loss at each of these conditions is presented. The batteries had not been discharged prior to the test. It is evident that high relative humidity is more detrimental to performance retention, leading to battery capacity degradation.

TABLE 13.6 Relative Degree of Environmental Tolerance

Relative humidity ±5%, %	Zinc/air battery type*		
	230-size	312-size	675-size
20	NA	−0.057	−0.084
60	−0.047	−0.036	−0.027
90	NA	NA	+0.018

* Storage periods 230-Size—21 days; 312-size—28 days; 675-size—30 days.

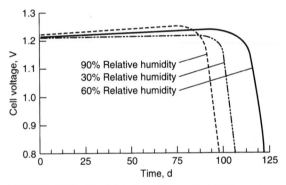

FIGURE 13.17 Zinc/air battery discharged at 21°C at various levels of relative humidity. (*Courtesy of Duracell, Inc.*)

TABLE 13.7 Weight and Capacity Changes as a Function of Storage at Normal and Extreme Relative Humidity Conditions—675-size Zinc/Air Battery

Relative humidity, %	Weight loss or gain, mg			Delivered capacity, mAh		
	0 days	10 days	30 days	0 days	10 days	30 days
20	0	−152	−2.51	500	465	283
60	0	0.99	1.56	500	490	410
90	0	3.28	3.54	500	176	30

The conditions of use of a battery also affect its performance. Batteries enclosed in a sealed battery compartment (with allowance for air access) or other enclosure undergo less performance loss due to water vapor transmission than batteries exposed in the open air. Batteries exposed to daily indoor and outdoor relative humidity conditions will undergo water vapor transmission cycling which tends to average out the effects of both environments. Indoor relative humidities tend to be closer to the conditions for recommended use (40 to 60% RH).

Note that the larger batteries such as the 675-size have greater tolerance to the detrimental effects of water gain or loss. This is typical in a comparison of batteries of different sizes. Larger batteries have more electrolyte and also a greater anode-free volume, making them more tolerant to water loss and gain conditions, respectively.

The operational life of zinc/air battery is most dependent on the control of gas transmission into and out of the cell. It is evident that water vapor transmission is the key factor in extending the service life for these batteries. A key area for research and development relating to zinc/air batteries is focused on membrane technology. A selectively permeable gas diffusion membrane, which allows air diffusion into the cell but excludes or greatly reduces water vapor transmission, will greatly broaden the range of application for zinc/air batteries. Papers and patent disclosures over the last few years have discussed a series of new materials under study.[18–20]

REFERENCES

1. G. W. Elmore and H. A. Tanner, U.S. Patent 3,419,900.

2. A. M. Moos, U.S. Patent 3,267,909.

3. R. G. Biddick, U.S. Patent 4,129,633.

4. E. Yeager, "Electrochemical Catalysis for Oxygen Electrodes," Rep. LBL-25817, Lawrence Berkeley Lab., Calif. 1988.

5. C. Warde and A. D. Glasser, U.S. Patent 3935027.

6. B. Szczesniak, et al., Abstract Number 280, Joint International Meeting of ECS and ISE, Paris, 1997.

7. M. Ohashi, H. Watabe, and H. Ogata, Japanese Patent Kokai 9-289045.

8. T. Saeki, T. Watabe, and S. Kobayashi, Japanese Patent Kokai 8-338836 and 8-315870.

9. K. Yoshida and M. Watabe, U.S. Patent 4,380,567.

10. R. Dopp, J. Ottman and J. Passanti, U.S. Patent 5,650,246.

11. A. Borbely and J. Molla, U.S. Patent 4,894,296.

12. A. Ohta, Y. Morita, et al., "Manganese Oxide as a Catalyst for Zinc-Air Cells," *Proc. Battery Material Symp.,* 1985.

13. J. Passanti and R. Dopp, U.S. Patent 5,308,711.

14. J. Ottman, B. Dopp and J. Burns, U.S. Patent 5,567,538.

15. H. Konishi and T. Yokoyama, "Air Cells for Pagers," National Tech. Rep., vol. 37, no. 1, JEC Press, Cleveland, Ohio, 1991.

16. J. W. Cretzmeyer, H. R. Espig, and J. C. Hall, "Commercial Zinc-Air Batteries," *Power Sources,* Vol. 6, Oriel Press, Newcastle-upon-Tyne, U.K., 1977.

17. S. Bender, J. W. Cretzmeyer, and J. C. Hall, "Long Life Zinc-Air Cells as a Power Source for Consumer Electronics, in *Progress in Batteries and Solar Cells,* vol. 2, JEC Press, Cleveland, Ohio, 1979.

18. T. Takamura, Y. Sato, M. Susuki, T. Nakamura, and K. Sasaki, "High Performance Zn-Air Cell Using Gas-Selective Membranes," *Proc. Int. Power Sources Symp.,* Brighton, U.K., The Paul Press, London, 1985.

19. M. Yoshino, S. Noya, and M. Yanagihara, Japanese patednt Kokai 2-216755 and 2-216756.

20. N. Yoshino, S. Noya, A. Hanabusa, and N. Yangihara, Japanese Patent Kokai 3-108255, 3-108256, and 3-108257.

CHAPTER 14
LITHIUM BATTERIES

David Linden and Thomas B. Reddy

14.1 GENERAL CHARACTERISTICS

Lithium metal is attractive as a battery anode material because of its light weight, high voltage, high electrochemical equivalence, and good conductivity. Because of these outstanding features, the use of lithium has dominated the development of high-performance primary and secondary batteries during the last two decades. (See Chap. 34 and 35 covering lithium secondary batteries.)[1]

Serious development of high-energy-density battery systems was started in the 1960s and concentrated on nonaqueous primary batteries using lithium as the anode. The lithium batteries were first used in the early 1970s in selected military applications, but their use was limited as suitable cell structures, formulations, and safety considerations had to be resolved. Lithium primary cells and batteries have since been designed, using a number of different chemistries, in a variety of sizes and configurations. Sizes range from less than 5 mAh to 10,000 Ah; configurations range from small coin and cylindrical cells for memory backup and portable applications to large prismatic cells for standby power in missile silos.

Lithium primary batteries, with their outstanding performance and characteristics, are being used in increasing quantities in a variety of applications, including cameras, memory backup circuits, security devices, calculators, watches, etc. Nevertheless, lithium primary batteries have not attained a major share of the market as was anticipated, because of their high initial cost, concerns with safety, the advances made with competitive systems and the cost-effectiveness of the alkaline/manganese battery. World-wide sales of lithium primary batteries for 1999 have been estimated at $1.1 billion.[2]

14.1.1 Advantages of Lithium Cells

Primary cells using lithium anodes have many advantages over conventional batteries. The advantageous features include the following:

1. *High voltage:* Lithium batteries have voltages up to about 4 V, depending on the cathode material, compared with 1.5 V for most other primary battery systems. The higher voltage reduces the number of cells in a battery pack by a factor of at least 2.

2. *High specific energy and energy density:* The energy output of a lithium battery (over 200 Wh/kg and 400 Wh/L) is 2 to 4 or more times better than that of conventional zinc anode batteries.

3. *Operation over a wide temperature range:* Many of the lithium batteries will perform over a temperature range from about 70 to −40°C, with some capable of performance to 150°C or as slow as −80°C.

4. *Good power density:* Some of the lithium batteries are designed with the capability to deliver their energy at high current and power levels.

5. *Flat discharge characteristics:* A flat discharge curve (constant voltage and resistance through most of the discharge) is typical for many lithium batteries.

6. *Superior shelf life:* Lithium batteries can be stored for long periods, even at elevated temperatures. Storage of up to 10 years at room temperature has been achieved and storage of 1 year at 70°C has also been demonstrated. Shelf lives over 20 years have been projected from reliability studies.

The performance advantages of several types of lithium batteries compared with conventional primary and secondary batteries, are shown in Secs. 6.4 and 7.3. The advantage of the lithium cell is shown graphically in Figs. 7.2 to 7.9, which compare the performance of the various primary cells. Only the zinc/air, zinc/mercuric oxide, and zinc/silver oxide cells, which are noted for their high energy density, approach the capability of the lithium systems at 20°C. The zinc/air cell, however, is very sensitive to atmospheric conditions; the others do not compare as favorably on a specific energy basis nor at lower temperatures.

14.1.2 Classification of Lithium Primary Cells

Lithium batteries use nonaqueous solvents for the electrolyte because of the reactivity of lithium in aqueous solutions. Organic solvents such as acetonitrile, propylene carbonate, and dimethoxyethane and inorganic solvents such as thionyl chloride are typically employed. A compatible solute is added to provide the necessary electrolyte conductivity. (Solid-state and molten-salt electrolytes are also used in some other primary and reserve lithium cells; see Chaps. 15, 20, and 21.) Many different materials were considered for the active cathode material; sulfur dioxide, manganese dioxide, iron disulfide, and carbon monofluoride are now in common use. The term "lithium battery," therefore, applies to many different types of chemistries, each using lithium as the anode but differing in cathode material, electrolyte, and chemistry as well as in design and other physical and mechanical features.

Lithium primary batteries can be classified into several categories, based on the type of electrolyte (or solvent) and cathode material used. These classifications, examples of materials that were considered or used, and the major characteristics of each are listed in Table 14.1.

Soluble-Cathode Cells. These use liquid or gaseous cathode materials, such as sulfur dioxide (SO_2) or thionyl chloride ($SOCl_2$), that dissolve in the electrolyte or are the electrolyte solvent. Their operation depends on the formation of a passive layer on the lithium anode resulting from a reaction between the lithium and the cathode material. This prevents further chemical reaction (self-discharge) between anode and cathode or reduces it to a very low rate. These cells are manufactured in many different configurations and designs (such as high and low rate) and with a very wide range of capacities. They are generally fabricated in cylindrical configuration in the smaller sizes, up to about 35 Ah, using a bobbin construction for the low-rate cells and a spirally wound (jelly-roll) structure for the high-rate designs. Prismatic containers, having flat parallel plates, are generally used for the larger cells up to 10,000 Ah in size. Flat or "pancake-shaped" configurations have also been designed. These soluble cathode lithium cells are used for low to high discharge rates. The high-rate designs, using large electrode surface areas, are noted for their high power density and are capable of delivering the highest current densities of any active primary cell.

TABLE 14.1 Classification of Lithium Primary Batteries*

Cell classification	Typical electrolyte	Power capability	Size, Ah	Operating range, °C	Shelf life, years	Typical cathodes	Nominal cell voltage, V	Key characteristics
Soluble cathode (liquid or gas)	Organic or inorganic (w/solute)	Moderate to high power, W	0.5 to 10,000	−80 to 70	5–20	SO_2 $SOCl_2$ SO_2Cl_2	3.0 3.6 3.9	High energy output, high power output, low-temperature operation, long shelf life
Solid cathode	Organic (w/solute)	Low to moderate power, mW-W	0.03 to 33	−40 to 50	5–8	V_2O_5 $AgV_2O_{5.5}$ Ag_2CrO_4 MnO_2 $Cu_4O(PO_4)_2$ $(CF)_n$ CuS FeS_2 FeS CuO	3.3 3.2 3.1 3.0 3.0 2.6 1.7 1.5 1.5 1.5	High energy output for moderate power requirements, nonpressurized cells
Solid electrolyte (see Chap. 15)	Solid state	Very low power, μW	0.003 to 0.5	0 to 100	10–25	$PbI_2/PbS/Pb$ I_2 (P2VP)	1.9 2.8	Excellent shelf life, solid state—no leakage, long-term microampere discharge

*For reserve batteries; see Chaps. 20 and 21.

Solid-Cathode Cells.   The second type of lithium anode primary cells uses solid rather than soluble gaseous or liquid materials for the cathode. With these solid cathode materials, the cells have the advantage of not being pressurized or necessarily requiring a hermetic-type seal, but they do not have the high-rate capability of the soluble-cathode systems. They are designed, generally, for low- to medium-rate applications such as memory backup, security devices, portable electronic equipment, photographic equipment, watches and calculators, and small lights. Button, flat, and cylindrical-shaped cells are available in low-rate and the moderate-rate jelly-roll configurations. A number of different solid cathodes are being used in lithium primary cells, as listed in Table 14.1. The discharge of the solid-cathode cells is not as flat as that of the soluble-cathode cells, but at the lower discharge rates and ambient temperature their capacity (energy density) may be higher than that of the lithium/sulfur dioxide cell.

Solid-Electrolyte Cells.   These cells are noted for their extremely long storage life, in excess of 20 years, but are capable of only low-rate discharge in the microampere range. They are used in applications such as memory backup, cardiac pacemakers, and similar equipment where current requirements are low but long life is critical (see Chap. 15).

In Fig. 14.1 the size or capacity of these three types of lithium cells (up to the 30-Ah size) is plotted against the current levels at which they are typically discharged. The approximate weight of lithium in each of these cells is also shown.

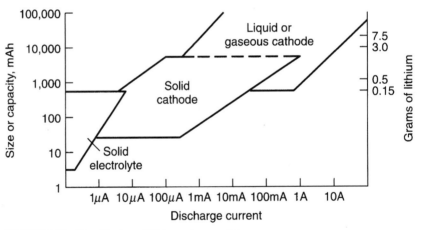

FIGURE 14.1   Classification of lithium primary cell types.

14.2 CHEMISTRY

14.2.1 Lithium

The main requirements for electrode materials used for high-performance (high specific energy and energy density) batteries are a high electrochemical equivalence (high coulombic output for a given weight of material) and a high electrode potential. It is apparent from Table 14.2, which lists the characteristics of metals used as battery anodes, that lithium is an outstanding candidate. Its standard potential and electrochemical equivalence are the highest of the metals; it excels in theoretical gravimetric energy density; and, with its high potential, it is inferior only to aluminum and magnesium on a volumetric energy basis (Watthours per liter). Aluminum, however, has not been used successfully as an anode except in reserve systems, and magnesium has a low practical operating voltage. Furthermore, lithium is preferred to the other alkali metals because of its better mechanical characteristics and lower reactivity. Calcium has been investigated as an anode, in place of lithium, because its higher melting point (838°C compared with 180.5°C for lithium) may result in safer operation, reducing the possibility of thermal runaway should high internal cell temperatures occur. To date, practical cells using calcium have not been produced.

Lithium is one of the alkali metals and it is the lightest of all the metallic elements, with a density about half that of water. When first made or freshly cut, lithium has the luster and color of bright silver, but it tarnishes rapidly in moist air. It is soft and malleable, can be readily extruded into thin foils, and is a good conductor of electricity. Table 14.3 lists some of the physical properties of lithium.[3,4]

TABLE 14.2 Characteristics of Anode Materials

Material	Atomic weight, g	Standard potential at 25°C, V	Density, g/cm³	Melting point, °C	Valence change	Electrochemical equivalence		
						Ah/g	g/Ah	Ah/cm³
Li	6.94	−3.05	0.534	180	1	3.86	0.259	2.08
Na	23.0	−2.7	0.97	97.8	1	1.16	0.858	1.12
Mg	24.3	−2.4	1.74	650	2	2.20	0.454	3.8
Al	26.9	−1.7	2.7	659	3	2.98	0.335	8.1
Ca	40.1	−2.87	1.54	851	2	1.34	0.748	2.06
Fe	55.8	−0.44	7.85	1528	2	0.96	1.04	7.5
Zn	65.4	−0.76	7.1	419	2	0.82	1.22	5.8
Cd	112	−0.40	8.65	321	2	0.48	2.10	4.1
Pb	207	−0.13	11.3	327	2	0.26	3.87	2.9

TABLE 14.3 Physical Properties of Lithium

Melting point	180.5°C
Boiling point	1347°C
Density	0.534 g/cm³ (25°C)
Specific heat	0.852 cal/g (25°C)
Specific resistance	$9.35 \times 10^6 \Omega \cdot cm$ (20°C)
Hardness	0.6 (Mohs scale)

Lithium reacts vigorously with water, releasing hydrogen and forming lithium hydroxide,

$$2Li + 2H_2O \rightarrow 2LiOH + H_2$$

This reaction is not as vigorous as that of sodium and water, probably due to the fairly low solubility and the adherence of LiOH to the metal surface under some conditions, however, the heat generated by this reactors may ignite the hydrogen which is formed and the lithium will then also burn. Because of this reactivity, however, lithium must be handled in a dry atmosphere and, in a battery, be used with nonaqueous electrolytes. (The lithium-water battery, described in Chap. 38, is an exception to this condition.)

14.2.2 Cathode Materials

A number of inorganic and organic materials have been examined for use as the cathode in primary lithium batteries.[1,5] The critical requirements for this material to achieve high performance are high battery voltage, high energy density, and compatibility with the electrolyte (that is, being essentially nonreactive or insoluble in the electrolyte). Preferably the cathode material should be conductive, although there are few such materials available and solid cathode materials are usually mixed with a conducting material, such as carbon, and applied to a conductive grid to provide the needed conductivity. If the cathode reaction products are a metal and a soluble salt (of the anode metal), this feature can improve cathode conductivity as the discharge proceeds. Other desirable properties are the cathode material are low cost, availability (noncritical material), and favorable physical properties, such as nontoxicity and nonflammability. Table 14.4 lists some of the cathode materials that have been studied for primary lithium batteries and gives their cell reaction mechanisms and the theoretical cell voltages and capacities.

14.2.3 Electrolytes

The reactivity of lithium in aqueous solutions requires the use of nonaqueous electrolytes for lithium anode batteries.[5] Polar organic liquids are the most common electrolyte solvents for the active primary cells, except for the thionyl chloride ($SOCl_2$) and sulfuryl chloride (SO_2Cl_2) cells, where these inorganic compounds serve as both the solvent and the active cathode material. The important properties of the electrolyte are:

1. It must be aprotic, that is, have no reactive protons or hydrogen atoms, although hydrogen atoms may be in the molecule.
2. It must have low reactivity with lithium (or form a protective coating on the lithium surface to prevent further reaction) and the cathode.
3. It must be capable of forming an electrolyte of good ionic conductivity.
4. It should be liquid over a broad temperature range.
5. It should have favorable physical characteristics, such as low vapor pressure, stability, nontoxicity, and nonflammability.

A listing of the organic solvents commonly used in lithium batteries is given in Table 14.5. These solvents are typically employed in binary or ternary combination. These organic electrolytes, as well as thionyl chloride (mp $-105°C$, bp $78.8°C$) and sulfuryl chloride (mp $-54°C$, bp $69.1°C$), are liquid over a wide temperature range with low freezing points. This characteristic provides the potential for operation over a wide temperature range, particularly low temperatures.

TABLE 14.4 Cathode Materials Used in Lithium Primary Batteries

Cathode material	Molecular weight	Valence change	Density, g/cm³	Theoretical faradic capacity (cathode only) Ah/g	Ah/cm³	g/Ah	Cell reaction mechanism (with lithium anode)	Theoretical cell Voltage, V	Specific Energy Wh/kg
SO_2	64	1	1.37	0.419	—	2.39	$2Li + 2SO_2 \rightarrow 2Li_2S_2O_4$	3.1	1170
$SOCl_2$	119	2	1.63	0.450	—	2.22	$4Li + 2SOCl_2 \rightarrow 4LiCl + S + SO_2$	3.65	1470
SO_2Cl_2	135	2	1.66	0.397	—	2.52	$2Li + SO_2Cl_2 \rightarrow 2LiCl + SO_2$	3.91	1405
Bi_2O_3	466	6	8.5	0.35	2.97	2.86	$6Li + Bi_2O_3 \rightarrow 3Li_2O + 2Bi$	2.0	640
$Bi_2Pb_2O_5$	912	10	9.0	0.29	2.64	2.41	$10Li + Bi_2Pb_2O_5 \rightarrow 5Li_2O + 2Bi + 2Pb$	2.0	544
$(CF)_n$	31	1	2.7	0.86	2.32	1.16	$nLi + (CF)_n \rightarrow nLiF + nC$	3.1	2180
$CuCl_2$	134.5	2	3.1	0.40	1.22	2.50	$2Li + CuCl_2 \rightarrow 2LiCl + Cu$	3.1	1125
CuF_2	101.6	2	2.9	0.53	1.52	1.87	$2Li + CuF_2 \rightarrow 2LiF + Cu$	3.54	1650
CuO	79.6	2	6.4	0.67	4.26	1.49	$2Li + CuO \rightarrow Li_2O + Cu$	2.24	1280
$Cu_4O(PO_4)_2$	458.3	8	—	0.468	—	2.1	$8Li + Cu_4O(PO_4)_2 \rightarrow Li_2O + 2Li_3PO_4 + Cu$	2.7	—
CuS	95.6	2	4.6	0.56	2.57	1.79	$2Li + CuS \rightarrow Li_2S + Cu$	2.15	1050
FeS	87.9	2	4.8	0.61	2.95	1.64	$2Li + FeS \rightarrow Li_2S + Fe$	1.75	920
FeS_2	119.9	4	4.9	0.89	4.35	1.12	$4Li + FeS_2 \rightarrow 2Li_2S + Fe$	1.8	1304
MnO_2	86.9	1	5.0	0.31	1.54	3.22	$Li + Mn^{IV}O_2 \rightarrow Mn^{III}O_2(Li^+)$	3.5	1005
MoO_3	143	1	4.5	0.19	0.84	5.26	$2Li + MoO_3 \rightarrow Li_2O + Mo_2O_5$	2.9	525
Ni_3S_2	240	4	—	0.47	—	2.12	$4Li + Ni_3S_2 \rightarrow 2Li_2S + 3Ni$	1.8	755
$AgCl$	143.3	1	5.6	0.19	1.04	5.26	$Li + AgCl \rightarrow LiCl + Ag$	2.85	515
Ag_2CrO_4	331.8	2	5.6	0.16	0.90	6.25	$2Li + Ag_2CrO_4 \rightarrow Li_2CrO_4 + 2Ag$	3.35	515
$AgV_2O_{5.5}$*	297.7	3.5	—	0.282	—	—	$3.5Li + AgV_2O_{5.5} \rightarrow Li_{3.5}AgV_2O_{5.5}$	3.24	655
V_2O_5	181.9	1	3.6	0.15	0.53	6.66	$Li + V_2O_5 \rightarrow LiV_2O_5$	3.4	490

*Multiple-step discharge; see Ref. 11 (Experimental values to +1.5 V cut-off).

TABLE 14.5 Properties of Organic Electrolyte Solvents for Lithium Primary Batteries

Solvent	Structure	Boiling point at 10^5 Pa, °C	Melting point, °C	Flash point, °C	Density at 25°C, g/cm³	Specific conductivity with 1M LiClO$_4$, S/cm^{-1}
Acetonitrile (AN)	$H_3C-C\equiv N$	81	−45	5	0.78	3.6×10^{-2}
γ-Butyrolactone (BL)	(cyclic ester)	204	−44	99	1.1	1.1×10^{-2}
Dimethylsulfoxide (DMSO)	$H_3C-S-CH_3$ ($\parallel O$)	189	18.5	95	1.1	1.4×10^{-2}
Dimethylsulfite (DMSI)	($O=S(OCH_3)_2$)	126	−141		1.2	
1,2-Dimethoxyethane (DME)	$H_2C-O-CH_3$ / $H_2C-O-CH_3$	83	−60	1	0.87	
Dioxolane (1,3-D)	(cyclic)	75	−26	2	1.07	
Methyl formate (MF)	$H-C-O-CH_3$ ($\parallel O$)	32	−100	−19	0.98	3.2×10^{-2}
Nitromethane (NM)	H_3C-NO_2	101	−29	35	1.13	1×10^{-2}
Propylene carbonate (PC)	(cyclic carbonate)	242	−49	135	1.2	7.3×10^{-3}
Tetrahydrofuran (THF)	$H_2C-CH_2-CH_2-CH_2-O$ (cyclic)	65	−109	−15	0.89	

The Jet Propulsion Laboratory (Pasadena, CA) has evaluated several types of lithium primary batteries to determine their ability to operate planetary probes at temperatures of −80°C and below.[6] Individual cells were evaluated by discharge tests and Electrochemical Impedance Spectroscopy. Of the five types considered ($Li/SOCl_2$, Li/SO_2, Li/MnO_2, Li-BCX and Li-CFn), lithium-thionyl chloride and lithium-sulfur dioxide were found to provide the best performance at −80°C. Lowering the electrolyte salt to ca. 0.5 molar was found to improve performance with these systems at very low temperatures. In the case of D-size Li/$SOCl_2$ batteries, lowering the $LiAlCl_4$ concentration from 1.5 to 0.5 molar led to a 60% increase in capacity on a baseline load of 118 ohms with periodic one-minute pulses at 5.1 ohms at −85 C.

Lithium salts, such as $LiClO_4$, $LiBr$, $LiCF_3SO_3$, and $LiAlCl_4$, are the electrolyte solutes most commonly used to provide ionic conductivity. The solute must be able to form a stable electrolyte which does not react with the active electrode materials. It must be soluble in the organic solvent and dissociate to form a conductive electrolyte solution. Maximum conductivity with organic solvents is normally obtained with a 1-Molar solute concentration, but generally the conductivity of these electrolytes is about one-tenth that of aqueous systems. To accommodate this lower conductivity, close electrode spacing and cells designed to minimize impedance and provide good power density are used.

14.2.4 Cells Couples and Reaction Mechanisms

The overall discharge reaction mechanism for various lithium primary batteries is shown in Table 14.4, which also lists the theoretical cell voltage. The mechanism for the discharge of the lithium anode is the oxidation of lithium to form lithium ions (Li^+) with the release of an electron,

$$Li \rightarrow Li^+ + e$$

The electron moves through the external circuit to the cathode, where it reacts with the cathode material, which is reduced. At the same time, the Li^+ ion, which is small (0.06 nm in radius) and mobile in both liquid and solid-state electrolytes, moves through the electrolyte to the cathode, where it reacts to form a lithium compound.

A more detailed description of the cell reaction mechanism for the different lithium primary batteries is given in the sections on those battery systems.[1,7]

14.3 CHARACTERISTICS OF LITHIUM PRIMARY BATTERIES

14.3.1 Summary of Design and Performance Characteristics

A listing of the major lithium primary batteries now in production or advanced development and a summary of their constructional features, key electrical characteristics, and available sizes are presented in Table 14.6. The types of batteries, their sizes, and some characteristics are subject to change depending on design, standardization, and market development. Manufacturers' data should be obtained for specific characteristics. The performance characteristics of these systems, under theoretical conditions, are given in Table 14.4. Comparisons of the performance of the lithium batteries with comparably sized conventional primary batteries are covered in Secs. 6.4 and 7.3. Detailed characteristics of some of these batteries are covered in Secs. 14.5 to 14.12.

TABLE 14.6 Characteristics of Lithium Primary Batteries

Soluble cathode batteries

System	Cathode	Electrolyte Solvent	Solute	Separator	Construction	Voltage, V Nominal	Working* (20°C)	Specific energy† Wh/kg	Energy density† Wh/L	Power density	Discharge profile	Available sizes
Lithium/sulfur dioxide (Li/SO₂)	SO₂ with carbon and binder on Al screen	AN	LiBr	Microporous Polypropylene	Spiral "jelly-roll" cylindrical construction; glass-to-metal seal	3.0	2.9–2.7	260	415	High	Very flat	Cylindrical batteries up to 35 Ah
Lithium/thionyl chloride (Li/SOCl₂) Low rate	SOCl₂ with carbon and binder on Ni or SS	SOCl₂	LiAlCl₄	Glass non-woven	Wafer construction	3.6	3.6–3.4	275	630	Low	Flat	0.4–1.7 Ah
					"Bobbin" in cylindrical construction	3.6	3.5–3.3	590	1100	Medium	Flat	Cylindrical batteries 1.2–19
High capacity					Prismatic with flat plates	3.6	3.5–3.3	480	950	Medium	Flat	12–10,000 Ah
High rate					Spiral "jelly-roll" cylindrical construction or flat disk	3.6	3.5–3.2	380	725	Medium to high	Flat	Cylindrical: 5–23 Ah Flat disk: up to 320 Ah
		SOCl₂ with halogen additives	LiAlCl₄	Glass mat	Spiral "jelly-roll" cylindrical construction	3.9	3.8–3.3	450	900	Medium	Flat	2–30 Ah
Lithium/sulfuryl chloride (Li/SO₂Cl₂)	SO₂Cl₂ with carbon and binder SS screen	SO₂Cl₂ (some with additives)	LiAlCl₄	Glass	Spiral "jelly-roll" cylindrical construction; glass-to-metal seal	3.95	3.5–3.1	450	900	Medium to high	Flat	7–30 Ah

TABLE 14.6 Characteristics of Lithium Primary Batteries (*Continued*)

Solid cathode batteries

System	Cathode	Electrolyte solvent	Salt	Separator	Construction	Nominal voltage, V	Working voltage, V	Energy density (Wh/kg)	Energy density (Wh/L)	Rate capability	Discharge profile	Sizes and comments
Lithium/carbon monofluoride (Li(CF)$_n$)	CF with carbon and binder on nickel collector	PC + DME or BL	LiBF$_4$ or LiAsF$_6$	Polypropylene	"Coin" construction crimped seal Pin type	3.0	2.7–2.5	215	550	Low to medium Low	Moderately flat Humped	Coin batteries to 500 mAh Small cylinders 25–50 mAh
					Spiral "jelly-roll" cylindrical construction crimped or glass-to-metal seal			250 (commercial) 590 (military)	635 (commercial) 1050 (military)			Cylindrical batteries to 5 Ah (commercial) and 1200 Ah (military)
Lithium/copper oxide (Li/CuO)	CuO pressed in cell can	1,3D	LiClO$_4$	Nonwoven glass	Rectangular with flat plates	1.5	1.5–1.4	440 (biomedical)	900	Low	High initial voltage drop, then moderatley flat	Rectangular batteries to 40 Ah
					"Bobbin" inside-out cylindrical construction:			280	650			Cylindrical batteries 500–3500 mAh
Lithium/iron disulfide (LiFeS$_2$)	FeS$_2$				"Jelly-roll" cylindrical construction: crimped seal	1.5	1.6–1.4	260	500	Medium to high	High initial drop, then flat	AA-size
Lithium/manganese dioxide (Li/MnO$_2$)	MnO$_2$ with carbon and binder on supporting grid	PC + DME	Li salt	Polypropylene	"Coin construction with flat electrodes	3.0	3.0–2.7	230	545	Low to medium	Moderately flat	Coin batteries 65–1000 mAh
		Organic solvent	Li salt	Polypropylene	"Jelly-roll" cylindrical construction; crimped and hermetic seals	3.0	2.8–2.5	230	535	Medium to high	Moderately flat	$\frac{2}{3}$A Cylindrical batteries typical, larger cells available to 33 Ah
		Organic solvent	Li salt	Polypropylene	"Bobbin" cylindrical construction	3.0	3.0–2.8	270	620	Low to medium	Moderately flat	Cylindrical batteries to 1.75 Ah
Lithium/silver vanadium oxide (Li/AgV$_4$O$_{11}$)	AgV$_2$O$_{5.5}$ with graphite and carbon	PC, DME	LiAsF$_6$	Microporous polypropylene	Rounded prismatic and D-shaped cross section	3.2	3.2–1.5	270	780	Low to mdium	Multiple plateaus	Special sizes for implantable medical devices

*Working voltages are typical for discharges at favorable loads.

†Energy densities are for 20°C, under favorable discharge conditions. See details in appropriate sections.

14.3.2 Soluble-Cathode Lithium Primary Batteries

Two types of soluble-cathode lithium primary batteries are currently available (Table 14.1). One uses SO_2 as the active cathode dissolved in an organic electrolyte solvent. The second type uses an inorganic solvent, such as the oxychlorides $SOCl_2$ and SO_2Cl_2, which serves as both the active cathode and the electrolyte solvent. These materials form a passivating layer or protective film of reaction products on the lithium surface, which inhibits further reaction. Even though the active cathode material is in contact with the lithium anode, self-discharge is inhibited by the protective film which proceeds at very low rates and the shelf life of these batteries is excellent. This film, however, may cause a voltage delay to occur: i.e. a time delay to break down the film and for the cell voltage to reach the operating level when the discharge load is applied. These lithium batteries have a high specific energy and, with proper design, such as the use of high-surface-area electrodes, are capable of delivering high specific energy at high specific power.

These cells generally require a hermetic-type seal. Sulfur dioxide is a gas at 20°C (bp −10°C), and the undischarged cell has an internal pressure of 3 to 4×10^5 Pa at 20°C. The oxychlorides are liquid at 20°C, but with boiling points of 78.8°C for $SOCl_2$ and 69.1°C for SO_2Cl_2, a moderate pressure can develop at high operating temperatures. In addition, as SO_2 is a discharge product in the oxychloride cells, the internal cell pressure increases as the cell is discharged.

The lithium/sulfur dioxide (Li/SO_2) battery is the most advanced of these lithium primary batteries. These batteries are typically manufactured in cylindrical configurations in capacities up to about 35 Ah. They are noted for their high specific power (about the highest of the lithium primary batteries, high energy density, and good low-temperature performance. They are used in military and specialized industrial, space and commercial applications where these performance characteristics are required.

The lithium/thionyl chloride ($Li/SOCl_2$) battery has one of the highest specific energies of all the practical battery systems. Figures 7.8 and 7.9 illustrate the advantages of the $Li/SOCl_2$ battery over a wide temperature range at moderate discharge rates. Figure 14.2 compares the discharge profile of the $Li/SOCl_2$ cell with the Li/SO_2 cell. At 20°C, at moderate discharge rates, the $Li/SOCl_2$ cell has a higher working voltage and about a 25% advantage in service life. The Li/SO_2 cell, however, does have better performance at low temperatures and high discharge rates and a lower voltage delay after storage. $Li/SOCl_2$ cells have been fabricated in many sizes and designs ranging from small button and cylindrical cells with

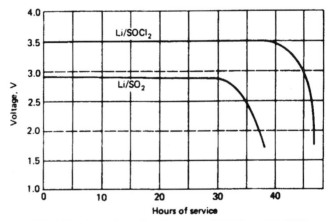

FIGURE 14.2   Comparison of performance of Li/SO_2 and $Li/SOCl_2$ C-size batteries; 100-mA discharge load at 20°C.

capacities below 1 Ah, to large prismatic cells with capacities as high as 10,000 Ah. Low-rate cells have been used successfully in many applications, especially as memory backup, for many years; high-rate cells are used in special applications.

The lithium/sulfuryl chloride (Li/SO_2Cl_2) battery has potential advantages because of its higher voltage (3.9 open-circuit voltage) and resultant higher specific energy. Suitable cathode electrode formulations and cell designs have been investigated to achieve the full capability of this electrochemical system. Figure 14.3 shows a comparison of the cathode polarization for Li/SO_2Cl_2 and $Li/SOCl_2$ batteries. Halogen additives have been used as a means to improve performance. Halogen additives are also used in some $SOCl_2$ cells.

Calcium has been investigated as an anode material in place of lithium in thionyl chloride cells. Safer operation was anticipated with calcium since its melting temperature of 838°C is not likely to be reached by any internally driven cell condition. While the discharge voltage is about 0.4 V lower than for the $Li/SOCl_2$ cell (open-circuit voltage is 3.25 V), the Ca/$SOCl_2$ cell has a flat discharge profile and about the same volumetric ampere-hour capacity. Shelf-life characteristics are also similar to those of the lithium anode cell.[9,10] However, calcium is significantly more difficult to process than lithium and passivation is a more significant factor. To date, no calcium-thionyl chloride batteries have been commercialized.

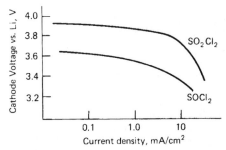

FIGURE 14.3 Comparison of cathode polarization curves, Li/SO_2Cl_2 vs. $Li/SOCl_2$ batteries. (*From Ref. 8.*)

14.3.3 Solid-Cathode Lithium Primary Cells

The solid-cathode lithium batteries are generally used in low- to moderate-drain applications and are manufactured mainly in small flat or cylindrical sizes ranging in capacity from 30 mAh to about 5 Ah, depending on the particular electrochemical system. Larger batteries have been produced in cylindrical and prismatic configurations. A comparison of the performance of solid-cathode lithium batteries and conventional batteries is presented in Chap. 7.

The solid-cathode batteries have the advantage, compared with the soluble-cathode lithium primary batteries, of being nonpressurized and thus not requiring a hermetic-type seal. A mechanically crimped seal with a polymeric gasket is satisfactory for most applications. On light discharge loads, the energy density of some of the solid-cathode systems is comparable to that of the soluble-cathode systems, and in smaller battery sizes may be greater. Their disadvantages, again compared with the soluble-cathode batteries, are a lower rate capability, poorer low-temperature performance, and a more sloping discharge profile.

To maximize their high-rate performance and compensate for the lower conductivity of the organic electrolytes, designs are used for these lithium cells to increase electrode area, such as a larger-diameter coin cell instead of button cells, multiple parallel electrodes, or the spirally wound jelly-roll construction for the cylindrical cells.

A number of different cathode materials have been used in the solid-cathode lithium cells. These are listed in Tables 14.4 and 14.6, which present some of the theoretical and practical performance data of these cells. The major features of the solid-cathode lithium cells are compared in Table 14.7. Many of the characteristics are similar, such as high specific energy and energy density and good shelf life. An important property is the 3-V cell voltage obtained with several of these cathodes. Some cathode materials have been used mainly in the coin or button cell designs while others, such as the manganese dioxide cathode, have been used in coin, cylindrical, and prismatic cells as well as in both high (spirally wound) and low (bobbin) rate designs.

Although a number of different solid-cathode lithium batteries have been developed and even manufactured, more recently the trend is toward reducing the number of different chemistries that are manufactured. The lithium/manganese dioxide (Li/MnO_2) battery was one of the first to be used commercially and is still the most popular system. It is relatively inexpensive, has excellent shelf life, good high-rate and low-temperature performance, and is available in coin and cylindrical cells. The lithium/carbon monofluoride ($Li\{CF\}_n$) battery

TABLE 14.7 Characteristics of Typical Lithium/Solid-Cathode Batteries

Type of battery	Operating voltage, V	Characteristics
Li/MnO_2	3.0	High specific energy and energy density; wide operating temperature range (-20 to 55°C); performance at relatively high discharge rates; minimal voltage delay; relatively low cost; available in flat (coin) and cylindrical batteries (high and low rates)
$Li/(CF)_n$	2.8	Highest theoretical specific energy, low- to moderate-rate capability; wide operating temperature range (-20 to 60°C); flat discharge profile; available in flat (coin), cylindrical and prismatic designs.
$Li/Cu_4O(PO_4)_2$	2.5	High specific energy; long storage life; operating temperature range up to 175°C; low- to moderate-rate capability; not currently available.
Li/CuO	1.5	Highest theoretical volumetric coulombic capacity (Ah/L); long storage life; low- to moderate-rate capability; operating temperature range up to 125 to 150°C; no apparent voltage delay. Potential replacement for alkaline-manganese but not currently available.
Li/FeS_2	1.5	Replacement for conventional zinc-carbon and alkaline-manganese dioxide batteries; higher power capability than conventional batteries and better low-temperature performance and storability. Currently available in AA size as a direct replacement for alkaline-manganese
Li/Ag_2CrO_4	3.1	High voltage, high specific energy and energy density; low-rate capability; high reliability; used in low-rate, long-term applications; high cost
$Li/AgV_2O_{5.5}$	3.2	High specific energy and energy density multiple-step discharge; good rate capability; used in implantable and other medical devices
Li/V_2O_5	3.3	High energy density; two-step discharge; used in reserve cells (Chap. 20).

is another of the early solid-cathode batteries and is attractive because of its high theoretical capacity and flat discharge characteristics. It is also manufactured in coin, cylindrical and prismatic configurations. The higher cost of polycarbon monofluoride has affected the commercial potential for this system but it is finding use in biomedical, military and space applications where cost is not a factor. The lithium/copper oxyphosphate $(Li/Cu_4O(PO_4)_2)$ battery was designed for high temperatures and special applications. It has a high specific energy and long shelf life under adverse environmental conditions but is not currently being manufactured.

The lithium/silver chromate (Li/Ag_2CrO_4) battery is noted for its high volumetric energy density for low-rate long-term applications. Its high cost has limited its use to special applications. The lithium/vanadium pentoxide (Li/V_2O_5) battery has a high volumetric energy density, but with a two-step discharge profile. Its main application has been in reserve batteries (Chap. 20). The lithium/silver vanadium oxide $(Li/AgV_2O_{5.5})$ battery is used in medical applications which have pulse load requirements as this battery is capable of relatively high-rate discharge.[11] The other solid-cathode lithium batteries operate in the range of 1.5 V and were developed to replace conventional 1.5-V button or cylindrical cells. The lithium/copper oxide (Li/CuO) cell is noted for its high coulombic energy density and has the advantage of higher capacity or lighter weight when compared with conventional cylindrical cells. It is capable of performance at high temperatures and, similar to the copper oxyphosphate cell, has a long shelf life under adverse conditions. The iron disulfide (Li/FeS_2) cell has similar advantages over the conventional cells, plus the advantage of high-rate performance. Once available in a button cell configuration, it is now being marketed commercially in a high-rate cylindrical AA-size battery as a replacement for zinc-carbon and alkaline-manganese dioxide batteries.

The remaining solid-cathode systems listed in Tables 14.1 and 14.4 are not currently commercially available.

Typical discharge curves for the major solid-cathode batteries are shown in Fig. 14.4. The discharge curves of the Li/SO_2 and $Li/SOCl_2$ batteries showing their flatter discharge profile, are also plotted for comparison purposes.

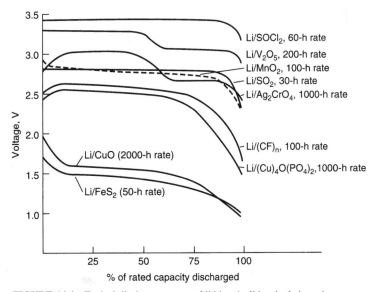

FIGURE 14.4 Typical discharge curves of lithium/solid-cathode batteries.

A comparison of the performance of several of the solid-cathode batteries in a low-rate button configuration and the higher-rate cylindrical configuration is presented in Secs. 6.4 and 7.3. In the button configuration (Table 7.4), the lithium batteries have an advantage in specific energy (Wh/kg) over many of the conventional batteries. This advantage may not be too important in these small battery sizes, but the lithium batteries have an advantage of lower cost, particularly when compared with the silver cells, and longer shelf life. In addition, the zinc/mercuric oxide battery, which once dominated the button battery market, and the cadmium/mercuric oxide battery are being phased out because of their use of hazardous mercury and cadmium.

In the larger cylindrical sizes (Table 7.5) the lithium cells have an advantage in both volumetric and gravimetric energy density. In some designs this advantage is even more significant at higher discharge loads. Figure 7.3 shows another comparison of the performance of solid-cathode and soluble-cathode lithium cells with aqueous cells.

It is important when making these comparisons, as discussed in Chap. 3, to identify the specific discharge conditions of the application since the comparative performance of each battery system can vary depending on the discharge conditions. For example, as shown in Fig. 14.5, the lithium/copper oxide battery, designed for optimum performance at low discharge rates, has a comparatively high energy output when discharged at these light discharge loads, but the output drops off considerably at high rates. The similarly sized high-rate spirally wound configuration for the lithium/manganese dioxide cell has a lower energy output at the low discharge rates, but can maintain this performance as the discharge rate is increased. The performance of each of the battery systems, under various discharge conditions, is presented in the sections discussing each specific system.

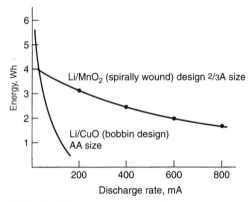

FIGURE 14.5   Comparison of Li/CuO and Li/MnO$_2$ batteries at 20°C. Batteries are equivalently sized.

The selection of a lithium versus a conventional cell thus becomes a tradeoff between the lower initial cost of most of the conventional cells, the performance advantages of the lithium cells, and the key requirements of the specific application.

14.4 SAFETY AND HANDLING OF LITHIUM BATTERIES

14.4.1 Factors Affecting Safety and Handling

Attention must be given to the design and use of lithium cells and batteries to ensure safe and reliable operation. As with most battery systems, precautions must be taken to avoid physical and electrical abuse because some batteries can be hazardous if not used properly. This is important in the case of lithium cells since some of the components are toxic or inflammable[12] and the relatively low melting point of lithium (180.5°C) indicates that cells must be prevented from reaching high internal temperatures.

Because of the variety of lithium cell chemistries, designs, sizes, and so on, the procedures for their use and handling are not the same for all cells and batteries and depend on a number of factors such as the following:

1. *Electrochemical system:* The characteristics of the specific chemicals and cell components influence operational safety.

2. *Size and capacity of cell and battery:* Safety is directly related to the size of the cell and the number of cells in a battery. Small cells and batteries, containing less material and, therefore, less total energy, are "safer" than larger cells of the same design and chemistry.

3. *Amount of lithium used:* The less lithium that is used, implying less energetic cells, the safer they should be.

4. *Cell design:* High-rate designs, capable of high discharge rates, versus low-power designs where discharge rate is limited, use of "balanced" cell chemistry, adequate intra- and intercell electrical connections, and other features affect cell performance and operating characteristics.

5. *Safety features:* The safety features incorporated in the cell and battery will obviously influence handling procedures. These features include cell-venting mechanisms to prevent excessive internal cell pressure, thermal cutoff devices to prevent excessive temperatures, electrical fuses, PTC devices and diode protection. Cells are hermetically or mechanically crimped-sealed, depending on the electrochemical system, to effectively contain cell contents if cell integrity is to be maintained.

6. *Cell and battery containers:* These should be designed so that cells and batteries will meet the mechanical and environmental conditions to which they will be exposed. High shock, vibration, extremes of temperature, or other adverse conditions may be encountered in use and handling, and the cell and battery integrity must be maintained. Container materials should also be chosen with regard to their flammability and the toxicity of combustion products in the event of fire. Container designs should also be optimized to dissipate the heat generated during discharge and to release pressure in the event of cell venting.

14.4.2 Safety Considerations

The electrical and physical abuses that may arise during the use of lithium cells are listed in Table 14.8 together with some generalized comments on corrective action. The behavior of specific cells is covered in the other sections of this chapter. The manufacturer's data should be consulted for more details on the performance of individual cells.

High-Rate Discharges or Short-Circuiting. Low-capacity batteries, or those designed as low-rate batteries, may be self-limiting and not capable of high-rate discharge. The temperature rise will thus be minimal and there will be no safety problems. Larger or high-rate cells can develop high internal temperatures if short-circuited or operated at excessively high

TABLE 14.8   Considerations for Use and Handling of Lithium Primary Batteries

Abusive condition	Corrective procedure
High-rate discharging or short-circuiting	Low-capacity or low-rate batteries may be self-limiting Electrical fusing, thermal protection Limit current drain; apply battery properly
Forced discharge (cell reversal)	Voltage cutoff Use low-voltage batteries Limit current drain Special designs ("balanced" cell) Use of diode in parallel across cell
Charging	Prohibit charging Diode protection to prevent or limit charging current
Overheating	Limit current drain Fusing, thermal cutoff, PTC devices Design battery properly Do not incinerate
Physical abuse	Avoid opening, puncturing, or mutilating cells Maintain battery integrity

rates. These cells are generally equipped with safety vent mechanisms to avoid more serious hazards. Such cells or batteries should be fuse-protected (to limit the discharge current). Thermal fuses or thermal switches should also be used to limit the maximum temperature rise. Positive temperature coefficient (PTC) devices are used in cells and batteries to provide this protection.

Forced Discharge or Voltage Reversal.   Voltage reversal can occur in a multicell series-connected battery when the better performing cells can drive the poorer cell below 0 V, into reversal, as the battery is discharged toward 0 V. In some types of lithium cells this forced discharge can result in cell venting or, in more extreme cases, cell rupture. Precautionary measures include the use of voltage cutoff circuits to prevent a battery from reaching a low voltage, the use of low-voltage batteries (since this phenomenon is unlikely to occur with a battery containing only a few cells in series), and limiting the current drain since the effect of forced discharge is more pronounced on high-rate discharges. Special designs, such as the "balanced" Li/SO_2 cell (see Sec. 14.5), also have been developed that are capable of withstanding this discharge condition. The use of a current collector in the anode maintains lithium integrity and may provide an internal shorting mechanism to limit the voltage in reversal.

Charging.   Lithium batteries, as well as the other primary batteries, are not designed to be recharged. If they are, they may vent or explode. Batteries which are connected in parallel or which may be exposed to a charging source (as in battery-backup CMOS memory circuits) should be diode-protected to prevent charging (see Sec. 5.2.4).

Overheating.   As discussed, overheating should be avoided. This can be accomplished by limiting the current drain, using safety devices such as fusing and thermal cutoffs, and designing the battery to provide necessary heat dissipation.

Incineration.   Lithium cells are either hermetically or mechanically sealed. They should not be incinerated without proper protection because they may rupture or explode at high temperatures.

Currently special procedures govern the transportation and shipment of lithium batteries and procedures for the use, storage, and handling of lithium batteries have been recommended.[13] Disposal of some types of lithium cells also is regulated. The latest issue of these regulations should be consulted for the most recent procedures (see Sec. 4.10 for details.) The U.S. Federal Aviation Agency has recently adapted technical standard order TSO-C142-Lithium Batteries, governing the installation and use of lithium primary batteries on commercial aircraft.[15]

14.5 LITHIUM/SULFUR DIOXIDE (Li/SO₂) BATTERIES

One of the more advanced lithium primary batteries, used mainly in military and in some industrial and space applications, is the lithium/sulfur dioxide (Li/SO_2) system. The battery has specific energy and energy density of up to 260 Wh/kg and 415 Wh/L respectively. The Li/SO_2 battery is particularly noted for its capability to handle high current and high power requirements, for its excellent low-temperature performance and for its long shelf life.

14.5.1 Chemistry

The Li/SO_2 cell uses lithium as the anode and a porous carbon cathode electrode with sulfur dioxide as the active cathode material. The cell reaction mechanism is

$$2Li + 2SO_2 \rightarrow Li_2S_2O_4\downarrow \text{ (lithium dithionite)}$$

As lithium reacts readily with water, a nonaqueous electrolyte, consisting of sulfur dioxide and an organic solvent, typically acetonitrile, with dissolved lithium bromide, is used. The specific conductivity of this electrolyte is relatively high and decreases only moderately with decreasing temperature (Fig. 14.6), thus providing a basis for good high-rate and low-temperature performance. About 70% of the weight of the electrolyte/depolarizer is SO_2. The internal cell pressure, in an undischarged cell, due to the vapor pressure of the liquid SO_2, is 3–4 $\times$ 10⁵ Pa at 20°C. The pressure at various temperatures is shown in Fig. 14.7. The mechanical features of the cell are designed to contain this pressure safely without leaking and to vent the electrolyte if excessively high temperatures and the resulting high internal pressures are encountered.

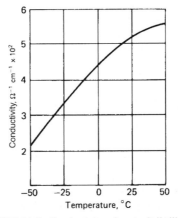

FIGURE 14.6 Conductivity of acetonitrile/lithium bromide/sulfur dioxide electrolyte (70% SO_2).

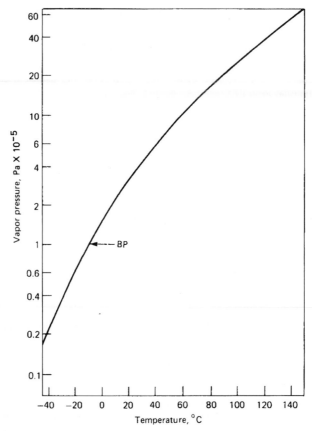

FIGURE 14.7 Vapor pressure of sulfur dioxide at various temperatures.

During discharge the SO_2 is used up and the cell pressure reduced somewhat. The discharge is generally terminated by the full use of available lithium, in designs where the lithium is the limiting electrode, or by blocking of the cathode by precipitation of the discharge product (cathode limited). Current designs are typically limited by the cathode so that some lithium remains at the end of discharge. The good shelf life of the Li/SO_2 cell results from the protective lithium dithionite film on the anode formed by the initial reaction of lithium and SO_2. It prevents further reaction and loss of capacity during storage.

Most Li/SO_2 cells are now fabricated in a balanced construction where the lithium:sulfur dioxide stoichiometric ratio is in the range of $Li:SO_2 = 0.9 - 1.05:1$. With the earlier designs, where the ratio was on the order of $Li:SO_2 = 1.5:1$, high temperatures, cell venting, or rupture and fires due to an exothermic reaction between residual lithium and acetonitrile, in the absence of SO_2, could occur on deep or forced discharge. Cyanides and methane can also be generated through this reaction. In the balanced cell the anode is protected by residual SO_2 and remains passivated. The conditions for the hazardous reaction are minimized since some protective SO_2 remains in the electrolyte.[16] A higher negative cell voltage, in reversal, of the balanced cell is also beneficial for diode protection, which is used in some designs to bypass the current through the cell and minimize the adverse effects of reversal.

The use of a current collector, typically an inlayed stripe of copper metal, also helps to maintain the integrity of the anode and leads to formation of a short-circuit mechanism since copper dissolution on cell reversal causes plated copper on the cathode to form an internal ohmic bridge.

14.5.2 Construction

The Li/SO$_2$ cell is typically fabricated in a cylindrical structure, as shown in Fig. 14.8. A jelly-roll construction is used, made by spirally winding rectangular strips of lithium foil, a microporous polypropylene separator, the cathode electrode (a Teflon-acetylene black mix pressed on an expanded aluminum screen), and a second separator layer. This design provides the high surface area and low cell resistance to obtain high-current and low-temperature performance. The roll is inserted in a nickel-plated steel can, with the positive cathode tab welded to the pin of a glass-to-metal seal and the anode tab welded to the cell case, the top is welded in place, and the electrolyte/depolarizer is added. The safety vent releases when the internal cell pressure reaches excessive levels, typically 2.41 MPa (350 psi) caused by inadvertent abusive use such as overheating or short-circuiting; and prevents cell rupture or explosion. The vent activates at approximately 95°C, well above the upper temperature limit for operation and storage, safely relieving the excess pressure and preventing possible cell rupture. Additional construction details have been previously described.[16] It is important to employ a corrosion-resistant glass or to coat the glass with a protective coating to prevent lithiation of the glass due to the potential difference between the cell case and the pin of the glass-to-metal seal.

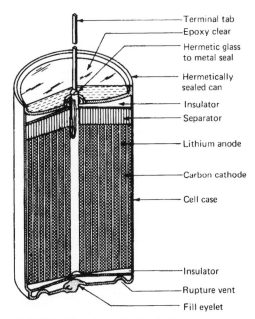

FIGURE 14.8 Lithium/sulfur dioxide batteries.

14.5.3 Performance

Voltage.   The open-circuit voltage of the Li/SO_2 battery is 2.95 V. The nominal voltage is usually specified as 3 V. The specific voltage on discharge is dependent on the discharge rate, discharge temperature, and state of charge; typical working voltages range between 2.7 and 2.9 V (see Figs. 14.9, 14.10, and 14.12). The cutoff voltage, the voltage by which most of the battery capacity has been exhausted, is typically 2 V.

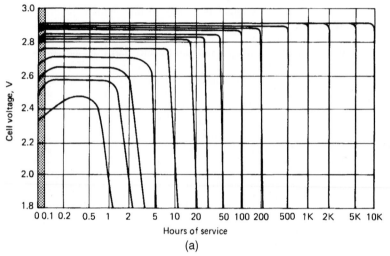

FIGURE 14.9(*a*)   Typical discharge characteristics of standard-rate Li/SO_2 battery at various loads at 20°C.

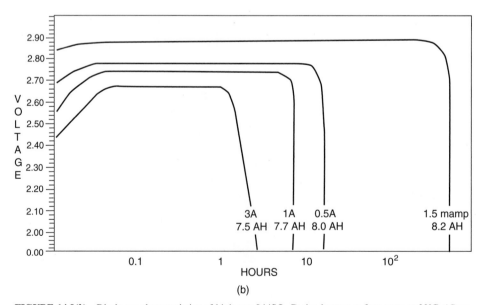

FIGURE 14.9(*b*)   Discharge characteristics of high-rate Li/SO_2 D-size battery at four rates at 23°C. (*Courtesy Hawker Eternacell, Inc.*).

Discharge.   Typical discharge curves for the standard-rate Li/SO$_2$ battery at 20°C are given in Fig. 14.9*a*. The high cell voltages and the flat discharge profile are characteristic of the Li/SO$_2$ battery. Another unique feature is the ability of the Li/SO$_2$ battery to be efficiently discharged over a wide range of current or power levels, from high-rate short-term or pulse loads to low-drain continuous discharges for periods of 5 years or longer. At least 90% of the battery's rated capacity may be expected on the long-term discharges. Figure 14.9*b* shows the discharge curves for a high-rate D-size battery at four rates up to 3 amps.

The high rate capability of the Li/SO$_2$ battery is illustrated in Chap. 6, Fig. 6.4, which compares the performance of various cylindrical primary and secondary batteries. The Li/SO$_2$ battery maintains a high capacity almost to the 1-h rate, whereas the performance of the zinc primary batteries begins to drop off significantly at the 20 to 50-h rate. The Li/SO$_2$ batteries are capable of higher-rate discharges on pulse loads. For example, a squat D cell designed in a high-rate construction can deliver pulse loads as high as 37.5 A producing 59 watts of power.[17] For high-rate designs, extended discharges, however, at rates above the 2-h rate may cause overheating. The actual heat rise depends on the battery design, type of discharge, temperature, and voltage. As discussed in Sec. 14.4, the design and use of the battery should be controlled to avoid overheating.

A recent study[17] has shown that the high-rate pulse output of the lithium/sulfur dioxide battery may be enhanced by a variety of design variables. Multiple tabbing (1 to 3) of both anode and cathode, optimizing the composition of the cathode mix and reducing the aspect ration (length/width) of the electrodes were all found to reduce polarization during high-rate, 10 s pulse discharge. D-size cells and thin D-size cells (1.1 in dia. × 2.20 in. high) with anodes and cathodes containing 2 tabs using an optimized cathode mix were found capable of producing 99 and 97 Watts respectively under 50 Amp, 10 s pulses. Ultimately, a 5/4 C-size cell without multiple tabbing but using the optimized cathode mix was selected for reasons of volumetric efficiency to produce a 74-cell, 110 V battery capable of providing 5500 Watt, 10 s pulses for a U.S. Navy application.

Effect of Temperature.   The Li/SO$_2$ battery is noted for its ability to perform over a wide temperature range, from −40 to 55°C. Discharge curves for a standard-rate Li/SO$_2$ battery at various temperatures are shown in Fig. 14.10. Significant, again, are the flat discharge curves over with wide temperature range, the good voltage regulation, and the high percentage of the 20°C performance available at the temperature extremes. As with all battery systems, the relative performance of the Li/SO$_2$ battery is dependent on the rate of discharge. In Fig. 14.11 the discharge performance of a standard-rate cell is plotted as a function of load and battery temperature.

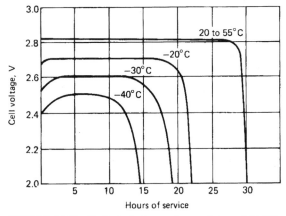

FIGURE 14.10   Typical discharge characteristics of Li/SO$_2$ battery at various temperatures, $C/30$ discharge rate.

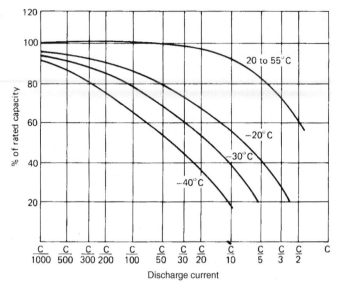

FIGURE 14.11 Performance of Li/SO$_2$ batteries as a function of discharge temperature and load.

Internal Resistance and Discharge Voltage. The Li/SO$_2$ battery has a relatively low internal resistance (about one-tenth that of conventional primary batteries) and good voltage regulation over a wide range of discharge loads and temperatures. The midprint voltage of the discharge of a standard-rate Li/SO$_2$ battery (to an end voltage of 2 V) at various discharge rates and temperatures is plotted in Fig. 14.12.

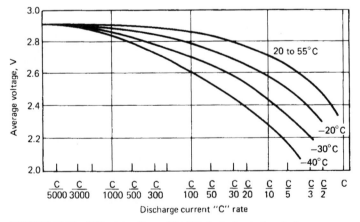

FIGURE 14.12 Midpoint voltage of Li/SO$_2$ batteries during discharge.

Service Life. The capacity or service life of the Li/SO_2 battery at various discharge rates given in Fig. 14.13. The data are normalized for a 1-kg or 1-L size battery and presented in terms of hours of service at various discharge rates. The linear shape of this curve is again indicative of the capability of the Li/SO_2 battery to be efficiently discharged at these extreme conditions. This data can be used in several ways to calculate the approximate performance of a given battery or to select a Li/SO_2 battery of suitable size for a particular application, recognizing that the specific energy of the larger-size batteries is higher than that of the smaller ones.

The service life of a battery at a given current load can be estimated by dividing the current (in Amperes) by the weight or volume of the battery. This value is located on the ordinate, and the service life, at a specific current and temperature, is read of the abscissa.

The weight or volume of a battery needed to deliver a required number of hours of service at a specified current load can be estimated by locating a point on the curve corresponding to the required service hours and discharge temperature. The battery weight or volume is calculated by dividing the value of the specified current (in Amperes) by the value of Amperes per kilogram or Amperes per liter obtained from the ordinate.

Shelf Life. The Li/SO_2 battery is noted for its excellent storage characteristics, even at temperatures as high as 70°C. Most primary batteries lose capacity while idle or on stand due to anode corrosion, side chemical reactions, or moisture loss. With the exception of the magnesium battery, most of the conventional primary batteries cannot withstand temperatures in excess of 50°C and should be refrigerated if stored for long periods. The Li/SO_2 battery, however, is hermetically sealed and protected during storage by the formation of a film on the anode surface. Capacity losses during stand are minimal.

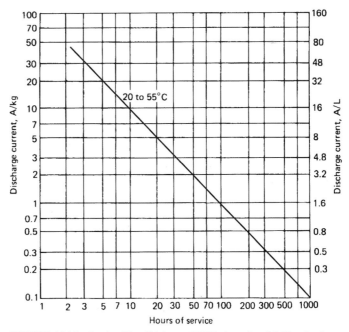

FIGURE 14.13 Service life of high-rate Li/SO_2 batteries; 2.0-V end voltage.

Recent data[18] on two-year old BA-5590 batteries consisting of 10 Li/SO$_2$ D-size cells discharged in series at 2 amps at +21 and −30°C showed a 6.5% capacity loss at the higher temperature but no loss at the lower temperature. Fourteen-year storage data was also obtained on BA-5598 batteries consisting of five "squat" D-size cells in series. These batteries showed only an 8% capacity loss when discharged at room temperature at 2 amps, but virtually no loss at cold temperature. In both cases, a lower operating voltage was observed after storage. Using multiple groups of batteries, stored for 4, 6 and 14 years under ambient conditions, the data shown in Fig. 14.14 was obtained. The capacity loss for the first two years is approximately 3%/year, but the rate of loss decreases significantly after that period. High-temperature storage of batteries was also carried out at +70 and +85°C, as shown in Fig. 14.15. At 70°C, these batteries showed 92% capacity retention after 1 month and 77% capacity retention after five months. At 85°C, 82% capacity retention was observed after one-month's storage. This study concluded that there was no obvious benefit to making long-term storage predictions based on accelerated aging tests at high temperature.

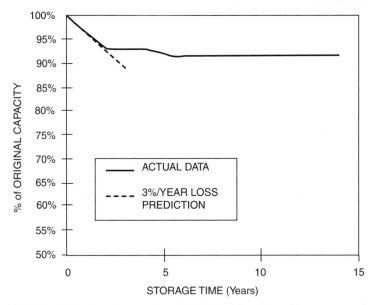

FIGURE 14.14   Capacity retention of Li/SO$_2$ batteries after ambient storage and discharge at the 2 amp rate.

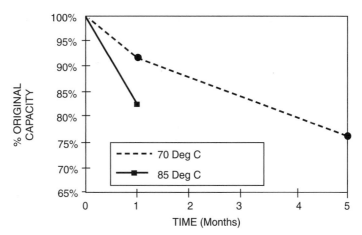

FIGURE 14.15   Effect of storage time/temperature on capacity of Li/SO$_2$ batteries.

Voltage Delay.   After extended long-term storage at elevated temperatures, the Li/SO$_2$ battery may exhibit a delay in reaching its normal operating voltage when placed on discharge, especially at high current loads and low temperatures. This start-up or voltage delay is caused by the protective film formed on the lithium anode, the characteristic responsible for the excellent shelf life of the cell. The specific delay time for a battery depends on such factors as the history of the battery, the specific cell design and components, the storage time and temperature, discharge load and temperature. Typically, the voltage delay is minimal or nonexistent for discharges at moderate to low rates at temperatures above $-20°C$. No delay is evident on discharge at 20°C, even after storage at 70°C for 1 year. On discharge at $-30°C$, the delay time is less than 200 ms after 8 weeks of storage at 70°C on discharges lower than the 40-h rate. At higher rates, the voltage delay increases with increasing storage temperature and time. At the 2-h discharge rate, for example, the maximum start-up time is about 80 s after 8 weeks of storage at 70°C; it is 7 s after 2 weeks of storage.[19] The start-up voltage delay can be eliminated by preconditioning with a short discharge at a higher rate to depassivate the anode until the operating voltage is reached since the delay will return only after another extended storage period.

14.5.4   Cell and Battery Types and Sizes

The Li/SO$_2$ batteries are manufactured in a number of cylindrical cell sizes, ranging in capacity to 35 Ah. Some of the cells are manufactured in standard ANSI cell sizes in dimensions of popular conventional zinc primary batteries. While these single batteries may be physically interchangeable, they are not electrically interchangeable because of the higher cell voltage of the lithium cell (3.0 V for lithium, 1.5 V for the conventional zinc cells). Table 14.9 lists some of the sizes and rated capacities of Li/SO$_2$ batteries that are currently manufactured.

TABLE 14.9 Typical Lithium/Sulfur Dioxide Cylindrical Cells (SAFT American, Inc.).*

Model number	Size	Nominal capacity (drain)	Maximum recommended continuous current	Pulse capability	Standard operating range	Max. OD	Max. H	Weight
LO 34 SX	1/3 R14 1/3C	0.86 Ah (30 mA)	0.25 A	0.50 A	−60/+70°C	25.6 mm	20.3 mm	16 g
LO 35 SX	2/3 R14 2/3 C	2.00 Ah (100 mA)	2.0 A	10 A	−60/+70°C	25.6 mm	35.9 mm	30 g
LO 26 SX	R20 D	7.5 Ah (240 mA)	3.0 A	10 A	−60/+70°C	33.8 mm	59.3 mm	85 g
LO 25 SX	"Fat" D	8.00 Ah (270 mA)	3.0 A	10 A	−60/+70°C	38.4 mm	50.3 mm	93 g
LO 52 SX	2 R20 DD	16.0 Ah (500 mA)	6.0 A	20 A	−60/+70°C	33.7 mm	115.6 mm	163 g
LO 38 SHX	"long" A	1.50 Ah (250 mA)	1.0 A	3.0 A	−60/+70°C	16.3 mm	57.2 mm	21 g
LO 29 SHX	R14 C	3.50 Ah (120 mA)	2.0 A	30 A	−60/+70°C	25.6 mm	50.4 mm	40 g
LO 43 SHX	5/4 R14 5/4 C	4.50 Ah (500 mA)	2.0 A	30 A	−60/+70°C	25.6 mm	59.3 mm	53 g
LO 40 SHX	2/3 "Thin" D	3.50 Ah (120 mA)	2.0 A	10 A	−60/+70°C	28.8 mm	41.6 mm	40 g
LO 30 SHX	"Thin" D	5.75 Ah (200 mA)	3.0 A	30 A	−60/+70°C	29.1 mm	59.8 mm	63 g
LO 26 SHX	R20 D	7.20 Ah (200 mA)	4.0 A	30 A	−60/+70°C	34.2 mm	59.8 mm	85 g
LO 39 SHX	F	11.0 Ah (200 mA)	8.0 A	60 A	−60/+70°C	30.7 mm	99.4 mm	125 g

*Pulse capability: 1 sec./min pulses over 2.0 volts.

14.5.5   Use and Handling of Li/SO$_2$ Cells and Batteries—Safety Considerations

The Li/SO$_2$ battery is designed as a high-performance system and is capable of delivering a high capacity at high discharge rates. The cell should not be physically or electrically abused, safety features should not be bypassed, and manufacturers' instructions should be followed.

Abusive conditions could adversely affect the performance of the Li/SO$_2$ battery and result in cell venting, rupture, explosion, or fire. Preventative measures are discussed in Sec. 14.4.

The Li/SO$_2$ battery is pressurized and contains materials that are toxic or flammable. Properly designed batteries are hermetically sealed so that there will be no leakage or outgassing, and they are equipped with safety vents which release if the batteries reach excessively high temperatures and pressures, thus preventing an explosive condition.

The Li/SO$_2$ batteries can deliver very high currents. Because high internal temperatures can develop from continuous high current drain and short circuit, batteries must be protected by electrical fusing and thermal cutoffs. Charging of Li/SO$_2$ batteries may result in venting, rupture, or even explosion and should never be attempted. Cells or groups of cells connected in parallel should be diode-protected to prevent one group from charging another. The balanced Li/SO$_2$ cell is designed to handle forced discharges or cell reversal and will perform safely within the specified bounds, but design limits should not be exceeded in any application.

Proper battery design, using the Li/SO$_2$ cell, should follow these guidelines:

1. Use electrical fusing and/or current-limiting devices to prevent high currents or short-circuits.
2. Protect with diodes if cells are paralleled or connected to a possible charging source.
3. Minimize heat buildup by adequate heat dissipation and protect with thermal cutoff devices.
4. Do not inhibit cell vents in battery construction.
5. Do not use flammable materials in the construction of batteries.
6. Allow for release of vented gases.
7. Incorporate resistor and switch to activate it to ensure complete depletion of active materials after normal discharge.
8. In certain cases, a diode is placed in parallel with the cell to limit the voltage excursion in reversal.

Currently special procedures govern the transportation, shipment, and disposal of Li/SO$_2$ batteries as well as other lithium batteries.[12–15] Procedures for the use, storage, and handling of these batteries also have been recommended. The latest issue of these regulations should be consulted for the most recent procedures.

14.5.6   Applications

The desirable characteristics of the Li/SO$_2$ battery and its ability to deliver a high energy output and operate over a wide range of temperatures, discharge loads, and storage conditions have opened up applications for this primary battery that, heretofore, were beyond the capability of primary battery systems (see Sec. 6.4).

Major applications for the Li/SO$_2$ battery are in military equipment, such as radio transceivers and portable surveillance devices, taking advantage of its light weight and wide-temperature operation. Table 14.10 lists the most common types of military Li/SO$_2$ batteries,

TABLE 14.10 U.S. Military Lithium/Sulfur Dioxide Batteries (Per MIL-B-49430)

Battery type	Open circuit voltage (series/parallel)(V)	Nominal voltage (series/parallel)(V)	Nominal capacity (series/parallel)(Ah)	Weight (g)	Typical applications
Ba-5112/U	12.0	11.2	1.8	180	Rescue radio/beacon
BA-5567/U	3.0	2.6	0.8	20.0	Night vision equipment
BA-5599/U	9.0	7.2	7.2	454	Test equipment
					Night vision equipment
BA-5600/U	9.0	8.4	7.2	363	Data terminals
BA-5800/U	6.0	5.6	7.2	220	Chemical agent monitors
					Global positioning equipment
BA-5847/U	6.0	5.6	7.2	240	Test equipment
					Antennas
					Night vision equipment
BA-5598	15.0 (with 3.0 Volt tap)	14.0	8.0	631	Radios (PRC-77, PRC-25), Scramblers
BA-5588	15.0	14.0	3.9	295	Handheld radios
					Gas masks
BA-5557	30.0/15.0	26.0/13.0	2.25/4.5	500	Digital message device
BA-5590	30.0/15.0	24.0/12.0	7.2/14.4	1021	Radios (SINCGARS)
					Satellite radios
					Scramblers
					Radar
					Loudspeakers
					UHF radios
					Range finders
					Counter measures
					Weather instruments
					Jammers
					Cooling systems

their characteristics and applications. Other military applications, such as sonobuoys and munitions, have long shelf-life requirements, and the active Li/SO₂ primary battery can replace reserve batteries used earlier. Some industrial applications have developed, particularly to replace secondary batteries and eliminate the need for recharging. Consumer applications have been limited to date because of restrictions in shipment and transportation and concern with its hazardous components.[20]

14.6 LITHIUM/THIONYL CHLORIDE (Li/SOCl₂) BATTERIES

The lithium/thionyl chloride (Li/SOCl₂) battery has one of the highest cell voltages (nominal voltage 3.6 V) and energy densities of the practical battery systems. Specific energy and energy densities range up to about 590 Wh/kg and 1100 Wh/L, the highest values being achieved with the low-rate batteries. Figures 7.8, 7.9, and 14.2 illustrate some of the advantageous characteristics of the Li/SOCl₂ cell.

Li/SOCl₂ batteries have been fabricated in a variety of sizes and designs, ranging from wafer or coin cells with capacities as low as 400 mAh, cylindrical cells in bobbin and spirally wound electrode structures, to large 10,000-Ah prismatic cells, plus a number of special sizes and configurations to meet particular requirements. The thionyl chloride system originally suffered from safety problems, especially on high-rate discharges and overdischarge, and a voltage delay that was most evident on low-temperature discharges after high-temperature storage.[21]

Low-rate batteries have been used commercially for a number of years for memory backup and other applications requiring a long operating life. The large prismatic batteries have been used in military applications as an emergency back-up power source. Medium- and high-rate batteries have also been developed as power sources for a variety of electric and electronic devices. Some of these batteries contain additives to the thionyl chloride and other oxyhalide electrolytes to enhance certain performance characteristics. These are covered in Sec. 14.7.

14.6.1 Chemistry

The Li/SOCl₂ cell consists of a lithium anode, a porous carbon cathode, and a nonaqueous SOCl₂:LiAlCl₄ electrolyte. Other electrolyte salts, such as LiGaCl₄ have been employed for specialized applications. Thionyl chloride is both the electrolyte solvent and the active cathode material. There are considerable differences in electrolyte formulations and electrode characteristics. The proportions of anode, cathode, and thionyl chloride will vary depending on the manufacturer and the desired performance characteristics. Significant controversy exists as to the relative safety of anode-limited vs. cathode-limited designs.[22] Some cells have one or more electrolyte additives. Catalysts, metallic powders, or other substances have been used in the carbon cathode or in the electrolyte to enhance performance.

The generally accepted overall reaction mechanism

$$4Li + 2SOCl_2 \rightarrow 4LiCl \downarrow + S + SO_2$$

The sulfur and sulfur dioxide are initially soluble in the excess thionyl chloride electrolyte, and there is a moderate buildup of pressure due to the generation of sulfur dioxide during the discharge. The lithium chloride, however, is not soluble and precipitates within the porous carbon black cathode as it is formed. Sulfur may precipitate in the cathode at the end of discharge. In most cell designs and discharge conditions, this blocking of the cathode is the factor that limits the cell's service or capacity. Formation of sulfur as a discharge product can also present a problem because of a possible reaction with lithium which may result in a thermal runaway condition.

The lithium anode is protected by reacting with the thionyl chloride electrolyte during stand forming a protective LiCl film on the anode as soon as it contacts the electrolyte. This passivating film, while contributing to the excellent shelf life of the cell, can cause a voltage delay at the start of a discharge, particularly on low-temperature discharges after long stands at elevated temperatures. The presence of trace qualities of moisture leads to the formation of HCl which increases passivation, as does the presence of ppm levels of iron. Some products have special anode treatments or electrolyte additives to overcome or lower this voltage delay.

The low freezing point of thionylchloride (below −110°C) and its relatively high boiling point (78.8°C) enable the cell to operate over a wide range of temperature. The electrical conductivity of the electrolyte decreases only slightly with decreasing temperature. Some of the components of the $Li/SOCl_2$ systems are toxic and flammable; thus exposure to open or vented cells or cell components should be avoided.

14.6.2 Bobbin-Type Cylindrical Batteries

$Li/SOCl_2$ bobbin batteries are manufactured in a cylindrical configuration, some in sizes conforming to ANSI standards. These batteries are designed for low- to moderate-rate discharge and are not typically subjected to continuous discharge at rates higher than the $C/100$ rate. They have a high energy density. For example, the D-size cell delivers 19.0-Ah at 3.5 V, compared with 15 Ah at 1.5 V for the conventional zinc-alkaline cells (see Tables 7.5 and 14.11).

Construction. Figure 14.16 shows the constructional features of the cylindrical $Li/SOCl_2$ cell, which is built as a bobbin-type construction. The anode is made of lithium foil which is swaged against the inner wall of a stainless or nickel-plated steel can; the separator is made of nonwoven glass fibers. The cylindrical, highly porous cathode, which takes up most of the cell volume, is made of Teflon-bonded acetylene black. The cathode also incorporates a current collector which is a metal cylinder in the case of the larger cells and a pin in the case of smaller cells which do not have an annular cavity.

TABLE 14.11 Characteristics of Typical High-Capacity and Wafer-Type Cylindrical Bobbin-Type $Li/SOCl_2$ Batteries

	$\frac{1}{2}$AA	$\frac{2}{3}$AA	AA	C	$\frac{1}{6}$D	D
Rated capacity at $C/1000$ rate, Ah	1.20	1.65	2.40	8.5	1.7	19.0
Dimensions (max)						
Diameter, mm	14.5	14.5	14.5	26.2	32.9	32.9
Height, mm	25.2	33.5	50.5	50	10.0	61.5
Volume, cm³	4.16	5.53	8.34	27.0	8.50	52.3
Weight, g	9.2	11.8	17.6	50.5	21.5	92.5
Maximum current for continuous use, mA	50	75	100	230		230
Specific Energy Wh/kg	456	490	475	590	275	720
Energy Density Wh/L	1010	1045	1010	1100	700	1270

Source: Tadiran, Ltd.

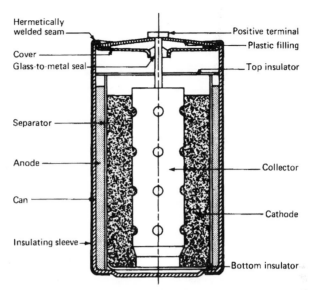

Hermetically welded seam

Positive terminal

Plastic filling

Cover

Glass-to-metal seal

Top insulator

Separator

Anode

Collector

Can

Cathode

Insulating sleeve

Bottom insulator

FIGURE 14.16   Cross section of bobbin-type Li/SOCl$_2$ battery. (*From Ref. 23.*)

Performance.   The open-circuit voltage of the Li/SOCl$_2$ cell is 3.65 V; typical operating voltages range between 3.3 and 3.6 V with an end voltage of 3.0 V. Typical discharge curves for the Li/SOCl$_2$ battery are shown in Fig. 14.17*a* for the D-size cell. The Li/SOCl$_2$ cell discharges are characterized by a flat profile with good performance over a wide range of temperatures and low- to moderate-rate discharges. Figure 14.17*b* shows the operating voltage of the bobbin D-cell at various drain rates and temperatures. The relationship of capacity with current is given in Fig. 14.18, showing the performance from −30 to 72°C. The Li/ SOCl$_2$ cell is capable of performance at unusually high temperatures. At 145°C, (Fig. 14.19) the cells deliver most of their capacity at high rates and up to 70% at low discharge rates (20 days of discharge).[22] Li/SOCl$_2$ cells are used to build battery packs that are employed in oil exploration and most withstand temperatures to 150°C as well as high levels of shock and vibration.

Figure 14.20 shows the behavior of AA cells on continuous low-rate discharge at 25°C. The discharge curve is very flat at these low-current drains, but capacity loss below the 2.4 Ah rating occurs below the 1000-hour rate due to parasitic self-discharge.

The capacity or service life of the high-capacity bobbin-type Li/SOCl$_2$ cell, normalized for a 1-kg and 1-L size cell, at various discharge temperatures and loads, is summarized in Fig. 14.21.

The long shelf life of the Li/SOCl$_2$ battery is due to the stability of the lithium anode in contact with the electrolyte, as a result of a protective LiCl film that forms on the lithium surface. The long shelf life can also be attributed to the stability of other cell components. For example, the can and cover are cathodically protected by the lithium, and the carbon, stainless-steel collector, and glass separator are all inert in the electrolyte. Figure 14.22 shows the loss of capacity after 3 years at 20°C, a loss of about 1 to 2% per year. Storage at 70°C results in a capacity loss of about 5% per year. Cells should preferably be stored in an upright position; storage on the side or upside down may result in higher capacity loss.

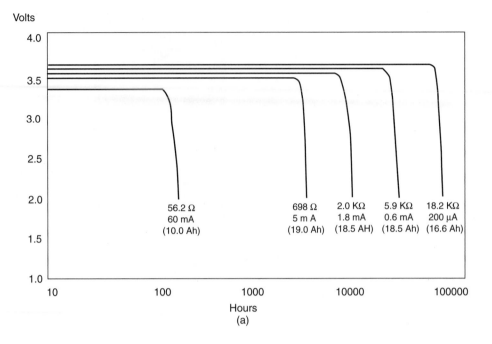

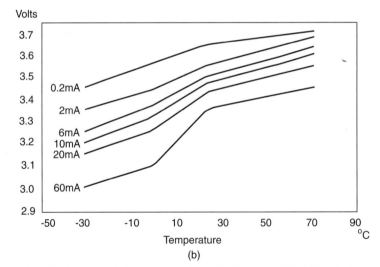

FIGURE 14.17   (*a*) Discharge characteristics of high-capacity $Li/SOCl_2$ cylindrical D-size bobbin battery at $+25°C$. (*b*) Operating (plateau) voltage of the same battery as a function of temperature at various drain rates. (*Ref. 23.*)

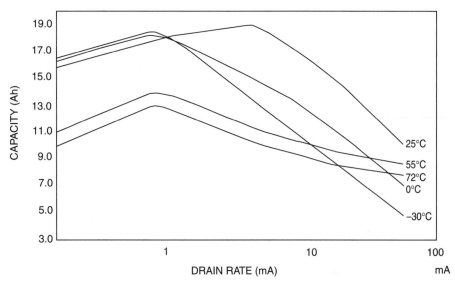

FIGURE 14.18 Performance characteristics of high-capacity cylindrical bobbin D-size batteries as a function of drain rate at various temperatures. (*Ref. 23.*)

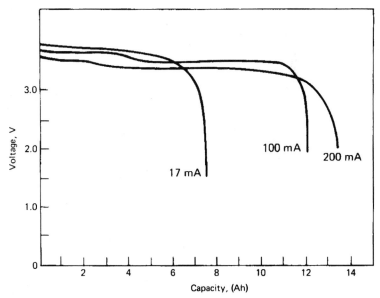

FIGURE 14.19 Discharge characteristics of Li/SOCl$_2$ cylindrical D-size bobbin battery at 145°C. (*From Ref. 24.*)

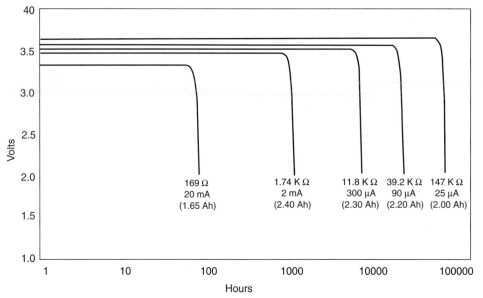

FIGURE 14.20 Discharge characteristics of high-capacity Li/SOCl₂ cylindrical AA-size bobbin battery on low-rate discharge at +25°C. (*Ref. 23.*)

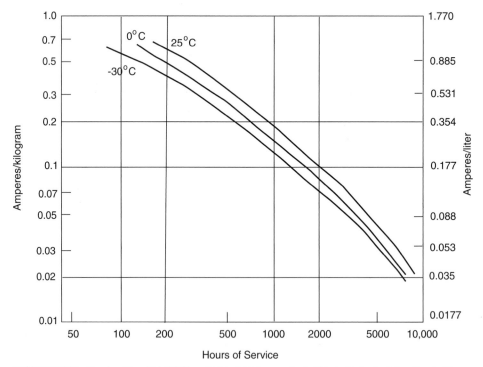

FIGURE 14.21 Service life of Li/SOCl₂ cylindrical high-capacity bobbin batteries to 2.0 volt cut-off.

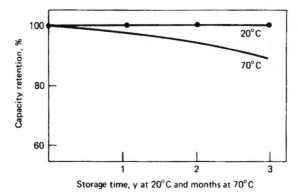

FIGURE 14.22 Capacity retention of Li/SOCl$_2$ cylindrical bobbin battery (*From Ref. 23.*)

After storage, the Li/SOCl$_2$, battery may exhibit a delay in reaching its operating voltage because of the formation of the LiCl film on the lithium surface. The voltage delay becomes more pronounced with a heavier discharge load and lower discharge temperature. The voltage delay of the Li/SOCl$_2$ cells can be improved by an in situ coating of the lithium anode with an ionic conductor-solid-electrolyte interface. The improvement is shown in Fig. 14.23 which compares the minimum voltage and the load after 2 years of storage for both the standard construction and the coated one. It shows the dependence of the closed-circuit voltage on the discharge current of AA cells after 2 years of storage at 25°C. Once the discharge is started, the passivation film is dissipated gradually, the internal resistance returns to its normal value, and the plateau voltage is reached. The passivation film may be removed more rapidly by the application of high-current pulses for a short period or, alternatively, by short-circuiting the batteries momentarily several times until the cell is activated. The use of a pulse provides more reproducible results.

Special Characteristics.   The bobbin batteries are designed to limit the possibility of hazardous operation and to eliminate the need (in some designs) for a safety vent. This is achieved by minimizing the reactive surface area and increasing the heat dissipation, thus limiting the short-circuit current and a hazardous temperature rise, respectively. These cells also are cathode-limited, a feature that was found safer than anode-limited cells for this design.[25] The batteries have withstood short circuits, forced discharge, and charging under certain conditions with no hazardous condition.[23,24,26] Batteries should not be disposed of in fire or subjected to long-term exposure at temperatures near 180°C because they may explode.

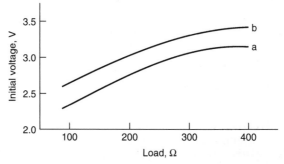

FIGURE 14.23 Li/SOCl$_2$ cylindrical AA-size bobbin batteries—minimum voltage vs. load after 2-year storage at 25°C; (*a*) standard construction; (*b*) with loading on lithium anode.

Battery Sizes. The bobbin-type $Li/SOCl_2$ batteries are manufactured in the standard ANSI cell sizes as well as in special cell and battery configurations. Although some of these batteries may be physically interchangeable with conventional zinc batteries, they are not electrically interchangeable because of their higher voltages.

Table 14.11 lists the properties of some of the typical bobbin-type batteries that are manufactured. These characteristics may vary with the manufacturer. Manufacturer's data should be consulted for specific data as well as for the characteristics of their other batteries.

14.6.3   Spirally Wound Cylindrical Batteries

Medium to moderately high-power $Li/SOCl_2$ batteries which are designed within spirally wound electrode structure are also available. These batteries were developed primarily to meet military specifications where high drains and low-temperature operation were required. They are now also used in selected industrial applications where these features are also needed.

A typical construction is shown in Fig. 14.24. The cell container is made of stainless steel, a corrosion-resistant glass-to-metal feed-through is used for the positive terminal, and the cell cover is laser sealed or welded to provide an hermetic closure. Safety devices, such as a vent and a fuse or a PTC device, are incorporated in the cell to protect against buildup of internal pressure or external short circuits.

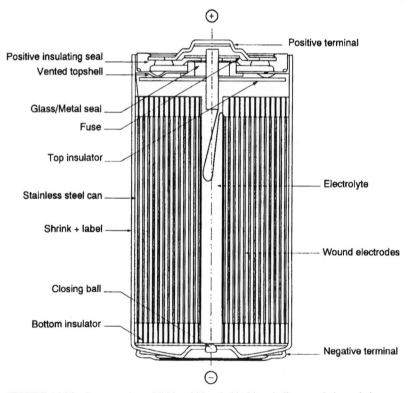

FIGURE 14.24   Cutaway view of lithium/thionyl chloride spirally wound electrode battery. (*Courtesy of SAFT America, Inc.*)

The discharge curves for a D-size battery are plotted in Fig. 14.25, showing the higher performance at the moderate drains compared to the bobbin cell (see Fig. 14.17). The inserts in Fig. 14.25 show the voltage response when the load is initially applied. The voltage of these cells recovers within 10 s as they incorporate an anode-coating technology, as discussed earlier.

Figure 14.26 summarizes the performance characteristics of the D-size battery, showing the relationship of voltage and capacity with current drain at several temperatures.

The capacity retention after storage at 20°C is shown in Fig. 14.27. Like the other Li/SOCl$_2$ batteries, these batteries have an excellent storage capability over a wide temperature range due to the buildup of a protective lithium chloride layer on the lithium. This passivation layer, however, may cause a voltage delay under some discharge conditions. Table 14.12 lists the characteristics of typical cylindrical spirally wound Li/SOCl$_2$ batteries.

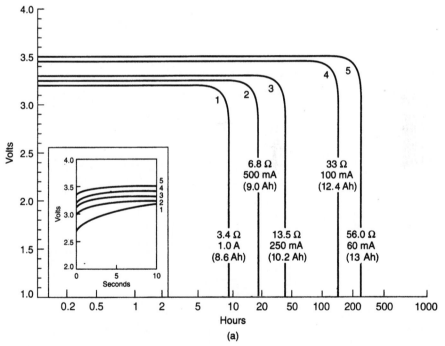

FIGURE 14.25 Discharge characteristics of spirally wound Li/SOCl$_2$ D-size battery, medium discharge rate at (*a*) 25°C. (*Courtesy of SAFT America, Inc.*)

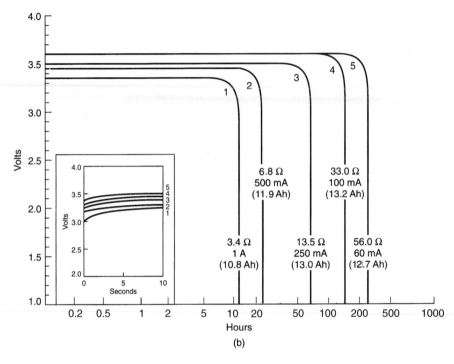

(b)

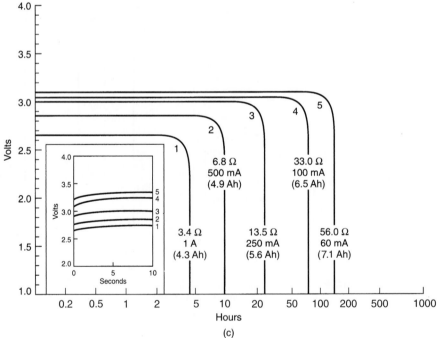

(c)

FIGURE 14.25 Discharge characteristics of spirally wound Li/SOCl$_2$ D-size battery, medium discharge rate at (b) 72°C, (c) −40°C. (*Courtesy of SAFT America, Inc.*) (*Continued*).

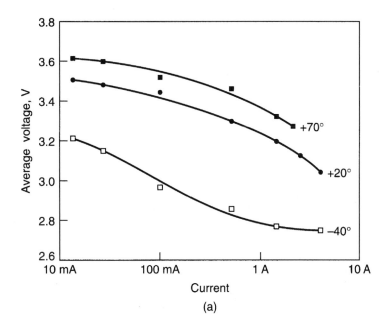

(a)

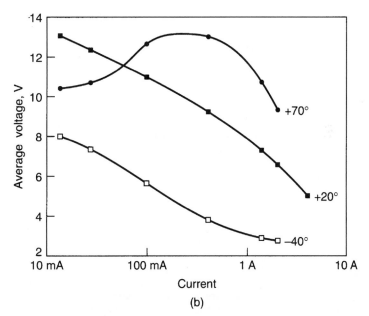

(b)

FIGURE 14.26   Discharge characteristics of spirally wound Li/SOCl₂ D-size battery, medium rate. (*a*) Voltage vs. current. (*b*) Capacity vs. current. (*Courtesy of SAFT America, Inc.*)

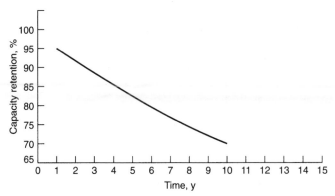

FIGURE 14.27 Capacity retention of spirally wound Li/SOCl₂ D-size cell, storage at 20°C. (*Courtesy of SAFT America, Inc.*)

TABLE 14.12 Characteristics of Typical Cylindrical Spirally Wound Li/SOCl₂ Batteries

	⅓C	C	D
Rated capacity at 20°C, Ah	1.15	5.5	13.0
Dimensions (max)			
Diameter, mm	26.2	26.0	33.1
Height, mm	18.9	49.9	61.4
Volume, cm³	10.2	26.5	52.8
Weight, g	24	51	100
Maximum current for continuous use, mA	400	800	1800
Specific energy/Energy density			
Wh/kg	168	377	455
Wh/L	395	726	860
Recommended operating temperature range, °C	−60 to 85		

Source: SAFT America, Inc.

14.6.4 Flat or Disk-Type Li/SOCl₂ Cells

The Li/SOCl₂ system was also designed in a flat or disk-shaped cell configuration with a moderate to high discharge rate capability. These batteries are hermetically sealed and incorporate a number of features to safely handle abusive conditions, such as short circuit, reversal, and overheating, within design limits.

The battery shown in Fig. 14.28 consists of a single or multiple assembly of disk-shaped lithium anodes, separators, and carbon cathodes sealed in a stainless-steel case containing a ceramic feed-through for the anode and insulation between the positive and negative terminals of the cell.[26]

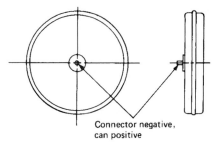

FIGURE 14.28 Disk-type Li/SOCl$_2$ cell.

The batteries have been manufactured in small and large diameter sizes. Originally developed by Altus Corp., they are currently being produced in large sizes only for U.S. Navy applications by HED Battery Corp., Santa Clara, CA. The characteristics of these batteries are summarized in Table 14.13. Discharge curves for large batteries are shown in Fig. 14.29. Typically the cells have a high energy density, flat discharge profiles, and the capability of performance over the temperature range of −40 to 70°C. On storage they can retain 90% of the capacity after storage of 5 years at 20°C, or 6 months at 45°C or of 1 month at 70°C.

TABLE 14.13 Characteristics of Disk-Type Li/SOCl$_2$ Batteries

Capacity, Ah	Diameter, mm	Height, mm	Weight, g	Maximum continuous current, A	Specific energy Wh/kg	Energy density Wh/L
500	432	127	7,270	7	240	915
1400	432	35	1,600	16	350	930
2000	432	51	17,700	25	385	910
8000	432	187	56,800	40	475	990

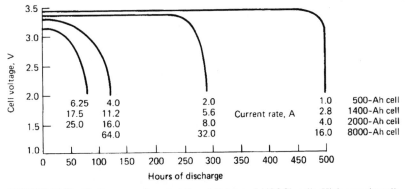

FIGURE 14.29 Performance characteristics of disk-type Li/SOCl$_2$ cells. High-capacity cell; typical performance at 0–25°C range to 2.5 V cut-off.

The cell design includes the following features:

1. *Short-circuit protection:* Structure of interconnects fuses at high currents, providing an open circuit.
2. *Reverse-voltage chemical switch:* Upon cell reversal, it allows cell to endure 100% capacity reversal, up to 10-h rate, without venting or pressure increase.
3. *Antipassivation (precoat lithium anode):* Reduces voltage delay by retarding growth of LiCl film; large cells stored for 2 years reach operating voltage within 20 s.
4. *Self-venting:* Ceramic seal is designed to vent cell at predetermined pressures.[27]

These cells are used as multicell batteries in naval applications.

Recent work on these designs[28,29] has involved 1000 and 1200 Amp-hour cells for application in a U.S. Navy Long-Range Mine Reconnaissance System (LMRS). These are scaled-down versions of 2350 Ah cells which had shown the ability to operate at the C/40 rate, providing a power density of 2.3 W/kg. Both 1000 and 1200 Ah cells were 20.3 cm in diameter with an annular cavity in the center of the disk. The former unit was 9.53 cm high, while the latter was 12.07 cm high. Both designs incorporate a ceramic-to-metal seal capable of carrying 60 amps and both were limited by the capacity of the carbon cathode with Li/SOCl$_2$ capacity ratio balanced. The 1000 Ah units were tested individually and as 4 and 12-cell batteries with 0.5 cm intercell insulators and compressed between 1.59 cm aluminum end-plates by tie-rods. The 12-cell battery consisted of three stacks of four cells with a diameter of 45.3 cm designed to fit within the hull of LRMS. Test data is summarized in Table 14.14. Based on the results of this testing, a 30-cell battery weighing about 205 kg would deliver 100 kWh at 100 Volts for operational power up to 5kW. Subsequently, the cell capacity was increased to 1200 Ah by increasing the cell height.[29] These cells were subjected to a series of safety tests as defined by NAVSEA INST 9310.1B (June 13, 1992) and U.S. Navy Technical Manual S9310-AQ-SAF-010. The 1200 Ah units were subjected to intermittent and sustained short circuits, forced discharge into voltage reversal, charging tolerance and high-temperature discharge and high-temperature exposure after low temperature (0°C) discharge. No cells produced venting, loss of material or case breach of any kind during these tests, nor were there indications of internal shorts or potentially violent conditions. The pulsed and sustained soft-shorts produced significant heating and pressure, but these were within the capability of the battery to operate safely. At sustained currents in excess of 110 Amps, the cathode appears to clog rapidly, limiting capacity. The exothermic response obtained when the battery was quickly heated to 75°C after cold discharge at 40 Amps at 0°C is a result of accelerated anode repassivation. The subsequent 55°C short-circuit

TABLE 14.14 Performance Characteristics of 1000 Ah LMRS Lithium/Thionyl Chloride Cells and Batteries (Number of Cells Tested Indicated in Parenthesis After Each Test)

Configuration	Rate	Ah	kWh	Wh/kg
Single (1)	C/22–C/67	931	3.12	108
Single (5)	C/25–C/67	913	3.00	105
Single (2)	C/40	927	3.09	111
4-cell	C/25–C/50	1053	3.58	125
4-cell	C/40	1075	3.67	126
4-cell	C/60	1004	3.41	119
12-cell	C/20–C/40	896	3.03	106
12-cell	C/20–C/40	1016	3.44	121

behavior, confirms this hypothesis. There was an indication that this response would have led to a thermal runaway. The 40 Amp, 55°C discharge demonstrated that the battery could operate safely in the absence of cooling for an extended period of time in a simulated vehicle structure. The tolerance to a moderate charging voltage indicated a margin level in potential failures of diodes. The forced reversal test demonstrated a moderate tolerance to these conditions. A fuse in the negative terminal assembly is being considered to withstand high-rate short circuits. This test program demonstrated the feasibility of using a large lithium/thionyl chloride propulsion battery for LMRS.

14.6.5   Large Prismatic Li/SOCl₂ Cells

The large high-capacity $Li/SOCl_2$ batteries were developed mainly as a standby power source for those military applications requiring a power source that is independent of commercial power and the need for recharging.[30–32] They generally were built in a prismatic configuration, as shown schematically in Fig. 14.30. The lithium anodes and Teflon-bonded carbon electrodes are made as rectangular plates with a supporting grid structure, separated by nonwoven glass separators, and housed in an hermetically sealed stainless-steel container. The terminals are brought to the outside by glass-to-metal feed-through or by a single feed-through isolated from the positive steel case. The cells are filled through an electrolyte filling tube.

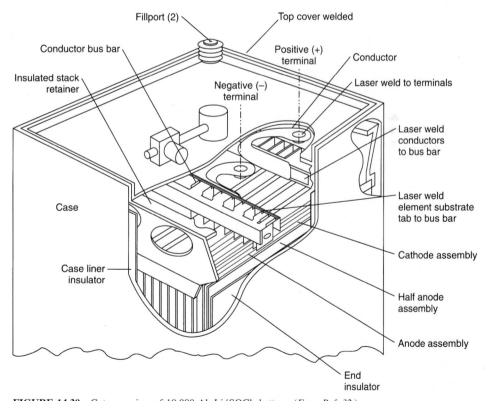

FIGURE 14.30   Cutaway view of 10,000-Ah $Li/SOCl_2$ battery. (*From Ref. 32.*)

The characteristics of several prismatic batteries are summarized in Table 14.15. These cells have a very high energy density. They are generally discharged continuously at relatively low rates (200–300-h rate), but are capable of heavier discharge loads. A typical discharge curve is shown in Fig. 14.31. The voltage profile is flat, and the cell operates just slightly above ambient temperature at this discharge load. During the course of the discharge there is a slight buildup of pressure, reaching a value of about 2×10^5 Pa at the end of the discharge. A higher-rate pulse discharge is shown in Fig. 14.32. The 2000-Ah cell was discharged continuously at a 5-A load, with 40-A pulses, 16 s in duration, superimposed once every day. A steady discharge voltage was obtained throughout most of the discharge, with only a slight reduction in voltage during the pulse. The batteries are capable of performance from −40 to 50°C; shelf-life losses are estimated at 1% per year.[32]

TABLE 14.15 Characteristics of Large Prismatic Li/SOCl$_2$ Batteries

Capacity, Ah	Height, mm	Length, mm	Width, mm	Weight, kg	Specific energy Wh/kg	Energy density Wh/L
2,000	448	316	53	15	460	910
10,000	448	316	255	71	480	950
16,500	387	387	387	113	495	970

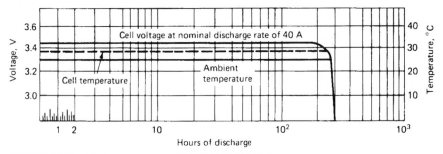

FIGURE 14.31 Discharge curves for 10,000-Ah Li/SOCl$_2$ battery.

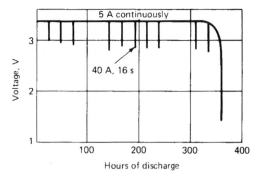

FIGURE 14.32 Discharge of high-capacity 2000-Ah Li/SOCl$_2$ battery.

14.6.6 Applications

The applications of the Li/SOCl$_2$ system take advantage of the high energy density and long shelf life of this battery system. The low-drain cylindrical batteries are used as a power source for CMOS memories, utility meters, and RFID tags such as the EZ Pass Toll collection system, programmable logic controllers and wireless security alarm system. Wide application in consumer-oriented applications is limited because of the relatively high cost and concern with the safety and handling of these types of lithium batteries.

The higher-rate cylindrical and the larger prismatic Li/SOCl$_2$ batteries are used mainly in military applications where high specific energy is needed to fulfill important mission requirements. A significant application for the large 10,000-Ah batteries was as standby power source as nine-cell batteries for the Missile Extended System Power in the event of loss of commercial or other power. These batteries are now being decommissioned.

A lithium/thionyl chloride battery was developed for use on the Mars Microprobe Mission, a secondary payload on the Mars 98 Lander Mission, which disappeared on entry into the Martian atmosphere in December, 1999.[33] The Microprobe power source is a four-cell lithium/thionyl chloride battery with a second redundant battery in parallel. The eight 2 Ah cells are arranged in a single-layer configuration in the aft-body of the microprobe. The lithium primary cells (and battery configuration) have been designed to survive the maximum landing impact that may reach 80,000 G and then be operational on the Martian surface to −80°C. Primary lithium-thionyl chloride batteries were selected for the Microprobes based on high specific energy and promising low temperature performance. A parallel plate design was selected as the best electrode configuration for surviving the impact without shorting during impact. A cross-section of the final 2 Ah Mars cell design showing the parallel plate electrode arrangement is shown in Fig. 14.33. For this cell, the cathodes are blanked from sheets of a Teflon®-bonded carbon composition attached to nickel-disc current collectors and connected in parallel. The ten full disc-anodes are also connected in parallel and are electrically isolated from the case and cover. The assembly fixture helps with component alignment and handling during stack assembly and during connection of the cathode and

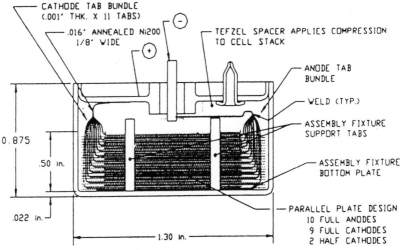

FIGURE 14.33 Vertical Cross-Section of the Final 2 Ah Cell Design. (*Courtesy of Yardney Technical Products, Inc.*)

anode substrate tabs to the cover and the glass-to-metal (GTM) seal anode terminal pin, respectably. The D-size diameter case is 2.22 cm height. The cover was redesigned after initial tests to minimize the chance of a GTM seal fracture during impact. The Tefzel® spacer, located between the cover and stack, helps in handling the stack during substrate tab connections and provides for the proper degree of cathode and separator compression once the cover is TIG welded to the case. The Mars cells are required to deliver 0.55 Ah of capacity at $-80°C$. A low temperature thionyl chloride electrolyte consisting of a 0.5 M LiGaCl$_4$ in SOCl$_2$ was developed during the initial phase of the program. As the result of this effort, the battery was able to operate at $-80°C$ on a 1 Amp discharge as shown in Fig. 14.34. The battery provided over 0.70 amp-hours at this extremely low temperature. The ability to withstand the 80,000 G impact was demonstrated by Air Gun tests performed at $-40°C$ into frozen desert sand followed by a simulated mission profile at $-60°C$ in an environmental chamber. The battery must supply power for a drill that provides a core sample of subsurface soil for water analysis. In addition, the power requirement for the 20 minutes water experiment was increased from 2.5 to ~6 Watts and increased power levels were required for telemetry at both -60 and $-80°C$. The total low temperature capacity and major tasks are listed in Table 14.16. The drill operation requires an initial current of 1A for 25 milliseconds, after which the current is in the range 75 to 85mA for the duration of the task. The soil sample heating operation lasted for 20 minutes with power in excess of 6 Watts required. The high rate transmission started out at 10.4 Watts (9.7 V), but the power level dropped off toward the end to 6.4 Watts (7.6 V) after nine minutes. The cell delivered a total of 0.724 Ah of low temperature capacity. Although the fate of the Microprobes is currently unknown, this program extended the state-of-the-art for lithium/thionyl chloride battery technology by demonstrating its ability to withstand 80,000 G impact and then operate at temperatures down to $-80°C$ and may be employed in future space missions.

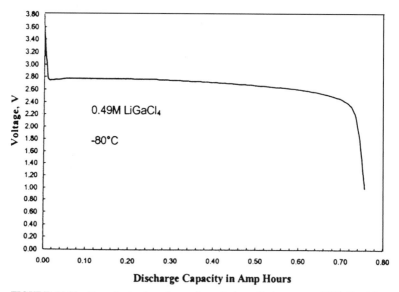

FIGURE 14.34   Capacity in Amp hours for a 1 Amp discharge at $-80°C$. For Mars Microprobe battery. (*Courtesy of Yardney Technical Products, Inc.*)

TABLE 14.16 Results of Air Gun Test on Mars Microprobe
Battery

Post impact battery discharge	
Output on profile	0.515 Ahr
Additional output at $-80°C$	0.157 Ahr
Total output	0.724 Ahr

Major tasks	
Calib. 9Ω, $-60°C$	9.5 V
Drill 136Ω, $-60°C$	11.7 V
H_2O 16Ω, $-60°C$	10.5 V
High X-mit 9Ω, $-60°C$	9.7 V
X-mit 59Ω, $-80°C$	7.6 V

14.7 LITHIUM/OXYCHLORIDE BATTERIES

The lithium/sulfuryl chloride (Li/SO_2Cl_2) battery is in addition to the lithium/thionyl chloride battery, the other oxychloride that has been used for primary lithium batteries. The Li/SO_2Cl_2 battery has two potential advantages over the $Li/SOCl_2$ battery:

1. A higher energy density as a result of a higher operating voltage (3.9-V open-circuit voltage) as shown in Fig. 14.3 and less solid product formation (which may block the cathode) during the discharge.

2. Inherently greater safety because sulfur, which is a possible cause of thermal runaway in the $Li/SOCl_2$ battery, is not formed during the discharge of the Li/SO_2Cl_2 battery.

3. A higher rate capability than the thionyl chloride battery as, during the discharge, more SO_2 is formed per mole of lithium, leading to a higher conductivity.

Nevertheless, the Li/SO_2Cl_2 system is not as widely used as the $Li/SOCl_2$ system because of several drawbacks:

(1) Cell voltage is sensitive to temperature variations; (2) higher self-discharge rate; (3) lower rate capability at low temperatures.

Another type of lithium/oxychloride battery involves the use of halogen additives to both the $SOCl_2$ and SO_2Cl_2 electrolytes. These additives given an increase in the cell voltage (3.9 V for the $Li/BrCl$ in the $SOCl_2$ system; 3.95 V for the Li/Cl_2 in the SO_2Cl_2 system), energy density and specific energy up to 1040 Wh/L and 480 Wh/kg, and safer operation under abusive conditions.

14.7.1 Lithium/Sulfuryl Chloride (Li/SO₂Cl₂) Batteries

The Li/SO_2Cl_2 battery is similar to the thionyl chloride battery using a lithium anode, a carbon cathode and the electrolyte/depolarizer of $LiAlCl_4$ in SO_2Cl_2. The discharge mechanism is:

Anode $$Li \rightarrow Li^+ + e$$

Cathode $$SO_2Cl_2 + 2e \rightarrow 2Cl^- + SO_2$$

Overall $$2Li + SO_2Cl_2 \rightarrow 2LiCl\downarrow + SO_2$$

The open-circuit voltage is 3.909 V.

Cylindrical, spirally wound Li/SO_2Cl_2 cells were developed experimentally but were never commercialized because of limitations with performance and storage. Bobbin-type cylindrical cells, using a sulfuryl chloride/$LiAlCl_4$ electrolyte and constructed similar to the design illustrated in Fig. 14.16, also showed a variation of voltage with temperature and a decrease of the voltage during storage. This may be attributed to reaction of chlorine which is present in the electrolyte and formed by the dissociation of sulfuryl chloride into Cl_2 and SO_2. This condition can be ameliorated by including additives in the electrolyte. Bobbin cells, made with the improved electrolyte, gave significantly higher capacities at moderate discharge currents, compared to the thionyl chloride cells.[34] This system has been employed for reserve lithium/sulfuryl chloride batteries, as well[35] (see Chap. 20).

14.7.2 Halogen-Additive Lithium/Oxychloride Cells

Another variation of the lithium/oxyhalide cell involves the use of halogen additives in both the $SOCl_2$ and the SO_2Cl_2 electrolytes to enhance the battery performance. These additives result in: (1) an increase in the cell voltage (3.9 V for BrCl in the $SOCl_2$ system (BCX), 3.95 V for Cl_2 in the SO_2Cl_2 system (CSC), and (2) an increase in energy density and specific energy to about 1040 Wh/L and 480 Wh/kg for the CSC system.

The lithium/oxyhalide cells with halogen additives offer among the highest energy density of primary battery systems. They can operate over a wide temperature range, including high temperatures, and have excellent shelf lives. They are used in a number of special applications—oceanographic and space applications, memory backup, and other communication and electronic equipment.

These lithium/oxychloride batteries are available in hermetically sealed, spirally wound electrode cylindrical configurations, ranging from AA to DD size in capacities up to 30 Ah. These batteries are also available in the AA size containing 0.5 g of Li and in flat disk-shaped cells. Figure 14.35 shows a cross section of a typical cell. Table 14.17 lists the different lithium-oxychloride batteries manufactured and their key characteristics. Two types of halogen-additive lithium/oxychloride batteries have been developed, as follows:

Li/SOCl$_2$ System with BrCl Additive (BCX).   This battery has an open-circuit voltage of 3.9 V and an energy density of up to 1070 Wh/L at 20°C. The BrCl additive is used to enhance the performance. The cells are fabricated by winding the lithium anode, the carbon cathode, and two layers of a separator of nonwoven glass into a cylindrical roll and packaging them in an hermetically sealed can with a glass-to-metal feed-through. The performance of the D-size battery at various temperatures and discharge rates is shown in Fig. 14.36. The discharge curves are relatively flat with a working voltage of about 3.5 V. The batteries are capable of performance over the temperature range of −55 to 72°C. The shelf-life characteristics of this battery are shown in Table 14.18. Capacity loss on storage is higher than with lithium systems using thionyl chloride only.

The addition of BrCl to the depolarizer may also prevent the formation of sulfur as a discharge product, at least in the early stage of the discharge, and minimize the hazards of the $Li/SOCl_2$ battery attributable to sulfur or discharge intermediates. The cells show abuse resistance when subjected to the typical tests, such as short circuit, forced discharge, and exposure to high temperatures.[36]

Li/SO$_2$Cl$_2$ with Cl$_2$ Additive (CSC).   This battery has an open-circuit voltage of 3.95 V and an energy density of up to 1040 Wh/L. The additive is used to decrease the voltage-delay characteristic of the lithium/oxyhalide cells. The typical operating temperature of these cells is −30 to 90°C. The cylindrical cells are designed in the same structure as those shown in Fig. 14.35.

Typical performance characteristics for this battery type are shown in Fig. 14.37. The cells show abuse resistance similar to the Li/BrCl in $SOCl_2$ cells when subjected to abuse tests.

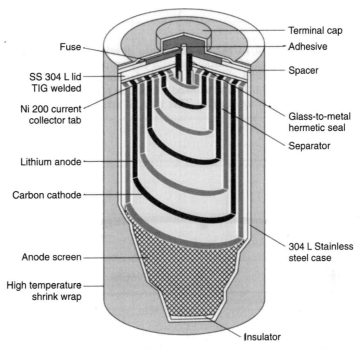

FIGURE 14.35   Cross section of lithium/oxychloride cell. (*Courtesy of Electrochem Industries.*)

TABLE 14.17   Typical Halogen Additive Oxychloride Batteries

	BrCl in $SOCl_2$				Cl_2 in SO_2Cl_2		
	AA	C	D	DD	C	D	DD
Voltage, V							
Open-circuit			3.9			3.95	
Average operating			3.4			3.3	
Rated capacity, 100-h rate, Ah	2.0	7.0	15.0	30.0	7.0	14.0	30.0
Dimensions							
Diameter, mm	13.7	25.6	33.5	33.5	25.6	33.5	33.5
Height, mm	48.9	48.4	59.3	111	48.4	59.3	111
Volume, cm³	7.21	24.9	52.3	98.2	24.9	52.3	98.2
Weight, g	16	55	115	216	52	116	213
Maximum current capability, mA	100	500	1000	3000	1000	2000	4000
Specific Energy/Energy Density							
Wh/kg	412	433	456	486	455	404	480
Wh/L	915	984	1000	1070	956	897	1040
Operating temperature range, °C			−55 to 72			−32 to 93°C	

Source: Electrochem Industries Div., Wilson Greatbatch Ltd.

TABLE 14.18   Storage Capacity Losses of Li/SOCl$_2$ with BrCl Additive Cells, D Size

Temperature, °C	1st year	Each following year
−40	2% loss	2% loss
21	7% loss	5% loss
72	20% loss	9% loss

Source: Electrochem Industries Div., Wilson Greatbatch, Ltd.

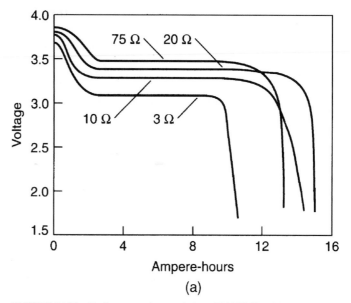

(a)

FIGURE 14.36   Performance characteristics of Li/SOCl$_2$ with BrCl-additive. D-size batteries. (*a*) Discharge characteristics at 20°C. (*Courtesy of Electrochem Industries, Div., Wilson Greatbatch, Ltd.*)

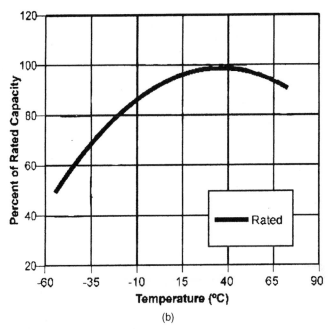

(b)

FIGURE 14.36(b) Capacity as a function of discharge temperature (100% represents rated capacity at room temperature)

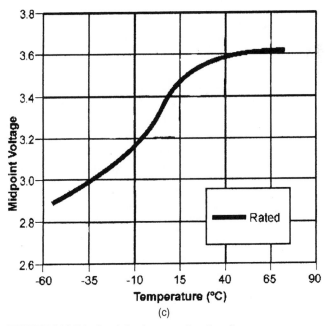

(c)

FIGURE 14.36(c) Loaded voltage as a function of temperature

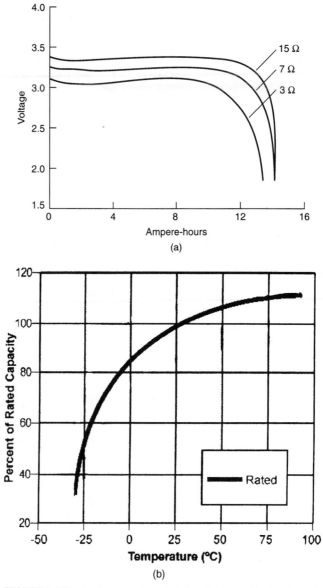

FIGURE 14.37 Performance characteristics of Li/SO_2Cl_2 with Cl_2^- additive in D-size batteries. (*a*) Discharge at 20°C. (*b*) Capacity vs. discharge temperature; 100% capacity delivered at 20°C. (*Courtesy of Electrochem Industries, Div., Wilson Greatbatch Ltd.*)

A recent study[37] has evaluated the effect of ambient temperature storage for up to 6 years. The interrelation of voltage stability, capacity retention, self-discharge and voltage delay has been studied. This source should be consulted to obtain detailed data on this system.

14.8 LITHIUM/MANGANESE DIOXIDE (Li/MnO₂) BATTERIES

The lithium/manganese dioxide (Li/MnO₂) battery was one of the first lithium/solid-cathode systems to be used commercially and is now the most widely used primary lithium battery. It is available in many configurations (including coin, bobbin, spirally wound cylindrical, and prismatic configurations in multicell batteries, and in designs for low, moderate, and moderately high drain applications. The capacity of batteries available commercially ranges up to 2.5 Ah. Larger sized batteries are available for special applications and have been introduced commercially. Its attractive properties include a high cell voltage (nominal voltage 3 V), specific energy above 230 Wh/kg and an energy density above 535 Wh/L, depending on design and application, good performance over a wide temperature range, long shelf life, storability even at elevated temperatures, and low cost.

The Li/MnO₂ battery is used in a wide variety of applications such as long-term memory backup, safety and security devices, cameras, many consumer devices and in military electronics. It has gained an excellent safety record during the period since it was introduced.

The performance of a Li/MnO₂ battery is compared with comparable mercury, silver oxide, and zinc batteries in Sec. 7.3 illustrating the higher energy output of the Li/MnO₂ battery.

14.8.1 Chemistry

The Li/MnO₂ cell uses lithium for the anode, and electrolyte containing lithium salts in a mixed organic solvent such as propylene carbonate and 1,2-dimethoxyethane, and a specially prepared heat-treated form of MnO₂ for the active cathode material.

The cell reactions for this system are

Anode
$$x\text{Li} \rightarrow \text{Li}^+ + e$$

Cathode
$$\underline{\text{Mn}^{\text{IV}}\text{O}_2 + x\text{Li}^+ + e \rightarrow \text{Li}_x\text{Mn}^{\text{III}}\text{O}_2}$$

Overall
$$x\text{Li} + \text{Mn}^{\text{IV}}\text{O}_2 \rightarrow \text{Li}_x\text{Mn}^{\text{III}}\text{O}_2$$

Manganese dioxide, an intercalation compound, is reduced from the tetravalent to the trivalent state producing Li_xMnO_2 as the Li^+ ion enters into the MnO_2 crystal lattice.[1,38]

The theoretical voltage of the total cell reaction is about 3.5 V, but an open circuit voltage of a new cell is typically 3.3 V. Cells are typically predischarged to lower the open circuit voltage to reduce corrosion.

14.8.2 Construction

The Li/MnO₂ electrochemical system is manufactured in several different designs and configurations to meet the range of requirements for small, lightweight, portable power sources.

Coin Cells. Figure 14.38 shows a cutaway illustration of a typical coin cell. The manganese dioxide pellet faces the lithium anode disk and is separated by a nonwoven polypropylene separator impregnated with the electrolyte. The cell is crimped-sealed, with the can serving as the positive terminal and the cap as the negative terminal.

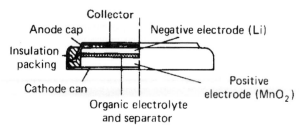

FIGURE 14.38   Cross-sectional view of Li/MnO$_2$ coin-type battery. (*Courtesy of Duracell, Inc.*)

Bobbin-Type Cylindrical Cells.   The bobbin-type cell is one of the two Li/MnO$_2$ cylindrical cells. The bobbin design maximizes the energy density due to the use of thick electrodes and the maximum amount of active materials, but at the expense of electrode surface area. This limits the rate capability of the cell and restricts its use to low-drain applications.

A cross section of a typical cell is shown in Fig. 14.39. The cells contain a central lithium anode core surrounded by the manganese dioxide cathode, separated by a polypropylene separator impregnated with the electrolyte. The cell top contains a safety vent to relieve pressure in the event of mechanical or electrical abuse. Welded-sealed cells are manufactured in addition to the crimped-seal design. These cells, which have a 10-year life, are used for memory backup and other low-rate applications.

Spirally Wound Cylindrical Cells.   The spirally wound cell, illustrated in Fig. 14.40, is designed for high-current pulse applications as well as continuous high-rate operation. The lithium anode and the cathode (a thin, pasted electrode on a supporting grid structure) are

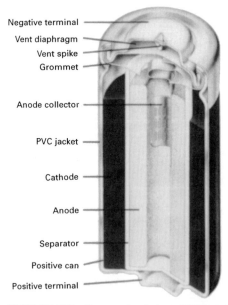

FIGURE 14.39   Cross-sectional view of Li/MnO$_2$ bobbin battery. (*Courtesy of Duracell, Inc.*)

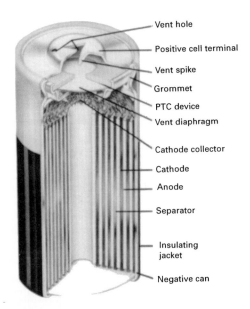

Vent hole

Positive cell terminal

Vent spike

Grommet

PTC device

Vent diaphragm

Cathode collector

Cathode

Anode

Separator

Insulating jacket

Negative can

FIGURE 14.40   Cross-sectional view of Li/MnO$_2$ spirally wound electrode battery. (*Courtesy of Duracell, Inc.*)

wound together with a microporous polypropylene separator interspaced between the two thin electrodes to form the jelly-roll construction. With this design a high electrode surface area is achieved and the rate capability increased.

High-rate spirally wound cells contain a safety vent to relieve internal pressure in the event the cell is abused. Many of these cells also contain a resettable positive temperature coefficient (PTC) device which limits the current and prevents the cell from overheating if short-circuited accidentally (see also Sec. 14.8.5).

Multicell 9-V Battery.   The Li/MnO$_2$ system has also been designed in a 9-V battery with 1200 mAh capacity in the ANSI 1604 configuration as a replacement for the conventional zinc battery. The battery contains three prismatic cells, using an electrode design that utilizes the entire interior volume, as shown in Fig. 14.41. An ultrasonically sealed plastic housing is used for the battery case.

Foil Cell Designs.   Other cell design concepts are being used to reduce the weight and cost of batteries by using lightweight cell packaging. One of these approaches is the use of heat-sealable thin foil laminates, in a prismatic cell configuration in place of metal containers. The design of a cell with a capacity of about 16 Ah is illustrated in Fig. 14.42. The cell contains 10 anode and 11 cathode plates in a parallel plate array.[39]

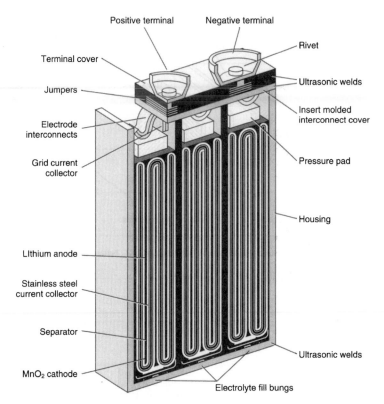

FIGURE 14.41 Cross-sectional view of three-cell 9-V Li/MnO$_2$ battery. (*Courtesy of Ultralife Batteries, Inc.*)

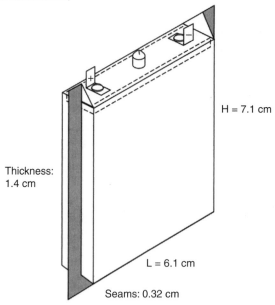

FIGURE 14.42 Foil cell design. (*From Ref. 39.*)

14.8.3 Performance

Voltage   The open-circuit voltage of the Li/MnO$_2$ battery is typically 3.1 to 3.3 V after pre-discharge. The nominal voltage is 3.0 V. The operating voltage during discharge ranges from about 3.1 to 2.0 V and is dependent on the cell design, state of charge, and the other discharge conditions. The end or cutoff voltage, the voltage by which most of the capacity has been expended, is 2.0 V, except under high-rate, low-temperature discharges, when a lower end voltage may be specified.

Discharge Characteristics of Coin-Type Batteries.   Typical discharge curves for the Li/MnO$_2$ coin cells are presented in Fig. 14.43. The discharge profile is fairly flat at these low to moderate discharge rates throughout most of the discharge, with a gradual drop near the end of life. This gradual drop in voltage can serve as a state-of-charge indicator to show when the battery is approaching the end of its useful life.

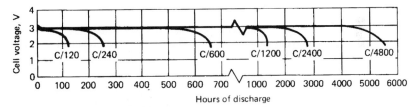

FIGURE 14.43   Typical discharge curves of Li/MnO$_2$ coin-type batteries. (*Courtesy of Duracell, Inc.*)

Some applications (such as an LED watch with backlight) require a high pulse load superimposed on a low background current. The performance of a coin-type battery under these conditions is shown in Fig. 14.44, illustrating good voltage regulation even at the higher discharge rates.

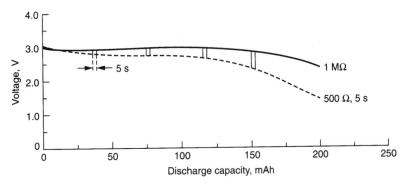

FIGURE 14.44   Pulse characteristics of Li/MnO$_2$ coin-type battery (190-mAh size) at 20°C. Test conditions: continuous load—1 MΩ ≈ 3 μA; pulse load—500 Ω ≈ 5.5 mA; duration—5 s; pulses—4; time between pulses—3 h. (*Courtesy of Duracell, Inc.*)

The Li/MnO$_2$ coin-type battery is capable of performing over a wide temperature range, from about -20 to $55°C$, as shown in Fig. 14.45. The midpoint voltages, when discharged at various loads and temperatures, are plotted in Fig. 14.46.

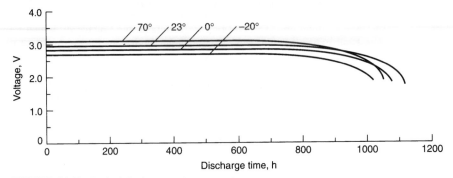

FIGURE 14.45   Typical discharge performance of Li/MnO$_2$ coin-type battery at various temperatures. (*Courtesy of Duracell, Inc.*)

The discharge characteristics of the Li/MnO$_2$ battery are summarized in Fig. 14.47, which shows the percent capacity delivered at various temperatures and discharge loads.

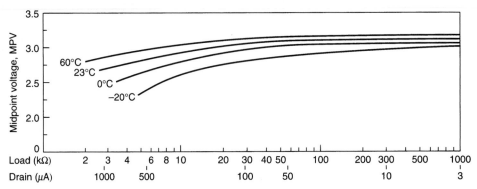

FIGURE 14.46   Midpoint voltage during discharge of Li/MnO$_2$ coin-type (280-mAh size). (*Courtesy of Duracell, Inc.*)

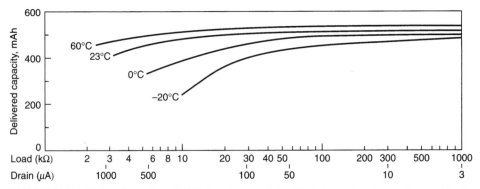

FIGURE 14.47   Delivered capacity of Li/MnO$_2$ coin-type (280-mAh size) at various temperatures and loads. (*Courtesy of Duracell, Inc.*)

Discharge Characteristics of Cylindrical Bobbin Batteries.   Typical discharge curves for the Li/MnO$_2$ cylindrical bobbin batteries are given in Fig. 14.48. These bobbin electrode batteries are designed for use at low to moderate discharge rates, delivering higher capacities at these discharge rates than the spirally wound electrode batteries of the same size (see Table 14.19). The discharge profile is fairly flat at these low rates throughout most of the discharge, with the typical gradual slope near the end of the discharge. The effect of a high pulse load superimposed on a low background current is shown in Fig. 14.49.

The performance of the Li/MnO$_2$ cylindrical bobbin battery at temperatures from −20 to 60°C is shown in Fig. 14.50. Operation of the coin-type and cylindrical bobbin electrode batteries at the lower temperatures is limited to the lower discharge rates.

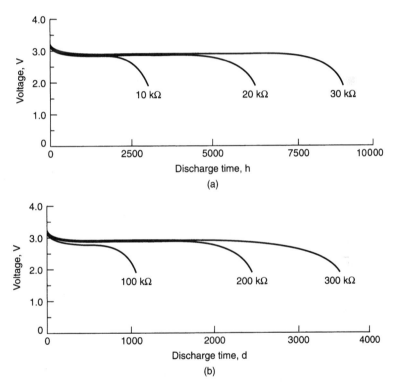

FIGURE 14.48   Discharge characteristics of Li/MnO$_2$ cylindrical bobbin battery (850-mAh size) at 20°C. (*a*) Discharge time in hours. (*b*) Discharge time in days. (*Courtesy of Duracell, Inc.*)

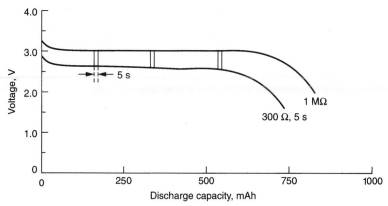

FIGURE 14.49 Pulse discharge characteristics of Li/MnO$_2$ cylindrical bobbin cell (850-mAh size) at 20°C. Test conditions: continuous load—1 M$\Omega \approx$ 2.9 μA; pulse load—300 $\Omega \approx$ 10 mA; duration—5 s; pulses—3; time between pulses—3 h. (*Courtesy of Duracell, Inc.*)

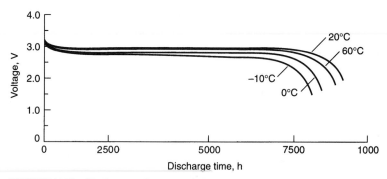

FIGURE 14.50 Discharge performance of Li/MnO$_2$ cylindrical bobbin cell (850-mAh size) at various temperature; 30-kΩ discharge rate. (*Courtesy of Duracell, Inc.*)

Discharge Characteristics of Cylindrical Spirally Wound Batteries.   Typical discharge curves for Li/MnO$_2$ cylindrical spirally wound batteries at various constant-current discharge loads and temperatures are given in Fig. 14.51. These batteries are designed for operation at fairly high rates and low temperatures. Their discharge profile is flat under most of these discharge conditions. The midpoint voltage when discharged at various loads and temperatures, is plotted in Fig. 14.52.

The characteristics of the batteries under constant power discharge are shown in Fig. 14.53. These data are expressed in terms of E-rate, which is calculated in a manner similar to calculating the C rate, but based on the rated Watt-hour capacity. For example, the $E/5$ rate for a cell rated at 4 Wh is 800 mW.

The discharge characteristics of the cylindrical spirally wound Li/MnO$_2$ battery at various temperatures and loads are summarized in Fig. 14.54. Figure 14.54a shows the percent capacity delivered on constant-resistance loads, Fig. 14.54b the percent capacity delivered on constant-current loads. The good performance of the Li/MnO$_2$ battery at the lower-rate discharges is evident, and it still delivers a higher percentage of its capacity at relatively high discharge rates compared to conventional aqueous primary cells.

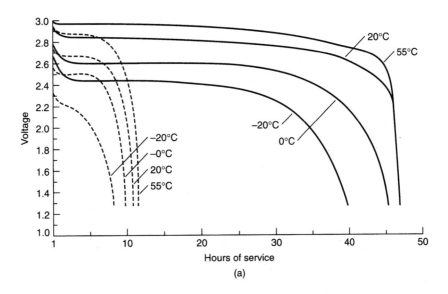

(a)

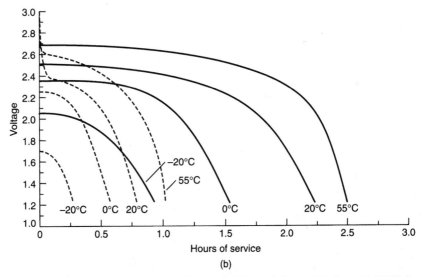

(b)

FIGURE 14.51 Discharge characteristics of cylindrical (spirally wound electrode) Li/MnO_2 battery (CR123A-size). (*a*) Discharge at 30 and 125 mA. Broken line—125 mA; solid line—30 mA. (*b*) Discharge at 500 and 1000 mA. Broken line—1 A; solid line—0.5 A. (*b*) Discharge at 500 and 1000 mA. Broken line—1 A; solid line—0.5 A.

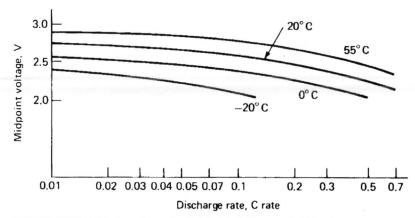

FIGURE 14.52 Midpoint voltage of cylindrical (spirally wound) Li/MnO$_2$ batteries during discharge; 2-V end voltage.

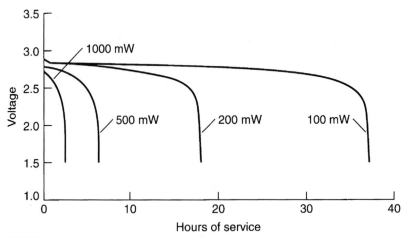

FIGURE 14.53 Discharge characteristics of cylindrical (spirally wound electrode) Li/MnO$_2$ cells (CR123A-size) under constant-power mode at 20°C.

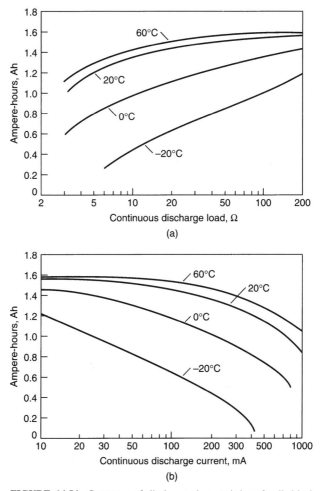

FIGURE 14.54 Summary of discharge characteristics of cylindrical (spirally wound) Li/MnO_2 battery (CR123A size); capacity vs. discharge load to 2.0 V per cell. (*a*) Constant-resistance loads. (*b*) Constant-current loads.

The discharge characteristics of a larger (D-size) spiral-wound Li/MnO_2 battery is shown in Fig. 14.55. The discharge characteristics indicate the fall-off in performance at higher rates (greater than 1 amp) and at lower temperatures.

Discharge Characteristics of Three-Cell 9 V Li/MnO_2 Battery. The performance of the 9-V, 1.2 Ah Li/MnO_2 battery is shown in Fig. 7.6. Typical discharge curves at 20°C and at temperatures from −40 to 71°C are shown in Fig. 14.56*a*. The voltage delay characteristics after 3 months' storage at 20°C are shown in Fig. 14.56*b*. The lithium battery has a higher voltage and delivers significantly more service than the comparable zinc anode batteries as shown in Fig. 7.6.

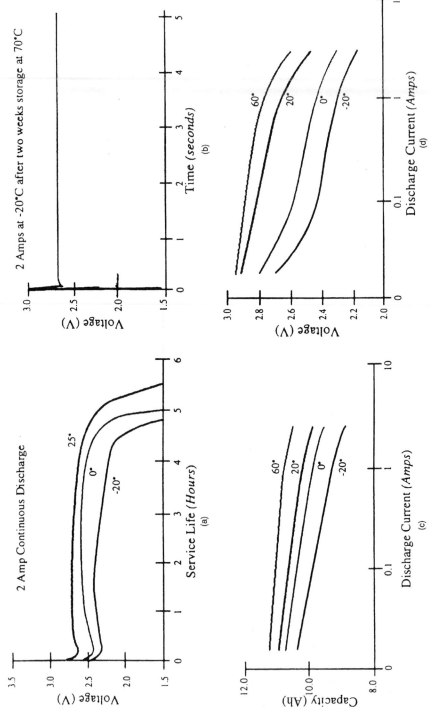

FIGURE 14.55 Discharge characteristics of spiral-wound Li/MnO$_2$ D-size battery. (*a*) Discharge curve at 2 amp rate at 3 temperatures; (*b*) start-up characteristic after 2-weeks storage at 70°C. (*c*) Capacity as a function of discharge rate at 4 temperatures. (*d*) Operating voltage vs discharge rate at 4 temperatures. (*Source: Manufacturer's Data Sheet.*)

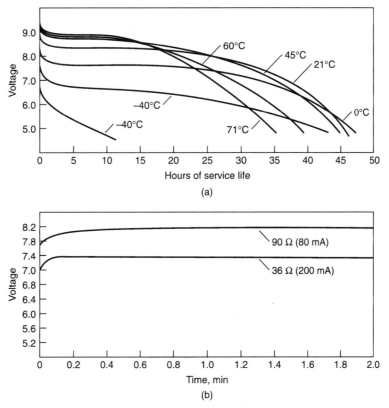

FIGURE 14.56 Discharge characteristics of 9-V Li/MnO$_2$ battery. (*a*) Discharge vs. temperature; 300-Ω continuous discharge. (*b*) Voltage delay after 3 months' storage at 20°C. (*Courtesy of Ultralife Batteries, Inc.*)

Internal Resistance. The internal resistance of the Li/MnO$_2$ battery, as with most battery systems, is dependent on the cell size, design, electrode, separator, as well as the chemistry. Inherently, the conductivity of the organic solvent-based electrolytes is lower than that of the aqueous electrolytes, and the Li/MnO$_2$ system, therefore, has a higher impedance than conventional cells of the same size and construction. Designs which increase electrode area and decrease electrode spacing, such as coin-shaped flat cells and spirally wound jelly-roll configurations, are used to reduce the resistance. Further, the lithium cells will perform relatively more efficiently at the lower temperatures because the conductivity of the organic solvents is less sensitive to temperature changes than it is for the aqueous solvents.

Figure 14.57 shows the change in internal resistance of a 280-mAh coin-type battery during a low-rate discharge at 20°C. Typically, the resistance is a mirror image of the voltage profile. It remains fairly constant for most of the discharge and increases at the end of life. The resistance values for typical Li/MnO$_2$ batteries are listed in Table 14.19.

Service Life. The capacity or service life of the different types of Li/MnO$_2$ cells, normalized for a 1-g and 1-cm^3 cell, at various discharge loads and temperatures, is summarized in Fig. 14.58. These data can be used to approximate the performance of a given cell or to determine the size and weight of a cell for a particular application.

Shelf Life. The storage characteristics of the Li/MnO$_2$ cell are shown in Fig. 14.59. This system is very stable in all of the configurations, with a loss of capacity of less than 1% annually. The cells do not have any noticeable voltage delay at the start of most discharges, except at low temperature on high discharge rates.

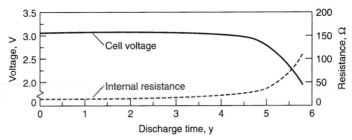

FIGURE 14.57 Internal resistance of Li/MnO$_2$ coin-type battery (280-mAh size), 5-μA drain, at 20°C.

TABLE 14.19 Typical Li/MnO$_2$ Batteries

International (IEC) type	Rated capacity, mAh*	Weight, g	Diameter, mm	Height, mm	Volume, cm^3	Specific† energy Wh/kg	Energy† density Wh/L	Impedance at 1 kHz, Ω
Low-rate flat or coin batteries, 3 V							
CR 1216	25	0.7	12.5	1.6	0.21	100	335	—
CR 1025	30	0.7	10.0	2.5	0.20	125	410	—
CR 1220	35	0.9	12.5	2.0	0.25	110	390	—
CR 1616	55	1.2	16.0	1.6	0.32	128	480	—
CR 2012	55	1.4	20.0	1.2	0.38	110	405	—
CR 1620	70	1.3	16.0	2.0	0.40	150	485	—
CR 2016	75	1.7	20.0	1.6	0.50	125	420	12–18
CR 2320	130	3.0	23.0	2.0	0.83	122	440	—
CR 2025	150	2.5	20.0	2.5	0.79	165	530	12–18
CR 2325	190	3.0	23.0	2.5	1.04	175	510	8–15
CR 2032	210	3.3	20.0	3.2	1.00	175	585	12–18
CR 2330	265	4.0	23.0	3.0	1.35	185	550	—
CR 2430	280	4.0	24.5	3.0	1.41	195	555	8–15
CR 3032	500	7.1	30.0	3.2	2.26	195	600	—
CR 2450	550	6.2	24.5	5.0	2.35	245	650	8–15
CR 2354	560	5.9	23.0	5.4	2.24	205	700	—
CR 2477	1000	10.5	24.5	7.7	3.62	260	770	—
Cylindrical batteries; spirally wound, 3 V							
CR 11108‡	160	3.3	11.6	10.8	1.14	155	395	3–5
CR 15H270‡	750	11.0	15.6	27.0	5.2	190	400	—
CR 17345‡	1400	17.0	17.	34.5	7.8	230	535	2–6
Cylindrical batteries, bobbin, 3 V							
CR 14250	850	9.5	14.5	25.0	11.1	250	580	9–13
CR 17335	1800	17.	17.	33.5	7.6	295	660	5–8

International (IEC) type	Rated capacity, mAh*	Weight, g	Batteries (dimensions in mm)		
Batteries (dimensions in mm)					
2 CR $\frac{1}{3}$N	160	9.4	13.0 (D) × 25.2 (H)	2 CR $\frac{1}{3}$N cells in series, 6 V	
2 CR 5	1300	37.	17 (T) × 34 (W) × 45 (H)	2 CR 17345 cells in series, 6 V	
CR-P2	1300	37.	19.5 (T) × 34 (W) × 36 (H)	2 CR 17345 cells in series, 6 V	
1604LE	1200	34.4	16.8 (T) × 25.8 (W) × 48.4 (H)	3 prismatic cells in series, 9 V	
CR-V6	1500	39.	29 (W) × 14.5 (T) × 52 (H)	2 "AA"-size cells in series, 6 V	
CR-V3	3000	34.	29 (W) × 14.5 (T) × 52 (H)	2 "AA"-size cells in parallel, 3 V	

*Low-rate batteries–C/200 rate; high-rate and cylindrical cells–C/30 rate.
†Based on average voltage of 2.8 V.
‡Common nomenclature: CR 11108 = CR $\frac{1}{3}$N; CR 15H270 = CR 2; CR 17345 = CR 123A.
Source: Manufacturers' data sheets.

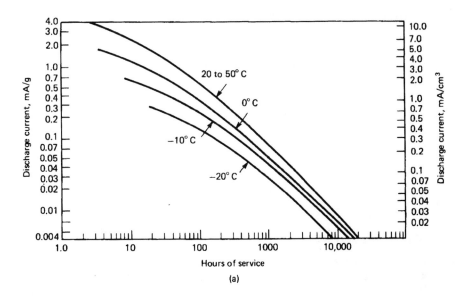

(a)

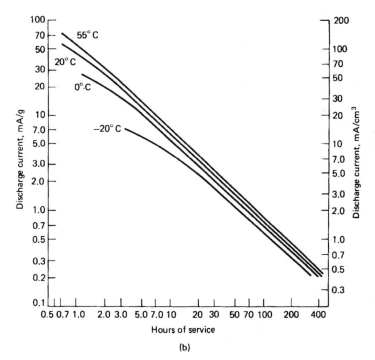

(b)

FIGURE 14.58   Service life of Li/MnO$_2$ cells to 2-V end voltage. (*a*) Low-rate coin-type batteries. (*b*) Small cylindrical batteries.

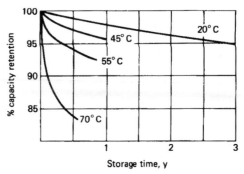

FIGURE 14.59 Shelf life of Li/MnO$_2$ small cylindrical batteries. (*Courtesy of Duracell, Inc.*)

14.8.4 Cell and Battery Sizes

The Li/MnO$_2$ cells are manufactured and commercially available in a number of flat and cylindrical batteries ranging in capacity from about 30 to 1400 mAh. Larger-size batteries have been developed in cylindrical and rectangular configurations. The physical and electrical characteristics of some of these are summarized in Tables 14.19, 14.20 and 14.21. In some instances interchangeability with other battery systems is provided by doubling the size of the battery to accommodate the 3-V output of the Li/MnO$_2$ cell compared with 1.5 V of the conventional primary batteries, for example battery type CR-V3.

TABLE 14.20 Characteristics of Larger Li/MnO$_2$ Batteries. Standard Type: C/20 Rate at Room Temperature to 2.0 V Cut-off. High-Rate Type: C/10 Rate at Room Temperature to a 2.0 V Cut-off

IEC cell size	Nominal capacity typical values (Ah)	Max. continuous current* (mA)	Dimensions dia. (mm)	ht. (mm)	Weight (g)
M03	0.2	40	14.5	25	8
M04	0.3	60	14.5	28	9
M49	1.6	300	22.5	32	24
M52	4.5	800	26	51	57
M52 HR	4.0	1200	26	51	59
M56	5.6	900	26	60	69
M56 HR	5.5	1500	26	60	71
M19 HR	9.0	2000	33.5	58	103
M20	10.5	2000	34	61	115
M20 HR	10.0	2500	34	61	117
M24 HR	20.0	4000	33.5	111	201
M58	11.0	2000	42	51	125
M25	33.0	4000	42	125	355
M62	33.0	5000	42	133	355

Source: FRIWO Silberkraft data sheets.

TABLE 14.21   Specifications for Commercially Available Foil Cells and Batteries

Manufacturer's model number	Dimension's thickness × L × W {mm}	Voltage (OCV) {volts}	Capacity {mAh}	Max. cont. current {mA}	Weight {g}
U3VF-A-T	0.8 × 38.6 × 30.0	3.2	27	20	1.1
U3VF-B-T	0.5 × 27.7 × 38.1	3.2	45	40	1.2
U3VF-G-T	1.1 × 36.3 × 25.7	3.2	120	12	1.1
U3VF-L-T	2.0 × 38.6 × 30.0	3.2	160	20	2.1
U6VF-K2	4.6 × 73.7 × 46.0	6.4	600	50	17.5
U3VF-K	2.3 × 73.7 × 46.0	3.2	800	55	8.5
U3VF-H	2.3 × 93.9 × 76.0	3.2	1800	60	18.0
U6VF-H2	3.9 × 93.2 × 78.2	6.4	1800	60	37.0
U3VF-D	2.3 × 92.0 × 92.0	3.2	2300	80	27.5

Source: Manufacturer's Data Sheets.

14.8.5   Applications and Handling

The main applications of the Li/MnO_2 system currently range up to several Ampere-hours in capacity, taking advantage of its higher energy density, better high-rate capability, and longer shelf life compared with the conventional primary batteries. The Li/MnO_2 batteries are used in memory applications, watches, calculators, cameras, and radio frequency identification (RFID) tags. At the higher drain rates, motor drives, automatic cameras, toys, personal digital assistants (PDAs), digital cameras and utility meters are excellent applications.

The low-capacity Li/MnO_2 batteries can generally be handled without hazard, but, as with the conventional primary battery systems, charging and incineration should be avoided as these conditions could cause a cell to explode.

The higher-capacity cylindrical batteries are generally equipped with a venting mechanism to prevent explosion, but the batteries, nevertheless, should be protected to avoid short circuits and cell reversal, as well as charging and incineration. Most of the high-rate batteries are also equipped with an internal resettable current and thermal protective system called a positive temperature coefficient (PTC) device. When a cell is short-circuited or discharged above design limits and the cell temperature increases, the resistance of the PTC device quickly increases significantly. This limits the amount of current which can be drawn from the cell and keeps the internal temperature of the cell within safe limits. Figure 14.60 shows the operation of the PTC device when a cell is short-circuited. After a short-circuit peak of about 10 A the current is abruptly limited and maintained at the depressed level. When the short-circuit is removed, the cell reverts to its normal operating condition. The short delay of several minutes before the PTC operates permits the cell to deliver pulse currents at higher values than the maximum permitted under continuous drain.

Larger spiral-wound cells and the foil-type shown in Fig. 14.42 are finding use in military electronics, including the U.S. Army's Twenty-First Century Land Warrior System, radios and the thermal weapons sight. Battery packs are also being employed for Emergency Positioning Indicating Radio Beacons (EPIRBs) and pipeline test vehicles. Smaller batteries are also available commercially in foil laminate packages for use in specialized applications such as toll collection transponders and RFID tags for shipping and inventory control.

The specific conditions for the use and handling of Li/MnO_2 batteries are dependent on the size as well as the specific design features. Manufacturers' recommendations should be consulted.

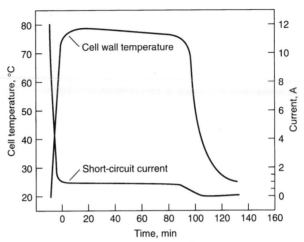

FIGURE 14.60   Short circuit of Duracell® XL™ CR123A battery.

14.9   *LITHIUM/CARBON MONOFLUORIDE {Li/(CF)ₙ} BATTERIES*

The lithium/carbon monofluoride $\{Li/(CF)_n\}$ battery was one of the first lithium/solid-cathode systems to be used commercially. It is attractive as its theoretical specifc energy (about 2180 Wh/kg) is among the highest of the solid-cathode systems. Its open-circuit voltage is 3.2 V, with an operating voltage of about 2.5–2.7 V. Its practical specific energy and energy density ranges up to 250 Wh/kg and 635 Wh/L in smaller sizes and 590 Wh/kg and 1050 Wh/L in larger sizes. The system is used primarily at low to medium discharge rates.

14.9.1   Chemistry

The active components of the cell are lithium for the anode and polycarbon monofluoride $(CF)_n$ for the cathode. The value of n is typically 0.9 to 1.2. Carbon monofluoride is an interstitial compound, formed by the reaction between carbon powder and fluorine gas. While electrochemically active, the material is chemically stable in the organic electrolyte and does not thermally decompose up to 400°C, resulting in a long storage life. Different electrolytes have been used; typical electrolytes are lithium hexafluorarsenate ($LiAsF_6$) in δ-butyrolactone (GBL) or lithium tetrafluoroborate ($LiBF_4$) in propylene carbonate (PC) and dimethoxyethane (DME).

The simplified discharge reactions of the cell are

Anode
$$n\mathrm{Li} \rightarrow n\mathrm{Li}^+ + ne$$

Cathode
$$(\mathrm{CF})_n + ne \rightarrow n\mathrm{C} + n\mathrm{F}^-$$

Overall
$$n\mathrm{Li} + (\mathrm{CF})_n \rightarrow n\mathrm{LiF} + n\mathrm{C}$$

The polycarbon monofluoride changes into amorphous carbon which is more conductive as the discharge progresses, thereby increasing the cell's conductivity, improving the regulation of the discharge voltage, and increasing the discharge efficiency. The crystalline LiF precipitates in the cathode structure.[1,40,41]

14.9.2 Construction

The Li/$(CF)_n$ system is adaptable to a variety of sizes and configurations. Batteries are available in flat coin or button, cylindrical, and rectangular shapes, ranging in capacity from 0.020 to 25 Ah; larger-sized batteries have been developed for specialized applications.

Figure 14.61 shows the construction of a coin-type battery. The Li/$(CF)_n$ cells are typically constructed with an anode of lithium foil rolled onto a collector and a cathode of Teflon-bonded polycarbon monofluoride and acetylene black on a nickel collector. Nickel-plated steel or stainless steel is used for the case material. The coin cells are crimped-sealed using a polypropylene gasket.

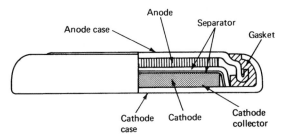

FIGURE 14.61   Cross-sectional view of Li/$(CF)_n$ coin-type battery. (*Courtesy of Panasonic, Division of Matsushita Electric Corp. of America.*)

The pin-type batteries use an inside-out design with a cylindrical cathode and a central anode in an aluminum case, as shown in Fig. 14.62.

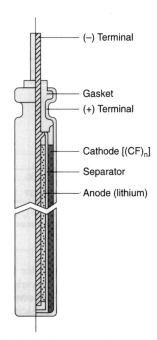

FIGURE 14.62   Cross-sectional view of Li/$(CF)_n$ pin-type battery. (*Courtesy of Panasonic, Division of Matsushita Electric Corp. of America.*)

The cylindrical batteries use a spirally wound (jelly-roll) electrode construction, and the batteries are either crimped or hermetically sealed. Their construction is similar to the cylindrical spiral-wound electrode design of the Li/MnO$_2$ battery shown in Fig. 14.40. The larger cells are provided with low-pressure safety vents.

14.9.3 Performance

Coin-Type Batteries. Figure 14.63 presents the discharge curves at 20°C for a typical Li/ (CF)$_n$ coin-type battery rated at 165 mAh. The voltage is constant throughout most of the discharge, and the coulombic utilization is close to 100% under low-rate discharge. Figure 14.64 presents the discharge curves for the same battery at various discharge temperatures. The behavior of the battery on a pulse discharge at 20°C is shown in Fig. 14.65.

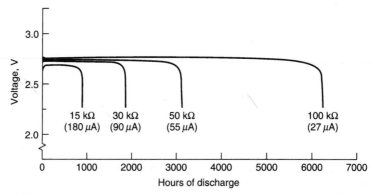

FIGURE 14.63 Typical discharge curves of Li/(CF)$_n$ coin-type battery at 20°C; rated capacity 165 mAh. (*Courtesy of Panasonic, Division of Matsushita Electric Corp. of America.*)

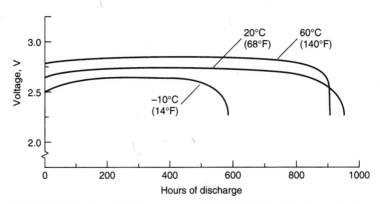

FIGURE 14.64 Typical discharge curves of Li/(CF)$_n$ 165 mAh coin-type battery at various temperatures; 15-kΩ discharge load; 180 μA. (*Courtesy of Panasonic, Division of Matsushita Electric Corp. of America.*)

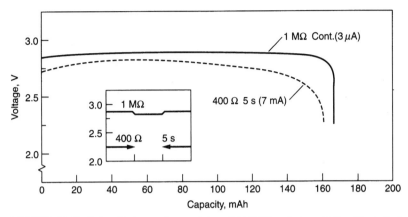

FIGURE 14.65 Pulse discharge characteristics of $Li/(CF)_n$ coin-type (165-mAh size) at 20°C. (*Courtesy of Panasonic, Division of Matsushita Electric Corp. of America.*)

The performance of the coin battery (165-mAh capacity) is summarized in Fig. 14.66. Figure 14.66*a* shows the average load voltage (plateau voltage during discharge) and Fig. 14.66*b* shows the capacity for discharges at various loads and temperatures.

Figure 14.67 summarizes the discharge performance data for $Li/(CF)_n$ coin-type batteries normalized for a 1-g and 1-mL battery. These data can be used to approximate the size or performance of a battery for a particular application.

Cylindrical Cells.   The cylindrical batteries are designed to operate at heavier discharge loads than the coin batteries. Figure 14.68 shows the discharge characteristics at 20°C of a cylindrical $\frac{2}{3}$A-size cell rated at 1.2 Ah; Fig. 14.69 presents the discharge curves at other discharge temperatures. In some cases an initial low voltage is observed with the $Li/(CF)_n$ battery; that is, the voltage drops initially below the operating level on load and recovers gradually as the discharge progresses. This is attributed to the fact that $(CF)_n$ is an insulator, but the resistance of the cathode decreased during the discharge as conductive carbon is produced.

The average load voltages for cylindrical batteries at various temperatures are given in Fig. 14.70. The performance data are then summarized in Fig. 14.71, which shows the effect of temperature and load on the service life of a battery normalized to unit weight (grams) and volume (cubic centimeters).

The internal impedance of these cells is relatively low. For example, the internal resistance of the BR $\frac{2}{3}$A battery ranges between 0.4 Ω at 25°C and 0.6 Ω at −20°C, but rises toward the end of the discharge, as shown in Fig. 14.72.

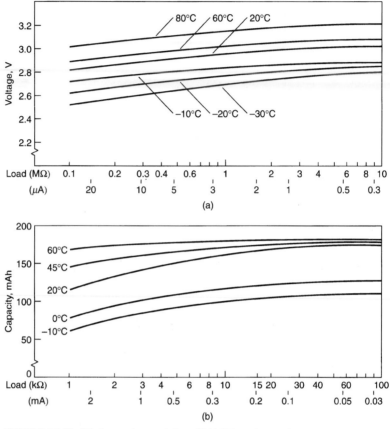

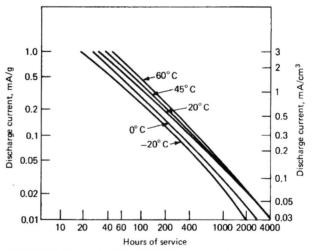

FIGURE 14.66 Discharge characteristics of $Li/(CF)_n$ coin-type battery (165-mAh size). (a) Operating voltage vs. discharge load; voltage at 50% discharge. (b) Capacity vs. discharge load; cutoff at 2.0 V. (*Courtesy of Panasonic, Division of Matsushita Electric Corp. of America.*)

FIGURE 14.67 Service life of $Li/(CF)_n$ coin-type batteries at various discharge rates and temperatures; 2.0-V end voltage.

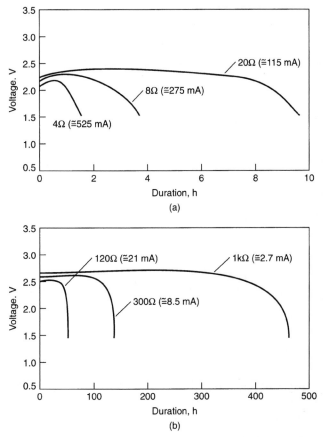

FIGURE 14.68   Typical discharge curves for $Li/(CF)_n$ cylindrical battery, BR $\frac{2}{3}$A size, at 20°C. (*Courtesy of Panasonic, Division of Matsushita Electric Corp. of America.*)

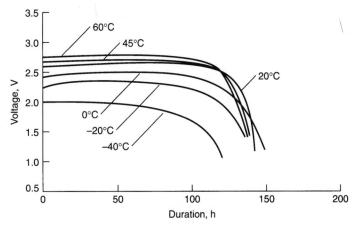

FIGURE 14.69   Typical discharge curves of $Li/(CF)_n$ cylindrical battery at various temperatures; rated capacity 1.2 Ah; 300-Ω discharge load. (*Courtesy of Panasonic, Division of Matsushita Electric Corp. of America.*)

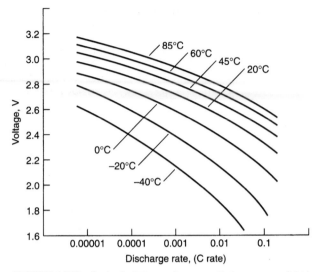

FIGURE 14.70   On-load plateau voltage vs. discharge rate of Li/ $(CF)_n$ cylindrical batteries. (*Courtesy of Panasonic, Division of Matsushita Electric Corp. of America.*)

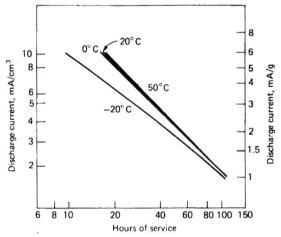

FIGURE 14.71   Service life of Li/$(CF)_n$ cylindrical batteries at various discharge rates and temperatures; 1.8-V end voltage.

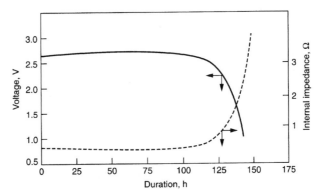

FIGURE 14.72 Internal impedance during discharge of a Li/(CF)$_n$ cylindrical battery BR $\frac{2}{3}$A size, at 20°C; 300-Ω discharge load. (*Courtesy of Panasonic, Division of Matsushita Electric Corp. of America.*)

High Temperature Coin-Type Batteries.   The long shelf life and flat discharge character-istics of the Li/(CF)$_n$ system are utilized in coin-type batteries designed for mounting on PC boards and used for long-term, low-drain or standby applications. The dimensions and per-formance characteristics of this battery are given in Table 14.22*a*. These units are rated for use from −40 to +150°C. The other Li/(CF)$_n$ cells are also capable of long-term discharge.

Shelf Life.   The Li/(CF)$_n$ cells have extremely good storage characteristics. Tests over more than 10 years of storage show a self-discharge rate of about 0.5% per year at 20°C and less than 4% per year at 70°C. These rates decrease on longer term storage.[42] This cell is also well-suited for applications requiring low current drain over an extended period of time. This is illustrated in Fig. 14.73, which shows the discharge characteristics of the $\frac{2}{3}$A-size cell at a 20-μA discharge rate over a period of 7 years. Voltage delay after storage is not apparent with these cells, except under stringent discharge conditions.

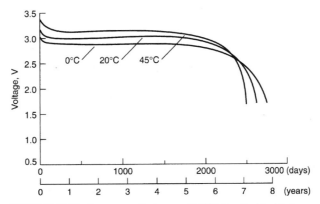

FIGURE 14.73 Long-term discharge of Li/(CF)$_n$ cylindrical battery. BR $\frac{2}{3}$A size; 150-kΩ discharge. (*Courtesy of Panasonic, Division of Matsushita Electric Corp. of America.*)

14.9.4 Cell and Battery Types

The Li/(CF)$_n$ batteries are manufactured in a number of coin, cylindrical, and pin configurations. The major electrical and physical characteristics of some of these batteries are listed in Table 14.22a and b. Manufacturers' specifications should be consulted for the most recent listings of commercially available cells.

Larger sizes of cells and batteries[43] have also been developed for military, governmental and space applications, as given in Table 14.22b. Spiral-wound and prismatic cells are used to build the multi-cell batteries given in this table. The smaller cylindrical cells employ a 0.030 cm thick steel case but larger units, such as the 1200 Ah cell, are reinforced with an epoxy-fiberglass cylinder to provide additional strength, with an increase in weight about half that of increasing the steel wall thickness. All these cells employ a Zeigler-type compression seal, a unique cutter vent mechanism and two layers of separator. The first is a microporous layer to prevent particulate migration, and the second is a non-woven polyphenylene sulfide material to provide high-strength, high-temperature stability and good electrolyte wicking action. These low-rate designs provide a specific energy of 600 Wh/kg and an energy density of 1000 Wh/L in the DD size and higher values for larger units. Capacity to a 2.0 volt cut-off as a function of temperature at four rates from 0.04 to 1.00 A is shown in Fig. 14.74 for the DD design. The capacity of this battery is relatively independent of temperature at the three lower currents and above 10°C, but decreases at the higher rate and lower temperatures. Discharge curves for the 1200 Ah reinforced cylindrical battery at the 2000 hours (ca. 500 mA) and the 1000-hour rate (ca. 1.0 Amp) are shown in Fig. 14.75. The trailing knee in these discharge curves has been attributed to electrolyte starvation at the end of the discharge. These batteries effectively demonstrate the ability of the lithium/carbon monofluoride system to provide very high specific energies and energy densities, in these low rate designs.

TABLE 14.22(a) Characteristics of Lithium/Carbon Monofluoride Li/(CF)$_n$ Batteries

				Coin batteries, 3V				
				Recommended drain		Dimensions		
Model no.	IEC	*Nominal capacity (mAh)	Pulse (mA)	Standard (mA)	Low (μA)	Diameter (mm)	Height (mm)	Weight (g)
BR1216	—	25	5	0.03	1	12.5	1.60	0.6
BR1220	—	35	5	0.03	1	12.5	2.00	0.7
BR1225	BR1225	48	5	0.03	1	12.5	2.50	0.8
BR1616	—	48	8	0.03	1	16.0	1.60	1.0
BR1632	—	120	8	0.03	1	16.0	3.20	1.5
BR2016	BR2016	75	10	0.03	1	20.0	1.60	1.5
BR2020	BR2020	100	10	0.03	2	20.0	2.00	2.0
BR2032	—	190	10	0.03	4	20.0	3.20	2.5
BR2320	BR2320	110	10	0.03	2	23.0	2.0	2.5
BR2325	BR2325	165	10	0.03	3	23.0	2.50	3.2
BR2330	—	255	10	0.03	5	23.0	3.00	3.2
BR3032	BR3032	500	10	0.03	10	30.0	3.20	5.5

TABLE 14.22(a)   Characteristics of Lithium/Carbon Monofluoride Li/(CF)$_n$ Batteries (*Continued*)

Coin type: High operating temperature, 3 V

Model no.	Dimensions (mm)		Nominal capacity (mAh)	Temperature range (°C)
	Diameter	Height		
†BR1225A	12.5	2.5	48	−40°C ~ 150°C
BR1632A	16.0	3.2	120	−40°C ~ 150°C
BR2330A	23.0	3.0	255	−40°C ~ 150°C
BR2477A	24.5	7.0	1000	−40°C ~ 125°C

Pin type, 3 V

Model no.	Nominal* capacity (mAh)	Recommended drain		Dimensions		Weight (g)	Operating temperature range
		Pulse (mA)	Standard (mA)	Diameter (mm)	Height (mm)		
BR425	25	4	0.5	4.2	25.9	0.55	−20°C ~ +60°C (−4°F ~ 140°F)
BR435	50	6	1	4.2	35.9	0.85	−20°C ~ +60°C (−4°F ~ 140°F)

Cylindrical type, 3 V

Model no.	Electrical characteristics (20°)				Dimensions (max.)		Weight (g)
	§Nominal capacity (mAh)	Standard drain (mA)	Continuous maximum drain (mA)	Max. pulse (mA)	Diameter (mm)	Approximate height (mm)	
BR-C	5,000	150	300	1,000	26.0	50.5	42.0
BR-A	1,800	2.5	250	1,000	17.0	45.5	18.0
BR-2/3A	1,200	2.5	250	1,000	17.0	33.5	13.5
‡BR-AH	2,000	2.5	250	1,000	17.0	45.5	18.0
‡BR-AG	2,200	2.5	250	1,000	17.0	45.5	18.0
‡BR-2/3AH	1,350	2.5	250	1,000	17.0	33.5	13.5
‡BR-2/3AG	1,450	2.5	250	1,000	17.0	33.5	13.5

*Nominal capacity shown is based on standard drain.
†Under development
‡H and G versions are higher capacity
§Nominal capacity is based on standard drain and cutoff voltage down to 2.0V at 20°C (68°F)

TABLE 14.22(b)   Characteristics of Large Lithium/Carbon Monofluoride Li/(CF)$_n$ Batteries

| | | Single-cell batteries | | |
| | | Dimensions (cm) | | Weight (grams) |
Part number	Capacity (Ah)	Diameter	Height	
LCF-111	240	6.62	16.51	880
LCF-112	35	3.02	13.84	170
LCF-117	1200	11.43	26.67	3950
LCF-119	400	11.43	9.53	1575
LCF-122	18	3.37	6.06	—
LCF-123	35	3.37	11.72	—
LCF-313	40	6.45(L) × 3.43(W) × 7.09(H)		230

| | | | Multicell batteries | | | | |
| | | | Dimensions (cm) | | | | |
Part number	Capacity (Ah)	Nominal voltage	H	L	W	Weight (grams)	Comments
MAP-9036	23.5	39	17.1	20.3	14.0	4586	Former shuttle range safety system
MAP-9046	3.74 (×2)	30 (×2)	15.9	17.3	7.6	3405	2 independent voltage sections
MAP-9225	240	15	24.9	30.7	6.5	6000	
MAP-9257	80	18	12.4	18.5	14.8	—	
MAP-9319	240	21	42.9	29.7	9.7	—	
MAP-9325	120/7.2	12/15	17.1	18.6	9.2	—	Optional casing
MAP-9334	80	6	16.8	7.6	4.8	—	Minuteman III GRP batteries
MAP-9381	70	39	31.3	20.0	9.7	—	Integrated capacitor bank
MAP-9382	80/70	33/12	20.1	17.6	14.1	—	2 independent voltage sections
MAP-9389	280	15	23.6	33.8	11		
MAP-9392	40	39	17.1	20.3	14.0		X-33 Range safety system

Source: Eagle-Picher Technologies.

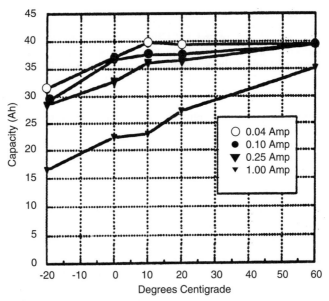

FIGURE 14.74 'DD' Li/(CF)$_n$ discharge performance as a function of rate and temperature.

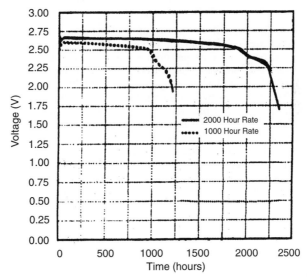

FIGURE 14.75 Typical discharge curves for 1200 Ah Li/(CF)$_n$ battery.

14.9.5 Applications and Handling

The applications of the Li/(CF)$_n$ battery are similar to those of the other lithium/solid-cathode batteries, again taking advantage of the high specific energy and energy density and long shelf life of these batteries. The Li/(CF)$_n$ coin batteries are used as a power source for watches, portable calculators, memory applications, and so on. The low-capacity miniature pin-type batteries have been used as an energy source for LEDs and for fishing lights and microphones. The larger cylindrical batteries can also be used in memory applications, but their higher drain capability also covers use in radio sets, for telemetry, and for photographic and similar general-purpose applications.

Handling considerations for the Li/(CF)$_n$ systems, too, are similar to those for the other lithium/solid-cathode systems. The limited current capability of the coin and low-capacity batteries restricts temperature rise during short circuit and reversal. These batteries can generally withstand this abusive use even though they are not provided with a safety vent mechanism. The larger batteries are provided with a venting device, but short circuit, high discharge rates, and reversal should be avoided as these conditions could cause the cell to vent. Charging and incineration likewise should be avoided for all batteries. The manufacturer's recommendations should be obtained for handling specific battery types.

14.10 LITHIUM/IRON DISULFIDE (Li/FeS$_2$) BATTERIES

Iron sulfide, in both the monosulfide (FeS) and the disulfide (FeS$_2$) forms, has been considered for use in solid-cathode lithium batteries. Only the disulfide battery has been commercialized because of its performance advantage due to its higher sulfur content and higher voltage. The monosulfide electrode has the advantage of reduced corrosion, longer life, and a single voltage plateau compared to the disulfide electrode, which discharges in two steps.

These batteries have a nominal voltage of 1.5 V* and can therefore be used as replacements for aqueous batteries having a similar voltage. Button-type Li/FeS$_2$ batteries were manufactured as a replacement for zinc/silver oxide batteries but are no longer marketed. They had a higher impedance and a slightly lower power capability but were lower in cost and had better low-temperature performance and storability.

Li/FeS$_2$ batteries are now manufactured in a cylindrical configuration. These batteries have better high-drain and low-temperature performance than the zinc/alkaline-manganese dioxide batteries. The performance of these two systems on constant-current discharge at various discharge rates, in the AA size, is compared in Fig. 14.76.

---

*ANSI Standard C18.3M, Part 1-1999.

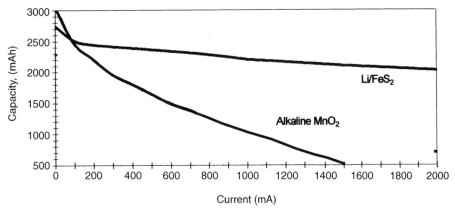

FIGURE 14.76   Discharge rate vs. capacity of Li/FeS$_2$ and Zn/alkaline/MnO$_2$ "AA"-size batteries at 21°C. (*Courtesy of Eveready Battery Co, Inc.*)

14.10.1   Chemistry

The Li/FeS$_2$ battery uses lithium for the anode, iron disulfide for the cathode, and lithium iodide in an organic solvent blend as the electrolyte. The cell reactions are

Anode $\qquad\qquad\qquad\qquad\qquad$ $4Li \rightarrow 4LI^+ + 4e$

Cathode $\qquad\qquad\qquad\qquad$ $\underline{FeS_2 + 4e \rightarrow Fe + 2S^{-2}}$

Overall $\qquad\qquad\qquad\qquad$ $4Li + FeS_2 \rightarrow Fe + 2Li_2S$

Intermediate species are formed during the discharge, but they have not been fully characterized. They are dependent on many variables, including discharge rate and temperature. Evidence of the intermediate species can be seen in the stepped discharge curve for Li/FeS$_2$ batteries on a light drain, as illustrated in Fig. 14.77.

The overall cell reaction for the monosulfide electrode is

$$2Li + FeS \rightarrow Fe + Li_2S$$

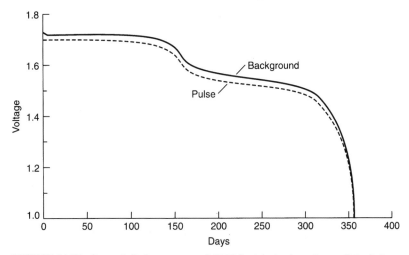

FIGURE 14.77   Stepped discharge curve of Li/FeS$_2$ AA-size batteries on light drain at 21°C. 5000-Ω background with 25-Ω, 1-s/week pulse. (*Courtesy of Eveready Battery Co., Inc.*)

14.10.2 Construction

Li/FeS$_2$ batteries may be manufactured in a variety of designs, including the button and both bobbin and spiral-wound-electrode cylindrical cells. A bobbin construction is most suitable for light-drain applications. The spiral-wound-electrode construction is needed for the heavier-drain applications, and it is this design that has been commercialized.

The construction of the spiral-wound cylindrical battery is shown in Fig. 14.78. These batteries typically have several safety devices incorporated in their design to provide protection against such abusive conditions as short circuit, charging, forced discharge, and over heating. Two safety devices are shown in the figure—a pressure relief vent and a resettable thermal switch, called a positive thermal coefficient (PTC) device. The safety relief vent is designed to release excessive internal pressure to prevent violent rupture if the battery is heated or abused electrically.

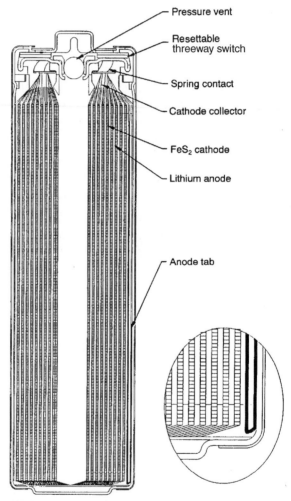

Pressure vent

Resettable threeway switch

Spring contact

Cathode collector

FeS$_2$ cathode

Lithium anode

Anode tab

FIGURE 14.78 Cross section and top assembly of Li/FeS$_2$ AA-size battery. (*Courtesy of Eveready Battery Co., Inc.*)

The primary purpose of the PTC is to protect against external short circuits, though it also offers protection under certain other electrical abuse conditions. It does so by limiting the current flow when the cell temperature reaches the PTC's designed activation temperature. When the PTC activates, its resistance increases sharply, with a corresponding reduction in the flow of current and, consequently, internal heat generation. When the battery (and the PTC) cools, the PTC resistance drops, allowing the battery to discharge again. The PTC will continue to operate in this manner for many cycles if an abusive condition continues or recurs. The PTC will not "reset" indefinitely, but when it ceases to do so, it will be in the high-resistance condition. The characteristics of PTCs (or any other current-limiting devices in the battery) may place some limitations on performance. These are discussed in more detail in Sec. 14.10.3.

14.10.3 Performance

Voltage.   The nominal voltage of the Li/FeS$_2$ system is given as 1.5 V and the open-circuit voltage of undischarged cells is approximately 1.8 V. The voltage on load drops within milliseconds, as shown in Fig. 14.79.

Discharge.   Li/FeS$_2$ batteries typically have a higher operating voltage and a flatter discharge profile than aqueous zinc/alkaline manganese dioxide 1.5-V batteries. This is illustrated in Fig. 14.80, which compares the performance of these two battery systems at relatively light and heavy constant-current discharge rates. These characteristics of the Li/FeS$_2$ battery result in higher energy and power output, especially on heavier drains were the operating voltage differences are greatest.

Typical discharge curves of the AA-size Li/FeS$_2$ battery under constant-resistance, constant-current, and constant-power discharge modes are shown in Figs. 14.81 to 14.83.

Operating Temperature.   Li/FeS$_2$ batteries are also suitable for use over a broad temperature range, generally -40 to 60°C. As can be seen in Fig. 14.84, service life is improved at elevated temperatures as it is at room temperature. In some applications there may be further limits on the maximum discharge temperature due to current limiting, which are part of the cell or battery device designs. Service life is reduced as the discharge temperature is lowered below room temperature, though the performance of the Li/FeS$_2$ battery is affected much less by low temperature than are aqueous systems.

Effects of Current-Limiting Devices.   Some current-limiting devices, such as fuses and PTCs, are designed to respond to high temperatures. Both the ambient temperature and internal cell heating can affect these devices, so any of the following factors may play a role:

Surrounding air temperature

Thermal insulating properties of battery container

Heat generated by equipment components during use

Cumulative heating effects of multicell batteries

Discharge rates and durations

Frequency and duration of rest periods

It may be necessary to consult the manufacturer or conduct testing to determine limitations in specific applications.

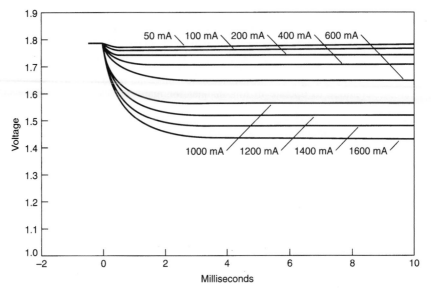

FIGURE 14.79 On-load voltage of Li/FeS$_2$ AA-size cells. (*Courtesy of Eveready Battery Co., Inc.*)

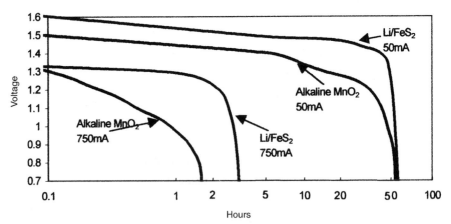

FIGURE 14.80 Performance of Li/FeS$_2$ and Zn/alkaline/MnO$_2$ "AA"-size batteries at 50 and 750 mA continuous discharge at 21°C. (*Courtesy of Eveready Battery Co, Inc.*)

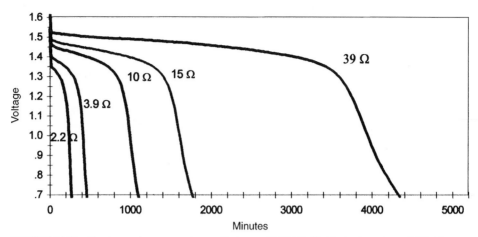

FIGURE 14.81 Constant resistance continuous discharge of Li/FeS$_2$ "AA"-size batteries at 21°C (*Courtesy of Eveready Battery Co, Inc.*)

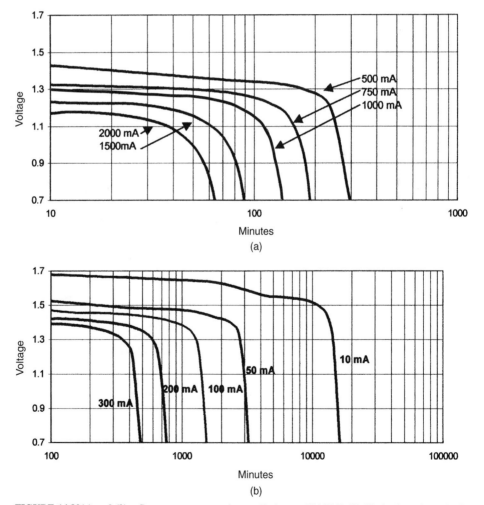

FIGURE 14.82(*a*) and (*b*) Constant current continuous discharge of Li/FeS$_2$ "AA"-size batteries at 21°C. (*Courtesy of Eveready Battery Co.*)

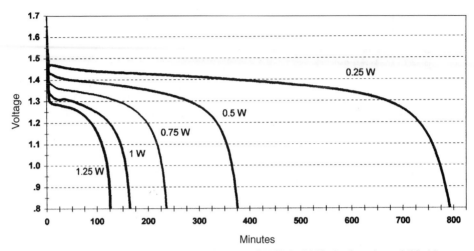

FIGURE 14.83   Constant power continuous discharge of Li/FeS$_2$ "AA"-size batteries at 21°C. (*Courtesy of Eveready Battery Co, Inc.*)

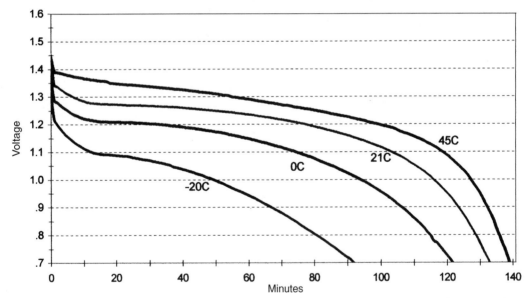

FIGURE 14.84   Performance of Li/FeS$_2$ "AA"-size batteries on a 1000 mA continuous discharge at various temperatures. (*Courtesy of Eveready Battery Co.*)

Impedance.   AC impedance is an electrical characteristic that is frequently used as an indicator of performance for aqueous batteries. The correlation is only poor at best with Li/FeS$_2$ batteries. There is a protective film that forms on the surface of the lithium anode. This film is an important factor in the excellent shelf life of the Li/FeS$_2$ cell. As the cell ages, this protective film increases with age. As the film increases, the impedance does as well. However, this film is easily disrupted when the battery is put on load, making impedance inappropriate as an indicator of expected Li/FeS$_2$ battery performance, especially after storage.

Storage Temperature.   Storage at high temperature will reduce the service life of Li/FeS$_2$ batteries, as it will with other systems. However, because of the very low levels off impurities in the materials used and the high degree of seal effectiveness required in lithium batteries, service maintenance of Li/FeS$_2$ batteries after high-temperature storage is better than expected with aqueous systems. The typical storage temperature range of Li/FeS$_2$ batteries is −40 to 60°C, though relatively short excursions above 60°C may be possible. In such cases the temperature at which the pressure relief vent will operate, and the effect of temperature on cell and battery cases and insulating materials should be understood.

14.10.4 Cell Types and Applications

Table 14.23 lists the characteristics of the cylindrical Li/FeS$_2$ AA-size battery that is currently available commercially. This battery has better high-drain and low-temperature performance than the conventional zinc cells and is intended to be used in applications that have a high current drain requirement, such as cameras, computers, and cellular phones. An example of the advantage of the lithium cell, compared to the Zn/alkaline/MnO$_2$ battery is shown in Fig. 14.85, which plots the number of flashes obtained with an autoflash SLR camera. The lithium battery outperforms the alkaline battery by delivering more flashes with a more rapid recycle time. See Fig. 6.12 for additional comparative data on the use of several battery types in photoflash use.

TABLE 14.23 Characteristics of Li/FeS$_2$ AA-Size Battery

Dimensions, mm	
Height	50
Diameter	14
Weight g	15
Voltage, V	
Nominal classification	1.5
Open-circuit	1.8
Capacity, at 20°C, Ah	
to 0.9 V	
1000 mA	2.8
300 mA	2.4
30 mA	2.6
0.3 mA	2.6
Energy density, Wh/L	
1000 mA	324
300 mA	436
30 mA	500
0.3 mA	540
1-kHz impedance, Ω	0.18

Source: Eveready Battery Co., Inc.

Button-type Li/FeS$_2$ batteries are no longer manufactured. Batteries using the Li/FeS system have not been manufactured for commercial use.

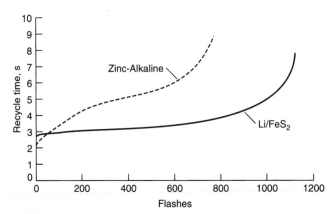

FIGURE 14.85   Comparison of number of flashes in automatic camera for zinc-alkaline and Li/FeS$_2$ AA-size batteries. One flash every 30 s with camera in manual mode. (*Courtesy of Eveready Battery Co., Inc.*)

14.11   LITHIUM/COPPER OXIDE (Li/CuO) AND LITHIUM/COPPER OXYPHOSPHATE [Li/Cu$_4$O(PO$_4$)$_2$] CELLS

The lithium/copper oxide (Li/CuO) system is characterized by a high specific energy and energy density (about 280 Wh/kg and 650 Wh/L) as copper oxide has one of the highest volumetric capacities of the practical cathode materials (4.16 Ah/cm^3). The battery has an open-circuit voltage of 2.25 V and an operating voltage of 1.2 to 1.5 V, which makes it interchangeable with some conventional batteries. The battery system also features long storability with a low self-discharge rate and operation over a wide temperature range.

Li/CuO batteries have been designed in button and cylindrical configurations up to about 3.5 Ah in size, mainly for use in low- and medium-drain, long-term applications for electronic devices and memory backup. Higher-rate designs as well as hermetically sealed batteries with glass-to-metal seals have also been manufactured.

Figure 14.86 compares the performance of a Li/CuO AA-size cylindrical battery with the zinc/alkaline/MnO$_2$ battery. The Li/CuO cell has a significant capacity advantage at low discharge rates, but loses this advantage at higher discharge rates.

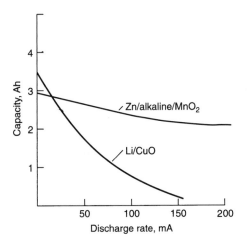

FIGURE 14.86 Comparison of Li/CuO and Zn/alkaline/MnO$_2$ AA-size batteries at 20°C.

14.11.1 Chemistry

The discharge reaction of the Li/CuO cell is

$$2Li + CuO \rightarrow Li_2O + Cu$$

The discharge proceeds stepwide, $CuO \rightarrow Cu_2O \rightarrow Cu$, but the detailed mechanism has not been clarified.[1,44] A double-plateau discharge has been observed at high-temperature (70°C) discharges at low rates, which blends into a single plateau under more normal discharge conditions.[45]

14.11.2 Construction

The construction of the Li/CuO button-type battery shown in Fig. 14.87*a* is similar to other conventional and lithium/solid-cathode cells. Copper oxide forms the positive electrode and lithium the negative. The electrolyte consists of lithium perchlorate in an organic solvent (dioxolane).

The cylindrical batteries (Fig. 14.87*b*) use an inside-out bobbin construction. A cylinder of pure porous nonwoven glass is used as the separator, nickel-plated steel for the case, and a polypropylene gasket for the cell seal. The can is connected to the cylindrical copper oxide cathode and the top to the lithium anode.

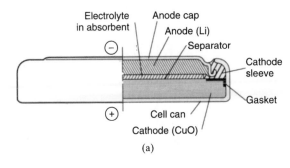

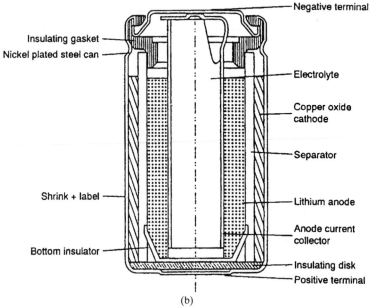

FIGURE 14.87 Lithium/copper oxide batteries (*a*) Button configuration. (*Courtesy of Panasonic, Division of Matsushita Electric Corp. of America.*) (*b*) Cylindrical battery, bobbin construction. (*Courtesy of SAFT America, Inc.*)

14.11.3 Performance

Button Battery. The performance of the 60-mAh Li/CuO button cell under various discharge conditions and temperatures is shown in Fig. 14.88.

Cylindrical Bobbin Li/CuO Battery. Typical discharge curves for this system are shown in Fig. 14.89. After a high initial load voltage, the discharge profile is flat at the relatively light loads. The bobbin construction does not lend itself to high-rate discharges and the battery capacity is significantly lowered with increasing discharge rates. The Li/CuO cylindrical battery operates over a wide temperature range, typically from −20 to 70°C, although the battery can operate outside these limits, but with changes in the discharge profile or load capability. Discharge curves at several different temperatures are shown in Fig. 14.90. The performance of the battery at temperatures from −40 to 70°C and at various loads is summarized in Fig. 14.91. The high capacity of the battery at the lighter loads falls off sharply with increasing load and decreasing temperatures.

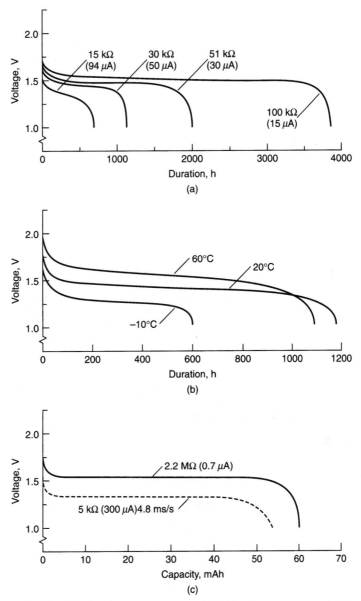

FIGURE 14.88 Discharge characteristics of Li/CuO button-type battery, 60-mAh size. (*a*) Load characteristics. (*b*) Temperature characteristics. (*c*) Pulse discharge characteristics. (*Courtesy of Panasonic, Division of Matsushita Electric Corp. of America.*)

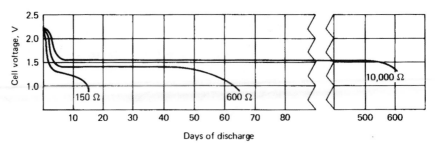

FIGURE 14.89 Typical discharge curves for Li/CuO AA-size battery at 20°C. (*Courtesy of SAFT America, Inc.*)

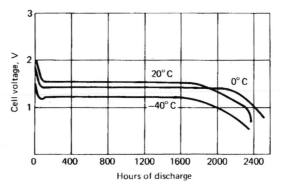

FIGURE 14.90 Effect of temperature on Li/CuO AA-size battery, 1-kΩ load. (*Courtesy of SAFT America, Inc.*)

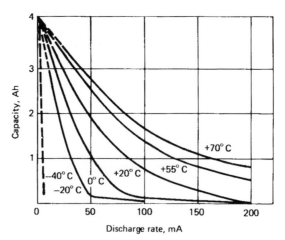

FIGURE 14.91 Capacity of Li/CuO AA-size battery as a function of discharge load and temperature. (*Courtesy of SAFT America, Inc.*)

The long-term storage capability of these Li/CuO cells is illustrated in Fig. 14.92. Figure 14.92*a* shows that there is only a minimum loss of capacity after 10 years of storage at room temperature, less than 0.5% per year. Performance after storage at high temperatures is plotted in Fig. 14.92*b*. The retention of residual capacity in partially discharged cells is said to be equivalent to that of fully charged cells.

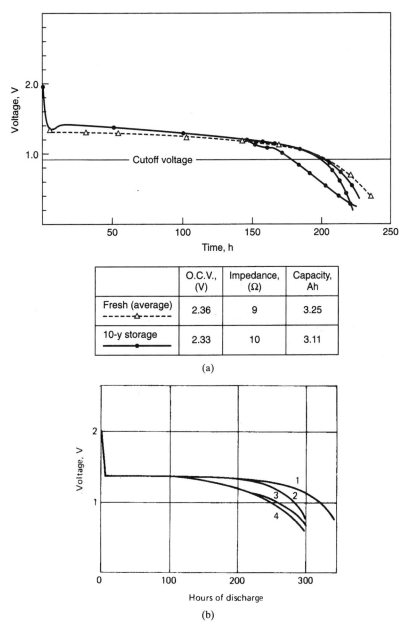

	O.C.V., (V)	Impedance, (Ω)	Capacity, Ah
Fresh (average) -----△-----	2.36	9	3.25
10-y storage ——•——	2.33	10	3.11

(a)

(b)

FIGURE 14.92 Effect of storage on performance on Li/CuO cylindrical batteries. (*a*) Discharges at 20°C before and after 10-year storage at 20°C. (*b*) Discharges at 20°C after 70°C storage. Curve 1—no storage; curve 2—6 months' storage; curve 3—12 months' storage; curve 4—18 months' storage. (*Courtesy of SAFT America, Inc.*)

High-Temperature Cells.   Specially designed hermetically sealed batteries have been developed for use at the high temperatures encountered, for example, by the oil-well logging industry, which uses down-hole tools operating to 150°C, the maximum temperature at which the Li/CuO cells can operate.

Spirally Wound Cells.   Cylindrical batteries in C and D sizes have been designed with spirally wound electrodes to meet higher drain requirements. Figure 14.93 shows the discharge performance of a Li/CuO D-size battery at various temperatures and at relatively low and high discharge rates.

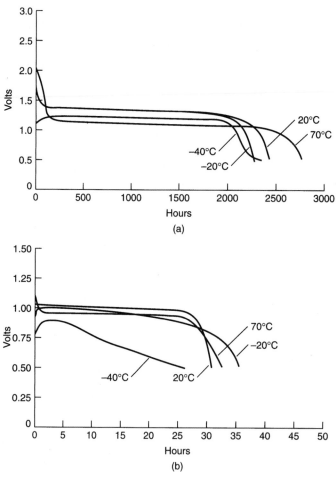

FIGURE 14.93   Performance of high- and low-rate Li/CuO D-size batteries. (*a*) Discharge at 147 Ω. (*b*) Discharge at 1.5 Ω.

14.11.4 Cell Types and Applications

The Li/CuO batteries that have been available in the button and small cylindrical (bobbin) configurations are listed in Table 14.24. Under the low-drain conditions these batteries have a significant capacity advantage over the conventional aqueous batteries. Combined with their excellent storability and operation over a wide temperature range, these batteries provide reliable power sources for applications such as memory backup, clocks, electric meters, and telemetry and, with high-temperature cells, in high-temperature environments. Specially designed units were also manufactured to meet higher drain applications. These batteries are no longer available commercially.

TABLE 14.24 Characteristics of Lithium/Copper Oxide Batteries

	Li/CuO		
	Button†	½AA	AA
Nominal voltage, V	1.5	1.5	1.5
Dimensions (max)			
Diameter, mm	9.5	14.5	14.5
Height, mm	2.7	26.0	50.5
Volume, cm^3	0.2	4.3	8.3
Weight, g	0.6	7.3	17.4
Rated capacity, Ah*	0.060	1.4	3.4
Specific energy/			
Energy density			
Wh/kg	150	285	290
Wh/L	450	485	610
Weight of lithium, g	—	0.4	0.9
Maximum current, mA	0.3	20	40

*At approximately $C/1000$ rate.
    Source: SAFT America, Inc. and Panasonic Div.,† Matsushiti Industrial Co.

14.12 *LITHIUM/SILVER VANADIUM OXIDE BATTERIES*

The lithium/silver vanadium oxide system has been developed for use in biomedical applications, such as cardiac defibrillators, neurostimulators and drug delivery devices. Electrochemical reduction of silver vanadium oxide (SVO) is a complex process and occurs in multiple steps from 3.2 to 2.0 V. This system is capable of high power, high energy density and high specific energy as is required for cardiac defibrillators, its principle application.

SVO with the stoichiometry of $AgV_2O_{5.5}$ is capable of providing 310 mAh/kg. Practical cells can achieve up to 780 Wh/L and 270 Wh/kg[46,47] and are capable of 1 to 2 Amp pulses as required by cardiac defibrillators.

14.12.1 Chemistry

SVO is produced by the thermal reaction of $AgNO_3$ and V_2O_5.[46] This material belongs to the class of vanadium bronzes and possesses semi-conducting properties. It is reduced chemically and electrochemically in a multi-step process in which V^{+5} is first reduced to V^{+4} in a two-step reaction, followed by reduction of Ag^{+1} to Ag^0 and then by partial reduction of V^{+4} to V^{+3}. These processes tend to overlap and plateaus are observed in the discharge curves at 3.2, 2.8, 2.2 and 1.8 V. When discharged to 1.5 V, 3.5 equivalents of lithium per mole of SVO are utilized based on the empirical formula $AgV_2O_{5.5}$. The following reactions have been proposed to account for this stoichiometry:

Anode: $3.5 \, Li \rightarrow 3.5 \, Li^+ + 3.5e$

Cathode: $\underline{3.5e + Ag(I)V_2(V)O_{5.5} \rightarrow Ag(0)V_2(IV,III)O_{5.5}}$

Overall: $3.5 \, Li + AgV_2O_{5.5} \rightarrow Li_{3.5}AgV_2O_{5.5}$

Electrolytes which have been employed with this system are 1 molar $LiBF_4$ in propylene carbonate (PC) and 1 Molar $LiAsF_6$ in 1:1 by volume PC-DME. The latter appears to be the electrolyte of choice at present. The addition of CO_2 or substances such as dibenzyl carbonate (DBC) or benzyl succinimidyl carbonate have been reported to reduce anode passivation which causes voltage delay and increased DC resistance between 40 and 70% depth of discharge.[48] These materials are believed to operate by lowering the impedance of the solid electrolyte interface (SEI) layer on the surface of the lithium anode.

14.12.2 Construction

Typical construction of a prismatic implantable Li/SVO cell designed for use in a cardiac defibrillator where high power capability is required is shown in Fig. 14.94.[49] The case and lid are made of 304L stainless steel and are the negative cell terminal. A glass-to-metal (GTM) seal employs TA-23 glass and a molybdenum pin. Although this GTM seal is intrinsically corrosion resistant in lithium primary cells, an elastomeric material and an insulating cap are added to the underside of the seal to provide an additional mechanical barrier and enhanced protection for the GTM. Two layers of insulating straps, one a fluropolymer and the other of mica are added to the underside of the lid to prevent contact of the cathode lead to the lid. The anode is constructed of two layers of lithium foil pressed on a nickel screen and heat sealed in a microporous polypropylene separator. The cathode consists of SVO with added carbon, graphite and PTFE binder. A typical composition is 94% SVO, 1% carbon, 2% graphite and 3% PTFE,[50] pressed on a finely meshed titanium screen under high pressure and then heat sealed in a microporous polypropylene separator. One anode assembly is wound around the individual cathode plates as shown in Fig. 14.94 and the anode tab welded to the case. Six to eight cathode plates are welded to a cathode lead bridge and an insulated tab connects the bridge to the Mo pin of the GTM seal. Following welding of the lid to the case, the cell is filled with electrolyte through the fill hole and a stainless steel ball inserted in the fill port, and an additional plug is then laser welded over the fill hole, providing an hermetic cell.

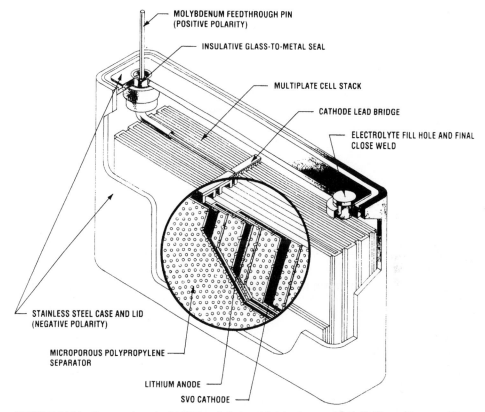

- MOLYBDENUM FEEDTHROUGH PIN (POSITIVE POLARITY)
- INSULATIVE GLASS-TO-METAL SEAL
- MULTIPLATE CELL STACK
- CATHODE LEAD BRIDGE
- ELECTROLYTE FILL HOLE AND FINAL CLOSE WELD
- STAINLESS STEEL CASE AND LID (NEGATIVE POLARITY)
- MICROPOROUS POLYPROPYLENE SEPARATOR
- LITHIUM ANODE
- SVO CATHODE

FIGURE 14.94 Construction of a Li/SVO cell designed for use in a cardiac defibrillator. (*Courtesy Wilson Greatbatch, Ltd.*)

14.12.3 Performance

A typical defibrillator operates by sensing fibrillation through two electrodes and delivering 25 to 40 Joule pulses to the heart, when necessary, to paralyze the heart which then recovers to a normal beating rhythm. The power source is required to deliver 10 to 30 microamps continuously for the sensing and/or pacing functions over many years. It must also be capable of providing the energy required to charge the capacitors in the defibrillator. This energy is typically delivered as 1 to 2 Amps for 10 seconds. A typical battery is therefore tested to provide four 2.0 Amp, 10 second pulses with 15 second intervals between pulses.

Figure 14.95 shows the continuous discharge curve for a resistive discharge at the two-year rate. Battery voltage is plotted against lithium equivalents per mole of SVO. Accelerated discharge testing consists of applying a series of four 2.0 Amp, 10 second pulses with 15 second intervals superimposed on a low-rate resistive discharge. Figure 14.96 shows a discharge curve for a 2.2 Ah Li/SVO battery subjected to a series of four pulses every 30 minutes. The voltage on the continuous background load is shown together with the minimum voltage on the first and fourth pulses to a 1.5 V cut-off. Additional testing is carried out with increased intervals between the pulse sequences. Figure 14.97 shows a discharge curve in which four 1.5 Amp, 10 second pulses superimposed on a 17.4 K ohm load are applied every two months to a 1.4 Ah battery. The continuous discharge voltage is plotted as is the fourth pulse minimum voltage which is above 2.0 volts after 45 months.[51]

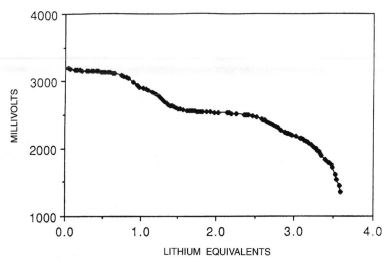

FIGURE 14.95   Discharge of lithium/silver vanadium oxide. Voltage is shown versus equivalents of lithium for a cathode stoichiometry of $AgV_2O_{5.5}$. (*Ref. 46.*)

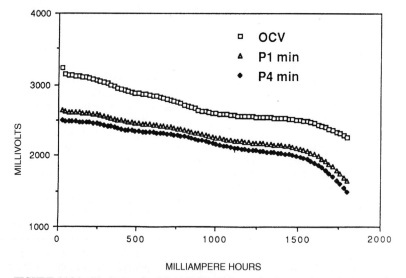

FIGURE 14.96   Discharge of a 2.2 Ah lithium/silver vanadium oxide defibrillator battery under a scheme of one pulse train of four pulses applied every 30 minutes. (*Ref. 46*).

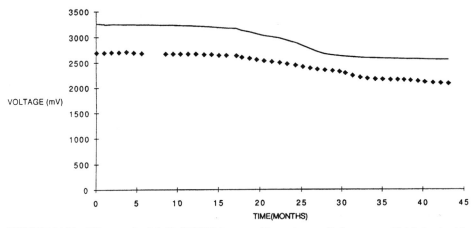

FIGURE 14.97 Life test of a 1.4 Ah Li/SVO battery with continuous discharge on a 17.4 kohm load is shown by the solid curve. The minimum voltage on the fourth 1.5 Amp pulse applied every two months is shown by the diamonds. (*Ref. 51*).

14.12.4 Cell and Battery Types

Single-cell batteries are produced in either prismatic or with a D-shaped cross-section similar to the design of lithium/iodine pacemaker batteries. High-rate models are produced in a prismatic design of 2.2 Ah capacity, weighing 27.9 g and occupying 10.3 cc. This unit provides 410 Wh/L and 150 Wh/kg. A similar unit in a D-shaped cross-section provides 1.18 Amp-hours capacity, a specific energy of 140 Wh/kg and an energy density of 400 Wh/L. These parameters were obtained under an accelerated pulse test regime to a 1.5 V cut-off.

Moderate rate units are also available. A rounded prismatic unit provides 2.5 Ah capacity, 270 Wh/kg and 780 Wh/L. A D-shaped moderate-rate battery product provides 1.6 Ah capacity with a specific energy of 250 Wh/kg and an energy density of 750 Wh/L. This data was obtained under a continuous 499 ohm discharge to a 2.0 V cut-off.

14.12.5 Applications

High-rate designs are employed in cardiac defibrillators, while moderate rate units find applications in implantable neurosimulators and drug infusion devices. Lithium/SVO batteries are employed for biomedical applications and as such must be produced under the Good Manufacturing Practices (GMP) for medical devices of the U.S. Food and Drug Administration.

Stability of the components and the electrochemical system provide self-discharge of 2%/year. Batteries are tested under both standard and abusive conditions. Standard tests are carried out on fresh, half-depleted and fully depleted units. Each battery is examined visually, dimensionally and radiographically before and after each test.[52] Qualification tests for implantable defibrillator batteries are: 1) high and low pressure exposure followed by mechanical shock and vibration and then followed by high temperature exposure; 2) thermal shock (70 to −40°C); 3) four-week temperature storage (55 and −40°C); 4) low-temperature storage; 5) forced overdischarge (single cell at C/10 rate and a two-cell battery); and 6) short circuit at 37°C. Cells bulge but do not vent on short circuit and pass other tests without failure. Abusive tests are conducted to determine the ability of the batteries to withstand severe conditions, but do not have survival requirements. These consist of crush, recharge, slow dent/puncture and high-rate forced discharge tests.

Traceability of materials and components to individual units is required and each is serialized. A comprehensive quality assurance program is also implemented. One to two percent of cells built are tested under conditions similar to those seen under actual use conditions as part of a quality assurance program.

REFERENCES

1. J. P. Gabano, *Lithium Batteries,* Academic, London, 1983.

2. D. MacArthur G. Blomgren, and R. Powers, "Lithium and Lithium Ion Batteries, 2000," Powers Associates, Westlake Ohio (2000).

3. Technical data, Foote Mineral Co., Exton, Pa; Lithium Corp. of America, Gastonia, N.C.

4. H. R. Grady, "Lithium Metal for the Battery Industry," *J. Power Sources* **5:**127 (1980), Elsevier Sequoia, Lausanne, Switzerland.

5. J. T. Nelson and C. F. Green, "Organic Electrolyte Battery Systems," U.S. Army Material Command Rep. HDL-TR-1588, Washington, D.C., Mar. 1972.

   J. O. Besenhard and G. Eichinger, "High Energy Density Lithium Cells, pt. I, Electrolytes and Anodes," *J. Electroanal. Chem.* **68:**1 (1976), Elsevier Sequoia, Lausanne, Switzerland.

   G. Eichinger and J. O. Besenhard, "High Energy Density Lithium Cells, pt. II, Cathodes and Complete Cells," *J. Electroanal. Chem.* **72:**1 (1980), Elsevier Sequoia, Lausanne, Switzerland.

6. F. Deligiannis, B. V. Ratnakumar, H. Frank, E. Davies, and S. Surampudi, *Proc. 38th Power Sources Conf.,* 373–377 (1996), Cherry Hill, N.J.

7. A. N. Dey, "Lithium Anode Film and Organic and Inorganic Electrolyte Batteries," in *Thin Solid Films,* vol. 43, Elsevier Sequoia S. A., Lausanne, Switzerland, 1977, p. 131.

8. S. Gilman and W. Wade, "The Reduction of Sulfuryl Chloride at Teflon-Bonded Carbon Cathodes," *J. Electrochem. Soc.* **127:**1427 (1980).

9. A. Meitav and E. Peled, "Calcium-Ca(AlCl$_4$)$_2$-Thionyl Chloride Cell: Performance and Safety," *J. Electrochem. Soc.* **129:**3 (1982).

10. R. L. Higgins and J. S. Cloyd, "Development of the Calcium-Thionyl Chloride Systems," *Proc. 29th Power Sources Conf.,* Electrochemical Society, Pennington, N.J., June 1980.

    M. Binder, S. Gilman, and W. Wade, "Calcium-Sulfuryl Chloride Primary Cell," *J. Electrochem. Soc.* **129:**4 (1982).

11. E. S. Takeuchi and W. C. Thiebolt, "The Reduction of Silver Vanadium Oxide in Lithium/Silver Vanadium Oxide Cells," *J. Electrochem. Soc.* **135,** No. 11, Nov. 1988.

    E. S. Takeuchi, "Lithium/Solid Cathode Cells for Medical Applications," *Proc. Int. Battery Seminar,* Boca Raton, Fla, 1993.

    A. Crespi, "The Characterization of Silver Vanadium Cathode Material by High-Resolution Electron Microscopy," *Proc. 7th Int. Meet. Lithium Batteries,* Boston, Mass., May 1994.

12. N. I. Sax, *Dangerous Properties of Industrial Materials,* Van Nostrand Reinhold, New York, N.Y.

13. *Transportation,* Code of Federal Regulations CFR 49, U.S. Government Printing Office, Washington, D.C.; Exemption DOT-E-7052, Department of Transportation, Washington, D.C.: "Technical Instructions for the Safe Transport of Dangerous Goods by Air," International Civil Aviation Organization, DOC 9284-AN/905, Montreal, Quebec, Canada.

14. E. H. Reiss, "Considerations in the Use and Handling of Lithium-Sulfur Dioxide Batteries," *Proc. 29th Power Sources Conf.,* Electrochemical Society, Pennington, N.J., June 1980.

15. Technical Standard Order: TSO-C142, Lithium Batteries, U.S. Dept. of Transportation, Federal Aviation Administration, Washington, D.C. (2000).

16. T. B. Reddy, *Modern Battery Technology,* Sec. 5.2, C. D. S. Tuck, ed., Ellis Horwood, N.Y. (1991).

17. M. Mathews, *Proc. 39th Power Sources Conf.,* pp. 77–80 (2000), Cherry Hill, N.J.

18. M. Sink, *Proc. 38th Power Sources Conf.,* pp. 187–190 (1998), Cherry Hill, N.J.

19. H. Taylor, "The Storability of Li/SO$_2$ Cells," *Proc. 12th Intersociety Energy Conversion Engineering Conf.,* American Nuclear Society, LaGrange Park, Ill., 1977.

20. D. Linden and B. McDonald, "The Lithium-Sulfur Dioxide Primary Battery—Its Characteristics, Performance and Applications," *J. Power Sources* **5:**35 (1980), Elsevier Sequoia, Lausanne, Switzerland.

21. R. C. McDonald et al., "Investigation of Lithium Thionyl Chloride Battery Safety Hazard," Tech. Rep. N60921-81-C0229, Naval Surface Weapons Center, Silver Spring, Md., Jan. 1983.

22. S. C. Levy and P. Bro, *Battery Hazards and Accident Prevention,* Sec. 10.3.2, Plenum Publishing Corp., New York, N.Y. (1994).

23. "Tadiran Lithium Inorganic Cells," Tadiran Electronics Industries, Inc. Port Washington, N.Y. 11050.

24. M. Babai and U. Zak, "Safety Aspects of Low-Rate Li/SOCl$_2$ Batteries," *Proc. 29th Power Sources Conf.,* Electrochemical Society, Pennington, N.J., June 1980.

25. K. M. Abraham and R. M. Mank, "Some Safety Related Chemistry of Li/SOCl$_2$ Cells," *Proc. 29th Power Sources Conf.,* Electrochemical Society, Pennington, N.J., June 1980.

26. R. L. Zupancic, "Performance and Safety Characteristics of Small Cylindrical Li/SOCl$_2$ Cells," *Proc. 29th Power Sources Conf.,* Electrochemical Society, Pennington, N.J., June 1980.

27. J. F. McCartney, A. H. Willis, and W. J. Sturgeon, "Development of a 200 kWh Li/SOCl$_2$ Battery for Undersea Applications," *Proc. 29th Power Sources Conf.,* Electrochemical Society, Pennington, N.J., June 1980.

28. A. Zolla, J. Westernberger, and D. Noll, *Proc. 39th Power Sources Conf.,* pp. 64–68 (2000), Cherry Hill, N.J.

29. C. Winchester, J. Banner, A. Zolla, J. Westenberger, D. Drozd and S. Drozd, *Proc. 39th Power Sources Conf.,* pp. 5–9 (2000), Cherry Hill, N.J.

30. K. F. Garoutte and D. L. Chua, "Safety Performance of Large Li/SOCl$_2$ Cells," *Proc. 29th Power Sources Conf.,* Electrochemical Society, Pennington, N.J., June 1980.

31. F. Goebel, R. C. McDonald, and N. Marincic, "Performance Characteristics of the Minuteman, Lithium Power Source," *Proc. 29th Power Sources Conf.,* Electrochemical Society, Pennington, N.J., June 1980.

32. D. V. Wiberg, "Non-Destructive Test Techniques for Large Scale Li/Thionyl Chloride Cells" *Proc. Int. Battery Seminar,* Boca Raton, Fla., 1993.

33. P. G. Russell, D. Carmen, C. Marsh, and T. B. Reddy, *Proc. 38th Power Sources Conf.,* pp. 207–210 (1998), Cherry Hill, N.J.

34. E. Elster, S. Luski, and H. Yamin, "Electrical Performance of Bobbin Type Li/SO$_2$Cl$_2$ Cells," *Proc. 11th Int. Seminar Batteries,* Boca Raton, Fla., March 1994.

35. S. McKay, M. Peabody, and J. Brazzell, *Proc. 39th Power Sources Conf.,* pp. 73–76 (2000), Cherry Hill, N.J.

36. C. C. Liang, P. W. Krehl, and D. A. Danner, "Bromine Chloride as a Cathode Component in Lithium Inorganic Cells," *J. Appl. Electrochem,* 1981.

37. D. M. Spillman and E. S. Takeuchi, *Proc. 38th Power Sources Conf.,* pp. 199–202 (1998), Cherry Hill, N.J.

38. H. Ikeda, S. Narukawa, and S. Nakaido, "Characteristics of Cylindrical and Rectangular Li/MnO$_2$ Batteries," *Proc. 29th Power Sources Conf.,* Electrochemical Society, Pennington, N.J., 1980.

39. T. B. Reddy and P. Rodriguez, "Lithium/Manganese Dioxide Foil-Cell Battery Development," *Proc. 36th Power Sources Conf.,* Cherry Hill, N.J., 1994.

40. A. Morita, T. Iijima, T., Fujii, and H. Ogawa, "Evaluation of Cathode Materials for the Lithium/Carbon Monofluoride Battery," *J. Power Sources* **5:**111 (1980), Elsevier Sequoia, Lausanne, Switzerland, 1980.

41. D. Eyre and C. D. S. Tuck, *Modern Battery Technology,* Sec. 5.3, C. D. S. Tuck, ed., Ellis Horwood, N.Y. (1991).

42. R. L. Higgins and L. R. Erisman, "Applications of the Lithium/Carbon Monofluoride Battery," *Proc. 28th Power Sources Symp.,* Electrochemical Society, Pennington, N.J., June 1978.

43. T. R. Counts, *Proc. 38th Power Sources Conf.,* 143–146 (1998), Cherry Hill, N.J.

44. T. Iijima, Y. Toyoguchi, J. Nishimura, and H. Ogawa, "Button-Type Lithium Battery Using Copper Oxide as a Cathode," *J. Power Sources* **5:**1 (1980), Elsevier Sequoia, Lausanne, Switzerland.

45. J. Tuner et al., "Further Studies on the High Energy Density Li/CuO Organic Electrolyte System," *Proc. 29th Power Sources Conf.,* Electrochemical Society, Pennington, N.J., June 1980.

46. E. S. Takeuchi, *Proc. 10th Int. Seminar on Primary and Secondary Battery Technology and Applications,* Deerfield Beach, Fla., March 1993.

47. E. S. Takeuchi and W. C. Thiebolt III, *J. Electrochem. Soc.* **135:**269 (1988).

48. H. Gan and E. S. Takeuchi, *Abstract No. 332,* Electrochemical Society Metting, Honolulu, HI, Oct. 17–22 (1999).

49. *A Guide to Implantable Medical Batteries,* Wilson Greatbatch Ltd., Clarence, N.Y.

50. R. A. Leising and E. S. Takeuchi, *Chem. of Materials,* **5:**738 (1993).

51. C. F. Holmes and M. Visby, *PACE* **14:**341 (1991).

52. E. S. Takeuchi, *J. Power Sources* **54:**115 (1995).

CHAPTER 15
SOLID-ELECTROLYTE BATTERIES

Boone B. Owens, Paul M. Skarstad, Darrel F. Untereker, and Stefano Passerini

15.1 GENERAL CHARACTERISTICS

Any battery consists of an electrochemically reactive couple separated by an ion-transport medium, or electrolyte. In most familiar batteries the electrolyte is a liquid. However, the availability of solids capable of being fabricated into electronically insulating elements with fairly low overall ionic resistance has stimulated the development of solid-electrolyte batteries. A few of these types of batteries have become available commercially and are important power sources at normal ambient temperatures ($\sim25°C$) for heart pacemakers, for preserving volatile computer memory, and for other low-power applications requiring long shelf and service lives. A chronology of various types of solid-electrolyte cells developed since 1950 is shown in Table 15.1. Two trends are indicated in this table. First, the low-energy silver-based systems have been replaced largely by developments of high-energy lithium anode-based batteries. Second, the batteries have gone from being pellet-based devices with relatively thick electrolyte layers (on the order of a millimeter) to thin film devices with electrolyte thicknesses on the order of micrometers. This reduction in the interelectrode distance has been necessary in order to overcome the ohmic losses due to the more resistive lithium-ion conducting electrolytes. Review articles which cover solid-electrolyte battery systems, including those not commercially available, are available in the literature.[1–10]

Commercially available solid-electrolyte batteries use lithium anodes. Lithium is attractive as an anode material for several reasons. First, it has a high specific capacity on both a weight and a volume basis. Second, it is strongly electropositive; this leads to high voltages when coupled with typical cathode materials. Finally, suitable lithium ion conductors are available for use as solid electrolytes. Table 15.2 compares theoretical values of the equivalent weight, equivalent volume, voltage, capacity density, and energy density for battery systems forming metal iodide as the discharge product. Among the alkali metals, the small equivalent volume of lithium more than compensates for the slightly higher voltages obtained with the heavier members of the group in determining the theoretical energy density. On the other hand, the high voltage with lithium compensates for the capacity advantage of the polyvalent metals. Only metal I_2 batteries of Ca, Sr, and Ba have theoretical specific energies approaching that of lithium. Moreover, no suitable conductive solid electrolytes are known for the polyvalent metals in Table 15.2. Therefore lithium appears to be the anode material of choice at the present time.

TABLE 15.1 Chronology of Solid-Electrolyte Batteries*

Date	Electrolyte	Log conductivity, $\Omega^{-1} \cdot cm^{-1}$	Typical cell system
1950–1960	AgI	-5	Ag/V_2O_5
1960–1965	Ag_3SI	-2	Ag/I_2
1965–1972	$RbAg_4I_5$	-0.5	Ag/Me_4NI_5
1965–1975	Beta-alumina	-1.5	$Na-Hg/I_2,PC$
1970–1975	$LiI(Al_2O_3)$	-5	Li/PbI_2
1970–1980	LiI	-7	$Li/I_2(P2VP)$
1978–1985	LiX-PEO	-7	Li/V_2O_5
1980–1986	$Li_{0.36}I_{0.14}O_{0.007}P_{0.11}S_{0.36}$	-3.3	Li/TiS_2
1983–1987	MEEP	-4	Li/TiS_2
1985–1992	Plasticized SPE	-3	Li/V_6O_{13}
1985–1992	$Li_{0.35}I_{0.12}O_{0.31}P_{0.12}S_{0.098}$	-4.7	Li/TiS_2
1990–2000	LiPON	-5.6	$Li/a-V_2O_5$
1992–2000	LiPON	-5.6	LiC_6/Li_xCoO_2 and LiC_6/Li_xNiO_2

PEO—polyethylene oxide: MEEP—poly(bis-(methoxy ethoxy ethoxide)): SPE—solid polymer electrolyte.[1]
LIPON = $Li_{0.39}N_{0.02}O_{0.47}P_{0.12}$.

Recent data on rechargeable systems with solid electrolytes will be found in chapters 34 and 35.

TABLE 15.2 Theoretical Values for Capacity Densities and Energy Densities of Balanced Metal/I_2-Batteries*

Anode metal	Anode equivalent weight, g/eq	Anode equivalent volume, cm^3/eq	E^0, V	Cell capacity density, Ah/cm^3	Cell energy density, Wh/cm^3
		Monovalent metals			
Li	6.9	13.0	2.8	0.69	1.9
Na	23.0	23.7	3.0	0.54	1.6
K	39.1	44.9	3.4	0.38	1.3
Rb	85.5	55.9	3.4	0.33	1.1
Cs	132.9	71.1	3.5	0.28	1.0
Cu	63.5	7.1	0.7	0.82	0.59
Ag	107.9	10.3	0.7	0.75	0.51
Tl	204.4	17.2	1.3	0.63	0.81
		Polyvalent metals			
Be	4.5	2.4	1.1	0.96	1.1
Mg	12.2	7.0	1.9	0.82	1.5
Ca	20.0	13.0	2.8	0.69	1.9
Sr	43.8	16.9	2.9	0.63	1.9
Ba	68.7	19.6	3.1	0.59	1.9
Zn	32.7	4.6	1.1	0.93	1.0
Cd	56.2	6.5	1.1	0.89	1.0
Al	9.0	3.3	1.0	0.83	0.9

*Equivalent weights, densities, and cell voltages were obtained from data in Weast.[11]

Complete solid-state batteries offer several advantages over those with fluid components. They generally exhibit high thermal stability, low rates of self-discharge (shelf life of 5 to 10 years or better), the ability to operate over a wide range of environmental conditions (temperature, pressure, and acceleration), and high energy densities (300 to 700 Wh/L). On the other hand, limitations associated with a complete solid-state battery include relatively low power capability (microWatt range) due to the high impedance of most solid electrolytes at normal ambient temperature, possible mechanical stresses due to volume changes associated with electrode discharge reactions, and reduced electrode efficiencies at high discharge rates (see Table 15.3). However, lithium polymer electrolyte batteries appear to have solved some of these deficiencies.

The only current commercial inorganic solid-electrolyte battery system is $Li/LiI/I_2(P2VP)$. The solid electrolyte LiI is formed in situ as the discharge product of the cell reaction. The cathode is a mixture of solid iodine and a saturated viscous liquid solution containing poly-2-vinylpyridine (P2VP) and iodine. The $Li/LiI/I_2 \cdot (P2VP)$ battery can be regarded as a quasi-solid-state system because of the high viscosity of the polymer-containing liquid phase and the preponderance of solid iodine in the material. However, the viscous liquid phase does impart a plasticity to the $I_2(P2VP)$ cathode, which makes these solid-state cells better able to adapt to volumetric changes during cell discharge.

TABLE 15.3 Major Advantages and Disadvantages of Lithium Solid-Electrolyte Batteries

Advantages	Disadvantages
Excellent storage stability—shelf life of 10 years or better	Low current drains (microamperes)
High energy densities	Power output reduced at low temperatures
Hermetically sealed—no gassing or leakage	Care must be exercised to prevent short-circuiting or
Wide operating temperature range, up to 200°C	shunting of cell (which could be a relatively high drain
Shock and vibration-resistant	on cell)

15.2 Li/LiI(Al₂O₃)/METAL SALT BATTERIES

Lithium batteries with the dispersed phase electrolyte of LiI and Al_2O_3 are no longer produced commercially, but the following description of these batteries, which were developed at Duracell in the 1970s, is included for its technological interest. The battery

$$Li/LiI(Al_2O_3)/metal\ salts$$

is a true solid-state battery. All components are solids, with no significant amount of liquid or vapor component coexisting. It was commercially available in the 1980s, but has since been withdrawn from the market.

15.2.1 General Characteristics

Solid-state batteries based on this system have the following properties: (1) long shelf life, (2) low power capability, (3) hermetically sealed, no gassing, (4) broad operating temperature range: −40 to 170°C for pure lithium anodes, up to and beyond 300°C with compound anodes, (5) high volumetric energy density, up to 0.6 Wh/cm³ in the system that were produced[12,13] and up to 1.0 Wh/cm³ in systems that were under development.[14,15]

These batteries were recommended for low-rate applications. They were particularly suited for applications requiring long life at low-drain or open-circuit conditions. Possible ambient-temperature applications included watches, pacemakers, and monitoring devices. The power characteristics and long shelf lives were well-suited to providing backup power for preserving volatile computer memory.

15.2.2 Cell Chemistry

Two different cathodes were used in these commercial solid-state cells. The original battery system used a mixture of PbI_2 and Pb for the cathode.[12] This was replaced by a mixture of PbI_2, PbS, and Pb.[13] A newer system, with increased energy density, used a mixture of TiS_2 and S as the cathode.[14] Properties of these battery systems are summarized in Table 15.4. Other cathode materials have also been investigated in cells of this type. They include As_2S_3 and various other metal sulfides.[15]

TABLE 15.4 Practical Solid-State Lithium Battery Systems

Cell	Practical energy density, Wh/cm^3	Cell reactions	E^0, V
$Li/LiI(Al_2O_3)/PbI_2$, Pb	0.1–0.2	$2Li + PbI_2 \rightarrow 2LiI + Pb$	1.9
$Li/LiI(Al_2O_3)/PbI_2$, PbS, Pb	0.3–0.6	$2Li + PbI_2 \rightarrow 2LiI + Pb$	1.9
		$2Li + PbS \rightarrow Li_2S + Pb$	1.8
$Li/LiI(Al_2O_2)/TiS_2$, S	0.9–1.0	$Li + TiS_2 \rightarrow Li_x TiS_2$	2.5–1.9
		$2Li + S \rightarrow Li_2S$	2.3

Source: Duracell, Inc.

The solid electrolyte used in these solid-state cells is a dispersion of lithium iodide and lithium hydroxide with alumina. Pure lithium iodide has a conductivity of about $10^{-7}\ \Omega^{-1} \cdot$ cm^{-1} at room temperature.[16–18] The conductivity of lithium iodide can be enhanced by several orders of magnitude by dispersal of substantial amounts (~50 mole percent) of high-surface-area alumina in the lithium iodide. Lithium ion conductivities as high as $10^{-4}\ \Omega^{-1} \cdot$ cm^{-1} have been reported in such dispersions at 25°C.[15,19]

The dispersed-phase lithium iodide with enhanced lithium ion conductivity remains a good electronic insulator with the partial electronic (hole) conductivity approximately $10^{-40}\ \Omega^{-1} \cdot$ cm^{-1}.[15] Thus this material is well-suited for application as a solid electrolyte. The mechanism by which the high-surface-area alumina, itself a poor conductor, enhances the conductivity of lithium iodide is not well understood. Enhanced interfacial conduction has been suggested.[20] The Arrhenius plot of conductivity for LiI (Al_2O_3) (Fig. 15.1) in the range −50 to 300°C is characterized by a single activation energy (0.4 eV).[15] At 300°C the conductivity reaches 0.1 $\Omega^{-1} \cdot$ cm^{-1}, allowing the construction of batteries with substantial rate capability. High-temperature solid-state secondary batteries using this electrolyte have been investigated for load-leveling and vehicle-propulsion applications.[15,21]

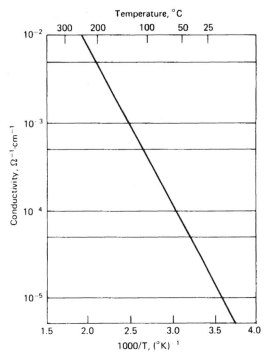

FIGURE 15.1 Ionic conductivity of $LiI(Al_2O_3)$ electrolyte. (*Courtesy of Duracell, Inc.*)

15.2.3 Cell Construction

The dispersed-phase ionic conductor $LiI(Al_2O_3)$ can be fabricated as a dense, conductive solid electrolyte element by uniaxial compression at room temperature.[13] No high-temperature sintering is required. The minimum thickness of an electrolyte element in a practical cell construction is about 0.2 mm. The cell is designed in a button or circular disk configuration, as shown in Fig. 15.2. The cell is fabricated by sequentially pressing the cell components (cathode, electrolyte, and anode) at appropriate pressures into the cell in the form of a thin circular disk. This cell can be incorporated in series-parallel combinations to give batteries with the desired voltage and current characteristics. The battery is fabricated by placing the cells in a stainless-steel container, holding the cells under spring pressure. There is no need to encapsulate or isolate each cell as required in conventional batteries. The energy density achieved in the cell and battery described in Table 15.5 is 0.4 Wh/cm³. Energy densities up to 0.6 Wh/cm³ have been achieved in practical cells with this chemical system.

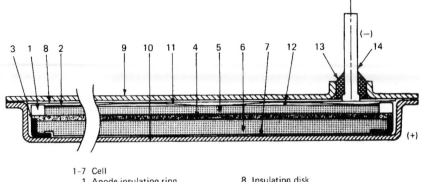

1-7 Cell
   1 Anode insulating ring
   2 Anode collector
   3 Cathode ring
   4 Electrolyte
   5 Anode
   6 Cathode
   7 Cathode collector

   8 Insulating disk
   9 Top-feed through assembly
  10 Can
  11 Wave spring
  12 Tab
  13 Loctite
  14 Glass-to-metal seal

FIGURE 15.2   Cross section of solid-state battery. (*Courtesy of Duracell, Inc.*)

TABLE 15.5   Properties of Commercial Li/LiI/PbI$_2$, PbS Batteries

E'', V	Capacity, mAh	Diameter, cm	Height, cm	Volume, cm^3	Weight, g	Energy density Wh/cm^3	Specific energy Wh/g
1.9	350	2.90	0.25	1.5	7.25	0.445	0.92
3.8	350	2.97	0.58	4.0	15.86	0.332	0.94

15.2.4 Performance Characteristics

The discharge properties of these solid-state batteries are characterized by high practical energy densities, low-rate capability, and low rates of self-discharge. From the linear polarization curve (Fig. 15.3) it is apparent that the resistance is essentially ohmic. Bridge measurements of cell resistance at 1 kHz are in agreement with values determined from polarization curves.

Figure 15.4 shows the discharge curves for the 350-mAh battery at various discharge rates at 20°C. At discharge rates below 6 μA, the discharge efficiency is close to 100% to a 1-V cutoff. The discharge curves for the same battery at various temperatures are shown in Fig. 15.5 for discharges at 3 μA (0.55 μA/cm^2 of electrolyte area) and 12 μA. The improved performance attained at the higher discharge temperatures is evident. These batteries can be discharged at higher temperatures than shown; however, for operation above 125°C, a modified electrode structure is required. At lower temperatures the current capability of the solid-state battery is reduced. In typical CMOS memory applications, however, the load current requirement usually is similarly reduced and the cell output tracks the CMOS power requirements. The service life delivered by the 350-mAh battery at various discharge temperatures and loads is given in Fig. 15.6. These data can be used to calculate the performance of the battery under various conditions of use.[22]

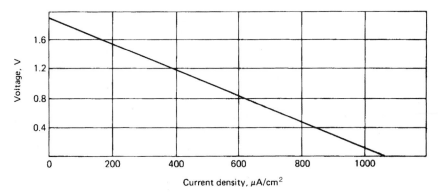

FIGURE 15.3   Typical polarization curve (IX battery. 25°C, cell surface area 1cm²). (*Courtesy of Duracell, Inc.*)

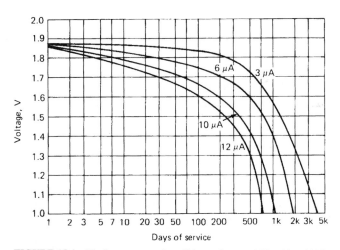

FIGURE 15.4   Discharge curves for solid-state battery. 350 mAh at 20°C. (*Courtesy of Duracell, Inc.*)

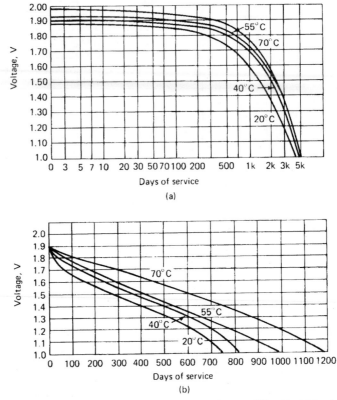

FIGURE 15.5 Discharge curves for solid-state battery. 350 mAh. (*a*) 3-μA discharge. (*b*) 12-μA discharge. (*Courtesy of Duracell, Inc.*)

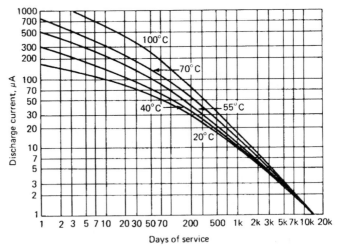

FIGURE 15.6 Service life of solid-state batteries. 350 mAh. to 1-V cutoff. (*Courtesy of Duracell, Inc.*)

15.2.5 Storage

Long-term storage tests show that there is no loss of capacity after storage periods up to 1 year at 20, 45, and 60°C (Fig. 15.7). The battery can withstand storage to 200°C, but for best performance, such high temperatures should be avoided. Temperatures above 200°C may cause bulging and failure of the seal. The long shelf life results from the chemical compatibility of the cell components, the absence of chemical reaction between the electrodes and the electrolyte, and the low electronic conductivity of the solid electrolyte which minimizes self-discharge. Longer-term tests show that there is no measurable loss of capacity after storage periods of 4 year at 20°C and up to 1 year at temperatures as high as 100°C. On the basis of these tests it is projected that the shelf life of these batteries is at least 15–20 years under normal storage conditions.[12]

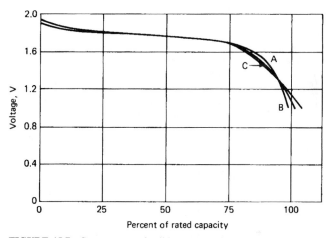

FIGURE 15.7   Storage tests of solid-state battery. Discharge after (*a*) no storage: (*b*) 1 year at 45°C: (*c*) 1 year at 60°. (*Courtesy of Duracell, Inc.*)

15.2.6 Handling

Solid-state batteries are designed primarily for low power and long service life. These batteries can withstand short circuit and voltage reversal, although these conditions should be avoided. No explosion due to pressure buildup or chemical reaction is known to occur under recommended operating temperatures. Prolonged short-circuiting will result in a separation between the electrode and the electrolyte, making the cell inoperative.

15.3 THE LITHIUM/IODINE BATTERY

15.3.1 General Characteristics

Lithium/iodine batteries are based on the reaction

$$Li + \tfrac{1}{2}I_2 \rightarrow LiI$$

The specific reactions for the cell using poly-2-vinylpyridine (P2VP) in the cathode are

Anode	$2Li \rightarrow 2Li^+ + 2e$
Cathode	$2Li^+ + 2e + P2VP \cdot nI_2 \rightarrow P2VP \cdot (n-1)I_2 + 2LiI$
Overall	$2Li + P2VP \cdot nI_2 \rightarrow P2VP \cdot (n-1)I_2 + 2LiI$

Lithium and iodine are consumed, and their reaction product, LiI, precipitates in the region between the two reactants. The LiI not only is the discharge product, but also serves as the cell separator and electrolyte. The theoretical energy density is 1.9 Wh/cm^3 (see Table 15.2). Practical values approaching 1 Wh/cm^3 can be obtained at discharge rates of 1 to 2 μA/cm^2. Commercially available lithium/iodine batteries have a solid anode of lithium and a polyphase cathode which is largely iodine. The iodine is made conductive by the addition of an organic material. Pyridine-containing polymers are most often used for this purpose, the additive in all present commercial batteries being P2VP. At ambient temperatures the iodine/P2VP mixtures are two-phase in undischarged batteries, liquid plus excess solid iodine.[23] The iodine content of the cathode decreases during discharge of the battery, and the remaining cathode material becomes hard as the battery nears depletion. As discharge proceeds, the layer of lithium iodide becomes thicker. The resistance of the battery also increases because of the growing amount of discharge product.

The volume change accompanying the cell discharge is negative. The theoretical value for this volume change is -15% for complete discharge of a balanced mixture of pure iodine and lithium.[23] It is somewhat less when the chemical cathode is not pure iodine. For example, a volume change of -12% is expected if the cathode is 91% iodine by weight.[24] The volume change may be accommodated by the formation of a porous discharge product or of macroscopic voids in the cell.

Cells are formed by contacting the iodine-containing cathode directly with the lithium anode. The chemical reaction between these two materials immediately forms a thin layer of lithium iodide between anode and cathode. This layer serves to separate the two electroactive materials electronically and prevents failure due to internal short-circuiting of the anode and cathode. This makes them especially suitable for applications requiring very high reliability.

Features of this system include low self-discharge, high reliability, and no gassing during discharge. Shelf life is 10 years or longer, and the cells can take a considerable amount of abuse without any catastrophic effects. Batteries of this type have found commercial applications powering various low-power devices such as cardiac pacemakers, solid-state memories, and digital watches. Power sources for portable monitoring and recording instruments and the like are also possible applications.

All the currently available Li/I$_2$ batteries have a nominal capacity of 15 Ah or less, and most have deliverable capacities under 5 Ah. All the Li/I$_2$ batteries intended for medical applications are designed to be cathode-limited.

15.3.2 Cell Construction

Several generic types of Li/I$_2$ cells have been produced, three of which were designed for medical applications such as cardiac pacemakers. Figure 15.8 shows the first type, which was phased out in the early 1980s. This unit had a case-neutral design and consisted of a stainless-steel housing with a plastic insulator that lined the inside of the case. A lithium envelope (the anode) fitted inside the plastic and contained the I$_2$(P2VP) depolarizer. The cathode current collector was located in the center of the cell. Current collector leads from both the anode and the cathode went through hermetic feed-throughs in the case. This cell was formed by pouring molten iodine depolarizer into the lithium envelope. After the cathode material solidified, the top of the lithium envelope was closed, the plastic cup added, and the final assembly completed. The construction used in this cell eliminates any contact between the case and the iodine depolarizer.

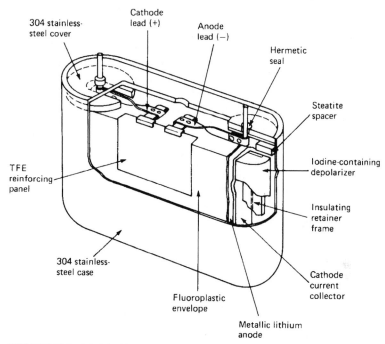

FIGURE 15.8 Model 802/35 Li/I$_2$ cell. (*Courtesy of Catalyst Research Corp.*)

A second construction uses a case-positive design of similar size. This cell is the type used today for most medical applications. A cutaway view is shown in Fig. 15.9. This cell is manufactured in a slightly different manner from that in Fig. 15.8. It contains a central lithium anode and uses the stainless-steel case of the cell as the positive-current collector. Most models are completely assembled with their header welded to the can before the cathode is added to the cell. Hot depolarizer is poured into the cell can through a small fill port, which is later welded shut. The anode current collector is brought out via a glass-to-metal feed-through.

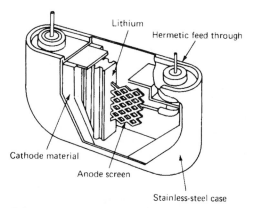

FIGURE 15.9 Cutaway view of typical can-positive Li/I$_2$ cell. (*Courtesy of Wilson Greatbatch, Ltd.*)

Manufacturers of these case-positive designs also precoat their anode assembly with a layer of pure P2VP prior to assembly. This coating is designed to protect the anode from the environment before assembly, but it also alters the electrical discharge behavior (see Sec. 15.3.4).

Another case-positive cell type has been used for medical applications. This unit is very similar to the other case-positive designs, but the cathode is not poured into the battery can. The iodine and P2VP are pelletized and then pressed onto the central anode assembly. After the pressing operation, the entire unit is slipped into a nickel can. An exploded view of this cell is shown in Fig. 15.10.

Case-neutral designs were developed to prevent corrosion of the exterior case and to minimize leakage to the feed-through by the iodine depolarizer. However, 5 years of real-time data have shown that no significant corrosion of stainless steel in contact with the cathode depolarizer or its vapor takes place in sealed can-positive cells. Tests show that corrosion occurs during the first few months after assembly and is limited to a 50-μm layer. Even at 60°C, corrosion of stainless steel by the iodine depolarizer has not proved to be a problem in the dry environment of the cell.

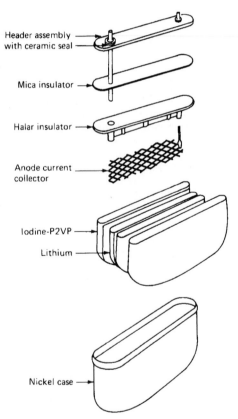

Header assembly with ceramic seal

Mica insulator

Halar insulator

Anode current collector

Iodine-P2VP

Lithium

Nickel case

FIGURE 15.10 Exploded view of case-positive Li/I$_2$ cell. (*Courtesy of Catalyst Research Corp.*)

Li/I_2 medical batteries are produced in a variety of sizes and shapes to meet specific applications. Their profiles range from rectangular to semicircular, or a combination. All of them are made quite thin and have flat sides because their primary application is in cardiac pacemakers. Cell thickness is typically 5 to 10 mm. The area of the lithium anodes ranges between 10 and 20 cm^2 in current batteries. Some cells use a ribbed anode (see Fig. 15.9) to increase the amount of active anode surface area in the battery.

The nonmedical batteries are made in more conventional button and cylindrical configurations. The hermetically sealed button-cell batteries are intended for powering digital watches and serving as backup power for computer memories. These batteries are made by pressing iodine cathode and lithium anode layers into a stainless-steel cup. The cup is the positive-current collector. A glass-to-metal feed-through brings the negative terminal to the exterior. Figure 15.11 shows a view of this cell type. The cylindrical (D-cell diameter) Li/I_2 battery is welded hermetically, like the other batteries described. It is case-positive; the negative connection is a button on the end of the cell. It is designed to withstand substantial shock, vibration, and abuse without venting, swelling, leaking, or exploding. These button-type batteries are no longer commercially available.

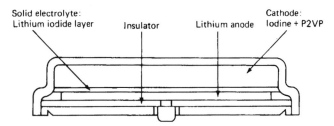

FIGURE 15.11   Button-type Li/I_2 battery. (*Courtesy of Catalyst Research Corp.*)

Manufacturing of all Li/I_2 batteries is done in a dry environment (typically less than 1% RH). In addition, all these batteries are sealed hermetically to prevent exchange of any material with the environment. Good sealing is required in order to maintain the desired electrical characteristics.

Connection to the medical-grade batteries is made by soldering or spot welding. The case-positive varieties usually have a pin or wire welded to the case to facilitate making the positive connection (see Figs. 15.9 and 15.10).

Manufacturers keep detailed records of the construction and manufacture of each battery intended for medical applications and each unit is individually serialized. This procedure allows the systematic tracing of the history and behavior of every battery, should the need arise.

15.3.3 Commercially Available Batteries

Table 15.6 summarizes the manufacturer's specifications for some typical pacemaker batteries. Batteries for medical use are relatively expensive because of the low manufacturing volume and demands for high reliability.

TABLE 15.6 WGL Specifications for Lithium/Iodine Pacemaker Batteries

Type	Manufacturer's rated capacity, Ah	Wt., g	Vol., cm³	Size, mm (length × width × height)	Energy density Wh/cm³	Specific energy Wh/g
9107	0.43	5.3	1.40	30 × 4.2 × 13	0.81	0.22
9114	0.56	7.0	1.82	28 × 6 × 14	0.82	0.21
9331	0.82	9.1	2.34	32 × 5 × 19	0.93	0.24
8711	0.86	9.5	2.52	27 × 5 × 22	0.90	0.24
8831	0.89	10.7	2.75	43 × 5 × 16	0.86	0.22
9412	0.92	10.2	2.20	32 × 5 × 21	1.11	0.24
9085	0.98	10.9	2.93	27 × 5 × 25	0.89	0.24
9438	0.98	10.7	2.93	27 × 5 × 25	0.89	0.24
9105	0.99	10.6	2.95	27 × 6 × 21	0.89	0.25
9236	1.00	11.0	2.85	24 × 6 × 25	0.93	0.24
8426	1.07	11.6	3.02	33 × 5 × 23	0.94	0.24
9074	1.15	12.9	3.28	43 × 5 × 18	0.93	0.24
8708	1.21	13.4	3.52	47 × 5 × 22	0.91	0.24
8431	1.28	13.8	3.67	27 × 5 × 31	0.92	0.25
8402	1.29	14.4	3.72	45 × 5 × 22	0.92	0.24
8843	1.36	13.6	3.73	32 × 6 × 23	0.97	0.27
8207	1.55	17.2	4.70	27 × 7.8 × 25	0.87	0.24
8950	1.58	15.9	4.45	27 × 6 × 31	0.94	0.26
8077	1.78	18.9	5.11	45 × 7 × 22	0.92	0.25
8206	1.98	20.4	5.54	27 × 7.8 × 31	0.95	0.26
8041	2.10	22.0	5.81	33 × 8.6 × 26	0.96	0.25
9111	2.36	22.8	6.24	33 × 7 × 31	1.00	0.27

Courtesy Wilson Greatbatch Ltd.

15.3.4 Discharge Performance Characteristics

The open-circuit voltage of a Li/I_2 battery is very near 2.80 V. This value is maintained through the useful life of the battery. The resistance of the lithium iodide discharge product controls the load voltage for most of the discharge. The morphology of the lithium iodide discharge product, in turn, depends on the way the battery is constructed. Only at the end of the discharge cycle is the resistance of the iodine depolarizer significant in the total internal resistance of the battery. Because the battery resistance is quite high and increases throughout life, the discharge curves are not flat, even at moderate current drains. Two characteristic types of discharge curves are observed.

Batteries made by putting the depolarizer against bare lithium anode metal discharge by growing a nearly planar layer of lithium iodide. This layer becomes thicker in proportion to the amount of discharge; hence the load voltage during this first phase of the discharge decreases linearly with the amount of charge removed from the battery, until all the available iodine is consumed. At that time the voltage drops at a steeper rate, indicative of the increasingly significant resistance of the iodine-deficient depolarizer. In the final, third phase the open-circuit as well as the load voltage drops as iodine is no longer available at unit thermodynamic activity. The cells for medical use are made cathode-limited so their end-of-life voltage change is much less abrupt than if they were lithium-limited. The normalized slope (Volts per Ampere-hour) for the straight-line portion of the discharge curves is approximated by:

$$\frac{\Delta E}{\Delta Q} = \frac{3600M}{F\sigma d} \cdot \frac{i}{A^2} = K\frac{i}{A^2}$$

where $\Delta E/\Delta Q$ = slope of discharge curve, V/Ah
      M = molecular weight of LiI
      d = density of LiI
      σ = lithium ion conductivity of LiI
      F = Faraday's constant
      i = discharge current
      A = area of anode, cm^2

Literature values of the Li^+ conductivity in LiI range from 0.2 to 0.8 $\times$ 10^{-6} $\Omega^{-1} \cdot cm^{-1}$ at 37°C.[16–18] K then lies between 1.5 and 6.0 $\times$ 10^6. Experimental values of K measured from discharge data have been found to be between 1.0 and 1.5 $\times$ 10^6. At high discharge rates, fracturing of the LiI layer apparently occurs, and this equation holds only over small portions of the discharge curve.

The resistance of these batteries behaves consistently with the load voltage curves. It builds up linearly with discharge until the cells are depleted of iodine. At discharge rates much above 10 $\mu A/cm^2$ a nonohmic component of the polarization begins to become increasingly important.[25] The nonohmic component appears to originate near the discharge product-cathode interface. The effects of this polarization may last a long time after high-rate discharge. Thus extrapolation of performance between rates is not simple.

A different discharge curve is obtained from lithium/iodine batteries which are constructed with a coating of pure P2VP applied to the anode before the iodine depolarizer is added to the battery. This coating alters the discharge behavior of these cells. The discharge product no longer grows in a planar fashion but rather in columnar-type groupings. It has been shown that a liquid breakdown reaction product of the P2VP anode coating has been observed in special experimental cells constructed on microscope slides.[26] This liquid is thought to be a liquid electrolyte which wets the lithium iodide discharge product and forms a lower resistance path in parallel to the lithium iodide.

Plots of discharge voltage versus state of discharge for these batteries are not linear. The battery resistance increases exponentially with the state of discharge. This dependence extends from the beginning of life throughout the region of discharge where the cathode is two-phase, consisting of crystalline iodine plus a conductive liquid, which is a product of a reaction between iodine and P2VP. During most of the discharge, the resistance is dominated by the accumulating discharge product and the charge-transfer reaction. Once the crystalline iodine is depleted, the resistivity of the cathode rises rapidly and soon dominates the resistance of the battery. The crossover from electrolyte to cathode domination of battery resistance gives the discharge curves of Li/I_2 batteries their characteristic shape.

Detailed studies of Li/I_2 batteries have shown that under conditions of mild discharge (≥ 5 $\mu A/cm^2$) the voltage loss below 2.8 V results from five physical processes: (1) a drop in open-circuit potential as the discharge proceeds into the single-phase region of cathode composition, (2) an exponential increase in electrolyte resistance throughout the discharge, (3) bulk ohmic cathode resistance, which decreases slightly as the discharge proceeds in the two-phase region of cathode composition, but which rises sharply after the crystalline iodine is depleted, (4) a charge-transfer resistance at the case-to-cathode interface, and (5) concentration polarization in the cathode. All of these processes have been characterized quantitatively in a general predictive steady-state discharge model of Li/I_2 battery discharge.[27]

Figure 15.12 shows a typical plot of load voltage versus state of discharge for both coated- and uncoated-anode groups of otherwise identical batteries at 37°C. The voltage data are shown between the initial voltage of 2.8 V and the cutoff voltage of 2.0 V. All data were obtained at a current density of 6.7 $\mu A/cm^2$. The voltage of the uncoated batteries decreases linearly until iodine depletion starts to occur. At this time the slope of the curve changes.

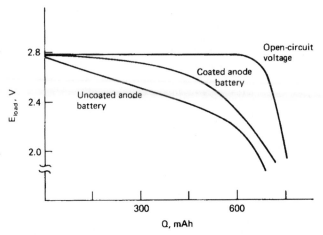

FIGURE 15.12   Load voltage vs. discharge state for uncoated and P2VP-coated anode Li/I_2 batteries discharged at 6.7 $\mu A/cm^2$. 37°C. (*Courtesy of Medtronic, Inc.*)

The same batteries with P2VP-coated anodes exhibit a higher load voltage at this current density. However, as the current density decreases, the differences between the discharge curves for the coated- and uncoated-anode batteries become smaller. Figure 15.13 shows the discharge voltage data for the coated-anode battery in Fig. 15.12 to 0 V with the corresponding resistance data. Log R versus Q is linear, and R varies between 100 and 8000 Ω during discharge. Typical discharge curves for a CRC 800 series battery, as well as changes in

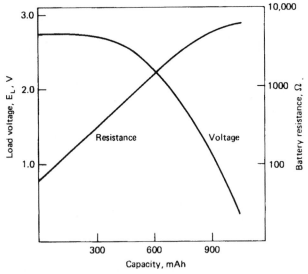

FIGURE 15.13   Load voltage and battery resistance vs. Q for coated-anode Li/I_2 battery discharged at 100 μA, 37°C. (*Courtesy of Medtronic, Inc.*)

battery resistance, are shown in Fig. 15.14. Catalyst Research Corporation reported that its 900 series batteries (manufactured by pressing the cathode depolarizer onto a central lithium anode assembly) exhibited discharge behavior very similar to that of the coated-anode batteries.

The button and D-size batteries made by Catalyst Research Corporation were reported to have discharge curves like the 900 series batteries. Figures 15.15 and 15.16 show projected discharge curves for two types of button cells. Figure 15.17 is a similar discharge plot for the D-size battery at 25°C. This cell delivered 7 Ah (0.45 Wh/cm^3) at the 2-month (5-mA) rate; capacities and energy densities at lower rates are expected to be larger. The projected discharge curves for the 1- and 2-mA rates are also shown.

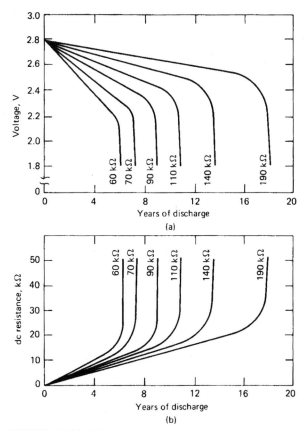

FIGURE 15.14   800 series Li/I$_2$ battery (*a*) Typical discharge curves. (*b*) Changes in cell resistance during discharge. (*Courtesy of Catalyst Research Corp.*)

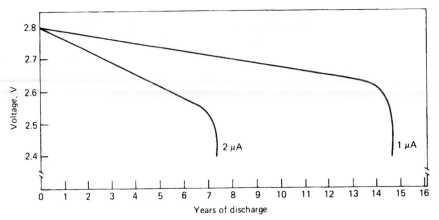

FIGURE 15.15 Projected discharge curves for S19P-20 button watch battery at 25°C. (*Courtesy of Catalyst Research Corp.*)

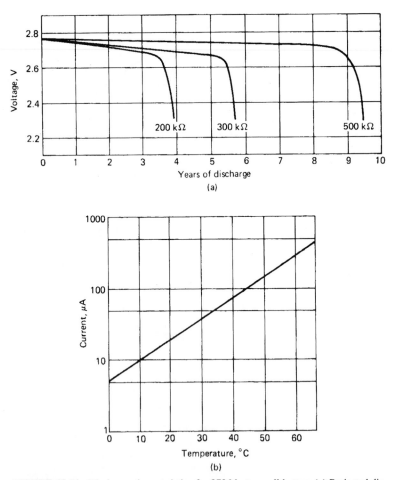

FIGURE 15.16 Discharge characteristics for 2736 button-cell battery (*a*) Projected discharge at 25°C. (*b*) Maximum continuous discharge current vs. temperature.[25] (*Courtesy of Catalyst Research Corp.*)

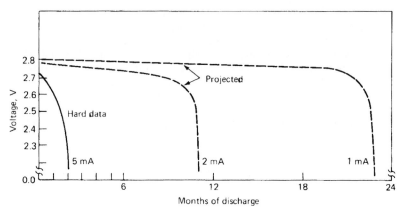

FIGURE 15.17 Discharge data for D-cell diameter Li/I$_2$ battery at 25°C. (*Courtesy of Catalyst Research Corp.*)

15.3.5 Self-Discharge

In addition to the ohmic and nonohmic polarization losses there are two other mechanisms for capacity loss in the Li/I$_2$ battery system. The first is self-discharge. It occurs by the direct combination of lithium and iodine which has diffused through the lithium iodide layer to reach the lithium anode. The amount of self-discharge observed in these batteries is dependent on the effective thickness of the lithium iodide discharge layer. For this reason the largest percentage of self-discharge occurs early in the battery's life when the lithium iodide layer is very thin. In fact, at a discharge rate of 1 to 2 μA/cm^2 virtually all self-discharge loss occurs by the time the battery is 25% depleted. Figure 15.18 shows a plot of power lost

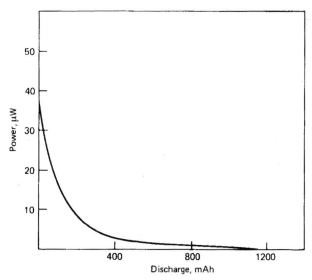

FIGURE 15.18 Power (heat) loss due to self-discharge vs. discharge state for typical Li/I$_2$ battery (calorimetric measurements made at open-circuit voltage). (*Courtesy of Medtronic, Inc.*)

to self-discharge versus state of discharge for a 2000-mAh stoichiometric capacity Li/I_2 medical battery at 37°C. These measurements were made using a microcalorimeter. The amount of self-discharge observed in the batteries precoated with a layer of P2VP is a little greater than in the batteries which do not receive this treatment. This is because the diffusion path between the cathode depolarizer and the lithium anode is thinner than in the uncoated batteries. However, the amount of self-discharge in any of these batteries is 10% or less (depending on the discharge rate) of the rated capacity of the cell over a 10 to 15-year lifetime and is therefore small compared to most battery systems. Calorimetric measurements have shown that self-discharge losses are smaller while current is being drawn from the battery. Most estimates (including that above) are made at open circuit. Therefore they over-estimate self-discharge losses by as much as 100%.

15.3.6 Other Performance Losses

The total iodine utilization in the Li/I_2 battery depends on the initial ratio of iodine to donor material in the cathode depolarizer as well as loss due to self-discharge. The iodine and amine nitrogen form a very strong 1:1 complex which is not dischargeable at a useful voltage. For this reason the amount of iodine that can be discharged is limited to that between the initial starting ratio of the battery material depolarizer and that of the 1:1 complex. In addition, the P2VP used as a donor material in these batteries is chemically attacked by the iodine in the battery. The result is that approximately another one-half mole of iodine per mole of nitrogen atoms in the donor materials becomes unavailable for discharge.[25] Thus the maximum amount of discharge capacity is equal to the iodine that lies between the initial cathode ratio and about a 1.5:1 final ratio. This value does not include the self-discharge and polarization losses discussed.

The effect of all losses upon the available capacity as a function of rate is summarized in Fig. 15.19 for a battery at 37°C. Here the deliverable capacity is estimated by starting with the stoichiometric capacity and subtracting the losses and limitations which arise from the various processes described. This battery had an initial cathode-to-donor ratio of 6.2 mol iodine per pyridine ring (15:1 weight ratio) in the cathode complex. The maximum utilization is predicted to be 60% of stoichiometric capacity for this ratio of iodine to P2VP. The mole ratio of iodine to donor material ranges from 6 to 8 in currently available batteries. The achievable utilization will depend on this ratio. (Most literature for commercial batteries gives the initial ratio in terms of weight ratio of iodine to donor material.)

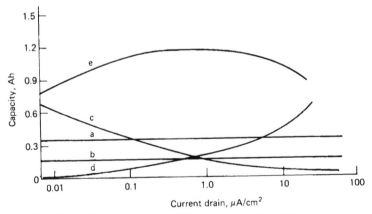

FIGURE 15.19 Capacity and loss projections for 2.0-Ah stoichiometric capacity Li/I_2 battery. *a*—loss to 1:1 adduct formation: *b*—loss to iodination reaction: *c*—projected loss to self-discharge: *d*—projected loss to polarization: *e*—net utilizable battery capacity. (*Courtesy of Medtronic, Inc.*)

15.3.7 Effect of Temperature

Since the majority of these batteries has been used as power sources for heart pacemakers, most experience with them is at body temperature (37°C). However, testing has been done at temperatures between -10 and 60°C. These results are more important for the nonmedical batteries. Figure 15.20 shows the variation of maximum continuous current for the CRC D battery as a function of temperature. Hard data exist between -10 and 50°C. Performance at the lower temperature is reduced due to the low conductivity of the depolarizer and LiI. At 60°C, self-discharge is increased along with the rate of other parasitic reactions, although the battery is usable at this temperature. Manufacturers give operating temperature ranges for nonmedical batteries between -20 or 0 and 50°C. The temperature at which best performance can be expected is between room temperature and 40°C.

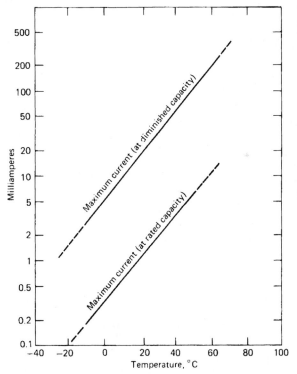

FIGURE 15.20 Maximum continuous-current capability of Li/I$_2$ battery vs. temperature. (*Courtesy of Catalyst Research Corp.*)

15.4   *Ag / RbAg₄I₅ / Me₄NI_n C BATTERIES*

These solid-electrolyte batteries are based on rubidium silver iodide ($RbAg_4I_5$) as the electrolyte. This material exhibits an unusually high ionic conductivity of 0.26 $\Omega^{-1} \cdot cm^{-1}$ at 25°C.[2] This permits battery discharge at much higher current drains than those available with LiI-based batteries. The initial characteristics of small 40-mAh five-cell batteries are shown in Table 15.7. Both tetramethylammonium pentaiodide and enneaiodide have been evaluated as active cathode agents.[25] These batteries demonstrated efficient discharge capability at temperatures ranging from −40 to 71°C. Constant-load discharge curves are shown in Figs. 15.21 and 15.22.[3] The two cathodes behave similarly. The Me_4NI_9 cathode has the higher iodine content and activity, which results in higher energy density and rate capability. The lower iodine activity of the Me_4NI_5 cathode gives rise to a lower rate of self-discharge and, therefore, better long-term storage characteristics. The batteries were tested after nearly 20 years of storage.[29] Discharge curves for the nominal 23°C stored batteries are shown in Figs. 15.23 and 15.24. These five-cell batteries delivered about 36 mAh of capacity to a 2-V cutoff, equal to 90% of their initial capacity as measured in 1972. The capacity is plotted in Fig. 15.25 as a function of storage time for the tests that were performed at a constant resistive load of 64.9 kΩ after storage times of 0, 10, and 20 years. Additional tests were conducted after 25 years of storage at 25°C and −15°C. The performance obtained this storage period was similar to that reported above after 20 years of storage. The average capacity loss rate was less than 0.5% per year after storage and discharge at room temperatures.[30]

TABLE 15.7   Solid-State Ag/I_2 Batteries*

	Ag/Me_4NI_9	Ag/Me_4NI_5
Single-cell voltage, V	0.662	0.650
Battery open-circuit voltage, V	3.31	3.25
Internal resistance, Ω	30	35

*1.27 cm diameter × 1.9 cm height; capacity = 40 mAh.

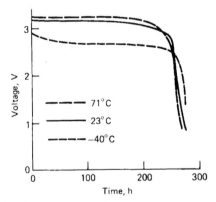

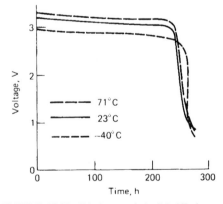

FIGURE 15.21   Discharge of Ag/Me_4NI_5 battery, 21.5-kΩ constant load. (*From Topics in Applied Physics, vol. 21, reprinted by permission of the publisher, Springer-Verlag.*)

FIGURE 15.22   Discharge of Ag/Me_4NI_9 battery. (*From Topics in Applied Physics, vol. 21, reprinted by permission of the publisher, Springer-Verlag.*)

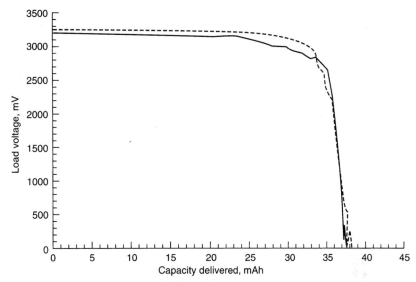

FIGURE 15.23   64.9-kΩ constant load discharge of $Ag/RbAg_4I_5/Me_4NI_5$ batteries after 20 years of storage at 25°C. Discharge temperature: solid curve—25°C, dashed curve—60°C. (*From Ref. 29.*)

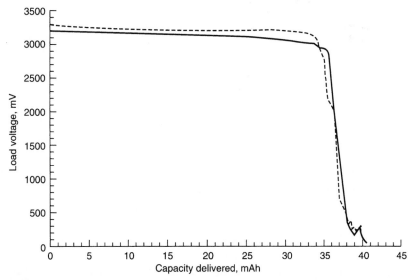

FIGURE 15.24   64.9-kΩ constant-load discharge of $Ag/RbAg_4I_5/Me_4NI_9$ batteries after 20 years of storage at 25°C. Discharge temperature: solid curve—25°C: dashed curve—60°C. (*From Ref. 29.*)

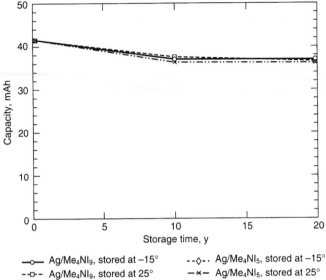

FIGURE 15.25   Capacity of $Ag/RbAg_4I_5/Me_4NI_n$ batteries as a function of storage time. (*From Ref. 29.*)

The results of the tests confirm the long-term stability of solid-electrolyte batteries with respect to self-discharge losses. However, the batteries stored at $-15°C$ suffered from electrolyte disproportionation and the concurrent resistance increase. This disproportionation was readily reversed by heating the batteries at 60°C for less than 30 min. Following this thermal treatment, the internal resistances returned to their normal values, and standard discharge behavior was observed.

The Ag/I_2 batteries were inherently deficient in two significant areas, voltage and size. The low cell voltage and high equivalent weights and volumes restricted these batteries to very low energy densities and specific energies (0.08 Wh/cm^3 or 13 Wh/kg). Consequently no commercial application resulted.

REFERENCES

1. B. B. Owens and P. M. Skarstad, "Ambient Temperature Solid State Batteries," *Solid State Ionics* **43:**665 (1992).

2. B. B. Owens, "Solid Electrolyte Batteries," in C. W. Tobias (ed.), *Advances in Electrochemistry and Electrochemical Engineering,* vol. 8, Wiley, New York, 1971, chap. 1, pp. 1–62.

3. B. B. Owens, J. E. Oxley, and A. F. Sammells, "Applications of Halogenide Solid Electrolytes," in S. Geller (ed.), *Solid Electrolytes, Topics in Applied Physics,* vol. 21, Springer-Verlag, New York, 1977, chap. 4, pp. 67–104.

4. C. C. Liang, "Solid-State Batteries," *Appl. Solid State Sci.* **4:**95–135 (1974).

5. T. Takahaski, *Denki Kagoku* **36:**402–412, 481–490 (1968).

6. B. B. Owens and P. M. Skarstad, "Ambient Temperature Solid-State Batteries," in P. Vashishta, J. N. Mundy, and G. K. Shenoy (eds.), *Fast Ion Transport in Solids,* Elsevier-North Holland, New York, 1979, pp. 61–68.

7. M. Gauthier, A. Belanger, B. Kapfer, and G. Vassort, "Solid Polymer Electrolyte Lithium Batteries," in J. R. MacCallum and C. A. Vincent (eds.), vol. 2, *Polymer Electrolyte Reviews,* Elsevier Applied Science, London, 1989.

8. C. F. Holmes, "Lithium/Halogen Batteries," in B. B. Owens (ed.) *Batteries for Implantable Biomedical Devices,* Plenum, New York, 1986.

9. C. A. C. Sequira and A. Hooper, (eds.), *Solid State Batteries,* NATO ASI ser. 101, Nijhoff Publishers, Dordrecht, The Netherlands, 1985.

10. B. V. R. Chowdari and S. Radhakrishna (eds.), *Materials for Solid State Batteries, Proc. Regional Workshop,* World Scientific Publ. Co., Singapore, 1986.

11. R. C. Weast (ed.), *Handbook of Chemistry and Physics,* 59th ed., Chemical Rubber Publishing Co., West Palm Beach, Fla., 1978–1979.

12. C. C. Liang, J. *Electrochem. Soc.* **120:**1289 (1973).

13. C. C. Liang and L. H. Barnette, J. *Electrochem. Soc.* **123:**453–458 (1976).

14. J. R. Rea, L. H. Barnette, C. C. Liang, and A. V. Joshi, "A New High Energy Density Battery System," in B. B. Owens and N. Margalit (eds.), *Proc. Symp. on Biomedical Implantable Applications and Ambient Temperature Lithium Batteries,* vol. 80-4, Electrochemical Society, Princeton, N.J., 1980, pp. 245–253.

15. C. C. Liang, A. V. Joshi, and N. E. Hamilton, J. *Appl. Electrochem.* **8:**445–454 (1978).

16. D. E. Ginnings and T. E. Phipps, J. *Am. Chem. Soc.* **52:**1340 (1930).

17. B. J. H. Jackson and D. A. Young, J. *Phys. Chem. Solids* **30:**1973–1976 (1969).

18. C. R. Schlaijker and C. C. Liang, "Solid-State Batteries and Devices," in W. Van Gool (ed.), *Fast Ion Transport in Solids,* North Holland, Amsterdam, 1973, pp. 689–694.

19. C. C. Liang, U.S. Patent 3,713,897 (1973).

20. A. M. Stoneham, E. Wade, and J. A. Kilner, *Mater. Res. Bull.* **14:**661–666 (1979).

21. J. R. Rea, *DOE* Battery and Electrochemical Contractors' Conf, Arlington, Va., vol. 11, session VII, Dec. 1979.

22. S. J. Garlock, "Characteristics of Lithium Solid State Batteries in Memory Circuits," Wescon 81, San Francisco, Calif., Professional Program Session Rec. 29, Electronic Conventions, Inc., El Segundo, Calif., Sept. 1981.

23. G. M. Phillips and D. F. Untereker, "Phase Diagram for the Poly-2-vinylpyridine and Iodine System," in B. B. Owens and N. Margalit (eds.), *Proc. Symp. on Biomedical Implantable Applications and Ambient Temperature Lithium Batteries,* vol. 80-4, Electrochemical Society, Princeton, N.J., 1980.

24. P. M. Skarstad, B. B. Owens, and D. F. Untereker, "Volume Changes on Discharge in Lithium-Iodine (Poly-2-vinylpyridine) Cells," *152d Electrochem. Soc. Meeting,* Atlanta, Oct. 1977, abstract 23.

25. K. R. Brennen and D. F. Untereker, "Iodine Utilization in Li/I_2 (Poly-2-vinylpyridine) Batteries," in B. B. Owens and N. Margalit (eds.), *Proc. Symp. on Biomedical Implantable Applications and Ambient Temperature Lithium Batteries,* vol. 80-4, Electrochemical Society, Princeton, N.J., 1980.

26. J. P. Phipps, T. G. Hayes, P. M. Skarstad, and D. Untereker, *Solid State Ionics* 18, **19:**1073 (1986).

27. C. L. Schmidt and P. M. Skarstad, "Development of a Physically-Based Model for the Lithium/Iodine Battery," in T. Keily and B. W. Baxter (eds.), *Power Sources,* vol. 13, International Power Sources Committee, Leatherhead, England, 1991, pp. 347–361.

28. D. L. Warburton, R. F. Bis, and B. B. Owens, "Five-Year Storage Tests of Solid-State Ag/I_2 Batteries," *28th Power Sources Symp.,* Atlantic City, N.J., June 12–15, 1978.

29. B. B. Owens and J. R. Bottelberghe, "Twenty Year Storage Test of Ag/$RbAg_4I_5$/I_2 Solid State Batteries," *Solid State Ionics* **62:**243 (1993).

30. D. L. Warburton, et al., "Twenty-five Year Shelf-life of an Active Primary Battery." *Proc. 38th Power Sources Conf.* pp. 89–92 (1998). Cherry Hill, N.J.

P · A · R · T · 3

RESERVE BATTERIES

CHAPTER 16
RESERVE BATTERIES— INTRODUCTION

David Linden

16.1 CLASSIFICATION OF RESERVE BATTERIES

Batteries, which use highly active component materials to obtain the required high energy, high power, and/or low-temperature performance, are often designed in a reserve construction to withstand deterioration in storage and to eliminate self-discharge prior to use. These batteries are used primarily to deliver high power for relatively short periods of time after activation in such applications as radiosondes, fuzes, missiles, torpedoes, and other weapon systems. The reserve design also is used for batteries required to meet extremely long or environmentally severe storage requirements.

In the reserve structure, one of the key components of the cell is separated from the remainder of the cell until activation. In this inert condition, chemical reaction between the cell components (self-discharge) is prevented, and the battery is capable of long-term storage. The electrolyte is the component that is usually isolated, although in some water-activated batteries the electrolyte solute is contained in the cell and only water is added.

The reserve batteries can be classified by the type of activating medium or mechanism that is involved in the activation:

Water-activated batteries: Activation by fresh- or seawater.

Electrolyte-activated batteries: Activation by the complete electrolyte or with the electrolyte solvent. The electrolyte solute is contained in or formed in the cell.

Gas-activated batteries: Activation by introducing a gas into the cell. The gas can be either the active cathode material or part of the electrolyte.

Heat-activated batteries: A solid salt electrolyte is heated to the molten condition and becomes ionically conductive, thus activating the cell. These are known as thermal batteries.

Activation of the reserve battery is accomplished by adding the missing component just prior to use. In the simplest designs, this is done by manually pouring or adding the electrolyte into the cell or placing the battery in the electrolyte (as in the case of seawater-activated batteries). In more sophisticated applications the electrolyte storage and the activation mechanism are contained within the overall battery structure, and the electrolyte is brought automatically to the active electrochemical components by remotely activating the activation mechanism. The trigger for activation can be a mechanical or electrical impulse, the shock and spin accompanying the firing of a shell or missile, and so on. Activation can be completed very rapidly if required, usually in less than one second. The penalty for automatic activation is a substantial reduction in the specific energy and/or energy density of the battery due to the volume and weight of the activating mechanism. It is therefore not general practice to rate these batteries in terms of specific energy or energy density.

The gas-activated batteries are a class of reserve batteries which are activated by introduction of a gas into the battery system. There are two types of gas-activated batteries: those in which the gas serves as the cathodic active material and those in which the gas serves to form the electrolyte. The gas-activated batteries were attractive because they offered the potential of a simple and positive means of activation. In addition, because the gas is nonconductive, it can be distributed through a multicell assembly without the danger of short-circuiting the battery through the distribution system. Gas-activated batteries are no longer in production, however, because of the more advantageous characteristics of other systems.

The thermal or heat-activated battery is another class of reserve battery. It employs a salt electrolyte, which is solid and, hence, nonconductive at the normal storage temperatures when the battery must be inactive. The battery is activated by heating it to a temperature sufficiently high to melt the electrolyte, thus making it ionically conductive and permitting the flow of current. The heat source and activating mechanism, which can be set off by electrical or mechanical means, can be built into the battery in a compact configuration to give very rapid activation. In the inactive stage the thermal battery can be stored for periods of 10 years or more.

16.2 CHARACTERISTICS OF RESERVE BATTERIES

Reserve batteries have been designed using a number of different electrochemical systems to take advantage of the long unactivated shelf life achieved by this type of battery design. Relatively few of these have achieved wide usage because of the lower capacity of the reserve structure (compared with a standard battery of the same system), poorer shelf life after activation, higher cost, and generally acceptable shelf life of active primary batteries for most applications. For the special applications that prompted their development, nevertheless, the reserve structure offers the needed advantageous characteristics.

In recent years, however, the use of reserve batteries has declined because of the improved storability of active primary batteries and the limited number of applications requiring extended storage. Most of these applications are for special military weapon systems.

The reserve batteries are usually designed for specific applications, each design optimized to meet the requirements of the application. A summary of the major types of reserve batteries, their major characteristics and advantages, disadvantages, and key areas of application is given in Table 16.1.

Conventional Systems. Reserve batteries employing the conventional electrochemical systems, such as the Leclanché zinc-carbon system, date back to the 1930–1940 period. This structure, in which the electrolyte is kept in a separate vial and introduced into the cell at the time of use, was employed as a means of extending the shelf life of these batteries, which was very poor at that time. Later similar structures were developed using the zinc-alkaline systems. Because of the subsequent improvement of the shelf life of these primary batteries and the higher cost and lower capacity of the reserve structure, batteries of this type never became popular.

TABLE 16.1 Characteristics of Reserve Batteries

System	Conventional system	Water-activated batteries	Metal/air batteries	Lithium/water batteries	Zinc/Silver oxide batteries	
					Manually activated	Automatically activated
General characteristics	Conventional cylindrical cells in reserve design (electrolyte separated in cell during storage)	Battery activated by adding or placing battery in water	Battery activated by adding electrolyte or placing battery in seawater	Primary reserve system, depending on controlled reaction of Li with H_2O	Battery activated by adding KOH electrolyte just prior to use	Electrolyte separately stored in battery; built-in device to automatically activate from remote or local position
Advantages	Reserve structure extends shelf life	High energy density; moderate to high rate capability; good low-temperature performance after activation; simple designs; easy activation	High energy density achieved by using oxygen from air or ocean environment	High energy density	Highest capacity of practical aqueous systems for high-rate use	High capacity, no maintenance; automatic activation. Excellent unactivated shelf life
Disadvantages/limitations	Lower capacity than conventional active cells; low to moderate discharge rates	Rapid self-discharge after activation; AgCl system is expensive	Self-discharge	Need to control Li reaction with H_2O: complex system and controls	Manual activation is inconvenient and undesirable for field use; low-temperature performance is poor	Activation device reduces energy density; costly, but warranted, for special applications
Chemistry: Anode	Zn	Mg, Zn	Zn, Mg, Al	Li	Zn	Zn
Cathode	MnO_2	AgCl, CuCl, MnO_2, $PbCl_2$, and others	Air or oxygen	H_2O, H_2O_2, O_2, AgO	AgO, Ag_2O	AgO, Ag_2O
Electrolyte	Salt	H_2O, seawater aqueous solutions	Seawater or alkaline solution	H_2O, LiOH	KOH	KOH
Nominal voltage, V	1.5	1.5–1.6	See Table 38.2	2.2	1.6	1.6
Performance characteristics Operating temperature, °C	0 to 50	−60 to 65 (after activation) Performance almost independent of ambient temperature after activation	See Chap. 38	0 to 30	0 to 60	0 to 60 (<0°C operation with heaters)
Specific energy and energy density, Wh/kg Wh/L	30 (at moderate rates) 60	AgCl 100–150 Others 45–80 50–200	See Chap. 38	160 (at 20-h rate) 135	60–60 (at high rates) 100–160	20–50 (at high rates) 100–200
Status	No longer in use	In limited use for special applications	In development, early production for special applications	In limited development	In production	In production
Major applications		Marine applications (torpedoes, sonobuoys); air-sea rescue; emergency lights	Multiple applications	Marine applications (torpedoes, sonobuoys, submersibles)	Special applications requiring high-rate, high-capacity batteries	Missile and torpedo applications
Reference	Section 16.2	Chapter 17	Chapter 38	Section 38.6.1	Chapter 18	Chapter 18

TABLE 16.1 Characteristics of Reserve Batteries (*Continued*)

System	Spin-dependent batteries	Lithium-nonaqueous batteries	Liquid ammonia batteries	Gas-activated batteries: Chlorine depolarized	Gas-activated batteries: Ammonia-vapor-activated (AVA)	Thermal batteries
General characteristics	Electrolyte stored separately in battery; activated by shock and spin of projectiles	Battery activated by introducing liquid electrolyte into battery system	Battery activated by introducing liquid NH_3 into battery system (NH_3 can be stored in ampul in battery)	Battery activated by introducing chlorine gas to act as the depolarizer	Battery activated by introducing ammonia gas to form the electrolyte with salt already in the battery	Battery activated by heating to a temperature sufficient to melt solid electrolyte, making it conductive
Advantages	Excellent unactivated shelf life; convenient, reliable, rapid "built-in" activation	High energy density; wide operating temperature range; flat discharge profile; excellent unactivated storage; high to low discharge rate capability	Wide operating temperature range; high applications	Potential for high-rate, high-capacity, good low-temperature performance; simple activation even at $-20°C$	Potential for good low-temperature performance; simple activation; excellent unactivated shelf life; high and moderate rate application	Performance independent of ambient temperature; rapid activation; excellent unactivated shelf life
Disadvantages/limitations	Activation device reduces energy density	Reserve structure has lower energy density than active primary systems	High pressure, poor wet stand	Short shelf life even in unactivated condition	Activation slow and nonuniform	Short lifetime; activation device reduces energy density. New designs, using Li or Li alloy anode, however, has higher energy density and longer lifetime
Chemistry:						
Anode	Pb · Zn · Li	Li · Li · Li	Mg	Zn	Zn	Ca · Mg · Li · Li
Cathode	PbO_2 · AgO · $SOCl_2$	V_2O_5 · SO_2 · $SOCl_2$	m · DNB	Cl_2	PbO_2	$CaCrO_4$ · V_2O_5 · LiCl/KCl · FeS_2
Electrolyte	HBF_4 · KOH · $SOCl_2$	Organic · Organic · $SOCl_2$	NH_4SCN, $KSCN(NH_3)$	Salt ($CaCl_2$, $ZnCl_2$)	$NH_4SCN(NH_3)$	LiCl/KCl · LiCl/KCl · LiCl/KCl
Nominal voltage, V	2.0 · 1.6 · 3.5	3.3 · 3.0 · 3.5	2.2	1.5	1.9	2.22–2.6 · 2.2–2.7 · 1.6–2.1
Performance characteristics:						
Operating temperature, °C	−40 to 60 (For HBF_4 system, other systems may require heating for low-temperature operation.)	−55 to 70	−55 to 70	−20 to 50	−55 to 75	−55 to 75
Specific energy and energy density						
Wh/kg	{ See Section 19.4 }	50–150 (depending on battery system)	45 (at high 100 rates) 60 (at low 130 rates)	40	25	10 up to 30 (for Ca batteries) 40 100 (for Li batteries)
Wh/L		100–300		60	50	
Status	In production	In production	Production terminated	Development effort terminated	Effort redirected to liquid ammonia batteries	In production, emphasis directed to newer lithium systems and longer lifetime
Major applications	Artillery and spin stabilized projectiles—fuzing control, or arming	Mine fuzing, missiles	Mine fuzing, missiles			Military ordnance (projectiles, rockets, missiles, fuzing)
Reference	Chapter 19	Chapter 20	Section 16.2	Section 16.2	Section 16.2	Chapter 21

Water-activated Batteries.   A reserve battery that was used widely is the water-activated type. This battery was developed in the 1940s for applications such as weather balloons, radiosondes, sonobuoys, and electric torpedoes requiring a low-temperature, high-rate, or high-capacity capability. These batteries use an energetic electrochemical system, generally a magnesium alloy, as the anode and a metal halide for the cathode. The battery is activated by introduction of water or an aqueous electrolyte. The batteries are used at moderate to high discharge rates for periods up to 24 h after activation.

These batteries may also be designed to be activated with seawater. They have been used for sonobuoys, other marine applications (lifejacket lights, etc.), and underwater propulsion. Activation can occur upon immersion into seawater or require the forced flow of seawater through the system. Many of these seawater batteries use a magnesium alloy anode with a metal salt cathode, as shown in Table 16.1.

Alloys of zinc, aluminum, and lithium have also been considered for special-purpose seawater batteries. Zinc can be used as the anode in low-current, low-power long-life batteries. It has the advantage of not sludging, but the disadvantage of being a low-power-density system. Zinc/silver chloride seawater batteries have been used as the power source for repeaters for submarine telephone cables (for example, 5 mA at 0.9–1.1 V for 1 year of operation).

Zinc and aluminum seawater batteries, using a silver oxide cathode, have higher energy densities than magnesium seawater batteries and can be discharged at high rates similar to the magnesium/silver chloride battery. The aluminum anode is subject to much higher corrosion rates than magnesium. Lithium is attractive because of its high energy and power density, and batteries using lithium as an anode were once in development using silver oxide or water as the cathode material. In general the combination of lithium with water is considered hazardous because of the high heat of reaction, but in the presence of hydroxyl ion concentrations greater than 1.5M a protective film is formed which exists in a dynamic steady state. Operation of these batteries requires very precise control of the electrolyte concentration, which requires sophisticated pumps and controls (also see Chapter 38).

Zinc, aluminum, or magnesium alloys are being used in reserve batteries using air as the cathode. With aluminum or magnesium, these batteries may be activated with saline electrolytes, and in some underwater application they may use oxygen dissolved in the seawater. Reserve or mechanically rechargeable air batteries, for higher-power applications such as for standby power or electric-vehicle propulsion, use zinc or aluminum alloys with alkaline electrolytes (also see Chapter 38).

Zinc/Silver Oxide Batteries.   Another important reserve battery uses the zinc/silver oxide system, which is noted for its high-rate capability and high specific energy. For missile and other high-rate applications, the cell is designed with thin plates and large-surface-area electrodes, which increase the high-rate and low-temperature capability of the battery and give a flatter discharge profile. This construction, however, reduces the activated shelf life of the battery, necessitating the use of a reserve battery design. The cells can be filled and activated manually, but for missile applications the zinc/silver oxide battery is used in an automatically activated design. This use requires a long period in a state of readiness (and storage), necessitating the reserve structure, a means for rapid activation, and an efficient high-rate discharge at the rate of approximately 2 to 20 min. Activation is accomplished within a second by electrically firing a gas squib which forces the stored electrolyte into the cells. Shelf life of the unactivated battery is 10 years or more at 25°C storage.

Spin Activated Batteries.   The spin-dependent design provides another means of activating reserve batteries using liquid electrolytes, taking advantage of the forces available during the firing of an artillery projectile. The electrolyte is stored in a container in the center of the battery. The shock of the firing breaks or opens the container, and the electrolyte is distributed into the annular-shaped cells by the centrifugal force of the spinning of the projectile.

Nonaqueous Electrolyte Batteries.   Nonaqueous electrolyte systems are also used in reserve batteries to take advantage of their lower freezing points and better performance at low temperatures. The liquid ammonia battery, using liquid ammonia as the electrolyte solvent, had been employed up to about 1990 as a power source for fuzes and low power ordnance devices which require a battery capable of performance over a wide temperature range and with an unactivated shelf life in excess of 10 years. The liquid ammonia battery is operable at cold as well as normal temperatures with little change in cell voltage and energy output. The battery typically uses a magnesium anode, a meta-dinitrobenzene-carbon cathode, and an electrolyte salt system based on ammonium and potassium thiocyanate. Activation is accomplished by introducing liquid ammonia into the battery cell where it combines with the thiocyanate salts to form the electrolyte:

$$NH_4SCN \xrightarrow{NH_3} NH_4^+ + SCN^-$$

$$KSCN \xrightarrow{NH_3} K^+ + SCN^-$$

Mechanically, this can be done by igniting a gas generator which forces the electrolyte into the cells or through an external force, such as a gun firing setback which breaks a glass ampule containing the liquid ammonia as shown in Fig. 16.1. Depending on the application, the battery can be designed for efficient discharge for several minutes or up to 50 or more hours of service. Typical performance characteristics are shown in Fig. 16.2. This battery is no longer in production; the only manufacturer was Alliant Techsystems, Power Sources Center, Horsham, PA.

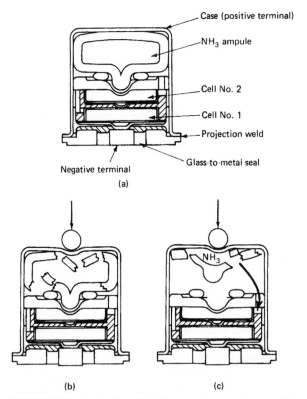

FIGURE 16.1   Activation of liquid ammonia reserve battery, Alliant model G2514. (*a*) Inactive cell. (*b*) Ampoule broken by external force. (*c*) Ammonia activates battery stack. (*Courtesy of Alliant Techsystems, Inc., Power Sources Center.*)

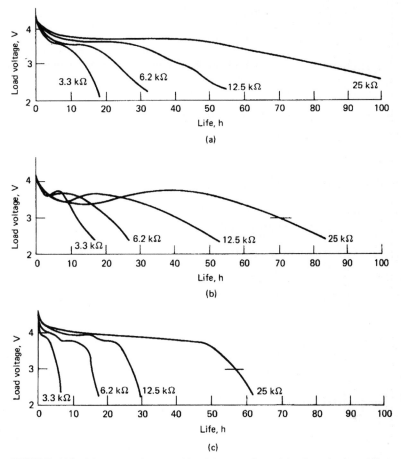

FIGURE 16.2 Discharge voltage profile of battery after axial spin activation. Alliant model G2514. (*a*) 20°C. (*b*) 70°C. (*c*) −55°C. (*Courtesy of Alliant Techsystems, Inc., Power Sources Center.*)

Lithium Anode Batteries.   The lithium anode electrochemical system is also being developed in reserve configurations to take advantage of its high energy density and good low-temperature performance. These batteries use either an organic electrolyte or a nonaqueous inorganic electrolyte because of the reactivity of lithium in aqueous electrolytes. Even though the active lithium primary batteries are noted for their excellent storability, the reserve structure is used to provide a capability of essentially no capacity loss even after storage periods in the inactive state of 10 years or more. The performance characteristics of the reserve battery, once activated, are similar to those of the active lithium batteries, but with a penalty of 50% or more in specific energy and energy density due to the need for the activation device and the electrolyte reservoir. Lithium is also being considered as an anode in aqueous reserve batteries for high-rate applications in a marine environment.

Gas-activated Batteries.   The gas-activated batteries were attractive because their activation was potentially simpler and more positive than liquid or heat activation. The ammonia vapor-activated (AVA) battery was representative of a system in which the gas served to form the electrolyte. (Solids such as ammonium thiocyanate will absorb ammonia rapidly to form electrolyte solutions of high conductivity.) In practice, ammonia vapor activation was found to be slow and nonuniform, and the development of the ammonia battery was directed to liquid ammonia activation which, in turn, was found to be inferior to newer developments. The chlorine-depolarized zinc/chlorine battery was representative of the gas depolarizer system. This battery used a zinc anode, a salt electrolyte, and chlorine, which was introduced into the cell, at the time of use, as the active cathode material. The battery was designed for very high rate discharge ranging from 1 to 5 min, but its poor shelf life while inactivated limited further development and use.

Thermal Batteries.   The thermal battery has been used extensively in fuzes, mines, missiles, and nuclear weapons which require an extremely reliable battery that has a very long shelf life, can withstand stress environments such as shock and spin, and has the ability to develop full voltage rapidly, regardless of temperature. The life of the battery after activation is short-the majority of applications are high-rate and require only 1–10 min of use—and is primarily dependent on the time the electrolyte can be maintained above its melting point. The energy density of the thermal battery is low; in this characteristic it does not compare favorably with other batteries except at the extremely high discharge rates. New designs, using lithium or lithium alloy anode, have resulted in a significant increase in the energy density as well as an increase in the discharge time to 1 to 2 hours.

CHAPTER 17
MAGNESIUM WATER-ACTIVATED BATTERIES

Ralph F. Koontz and R. David Lucero

17.1 GENERAL CHARACTERISTICS

The water-activated battery was first developed in the 1940s to meet a need for a high-energy-density, long-shelf-life battery, with good low-temperature performance, for military applications.

The battery is constructed dry, stored in the dry condition, and activated at the time of use by the addition of water or an aqueous electrolyte. Most of the water-activated batteries use magnesium as the anode material. Several cathode materials have been used successfully in different types of designs and applications.

The magnesium/silver chloride seawater-activated battery was developed by Bell Telephone Laboratories as the power source for electric torpedoes.[1] This work resulted in the development of small high-energy-density batteries readily adaptable for use as power sources for sonobuoys, electric torpedoes, weather balloons, air-sea rescue equipment, pyrotechnic devices, marine markers, and emergency lights.

The magnesium/cuprous chloride system became commercially available in 1949.[2,3] Compared with the magnesium/silver chloride battery, this system has lower energy density, lower rate capability, and less resistance to storage at high humidities, but its cost is significantly lower. Although the magnesium/cuprous chloride system can be used for the same purposes as the magnesium/silver chloride battery, its major application was in airborne meteorological equipment, where the use of the more expensive silver chloride system was not warranted. The cuprous chloride system does not have the physical or electrical characteristics required for use as the power source for electric torpedoes. Recently, the magnesium/cuprous chloride chemistry has been developed for aviation and marine lifejacket lights (see Sec. 17.5.5).

Because of the high cost of silver, the impracticality of recovering it after use, other nonsilver water-activated batteries were developed, primarily as the power source for antisubmarine warfare (ASW) equipment.

The systems which have been developed and used successfully are magnesium/lead chloride,[4] magnesium/cuprous iodide-sulfur,[5–7] magnesium/cuprous thiocyanate-sulfur,[8] and magnesium/manganese dioxide utilizing an aqueous magnesium perchlorate electrolyte.[9–11] None of these systems can compete with the magnesium/silver chloride system in almost every attribute except cost.

Magnesium seawater-activated batteries, using dissolved oxygen in the seawater as the cathode reactant, also have been developed for application in buoys, communications, and underwater propulsion. These batteries, as well as the use of other metals as anodes for water-activated batteries, are covered in Chaps. 16 and 38.

Another seawater battery system being investigated for low rate long duration undersea vehicle applications consists of a magnesium anode, a palladium and iridium catalyzed carbon paper cathode and a solution-phased catholyte of seawater, acid and hydrogen peroxide. The magnesium/hydrogen peroxide system has a voltage of 2.12 volts and is expected to be capable of more than 500 Wh/kg when configured for large scale unmanned undersea vehicles.[12]

The advantages and disadvantages of the various water-activated magnesium batteries are given in Table 17.1.

TABLE 17.1   Comparison of Silver and Nonsilver Cathode Batteries

Advantages	Disadvantages
Silver chloride cathodes	
Reliable	High raw material costs
Safe	High rate of self-discharge after activation
High power density	
High energy density	
Good response to pulse loading	
Instantaneous activation	
Long unactivated shelf-life	
No maintenance	
Nonsilver cathodes	
Abundant domestic supply	Requires supporting conductive grid
Low raw-material cost	Operates at low current densities
Instantaneous activation	Low energy density compared to silver
Reliable, safe	High rate of self-discharge after activation
Long unactivated shelf life	
No maintenance	

17.2 CHEMISTRY

The principal overall and current-producing reactions for the water-activated magnesium batteries are as follows:

1. Magnesium/silver chloride

Anode	$Mg - 2e \rightarrow Mg^{2+}$
Cathode	$2AgCl + 2e \rightarrow 2Ag + 2Cl^-$
Overall	$Mg + 2AgCl \rightarrow MgCl_2 + 2Ag$

2. Magnesium/cuprous chloride

Anode	$Mg - 2e \rightarrow Mg^{2+}$
Cathode	$2CuCl + 2e \rightarrow 2Cu + 2Cl^-$
Overall	$Mg + 2CuCl \rightarrow MgCl_2 + 2Cu$

3. Magnesium/lead chloride

Anode	$Mg - 2e \rightarrow Mg^{2+}$
Cathode	$PbCl_2 + 2e \rightarrow Pb + 2Cl^-$
Overall	$Mg + PbCl_2 \rightarrow MgCl_2 + Pb$

4. Magnesium/cuprous iodide, sulfur

Anode	$Mg - 2e \rightarrow Mg^{2+}$
Cathode	$Cu_2I_2 + 2e \rightarrow 2Cu + 2I^-$
Overall	$Mg + Cu_2I_2 \rightarrow MgI_2 + 2Cu$

5. Magnesium/cuprous thiocyanate, sulfur

Anode	$Mg - 2e \rightarrow Mg^{2+}$
Cathode	$2CuSCN + 2e \rightarrow 2Cu + 2SCN^-$
Overall	$Mg + 2CuSCN \rightarrow Mg(SCN)_2 + 2Cu$

6. Magnesium/manganese dioxide

Anode	$Mg - 2e \rightarrow Mg^{2+}$
Cathode	$2MnO_2 + H_2O + 2e \rightarrow Mn_2O_3 + 2OH^-$
Overall	$Mg + 2MnO_2 + H_2O \rightarrow Mn_2O_3 + Mg(OH)_2$

A side reaction also occurs between the magnesium anode and the aqueous electrolyte, resulting in the formation of magnesium hydroxide, hydrogen gas, and heat.

$$Mg + 2H_2O \rightarrow Mg(OH)_2 + H_2$$

In immersion-type batteries the hydrogen evolved creates a pumping action which helps purge the insoluble magnesium hydroxide from the battery. Magnesium hydroxide remaining within a cell can fill the space between the electrodes which can become devoid of electrolyte, prevent ionic flow, and cause premature cell and battery failure.

The heat evolved improves the performance of immersion-type batteries; it enables dunk-type batteries to operate at low ambient temperatures and forced-flow batteries to operate at high current densities.

Those cathodes containing sulfur exhibit a higher potential versus magnesium than cathodes possessing only the prime depolarizer. During discharge the sulfur probably reacts with the highly active copper formed when the prime depolarizer is reduced producing a copper sulfide, thus accounting for the fact that no copper is observed at end of discharge. This reaction may also prevent copper from plating out on the magnesium, thus deterring premature voltage drop. In those cases where the battery is allowed to discharge past the point where all prime depolarizer is gone and magnesium is present, hydrogen sulfide can be produced. Hydrogen sulfide can also result if the cell is short-circuited.

17.3 TYPES OF WATER-ACTIVATED BATTERIES

Water-activated batteries are manufactured in the following basic types:

1. *Immersion* batteries are designed to be activated by immersion in the electrolyte. They have been constructed in sizes to produce from 1.0 V to several hundred volts at currents up to 50 A. Discharge times can vary from a few seconds to several days. A typical immersion-type water-activated battery is shown in Fig. 17.1.

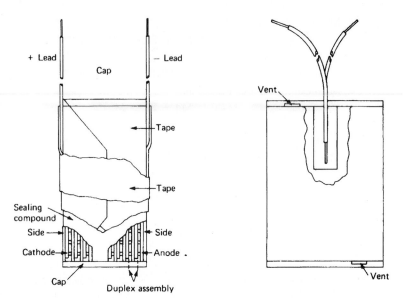

FIGURE 17.1   Seawater battery, immersion type.

2. *Forced-flow* batteries are designed for use as the power source for electric torpedoes. The name is derived from the fact that seawater is forced through the battery as the torpedo is driven through the water. Because of the heat generated during discharge and electrolyte recirculation, these systems can perform at current densities up to 500 mA/cm² of cathode surface area. Batteries containing from 118 to 460 cells which will produce from 25 to 460 kW of power have been developed. Discharge times are about 10–15 min. A diagrammatic representation of a torpedo battery and a torpedo battery with recirculation voltage control is shown in Fig. 17.2.

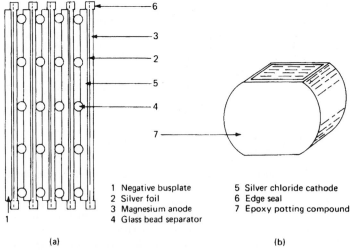

1 Negative busplate	5 Silver chloride cathode
2 Silver foil	6 Edge seal
3 Magnesium anode	7 Epoxy potting compound
4 Glass bead separator	

(a)   (b)

FIGURE 17.2   Diagrammatic representation of torpedo battery construction. (*a*) Cell construction. (*b*) Battery configuration.

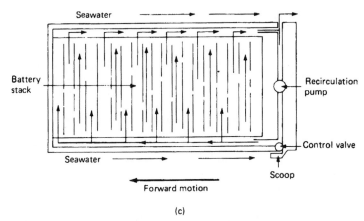

(c)

FIGURE 17.2 (*c*) Recirculation voltage control (*Continued*).

3. *Dunk-type* batteries are designed with an absorbent separator between the electrodes and are activated by pouring the electrolyte into the battery, where it is absorbed by the separator. Batteries of this type have been designed to produce from 1.5 to 130 V at currents up to about 10 A. Lengths of discharge vary from about 0.5 to 15 h. Figure 17.3 is a diagrammatic representation of a magnesium/cuprous chloride battery used in radiosonde applications. A pile-type construction is used. A sheet of magnesium is separated from the cuprous chloride cathode by a porous separator which also serves to retain the electrolyte. The cathode is a pasted type made by applying a paste of powdered cuprous chloride and a liquid binder onto a copper grid or screen. The assembly is taped together to form the battery. The batteries are also made in spiral or jelly-roll design. (See Figure 17.22 for an illustration of this battery.)

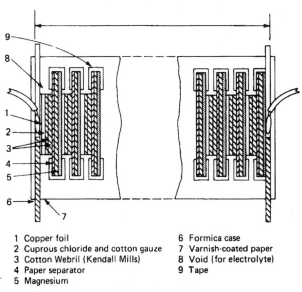

1 Copper foil	6 Formica case
2 Cuprous chloride and cotton gauze	7 Varnish-coated paper
3 Cotton Webril (Kendall Mills)	8 Void (for electrolyte)
4 Paper separator	9 Tape
5 Magnesium	

FIGURE 17.3 Diagrammatic representation of magnesium/cuprous chloride dunk-type battery. (*Courtesy of Eagle-Picher Industries.*)

17.4 CONSTRUCTION

Water-activated cells consist of an anode, a cathode, a separator, terminations, and some form of encasement. A battery consists of a multiplicity of cells connected in series or series-parallel. Such an assembly requires a method to connect the cells in the desired configuration plus a method to control leakage currents. The voltage of a cell depends primarily on the electrochemical system involved. To increase voltage, a number of cells must be connected in series. The capacity of a cell in ampere-hours is primarily dependent on the quantity of active material in the electrodes. The ability of a cell to produce a given current at a usable voltage depends on the area of the electrode. To decrease current density so as to increase load voltage, the electrode area must be increased. Power output depends on the temperature and salinity of the electrolyte. Power output can be increased by increasing the temperature or the salinity of the electrolyte.

The basic components of a single cell, a duplex assembly for connecting cells in series, and a finished battery are illustrated in Figs. 17.4, 17.5, and 17.1, respectively.[12–16] The illustrations represent batteries designed for use by immersion in the electrolyte as contrasted to a dunk-type (radiosonde) battery, which is activated by pouring the electrolyte into the battery, or a forced-flow electric torpedo battery. The construction principles with slight variations are similar in all cases.

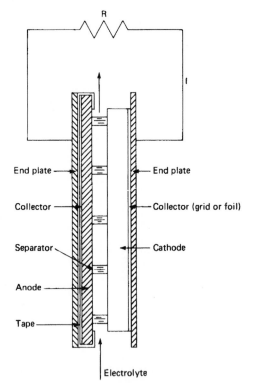

FIGURE 17.4 Basic water-activated cell.

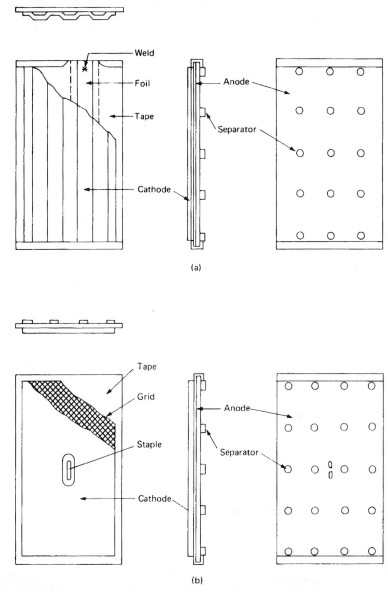

FIGURE 17.5 Duplex electrode assemblies. (*a*) Silver. (*b*) Nonsilver.

17.4.1 Components

A more detailed description of the various cell and battery components and construction elements follows.

Anode (*Negative Plate*). The anode is made from sheet magnesium. Magnesium AZ61A is preferred because it tends to sludge and polarize less. In some cases AZ31B alloy is used; however, this alloy gives slightly lower voltage, polarizes at high current densities, and sludges more. In recent years magnesium alloys AP65 and MTA75 have been developed and evaluated. These are high-voltage alloys giving load voltages of 0.1–0.3 V higher than AZ61A. MTA75 is a higher-voltage alloy than AP65. These alloys sludge more; however, under some forced-flow discharge conditions, the sludging problem may be controlled. These alloys are not used extensively in the United States; however, they are used in the United Kingdom and Europe in electric torpedo batteries. Composition ranges of these alloys are shown in Table 17.2.

TABLE 17.2   Composition Range for Battery Plate Alloys

Element	AZ31		AZ61		AP65		MELMAG 75	
	% Min.	% Max.	% Min.	% Max.	% Min.	% Max.	% Min.	% Max.
Al	2.5	3.5	5.8	7.2	6.0	6.7	4.6	5.6
Zn	0.6	1.4	0.4	1.5	0.5	1.5	—	0.3
Pb	—	—	—	—	4.4	5.0	—	—
Tl	—	—	—	—	—	—	6.6	7.6
Mn	0.15	0.7	0.15	0.25	0.15	0.30	—	0.25
Si	—	0.1	—	0.05	—	0.3	—	0.3
Ca	—	0.04	—	0.3	—	0.3	0.3	—
Cu	—	0.05	0.05	0.05	0.005	—	—	—
Ni	—	0.005	—	0.005	—	0.005	—	0.005
Fe	—	0.006	—	0.006	—	0.010	—	0.006

Cathode (***Positive Plate***).   The cathode consists of a depolarizer and a current collector. These depolarizers are powders and are nonconductive. In order for the depolarizer to function, a form of carbon is added to impart conductivity; a binder is added for cohesion, and a metal grid is used as a current collector, a base for the cathode, to facilitate intercel connections and battery terminations. Possible cathode formulations are shown in Table 17.3[1,3–5,8]

TABLE 17.3   Cathode Compositions

	Silver chloride[1]	Cuprous iodide[5,6]	Cuprous thiocyanate[8]	Lead chloride[4]	Cuprous chloride
Depolarizer, %/w	100	73	75–80	80.7–82.5	95–100
Sulfur, %/w	—	20	10–12	—	—
Additive, %/w	—	—	0–4	2.3–4.4	—
Carbon, %/w	—	7	7–10	9.6–9.8	—
Binder, %/w	—	—	0–2	1.5–1.6	0–5
Wax, %/w	—	—	—	3.8	—

Silver chloride is a special case. Silver chloride can be melted, cast into ingots, and rolled into sheet stock in thicknesses from about 0.08 mm up. Since this material is malleable and ductile, it can be used in almost any configuration. Silver chloride is nonconductive and is made conductive by superficially reducing the surface to silver by immersion in a photographic developing solution. No base grid need be used with silver chloride.

Nonsilver cathodes are usually prismatic in shape and are flat. Silver chloride cathodes are used flat and corrugated in many configurations.

Separators.   Separators are nonconductive spacers placed between the electrodes of immersion- and forced-flow-type batteries to form a space for free ingress of electrolyte and egress of corrosion products. Separators in the form of disks, rods, glass beads, or woven fabrics may be used.[13–14]

Dunk-type batteries utilize a nonwoven, absorbent, nonconductive material for the dual purpose of separating the electrodes and absorbing the electrolyte.

Intercell Connections.   In a series-arranged battery of pile construction, the anode of one cell is connected to the cathode of the adjacent cell. To accomplish this without producing a short-circuited cell, an insulating tape or film is placed between the electrodes on nonsilver batteries. For silver batteries silver foil is used alone or in conjunction with an insulating tape.

For nonsilver cells the connection is made by stapling the electrodes together through the insulator.[15] For silver cells, the silver chloride, surface-reduced to silver, is heat-sealed to silver foil, which has been previously welded to the anode. Where large surface areas are involved, contact between silver and silver foil can be made by pressure alone.

Terminations.   For silver chloride cathodes the lead is soldered directly to silver foil, which has been heat-sealed to one surface of the silver chloride. Leads are soldered directly to the collector grid of nonsilver cathodes or soldered to a piece of copper foil, which has been stapled to the collector grid.

The anode connection is made by soldering the lead to silver foil, which has been welded to the anode, or by welding directly to the anode.

Encasement.   The battery encasement must effectively rigidize the battery and provide openings at opposite ends to allow free ingress and egress of electrolyte and corrosion products.

The periphery of the battery must be sealed in such a manner that the cells contact the external electrolyte only at the openings provided at the top and bottom of the battery. The encasement can be accomplished by using premolded pieces, caulking compounds, epoxy resins, an insulating sheet, or hot-melt resins.[13–16] For single-batteries these precautions are not necessary.

17.4.2 Leakage Current

All the cells in the immersion- and forced-flow-type batteries operate in a common electrolyte. Since the electrolyte is conductive and continuous from cell to cell, conductive paths exist from each point in a battery to every other point. Current will flow through these conductive paths to points of different potential. This current is referred to as "leakage current" and is in addition to the current flowing through the load. Electrodes must be designed to compensate for these leakage currents.

Leakage currents for a small number of cells can be reduced by increasing the resistance path from a cell to the common electrolyte or that of the common electrolyte between adjacent cells. Leakage currents for a large number of cells can be reduced by increasing the resistance of the common electrolyte external to the individual cells.

By construction the conducting paths from cell to cell are made as long as possible. In many instances the negative or positive of the battery is connected to an external metal surface. Leakage currents flow from the battery to this surface. These leakage currents are controlled by placing a cap containing a slot over the battery openings. If one terminal is connected to an external conductive surface, the slot in the cap is opened to the electrolyte only on that side of the battery. Where neither terminal is connected to an external conductive surface, either end of the cap may be opened; however only that on one side of the battery should be opened.

The resistance (ohms) of the slot in the cap may be calculated using the formula

$$R = p\,\frac{l}{a}$$

where R = resistance, Ω
l = length of slot, cm
a = cross-sectional area of slot, cm^2
p = resistance of electrolyte for temperature and salinity in which battery is operating, $\Omega \cdot$ cm

For dunk-type batteries the electrolyte continuity from cell to cell is broken when the electrolyte is absorbed in the separator. The excess is poured off the battery or spun away from the cells by some external force applied to the battery.

17.4.3 Electrolyte

Seawater-activated batteries are designed to operate in an infinite electrolyte, namely, the oceans of the world. However, for design, development, and quality control purposes, it is not practical to use ocean water. Thus it is common practice throughout the industry to use a simulated ocean water. A commercial product, composed of a blend of all the ingredients required, simplifies the manufacture of simulated ocean water test solutions.

Dunk-type batteries, activated by pouring the electrolyte into the battery where it is absorbed by the separator, can utilize water or seawater when the temperature is above freezing. At lower temperatures special electrolytes can be used. The use of a conducting aqueous electrolyte will result in faster voltage buildup. However, the introduction of salts in the electrolyte will increase the rate of self-discharge.

17.5   PERFORMANCE CHARACTERISTICS

17.5.1 General

A summary of the performance characteristics of the major water-activated batteries currently available is given in Table 17.4.

Voltage versus Current Density.   Figures 17.6 and 17.7 are representative voltage versus current density curves for several water-activated battery systems at 35 and 0°C, respectively, using a simulated ocean water electrolyte.

Discharge Curves.   Discharge curves of the magnesium/silver chloride, magnesium/cuprous thiocyanate-sulfur, magnesium/cuprous iodide-sulfur, and magnesium/lead chloride electrochemical systems, discharged continuously through various resistances in simulated ocean water and high and low temperatures and salinities, are shown in Figs. 17.8 to 17.15. These data show the advantageous performance of the silver chloride system.

Service Life.   The capacities per unit of weight versus the average power output of these same electrochemical systems, at high and low temperatures and salinities, are shown in Fig. 17.16 and 17.17.

TABLE 17.4 Performance Characteristics of Water-Activated Batteries

Cathode	Silver chloride	Lead chloride	Cuprous iodide	Cuprous thiocyanate	Cuprous chloride[a]
Anode			Magnesium		
Electrolyte		Tapwater, seawater, or other conductive aqueous solutions			
Open-circuit voltage, V	1.6–17	1.1–1.2	1.5–1.6	1.5–1.6	1.5–1.6
V per cell at 5 mA/cm^2[b]	1.42–1.52	0.90–1.06	1.33–1.49	1.24–1.43	1.2–1.4
Activation, s:					
35°C[c]	<1	<1	<1	<1	
RT[d]	—	—	—	—	1–10
0°C[e]	45–90	45–90	45–90	45–90	
Internal resistance, Ω[f]	0.1–2	1–4	1–4	1–4	2
Ah/g cath. theor.[g]	0.187	0.193	0.141	0.220	0.271
Usable capacity, % of theoretical	60–75	60–75	60–75	60–75	60–75
Wh/kg	100–150	50–80	50–80	50–80	50–80
Wh/L	180–300	50–120	50–120	50–120	20–200
Operating temperatures, °C[h]			−60 to +65		

[a] All but cuprous chloride are immersion type. Cuprous chloride is dunk type.
[b] See voltage vs. current density curves.
[c] Battery preconditioned at +55°C, then immersed in simulated ocean water of 3.6 wt.%
[d] Electrolyte at room temperature poured into battery and absorbed by separator.
[e] Battery preconditioned at −20°C, then immersed in simulated ocean water of 1.5 wt. %.
[f] Depends on battery design.
[g] 100% active material.
[h] Following activation at room temperature.

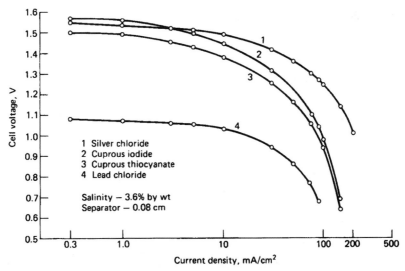

FIGURE 17.6 Representative cell voltages vs. current density at 35°C.

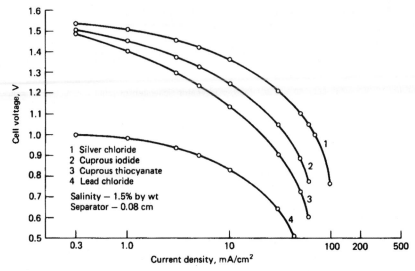

FIGURE 17.7   Representative cell voltages vs. current density at 0°C.

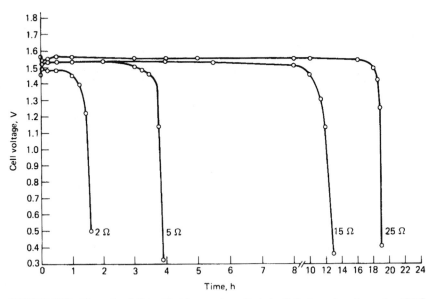

FIGURE 17.8   Magnesium/silver chloride seawater-activated cell discharged continuously at 35°C in simulated ocean water, 3.6% salinity.

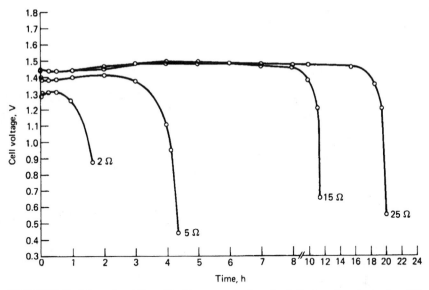

FIGURE 17.9 Magnesium/silver chloride seawater-activated cell discharged continuously at 0°C in simulated ocean water, 1.5% salinity.

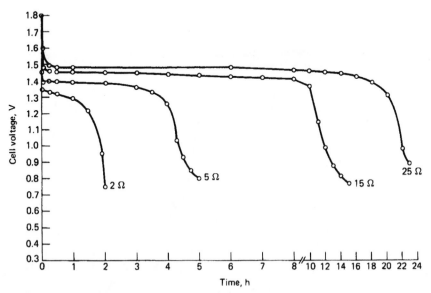

FIGURE 17.10 Magnesium/cuprous thiocyanate seawater-activated cell discharged continuously at 35°C in simulated ocean water, 3.6% salinity.

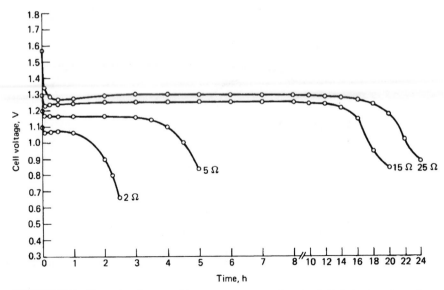

FIGURE 17.11   Magnesium/cuprous thiocyanate seawater-activated cell discharged continuously at 0°C in simulated ocean water, 1.5% salinity.

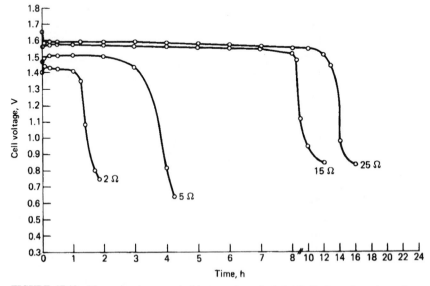

FIGURE 17.12   Magnesium/cuprous iodide seawater-activated cell discharged continuously at 35°C in simulated ocean water, 3.6% salinity.

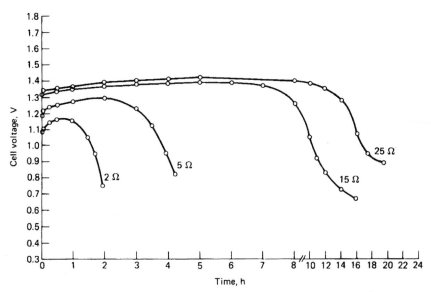

FIGURE 17.13 Magnesium/cuprous iodide seawater-activated cell discharged continuously at 0°C in simulated ocean water, 1.5% salinity.

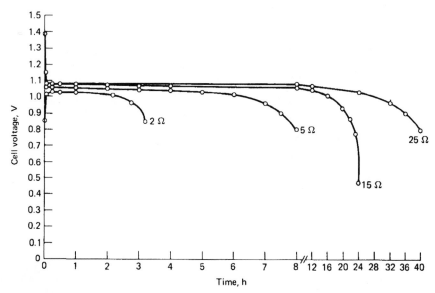

FIGURE 17.14 Magnesium/lead chloride seawater-activated cell discharged continuously at 35°C in simulated ocean water, 3.6% salinity.

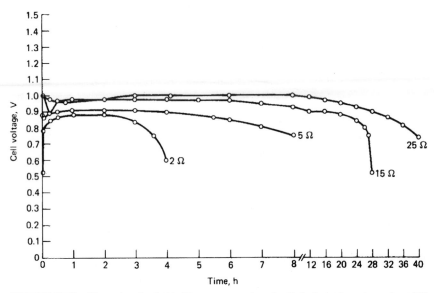

FIGURE 17.15 Magnesium/lead chloride seawater-activated cell discharged continuously at 0°C in simulated ocean water, 1.5% salinity.

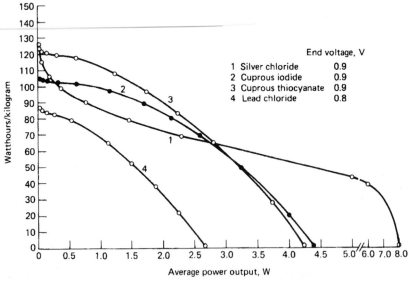

FIGURE 17.16 Capacity vs. power output of seawater-activated cells discharged continuously at 35°C in simulated ocean water, 3.6% salinity.

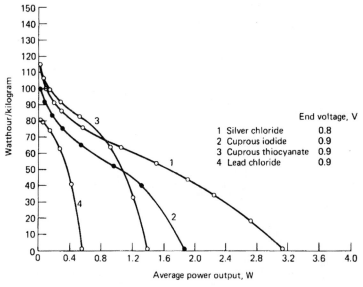

FIGURE 17.17 Capacity vs. power output of seawater-activated cells discharged continuously at 0°C in simulated ocean water, 1.5% salinity.

17.5.2 Immersion-Type Batteries

The performance of these same systems, designed as immersion-type batteries to meet the physical, electrical, and environmental specifications listed in Table 17.5, is shown in Figs. 17.18 to 17.20. The performance characteristics are summarized in Table 17.6.

TABLE 17.5 Performance Specifications for Seawater-Activated Battery

Load	$80 \pm 2 \, \Omega$	
Life	9 h	
Voltage	15.0 V in. from 90 s to 9 h	
	19.0 V max.	
Activation*	60 s to 13.5 V	
	90 s to 15.0 V	
Battery size:	Silver	Nonsilver
Height, cm	7.7 max.	10.6 max.
Width, cm	5.7 max.	7.6 max.
Thickness, cm	4.2 max	5.7 max.
Weight, g	255 ± 14	482 ± 85
Environmental:		
Storage	From -60 to $+70°C$ for 5 years†	
	90 days at -50 to $+40°C$ at 90% RH (see Table 17.5)	
	10 days per MIL-T-5422E (see Table 17.5)	
Vibration, Hz	5–500	
Electrolyte:		
Low temperature	Ocean water of 1.5% salinity by weight at $0 \pm 1°C$	
High temperature	Ocean water of 3.6% salinity by weight at $+34 \pm 1°C$	

*Battery preconditioned at $-20°C$ prior to immersion in ocean water of 1.5% salinity by weight at $0 \pm 1°C$.

†In equipment packed in sealed plastic container with appropriate desiccant.

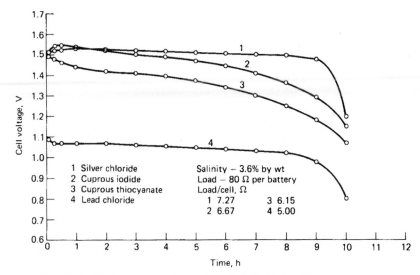

FIGURE 17.18   Discharge curves of seawater-activated batteries at 35°C.

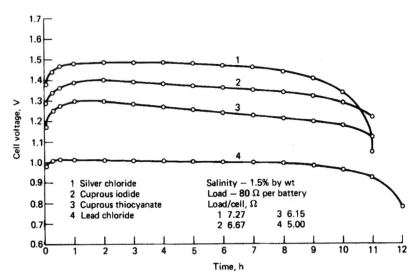

FIGURE 17.19   Discharge curves of seawater-activated batteries at 0°C.

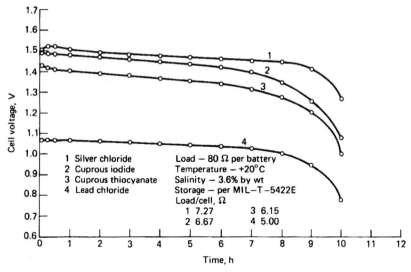

FIGURE 17.20   Discharge curves of seawater-activated batteries, 10-day humidity.

TABLE 17.6   Performance Summary of Seawater-Activated Batteries

	Silver chloride	Cuprous iodide	Cuprous thiocyanate	Lead chloride
Number of cells	11	12	13	16
Battery dimensions:				
Height, cm	7.5	9.8	10.2	10.5
Width, cm	5.5	7.6	7.4	7.5
Thickness, cm	3.9	4.4	5.7	4.5
Weight, g	252	516	478	458
Activation:				
Low temp.:				
To 13.5 V, s	<15	<15	<15	<15
To 15.0 V, s	60	60	60	15
High temp.:				
To 15.0 V, s	<1	<1	<1	<1
Life:				
High temp., h	9.67	9.4	9.3	9.5
Low temp., h	9.80	10.3	10.3	10.7
Load resistance (per cell), Ω*	7.27	6.67	6.15	5.0
Cutoff voltage (per cell), V*	1.364	1.25	1.154	0.9375
Average current, A	0.206	0.220	0.236	0.219
Average volts per cell, V*	1.497	1.463	1.378	1.048
Wh/L	204	110	90	100
Wh/kg	130	70	79	75

   *As each battery system contains a different number of cells, cell load resistances and cell voltages are different for each battery.

17.5.3 Forced-Flow Batteries

With the development of the recirculation system in which the inflow of fresh electrolyte can be controlled, thereby maintaining the temperature and conductivity of the electrolyte, the performance of electric torpedo batteries has been improved markedly. With recirculation and flow control, a recirculation pump (see Fig. 17.2) and a voltage-sensing mechanism are added to the battery system. By this method the temperature of the battery and the conductivity of the seawater electrolyte increase. Since battery voltage increases directly with temperature and conductivity, it is possible to control the output of the battery by controlling the intake of electrolyte by means of the voltage-sensing mechanism.

The performance of one type of torpedo battery with and without recirculation voltage control is shown in Fig. 17.21.[17] The blocked-in area represents the limits within which an electric torpedo battery with recirculation and flow control will perform when discharged under any of the conditions shown by the three individual curves. All voltages pertinent to the start and finish of the battery are shown by the three individual curves.

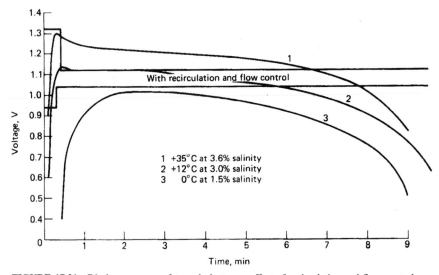

FIGURE 17.21   Discharge curves of torpedo battery—effect of recirculation and flow control.

17.5.4 Dunk-Type Batteries

Magnesium/Cuprous Chloride Batteries.   The magnesium/cuprous chloride battery was widely used in applications requiring low-temperature performance, such as radiosondes, having replaced the more expensive magnesium/silver chloride system in applications where weight and volume are not critical. Figure 17.22 illustrates a typical magnesium/cuprous chloride battery. The pile-type construction shown in Fig. 17.3 is used.

The battery is activated by filling it with water, and full voltage is reached within 1 to 10 min. The battery is best suited for discharge at about the 1–3-h rate at temperatures from +60 to −50°C after activation at room temperature. Overheating and dry-out will occur on high current drains, and self-discharge limits the life after activation. For best service these batteries should be put into use soon after activation. The heat that is developed during discharge can be used to advantage in batteries which are operated at low temperatures; therefore, the energy output varies little with decreasing temperature. Figure 17.23 shows the discharge curve for this battery at various temperatures. Figure 17.24 gives some typical discharge curves for this type of battery with a similar design at various discharge loads.

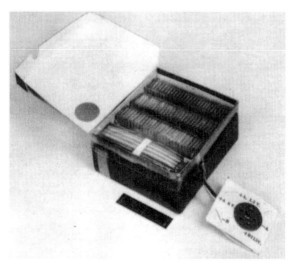

FIGURE 17.22 Magnesium/cuprous chloride radiosonde battery. Size: $10.2 \times 11.7 \times 1.9$ cm; weight: 450 g, rated capacity: A_1 section—1.5 V, 0.3 Ah; A_2 section—6.0 V, 0.4 Ah; B section—115 V, 0.08 Ah.

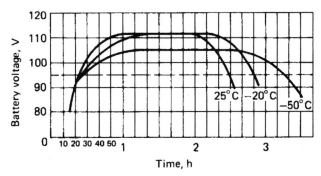

FIGURE 17.23 Discharge curves of magnesium/cuprous chloride radiosonde battery, 115-V section; discharge load: 3050 Ω.

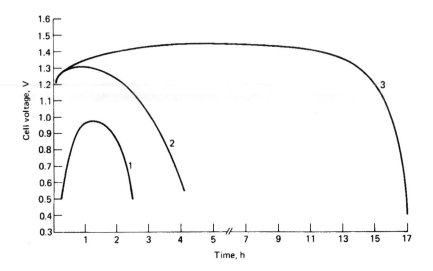

Cell no.	Load, Ω	Dimensions, cm			
		Volume	Length	Height	Thickness
1	2.5	10.2	8.2	2.5	0.5
2	8.0	2.5	2.2	3.8	0.3
3	125	1.3	2.0	2.0	0.3

FIGURE 17.24 Discharge curves of magnesium/cuprous chloride water-activated batteries at 20°C; electrolyte: tapwater.

Magnesium/Manganese Dioxide Battery.   This reserve battery consists of a magnesium anode and a manganese dioxide cathode.[10,18] It is activated by pouring an aqueous magnesium perchlorate electrolyte into the cells of the battery, where it is absorbed by the separators. Electrolyte absorption occurs within a few seconds at 0°C or above, but 3 min or more are required at −40°C due to the viscosity of the electrolyte.

The battery can deliver between 80 and 100 Wh/kg over the temperature range of −40 to +45°C at the 10–20-h discharge rate. Over 75% of the battery's fresh capacity is available after 7 days' activated stand at 20°C and 4 days' storage at 45°C. Typical discharge curves are shown in Fig. 17.25 for a five-cell 10-Ah battery, weighing about 1 kg and being 655 cm³ in size.

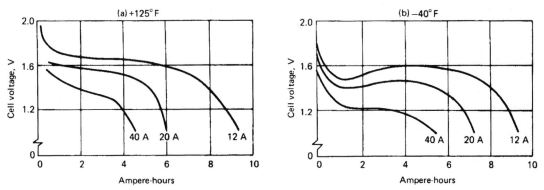

FIGURE 17.25 Typical discharge curves of magnesium/manganese dioxide cell, 10-Ah size. (*Courtesy of Eagle-Picher Industries.*)

17.6 BATTERY APPLICATIONS

Water-activated batteries can be viable candidates as the power source for many types of equipment. The choice of which battery to use becomes one of economics. By proper design all will perform similarly. Where high current densities are required and cost is secondary, the magnesium/silver chloride system is best. All can be used as immersion or dunk-type batteries; however, all but the magnesium/cuprous chloride system will withstand long storage times at high temperatures and high humidities. At the present state of the art only the magnesium/silver chloride system is suitable for use in forced-flow batteries.

17.6.1 Water-activated Batteries for Aviation and Marine Lifejacket Lights

The magnesium/cuprous chloride water-activated battery system is being used in FAA and U.S. Coast Guard approved aviation and marine lifejacket lifts. A typical light is shown in Fig. 17.26.

The single cell battery has a cathode approximately 5 mm thick with a footprint of 7.25 by 2 mm. Table salt is added to the cathode mix[19] to obtain an adequate voltage in freshwater. (The holes in the battery case are optimized to maintain electrolyte salinity while allowing flushing of discharge products). After being mixed while heated, and then cooled and re-chopped, the powder is pressed and reheated in an automatic hydraulic press. The cathode is pressed with a titanium wire current collector, which is wire brushed before manufacture to remove oxide buildup.

The cell is constructed with two anodes, each with the same footprint as the cathode, connected in parallel and placed on either side of the cathode. The anodes are AZ61 electrochemical magnesium sheet.

Typical cell voltage at a 220 to 240 mA (C/12) discharge (against a miniature incandescent lamp) starts at 1.77V in salt water and goes down gradually to about 1.65V before a sharp voltage drop signaling the end of discharge. Voltages in fresh water are about 0.1V lower. Total capacity is about 3000 mAh.

A battery, with two cells wired in series for international marine use, uses a AT61 sheet because of the requirement for higher voltage. In salt water, the cell voltage at a 340 mA (C/8) discharge (against a highly efficient-gas-filled miniature lamp) is as high as 1.87V early in the discharge and drops to about 1.8V after 8 hours. Again, the voltage is fresh water voltage is about 100 mV less per cell. This discharge is shown in Fig. 17.27.

FIGURE   17.26 Lifejacket   light,   using magnesium/cuprous chloride water-activated battery. (*Courtesy of Electric Fuel Ltd.*[20])

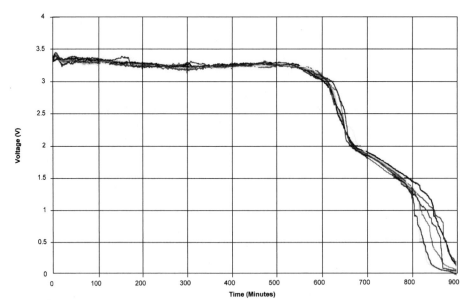

FIGURE 17.27   Typical discharge of 6 WAB-MX8 batteries in fresh tap water at 330 mA (*Courtesy of Electric Fuel Ltd.*[20])

Because the salt added to the cathode makes the cathode even more hygroscopic than it would otherwise be, the batteries are preferably stored with a removable pull-plug used to seal the holes in the battery case.

The characteristics of the lifejacket lights are given in Table 17.7.

TABLE 17.7 Characteristics of Lifejacket Lights

Electric fuel model no.	Nominal voltage, V	Nominal size, cm			Nominal discharge capacity		
		Length	Width	Height	Time	Wh	Normal usage mode
WAB-H12	1.7	2.9	1.6	9.3	12h	4.4	Aviation/Marine Lifejacket light
WAB-H18	1.7	2.9	1.6	9.3	8h+	3.3	Aviation Lifejacket light
WAB-MX8	3.6	3.1	3.3	9.5	8h+	10.7	Marine Lifejacket light

Source: Electric Fuel Ltd.[20]

17.6.2 Magnesium/Silver Chloride Batteries

Figure 17.28 illustrates two of the magnesium/silver chloride batteries currently manufactured. These batteries are used in the following types of applications:

Lifeboat emergency equipment on commercial airlines

Sonobuoys

Radio and light beacons

Underwater Ordnance

Radiosonde units—balloon transport equipment, high altitude low ambient temperature operation.

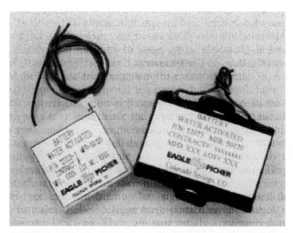

FIGURE 17.28 Magnesium/silver chloride batteries; 12023-1 and 12073.

17.7   BATTERY TYPES AND SIZES

Although "standard lines" of water-activated batteries were once manufactured, most batteries now are designed and manufactured for specific applications. Tables 17.8 and 17.9 list some of the standard and special purpose magnesium/cuprous chloride and magnesium/silver chloride batteries that were manufactured. Of these, only the two batteries illustrated in Fig. 17.28 are currently manufactured.

TABLE 17.8   Magnesium/Cuprous Chloride Water-Activated Batteries

E-P number	Other designation	Nominal voltage/ selection, V	Nominal size, cm			Nominal discharge capacity		Normal usage mode
			Length	Width	Height	Time	Wh	
MAP-12037	PIBAL	3.0	1.3	3.2	5.1	30 min	0.8	Airborne, lighting type
MAP-12051	—	18.0	6.8	3.8	5.7	120 min	2.16	Airborne, radiosonde
MAP-12053	BA-259	A-1.5 B-6.0 C-115.0	11.7	10.2	6.0	A-90 min B-90 min C-90 min	0.34 1.89 650.4	Airborne, radiosonde
MAP-12060	—	18.0	5.1	5.4	5.1	120 min	5.4	Airborne, radiosonde
MAP-12061	—	22.5	5.1	7.0	5.1	90 min	7.59	Airborne, radiosonde
MAP-12064	BA-253	6.0	10.2	3.8	3.8	45 min	2.25	Airborne, lighting, type
MAP-12071	—	20.0	6.3	7.6	16.0	8.1 h	53.46	Submerged, buoy system

Source: Eagle-Picher Technologies, LLC[21]

TABLE 17.9   Magnesium/Silver Chloride Water-Activated Batteries

E-P number	Other designation	Nominal voltage, V	Approximate size, cm			Nominal discharge capacity		
			Length	Width	Height	Time	Wh	Ah
MAP-2023-1	Squib firing battery	5.5	5.1	2.5	5.4	1 min	0.315	0.0572
MAP-12062	—	48	12.1	dia.	33	20 min	400	8.33
MAP-12065	—	4.5	6.3	6.7	13.9	50 h	157.5	35
MAP-12066	—	7.5	5.1	5.1	16.5	14 h	138	18.4
MAP-12067	MK-72 Squib firing battery	0.75	2.8	dia.	2.5	13 s	0.0010	0.0014
MAP-12069	—	10	7.6	2.5	8.9	6 h	1.5	0.15
MAP-12070	—	12	5.1	2.6	10	9 h	53.2	4.44
MAP-12073	—	14.5	7.6	2.8	5.1	15 h	14.55	1
MAP-12074	—	10.5	4.1	5.1	25	48 h	95.35	9

Source: Eagle-Picher Technologies, LLC[21]

REFERENCES

1. National Defense Research Committee, *Final Report on Seawater Batteries,* Bell Telephone Laboratories, New York, 1945.

2. L. Pucher, "Cuprous Chloride-Magnesium Reserve Battery," *J. Electrochem. Soc.* **99:**203C (1952).

3. B. N. Adams, "Batteries," U.S. Patent 2,322,210, 1943.

4. H. N. Honer, F. P. Malaspina, and W. J. Martini, "Lead Chloride Electrode for Seawater Batteries," U.S. Patent 3,943,004, 1976.

5. H. N. Honor, "Deferred Action Battery," U.S. Patent 3,205,896, 1965.

6. N. Margalit, "Cathodes for Seawater Activated Cells," *J. Electrochem. Soc.* **122:**1005 (1975).

7. J. Root, "Method of Producing Semi-Conductive Electronegative Element of a Battery," U.S. Patent 3,450,570, 1969.

8. R. F. Koontz and L. E. Klein, "Deferred Action Battery Having an Improved Depolarizer," U.S. Patent 4,192,913, 1980.

9. E. P. Cupp, "Magnesium Perchlorate Batteries for Low Temperature Operation," *Proc. 23d Annual Power Sources Conf.,* Electrochemical Society, Pennington, N.J., 1969, p. 90.

10. N. T. Wilburn, "Magnesium Perchlorate Reserve Battery," *Proc. 21st Annual Power Sources Conf.,* Electrochemical Society, Pennington, N.J., 1967, p. 113.

11. W. A. West-Freeman and J. A. Barnes, "Snake Battery; Power Source Selection Alternatives," NAVSWX TR 90-366, Naval Surface Warfare Center, Silver Spring, Md., 1990.

12. M. G. Medeiros and R. R. Bessette, "Magnesium-Solution Phase Catholyte Seawater Electrochemical System," *Proc. 39th Power Sources Conf.,* Cherry Hill, N.J., June 2000, p. 453.

13. M. E. Wilkie and T. H. Loverude, "Reserve Electric Battery with Combined Electrode and Separator Member," U.S. Patent 3,061,659, 1962.

14. K. R. Jones, J. L. Burant, and D. R. Wolter, "Deferred Action Battery," U.S. Patent 3,451,855, 1969.

15. H. N. Honor, "Seawater Battery," U.S. Patent 3,966,497, 1976.

16. H. N. Honer, "Multicell Seawater Battery," U.S. Patent 2,953,238, 1976.

17. J. F. Donahue and S. D. Pierce, "A Discussion of Silver Chloride Seawater Batteries," Winter Meeting, American Institute of Electrical Engineers, New York, 1963.

18. H. R. Knapp and A. L. Almerini, "Perchlorate Reserve Batteries," *Proc. 17th Annual Power Sources Conf.,* Electrochemical Society, Pennington, N.J., 1963, p. 125.

19. U.S. Patent No. 5,424,147.

20. Electric Fuel, Ltd., Beit Shemesh, Israel.

21. Eagle-Picher Technologies, LLC, Power Systems Dept., Colorado Springs, CO.

CHAPTER 18
ZINC/SILVER OXIDE RESERVE BATTERIES

James M. Dines and Elliott M. Morse
Revised by Curtis Brown

18.1 GENERAL CHARACTERISTICS

An important reserve battery, particularly for missile and aerospace applications, is the zinc/silver oxide electrochemical system, which is noted for its high-rate capability and high energy density. The cell is designed with thin plates and large-surface-area electrodes, which augment its high-rate and low-temperature capability and provide a flat discharge characteristic. This design, however, reduces the activated or wet shelf life of the battery, necessitating the use of a reserve-battery design to meet storage requirements.

The zinc/silver oxide electrochemical system was the metallic couple with which Volta demonstrated the possibility of using dissimilar metals in a "pile-type" multicell construction to obtain a substantial electric voltage. The system existed somewhat as a laboratory device until Professor André designed a practical secondary cell early in World War II.

Subsequent to World War II, the U.S. military became interested in a dry-charged primary version for use in airborne electronics and missiles because of its very high energy output per unit weight and volume and high-rate capability. The ultimate result of this interest was the development of lightweight batteries for the aerospace industry, both military and civilian. The entire manned space program was keyed to zinc/silver oxide reserve batteries as the power sources for the various flight vehicles.

Zinc/silver oxide reserve batteries are divided into two classes, manually activated and remotely activated. In general the manually activated types are used for space systems and accessible terrestrial applications and are usually packaged in more conventional configurations. The remote or automatically activated types are used principally for weapon and missile systems. This use requires a long period of readiness (in storage), a means for rapid remote activation, and an efficient discharge at high discharge rates, typically in the 2 to 20-min. range. The performance of manually activated types ranges from about 60 to 220 Wh/kg and 120 to 550 Wh/L. For remotely activated types, the specific energy and energy density are reduced because of the self-contained activating device and ranges from about 11 to 88 Wh/kg and 24 to 320 Wh/L.

18.2   *CHEMISTRY*

The electrochemical reactions associated with the discharge of a zinc/silver oxide battery as a primary system are generally considered to proceed as follows. The cathode or positive electrode is silver oxide and may be either Ag_2O (monovalent), AgO (divalent), or a mixture of the two. The anode or negative electrode is metallic zinc, and the electrolyte is an aqueous solution of potassium hydroxide. The chemical reactions and the associated voltages at standard conditions are

$$Zn + 2AgO + H_2O \rightarrow Zn(OH)_2 + Ag_2O \qquad E^0 = 1.815 \text{ V}$$

$$Zn + Ag_2O + H_2O \rightarrow Zn(OH)_2 + 2Ag \qquad E^0 = 1.589 \text{ V}$$

The total cell reaction with 31% KOH electrolyte at 25°C is

$$Zn + AgO + H_2O \rightarrow Zn(OH)_2 + Ag \qquad E = 1.852 \text{ V}$$

18.3   *CONSTRUCTION*

A typical assembly of the manually activated reserve zinc/silver oxide cell is shown in Fig. 18.1. These batteries are designed to be filled with electrolyte, just before use. The conventional cell design is a prismatic container with positive and negative terminals and a combination fill/vent cap. Batteries are formed by connecting single cells in series and packaging them in a unit container. Batteries used in space programs utilize thin-gauge stainless-steel, titanium, magnesium or composite containers to minimize weight.

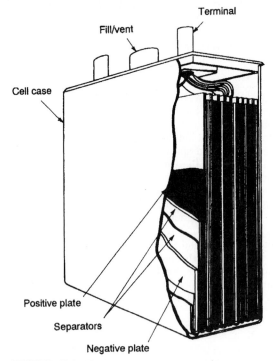

FIGURE 18.1   Typical construction of primary reserve zinc/silver oxide cell. (*Courtesy of Eagle-Picher Industries.*)

18.3.1 Cell Components

The components of a reserve zinc/silver oxide cell consist of the positive plates, the negative plates, and the separators. The components are assembled such that each negative plate is protected from direct contact with the adjacent positive plate by a separator. The cell components are assembled and packaged in a container; the plates can be prepared in either a dry and charged condition or dry and uncharged condition.

Positive Plates.   The positive plates are prepared by applying silver or silver oxide powder to a metallic grid. Copper, nickel, and silver have all been used for grid material, with silver the most prevalent for reasons of electrochemical stability and conductivity. After the silver powder is pressed or sintered to the grid, the plates are electroformed in an alkaline solution, then washed thoroughly and air-dried at a moderate temperature (usually 20 to 50°C). The nominally divalent oxide thus formed is relatively stable at ambient temperatures but tends to lose oxygen and degrade to the monovalent state with increasing temperatures and time. Continuous exposure to high temperatures (70°C) causes reduction to the monovalent oxide in a few months.

Negative Plates.   The negative plates may be prepared by pasting or pressing zinc powder or zinc oxide onto a grid or by electroplating zinc from an alkaline bath to form a very active spongy zinc deposit.

Both positive and negative electrodes may vary in thickness from 0.12 mm as a practical minimum to 2.5 mm maximum for positives and 2 mm maximum for negatives. The extremely thin plates are utilized for very short-life, high-discharge-rate automatically activated batteries; the thick plates are employed in manually activated batteries designed for continuous discharge over several months at very low currents.

Separator Materials.   Typical separator materials used in zinc/silver oxide cells include regenerated cellulose films (cellophane, fiber-reinforced or silver-treated cellophane), nonwoven synthetic fiber mats of nylon, Dacron, and polypropylene, and nonwoven rayon fiber mats. The synthetic fiber mats are frequently placed adjacent to the positives to protect the cellophane from the highly oxidizing influence of that material. The cellophane, a semipermeable film, prevents buildup of particles between plates (while allowing ionic transfer), thus preventing interplate short circuits. The rayon mat absorbs the electrolyte solution and distributes it over the plate surfaces. Cells intended for automatic activation normally are not designed with the film separators for they require too long for complete wetting. The open-mat separators provide sufficient protection from interplate short-circuiting for several hours.

Separator materials are necessary for cell operation because they prevent short circuits, but they also impede current flow, causing an *IR* drop within the cell. Very high discharge rate cells must have very low internal impedance, hence a minimum of separator material. As a result this type of cell is restricted to very short wet-life applications. The semipermeable film is the separator which contributes most to *IR* drop and also to protection against short circuits. Long-life cells may contain five or six layers of cellophane. They are therefore better suited to medium or low discharge rates.

Electrolyte.   The electrolyte used for reserve zinc/silver oxide cells is an aqueous solution of potassium hydroxide. High and medium discharge rate cells use a 31% by weight electrolyte solution because this composition has the lowest freezing point and is close to the minimum resistance which occurs at 28 wt. %. Low-rate cells may use a 40–45% solution since lower rates of hydrolysis of cellulosic separators occur with the higher KOH concentrations.

18.3.2 High- and Low-Rate Designs

A battery intended to be discharged at a 5 to 60 min. rate is considered a high-rate design. These cells are designed primarily to deliver high current and require a large plate surface area. They therefore contain many very thin plates. The separators also have as low an impedance as possible, that is, one or two layers of cellophane versus five or six layers for low-rate cells. The 31% potassium hydroxide electrolyte has a high conductivity and is therefore employed in high-rate cells.

Low-rate batteries are in a class intended for discharge at rates ranging from 10 to 1000 hours with emphasis on high specific energy and energy density. The plates are thick (2 mm), and relatively high impedance separator wraps are used. A higher concentration of electrolyte (40%), which permits a greater Ampere-hour capacity, can also be used. This design configuration also gives a substantial improvement in the activated or wet stand capability of the cell.

18.3.3 Automatically Activated Types

The automatically activated type battery is a class of reserve battery intended for quick preparation for use after an undetermined period subsequent to installation. The very high energy output of the primary zinc/silver oxide system and the use of an integrally designed system for injecting the electrolyte into the cells combine to provide an efficient power source for weapons and other systems requiring a long-term ready state. Figure 18.2 shows a typical automatically activated battery for a missile application.

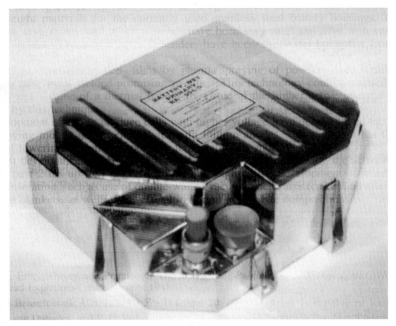

FIGURE 18.2 Zinc/silver oxide primary reserve battery designed for automatic activation. (*Courtesy of Eagle-Picher Industries.*)

Four kinds of activation systems have been utilized in this type of battery for transferring electrolyte from a reservoir to the cells. All the systems depend on gas pressure to move the electrolyte, and the most conventional source of gas is a pyrotechnic device.

The "gas generator" is a small cartridge which contains an ignitable propellant material and an electrically fired ignitor or "match." Figure 18.3 shows the four types of battery design.

The tubular reservoir (Fig. 18.3a) can assume many forms. It is usually coiled around the battery, as shown in Fig. 18.4 (an assembly of a battery with a tubular reservoir), but it can also be formed with 180° bends into a flat shape, or it can be configured to fit into available nonstandard volumes into which a missile battery is often mounted. The tubular reservoir is fitted with foil diaphragms at each end. For activation, the gas generator located at one end can be electrically ignited; the gas causes the diaphragms to break, and the electrolyte is forced into a manifold which distributes it to the cells of the battery. The piston activator (Fig. 18.3b) operates by pushing the electrolyte out of a cylindrical reservoir when a gas generator is fired behind it. The tank activator contains the electrolyte in a variable-geometry tank with a gas generator located at the top. When the gas enters at the top, the electrolyte is forced out through an aperture at the bottom. The system is position-sensitive and will operate properly only when in an upright position relative to the components. The tank-diaphragm activator (Fig. 18.3d) uses a sphere or spheroid tank with a diaphragm attached internally at the major circumference. When the gas generator is fired, the diaphragm moves to the opposite side, forcing the electrolyte out through an aperture in the reservoir side of the tank.

Of the four systems, the tubular system is the most versatile, but in simple battery shapes may be heavier. The piston and diaphragm systems have moving parts and thus can be less reliable; they are also less adaptable to special shapes. The tank is efficient but position-sensitive.

The operating sequence of an automatically activated battery involves: (1) application of ignition current, (2) gas generator burning and associated gas production, (3) rupture of a diaphragm, (4) movement of electrolyte out of the reservoir into the distribution manifold, and (5) filling of the cells with electrolyte. In a typical operation the total sequence involves less than one second. In many applications the electrical load is wired directly to the battery, and so the battery activates under load. Figure 18.5 shows the rise times of the voltage under load (A) and under no-load (B) condition for a battery not used until 6 hours later. The delayed-use battery has film separators, and the slower wetting is reflected by the longer rise time.

Automatically activated batteries suffer a weight and volume penalty compared with manually activated types, but the design permits the use of a high-performance battery when there is no time available to activate manually or the unit is inaccessible. In many applications both conditions exist. The volume penalty is usually about 2 times, and the weight penalty about 1.6 times the basic battery. Most automatically activated battery designs utilize an integral electric heater. The heater maintains the electrolyte at about 40°C or at a temperature which will raise a cold battery to 40°C when activation occurs. The use of heaters permits the design of batteries which can meet close voltage tolerances, thus improving the capability of the weapons' electric and electronic systems when operating over a wide temperature range.

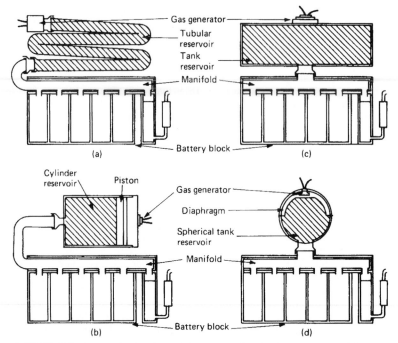

FIGURE 18.3   Schematic drawings of four types of activation systems used in automatically activated batteries. (*a*) Tubular reservoir. (*b*) Piston activator. (*c*) Tank activator. (*d*) Tank-diaphragm activator. (*Courtesy of Eagle-Picher Industries.*)

FIGURE 18.4   Assembly of automatically activated zinc/silver oxide primary battery with tubular coil reservoir. (*Courtesy of Eagle-Picher Industries.*)

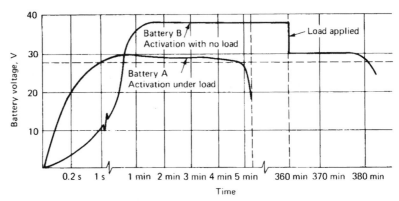

FIGURE 18.5   Voltage rise time for automatically activated zinc/silver oxide batteries at 25°C. (*Courtesy of Eagle-Picher Industries.*)

18.4   *PERFORMANCE CHARACTERISTICS*

Zinc/silver oxide reserve batteries as a class are somewhat unique in that they are almost entirely committed to specific applications. These applications require the flat voltage profile and the high specific energy and energy density available from this system, and they often demand a special design for each requirement. If a low-temperature environment is involved, battery heaters are used. If the discharge requires a wide range of current with only a small voltage variation, many very thin plates are used. A very high capacity requirement at low rates requires the use of thick plates and more concentrated electrolyte. There is no standard design or size because there is no typical application. The applications always demand the maximum from the battery design in capacity and voltage regulation at a minimum weight and volume. Multiple batteries are commonly packaged as a single unit, providing a range of load current and Ampere hour capacity in one convenient package.

18.4.1   Voltage

The open-circuit voltage of the zinc/silver oxide cell will range from 1.6 to 1.85 V per cell. The nominal load voltage is 1.5 V, and typical end voltages are 1.4 V for low-rate cells and 1.2 V for high-rate cells. At high rates, such as a 5 to 10-min. discharge rate, the output voltage would be about 1.3 to 1.4 V per cell, whereas the 2-hour rate discharge voltage would be slightly above 1.5 V. Figure 18.6 shows a family of discharge curves at four different current densities. The voltage level is inversely related to the current density (calculated from the area of the active plate surface). Thus based on 100 cm^2 of positive-plate surface, a 10 A discharge rate would be a 0.1 A/cm^2 current density. If the discharge rate is doubled to 0.2 A/cm^2, the voltage level would decrease, and if the rate is lowered to 0.05 A/cm^2, the voltage level would increase.

In cell design, the Ampere-hour capacity of the cell is determined by the amount of silver oxide active material present (zinc active material is provided in excess because of the cost relationship to silver), but the voltage is determined by the current density. In a fixed volume, higher discharge rates can be obtained without lowering the battery voltage by using thinner plates (thus providing more plates per cell element and lowering the current density), but with a reduction of capacity. The lower the current density at which a battery can operate, the better the voltage regulation with changing rates of discharge.

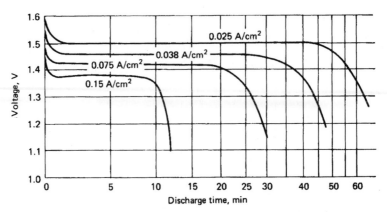

FIGURE 18.6   Effect of changing current density on battery voltage at 25°C. (*Courtesy of Eagle-Picher Industries.*)

18.4.2 Discharge Curves

A set of discharge curves for high-rate batteries is shown in Fig. 18.7 and for low-rate batteries in Fig. 18.8. The designs for these two types of batteries are quite different, with the principal difference being the thickness of the plates. The thin plates used in high-rate cells provide more surface area for lower current density, thus better voltage control and also more efficient utilization of the active material. At lower rates of discharge, as in Fig. 18.8, the voltage level is higher and active material utilization is also excellent, both because of lower current density. It will be noted that the low-rate discharge curves are above 1.6 V for a period of time. This is the effect of the divalent oxide, which affects voltage only at low rates. Most of the divalent capacity is obtained at high rates, but its voltage is decreased by the higher current density imposed.

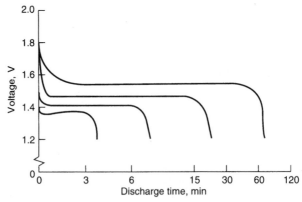

FIGURE 18.7   Discharge curves for high-rate zinc/silver oxide batteries at 25°C. (*Courtesy of Eagle-Picher Industries.*)

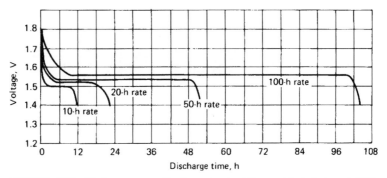

FIGURE 18.8 Discharge curves for low-rate zinc/silver oxide batteries at 25°C. (*Courtesy of Eagle-Picher Industries.*)

18.4.3 Effect of Temperature

The family of curves shown in Fig. 18.9 illustrates the performance obtained from a high-rate battery when discharged over a range of temperatures. It should be understood that the change in voltage levels caused by temperature is closely related to the changes caused by current density. Thus the adverse effect of cold temperature can be improved by reducing the current density of the cell, and the voltage and capacity of batteries discharged at high current densities can be improved by increasing their operating temperature. Figure 18.10 shows a family of curves for low-rate batteries discharged at various temperatures. The two sets of curves show that the zinc/silver oxide system is significantly affected at temperatures below 0°C and thus is not recommended for applications in this environment without heaters.

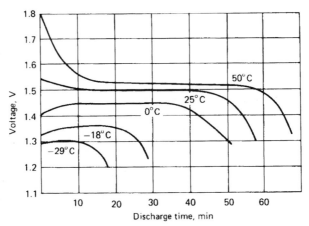

FIGURE 18.9 Effects of temperature on high-rate zinc/silver oxide primary batteries discharged at the 1-h rate. (*Courtesy of Eagle-Picher Industries.*)

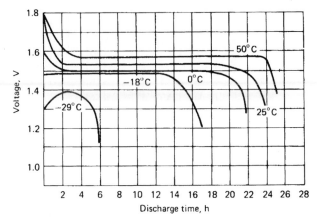

FIGURE 18.10   Effects of temperature on low-rate zinc/silver oxide primary batteries discharged at the 24-h rate. (*Courtesy of Eagle-Picher Industries.*)

18.4.4   Impedance

Figure 18.11 shows the dynamic internal resistance (DIR) of a high-rate cell at various stages of discharge and temperature. These curves show a declining ($\Delta V/\Delta I$) ratio until the end of discharge, at which time the dynamic resistance rises rapidly. The declining impedance is caused by an improvement in the positive-plate conductivity and a temperature rise during the discharge. This feature can vary considerably, depending on cell design, ambient temperature of the discharge, and the point in time after the change of discharge rate when the voltage change is observed.

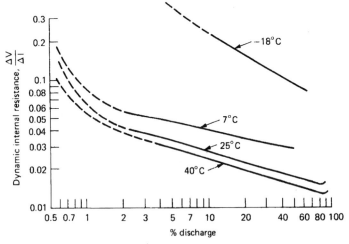

FIGURE 18.11   Dynamic internal resistance of zinc/silver oxide primary batteries. (*Courtesy of Eagle-Picher Industries.*)

18.4.5 Service

The performance of zinc/silver oxide batteries in Amperes per unit weight and volume versus service time is given in Fig. 18.12. It can be noted, again, that this battery system is particularly sensitive to temperatures below 0°C. These data are applicable, within reasonable accuracy, for both high- and low-rate designs.

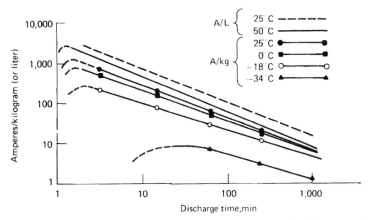

FIGURE 18.12 Service life of zinc/silver oxide primary batteries. (*Courtesy of Eagle-Picher Industries.*)

18.4.6 Shelf Life

The dry shelf life of the zinc/silver oxide battery is shown in Fig. 18.13, which gives storage data at 25, 50, and 74°C for periods of up to 2 years. The losses shown are based on the assumption that the positive active material is divalent silver oxide, which slowly degrades to monovalent oxide at temperatures above about 20°C. Degradation of the negative plate is minimal. It is expected that the monovalent oxide level would be reached in about 30 months when the storage temperature is 50°C. Experience has shown that batteries stored at average ambient temperatures of 25°C or lower retain capacity at or above the monovalent oxide level for a period of 25 years or longer.

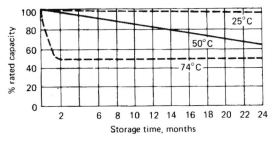

FIGURE 18.13 Dry storage of zinc/silver oxide primary batteries. (*Courtesy of Eagle-Picher Industries.*)

The wet shelf life of the zinc/silver oxide battery varies considerably with design and method of manufacture. Figure 18.14 provides a guide to the expected performance of most designs. The wet shelf-life degradation is caused principally by loss of negative-plate capacity (dissolution of the sponge zinc in the electrolyte) or development of short circuits through the cellulosic separators.

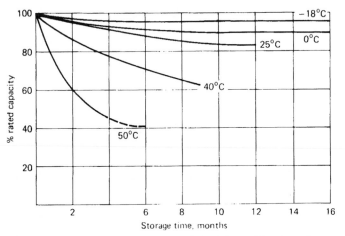

FIGURE 18.14   Wet (activated) storage of zinc/silver oxide primary batteries. (*Courtesy of Eagle-Picher Industries.*)

18.5   *CELL AND BATTERY TYPES AND SIZES*

Single-cell units of the reserve zinc/silver oxide type are available in sizes from about 1 Ah as a minimum up to about 775 Ah. Table 18.1*a* and 18.1*b* provides the specifications for a series of high-rate cells ranging in capacity from 1 to 250 Ah and a series of low-rate cells ranging in capacity from about 2 to 2680 Ah. These are all manually activated.

Table 18.2 lists a number of automatically activated batteries which have been designed to meet various specific applications. Most of these batteries are high-rate with a short wet-life. The weight and volume of this type are more a function of the load requirements and the space envelope provided than voltage and capacity.

TABLE 18.1a Zinc/Silver Oxide Manually Activated Batteries

High-rate cells, 15-mm rate					Low-rate cells, 20-h rate					Physical dimensions, cm		
Cell type*	Cap., Ah	Specific Energy Wh/kg	Energy Density Wh/L	Wt, g	Cell type*	Cap., Ah	Specific Energy Wh/kg	Energy Density Wh/L	Wt, g	Length	Width	Height
SZH 1.0	1.0	57	104	25	SZL 1.7	1.7	84	171	30	1.09	2.69	5.16
SZH 1.6	1.6	66	110	35	SZL 2.8	2.8	88	201	50	1.25	3.07	5.72
SZH 2.4	2.4	66	116	55	SZL 4.5	4.5	92	220	75	1.42	3.50	6.32
SZH 4.0	4.0	66	128	90	SZL 7.5	7.5	97	250	120	1.63	4.00	7.09
SZH 7.0	7.0	66	134	160	SZL 16.8	16.8	106	305	240	2.00	4.95	8.48
SZH 16.0	16.0	66	140	370	SZL 43.2	43.2	125	397	520	2.54	6.27	10.39
SZH 68.0	68.0	80	196	1290	SZL 160.0	160.0	187	470	1330	3.73	9.27	15.09
SZH 250.0	250.0	154	410	2450	—	—	—	—	—	4.32	9.45	22.43
—	—	—	—	—	SZL 410.0	410.0	210	560	3000	4.22	13.84	19.35
—	—	—	—	—	SZL 775.0	775.0	276	957	4380	6.96	8.36	21.70

* Eagle-Picher Industries.

TABLE 18.1b Zinc/Silver Oxide Manually Activated Primary Batteries*

Cell Model†	Type‡	Voltage§	Capacity Ah	Specific Energy Wh/kg	Energy Density Wh/L	Weight g	Dimensions, cm			Volume L
							Height	Width	Depth	
PM1	HR	1.42	2.0	92	147	31	5.13	2.74	1.37	0.019
PMV2	HR	1.48	5.3	103	184	76	6.42	4.37	1.52	0.043
PM3	HR	1.41	6.4	106	187	85	7.26	4.37	1.52	0.048
PML4	HR	1.42	8.3	113	208	104	8.53	4.37	1.52	0.057
PM5	MR	1.49	9.9	119	187	124	7.36	5.28	2.03	0.079
PMC5	MR	1.48	12.3	141	231	129	7.36	5.28	2.03	0.079
PMC10	MR	1.48	28	152	312	272	12.00	5.89	1.88	0.133
PM15	HR	1.42	19	92	180	292	12.55	5.89	2.03	0.150
PMV16	HR	1.47	18	72	141	365	15.57	5.84	2.06	0.187
PM30	HR	1.44	41	98	169	600	16.64	8.28	2.54	0.350
PM58	HR	1.42	56	85	162	938	18.42	8.26	3.23	0.491
PML100	LR	1.50	118	180	376	982	13.74	9.70	3.53	0.470
PML140	LR	1.49	165	197	439	1,250	16.36	9.70	3.53	0.560
PML170	LR	1.48	200	197	469	1,500	18.44	9.70	3.53	0.631
PML400	LR	1.47	375	218	566	2,525	16.10	15.27	3.96	0.974
PML2500	LR	1.48	2680	221	721	17,960	47.90	10.72	10.72	5.505

*These batteries are normally used as primaries. However, they all can be recharged (typically 3 to 10 cycles).
† Yardney Technical Products.
‡ HR = High rate, MR = Medium rate, LR = Low rate.
§ HR = 15-minute rate, MR = 1-h rate, LR = 5-h rate.

TABLE 18.2 Zinc/Silver Oxide Automatically Activated Batteries

Part Number*	Application	Weight, kg	Volume, L	Voltage, V	Current, A	Capacity, Ah	Energy density	
							Wh/kg	Wh/L
EPI 4331	AIM-7	1.0	0.45	26	10.0	0.8	21	46
EPI 4568	Peacekeeper	3.3	1.89	30	2.0	3.8	35	60
EPI 4500	Pariot	3.6	1.61	51	18.0	1.5	21	48
YTP 15148	Trident I	5.0	1.20	28	6.0	12.0	65	284
EPI 4567	Peacekeeper	6.2	3.46	30	11.0	16.0	77	139
EPI 4470	Harpoon	8.6	3.5	28	27, 40	8, 12	65	160
EPI 4445	Torpedo	9.3	4.8	28	30	20	60	117
YTP 15066	Trident I	14.5	3.8	30, 31	15, 23	4, 10	30	112
EPI 4569	Trident II	30.4	19.9	30, 31	18, 42	20, 10	30	46
YTP P-530	Minuteman	0.77	0.36	30	10.0	0.46	17.9	38.3
YTP P-515	Sparrow	0.99	0.45	24	11.0	0.45	10.9	24.0
YTP P-512	NMD	0.86	0.30	30	13.0	0.30	10.5	30.0
YTP P-468	AGM130	7.03	2.70	28	30	12.08	45.0	117.0
YTP P-471	Peacekeper	19.5	12.1	76, 31	16.7, 40	5.46, 40.90	86.3	139.3

*EPI—Eagle-Picher Industries; YTP—Yardney Technical Products.

18.15

18.6 SPECIAL FEATURES AND HANDLING

Both manually and automatically activated zinc/silver oxide batteries were developed to meet highly stringent requirements with regard to performance and reliability. The time and temperature of storage prior to use are of importance, and records should be maintained to ensure use within allowable limits. Special care must be exercised to ensure that the proper amount of the specified type of electrolyte is added to each cell of a manual-type battery and that, after activation, the unit is discharged within the shelf-life limitation at the proper temperature. Some battery containers have pressure-relief valves or heaters, or both, and these must be carefully maintained and monitored.

Automatically activated batteries require special preinstallation check out of gas generator ignitor circuits, heater circuits, and vent fittings. For long-term installations, there should be monitoring of the ambient temperature to prevent degradation caused by exposure to high temperatures. Periodic checks should be made to ensure that the ignitor circuits are intact because some circuits are sensitive to electromagnetic fields. After activation, if the battery is not discharged within the specified time, it must be replaced.

The proper electrical performance of these batteries is best ensured by operating them at temperatures at or slightly above room temperature. Temperatures below 15°C can adversely affect the voltage regulation of high-rate batteries, and below 0°C there is also considerable loss of capacity for both types.

18.7 COST

The cost of high-performance primary zinc/silver oxide batteries is dependent on the specifications to which they are built and the quantity involved. Manual-type batteries may cost anywhere from $5 to $15 per Watthour; remote-activated types will cost about $15 to $20 per Watthour. When the price of silver is high, material cost becomes one of the chief disadvantages of these batteries. There are many applications, however, in which no other technology can meet the high energy density of the zinc/silver oxide primary system.

BIBLIOGRAPHY

Bauer, P.: *Batteries far Space Power Systems,* U.S. Government Printing Office, Washington, D.C., 1968.

Cahoon, N. C., and G. W. Heise: *The Primary Battery,* Wiley, New York, 1969.

Chubb, M. F., and J. M. Dines: "Electric Battery," U.S. Patent 3,022,364.

Eagle-Picher Industries, Joplin, MO, brochure.

Fleiseher, A., and J. J. Lander: *Zinc Silver Oxide Batteries,* Wiley, New York, 1971.

Hollman, F. G., et al.: "Silver Peroxide Battery and Method of Making," U.S. Patent 2,727,083.

Jasinski, R.: *High Energy Batteries,* Plenum, New York, 1967.

Yardney Technical Products, Pawcatuck, CT, brochure.

CHAPTER 19
SPIN-DEPENDENT RESERVE BATTERIES

Asaf A. Benderly
Revised by Allan B. Goldberg

19.1 GENERAL CHARACTERISTICS

Various military, and a few civilian, applications with long shelf-life requirements must turn to reserve batteries for their electric power. This is particularly true when the system requires that the power supply be integrally packaged with the electronics and not replaced throughout the storage life of the system. Typical of such applications are fuzing, control, and arming systems for artillery and other spin-stabilized projectiles.

High spin forces, such as those encountered in artillery projectiles, may produce a difficult environment for many battery designs. However, special designs for liquid-electrolyte reserve batteries have evolved that take advantage of spin to bring about their activation and then keep the electrolyte within the cell structure.

A typical spin-dependent reserve battery is illustrated in Figs. 19.1 and 19.2. The electrode stack consists of electrodes and cell spacers of an annular configuration packaged dry and therefore capable of long-term storage. A metal ampoule, inserted in the center hole of the stack, houses the electrolyte. Upon firing of the gun, the ampoule opens; the electrolyte is released and is then distributed into the annular-shaped cells centrifugally, thereby causing the battery to become active.

FIGURE 19.1 Cross-section of lead/fluoboric acid/lead dioxide multi-cell reserve battery showing "dashpot" cutter for copper ampoule. (*Courtesy of U.S. Department of the Army.*)

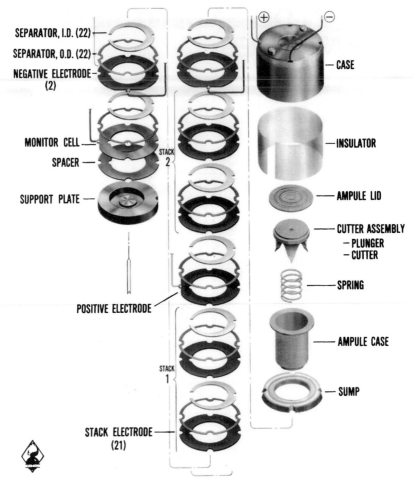

FIGURE 19.2   Component parts of lead/fluoboric acid/lead dioxide multicell reserve battery, PS 416 power supply. (*Courtesy of U.S. Department of the Army.*)

19.2   *CHEMISTRY*

The chemistry most commonly employed in spin-dependent liquid-electrolyte reserve batteries has been the lead/fluoboric acid/lead dioxide cell represented by the following simplified reaction:

$$Pb + PbO_2 + 4HBF_4 \rightarrow 2Pb(BF_4)_2 + 2H_2O$$

Fluoboric acid, rather than the more common sulfuric acid electrolyte, is used for these applications because it performs better at the very low temperatures required for these military applications. This low-temperature performance is due in part to the absence of insoluble reaction products as the reserve battery discharges.

More recently, spin-dependent liquid-electrolyte reserve batteries employing lithium anodes have been developed. The most promising system is that in which thionyl chloride serves in the dual role of electrolyte carrier and active cathodic depolarizer (see Chap. 20). The accepted cell reaction for this system is

$$4Li + 2SOCl_2 \rightarrow 4LiCl + S + SO_2.$$

At one time, the zinc/potassium hydroxide/silver oxide system was also employed in spin-dependent reserve batteries. More frequently, this reserve system has been used in nonspin applications, such as missiles, where the electrolyte is driven into place by a gas generator or other activation method (Chap. 18). This system is again finding favor in some applications where the potential hazards of lithium-based systems can create safety problems. The chemistry of the zinc/silver oxide couple can be represented by either of two reactions, depending on the oxidation state of the silver oxide:

$$2AgO + Zn \rightarrow Ag_2O + ZnO$$

$$Ag_2O + Zn \rightarrow 2Ag + ZnO$$

Within the past five years, thermal batteries that can operate at high spin rates (300 rps) have been developed and successfully demonstrated. These batteries have been based on the now standard lithium (alloy)/iron disulfide couple employed in most thermal batteries (see Chap. 21 for a detailed discussion of these chemistries).

19.3 DESIGN CONSIDERATIONS

19.3.1 Electrode-Stack Arrangement

The electrode stack may be arranged in two ways. One favors a high-voltage output, and the other a high-current output. The former generally uses bipolar electrodes; that is, electrodes wherein anodic and cathodic materials, respectively, are applied to the opposite sides of a metal substrate. Such bipolar electrode plates are stacked in a pile or series configuration, making automatic contact from one cell to the next. The voltage output of such a stack is the sum of all the cells. In the high-current configuration, electrode plates coated with anodic material on both sides of the substrate are stacked alternately with plates coated with cathodic material on both sides. All anodic plates are connected electrically in parallel through tabs. All cathodic plates are similarly connected. The two electrical connections constitute the effective terminals of the battery. This type of parallel stack is, in effect, a single electrochemical cell with a larger electrode area (Fig. 19.3). Where required by the application, multiples of series stacks can be connected in parallel, thereby yielding both high-voltage and high-current outputs.

FIGURE 19.3   Electrode stack and case of lead/fluoboric acid/lead dioxide parallel-construction (single cell) battery. (*Courtesy of U.S. Department of the Army.*)

19.3.2 Electrolyte Volume Optimization

The electrolyte capacity of the ampoule must be matched to the composite volume of all the cells in the battery. A parallel construction battery is reasonably tolerant of electrolyte flooding or starvation since it is a single cell. A series configuration, however, can tolerate no flooding since that condition produces intercell short circuits in the electrolyte fill channel or manifold. The opposite condition, that is, insufficient electrolyte, may leave one or more cells empty and therefore fail to provide continuity throughout the cell stack.

Since temperature extremes have a greater effect on the expansion and contraction of the liquid electrolyte than on the volume of the cells, an electrolyte volume that ensures that the cells will be reasonably full at low temperatures usually leads to an excess of electrolyte at higher temperatures. This excess must be accommodated in the design of the battery by the use of a "sump." In some short-life batteries, a match is established at high temperature with the recognition that cells will be less than full at lower temperatures. To ensure that some electrolyte enters each cell (so that continuity can be maintained), leveling holes may be provided from cell to cell. Though kept very small to reduce the effect of inevitable intercell short circuits, these holes do dissipate some of the capacity of the battery.

19.3.3 Cell Sealing

Since the individual cells of a spin-dependent liquid-electrolyte reserve battery are generally annular in shape and are filled by centrifugal force, the periphery of the cell must be sealed to keep electrolyte from leaking out. This sealing is typically accomplished by a plastic barrier formed around the outside of the electrode-spacer stack. For lead/fluoboric acid/lead dioxide batteries, this barrier is formed by fish paper (a dense, impervious paper) coated with polyethylene that melts at a relatively low temperature (similar to that used on milk cartons). Cell spacers are punched from the coated fish paper and placed between the electrodes. The stack is then clamped together and heated in an oven at a temperature sufficient to fuse the polyethylene, which then acts as an adhesive and sealer between the electrodes.

19.3.4 Ampoules

Early designs of liquid-electrolyte reserve batteries used glass ampoules to house the electrolyte, and in fact, some modern batteries still use such ampoules. These ampoules are generally smashed by the acceleration force of gunfire or by the explosive output of a primer or squib. Although these forces are ample, there is also a tendency for rough handling or a drop on a hard surface to cause inadvertent glass ampoule breakage. This would destroy the battery due to the premature leakage of electrolyte into the cells.

A major advance in battery ruggedness resulted from the design of metal, usually copper, ampoules with internal cutting mechanisms. One version employs a cutter that is activated by a combination of spin and acceleration (Fig. 19.4), both provided by the act of gunfire. Another version relies on a dashpot cutter mechanism (Figs. 19.1 and 19.2). This mechanism requires a sustained acceleration (several milliseconds experienced in gunfire), but will not function when subjected to the much shorter (a portion of a millisecond) shock pulse resulting from being dropped on a hard surface. The use of these "intelligent" ampoules, which are capable of discriminating between the forces of gunfire and those of rough handling, has resulted in a substantial improvement in battery reliability and safety.

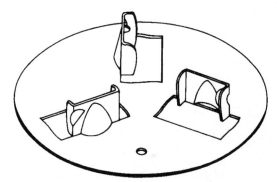

FIGURE 19.4 Three-bladed cutter for copper ampoule requiring spin and acceleration for activation. (*Courtesy of U.S. Department of the Army.*)

19.3.5 Safety in Lithium-Based Batteries

For at least the past ten years, reserve single-cell batteries that use various lithium-based electrochemistries have been employed in a number of fuzing applications. These cells are normally constructed with an anode-separator-cathode assembly spirally wound around a centrally located glass electrolyte ampoule. The electrolyte ampoule is normally broken upon gunfire or as the result of the bottom of the cell case being struck with a squib- or spring-driven device. Since these are single-cell devices, there is no chance for intercell short-circuiting and subsequent safety problems.

However, in multicell pile configuration reserve batteries, there is a considerable chance for intercell short-circuiting in the common electrolyte manifold. This intercell shortcircuiting not only dissipates the capacity of the cells, but it can also allow for dendritic growth, which can lead to electronic short-circuiting of the cells with catastrophic results. Experience has shown that such dendritic growth can be minimized or eliminated if all interior metallic surfaces of the battery have a nonconductive (usually Teflon-based) coating.

19.4   PERFORMANCE CHARACTERISTICS

19.4.1   General

Energy and Power Density.   Liquid-electrolyte reserve batteries are not normally rated in terms of energy or power per unit weight or per unit volume. Because of the need to provide double the volume for the electrolyte (one volume in an ampoule, the other in the cells themselves), such batteries are not highly space efficient. Space is also consumed by the ampoule-opening mechanism and the cell-sealing material. Furthermore, the cell area is sometimes not exposed to the electrolyte because of the spin eccentricity of the projectile, which houses the battery. Finally, such batteries are generally designed for short-lifetime applications, such as the flight time of an artillery projectile (approximately 3 min).

Operating Temperature Limits.   Like most other batteries, the performance of liquid-electrolyte reserve batteries is affected by temperature. Military applications frequently demand battery operations at all temperatures between −40 and 60°C, with storage limits of −55 to 70°C. These requirements are routinely met by the lead/fluoboric acid/lead dioxide systems and, with some difficulty at the low-temperature end, by the lithium/thionyl chloride and zinc/potassium hydroxide/silver oxide systems. Provision is occasionally made to warm the electrolyte prior to the activation of the two latter systems.

Voltage Regulation.   Since the voltage sustained by a liquid-electrolyte reserve battery at low temperatures and under heavy electric loading is much lower than that which it delivers at high temperatures, a serious problem of voltage regulation frequently results. In some situations, the ratio of high- to low-temperature voltage may be as much as 2:1. This problem may be avoided by the use of thermal batteries (Chap. 21), which provide their own pyrotechnically induced operating temperature, irrespective of the ambient temperature. Until recently, thermal batteries were extremely ineffective at high spin rates, but progress has been made in this field and thermal batteries capable of withstanding spin rates of 300 rps are now available.

Shelf Life.   The shelf life of liquid-electrolyte reserve batteries is highly dependent on the storage temperature, with high temperatures being the more deleterious. Zinc/silver oxide cells are probably the most vulnerable of the generally used systems because of the reduction of silver oxide and the passivation of zinc. Ten-year storage life is probably the best that can be expected unless the battery is substantially overdesigned. Lead/fluorboric acid/lead dioxide batteries also degrade with time, in both the loss of capacity and the lengthening of activation time. However, if objectionable organic materials are avoided in battery construction and the battery is designed with some safety factors, 20 to 25 years of shelf life may be realized. Lithium/thionyl chloride reserve systems are still too new to have a documented storage history; however, a long storage capability is projected for a properly (dry) built and sealed battery.

Linear and Angular Acceleration Limits.   Since spin-activated batteries are normally expected to be used in environments where guns are used, they must be built to withstand the forces of gunfire. With the development of the ampoules and the construction methods described, such batteries can withstand linear acceleration to the 20,000 to 30,000-g level and spin rates as great as 30,000 rpm. The sizes intended for small-caliber (20 to 40-mm) projectiles will withstand linear g levels 2 to 5 times that high.
   As an assist in withstanding these forces, the battery assembly is sometimes encapsulated in a supporting plastic. A popular design involved a molded plastic cup to house the stack and ampoule assembly, which was locked in place with an epoxy resin. More recently, the stack and ampoule assembly has been encapsulated in situ in a RIM (reaction impingement molding) process using a high-impact polyurethane foam, a process that allows demolding in just minutes. These two types of support are shown in Fig. 19.5.

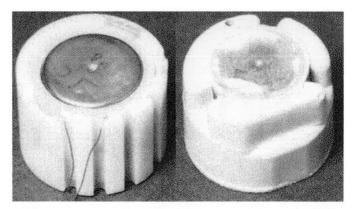

FIGURE 19.5 Stack and ampoule assembly of a lead/fluoboric acid/lead dioxide reserve battery supported by potting in epoxy in a molded case (left) and by in situ molding using a reaction impingement molded polyurethane foam (right). (*Courtesy of U.S. Department of the Army.*)

Activation Time. The time from initiation of the battery to the point at which it delivers and sustains a requisite level of voltage across a specified electric load is defined as the activation time. For a spin-dependent liquid-electrolyte reserve battery, this time would include the times for ampoule opening, electrolyte distribution, clearing of electrolyte short circuits in the filling manifold, depassivation of electrodes, and elimination of any form of polarization. Activation times are usually longest at low temperatures, where increased viscosity of the electrolyte and decreased ion mobility are most significant.

The application normally establishes the maximum allowable activation time, and reserve batteries are frequently designed to reach 75 or 80% of their peak voltage within this required time. A typical application requiring a very short activation time, perhaps less than 100 ms, would be a time fuze for an artillery projectile. Battery power is required to start the timer. Hence a stretch-out or uncertainty of time to reach timer voltage could result in a serious timing error, with a corresponding ineffectiveness of gunfire. In some cases, safety can be adversely affected by a timing error. In less critical situations, 0.5 to 1.0 s is allowed for activation.

19.4.2 Performance of Specific Electrochemical Systems

The physical and electrical characteristics of several typical spin-dependent reserve batteries are presented in Table 19.1.

Lead/Fluoboric Acid/Lead Dioxide Battery. Discharge curves for a typical lead/fluoboric acid/lead dioxide liquid-electrolyte reserve battery employed to power the proximity fuze of an artillery shell are given in Fig. 19.6. The slight rise in the low-temperature curve is due to its gradual rise in temperature in a room-temperature spinning tester. Similarly, the high-temperature curve is falling faster than it would in a true isothermal situation.

TABLE 19.1   Typical Spin-Dependent Reserve Batteries

Reference	Electro-chemical system	Height, cm	Diameter, cm	Weight, g	Nominal voltage, V	Nominal capacity, Wh
Fig. 19.1	$Pb/HBF_4/PbO_2$	4.1	5.7	280	35	0.5
Fig. 19.3	$Pb/HBF_4/PbO_2$	2.5	3.8	75	1.5	0.05
Fig. 19.8	$Li/SOCl_2$	1.67	3.8	70	12	0.37
	$Zn/KOH/AgO$	1.3	5.1	80	1.4	0.65

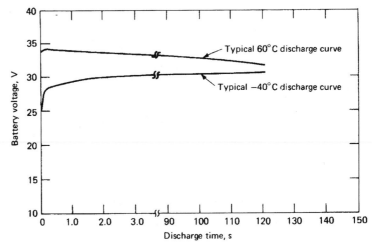

FIGURE 19.6   Discharge curves of a spinning lead/fluoboric acid/lead dioxide series-configuration reserve battery. Current density: 100 mA/cm². (*Courtesy of U.S. Department of the Army.*)

Lithium/Thionyl Chloride Battery.   Until recently, spin-dependent batteries were expected to function for short periods of time and only under sustained spin (necessary to keep the electrolyte within the cells). New applications have arisen that require a battery capable of withstanding artillery fire and spinning for a short time followed by some substantial operating time in a nonspin mode. Such applications include artillery delivery of mines or communication jammers intended to function after impact with the ground, or projectiles and submunitions that are operative while being slowed down by parachute.

The lithium-based liquid-electrolyte reserve battery holds promise of fulfilling this difficult combination of requirements. A typical cell, as illustrated in Fig. 19.7, incorporates an absorbing separator such as a nonwoven glass mat between the electrodes, and a long, high-resistance electrolyte filling path. After cell filling under spin, the absorbing material causes the electrolyte to be retracted away from the manifold and retained within the cell after cessation of spin. These design features, coupled with the long wet-stand capability of the lithium/thionyl chloride system, have paved the way for reserve batteries in applications that previously had to depend on the use of active batteries with relatively shorter storage capability. A discharge curve for a multicell, liquid-electrolyte reserve battery is given in Fig. 19.8 (also see Chap. 20).

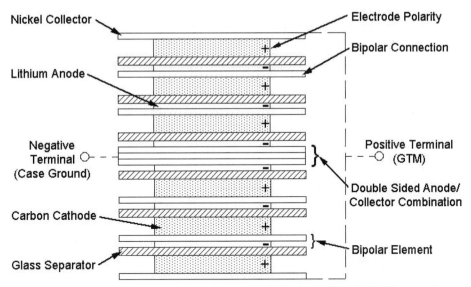

FIGURE 19.7   Quarter cross-sectional view of the cell stack of a lithium/thionyl chloride reserve battery. (*Courtesy of Alliant Power Sources Company.*)

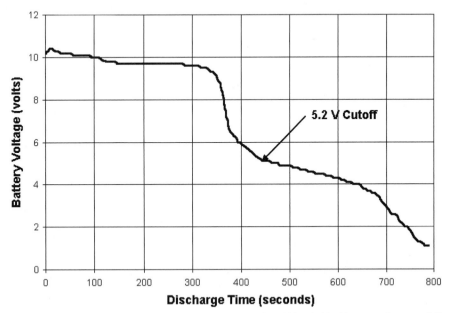

FIGURE 19.8   Discharge curve of a spinning (45 rps) lithium/thionyl chloride reserve battery at 25° C. Current density: 35 mA/cm². (*Courtesy of Alliant Power Sources Company.*)

Spin-Capable Thermal Batteries.   Because of their lower susceptibility to temperature extremes and known superior shelf life (without degradation), thermal batteries have been desired as an alternative to liquid-electrolyte reserve batteries for some time. The primary failure mode for thermal batteries in a high-spin environment had always been intercell short-circuiting at the cell stack edges due to the leakage of molten conductive materials at battery operating temperatures. New construction techniques, new electrochemistries that allow for higher electrolyte binder contents, and lithium alloy anodes that prevent migration of the anode material have made spin-capable thermal batteries practical.

BIBLIOGRAPHY

Benderly, A. A.: "Power for Ordnance Fuzing," *National Defense,* Mar.–Apr. 1974.

Biggar, A. M.: "Reserve Battery Requiring Two Simultaneous Forces for Activation," *Proc. 24th Annual Power Sources Symp.,* Electrochemical Society, Pennington, NJ, pp. 39–41, 1970.

Biggar, A. M., R. C. Proestel, and W. H. Steuernagel: "A 48-Hour Reserve Power Supply for a Scatterable Mine," *Proc. 26th Annual Power Sources Symp.,* Electrochemical Society, Pennington, NJ, pp. 126–129, 1974.

Doddapaneni, H., D. L. Chua, and J. Nelson: "Development of a Spin Activated, High Rate, Li/SOCl$_2$ Bipolar Reserve Battery," *Proc. 30th Annual Power Sources Symp.,* Electrochemical Society, Pennington, NJ, pp. 201–204, 1982.

Krieger, F. C.: "Miniaturized Thermal Reserve Battery," *Proc. 38th Annual Power Sources Conference,* U.S. Army CECOM/ARL, Cherry Hill, NJ, pp. 231–234, 1998.

Morganstein, M., and A. B. Goldberg: "Reaction Impingement Molding (RIM) Encapsulation of a Fuze Power Supply," *Proc. of the 4th International SAMPE Electronics Conference,* Society for the Advancement of Material and Process Engineering, Covina, CA, pp. 753–764, 1990.

Schisselbauer, P. F., and D. P. Roller, "Reserve g-Activated, Li/SOCl$_2$ Primary Battery for Artillery Applications," *Proc. 37th Annual Power Sources Conference,* U.S. Army CECOM/ARL, Cherry Hill, NJ, pp. 357–360, 1996.

Turrill, F. G., and W. C. Kirchberger: "A One-Dollar Power Supply for Proximity Fuzes," *Proc. 24th Annual Power Sources Symp.,* Electrochemical Society, Pennington, NJ, pp. 36–39, 1970.

CHAPTER 20
AMBIENT-TEMPERATURE LITHIUM ANODE RESERVE BATTERIES

David L. Chua, William J. Eppley, and Robert J. Horning

20.1 GENERAL CHARACTERISTICS

The use of lithium metal as an anode in reserve batteries provides a significant energy advantage over the traditional reserve batteries because of the high potential and low equivalent weight (3.86 Ah/g) of lithium. A lithium reserve battery can operate at a voltage close to twice that of the conventional aqueous types. Due to the reactivity of lithium in aqueous electrolytes, with the exception of the special lithium-water and lithium-air batteries (see Sec. 38.6), lithium batteries must use a nonaqueous electrolyte with which lithium is nonreactive.

The various ambient-temperature active (non-reserve) lithium batteries are covered in Chap. 14. Of these systems, the ones demonstrating the higher energy densities and rate capabilities are Li/SO_2, Li/V_2O_5, $Li/SOCl_2$, and Li/Li_xCoO_2. The discharge characteristics of these batteries are shown in Fig. 20.1. These are the electrochemical systems that are predominately employed in the reserve-type configurations.

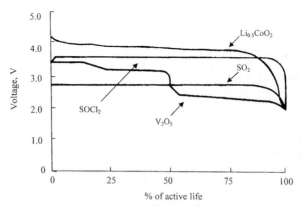

FIGURE 20.1 Performance comparison of lithium anode primary systems at 20°C. Thionyl chloride ($SOCl_2$)—3.6 V; vanadium pentoxide (V_2O_5)—3.4 V; sulfur dioxide (SO_2)—2.9 V; precharged lithiated cobalt oxide (Li_xCoO_2, $x = 0.4 - 0.5$) − 4.0 V.

In the reserve construction, the electrolyte is physically separated from the electrode active materials until the battery is used and it is stored in a reservoir prior to activation. This design feature provides a capability of essentially undiminished output even after storage periods, in the inactive state, of over 14 years. The reserve feature, however, results in an energy density penalty of as much as 50% compared with the active lithium primary batteries. Key contributors to this penalty are the activation device and the electrolyte reservoir.

In the selection of a lithium anode electrochemical system for packaging into a reserve battery, besides such important considerations as physical properties of the electrolyte solution and performance as a function of the discharge conditions, factors such as the stability of the electrolyte and the compatibility of the electrolyte with the materials of construction of the electrolyte reservoir are of special importance.

20.2   CHEMISTRY

20.2.1   Lithium / Vanadium Pentoxide (Li / V_2O_5) Cell

The basic cell structure of this system consists of a lithium anode, a microporous polypropylene film separator, and a cathode that is usually composed of 90% V_2O_5 and 10% graphite, on a weight basis. When it is used in a reserve battery, the prevalent electrolyte is $2M$ $LiAsF_6$ + $0.4M$ $LiBF_4$ in methyl formate (MF) because of its excellent stability during long-term storage.

As shown in Fig. 20.1, the Li/V_2O_5 system has a two-plateau discharge characteristic. A net cell reaction, involving the incorporation of lithium in V_2O_5, has been postulated to account for the first plateau,

$$Li + V_2O_5 \rightarrow LiV_2O_5$$

The initial voltage level ranges from 3.4 to 3.3 V, decreases to 3.3 to 3.2 V for approximately 50% of the active life of the first discharge plateau, at which point the range again decreases to a level of 3.2 to 3.1 V, which is maintained for the balance of the first plateau of discharge. After completion of the first plateau, the Li/V_2O_5 system undergoes a rapid change in voltage to the second discharge plateau around a voltage range of 2.4 to 2.3 V. This step involves the formation of reduced forms of V_2O_5, although specific mechanisms remain unclear.[1] This second plateau is relatively more sensitive to temperature and discharge rate, and it is for this reason that most Li/V_2O_5 cells (active and reserve) are designed to operate at only the first discharge plateau level.[2]

The long-term storage capability of Li/V_2O_5 reserve cells is heavily dependent on the stability of the electrolyte solution. $LiAsF_6$ in MF electrolyte is unstable due to the decomposition reactions involving the hydrolysis of methyl formate followed by the dehydration of the hydrolysis product(s).[3] These reactions result in a premature fracture of the glass ampoule used as the electrolyte reservoir. The stability of the $LiAsF_6$:MF electrolyte solution was achieved by making the solution either neutral or alkaline. In practiced, this is accomplished by using two electrolyte salts ($LiAsF_6$ + $LiBF_4$:MF) and by incorporating lithium metals to scavenge water in the glass ampoule.

20.2.2   Lithium / Thionyl Chloride ($Li / SOCl_2$)

The basic cell structure that is generally used for this system consists of a lithium anode, a nonwoven glass separator and a Teflon®-bonded carbon cathode which serves only as the reaction site medium. One unique feature of this chemistry is the fact that thionyl chloride ($SOCl_2$) serves two functions—as the solvent of the commonly used $LiAlCl_4$ in $SOCl_2$ electrolyte solution and as the active cathode material (see Sec. 14.6).

Figure 20.1 shows the marked advantage in discharge performance of the $Li/SOCl_2$ system. The accepted net cell reaction for this system is

$$4Li + 2SOCl_2 \rightarrow 4LiCl + S + SO_2$$

Most of the sulfur dioxide formed during discharge is dissolved in the electrolyte and practically no gas pressure is generated.[4] Depending on the discharge rate and temperature the $Li/SOCl_2$ system normally exhibits a working voltage range of between 3.0 and 3.6 V with a flat discharge characteristic. These excellent discharge characteristics—high voltage and flat discharge—are best attained in a reserve cell, especially is the discharge current density is high. In an active primary $Li/SOCl_2$ cell, the lithium anode is coated with a passive LiCl film. Under sustained storage coupled with high-temperature exposure the passivating film will limit the current-handling capability of the anode as well as increasing the time required to reach operating voltage.[4]

The conventional $LiAlCl_4$ in $SOCl_2$ electrolyte solution has proved to have excellent stability. Electrolyte glass ampoules exposed to +74°C did not show any sign of apparent degradation up to at least 12 years of storage. Because of this and its overall performance superiority, the $Li/SOCl_2$ reserve cell has become the system of choice for high-energy reserve batteries.

Recently the use of excess $AlCl_3$, a Lewis acid, in the conventional $LiAlCl_4$ in $SOCl_2$ electrolyte solution was shown to improve the rate capability of the $Li/SOCl_2$ system.[5,6] It should be noted, however, that the inherent stability of this high-rate electrolyte has yet to be established so as to ensure its application in reserve cells.

20.2.3   Lithium/Sulfur Dioxide (Li/SO₂)

The Li/SO_2 system uses a basic cell structure consisting of a lithium anode, a separator, and a Teflonated® carbon cathode, similar to one used in the $Li/SOCl_2$ system, which serves as the reaction site. The electrolyte solution commonly employed contains a mixture of lithium bromide (LiBr), acetonitrile (AN), and sulfur dioxide (SO_2), which also serves as the active cathode material.

One serious problem in using the LiBr in $AN–SO_2$ electrolyte solution for reserve cells is its instability during storage. Although this electrolyte solution is commonly employed in active primary cells, it is unsuitable for reserve battery applications because it decomposes to form highly reactive and solid products when stored in the absence of cell components. Replacing LiBr with lithium hexafluoroarsenate ($LiAsF_6$) results in an electrolyte solution with good stability. The functional performance of the $LiAsF_6$ electrolyte is equivalent or superior to that of the LiBr solution for low to moderate rate.[7–9].

The Li/SO_2 reserve battery, using the stable electrolyte solution ($LiAsF_6$ in $AN-SO_2$), follows the same net cell reaction

$$2Li + 2SO_2 \rightarrow Li_2S_2O_4 \downarrow$$

as in the active primary cell. It should be noted, however, that $LiAsF_6$ in $AN–SO_2$ electrolyte solution is limited to moderate- or lower-rate applications due to poorer electrolyte conductivity. Because of this, the emphasis on using the Li/SO_2 system for reserve applications has been shifted to the higher-performance $Li/SOCl_2$ system.

20.2.4   Lithium / Pre-Charged Li$_x$CoO$_2$ (0.5 ≤ x < 1) Cell

The use of pre-charged Li$_x$CoO$_2$ is a new approach to high-voltage and high-energy cathode systems for reserve battery applications. Key criteria are the ability to process the cathode materials into a stable raw material for the reserve cells. In the case of the Li$_x$CoO$_2$ cathode, lithium can be extracted electrochemically to where $x \geq 0.5$ and then processed successfully as a raw material for primary active cell applications.[10] The Li/pre-charged Li$_x$CoO$_2$ (0.5 ≤ x < 1) cell structure is very similar to the Li/V$_2$O$_5$ cell. Except for the difference in cathode material, all other design features are essentially the same.

Both the Li/V$_2$O$_5$ and Li/pre-charged Li$_x$CoO$_2$ (0.5 ≤ x < 1) cells offer an unique design feature, permitting the cells to be recharged when demanded by a specific application. One application that used the pre-charged Li$_x$CoO$_2$ cathode technology is described in Sec. 20.3.2.

20.3   CONSTRUCTION

20.3.1   General Considerations

Lithium anode reserve batteries are basically composed of three major components:

1. Activation and electrolyte delivery system
2. Electrolyte reservoir
3. Cell and/or battery unit

However, the actual design can vary widely, depending on the application. The design can vary from a simple, small, single cell with an ampoule manually activated, to a very large, complex, multicell battery with an automatic electric initiation mechanism to transfer the electrolyte from the reservoir chamber to a high-voltage multicell battery stack. Both the electrodes and the hardware components are essentially the same as the primary active units, but with allowances made for electrolyte storage and electrolyte delivery into the cells at the time of activation. In addition, the electrochemical and hardware components must be constructed of a rugged maintenance-free design to survive severe environmental and performance requirements as most are used in military or special applications. For example, Table 20.1 lists typical requirements of lithium reserve batteries and illustrates the reason for many of their unique construction and design features.

TABLE 20.1   Typical Characteristics of Lithium Anode Reserve Batteries

Operating temperature range −55 to 70°C
10- to 20-year unactivated storage life
Hermetically sealed
High energy density
High reliability
Low electrical noise
Flat discharge voltage profile
Rapid voltage rise after initiation
Mechanical environmental capability:
   Acceleration shocks up to 20,000 g
   High spin up to 20,000 rpm
   Transportation and deployment vibration levels
Operating life from several seconds up to 1 year

Some common construction features are used in the design of lithium reserve batteries. The outer case is generally made of a 300-series stainless steel since it offers the corrosion resistance against both the internal system and the external environment during its long-term use. Various welding techniques such as laser, tungsten inert gas (TIG), resistance, and electron beam can be applied to the 300-series stainless steels. Thus the outer case provides a true 20-year reliability, capable of maintaining the hermeticity required for reserve lithium batteries. The electrical terminals used are generally glass-to-metal types, which also provide the hermeticity required for long-term storage.

20.3.2 Types of Lithium Anode Reserve Batteries

Three basic lithium reserve battery types are being manufactured at the present time:

1. Single-cell battery with electrolyte stored in a glass ampoule
2. Multiple single cells using bellows for the electrolyte storage reservoir
3. Multicells of bipolar construction with either a glass ampoule or a reservoir for electrolyte storage

Ampoule Type.   Single-cell reserve types using an ampoule as the electrolyte storage reservoir are the most reliable of the reserve designs due to their simple construction and lack of intercell leakage problems associated with multicell batteries. One group of these cells is sized to the ANSI standard specifications, and the other group consists of those cells built for special-purpose applications which are not sized to the ANSI specifications. Both groups, however, are very similar in construction.

Figure 20.2 shows the cross section of a reserve lithium anode cell in an A-size configuration of about 1 Ah, using the $Li/SOCl_2$ system.[6] The cell consists of concentrically arranged components. A lithium anode is swaged against the inner wall of a stainless-steel cylindrical can. A nonwoven glass separator is located adjacent to the anode. The Teflon®-bonded carbon cathode is inserted against the separator. A cylindrical nickel current collector provides the electrical contact to the positive terminal and houses the hermetically sealed glass ampoule. The ampoule is held firmly in place by upper and lower insulating supports, which protect it from premature breakage while permitting transmission of a direct force at the bottom of the case to shatter the ampoule at the time of activation. The unit is sealed hermetically to ensure long shelf life in the unactivated condition. Activation is achieved by applying a sharply directed force at the bottom of the cell case to shatter the glass ampoule. The electrolyte is absorbed by the porous cathode and the glass separator, thereby activating the battery.

Another design has been developed for mine and fuze applications, using both the Li/V_2O_5 and the $Li/SOCl_2$ systems, in the capacity range of 100 to 500 mAh.[11] The cross sections of these two cells are shown in Figs. 20.3 and 20.4 respectively. Both cells are similar with respect to the external hardware and the internal arrangement of the components. The case and header assembly are projection-welded together at the case flange. The header serves as the cover for the cells and incorporates a glass-to-metal seal for the center terminal pin made of nickel-iron Alloy52. The terminal pin has negative polarity (both cell designs), and the balance of the header and case surface have positive polarity. The hermetically sealed hardware in conjunction with the reserve feature of the design makes it possible to achieve storage times in excess of 20 years.

The internal arrangement of the components consists of annularly located electrodes about a central glass ampoule used as the electrolyte solution reservoir. In addition there are various insulating components in the upper and lower portions of the cell, used to prevent internal short-circuiting.

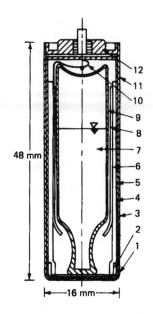

48 mm

16 mm

1 Insulator	7 Electrolyte
2 Bottom separator	8 Current collector
3 Cell can	9 Glass ampul
4 Lithium anode	10 Positive terminal tab
5 Separator	11 Top spacer
6 Carbon cathode	12 Cell cover

FIGURE 20.2 Cross section of Li/SOCl$_2$ A-size reserve cell. (*Courtesy of Tadiran Industries, Ltd.*)

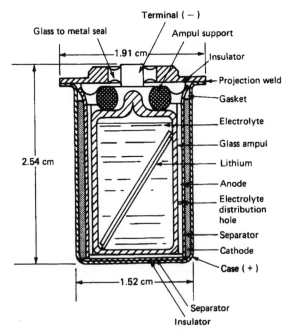

FIGURE 20.3 Cross section of Li/V$_2$O$_5$ reserve cell. Alliant model G2659. (*Courtesy of Alliant Techsystems, Inc.*)

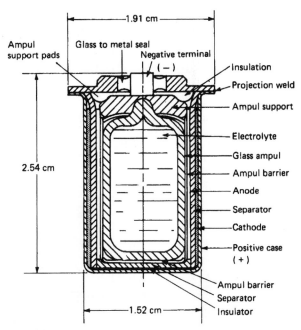

FIGURE 20.4 Cross section of LiSOCl$_2$ reserve cell. Alliant model G2659B1. (*Courtesy of Alliant Techsystems, Inc.*)

Several features account for most of the design differences between these two cells. In the Li/SOCl$_2$ reserve cell, the glass ampoule also contains the cathode oxidant, SOCl$_2$, while the cathode oxidant of the Li/V$_2$O$_5$ reserve cells is contained in the cathode structure. Directly adjacent to the Li/SOCl$_2$ cell case is the Teflonated® carbon, while in the case of the Li/V$_2$O$_5$ cell, the cathode is molded from a dry mixture of V$_2$O$_5$ and graphite. The Teflonated®-carbon cathode for the reduction of SOCl$_2$ is made in sheet form and is attached to a metal grid rolled to shape and inserted against the inside wall of the case. Another difference is the way the electrical connection is made for the two cathodes. The V$_2$O$_5$ connection is made by the direct-pressure contact of the molded cathode, whereas with the SOCl$_2$ system the cathode lead is welded to the case at the time the cover is welded. The lithium anode structure consists of pure lithium metal, which is pressed onto an expanded metal grid of 316L stainless steel. One end of a flat 316L stainless-steel lead is spot-welded to the pin of the glass-to-metal seal. Rolled into a cylinder, the anode is inserted into the cell next to the separator. Both cells are provided with an ampoule support in order to survive the shock environment specified. In the Li/SOCl$_2$ system, Tefzel and glass have been found to be chemically stable for use as insulators, separators, and supports. The Li/V$_2$O$_5$ system allows more flexibility because many rubbers and plastics can be used.

Multicell Single-Activator Design. For those applications where higher than single-cell voltages are required, a battery is constructed of two or more cells, depending, of course, on the voltage needed. Typical voltages are 12 and 28 V, and for lithium anode cells with a 2.7 to 3.3-V operating voltage, this would require anywhere from 4 to 10 cells for each battery. This family of batteries is unique with respect to the method of cell activation and the containment of electrolyte in multiple cells initiated from a single self-contained reservoir of electrolyte. Batteries of this design are used in preference ot the bipolar type to achieve higher cell capacities and to allow discharge times up to 1 year or more, through the tight control of intercell leakage. The leakage currents are controlled and limited to usually less than several percent of the discharge current. This feature, however, limits these batteries from being miniaturized, which is possible with many bipolar designs.

An example of this design approach is the Li/SO$_2$ reserve battery illustrated in Fig. 20.5. The battery is cylindrical and contains three main components: (1) the electrolyte storage reservoir section, (2) the electrolyte manifold and activation system, and (3) the reserve cell compartment. About one-half of the internal battery volume contains the electrolyte reservoir. The reservoir section consists primarily of a collapsible bellows in which the electrolyte solution is stored. Surrounding the bellows, between it and the outer battery case, is a space that holds a specific amount of gas/liquid. The gas is selected such that its vapor pressure always exceeds that of the electrolyte, thereby providing the driving force for eventual liquid transfer into the cell chamber section once the battery has been activated.

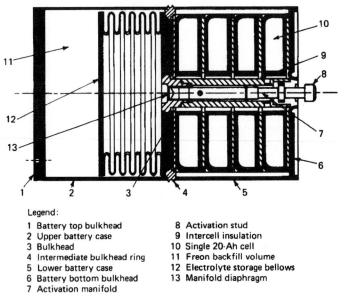

FIGURE 20.5   Cross section of 20-Ah Li/SO$_2$ multicell battery.

Legend:

1 Battery top bulkhead	8 Activation stud
2 Upper battery case	9 Intercell insulation
3 Bulkhead	10 Single 20-Ah cell
4 Intermediate bulkhead ring	11 Freon backfill volume
5 Lower battery case	12 Electrolyte storage bellows
6 Battery bottom bulkhead	13 Manifold diaphragm
7 Activation manifold	

In the remaining half of the battery volume there is the centrally located electrolyte manifold and activation system housed in a 1.588-cm-diameter tubular structure plus the series stack of four toroidally shaped cells that surround the manifold/activation system.

The manifold and cells are separated from the reservoir by an intermediate bulkhead. In the bulkhead there is a centrally positioned diaphragm of thin section to be pierced by the cutter contained within the manifold. In fabrication, the diaphragm is assembled as part of the tubular manifold which, in turn, is welded as a subassembly to the intermediate bulkhead. Figure 20.6 is a more detailed cross-sectional view of the electrolyte manifold and activation system with the major components identified.

The activating mechanism consists of a cutter that is manually moved into the diaphragm, cutting it and thereby allowing electrolyte to flow. The movement of the cutter is accomplished by the turning of an external screw that is accessible in the bottom base of the battery. The cutter section and the screw mechanism are isolated from one another by a small collapsible metal cup that is sealed hermetically between the two sections. This prevents external electrolyte leakage. The manifold section is a series of small nonconductive plastic tubes connected to one end of the central cylinder and to each of the individual cells at the other end. The long length and small cross-sectional area of the tubes minimize intercell leakage losses during the period of time that electrolyte is present in the manifold structure.

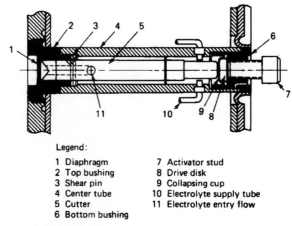

Legend:

1 Diaphragm
2 Top bushing
3 Shear pin
4 Center tube
5 Cutter
6 Bottom bushing
7 Activator stud
8 Drive disk
9 Collapsing cup
10 Electrolyte supply tube
11 Electrolyte entry flow

FIGURE 20.6 Cross section of electrolyte manifold and activation system.

In this application four individual cells are required to meet the voltage requirement. (The number of cells is, of course, adjustable with minor modification to meet a wide range of voltage needs.) Each cell contains flat circular anodes and cathodes that are separately wired in parallel to achieve the individual cell capacity and plate area needed for a given set of requirements. To fabricate, the components, with intervening separators, are alternately stacked around the cell center tube, after which the parallel connections are made. The cells are individually welded about the inner tube and outer perimeter to form hermetic units ready for series stacking within the battery. Connections from the cells are made to external terminals which are located in the bottom bulkhead of the battery.

Figure 20.7 shows the major battery components prior to assembly. The components shown are fabricated primarily from 321 stainless steel, and the construction is accomplished with a series of TIG welds. The hardware shown is designed specifically for use with the lithium/sulfur dioxide electrochemical system: however, it is adaptable, with minor modifications to other liquid and solid oxidant systems. The battery can also be adapted to electrical rather than manual activation.

An example of this reserve design approach being used with the lithium/thionyl chloride chemistry is shown in Fig. 20.8a. This high-power reserve battery, designated by the U.S. Navy as battery BA-6511 SLQ, was developed to provide electric power for a family of

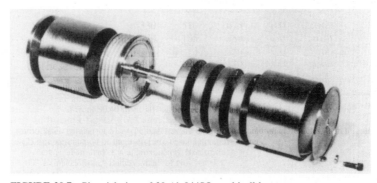

FIGURE 20.7 Pictorial view of 20-Ah Li/SO$_2$ multicell battery.

(a)

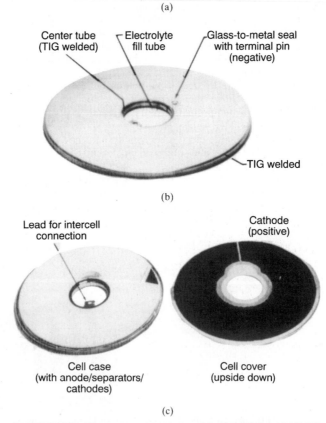

Center tube (TIG welded)
Electrolyte fill tube
Glass-to-metal seal with terminal pin (negative)
TIG welded

(b)

Lead for intercell connection
Cathode (positive)
Cell case (with anode/separators/cathodes)
Cell cover (upside down)

(c)

FIGURE 20.8   High-power reserve battery BA-6511/SLQ. (*a*) Li/SOCl$_2$ reserve battery. (*b*) High-power cells. (*c*) High-power cell case and electrode assembly. (*Courtesy of Alliant Techsystems, Inc.*)

ocean buoys.[12] The reserve battery was selected for this application to eliminate the problems of self-discharge and passivation associated with extended stand of an active battery, but also for safety as the electrolyte is stored separately from the battery until activation.

The battery weighs about 145 lb and is contained in a package that is 29.2 cm in diameter and 43.2 cm long. The battery contains 21 cells; 18 cells compose a 56-V section, delivering 4 kW, and rated at 65 Ah; 3 cells are in a 10-V section, delivering 7 A, and rated at 57 Ah. The electrolyte is stored in a reservoir and is distributed to the 21 cells via a unique mani-folding system. Activation is initiated by an explosive squib, and a stored energy system within the reservoir provides the motive power. The cell design used for this battery (Fig. 20.8b) is a circular wafer with a hole through the center to provide a channel for electrical and tubing connections. The two types of cells used in the battery are physically similar, differing only in height and capacity as a result of one less set of electrodes. The cells used in the high-voltage, high-rate section contain five anodes and six cathodes. Anodes are single-sided with lithium pressed onto expanded nickel grids. The cathode is cut from coated stock of Teflonated® carbon on a nickel screen, as shown in Fig. 20.8c. Nonwoven glass separators are used. The specifications for the two cells are listed in Table 20.2.

TABLE 20.2   Characteristics of Li/SOCl$_2$ Reserve Cells Model G3070A2

	Low-rate reserve cell	High-rate reserve cell
Performance		
Open-circuit voltage (activated)	3.67 V	3.67 V
Voltage under load	3.40 V, 7 A at 20°C	3.10 V, 72 A at 20°C
Rated capacity	57 Ah at 7 A to 2.67 V at 20°C	65 Ah at 72 A to 2.63 V at 20°C
Physical characteristics:		
Max diameter, OD	28.5 cm	28.5 cm
Max diameter, ID	6.7 cm	6.7 cm
Max height	0.89 cm	1.04 cm
Cell weight with electrolyte	1310 g	1485 g
Case material	Stainless steel	Stainless steel

Source: Alliant Techsystems, Inc.

Another more recent example of this reserve design using a pre-charged Li$_x$CoO$_2$ ($0.5 \leq x < 1$) chemistry[13] is shown in Fig. 20.9. This reserve battery consists of three hermetically-welded cell cases with a central reservoir enclosed in a stainless steel battery housing. Such a battery was developed to power the Hand-Emplaced Wide Area Munitions (HWAM).

Figure 20.10 shows another multicell battery designed for light weight missile applications.[14] Based on the Li/oxyhalide technology, it used advanced thin electrode technology that has high energy utilization and low electrical impedance. A composite separator was also used that combines high electrolyte absorption and mechanical integrity. This type of high power design sees applications such as the Theater High Altitude Area Defense (THAAD) in a Kill Vehicle for the Ground Based Interceptor (GBI) program. Other advantages of this light weight, high power battery are: (1) reduction in battery weight over the thermal or silver zinc system; (2) specific energy greater than 250 Wh/kg; (3) gain in weight advantage as the mission time increases or as the energy to power ratio increases; (4) high power delivery even after 10 years of storage at temperature below −32°C; and (5) low operating temperature allowing locations near heat sensitive electronics.

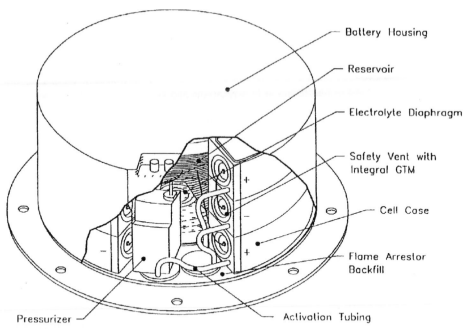

FIGURE 20.9 Design of reserve Li/Li_xCoO_2 battery for Hand-emplaced Wide Area Munitions (HWAM). (*Courtesy of Alliant Techsystems, Inc.*)

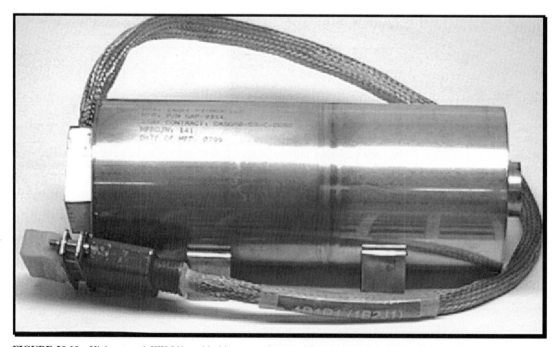

FIGURE 20.10 High-power 1 KW Li/oxychloride reserve battery. (*Courtesy of Eagle-Picher Technologies*)

Multicell Bipolar Construction with Single-Activator Reservoir. Lithium anode reserve batteries, using bipolar construction, are relatively few in number and always developed for specific applications. The bipolar construction—one component used as both the anode collector of one cell and the cathode collector of the next cell in the stack—is not unique to the lithium reserve battery, but an adaptation of techniques used in other types of batteries. There are several advantages of the bipolar construction:

Very high energy and power density for high-voltage batteries

Rugged construction to withstand spin and setback forces from artillery firing

Flexibility to adjust voltages in the cell stack

Adaptability to varying energy and power requirements

Figure 20.11 is an illustration of a reserve lithium/thionyl chloride battery using a bipolar plate construction. This battery weighs approximately 5.4 kg and has a volume of 2000 cm.[3]

Activation of the reserve battery is accomplished by supplying an electric pulse to the battery by firing an electric squib or actuator or by some mechanical means. This type of reserve battery has been used chiefly in artillery shells for electronic fuze power supplies and in missiles for the electronic power supply. Therefore the electric pulse can be supplied prior to firing or at the time of launch. However, for artillery fuze power supplies, the battery is usually activated by the launch acceleration (set back) and/or the spin forces. The accel-

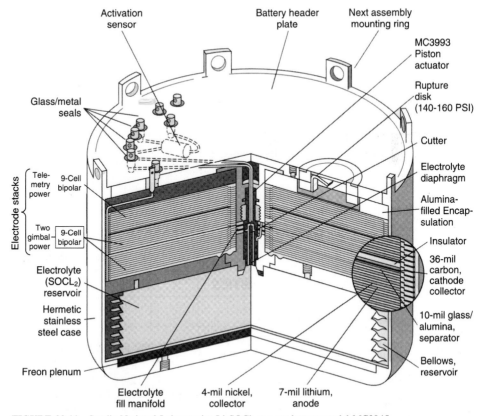

FIGURE 20.11 Sandia National Laboratories Li-SOCl$_2$ reserve battery model MC3945.

eration force of the artillery shell releases a firing pin which strikes and fires a primer. The primer can ignite a gas generator or directly release a stored gas by breaking open a metal diaphragm.

Once the battery has been initiated as described, the gas pressure (such as from a gas generator, stored gas/liquid, or CO_2) forces the electrolyte into each of the cells through a manifold (electrolyte distribution network).

The electrolyte reservoir is generally made using a collapsible cup, a bellows, or a wound tubing design. These serve to hold the electrolyte during the long inactive storage period and act as the delivery mechanism during activation. Each reservoir has some type of diaphragm which is broken with high pressure or mechanical means to allow the electrolyte to enter the cell-stack part of the battery hardware.

The bipolar cell stack with the electrolyte distribution manifold in the center comprises the battery section. When electrolyte enters the center manifold, it is distributed to each cell through holes or passageways in the housing encompassing the battery. The design of the manifold is the key to controlling intercell leakage. For bipolar batteries, life requirements are relatively short (seconds to several hours); therefore the manifolding is relatively simple. But when longer life is needed, the parasitic leakage currents are controlled by the length and area of the leakage path.

Another battery of this design was developed as a power source for the Extended Range Guided Munition (ERGM). Ultimately, this design would be similar to the design shown in Fig. 20.10. However, the development test fixture used is shown in Fig. 20.12. It uses the $LI/SOCl_2$ chemistry but with a special lithium tetrachlorogallate electrolyte optimized for performance. The test fixture employs gas pressure to compress the bellows, rupturing the diaphragm and activating the battery. In actual use, set-back forces on launch perform this function.

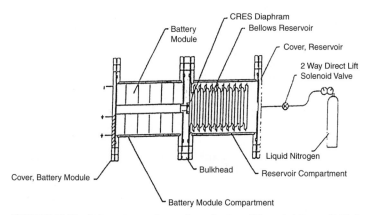

FIGURE 20.12   Laboratory test fixture for activation of Extended Range Guided Missile (ERGM) battery using $Li/SOCL_2$ system. (*Courtesy Yardney Technical Products, Inc.*)

TABLE 20.3   Specifications for ERGM Battery

Description	Specifications
12 V Section Regulation	• Load voltage Range: 9.5 to 16.0 V • 60W Applied Continuously
28 V Section Regulation	• Load Voltage Range: 24 to 40V • 15 W applied continuously • 125 Pulses, 8 Amperes, 0.1 Sec. Duration, Evenly Distributed
Operating Life	• 480 Seconds Minimum

20.4 PERFORMANCE CHARACTERISTICS

20.4.1 Ampoule-Type Batteries

Voltage characteristics at the "time of activation" are unique and an important feature of reserve batteries. This is especially true for military applications, where reserve batteries must normally be designed to meet operational voltage in less than 1 s and in many cases even less than $\frac{1}{2}$ s. For nonmilitary use, activation times to operating voltage level are less critical. However, for a given reserve battery design and the electrochemical couple used, the activation time is dependent on the discharge rate and temperature.

In general, the voltage rise times for both $Li/SOCl_2$ and Li/V_2O_5 have similar characteristics. Figure 20.13 shows the rise-time characteristics for the $Li/SOCl_2$ battery (illustrated in Fig. 20.4) at five temperatures at a current density of 0.1 mA/cm^2 (approximately a $C/500$ rate). Rise times are typically below 20 ms at ambient (24°C) and higher temperatures but increase up to 500 ms at the lower temperatures. The ability to activate rapidly is primarily due to the cell design, which allows the electrolyte to penetrate and wick into the porous electrode and separator at the instant of ampoule breakage.

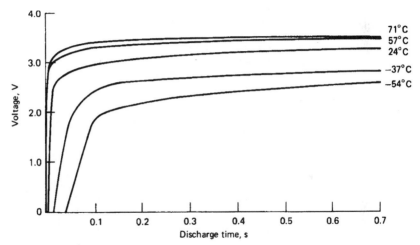

FIGURE 20.13   Rise-time characteristics after activation of $Li/SOCl_2$ reserve battery, Alliant model G2659B1; load = 4.35 kΩ. (*Courtesy of Alliant Techsystems, Inc.*)

The voltage levels of both the Li/V_2O_5 and the $Li/SOCl_2$ systems (batteries illustrated in Figs. 20.3 and 20.4) under steady-state discharge conditions are shown in Fig. 20.14. These two systems are very close in voltage at the lower temperatures, ranging from 3.3 to 3.0 V at current densities of less than 1 mA/cm^2. At higher temperatures, ambient and above up to 74°C, the $Li/SOCl_2$ battery operates above 3.5 V, whereas the Li/V_2O_5 batteries normally operate between 3.2 and 3.4 V. The higher voltage and increased capacity account for the significant increase in energy density of the $SOCl_2$ over the V_2O_5 battery. As shown in Fig. 20.14, the V_2O_5 system has very little change in capacity over the wide temperature range but is still much lower in capacity than the $SOCl_2$ battery when discharged at the same rate, namely 0.1 mA/cm^2. Although the capacity and voltage of the $Li/SOCl_2$ battery are lower at cold temperatures, its output is still higher than that of most other systems and its voltage profile is characterized by a flat single-step plateau. The high-temperature curve is also extremely flat and typically discharges above 3.6 V at a current density of 0.1 mA/cm^2. The voltage characteristics on ambient temperature discharges are similar to those at high temperature except for a slightly lower load voltage when discharged at the same rate,

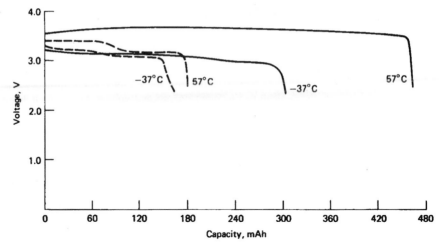

FIGURE 20.14 Comparison of discharge profiles for reserve-type Li/V$_2$O$_5$ (---) and Li/SOCl$_2$ (——) batteries. Current density = 0.1 mA/cm^2.

averaging 3.5 V. Table 20.4 compares the output parameters of the two systems with identical hardware and shows the superior performance of the Li/SOCl$_2$ battery. The similarity in voltage and the fact that the same hardware is used for both systems permits a one-for-one replacement.

Figure 20.15 shows the effect of inactive storage of up to 12 months at 71°C on the Li/SOCl$_2$ battery performance over the temperature range of −54 to 71°C. No significant effect on performance was found as a result of the storage. The slightly lower voltages during discharge or the voltage delays when the load is first applied on active (nonreserve) batteries were not present with the reserve batteries. Figure 20.15 also gives a summary of the performance of fresh batteries at various discharge loads and temperatures.

Figure 20.16 shows the discharge curves of the Li/SOCl$_2$ reserve A-size battery (illustrated in Fig. 20.2). The current drain capability of a reserve system significantly exceeds that of the corresponding active primary battery. Figure 20.16 shows the discharge characteristics at 1.25 kΩ (about 3 mA) or 0.15 mA/cm^2. Currents higher than 1.5 A (current density of 100 mA/cm^2) can be obtained at voltages higher than 2.0 V for several minutes at −10°C. The specific energy of the cells, to a cutoff voltage of 2.0 V, as a function of the discharge current at various temperatures is shown in Fig. 20.17. The performance at higher temperatures is close to that for 25°C.

TABLE 20.4 Performance Comparison between Li/SOCl$_2$ and Li/V$_2$O$_5$ Systems

System	Temperature, °C	Cell voltage, V	Capacity, mAh	Cell volume, cm^3	Cell weight, g	Specific energy Wh/kg	Energy density Wh/L
Li/V$_2$O$_5$*	−37	3.15	160	5.1	10	50.4	98.8
	57	3.30	180	5.1	10	59.4	116.5
Li/SOCl$_2$†	−37	3.05	300	5.1	10.5	87.1	179.4
	57	3.60	450	5.1	10.5	154.3	317.6

* Alliant model G2659.
† Alliant model G2659B1.

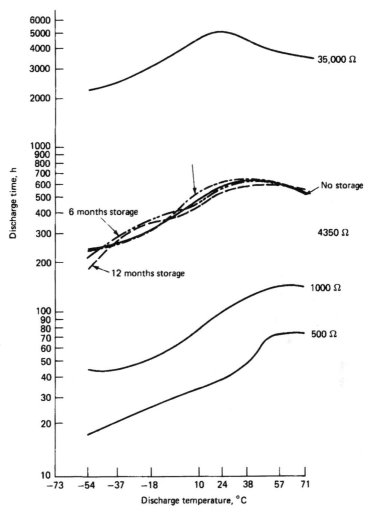

FIGURE 20.15   Effect of discharge rate and inactive storage on Li/SOCl$_2$ reserve battery, Alliant model G2659B1. (*Courtesy of Alliant Techsystems, Inc.*)

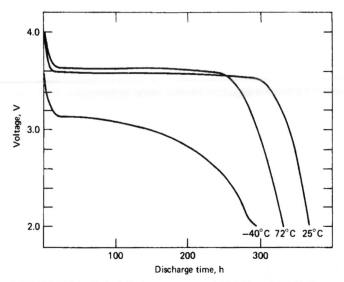

FIGURE 20.16 Typical discharge curves on 1.25 kOhm of Li/SOCl₂ reserve battery, Tadiran model TL-5160. (*Courtesy of Tadiran Industries, Ltd*).

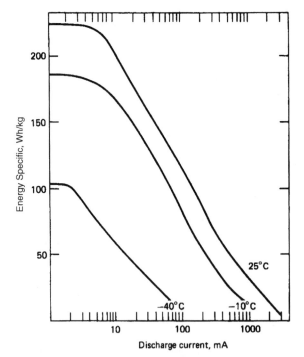

FIGURE 20.17 Specific energy as a function of discharge current and temperature for Li/SOCl₂ reserve battery, Tadiran model TL-5160. (*Courtesy of Tadiran Industries, Ltd.*)

20.4.2 Multicell Battery Design

The performance characteristics of the Li/SO_2 multicell single-activation design battery are shown in Fig. 20.18. The activation and discharge profiles for a 12-V, 100-Ah battery using an $LiAsF_6$ in $AN–SO_2$ electrolyte are illustrated. Because the battery is activated manually, the slower voltage rise time is expected because the cutting of the diaphragm in the center bulkhead requires several turns on the activation bolt. The battery could easily be activated with an electric or mechanical input to a piston actuator or squib to cut the diaphragm to improve the voltage rise time. Although the operational life of the battery can be very short, the data illustrate its capability for a long-term discharge at low discharge rates if a stable electrolyte is used.

Typical discharge curves at several temperatures for the two types of single-cells used in the $Li/SOCl_2$ reserve battery illustrated in Fig. 20.8 are shown in Fig. 20.19.

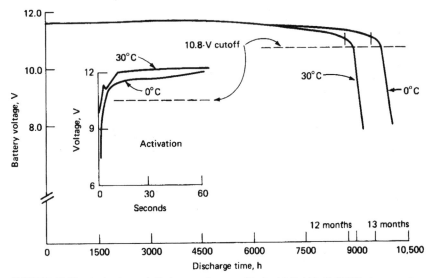

FIGURE 20.18 Activation and discharge voltage profile for 12-V, 100-Ah Li/SO_2 battery; electrolyte: $LiAsF_6$ in $AN–SO_2$.

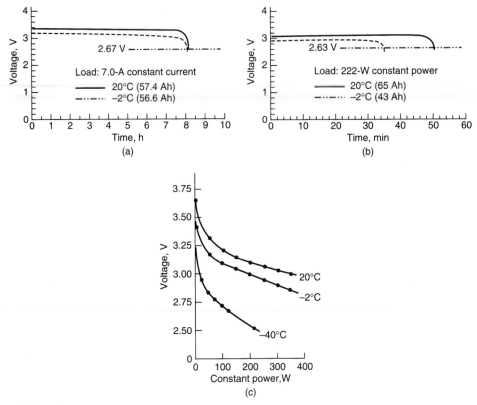

FIGURE 20.19 Discharge characteristics of Li/SOCl₂ reserve batteries, Alliant model G3070A2. (*a*) Low-rate-cell discharge profile: (*b*) High-rate-cell discharge profile. (*c*) Voltage characteristics of high-rate cell at various temperatures and power levels. (*Courtesy of Alliant Techsystems, Inc.*)

REFERENCES

1. A. N. Dey, "Lithium Anode Film and Organic and Inorganic Electrolyte Batteries," in *Thin Solid Films,* vol. 43, Elsevier Sequoia, Lausanne, Switzerland, 1977, p. 131.

2. R. J. Horning, "Small Lithium/Vanadium Pentoxide Reserve Cells," *Proc. 10th Intersoc. Energy Convers. Eng. Conf.,* 1975.

3. W. B. Ebner and C. R. Walk, "Stability of LiAsF₆-Methyl Formate Electrolyte Solutions," *Proc. 27th Power Sources Conf.,* 1976.

4. B. Ravid, *A Reserve-Type Lithium-Thionyl Chloride Battery,* Tadiran Israel Electronics Industries, 1979.

5. M. J. Domenicomi and F. G. Murphy, "High Discharge Rate Reserve Cell and Electrolyte," U.S. Patent 4,150,198, Apr. 17, 1979.

6. M. Babai, U. Meishar, and B. Ravid, "Modified Li/SOCl₂ Reserve Cells with Improved Performance," *Proc. 29th Power Sources Conf.,* June 1980.

7. P. M. Shah, "A Stable Electrolyte for Li/SO₂ Reserve Cells," *Proc. 27th Power Sources Symp.,* 1976.

8. P. M. Shah and W. J. Eppley, "Stability of the LiAsF₆:An:SO₂ Electrolyte," *Proc. 28th Power Sources Symp.,* 1978.

9. R. J. Horning and K. F. Garoutte, "Li/SO$_2$ Multicell Reserve Structure," *Proc. 27th Power Sources Symp.,* 1976.

10. Hsiu-Ping Lin and K. Burgess, "Synthesis of Charged Li$_x$CoO$_2$ ($0 < x < 1$) For Primary and Secondary Batteries," U.S. Patent 5,667,660 (1997).

11. W. J. Eppley and R. J. Horning, "Lithium/Thionyl Chloride Reserve Cell Development," *Proc. 28th Power Sources Symp.,* 1978.

12. J. Nolting and N. A. Remer, "Development and Manufacture of a Large Multicell Lithium-Thionyl Chloride Reserve Battery," *Proc. 35th International Power Sources Symp.,* 1992.

13. C. Kelly, "Development of HWAM Li$_x$CoO$_2$ Reserve Battery," Report No. NSWCCD-TR-98/005, April 1997.

14. S. McKay, M. Peabody and J. Brazzell, *Proc. 39th Power Sources Conf.,* pp. 73–76, 2000.

15. P. G. Russell, D. C. Williams, C. Marsh, and T. B. Reddy, *Proc. 6th Workshop for Battery Exploratory Development,* pp. 277–281, 1999.

CHAPTER 21
THERMAL BATTERIES

Visvaldis Klasons and Charles M. Lamb

21.1   GENERAL CHARACTERISTICS

Thermal batteries are primary reserve batteries that employ inorganic salt electrolytes. These electrolytes are relatively nonconductive solids at ambient temperatures. Integral to the thermal battery are pyrotechnic materials scaled to supply sufficient thermal energy to melt the electrolyte. The molten electrolyte is highly conductive, and high currents may then be drawn from the cells.

The activated life of a thermal battery depends on several factors involving cell chemistry and construction. Once activated, and as long as the electrolyte remains molten, thermal batteries may supply current, discharging the active materials to the point of functional exhaustion. On the other hand, even with excess active materials present, the batteries will eventually cease functioning due to the loss of internal heat and subsequent re-solidification of the electrolyte. Hence, two of the primary factors behind thermal battery active life are:

1. Compositions and masses of the active cell stack materials (i.e. anodes and cathodes), and

2. Other construction details, including the overall battery shape and the types and amounts of thermal insulation.

Depending on the battery design, which is ultimately determined by the specific requirements of the application, the activated thermal battery may supply electric power for only a few seconds, or may function for over an hour.

Initiation of a thermal battery is normally provided by an energy impulse from an external source to a built-in initiator. The initiator, typically an electric match, an electro-explosive device (squib), or a percussion primer, ignites the cell stack pyrotechnics. Rise time, the time interval between the initiation impulse and that time at which the battery can sustain a current at voltage, varies as a function of battery size, design, and chemistry. Rise times of several hundred milliseconds are not uncommon for large units. Small batteries have been designed to reliably achieve operating conditions within 10 to 20 milliseconds.

The shelf life of an unactivated thermal battery is typically 10 to 25 years, depending upon design. Once activated and discharged, though, they are not reusable or rechargeable.

Current developments in extending the activated life capabilities of thermal batteries have widened their suitability and application potential in new military as well as industrial/civilian systems.

Thermal batteries were first developed in Germany in the 1940s, and were used primarily for weapons applications.[1-3] Batteries containing multiple cells and integral pyrotechnic materials have been produced since 1947.[4] Because of their high reliability and long shelf life, thermal batteries are ideally suited for military ordnance purposes. Consequently, they have been widely used in missiles, bombs, mines, decoys, jammers, torpedoes, space exploration systems, emergency escape systems, and similar applications. Figure 21.1 illustrates typical thermal battery configurations.

FIGURE 21.1   Typical thermal batteries. (*Courtesy of Catalyst Research Corp.*)

Some of the advantages of thermal batteries include:

1. Very long shelf life (up to 25 years) in a "ready" state without degradation in performance.
2. Almost "instant" activation; fast start designs can provide useful power in hundredths of a second.
3. Peak-power densities can exceed 11 W/cm^2.
4. Very high demonstrated reliability and ruggedness following long-term storage over a wide temperature range and severe dynamic environments.
5. No maintenance required; they can be permanently installed in equipment.
6. Self-discharge is generally negligible. An unactivated battery can support almost no current.
7. Wide operating temperature range.
8. No outgassing; the batteries are hermetically sealed.
9. Custom designed for specific voltage, start time, current, and physical configuration requirements.

The disadvantages of thermal batteries include:

1. Generally short activated lives (typically less than 10 min), but they can be designed to operate for more than 2 hours.
2. Low to moderate energy densities and specific energies.
3. Surface temperatures can typically reach 230°C or higher.
4. Voltage output is nonlinear, and decreases with life.
5. One time use. Once activated, they cannot be turned off or reused (recharged).

21.2 DESCRIPTION OF ELECTROCHEMICAL SYSTEMS

A number of electrochemical systems have been used in thermal batteries. As materials and techniques have improved the state-of-the-art (SOA) performance of these batteries, older designs have gradually disappeared. Battery designs with older technologies still exist, however, and continue to be manufactured. In some cases, the continuing production of an "antiquated" system is driven by economics. Redesign and requalification of an existing battery with a newer technology is often economically unacceptable. Table 21.1 lists some of the more common types of electrochemical systems that have been used over the years.

All thermal battery cells consist of an alkali or alkaline earth metal anode, a fusible salt electrolyte, and a metal salt cathode. The pyrotechnic heat source is usually inserted between cells in a series cell-stack configuration.

TABLE 21.1 Types of Thermal Batteries

Electrochemical system: anode/electrolyte/cathode	Operating cell voltage	Characteristics and/or applications
$Ca/LiCl\text{-}KCl/K_2Cr_2O_7$	2.8–3.3	Very fast activation times; short lives; used in "pulse" applications
$Ca/LiCl\text{-}KCl/WO_3$	2.4–2.6	Medium-short lives; low electrical noise; not severe physical environments
$Ca/LiCl\text{-}KCl/CaCrO_4$	2.2–2.6	Medium lives; severe dynamic environments
$Mg/LiCl\text{-}KCl/V_2O_5$	2.2–2.7	Medium-short lives; severe physical environments
$Ca/LiCl\text{-}KCl/PbCrO_4$	2.0–2.7	Fast activation; short lives
$Ca/LiBr\text{-}KBr/K_2Cr_2O_4$	2.0–2.5	Short lives; used in high-voltage, low-current applications
$Li(alloy)/LiCl\text{-}LiBr\text{-}LiF/FeS_2$	1.6–2.1	Short to medium lives, high current capacity; severe physical environments
$Li(metal)/LiCl\text{-}KCl/FeS_2$	1.6–2.2	Long lives, high current capacity; severe physical environments
$Li(alloy)/LiBr\text{-}KBr\text{-}LiF/CoS_2$	1.6–2.1	Long lives (past 1 h), high current capacity; severe physical environments

21.2.1   Anode Materials

Until the 1980s, most thermal battery designs employed a calcium metal anode—with calcium foil generally attached to an iron, stainless steel, or nickel foil current collector or backing. A "bimetal" anode is manufactured by vapor depositing the calcium on the backing material. Here, the calcium anode thickness usually ranges between 0.03 and 0.25 mm. In other designs, calcium foil is either pressed onto a perforated "cheese grater"-type backing sheet, or is spot-welded to the backing. Magnesium metal is another anode material that has been widely used, both in "bimetal"-form and in pressed or spot-welded anode configurations.

Introduced in the mid-1970s, lithium has become the most widely used anode material in thermal batteries. There are two major configurations of lithium anodes: lithium alloy and lithium metal. The most commonly used alloys are lithium aluminum, with about 20 weight percent lithium and lithium (silicon), with about 44 weight percent lithium. Lithium-boron alloy has also been evaluated, but has not been used widely because of its higher cost.

LiAl and Li(Si) alloys are processed into powders, which are cold-pressed into anode wafers or pellets that range in thickness from 0.75 to 2.0 mm. In the cell, the alloy pellet is backed with an iron, stainless steel, or nickel current collector. Lithium alloy anodes function in activated cells as solid anodes, and must be maintained below melt or partial melt temperatures. Forty-four weight percent Li(Si) alloy will partially melt at 709°C, while α, β-LiAl will exhibit partial melting at 600°C. If these melting temperatures are exceeded, the melted anode may come in contact with cathode material, allowing a direct, highly exothermic chemical reaction and cell short-circuiting.

Lithium metal anodes function in activated cells at temperatures above the melting temperature of lithium, 181°C. To prevent the molten lithium from flowing out of the cells and short-circuiting the battery, it is combined with a high-surface-area binder of metal powder or metal foam. The binder holds the lithium in place by surface tension.

Lithium metal anodes are prepared by combining the binder material with molten lithium, followed by pressing the solidified mixture into thin foil, typically 0.07 to 0.65 mm thick. The foil is then cut into cell-sized parts. The anode foil parts are enclosed in iron-foil cups, which provide added protection against the migration of any free lithium (which can result in cell shorting) and also serve as electron collectors (electrical connections). Such anodes can function at cell temperatures greater than 700°C without significant loss of performance.[5] Each thermal battery designer or manufacturer has developed a number of anode configurations, from which the most suitable may be selected, depending upon specific battery performance requirements.

21.2.2   Electrolytes

Historically, most thermal battery designs have used a molten eutectic mixture of lithium chloride and potassium chloride as the electrolyte (45:55 LiCl:KCl by weight, mp = 352°C). Halide mixtures containing lithium have been preferred because of their high conductivities and general overall compatibility with the anodes and cathodes. Compared to most lower-melting oxygen-containing salts, the halide mixtures are less susceptible to gas generation via thermal decomposition or other side reactions. More recent electrolyte variations, containing bromides, have been developed for thermal batteries to achieve a lower melting point (and thus extend the operating life) or to reduce the internal resistance (and raise the current capability) of the batteries. These include LiBr-KBr-LiF (mp = 320°C), LiCl-LiBr-KBr (mp = 321°C), and the all-Li^+ electrolyte LiCl-LiBr-LiF (mp = 430°C).[6] Electrolytes with mixed-cations (*e.g.*, Li^+ and K^+, instead of all-Li^+) are subject to the establishment of Li^+ concentration gradients during discharge. These concentration gradients can give rise to localized freezing out of salts, especially during high current draw.[7]

At battery operating temperatures, the viscosity of molten salt electrolytes is very low (*ca.* 1 centiPoise). In order to immobilize the molten electrolytes, binders are added to the salts during compounding. Earlier blends, originally developed for $Ca/CaCrO_4$ systems and the original $LiAl/FeS_2$ batteries, employed clays, such as kaolin, and fumed silica as effective binders for the salts. These siliceous materials will react with Li(Si) and lithium metal anodes, however. High surface area MgO is sufficiently inert for the more reactive anodes, and is presently the binder of choice in most systems.

21.2.3 Cathode Materials

A wide variety of cathode materials has been used for thermal batteries. These include calcium chromate ($CaCrO_4$), potassium dichromate ($K_2Cr_2O_7$), potassium chromate (K_2CrO_4), lead chromate ($PbCrO_4$), metal oxides (V_2O_5, WO_3), and sulfides (CuS, FeS_2, CoS_2). The criteria for suitable cathodes include high voltage against a suitable anode, compatibility with halide melts, and thermal stability to approximately 600°C. Calcium chromate has been most often used with calcium anodes because of its high potential (at 500°C, V = 2.7) and its thermal stability at 600°C. FeS_2 and (more recently) CoS_2 are used with modern lithium-containing anodes (FeS_2 to 550°C and CoS_2 to 650°C).

21.2.4 Pyrotechnic Heat Sources

The two principal heat sources that have been used in thermal batteries are heat paper and heat pellets. Heat paper is a paper-like composition of zirconium and barium chromate powders supported in an inorganic fiber mat. Heat pellets are pressed tablets or pellets consisting of a mixture of iron powder and potassium perchlorate.

The Zr-$BaCrO_4$ heat paper is manufactured from pyrotechnic-grade zirconium powder and $BaCrO_4$, both with particle sizes below 10 microns. Inorganic fibers, such as ceramic and glass fibers, are used as a structure for the mat.[8] The mix, together with water, is formed into a paper—either as individual sheets by use of a mold or continuously through a paper-making process. The resultant sheets are cut into parts and dried. Once dried, the material must be handled very carefully since it is very susceptible to ignition by static charge and friction. Heat paper has a burning rate of 10 to 300 cm/s and a usual heat content of about 1675 J/g (400 cal/g). Heat paper combusts to an inorganic ash with electrical resistivity. If inserted between cells, it must be used in combination with highly conductive inter-cell connectors. In some battery designs, combusted heat paper serves as an electrical insulator between cells. In those applications it may have an additional layer of ceramic fibers only, known as *base sheet,* to enhance its dielectric properties. In most modern pellet-type batteries, heat paper is used only as an ignition or fuse train, if at all. In this application, the heat paper fuse, which is ignited by the initiator, in turn ignites the heat pellets, which are the primary heat source in these batteries.

Heat pellets are manufactured by cold-pressing a dry blend of fine iron powder (1 to 10 micron) and potassium perchlorate. The iron content ranges from 80 to 88% by weight, and is considerably in excess of stoichiometry. Excess iron provides the combusted pellet with sufficient electronic conductivity, eliminating the need for separate inter-cell connectors. The heat content of Fe-$KClO_4$ pellets ranges from 920 J/g for 88% iron to 1420 J/g for 80% iron. Burning rates are generally slower than those of heat paper, and the energy required to ignite them is greater. For that reason, the heat pellet is less susceptible to inadvertent ignition during battery manufacture. Heat pellets (and especially unpelletized heat powder) must, nevertheless, be handled with extreme care and protected from potential ignition sources.

After combustion, the heat pellet is an electronic conductor, simplifying inter-cell connection and battery design. It also retains its physical shape and is very stable under dynamic environments (such as shock vibration and spin). This contributes greatly to the general ruggedness of battery designs that incorporate heat pellets. Another major advantage of heat pellets is that their enthalpy of reaction is much higher than that of heat paper ash. Thus, they serve as heat reservoirs, retaining considerable heat after combustion, and tend to extend the battery active life by virtue of their greater thermal mass.

21.2.5 Methods of Activation

Thermal batteries are activated by applying an external signal to an initiation device that is incorporated in the battery. There are four generally used methods of activation: electric signal to an electric igniter; mechanical impulse to a percussion primer; mechanical shock to an inertial activator; and optical energy (laser) signal to a pyrotechnic material.

Electric igniters typically contain one or more bridge wires and a heat-sensitive pyrotechnic material. Upon application of an electric current, the bridge wire ignites the pyrotechnic, which in turn ignites the heat source in the thermal battery. Igniters generally fall into two categories: squibs and electric matches. A typical squib is enclosed in a sealed metal or ceramic enclosure and contains one or two bridge wires. The most commonly used types require a minimum activation current of 3.5 A and have a maximum no-fire limit of 1 A or 1 W (whichever is greater). Electric matches do not have a sealed enclosure and typically contain only one bridge wire. They require an activation current of 500 mA to 5 A and should not be subjected to a no-fire test current of more than 20 mA. Squibs are 4 to 10 times more expensive than electric matches, but are required for applications that may encounter environments with electromagnetic radiation.

Percussion primers are pyrotechnic devices that are activated by impact from a mechanical striking device. Typically, a primer is activated by an impact of 2016 to 2880 g · cm applied with a 0.6 to 1.1 mm spherical radius firing pin. Primers are installed in primer holders that are integral parts of the battery enclosure.

Inertial or setback activators are devices that are activated by a large-magnitude shock or rapid acceleration, as is generated upon firing of a mortar or artillery round. They are designed to react to a predetermined combination of g force and its duration. Inertial activators are typically firmly mounted inside the battery structure in order to withstand severe dynamic environments.

Optical energy (laser) activation of thermal batteries is accomplished by "firing" a laser beam through an optical "window" installed in the outer enclosure of the battery and igniting a suitable pyrotechnic material inside the unit. This method has found utility in applications where severe electromagnetic interference would be disruptive to an electrical firing method.

Thermal batteries can be equipped with more than one activation device. The multiple activators can be of the same type or of any combination required by the application.

21.2.6 Insulation Materials

Thermal batteries are designed to maintain hermeticity throughout their service lives, even though their internal temperatures reach or even exceed 600°C. The thermal insulation used to retard heat loss from the cell stack and minimize peak surface temperatures must be anhydrous and must have high thermal stability. Ceramic fibers, glass fibers, certain high-temperature polymers, and their combinations have been used as thermal insulators. Older battery designs may still have asbestos insulation, which was widely used before the 1980s.

Electrical insulation materials for conductors, terminals, initiators, and other electrically conductive components are typically made of mica, glass or ceramic fiber cloths, and high temperature-resistant polymers.

Thermal insulation is located around the periphery and at both ends of the cell stack. Some designs also incorporate high-temperature epoxy potting materials as insulation and structural support for the initiators and electric conductors on the terminal end (header) of the batteries. Long-life batteries (20+ min.) usually incorporate high-efficiency thermal insulation materials such as Min-K (Johns Manville Co.) or Micro-Therm (Constantine Wingate, Ltd.). Extended life batteries (1 hr and longer) may incorporate vacuum blankets and / or double cases with vacuum space between them to retard heat loss. Special high-thermal-capacity pellets and extra "dummy" cells are also used at the ends of cell stacks to retard heat loss and thus prolong the activated life of some batteries.[9] Figure 21.2 shows a typical arrangement of thermal insulation and an encapsulated header assembly with initiator (squib).

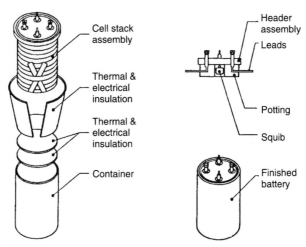

FIGURE 21.2 Typical thermal battery assembly. (*Courtesy of Eagle-Picher Technologies, LLC.*)

A very effective method for extending the activated battery life and reducing the effects of heat on thermally sensitive components located near the battery is to use an external thermal blanket. Provided that it is protected from external contamination, a thermal blanket is more effective than internal insulation, primarily because the hot gasses that are generated inside the battery during activation cannot penetrate it. External insulation, mounting methods, and the surrounding environment have a significant effect on the heat loss from the battery and all of these must be taken into consideration in the design of thermal batteries.

21.3 CELL CHEMISTRY

A wide variety of different cell chemistries have been developed and used in thermal batteries. At this time, the most widely used chemistry is lithium/iron disulfide (Li/FeS_2), with calcium/calcium chromate ($Ca/CaCrO_4$) as a distant second. There are special applications, though, where one of the other less used chemistries could offer special advantages. As an example, the requirement for a very fast activation time with a relatively short activated life would be provided by the calcium/potassium dichromate ($Ca/LiCl$-$KCl/K_2Cr_2O_7$) system or the calcium/lead chromate ($Ca/LiCl$-$KCl/PbCrO_4$) system. For a very general overview, Table 21.2 lists some example-specific performance characteristics of various thermal battery chemistries and designs.

TABLE 21.2 Characteristics of Various Thermal Batteries

Cell type	Volume, cm³	Weight, g	Nominal voltage, V	Current, A	Peak power, W	Average life, s	Specific energy (Wh/kg)	Energy density (Wh/L)
Cup/WO₃	450	850	7	5.8	41	70	2.3	4.3
Open cell/tape/WO₃	100	385	50	0.36	15	70	1.3	5.0
Open cell/tape/dichromate	44	148	18	26.0	462	1.2	1.0	3.5
Open cell/tape/dichromate	1	5.5	10	5.0	50	0.15	0.4	2.1
Open cell/tape/bromide	81	225	203	0.02	4	45	0.2	0.7
Pellet/CaCrO₄/heat paper	123	310	42	2.9	125	25	2.8	6.8
Pellet/CaCrO₄/heat pellet	105	307	28	1.2	34	150	4.6	13.4
Pellet/CaCrO₄/heat pellet	105	307	28	2.5	75	60	3.8	11.1
Pellet/LiM*/FeS₂	92.3	320	25	15.0	420	35	11.4	39.0
Pellet/LiM*/FeS₂	170	505	28	12.0	378	140	26.2	82.0
Pellet/LiM*/FeS₂	208	544	138	1.0	138	250	32.2	84.1
Pellet/LiM*/FeS₂	3120	6620	315	10.0	3600	250	33.1	77.0
Pellet/LiM*/FeS₂	334	907	65	7.95	541	320	43.0	116.0
Pellet/LiM*/FeS₂	552	1400	27	12.0	372	600	38.7	111.8
Pellet/LiM*/FeS₂	1177	270	27	17.0	459	900	35.1	97.5

M*—either alloy or metal.

21.3.1 Lithium/Iron Disulfide

There are three common lithium anode configurations: Li(Si) alloy, LiAl alloy, and Li metal in metal matrix, Li(M), where the matrix is usually iron powder. With the difference that the alloy anodes remain solids and the lithium in the Li(Fe) mix is molten in an activated cell, all three anodes participate in the cell reaction similarly. All may be used with the same FeS_2 cathode and the same electrolytes. These electrolytes may be the basic LiCl-KCl eutectic electrolyte, LiCl-LiBr-LiF electrolyte for best ionic conductivity, or a lower-melting-point electrolyte such as LiBr-KBr-LiF for extended activated life. Since the FeS_2 is a good electronic conductor, the electrolyte layer is necessary in order to prevent direct anode-to-cathode contact and cell short-circuiting. When molten, the electrolyte between the anode and the cathode is held in place by capillary action through the use of a chemically compatible (inert) binder material. MgO is the preferred material for this application.[10]

The Li/FeS_2 electrochemical system has become the preferred system because it does not contain any parasitic chemical reactions. The extent of self-discharge depends on the type of electrolyte used and the cell temperature.[11] The predominant discharge path for cathodes is:

$$3Li + 2FeS_2 \rightarrow Li_3Fe_2S_4 \ (2.1 \text{ V})$$

$$Li_3Fe_2S_4 + Li \rightarrow 2Li_2FeS_2 \ (1.9 \text{ V})$$

$$Li_2FeS_2 + 2Li \rightarrow Fe + 2Li_2S \ (1.6 \text{ V})$$

Most batteries are designed to use only the first and sometimes the second cathode transitions to avoid changes in cell voltage.

The transitions that occur at the anode depend on the alloy used. For LiAl:

$$\beta\text{-LiAl} \ (ca. \ 20 \text{ wt } \% \text{ Li}) \rightarrow \alpha\text{-Al (solid solution)}$$

Below approximately 18.4 weight percent lithium (lower limit for all β-LiAl) and above 10 weight percent lithium (upper limit for α-Al), the alloy is two-phase α,β-LiAl. This fixes the alloy voltage on a plateau. This plateau is about 300 mV less than the voltage afforded by pure lithium metal.

The composition transitions for Li(Si) are:

$$Li_{22}Si_5 \rightarrow Li_{13}Si_4 \rightarrow Li_7Si_3 \rightarrow Li_{12}Si_7$$

An anode voltage plateau is defined for compositions falling between each adjacent pair of alloys. That is, the first plateau occurs between $Li_{22}Si_5$ and $Li_{13}Si_4$. The 44 weight percent Li(Si) composition falls here, and begins its discharge approximately 150 mV less than that of pure lithium.

The use of FeS_2 as a cathode material can cause a large voltage transient or "spike" of 0.2 V or more per cell, which is evident immediately after activation and lasts from milliseconds to a few seconds. This phenomenon is related to the impact of temperature, the amounts of electroactive impurities in the raw material (iron oxides and sulfates), elemental sulfur from FeS_2 decomposition, and the activity of lithium not being fixed in the cathode. In applications where the voltage has to be well regulated, this "spike" is not acceptable. The voltage transient can be virtually eliminated by the addition of small amounts of Li_2O or Li_2S (typically 0.16 mol Li per mol FeS_2) to the catholyte (FeS_2 and electrolyte blend), a method known as *multiphase lithiation*.[12] The spike can also be reduced (but not eliminated) by thoroughly washing or vacuum treating the FeS_2 to remove acid-soluble impurities and elemental sulfur.

The Li/FeS_2 electrochemical system has a number of important advantages over other systems, including $Ca/CaCrO_4$. These advantages include:

- Tolerance of a wide range of discharge conditions, from open circuit to high current densities
- High current capabilities; 3 to 5 times that of $Ca/CaCrO_4$
- Highly predictable performance
- Simplicity of construction
- Tolerance to processing variations
- Stability in extreme dynamic environments

As a result of these advantages, this system has become the predominant choice for a wide range of high-reliability military and space applications.

21.3.2 Lithium / Cobalt Disulfide

As a cathode *vs.* lithium in molten salt electrolyte cells, cobalt disulfide exhibits a slightly lower voltage than does iron disulfide. Cobalt disulfide has a greater thermal stability with respect to loss of sulfur, however. The decomposition reactions for cobalt disulfide at elevated temperatures are:

$$3CoS_2 \rightarrow Co_3S_4 + S_2 \text{ (g)}$$

$$3Co_3S_4 \rightarrow Co_9S_8 + 2S_2 \text{ (g)}$$

For iron disulfide at elevated temperatures:

$$2FeS_2 \rightarrow 2FeS + S_2 \text{ (g)}$$

As a rough indicator of the relative stabilities, FeS_2 will have a sulfur vapor pressure (p_{S2}) of 1 atm in equilibrium with it at 700°C, whereas p_{S2} = 1 atm for CoS_2 at 800°C. It is, therefore, no surprise that the substitution of CoS_2 for FeS_2 can yield a more high-temperature-stable cell, and is therefore useful in batteries with activated lives of over 1 hour.[13] In an active battery, the decomposition of FeS_2 to FeS and elemental sulfur becomes significant above approximately 550°C. The free sulfur can combine directly with the Li anode in a highly exothermic reaction. Not only would this reduce available anode capacity, but the extra heat can cause even more thermal decomposition of the cathode. CoS_2, which is stable up to 650°C, allows higher initial stack temperatures to be sustained without excessive degradation of the cathode. It has also been demonstrated that cells with CoS_2 cathodes have a lower internal resistance later in activated life than do FeS_2 cathodes.

21.3.3 Calcium / Calcium Chromate

The reactions that take place in a $Ca/CaCrO_4$ thermal cell during activation must be in critical balance for the cell to function properly. Upon activation (application of heat), the calcium anode reacts with lithium ions in the LiCl-KCl eutectic electrolyte to form liquid beads of Ca-Li alloy. This alloy becomes the operational anode in the subsequent electrochemical reaction. The anodic half-cell reaction is:

$$CaLi_x \rightarrow CaLi_{x-y} + yLi^+ + ye^-$$

The Ca-Li alloy beads also react with dissolved $CaCrO_4$, forming a coating of $Ca_5(CrO_4)_3Cl$.[15,16] This Cr(V) compound is the same species that is formed in the cathodic half-cell reaction:

$$3CrO_4^{-2} + 5Ca^{+2} + Cl^- + 3e^- \rightarrow Ca_5(CrO_4)_3Cl$$

This "product" acts as a separator or mass transport barrier between the cathode and the anode to limit electrochemical self-discharge. If the integrity of this separator is breached, the battery can experience a "thermal runaway" condition, whereby the active electrochemical components are chemically consumed with accompanying generation of large amounts of excess heat. At the same time, if battery conditions are such that alloy formation exceeds usage, the excess alloy can cause periodic shorting, the "alloy noise" sometimes seen in cold-stored batteries.

The balance between chemical and electrochemical reactions in this system is dependent on the source of materials (particularly $CaCrO_4$), processing variations, density of compression-formed pellets, operating temperature of the cell, rate of current drain, and other variables.[17] Consequently, this system has been gradually phased-out in favor of the more stable and predictable lithium/iron disulfide cell chemistry which also has a higher energy density.

21.4 CELL CONSTRUCTION

A number of factors, including the cell chemistry used, the operating environments of the battery and the preferences of the designer, determine the choice of cell design. Basically, all cell designs fall into three categories: cup cells, open cells, and pelletized cells. To meet specific performance requirements, some designs may incorporate aspects of more than one cell category. Figure 21.3 illustrates the relative thickness ranges of the different cell designs.

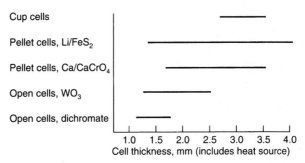

FIGURE 21.3 Thicknesses of thermal battery cells.

21.4.1 Cup Cells

The typical cup cell features a two layer anode (calcium or magnesium) having active anode material on both sides of a central current collector. On either side of the anode is an electrolyte pad made of glass tape impregnated with eutectic electrolyte. Next to each electrolyte are depolarizer pads consisting of cathode material ($CaCrO_4$ or WO_3) in an inorganic fiber matrix (paper). The cell is enclosed in a nickel foil cup and cover that are tightly crimped (Fig. 21.4a). Some designs also incorporate inorganic fiber mat gaskets and nickel "eyelets" to help prevent the molten electrolyte from leaking out of the activated cell. Zr/ $BaCrO_4$ heat paper pads located on either side provide heat to the cup cell.

Cup cells have the advantage of large reactive surfaces (they are two-sided or bipolar), and can contain relatively large amounts of reactive materials. Their disadvantages are that they are difficult to seal against electrolyte leakage and they have low heat capacity. The $Ca/CaCrO_4$ cell chemistry is also prone to "alloying" (producing excess molten Ca-Li alloy), which can short-circuit the cells. In order to obtain required short activation times, cup cells typically have to be pre-melted or "pre-fused" prior to assembly into cell stacks. Inter-cell electrical connections are accomplished by spot-welding the cell output leads between each cell, which presents a potential reliability problem.

Currently, cup cells have limited application, and are found primarily in older battery designs.

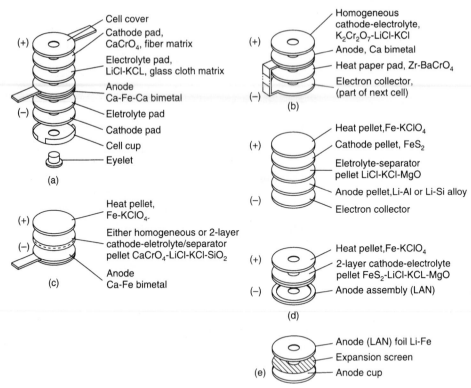

FIGURE 21.4   Variations in cell configurations: (*a*) Cup cell. (*b*) Open cell. (*c*) Ca/CaCrO$_4$ pellet cell. (*d*) Li alloy/FeS$_2$ and Li/FeS$_2$ pellet cell. (*e*) Li metal/FeS$_2$, (LAN) anode assembly.

21.4.2   Open Cells

The open-cell design is similar in construction to the cup cell, except that it is not enclosed in a cup (Fig. 21.4*b*). Elimination of the cup is possible because the amount of electrolyte is reduced to the extent that practically all of it is bound to the glass fiber cloth matrix by surface tension. Some designs use homogeneous electrolyte-depolarizer pads; others have discrete parts. Open-cell designs typically incorporate a combination anode-electron collector, usually in the shape of a "dumbbell." This combination part has anode material vacuum-deposited on one end (which serves as anode in one cell), while the other end is an electron collector in the next, series-connected cell. A narrow bridge connects the two ends of the dumbbell. The bridge serves as an inter-cell connector, eliminating the need for spot welds. Zr/BaCrO$_4$ heat paper pads heat the open cells, which are assembled between the folded dumbbells.

The open-cell design is used in relatively short-life applications and in pulse batteries. Their parts can be made very thin to promote very rapid heat transfer and obtain short activation times.

21.4.3 Pellet Cells

In pellet cells, the electrolyte, cathode, and heat source are in pellet (wafer) form. Anodes can be of different configurations, depending on which electrochemical system is used. For pellet production, the cell component chemicals are processed into powders, and the powders are uniaxially pressed into the parts. Electrolytes, which melt at cell operating temperatures, are combined with inert binders, which hold the molten salts in place by capillary action or surface tension, or both.

A typical pelletized $Ca/CaCrO_4$ cell, as shown in Fig. 21.4c, is made up of the following:

1. A calcium anode—either calcium foil (on nickel or iron foil collector) or calcium bimetal (deposited on either iron or nickel collector)
2. A pelletized electrolyte powder blend—consisting of LiCl-KCl eutectic salts and either SiO_2 or kaolin as binders
3. A pelletized cathode powder blend—consisting of $CaCrO_4$, LiCl-KCl eutectic salts, and SiO_2 or kaolin binder
4. A pelletized heat source—a blend of iron powder and $KClO_4$. (Alternatively, this may be a non-pelletized heat source assembly made up of Zr-$BaCrO_4$ heat paper in a nickel or iron foil dumbbell with the anode of the next cell on the outside—similar to the anode and heat source in open-cell designs.)

Variations of this cell design include 1) the use of a two-layer pellet with discrete electrolyte and cathode layers formed into one part; and 2) the use of a homogeneous pellet that has the electrolyte and cathode powders blended together (depolarizer-electrolyte-binder or DEB pellet).[18] A typical Li/FeS_2 cell, as illustrated in Fig. 21.4d, is made up of the following:

1. A lithium anode—of either pelletized lithium alloy powder or a lithium metal anode assembly
2. A pelletized electrolyte powder blend—consisting of a salt mixture and MgO binder. The salts may include mixtures such as LiCl-KCl eutectic, LiBr-KBr-LiF, or LiCl-LiBr-KBr
3. A pelletized cathode powder blend—of FeS_2 and electrolyte with either MgO or SiO_2 binder
4. A pelletized heat powder blend—of pyrotechnic-grade iron powder and $KClO_4$
5. An electrical collector—of iron or stainless steel foil, located between the heat pellet and the lithium alloy anode pellet. This part is not used with a lithium metal anode assembly, which has an integral metal foil cup. In some cases, especially in longer-life batteries, a second metal foil "collector" is placed between the FeS_2 cathode and the heat pellet to buffer or prevent the cathode from exposure to excessive heat.

The pressure used for pelletizing the cell components is critical. In the case of $Ca/CaCrO_4$ designs, the forming pressures, and hence the resultant densities of the electrolyte and cathode pellets, have a profound effect on the reactivities of the cells. The components of the Li/FeS_2 systems, except for the heat pellets, are less sensitive to variations in density. Heat pellet ignition sensitivity and burning rate are significantly affected by changes in density, however, with high density decreasing ignition sensitivity and rate. The design parameters of a representative $Li(Si)/FeS_2$ cell are shown in Table 21.3.[19]

The use of pellet-type cell construction has significantly increased the performance capability of thermal batteries. Pellet designs have particular advantages in longer-activated-life, high-current-drain applications. They are structurally very rugged, can operate reliably over wider ambient temperature ranges, and are generally less expensive to manufacture than older designs. There are applications, however, such as those requiring fast activation times and high-voltage pulses, where open-cell designs with $Ca/LiCl$-$KCl/K_2Cr_2O_7$ or $Ca/LiCl$-$KCl/PbCrO_4$ cell chemistries and heat paper are more suitable.

TABLE 21.3   Cell Components of 3400-A/s, Li-Si/FeS$_2$, Thermal Battery Cell (*From Street*[19])

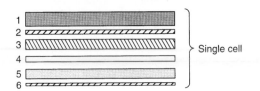

	Component	Chemical composition	Chemical ratio w/% ± 1	Density (g/cm^2) ± 0.05	Forming force, tons	Pellet Thickness, cm	Weight, g ± 0.1
1	Heat pellet	Fe/KClO$_4$	88/12	3.40	60	0.14	22
2	Cathode current collector	SST-304	—	7.75	—	0.013	4.6
3	Cathode pellet	FeS$_2$/LiCl-KCl/SiO$_2$	64/16/20	2.9	200	0.06	8.5
4	Separator pellet	LiCl-KCl-Li$_2$O/MgO	65/35	1.75	90	0.06	4.5
5	Anode pellet	Li/Si	44/56	1.0	115	0.1	4.5
6	Anode current collector	SST-304	—	7.75	—	0.013	4.6

21.5   CELL-STACK DESIGNS

All thermal batteries are designed to satisfy a specific set of performance requirements, each of which includes output voltage, current drain, and activated life. In designing a battery, the output voltage determines the number of cells that must be connected in series. Since each cell produces a fixed maximum voltage (from 1.6 to 3.3 V on open circuit, depending on the cell chemistry used), the battery output will be in multiples of discrete cell voltages. Batteries containing over 180 series-connected cells with an overall output voltage near 400 V have been successfully manufactured. Typical batteries contain 14 to 80 cells, and have an output voltage of 28 to 140 V. Figure 21.5 illustrates two different cell-stack configurations, one with cup cells and the other with pellet-type cells.

The current-carrying capacity of each cell is determined by the reactive surface area of the cell, which is directly related to the cell size (diameter). As with cell voltages, the maximum useful current densities (Amperes per unit area) differ greatly among cell chemistries (see Tables 21.4 and 21.5). The effective cell area, and hence the current-carrying capacity of a battery, can be adjusted by electrically connecting any number of cells in parallel.

Thermal batteries can be designed to provide multiple output voltages by electrically connecting the required number of cells in series. The multiple-voltage outputs can be drawn either from cells that are common to more than one output or from isolated cells whose output is not shared. An electrically isolated group of cells must be used for circuits that cannot tolerate "crosstalk" from other circuits in a system. It is also possible to combine cell-stack sections with different cell chemistries in the same battery. Such combinations yield the specific performance characteristics of both chemistries from a common unit. An example of this is a battery that combines a cell stack with a chemistry that has a very short

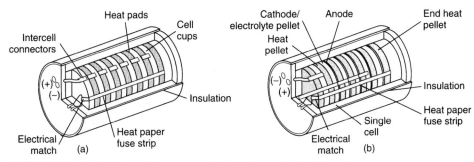

FIGURE 21.5 Typical thermal battery assemblies. (*a*) Cup cells. (*b*) Pellet cells.

TABLE 21.4 Attainable Current Density of Different Cell Designs

| | Current density, mA/cm^2 | | |
Cell design	10-s rate	100-s rate	1000-s rate
Cup cell	620	35	
Open cell/dichromate	54		
Pellet cell/two-layer Ca/CaCrO$_4$	790	46	
Pellet cell/DEB Ca/CaCrO$_4$	930	122	
Pellet cell/Li/FeS$_2$	>2500	610	150

TABLE 21.5 Typical Power and Energy Densities of Li/FeS$_2$ Thermal Batteries

Battery volume, cm^3	Power density, W/cm^2	Energy density, Wh/L	Activated life, s
20	11.25	46.87	15
29	1.44	34.20	85
70	2.59	35.97	50
108	0.65	32.41	180
170	1.98	109.80	200
171	10.64	118.26	40
183	2.29	63.75	100
306	0.51	39.65	280
311	2.25	75.03	700
552	0.15	67.63	1600
1176	0.40	101.19	900
1312	0.17	85.37	1800
3120	1.11	83.30	270

start time with a different cell stack that can provide a high current over a long activated life. Where such combinations are used, the outputs from the different cell-stack sections are often diode-isolated to prevent one section from charging the other. Some thermal battery designs combine two or more discrete batteries into an assembly that may have a number of different, mutually isolated voltage outputs with widely varying current capabilities.

Cells comprising a cell stack are typically held in place by the closing compression applied when the battery cover is secured by welding it to the battery case. Some battery designs incorporate an inner case to maintain compression on the cell stack while the outer case and cover combination provides hermetic enclosure for the unit. Figure 21.6 pictures a battery design that employs an inner cell stack case.

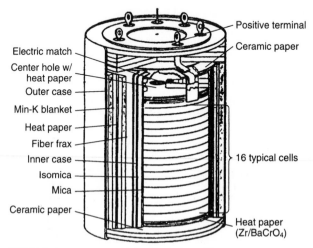

FIGURE 21.6 Typical thermal battery assembly with inner case. (*Courtesy of SAFT America, Inc.*)

21.6 PERFORMANCE CHARACTERISTICS

Thermal batteries are custom-designed to satisfy a specific set of performance requirements. These include not only output voltage, current, activated life, and voltage rise time (start), but also storage and activated-life environments, mounting, surface temperature, activation method and energy, and others. For this reason it is very important that the user or systems designer have a close technical interface with the battery designer during the design and development phases of the battery.

21.6.1 Voltage Regulation

Thermal battery output voltages are not linear. After reaching a peak level, typically within 1 second after activation, the voltage starts to decay until it eventually drops below the minimum useful level. Voltage regulation is the range between the specified minimum and maximum limits. Typically, the minimum voltage limit is 75% of the peak voltage. The battery output profile (consisting of the rise time, peak voltage, and rate of decay) depends on the cell chemistry, and is strongly affected by the operating temperature and applied load. Figure 21.7 illustrates the effects of discharge load on a typical battery output profile.

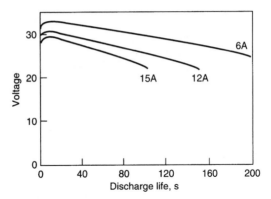

FIGURE 21.7   Discharge voltage curves of typical Li/FeS$_2$ thermal battery at three different current drain rates.

21.6.2 Activation Time

The activation time (rise time) is the time interval from the application of energy to the initiation device until the battery output voltage reaches the minimum specified limit. The activation time is affected by the operating temperature, applied load, and cell chemistry used. Lowering the operating temperature or increasing the load typically increases the activation time. Typical Li/FeS$_2$ batteries have activation times from 0.35 to 1.00 s. Large, high-capacity batteries can have activation times as long as 3 s. (Large diameter heat pellets take longer times to burn.) On the other hand, fast-activating chemistries such as Ca/K$_2$Cr$_2$O$_7$ can yield activation times as short as 12 ms. Figure 21.8 shows activation time ranges for various cell chemistries and Fig. 21.9 illustrates the effects of ambient temperature.

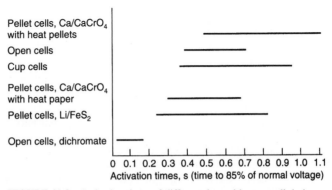

FIGURE 21.8   Activation times of different thermal battery cell designs.

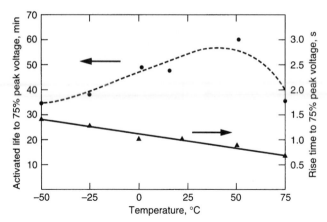

FIGURE 21.9   Activated life and rise time of Li/FeS$_2$ battery. (*From Quinn and Baldwin.*[20])

21.6.3 Activated Life

The activated (operating) life is typically specified as the time from the initial application of the activation energy until the battery voltage drops below the minimum specified limit. Activated life is affected by the cell chemistry used, the operating temperature environment, and the current drain. Typically, thermal batteries are thermally balanced (total cell mass *vs.* caloric input) to have the longest activated lives between the high and low operating temperature limits, or near room ambient. Lives will get shorter near each temperature limit. At the low limit, the electrolyte will start freezing sooner, whereas at the high limit the thermal degradation of FeS$_2$ occurs at a faster rate, depleting active materials.

21.6.4 Interface Considerations

The following performance and design characteristics must be noted when designing a system that interfaces with a thermal battery:

1. An unactivated battery has a very high internal resistance (megOhms). Once activated, an individual cell's resistance is between 0.003 and 0.02 Ohm, depending on the cell design. The internal resistance of the battery is equal to the sum of the resistances of all series-connected cells.
2. Some cell chemistries, such as Li/FeS$_2$, are tolerant of backcharging from an external power source. Others, however, such as Ca/CaCrO$_4$ must not be subjected to backcharging at all.
3. Electric actuators contain bridge wires that may not burn through during activation and, if not disconnected, may act as a parasitic load on the external ignition circuit.
4. Leakage paths that can adversely load the battery may develop in an activated battery between electrically "live" components and the battery case or activator circuits. System requirements, such as case grounding, cell-stack common output, and activator circuit grounding must be specified so that special insulation provisions can be incorporated into the battery design.

5. The surface temperature of an activated battery may reach 400°C. The type of battery mounting, the heat transfer properties of the mounting, the effects of high temperature on the surrounding components, and the proximity of combustible materials must be considered. The battery surface temperature can usually be reduced significantly by incorporating added (or more efficient) thermal insulation. This is achieved, however, at considerable cost and increase in battery volume. Figures 21.10 and 21.11 illustrate typical surface temperatures of thermal batteries.

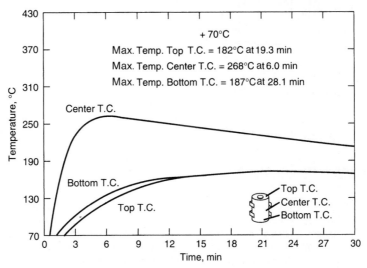

FIGURE 21.10   Surface temperature profiles for long-life thermal battery. (*From Street.*[19])

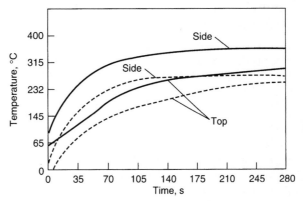

FIGURE 21.11   Surface temperature profiles for a medium-life thermal battery. Solid line-tested at 71°C; broken line-tested at −53°C.

21.7 TESTING AND SURVEILLANCE

The safety and reliability of thermal batteries has been a matter of continuing study since they were first developed. To identify defective units, most designs are 100% tested for hermeticity, polarity, electrical insulation resistance, and activation circuit resistance (if applicable) on manufacture. Most units are also radiographed. Prior to commencement of production, a sample group of 10 to as many as 500 batteries is subjected to qualification tests. This series of tests includes the most severe environmental and discharge conditions to which the particular battery design will be exposed in actual field use. Almost all thermal batteries are fabricated in homogeneous groups or lots, and samples from each lot are discharged to demonstrate compliance with the performance requirements. Usually the samples are discharged at maximum specified loads, often with concurrently imposed environmental forces. By using such test programs, reliability values greater than 99% and safety values greater than 99.9% have been demonstrated innumerable times in the last five decades.

Lithium thermal batteries designed for use in U.S. Navy systems are subject to safety tests per Navy technical manual S9310-AQ-SAF, "Battery, Navy Lithium Safety Program Responsibilities and Procedures." These tests are designed to assure that the battery design is safe not only in proper storage and use, but also when subjected to inadvertent misuse and conditions caused by accidents, such as backcharging, short circuits, and fires.

21.8 NEW DEVELOPMENTS

The primary aim of new research and development in the thermal battery area has been to increase the energy density and specific energy of the practical unit. Two possible approaches to this goal are to: (1) decrease the total volume and mass of the battery; and (2) increase the voltage or the current-carrying capacity per cell volume and mass.

In the area of decreasing battery mass, investigations are being conducted in substituting lighter-weight materials for the currently used stainless steel battery housings. Titanium, aluminum, composites, and other materials have been suggested and tried with varying degrees of success. Titanium cases and headers have been successful but suffer from higher cost.

Efforts to deposit thin FeS_2 films by plasma-spraying of powders onto stainless steel substrates have been yielded promising results.[21] This technology could potentially reduce the mass and volume of a thermal battery by virtue of the higher active material densities afforded by this technique.

Development efforts to produce cells with higher voltage have demonstrated the potential of employing molten nitrate electrolytes with lithium anodes.[22] This system has the added benefit of lowering the battery operating temperature by more than 200°C.

Recent development effort has also been directed toward increasing the activated life of batteries past 2 hrs up to 4 hrs. This effort has required the development of more efficient thermal insulation, such as use of double-walled vacuum enclosures (cases) and multi-layered insulating blankets, as well as lower melting point electrolyte compositions.

REFERENCES

1. G. O. Erb, "Theory and Practice of Thermal Cells," *Publication BIOS/Gp 2/HEC 182 Part II,* Halstead Exploiting Centre, June 6, 1945.
2. O. G. Bennett et al., U.S. Patent 3,575,714, Apr. 20, 1971.
3. B. H. van Domelen, and R. D. Wehrle, "A Review of Thermal Battery Technology," *Intersoc Energy Convers. Conf.,* 1974.
4. F. Tepper, "A Survey of Thermal Battery Designs and Their Performance Characteristics," *Intersoc. Energy Convers. Conf.,* 1974.

5. G. C. Bowser, D. E. Harney, and F. Tepper, "A High Energy Density Molten Anode Thermal Battery," *Power Sources* **6** (1976).

6. R. A. Guidotti, and F. W. Reinhardt, "Evaluation of Alternate Electrolytes for Use in Li(Si)/FeS$_2$ Thermal Batteries," *Proc. 33rd Power Sources Symp.,* 1988, pp. 369–376.

7. L. Redey, J. A. Smaga, J. E. Battles, and R. Guidotti, "Investigation of Primary Li-Si/FeS$_2$ Cells", *ANL-87-6*, Argonne National Laboratory, Argonne, IL, June 1987.

8. W. H. Collins, U.S. Patent 4,053,337, Oct. 11, 1977.

9. C. S. Winchester, "The LAN/FeS$_2$ Thermal Battery System," *Power Sources* **13** (1982).

10. Z. Tomczuk, T. Tani, N. C. Otto, M. F. Roche, and D. R. Vissers, *J. Electrochem. Soc.* **129(5):**925–932 (1992).

11. R. A. Guidotti, R. M. Reinhardt, and J. A. Smaga, "Self-Discharge Study of Li-Alloy/FeS$_2$ Thermal Cells," *Proc. 34th Int. Power Sources Symp.,* 1990, pp. 132–135.

12. R. A. Guidotti, "Methods of Achieving the Equilibrium Number of Phases in Mixtures Suitable for Use in Battery Electrodes, e.g., for Lithiating FeS$_2$," U.S. Patent 4,731,307, Mar. 15, 1988.

13. R. A. Guidotti, and F. W. Reinhardt, "The Relative Performance of FeS$_2$ and CoS$_2$ in Long-Life Thermal-Battery Applications," *Proc. 9th Int. Symp. Molten Salts,* 1994.

14. R. A. Guidotti, and F. W. Reinhardt, "Characterization of the Li(Si)/CoS$_2$ Couple for a High-Voltage, High-Power Thermal Battery," *SAND2000-0396*, 2000.

15. R. A. Guidotti, and F. W. Reinhardt, "Anodic Reactions in the Ca/CaCrO$_4$ Thermal Battery," *SAND83-2271*, 1985.

16. R. A. Guidotti, and W. N. Cathey, "Characterization of Cathodic Reaction Products in the Ca/CaCrO$_4$ Thermal Battery," *SAND84-1098*, 1985.

17. R. A. Guidotti, F. W. Reinhardt, D. R. Tallant, and K. L. Higgins, "Dissolution of CaCrO$_4$ in Molten LiCl-KCl Eutectic," *SAND83-2272*, 1984.

18. D. M. Bush et al., U.S. Patent 3,898,101, Aug. 3, 1975.

19. H. K. Street, "Characteristics and Development Report of the MC3573 Thermal Battery," *SAND82-0695*, 1983.

20. R. K. Quinn, and A. R. Baldwin, "Performance Data for Lithium-Silicon/Iron Disulfide Long Life Primary Thermal Battery," *Proc. 29th Power Sources Symp.,* 1980.

21. H. Ye et al, "Novel Design and Fabrication of Thermal Battery Cathodes Using Thermal Spray," Spring Meeting of the Materials Research Society, San Francisco, CA, April 5–9, 1999.

22. M. H. Miles, "Lithium Batteries Using Molten Nitrate Electrolytes," *Proc. 14th Annual Battery Conf.,* Long Beach, 1999.

BIBLIOGRAPHY

Askew, B. A., and R. Holland: "A High Rate Primary Lithium-Sulfur Battery," *Power Sources* 4 (1972).

Birt, D., C. Feltham, G. Hazzard, and L. Pearce: "The Electrochemical Characteristics of Iron Sulfide and Immobilized Salt Electrolytes," *Power Sources* **7** (1978).

Baird, M. D., A. J. Clark, C. R. Feltham, and L. H. Pearce: "Recent Advances in High Temperature Primary Lithium Batteries," *Power Sources* **7** (1978).

Bowser, G. C., et al.: U.S. Patent 3,891,460, June 24, 1975.

Bowser, G. C., et al.: U.S. Patent 3,930,888, Jan. 1976.

Bush, D. M., and D. A. Nissen: "Thermal Cells and Batteries Using the Mg/FeS$_2$ and LiAl/FeS$_2$ Systems," *Proc. 28th Power Sources Symp.,* 1978.

Collins, W. H.: U.S. Patent 1,482,738, Aug. 10, 1977.

De Gruson, J. A.: "Improved Thermal Battery Performance," *AFAPL-TR-79-2042*, Eagle Picher Industries, 1979.

Delnick, F. M., R. A. Guidotti, and D. K. McCarthy: "Chromium (V) Compounds as Cathode Materials in Electrochemical Power Sources," U.S. Patent 4,508,796, Apr. 2, 1985.

Guidotti, R. A., and F. W. Reinhardt: "Lithiation of FeS$_2$ for Use in Thermal Batteries," *Proc. 2nd Annual Battery Conf. on Applications and Advances,* 1987, paper 87DS-3.

Guidotti, R. A., F. M. Reinhardt, and W. F. Hammeter: "Screening Study of Lithiated Catholyte Mix for a Long-Life Li(Si)/FeS$_2$ Thermal Battery," *SAND 85-1737,* 1988.

Hansen, M.: *Constitution of Binary Alloys,* McGraw-Hill, New York, 1958.

Harney, D. E.: U.S. Patent 4,221,849, Sept. 9, 1980.

Kuper, W. E.: "A Brief History of Thermal Batteries," *Proc. 36th Power Sources Conf.,* Cherry Hill, N.J., June 1994.

Quinn, R. K., et al.: "Development of a Lithium Alloy/Iron Disulfide 60-Minute Primary Thermal Battery," *SAND79-0814,* (1979).

Schneider, A. A., et al.: U.S. Patent 4,119,796, Oct. 10, 1978.

Searcey, J. Q., et al.: "Improvements in Li(Si)/FeS$_2$ Thermal Battery Technology," *SAND82-0565,* 1982.

Szwarc, R.: "Study of Li-B Alloy in LiCl-KCl Eutectic Thermal Cells Utilizing Chromate and Iron Disulfide Depolarizer," Gepp-TM-426, General Electric Co., Neut. Dev. Dept., 1979.

SECONDARY BATTERIES

CHAPTER 22
SECONDARY BATTERIES— INTRODUCTION

David Linden and Thomas B. Reddy

22.1 GENERAL CHARACTERISTICS AND APPLICATIONS OF SECONDARY BATTERIES

Secondary or rechargeable batteries are widely used in many applications. The most familiar are starting, lighting, and ignition (SLI) automotive applications; industrial truck materials-handling equipment; and emergency and standby power. Small, secondary batteries are also being used in increasing numbers to power portable devices such as tools, toys, lighting, and photographic, radio, and more significantly, consumer electronic devices (computers, camcorders, cellular phones). More recently, secondary batteries have received renewed interest as a power source for electric and hybrid electric vehicles. Major development programs have been initiated toward improving the performance of existing battery systems and developing new systems to meet the stringent specifications of these new applications.

The applications of secondary batteries fall into two major categories:

1. Those applications in which the secondary battery is used as an energy storage device, being charged by a prime energy source and delivering its energy to the load on demand, when the prime energy source is not available or is inadequate to handle the load requirement. Examples are automotive and aircraft systems, uninterruptible power supplies and standby power sources, and hybrid applications.

2. Those applications in which the secondary battery is discharged (similar in use to a primary battery) and recharged after use, either in the equipment in which it was discharged or separately. Secondary batteries are used in this manner for convenience, for cost savings (as they can be recharged rather than replaced), or for power drains beyond the capability of primary batteries. Most consumer electronics, electric-vehicle, traction, industrial truck, and some stationary battery applications fall in this category.

Conventional aqueous secondary batteries are characterized, in addition to their ability to be recharged, by high power density, flat discharge profiles, and good low-temperature performance. Their energy densities and specific energies, however, are usually lower, and their charge retention is poorer than those of primary battery systems (see Chap. 6). Rechargeable batteries, such as lithium ion technologies, however, have higher energy densities, better charge retention, and other performance enhancements characterized by the use of higher energy materials. Power density may be adversely affected because of the use of aprotic solvents in the electrolyte, which have lower conductivity than the aqueous electrolyte. This has been compensated for by using high surface area electrodes.

Secondary batteries have been in existence for over 100 years. The lead-acid battery was developed in 1859 by Planté. It is still the most widely used battery, albeit with many design changes and improvements, with the automotive SLI battery by far the dominant one. The nickel-iron alkaline battery was introduced by Edison in 1908 as a power source for the early electric automobile. It eventually saw service in industrial trucks, underground work vehicles, railway cars, and stationary applications. Its advantages were durability and long life, but it gradually lost its market share because of its high cost, maintenance requirements, and lower specific energy.[1]

The pocket-plate nickel-cadmium battery has been manufactured since 1909 and was used primarily for heavy-duty industrial applications. The sintered-plate designs, which led to increased power capability and energy density, opened the market for aircraft engine starting and communications applications during the 1950s. Later the development of the sealed nickel-cadmium battery led to its widespread use in portable and other applications. The dominance of this technology in the portable rechargeable market has been surplanted initially by nickel-metal hydride and more recently by lithium-ion batteries which provide higher specific energy and energy density.

As with the primary battery systems, significant performance improvements have been made with the older secondary battery systems, and a number of newer types, such as the silver-zinc, the nickel-zinc, nickel-hydrogen, and lithium ion batteries, and the high-temperature system, have been introduced into commercial use or are under advanced development. Much of the development work on new systems has been supported by the need for high-performance batteries for portable consumer electronic applications and electric vehicles. Figure 22.1 illustrates the advances achieved in and the projections of the performance of rechargeable batteries for portable applications.

The specific energy and energy density of portable rechargeable nickel-cadmium batteries have not improved significantly in the past decade, and now stand at 35 Wh/kg and 100 Wh/L, respectively. Through the use of new hydrogen-storage alloys, improved performance in nickel-metal hydride batteries has been achieved and that system now provides 75 Wh/kg and 240 Wh/L. A major increase in performance of lithium ion systems was seen in the late 1990s due to the use of carbon materials in the negative electrode with much higher specific capacity. These batteries now provide a specific energy of 150 Wh/kg and an energy density of 400 Wh/L in the small cylindrical sizes employed for consumer electronics applications. The lithium/lithiated manganese dioxide rechargeable AA cell was withdrawn from the market in the late 1990s (see Chap. 34) and, although significant research and development with lithium metal continue in this field, no products are commercially available at the present time.

The worldwide secondary battery market is now approximately $20 billion annually. A world perspective of the use of secondary batteries by application is presented in Table 22.1. The lead-acid battery is by far the most popular, with the SLI battery accounting for a major share of the market. This share is declining gradually, due to increasing applications for other types of batteries. The market share of the alkaline battery systems is about 25%. A major growth area has been the non-automotive consumer applications for small secondary batteries. Lithium ion batteries have emerged in the last decade to capture a 50% share of the market for small sealed consumer batteries, as indicated in Table 22.1. The typical characteristics and applications of secondary batteries are summarized in Table 22.2.

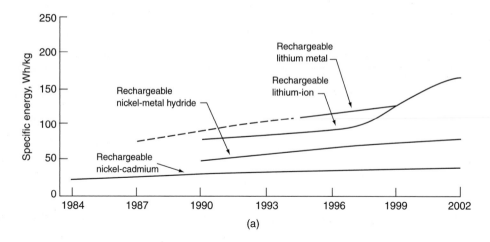

(a)

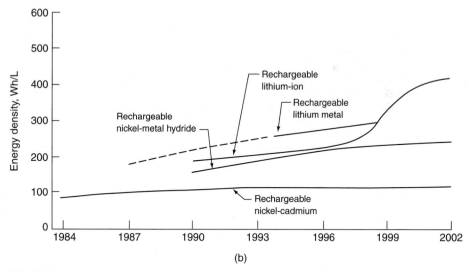

(b)

FIGURE 22.1 Advances in performance of portable rechargeable batteries. (*a*) Specific energy (Wh/kg). (*b*) Energy density (Wh/L).

TABLE 22.1   Worldwide Secondary Battery Market at Manufacturers' Prices—1999 (in millions of dollars)*†

Market segment	Battery system		
	Lead-acid	Alkaline	Lithium ion
Vehicle SLI	9600	—	—
Industrial:			
Standby and UPS	1500	400	—
Tractions incl. forklift trucks	1200	200	—
Consumer and instruments, small sealed cells	200	2430	2500
Energy storage			
Solar and load levelling	130	30	—
Military, aircraft, and space, incl. submarines	70	400	—
Vehicular propulsion			
Golf carts	200	—	—
Electric vehicles and hybrid electric vehicles	40	200	—
Total	12,940	3460	2500

*Values do not include complete data from former Soviet Union and China.
†Retail values can be as much as three times manufacturers' prices.
Source: From Refs. 2 and 3.

TABLE 22.2   Major Characteristics and Applications of Secondary Batteries

System	Characteristics	Applications
Lead-acid:		
Automotive	Popular, low-cost secondary battery, moderate specific-energy, high-rate, and low-temperature performance; maintenance-free designs	Automotive SLI, golf carts, lawn mowers, tractors, aircraft, marine
Traction (motive power)	Designed for deep 6–9 h discharge, cycling service	Industrial trucks, materials handling, electric and hybrid electric vehicles, special types for submarine power
Stationary	Designed for standby float service, long life, VRLA designs	Emergency power, utilities, telephone, UPS, load leveling, energy storage, emergency lighting
Portable	Sealed, maintenance-free, low cost, good float capability, moderate cycle life	Portable tools, small appliances and devices, TV and portable electronic equipment
Nickel-cadmium:		
Industrial and FNC	Good high-rate, low-temperature capability, flat voltage, excellent cycle life	Aircraft batteries, industrial and emergency power applications, communication equipment
Portable	Sealed, maintenance-free, good high-rate low-temperature performance, excellent cycle life	Railroad equipment, consumer electronics, portable tools, pagers, appliances, and photographic equipment, standby power, memory backup
Nickel-metal hydride	Sealed, maintenance-free, higher capacity than nickel-cadmium batteries	Consumer electronics and other portable applications; electric and hybrid electric vehicles
Nickel-iron	Durable, rugged construction, long life, low specific energy	Materials handling, stationary applications, railroad cars
Nickel-zinc	High specific energy, extended cycle life and rate capability	Bicycles, scooters, trolling motors
Silver-zinc	Highest specific energy, very good high-rate capability, low cycle life, high cost	Lightweight portable electronic and other equipment; training targets, drones, submarines, other military equipment, launch vehicles and space probes
Silver-cadmium	High specific energy, good charge retention, moderate cycle life, high cost	Portable equipment requiring a lightweight, high-capacity battery; space satellites
Nickel-hydrogen	Long cycle life under shallow discharge, long life	Primarily for aerospace applications such as LEO and GEO satellites
Ambient-temperature rechargeable "primary" types [Zn/ MnO_2]	Low cost, good capacity retention, sealed and maintenance-free, limited cycle life and rate capability	Cylindrical cell applications, rechargeable replacement for zinc-carbon and alkaline primary batteries, consumer electronics (ambient-temperature systems)
Lithium ion	High specific energy and energy density, long cycle life	Portable and consumer electronic equipment, electric vehicles, and space applications

22.2   TYPES AND CHARACTERISTICS OF SECONDARY BATTERIES

The important characteristics of secondary or rechargeable batteries are that the charge and discharge—the transformation of electric energy to chemical energy and back again to electric energy—should proceed nearly reversibly, should be energy efficient, and should have minimal physical changes that can limit cycle life. Chemical action, which may cause deterioration of the cell's components, loss of life, or loss of energy, should be absent, and the cell should possess the usual characteristics desired of a battery such as high specific energy, low resistance, and good performance over a wide temperature range. These requirements limit the number of materials that can be employed successfully in a rechargeable battery system.

22.2.1   Lead-Acid Batteries

The lead-acid battery system has many of these characteristics. The charge-discharge process is essentially reversible, the system does not suffer from deleterious chemical action, and while its energy density and specific energy are low, the lead-acid battery performs reliably over a wide temperature range. A key factor for its popularity and dominant position is its low cost with good performance and cycle-life.

The lead-acid battery is designed in many configurations, as listed in Table 22.2 (see also Chaps. 23 and 24), from small sealed cells with a capacity of 1 Ah to large cells, up to 12,000 Ah. The automotive SLI battery is by far the most popular and the one in widest use. Most significant of the advances in SLI battery design are the use of lighter-weight plastic containers, the improvement in shelf life, the "dry-charge" process, and the "maintenance-free" design. The latter, using calcium-lead or low-antimony grids, has greatly reduced water loss during charging (minimizing the need to add water) and has reduced the self-discharge rate so that batteries can be shipped or stored in a wet, charged state for relatively long periods.

The lead-acid industrial storage batteries are generally larger than the SLI batteries, with a stronger, higher-quality construction. Applications of the industrial batteries fall in several categories. The motive power traction types are used in materials-handling trucks, tractors, mining vehicles, and, to a limited extent, golf carts and personnel carriers, although the majority in use are automotive-type batteries. A second category is diesel locomotive engine starting and the rapid-transit batteries, replacing the nickel-iron battery in the latter application. Significant advances are the use of lighter-weight plastic containers in place of the hard-rubber containers, better seals, and changes in the tubular positive-plate designs. Another category is stationary service: telecommunications systems, electric utilities for operating power distribution controls, emergency and standby power systems, uninterruptible power systems (UPS), and in railroads, signaling and car power systems.

The industrial batteries use three different types of positive plates: tubular and pasted flat plates for motive power, diesel engine cranking, and stationary applications, and Planté designs, forming the active materials from pure lead, mainly in the stationary batteries. The flat-plate batteries use either lead-antimony or lead-calcium grid alloys. A relatively recent development for the telephone industry has been the "round cell," designed for trouble-free long-life service. This battery uses plates, conical in shape with pure lead grids, which are stacked one above the other in a cylindrical cell container, rather than the normal prismatic structure with flat, parallel plates.

An important development in lead-acid battery technology is the Valve-Regulated Lead-Acid battery (VRLA) (see Chap. 24). These batteries operate on the principle of oxygen recombination, using a "starved" or immobilized electrolyte. The oxygen generated at the positive electrode during charge can, in these battery designs, diffuse to the negative electrode, where it can react, in the presence of sulfuric acid, with the freshly formed lead. The VRLA design reduces gas emission by over 95% as the generation of hydrogen is also suppressed. Oxygen recombination is facilitated by the use of a pressure-relief valve, which is closed during normal operation. When pressure builds up, the valve opens at a predetermined value, venting the gases. The valve reseals before the cell pressure decreases to atmospheric pressure. The VRLA battery is now used in about 70% of the telecommunication batteries and in about 80% of the uninterrupted power source (UPS) applications.

Smaller sealed lead-acid cells are used in emergency lighting and similar devices requiring backup power in the event of a utility power failure, portable instruments and tools, and various consumer-type applications. These small sealed lead-acid batteries are constructed in two configurations, prismatic cells with parallel plates, ranging in capacity from 1 to 30 Ah, and cylindrical cells similar in appearance to the popular primary alkaline cells and ranging in capacity up to 25 Ah. The acid electrolyte in these cells is either gelled or absorbed in the plates and in highly porous separators so they can be operated virtually in any position without the danger of leakage. The grids generally are of lead-calcium-tin alloy; some use grids of pure lead or a lead-tin alloy. The cells also include the features for oxygen recombination and are considered to be VRLA batteries (see Chap. 24).

Lead-acid batteries also are used in other types of applications, such as in submarine service, for reserve power in marine applications, and in areas where engine-generators cannot be used, such as indoors and in mining equipment. New applications, to take advantage of the cost effectiveness of this battery, include load leveling for utilities and solar photovoltaic systems. These applications will require improvements in the energy and power density of the lead-acid battery.

22.2.2 Alkaline Secondary Batteries

Most of the other conventional types of secondary batteries use an aqueous alkaline solution (KOH or NaOH) as the electrolyte. Electrode materials are less reactive with alkaline electrolytes than with acid electrolytes. Furthermore, the charge-discharge mechanism in the alkaline electrolyte involves only the transport of oxygen or hydroxyl ions from one electrode to the other; hence the composition or concentration of the electrolyte does not change during charge and discharge.

Nickel-Cadmium Batteries. The nickel-cadmium secondary battery is the most popular alkaline secondary battery and is available in several cell designs and in a wide range of sizes. The original cell design used the pocket-plate construction. The vented pocket-type cells are very rugged and can withstand both electrical and mechanical abuse. They have very long lives and require little maintenance beyond occasional topping with water. This type of battery is used in heavy-duty industrial applications, such as materials-handling trucks, mining vehicles, railway signaling, emergency or standby power, and diesel engine starting. The sintered-plate construction is a more recent development, having higher energy density. It gives better performance than the pocket-plate type at high discharge rates and low temperatures but is more expensive. It is used in applications, such as aircraft engine starting and communications and electronics equipment, where the lighter weight and superior performance are required. Higher energy and power densities can be obtained by using nickel foam, nickel fiber, or plastic-bonded (pressed-plate) electrodes. The sealed cell is a third design. It uses an oxygen-recombination feature similar to the one used in sealed lead-acid batteries to prevent the buildup of pressure caused by gassing during charge. Sealed cells are available in prismatic, button, and cylindrical configurations and are used in consumer and small industrial applications.

Nickel-Iron Batteries.   The nickel-iron battery was important from its introduction in 1908 until the 1970s, when it lost its market share to the industrial lead-acid battery. It was used in materials-handling trucks, mining and underground vehicles, railroad and rapid-transit cars, and in stationary applications. The main advantages of the nickel-iron battery, with major cell components of nickel-plated steel, are extremely rugged construction, long life, and durability. Its limitations, namely, low specific energy, poor charge retention, and poor low-temperature performance, and its high cost of manufacture compared with the lead-acid battery led to a decline in usage.

Silver Oxide Batteries.   The silver-zinc (zinc/silver oxide) battery is noted for its high energy density, low internal resistance desirable for high-rate discharge, and a flat second discharge plateau. This battery system is useful in applications where high energy density is a prime requisite, such as electronic news gathering equipment, submarine and training target propulsion, and other military and space uses. It is not employed for general storage battery applications because its cost is high, its cycle life and activated life are limited, and its performance at low temperatures falls off more markedly than with other secondary battery systems.

The silver-cadmium (cadmium/silver oxide) battery has significantly longer cycle life and better low-temperature performance than the silver-zinc battery but is inferior in these characteristics compared with the nickel-cadmium battery. Its energy density, too, is between that of the nickel-cadmium and the silver-zinc batteries. The battery is also very expensive, using two of the more costly electrode materials. As a result, the silver-cadmium battery was never developed commercially but is used in special applications, such as nonmagnetic batteries and space applications. Other silver battery systems, such as silver-hydrogen and silver-metal hydride couples, have been the subject of development activity but have not reached commercial viability.

Nickel-Zinc Batteries.   The nickel-zinc (zinc/nickel oxide) battery has characteristics midway between those of the nickel-cadmium and the silver-zinc battery systems. Its energy density is about twice that of the nickel-cadmium battery, but the cycle life previously has been limited due to the tendency of the zinc electrode toward shape change which reduces capacity and dendrite formations, which cause internal short-circuiting.

Recent development work has extended the cycle life of nickel-zinc batteries through the use of additives in the negative electrode in conjunction with the use of a reduced concentration of KOH to repress zinc solubility in the electrolyte. Both of these modifications have extended the cycle life of this system so that it is now being marketed for use in electric bicycles, scooters and trolling motors in the United States and Asia.

Hydrogen Electrode Batteries.   Another secondary battery system uses hydrogen for the active negative material (with a fuel-cell-type electrode) and a conventional positive electrode, such as nickel oxide. These batteries are being used exclusively for the aerospace programs which require long cycle life at low depth of discharge. The high cost of these batteries is a disadvantage which limits their application. A further extension is the sealed nickel/metal hydride battery where the hydrogen is absorbed, during charge, by a metal alloy forming a metal hydride. This metal alloy is capable of undergoing a reversible hydrogen absorption-desorption reaction as the battery is charged and discharged respectively. The advantage of this battery is that its specific energy and energy density are significantly higher than that of the nickel-cadmium battery. The sealed nickel-metal hydride battery, manufactured in small prismatic and cylindrical cells, is being used for portable electronic applications and are being employed for other applications including hybrid electric vehicles. Larger sizes are finding use in electric vehicles (see Chap. 30).

Zinc/Manganese Dioxide Batteries.   Several of the conventional primary battery systems have been manufactured as rechargeable batteries, but the only one currently being manufactured is the cylindrical cell using the zinc/alkaline-manganese dioxide chemistry. Its major advantage is a higher capacity than the conventional secondary batteries and a lower initial cost, but its cycle life and rate capability are limited.

Lithium Ion Batteries.   Lithium ion batteries have emerged in the last decade to capture over half of the sales value of the secondary consumer market, with applications such as laptop computers, cell phones and camcorders (known as the "Three-C" market). Production capacity has recently been estimated to be 75 million/cells per month,[3] These cells provide high energy density and specific energy (see Fig. 22.1) and long cycle life, typically greater than 1000 cycles @ 80% depth of discharge. When built into batteries, battery management circuitry is required to prevent over charge and over discharge, both of which are deleterious to performance. The circuits may also provide an indication of state-of-charge and safety features in the case of an over-current or an over-heating condition (see Chap. 5). More detailed information on this subject may be found in Chap. 35.

22.3 COMPARISON OF PERFORMANCE CHARACTERISTICS FOR SECONDARY BATTERY SYSTEMS

22.3.1 General

The characteristics of the major secondary systems are summarized in Table 22.3. This table is supplemented by Table 1.2, which lists several theoretical and practical electrical characteristics of these secondary battery systems. A graphic comparison of the theoretical and practical performances of various battery systems is given in Fig. 3.3. This shows that up to only about 20 to 30% of the theoretical capacity of a battery system is attained under practical conditions as a result of design and the discharge requirements.

It should be noted, as discussed in detail in Chaps. 3 and 6, that these types of data and comparisons (as well as the performance characteristics shown in this section) are necessarily approximations, with each system being presented under favorable discharge conditions. The specific performance of a battery system is very dependent on the cell design and all the detailed and specific conditions of the use and discharge-charge of the battery.

A qualitative comparison of the various secondary battery systems is presented in Table 22.4. The different ratings given to the various designs of the same electrochemical system are an indication of the effects of the design on the performance characteristics of a battery.

TABLE 22.3 Characteristics of the Major Secondary Battery Systems

Common name	Lead-acid				Nickel-cadmium			
	SLI	Traction	Stationary	Portable	Vented pocket plate	Vented sintered plate	Sealed	FNC
Chemistry:								
Anode	Pb	Pb	Pb	Pb	Cd	Cd	Cd	Cd
Cathode	PbO_2	PbO_2	PbO_2	PbO_2	NiOOH	NiOOH	NiOOH	NiOOH
Electrolyte	H_2SO_4 (aqueous solution)	H_2SO_4 (aqueous solution)	H_2SO_4 (aqueous solution)	H_2SO_4 (aqueous solution)	KOH (aqueous solution)	KOH (aqueous solution)	KOH (aqueous solution)	KOH (aqueous solution)
Cell voltage (typical), V:								
Nominal	2.0	2.0	2.0	2.0	1.2	1.2	1.2	1.2
Open circuit	2.1	2.1	2.1	2.1	1.29	1.29	1.29	1.35
Operating	2.0–1.8	2.0–1.8	2.0–1.8	2.0–1.8	1.25–1.00	1.25–1.00	1.25–1.00	1.25–0.85
End	1.75 (lower operating and end voltage during cranking operation)	1.75	1.75 (except when on float service)	1.75 (where cycled)	1.0	1.0	1.0	1.00–0.65
Operating temperatures, °C	−40 to 55	−20 to 40	−10 to 40[c]	−40 to 60	−20 to 45	−40 to 50	−40 to 45	−50 to 60
Specific energy and energy density (at 20°C)								
Wh/kg	35	25	10–20	30	20	30–37	35	10–40
Wh/L	70	80	50–70	90	40	58–96	100	15–80
Discharge profile (relative)	Flat	Flat	Flat	Flat	Flat	Very flat	Very flat	Flat
Power density	High	Moderately high	Moderately high	High	High	High	Moderate to high	Very high
Self-discharge rate (at 20°C), % loss per month[b]	20–30 (Sb-Pb) 2–3 (maintenance-free)	4–6	—	4–8	5	10	15–20	10–15
Calendar life, years	3–6	6	18–25	2–8	8–25	3–10	2–5	5–20
Cycle life, cycles[c]	200–700	1500	—	250–500	500–2000	500–2000	300–700	500–10,000
Advantages	Low cost, ready availability, good high-rate, high- and low-temperature operation (good cranking service), good float service, new maintenance-free designs	Lowest cost of competitive systems (also see SLI)	Designed for "float" service lowest cost of competitive systems (also see SLI)	Maintenance-free: long life on float service; low- and high-temperature performance; no "memory" effect; operates in any position	Very rugged, can withstand physical and electrical abuse; good charge retention, storage and cycle life lowest cost of alkaline batteries	Rugged; excellent storage; good specific energy and high-rate and low-temperature performance	Sealed, no maintenance; good low-temperature and high-rate pacformance; long life cycle; operates in any position	Sealed, no maintenance, high power capability even at low temperature, long cycle life at low depth of discharge, fast charging
Limitations	Relatively low cycle life; limited energy density; poor charge retention and storability; hydrogen evolution	Low energy density; less rugged than competitive systems; hydrogen evolution	Hydrogen evolution	Cannot be stored in discharged condition; lower cycle life than sealed nickel-cadmium; difficult to manufacture in very small sizes	Low energy density	High cost; "memory" effect; thermal runaway problem	Sealed lead-acid battery better at high temperature and float service, "memory" effect	Lower energy density than sintered plate design
Major battery types available	Prismatic cells; 30–200 Ah at 20-h rate	Based on positive plate design; 45–200 Ah per positive plate	Based on positive plate design: 5–400 Ah per positive to 1440 Ah plate	Sealed cylindrical cells; 2.5–25 Ah; prismatic cells; to 1440 Ah	Prismatic cells; 5–1300 Ah	Prismatic cells; 1.5–100 Ah	Button cells to 0.5 Ah; cylindrical cells to 10 Ah	Prismatic designs to 450 Ah

[a] Based on $C/LiCoO_2$ lithium-ion battery (see Chap. 35) (characteristics vary with battery system and design).
[b] Self-discharge rate usually decreases with increasing storage time.
[c] Dependent on depth of discharge.
[d] High rate Zn/AgO battery.
[e] Low rate Zn/AgO battery.

Nickel-iron (conventional)	Nickel-zinc	Zinc/silver oxide (silver-zinc)	Cadmium/silver oxide (silver-cadmium)	Nickel-hydrogen	Nickel-metal hydride	Rechargeable "primary" types, Zn/MnO$_2$	Lithium ion systems[a]
Fe NiOOH KOH (aqueous solution)	Zn NiOOH KOH (aqueous solution)	Zn AgO KOH (aqueous solution)	Cd AgO KOH (aqueous solution)	H$_2$ NiOOH KOH (aqueous solution)	MH NiOOH KOH (aqueous solution)	Zn MnO$_2$ KOH (aqueous solution)	C LiCoO$_2$ Organic solvent
1.2 1.37 1.25–1.05 1.0	1.65 1.73 1.6–1.4 1.2	1.5 1.86 1.7–1.3 1.0	1.1 1.41 1.4–1.0 0.7	1.4 1.32 1.3–1.15 1.0	1.2 1.4 1.25–1.10 1.0	1.5 1.5 1.3–1.0 1.0	4.0 4.1 4.0–3.0 3.0
−10 to 45	−10 to 50	−20 to 60	−25 to 70	0 to 50	−20 to 50	−20 to 40	−20 to 50
30 55 Moderately flat	50–60 80–120 Flat	105[d] 180 Double plateau	70 120 Double plateau	64 (CPV) 105 (CPV) Moderately flat	75 240 Flat	85 250 Sloping	150 400 Sloping
Moderate to low	High	Very high (for high rate-design)	Moderate to high	Moderate	Moderate to high	Moderate	Moderate; high in prismatic designs
20–40	<20	5	5	Very high except at low temp.	15–25		2
8–25	—	2	3 (vented) 4 (sealed)	—	2–5		
2000–4000	500	50–100	300–800	1500–6000 40,000 at 40% (DOD)	300–600	15–25	1000+
Very rugged, can withstand physical and electrical abuse; long life (cycling or stand)	High energy density; relatively low cost; good low-temperature performance	High energy density; high discharge rate; low self-discharge	High-energy density; low self-discharge, good cycle life	High energy density; long cycle life at low DOD; can tolerate overcharge	High energy density; sealed, good cycle life	Good shelf life; low cost	High specific energy and energy density; low self-discharge; long cycle life
Low power and energy density; high self-discharge; hydrogen evolution; high cost and high maintenance cost	Improved cycle life at reduced specific energy	High cost; low cycle life; decreased performances at low temperatures	High cost; decreased performance at low temperatures	High initial cost; self-discharge proportional to H$_2$ presure and temperature	Intermediate in cost between NiCad and Li Ion	Limited cycle life; low drain applications; small size only	Lower rate (compared to aqueous systems)
Decreasing significance in developed countries	In production for electric bicycles and scooters and trolling motors: 2–100 Ah sizes	Prismatic cells: <1 to 1000 Ah; special types to 5000 Ah	Prismatic cells: <1 to 1000 Ah	Aerospace applications (up to 100 Ah)	Button and cylindrical cells to 4.1 Ah, large prismatics to 100 Ah	Cylindrical cells to 10 Ah	Cylindrical and prismatic cells to 100 Ah

TABLE 22.4   Comparison of Secondary Batteries*

System	Energy density	Power density	Flat discharge profile	Low-temperature operation	Charge retention	Charge acceptance	Efficiency	Life	Mechanical properties	Cost
Lead-acid:										
Pasted	4	4	3	3	4	3	2	3	5	1
Tubular	4	5	4	3	3	3	2	2	3	2
Planté	5	5	4	3	3	3	2	2	4	2
Sealed	4	3	3	2	3	3	2	3	5	2
Lithium-metal	1	3	3	2	1	3	3	4	3	4
Lithium-ion	1	2	3	2	2	1	1	1	3	2
Nickel-cadmium:										
Pocket	5	3	2	1	2	1	4	2	1	3
Sintered	4	1	1	1	4	1	3	2	1	3
Sealed	4	1	2	1	4	2	3	3	2	2
Nickel-iron	5	5	4	5	5	2	5	1	1	3
Nickel-metal hydride	3	2	2	2	4	2	3	3	2	3
Nickel-zinc	2	3	2	3	4	3	3	4	3	3
Silver-zinc	1	1	4	3	1	3	2	5	2	4
Silver-cadmium	2	3	5	4	1	5	1	4	3	4
Nickel-hydrogen	2	3	3	4	5	3	5	2	3	5
Silver-hydrogen	2	3	4	4	5	3	5	2	3	5
Zinc-manganese dioxide	2	4	5	3	1	4	4	5	4	2

*Rating: 1 to 5, best to poorest.

22.3.2 Voltage and Discharge Profiles

The discharge curves of the conventional secondary battery systems, at the $C/5$ rate, are compared in Fig. 22.2. The lead-acid battery has the highest cell voltage of the aqueous systems. The average voltage of the alkaline systems ranges from about 1.65 V for the nickel-zinc system to about 1.1 V. At the $C/5$ discharge rate at 20°C there is relatively little difference in the shape of the discharge curve for the various designs of a given system. However, at higher discharge rates and at lower temperatures, these differences could be significant, depending mainly on the internal resistance of the cell.

Most of the conventional rechargeable battery systems have a flat discharge profile, except for the silver oxide systems, which show the double plateau due to the two-stage discharge of the silver oxide electrode, and the rechargeable zinc/manganese dioxide battery.

The discharge curve of a lithium ion battery, the carbon/lithiated cobalt oxide system, is shown for comparison. The cell voltages of the lithium ion batteries are higher than those of the conventional aqueous cells because of the characteristics of these systems. The discharge profile of the lithium ion batteries are usually not as flat due to the lower conductivity of the nonaqueous electrolytes that must be used and to the thermodynamics of intercalation electrode reactions (see Chapter 35). The average discharge voltage for a lithium ion cell is 3.6 V, which allows one unit to replace three Nicad or NiMH cells in a battery configuration.

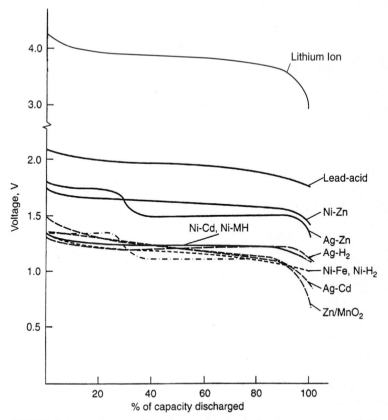

FIGURE 22.2 Discharge profiles of conventional secondary battery systems and rechargeable lithium ion battery at approximately $C/5$ discharge rate.

22.3.3 Effect of Discharge Rate on Performance

The effects of the discharge rate on the performance of these secondary battery systems are compared again in Fig. 22.3. This figure is similar to a Ragone plot, except that the abscissa is expressed in hours of service instead of specific energy (Wh/kg). This figure shows that hours of service each battery type (unitized to 1-kg battery weight) will deliver at various power (discharge current $\times$ midpoint voltage) levels. The higher slope is indicative of superior retention of capacity with increasing discharge load. The specific energy can be calculated by the following equation:

$$\text{specific energy} = \text{specific power} \times \text{hours of service}$$

or

$$\text{Wh/kg} = \text{W/kg} \times \text{h} = \frac{A \times V \times h}{kg}$$

Figure 22.4 is a Ragone plot on a semi-log scale comparing the performance of the nickel-cadmium and sealed nickel-metal hydride in AA size and the new lithium ion battery in a 14500 cylindrical configuration, on a gravimetric and volumetric basis at 20°C.

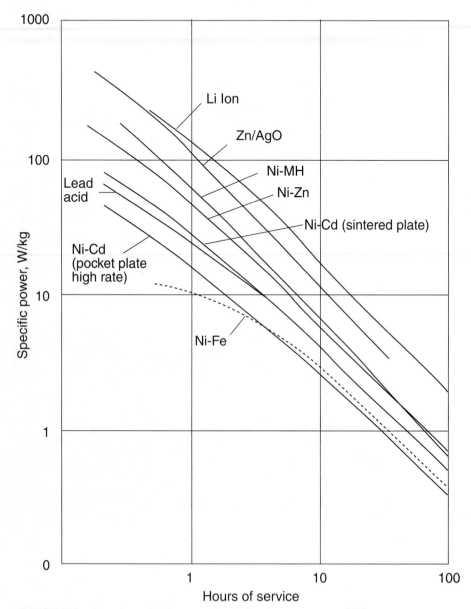

FIGURE 22.3 Comparison of performance of secondary battery systems at 20°C.

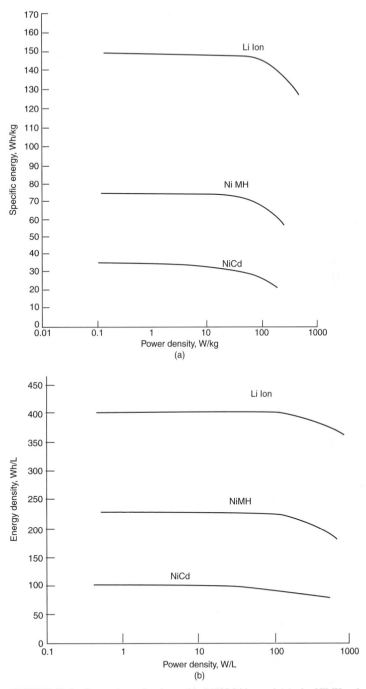

FIGURE 22.4 Comparison of rechargeable 14500 Li ion and AA-size NiMH and NiCd batteries at 20°C. (*a*) Specific energy vs. power density. (*b*) Energy density vs. power density.

22.3.4 Effect of Temperature

The performance of the various secondary batteries over a wide temperature range is shown in Fig. 22.5 on a gravimetric basis. In this figure, the specific energy for each battery system is plotted from -40 to $60°C$ at about the $C/5$ discharge rate. The lithium ion system has the highest energy density to $-20°C$. The sintered-plate nickel-cadmium and nickel-metal hydride batteries show higher percentage retention. In general the low-temperature performance of the alkaline batteries is better than the performance of the lead-acid batteries, again with the exception of the nickel-iron system. The lead-acid system shows better characteristics at the higher temperatures. These data are necessarily generalized for the purposes of comparison and present each system under favorable discharge conditions. Performance is strongly influenced by the specific discharge conditions.

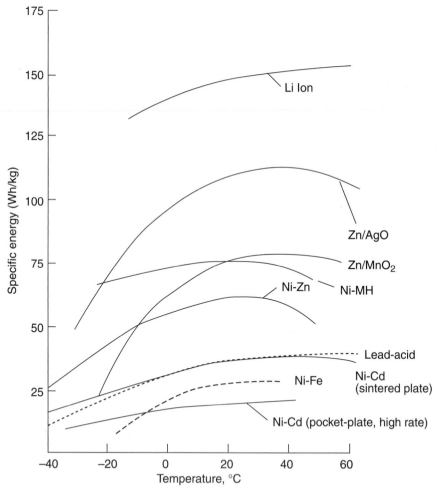

FIGURE 22.5 Effect of temperature on specific energy of secondary battery systems at approximately $C/5$ discharge rate.

22.3.5 Charge Retention

The charge retention of most of the conventional secondary batteries is poor compared with that of primary battery systems (see Fig. 6.7). Normally, secondary batteries are recharged on a periodic basis or maintained on "float" charge if they are to be in a state of readiness. Most alkaline secondary batteries, especially the nickel oxide batteries, can be stored for long periods of time even in a discharged state without permanent damage and can be recharged when required for use. The lead-acid batteries, however, cannot be stored in a discharged state because sulfation of the plates, which is detrimental to battery performance, will occur.

Figure 22.6 shows the charge retention properties of several different secondary battery systems. These data are also generalized for the purpose of comparison. There are wide variations of performance depending on design and many other factors, with the variability increasing with increasing storage temperature. Typically, the rate of loss of capacity decreases with increasing storage time.

The silver secondary batteries, the Zn/MnO_2 rechargeable battery, and lithium-ion systems have the best charge retention characteristics of the secondary battery systems with typical lithium ion batteries, self discharge is typically 2% per month at ambient temperature. Low-rate silver cells may lose as little at 10 to 20% per year, but the loss with high-rate cells with large surface areas could be 5 to 10 times higher. The vented pocket- and sintered-plate nickel-cadmium batteries and the nickel-zinc systems are next; the sealed cells and the nickel-iron batteries have the poorest charge retention properties of the alkaline systems.

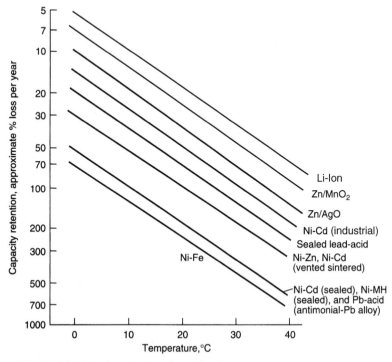

FIGURE 22.6 Capacity retention of secondary battery systems.

The charge retention of the lead-acid batteries is dependent on the design, electrolyte concentration, and formulation of the grid alloy as well as other factors. The charge retention of the standard automotive SLI batteries, using the standard antimonial-lead grid, is poor, and these batteries have little capacity remaining after six-months' storage at room temperature. The low antimonial-lead designs and the maintenance-free batteries have much better charge retention with losses on the order of 20 to 40% per year.

One of the potential advantages of the lithium metal rechargeable batteries is their good charge retention which, in many cases, should be similar to the charge retention characteristics of the lithium primary batteries.

22.3.6 Life

The cycle life and calendar life of the different secondary battery systems are also listed in Table 22.3. Again, these data are approximate because specific performance is dependent on the particular design and the conditions under which the battery is used. The depth of discharge (DOD), for example, as illustrated in Fig. 22.7, and the charging regime strongly influences the battery's life.[4]

Of the conventional secondary systems, the nickel-iron and the vented pocket-type nickel-cadmium batteries are best with regard to cycle life and total lifetime. The nickel-hydrogen battery developed mainly for aerospace applications, has demonstrated very long cycle life under shallow depth of discharge. The lead-acid batteries do not match the performance of the best alkaline batteries. The pasted cells have the shortest life of the lead-acid cells; the best cycle life is obtained with the tubular design, and the Planté design has the best lifetime.

One of the disadvantages of using zinc, lithium, and other metals with high negative standard potentials in rechargeable batteries is the difficulty of successful recharging and obtaining good cycle and calendar lives. The nickel-zinc battery has recently been improved to provide extended cycle life as seen in Fig. 22.7. The lithium-ion system, however, has also been shown to have good cycle life.

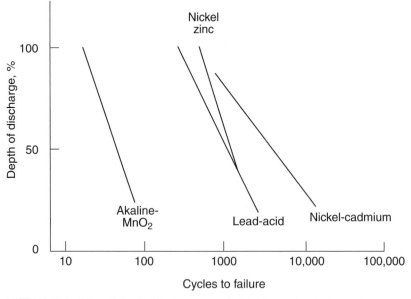

FIGURE 22.7   Effect of depth of discharge on cycle life of secondary battery systems.

22.3.7 Charge Characteristics

The typical charge curves of the various secondary aqueous-systems at normal constant-current charge rates are shown in Fig. 22.8. Most of the batteries can be charged under constant-current conditions, which is usually the preferred method of charging, although, in practice, constant-voltage or modified constant-voltage methods are used. Some of the sealed batteries, however, may not be charged by constant-voltage methods because of the possibility of thermal runaway. Generally the vented nickel-cadmium battery has the most favorable charge properties and can be charged by a number of methods and in a short time. These batteries can be charged over a wide temperature range and can be overcharged to some degree without damage. Nickel-iron batteries, sealed nickel-cadmium batteries, and sealed nickel/metal hydride batteries have good charge characteristics, but the temperature range is narrower for these systems. The nickel/metal hydride battery is more sensitive to overcharge, and charge control to prevent overheating is advisable. The lead-acid battery also has good charge characteristics, but care must be taken to prevent excessive overcharging. The zinc/manganese dioxide and zinc/silver oxide batteries are most sensitive with regard to charging; overcharging is very detrimental to battery life. Figure 22.9 shows typical constant current–constant voltage charging characteristics of an 18650 lithium ion battery.

Table 22.5 summarizes the typical conditions for charging the different systems. However, the chapters on individual battery systems and manufacturers' recommendations should be consulted because of the different procedures used.

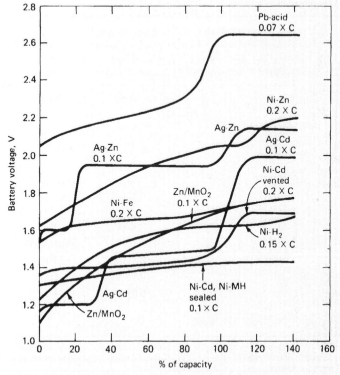

FIGURE 22.8 Typical charge characteristics of secondary battery systems, constant-current charge at 20°C. (*Adapted from Falk and Salkind. Ref. 5.*)

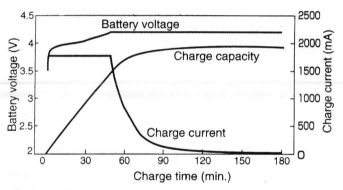

FIGURE 22.9   Charging characteristics of a typical cylindrical 18650 lithium ion battery at 20°C. Battery is charged at constant current of 1.8 Amps (nominal C-rate) to 4.2 Volts followed by a taper charge at this voltage for a total time of 2 hours.

TABLE 22.5   Charging Characteristics of Secondary Batteries

System	Charged methods* Preferred	Not recommended	Recommended constant-current charge rate, C (A)	Over-charge-ability	Temperature range for charging, °C	Efficiencies† Ah, %	Wh, %
Lithium ion	cc, cv		0.20	None	−20 to 50	99	95
Lead-acid							
Pasted, Planté	cc, cv		0.07	Fair	−40 to 50	90	75
Tubular	cc, cv		0.07	Fair	−40 to 50	80	70
Nickel-cadmium:							
Industrial vented	cc, cv		0.2	Very good	−50 to 40	70	60
Sintered vented	cc, cv		0.2	Very good	−55 to 75	70–80	60–70
Sealed	cc	cv	0.1–0.3‡	Very good	0 to 40	65–70	55–65
Nickel/metal hydride	cc	cv	0.1‡	Fair	0 to 40	65–70	55–65
Nickel-iron	cc	cv	0.2	Very good	0 to 45	80	60
Nickel-zinc	cc, cv		0.1–0.4	Fair	−20 to 40	85	70
Silver-zinc	cc		0.05–0.1	Poor	0 to 50	90	75
Silver-cadmium	cc		0.01–0.2	Fair	−40 to 50	90	70
Zn/MnO$_2$	cv	cc w/o v. limit		Fair	10–30		55–65

*Constant current (cc) includes two-rate charging, and constant voltage (cv) includes modified constant-voltage charging.
†All data are related to normal rates of charge and discharge and room-temperature operation.
‡Fast charge procedures can be used with charge control.
Source: Based on Falk and Salkind Ref. 5.

Many manufacturers are now recommending "fast" charge methods to meet consumer and application demand for recharging in less than 2 to 3 h. These methods require control to cut off the charge before an excessive rise in gassing, pressure, or temperature occurs. These could cause venting or a more serious safety hazard, or they could result in a deleterious effect on the battery's performance or life. Pulse charging is also being employed with some systems to provide higher charge rates.

In general, control techniques are useful for recharging most secondary batteries. They can be employed in several ways: to prevent overcharging, to facilitate "fast" charging, to sense when a potentially deleterious or unsafe condition may arise and cut off the charge or reduce the charging rate to safe levels. Similarly, discharge controls are also being used to maintain cell balance and to prevent overdischarge.

Another approach is the "smart" battery. These batteries incorporate features:

1. To control the charge so that the battery can be charged optimally and safely
2. For fuel gauging to indicate the remaining charge left in the battery
3. Safety devices to alert the user to unsafe or undesirable operation or to cut off the battery from the circuit when these occur.

See Sec. 5.6

22.3.8 Cost

The cost of a secondary battery may be evaluated on several bases, depending on the mode of operation. The initial cost is one of the bases for consideration. Other factors are the number of charge-discharge cycles that are available, or the number delivered in an application, during a battery's lifetime, or the cost determined on a dollar-per-cycle or dollar-per-total-kilowatt-hour basis. The cost of charging, maintenance, and associated equipment may also have to be considered in this evaluation. See Sec. 6.4.3. In an emergency standby service or SLI-type application, the important factors may be the calendar life of the battery (rather than as cycle life) and the cost is evaluated on a dollar-per-operating-year basis.

The lead-acid battery system is by far the least costly of the secondary batteries, particularly the SLI type. The lead-acid traction and stationary batteries, having more expensive constructional features and not as broad a production base, are several times more costly, but are still less expensive than the other secondary batteries. The nickel-cadmium and the rechargeable zinc/manganese dioxide batteries are next lowest in cost, followed by the nickel/metal hydride battery. The cost is very dependent on the cell size or capacity, the smaller button cells being considerably more expensive than the larger cylindrical and prismatic cells. The nickel-iron battery is more expensive and, for this reason among others, lost out to the less expensive battery system.

The most expensive of the conventional-type secondary batteries are the silver batteries. Their higher cost and low cycle life have limited their use to special applications, mostly in the military and space applications, which require their high energy density. The nickel-hydrogen system is more expensive due to its pressurized design and a relatively limited production. However, their excellent cycle life under conditions of shallow discharge make them attractive for aerospace applications. The cost of cylindrical lithium ion batteries has been decreasing rapidly as production rates have increased and has recently been stated to be $1.22/Wh.[6]

An important objective of the program for the development of secondary batteries for electric vehicles and energy storage is to reduce the cost of these battery systems. Cost factors are discussed in Chap. 37.

ACKNOWLEDGMENT

The contributions of Prof. Alvin J. Salkind and Mr. Anthony G. Cannone to this chapter are gratefully acknowledged.

REFERENCES

1. A. J. Salkind, D. T. Ferrell, and A. J. Hedges, "Secondary Batteries 1952–1977," *J. Electrochem. Soc.* **125**(8), Aug. 1978.

2. A. J. Salkind, Director, Battery Laboratory, Rutgers University, NJ, 2001.

3. H. Takeshita, *Proc. 18th Int. Seminar on Primary & Secondary Batteries,* Fort Lauderdale, FL (March 5–8, 2001).

4. L. H. Thaller, "Expected Cycle Life vs. Depth of Discharge Relationships of Well-Behaved Single Cells and Cell Strings," *J. Electrochem. Soc.* **130**(5), May 1983.

5. S. U. Falk and A. J. Salkind, *Alkaline Storage Batteries,* Wiley, New York, 1969.

6. D. MacArthur, G. Blomgren, and R. Powers, *Lithium and Lithium Ion Batteries, 2000,* Powers Associates, Sept. 30, 2000.

CHAPTER 23
LEAD-ACID BATTERIES

Alvin J. Salkind, Anthony G. Cannone, and Forrest A. Trumbure

23.1 GENERAL CHARACTERISTICS

The lead-acid battery has been a successful article of commerce for over a century. Its production and use continue to grow because of new applications for battery power in energy storage, emergency power, and electric and hybrid vehicles (including off-road vehicles) and because of the increased number of vehicles for which it provides the energy for engine starting, vehicle lighting, and engine ignition (SLI). Its sales represent approximately 40 to 45% of the sales value of all batteries in the world, or a market value, in 1999, of about $15 billion at manufacturers' levels and 2 to 3 times this value at retail levels. (These values do not include some countries such as Russia and China, for which complete market data are not available.) This battery system is also used extensively in telephone systems, power tools, communication devices, emergency lighting systems, and as the power source for mining and material-handling equipment. The wide use of the lead-acid battery in many designs, sizes, and system voltages is accounted for by the low price and the ease of manufacture on a local geographic basis of this battery system. The lead-acid battery is almost always the least expensive storage battery for any application, while still providing good performance and life characteristics.

New uses, designs, and fabrication processes are still being introduced at significant rates. Some of the new designs are for modern electric-vehicle, energy-storage, and electronics applications. There have been many improvements in lead-acid battery design and charger system logic to make high-voltage batteries more uniform in performance. Electric-vehicle batteries are typically 100 to 300 V systems. The lead-acid battery has a high electrical turnaround efficiency, 75 to 80%, which makes the system attractive for electric-vehicle and energy-storage use. Traditional vertical-plate batteries are capable of energy densities greater than 40 Wh/kg, and a horizontal-plate design with higher energy and power densities have found use in traction and fork lift applications. Modified lead-acid batteries are being investigated for both electric and hybrid-drive vehicles. The world's largest energy-storage battery system was finished in late 1988. This 40-MWh battery, located in Chino, Calif., uses individual industrial-size lead-acid cells in series and parallel connection to make a 10-MW system delivering energy into the utility grid at 2000 V and 8000 A for 4 hours. AC to DC conversion is built into the system. This battery operated for more than a decade as a demonstration project. At the other extreme, small individual lead-acid cells and batteries are now available with quick connects for use in small electric appliances and electronics applications. Many of these newer applications require low-maintenance or maintenance-free designs. Thin film capacitor-like lead-acid batteries have become commercially available in the past few years, for consumer and electronic applications. These are discussed in detail in Chap. 24. Some larger industrial cells are often virtually maintenance-free using the oxygen-recombination principle and a resealable Bunsen vent. An approach to high-energy-density, high-power-density, high-cycle-life lead-acid battery design is the bipolar design, a design which is still being pursued. The problems which prevent this design from larger scale commercial use relate to the availability of a bipolar material which is electronically conductive, nonporous to ions, low cost, and stable against both positive and negative active materials. Conductive plastics, which are used in some battery systems, have not been successful in lead batteries. Experiments have been carried out with a bipole made from tin oxide coated glass encapsulated in a plastic matrix, and with multilayers of different lead alloys to slow the penetration of the bipole by corrosion.

The overall advantages and disadvantages of the lead-acid battery, compared with other systems, are listed in Table 23.1.

TABLE 23.1 Major Advantages and Disadvantages of Lead-Acid Batteries

Advantages	Disadvantages
Popular low-cost secondary battery—capable of manufacture on a local basis, worldwide, from low to high rates of production	Relatively low cycle life (50–500 cycles)*
	Limited energy density—typically 30–40 Wh/kg
Available in large quantities and in a variety of sizes and designs—manufactured in sizes from smaller than 1 Ah to several thousand Ampere-hours	Long-term storage in a discharged condition can lead to irreversible polarization of electrodes (sulfation)
Good high-rate performance—suitable for engine starting (but outperformed by some nickel-cadmium and nickel metal-hydride batteries)	Difficult to manufacture in very small sizes (it is easier to make nickel-cadmium button cells in the smaller than 500-mAh size)
Moderately good low- and high-temperature performance	Hydrogen evolution in some designs can be an explosion hazard (flame arrestors are installed to prevent this hazard)
Electrically efficient—turnaround efficiency of over 70%, comparing discharge energy out with charge energy in	Stibene and arsine evolution in designs with antimony and arsenic in grid alloys can be a health hazard
High cell voltage—open-circuit voltage of >2.0 V is the highest of all aqueous-electrolyte battery systems	Thermal runaway in improperly designed batteries or charging equipment
Good float service	Positive post blister corrosion with some designs
Easy state-of-charge indication	
Good charge retention for intermittent charge applications (if grids are made with high-overvoltage alloys)	
Available in maintenance-free designs	
Low cost compared with other secondary batteries	
Cell components are easily recycled.	

*Up to 2000 cycles can be attained with special designs.

The lead-acid battery is manufactured in a variety of sizes and designs, ranging from less than 1 to over 10,000 Ah. Table 23.2 lists many of the various types of lead-acid batteries that are available.

TABLE 23.2 Types and Characteristics of Lead-Acid Batteries

Type	Construction	Typical applications
SLI (starting, lighting, ignition)	Flat-pasted plates (option: maintenance-free construction)	Automotive, marine, aircraft, diesel engines in vehicles and for stationary power
Traction	Flat-pasted plates; tubular and gauntlet plates	Industrial trucks (material handling)
Vehicular propulsion	Flat-pasted plates; tubular and gauntlet plates; also composite construction	Electric vehicles, golf carts, hybrid vehicles, mine cars, personnel carriers
Submarine	Tubular plates; flat-pasted plates	Submarines
Stationary (including energy-storage types such as charge retention, solar photovoltaic, load leveling)	Planté;* Manchester;* tubular and gauntlet plates; flat-pasted plates; circular conical plates	Standby emergency power: telephone exchange, uninterruptible power systems (UPS), load leveling, signaling
Portable (see chap. 24)	Flat-pasted plates (gelled electrolyte, electrolyte absorbed in separator); spirally wound electrodes; tubular plates	Consumer and instrument applications: portable tools, appliances, lighting, emergency lighting, radio, TV, alarm systems

*Now rarely used.

23.1.1 History

Practical lead-acid batteries began with the research and inventions of Raymond Gaston Planté in 1860, although batteries containing sulfuric acid or lead components were discussed earlier.[1] Table 23.3 lists the events in the technical development of the lead-acid battery. In Planté's fabrication method, two long strips of lead foil and intermediate layers of coarse cloth were spirally wound and immersed in a solution of about 10% sulfuric acid. The early Planté cells had little capacity, since the amount of stored energy depended on the corrosion of a lead foil to lead dioxide to form the positive active material, and similarly the negative electrode was formed by roughening of another foil (on cycling) to form an extended surface. Primary cells were used as the power sources for this formation. The capacity of Planté cells was increased on repeated cycling as corrosion of the substrate foils created more active material and increased the surface area. In the 1870s magnetoelectric generators became available to Planté, and about this time the Siemens dynamo began to be installed in central electric plants. Lead-acid batteries found an early market to provide load leveling and to average out the demand peaks. They were charged at night, similar to the procedure now planned for modern load-leveling energy-storage systems.

Subsequent to Planté's first developments, numerous experiments were done on accelerating the formation process and coating lead foil with lead oxides on a lead plate pretreated by the Planté method. Attention then turned to other methods for retaining active material, and two main technological paths evolved.

1. Coating a lead oxide paste on cast or expanded grids, rather than foil, in which the active material developed structural strength and retention properties by a "cementation" process (interlocked crystalline lattice) through the grid and active mass. This is generally referred to as flat-plate design.

2. The tubular electrode design, in which a central conducting wire or rod is surrounded by active material and the assembly encased in an electrolyte porous insulated tube, which can be either square, round, or oval.

TABLE 23.3   Events in Technical Development of Lead-Acid Battery

		Precursor systems
1836	Daniell	Two-fluid cell; copper/copper sulfate/sulfuric acid/zinc
1840	Grove	Two-fluid cell; carbon/fuming nitric acid/sulfuric acid/zinc
1854	Sindesten	Polarized lead electrodes with external source

		Lead-acid battery developments
1860	Planté	First practical lead-acid battery, corroded lead foils to form active material
1881	Faure	Pasted lead foils with lead oxide-sulfuric acid pastes for positive electrode, to increase capacity
1881	Sellon	Lead-antimony alloy grid
1881	Volckmar	Perforated lead plates to provide pockets for support of oxide
1882	Brush	Mechanically bonded lead oxide to lead plates
1882	Gladstone and Tribs	Double sulfate theory of reaction in lead-acid battery: $$PbO_2 + Pb + 2H_2SO_4 \rightleftharpoons 2PbSO_4 + 2H_2O$$
1883	Tudor	Pasted mixture of lead oxides on grid pretreated by Planté method
1886	Lucas	Formed lead plates in solutions of chlorates and perchlorates
1890	Phillipart	Early tubular construction—individual rings
1890	Woodward	Early tubular construction
1910	Smith	Slotted rubber tube, Exide tubular construction
1920 to present		Materials and equipment research, especially expanders, oxides, and fabrication techniques
1935	Haring and Thomas	Lead-calcium alloy grid
1935	Hamer and Harned	Experimental proof of double sulfate theory of reaction
1956– 1960	Bode and Voss Ruetschi and Cahan Burbank Feitknecht	Clarification of properties of two crystalline forms of PbO_2 (alpha and beta)
1970s	McClelland and Devit	Commercial spiral-wound sealed lead acid battery
		Expanded metal grid technology; composite plastic/metal grids; sealed and maintenance-free lead-acid batteries; glass fiber and improved separators; through-the-partition intercell connectors; heat-sealed plastic case-to-cover assemblies; high-energy-density batteries (above 40 Wh/kg); conical grid (round) cell for long-life float service in telecommunications facilities
1980s		Sealed valve-regulated batteries; quasi-bipolar engine starter batteries; improved low-temp. performance; world's largest battery installed (Chino, Calif.); 40-MWh lead-acid load leveling
1990s		Electric-vehicle interest reemerges; bipolar battery designs for highpower use in uninterruptible power supplies, power tool market and electronic back-up. Thin foil cells, small cells for consumer and current road applications.
2000s		Planned introduction of 36 Volt SLI battery system for automobiles

Simultaneous with the advances in developing and retaining active material was work in strengthening the grid by casting it from lead alloys such as lead-antimony (e.g., Sellon, 1881) or lead-calcium (e.g., Haring and Thomas, 1935).[2] The technical knowledge for an economical manufacture of reliable lead-acid batteries was in place by the end of the nineteenth century, and subsequent growth of the industry was rapid. Improvements in design, manufacturing equipment and methods, recovery methods, active material utilization and production, supporting structures and components, and nonactive components such as separators, cases, and seals continue to improve the economic and performance characteristics of lead-acid batteries. Intense development work continues on applications for energy storage, electric bicycles and electric vehicles.

23.1.2 Manufacturing Statistics and Lead Use

The major present-day use of lead-acid storage batteries is in vehicle starting (SLI) applications. Most vehicles use a 12-V battery with a capacity in the range of 40 to 60 Ah. This battery, weighing about 14.5 kg, has sufficient high-rate capacity to deliver the 450 to 800 A necessary to start an automobile engine. The power demand for the automobiles in the year 2003 is expected to be greater than that supplied by the present 12-V battery systems. With all of the new and planned onboard systems, the battery is expected to be a nominal 36 Volts which corresponds to a 42 V charged voltage.

Approximately 60% of the battery weight is lead or lead components. Batteries represent the most important use of lead in the world. The yearly consumption of lead for the manufacture of batteries was about 3 million metric tonnes in 1992, approximately half of the total world output of primary and secondary lead. The percentage of lead used in batteries is increasing as other uses disappear because of environmental and health concerns. There is a direct correlation between the number of motor vehicles registered and the number of SLI battery units sold annually, as shown in Table 23.4a.[3] Although SLI units represent over 70% of the monetary value of the total lead battery market at present, this percentage is declining as the other applications for storage batteries continue to grow. A world perspective on the manufacture and use of SLI lead-acid batteries is presented in Table 23.4b.

TABLE 23.4a Market Growth of Lead-Acid Battery in United States*

	1960	1969	1980	1991	1999**	2003 est
MV registration	74	105	158	190	250	260
SLI units (original equipment and replacement)	34	47	62	76	100	110
Car-battery ratio‡	2.2	2.2	2.5	2.5	2.5	2.4
SLI sales, $	330	510	1675	2100	2700	2800
Industrial, $	70	105	380	550	1015	1500
Consumer, $	1	3	55	100	150	160
Total, $	400	620	2110	2750	3965	4460

*All units in millions, except car-battery ratio, values are at manufacturers' pricing.
Battery prices are affected by the price of lead.
   Lead prices varied from $0.40 to $1.40/kg between 1978 and 1999, (lead price Dec. 1999 = $0.91/kg.)
   ‡A considerable number of SLI-type batteries are used in boats and other vehicles, therefore this ratio is not an exact representation of the useful life of an automotive battery.
   NOTE: The Canadian market is approximately 10% of the U.S. market.
   Year 2003 estimates from "Battery Man" of Nov. 1999.
   **A large growth in industrial battery sales began in 1995.

TABLE 23.4b World Battery Trends-SLI Shipment*

	1990	1991[4]	% of total (1991)	1999
North America	84.9	81.4	35	120
Europe	67.3	69.6	29	79
Asia/Pacific	53.7	58.8	24	—
Latin American	17.9	18.9	8	—
Africa/Middle East	10.0	9.6	4	—
Total	233.8	238.3	100	320 est.

*In millions of units.
Source: From Amistadl.[4]

An important aspect of lead is that it is a recoverable resource. It has been estimated that more than 95% of the batteries sold in the United States are ultimately recycled, and it takes considerably less energy to recycle lead, a low-melting metal (mp 327.4°C), than to produce the metals used in other storage battery systems (nickel, iron, zinc, silver, and cadmium) in battery-grade quality.

23.2 CHEMISTRY

23.2.1 General Characteristics

The lead-acid battery uses lead dioxide as the active material of the positive electrode and metallic lead, in a high-surface-area porous structure, as the negative active material. The physical and chemical properties of these materials are listed in Table 23.5.[5] Typically, a charged positive electrode contains both α-PbO_2 (orthorhombic) and β-PbO_2 (tetragonal). The equilibrium potential of the α-PbO_2 is more positive than that of β-PbO_2 by 0.01 V. The α form also has a larger, more compact crystal morphology which is less active electrochemically and slightly lower in capacity per unit weight; it does, however, promote longer cycle life. Neither of the two forms is fully stoichiometric. Their composition can be represented by PbO_x, with x varying between 1.85 and 2.05. The introduction of antimony, even at low concentrations, in the preparation or cycling of these species leads to a considerable increase in their performance. The preparation of the active material precursor consists of a series of mixing and curing operations using leady lead oxide (PbO + Pb), sulfuric acid, and water. The ratios of the reactants and curing conditions (temperature, humidity, and time) affect the development of crystallinity and pore structure. The cured plate consists of lead sulfate, lead oxide, and some residual lead (<5%). The positive active material, which is formed electrochemically from the cured plate, is a major factor influencing the performance and life of the lead-acid battery. In general the negative, or lead, electrode controls cold-temperature performance (such as engine starting).

The electrolyte is a sulfuric acid solution, typically about 1.28 specific gravity or 37% acid by weight in a fully charged condition.

TABLE 23.5 Physical and Chemical Properties of Lead and Lead Oxides (PbO_2)

Property	Lead	α-PbO_2	β-PbO_2
Molecular weight, g/mol	207.2	239.19	239.19
Composition		$PbO_{1.94-2.03}$	$PbO_{1.87-2.03}$
Crystalline form	Face-centered cubic	Rhombic (columbite)	Tetragonal (rutile)
Lattice parameters, nm	$a = 0.4949$	$a = 0.4977$	$a = 0.491-0.497$
		$b = 0.5948$	$c = 0.337-0.340$
		$c = 0.5444$	
X-ray density, g/cm^3	11.34	9.80	~9.80
Practical density at 20°C (depends on purity), g/cm^3	11.34	9.1–9.4	9.1–9.4
Heat capacity, cal/deg·mol	6.80	14.87	14.87
Specific heat, cal/g	0.0306	0.062	0.062
Electrical resistivity, at 20°C, $\mu\Omega$/cm	20	~100×10^3	
Electrochemical potential in 4.4M H_2SO_4 at 31.8°C, V	0.356	~1.709	~1.692
Melting point, °C	327.4		

Source: Ref. 5 and others.

As the cell discharges, both electrodes are converted to lead sulfate. The process reverses on charge:

Negative electrode
$$Pb \underset{charge}{\overset{discharge}{\rightleftharpoons}} Pb^{2+} + 2e$$

$$Pb^{2+} + SO_4^{2-} \underset{charge}{\overset{discharge}{\rightleftharpoons}} PbSO_4$$

Positive electrode
$$PbO_2 + 4H^+ + 2e \underset{charge}{\overset{discharge}{\rightleftharpoons}} Pb^{2+} + 2H_2O$$

$$Pb^{2+} + SO_4^{2-} \underset{charge}{\overset{discharge}{\rightleftharpoons}} PbSO_4$$

Overall reaction
$$Pb + PbO_2 + 2H_2SO_4 \underset{charge}{\overset{discharge}{\rightleftharpoons}} 2PbSO_4 + 2H_2O$$

As shown, the basic electrode processes in the positive and negative electrodes involve a dissolution-precipitation mechanism and not a solid-state ion transport or film formation mechanism.[5] The discharge-charge mechanism, known as double-sulfate reaction, is also shown graphically in Fig. 23.1.[6] As the sulfuric acid in the electrolyte is consumed during discharge, producing water, the electrolyte is an "active" material and in certain battery designs can be the capacity-limiting material.

As the cell approaches full charge and the majority of the $PbSO_4$ has been converted to Pb or PbO_2, the cell voltage on charge becomes greater than the gassing voltage (about 2.39 V per cell) and the overcharge reactions begin, resulting in the production of hydrogen and oxygen (gassing) and the resultant loss of water.

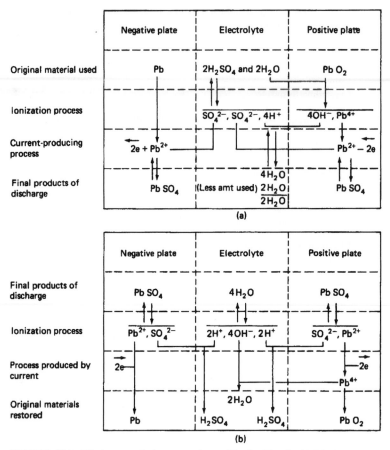

FIGURE 23.1 Discharge and charge reactions of lead-acid cell. (*a*) Discharge reactions. (*b*) Charge reactions. (*From Mantell.*[6])

Negative electrode	$2H^+ + 2e \rightarrow H_2$
Positive electrode	$\underline{H_2O - 2e \rightarrow \frac{1}{2}O_2 + 2H^+}$
Overall reaction	$H_2O \rightarrow H_2 + \frac{1}{2}O_2$

In sealed lead-acid cells this reaction is controlled to minimize hydrogen evolution and the loss of water by recombination of the evolved oxygen with the negative plate (see Sec. 24.2).

The general performance characteristics of the lead-acid cell, during charge and discharge, are shown in Fig. 23.2. As the cell is discharged, the voltage decreases due to depletion of material, internal resistance losses, and polarization. If the discharge current is constant, the voltage under load decreases smoothly to the cutoff voltage and the specific gravity decreases in proportion to the Ampere-hours discharged.

An analysis of the behavior of the positive and negative plates can be done by measuring the voltage between each electrode and a reference electrode, the "half-cell" voltage. Figure 23.3 illustrates this analysis, using a cadmium reference electrode. The industry is, however, shifting away from cadmium reference electrodes to more stable materials as discussed later (Sec. 23.2.3).

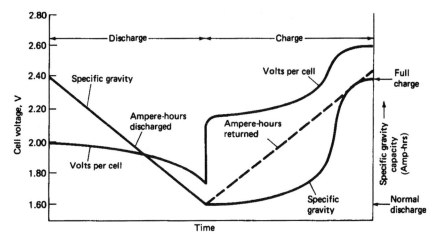

FIGURE 23.2 Typical voltage and specific gravity characteristics of lead-acid cell at constant-rate discharge and charge.

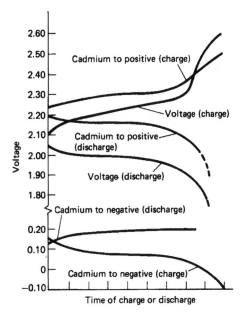

FIGURE 23.3 Typical charge-discharge curves of lead-acid cell. (*From Mantell.*[6])

Voltage.   The nominal voltage of the lead-acid cell is 2 V; the voltage on open circuit is a direct function of the electrolyte concentration, ranging from 2.125 V for a cell with 1.28 specific gravity electrolyte to 2.05 V with 1.21 specific gravity (see Sec. 23.2.2). The end or cutoff voltage on moderate-rate discharges is 1.75 V per cell, but may range to as low as 1.0 V per cell at extremely high discharge rates at low temperatures.

Specific Gravity.   The selection of specific gravity used for the electrolyte depends on the application and service requirements (see Table 23.12). The electrolyte concentration must be high enough for good ionic conductivity and to fulfill electrochemical requirements, but not so high as to cause separator deterioration or corrosion of other parts of the cell, which would shorten life and increase self-discharge. The electrolyte concentration is deliberately reduced in high-temperature climates. During discharge, the specific gravity decreases in proportion to the Ampere-hours discharged (Table 23.6). The specific gravity is thus a means for checking the state of charge of the battery. On charge, the change in specific gravity should similarly be proportional to the Ampere-hour charge accepted by the cell. However, there is a lag because complete mixing of the electrolyte does not occur until gassing commences near the end of the charge.

TABLE 23.6   Specific Gravity of Lead-Acid Battery Electrolytes at Different States of Charge for Various Designs.*

	Specific gravity			
State of charge	A	B	C	D
100% (full charge)	1.330	1.280	1.265	1.225
75%	1.300	1.250	1.225	1.185
50%	1.270	1.220	1.190	1.150
25%	1.240	1.190	1.155	1.115
Discharged	1.210	1.160	1.120	1.0

Assumes flooded cell design.
*Specific gravity may range from 100 to 150 points between full charge and discharge depending on cell design: A—electric vehicle battery; B—traction battery; C—SLI battery; D—stationary battery.

23.2.2 Open-Circuit Voltage Characteristics

The open-circuit voltage for a battery system is a function of temperature and electrolyte concentration as expressed in the Nernst equation for the lead-acid cell (see also Chap. 2).

$$E = 2.047 + \frac{RT}{F} \ln \left(\frac{\alpha H_2SO_4}{\alpha H_2O} \right)$$

Since the concentration of the electrolyte varies, the relative activities of H_2SO_4 and H_2O in the Nernst equation change. A graph of the open-circuit voltage versus electrolyte concentration at 25°C is given in Fig. 23.4. The plot is fairly linear above 1.10 specific gravity, but shows strong deviations at lower concentrations. The open-circuit voltage is also affected by temperature. The temperature coefficient of the open-circuit voltage of the lead-acid battery is shown in Fig. 23.5. Where dE/dT is positive, such as above 0.5 Molar H_2SO_4, the reversible potential of the system increases with increasing temperature. Below 0.5 M, the temperature coefficient is negative. Most lead-acid batteries operate above 2 Molar H_2SO_4 (1.120 specific gravity) and have a thermal coefficient of about +0.2 mV/°C.

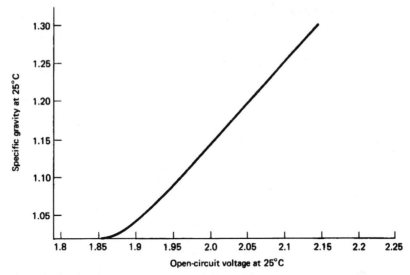

FIGURE 23.4 Open-circuit voltage of lead-acid cell as a function of electrolyte specific gravity.

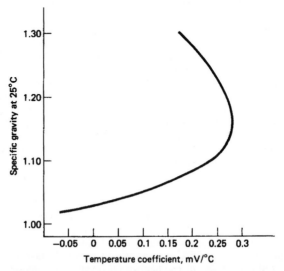

FIGURE 23.5 Temperature coefficient of open-circuit voltage of lead-acid cell as a function of electrolyte specific gravity.

23.2.3 Polarization and Resistive Losses

When a battery is being discharged, the voltage under load is lower than the open-circuit voltage at the same concentrations of H_2SO_4 and H_2O in the electrolyte and Pb or PbO_2 and $PbSO_4$ in the plates. The thermodynamically stable state for batteries is the discharged state. Work (charging) must be done to cause the equilibria of the electrochemical reactions to go toward PbO_2 in the positive and Pb in the negative. Thus the voltage of the power source for recharging the lead-acid battery must be higher than the Nernst voltage of the battery on open circuit.

These deviations from the open-circuit voltage during charge or discharge are due, in part, to resistive losses in the battery and, in part, to polarization. These losses can be measured by use of an interrupted discharge, where the IR losses can be estimated by Ohm's law ($\Delta E / \Delta I = R$) within a few seconds to a few minutes after the discharge is stopped. The effect of polarization can take several hours to measure in order for diffusion to allow the plate interiors to reequilibrate. AC impedance spectroscopy techniques are also of value. Polarization is more easily measured by use of a reference electrode. The standard reference of hydrogen on platinum is not practical for most measurements on lead-acid batteries, and several other sulfate-based reference electrode systems are used. A review of reference electrodes[7] neglects several very practical sulfate electrodes. Still, a commonly used electrode for battery maintenance is the cadmium "stick," but it is not especially stable (± 20 mV/day). Mercury-mercurous sulfate reference electrodes are stable and are available from several vendors. A novel $Pb/H_2SO_4/PbO_2$ reference electrode has been patented.[8] This electrode measures the polarization on charge or discharge directly, without need for a correction of different thermal coefficients of EMF. The change in polarization between the start and the end of discharge is typically 50 to several hundred mV, and the cell capacity is limited by the plate group (positive or negative) that has the largest change in polarization during discharge. When both groups in a cell change about equally, the capacity limitation is more likely depletion of H_2SO_4 in the electrolyte than depletion of Pb or PbO_2 in the plates. On charge, the polarization is a good measure that both positives and negatives have been recharged: the plate polarizations change by more than 60 mV between start and end of recharge. Polarization voltages stabilize at some value when plates are recharged and are gassing freely.

23.2.4 Self-Discharge

The equilibria of the electrode reactions are normally in the discharge direction since, thermodynamically, the discharged state is most stable. The rate of self-discharge [loss of capacity (charge) when no external load is applied] of the lead-acid cell is fairly rapid, but it can be reduced significantly by incorporating certain design features.

The rate of self-discharge depends on several factors. Lead and lead dioxide are thermodynamically unstable in sulfuric acid solutions, and on open circuit, they react with the electrolyte. Oxygen is evolved at the positive electrode and hydrogen at the negative, at a rate dependent on temperature and acid concentration (the gassing rate increases with increasing acid concentration) as follows:

$$PbO_2 + H_2SO_4 \rightarrow PbSO_4 + H_2O + \tfrac{1}{2}O_2$$

$$Pb + H_2SO_4 \rightarrow PbSO_4 + H_2$$

For most positives, the formation of $PbSO_4$ by self-discharge is slow, typically much less than 0.5%/day at 25°C. (Some positives which have been made with nonantimonial grids can fail by a different mechanism on open circuit, namely, the development of a grid-active material barrier layer.) The self-discharge of the negative is generally more rapid, especially if the cell is contaminated with various catalytic metallic ions. For example, antimony lost from the positive grids by corrosion can diffuse to the negative, where it is deposited, resulting in a "local action" discharge cell which converts some lead active material to $PbSO_4$. New batteries with lead-antimony grids lose about 1% of charge per day at 25°C, but the charge loss increases by a factor of 2 to 5 as the battery ages. Batteries with nonantimonial lead grids lose less than 0.5% of charge per day regardless of age. This is illustrated in Fig. 23.6a.[9] Maintenance-free and charge-retention-type batteries, where the self-discharge rate must be minimized, use low-antimony or antimony-free alloy (such as calcium-lead) grids. However, because of other beneficial effects of antimony, its complete elimination may not be desirable, and low-antimony–lead alloys are a useful compromise.

Self-discharge is temperature-dependent, as shown in Fig. 23.6b.[10] The graph shows the fall in specific gravity per day of a new fully charged battery with 6% antimonial lead grids. Self-discharge can thus be minimized by storing batteries at temperatures between 5 and 15°C.

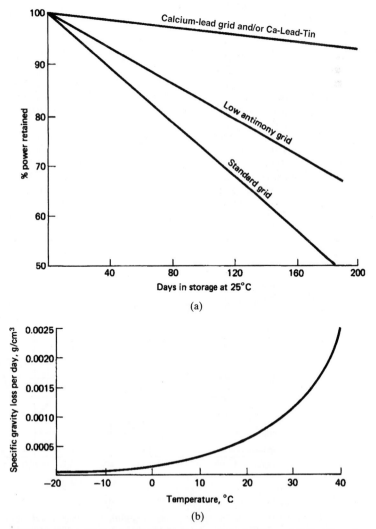

(a)

(b)

FIGURE 23.6 (a) Capacity retention during stand or storage at 25°C. (*From Sabatino.*[9]) (b) Loss of specific gravity per day with temperature of a new, fully charged lead-acid battery with 6% antimonial lead grids. (*From Ref. 10.*)

23.2.5 Characteristics and Properties of Sulfuric Acid

The major characteristics and properties of the sulfuric acid electrolyte, as they apply to the operation of the lead-acid battery, are listed in Table 23.7. The freezing points of sulfuric acid solutions at various concentrations are also plotted in Fig. 23.7a. The freezing point of aqueous sulfuric acid solutions varies significantly with concentration. Batteries must therefore be designed so that the electrolyte concentration is above the value at which the electrolyte would freeze when exposed to the anticipated cold. Alternatively, the battery can be insulated or heated so that it remains above the electrolyte freezing temperature.

Figure 23.7b shows the specific resistivity of sulfuric acid solutions at various specific gravities as a function of temperature from -40 to $40°C$.

TABLE 23.7   Properties of Sulfuric Acid Solutions*

Specific gravity		Temperature coeff. α	H_2SO_4			Freezing point, °C	Electrochemical equivalent (per liter of acid), Ah
At 15°C	At 25°C		Wt., %	Vol., %	Mol/L		
1.00	1.000	—	0	0	0	0	0
1.05	1.049	33	7.3	4.2	0.82	−3.3	22
1.10	1.097	48	14.3	8.5	1.65	−7.7	44
1.15	1.146	60	20.9	13.0	2.51	−15	67
1.20	1.196	68	27.2	17.7	3.39	−27	90
1.25	1.245	72	33.2	22.6	4.31	−52	115
1.30	1.295	75	39.1	27.6	5.26	−70	141
1.35	1.345	77	44.7	32.8	6.23	−49	167
1.40	1.395	79	50.0	38.0	7.21	−36	
1.45	1.445	82	55.0	43.3	8.2	−29	
1.50	1.495	85	59.7	48.7	9.2	−29	

*To calculate the specific gravity for any temperature, °C, SG (t) = SG (15°C) + $\alpha \times 10^{-5}$ (15 − t).

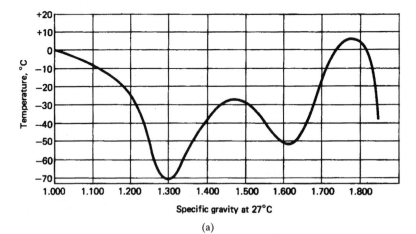

(a)

FIGURE 23.7a Freezing points of sulfuric acid solutions at various specific gravities. The inflection points result from the different water to SO_3 hydration ratios.

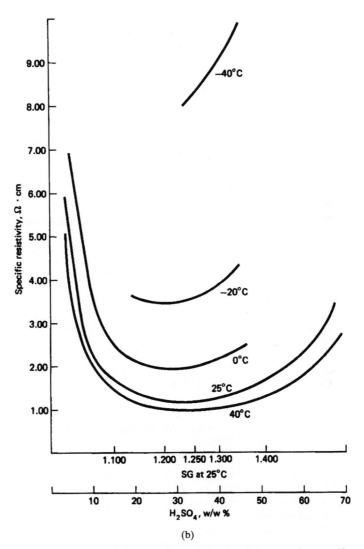

FIGURE 23.7*b*   Specific resistivity of sulfuric acid solutions at various specific
gravities and temperatures.

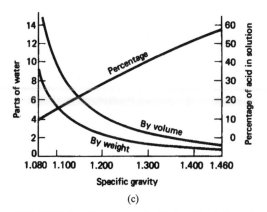

FIGURE 23.7c   Preparation of sulfuric acid solutions of any specific gravity from concentrated sulfuric acid. (*From G. W. Vinal, Storage Batteries, Wiley, New York, 1955, p. 129.*)

The specific gravities for several types of lead-acid battery designs are given in Table 23.12; the change in specific gravity at different states of charge is shown in Table 23.6. A comparison with freezing-point data will show that battery type *A* will freeze at −30°C when fully discharged while battery type D will freeze at about −5°C, a factor which must be considered in the design of the battery and the battery housing. The acid concentration for most lead-acid batteries for use in temperate climates is usually between 1.26 and 1.28 specific gravity. Higher-concentration electrolytes tend to attack some separators and other components; lower concentrations tend to be insufficiently conductive in a partially charged cell and freeze at low temperatures. In high-temperature climates, a lower concentration is used, and in stationary cells with larger proportional electrolyte volumes and no high-rate demands, electrolytes with specific gravity as low as 1.21 are used (see Table 23.12).

Figure 23.7c indicates the method of preparing sulfuric acid solutions of any specific gravity from concentrated sulfuric acid.

23.3 CONSTRUCTIONAL FEATURES, MATERIALS, AND MANUFACTURING METHODS

Lead-acid batteries consist of several major components, as shown in a cutaway view of Fig. 23.8. This figure shows the construction of an automotive SLI battery. Batteries for other applications have analogous components, as illustrated and described in Secs. 23.4–23.6. The applications of the various cells and batteries dictate the design, size, quantities, and types of materials that are used.

The active components of a typical lead-acid battery constitute less than one-half of its total weight. A breakout of the weights of the components of several types of lead-acid batteries is shown in Fig. 23.9.

The battery components are fabricated and processed as shown in the flowsheets of Fig. 23.10. The major starting material is highly purified lead.[11] The lead is used for the production of alloys (for subsequent conversion to grids) and for the production of lead oxides [for subsequent conversion first to paste and ultimately to the lead dioxide positive active material (Fig. 23.10a) and the sponge lead negative active material].

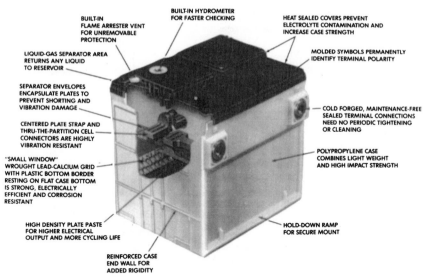

FIGURE 23.8   Maintenance-free lead-acid SLI battery. (*Courtesy of Delphi Energy Systems.*)

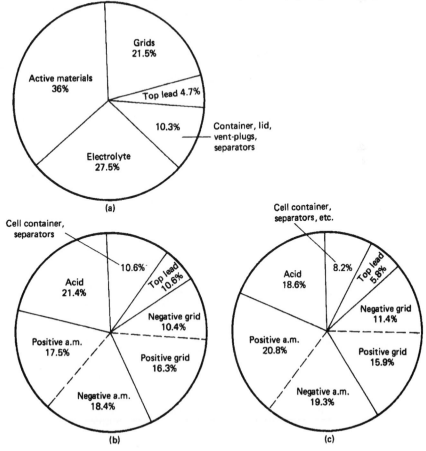

FIGURE 23.9   Weight analysis of typical lead-acid batteries. (*a*) SLI battery. (*b*) Tubular industrial battery. (*c*) Flat-plate traction battery. (*From Ref. 10.*)

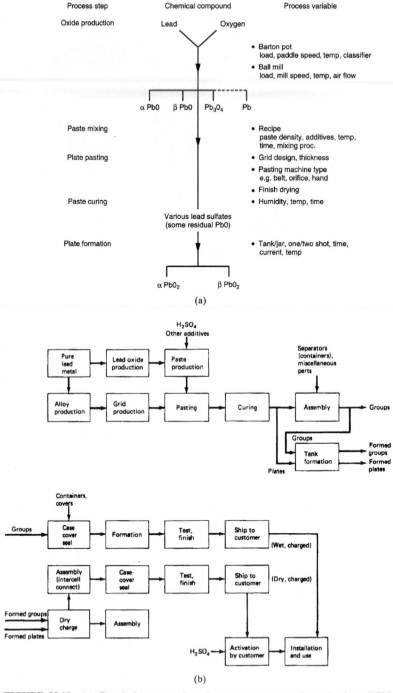

FIGURE 23.10   (*a*) Chemical compounds and process parameters in production of SLI batteries. (*b*) Production flow sheet for lead-acid batteries.

Automotive lead-acid batteries (SLI) are produced mainly in high-volume plants with a great deal of automation. Many modern factories are capable of producing quantities on the order of 20,000 to 30,000 batteries per day. On average, an automated facility might require less than 500 employees, including all staffing levels. The automation has been prompted by environmental, reliability, and cost considerations.

23.3.1 Alloy Production

Pure lead is generally too soft to be used as a grid material. Exceptions that use pure lead plates are some special, very thick plate Planté or pasted-plate batteries, some small spiral wound batteries, some valve regulated cells and batteries (see Fig. 23.12c) and a cylindrical cell. The latter were developed by Bell Laboratories (now part of Lucent Technologies) (see Fig. 23.36).[12]

The pure lead has been hardened, traditionally, by the addition of antimony metal. The amount of antimony has varied between 5 and 12% by weight, generally dependent on the availability and cost of antimony. Typical modern alloys, especially for deep-cycling applications, contain 4 to 6% antimony. The trend in grid alloys is to go to even lower antimony contents, in the range of 1.5 to 2% antimony, in order to reduce the maintenance (water addition) that the battery will require. As the antimony content goes below 4%, the addition of small amounts of other elements is necessary to prevent grid fabrication defects and grid brittleness. These elements, such as sulfur, copper, arsenic, selenium, tellurium and various combinations of these elements act as grain refiners to decrease the lead grain size.[13–15]

Some of the alloying elements, not previously described as grain refiners, fall into two broad classes of elements that are beneficial or detrimental to grid production or battery performance. Beneficial elements include tin, which operates synergistically with antimony and arsenic to improve metal fluidity and castability. Silver and cobalt are also thought to improve corrosion resistance. Detrimental elements include iron, which increases drossing;[1] nickel, which affects battery operation; and manganese, which attacks paper separators. Cadmium has been used in grid alloys to enhance processability in antimonial alloys to minimize the detrimental effects of antimony. Cadmium, however, is not popular because of its toxicity and difficulty of removal during lead recovery (recycling) operations. Bismuth exists in many lead ore feedstocks and has been reported to both increase and decrease grid corrosion rates.

A second class of lead alloys has been developed which uses calcium or other alkaline earth elements for stiffening. These alloys were developed originally for telephone service applications.[2,16] Antimony from the grids is dissolved during battery operation and migrates to the negative plates where it redeposits, which results in increased hydrogen evolution and water loss. For telephone applications, more stable battery operation and less frequent watering were desired. The composition of the alloy depends somewhat on the grid manufacturing process. Calcium is used in the range of 0.03 to 0.20% but for corrosion resistance the preferred range is 0.03 to 0.05%. A variation has been to substitute strontium for calcium. Barium has been investigated but is generally felt to be detrimental to performance. Tin has been used to enhance the mechanical and corrosion-resistant properties of the Pb-Ca alloys and is usually used in the range of 0.25 to 2.0% by weight. The trend in nonantimonial alloy development is toward ternary alloys (Pb-Ca-Sn) containing a minimal amount of tin because of the expense of this element. Some batteries are produced with a quaternary alloy—the fourth element being aluminum—to stabilize the drossing loss of the alkaline earth element (calcium or strontium) from the molten alloy. Grain refining is done by the alkaline earth metal, and no other elements (impurities) are desired. The properties of the alloys are summarized in Table 23.8.[13]

TABLE 23.8 Properties of Lead Alloys

Alloys of the 1970s

Property	Conventional antimony	Low antimony	Cast lead-calcium-tin		Lead-strontium-tin-aluminum	Lead-cadmium-antimony	Wrought-lead-calcium-tin
			0.1Ca 0.3Sn	0.1Ca 0.7Sn			0.065Ca 0.7Sn
Ultimate tensile strength, Pa $\times 10^{-6}$	38–46	33–40	40–43	47–50	53	33–40	60
Percent elongation	20–25	10–15	25–35	20–30	15	25	10–15

Property	Cast conventional antimony	Cast low-antimony	Cast lead-calcium	Cast lead-strontium	Cast lead-cadmium antimony	Wrought lead-calcium-tin (1st generation)
Ease of grid manufacture	Good	Fair	Fair	Fair to good	Fair	Good
Mechanical	Good	Fair	Fair to good	Fair to good	Fair	Good
Corrosion	Fair	Fair	Good	Good	Fair	Good
Battery performance	Poor	Fair	Good	Good	Good	Fair to good
Economics	Good	Good	Fair	Poor	Fair	Fair to good

Alloys of the 1980s and 1990s

	Cast alloys				Wrought alloys				Cast and wrought
Property	Lead-calcium-tin 0.1Ca 0.3Sn	Lead-calcium 0.1Ca	Lead-calcium-tin with aluminum	Lead-calcium with aluminum	Lead-calcium-tin 0.065Ca 0.3Sn	Lead-calcium-tin 0.065Ca 0.5Sn	Lead-calcium 0.075Ca	Low antimony 2.5–3.0% Sb	Lead 0.01–1.5Sn
Ultimate tensile strength, Pa $\times$ 10^{-6}	40–43	37–39	40–43	37–39	43–47	47	43	37–40	
Percent elongation	25–35	30–45	25–35	30–45	15	15	25	25–40	

	Cast alloys		Wrought alloys	
Property	Low antimony	Lead-calcium	Wrought lead-calcium-tin (2d generation)	Wrought low antimony
Ease of grid manufacture	Fair to good	Good (aluminum)	Good	Good
Mechanical	Fair	Fair to good	Good	Good
Corrosion	Fair	Good	Good	Fair to good
Battery performance	Fair to good	Good	Fair to good	Fair
Economics	Good	Good (lower tin)	Good	Good

NOTE: Alloy constituents given in weight percent.

23.3.2 Grid Production

Two general classes of grid production methods virtually describe all modern production, but two other classes of production techniques might become widespread in the future. These are listed in Table 23.9.

TABLE 23.9   Grid Production Methods

Book mold cast
  Gravity cast
  Injection molded (die cast)
Mechanically worked (Planté, Manchester)
Continuous cast, drum cast
Continuous cast, wrought-expanded, cast-expended
  Casting
  Working
  Expansion
    Progressive die expansion
    Precision expanded
    Rotary expanded
    Rotary expansion
    Diagonal/slit expansion
  Punching
Composite

The purposes of the grid are to hold the active material mechanically and conduct electricity between the active material and the cell terminals. The mechanical support can be provided by nonmetallic materials (polymer, ceramic, rubber, etc.) inside the plate, but these are not electrically conductive. Additional mechanical support is sometimes gained by the construction method or by various wrappings on the outside of the plate. Metals other than lead alloys have been investigated to provide electrical conductivity, and some (copper, aluminum, silver) are more conductive than lead. These alternate conductors are not corrosion-resistant in the sulfuric acid electrolyte and are often more expensive than lead alloys. Titanium has been evaluated as a grid material; it is not corroded after special surface treatments but is very expensive. Copper grids are used in the negatives of some submarine batteries.

The grid design is generally a rectangular framework with a tab or lug for connection to the post strap. For cast grids, the framework features a heavy external frame and a lighter internal structure of horizontal and vertical bars. In some grid designs the frame tapers with the greater width near the lug; the internal bars may also be tapered. A recent advance in grid design is the "radial" grid, with the vertical wires displaced along the frame, pointing directly toward the tab area in order to increase grid conductivity (Fig. 23.11). The radial design has been further refined to a composite of lead alloy radial conductor arrangement cast into a rectangular plastic frame. An example of this composite grid is shown in Fig. 23.12a. The grids used in the cylindrical-cell design (Fig. 23.12b) incorporate both concentric and radial members. This system has been in commercial production since 1970, with most of the original cells still in use. An example of a balanced positive grid design is shown in Fig. 23.12c.

(a)   (b)

FIGURE 23.11 Examples of lead-acid battery cast grids. (*a*) Conventional cast flat grid. (*b*) Radial-design grid.

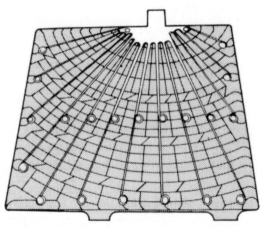

FIGURE 23.12*a* Composite grid, radial conductor. Grid combines diagonal conducting members with light robust plastic frame.

"Book-mold" casting historically accounted for most grid production. Permanent molds are made from steel (Meehanite) blocks by machining grooves to form the grid frames and internal lattice structure. The molds are filled when closed with an amount of lead sufficient to form the grid and leave an excess gate or sprue which is subsequently trimmed off by a cutting or stamping operation. The grid alloy is put into the mold from a ladle in a recirculation lead alloy stream, from a metering valve in a nonrecirculation lead stream, or from a hand-filled ladle. A variation on book-mold casting is injection molding or die casting of battery grids. Here the lead alloy is forced into a clamped mold by high injection pressure. Depending on the alloy characteristics, injection molding can be capable of very high production rates.

Another method of grid manufacture is via mechanical treatment of a strip or slab of lead alloy. The traditional procedures (Planté-type plate) have been either to cut grooves into a thick lead plate, thereby increasing its surface area, or to crimp and roll up lead strips into rosettes which are inserted into round holes in a cast plate. The resultant plates are formed electrolytically into positives in the classic Planté and Manchester designs (Fig. 23.13).

FIGURE 23.12*b* Balanced positive design[27] takes into account grid corrosion and growth and promotes the maintenance of contact of the grid with the active material, while maintaining the shape of the plate and its angle with the horizontal. This concept has also been carried into the prismatic grid structure. (*Courtesy of AT&T.*)

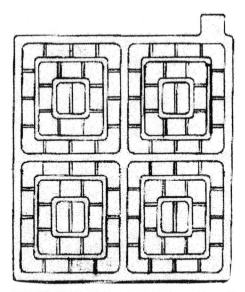

FIGURE 23.12*c* Balanced rectangular positive grid design. This design promotes active material contact and accounts for grid corrosion and growth in a prismatic cell. (*From Ref. 36.*)

(a)

(b)

FIGURE 23.13 Planté and Manchester plates. (*a*) Planté. (*b*) Manchester.

The third major grid production method is circumferential continuous casting onto a mold cut into the surface of a drum. Successful high-speed production of up to 150 grids per minute has been reported. Continuous-cast grids are not symmetrical about a central planar axis and need to be overpasted to hold the active material in place.

A fourth major grid production method, expansion from wrought or cast lead alloy strip, is rapidly supplanting book-mold casting as the preferred method for the manufacture of SLI battery grids. The advantages of this method are lower grid weight per unit of battery electrical performance, the capability to manufacture a wide variety of sizes with a minimum investment in tooling, a very high-rate production capability (up to 600 plates per minute), and very uniform grid and plate sizes. Most development and commercialization have been done on nonantimonial lead-calcium tin alloys (Fig. 23.14). Strip is produced from cast slabs by a variety of proprietary metal-working processes, and the thin strip is slit to the width specified by the battery manufacturer. The worked metal increases in strength as it decreases in thickness during this processing.

FIGURE 23.14   Expanded wrought grid for lead-acid batteries.

Machinery to produce grids from wrought strip has been developed and put into production by several manufacturers. Four types of machinery are involved: progressive die expansion, precision expansion, rotary expansion, and diagonal slit expansion. Progressive die expansion has been the most extensively utilized of the four methods, but rotary expansion is of increasing importance. Continuous drum casting of automotive grids is also challenging the expansion processes as the dominant manufacturing method for calcium alloy grids used in negative plates in automotive batteries. Positive-plate grids are most often produced of low-antimony lead cast in book molds.

Whatever grid production method is used, there is often the need for small cast parts for plate and cell interconnections and connection to external equipment. These parts have traditionally been cast in fixed molds, sometimes with mold inserts to allow a variety of similar parts to be made in each mold. Newer battery production methods often produce these various interconnections automatically in the course of battery assembly.

23.3.3 Lead Oxide Production

Lead is used to make the active materials as well as the grids. The lead must be highly refined (usually virgin or primary lead) to preclude contamination of the battery. It is described as corroding-grade lead in ASTM specification B29.[11] Lead is oxidized by either of two processes—the Barton pot or the ball mill.[17] In the Barton pot process, a fine stream of molten lead is swept around inside a heated pot-shaped vessel, and oxygen from the air reacts with fine droplets or particles to produce an oxide coating around each droplet. Typical Barton pot oxides contain 15 to 30% free lead, which usually exists as the core of each fine leady oxide spherically shaped particle. Barton pots are available in a variety of sizes up to 1000 kg/h output.

Ball milling describes a larger variety of processes. Lead pieces are put into a rotary mechanical mill, and the attrition of the pieces causes fine metallic flakes to form. These are oxidized by an airflow, and the airflow also serves to remove the leady oxide particles to collection in a baghouse. The feedstock for ball mills can range from small cast slugs weighing less than 30 g to full pigs of lead weighing approximately 30 kg. Typical ball mill oxides also contain 15 to 30% free lead in the shape of a flattened platelet core surrounded by an oxide coating.

Some battery positive plates use an additive of red lead (Pb_3O_4), which is more conductive than PbO, to facilitate the electrochemical formation of PbO_2, Red lead is produced from leady oxide by roasting this material in an airflow until the desired conversion is complete. Such processing reduces the free lead content and generally increases the oxide particle size. A variety of other oxides and lead-containing materials have been used to produce battery plates but are of only historical interest.[17] Positive plates for the Lucent Technologies batteries (formerly Bell Laboratories) were initially made with tetrabasic lead sulfate ($4PbO \cdot PbSO_4$), which is a precursor for α-PbO_2. These plates now contain up to 25% red lead (Pb_3O_4) in order to facilitate the electrochemical formation process.

23.3.4 Paste Production

Lead oxide is converted to a plastic doughlike material so that it can be affixed to the grids. Leady oxide is mixed with water and sulfuric acid in a mechanical mixer. Three types of mixers are commonly used—the change can or pony mixer, the muller, and a vertical muller.

The pony mixer is the traditional unit. A preweighed amount of leady oxide is placed into the mixing tub, and this is wetted first with water and then with sulfuric acid solution. Dry paste additives, if any, are premixed into the leady oxide before water addition. These additives can be plastic fibers to enhance the mechanical strength of the dried paste, expanders to maintain negative-plate porosity in operation, and various other proprietary additives which ease processing or are believed to improve battery performance. Muller mixers are usually filled first with the water component, then the oxide, then the acid.

As mixing proceeds, the paste viscosity increases, then decreases, as measured by the amount of power consumed by the mixer motor. The paste becomes hot from the mechanical mixing and from the reaction of H_2SO_4 with the leady oxide. Paste temperature is controlled by cooling jackets on the mixer or by evaporation of water from the paste. The amounts of water and acid for a given amount of oxide will be different for the two mixer types and will also depend on the intended use of the plates: SLI plates are generally made at a low $PbO:H_2SO_4$ ratio and deep-cycling plates at a high $PbO:H_2SO_4$ ratio. Sulfuric acid acts as a bulking agent—the more acid used, the lower the plate density will be. The total amount of liquids and the type of mixer used will affect final paste consistency (viscosity). Paste mixing is controlled by the measurement of paste density using a cup with a hemispherical cavity and by the measurement of paste consistency with a penetrometer.

23.3.5 Pasting

Pasting is the process by which the paste is actually integrated with the grid to produce a battery plate. This process is a form of extrusion, and the paste is pressed by hand trowel or by machine into the grid interstices. Two types of pasting machines are used: a fixed-orifice paster that pushes paste into both sides of the plate simultaneously and a belt paster in which paste is pressed into the open side of a grid that is being conveyed past a paste hopper on a porous belt. The amount of paste applied to a plate by a belt paster is regulated by the spacing of the hopper above the grid on the belt and the type of troweling (roller or rubber squeegee) used at the hopper exit. Using identical paste and grids, a trowel roller machine packs the paste both thicker and more densely than a rubber squeegee machine. As plates are pasted on either belt pasting machine, water is forced out of the paste, into the belt, and ultimately to a sump on or near the machine. The sump material can be used in place of some of the liquids for subsequent batches of negative paste.

Grids are automatically or manually placed onto the belt before being moved under the paste hopper. Most smaller-sized plates are made as "doubles" joined at the feet (cast) or at the tab edge (wrought expanded), or as panels of varying number of plates. Typically the belt is 35 to 50 cm wide and can handle such doubles. Industrial stationary or traction plates (being larger) are pasted by lengthwise feed into the machine or are hand-pasted.

After pasting, plates are racked or stacked for curing. Stacked plates contain enough moisture to stick together, and so before stacking, the plate surfaces are dried somewhat by a rapid passage through a high-temperature drier or over heated platens. Some carbon dioxide from the combustion process might be absorbed on the surface such that the surface is made "harder." The flash drying process may also help start the curing reactions. Thicker plates are usually placed the long edge upward in racks rather than being stacked horizontally on pallets after flash drying.

Wrought expanded plates and some cast plates are cut into discrete plate portions by a slitter machine in the pasting line. Some manufacturers also have the plate lugs brushed clean of paste and surface oxidation on the same machinery.

In Europe, and less commonly in the United States, many of the heavy-duty battery positive plates are made in porous tubular sheaths. The grid is cast or injection-molded of lead, with long-finned spines attached to a header bar and a connection lug. Individual woven fiberglass plastic sheaths or a multitube gauntlet are placed on the spines. These plates are filled with powder or with a slurried paste until the tubes are full. A plastic cap plugs the open sheath ends and becomes the bottom of the plate (Fig. 23.15).

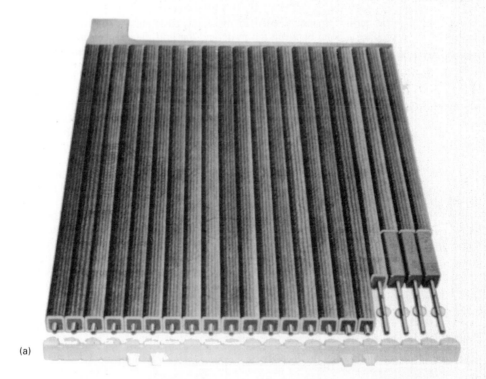

(a)

(b)

FIGURE 23.15 Tubular and gauntlet plates. (*a*) Tubular. (*b*) Gauntlet.

23.3.6 Curing

The curing process is used to make the paste into a cohesive, porous mass and to help produce a bond between the paste and the grid. Several different curing processes are used for lead oxides, depending on paste formulation and the intended use of the battery.[15]

Typical cure for SLI plates is "hydroset," at low temperature and low humidity for 24 to 72 h. The temperature is preferably between 25 and 40°C; the humidity is that contained in the flash-dried plates, typically 8 to 20% H_2O by weight. The plates are usually covered by canvas, plastic, or other materials to help retain both temperature and moisture. Some manufacturers use enclosed rooms for the hydroset, and these rooms may be heated where required by climatic conditions. As the plates cure, they reach a peak temperature, and temperature and humidity decrease. Hydroset typically produces tribasic lead sulfate, which gives high energy density.

More and more manufacturers are using curing ovens, where temperature and humidity can be precisely controlled. This controls the peak temperature and makes sure that sufficient moisture is available to oxidize the remaining free lead in the paste. Peak temperatures in the range of 65 to 90°C are used; at higher temperatures or longer times, significant amounts of tetrabasic lead sulfate are produced in the plate, and the plates generally have lower energy density. At the end of curing, the free lead content of the paste should be below 5%, preferably as low as possible. (Insufficient paste cure is observed as "soft," easily broken pasted plates, usually pale in color.) If the plates have not cured, they can be rewetted and reheated to force the cure. Another process to force completion of curing is to dip the partially cured plates into dilute sulfuric acid. This latter process ("pickling") is also used for cure of powder-filled tubular positive plates.

Cured plates are stored until use. Shelf life is not critical, but the high cost of inventory usually makes storage time minimal.

23.3.7 Assembly

The simplest cell consists of one negative, one positive, and one separator between them. Most practical cells contain about 3 to 30 plates with separators in between. Individual or leaf separators are generally used. The use of "envelope" separators, which surround either the positive or negative plate, or both, is becoming more popular in small, sealed cells, SLI, motive power, and standby batteries to facilitate production and to control lead contamination during manufacture.

Separators are used to electrically insulate each plate from its nearest counterelectrode neighbors but must be porous enough to allow acid transport into or out of the plates. The properties of typical separators are given in Table 23.10.

SLI batteries use either phenolic/cellulosic separators, sintered PVC separators or more recently a trend toward microporous polyethylene in either "leaf" or "envelope" form. The average pore diameters are in the 5 to 30-μm range. Heavy-duty batteries usually have microporous rubber or polyolefin (mainly polyethylene) separators which have smaller pore sizes and give longer life at a higher cost. Glass fiber scrim is added to some separator materials to improve acid retention.

Industrial traction-pasted positive plates are usually wrapped with several layers of fibrous glass matting to help retain the active material during use. The inner layer consists of usually very fine parallel strands of "slyver" glass, and subsequent layers are randomly oriented glass fibers held together by a plastic (styrene or acrylic) binder. The matting is held in place by a heat-sealed, perforated rubber retainer. In lieu of plate wrapping, many heavy-duty batteries use a layer of glass matting on the ribs of the separator to help retain the positive active material.

TABLE 23.10 Comparison of Various Types of Battery Separator Materials

Type	Rubber	Cellulose	PVC	PE	Glass fiber	Microglass
Year available	1930	1945	1950	1970	1980	1985
Backweb, mils	20+	17–30	12–20	7–30	22–26	10–150
Porosity, %	60 ± 5	65 ± 5	40+	60 ± 5	85 ± 5	90+
Maximum pore size, μm	>5	35	25	<1	45	15–30
Average pore size, μm	3	25	15	<0.1	80	10–15
Electrical resistance, (in acid) mΩ·in^2	30–50	25–30	15–30	8–40	10	<5
Purity	Good	Fair	Good	Good	Good	Excellent
Corrosion resistance	Good	Fair	Excellent	Excellent	Excellent	Excellent
Flexibility	Brittle	Brittle	Brittle	Excellent	Good	Fair to good

Source: *Battery Man.* (Mar. 1993).

Plates and separators are stacked manually or by a stacking machine. Stacked elements are staged on roller conveyors or carts as input to the interplate welding operations. Welding is done by two general methods: melting of the lugs in a mold with the lugs facing upward, or immersion of lugs facing downward into pools of molten lead alloy contained in a pre-heated mold. The first method is the traditional assembly method for lead-acid batteries. In this method the plate lugs fit up through slots in a mold "comb"; the shape and the size of the group strap are delineated by the "dam" and "back iron" portions of the tooling. Some battery manufacturers use slotted "crowfoot" posts to fit over the plate lugs to speed the welding process. The second welding method is called the "cast-on strap" process and is typically used for SLI cells. Stacked elements are loaded into slots of the cast-on machine. A mold which has cutouts corresponding to the desired straps and posts is preheated and filled with the appropriate molten lead alloy, making sure not to join lead or lead-calcium alloys with antimony alloys. The mold and the stacked elements are moved until the plate lugs are immersed in the strap cutouts. External cooling solidifies the strap onto and around each lug, and the elements are moved to a point where they can be dropped into a battery case. Visual examination can differentiate between the two welding methods: fixture-welded plate straps are usually thicker and smoother than cast-on straps; cast-on straps also will usually show a convex meniscus of metal between adjacent plate lugs on the underside of the strap if the lug is properly cleaned of paste. A good weld is required between each plate lug and the strap so that high-rate discharge performance is maximized. The resultant assemblage of plates and separators is known as an element, and the welded subelements are known as groups. Electrical testing for short circuits is usually done on elements before further assembly.

Cast-on battery elements are either continuously connected or made in discrete one-cell modules. The first method requires that long intercell connections be used, which travel over the intercell partition and are seated in a slot in this partition; this is known as the loop-over-partition design. In the second cast-on method, tabs on the ends of the plate straps are positioned over holes that have been prepunched into the intercell partitions of a battery case. These tabs are welded together manually with a very small torch or automatically by a resistance welding machine. The latter also squeezes the tabs and the intercell partition to provide a leak-proof seal.

Industrial traction cells and old-style SLI cells have been connected into batteries after the cell cases and covers are sealed together. Traction batteries are needed in thousands of different sizes for various applications, and the standard unit of construction is a cell, not a quantity of plates and separators. A heavy steel tray is fabricated and coated with an acid-resistant coating (urethane, epoxy, etc.). Traction cells are placed into the tray and shimmed as necessary, and intercell connections are welded on. Heavy flexible wires (made from welding cable) are welded to the end cells for connection to the external circuit.

23.3.8 Case-to-Cover Seal

Four different processes have been used to seal battery cases and covers together. Enclosed cells are necessary to minimize safety hazards related to the acidic electrolyte, to the potentially explosive gases produced on overcharge, and to electrical shock. Most SLI batteries and many modern traction cells are sealed with fusion of the case and cover. The fusion comes from preheating each on a platen, then forcing the two together mechanically, or from ultrasonic welding of the case and cover. Fusion-sealed batteries are virtually impossible to repair. At best the elements can be salvaged, but the cover and usually the case are discarded and replaced. A few SLI batteries are sealed using an epoxy cement which fills a groove in the cover; the battery is inserted and positioned so that the case and intercell partition lips fit into the epoxy-filled groove. Heat is used to activate the catalyst to set the epoxy.

Some small deep-cycling batteries feature tar (asphalt)-sealed cases and individual cell covers. Here the tar seal allows easy repair to the battery. Traditionally all batteries were made this way before about 1960, but heat seals are typical for SLI batteries today. Molten tar is dispensed from a heated kettle to fill a groove between the cover and the case. The tar must be hot enough to flow easily but cool and viscous enough to solidify before running down into the cell.

Stationary batteries in plastic cases are sealed with epoxy glues, with solvent cement, or (for PVC copolymer cases and covers) with a thermal seal. Terminals are cast or welded on. Some very large stationary and traction cells are made so that coolant can be circulated through the terminals, and others are made with terminals with copper inserts for increased conductivity and mechanical strength.

23.3.9 Tank Formation

Plates or assembled groups can be electrically formed or charged before assembly into the case. When SLI plates are formed, these are usually formed as "doubles," with two to five panels stacked together in a slotted plastic formation tank, spaced an inch or less from stacks of counterelectrode-pasted panels in adjacent slots. The stacks are arranged so that all positive lugs protrude out of one side of the tank top and all negative lugs protrude out of the other side of the tank top. All lugs with the same polarity are connected by welding to a heavy lead bar, and the two bars are connected to a low-voltage, constant-current power supply. The tank is filled with electrolyte, and current is passed until the plates have been formed: the positives are converted to a deep brownish black and the negatives to a soft gray which shows a bright metallic streak when scratched. Industrial plates are usually formed singly. Sometimes these are also formed against dummy plates or grids. A variety of tank materials have been used, but the most common are PVC, polyethylene, or lead. The tanks are arranged so that the acid can be drained and refilled because formation increases the electrolyte concentration.

A variety of formation conditions are used, with variations in electrolyte density, charging rate (current), and temperature. Electrolyte is typically dilute, in the range of 1.050–1.150 specific gravity. The charging rate is usually fixed, but some manufacturers use a sequence of two or three different charging rates for different periods of time.

Tank-formed groups or plates are somewhat unstable (negatives will spontaneously oxidize in air) and therefore are "dry charged" before use (see Sec. 23.3.11).

Modern electronic chargers for formation are made to operate in a constant-current mode, either by a saturable reactor control or by use of silicon controlled rectifiers (SCRs). The most recent development in formation chargers is to control the current and time by use of a microcomputer. Some formation schedules include charge at three or more different currents which start at low currents, go to higher currents, and then revert to lower currents. Current adjustment during formation minimizes damage to the cells by high temperatures and the need for cell cooling by water spray or forced air.

23.3.10 Case Formation

The more usual method of formation is to completely assemble the battery, fill it with electrolyte, and then apply the formation charge. This method is used for SLI and most stationary and traction batteries. A variety of formation conditions are used, similar to those for tank formation. The two major formation processes are the two-shot formation process (used for stationary and traction batteries) and the one-shot formation process (used for most SLI batteries). In the two-shot formation, the electrolyte is dumped to remove the low-density initial electrolyte and refilled with more concentrated electrolyte, chosen so that when this is mixed with the dilute initial acid residue which is absorbed in the elements or trapped in the case, the cell electrolyte will equilibrate at the desired density (Table 23.11). Typical values of the electrolyte specific gravity at full charge after formation are given in Table 23.12.

TABLE 23.11 Formation Processes

	One-shot	Two-shot
Typical application	SLI	All others, some SLI
Electrolyte concentrations, sp gr.:		
Initial	1.200	1.005–1.150
Final	1.280	1.150–1.230
Subsequent processing	None	Dump and refill with 1.280–1.330 sp gr electrolyte; continue charge for several hours

TABLE 23.12 Specific Gravity of Electrolytes at Full Charge at 25°C

	Specific gravity	
Type of battery	Temperate climates	Tropical climates
SLI	1.260–1.290	1.210–1.230
Heavy duty	1.260–1.290	1.210–1.240
Golf cart	1.260–1.290	1.240–1.260
Golf cart (electric vehicle)	1.275–1.325	1.240–1.275
Traction	1.275–1.325	1.240–1.275
Stationary	1.210–1.225	1.200–1.220
Diesel starting (raiload)	1.250	
Aircraft	1.260–1.285	1.260–1.285

23.3.11 Dry Charge

The performance of wet batteries degrades with long periods of inactivity, especially when stored at warm temperatures. A loss of 1 to 3% of capacity each day is possible with SLI batteries that contain antimonial lead grids. The loss on stand can be much lower for maintenance-free batteries (0.1 to 3%/day). When lead-acid batteries must be stored for a long time, especially in high ambient temperatures, or when batteries are shipped for export, their performance can be stabilized by removal of the electrolyte by one of several methods.

When the electrolyte is removed, the battery is termed "dry-charged" (that is, charged and dry) or "charged and moist." The first process is done before the battery elements are assembled inside the case and cover. The plates can be tank-formed, water-washed, then dried in an inert gas before the element-welding portion of assembly. Alternately the welded element can be tank-formed, washed, and then dried in an inert gas. The latter process is simpler to carry out, but it is necessary that the separators can be rewetted easily after being washed and dried. The assembly (case, elements, cover) is completed and the battery is sealed. The battery can be stored in this dry-charged state for up to several years before reactivation and use.

Several processing innovations have been commercialized in the past 10 years to convert wet-charged batteries into moist or semidry batteries. In one process, most of the electrolyte is removed by centrifugation. Another process uses an inorganic salt (sodium sulfate) in the electrolyte, which minimizes degradation during storage and assists in an eventual reactivation. A battery is formed, dumped, refilled with electrolyte which contains the additive, high-rate-discharge tested, and then finally dumped. The high-rate electrical discharge (to simulate engine cranking) allows testing of an assembled "damp-dry" battery, but this also probably blocks the plate surfaces with a thin layer of small lead sulfate crystals. These crystals then minimize plate or separator degradation during storage when the battery is sealed.

23.3.12 Testing and Finishing

Electrical tests are used to check the performance of batteries before they are sold and often before they are put into use. The type of test employed depends on the intended use for the battery. SLI batteries are tested by brief discharges at very high currents (200 to 1500 A) to simulate engine-cranking performance. Stationary and traction cells are discharged at a rate specified by the user, usually in the range of 1 to 10 h if stationary and 4 to 8 h for traction. The discharge for SLI batteries is usually done by dissipation through a fixed low-value resistor or by a brief, high-rate electrical discharge driven by a power supply. Heavy-duty batteries are discharged through a resistor, a transistorized load, or an inverter.

The final manufacturing steps consist of improving the battery appearance by washing, drying, painting, installing vent plugs, and labeling as desired. Rubber-case batteries are usually painted; plastic-case batteries are not. A large variety of plaques and labels are available which can describe the battery, its performance, and use. Product liability requirements in many countries mandate that the user be warned of the hazardous nature of the battery, especially that the electrolyte is corrosive and that gases are formed which can be explosive.

Traction batteries are physically sized to fit a myriad of different forklift trucks, and so the final assembly for a traction battery consists of inserting preformed and pretested cells into a sturdy metal box (tray), making intercell connections, making cable connections, and sometimes adding a tar or plastic (urethane) material in the spaces between cell covers.

23.3.13 Shipping

Small batteries (SLI and golf-cart types) are palletized several layers high for long-distance shipment. The batteries are cushioned by five-sided (slipover) or six-sided cardboard boxes with cardboard or wood sheet between layers. The batteries are held laterally by banding or by plastic sheet which is shrink- or stretch-wrapped around a full pallet. Pallets need to be very sturdy to withstand the battery weight and handling abuse. Large batteries are palletized, banded, and cushioned as appropriate.

Batteries have traditionally been shipped only minimal distances because of their fragile nature, their weight, and their corrosive contents. The latter cause common carriers to charge a significant premium for battery shipment and usually preclude shipment by air. As the number of small, localized-sales battery manufacturers continues to decrease in the United States, the remaining large manufacturers now usually ship to their distribution chain on their own trucks.

23.3.14 Activation of Dry-Charged Batteries

When batteries have been dry-charged, they have to be reactivated before use. Activation consists of unpacking the battery, filling the cells with electrolyte (which sometimes is shipped with the battery in a separate package), charging the battery (if time is available), and testing the battery performance. When dry-charged batteries are activated, the materials which had been used to seal the vent holes must be removed and discarded.

23.4 SLI (AUTOMOTIVE) BATTERIES: CONSTRUCTION AND PERFORMANCE

23.4.1 General Characteristics

The design of lead-acid batteries is varied in order to maximize the desired type of performance. Tradeoffs exist for optimization among such parameters as power density, energy density, cycle life, "float-service" life, and cost.

High power density requires that the internal resistance of the battery be minimal. This affects grid design, the porosity, thickness, and type of separator, and the method of intercell connection. High power and energy densities also require that plates and separators be thin and very porous and, usually, that paste density be very low. High cycle life requires premium separators, high paste density, the presence of α-PbO_2 or another bonding agent, modest depth of discharge, good maintenance, and, usually, the use of high-antimony (5 to 7% Sb) grid alloy. Low cost requires both minimum fixed and variable costs, high-speed automated processing, and no premium materials for the grid, paste, separator, and other cell and battery components.[1,14,18–20]

The automotive industry is planning to introduce a 36/42 volt battery system in the first decade of the new millennium (see Sec. 23.9.1).

23.4.2 Construction

An SLI battery whose main function is to start an internal combustion engine, discharges briefly but at a high current. Once the engine is running, a generator or alternator system recharges the battery and then maintains it on "float" at full charge or slight overcharge. In recent automobile designs the parasitic electrical load of lights, motors, and electronics causes a gradual discharge of the battery when the engine is not in operation. This factor, coupled with normal self-discharge, introduces a significant cycling component into the normal cranking/floating duty cycles. Studies of SLI battery life and failure modes are presented in Secs. 23.4.3 and 23.8.4.

The cranking ability of an SLI battery is directly proportional to the geometric area of plate surface, with the proportionality factor typically 0.155 to 0.186 cold-crank Amperes (CCA) at $-17.8°C$ (0°F) per square centimeter of positive-plate surface. Cranking performance is generally limited by the positive plate at higher temperatures ($>18°C$) and by the negative plate at lower temperatures ($<5°C$). The ratio of positive surface to negative surface is fixed by design. To maximize the cranking capacity, SLI battery designs emphasize grids with minimum electrical resistance (using a variety of radial and expanded grid designs), thin plates, and a higher concentration of electrolyte than motive-power or stationary batteries.

Usually an "outside-negative" ($n + 1$ negative plates interspersed with $2n$ separators and n positive plates) design is used. However, in order to balance the cranking rating with the requirement or electrical load, as well as to facilitate automatic assembly, SLI batteries with an even number of plates, or "outside-positive" designs, are widely produced in the United States.

A major advance is the maintenance-free SLI battery, which has several characteristics that distinguish it from the conventional battery. It requires no addition of water during its life, it has significantly improved capacity retention during storage, and it has minimal terminal corrosion. The construction of a typical maintenance-free SLI battery was illustrated in Fig. 23.8. This type of battery relies mainly on charge control to prevent electrolysis of water and dry-out as compared to the small sealed consumer designs which rely on oxygen recombination (see Chap. 24).

The SLI maintenance-free battery has a large acid reservoir, made possibly by the use of smaller plates and placement of the element directly on the bottom of the container, eliminating the sludge space. The positive plates are usually enveloped in a microporous separator that prevents active material from falling to the bottom of the container and creating a short circuit. An important feature of the maintenance-free battery is the use of nonantimonial (such as calcium-lead) or low-antimonial lead grids. The use of these grids reduces the overcharge current significantly, reducing the rate at which water is lost during overcharge, as well as improving the stand characteristics (see Sec. 23.7). The use of the expanded grid, produced from wrought lead-calcium strip, is also shown in the figure. Most SLI batteries, so called hybrid, are built using lead-calcium-tin grids for the negative and low-antimony lead grids for the positive electrode.

Another refinement of the SLI battery is illustrated in Fig. 23.16. In this design the plates are approximately one-fifth the width of conventional SLI battery plates and are inserted parallel, rather than perpendicular, to the length of the battery case. This design reduces the internal impedance of the cell and gives very high CCA ratings.

Heavy-duty SLI batteries for trucks, buses, and construction equipment are designed similar to the passenger vehicle SLI batteries but use heavier and thicker plates with high-density paste, premium separators often with glass mats, anchor-bonding of the element to the bottom of the case, rubber cases, and other such features to enhance longer life. This is necessary to provide maximum mechanical strength for the physically large (up to 530 by 285 mm) case dimensions. Because the thick plates provide less cranking current than the thinner plates (since fewer can be included in a cell of a given size), series or series-parallel connections of batteries are used. Typically, the 12-V monoblocks are connected in series for cranking at 24 V and in parallel for running and recharging at 12 V. A few sizes of maintenance-free heavy-duty batteries have also been produced.

SLI-type batteries are also used on motorcycles and boats. Batteries for recreational marine use generally have thicker plates (to give more capacity) and higher-density paste. They have the same Battery Council International type designations as automotive batteries. See Sec. 4.10 for a listing of BCI battery types. Marine batteries are also manufactured in four-cell 8-V monoblocs.

Aircraft use SLI-type batteries with special spill-proof vent caps which preclude loss of electrolyte during flight.

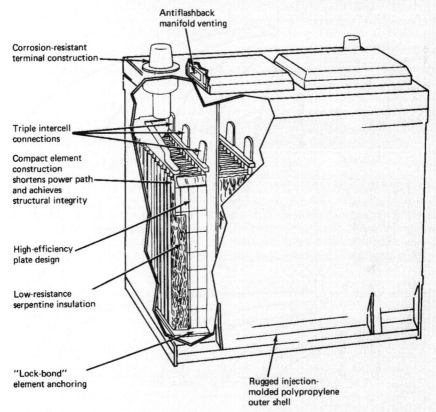

Antiflashback
manifold venting

Corrosion-resistant
terminal construction

Triple intercell
connections

Compact element
construction
shortens power path
and achieves
structural integrity

High-efficiency
plate design

Low-resistance
serpentine insulation

"Lock-bond"
element anchoring

Rugged injection-
molded polypropylene
outer shell

FIGURE 23.16 Cathanode lead-acid SLI battery.

23.4.3 Performance Characteristics

Discharge Performance.   Discharge curves, showing the discharge profile of SLI-type batteries at several constant-current discharge rates, are presented in Fig. 23.17. The typical
final or end voltages at these discharge rates are also shown. Higher service capacity is
obtained at the lower discharge rates. At the higher discharge rates, the electrolyte in the
pore structure of the plates becomes depleted and the electrolyte cannot diffuse rapidly
enough to maintain the cell voltage. Intermittent discharge, which allows time for the electrolyte to recirculate, or forced circulation of the electrolyte will improve high-rate performance. In general the lead-acid cell may be discharged without harm at any rate of current it
will deliver, but the discharge should not be continued beyond the point where the cell
approaches exhaustion or where the voltage falls below a useful value.

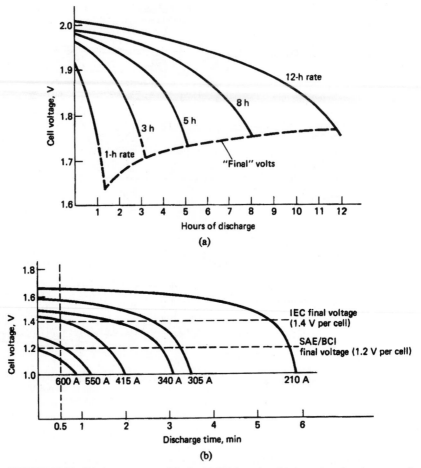

FIGURE 23.17   Discharge curves of lead-acid SLI batteries. (*a*) At various hourly rates and 25°C. (*b*) At various high rates and −17.8°C. Battery rated at 70 Ah, 20-h rate at 25°C.

Effect of Temperature on Performance.   Figure 23.18(a) shows typical discharge curves for a lead-acid single-cell battery at several discharge temperatures. Fig. 23.18(b) shows the discharge characteristics of a 12 V, 60 Ah battery when discharged at 340 A at temperatures from −30 to 25°C. Higher discharge voltages and capacities are obtained at the higher temperatures and lower discharge rates.

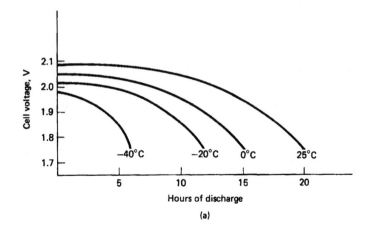

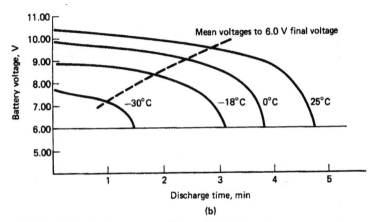

FIGURE 23.18 Discharge curves of lead-acid SLI batteries at various temperatures. (*a*) At *C*/20 rate. (*b*) At 340 A. 12-V battery, nominal capacity 60 Ah, 20-h rate at 25°C.

The effect of discharge rate and temperature on the capacity of the lead-acid battery is summarized in Fig. 23.19, which shows the percentage of the 20-h rate capacity delivered under different discharge conditions. Although the battery will operate over a wide temperature range, continuous operation at high temperatures may reduce life as a result of an increase in the rate of corrosion (see Sec. 23.8.1).

The performance of the lead-acid SLI-type cell at different temperatures and loads is given in another form in Fig. 23.20. The logarithm of the current drain is plotted against the logarithm of the service hours, in accordance with Peukert's relationship (Chap. 3, Sec. 3.2.6). The linear relationship is maintained over a wide range, with divergences appearing on the high-rate and low-temperature discharges. In this figure, the data have been normalized to unit cell weight (kilograms) and unit cell volume (liters). Figure 23.20 can be used to approximate the performance of various size cells over the operating conditions shown or to determine the size and weight of a battery to meet a particular service requirement.

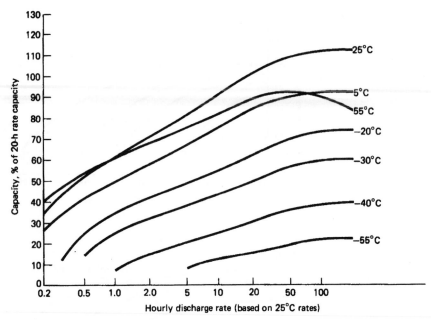

FIGURE 23.19 Performance of lead-acid SLI batteries at various temperatures and discharge rates to 1.75-V per cell end voltage.

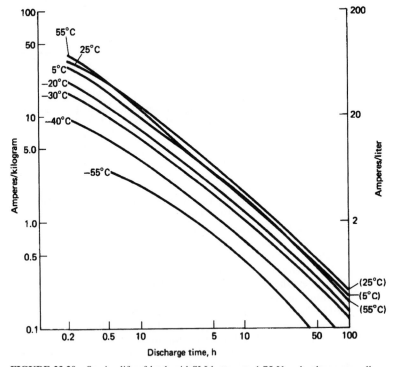

FIGURE 23.20 Service life of lead-acid SLI battery to 1.75-V end voltage per cell.

Internal Impedance.   The high current requirement for engine cranking demands that the batteries be designed with low resistance; for example, that conductors have large cross sections and minimal lengths, that separators have maximum porosity and minimum backweb thickness, and that the electrolyte be in the range of low resistance. The relationship between plate surface area and CCA suggests the involvement of the electrochemical double layer of the porous active materials. Low frequencies are generally necessary to evaluate the capacitive reactance component of battery impedance—strictly the resistance impedance components can be evaluated by Ohm's law by determining the voltage difference at two levels of discharge current. The resistance of a lead-acid battery increases during a discharge almost linearly with the decrease of the specific gravity of the electrolyte. The difference in resistance between full charge and discharge is in the order of 40%. The effect of temperature on the resistance of the battery is shown in Fig. 23.21; the battery resistance increases by about 50% between 30 and $-18°C$.

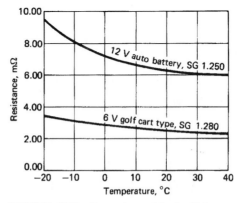

FIGURE 23.21   Comparison of lead-acid battery designs. Effect of temperature on battery resistance.

Self-Discharge.   A lead-acid battery loses capacity during open-circuit stand (self-discharge). This loss is more severe with batteries which use antimonial lead grid alloys in the positive plates. A comparison of the open-circuit stand loss of conventional antimonial lead (>4% Sb), low antimonial lead (<3% Sb), and nonantimonial lead grids is shown in Fig. 23.6. This loss is most easily detected by a drop in the terminal voltage of the battery and/or the specific gravity. The sulfuric acid reacts, primarily on the surface of the negative plate, in small local self-discharge "cells" where antimony and lead are in contact, becoming a small particle of lead sulfate. In batteries using calcium-lead nonantimonial lead negative and antimonial lead positive grids, self-discharge loss is minimized until antimony diffuses to the negative. This is especially true for the low-antimonial alloy positives.

Life and Failure Modes.   The life of SLI batteries is affected by the design, the processing, and the operational environment of the battery. Because of the automated assembly methods used today, SLI batteries are fairly consistent in life under ideal operating conditions, but the wide variety of operating conditions tends to spread the failure distribution. Warranty coverage for a failed battery is often more dependent on marketing strategy than on the statistical expectations of the failure rate.

SLI battery design, materials, and operation have changed markedly in the past two decades; life and failure mode distribution have also changed. In Fig. 23.22*a* the average age of failed batteries is plotted. Possible explanations for the shorter life in 1982 may be a

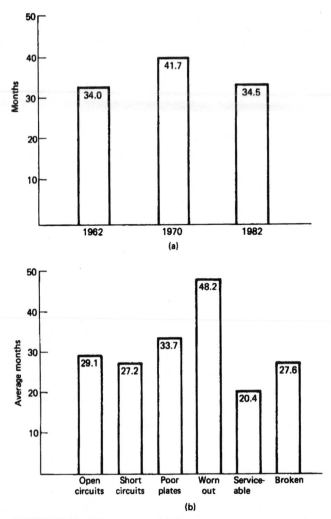

FIGURE 23.22   Failure modes of SLI batteries. (*a*) Average age of returned batteries. (*b*) Failure mode of returned batteries. (*From Amistadl.*[4])

reduction in battery size and more demanding performance requirements. These averages include taxis, police cars, and other heavy-duty users, which account for the relatively low age for failed batteries. Figure 23.22*b* shows the failure modes for these batteries, which are described in more detail in Table 23.13. A higher incidence of short-circuited batteries used in warmer climates suggests that grid corrosion is still a major failure mode. The "worn-out" category includes low electrolyte level, and it should be noted that many maintenance-free SLI batteries are sealed so that water, lost to evaporation and electrolysis, cannot be replaced.

   SLI batteries are not designed for deep-cycling service, and very short lives are generally obtained with such operation. The deep-cycling capability of SLI batteries is covered in Sec. 23.5.2.

TABLE 23.13   Summary of Failure Modes of Lead-Acid SLI Batteries

1. Open circuits
   a. Terminal
   b. Cell to terminal
   c. Cell to cell
   d. Broken straps
   e. Plates off
	4. Worn out
	   a. Worn out
	   b. Undercharge
	   c. Low level (electrolyte)
	   d. Terminal corrosion
	   e. Underformed

2. Short circuits
   a. Plate to strap
   b. Plate to plate (plate fault)
   c. Plate to plate (separator fault)
   d. Plate to plate (sediment/moss)
   e. Vibration short circuit
	5. Serviceable
	   a. Serviceable
	   b. Discharged only
	6. Broken
	   a. Broken container
	   b. Broken cover

3. Poor plates
   a. Overcharge/overheat
   b. Grid corrosion
   c. Paste adhesion
   d. Paste sulfation
   e. Paste undeformed
	c. Damaged terminal
	   d. Internal damage
	   e. Other

Standard Tests for Rating SLI Batteries.   Several standard tests have been devised to evaluate and rate the performance of SLI batteries under conditions simulating the major requirements of their applications. The cold-cranking Amperes (CCA) test evaluates the capability of the battery to deliver power to crank an engine at cold temperatures. The cranking-test rating is the current that a fully charged battery can deliver at $-17.8°C$ for 30 s to a voltage of 1.2 V per cell. If the measured voltage is above or below this value at 30 s, the CCA value can be calculated by multiplying the discharge current by the correction factor shown in Fig. 23.23. Figure 23.17*b* illustrates the performance of a 70-Ah cell with a CCA rating of 550 A.

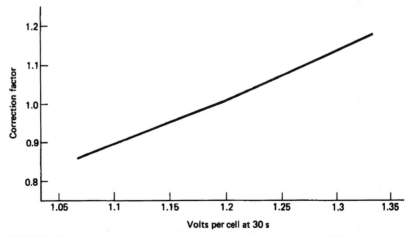

FIGURE 23.23   Correction factor for calculating cold-cranking Ampere (CCA) rating.

Reserve capacity is measured in a test of the battery's ability to provide power for lights, ignition, and the auxiliaries. The reserve capacity is defined as the number of minutes a fully charged battery can maintain a current of 25 A to 1.75 V per cell at 25°C.

Other SLI tests are included in the SAE battery test standard J 537 on charge rate acceptance, overcharge life, and vibration resistance. A standard SLI life test is specified in SAE J240A. This test consists of a shallow discharge at 25 A followed by a brief charge at voltage and current limits for 10 min.

23.4.4 Cell and Battery Types and Sizes

SLI battery sizes have been standardized by both the automotive industry through the Society of Automotive Engineers (SAE), Warrendale, Pa., and the battery industry through the Battery Council International (BCI), Chicago.[21] The BCI nomenclature follows the standards adopted by its predecessor, the American Association of Battery Manufacturers (AABM). The latest standards are published annually by the Battery Council International (BCI).[21,22] Internationally, standardization is handled by the International Electrotechnical Commission (IEC). More detailed information on these standards is found in Chapter 4, Sec. 4.10 and Table 4.10.

23.5 DEEP-CYCLE AND TRACTION BATTERIES: CONSTRUCTION AND PERFORMANCE

23.5.1 Construction

The prime requirement for deep-cycling batteries for traction applications is maximum cycle life, then, if possible, high energy density and low cost. In an electric forklift application, in fact, light weight may not be advantageous because the battery's weight usually is needed to counterbalance the payload. The life of these batteries is improved by the use of thick plates with high paste density, usually a high-temperature and high-humidity cure, low electrolyte density formation, premium separators, and one or more layers of glass fiber matting (to retain the active material in the positive plates). The major modes of failure are disintegration of the PbO_2 positive active mass and corrosion of the positive grids. The deep-cycling battery is usually designed to be capacity-limited when new by the amount of electrolyte and not by the material in the plates. This serves to protect the plates and maximize their life. Both negatives and positives are degraded during use, but at end of life, battery capacity is generally limited by the positive plate. Battery failure, for cycle life rating purposes, is considered to occur when the battery will no longer produce 60–80% of its initial or rated discharge capacity.

A typical traction battery, using flat-pasted plates, is shown in Fig. 23.24. Cells are always made with an outside-negative design (e.g., n positive plates, $n + 1$ negative plates). Deep-cycling traction batteries are built as an assemblage of individual cells. If the battery's performance is limited by a catastrophic failure of one (or a few) cell(s), then those cells can be repaired or replaced in a cost-effective manner. Power requirements vary widely with the load, distance traveled, and lifting or climbing requirements. Battery sizes are determined by the forklift truck manufacturer and can be "calibrated" in the actual application by the use of an Ampere-hour meter. A rough indication of the suitability of a traction battery for an application is the change in the specific gravity of the electrolyte during use. A larger battery size (or battery replacement or repair) is indicated when full operation cannot be achieved.

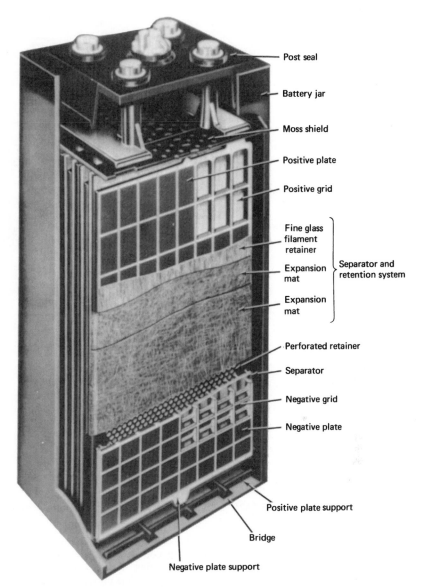

Post seal

Battery jar

Moss shield

Positive plate

Positive grid

Fine glass
filament
retainer

Expansion
mat

Separator and
retention system

Expansion
mat

Perforated retainer

Separator

Negative grid

Negative plate

Positive plate support

Bridge

Negative plate support

FIGURE 23.24   Flat-pasted-plate lead-acid traction battery. (*Courtesy of C&D Technologies.*)

Although the flat-pasted (Faure) positive plate is typical for deep-cycling batteries in the United States, some cycling batteries in the United States and most cycling batteries in the rest of the world are built with tubular or gauntlet-type positives (Fig. 23.25). The tubular construction minimizes both grid corrosion and shedding, and long life is characteristic of these designs, but at a higher initial cost. Flat-pasted negative plates are used in conjunction with these positives and the cells are of the outside-negative design.

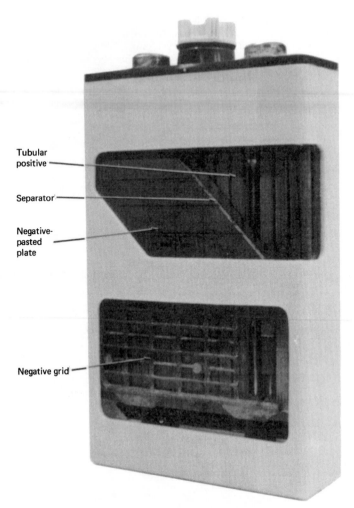

FIGURE 23.25   Lead-acid cell with tubular positive plates. (*Courtesy of Enersys, Inc.*)

Small traction batteries (such as for golf carts) are designed to be intermediate between full-sized traction batteries and SLI batteries. Traction design concepts sometimes utilized include high paste density, careful control of plate curing and formation, to maximize content of the positive plates, glass matting against the positive, and tubular positives. SLI concepts sometimes utilized for golf-cart and other electric-vehicle batteries include thin cast radial grids, minimum separator resistance, through-the-partition intercell connection, and heat or epoxy-sealed plastic cases and covers. Cost is also an important factor.

For on-the-road electric-vehicle applications, the major criterion has been high energy density, which results in maximum range, and SLI battery design has prevailed over traction battery design. In traditional cycling batteries with a few widely spaced plates, good electrolyte homogeneity occurs by convective flow. When plates are made thinner and more closely spaced for high-discharge-rate applications, such as for electric vehicle propulsion, the electrolyte has been found to become stratified during operation. A variety of electrolyte mixing devices has been designed not only to offset stratification but also to increase the discharge efficiency of the battery.

Military submarines of the diesel-electric type require cycling batteries for propulsion. These batteries are made with nonantimonial lead grids because stibine and arsine produced on charge are unacceptable for personnel health in a closed environment. The plates are much larger than most traction cells—up to 600 cm wide and up to 1500 cm tall. Both flat-pasted and tubular positive plates are used.

23.5.2 Performance Characteristics

Batteries for traction and deep-cycle applications use cells with either pasted or tubular positive plates. In general, the performance of the two types of plates is similar, but the tubular or gauntlet plates show lower polarization losses because of the larger active surface area, better retention of the positive active material, and reduced loss on stand. The loss of capacity on stand at room temperature for the two plate structures, as measured by the drop in specific gravity, is shown in Fig. 23.26.

Typical discharge curves for the two types of traction cells are shown in Fig. 23.27. The relationship of discharge current to Ampere-hour capacity, up to various end voltages, is shown in Fig. 23.28. These data are presented on the basis of the positive plate since cell design and performance data of traction batteries are generally based on the number and size of positive plates that are in the cell. As is typical with most batteries, the capacity decreases with increasing discharge load and increasing end voltage.

The same relationship and comparison of the performance of the pasted and tubular plates are plotted in Fig. 23.29. These data show the superiority of the tubular plate as the discharge rate is increased.

Figure 23.30 shows the increase in available service on intermittent discharge, carried out over different periods, as compared with a continuous discharge. The gain is more pronounced at the heavier discharge loads and when the intermittent discharge is spread out over a longer period, thus allowing more time for recovery.

The effect of temperature on the discharge performance of traction-type batteries is illustrated in Fig. 23.31.

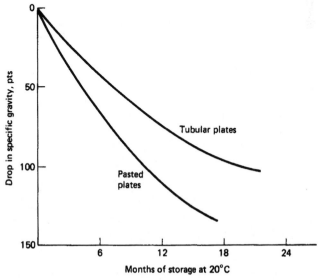

FIGURE 23.26   Retention of charge of pasted- vs. tubular-plate traction batteries.

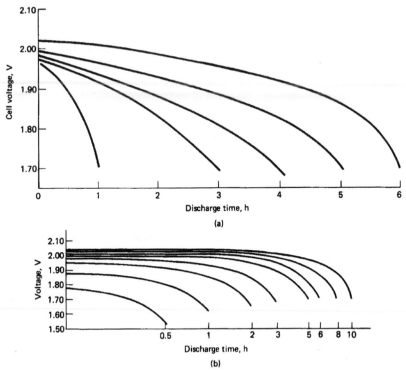

FIGURE 23.27   Discharge characteristics of traction batteries at 25°C. (*a*) Flat-pasted-plate batteries. (*b*) Tubular positive batteries.

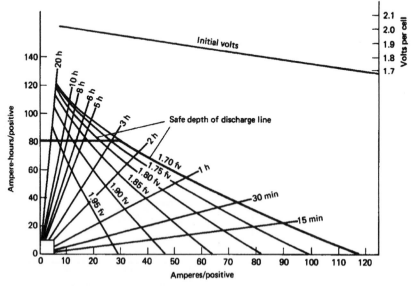

FIGURE 23.28   Performance characteristics of industrial flat-pasted plate traction battery to various final voltages (fv), at 25°C based on positive plate rated at 100 Ah at 6-h rate.

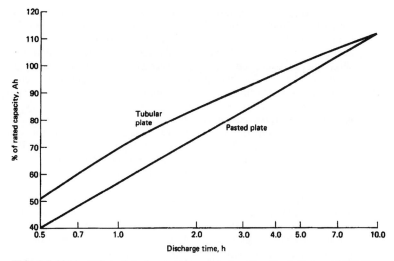

FIGURE 23.29 Effect of discharge rate on capacity of traction batteries at 25°C. Comparison of performance of flat-pasted-plate vs. tubular-plate batteries.

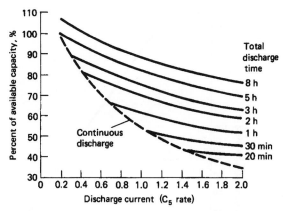

FIGURE 23.30 Capacity available on intermittent discharge of traction batteries at 25°C.

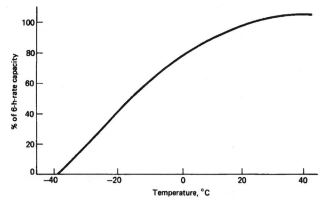

FIGURE 23.31 Effect of temperature on capacity of traction batteries, typical flat-plate design. (*From Ref. 23.*)

The cycle life characteristics of traction batteries are presented in Fig. 23.32. This figure shows the relationship of cycle life to depth of discharge at the 6-h discharge rate, cycle life being defined as the number of cycles of 80% of rated capacity. It is quite evident that the deeper the cells are discharged, the shorter their useful life, and that 80% depth of discharge should not be exceeded if full life expectancy is to be attained. Figure 23.28 shows the safe depth of discharge for other discharge rates. At low rates the discharge should be terminated at the higher end voltages as shown, until the 1.70-V line is intercepted; then the discharges at the higher rates can be run to 1.7-V per cell final voltage. Typical cycle life expectancy is 1500 cycles (approximately 6 years).

The relationship of discharge current and service time for several small deep-cycling batteries is shown in Fig. 23.33. At very high discharge rates, Peukert's relationship does not hold as well as for the SLI types, and the performance deviates at shorter discharge times. Nevertheless, such deep-cycling batteries can be used for cranking service and may be preferred if the battery will be deeply or repeatedly discharged in operation. Conversely, an SLI battery generally makes a poor deep-cycling battery; SLI batteries are usually made with nonantimonial lead grids (U.S. practice), and cycling generally causes the development of a grid-active material barrier layer which shortens cycle life. A comparison of the cycle life at a low discharge rate (25 A) of an SLI battery with a deep-cycle design of the same physical size is shown in Fig. 23.34.

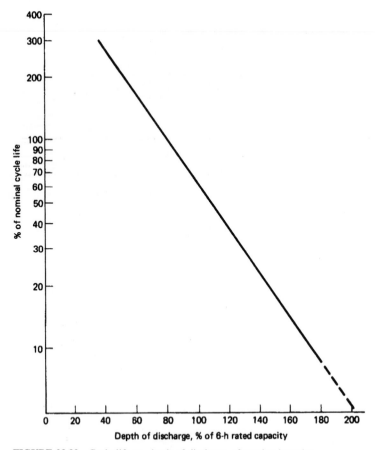

FIGURE 23.32   Cycle life vs. depth of discharge of traction batteries.

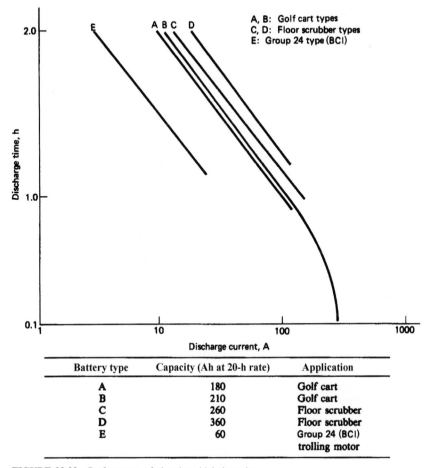

Battery type	Capacity (Ah at 20-h rate)	Application
A	180	Golf cart
B	210	Golf cart
C	260	Floor scrubber
D	360	Floor scrubber
E	60	Group 24 (BCI) trolling motor

FIGURE 23.33   Performance of electric-vehicle batteries.

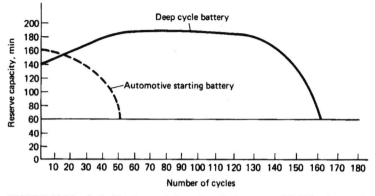

FIGURE 23.34   Cycle life characteristics at low discharge rate (25 A) for deep-cycle vs. SLI-type batteries.

23.5.3 Cell and Battery Types and Sizes

Traction or motive-power batteries are made in many different sizes, limited only by the battery compartment size and the required electrical service. The basic rating unit is the positive-plate capacity, given in Ampere-hours at the 5- or 6-h rate. Table 23.14a lists the typical U.S. traction plate sizes using flat-pasted plates; between 5 and 33 plates are used to assemble traction cells, as also shown in the table. Ratings of the cell are the product of the capacity of a single positive plate multiplied by the number of positive plates. The cells, in turn, are assembled in a variety of battery layouts, with typical voltage in 6-V increments (e.g., 6, 12, 18 to 96 V) resulting in almost 1000 battery sizes. Popular traction battery sizes are the 6-cell, 11-plates-per-cell, 75-Ah positive-plate (375-Ah cell) and the 6-cell, 13 plates-per-cell, 85-Ah positive-plate (510-Ah cell) batteries. Table 23.14b presents similar information on the tubular positive-plate batteries.

Several SLI group sizes have been used for deep-cycling applications, especially taller versions of otherwise SLI lengths and widths. Some of these are listed in Table 23.15.

A variation of the forklift battery design is used for some on-the-road electric vehicles. Table 23.16 lists the characteristics of typical electric-vehicle batteries. Manufacturers' catalogs and data should be consulted for specific information on sizes and performance.

TABLE 23.14a Typical Traction Batteries (United States), Flat-Pasted Plates

Positive-plate capacity, Ah at 6-h rate	Plate dimensions, mm					Cell size, *† (positive plates per cell)
		Width		Thickness		
	Height	Positive	Negative	Positive	Negative	
45	275	143	138	6.5	4.6	5–16
55	311	143	138	6.5	4.6	5–16
60	330	143	138	6.5	4.6	5–16
75	418	143	138	6.5	4.6	2–16
85	438	146	146	7.4	4.6	3–16
90	489	138	143	6.5	4.6	3–16
110	610	143	143	7.4	4.6	4–12
145	599	200	200	6.5	4.7	4–10,12,15
160	610	203	203	7.2	4.7	4–10,12,15

*All cells have n positive plates and $n + 1$ outside negative plates.

†Typical cell characteristics: Six positive, 85-Ah plates (510-Ah cell); weight: 45 kg; size: length, 127 mm; width, 159 mm; height, 616 mm.

Source: C & D Technologies.

TABLE 23.14b Typical Traction Batteries, Tubular Plates

Positive-plate capacity, Ah at 6-h rate	Dimensions, mm					Cell size, †,‡ (positive plates per cell)
	Height	Width		Thickness		
		Positive	Negative	Positive	Negative	
49	249	147	144	9.1	*	4–10
55	258	147	144	9.1	*	4–10
57	300	147	144	9.1	*	4–10
75	344	147	144	9.1	*	4–10
85	418	147	144	9.1	*	4–10
100	445	147	144	9.1	*	4–10
110	565	147	144	9.1	*	4–10
120	560	147	144	9.1	*	4–10
170	560	204	203	9.1	*	3–8

*Varies from 5 to 8 mm depending on manufacturer.
†All cells have n positive and $n + 1$ outside negative plates. Negatives are flat-pasted plates.
‡Typical cell characteristics; six positive, 85-Ah plates (510-Ah cell); weight: 36 kg; size: length, 127 mm; width, 157 mm; height, 549 mm.
Source: Enersys, Inc.

TABLE 23.15 Small Deep-Cycling Batteries

BCI type	Volts	Dimensions, mm			Ratings	Typical operational current, A	Applications
		L	W	H			
U1	12	197	132	186	30–45 Ah at 20 h	25 ⎫	Trolling
24	12	260	173	225	75–90 Ah at 20 h	25 ⎬	motors
27	12	306	173	225	90–105 Ah at 20 h	25 ⎭	Wheelchairs
GC2	6	264	183	260	75 min at 75 A	75 (GC) ⎫	Golf carts
(GC2H)	6	264	183	260	95–90 min at 75 A	300 (EV) ⎪	Electric
Not assigned	6	264	183	260	100–100 min at 75 A	300 (EV) ⎬	vehicles
Not assigned	12	261	181	279	105 Ah at 20 h	150 (EV) ⎭	
Not assigned	6	295	178	276	200–230 Ah at 20 h	50–75 ⎫	Floor
Not assigned	12	241	166	239	50–70 Ah at 20 h	50–75 ⎭	maintenance machinery
Not assigned	12	518	276	445	350–400 Ah at 20 h	30–50	Mine cars

Source: BCI Technical Committee, Battery Council International.

TABLE 23.16 Typical Electric Vehicle (EV) Batteries

BCI group	Volts	Plates per battery	Maximum overall dimensions, mm			Weight, kg	Ah at 2 h	Ah at 3 h	75 A (min)	Wh/kg at 3-h rate
			Length	Width	Height					
U1	12	54	197	132	186	95	20	22	15	26
24	12	78	260	173	225	22	55	59	39	31
GC2	6	57	264	183	270	26	126	135	100	29
27	12	90	306	173	225	24	62	68	45	32
GC2	6	57	264	183	270	30	150	171	120	33
GC2	6	39	264	183	280	27	158	174	140	37

23.6 STATIONARY BATTERIES: CONSTRUCTION AND PERFORMANCE

23.6.1 Construction

Designs for stationary batteries have changed much more slowly than those for SLI and traction batteries. This is not surprising in light of the much longer service life of the stationary batteries. Heavy, thick plates (including Planté as well as pasted Faure and tubular positives) with high paste density are generally used.[24] Curing is very important, and pasted plates are usually carefully dried to prevent cracks and degradation of the grid-paste interface.

The stationary battery is designed with excess electrolyte (highly flooded) to minimize maintenance and the watering interval and is generally positive-plate-limited in capacity (compared with traction batteries, for example, which are electrolyte- or acid-limited). The stationary batteries are capable of being floated and moderately overcharged.

The thick-plate design of stationary batteries reflects the fact that high energy and power densities are not as necessary as is the case for SLI and traction batteries. The overcharge operation of stationary batteries requires a large electrolyte volume (which can be accommodated as the batteries are mounted in fixed positions) and usually nonantimonial lead grids (to maximize the intervals between watering). The overcharge causes some positive-grid corrosion, and this is manifested as "growth" or expansion of the grid. The dimensions from the positive plate to the inside container are scaled so that the positives can grow by up to 10% before the plates touch the container walls. If the growth is greater than 10%, the active material is sufficiently loose on the grid so that the capacity becomes severely positive-limited and the battery must be replaced.

The positives are usually supported from hanging lugs or nonconductive rods, which are borne by the tops of the negatives. The containers are usually transparent thermoplastics (acrylonitrile-butadiene-styrene, styrene-acrylonitrile resin, polycarbonate, PVC), but some small stationary batteries are built in translucent polyolefin containers similar to those used for SLI batteries. The stationary batteries were the first application for flame-retardant vent caps which are now also standard on most SLI batteries.

The positive plate has the greater influence on the performance and life of the battery. A variety of positives are used, depending as much on tradition and custom as on the performance characteristics. The flat-pasted stationary batteries are popular in the United States because of their lower costs, lower maintenance, and lower generation of hydrogen. Lead-calcium tubular plates are now being introduced. For most of the standby emergency applications the grids are cast in lead-calcium alloy. Planté and tubular designs are popular in Europe because of their longer life. All stationary batteries today use pasted negative plates generally with n positive plates and $n + 1$ negative plates (outside-negative design). This is done because of the need for proper support of the positives, which tend to grow or expand during their life. One design used by some manufacturers is to make the two outermost negative plates thinner than the inside negative plates because the outermost surfaces are not easily recharged. Figure 23.35 is an illustration of a stationary battery system installation.

A significantly different approach to stationary battery design is the cylindrically shaped battery of Lucent Technologies (formerly AT&T Bell Laboratories).[25,26] Traditionally, prismatic-shaped stationary batteries had failed after 5 to 20 years of service in telephone systems. The battery, illustrated in Fig. 23.36, incorporates a number of innovations in order to achieve a battery life initially predicted to be 30 years or longer. These include lattice-type circular-shaped pure lead grids (cupped at a 10° angle), plates stacked horizontally one above the other instead of in the conventional vertical construction, chemically produced tetrabasic lead sulfate (TTB) positive paste, positives welded around the external plate circumference, negatives welded to a central conductor core, and heat-sealed copolymer container and cover. The use of pure lead in place of lead-calcium is to reduce positive-plate grid growth; the

FIGURE 23.35 Stationary battery system installation.

circular and slightly concave shape of the plates is to counter the effect of growth and ensure good contact of the active material and grid during the life of the battery. An alloy of Pb/Sn[28,35] prevents positive terminal post corrosion (nodular) and postseal leakage for the life of the cell. Twenty-year-old batteries with this alloy have shown no nodular corrosion and no post-seal leakage. Grid growth is caused by the conversion of lead into PbO_2 on the grid surfaces and at the grain boundaries. The formation of this PbO_2 on the grid adds to the amount of active paste material of the plate and, in the case of concentric grid,[27] increases the plate capacity over time. The Bell Labs researchers[12] found that the growth of the positive-grid members is proportional to their surface area and inversely proportional to their cross-sectional area. Maintaining this ratio of surface area to cross-sectional area for the concentric members in the grid is accomplished by varying the cross section and the surface areas. Thus the shape of the grid remains constant and the only change is caused by the formation of the lead dioxide on the lead grid surfaces.

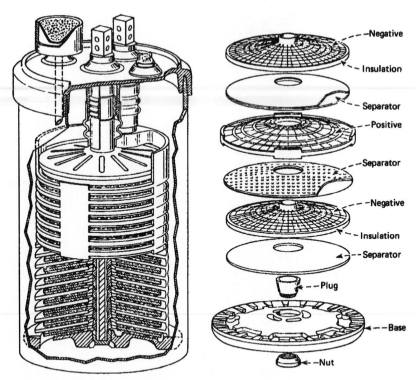

FIGURE 23.36 Cutaway and exploded views of Bell System lead-acid battery. (*From Ref. 25.*)

Accelerated tests projected a capacity increase to accompany the grid growth. In 1988 verification tests were conducted at a site which had cells manufactured in 1973, that is, 15 years earlier. A 24-V string, selected at random out of 11 strings, was discharged at the same 5-h rate used at the time they were manufactured. The capacity behavior on aging is compared to the predicted capacity behavior at 22°C, which was the average yearly temperature of this particular location (Fig. 23.37*a*). The plate growth of the 54 plates measured is summarized in Fig. 23.37*b*, where the findings were compared to that predicted from accelerated testing. The expected capacity and plate growth increases at 22°C and 15 years were 0.25% per year or 3.8% total capacity and 0.027% per year or 0.4% plate growth, respectively. More recent 23-year corrosion data projects a life of 68 to 69 years for round cell batteries.

Some stationary batteries are designed to cycle rather than "float." For these applications the design criteria of the traction batteries are applicable (see Sec. 23.5.1). Applications of cycling stationary batteries include load leveling, utility peak-power shaving, and photovoltaic energy-storage systems.

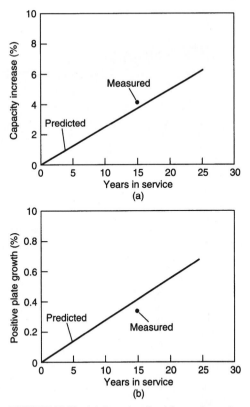

FIGURE 23.37 (*a*) Capacity after 15 years in service vs. prediction from accelerated testing. (*From Biagetti.*[28]) (*b*) Positive-plate growth after 15 years in service vs. prediction from accelerated testing.

23.6.2 Performance Characteristics

Batteries for stationary applications may use cells with flat-pasted, tubular, Planté, or Manchester positive plates. Typical discharge curves for the flat-pasted-type stationary cell at various discharge rates at 25°C are shown in Fig. 23.38, and the effect of the discharge rate on the capacity of the cell is summarized in Fig. 23.39. Generally, the discharge rate for stationary cells is identified as the hourly rate (the current in Amperes that the battery will deliver or the rate hours) rather than the C rate used for other types of batteries.

Figure 23.40(*a*)–(*d*) is a series of curves showing the specific performance characteristics of the four types of stationary batteries at 25°C based on positive-plate design. An electrolyte with a specific gravity of 1.215 is used in all these batteries. The format used in these figures consists of two sections. The lower log-log section shows the capacity (expressed in discharge time) the particular positive plate will deliver at the specified current (expressed in Amperes per positive plate) to various voltages including a final voltage. The upper semilog section shows the cell voltage at various stages of the discharge at various discharge rates (also expressed in Amperes per positive plate). The final voltage is the voltage at which the cell can no longer supply useful energy.

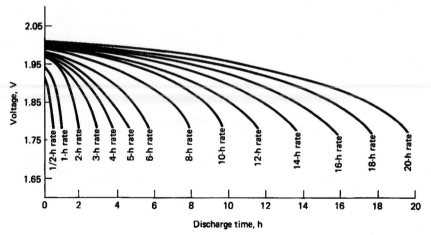

FIGURE 23.38 Discharge curves of flat-pasted lead-acid stationary batteries (specific gravity 1.215) at various discharge rates at 25°C.

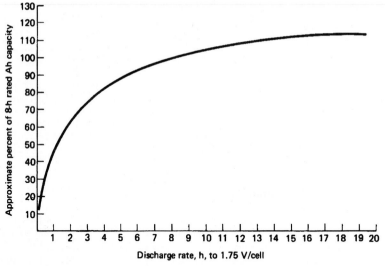

FIGURE 23.39 Effect of discharge rate on cell capacity at 25°C for flat-pasted lead-acid stationary batteries (specific gravity 1.215) to 1.75-V end voltage.

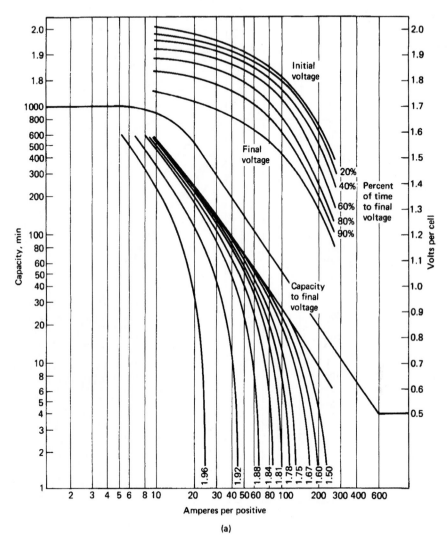

FIGURE 23.40 Performance curves of lead-acid stationary batteries at 25°C (S-shaped curves, based on positive-plate performance). (*a*) Antimony flat-pasted plate, 125 Ah at 8-h rate; 290-mm height, 239-mm width, 8.6-mm thickness. (*Courtesy of Enersys, Inc.*)

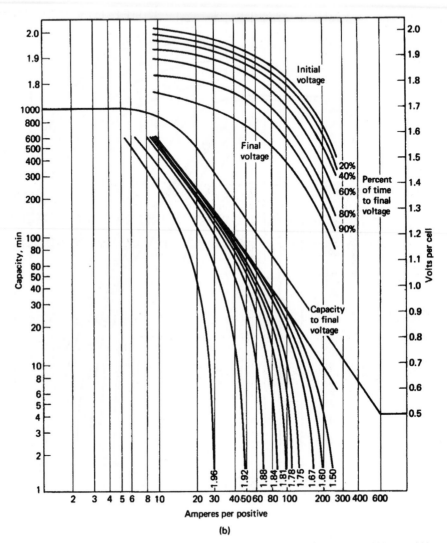

FIGURE 23.40   (*b*) Calcium flat-pasted plate, 125 Ah at 8-h rate; 290-mm height, 239-mm width, 8.6-mm thickness. (*Courtesy of Enersys, Inc.*) (*Continued*).

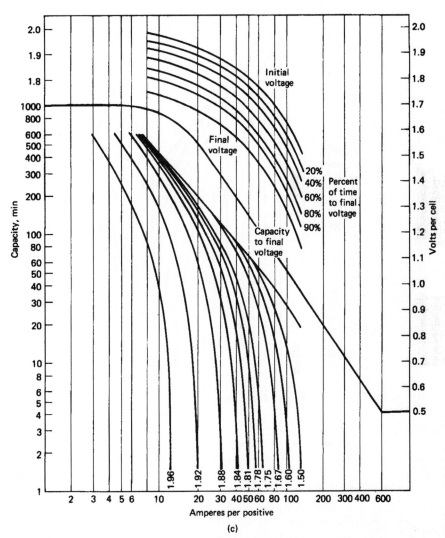

FIGURE 23.40 (*c*) Ironclad tubular plate, 70 Ah at 8-h rate; 274-mm height, 203-mm width, 8.9-mm thickness. (*Courtesy of Enersys, Inc.*) (*Continued*).

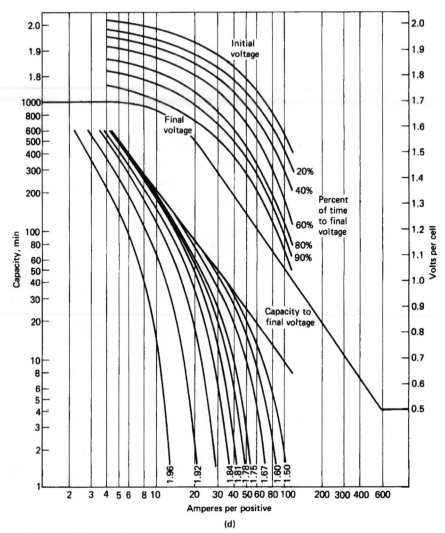

FIGURE 23.40   (*d*) Manchex plate, 40 Ah at 8-h rate; 197-mm height, 197-mm width, 11.2-mm thickness. (*Courtesy of Enersys, Inc.*) (*Continued*).

The energy density of the flat-pasted positive-plate and the tubular positive-plate batteries is similar. It is lower for the Planté positive-plate batteries. The high-rate performance of the flat-pasted positive cells is better because these plates can be made thinner than the tubular or Planté plates.

The optimal temperature for the use of stationary batteries ranges from 20 to 30°C, although temperatures from −40 to 50°C can be tolerated. The effect of temperature on the capacity of stationary batteries at different discharge loads is shown in Fig. 23.41. High-temperature operation, however, increases self-discharge, reduces cycle life, and causes other adverse effects, as discussed in Sec. 23.8.1.

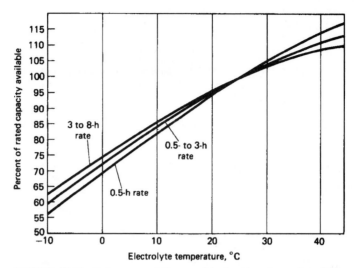

FIGURE 23.41 Performance of flat-pasted lead-acid stationary batteries at various temperatures and discharge rates. (*Courtesy of C&D Technologies, Dynasty Div.*)

The rates of self-discharge of the various types of stationary batteries are compared in Fig. 23.42, which shows the relative float current at a specified float voltage. The float current under these conditions is a measure of self-discharge or local action. It is lowest for the calcium-lead grid pasted positives and remains low throughout the life. The float current is progressively higher for the tubular antimony-lead positives, the pasted antimony-lead positives, and the Manchester-type positives—at the beginning and throughout the battery's life. If the float current is not increased periodically, the antimonial cells will all become progressively self-discharged and sulfated.

For fully charged batteries, the self-discharge rate at 25°C for the calcium-lead positive-plate cells is about 1% per month, 3% for the Planté, and about 7 to 15% for the antimonial lead positive cells. At higher temperatures, the self-discharge rate increases significantly, doubling for each 10°C rise in temperature.

The float current for the calcium-lead and antimonial lead batteries is shown in Fig. 23.43 under float charge at voltages between 2.15 and 2.40 V per cell. It has been found that more than 50 mV positive and negative overpotential is necessary to prevent self-discharge so that 0.005 A float current per 100 Ah of battery capacity is required for the lead-calcium batteries. Antimonial lead batteries initially require at least 0.06 A per 100 Ah, but this increases to 0.6 A per 100 Ah as the battery ages. The higher float current also increases the rate of water consumption and evolution of hydrogen gas.

Various, and at times conflicting, claims about the life of stationary battery designs are made by the different manufacturers worldwide. Generally, the flat-pasted antimonial lead batteries have the shortest life (5 to 18 years), followed by the flat-pasted calcium-lead batteries (15 to 25 years), the tubular batteries (20 to 25 years), and the Planté batteries (25 years).

Life on float service has been found to be related to temperature (Arrhenius-type behavior), as plotted in Fig. 23.44. The growth rate constant k is plotted for several different types of grid alloys used for the telephone system. At 25°C the time to reach 4% growth, an upper limit before the battery's integrity is impaired, is calculated to be 13.8 years for PbSb, 16.8 years for PbCa, and 82 years for pure lead.[25]

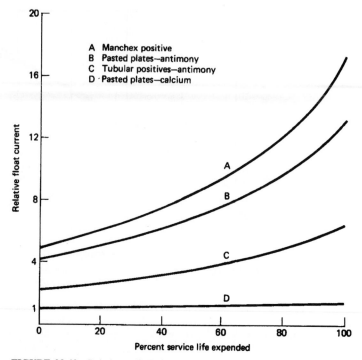

FIGURE 23.42 Relative self-discharge of lead-acid stationary batteries of different construction.

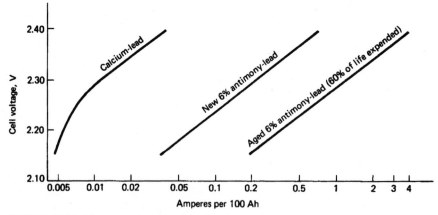

FIGURE 23.43 Float current characteristics of stationary batteries at 25°C, 100-Ah cells, fully charged, 1.210 specific gravity.

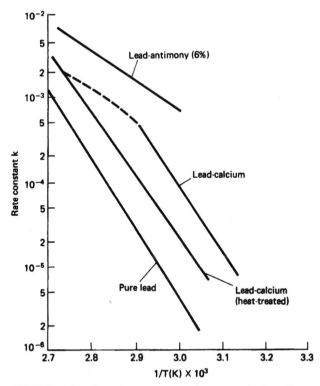

FIGURE 23.44 Corrosion rate constant log k vs. $1/T$ for different lead-alloy grids. (*From Ref. 25.*)

23.6.3 Cell and Battery Types and Sizes

Stationary batteries, like traction batteries, are available in a variety of plate and cell sizes. Stationary battery systems are assembled on insulated metal racks with groups of cells in series, having nominal system voltages of between 12 and 160 V. Some battery installations are made with several such series cell strings connected in parallel for greater storage capacity. Most stationary batteries in the United States are made with pasted positive plates. In Europe, Planté and tubular positives are more popular.

Listings of stationary batteries with flat-pasted, tubular, and Planté plates are given in Tables 23.15 to 23.17. The basic rating unit is the positive plate, given in Ampere-hours at the 8-h rate unless specific otherwise, ratings of the cell are the product of the capacity of a single positive plate multiplied by the number of positive plates. A popular stationary battery size is the 1680-Ah cell (168-Ah positive, 10 positive or 21 plates per cell) for use in telephone exchanges. The former Bell System has designed a special cylindrical cell in this capacity rating (Lucent Technologies specification KS-20472, list 1).

TABLE 23.17a   Typical Stationary Batteries, Flat-Pasted Plates

Positive-plate capacity, Ah at 8-h rate	Plate dimensions, mm				Cell size*† (positive plates per cell)
			Thickness		
	Height	Width	Positive	Negative	
5	89	63.5	6.6	4.3	2,4
25	149	143	6.6	4.3	1–8
90–95	290	222	7.9	5.3	2–12
150–155	381	304	6.4	4.6	2–17
166	381	304	7.9	5.3	5–16
195	457	338	6.9	5.3	13–18
412	1816	338	7.6	5.5	17–19

*Typical cell construction: n positive and $n + 1$ outside negative plates per cell. Some smaller cell sizes are assembled in multiples of two, three, or four cells in a monolithic container.

†Typical cell characteristics: 10 positive 168-Ah plates (1680-Ah cell); weight: 140 kg; size: length, 270 mm; width, 359 mm; height, 575 mm.

Source: C & D Technologies, Inc., Blue Bell, Pa.

TABLE 23.17b   Typical Stationary Batteries, Tubular Plates

Positive-plate capacity, Ah		Plate dimensions, mm				Cell size*† (positive plates per cell)
At 4-h rate	At 8-h rate			Thickness		
		Height	Width	Positive	Negative	
26	31.25	157	203	8.9	5.6	4
76	96	277P	234P	8.9		3–10
		290N	239N		6.1	
88	105	277P	234P	8.9		3–10
		290N	239N		6.1	
124	152	366	307	8.9	4.8	5–14

*Typical cell construction: n positive and $n + 1$ outside negative plates per cell; used with flat-pasted negative plates.

†Typical cell characteristics: 11 positive 152-Ah plates (1672-Ah cell); weight: 128 kg; size: length, 272 mm; width, 368 mm; height, 577 mm.

Source: Enersys, Inc., and Tudor AB.

TABLE 23.17c Typical Stationary Batteries, Planté Plates

| Positive-plate capacity, Ah at 8-h rate | Plate dimensions, mm | | | | Cell size* (positive plates per cell) |
| | Height | Width | Thickness | | |
			Positive	Negative	
		Planté type†‡			
8	140	140	9.5	4.7	3, 5, 7
20			9.5	4.7	5–17
40			11.1		9–25
80	286	233	9.5	4.7	2–7
83			11.1		13–25
		Manchester type†§			
20	155	149	9.7	4.6	2, 3, 4
40	197	197	11.2	4.6	2–9
83	282	292	11.2	4.6	5–12

*Typical cell construction: n positive and $n + 1$ negative outside plates per cell.
†Used with flat-pasted negative plates.
‡Typical Planté cell characteristics: 2 positive 80-Ah plates (160-Ah cell). Two-cell battery size: L 283 mm, W 159 mm, H 463 mm.
Source: Gould, Inc.
§Typical Manchester cell characteristics: 4 positive 40-Ah plates (160-Ah cell). One-cell battery weight: 40 kg; size: length, 131 mm; width, 257 mm; height, 455 mm.
Source: Enersys, Inc.

23.7 CHARGING AND CHARGING EQUIPMENT

23.7.1 General Considerations

In the charging process, DC electric power is used to reform the active chemicals of the battery system to their high-energy, charged state. In the case of the lead-acid battery, this involves, as shown in Sec. 23.2, the conversion of lead sulfate in the positive electrodes to lead oxide (PbO_2), the conversion of lead sulfate of the negative electrode to metallic lead (sponge lead), and the restoration of the electrolyte from a low-concentration sulfuric acid solution to the higher concentration of approximately 1.21 to 1.30 specific gravity. Since a change of phase from solid to solution is involved with the sulfate ion, charging lead-acid batteries has special diffusional considerations and is temperature-sensitive. During charge and discharge the solid materials which go into solution as ions are reprecipitated as a different solid compound. This also causes a redistribution of the active material. The rearrangement will tend to make the active material contain a crystal structure with fewer defects, which results in less chemical and electrochemical activity. Therefore the lead-acid battery is not as reversible physically as it is chemically.[29] This physical degradation can be minimized by proper charging, and often batteries discarded as dead can be restored with a long, slow recharge (3 to 4 days at 2 to 3 A for SLI batteries).

A lead-acid battery can generally be charged at any rate that does not produce excessive gassing, overcharging, or high temperatures. The battery can absorb a very high current during the early part of the charge, but there is a limit to the safe current as the battery becomes charged. This is shown in Fig. 23.45, which is a graphic representation of the Ampere-hour rule

$$I = Ae^{-t}$$

where I is the charging current, A is the number of Ampere-hours previously discharged from the battery, and t equals time. Because there is considerable latitude, there are a number of charging regimes, and the selection of the appropriate method depends on a number of considerations, such as the type and design of the battery, service conditions, time available for charging, number of cells or batteries to be charged, and charging facilities. Figure 23.46 shows the relation of cell voltage to the state of charge and the charging current. The figure shows that a fully discharged battery can absorb high currents with the charging voltage remaining relatively low. However, as the battery becomes charged, the voltage increases to excessively high values if the charge is maintained at the high rate, leading to overcharge and gassing. The charge current should be reduced to reasonable values at the battery reaches full charge.

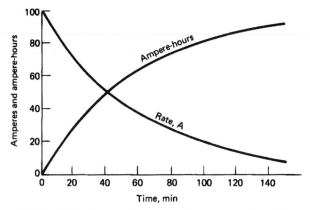

FIGURE 23.45   Graphic illustration of ampere-hour law. (*From Vinal.*[18])

In automotive, marine, and other vehicle applications, the DC electric power is usually provided by an on-board generator or alternator driven from the prime engine. These devices have a voltage and current limiter to prevent overcharging. The proper limit is dependent on the chemistry and physical construction of the cell or battery. For the traditional automotive batteries which use antimonial lead alloy as grid material, voltage limits in the range of 14.1 to 14.6 V for a nominal 12-V battery are usual. With the newer maintenance-free batteries, which use a calcium-lead alloy grid or other grid material with high hydrogen overvoltage, higher charging voltages, in the range of 14.5 to 15.0 V, can be used without danger of overcharge. Batteries in automobile and similar applications today see what is close to cycling rather than float service, but the charging controls are such that very little gas is evolved on charge. This minimizes the requirement for watering, but makes accurate control of the charge necessary. The charging rates for the different types of SLI batteries are compared in Fig. 23.47.[9] The calcium-lead maintenance-free battery is less affected by high settings in the voltage regulator than the other batteries.

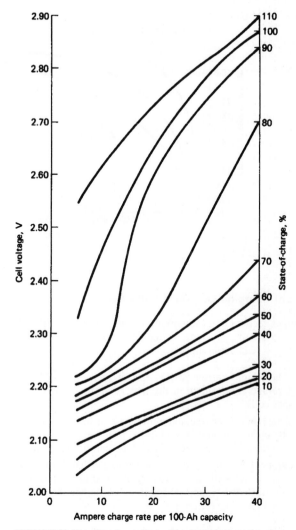

FIGURE 23.46   Charging voltage of lead-acid battery at various states of charge. (*From Ref. 23.*)

In many nonautomotive applications, charging is done separately from the system using the battery. The direct current necessary for charging is usually obtained by rectifying alternating current. These chargers include wall-hung units and mobile units as well as floor-mounted units. Newer charger designs have microprocessor controls, can sense battery condition, temperature, voltage, charge current, and so on, and are capable of changing charging rates during the charge. Most rectifiers produce some AC ripple with the direct current, which causes additional heating of the battery. This should be minimized, especially near the end of the charge when batteries tend to get hot. Pulse charging and the use of asymmetric alternating current have been proposed as a means to overcome this problem, but practical lead-acid batteries have such large capacitances that the pulses are smoothed out and the effects minimized.[29]

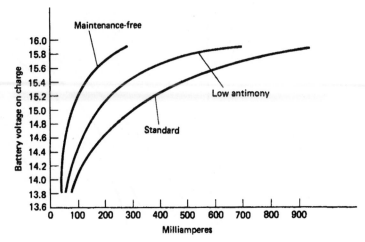

FIGURE 23.47 Charging characteristics of lead-acid SLI batteries at 25°C.

23.7.2 Methods of Charging Lead-Acid Batteries

Proper recharging is important to obtain optimum life from any lead-acid battery under any conditions of use. Some of the rules for proper charging are given below and apply to all types of lead-acid batteries.

1. The charge current at the start of recharge can be any value that does not produce an average cell voltage in the battery string greater than the gassing voltage (about 2.4 V per cell).

2. During the recharge and until 100% of the previous discharge capacity has been returned, the current should be controlled to maintain a voltage lower than the gassing voltage. To minimize charge time, this voltage can be just below the gassing voltage.

3. When 100% of the discharged capacity has been returned under this voltage control, the charge rate will have normally decayed to the charge "finishing" rate. The charge should be finished at a constant current no higher than this rate, normally 5 A per 100 Ah of rated capacity (referred to as the 20-h rate).

A number of methods for charging lead-acid batteries have evolved to meet these conditions. These charging methods are commonly known as:

1. Constant-current, one-current rate
2. Constant-current, multiple decreasing-current steps
3. Modified constant current
4. Constant potential
5. Modified constant potential with constant initial current
6. Modified constant potential with a constant finish rate
7. Modified constant potential with a constant start and finish rate
8. Taper charge
9. Pulse charging
10. Trickle charging
11. Float charging
12. Rapid charging

Constant-Current Charging.   Constant-current recharging, at one or more current rates, is not widely used for lead-acid batteries. This is because of the need for current adjustment unless the charging current is kept at a low level throughout the charge (Ampere-hour rule), which will result in long charge times of 12 h or longer. Typical charger and battery characteristics for the constant-current charge, for single and two-step charging, are shown in Fig. 23.48.

Constant-current charging is used for some small lead-acid batteries (see Chap. 24). The use of a constant-current charge during the initial battery "formation" charge has been described in Sec. 23.3. Constant-current charging is also used at times in the laboratory because of the convenience of calculating Ampere-hour input and because constant-current charging can be done with simple, inexpensive equipment. Constant-current charging at half the 20-h rate can be used in the field to decrease the sulfation in batteries which have been over-discharged or undercharged. This treatment, however, may diminish battery life and should be used only with the advice of the battery manufacturer.

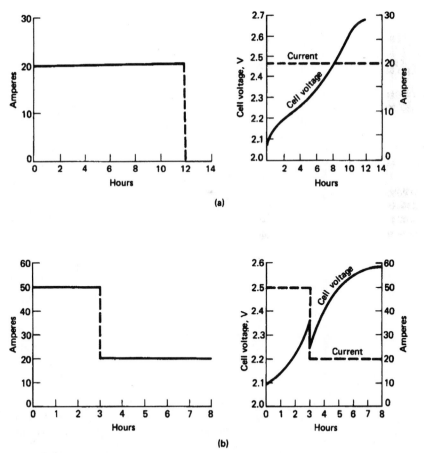

FIGURE 23.48 Typical charger and battery characteristics for constant-current charging of lead-acid batteries. (*a*) Single-step constant-current charging. (*b*) Two-step constant-current charging. (*From Ref. 10.*)

Constant-Potential Charging.   The characteristics of constant-potential and modified constant-potential charging are illustrated in Fig. 23.49. In normal industrial applications, modified constant-potential charging methods are used (methods 5, 7, and 8). Modified constant-potential charging (method 5) is used for on-the-road vehicles and utility, telephone, and uninterruptible power system applications where the charging circuit is tied to the battery. In this case the charging circuit has a current limit, and this value is maintained until a predetermined voltage is reached. Then the voltage is maintained constant until the battery is called on to discharge. Decisions must be made regarding the current limit and the constant-voltage value. This is influenced by the time interval when the battery is at the constant voltage and in a 100% state of charge. For this "float"-type operation with the battery always on charge, a low charge current is desirable to minimize overcharge, grid corrosion associated with overcharge, water loss by electrolysis of the electrolyte, and maintenance to replace this water. To achieve a full recharge with a low constant potential requires the proper selection of the starting current, which is based on the manufacturer's specifications.

The modified constant-potential charge, with constant start and finish rates, is common for deep-cycling batteries which are typically discharged at the 6-h rate to a depth of 80%; the recharge is normally completed in an 8-h period. The charger is set for the constant potential of 2.39 V per cell (the gassing voltage), and the starting current is limited to 16 to

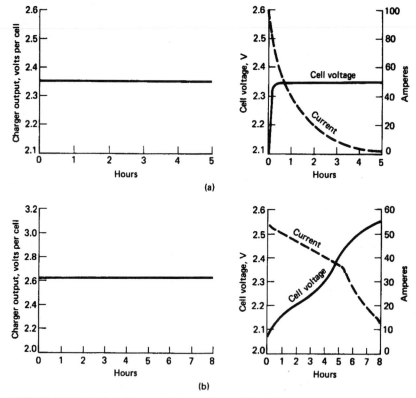

FIGURE 23.49   Typical charger and battery characteristics for constant-potential charging of lead-acid batteries. (*a*) Constant-potential charging. (*b*) Modified constant-potential charging. (*From Ref. 10.*)

20 A per 100 Ah of the rated 6-h Ampere-hour capacity by means of a series resistor in the charger circuit. This initial current is maintained constant until the average cell voltage in the battery reaches 2.39 V. The current decays at constant voltage to the finishing rate of 4.5 to 5 A per 100 Ah, which is then maintained to the end of the charge. Total charge time is controlled by a timer. The time of charge is selected to ensure a recharge input capacity of a predetermined percent of the Ampere-hour output of the previous discharge, normally 110 to 120%, or 10 to 20% overcharge. The 8-h charging time can be reduced by increasing the initial current limit rate.

Taper Charging.   Taper charging is a variation of the modified constant-potential method, using less sophisticated controls to reduce equipment cost. The characteristics of taper charging are illustrated in Fig. 23.50. The initial rate is limited, but the taper of voltage and current is such that the 2.39 V per cell at 25°C is exceeded prior to the 100% return of the discharge ampere-hours. This method does result in gassing at the critical point of recharge, and the cell temperature is increased. The degree of gassing and temperature rise is a variable depending on the charger design, and battery life can be degraded from excessive battery temperature and overcharge gassing (see Sec. 23.8.3). The gassing voltage decreases with increasing temperature; correction factors given in Table 23.18 provide the voltage correction factors at temperatures other than 25°C.

The end of the charge is often controlled by a fixed voltage rather than a fixed current. Therefore when a new battery has a high counter-EMF, this final charge rate is low and the battery often does not receive sufficient charge within the time period allotted to maintain the optimum charge state. During the latter part of life when the counter-EMF is low, the charging rate is higher than the normal finishing rate, and so the battery receives excessive charge, which degrades life. Thus the taper charge does degrade battery life, which must be justified by the use of less expensive equipment.

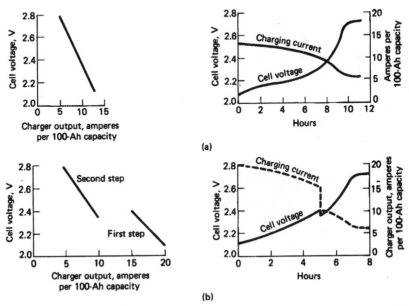

FIGURE 23.50 Typical charger and battery characteristics for taper charging of lead-acid batteries. (*a*) Single-step taper charge. (*b*) Two-step taper charge. (*From Ref. 10.*)

TABLE 23.18 Correction Factors for Cell Gassing Voltage

Electrolyte temperature, °C	Cell gassing voltage, V	Correction factor, V
50	2.300	−0.090
40	2.330	−0.060
30	2.365	−0.025
25	2.390	0
20	2.415	+0.025
10	2.470	+0.080
0	2.540	+0.150
−10	2.650	+0.260
−20	2.970	+0.508

For photovoltaic battery systems and other systems designed for optimum life, charging control and regulation circuits should produce a pattern of voltage and current equivalent to the best industrial circuits. Modified constant-potential charging methods with constant initial current (methods 5 and 7) are preferred. Optimum control to maximize life and energy output from the battery is best achieved when the depth of discharge and the time for recharge are predetermined and repetitive, a condition not always realized in solar photovoltaic applications.

Pulse Charging.   Pulse charging is also used for traction applications, particularly in Europe. In this case the charger is periodically isolated from the battery terminals and the open-circuit voltage of the battery is automatically measured (an impedance-free measurement of the battery voltage). If the open-circuit voltage is above a preset value, depending on a reference temperature, the charger does not deliver energy. When the open-circuit voltage decays below that limit, the charger delivers a DC pulse for a fixed time period. When the battery state of charge is very low, charging current is connected almost 100% of the time because the open-circuit voltage is below the present level or rapidly decays to it. The duration of the open-circuit and the charge pulses are chosen so that when the battery is fully charged, the time for the open-circuit voltage to decay is exactly the same as the pulse duration. When the charger controls sense this condition, the charger is automatically switched over to the finish rate current and short charging pulses are delivered periodically to the battery to maintain it at full charge. In many industrial applications high-voltage batteries may be used and difficulty can be encountered in keeping the cells in a balanced condition. This is particularly true when the cells have long periods of standby use with different rates of self-decay. In these applications the batteries are completely discharged and recharged periodically (usually semiannually) in what is called an equalizing charge, which brings the whole string of cells back to the complete charge state. On completion of this process, the liquid levels in the cells must be checked and water added to depleted cells as required. With the newer types of maintenance-free cells, which are semisealed, such equalizing charges and differential watering of the cells may not be possible, and special precautions are taken in the charger design to keep the cells at an even state of charge.

Trickle Charging.   A trickle charge is a continuous constant-current charge at a low (about $C/100$) rate, which is used to maintain the battery in a fully charged condition, recharging it for losses due to self-discharge as well as to restore the energy discharged during intermittent use of the battery. This method is typically used for SLI and similar type batteries when the battery is removed from the vehicle or its regular source of charging energy for charging.

Float Charging.   Float charging is a low-rate constant-potential charge also used to maintain the battery in a fully charged condition. This method is used mainly for stationary batteries which can be charged from a DC bus. The float voltage for a non-antimonial grid battery containing 1.210 specific gravity electrolyte and having an open-circuit voltage of 2.059 V per cell is 2.17 to 2.25 V per cell.

Rapid Charging.   In many applications, it is desirable to be able to rapidly recharge the battery within an hour or less. As is the case under any charging condition, it is important to control the charge to maintain the morphology of the electrode, to prevent a rise in the temperature, particularly to a point where deleterious side reactions (corrosion, conversion to nonconducting oxides, high solubility of materials, decomposition) take place, and to limit overcharge and gassing. As these conditions are more prone to occur during high-rate charging, charge control under these conditions is critical.

The availability of small, low-cost but sophisticated semiconductor chips has made effective methods of controlling the charging voltage-current-profile feasible. These devices can be used to either terminate the charge, limit the charge current, or switch between charge regimes when potentially damaging conditions arise during the charge.

A number of different techniques have been developed for effective rapid recharge. In one method, referred to as "reflex" charging, a brief discharge pulse of a fraction of a second, is incorporated into the charging regime. This technique has been found to be effective in preventing an excessive rise in temperature during rapid (15-min) high-rate recharging.

23.8 MAINTENANCE SAFETY, AND OPERATIONAL FEATURES

23.8.1 Maintenance

It is common for industrial lead-acid batteries to function for periods of 10 years or longer. Proper maintenance can ensure this extended useful life. Five basic rules of proper maintenance are:

1. Match the charger to the battery charging requirements.
2. Avoid overdischarging the battery.
3. Maintain the electrolyte at the proper level (add water as required).
4. Keep the battery clean.
5. Avoid overheating the battery.

In addition to these basic rules, as the battery is made of individual cells connected in series, the cells must be properly balanced periodically.

Charging Practice.   Poor charging practice is responsible for short battery life more than any other cause. Fortunately the inherent physical and chemical characteristics of lead-acid batteries make control of charging quite simple. If the battery is supplied with DC energy at the proper charging voltage, the battery will draw only the amount of current that it can accept efficiently, and this current will reduce as the battery approaches full charge. Several types of devices can be used to ensure that the charge will terminate at the proper time. The specific gravity of the electrolyte should also be checked periodically for those batteries that have a removable vent and adjusted to the specified value (see Tables 23.7 and 23.12).

Overdischarge. Overdischarging the battery should be avoided. The capacity of large batteries, such as those used in industrial trucks, is generally rated in Ampere-hours at the 6-h discharge rate to a final voltage of 1.75 V per cell. These batteries can usually deliver more than rated capacity, but this should be done only in an emergency and not on a regular basis. Discharging cells below the specified voltage reduces the electrolyte to a low concentration, which has a deleterious effect on the pore structure of the battery. Battery life has been shown to be a direct function of the depth of discharge, as illustrated in Fig. 23.51.[30]

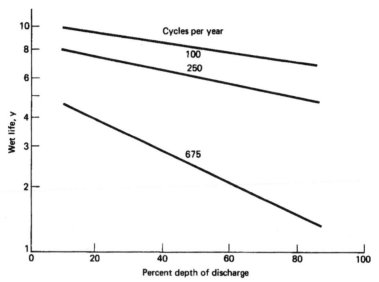

FIGURE 23.51 Effect of depth of discharge and number of cycles per year on wet life at 25°C. (*From Ref. 30.*)

Electrolyte Level. During normal operation, water is lost from a battery as the result of evaporation and electrolysis into hydrogen and oxygen, which escape into the atmosphere. Evaporation is a relatively small part of the loss, except in very hot, dry climates. With a fully charged battery, electrolysis consumes water at a rate of 0.336 cm^3 per Ampere-hour overcharge. A 500-Ah cell overcharged 10% can thus lose 16.8 cm^3, or about 0.3% of its water each cycle. It is important that the electrolyte be maintained at the proper level in the battery. The electrolyte not only serves as the conductor of electricity but is a major factor in the transfer of heat from the plates. If the electrolyte is below the plate level, then an area of the plate is not electrochemically active; this causes a concentration of heat in other parts of the cell. Periodic checking of water consumption can also serve as a rough check on charging efficiency and may warn when adjustment of the charger is required.

Since replacing water can be a major maintenance cost, water loss can be reduced by controlling the amount of overcharge and by using hydrogen and oxygen recombining devices in each cell where possible. Addition of water is best accomplished after recharge and before an equalization charge. Water is added at the end of the charge to reach the high acid level line. Gassing during charge will stir the water into the acid uniformly. In freezing weather, water should not be added without mixing, as it may freeze before gassing occurs. Water added must be either distilled water, demineralized water, or local water which has been approved for use in batteries. Automatic watering devices and reliability testing can reduce maintenance labor costs further. Overfilling must be avoided because the resultant overflow of acid electrolyte will cause tray corrosion, ground paths, and loss of cell capacity. A final check of specific gravity should be made after water has been added to ensure correct acid concentration at the end of charge. A helpful approximation is

$$\text{Specific gravity} = \text{cell open-circuit voltage} - 0.845$$

which permits electrical monitoring of specific gravity on an occasional basis (see also Fig. 23.4). Although distilled water is no longer specified by most battery manufacturers, good-quality water, low in minerals and heavy-metal ions such as iron, will help prolong battery life.

Cleanliness.   Keeping the battery clean will minimize corrosion of cell post connectors and steel trays and avoid expensive repairs. Batteries commonly pick up dry dirt, which can be readily blown off or brushed away. This dirt should be removed before moisture makes it a conductor of stray currents. One problem is that the top of the battery can become wet with electrolyte any time a cell is overfilled. The acid in this electrolyte does not evaporate and should be neutralized by washing the battery with a solution of baking soda and hot water, approximately 1 kg of baking soda to 4 L of water. After application of such a solution, the area should be rinsed thoroughly with water.

High Temperature—Overheating.   One of the most detrimental conditions for a battery is high temperature, particularly above 55°C, because the rates of corrosion, solubility of metal components, and self-discharge increase with increasing temperature. High operating temperature during cycle service requires higher charge input to restore discharge capacity and local action (self-discharge) losses. More of the charge input is consumed by the electrolysis reaction because of the reduction in the gassing voltage at the higher temperature (see Table 23.19). While a 10% overcharge per cycle maintains the state of charge at 25 to 35°C, 35 to 40% overcharge may be required to maintain the state of charge at the higher (60 to 70°C) operating temperatures. On float service, float currents increase at the higher temperatures, resulting in reduced life. Eleven days float at 75°C is equivalent in life to 365 days at 25°C.

Batteries intended for high-temperature applications should use a lower initial specific gravity electrolyte than those intended for use at normal temperatures (see Table 23.12). Other design features, such as the use of more expander in the negative plate, are also important to improve operation at high temperatures.

Cell Balancing.   During cycling, a high-voltage battery having many cells in a series string can become unbalanced, with certain cells limiting charge and discharge. Limiting cells receive more overcharge than other cells in the string, have greater water consumption, and thus require more maintenance. The equalization charge has the function of balancing cells in the string at the top of charge. In an equalization charge, the normal recharge is extended for 3 to 6 h at the finishing rate of 5 A per 100 Ah, 5-h rated capacity, allowing the battery voltage to rise uncontrolled. The equalization charge should be continued until cell voltages and specific gravities rise to a constant, acceptable value. Frequency of equalization charge is normally a function of the accumulative discharge output and will be specified by the manufacturer for each battery design and application.

23.8.2 Safety

Safety problems associated with lead-acid batteries include spills of sulfuric acid, potential explosions from the generation of hydrogen and oxygen, and the generation of toxic gases such as arsine and stibine. All these problems can be satisfactorily handled with proper precautions. Wearing of face shields and plastic or rubber aprons and gloves when handling acid is recommended to avoid chemical burns from sulfuric acid. Flush immediately and thoroughly with clean water if acid gets into the eyes, skin, or clothing and obtain medical attention when eyes are affected. A bicarbonate of soda solution (100 g per liter of water) is commonly used to neutralize any acid accidentally spilled. After neutralization the area should be rinsed with clear water.

Precautions must be routinely practiced to prevent explosions from ignition of the flammable gas mixture of hydrogen and oxygen formed during overcharge of lead-acid batteries. The maximum rate of formation is 0.42 L of hydrogen and 0.21 L of oxygen per Ampere-hour overcharge at standard temperature and pressure. The gas mixture is explosive when hydrogen in air exceeds 4% by volume. A standard practice is to set warning devices to ring alarms at 20 to 25% of this lower explosive limit (LEL). Low-cost hydrogen detectors are available commercially for this purpose.

With good air circulation around a battery, hydrogen accumulation is normally not a problem. However, if relatively large batteries are confined in small rooms, exhaust fans should be installed to vent the room constantly or to be turned on automatically when hydrogen accumulation exceeds 20% of the lower explosive limit. Battery boxes should also be vented to the atmosphere. Sparks or flame can ignite these hydrogen atmospheres above the LEL. To prevent ignition, electrical sources of arcs, sparks, or flame must be mounted in explosion-proof metal boxes. Battery cells can similarly be equipped with flame arrestors in the vents to prevent outside sparks from igniting explosive gases inside the cell cases. It is good practice to refrain from smoking, using open flames, or creating sparks in the vicinity of the battery. A considerable number of the reported explosions of batteries come from uncontrolled charging in nonautomotive applications. Often batteries will be charged, off the vehicle, for long periods of time with an unregulated charger. In spite of the fact that the charge currents can be low, fair volumes of gas can accumulate. When the battery is then moved, this gas vents, and if a spark is present, explosions have been known to occur. The introduction of the calcium alloy grids has minimized this problem, but the possibility of explosion is still present.

Some types of batteries can release small quantities of the toxic gases stibine and arsine. These batteries have positive or negative plates which contain small quantities of the metals antimony and arsenic in the grid alloy to harden the grid and to reduce the rate of corrosion of the grid during cycling. Arsine (AsH_3) and stibine (SbH_3) are generally formed when the arsenic or antimony alloy material comes into contact with nascent hydrogen, usually during overcharge of the battery, which then combines to form these colorless and essentially odorless gases. They are extremely dangerous and can cause serious illness and death. The OSHA 1978 concentration limits for SbH_3 and AsH_3 are 0.1 and 0.05 ppm, respectively, as a maximum allowable weighted average for any 8-h period. Ventilation of the battery area is very important. Indications are that ventilation designed to maintain hydrogen below 20% LEL (approximately 1% hydrogen) will also maintain stibine and arsine below their toxic limits.

The ordinary 12-V SLI automotive battery is a minor shock hazard. The hazard level increases with higher-voltage systems, and systems in the range of 84 to 360 V are being used for electric vehicles. Systems as high as 1000 V are under study for fixed-location energy-storage systems for load leveling. Batteries are electrically alive even in the discharged state, and the following precautions should be practiced:

1. Keep the top of the battery clean and dry to prevent ground short circuits and corrosion.

2. Do not lay metallic objects on the battery. Insulate all tools used in working on batteries.

3. Remove jewelry and any other electrical conductor before inspecting or servicing batteries.

4. When lifting batteries, use insulated lifting tools to avoid risks or short circuits between cell terminals by lifting chains or hooks.

5. Make sure gases do not accumulate in batteries before they are moved.

23.8.3 Effect of Operating Parameters on Battery Life

Operating parameters which have a strong influence on battery life are depth of discharge, number of cycles used each year, charging control, type of storage, and operating temperature. In some cases the battery design features which increase life tend to decrease the initial capacity, power, and energy output. It is important, therefore, that the design features of the battery be selected to match the operating and life requirements of the application.

1. Increasing the depth of discharge decreases cycle life, as illustrated in Figs. 23.52[30] and 23.32.
2. Increasing the number of cycles performed per year decreases the wet life (Sec. 23.8.1 and Fig. 23.51).
3. Excessive overcharging leads to increasing positive grid corrosion, active material shedding, and shorter wet life.

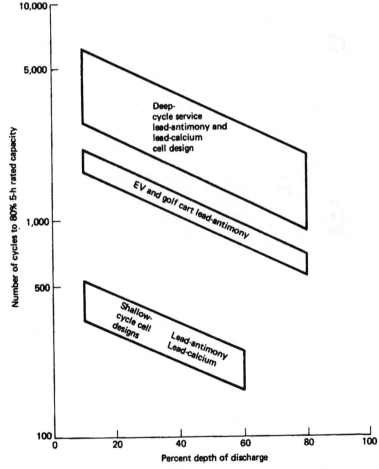

FIGURE 23.52 Effect of cell design and depth of discharge on cycle life of various types of lead-acid batteries at 25°C. (*From Ref. 30.*)

4. Storing wet cells in a discharged condition promotes sulfation and decreases capacity and life.

5. Proper charging operations with good equipment maintain the desired state of charge with a minimum of overcharge and lead to optimum battery life.

6. Stratification of the electrolyte in large cells into levels of varying concentration can limit charge acceptance, discharge output, and life unless controlled during the charge process. During a recharge, sulfuric acid of higher concentration than the bulk electrolyte forms in the pores of the plates. This higher-density acid settles to the bottom of the cell, giving higher specific gravity acid near the bottom of the plates and lower specific gravity acid near the top of the plates. This stratification accumulates during the nongassing periods of charge. During the gassing periods of overcharge, partial stirring is accomplished by gas bubbles formed at and rising along the surfaces of the plates and in the separator system. During discharge, acid in the pores of the plates and near their surface is diluted; however, concentration gradients set up by longer charge periods are seldom compensated entirely, particularly if the discharge periods are shorter, as is usually the case. Diffusion processes to eliminate these concentration gradients are very slow, and stratification during repetitive cycling can become progressively greater. Two methods for stratification control are by deliberate gassing of the plates during overcharge at the finishing rate and by stirring of cell electrolyte by pumps (usually airlift pumps). The degree of success in eliminating stratification is a function of cell design, the design of the pump accessory system, and cell operating procedures.

23.8.4 Failure Modes

The failure modes of lead-acid batteries depend on the type of application and the particular battery design. This is the rationale for the manufacture of different batteries since each one is designed to give optimum performance in a specific type of use. The more prevalent failure modes for the different types of lead-acid batteries are listed in Table 23.19.[31] Significantly, if a battery is properly maintained, most of the inherent failures are due to the degradation of the positive plate through either grid corrosion or paste shedding. These failures are irreversible, and when they occur, the battery must be replaced. Details of the failure modes of SLI batteries are given in Sec. 23.4.3.

The failure mode and the time to failure can be modified by changes in the inner parameters (I), such as battery materials, processing, and design, or by the conditions of use, designated as the outer parameters (O). Some of these are listed in Table 23.20.[31]

TABLE 23.19 Failure Modes of Lead-Acid Batteries

Battery type	Normal life	Normal failure mode
SLI	Several years	Grid corrosion
SLI (maintenance-free)	Several years	Lack of water, damage to positive plates
Golf cart	300–600 cycles	Positive shedding and grid corrosion, sulfation
Stationary (industrial)	6–25 years	Grid corosion
Traction (industrial)	Minimum 1500 cycles	Shedding, grid corrosion

TABLE 23.20 Modification of Lead-Acid Battery Failure Rate

Failure mechanism	Rate of failure modification*
Shedding positives	I: active mass structure, battery design O: number of cycles, depth of discharge, charge factor
Sulfation/leading of negatives	I: active mass additives O: temperature, charge factor, maintenance
Positive grid corrosion (overall, localized, or positive grid growth)	I: grid alloy, casting conditions, active mass
Separators	I: electrolyte concentrations, battery design O: temperature, charge factor, maintenance
Case, cover, vents, external battery connections	I: battery materials and design O: maintenance, abuse

*I—inner parameters; O—outer parameters.

23.9 APPLICATIONS AND MARKETS

The lead-acid battery is used in a wide variety of applications, and in the past few years many new applications have arisen. The various types of lead-acid batteries and their applications are listed in Table 23.2. The new uses of lead-acid batteries are mainly associated with the smaller sealed maintenance-free cells used in electronic and portable devices and with the advanced designs for energy storage and electric vehicles.

23.9.1 Automotive Applications

Traditionally the most common use of the lead-acid battery is for starting, lighting, and ignition in automobiles and other vehicles with internal combustion engines. Almost all of these now have 12-V nominal electric systems. Most of the earlier generators have been replaced by alternators and electromechanical regulators by electronic/solid-state controls. High cranking ability at low temperatures is still the major design factor, but SLI batteries today see more cycling-type service (compared with float service) because of the electrical load of the auxiliaries. Size and weight reduction have also become important as well as the battery geometry. Batteries are normally located in the cool air stream ahead of the engine to prevent their overheating; thus their geometry can affect the profile of the vehicle. These factors have led to the redesign of the lead-acid battery for SLI applications. The most important changes were:

- Change from high-antimony (4 to 5%) lead alloy grid to a low-antimony (1 to 2%) or nonantimonial lead alloy grid, thus reducing hydrogen evolution
- Use of thinner electrodes
- Better separators with lower electrical resistance
- Plate tabs located in from corners, and grids redesigned for high conductivity
- Semisealed, maintenance-free construction

Automotive-type batteries are also used on trucks, aircraft, industrial equipment, and motorcycles as well as in many other applications. They are used in off-road vehicles, such as snowmobiles, in boats to crank inboard and outboard engines, and in various farm and construction equipment. Military vehicles in the United States and NATO countries have standardized on a 24-V electric system that is provided by a series connection of two 12-V batteries.

The term "SLI battery" has evolved into something of a misnomer. In addition to starting, lighting and ignition, the automotive SLI battery may provide the power for many other functions. Although these features may not pose much of a burden individually, collectively they add up to a significant drain on the SLI battery. Some of today's automobiles require up to 2 to 3 kW of power. This could more than double in the next few years. Table 23.21 shows some current or anticipated features requiring power exclusive of the SLI functions. The typical SLI battery is not designed to handle the cycling demands prompted by some of these items. As a result, the automotive industry is planning to go to a 36/42 volt system in the near future.[32,33] The 42-V figure represents the nominal charge voltage for an eighteen cell battery, while 36 V is the minimal operating voltage. Generally, only one of these figures is mentioned as characterizing these new systems. A major advantage of going to the higher voltage is, that at a given power level, the current required would be proportionately less than with the conventional 12-V battery. The higher voltage will result in a substantial saving in weight of the current-carrying distribution system in the car, i.e., less copper will be required. In turn, this saving is projected to translate into a 5 to 10% increase in gas mileage.

Simply scaling up the current SLI lead-acid battery from 6 to 18 cells leads to a number of problems, not the least of which is the significant increase in the weight of the battery. In addition, with all the added drains on the battery and the possibility of automatic start-stop, the battery will experience deeper and more numerous discharges than current SLI batteries are designed to handle. Like VRLA batteries, today's SLI batteries are maintenance free in that they don't require addition of water during their operating life. However, unlike the starved electrolyte VRLA batteries, SLI batteries contain flooded cells. To address the weight problem, VRLA batteries are an obvious choice. Unfortunately, as discussed in Chap. 24, the wide temperature swings encountered under the hood of an automobile present a significant challenge to VRLA technology.

TABLE 23.21   Features Requiring Power for Present and Future Automobile Designs (Exclusive of SLI Function).

Alarms (may include flashing LEDs)	Audio-radio, tape or CD players
Computer	Global positioning features (maps, routing, emergency location)
Electric suspension	Electromagnetic valve trains
Automatic start-stop of engine	Electric heating of catalysts
Air conditioning	Sensing (e.g., for airbag deployment)
Electric heating of seats	Anti-lock braking
Electric steering	Electrochromic mirrors
Power windows	Rear window deicer/defogger
Rear seat entertainment center	Cigarette lighter (other functions)
Electric door locks	Cruise control
Clock	

Other potential problems associated with the higher voltage platform include such items as electrical noise and more parasitic drains due to the higher voltage. Higher voltages are undesirable due to potential shock hazard if the vehicle chasis is grounded. The higher costs and the need for the semiconductor industry to modify their electronic circuits and devices also present challenges. Counteracting the increased gas mileage anticipated from the reduced amount of copper, the increased power requirements in future automobiles require more fuel. An increase from a 2-kw to a 4-kw system will result in a reduction in gas mileage of up to 6 miles per U.S. gallon. The timeframe for the introduction of a higher voltage automotive battery system in volume production is currently a matter of controversy. Estimates range from 2003 up to 2010 or later. A prime mover in the trend to higher voltage systems is the MIT Consortium on Advanced Automotive Electrical/Electronic Components and Systems, a group that includes various auto manufacturers and suppliers.

Suggested approaches to a 36/42-V platform include a dual battery system combining lead-acid with either a nickel-metal hydride or a lithium-ion battery. On the other hand, some suggest abandoning the lead-acid battery altogether and going to a nickel-metal hydride battery. Two commercially available Hybrid Electric Vehicles on the market in 2000 are using the latter type of battery exclusively. The major advantage of lead-acid over the other two systems remains cost. At this point, it seems likely that there will be a variety of approaches to this 36/42-V regime. The role of lead-acid in this mix may depend critically on the progress in VRLA technology, especially in the area of subduing the tendency for thermal runaway under adverse conditions (see Chap. 24).

One example of a dual battery now being marketed is the so-called Gemini Twinpower™ battery manufactured in China. This system comprises two 12-V batteries in a single case with an associated "Energy Management Controller" (EMC). One 12-V battery has thin plates for the starter function; the other 12-V battery has thick plates and glass mat separators designed for cycle life. The EMC maintains the starter battery fully charged and, during starting, combines the two batteries for starting, while isolating the batteries when the engine is turned off. This "smart" battery could be considered an intermediate step towards the 36/42-V system.

23.9.2 Small Sealed Lead-Acid Cells

In recent years there has been a significant increase in the use of battery-operated consumer equipment such as portable tools, lighting devices, instruments, photographic equipment, calculators, radio and television, toys, and appliances. Batteries for these applications are generally of low capacity, up to 25 Ah. Storage batteries are used frequently because of their high power capability and rechargeability, but they have to be sealed or of the nonspill type in order to function in all positions. Vented lead-acid batteries of the electrolyte-retaining (ER) type and cylindrical (see Chap. 24) or prismatic cells are used in competition with sealed nickel-cadmium cells. The lead-acid batteries offer lower initial cost, better float service, higher cell voltage, and the absence of memory effect (loss of capacity on shallow cycling). The nickel-cadmium cell has longer life and better cycling service.

The small sealed or semisealed lead-acid cells are available as single 2-V units or as multiple-cell units, usually in 6-V monobloc constructions. They are an outgrowth of the earlier ER-type batteries in which the electrolyte was absorbed in wood pulp separators. The ER-type cells, while spill-proof, contained more electrolyte, did not recombine oxygen on overcharge, and were vented.

A related small deep-discharge lead-acid battery is the one used for miner's lamps and similar equipment. These are 4-V units which are vented and can be watered. They are designed to deliver 1 A for 12 h between charges.

23.9.3 Industrial Applications

Applications for lead-acid batteries, other than the SLI and small sealed power units, fall into two categories, as shown in Table 23.22—those based on automotive-type constructions and those based on industrial-type constructions. Often several designs can be used for a single type of application.

TABLE 23.22 Major Applications of Lead-Acid Batteries (Non-SLI Types)

Automotive and small energy storage designs		Industrial designs		
Traction	Special	Stationary	Traction (motive power)	Special
Golf cart	Emergency lighting	Switch gear	Mine locomotives	Submarines
Off-road vehicles	Alarm signals	Emergency lighting	Industrial trucks	Ocean buoys
On-road vehicles	Photovoltaic	Telecommunication facilities	Large electric vehicles	
	Sealed cells (for tools, instruments, electronic devices, etc.)	Railway signals		
		Uninterrupted power supply		
		Photovoltaics		
		Load leveling and energy storage		

23.9.4 Electric Vehicle

Lead-acid batteries have powered off-the-road electric vehicles such as golf carts, forklift trucks and airport baggage carriers for many decades. Most of these vehicles employ 36-V systems utilizing six 6-V batteries in series. Many on-the-road designs have evolved based on enhancing the mechanical properties of golf carts. These are traditionally used in retirement villages and non-city applications. Lead-acid batteries have also been used for true on-the-road vehicle designs, such as the General Motors EV1, and a number of delivery vans. Small battery powered vans have been used in the UK for milk delivery for the past century and 10–15,000 such vehicles are estimated to be in current use. Many of the modern electric vehicles have utilized high voltage (AC or DC) motors, with batteries operating in the range of 200 to 300 V.

Such high voltages complicate the battery design and also compromise life. One problem is simply the large number of cells in series needed to attain the high voltage. Reliability analyses in the battery industry have shown that statistically, the life of a battery decreases as the number of cells in series increases. This is not a surprising conclusion but rather an expected result of having units with individual failure rates in a series of such units. To anticipate and accommodate the failure of individual cells without compromising the whole battery requires the added complication of electronic circuitry to switch bad cells out of the series connection. If full voltage is crucial, one might require standby cells to be switched into the circuit at the same time. Additional problems associated with the high voltage include higher leakage currents, ground short problems and corrosion. With the larger battery, there is also a safety concern associated with the possible accumulation of greater amounts of hydrogen than with conventional SLI batteries.

The EV1, now in its second generation, is a limited production vehicle designed from the ground up as an electric vehicle. It has an aerodynamic teardrop shape, regenerative braking to charge the battery, together with an aluminum structure and composite body panels for reduced weight. A battery of twenty-six 12-V valve-regulated lead-acid batteries powers a 137 horsepower, 3-phase AC induction motor. The estimated driving range is between 55 and 95 miles between charges, depending on driving conditions and driver habits. An optional nickel-metal hydride battery is estimated to extend the range from 75 to 130 miles. With air conditioning, traction control, cruise control, anti-lock braking, speeds up to 80 miles per hour, and other features, the EV1 is not a stripped-down vehicle. Although such a zero-emissions electric vehicle would appear to be the answer to environmental concerns and the benefits have been widely publicized, consumer acceptance of a short range, more expensive vehicle has not been widespread. On the other hand, the introduction of low-emissions hybrid electric-internal combustion vehicles, notably the Toyota Prius, generated immediate enthusiasm and sales, with favorable reviews in the automotive press. The Prius employs an Ni-MH battery, as does the Honda Insight. In 1999, the costs for the major EV candidate battery systems were roughly $200 to 400/kWh for lead-acid, $500 to 1,000/kWh for Ni-MH and >$1,000/kWh for lithium-ion batteries.[34] It is clear that mass production of pure on-the-road electric vehicles will require lower cost batteries, increased range between charges and an infrastructure of charging facilities to support the electric vehicles. Lead-acid technology seems suitable only for those willing to accept a truly low-range, pure-electric vehicle.

23.9.5 Energy-Storage Systems

Secondary batteries are now being considered for load leveling in electric utility systems as an alternative to meet peak power demands currently provided with energy-expensive oil- or gas-fueled turbines (see Chap. 37). Large batteries, on the order of 50 MWh at 1000 V, are required. The lead-acid battery, again, is a major candidate for a near-term solution for this application. The goal is to obtain in excess of 2000 cycles or 10 year of operation at a cost of about $90 per kiloWatthour.

Smaller-sized batteries are used for energy storage in systems employing renewable but interruptible energy sources, such as wind and solar (photovoltaic) energy. These systems are usually located on the customer side of the utility power grid. The system generally handles the following functions:

1. Converts solar, wind, or other such prime energy source to direct electric power
2. Regulates the electric power output
3. Feeds the electric energy into an external load circuit to perform work
4. Stores the electric energy in a battery subsystem for later use

A block diagram of a typical photovoltaic system is shown in Fig. 23.53.[30]

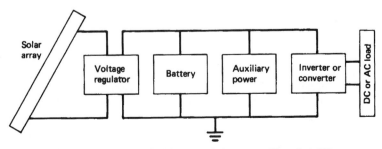

FIGURE 23.53 Components of solar photovoltaic system. (*From Ref. 30.*)

The selection of the proper battery for these types of applications requires a complete analysis of the battery's charge and discharge requirements, including the load, the output and pattern of the solar or alternative energy source, the operating temperature, and the efficiency of the charger and other system components. Golf-cart-type lead-acid batteries and modified electric-vehicle designs are widely used in these small stationary energy-storage systems because they are the least expensive design in commercial production. Maintenance-free batteries are also being developed for these applications. These batteries (100-Ah size) can give maintenance-free unattended operation, a self-discharge rate less than 5% per month, a recharge in less than 8 h, and 1000 to 2000 cycles to an 80% depth of discharge.

23.9.6 Power Conditioning and Uninterrupted Power Systems

DC Power Systems.   A new concept in standby power is the DC power system with battery backup. These power systems include a battery charger (rectifier/charger) which has a sufficient capacity to recharge the batteries at the proper voltage while simultaneously supplying power to the DC load. In addition, equipment protection and isolation of the line voltage from the secondary windings of the special power transformer are designed into the system.

Static Uninterruptible AC Power System (UPS).   In this power system, a storage battery is linked to the utility power to provide a continuity of service in the event of an interruption of the utility power. The continuous UPS system (Fig. 23.54) is illustrative of this type of device. During normal operation, the AC line supplies power to the static battery charger (rectifier-charger) which, in turn, "float" charges the battery and, at the same time, supplies DC power to the static inverter. The inverter, in turn, supplies power to the AC load. A synchronizing signal (if used) from the AC power line can maintain the phase and frequency of the inverter output the same as in the power line. This maintains the accuracy of timing devices such as clocks and recorder charts.

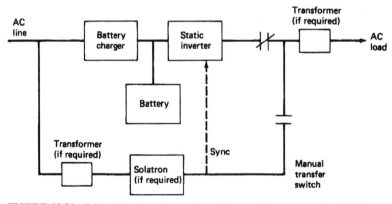

FIGURE 23.54   Schematic of continuous-type uninterruptible power system (UPS).

The voltage regulator within the inverter maintains the AC load voltage constant throughout the load range as well as during periods of equalizing charge on the storage battery. Transient and steady-state power-line variations are isolated from the load by the regulating action of the battery charger in conjunction with the filtering action of the battery and the inverter.

Upon AC line power failure, the battery charger ceases to operate and the battery instantly supplies power to the inverter and sustains the AC load without interruption. The synchronizing signal is also lost during the power failure. The inverter, therefore, continues to operate on its own internal frequency reference ($\pm 1\%$ for standard units). The Ampere-hour capacity of the battery determines the operating time of the system. The inverter is designed to maintain a constant voltage output as the battery voltage drops during the discharge. When AC power returns, the charger restores the battery energy and at the same time supplies power to the inverter. A more detailed description of other types of UPS systems employing VRLA Lead Acid Batteries is found in Sec. 24.8.

23.9.7 Marine Batteries

The marine battery market includes small and large leisure craft used for fishing, sailing, and travel as well as larger commercial vessels engaged in towing, passenger transportation, and workboat activities. In general lead-acid battery systems are used with system voltages ranging from 6 to 220 V where recharging is accomplished by an engine generator or alternator. Three types of designs are common: conventional flooded cells (mainly in the larger cell designs), absorbed electrolyte designs, and gelled electrolyte designs. (The latter two are VRLA batteries and are discussed in Chapter 24.)

Marine service differs from automotive in several aspects. Lights, refrigeration, blowers, motors, radio, and other electrical equipment results in cycling service, with often a delay between discharge and charge. In marine service, the batteries should have much greater capacity than would normally be specified for the same horsepower equipment in a shore-based application.

The key engineering features utilized by one prominent manufacturer of marine batteries[37] include the following:

- A special grid design with heavier vertical and horizontal members.
- High density active materials, both positive and negative.
- Positive plates double insulated with thick woven glass matting and then sealed in a microporous polyethylene envelope.

In some designs, individual cells can be replaced while outer plastic cases provide high impact and environmental protection.

REFERENCES

1. H. Bode, *Lead-Acid Batteries,* Wiley, New York, 1977.
2. H. E. Haring and U. B. Thomas, *Trans. Electrochem. Soc.* **68:**293 (1935).
3. A. J. Salkind, D. T. Ferrell, and A. J. Hedges, "Secondary Batteries: 1952–1977," *J. Electrochem. Soc.* **125:**311C (1978).
4. Battery Council International Annual Report.
5. P. Ruetschi, "Review of the Lead-Acid Battery Science and Technology," *J. Power Sources* **2:**3 (1977/1978).
6. C. Mantell, *Batteries and Energy Systems,* 2d ed., McGraw-Hill, New York, 1983.
7. D. J. G. Ives and G. J. Janz, *Reference Electrodes,* Academic, New York, 1961.
8. E. A. Willihnganz, U.S. Patent 3,657,639.
9. A. Sabatino, *Maintenance-free Batteries, Heavy Duty Equipment Maintenance,* Irving-Cloud Publishing, Lincolnwood, Ill., 1976.

10. Special Issue on Lead-Acid Batteries, *J. Power Sources* **2**(1) (1977/1978).

11. ASTM Specification B29, "Pig Lead Specifications," American Society for Testing and Materials, Philadelphia, Pa., 1959.

12. A. G. Cannone, D. O. Feder, and R. V. Biagetti, *Bell Sys. Tech. J.* **19**:1279 (1970).

13. A. T. Balcerzak, *Alloys for the 1980's,* St. Joe Lead Co., Clayton, Mo., 1980.

14. *Grid Metal Manual,* Independent Battery Manufacturers Association (IBMA), Largo, Fla., 1973.

15. N. E. Hehner, *Storage Battery Manufacturing Manual,* Independent Battery Manufacturers Association (IBMA), Largo, Fla., 1976.

16. U. B. Thomas, F. T. Foster, and H. E. Haring, *Trans. Electrochem. Soc.* **92**:313 (1947).

17. N. E. Hehner and E. Ritchie, *Lead Oxides,* Independent Battery Manufacturers Association (IBMA), Largo, Fla.

18. G. W. Vinal, *Storage Batteries,* 4th ed., Wiley, New York, 1955.

19. M. Barak, *Electrochemical Power Sources,* Peter Peregrinus, Stevanage, U.K., 1980.

20. *Battery Service Manual,* 9th ed., Battery Council International, Chicago, 1982.

21. Battery Council International, 401 N. Michigan Ave., Chicago, Ill. 60611.

22. Battery Council International, *Battery Replacement Data Book,* 2000.

23. *Gould Battery Handbook,* Gould Inc., Mendota Heights, Minn., 1973.

24. E. J. Friedman et al., *Electrotechnology,* vol. 3: *Stationary Lead-Acid Batteries.* Ann Arbor Science Publishers, Ann Arbor, Mich., 1980.

25. *Bell Sys. Tech. J.* **49**(7) (Sept. 1970).

26. R. V. Biagetti and H. J. Luer, "A Cylindrical, Pure Lead, Lead-Acid Cell for Float Service," *J. Power Sources* **4** (1979).

27. A. G. Cannone, U.S. Patent 3,556,853, Jan. 19, 1971.

28. R. V. Biagetti, "The AT&T Lineage 2000 Round Cell Revisited: Lessons Learned; Significant Design Changes; Actual Field Performance v. Expectations," *INTELECT—Int. Telecommunications Energy Conf.,* Kyoto, Japan, Nov. 5–8, 1991.

29. E. Ritchie, International Lead-Zinc Research Organization Project LE-82-84, Final Rep., New York, Dec. 1971.

30. "Handbook of Secondary Storage Batteries and Charge Regulators in Photovoltaic Systems." Exide Management and Technology Co., Rep. 1-7135, Sandia National Laboratories, Albuquerque, N.M., Aug. 1981.

31. G. E. Mayer, "Critical Review of Battery Cycle Life Testing Methods," *Proc. 5th Int. Electric Vehicle Symp.,* Philadelphia, Pa., Oct. 1978.

32. C. J. Murray, *Electrical Engineering Times,* Issue 1128, Aug. 20, 2000.

33. *Batteries International,* Issue 42, pp. 71–76, Jan. 2000.

34. D. M. MacArthur, G. E. Blomgren, and R. A. Powers, *1999 Battery Industry Developments,* Powers Associates, Westlake, OH. (2000).

35. A. G. Cannone, U.S. Patent 4,605,605, Aug. 12, 1986.

36. A. G. Cannone, U.S. Patent 4,980,252, Dec. 25, 1980.

37. Product Literature, Rolls Battery Engineering, Salem, MA, 01970, USA.

CHAPTER 24
VALVE REGULATED LEAD-ACID BATTERIES

Alvin J. Salkind, Ronald O. Hammel, Anthony G. Cannone, and Forrest A. Trumbore

24.1 GENERAL CHARACTERISTICS

Lead-acid battery designs for many small portable, and some larger fixed applications, have often been referred to as sealed and/or maintenance free. This has been an accurate description to the extent that there was no need or opportunity to replace electrolyte. However, virtually no design included a true hermetic seal, and only a pressure release valve which limited inflow or outflow of gas to the cell. The valves released internal gasses at pressures ranging from a few tenths of an atmosphere to a few atmospheres. In addition, most lead-acid battery cases are molded of plastics, which are hydrogen permeable.

The newer designation for these designs is as a "Valve Regulated Lead-Acid Battery" or VRLA. They differ from the conventional flooded lead-acid battery designs by containing only a limited amount of electrolyte ("starved" electrolyte) absorbed in a separator or immobilized in a gel. In most designs, the cell capacity is limited by the amount of positive active material. The starved electrolyte and excess of negative active material facilitate the recombination of oxygen produced during overcharge or "float" charge with the negative active material. The resealable valves are normally closed to prevent the entrance of oxygen from the outside air. The vent pressure design depends on the manufacturer and predominantly by the case shape and material. VRLA designs have two usual shapes, one with spirally-wound electrodes (jelly-roll construction) in a cylindrical container, and the second with flat plates in a prismatic container. The cylindrical containers can maintain higher internal pressures without deformation and can be designed to have a higher release pressure than the prismatic cells. In some designs, an outer metal container is used to prevent deformation of the plastic cases at higher temperatures and internal cell pressures. The range of venting pressures includes a high of 25 to 40 psi for a metal-sheathed, spirally wound cell to 1 to 2 psi for a large prismatic battery.

The electrolyte is commonly immobilized in two ways:

- *Absorbed electrolyte:* A highly porous and absorbent mat, fabricated from microglass fibers, is partially filled with electrolyte, and acts as the separator/electrolyte reservoir.

- *Gelled electrolyte:* Fumed silica is added to the electrolyte, causing it to harden into a gel. On subsequent charges, some water is lost, drying the gel until a network of cracks and fissure develops between the positive and negative electrodes. The openings created provide the path for the oxygen recombination reaction.

It should be noted that silica reacts with sulfuric acid and the absorption or gelation is chemical as well as physical in nature. The immobilization of the electrolyte allows batteries to operate in different orientations without spillage. In larger industrial applications, the batteries can be installed on their sides, permitting compact installations that use up to 40% less floor space and volume. An orientation in which the plates are horizontal to the ground is referred to as "pancake style."[1] Significant improvements in cycle life have been reported, in some circumstances, in cells cycled with their electrodes parallel to the ground.[2]

The use of VRLA designs is becoming more popular and accounted for over 75% of telecommunication and UPS applications in 1999. The development of advanced charging techniques has also increased the use of VRLA batteries in cycling applications such as forklift service. New market opportunities in portable electronics, power tools, and hybrid electric vehicles have stimulated the development of new designs of lead-acid batteries. However, the VRLA designs do not have the capability of handling certain types of abuse as well as conventional flooded batteries. The electrolyte, which provides the major internal heat sink in cells, is much more limited in VRLA cells. As a result, VRLA designs are more prone to thermal runaway under abusive conditions that pose little hazard for flooded cells. This is particularly true when VRLA batteries are subjected to operations at elevated temperatures. Recent discussion of high temperature (over 40°C) effects as well as a discussion of advantages and disadvantages have been given.[3,4] A detailed review of VRLA applications has been given in a series of articles.[5] To insure long battery life under conditions of higher temperature operations or overheating due to oxygen recombination, precautions must be taken.

A comparison of gel designs with absorbed electrolyte designs in fork lift truck applications was presented[6] for one particular set of applications. However, new materials and designs are evolving and the comparisons should be ongoing. Much activity is being reported in the area of special fibers, blends, and surface treatments for the absorbing mat separators.

The major advantages and disadvantages of VRLA batteries are listed in Table 24.1.

TABLE 24.1 Major Advantages and Disadvantages of VRLA Batteries

Advantages	Disadvantages
Maintenance-free	Should not be stored in discharged condition
Moderate life on float service	Relatively low energy density
High-rate capability	Lower cycle life than sealed nickel-cadmium battery
High charge efficiency	Thermal runaway can occur with incorrect charging or improper thermal management
No "memory" effect (compared to nickel-cadmium battery)	
"State of charge" can usually be determined by measuring voltage	More sensitive to higher temperature environment than conventional lead-acid batteries
Relatively low cost	
Available from small single-cell units (2 V) to large 48 V batteries	

24.2 CHEMISTRY

Although the design and the construction of the VRLA battery are different, its chemistry is that of the traditional lead-acid battery. The basic "double sulfate" reaction applies to the overall reaction:

$$PbO_2 + Pb + 2H_2SO_4 \underset{\text{charge}}{\overset{\text{discharge}}{\rightleftharpoons}} 2PbSO_4 + 2H_2O$$

The reaction at the positive electrode is

$$PbO_2 + 3H^+ + HSO_4^- + 2e \underset{\text{charge}}{\overset{\text{discharge}}{\rightleftharpoons}} 2H_2O + PbSO_4$$

and at the negative electrode,

$$Pb + HSO_4^- \underset{\text{charge}}{\overset{\text{discharge}}{\rightleftharpoons}} PbSO_4 + H^+ + 2e$$

When the cell is recharged, finely divided particles of $PbSO_4$ are electrochemically converted to sponge lead at the negative electrode and PbO_2 at the positive electrode. As the cell approaches complete recharge, when the majority of the $PbSO_4$ has been converted to Pb and PbO_2, the overcharge reactions begin. For conventional flooded lead-acid batteries, the result of these reactions is the production of hydrogen and oxygen gas and subsequent loss of water.

A unique aspect of the VRLA design is that the majority of the oxygen generated within the cells at normal overcharge rates is recombined within the cell. The pure lead-tin grids used in the construction minimize the evolution of hydrogen on overcharge. Most of the hydrogen generated within the cell is released to the atmosphere through the vent or through the plastic container.

Oxygen will react with lead at the negative plate in the presence of H_2SO_4 as quickly it can diffuse to the lead surface,

$$Pb + HSO_4^- + H^+ + \tfrac{1}{2}O_2 \underset{\text{charge}}{\overset{\text{discharge}}{\rightleftharpoons}} PbSO_4 + H_2O$$

In a flooded lead-acid battery this diffusion of gases is a slow process, and virtually all the H_2 and O_2 escape from the cell rather than recombine. In the VRLA battery the closely spaced plates are separated by a glass mat which is composed of fine glass strands in a porous structure. The cell is filled with only enough electrolyte to coat the surfaces of the plates and the individual glass strands in the separator, thus creating the starved-electrolyte condition. This condition allows for the homogeneous gas transfer between the plates, necessary to promote the recombination reactions. Additional discussion of VRLA chemical kinetics can be found in the literature.[4]

The pressure release valve maintains an internal pressure and this condition aids recombination by retaining the gases long enough within the cell for diffusion to take place. The net result is that water, rather than being released from the cell, is cycled electrochemically to take up the excess overcharge current beyond that used for conversion of active material. Thus the cell can be overcharged sufficiently to convert virtually all the active material without loss of water, particularly at the recommended recharge rates. At continuous high overcharge rates (such as $C/3$ and above), gas buildup becomes so rapid that the recombination process is not as highly efficient, and O_2 as well as H_2 gas is released from the cell as the cell vents. Charging at these higher rates should therefore be avoided.

24.3   CELL CONSTRUCTION

24.3.1   VRLA Cylindrical Cells[7]

A cross section of a VRLA cell and a breakdown of the basic components contained in the cell are shown in Figs. 24.1 and 24.2. Both the positive and the negative grids are made from 99.99% pure lead with 0.6% tin added for deep discharge recovery. The lead grid is relatively thin, 0.6 to 0.9 mm, to provide for a high plate surface area and resultant high discharge rates. The plates are pasted with lead oxides, separated by an absorbing glass mat, and spirally wound to form the basic cell element. Lead posts are then welded to the exposed positive- and negative-plate tabs. The terminals are inserted through the polypropylene inner top and are effectively sealed by expansion into the lead posts. The element is then stuffed into the liner and the top and liner are bonded together. At this state of construction, the cell is sealed except for the open vent hole. Sulfuric acid is then added and the relief valve is placed over the vent hole. The sealed element is then inserted into the metal can, the outer plastic top added, and crimping completes the assembly. The metal case provides mechanical strength and does not affect the operation of the resealable vent. The cell is then formed electrochemically.

Monobloc batteries, using the cylindrical cell, are produced with two to six cells interconnected in a single plastic container. These 4-, 6-, and 12-V batteries have performance characteristics similar to those of the single cell. The monobloc design is illustrated in Fig. 24.3. A newer type of small cylindrical lead-acid cell made within film electrodes has been introduced.[8] This design has been characterized by Atwater et al.[9] for use in portable electronic and communication applications.

FIGURE 24.1 Cross section of the VRLA cell. Components identified in Fig. 24.2. (*Courtesy of Hawker Energy Products, Inc.*)

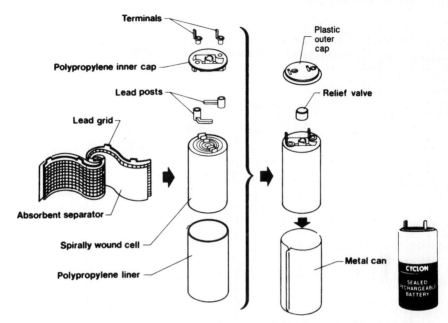

FIGURE 24.2   Components of VRLA cell. (*Courtesy of Hawker Energy Products, Inc.*)

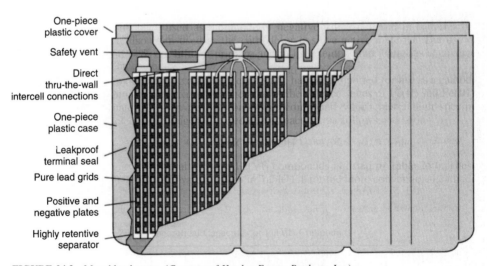

FIGURE 24.3   Monobloc battery. (*Courtesy of Hawker Energy Products, Inc.*)

24.3.2 VRLA Prismatic Cells

A cutaway view of a prismatic lead-acid cell is shown in Fig. 24.4*a* An exploded view of a three-cell monobloc battery, using this prismatic cell, is given in Fig. 24.4*b*.

Nonantimonial calcium-lead alloys are usually used for the grids for the prismatic cell because these grids have a lower self-discharge rate and reduced gassing. However, the

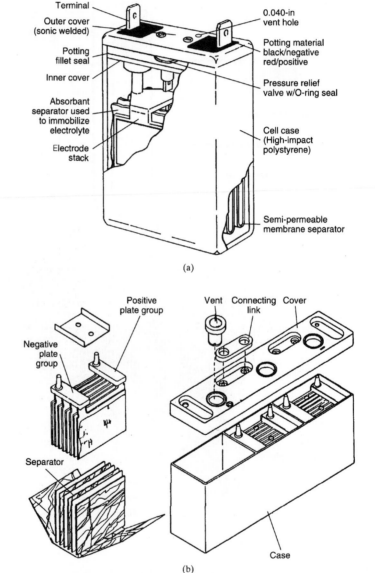

(a)

(b)

FIGURE 24.4   Typical prismatic lead-acid cell and monobloc battery (*a*) Cutaway view (*Courtesy of Eagle-Picher Industries, Inc.*) (*b*) Exploded view. (*Courtesy of Johnson Controls, Inc.*)

nonantimonial lead grids tend to reduce the deep-cyclability of the batteries unless additives are incorporated into the cell. Most of these are proprietary, the most common being phosphoric acid. The balance between cyclability, capacity, and float life is controlled by the ratio of α-PbO$_2$ to β-PbO$_2$, the paste density, the amount and concentration of the electrolyte, the composition of the grid alloy, and the type and amount of additive.

The electrolyte is absorbed in blotterlike glass-fiber separators or is gelled. Many acid-resistant materials, such as burnt clay, pumice, sand, Fuller's earth, plaster of paris, and asbestos, has been used for gelling. Present practice is to use fumed silica. When gelling agents are used, the electrodes are usually formed with dilute acid for the first step of a two-step formation, often in tanks. The gellation process is introduced in the final battery assembly.

24.3.3 Thin Prismatic Cells

Thin flat prismatic sealed lead-acid cells have been designed for portable applications as they offer more flexibility in the design of the battery. They use space more efficiently than cylindrical cells, resulting in a higher volumetric energy density, and the slim design is adaptable to small-footprint equipment.[10,11] An exploded view of a typical flat cell is shown in Fig. 24.5.

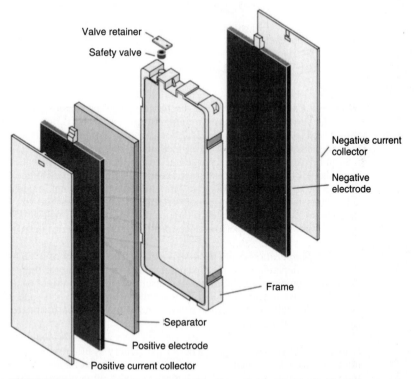

FIGURE 24.5   Exploded view of typical thin, flat sealed lead-acid cell. Metal plates, each approximately 0.3 mm thick, are laminated with special resin films to negative and positive electrode plates, both physically and electrically. They are then fused to a resin frame by a pressurized thermal bonding process. (*Courtesy of Panasonic Industrial Co.*)

24.4 *PERFORMANCE CHARACTERISTICS*

24.4.1 VRLA Cylindrical Cells

Voltage.   The nominal voltage of a VRLA single-cell battery is 2.0 V, and it is typically discharged to 1.75 V per cell under load. The open-circuit voltage depends on the state of charge, as plotted in Fig. 24.6, based on a $C/10$ discharge rate. The open-circuit voltage can therefore be used to approximate the state of charge. The curve is accurate to within 20% if the battery has not been charged or discharged within 24 h; it is accurate to within 5% if the battery has not been used for 5 days. The measurement of the open-circuit voltage to determine the state of charge is based on the relationship between the electromotive force (OCV) and the concentration of the sulfuric acid in the battery.

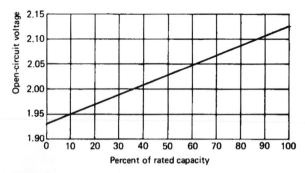

FIGURE 24.6   Open-circuit voltage vs. state of charge.

Discharge Characteristics.   The discharge voltage profiles of typical VRLA single-cell batteries, at various temperatures ranging from −40 to 65°C for various discharge rates, are shown in Fig. 24.7 (see Table 24.2 for capacity data on various size cells). The discharge voltage curves are relatively flat at medium to low rates. These curves are based on smaller 2.5- and 5-Ah batteries. Discharge curves for the larger 25-Ah battery are slightly different from those of the smaller batteries because of the greater distance from the center of the plate to the external stud. This gives a higher effective internal impedance per unit of capacity and results in a slightly lower performance at higher rates and lower temperatures. Figure 24.8 shows a set of discharge voltage curves for a 2.5-Ah cell at 25°C, which further illustrates the good voltage performance of the cell even at high rates of discharge.

Effect of Temperature and Discharge Rate.   The capacity of a VRLA battery, as with most batteries, is dependent on the discharge rate and temperature, the capacity decreasing with decreasing temperature and increasing discharge rate. The effect of temperature at the $C/10$, C, and $5C$ rates is shown in Fig. 24.9 for the cylindrical type D and X for the batteries. The larger 25-Ah cell gives a lower percentage of the 25°C performance at lower temperatures and higher discharge rates.

High-Rate Pulse Discharge.   The VRLA battery is effective in applications which require high-rate discharge, such as in engine-starting. The voltage-time curves for the battery at room temperature at the $10C$ discharge rate, both on continuous discharge and for a 16.7% duty cycle (10-s pulse, 50-s rest), are shown in Fig. 24.10 for 25 and −20°C.
    It is apparent from these data that the capacity of the VRLA battery is increased greatly when an intermittent pulse discharge is used. This is true because of the phenomenon known as "concentration polarization." As a discharge current is drawn from the cell, the sulfuric acid in the electrolyte reacts with the active materials in the electrodes. This reaction reduces the concentration of the acid at the electrode-electrolyte interfaces. Consequently the cell

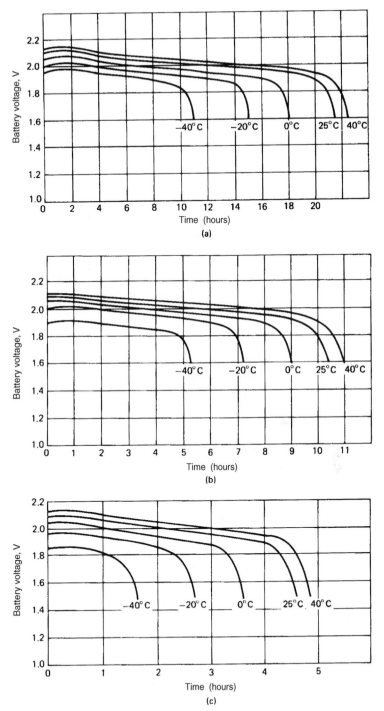

FIGURE 24.7 Discharge curves of cylindrical VRLA D and X single-cell batteries. Discharge rate (*a*) at *C*/20, (*b*) at *C*/10, (*c*) at *C*/5.

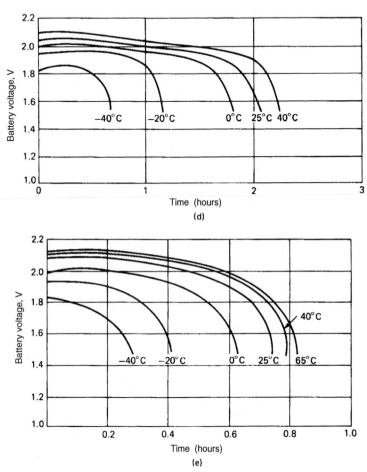

FIGURE 24.7 (*d*) at *C*/2.5, (*e*) at 1*C* (see Table 24.2 for capacity data) (*Continued*).

TABLE 24.2 VRLA Cylindrical Batteries

Model	Capacity, Ah $C/10$	$C/20$	$1C$	Dimensions, mm Height	Diameter, width	Length	Weight (typical), g	Specific energy @$C/20$ Wh/kg	Maximum discharge, A
				Single cells					
D	2.5	2.7	1.8	67.3	34.3	N/A	180	30.0	40
X	5.0	5.4	3.2	80.3	49.5	N/A	390	27.6	40
J	12.5	13.0	9.0	135.7	51.8	N/A	840	30.8	60
BC	25.0	26.0	17.5	172.3	65.3	N/A	1580	32.9	250
DT	4.5	4.8	3.7	102.9	34.3	N/A	272	35.3	40
E	8.1	8.4	6.2	108.7	44.5	N/A	549	30.6	40
				Monobloc batteries (preassembled batteries in common sizes)					
D, 4 V	2.5	2.7	1.8	70	45	78	360		40
D, 6 V	2.5	2.7	1.8	70	45	113	540		40
X, 4 V	5.0	5.4	3.2	77	54	96	740		40
X, 6 V	5.0	5.4	3.2	77	54	139	1110		40
E, 4 V	8.0	8.6	5.8	102	54	96	1110		40
E, 6 V	8.0	8.6	5.8	102	54	139	1670		40

Source: Hawker Energy Products, Inc.

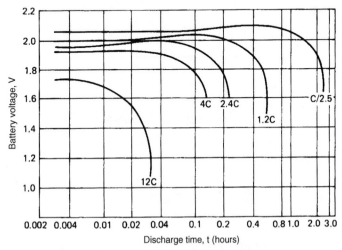

FIGURE 24.8 Discharge curves of VRLA cylindrical D and X batteries at 25°C, high discharge rates.

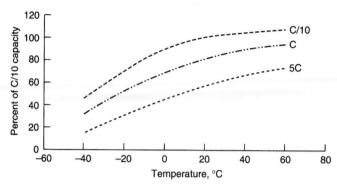

FIGURE 24.9 Effect of temperature on capacity for D and X size cylindrical units.

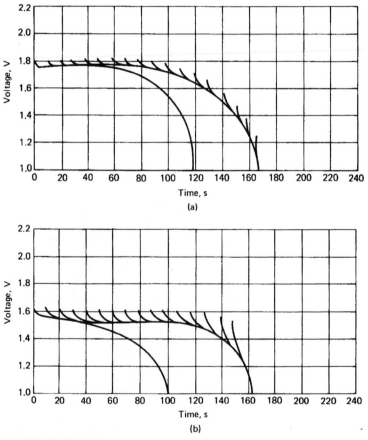

FIGURE 24.10 High-rate pulse performance at 10*C* discharge rate for VRLA cylindrical units. (*a*) 25°C. (*b*) −20°C. The upper curves are pulsed, the lower curves are continuous discharges.

voltage drops. During the rest period, the acid in the bulk of the solution diffuses into the electrode pores to replace the acid which has been used up. The cell voltage then increases as acid equilibrium is established. During a pulse discharge the acid is allowed to equilibrate between pulses, it is not depleted as quickly and the total cell capacity is increased.

The ability of the cell to provide high discharge currents and maintain usable voltage is illustrated in Fig. 24.11. The effect of the discharge rate on the midpoint voltage is shown at both 22 and $-20°C$. (The voltage indicated was measured midway through the discharge.)

The curves in Fig. 24.12 illustrate the maximum power that can be delivered as a function of the discharge rate at room temperature and at $-20°C$. The maximum power obtainable increases as the temperature increases.

Discharge Level.   As with all rechargeable batteries, discharging the VRLA battery beyond the point at which 100% of the capacity has been removed can shorten the life of the battery or impair its ability to accept a charge.

The voltage point at which 100% of the usable capacity of the cell has been removed is a function of the discharge rate, as shown in the upper envelope of the curve of Fig. 24.13. The lower curve shows the minimum voltage level to which the battery may be discharged with no effect on recharging capability. For optimum life and charge capability, the cell should be disconnected from the load at the voltages within the gray area between the two curves.

Under these "overdischarge" conditions, the sulfuric acid electrolyte can be depleted of the sulfate ion and become mainly water, which can create several problems. A lack of sulfate ions as charge conductors will cause the cell impedance to appear high and little charge current to flow. Longer charge times or alteration of the charge voltage may be required before normal charging may resume.

Another potential problem is the solubility of lead sulfate in water. In a severe deep-discharge condition, the lead sulfate present at the plate surfaces can go into solution in the aqueous electrolyte. Upon recharge, the water and the sulfate ion in the lead sulfate convert to sulfuric acid, leaving a precipitate of lead metal in the separator. This lead metal can result in dendritic short circuits between the plates and subsequently cell failure.

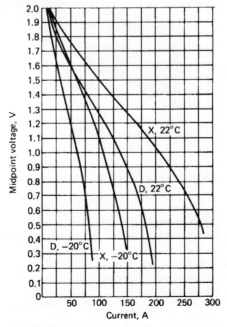

FIGURE 24.11   Effect of discharge rate on cell midpoint voltage of VRLA batteries.

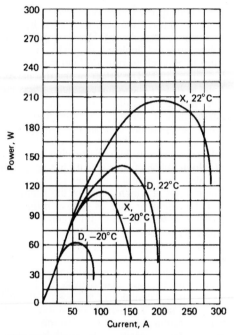

FIGURE 24.12 Instantaneous maximum peak power of VRLA batteries at 22 and −20°C.

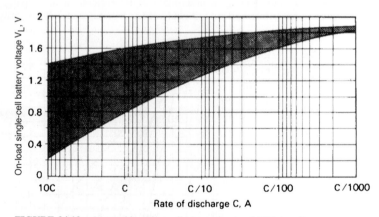

FIGURE 24.13 Acceptable voltage discharge levels of VRLA cells.

Storage Characteristics. Most batteries lose their stored energy when allowed to stand on open circuit due to the fact that the active materials are in a thermodynamically unstable state.[12] The rate of self-discharge is dependent on the chemistry of the system and the temperature at which it is stored. The cylindrical VRLA battery is capable of long storage without damage, as can be seen in Fig. 24.14, which plots the maximum storage time against the storage temperature. This curve shows the maximum number of days at any given temperature from 0 to 70°C for the cell to self-discharge from 2.18 down to 1.81 V open circuit.

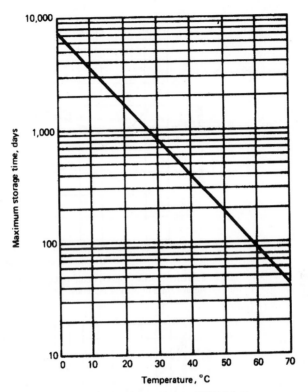

FIGURE 24.14 Storage characteristics of VRLA battery.

The cell should not be allowed to self-discharge below 1.76 V, because the recharge characteristics of the cell change appreciably and the cycle life cannot be predicted accurately. The capacity of these cells can be restored by recharging; however, the first charge on a cell that has been allowed to self-discharge down to 1.76 V will take longer than normal, and the first discharge will generally not deliver rated capacity. Subsequent cycles, however, will show an increase in the cell capacity to rated value.

It is important to recognize that the self-discharge rate of the VRLA battery is nonlinear; thus the rate of self-discharge changes as the state of charge of the cell changes. When the cell is in a high state of charge, that is, 80% or greater, the self-discharge is very rapid. The cell may discharge from 100 to 90% at room temperature in a matter of a week or two. Conversely, at the same temperature it may take 10 weeks or longer for the same cell to self-discharge from 20% state of charge down to 10% state of charge. Figure 24.15 is a curve of open-circuit voltage versus percent remaining storage time, which shows the nonlinearity of the self-discharge reaction.

By the use of Figs. 24.14 and 24.15, the number of days of storage which remain before a battery must be recharged can be calculated. As an example, if a battery has an open circuit voltage of 2.00 V, the state of charge, as determined from Fig. 24.15, is 37%. From Fig. 24.15 (again at an open-circuit voltage of 2.00 V), the remaining storage time is 82%. Figure 24.16 shows that at 20°C, the battery can be stored for a total of 1200 days before it must be recharged. Therefore the remaining storage time is 82% of 1200 or 984 days. This is the number of days at 2.00 V open-circuit voltage that a battery can be stored before it will reach 1.76 V and must be recharged.

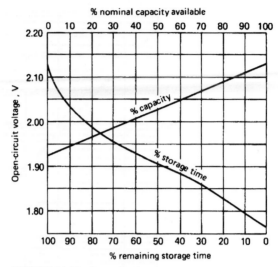

FIGURE 24.15   Open-circuit voltage vs. percent remaining storage time of VRLA batteries.

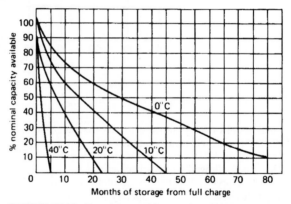

FIGURE 24.16   Remaining usable capacity of VRLA batteries after storage.

Figure 24.16 is a curve of the remaining usable capacity in a VRLA battery versus months of storage at various temperatures. This curve is convenient in determining the approximate remaining capacity after a given storage time at a particular temperature.

Life.   The life of all rechargeable battery systems is variable, depending on the type of use, environment, cycling, and charge to which the battery is subjected during its life.

1. *Cycle life:* Figures 24.17 and 24.18 illustrate the effect of several of the factors which control the cycle life. Figure 24.17 shows the effect of the charging voltage and demonstrates the need to select the proper charging voltage for a particular cycle regime. The figure is somewhat misleading, however, in that it would indicate that a low charging voltage, say 2.35 V, would yield a low cycle life, and this is not true. For example, for an application

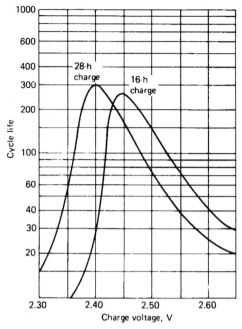

FIGURE 24.17   Effect of charge voltage on cycle life of VRLA batteries at various charge times at 25°C and approximately 100% DOD. End of cycle life—80% of rated capacity; discharge rate—$C/5$ to 1.6 V.

where the cells would be used about three times a week and left on charge the rest of the time, 2.35 V would be quite adequate. Most of the capacity can be returned to the cells within 16 h. The occasional long charge periods would maintain capacity and thus optimize the total cycle life. Generally, 2.45 V per cell is a better charging voltage for regimes of about one cycle per day.

Figure 24.18 shows the general effect of the depth of discharge (DOD) on the cycle life; typically at 100% DOD, 200 cycles are characteristic. It demonstrates that high cycle life can be achieved by slightly oversizing the battery for the application to reduce the depth of discharge. Figure 24.19 is a curve of capacity versus cycles for a 2.5-Ah cell (2.35 Ah at 5-h rate) cycled at 1 cycle per day at a $C/5$ discharge rate to 1.6 V per cell and an 18-h charge at 2.5-V constant voltage. The cell takes from 20 to 25 cycles to achieve rated capacity, exceeds rated capacity, and then begins to fall off slowly. The initial increase in capacity is a function of forming the cell.

   2. *Float life:* The expected float life of the VRLA battery is greater than 8 years at room temperature, arrived at by using accelerated testing methods, specifically, at high temperatures.

The primary failure mode of the VRLA battery can be defined as growth of the positive plate. Because this growth is the result of chemical reactions within the cell, the rate of growth increases with increasing temperature. In Fig. 24.20 the float life is plotted against temperature. The solid lines represent data from float-life tests performed at two float voltages, 2.3 and 2.4 V per cell. This graph can be used to determine the expected float life at various temperatures. End of life is defined as the failure of the cell to deliver 80% of rated capacity.

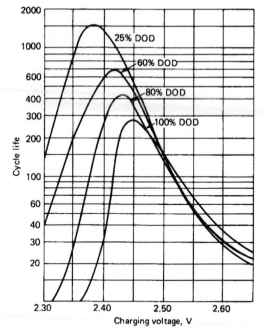

FIGURE 24.18 Effect of depth of discharge on cycle life of VRLA batteries as a function of charging voltage at 25°C, 16-h charge.

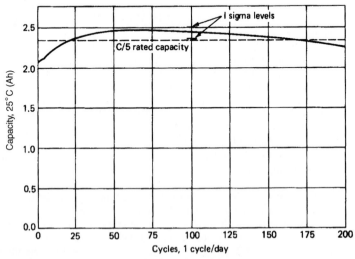

FIGURE 24.19 Effect of cycling on cell capacity, VRLA cell, $C/5$ discharge rate. Rated at 2.35 Ah at $C/5$ rate.

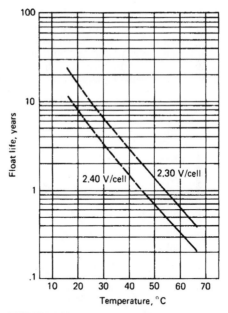

FIGURE 24.20 Float life of VRLA batteries.

24.4.2 Performance Characteristics—VRLA Prismatic Batteries

Typical discharge curves of the prismatic lead-acid battery at 20°C are shown in Fig. 24.21 which illustrates the high-rate capability and the flat discharge profile of this battery system. The cells are usually rated at the 20-h or .05C rate, and the figure shows the performance of the battery at discharge currents, expressed in terms of the $C/20$ rate, to various end voltages. For example, for a cell rated at 5 Ah at the 20-h rate, the $C/20/5$ rate is

$$\frac{C/20}{5} = 0.2 \; C/20 = (0.2)(5) = 1 \text{ A}$$

The $C/20/5$ rate is 1.0 A. The $C/20$ rate is 0.25 A.

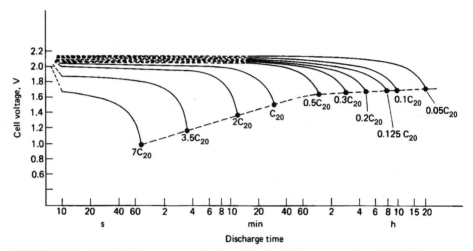

FIGURE 24.21 Typical discharge curves of prismatic lead-acid battery at various $C/20$ discharge rates. 20°C. (*Courtesy of Johnson Controls.*)

Figure 24.22 shows the effect of discharge rate and temperature on the delivered capacity of the battery. A fully charged battery can operate over a very wide temperature range. These data are summarized in Fig. 24.23 which plots the service life of the sealed prismatic lead-acid battery at various temperatures and discharge rates (C-rates). Manufacturers' data should be obtained for specific performance characteristics as the characteristics may vary depending on the size and design of the battery.

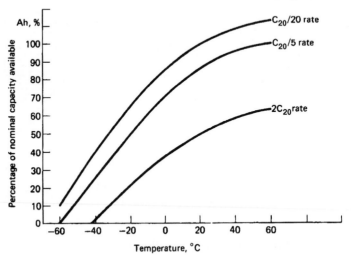

FIGURE 24.22   Effect of temperature and discharge rate on capacity of prismatic lead-acid battery.

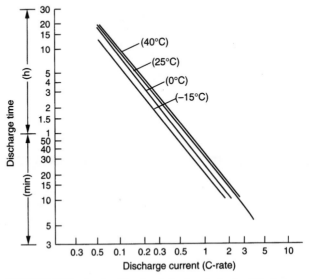

FIGURE 24.23   Service life of sealed prismatic lead-acid batteries.

The typical self-discharge characteristics of the sealed prismatic lead-acid battery are shown in Fig. 24.24. Figure 24.24*a* shows the capacity retention after storage for different periods of time at several temperatures. Figure 24.24*b* shows the time it takes for the capacity to decrease to 50% of the rated capacity throughout the temperature range. It loses relatively little capacity in storage compared with the more conventional antimonial-lead grid battery. The rate of self-discharge is about 4% per month. Once the battery has self-discharged to the level where it has about a 50% state of charge, recharging the battery is advisable. The residual capacity can be estimated by measuring the open-circuit voltage, as shown in Fig. 24.25.

The service-life characteristics of the gelled lead-acid battery on float service are shown in Fig. 24.26. The cycle-life characteristics, to different depths of discharge (DOD), are shown in Fig. 24.27. A characteristic of the battery is that the capacity increases in the early stages of life and reaches a maximum at about 10 to 30 cycles. It will also gain in capacity

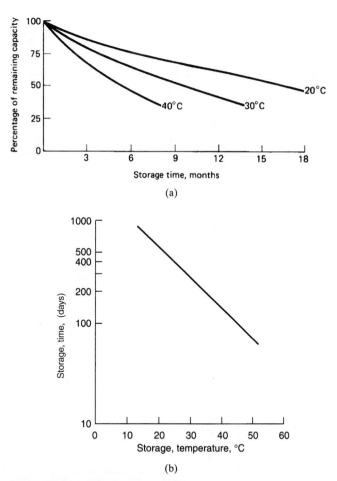

FIGURE 24.24 Characteristics of prismatic lead-acid battery. (*a*) Self-discharge at various temperatures. (*b*) Storage time and temperature to 50% of nominal capacity.

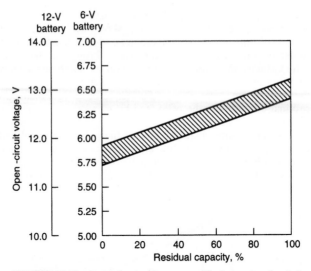

FIGURE 24.25 Open-circuit voltage vs. residual capacity of sealed prismatic lead-acid batteries at 25°C.

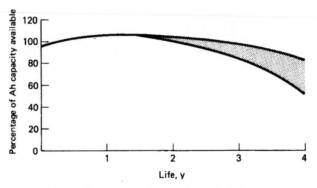

FIGURE 24.26 Performance of prismatic lead-acid battery on float service at 20°C. Float voltage = 2.25 − 2.3 V per cell.

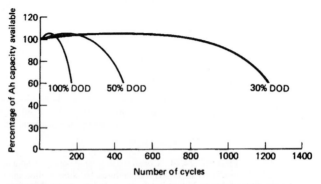

FIGURE 24.27 Cycle service life of prismatic lead-acid battery in relation to depth of discharge (DOD) at 20°C.

while on extended charge, as in float service where discharges are infrequent. The battery also does not exhibit a "memory" effect, as is the case with the sealed nickel-cadmium battery which may become conditioned when used for short periods, and may not be able to deliver full capacity when required.

The following are the recommended temperature ranges for the operation and storage of the prismatic lead-acid battery: discharge, -15 to $50°C$; charge, 0 to $40°C$; and storage, -15 to $40°C$.

24.4.3 Performance Characteristics—Thin Prismatic Batteries

The performance characteristics of the thin prismatic batteries are similar to those of the other sealed units. Their advantage becomes more apparent in the fabrication of battery packs as the minimum of void space required in the pack enhances the energy density.

The characteristic discharge curves for the thin sealed lead-acid batteries are shown in Fig. 24.28 and the effect of temperature on discharge capacity at various current drains is plotted in Fig. 24.29. These data are summarized in Fig. 24.30, which plots the service hours of a typical battery. The performance can be approximated from this monograph.

The loss of capacity with storage at different temperatures is also shown in Fig. 24.30. As with all lead-acid batteries, they should be recharged when the charge retention is below 50%. The cycle life of the thin prismatic batteries is similar to the one shown for the VRLA cells in Fig. 24.27.

The polarization characteristics of the thin prismatic VRLA batteries referred to in Fig. 24.28 are shown in Fig. 24.31a and b.[9] The data were taken with batteries in the horizontal (pancake) position. This data illustrates the effect of state of charge on the current-voltage relationship for both charge and discharge conditions.

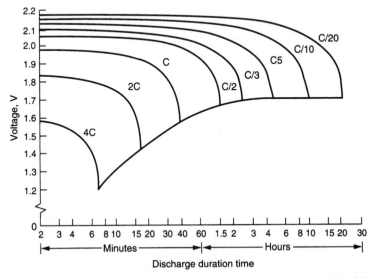

FIGURE 24.28 Characteristic discharge curves for thin prismatic single-cell sealed lead-acid batteries at 20°C. (*Courtesy of Portable Energy Products, Inc.*)

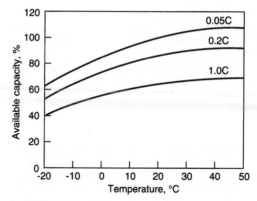

FIGURE 24.29 Effect of temperature on discharge performance of thin prismatic lead-acid batteries.

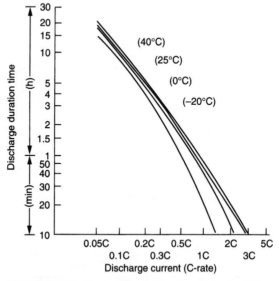

FIGURE 24.30 Service life of thin prismatic lead-acid batteries.

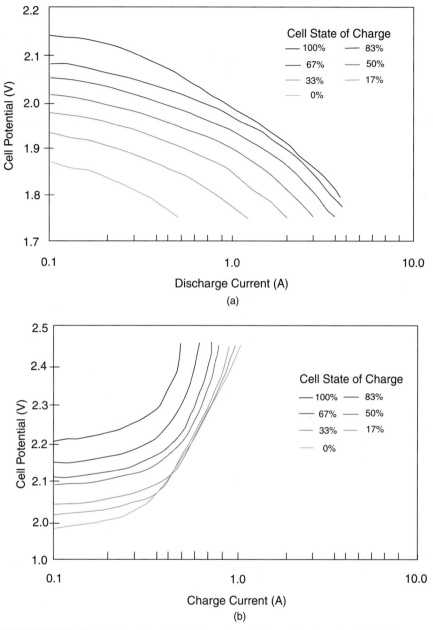

FIGURE 24.31 Polarization characteristics vs. state of charge for prismatic lead acid battery. (*a*) during discharge; (*b*) during charge.

24.4.4 Other Cell Designs

The lead-acid battery system still has a significant share of the portable rechargeable battery market. Its low cost is one of its advantages, but its energy density and power density and cycle life are inferior to the potential of the newer advanced systems.

New designs are being investigated to improve these characteristics. This work is directed towards high surface area electrodes with thin active material layers using such materials and designs as light-weight fiber-glass reinforced lead wire grids, thin metal foils, bipolar plates, forced flow electrolyte systems, and unique cell assemblies. Improvement in specific energy and energy density to 40 Wh/kg and 100 Wh/L with the possibility of rapid recharge have been attained.[13,14] Figure 24.32a compares the specific energy and specific power on a cell basis, of the newer designs with conventional technology. Figure 24.32b shows the peak power capability (W/kg) as a function of the depth-of-discharge.

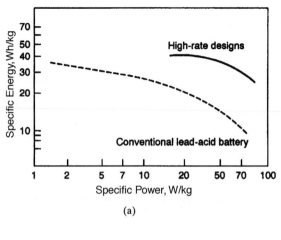

(a)

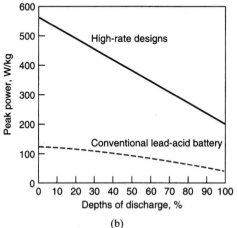

(b)

FIGURE 24.32 Comparison of characteristics of conventional lead-acid and experimental high-rate design batteries. (*a*) Specific energy vs. specific power. (*b*) Peak power vs. depth of discharge. (*Proceedings of International Seminar on Primary and Secondary Batteries, Boca Raton, Fla., 1994.*)

24.5 *CHARGING CHARACTERISTICS*

24.5.1 General Considerations[15]

Charging the VRLA battery, like charging other secondary systems, is a matter of replacing the energy depleted during discharge. Because this process is inefficient, it is necessary to return more than 100% of the energy removed during discharge. The amount of energy necessary for recharge depends on how deeply the battery was discharged, the method of recharge, the recharge time, and the temperature. The overcharge required in the lead-acid battery is associated with the generation of gases and corrosion of the positive-grid materials. In conventional flooded lead-acid batteries the gases generated are released from the system, resulting in a loss of water, which is replenished by maintenance. The VRLA battery incorporates the gas recombination principle which allows the oxygen generated at normal overcharge rates to be reduced at the negative plate, eliminating oxygen outgassing. Hydrogen gas generation has been substantially reduced by the use of nonantimonial lead grid material. The corrosion of the positive grid has been reduced by the use of pure or special alloy lead. Also, the effects of corrosion of the positive grid have been minimized by the element construction.

Charging can be accomplished by various methods. Constant-voltage charging is the conventional method for lead-acid batteries and is also preferred for VRLA batteries. However, constant current, taper current, and variations thereof can also be used.

VRLA batteries during cycling and float charge can become unbalanced in that the corrosion rate of the positive becomes unequal to the self-discharge rate of the negative. A suggested remedy[16] is to utilize a catalyst in the vent space of the cell to restore balance by recombining hydrogen and oxygen.

24.5.2 Constant-Voltage Charging

Constant-voltage charging is the most efficient and fastest method of charging a VRLA battery. Figure 24.33 shows the recharge times at various charge voltages for a cell discharged to 100% of capacity. The charger required to achieve these times at given voltages must be capable of at least the $2C$ rate. If the constant-voltage charger used has less than the $2C$ rate of charge capability, the charge times should be lengthened by the hourly rate at which the charger is limited; that is, if the charger is limited to the $C/10$ rate, then 10 h should be added to each of the charge voltage-time relationships; if the charger is limited to the $C/5$ rate, then 5 h should be added, and so on. There are no limitations on the maximum current imposed by the charging characteristics of the battery.

Figure 24.34 is a set of curves of charge current versus time for 2.5-Ah batteries charged by a constant voltage of 2.45 V with chargers limited to 2-, 1-, and 0.3-A currents. As shown, the only difference in these three charges is the length of time necessary to recharge the battery.

Figure 24.35a shows the charge rate versus time and the percent of previous discharge capacity returned versus time at 2.35-V constant voltage for a battery discharged at the $C/2.5$ rate to 80% DOD. The time necessary to recharge the battery to 100% of the previous discharge capacity is 1.5 h. The charger used was capable of an output in the $4C$ range and of voltage regulation at the terminals of better than 0.1%. The initial high inrush of current caused internal cell heating, which enhanced charge acceptance. If the battery had been more deeply discharged, the length of time to reach 100% capacity would have increased. The current decays exponentially with time, and after 3 h on charge, the current has decayed to a low level. At 2.35 V, the cell will accept whatever current is necessary to maintain capacity.

Figure 24.35b is a similar plot but for a battery charged at 2.50 V constant voltage. All of the capacity taken out on the previous discharge is returned in approximately one-half hour.

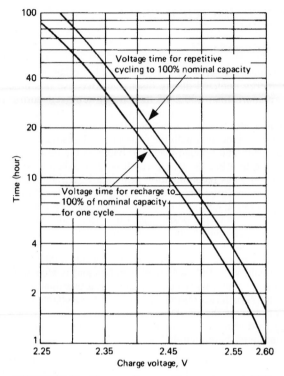

FIGURE 24.33 Charge voltage vs. time on charge at 25°C.

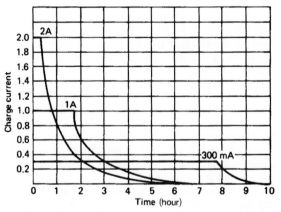

FIGURE 24.34 Charge current vs. time at 2.45-V constant voltage with various current limits (2.5-Ah battery, $C/10$ rate).

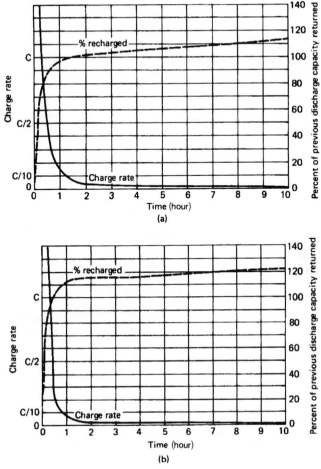

FIGURE 24.35 Charge rate and percent recharged vs. time during constant-voltage charge at 25°C. (*a*) 2.35 V. (*b*) 2.50 V.

24.5.3 Fast Charging[17]

A fast charge is defined as a method of charge that will return the full capacity in less than 4 h. However, many applications require 1 h or less.

Unlike conventional flooded lead-acid batteries, the VRLA design uses a starved electrolyte system where most of the electrolyte is contained within a highly retentive separator, which then creates the starved plates necessary for homogeneous gas-phase transfer. The gassing problem inherent in conventional lead-acid cells is not evident with this system, as the oxygen given off on overcharge is able to recombine within the VRLA battery. The large surface area of the thin plates used in some VRLA batteries reduces the current density to a level far lower than normally seen in the fast charge of conventional lead-acid batteries, thereby enhancing the fast-charge capabilities.

Figure 24.36 shows the charge rate or the current the VRLA battery can accept for a 1-h charge at three different voltages. The charger has a capability of delivering up to a $5C$ charge rate. The battery has a high charge acceptance at the beginning of the charge time: in fact, in the case of the 2.55-V-per-cell charge, the cell accepted the full current capability of the charger for the first 3 to 4 min. In the case of the 2.7-V-per-cell charge, there was a considerable amount of overcharging starting at 30 min, which caused internal heating and a consequent increase in charge current.

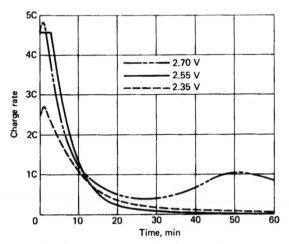

FIGURE 24.36   Charge rate vs. time for three charge voltages.

Figure 24.37 is a set of curves of normalized charge efficiency versus time in minutes for the three different voltages. This efficiency figure was calculated by dividing the total Ampere-hour capacity returned by the previous discharge capacity removed. On the 2.55 V charge, 100% of the capacity removed on the previous cycle was returned in 15 min. With the 2.7 V charge, a 60% overcharge was returned at the end of the 60-min charge.

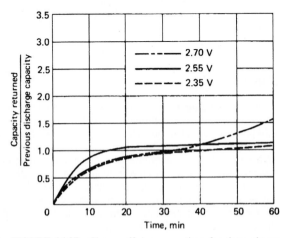

FIGURE 24.37   Charge efficiency vs. time for three charge voltages.

Figure 24.38 plots the discharge time in minutes vs. cycle number for the three charge voltages. Also, a set of reference batteries was charged at 2.5 V constant voltage for 16 h and discharged at the 1*C* rate. This reference curve is displayed to show the expected capacity at the 1*C* rate. It can be seen from these data that the 2.55-V-per cell curve most closely approximates the reference line. The battery charged at 2.7 V per cell received too much overcharge and, therefore, the capacity degraded after 15 cycles. The battery charged at 2.25 V achieved a value of approximately 75% of the reference and continued to cycle at that level.

These data show that the thin-plate VRLA battery can be fast-charged to 100% of the rated capacity in less than 1 h. A constant-voltage charger set at 2.5 to 2.55 V per cell and capable of the 3*C* to 4*C* rate of charge is preferred. It should be noted, as discussed, that charging at 2.7 V per cell for prolonged periods will damage the battery.

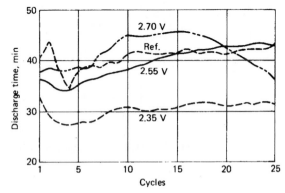

FIGURE 24.38   Effect of cycling on discharge time for three charge voltages.

24.5.4 Float Charging

When the VRLA battery is to be float-charged as in a standby application, the constant-voltage charger should be maintained between 2.3 and 2.4 V for maximum life. Continuous charging at greater than 2.4 V per cell is not recommended because of accelerated grid corrosion. Figure 24.39 gives the approximate values of voltage a battery will attain when float-charged at 25 and 65°C, or the charge rate a battery will accept if it has been charged for a sufficient period of time so that it is in a state of overcharge equilibrium. These curves can also be used to determine the approximate value of continuous constant current (trickle charge) that will maintain the proper float voltage. As an example, if a battery were trickle-charged at the 0.001*C* rate, its average voltage per cell on overcharge would be 2.35 V at 25°C. Conversely, if a cell were constant-voltage-charged at 2.35 V, its overcharge rate would be 0.001*C*.

High temperatures accelerate the rate of the reactions which reduce the life of a battery. At increased temperatures, the voltage necessary for returning full capacity in a given time is reduced because of the increased reaction rates within the battery. To maximize life, a negative charging temperature coefficient of approximately −2.5 mV/°C per cell is used at temperatures significantly different from 25°C. Figure 24.40 shows the recommended charging voltage at various temperatures for a sealed battery float-charged at 2.35 V per cell at 25°C. It is obvious from this curve that at extremely low temperatures, a significantly greater temperature coefficient than −2.5 mV/°C is required to achieve full recharge of the cell. Figure 24.40 also shows the voltage compensation under cycling service. The voltage compensation keeps the charging current at about the same value that it would be at 25°C when

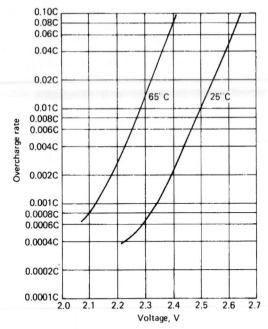

FIGURE 24.39 Overcharge current and voltage.

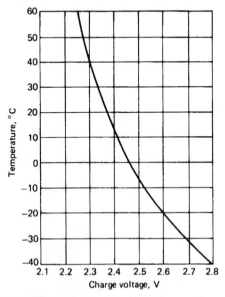

FIGURE 24.40 Recommended charge voltage at various temperatures (temperature compensation.)

the battery temperature is different. Temperature compensation of the charging voltage prevents thermal runaway of the batteries when they are used at high temperatures and assures adequate charging if the battery temperature is low.

When trickle-charging, it may be necessary to increase the charge rate at higher temperatures to maintain the proper float voltage. From Fig 24.39, it can be seen that for a battery trickle-charged at the $0.001C$ rate at 25°C, the float voltage would be 2.34 V. However, at the same rate at 65°C, the float voltage would be approximately 2.12 V, which is below the open-circuit voltage of the cell. At 65°C the trickle-charge current would need to be increased to approximately $0.01C$ to maintain the proper float voltage.

24.5.5 Constant-Current Charging

Constant current is another efficient method of charging the VRLA battery. Constant-current charging is accomplished by the application of a nonvarying constant-current source. This charge method is especially effective when several cells are charged in series, since it tends to eliminate any charge imbalance in a battery. Constant-current charging charges all cells equally because it is independent of the charging voltage of each cell in the battery. Figure 24.41 shows a family of curves of battery voltage versus percent capacity of previous discharges returned at different constant-current charging rates. As shown by these curves at different charge rates, the voltage increases sharply as the full charge state is approached. This increase in voltage is caused by the plates going into overcharge when most of the

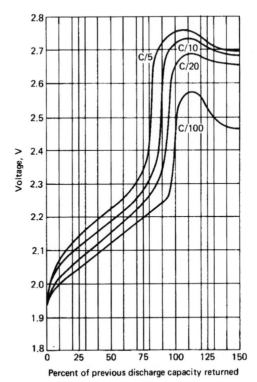

FIGURE 24.41 Voltage curves for batteries charged at various constant-current rates at 25°C.

active material on the plates has been converted from lead sulfate to lead on the negative plate and to lead dioxide on the positive plate. The voltage increase will occur at lower states of charge when the cell is being charged at higher rates. This is because at the higher constant-current charge rates the charging efficiency is reduced. The voltage curves in Fig. 24.41 are somewhat different from those for a conventional lead-acid battery due to the effect of the recombination of gases on overcharge within the system. The VRLA battery is capable of recombining the oxygen produced on overcharge up to the $C/3$ rate of constant-current charge. At higher rates the recombination reaction is exceeded by the rate of gas generation.

While constant-current charging is an efficient method, continued application at rates above $C/500$, after the battery is fully charged, can be detrimental to life. At overnight charge rates ($C/10$ to $C/20$), the large increase in voltage at the nearly fully charged state is a useful indicator for terminating or reducing the rates for a constant-current charger. If the rate is reduced to $C/500$, the battery can be left connected continuously and give 8 to 10 years of life at 25°C. Figure 24.42 is a plot of voltage versus time for a battery charging at the $C/15$ rate of constant current at 25°C. This battery had previously been discharged to 100% depth of discharge at the $C/5$ rate. This curve shows that the battery is not fully charged at the time the voltage increase occurs and must receive additional charging. If a battery is to be charged with constant current at higher than room temperature, then some temperature compensation must be built into the voltage-sensing network. As explained in Sec. 24.5.4, at higher temperatures and given charging rates, the battery voltage on overcharge is reduced. Therefore the rise in voltage at close to full charge will be somewhat depressed.

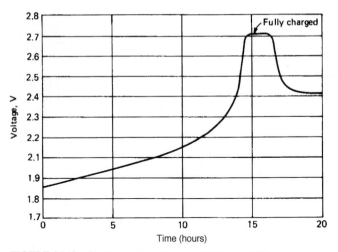

FIGURE 24.42   Constant-current charge at $C/15$ rate, 25°C.

24.5.6 Taper-Current Charging

Although taper-current chargers are among the least expensive types of chargers, their lack of voltage regulation can be detrimental to the cycle life of any type of battery. The VRLA battery can withstand charge voltage variations, but some caution in using taper-current chargers is recommended. A taper-current charger contains a transformer for voltage reduction and a half- or full-wave rectifier for converting from alternating to direct current. The output characteristics are such that as the voltage of the battery increases during charge, the charging current decreases. This effect is achieved by use of the proper wire size and the turns ratio. Basically, the turns ratio from primary to secondary determines the output voltage at no load, and the wire size in the secondary determines the current at a given voltage. The transformer is essentially a constant-voltage transformer which depends entirely on the AC line voltage regulation for its output-voltage regulation. Because of this method of voltage regulation, any changes in input line voltage directly affect the output of the charger. Depending on the charger design, the output-to-input voltage change can be more than a direct ratio; for example, a 10% line-voltage change can produce a 13% output-voltage change.

When considering the cost advantage of using a half-wave rectifier versus a full-wave rectifier in a taper-current charger, it should be noted that the half-wave rectifier supplies a 50% higher peak-to-average-voltage ratio than the full-wave rectifier. Therefore the total life of the battery for a given average charge voltage can be reduced for the half-wave type of charger because of the higher peak voltages. A DC ripple can lead in time to decreased performance through degradation of the active material and the grid. An AC ripple can be a more significant factor in premature battery failure, especially in float or uninterruptible power systems. The repeated charging and discharging of the battery shortens the battery life through heat generation and corrosion.

There are several charging parameters which must be met. The parameter of main concern is the recharge time to 100% nominal capacity for cycling for applications. This parameter can primarily be defined as the charge rate available to the battery when each cell is at 2.2 and 2.5 V. The charge voltage at which approximately 50% of the charge has been returned to the battery at normal charge rates of $C/10$ to $C/20$ is 2.2-V-per-cell; the 2.5 V per cell point represents the voltage at which the battery is in overcharge. Given the charge rate at 2.2 V, the recharge time for a taper-current charger can be defined by

$$\text{Recharge time} = \frac{1.2 \times \text{capacity discharged previously}}{\text{charge rate at 2.2 V}}$$

It is recommended that the charge rate at 2.5 V be between $C/50$ maximum and $C/100$ minimum to ensure that the battery will be recharged at normal rates and will not be severely overcharged if the charger is left connected for extended time periods.

Figure 24.43 is a set of output voltage versus current curves for a typical 2.5-Ah-battery taper-current charger. The three curves show the change in output with a variation in input voltage from 105 to 130 V AC. This particular charger, at 120-V AC input, will charge a three-cell D-sized battery (rated at 2.5 Ah) which had been previously discharged to 100% depth of discharge in 30 h by the following equation:

$$\text{Recharge time} = \frac{1.2 \times 2.5}{0.100 \text{ A}} = 30 \text{ h}$$

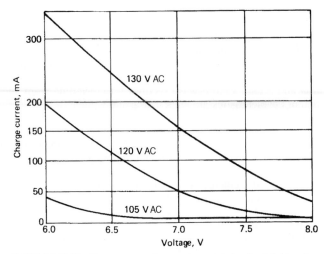

FIGURE 24.43   Taper-current charger characteristics for 2.5-Ah cylindrical three-cell VRLA battery.

24.5.7 Parallel/Series Charging

VRLA batteries can be charged or discharged in parallel. When more than four strings of cells are paralleled, it is advisable to use diodes in both the charge and the discharge path in the circuit. The discharge diodes prevent a battery from discharging into a paralleled battery should a cell short-circuit in the battery. The charge diodes, in conjunction with the fuse, will prevent a battery with a short-circuited cell from accepting all the charge current from the charger and subsequently prevent the other paralleled batteries from being fully charged. The fuse should be sized by dividing the maximum charge current by the number of batteries in parallel and multiplying this value times 2. This should result in the fuse opening on charge in a parallel string which has a short-circuited cell.

When float-charging many cells in series, 12 or more, for example, it is advantageous to use a trickle charge of $C/500$ maximum in parallel with the float charger. This trickle charge will tend to balance all cells in the battery by driving a continuous trickle charge equally through all cells.

24.5.8 Charge-Current Efficiency

Charge-current efficiency is the ratio of the current that is actually used for electrochemical conversion of the active material from lead sulfate to lead and lead dioxide to the total current supplied to the cell on recharge. The current which is not used for charging is consumed in parasitic reactions within the cell such as corrosion and gas production.

The charging efficiency is high for a VRLA battery. The distinctly high ratio of plate surface area to Ampere-hour capacity allows for higher charging rates and, therefore, efficient charging.

Charge-current efficiency is a direct function of the state of charge. The charge efficiency of a battery is high until it approaches full charge, at which time the overcharge reactions begin and the charge efficiency decreases. Obviously, past the point of full recharge, the efficiency falls to zero.

Figure 24.44 is a curve of charge current efficiency versus voltage at various constant voltages. Increasing voltage decreases the efficiency because of increased parasitic currents. The efficiency shows a marked decrease below the open-circuit voltage, typically 2.15 to 2.18 V, because the charge voltage is not high enough to support the charging reaction.

Figure 24.45 is a curve of efficiency versus log rate at various constant-current charge rates. As can be seen from the curve, at rates up to $C/10$ the efficiency approaches 100%. At higher rates there is a decrease in efficiency because, as the cell approaches the fully charged state, the surfaces of the plates become fully charged. This increases the charging reaction rates and results in increased voltages and gassing. At low charge rates, the efficiency drops because the charge current is equivalent to the parasitic currents and the battery voltage approaches the open-circuit value.

Figure 24.46 shows the charge acceptance during charge at various temperatures and charge rates.

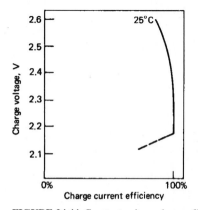

FIGURE 24.44 Constant-voltage charge efficiency.

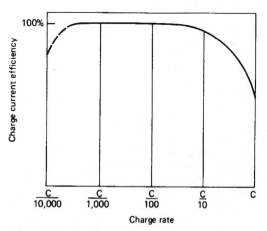

FIGURE 24.45 Constant-current charge efficiency.

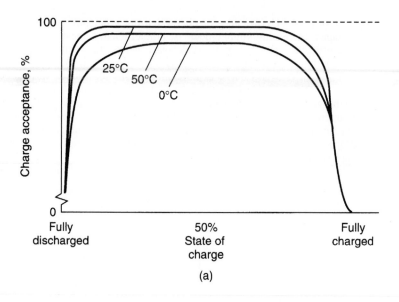

(a)

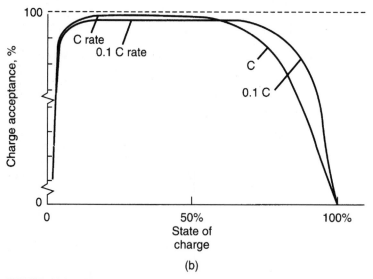

(b)

FIGURE 24.46 Charge acceptance of sealed lead-acid batteries. (*a*) At various temperatures. (*b*) At various charge rates.

24.6 SAFETY AND HANDLING

Two primary considerations relative to the application of VRLA batteries should be recognized to ensure that their usage is safe and proper: gassing and short-circuiting.

24.6.1 Gassing

Lead-acid batteries produce hydrogen and oxygen gases internally during charging and overcharging. These gases are released in an explosive mixture from conventional lead-acid batteries and therefore must not be allowed to accumulate in a confined space. An explosion could occur if a spark were introduced.

The VRLA battery, however, operates on 100% recombination of the oxygen gas produced at recommended rates of charging and overcharging, and so there is no oxygen outgassing. During normal operation, some hydrogen gas and also some carbon dioxide gas are given off. The hydrogen outgassing is essential with each cycle to ensure continued internal chemical balance. The lead grid construction of the VRLA battery minimizes the amount of hydrogen gas produced. Carbon dioxide is produced by oxidation of organic compounds in the cell.

The minute quantities of gases which are released from the VRLA battery with recommended rates of charge and overcharge will normally dissipate rapidly into the atmosphere. Hydrogen gas is difficult to contain in anything but a metal or glass enclosure, that is, it can permeate a plastic container at a relatively rapid rate. Because of the characteristics of gases and the relative difficulty in containing them, most applications will allow for their release into the atmosphere. However, if the VRLA battery is being designed into a gastight container, precautions must be taken so that the gases produced can be released to the atmosphere. If hydrogen is allowed to accumulate and mix with the atmosphere at a concentration of between 4 and 79% (by volume at standard temperature and pressure), an explosive mixture would be present which would be ignited in the presence of a spark or flame.

Another consideration is the potential failure of the charger. If the charger malfunctions, causing higher-than-recommended charge rates, substantial volumes of hydrogen and oxygen will be vented from the battery. This mixture is explosive and should not be allowed to accumulate. Adequate ventilation is required. Therefore the VRLA battery should never be operated in a gastight container. The batteries should never be totally encased in a potting compound since this prevents the proper operation of the venting mechanism and free release of gas. Furthermore, considerable pressure can build up in a gas-tight container. This can occur during storage because of the continuous generation of carbon dioxide gas. Such pressure is further compounded during charging by the generation of hydrogen.

24.6.2 Short-circuiting

These batteries have low internal impedance and thus are capable of delivering high currents if externally short-circuited. The resultant heat can cause severe burns and is a potential fire hazard. Particular caution should be used when any person working near the open terminals of cells or batteries is wearing metal rings or watchbands. Inadvertently placing these metal articles or tools across the terminals could result in severe skin burns.

24.7 BATTERY TYPES AND SIZES

A listing of VRLA cylindrical batteries is given in Table 24.2. The performance of these units at various current drains at 25°C is given in Fig. 24.47. A number of multicell batteries are available that use these cells in various series/parallel configurations.

Table 24.3 and 24.4 list some of the typical VRLA prismatic lead-acid batteries that are manufactured. Information on other manufacturers' products can be obtained by consulting their websites, such as yuasastationary.com, panasonic.com and www.hoppecke.com. Unlike some of the other types of lead acid batteries, there is no standardized list of sizes. Hence sizes, weights, and capacity ratings may vary from manufacturer to manufacturer.

Table 24.5 lists some of the thin prismatic VRLA batteries. These are generally used in multicell battery packs which can be built in a variety of configurations, depending on the equipment requirements. Typical discharge performance characteristics are shown in Fig. 24.30.

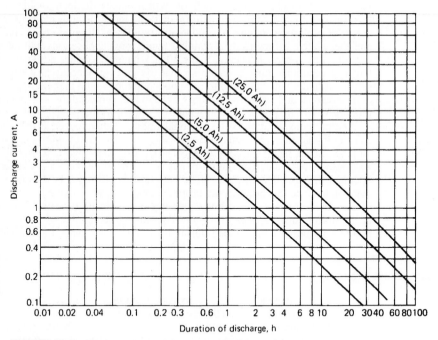

FIGURE 24.47 Discharge times of four types of VRLA cylindrical batteries at 25°C.

TABLE 24.3 Specifications for Typical VRLA Lead-Calcium Batteries

	LS 12-25	LS 6-50	LS 12-55	LS 12-80	LS 12-100	LS 6-125	LS 6-200	LS 4-300	LS 2-600
Nominal voltage	12 volts	6 volts	12 volts	12 volts	12 volts	6 volts	6 volts	4 volts	2 volts
Number of cells	6/unit	3/unit	6/unit	6/unit	6/unit	3/unit	3/unit	2/unit	1/unit
Rated 8-hr capacity (Ampere-hours to 1.75 Vpc)	25 Ah to 10.5 volts	50 Ah to 5.25 volts	52 Ah to 10.5 volts	80 Ah to 10.5 volts	100 Ah to 10.5 volts	123 Ah to 1.75 volts	200 Ah to 5.25 volts	300 Ah to 3.5 volts	600 Ah to 1.75 volts
Rated 15-min capacity (kiloWatts to 1.67 Vpc) per cell	0.092	0.185	0.172	0.275	0.344	0.430	0.688	1.032	2.063
Internal resistance per cell	0.10 Ohms	0.003 Ohms	0.00157 Ohms	0.00094 Ohms	0.000786 Ohms	0.00062 Ohms	0.000393 Ohms	0.000262 Ohms	0.000131 Ohms
Short circuit current	1155 A	2310 A	1274 A	2128 A	2545 A	3180 A	5089 A	7634 A	15267 A
Unit height	7.11 in (181 mm)	7.11 in (181 mm)	9.20 in (234 mm)	9.20 in (234 mm)	9.20 in* (234 mm)	9.95 in (253 mm)	9.20 in* (234 mm)	9.20 in* (234 mm)	9.20 in* (234 mm)
Unit length (includes handles)	7.64 in (194 mm)	7.64 in (194 mm)	10.20 in (259 mm)	13.94 in (354 mm)	16.58 in (421 mm)	16.60 in (422 mm)	16.58 in (421 mm)	16.58 in (421 mm)	16.58 in (421 mm)
Unit width	5.20 in (132 mm)	5.20 in (132 mm)	6.80 in (173 mm)	6.80 in (173 mm)	6.84 in (174 mm)	3.40 in (86 mm)	6.84 in (174 mm)	6.84 in (174 mm)	6.84 in (174 mm)
Weight	23 lbs (10 kg)		56 lbs (25 kg)	79 lbs (36 kg)	95 lbs (43 kg)	59 lbs (27 kg)	95 lbs (43 kg)		
Plate thickness positive negative	0.124 in (3.15 mm) 0.065 in (1.65 mm)			0.140 in (3.56 mm) 0.085 in (2.16 mm)		0.150 in (3.81 mm) 0.080 in (2.03 mm)	0.140 in (3.56 mm) 0.085 in (2.16 mm)		
Terminal characteristics	0.55 in diameter, threaded-brass insert, 0.50 in deep. Fasten with 10-32 stainless steel hex bolt/washer.			1.00 in diameter, threaded copper-brass insert, 0.75 in deep. Fasten with 1/4-20 stainless steel hex bolt/washer.		0.65 mm in diameter, threaded-brass insert, 0.50 in deep. Fasten with 1/4-20 stainless steel hex bolt/washer.	1.00 in diameter, threaded copper-brass insert, 0.75 in deep. Fasten with 1/4-20 stainless steel hex bolt/washer.		

Source: C and D Technologies Product Brochure 12-373. See website:cdpowercom.com.

TABLE 24.4   Specifications for Typical Pure Lead-Tin VRLA Batteries

Product (capacity)	Internal res. of fully charged battery mΩ @ 25°C	Nominal short circuit current for charged battery	Dimensions			Weight lb. (kg)
			Length in. (mm)	Width in. (mm)	Height in. (mm)	
G13EP (13Ah)	8.5	1,400A	6.910 (175.51)	3.282 (83.36)	5.113 (129.87)	10.8 (4.9)
G13EPX (13Ah)	8.5	1,400A	6.998 (177.75)	3.368 (85.55)	5.165 (131.19)	12.0 (5.4)
G16EP (16Ah)	7.5	1,600A	7.150 (181.61)	3.005 (76.33)	6.605 (167.77)	35.5 (6.1)
G16EPX (16Ah)	7.5	1,600A	7.267 (184.58)	3.107 (78.92)	6.666 (169.32)	14.7 (6.7)
G26EP (26Ah)	5.0	2,400A	6.565 (166.75)	6.920 (175.77)	4.957 (125.91)	22.3 (10.1)
G26EPX (26Ah)	5.0	2,400A	6.636 (168.55)	7.049 (179.04)	5.040 (128.02)	23.8 (10.8)
G42EP (42Ah)	4.5	2,600A	7.775 (197.49)	6.525 (165.74)	6.715 (170.56)	32.9 (14.9)
G42EPX (42Ah)	4.5	2,600A	7.866 (199.80)	6.659 (169.14)	6.803 (172.80)	35.1 (15.9)
G70EP (70Ah)	3.5	3,500A	13.020 (330.71)	6.620 (168.15)	6.930 (176.02)	53.5 (24.3)
G70EPX (70Ah)	3.5	3,500A	13.020 (330.71)	6.620 (168.15)	6.930 (176.02)	56.0 (25.4)
G26VP (26Ah)	5.0	2,400A	6.565 (166.75)	6.920 (175.77)	4.957 (125.91)	22.3 (10.1)
G26VPX (26Ah)	5.0	2,400A	6.636 (168.55)	7.049 (179.04)	5.040 (128.02)	23.8 (10.8)
G42VP (42Ah)	4.5	2,600A	7.775 (197.49)	6.525 (165.74)	6.715 (170.56)	32.9 (14.9)
G42VPX (42Ah)	4.5	2,600A	7.866 (199.80)	6.559 (169.14)	6.803 (172.80)	35.1 (15.9)

Source: Hawker Energy Products, Inc. Brochure. See website:www.hawker.invensys.com.

TABLE 24.5   Thin VRLA Prismatic Batteries

Capacity, Ah			Dimensions, mm			Weight (typical), g	Specific Energy Wh/Kg	Energy Density Wh/L
C/10	C/20	1C	Thickness	Width	Length			
1.40	1.46	1.02	8.7	33.0	97	90	32.4	105.0
1.80	1.88	1.31	8.7	41.7	97	115	32.6	107.0
	3.0	1.9	8.4	66.5	99.3	140	42.8	108.0
	5.0	3.2	8.4	79.2	135.1	226	44.2	111.0

Source: Top two batteries: Panasonic, Division of Matsushita Electric Corp. of America (see: panasonic.com). Bottom two batteries: Portable Energy Products, Inc.

24.8 APPLICATIONS OF VRLA BATTERIES TO UNINTERRUPTIBLE POWER SUPPLIES

The major application of the VRLA battery is in the standby power market, ranging from low-power (generally less than 5 KVA) applications such as emergency lighting or uninterruptible power supplies (UPSs) for individual computers or work stations to high-power UPSs in telecommunications facilities. A continuous supply of power is also critical in areas such as banking, stock exchanges, hospitals, air traffic control centers etc., where brief interruptions pose the risk of loss of critical data or hazards to health and safety. The low-power UPS systems are generally used where a power loss is acceptable as long as there is sufficient power to allow time for a safe shutdown of equipment. In a high-power application, the UPS is typically required to provide power until a generator can be brought on line.

UPS systems are generally one of three basic designs: (1) standby or off-line; (2) on-line; or (3) line interactive/hybrid. In most UPSs, AC power is fed to a battery charger/rectifier to provide the DC power to float charge the battery. The output of the battery is connected to an inverter that converts the DC output of the battery and/or the battery charger/rectifier to provide AC power needed to run the load. In the standby mode off-line UPS system, the battery and inverter only come into play when the normal AC power fails. In the on-line UPS system, the battery and inverter are always in the circuit. When the AC power fails, the battery is already on-line supplying power to the inverter and no voltage dropout occurs, as opposed to the off-line case where the voltage may drop out, typically for milliseconds before the battery/inverter duo is switched into service. The on-line UPS system serves also to smooth out any voltage fluctuations with the battery/inverter a continuously active element in the circuit (see Sec. 23.9.6). The line interactive hybrid UPS system utilizes an automatic voltage regulator and a special transformer to smooth any under- or over-voltages and to ease the transition to complete battery backup only during outages of the input AC power. A comprehensive treatment of UPS systems, markets and alternative UPS systems (e.g., flywheels) is given in Ref 18.

The experiences with VRLA batteries in the high-power UPS arena have served to demonstrate the complex problems associated with these batteries compared to their flooded counterparts. For most of the 20th century, lead-acid batteries for high-power UPS systems, notably those employed in telecommunications facilities, were of the flooded type. Flooded batteries have relatively large footprints, can spill acid and require periodic maintenance in the form of watering, a costly operation for a UPS facility involving a large number of cells. In the mid 1980s, with the advent of the VRLA battery and its maintenance-free feature, there was immediate interest in replacing flooded with VRLA batteries. Initially, expectations were that VRLA batteries would provide the roughly 20-year life found with flooded batteries. Failures of VRLA batteries occurred after only a few years of service in many cases. One unanticipated problem, negative plate self-discharge is considered below. In the year 2000, more realistic claims of 5–10 year battery life were the rule and there was a trend back to flooded lead-acid for the high-power UPS applications. Some switching from lead-acid to nickel-cadmium or nickel-metal hydride batteries for low-power UPS applications is also taking place or being considered.

The life of a VRLA battery in a UPS application depends not only on the design of the battery and the quality of its manufacture, but also very strongly on the usage. Most of the performance figures quoted in the manufacturers' specifications are for operating temperatures of 25°C. Any significant deviation from that temperature, higher or lower, can result in poor performance, especially in the hands of a customer without knowledge of the proper handling of VRLA batteries. For example, the optimal charging regime for a VRLA battery is quite dependent on the temperature and must be modified for either higher or lower temperatures (see Sec. 24.5.4). VRLA batteries that perform perfectly well in a constant temperature environment may perform quite poorly, even exploding or catching fire, in an outdoor cabinet in a telecommunications application in a variable climate.

There is, however, a considerable amount of research going into various ways to improve VRLA technology to overcome some of its deficiencies. These deficiencies are related generally to the oxygen recombination feature of the VRLA battery. One problem is negative plate self-discharge, a problem which has been found even in batteries that have been operated under conditions conforming to the battery manufacturers' recommendations. This negative self-discharge leads to the negative being in a significantly reduced state of charge and a reduced battery capacity results. In flooded batteries, the float current is more than ample to keep the negative plate fully charged. In VRLA batteries this is not necessarily the case and at least three approaches have been suggested for improving the situation. One is to increase the purity of the negative plate. Certain impurities tend to lower the overvoltage for hydrogen evolution. A second approach is to increase the corrosion rate of the positive grid, not an attractive alternative for long-life batteries. Another approach is the addition of a catalyst to promote the recombination of hydrogen and oxygen.[19] This catalytic approach should compensate for impure active materials as well as for air leaks into the cell. The addition of a catalyst may also increase the negative polarization, thus lowering the positive polarization, which in turn would lessen positive grid corrosion. A combination of higher purity negative plates and a catalyst could be an even better solution to negative plate polarization but the purity problem is complicated by the desire to use recycled lead, with attendant needs for improved refining processes. One company is manufacturing catalytic devices for use in new batteries or in some cases, retrofitting batteries already in service.

Whether any of the above remedies would resolve the safety issues associated with thermal runaway, especially under conditions of operation and improper charging outside recommended temperature limits remains an open question. An overview of the controversies that have surrounded VRLA batteries and their performance can be gleaned from proceedings of recent INTELEC conferences.[19] VRLA batteries with catalysts and other improvements have not yet been in the field long enough to determine whether the newer designs and additions will indeed prove the solutions to some of the VRLA problems. Controversy is still rampant over the current and future prospects for VRLA batteries in UPS applications.

Data on two examples of VRLA batteries now available are given in Tables 24.6*a*, *b* and *c*. These cells employ lead-calcium-tin positive grids and lead-calcium negative grids, a self-resealing safety vent releasing at 2 psi. Lead-calcium alloy grids are also used by other manufacturers. Batteries are floated at 2.23 to 2.35 volts per cell at 25°C.

The discharge rate data for a small 6-cell 12 V, 25 Ah (at $C/8$ to 10.5 V) VRLA battery are presented in Table 24.7. The dimensions of this battery are: height 181 mm; width 132 mm; length 194 mm with a weight of 10 kg.

The discharge rate data are presented in Table 24.8 for a 8 volt, 427 kg, 1360 Ah battery model #4DDV85-33 manufactured by Enersys, Inc.

The above data are for randomly selected batteries out of many marketed for UPS applications, and are believed to be representative of the products currently on the market for UPS applications.

TABLE 24.6*a*   Prismatic VRLA Sincle-cell Battery Characteristics at 25°C

Voltage	Capacity, Ah	Height, mm	Width, mm	Length, mm	Weight, kg	Max. amps*
2 V/cell	500 (C/8) 346 (C/2)	368	182	228	32	1000
2 V/cell	1440 (C/8) 968 (C/2)	580	328	183	88	2880

*Maximum current for 1 minute duration.
Source: Panasonic Website: www.panasonic.com.

TABLE 24.6b   Self-discharge Data for Cells in Table 24.6(a) at 25°C

Storage time	3 months	6 months	12 months	18 months
% Initial Capacity	91	82	64	50

TABLE 25.6c   Discharge Rate (Amperes) for 1440 Ah Battery in Table 24.6(a)

Hours	20	10	8	4	1
Cut-off voltage:					
1.60	76	151	187	319	894
1.75	72	143	180	300	731
1.90	64	128	151	259	540

TABLE 24.7   Discharge Rate (Amps) for Six-cell, 25 Ah Battery

Time	20 h	10 h	30 min	10 min	1 min
Cut-off voltage:					
1.75 V	1.4	2.6	28.4	57.2	113.1
1.90 V	1.2	2.2	22.4	39.8	57.9

Source: C&D Technologies Web site: www.cdpowercom.com

TABLE 24.8   Discharge Rate (Amps) for 8 Volt, 1360 Ah Battery

Time	24 h	10 h	1 h	15 min	1 min
Cut-off voltage: 1.75 V/cell	61	145	672	1248	1472

Source: Enersys, Inc.

REFERENCES

1. S. Takahashi, K. Hirakawa, M. Morimitsu, Y. Yamagachi, and Y. Nakayama, "Development of a Long Life VRLA Battery for Load Leveling-2," Yuasa-JIHO, No. 88, p. 34–38, April 2000.

2. A. G. Cannone, A. J. Salkind, and F. A. Trumbore. *Proc. 13th Annual Battery Conf.* Long Beach, CA, Jan. 1998.

3. M. Pavlov, Conference on Oxygen Cycle in Lead- Acid Batteries, 7th ELBC, Dublin, Ireland, 19 September 2000.

4. D. Berndt, "Valve-Regulated Lead-Acid Battery," and "Lead-Acid Batteries." (Conference on Oxygen Cycle in Lead and Batteries, 7th ELBC, Dublin, Ireland (Sept. 2000)

5. P. Moseley, "Improving the Valve Regulated Lead-Acid Battery," Proc. 1999 IBMA Conf., Battery Man, Feb. 2000, p. 16–29.

6. W. W. McGill III, "Gel vs. VRLA Lift Truck Batteries," Battery Man, Feb. 2000, p. 34–36.

7. D. H. McClelland et al., U.S. Patent 3,704,173 and U.S. Patent 3,862,861.

8. Bolder Battery Co. Literature, Bolder, CO.

9. T. B. Atwater, L. P. Jarvis, P. J. Cygun, and A. J. Salkind, *7th ELBC,* Dublin, Ireland (2000).

10. *Sealed Lead-Acid Batteries Technical Handbook,* Panasonic Industrial Co., Secaucus, N.J.

11. G. A. Moneypenny, "Thinline Batteries for Portable Applications," *Proc. 11th Int. Seminar on Batteries,* Boca Raton, Fla., March 1994.

12. K. R. Bullock and D. H. McClelland, "The Kinetics of the Self-Discharge Reaction in a Sealed Lead-Acid Cell." *J. Electrochem. Soc.* 123:327 (Mar. 1976).

13. J. Arias, "Advanced Bipolar Lead-Acid Batteries," *Proc. 11th Int. Seminar on Batteries,* Boca Raton, Fla., March 1994 and T. Juergens et al., "A New Sealed High-Power Lead-Acid Battery." Proc. 11th Int. Seminar on Batteries, Boca Raton, FL, March, 1994.

14. T. Juergens et al., "A New High Rate Sealed Lead-Acid Battery," *Proc. 36th Power Sources Conf.,* Cherry Hill, N.J., June 1994.

15. R. O. Hammel, "Charging Sealed Lead Acid Batteries," *Proc 27th Annual Power Sources Symp.,* 1976.

16. W. Jones and D. O. Feder, Batteries International, pp. 77–83 (1997).

17. R. O. Hammel, "Fast Charging Sealed Lead Acid Batteries," extended abstracts, pp. 34–36, Electrochem. Soc. Meeting, Las Vegas, Nev., 1976.

18. J. Plante in *Power 2000,* pp. 30–94 (Supplement to *EE Times* (2000)).

19. E. Jones, INTELEC 2000.

CHAPTER 25
IRON ELECTRODE BATTERIES

John F. Jackovitz and Gary A. Bayles*

25.1 GENERAL CHARACTERISTICS

Iron electrodes have been used as anodes in rechargeable battery systems since the introduction of the nickel-iron rechargeable battery at the turn of the century by Junger in Europe and Edison in the United States.[1] Even today the batteries are produced in a fashion similar to the original construction. New constructions have been developed which give better high-rate performance and have lower manufacturing costs. Today the nickel-iron battery is the most common rechargeable system using iron electrodes. Iron-silver batteries have been tested in special electronic applications, and iron/air batteries have shown promise as motive power systems. The characteristics of the iron battery systems are summarized in Tables 25.1 and 25.2.

As designed by Edison, the nickel-iron battery was and is almost indestructible. It has a very rugged physical structure and can withstand electrical abuse such as overcharge, over-discharge, discharged stand for extended periods, and short-circuiting. The battery is best applied where high cycle life at repeated deep discharges is required (such as traction applications) and as a standby power source with a 10- to 20-year life. Its limitations are low power density, poor low-temperature performance, poor charge retention, and gas evolution on stand. The cost of the nickel-iron battery lies between the lower-cost lead-acid and the higher-cost nickel-cadmium battery in most applications, with the exception of limited use applications in electric vehicles and mobile industrial equipment.

Most recently, iron electrodes have been considered and tested as cathodes too. Based upon high valence state iron, Fe(VI), these cathodes have shown promise in experimental cells when coupled with zinc or metal hydride anodes for portable primary and secondary batteries. The results are discussed in Sec. 25.7.

---

*Ralph J. Brodd was the original author for this chapter.

TABLE 25.1   Iron Electrode Battery Systems

System	Uses	Advantages	Disadvantages
Iron/nickel oxide (tubular)	Material handling vehicles, underground mining vehicles, miners' lamps, railway cars and signal systems, emergency lighting	Physically almost indestructible Not damaged by discharged stand Long life, cycling or stand Withstands electrical abuse: overcharge, overdischarged, short-circuiting	High self-discharge Hydrogen evolution on charge and discharge Low power density Lower energy density than competitive systems Poor low-temperature performance Damaged by high temperatures Higher cost than lead-acid Low cell voltage
Iron/air	Motive power	Good energy density Uses readily available materials Low self-discharge	Low efficiency Hydrogen evolution on charge Poor low-temperature performance Low cell voltage
Iron/silver oxide	Electronics	High energy density High cycle life	High cost Hydrogen evolution on charge

TABLE 25.2   System Characteristics

System	Nominal voltage, V		Specific energy Wh/kg	Energy density Wh/L	Specific power W/kg	Cycle life, 100% DOD
	Open-circuit	Discharge				
Iron/nickel oxide						
Tubular	1.4	1.2	30	60	25	4000
Developmental	1.4	1.2	55	110	110	>1200
Iron air	1.2	0.75	80		60	1000
Iron/silver oxide	1.48	1.1	105	160	—	>300

25.2   CHEMISTRY OF NICKEL-IRON BATTERIES

The active materials of the nickel-iron battery are metallic iron for the negative electrode, nickel oxide for the positive, and a potassium hydroxide solution with lithium hydroxide for the electrolyte. The nickel-iron battery is unique in many respects. The overall electrode reactions result in the transfer of oxygen from one electrode to the other. The exact details of the reaction can be very complex and include many species of transitory existence.[2–4] The electrolyte apparently plays no part in the overall reaction, as noted in the following reactions:

$$Fe + 2NiOOH + 2H_2O \underset{\text{charge}}{\overset{\text{discharge}}{\rightleftharpoons}} 2Ni(OH)_2 + Fe(OH)_2 \qquad \text{(first plateau)}$$

$$3Fe(OH)_2 + 2NiOOH \underset{\text{charge}}{\overset{\text{discharge}}{\rightleftharpoons}} 2Ni(OH)_2 + Fe_3O_4 + 2H_2O \qquad \text{(second plateau)}$$

The overall reaction is

$$3Fe + 8NiOOH + 4H_2O \underset{\text{charge}}{\overset{\text{discharge}}{\rightleftharpoons}} 8Ni(OH)_2 + Fe_3O_4$$

The electrolyte remains essentially invariant during charge and discharge. It is not possible to use the specific gravity of the electrolyte to determine the state of charge as for the lead-acid battery. However, the individual electrode reactions do involve an intimate reaction with the electrolyte.

A typical charge-discharge curve of an iron electrode is shown in Fig. 25.1. The two plateaus on charge correspond to the formation of the stable +2 and +3 valent states of the iron reaction products. The reaction of the iron electrode can be written as

$$Fe + nOH^- \rightarrow FE(OH)_n^{2-n} + 2e \qquad \text{(first plateau)}$$

and

$$Fe(OH)_n^{2-n} \rightarrow Fe(OH)_2 + nOH^-$$

$$Fe(OH)_2 + OH^- \rightarrow Fe(OH)_3 + e \qquad \text{(second plateau)}$$

Then

$$2Fe(OH)_3 + Fe(OH)_2 \rightarrow Fe_3O_4 + 4H_2O$$

Iron dissolved initially as the +2 species in alkaline media. The divalent iron complexes with the electrolyte to form the $Fe(OH)_n^{2-n}$ complex of low solubility. The tendency to supersaturate plays an important role in the operation of the electrode and accounts for many important aspects of the electrode performance characteristics. Continued charge forms the +3 valent iron which, in turn, interacts with +2 valent iron to form Fe_3O_4.

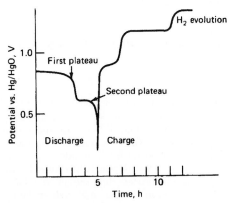

FIGURE 25.1 Discharge-charge curve of an iron electrode. (*From Ref. 5.*)

The superior life-cycling characteristics of the iron electrode result from the low solubility of the reaction intermediates and oxidized species. The supersaturation on discharge results in the oxidized material forming small crystallites near the reaction site. On charge, the low solubility also slows the crystal growth of the iron, thereby helping to ensure formation of the original active high-surface-area structure. The low solubility also accounts for poor high-rate and low-temperature performance as the discharged (oxidized) species precipitate at or near the reaction site and block the active surface. The performance characteristics are substantially improved, however, in the advanced nickel-iron batteries by the use of a superior electrode grid structure, such as fiber-metal, which provides intimate contact with the iron active material throughout the volume of porous structure.

Sulfide addition to the iron electrode radically changes the electrocrystallization kinetics. It increases the supersaturation and makes reaction more reversible. Sulfide also absorbs on the surface to block crystallization sites and raises the hydrogen evolution reaction on charge. Lithium salt additions seem to make the electrode perform more reversibly, perhaps by enhancing the solubility of the reaction intermediates.

Nickel electrode reactions[6,7] are generally thought to be solid-state-type reactions wherein a proton is injected or rejected from the lattice reversibly on discharge and on charge, respectively.

$$\beta\text{-Ni(OH)}_2 \xleftarrow[\text{in KOH}]{\text{transformation}} \alpha\text{-Ni(OH)}_2$$

$$\text{reduction} \atop (\text{discharge}) \updownarrow \text{oxidation} \atop (\text{charge}) \qquad \text{oxidation} \atop (\text{charge}) \updownarrow \text{reduction} \atop (\text{discharge})$$

$$\beta\text{-NiOOH} \xrightarrow[\text{in KOH}]{\text{overcharge}} \gamma\text{-NiOOH}$$

The oxidation (charge) voltage for the α and β materials is more positive than the discharge voltage by 60 mV and 100 mV, respectively. The β-Ni(OH)$_2$ is the usual electrode material. It is converted on charge to β-NiOOH with about the same molar volume. On overcharge the γ structure can form. This form also incorporates water and potassium (and lithium) into the structure. Its molar volume is about 1.5 times the β form. This is thought to be responsible in large part for the volume expansion (swelling) which occurs on charging the battery. The α form then results on discharge of the γ form. Its molar volume is about 1.8 times the β form, and the electrode can swell further on discharge. On discharge stand in concentrated electrolyte, the α form converts to the β form. Cobalt additions (2 to 5%) improve the charge acceptance (reversibility) of the nickel electrode.

25.3 CONVENTIONAL NICKEL-IRON BATTERIES

25.3.1 Construction

The construction of a tubular or pocket plate nickel-iron cell is shown in Fig. 25.2. The active materials are filled in nickel-plate perforated steel tubes or pockets. The tubes are fastened into plates of desired dimensions and assembled into cells by interleaving the positive and negative plates. The container is fabricated from nickel-plated sheet steel. The cells may be assembled into batteries in molded nylon cases or mounted into wooden traps. The steel cases may be coated with plastic or rubber for insulation or spaced by insulating buttons.

The manufacturing process has remained relatively unchanged for over 50 years. The processes are designed to produce materials of highest purity and with special particle characteristics for good electrochemical performance.

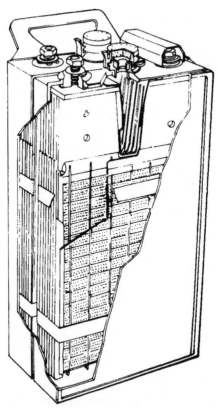

FIGURE 25.2   Cross section of typical nickel-iron battery. (*Courtesy of SAFT America, Inc.*)

Negative Electrode.   To produce the anode active material, pure iron is dissolved in sulfuric acid. The $FeSO_4$ is recrystallized, dried, and roasted (815 to 915°C) to Fe_2O_3. The material is washed free of sulfate, dried, and partially reduced in hydrogen. The resulting material (Fe_3O_4 and Fe) is partially oxidized, dried, ground, and blended. Small amounts of additives, such as sulfur, FeS, and HgO, are blended in to increase battery life by acting as depassivators, reducing gas evolution, or improving conductivity.

To make the anode current collector, steel strips or ribbon are perforated and nickel-plated. After drying and annealing, the strip is formed into a pocket, about 13 mm wide and 7.6 mm long. One end is left open and filled with the iron active material. A machine automatically introduces the active material and tamps it into the pockets. After filling, the negative pockets are crimped and pressed into openings in a nickel-plated steel frame.

Positive Electrode.   The positive active material consists of nickel hydroxide in alternate layers with nickel flake. High-purity nickel powder or shot is dissolved in sulfuric acid. The hydrogen evolved is used in making the iron active material. The acidity of the resulting solution is adjusted to pH 3 to 4 and filtered to remove ferric iron and other insoluble materials. If needed, the solution may be further purified to remove traces of ferrous iron and copper. Cobalt sulfate may be added in the proportion of 1.5% to improve nickel electrode performance. The nickel sulfate solution is sprayed into hot 25 to 50 NaOH solution. The resulting slurry is filtered, washed, dried, crushed, and screened to yield particles which pass 20- but not 200-mesh screens.

Special nickel flake (1.6×0.01 mm) is produced by electrodepositing alternate layers of nickel and copper on stainless steel. The electroplate is stripped and cut into squares. The copper is dissolved out in hot sulfuric acid, and the resulting nickel flakes are washed free of copper and dried at low temperature to prevent nickel oxide formation. With a modified process[8] the flakes of proper shape and size can be produced as a single layer, eliminating the need for deposition of alternate copper layers. As in the negative electrode, the positive-electrode process starts with perforated steel ribbon which is nickel-plated and annealed. The ribbon is wound into tubes with an interlocked seam. Two types, right- and left-wound tubes, are produced, typically of 6.3-mm diameter. The tubes are filled with alternate layers of nickel hydroxide and nickel flakes. Each layer is tamped (144 kg/cm^2) to ensure good contact. There are 32 layers of flake per centimeter. To prevent the seam from opening during the rigors of charge and discharge, rings are placed around the tubes at uniform intervals of about 1 cm. The tubes are enclosed, and the pinched ends are locked into the nickel-plated steel grid frame. The "rights" and "lefts" are alternated so that any tendency to distort on the part of one tube is counteracted by the next one. The positive electrode can also be made in the pocket plate construction, as described above under "Negative Electrode."

Cell Assembly.   The configuration and size of the tubes and pockets determine the capacity for each plate. The plates are then assembled into electrodes to meet the capacity requirements of each cell.

Each plate group is assembled by bolting a terminal pole and a selected number of plates, depending on capacity, to a steel rod which passes through the grid at the top of the plates. Groups of positive and negative plates are intermeshed to form the element. A cell usually contains one or more negative than positive plates. The cells are made positive limiting for best cycle life.

The positive and negative plates are separated by hard rubber or plastic pins called "hair pins" or "hook pins," which fit into spaces formed by the tubular positive and flat negative electrodes.

Electrolyte.   The electrolyte is a 25 to 30% KOH solution with up to 50 g/L of LiOH added. The composition of the replacement electrolyte to compensate for losses due to spray from the vent cap is about 23% caustic with about 25 g/L LiOH. Occasionally the electrolyte is replaced completely to rejuvenate the cell performance. The renewal electrolyte is about 30% KOH with 15 g/L LiOH.

Lithium additions to the electrolyte are important but not completely understood. Lithium hydroxide improves cell capacity and prevents capacity loss on cycling and also seems to facilitate nickel electrode kinetics. It expands the working plateau on charge and delays oxygen evolution. Some evidence exists for the formation of Ni^{4+}, which improves electrode capacity. Lithium also decreases the carbonate content in the electrolyte since Li_2CO_3 is not very soluble. It also decreases the tendency for swelling of the positive active material but increases the resistivity of the cell electrolyte.

Shortly after initiation of charge, hydrogen evolution begins on the iron electrode. The considerable hydrogen evolution on charge presumably helps counteract iron passivation in alkaline solution. Mercury additions also have a similar effect, but only in the early formation cycles.

25.3.2 Performance Characteristics of Nickel-Iron Battery

Voltage.   A typical discharge-charge curve of a commercial iron/nickel oxide battery is shown in Fig. 25.3. The battery's open-circuit voltage is 1.4 V; its nominal voltage is 1.2 V. On charge, at rates most commonly used, the maximum voltage is 1.7 to 1.8 V.

Capacity.   The capacity of the nickel-iron battery is limited by the capacity of the positive electrode and, hence, is determined by the length and number of positive tubes in each plate. The diameter of the tubes generally is held constant by each manufacturer. The 5-h discharge rate is commonly used as the reference for rating its capacity.

The conventional nickel-iron battery has moderate power and energy density and is designed primarily for moderate to low discharge rates. It is not recommended for high-rate applications such as engine starting. The high internal resistance of the battery lowers the terminal voltage significantly when high rates are required. The relationship between capacity and rate of discharge is shown in Fig. 25.4.

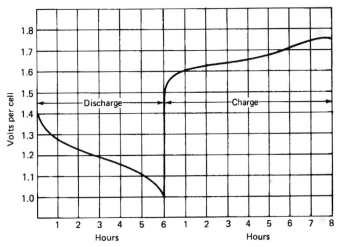

FIGURE 25.3   Typical voltage characteristics during constant-rate discharge and recharge. (*From Ref. 9.*)

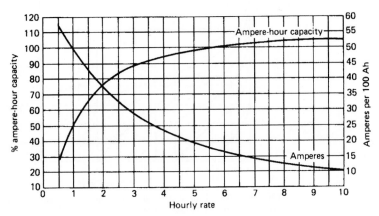

FIGURE 25.4   Curves of capacity vs. discharge rate at 25°C; end voltage 1.0 V per cell. (*From Ref. 9.*)

If a battery is discharged at a high rate and then at a lower rate, the sum of the capacities delivered at the high and low rates nearly equals the capacity that would have been obtained at the single discharge rate. This is illustrated in Fig. 25.5.

Discharge Characteristics.   The nickel-iron battery may be discharged at any current rate it will deliver, but the discharge should not be continued beyond the point where the battery nears exhaustion. It is best adapted to low or moderate rates of discharge (1- to 8-h rate). Figure 25.6 shows the discharge curves at different rates of discharge at 25°C.

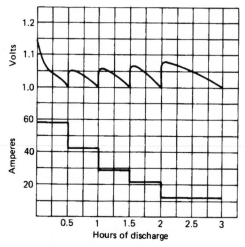

FIGURE 25.5   Effect of decreasing rate on battery voltage of nickel-iron cell.

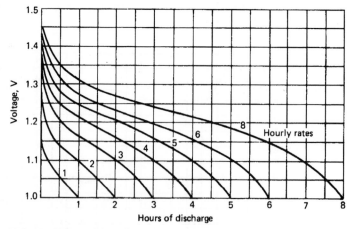

FIGURE 25.6   Time-voltage discharge curves of nickel-iron battery; end voltage 1.0 V per cell. (*From Ref. 9.*)

Effect of Temperature.   Figure 25.7 shows the effects of temperature on the discharge. The capacity at 25°C is normally taken as the standard reference value. The decrease in performance is generally attributed to passivity of the iron electrode and decreased solubility of the reaction intermediate. At low temperature, increased resistivity and viscosity of the electrolyte along with slower nickel electrode kinetics contribute to the fall-off of capacity. Care must be exercised to keep the temperature from exceeding about 50°C as the self-discharge of the nickel positive electrode is accelerated. Also, the increased solubility of iron at high temperature can adversely affect operation of the nickel electrode by incorporating soluble iron into the nickel hydroxide crystal lattice. The battery is seldom used below −15°C.

Hours of Service.   The hours of service on discharge that a typical nickel-iron battery, normalized to unit weight (kilograms) and volume (liters), will deliver at various discharge rates and temperatures is summarized in Fig. 25.8.

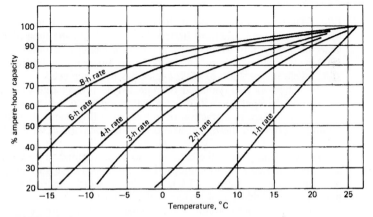

FIGURE 25.7   Effect of temperature on capacity at various rates. (*From Ref. 9.*)

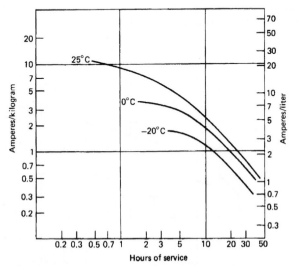

FIGURE 25.8   Hours of service of nickel-iron battery at various discharge rates and temperatures; end voltage 1.0 V cell.

Self-Discharge.   The self-discharge rate, charge retention, or stand characteristic of the nickel-iron battery is poor. At 25°C a cell will lose 15% of its capacity in the first 10 days and 20 to 40% in a month. At lower temperatures the self-discharge rate is lower. For example, at 0°C the losses are less than one-half of those experienced at 25°C.

Internal Resistance.   To a rough approximation, the internal resistance R_i can be estimated for tubular Ni-Fe from the equation

$$R_i \times C = 0.4$$

where R_i = internal resistance, Ω
C = battery capacity, Ah

For example, $R_i = 0.004 \ \Omega$ for a 100-Ah battery. The value of R_i remains constant through the first half of the discharge, then increases about 50% during the latter half of the discharge.

Life.   The main advantages of the tubular-type nickel-iron battery are its extremely long life and rugged construction. Battery life varies with the type of service but ranges from 8 years for heavy duty to 25 years or more for standby or float service. With moderate care, 2000 cycles can be expected; with good care, for example, by limiting temperatures to below 35°C, cycle life of 3000 to 4000 cycles has been achieved.

   The battery is less damaged by repeated deep discharge than any other battery system. In practice, an operator will drive a battery-operated vehicle until it stalls, at which point the battery voltage is a fraction of a volt per cell (some cells may be in reverse). This has a minimal effect on the nickel-iron battery in comparison with other systems.

Charging.   Charging of the batteries can be accomplished by a variety of schemes. As long as the charging current does not produce either excessive gassing (spray out of the vent cap) or temperature rise (above 45°C), any current can be used. Excessive gassing will require more frequent addition of water. If the cell voltage is limited to 1.7 V, these conditions should not be a consideration. Typical charging curves are given in Fig. 25.9. The Ampere-hour input should return 25 to 40% excess of the previous discharge to ensure complete charging. The suggested charge rate is normally between 15 and 20 A per 100 Ah of battery capacity. This rate would return the capacity in the 6- to 8-h time frame. The effect of temperature on charging is shown in Fig. 25.10.

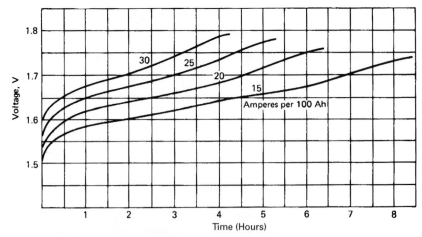

FIGURE 25.9   Typical charging voltage for nickel-iron battery at various rates. (*From Ref. 9.*)

Constant current and modified constant potential (taper), shown in Fig. 25.11, are common recharging techniques. The charging circuit should contain a current-limiting device to avoid thermal runaway on charge. Recharging each night after use (cycle charging) is the normal procedure. The batteries can be trickle-charged to maintain them at full capacity for emergency use. A trickle charge rate of 0.004 to 0.006 A/Ah of battery capacity overcomes the internal self-discharge and maintains the battery at full charge. Following an emergency discharge, a separate recharge is needed. For applications such as railroad signals, charging at a continuous average current may be the most economic method. Here a modest drain is required when no trains are passing but quite a heavy drain when a train passes, yet the total Ampere-hours over a period of 24 h remains fairly constant. For this situation, a constant current equal to that required to maintain the battery can be used.

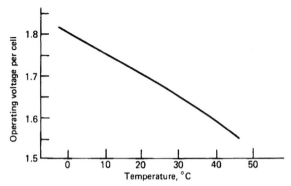

FIGURE 25.10   Variation of relay operating voltage with temperature. (*From Ref. 9.*)

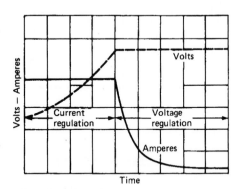

FIGURE 25.11   Effects of "regulators" with voltage and current regulation. (*From Ref. 9.*)

25.3.3   Sizes of Nickel-Iron Batteries

Nickel-iron batteries have been available in sizes ranging from about 5 to 1250 Ah. In recent years they have become less popular, giving way to the lead-acid and nickel-cadmium batteries, and are no longer manufactured by many of the original manufacturers. Table 25.3 lists the physical and electrical characteristics of typical nickel-iron batteries.

TABLE 25.3 Typical Nickel-Iron Batteries

Nominal capacity, Ah	169	225	280	337	395	450	560	675
Nominal current, A:								
5-h discharge	34	45	56	67	79	90	112	135
Cell weight, filled, kg	8.8	10.8	12.9	15.3	17.4	19.5	24.3	28.6
Installed weight, kg	9.8	12.0	14.3	16.9	19.3	21.7	26.5	31.2
Electrolyte (1.17 kg/L), kg	1.8	2.2	2.6	3.0	3.4	3.8	4.9	5.9
Cell dimensions, mm*								
Length	52	66	82	96	111	125	156	186
Width	130	130	130	130	130	130	135	135
Height	534	534	534	534	534	534	534	534
Battery dimensions, mm†:								
Length:								
2 cells	—	—	—	265	295	321	343	343
3 cells	—	—	—	376	421	460		
4 cells	284	367	431	487				
5 cells	346	448	545					
6 cells	408	546						
Width	161	161	161	161	161	161	197	228
Height	568	573	582	582	582	582	590	590

* See drawing (*a*).
† See drawing (*b*).
Source: Varta Batteries AG, Hanover, Germany.

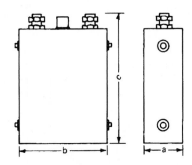

(*a*) Cell showing dimensions used in Table 25.3.

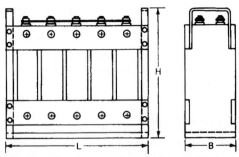

(*b*) Multicell battery showing dimensions used in Table 25.3. Tolerances are 5, 3, and 3 mm for dimensions *L*, *B*, and *H*, respectively.

25.3.4 Special Handling and Use of Nickel-Iron Batteries

The battery should be operated in a well-ventilated area to prevent the accumulation of hydrogen. Under certain circumstances, hydrogen can be ignited by a spark to cause an explosion with a resulting fire. In multicell batteries the usual precautions in dealing with high voltages should be taken.

If the battery is to be out of service for more than a month, it should be stored in the discharged condition. It should be discharged and short-circuited, then left in that condition for the storage period. Filling caps must be kept closed. If this procedure is not followed, several cycles are required to restore the capacity upon reactivation.

Constant-voltage charging is not recommended for conventional nickel-iron batteries. It may lead to a thermal runaway condition which results in dangerous conditions and can severely damage the battery. When the battery nears full charge, the gassing reactions produce heat and the temperature rises, lowering the internal resistance and the cell EMF. Accordingly, the charge current increases under constant-voltage charge. This increased current further increases the temperature, and a vicious cycle is started. A modified constant-voltage charging with current limiting is, however, acceptable.

25.4 *ADVANCED NICKEL-IRON BATTERIES*

The desire to use the attractive features of the nickel-iron couple, such as ruggedness and long life, in applications requiring high-rate performance and low manufacturing costs has led to the development of advanced nickel-iron batteries with performance characteristics suitable for electric automobiles and other mobile traction applications. The capability of these batteries permits an electric vehicle a range of at least 150 km between charges, acceleration rapid enough to merge into highway traffic, and a cycling life equivalent to 10 or more years of on-the-road service. The advanced battery utilizes sintered-fiber metal (steel wool) plaques, impregnated with active material, for both the positive and the negative electrodes. Nonwoven polypropylene sheets are used as separators between electrodes. The techniques for making plaques, impregnation and activation, stacking, and assembly are all amenable to high-volume production methods similar to those used in lead-acid battery manufacture.

The battery system design incorporates an electrolyte management system to minimize the maintenance problems associated with its widespread deployment in the public sector. This system, shown schematically in Fig. 25.12, provides for semiautomatic watering of the cells by utilizing a single-point watering port. The flow of electrolyte through the cells during the charging cycle permits heat removal and effective management of gas evolved during the charge. Uniform specific gravity for all cells is ensured and specific gravity maintenance is easily achieved by use of the system.

Both the positive and the negative plates of the Westinghouse nickel-iron battery used fiber metal plaques as the substrate. Two methods of active nickel impregnation were developed and used in demonstration batteries. An electroprecipitation process (EPP), developed in the mid-1960s, demonstrated good performance, ruggedness, and long cycle life. The EPP deposits nickel hydroxide electrochemically into the porous substrate. Efficient use of the nickel material is achieved, with active material utilization of 0.14 Ah/g of total electrode. An alternate nickel electrode manufacturing process was also developed which entails the preparation of a nickel hydroxide paste that is then loaded into the fiber metal substrate by roll pasting methods. Pasted nickel electrodes demonstrated performance equivalent to EPP electrodes (0.14 Ah/g of total electrode) while demonstrating a less expensive manufacturing process. The iron electrodes were also produced by a pasting process. Iron oxide, Fe_2O_3, was paste-loaded into the fiber metal electrode substrate and then furnace-reduced in a hydrogen atmosphere. These electrodes demonstrated 0.26 Ah/g of total electrode or, better, at $C/3$ discharge rates.

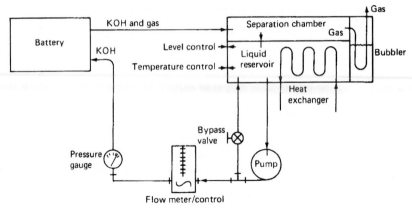

FIGURE 25.12   Schematic of electrolyte circulation system.

The performance characteristics of the advanced nickel-iron batteries, as typified by the Westinghouse system, are summarized in Table 25.4. Figure 25.13 shows a typical discharge curve for a 90-cell electric-vehicle battery at the $C/3$ rate. The battery capacity and energy as a function of discharge rate are shown in Fig. 25.14. Cell power and voltage characteristics, as a function of discharge rate and state of discharge, are presented in Fig. 25.15. The variation in battery capability with temperature, based on tests on five-cell modules, is shown in Fig. 25.16.[10-12] The Eagle-Picher Company also developed a nickel-iron battery using sintered-nickel electrodes similar to those described in Chap. 27. The iron electrode is similar to the Swedish National Development Corporation iron electrode discussed in Sec. 25.5. The performance of this battery is similar to that given in Figs. 25.13 to 25.16.[13]

TABLE 25.4   Advanced Nickel-Iron Battery Performance Characteristics*
Demonstrated as of December 1991

Capacity, † Ah	210
Specific energy, † Wh/kg	55
Energy density, † Wh/L	110
Specific power, ‡ W/kg	100
Cycle life§	>900
Urban range, km	
with regenerative braking	154
Without regenerative braking	125
Projected production cost, $/kWh (1990 $)	200–250

* Based on the Westinghouse nickel-iron battery.
† At the $C/3$ rate.
‡ 30-s average at 50% state of charge.
§ Cycle to 100% depth of discharge; life to 75% of rated energy.

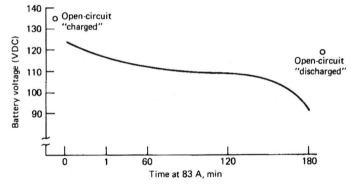

FIGURE 25.13   Battery voltage at $C/3$ (83-A) discharge rate.

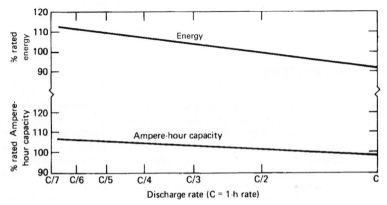

FIGURE 25.14   Capacity as a function of discharge rate. (*Courtesy of Westinghouse Electric Corp.*)

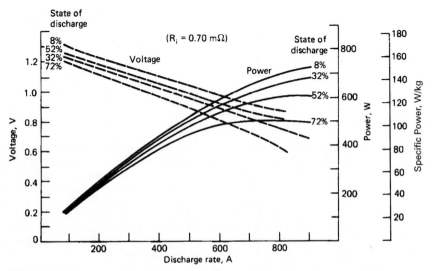

FIGURE 25.15   Power characteristics of 210-Ah nickel-iron battery. (*Courtesy of Westinghouse Electric Corp.*)

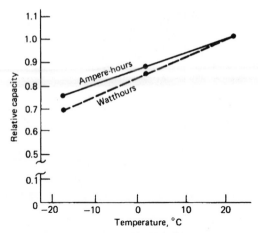

FIGURE 25.16   Effect of temperature on capacity and energy of nickel-iron battery ($C/3$ rate).

25.5   IRON/AIR BATTERIES

Rechargeable metal/air batteries have an advantage over conventional systems as only one reactant (the anode material) need be contained within the battery. The electrically rechargeable iron/air cell has a lower specific energy than the mechanically rechargeable cells (see Chap. 38) but has the advantage of potentially lower life-cycle costs. Unlike zinc, the iron electrodes do not suffer a severe redistribution of active materials or gross shape change upon prolonged electrical cycling. The iron/air cell is another candidate as a motive power source, especially for electric vehicles. The cell reactions are

$$O_2 + 2Fe + 2H_2O \underset{\text{charge}}{\overset{\text{discharge}}{\rightleftharpoons}} 2Fe(OH)_2 \qquad \text{(first plateau)}$$

$$3Fe(OH)_2 + \tfrac{1}{2}O_2 \underset{\text{charge}}{\overset{\text{discharge}}{\rightleftharpoons}} Fe_3O_4 + 3H_2O \qquad \text{(second plateau)}$$

The iron electrode kinetics are covered in Sec. 25.2. The oxygen electrode reactions follow the kinetic path with peroxide as an intermediate. The oxygen electrode reactions in simple form are

$$O_2 + 2H_2O + 2e \rightarrow H_2O_2 + 2OH^-$$

$$H_2O_2 + 2e \rightarrow 2OH^-$$

The single most important life-limiting factor in this battery system is the stability of the air electrode, which loses its ability to function reversibly as it undergoes repeated charges and discharges. The oxygen and peroxide evolved on charge and discharge may attack the substrate, alter the activity of the catalyst, and delaminate the wetproofing film. Separate air (oxygen) electrodes and circuits can be employed in the charge and discharge modes; however, considerations of system weight and volume favor the use of a bifunctional electrode, that is, a single electrode capable of sustaining either oxygen reduction or evolution. These electrodes must be stable over the potential range of both reactions, a fact which poses constraints on material stability and electrode design.

Several designs were developed, but most work on iron/air has been discontinued in favor of zinc/air.

The Swedish National Development Corporation's iron/air cell used the sintered-iron-mesh anode construction.[5,14,15] A pore-forming material could be included to control the development of the optimum electrode structure. The resulting pressed matrix was treated in H_2 at 650°C. The pore-forming material could be leached out after treatment. The active material utilization approached 65%. The air electrode was a porous-nickel double-layer structure (0.6 mm thick) composed of sintered nickel of coarse and fine porosities. The coarse layer on the electrolyte side was catalyzed with silver and impregnated with hydrophobic agents. The electrodes were welded into a polymer frame and formed into cells, as shown in Fig. 25.17. There were two air electrodes for each iron electrode. Air was forced past the electrode at about 2 times the stoichiometric requirement during operation. A schematic and a photo of a 30-kWh battery are shown in Fig. 25.18 and Fig. 25.19, respectively. Electrolyte circulation was used to control heat balance and remove gases generated during operation. Carbon dioxide was scrubbed from the incoming air using NaOH. The air was then humidified to minimize electrolyte loss. Overall the auxiliary systems require less than 10% of the system output.

Typical charge-discharge curves for an average battery in the iron/air battery are shown in Fig. 25.20. The marked difference in charge and discharge voltages accounts largely for the low overall system efficiency. Figure 25.21 shows the power-producing characteristics. The system is capable of over 1000 cycles, limited by the gradual deterioration of the air electrode.

The Westinghouse iron/air battery used a construction similar to that described for the Swedish National Development Corporation's system.[16] The sintered-iron electrode was somewhat similar to that described before. The iron electrode for this iron/air battery had a high active iron content and lower cycle life compared with the electrode described previously. Particles of iron powder were sintered to form a structure without the steel fiber

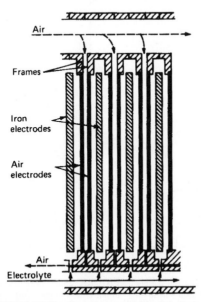

FIGURE 25.17 Cross section of Swedish National Development Corporation's iron/air battery pile. (*From Ref. 5.*)

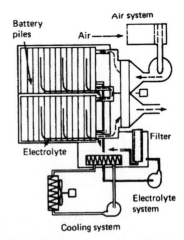

FIGURE 25.18 Schematic cross section of Swedish National Development Corporation's iron/air battery including auxiliary system. (*From Ref. 5.*)

FIGURE 25.19 Swedish National Development Corporation's 30-kWh iron/air battery system. (*Courtesy of Swedish National Development Corp.*)

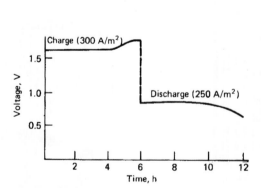

FIGURE 25.20 Charge-discharge voltages for battery in Swedish National Development Corporation's iron/air battery. (*From Ref. 5.*)

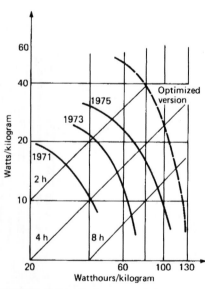

FIGURE 25.21 Performance of Swedish National Development Corporation's iron/air batteries. (*From Ref. 5.*)

substrate. Electrodes with this construction demonstrated up to 0.44 Ah/g. The air electrode was bifunctional and used a Teflon-bonded carbon-based structure with complex silver catalysts (silver content was less than 2 mg/cm^2) supported on a silver-plated nickel screen. The Westinghouse system used a horizontal flow concept to improve performance and control gas and heat. Good life was demonstrated for over 300 cycles with an air electrode of potentially very low cost. A summary of the projected characteristics of the 40-kWh battery is given in Table 25.5.

The Siemens cell was similar except that the air electrode was fabricated with two layers: a hydrophilic layer of porous nickel on the electrolyte side for oxygen evolution and a hydrophobic layer [carbon black bonded with Teflon® (PTFE) and catalyzed with silver] on the air side for oxygen reduction. The dual porosity helped to shield the silver catalyst from oxidation. As many as 200 cycles were achieved.[17]

TABLE 25.5   Characteristics of Westinghouse Iron/Air Electric-Vehicle Battery

Electric vehicle:	
Weight	900 kg curb weight
Range	240 km
Battery:	
Capacity	40 kWh
Power	10 kW continuous power
Weight	530 kg
Volume	0.04 m^3
Cost	$150/kWh

25.6 SILVER-IRON BATTERY

The silver-iron battery has been limited in use because of its high cost. Its theoretical energy density is essentially equal to that of the more popular silver-zinc system. The silver-iron battery has good cycle life compared with the silver-zinc and provides a battery of high reliability, long life, and better durability where high specific energy content is essential.[18–23] Figure 25.22 shows a 3.5-kWh battery designed for telecommunication use, whereas Fig. 25.23 shows a 9.5-kWh battery designed for use in a submersible.

The cell reactions are

$$Fe + 2AgO + H_2O \underset{\text{charge}}{\overset{\text{discharge}}{\rightleftharpoons}} Fe(OH)_2 + Ag_2O \quad \text{(first plateau)}$$

$$Fe + Ag_2O + H_2O \underset{\text{charge}}{\overset{\text{discharge}}{\rightleftharpoons}} Fe(OH_2) + 2Ag \quad \text{(second plateau)}$$

$$3Fe(OH)_2 + Ag_2O \underset{\text{charge}}{\overset{\text{discharge}}{\rightleftharpoons}} Fe_3O_4 + 3H_2O + 2Ag \quad \text{(third plateau)}$$

In practice, only the first and second discharge plateaus are used. The overall reaction is

$$2Fe + 2AgO + 2H_2O \underset{\text{charge}}{\overset{\text{discharge}}{\rightleftharpoons}} 2Fe(OH)_2 + 2Ag \quad E^0 = 1.34 \text{ V}$$

FIGURE 25.22   3.5-kWh telecommunications iron/silver oxide battery. (*Courtesy of Westinghouse Electric Corp.*)

FIGURE 25.23   9.5-kWh iron/silver oxide battery for submersible vehicle. (*Courtesy of Westinghouse Electric Corp.*)

Charge/Discharge Characteristics.   Typical charge-discharge curves for a silver-iron battery of the type shown in Fig. 25.22 are given in Fig. 25.24. The electrolyte is KOH of 1.31 specific gravity with 15 g/L LiOH added. The batteries can withstand several complete reversals without appreciable adverse effect on capacity.

Separator System and Cycle Life.   Multilayer microporous polyethylene, nonwoven felt polypropylene, and cellophanes are some of the materials typically used as separators in this system. It is important to note that the choice of separator has very little to do with the iron electrode, which is extremely stable in KOH and does not react with the separator system. Rather, the separator system must be selected to retard the migration of silver to the iron electrode and to withstand the oxidative effects of the silver electrode itself. The particular separator system chosen will therefore determine the cycle life, the shelf life, and the power capabilities. Consequently the separator system is usually application-specific. The typical cycle life at 100% depth of discharge and 10% overcharge is shown in Fig. 25.25.

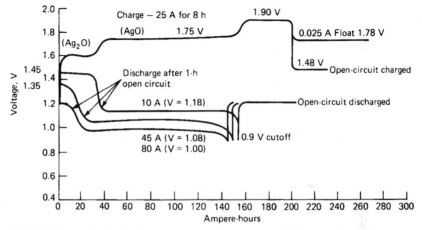

FIGURE 25.24   Charge-discharge characteristics of nominal 140-Ah iron/silver oxide battery. (*From Ref. 20.*)

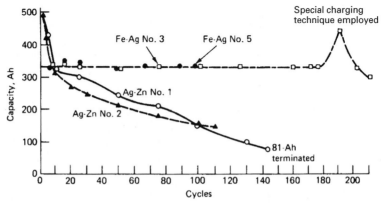

FIGURE 25.25   Cyclic life performance of zinc/silver oxide and iron/silver oxide prototype batteries. (*From Ref. 20.*)

Temperature Effects.   As with other alkaline battery systems, silver-iron performance can depend on the operating temperature. Cells designed for long-life and low-rate applications normally have a higher internal resistance than those designed for high rate and shorter life. The two designs therefore behave differently when the discharge temperature is decreased, as shown in Figs. 25.26 and 25.27.

Experimental Designs.   Experimental tests were conducted on designs other than the monopolar, prismatic designs discussed so far. Hand-assembled bipolar and jelly-roll cells were tested at the laboratory demonstration level. Results of voltage polarization tests on those designs are compared to the prismatic design in Fig. 25.28. The voltage characteristics of the jelly-roll design would be expected to improve considerably as assembly is refined and automated. Designs such as these would be suited for use in smaller portable systems such as communication devices that may require high power and energy density.

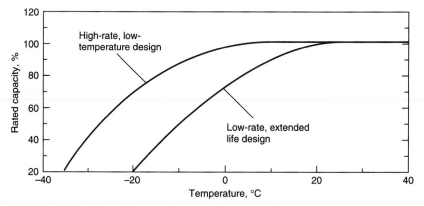

FIGURE 25.26   Effect of temperature on discharge capacity for different battery designs, $C/10$ discharge rate. (*Courtesy of Westinghouse Electric Corp.*)

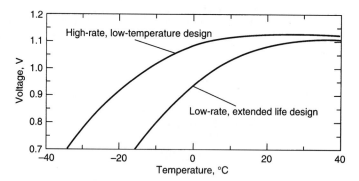

FIGURE 25.27   Effect of temperature on discharge voltage for different battery designs; $C/10$ discharge rate. (*Courtesy of Westinghouse Electric Corp.*)

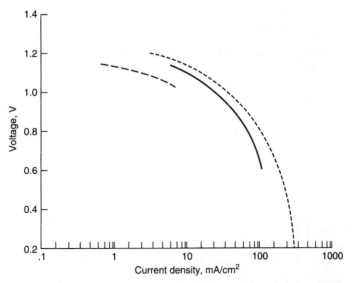

FIGURE 25.28 Voltage polarization characteristics for silver-iron system in experimental tests on three conventional types of cell designs. - - -, bipolar; ——, prismatic; – – –, jelly roll. (*Courtesy of Westinghouse Electric Corp.*)

25.7 IRON MATERIALS AS CATHODES

Iron has conventionally been used as the anode or negative active material in batteries but iron compounds have also been used as the cathode or positive active material. The use of iron sulfides (FeS and FeS_2) in lithium primary and in high temperature batteries is covered in Chapters 14 and 41.

Recently, an iron oxide, having a high valence state, has been reported for use as a cathode active material.[24] Iron normally exists as a metal or in the valence states of Fe(II) and Fe(III). The new cathode material is an Fe(VI)-containing compound which has a high specific capacity due to a 3-electron change in its reduction reaction, as follows:

$$FeO_4^{2-} + 3H_2O + 3e \rightarrow FeOOH + 5OH^- \qquad E^0 = \sim 0.9 \text{ Volt}$$

The theoretical capacity of several of these Fe(VI) compounds is listed in Table 25.6. These values can be compared with the values given in Table 1.1 for the more conventional cathode materials.

TABLE 25.6 Theoretical Capacities of Fe(VI) Compounds

Material	Molecular weight (g.)	Valence change	Electrochemical Equivalence	
			(mAh/g)	(g/Ah)
Li_2FeO_4	133.7	3	601	1.66
Na_2FeO_4	165.8	3	485	2.06
K_2FeO_4	198.1	3	406	2.46
$BaFeO_4$	257.2	3	313	3.19

The characteristics of Fe(VI) compounds have not been studied extensively in the past mainly because of the perception that these materials are highly unstable. While Li_2FeO_4 and Na_2FeO_4 are soluble in alkaline hydroxide, $BaFeO_4$ and K_2FeO_4 show evidence of low alkaline solubility and high stability as shown in Fig. 25.29. Further, their stability is greater in the more concentrated alkaline solutions that are used as battery electrolytes. This data has been extrapolated to suggest that, over a 10-year period, there will be less than a 10% loss of Fe(VI) in concentrated potassium hydroxide solutions using highly purified materials.

Electrochemically, the FeO_4^{2-} species have a high reduction potential, on the order of 0.9 V. Against an anode of zinc, the open circuit potential was found to be 1.75 V and 1.85 V for the K_2FeO_4 and $BaFeO_4$ cell, respectively. The proposed discharge reaction mechanism is as follows:

$$MFe(VI)O_4 + \tfrac{3}{2}Zn \rightarrow \tfrac{1}{2}Fe(III)_2O_3 + \tfrac{1}{2}ZnO + MZnO_2$$

where $M = K_2$ or Ba

The theoretical capacities and specific energy for the two batteries are given in Table 25.7. These values can be compared with those of other batteries listed in Table 1.2. The values for the zinc/iron oxide cells are higher than most of the other batteries with the exception of the lithium and air-breathing systems.

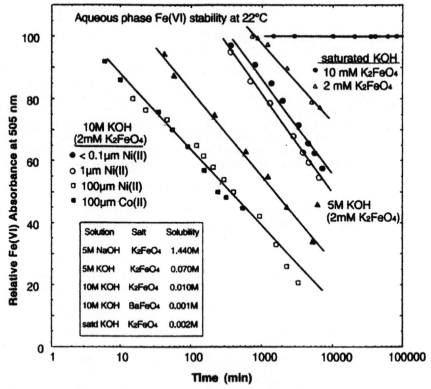

FIGURE 25.29 Stability of Fe(VI) in alkaline electrolyte with various concentrations of OH^-, K_2FeO_4 salts, and Co(II) and Ni(II) impurities. (*From Ref. 24.*)

TABLE 25.7 Theoretical Capacity and Specific Energy for $MFeO_4$ batteries

Couple	Open circuit voltage (V)	Theoretical specific capacity (g/Ah)	Theoretical specific capacity (Ah/g)	Theoretical specific energy (Wh/kg)
Zn/K_2FeO_4	1.75	3.68	0.271	475
$Zn/BaFeO_4$	1.85	4.41	0.226	419

The discharge characteristics and specific energy of experimental primary alkaline button cells, with zinc anodes and Fe(VI) compounds cathodes, were measured and compared to those with MnO_2 cathodes. These data are plotted in Figure 25.30 and illustrate the higher energy output of the cells fabricated with the Fe(VI) cathodes. Similar results were obtained with cells fabricated in the conventional cylindrical construction.

The Fe(VI) compounds were also shown to be rechargeable. A button cell, using a metal hydride anode and a capacity limited K_2FeO_4 cathode, was cycled for several cycles to a 75% depth of discharge and for more than 400 cycles to a 30% depth of discharge. The open circuit voltage of the cell was 1.3 V and the midpoint voltage was 1.1 V, similar to the voltage characteristics of the nickel-metal hydride cell.

The Fe(VI) compounds are promising candidates for cathode materials for both primary and rechargeable alkaline batteries. The reported results demonstrate their higher specific energy compared to other cathode materials now used in alkaline batteries. The questions of long-term stability, shelf life and other critical performance characteristics, large scale manufacture, materials cost, and so on, still have to be resolved and are subject to further evaluation.

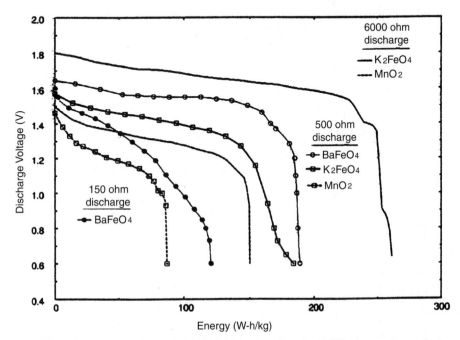

FIGURE 25.30 Capacity of several experimental button cells using Fe(VI) compounds as cathodes and Zn as anodes, compared to conventional Zn/MnO_2 button cells. (*From Ref. 24.*)

REFERENCES

1. S. U. Falk and A. J. Salkind, *Alkaline Storage Batteries,* Wiley, New York, 1969.

2. A. J. Salkind, C. J. Venuto, and S. U. Falk, "The Reaction at the Iron Alkaline Electrode," *J. Electrochem. Soc.* **111:**493 (1964).

3. R. Bonnaterre, R. Doisneau, M. C. Petit, and J. P. Stervinou, in J. H. Thompson (ed.), *Power Sources,* vol. 7, Academic, London, 1979, p. 249.

4. L. Ojefors, "SEM Studies of Discharge Products from Alkaline Iron Electrodes," *J. Electrochem. Soc.* **123:**1691 (1976).

5. B. Anderson and L. Ojefors, in J. H. Thompson (ed.), *Power Sources,* vol. 7, Academic, London, 1979, p. 329.

6. J. L. Weininger, in R. G. Gunther and S. Gross (eds.), *The Nickel Electrode,* vol. 82-4, Electrochemical Society, Pennington, N.J., 1982, pp. 1–19.

7. D. Tuomi, "The Forming Process in Nickel Positive Electrodes," *J. Electrochem. Soc.* **123:**1691 (1976).

8. INCO ElectroEnergy Corp. (formerly ESB, Inc.), Philadelphia, Pa.

9. "Nickel Iron Industrial Storage Batteries," Exide Industrial Marketing Divisions of ESB, Inc., 1966.

10. F. E. Hill, R. Rosey, and R. E. Vaill, "Performance Characteristics of Iron Nickel Batteries," *Proc. 28th Power Sources Symp.,* Electrochemical Society, Pennington, N.J., 1978, p. 149.

11. R. Rosey and B. E. Tabor, "Westinghouse Nickel-Iron Battery Design and Performance," EV Expo 80, EVC #8030, May 1980.

12. W. Feduska and R. Rosey, "An Advanced Technology Iron-Nickel Battery for Electric Vehicle Propulsion," *Proc. 15th IECEC,* Seattle, Aug. 1980, p. 1192.

13. R. Hudson and E. Broglio, "Development of the Nickel-Iron Battery System for Electric Vehicle Propulsion," *Proc. 29th Power Sources Conf.,* Electrochemical Society, Pennington, N.J., 1980.

14. L. Carlsson and L. Ojefors, "Bifunctional Air Electrode for Metal-Air Batteries," *J. Electrochem. Soc.* **127:**525 (1980).

15. L. Ojefors and L. Carlson, "An Iron-Air Vehicle Battery," *J. Power Sources,* **2:**287 (1977/78).

16. J. F. Jackovitz and C. T. Liu, *Extended Abstracts: 9th Battery and Electrochemical Contractors' Conf.,* USDOE, Alexandria, Va., Nov. 12–16, 1989, pp. 319–324.

17. H. Cnoblock, D. Groppel, D. Kahl, W. Nippe, and G. Siemsen, in D. H. Collins (ed.), *Power Sources,* vol. 5, Academic, London, 1975, p. 261.

18. O. Lindstrom, in D. H. Collins (ed.), *Power Sources,* vol. 5, Academic, London, 1975, p. 283.

19. *The Silver Institute Letter,* vol. 7, no. 3 (1977).

20. J. T. Brown, Extended Abstract No. 28, Battery Div. The Electrochemical Society, Las Vegas (NV), pp. 76–77, (1977).

21. E. Buzzelli, "Silver-Iron Battery Performance Characteristics," *Proc. 28th Power Sources Symp.,* Electrochemical Society, Pennintgon, N.J., 1978, p. 160.

22. G. A. Bayles, E. S. Buzzelli, and J. S. Lauer, "Progress in the Development of a Silver-Iron Communications Battery," *Proc. 34th Int. Power Sources Symp.,* Cherry Hill, N.J., June 1990.

23. G. A. Bayles, J. S. Lauer, E. S. Buzzelli, and J. F. Jackovitz, "Silver-Iron Batteries for Submersible Applications," *Proc. 3rd Annual Underwater Vehicle Conf.,* Baltimore, Md., June 1989.

24. S. Licht, B. Wang, and S. Ghosh, *Science* **128:**1039–1042 (1999).

CHAPTER 26
INDUSTRIAL AND AEROSPACE NICKEL-CADMIUM BATTERIES

Arne O. Nilsson and Christopher A. Baker

26.1 INTRODUCTION

The vented pocket-plate battery is the oldest and most mature of the various designs of nickel-cadmium batteries available. It is a very reliable, sturdy, long-life battery, which can be operated effectively at relatively high discharge rates and over a wide temperature range. It has very good charge retention properties, and it can be stored for long periods of time in any condition without deterioration. The pocket-plate battery can stand both severe mechanical abuse and electrical maltreatment such as overcharging, reversal, and short-circuiting. Little maintenance is needed on this battery. The cost is lower than for any other kind of alkaline storage battery; still, it is higher than that of a lead-acid battery on a per Watthour basis. The major advantages and disadvantages of this type of battery are listed in Table 26.1.

The pocket plate battery is manufactured in a wide capacity range, 5 to more than 1200 Ah, and it is used in a number of applications. Most of these are of an industrial nature, such as railroad service, switchgear operation, telecommunications, uninterruptible power supply, and emergency lighting. The pocket plate battery was also used in military and space applications.

The pocket plate batteries are available in three plate thicknesses to suit the variety of applications. The high-rate designs use thin plates for maximum exposed plate surface per volume of active material. They are used for the highest-rate discharge. The low-rate designs

TABLE 26.1 Major Advantages and Disadvantages of Industrial and Aerospace Nickel-Cadmium Batteries

Advantages	Disadvantages
Long cycle life	Low energy density
Rugged; can withstand electrical and physical abuse	Higher cost than lead-acid batteries
Reliable; no sudden death	Contains cadmium
Good charge retention	
Excellent long-term storage	
Low maintenance	

use thick plates to obtain maximum volume of active material per exposed plate surface. These types are used for long-term discharge. The medium-rate designs use plates of middle thickness and are suited for applications between, or combinations of, high-rate and long-term discharge.

Developmental work has been conducted almost continuously since the introduction of the pocket-plate nickel-cadmium battery to improve the performance characteristics of the battery and reduce weight. The sintered plate, which can be constructed in a thinner form than the pocket plate, was developed during the 1940s and has a lower internal resistance and gives superior high rate and low temperature performance compared to the pocket plate. It is used in high-power applications, such as engine starting, and in low-temperature environments. The sintered-plate battery is covered in Chap. 27. Further development of the sintered plate led to the design of smaller batteries for portable equipment and subsequently to the sealed, maintenance-free nickel-cadmium battery covered in Chap. 28.

The sintered-plate battery was found to be too expensive and complex to manufacture. It used a large amount of nickel and was impractical for medium-rate thick electrodes or for cells larger than 100 Ah. The pocket plate battery was too heavy for many applications. Recent developmental work has been directed to more effectively use the costly materials—nickel and cadmium—and toward simplified manufacturing processes. The design philosophy was to develop a high-surface-area, conductive-plate structure that would be light, easy to manufacture, inexpensive and to eliminate the troublesome aspects of sintered plate technology, namely, the sintering process and the chemical impregnation of the active materials. Taking advantage of new polymer materials and plating techniques, this work has resulted in a new electrode structure, the fiber-structured electrode (the Fiber Nickel Cadmium Battery—FNC) developed by Deutsche Automobilgesellschaft mbH (DAUG).

The fiber plates are manufactured from either a mat of pure nickel fibers or more commonly nickel-plated plastic fibers. To make the plastic fiber conductive, a thin layer of nickel is applied by electroless plating and thereafter a sufficiently thick layer of nickel for good conductivity is applied by electroplating. The plastic is then burned off, leaving a mat of hollow nickel fibers. The nickel fiber plaque is welded to a nickel-plated steel tab. Figure 26.1a shows the structure of a nickel fiber plaque before impregnation and Fig. 26.1b shows an unformed positive electrode.

This fiber electrode technology, while originally developed for EV applications, was first used for industrial low and medium rate vented cells. It is now being used in all types of nickel-cadmium as well as nickel-metal hydride batteries, including high-rate batteries for engine-cranking and sealed cells with oxygen recombination. Details of this technology are covered in Sec. 26.6.

A more recent design that has shown significantly improved performance characteristics is the plastic-bonded or pressed-plate electrode. This new development of electrode materials in industrial batteries is a spin-off from the development of electrode materials for use in aircraft and sealed portable consumer batteries. In the plastic-bonded plate, which is mainly used in the cadmium electrode, the active material cadmium oxide is mixed with a plastic powder, normally PTFE, and a solvent to produce a paste. The paste is isotropic and the materials are manufactured at the final density for the active material. As a result, dust problems are eliminated during manufacturing. The paste is extruded, rolled, or pasted onto a center current collector normally made of nickel-plated perforated steel. The plate structure is welded to nickel-plated steel tabs.

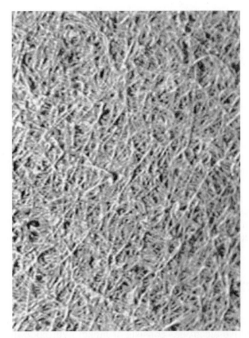

FIGURE 26.1a Nickel-fiber electrode structure before impregnation. (*Courtesy Acme Electric Corp.*)

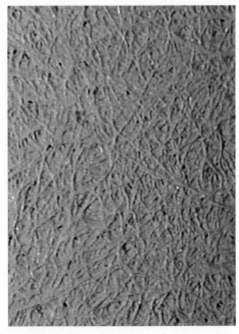

FIGURE 26.1b Unformed, pasted nickel positive electrode. (*Courtesy Acme Electric Corp.*)

26.2 CHEMISTRY

The basic electrochemistry is the same for the vented pocket plate, sintered plate, fiber and plastic-bonded plate types as well as for other variations of the nickel-cadmium system. The reactions of charge and discharge can be illustrated by the following simplified equation:

$$2NiOOH + 2H_2O + Cd \underset{\text{charge}}{\overset{\text{discharge}}{\rightleftharpoons}} 2Ni(OH)_2 + Cd(OH)_2$$

On discharge, trivalent nickel oxy-hydroxide is reduced to divalent nickel hydroxide with consumption of water. Metallic cadmium is oxidized to form cadmium hydroxide. On charge, the opposite reactions take place. The electromotive force (EMF) is 1.29 V.

The potassium hydroxide electrolyte is not significantly changed with regard to density or composition during charge and discharge, in contrast to the sulfuric acid in lead-acid batteries. The electrolyte density is generally approximately 1.2 g/mL. Lithium hydroxide is often added to the electrolyte for improved cycle life and high-temperature operation. A more detailed description of the cell reaction on overcharge is found in Sec. 26.6.

26.3 CONSTRUCTION

A cutaway view of a modern pocket-plate cell is shown in Fig. 26.2. The active material for the positive electrodes consists of nickel hydroxide mixed with graphite for conductivity and

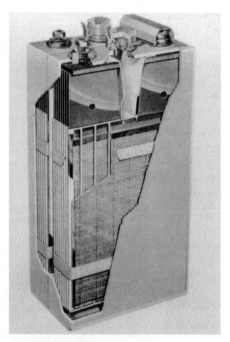

FIGURE 26.2   Pocket plate cells.

additives such as barium or cobalt compounds for improved life and capacity. The active material for the negative electrodes is prepared from cadmium hydroxide or cadmium oxide mixed with iron or iron compounds and sometimes also with nickel. The iron and nickel materials are added to stabilize the cadmium, prevent crystal growth and agglomeration, and improve conductivity. Typical active material compositions are shown in Table 26.2.

TABLE 26.2 Typical Composition of Active Materials for Pocket Plate Cells in the Discharged State

Positive active material		Negative active material	
Substance	Weight %	Substance	Weight %
Nickel (II) hydroxide	80	Cadmium hydroxide	78
Cobalt (II) hydroxide	2	Iron	18
Graphite	18	Nickel	1
		Graphite	3

The positive and negative electrodes of pocket-plate nickel-cadmium batteries are made using the same basic design to hold the active materials. The pocket plates are built up of flat pockets of perforated steel strips holding the active materials. The thin steel strips are perforated by hardened steel needles or by a technique using profiled roller dies. The specific hole area is between 15 and 30%. The strips are nickel-plated to prevent "iron poisoning" of the positive active material.

The active mass is either pressed into briquettes, which are fed into the preshaped perforated strip, or fed into the preshaped strip as a powder. The upper and lower steel strips are folded together by rollers. A number of these folded strips are arranged to interlock with each other to form long electrode sheets, which are then cut to electrode blanks. Electrodes are made from these blanks by providing them with steel frames for mechanical stability and for current takeoff.

The electrodes are made with different thicknesses (1.5 to 5 mm) to provide cells for high, medium and low-rate discharge rates. The negative plate is always thinner (30 to 40%) than the positive.

The electrodes are bolted or welded to electrode groups. Plate groups of opposite polarity are intermeshed and electrically separated from each other by plastic pins and plate edge insulators, Sometimes separators or perforated plastic sheets or plastic ladders are used between the electrodes. The distance between plates of different polarity in an element may vary from less than 1 mm for high-rate cells to 3 mm for low-rate cells.

The elements are inserted into cell containers of plastic or stainless steel. Plastic containers are made from polystyrene, polypropylene, or flame-retardant plastics. Important advantages of plastic containers over steel containers are that they allow visual control of the electrolyte level and they require no protection against corrosion. Also, they have lower weight and they can be more closely packed in the battery. The main drawbacks are that they are more sensitive to high temperatures and they require somewhat more space than steel containers. A plastic-bonded plate cell in a plastic container is shown in Fig. 26.3.

FIGURE 26.3 Plastic-bonded plate cell.

Figure 26.4 shows a partial cut-away view of a fiber nickel-cadmium (FNC) cell. The case and cover are polypropylene and are welded together. The electrode assembly shows a negative electrode, a corrugated separator and a positive electrode. O-ring seals are employed in the bushings for the terminals to ensure gas retention and a vent valve is seen in the case cover between the terminals. A catalytic gas recombination plug in the vent valve is employed for some applications. The terminals are nickel-plated copper, and a 1.19 kg/L KOH electrolyte is typically employed.

Cells are assembled into batteries in many different ways. Often 2 to 10 cells are mounted in a separate battery unit, several of which may be used to form the complete battery. A typical battery is shown in Fig. 26.5. Cells in plastic containers are also assembled into batteries by putting the single cells close together on a rack or a stand and connecting them with intercell connectors. This is especially the case for stationary applications (Fig. 26.6). Cells in steel containers can be assembled in a similar way; however, here the cells must be spaced from one another and insulated from the rack.

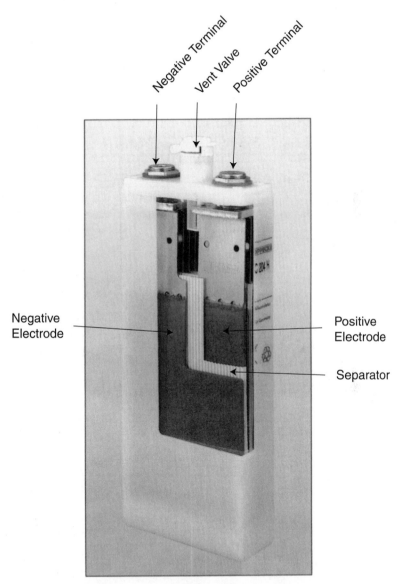

FIGURE 26.4 Partial cut-away view of fiber nickel-cadmium (FNC) cell. (*Source: Hoppecke Batteries.*)

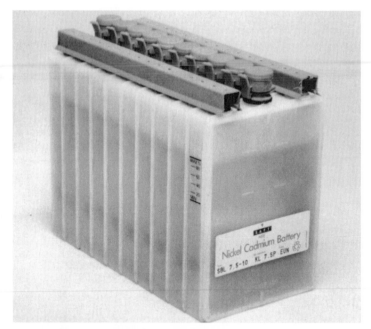

FIGURE 26.5 Ten-cell welded polypropylene unit.

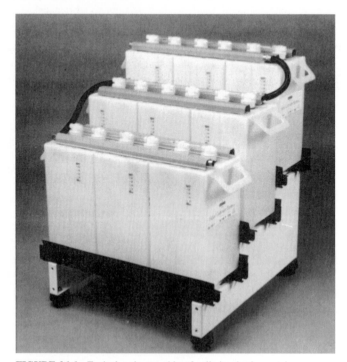

FIGURE 26.6 Typical rack assembly of cells in plastic containers.

26.4 PERFORMANCE CHARACTERISTICS

26.4.1 Energy Density and Specific Energy

Typical specific energy and energy density values for pocket plate single-cell batteries are 20 Wh/kg and 40 Wh/L, with the best values for commercially available units reaching 27 Wh/kg and 55 Wh/L. The corresponding values for complete pocket plate batteries are 19 Wh/kg and 32 Wh/L and 27 Wh/kg and 44 Wh/L, respectively. These data are based on the nominal capacity and the average discharge voltage at the 5-h rate. The specific energy and energy density of larger fiber plate batteries approach 40 Wh/kg and 80 Wh/L. Batteries with plastic-bonded plates approach 56 Wh/kg and 110 Wh/L. This compares to a specific energy of 30 to 37 Wh/kg and an energy density of 58 to 96 Wh/L for sintered-plate designs. See Chap. 27.

26.4.2 Discharge Properties

The nominal voltage of a nickel-cadmium battery is 1.2 V. Although discharge rate and temperature are of importance for the discharge characteristics of all electrochemical systems, these parameters have a much smaller effect on the nickel-cadmium battery than on, for instance, the lead-acid battery. Thus pocket plate nickel-cadmium batteries can be effectively discharged at high discharge rates without losing much of the rated capacity. They can also be operated over a wide temperature range.

Typical discharge curves at room temperature for pocket plate and plastic-bonded plate batteries at various constant discharge rates are shown in Fig. 26.7. Even at a discharge current as high as $5C$ (where C is the numerical value of the capacity in Ah), a high-rate pocket plate battery can deliver 60% of the rated capacity and a plastic-bonded battery as much as 80%. Battery capacities as a function of discharge rate and cutoff voltage are given in Fig. 26.8.

Pocket-plate nickel-cadmium batteries can be used at temperatures down to $-20°C$ with the standard electrolyte. Cells filled with a more concentrated electrolyte can be used down to $-50°C$. Figure 26.9 shows the effect of temperature on the relative performance of a nickel-cadmium medium-rate battery with standard electrolyte.

Batteries can also be used at elevated temperatures. Although occasional operation at very high temperatures is not detrimental, 45 to 50°C is generally considered as the maximum permissible temperature for extended periods of operation.

Figure 26.10 shows typical so-called starter curves for high-rate batteries. The batteries can deliver as much current as $20C$ A during 1 s to a final voltage of 0.6 V.

Occasional overdischarge or reversal of nickel-cadmium batteries is not detrimental, nor is complete freezing of the cells. After warming up, they will function normally again.

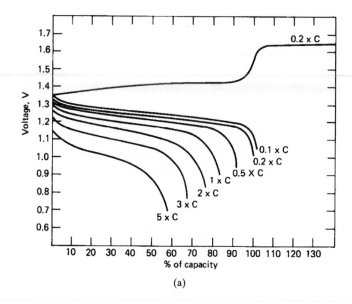

(a)

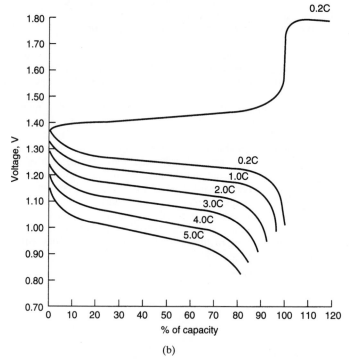

(b)

FIGURE 26.7 Charge and discharge characteristics of nickel-cadmium batteries at 25°C. (*a*) Pocket plate battery, high rate. (*b*) Plastic-bonded plate battery, high rate.

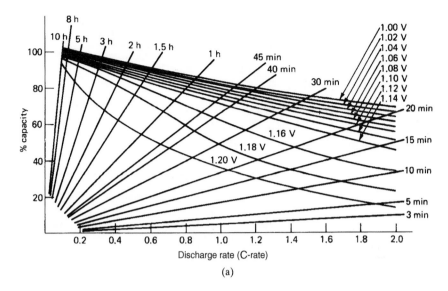

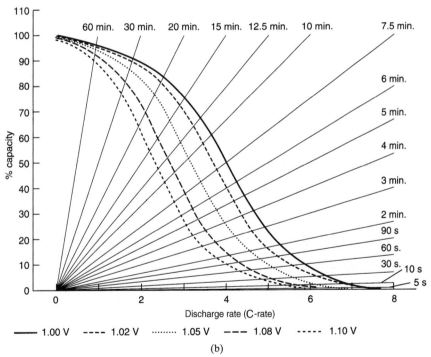

FIGURE 26.8   Discharge characterstics of nickel-cadmium batteries at 25°C; capacity as a function of discharge rate and cutoff voltage. (*a*) Pocket plate battery, high rate. (*b*) Plastic-bonded plate battery, high rate.

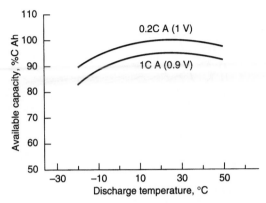

FIGURE 26.9   Typical available capacity at different temperatures for nickel-cadmium medium-rate batteries with standard electrolyte, fully charged at 25°c.

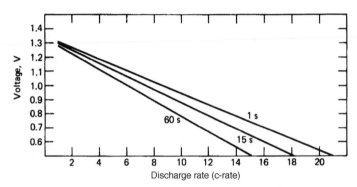

FIGURE 26.10   Voltage-current curves for high-rate pocket plate batteries at 25°C.

26.4.3   Internal Resistance

Nickel-cadmium batteries generally have a low internal resistance. Typical DC resistance values are 0.4, 1, and 2 mΩ, respectively, for a charged 100-Ah high-, medium-, and low-rate pocket plate single-cell battery. The internal resistance is largely inversely proportional to the battery size in a given series. Decreasing temperature and decreasing state of charge of a battery will result in an increase of the internal resistance. The internal resistance of fiber-plate batteries is 0.3 mΩ for a high-rate design and 0.9 mΩ for a low-rate design. Plastic-bonded plate batteries have an internal resistance as low as 0.15 mΩ.

26.4.4   Charge Retention

Charge retention characteristics of vented pocket-plate batteries at 25°C are shown in Fig. 26.11. Charge retention is temperature-dependent, the capacity loss at 45°C being about three times higher than at 25°C. There is virtually no self-discharge at temperatures lower than −20°C. Charge retention for fiber and plastic-bonded plate batteries has similar characteristics; their charge retention corresponds to that shown in Fig. 26.11 for high-rate batteries.

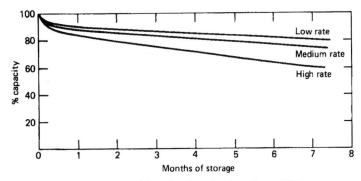

FIGURE 26.11   Charge retention of pocket plate batteries at 25°C.

26.4.5   Life

The life of a battery can be given either as the number of charge and discharge cycles that can be delivered or as the total lifetime in years. Under normal conditions a nickel-cadmium battery can reach more than 2000 cycles. The total lifetime may vary between 8 and 25 years or more, depending on the design and application and on the operating conditions. Batteries for diesel engine cranking normally last about 15 years, batteries for train lighting have normal lives of 10 to 15 years, and stationary standby batteries have lives of 15 to 25 years.

Factors which affect the battery life are the operating temperature, the discharge depth, and the charging regime. Low or moderate operating temperatures should always be preferred. Batteries operating at elevated temperatures or in cycling applications should be filled with electrolyte to which lithium hydroxide has been added.

The factors behind the excellent reliability and very long life of the nickel-cadmium batteries are the mechanically strong design, the absence of corrosive attack of the electrolyte on the electrodes and other components in the cell, and, furthermore, the ability of the battery to withstand electrical abuse such as reversal or overcharging and to stand long-time storage in any state of charge.

26.4.6   Mechanical and Thermal Stability

Nickel-cadmium cells and batteries are mechanically very robust and can withstand severe mechanical abuse and rough handling in general. The electrode groups are carefully bolted or, in more recent designs such as FNC, welded together. The cell containers are made of steel or high-impact plastics.

The electrolyte does not attack any of the components in the cell, and, accordingly, there is no risk of decreased strength during the lifetime of the battery. Cases of so-called sudden death due to corroded lugs or terminals cannot occur.

The thermal resistance of the nickel-cadmium batteries is also very good. These batteries can withstand temperatures up to 70°C or more without mechanical damage. Cells in polypropylene or steel containers are the best in this respect. Saline or corrosive environments present no problems for cells in plastic containers.

26.4.7 Memory Effect

The memory effect—the tendency of a battery to adjust its electrical properties to a certain duty cycle to which it has been subjected for an extended period of time—has been a problem with nickel-cadmium batteries in some applications. Pocket, fiber, and plastic-bonded plate cells do not show this tendency. See Sec. 27.7.2 for a description of the memory effect with sintered-plate nickel-cadmium batteries.

26.5 *CHARGING CHARACTERISTICS*

Pocket-plate nickel-cadmium batteries may be charged at constant current, constant voltage, or modified constant voltage. Constant-current charge characteristics are shown in Fig. 26.7. Charging is normally carried out at the 5-h rate for 7 h for a fully discharged battery. Overcharging is not detrimental but should be avoided as it leads to increased gassing and decomposition of water. Charging can be carried out in the temperature range of -50 to $45°C$. However, at the extreme temperatures, the charging efficiency is lower.

Constant-voltage charging characteristics with current limitations are shown in Fig. 26.12. The current is often limited to 0.1 to $0.4C$ A, and charging is normally carried out in the voltage range of 1.50 to 1.65 V per cell. The charging time may vary from 5 to more 25 h, depending on current limitation value and cell type.

In some applications such as emergency and standby, it is necessary to keep the battery in a high state of charge. A convenient way is to connect the battery in parallel with the ordinary current source and the load and to float the battery at 1.40 to 1.45 V per cell. The floating may be combined with a supplementary charge at fixed intervals or after each discharge.

The Ampere-hour efficiency of the pocket plate battery is 72% when going from the discharged to the fully charged state. The Watt-hour efficiency is approximately 60%. The best plastic-bonded plate batteries have an Ampere-hour efficiency of 85% and a Watt-hour efficiency of 73%.

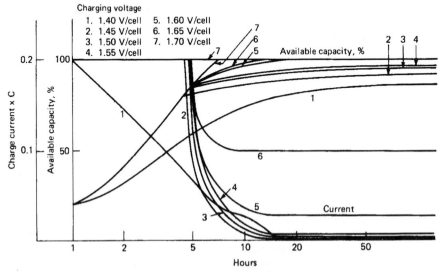

FIGURE 26.12 Constant-voltage charging with current limitation $0.2C$ of medium-rate pocket plate batteries at 25°C.

26.6 FIBER NICKEL-CADMIUM (FNC) BATTERY TECHNOLOGY

26.6.1 FNC Electrode Technology

An ideal electrode will feature the following characteristics:

- Provide a high surface area conductive matrix to contact the active material
- Provide sufficient porosity for high active material loading and an open structure for good electrolyte penetration.
- Have sufficient electrical conductivity to carry the current to the tab with minimal voltage drop, yet will still be light
- Fully contain the active material
- Able to accommodate the dimensional changes during battery charge and discharge without fatigue cracking
- Tolerate mechanical shock and vibration
- Chemically and thermally inert to the battery environment. Will not introduce any undesirable impurities into the cell.
- Utilize a simple process for the loading of the active material
- Strong enough to tolerate the cell manufacturing processes
- Versatile enough to allow manufacturing of various sizes, thickness, conductivity, and porosity
- Economical

It is in this area of plate design that the FNC technology has made a significant step forward in comparison to older technologies. The core of the FNC technology is the three-dimensional nickel-plated fiber matrix. The nickel coating is optimized to the expected current density of the battery. Thus, there is no excess nickel. Electrodes of thickness ranging from 0.5 to 10 mm targeted at ultra high (XX), high (X and H), medium (M), and low rate (L) designs are fabricated in a common process. The nickel fibers are very compact with one cubic centimeter of electrode volume nominally containing 300 meters of conducting filament. This current collecting matrix is 90% porous, allowing excellent utilization of the active material. The result is improved low temperature performance, a lower charge coefficient, and significantly higher power capability (see Fig. 26.1a).

The structure is highly porous and thus allows for high loading of active materials as well as excellent penetration of electrolyte. No conductive diluents like graphite or iron are required. Yet, due to the very high surface area of the fibers, the contact between the current carrying fiber matrix and the paste is very good. Because of this, losses are low, resulting in improved efficiency. The paste is loaded into the electrodes mechanically, and no impurities are introduced in the process. Active material (nickel hydroxide in the positive plate and cadmium hydroxide in the negative) is mechanically imbedded directly into the fiber plate. The pure active material contributes to longer life, lower self discharge and a more consistent and reliable product (see Fig. 26.1b). Consequently, a high surface area plate capable of high current loads and very long life has been realized.

The processes and cell design associated with FNC technology have resulted in improved battery performance. Improved charging efficiency has reduced the gassing on overcharge, and with it, the frequency of water topping for the vented cells. The design of the FNC plate allows elastic expansion and contraction during charge and discharge, eliminating one of the main causes of nickel cadmium plate degradation. This plate flexibility also provides increased shock and vibration tolerances. The flexibility of the electrode structure eliminates mechanical cracks and associated plate degradation resulting in increased battery life.

26.6.2 Manufacturing Flexibility

The power capability of a battery will affect its potential applications. A high power battery will be capable of delivering most of its capacity in a few minutes. To maximize battery power, it is necessary to minimize cell resistance. To that end, high power cells are designed with high surface areas, thin electrodes and with high metallic contents. However, the above measures will increase battery weight, volume, and cost. Given these tradeoffs, it is desirable to optimize the cell design for the application. Thus, it is more efficient and economical to build a cell for lower power applications utilizing thick electrodes with low metallic contents.

Practical manufacturing constraints inhibited the development of high-capacity low-rate sintered-plate batteries. In contrast, the FNC technology covers a wide range of power capabilities. The thickness of the fiber electrode, and the amount of conductive metallic nickel on it, are varied within an order of magnitude. This results in the capability to produce high-capacity, low-weight, low-cost batteries or ultra-high power, higher weight, and higher cost cells using the same processes and equipment. For the user, this means that the characteristics of sinter foils and the various types of pocket or foam-plate batteries no longer have to be considered separately. The FNC system has the same properties and basic characteristics over the entire range of applications.

26.6.3 Sealed Versus Vented Designs

The charging of aqueous-nickel batteries always occurs in competition with water electrolysis. Toward the end of the charging cycle, oxygen is typically evolved at the positive electrode and hydrogen may be evolved at the negative electrode. The way that the cell deals with these evolved gases will determine whether the cell can be sealed. In sealed cells, the gases are recombined internally. In an open cell, the gases are allowed to vent, hence the name vented cell.

The reactions that produce the gases are called the overcharge reactions. These reactions differ depending on whether one deals with a sealed or vented cell.

Overcharge Reactions:

Vented (open) cell:

Positive:
$$4OH^- \rightarrow 2H_2O + O_2 + 4e^-$$

Negative:
$$\underline{4H_2O + 4e^- \rightarrow 2H_2 + 4OH^-}$$

Net:
$$2H_2O \rightarrow 2H_2 + O_2$$

The net result here is the electrolysis of water to hydrogen and oxygen.

Sealed Cell:

Positive:
$$4OH^- \rightarrow 2H_2O + O_2 + 4e^-$$

Negative:
$$\underline{2Cd(OH)_2 + 4e^- \rightarrow 2Cd + 4OH^-}$$

Net Electrochemical:
$$2Cd(OH)_2 \rightarrow 2Cd + 2H_2O + O_2$$

Chemical
Recombination on Negative:
$$2Cd + O_2 + 2H_2O \rightarrow 2Cd(OH)_2$$

The result in a sealed cell is that electricity is converted into heat without any net chemical change in the cell. The overcharge reaction is exothermic, particularly the chemical recombination reaction in the sealed cell.

The overcharge reaction on the positive plate starts before the cell is fully charged, so that some oxygen evolution on charging is unavoidable. At higher temperatures, the oxygen evolution starts at a lower voltage. This results in lower charging efficiencies at higher temperatures and, in the case of the vented cell, an increased need for the addition of water. Additionally, this leaves the positive plate undercharged while the negative plate continues toward a full charged state. The resulting plate imbalance reduces battery capacity. To regain the lost capacity, vented batteries require deep discharge conditioning with each cell being clipped out (shorted).

26.6.4 Sealed FNC Maintenance-Free Batteries

With the development of sealed FNC technology, the first maintenance-free high rate prismatic nickel cadmium battery was introduced. In the sealed cell, an unfilled nickel-coated fiber plate is placed between two cadmium filled negative plates. This effectively results in a split negative with an unfilled central region. This inactivated section serves as a catalytic site for rapid oxygen reduction. The main oxygen pathway to the recombination site is through the plate pores, which, in the FNC plate construction, are relatively large. This provides the oxygen with easy access to the large recombination reaction surfaces, as shown in Fig. 26.13.

Because rapid oxygen recombination eliminates the pressure buildup normally associated with sealed nickel cadmium cells, high charging rates can be sustained even in the overcharge mode. Also, it is possible to use conventional nylon cell construction to produce prismatic sealed cells rather than the cylindrical design required for high pressure cells. The sealed

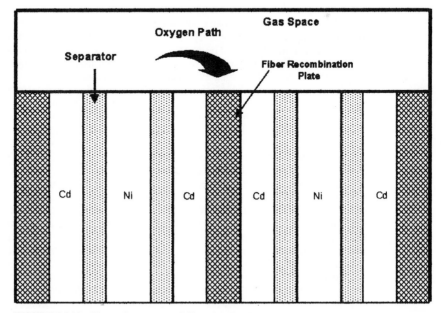

FIGURE 26.13   Electrode structure of fiber nickel-cadmium cell.

FNC prismatic cell case is made of either polyamide (nylon) or stainless steel. The negative pressure within the cell (approximately 0.1 bar absolute) allows for pressure change due to oxygen generation during overcharge without causing the cell walls to expand. In addition, the external air pressure presses the vessel walls and plate assembly together, enhancing the hydraulic contact between the electrodes and providing mechanical rigidity. The plastic cell cases are covered with an aluminum foil wrap which serves as a gas barrier to block the migration of gases and water through the plastic. Many years of operation are possible without gasses penetrating the cell, assuring the electrode materials remain active.

All nickel cadmium batteries must be overcharged to achieve a 100% state of charge. During the overcharge stage, excessively charged portions evolve oxygen and hydrogen. In vented batteries these gasses, along with water vapor, are vented outside the cell. The lost liquid must be replaced. The sealed FNC battery completely eliminates the loss of any gases from within the cell. Oxygen generated is rapidly recombined on the negative electrode. Hydrogen evolution is avoided by excess discharged cadmium on the negative electrode. This recombination process also keeps the plates in balance, eliminating the capacity loss that would otherwise result. Electrolyte spillage and corrosion is completely eliminated. In the absence of free KOH, even aluminum battery boxes may be used to reduce weight.

In the event of cell reversal or a failure to control charging voltage, hydrogen gas will be produced. A recombination plate of Pt/Pd-catalyzed plaque located within the cell provides for hydrogen recombination. The oxygen source, for hydrogen recombination, is provided by the self-discharge reaction on the positive electrode or the overcharge reaction on the following charge.

A safety valve at the top of the cell is provided to allow excessive pressure to escape should the battery be abused to the point that the electrolyte boils. This condition might be caused by a severe overcharge where adequate heat dissipation is not provided. Under high temperature abuse ($+100°C$ and above), the safety valve will open at approximately 45 psia over pressure allowing water vapor to escape. Electrolyte will not be expelled, even with the cell in an inverted position. When the cell is allowed to cool, the valve will reseal and the negative pressure cell will return to a normal operating condition. A reduction in cell capacity may be anticipated due to the loss of water from within the cell.

A high negative to positive capacity ratio (in excess of 2:1) is employed in the cell design. This high excess of negative capacity allows for fast charge without hydrogen evolution over a wide range of temperatures. The excess negative capacity also provides for high power at low temperature. Forming the electrodes prior to stack assembly in order to minimize the carbonate level within the sealed FNC cell further increases low temperature performance.

Positive and negative plates are connected to their respective terminal posts by nickel tabs. The tabs are attached directly to the fiber plates by a patented welding process and then fastened directly to nickel-plated, solid-copper terminal posts. The electrical path of each cell type is designed for maximum electrical performance.

Plate stacking is the same as previously discussed. Single positive plates are separated from the negative cadmium electrode by an electrolyte wet separator. The cadmium electrode is in three parts: two fiber frameworks carrying the negative active material, and an unfilled fiber recombination electrode placed between them. The large recombination surface area is sufficient to handle a 2 C rate charge on a fully charged battery. With the unfilled recombination plate being the primary gas path for oxygen recombination, small pore size separator can be used. The separator is designed to be completely filled with electrolyte, thus contributing to improved high rate performance. Additionally, the recombination plate acts as a reservoir for electrolyte, allowing for volumes in excess of 4 ml/Ah. This prevents stack dry-out as a possibility for premature cell failure.

Sealed FNC batteries are fail-safe. Even if subjected to extreme overcharge to the point at which the electrolyte boils, the battery will not go into thermal runaway. Instead, the hot cells dry out, with the loss of water vapor causing the battery impedance to increase. As the impedance increases, the current will decrease. After a time, the battery will no longer accept the charge current and it will cool down.

26.6.5 Performance

The high current performance capability of the sealed FNC battery design is exemplified by a short circuit test performed on a model KCF XX47 battery which produced currents approaching 4000 amps (see Fig. 26.14). The KCF XX47 unit was designed to meet the requirements for large auxillary power unit (APU) and direct engine starting. The constant voltage discharge of 12 V demonstrates the extraordinary high power capability of the battery (see Fig. 26.15). The KCF XX47 battery's high power performance is also demonstrated with start curves for a large, wide-body aircraft APU. The first two discharge sequences represent unsuccessful start attempts with the third showing a successful start. The minimum voltage required for this particular specification is 13 V, while the FNC battery provides over 16 V (see Fig. 26.16).

Cold temperature performance available with the sealed FNC cells is also impressive. Fig. 26.17 shows the capacity of a 28 V 47 Ah battery for four different discharge rates with the battery soaked at a temperature of −18°C.

Sealed FNC batteries have shown outstanding cycle life at both low and high rates of discharge. Figure 26.18 provides cycle life data for Low Earth Orbit (LEO) cycle testing. The data demonstrates a cycle life in excess of 10,000 cycles for 35% DOD, 10°C, C/2 cycling. By maintaining the stable end of discharge voltage, the low recharge coefficient (approximately 3%) demonstrates the superior charge efficiency of the sealed FNC cell.

Deep discharge cycling is not necessary, eliminating the necessity for battery removal from the aircraft during maintenance. Capacity checks, if desired, can be accomplished with the battery installed by using a portable discharge/charger unit because the sealed FNC design does not exhibit the memory effect typically found in other NiCd's. This is illustrated in Fig. 26.19 which consists of three capacity checks taken after 7845 60% DOD LEO cycles. The first capacity is directly out of orbital cycling at approximately the C rate. The second capacity value is after a full recharge (C/10 for 16 hours) and C/2 discharge. The third capacity value is after a resistive let down and subsequent full recharge. Of particular importance is the shallow voltage depression that occurs around 48 Amp-hours into the

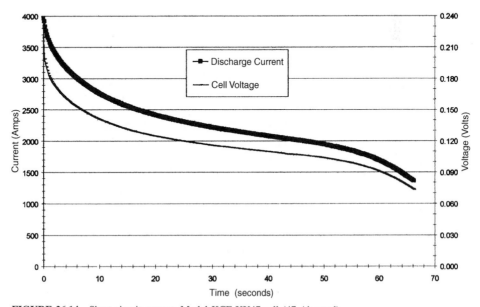

FIGURE 26.14 Short circuit current. Model KCF XX47 cell (47 Ah rated).

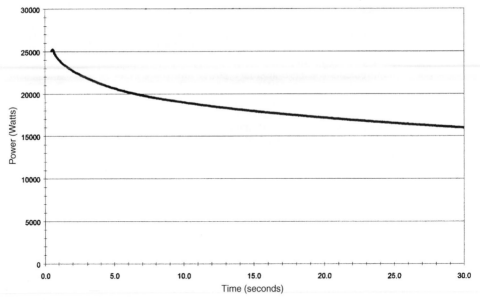

FIGURE 26.15   Constant voltage (12.0 Volts) discharge. Room temperature. Model XX47 battery (47 Ah rated).

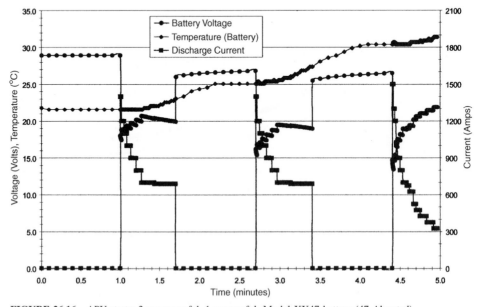

FIGURE 26.16   APU starts: 2 unsuccessful, 1 successful. Model XX47 battery (47 Ah rated).

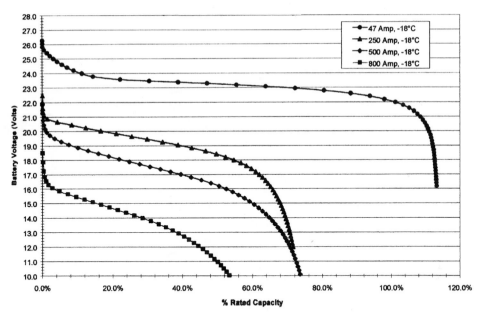

FIGURE 26.17   Constant current discharge. Battery charged at 25°C. Battery discharged at −18°C. Model XX47 battery (47 Ah rated).

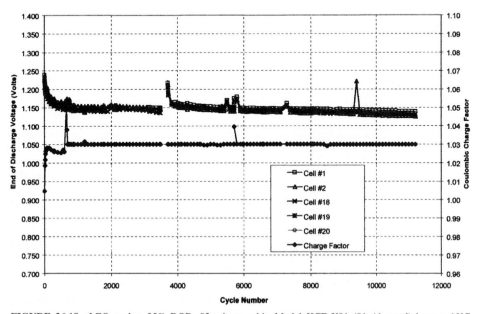

FIGURE 26.18   LEO cycles, 35% DOD, 92 minute orbit. Model KCF X81 (81 Ah rated) battery. 10°C ambient temperature. End of discharge voltage and coulombic charge factor.

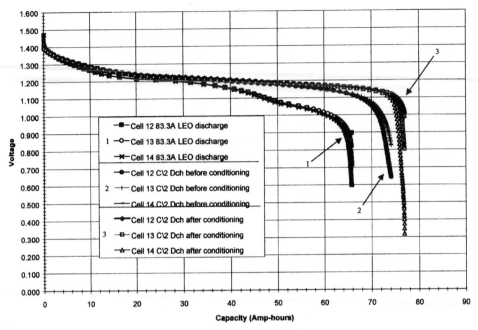

FIGURE 26.19 Capacity checks after 7845 LEO cycles, 60% DOD. Out of orbital capacity, before conditioning, after conditioning. Model KCF X81 (81 Ah rated). 10°C ambient temperature.

discharge, the point where the prior cycles discharge had ended. After 7845 cycles, the cell continued to discharge pasts this point to an out of orbital capacity of 68 Amp hours.

Charging characteristics of the sealed FNC cells are simple, yet different from vented NiCds. Because of the recombination that takes place during overcharge, the normal dV/dt behavior is not always observed. In addition, heat is generated during the overcharge from the recombination reaction, providing a reliable parameter for charge control. Changing from main mode to topping mode and charge termination is determined by battery temperature rise (Δ T).

The preferred charge is at constant current with a voltage clip (maximum voltage) of 1.55 V per cell. This provides an excellent charge rate while keeping the voltage below the level where both hydrogen and oxygen would be generated. This voltage is also sufficient to charge the battery at temperatures as low as −40°C. For many applications, this means that a heater blanket is not required.

A completely charged FNC battery has sufficient recombination to continue to accommodate a 2 C overcharge rate. Even higher rates are possible with the battery in a lower state of charge. A 200 Amp, 47 Ah cell charge profile is shown in Fig. 26.20. Voltage, current and temperature curves are shown for the charge sequence. The discharge voltage curve indicates the cell capacity following the 200 Amp charge.

The superior recombination rate of the oxygen produced in overcharge is demonstrated in Fig. 26.21. Internal cell pressure was taken at three different temperatures for a C rate charge. Note that at the end of charge and into overcharge, the internal pressure is well below atmospheric pressure.

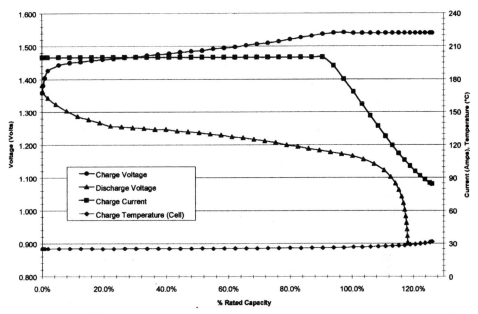

FIGURE 26.20   4C (200 amp) Charge. 1 C discharge. Model KCF XX47 Battery (81 Ah rated).

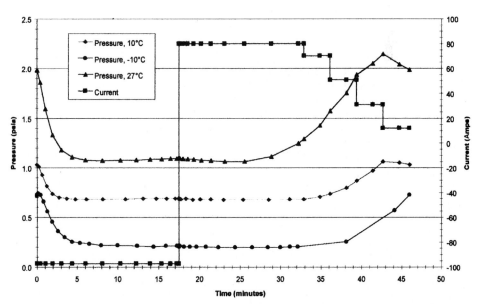

FIGURE 26.21   Cell internal pressure and current for 35% DOD accelerated LEO cycle. Model KCF X81 battery (81 Ah rated).

26.7   MANUFACTURERS AND MARKET SEGMENTS

Table 26.3 contains data regarding prominent manufacturers of industrial nickel-cadmium batteries, Table 26.4 lists the market segments and applications for these batteries.

TABLE 26.3   Major Manufacturers of Industrial and Aerospace Nickel-Cadmium Batteries (Does Not Include Sintered-Plate Designs—See Chap. 27)

Manufacturer/country	Trademark	Product range		
		Pocket-plate	Fiber plate	Plastic-bonded plate
Acme Electric, U.S.A	Acme		X	
Alcad Ltd., U.K.	Alcad	X		
HBL-NIFE Power Systems, India	HBL	X		
Hoppecke Batterien, Germany	Hoppecke		X	
Japan Storage Battery, U.S.A.	GS	X		X
Marathon Battery, U.S.A.	Marathon	X		X
SAFT, S.A., France	SAFT	X		X
Tudor S.A., Spain	Tudor	X	X	
Varta, Germany	Varta	X		
Yuasa, Japan	Yuasa	X		

Table 26.5 lists parameters of one line of standard sealed FNC cells currently manufactured. As detailed in the prior text, the manufacturing processes and cell design have the flexibility to develop capacities and packaging to fit a variety of application.

Figure 26.22 is a picture of a 28 V, 47 Ah airborne battery system with battery and dedicated charger which uses the model XX47 unit described in Table 26.5.

Table 26.6 lists the characteristics of a second line of sealed FNC batteries which are available in X, H, M and L designs. These operate with lower pressure vents than the batteries listed in Table 26.5.

TABLE 26.4 Market Segments and Applications for Vented Industrial Nickel-Cadmium Batteries

	Pocket plate			Fiber plate				Plastic-bonded plate	
Cell range*	H	M	L	XX	H or X	M	L	H	M
Capacity, Ah	10–1000	10–1250	10–1450	23–47	10–220	20–450	20–490	11–190	20–200
Applications	UPS, starting, switchgear	UPS, switchgear, auxiliary power, emergency power	Lighting, alarms, signalling, communications, standby power	Aircraft	UPS, satellites, starting, switchgear, traction, power stations and substations	UPS, switchgear, auxiliary power, emergency power	Lighting, UPS, alarms, signalling, telecommunications, standby power	UPS, starting, switchgear, traction, aircraft	Lighting, auxiliary power, traction
Railroad	X	X	X		X	X	X	X	X
Mass transit	X	X	X		X	X	X	X	X
Industry	X	X	X		X	X	X	X	
Buildings	X	X	X		X	X	X	X	
Hospitals	X	X	X		X	X	X	X	
Oil and gas	X	X	X		X	X	X	X	
Airports	X	X	X		X	X	X	X	
Marine	X	X	X		X	X	X	X	X
Military	X	X	X		X	X	X	X	X
Telecommunications	X	X	X		X	X	X	X	
Photovoltaics			X				X	X	
AGV/hybrid vehicles					X				

X–H—high rate; M—medium rate; L—low rate.
XX—ultra high-rate.

TABLE 26.5 Characteristics of Typical Fiber Nickel Cadmium Batteries

Battery type	Rated capacity (Ah)	Resistance mΩ (fully charged) 25°C	Weight (kg)	Length (mm)	Width (mm)	Height (mm)
X7	6.5	2.6	0.28	57	24	105
X15	15	1.4	0.57	61	29	178
X18	17	1.1	0.73	67	36	147
X26	26	0.8	1.13	115	25	168
X44	44	0.6	1.72	115	41	168
X55	55	0.5	2.05	115	54	166
X68	68	0.4	2.59	115	59	166
X81	81	0.5	3.13	115	54	224
XX23	23	0.7	1.13	115	25	168
XX40	40	0.4	1.72	115	41	168
XX47	47	.035	2.05	115	54	166
H8	8		0.29	57	24	100
H85	78		2.45	115	59	166

Source: Courtesy Acme Electric Corp.

FIGURE 26.22 FNC airborne battery system. Battery (left); charger (right). 28 V, 47 Ah battery.

TABLE 26.6 Standard Line of Sealed FNC Batteries in X, H, M and L Models

Capacities, dimensions and weights

Type	Cs capacity in Ah	Length	Width	Height	Cell weight with electrolyte* max. kg	Cell weight without electrolyte** max. kg	Weight of solid electrolyte in kg	Amount of electrolyte in litres
FNC 103 X	10	30	122	250	1.38	1.08	0.06	0.25
FNC 106 X	19	47	122	250	2.46	1.89	0.12	0.48
FNC 110 X	33	72	122	250	3.74	2.85	0.18	0.75
FNC 114 X	45	92	122	250	5.04	3.86	0.24	0.99
FNC 118 X	58	115	122	250	6.24	4.83	0.29	1.18
FNC 203 X	13	30	122	309	1.80	1.32	0.10	0.40
FNC 206 X	25	47	122	309	3.16	2.35	0.23	0.93
FNC 210 X	43	72	122	309	4.83	3.56	0.26	1.06
FNC 214 X	60	92	122	309	6.31	4.83	0.30	1.25
FNC 218 X	77	115	122	309	7.84	5.44	0.49	2.02
FNC 201 H	12	30	122	309	1.52	0.95	0.12	0.48
FNC 202 H	23	30	122	309	1.77	1.33	0.09	0.37
FNC 203 H	35	47	122	309	2.68	1.84	0.17	0.71
FNC 204 H	46	47	122	309	3.00	2.26	0.15	0.63
FNC 205 H	58	72	122	309	4.17	2.78	0.29	1.17
FNC 206 H	69	72	122	309	4.36	3.05	0.27	1.10
FNC 207 H	80	72	122	309	4.58	3.48	0.23	0.93
FNC 208 H	93	92	122	309	5.58	3.98	0.33	1.34
FNC 209 H	104	92	122	309	5.78	4.38	0.29	1.18
FNC 210 H	115	115	122	309	6.91	4.85	0.42	1.73
FNC 211 H	125	115	122	309	7.21	5.31	0.39	1.60
FNC 307 H	140	92	194	309	8.58	6.04	0.52	2.14
FNC 308 H	160	92	194	309	9.16	6.71	0.50	2.06
FNC 309 H	180	92	194	309	9.24	7.15	0.43	1.76
FNC 310 H	200	115	194	309	11.11	8.11	0.62	2.52
FNC 311 H	220	115	194	309	11.61	8.71	0.60	2.44

Type	Cs capacity in Ah	Length	Width	Height	Cell weight with electrolyte* max. kg	Cell weight without electrolyte** max. kg	Weight of solid electrolyte in kg	Amount of electrolyte in litres
FNC 201 L	20	30	122	309	1.60	1.01	0.12	0.49
FNC 202 L	40	47	122	309	2.59	1.64	0.19	0.80
FNC 203 L	60	47	122	309	2.76	2.07	0.15	0.58
FNC 204 L	80	72	122	309	4.05	2.69	0.30	1.15
FNC 205 L	100	72	122	309	4.28	3.15	0.25	0.95
FNC 206 L	120	92	122	309	5.24	3.70	0.34	1.30
FNC 207 L	140	92	122	309	5.61	4.21	0.31	1.18
FNC 208 L	160	115	122	309	6.83	4.76	0.45	1.74
FNC 209 L	180	115	122	309	6.92	5.03	0.41	1.58
FNC 306 L	200	92	194	309	8.47	5.99	0.54	2.08
FNC 307 L	233	92	194	309	9.17	7.10	0.45	1.74
FNC 308 L	266	115	194	309	10.68	7.66	0.65	2.53
FNC 309 L	300	115	194	309	10.90	8.23	0.58	2.25
FNC 404 L	150	77	157	405	7.33	4.66	0.58	2.25
FNC 405 L	185	77	157	405	7.79	5.42	0.51	1.99
FNC 406 L	225	109	157	405	10.31	6.61	0.80	3.11
FNC 407 L	265	109	157	405	10.91	7.36	0.77	2.98
FNC 408 L	300	109	157	405	12.00	8.86	0.68	2.65
FNC 409 L	340	109	157	405	13.10	10.36	0.59	2.30
FNC 410 L	375	157	157	405	15.31	10.00	1.15	4.45
FNC 411 L	415	157	157	405	15.81	10.61	1.13	4.37
FNC 412 L	450	157	157	405	16.51	11.66	1.05	4.08
FNC 413 L	490	157	157	405	17.11	12.76	0.94	3.66
FNC 201 M	20	30	122	309	1.54	1.01	0.11	0.44
FNC 202 M	40	47	122	309	2.58	1.68	0.20	0.76
FNC 203 M	60	47	122	309	2.84	2.19	0.14	0.54
FNC 204 M	80	72	122	309	4.16	2.86	0.28	1.09
FNC 205 M	100	72	122	309	4.46	3.51	0.21	0.80
FNC 206 M	120	92	122	309	5.56	4.09	0.32	1.24
FNC 207 M	140	92	122	309	5.90	4.50	0.30	1.18
FNC 208 M	160	115	122	309	7.11	5.15	0.42	1.64
FNC 209 M	180	115	122	309	7.41	5.81	0.35	1.34
FNC 306 M	200	92	194	309	8.51	6.00	0.54	2.10
FNC 307 M	233	92	194	309	9.23	7.38	0.40	1.56
FNC 308 M	266	115	194	309	10.00	7.11	0.63	2.44
FNC 309 M	300	115	194	309	10.94	8.40	0.53	2.05
FNC 404 M	150	77	157	405	7.51	4.20	0.72	2.78
FNC 405 M	185	77	157	405	7.81	4.62	0.69	2.68
FNC 406 M	225	109	157	405	10.51	7.00	0.76	2.94
FNC 407 M	265	109	157	405	10.51	7.51	0.65	2.52
FNC 408 M	300	109	157	405	12.00	8.46	0.57	2.20
FNC 409 M	340	157	157	405	15.51	10.00	1.19	4.62
FNC 410 M	375	157	157	405	16.00	11.00	1.09	4.20
FNC 411 M	415	157	157	405	16.21	11.61	1.00	3.87
FNC 412 M	450	157	157	405	16.81	12.61	0.91	3.53

* filled and charged
** unfilled and uncharged

Source: Hoppecke Batterien.

26.8 APPLICATIONS

Because of their favorable electrical properties, excellent reliability, low maintenance, rugged design, and long life, nickel-cadmium batteries are used in a large variety of applications, as indicated in Table 26.4. Most of these are of an industrial nature, but this type of battery is also used in many commercial, military and space applications.

The nickel-cadmium battery was originally developed for traction applications, and since the early years of this century, it has been used extensively in railroad applications. Today the nickel-cadmium battery is the system of choice in a variety of railroad and mass-transit installations around the world. Approximately 40% of all industrial nickel-cadmium batteries produced are used in train lighting and air-conditioning for rail cars, emergency and standby systems such as emergency brakes, door openers, and lighting in mass-transit and subway cars, diesel-engine cranking in locomotives and commuter cars, railroad signaling, communication along tracks, as well as standby power in rail stations and traffic control systems. The pocket-plate battery has traditionally dominated this market segment, but in recent years, with demands for higher energy per unit weight and volume, plastic-bonded and fiber-plate batteries have penetrated this market, particularly for high-speed trains, mass-transit cars, subway cars, and light rail vehicles. Where ruggedness and long durability are the main requirements, the pocket-plate battery still maintains its position.

In stationary applications where reliability is a must, nickel-cadmium batteries are used in standby and emergency installations where life and great economic values would be endangered by a power failure. Examples of such installations are emergency power in hospital operating theaters, standby power for all vital functions on off-shore oil rigs, uninterruptible power supplies (UPSs) for large computer systems in banks and insurance companies, standby power in process industries, and emergency lighting and landing systems in airports.

The nickel-cadmium battery is also used in power-generating stations and power distribution networks where power supplies must not break down. The batteries are used in switchgear applications and for control and monitoring functions. In centralized emergency lighting systems in hospitals, public buildings, sports arenas, and schools, nickel-cadmium batteries are often specified in building codes and by consultants in many industrialized countries.

In case of failure of the primary power supply, diesel generators or gas turbines are installed to take over the power supply. For reliable and fast-acting start-up of these engines nickel-cadmium batteries have proven to be the best emergency power source.

In portable applications were batteries are exposed to temperature extremes or rough handling, nickel-cadmium batteries are used for signal lamps, hand lamps, search lights, and portable instruments. Vented spillproof batteries are used in large devices, whereas sealed nickel-cadmium batteries dominate the smaller ones (see Chap. 27).

The industrial battery market is dominated by the lead-acid battery, and the nickel-cadmium battery is a niche-market product. The reason for this is the higher capital cost for nickel-cadmium batteries compared to lead-acid batteries. Where only energy is required, the lead-acid battery is the least expensive, as its cost per Watt-hour is lower than that for nickel-cadmium batteries. However, in cost per Watt or life cycle cost, nickel-cadmium batteries can compete with lead-acid batteries due to much better high-rate performance and longer life combined with low maintenance costs. A typical example is locomotive diesel-engine cranking, where a nickel-cadmium battery with only one-third of the Ampere-hour capacity and a life four times that of the lead-acid battery can do the job. In applications with short-duration discharges—standby and emergency equipment are usually used for less than a half-hour—the rated capacity of a battery is of little importance. The size of the battery is chiefly determined by the power need. The nickel-cadmium battery is unmatched in industrial applications when reliability and durability are considered in a life-cycle cost calculation.

Fiber nickel-cadmium batteries of the ultra high-rate (XX) and high-rate (X) design are employed primarily in aircraft, military and space applications. They have been used for the Boeing 777 main and APU batteries, for the MacDonald-Douglas (now Boeing) MD-90 APU and emergency position batteries and for the Apache helicopter main battery. They have also been employed as the training battery for a portable missile system, and are qualified for use in a Low Earth Orbit (LEO) satellite scheduled for launch in 2002 on a two-year mission.

Because of the variety of applications for nickel-cadmium batteries, it is important to select the best technology for the application. The characteristics are somewhat different for the three technologies available today for industrial use.

The pocket-plate battery has the lowest cost of the three technologies and is known for high reliability and fail-safe operation. However, the energy and power density limit its use in some areas. The fiber plate battery has lower internal resistance than the pocket-plate battery and is also available in ultra-high, high-, medium- and low-rate cells. Where very high energy and power density are required, the plastic-bonded plate may be the choice. The plastic-bonded and fiber plate batteries are the only technologies possible for use in automated guided vehicles (AGVs). They have also cost and performance advantages in some traditional pocket-plate battery applications such as engine cranking, switchgear, and uninterruptible power supplies (UPSs) where only very short duration discharge is required.

BIBLIOGRAPHY

General:

Barak, M. (ed.) *Electrochemical Power Sources,* Peter Peregrinus, London, 1980.

Falk, S. U., and A. J. Salkind, *Alkaline Storage Batteries,* Wiley, New York, 1969.

Jacksch, H.-D., *Batterie Lexikon,* pp. 348–394, Pflaum Verlag, Munich (1993).

Kinzelbach, R., *Stahlakkumulatoren,* Varta, Hannover, Germany, 1968.

Miyake, Y., and A. Kozawa, *Rechargeable Batteries in Japan,* JEC Press, Cleveland, Ohio, 1977.

Plastic-Bonded Electrode Technology:

McRae, B., and D. Nary, *Proc. of the 38th Power Sources Conf.,* pp. 123–126 (1998).

FNC Technology:

Anderman M., C. Baker, and F. Cohen, *Proc. of the 32nd Intersociety Energy Conversion Conf.,* Vol. 1, p. 97465 (1997).

Baker, C., *Proceedings of the SAE Power Systems Conference,* Williamsbury, VA (1997). See *Advanced Battery Technology,* April, 1997.

Baker C., and M. Barekatien, *Proc. SAE Power Systems Conf.,* San Diego, CA, Oct 31–Nov. 2, 2000.

FNC Vented Nickel-Cadmium Batteries, Hoppecke Batterien.

CHAPTER 27
VENTED SINTERED-PLATE NICKEL-CADMIUM BATTERIES

John M. Evjen and Arthur J. Catotti

Revised by: R. David Lucero, David F. Pickett Jr., and Timothy M. Kulin

27.1 GENERAL CHARACTERISTICS

The sintered-plate nickel-cadmium battery is a mature development of the nickel-cadmium system, having a higher energy density, up to 50% greater than its predecessor, the pocket-type construction. The sintered plate can be constructed in a much thinner form than the pocket plate, and the cell has a much lower internal resistance and gives superior high-rate and low-temperature performance. A flat discharge curve is characteristic of the battery, and its performance is less sensitive than other battery systems to changes in discharge load and temperature. The sintered-plate battery has most of the favorable characteristics of the pocket-type battery, although it is generally more expensive. It is electrically and mechanically rugged, is very reliable, requires little maintenance, can be stored for long periods of time in a charged or uncharged condition, and has good charge retention. Batteries losing capacity through self-discharge can be restored to full service with a normal charge. The major advantages and disadvantages of this battery type are given in Table 27.1.

TABLE 27.1 Major Advantages and Disadvantages of Vented Sintered-Plate Nickel-Cadmium Batteries

Advantages	Disadvantages
Flat discharge profile	Higher cost
Higher energy density (50% greater than pocket plate)	Memory effect (voltage depression)
Superior high-rate and low-temperature performance	Temperature controlled charging system required to enhance life
Excellent long-term storage	
Good capacity retention; capacity can be restored by recharge	

For these reasons, vented sintered-plate nickel-cadmium batteries are used in applications requiring high-power discharge service such as aircraft turbine engine and diesel engine starting as well as other mobile and military equipment. The battery provides outstanding performance where high peak power and fast recharging are required. In many applications the vented sintered-plate battery is used because it leads to a reduction in size, weight and maintenance as compared with other battery systems. This is particularly true in systems subject to low-temperature operation. The rise in terminal voltage of the vented cell at the end of charge also provides a useful characteristic for controlling the charge.

27.2 CHEMISTRY

Vented sintered-plate nickel-cadmium cells, in the discharged state, consist of flat positive nickel hydroxide and negative cadmium hydroxide plates, separated by materials that act as a gas barrier and electrical separator. The electrolyte, normally a 31% potassium hydroxide solution completely covers the plates and separators: for this reason vented cells are referred to as "flooded cells."

In the sintered-plate design, the active materials are held within the pores of a sintered-nickel structure. Nickel hydroxide with 3% to 10% cobalt hydroxide is the active material of the positive plate while cadmium hydroxide is the active material of the negative plate.

The electrochemistry of the charge and discharge of the positive electrode is quite complex and not well understood,[1] especially, the role that cobalt plays in the active material.[2] For simplicity, lets consider the role of nickel hydroxide in the charge-discharge reaction.

During charge the nickel hydroxide in the positive electrode is oxidized to nickel oxyhydroxide (NiOOH) and higher valence states of nickel. Potassium and water are also incorporated into the active material as potassium hydroxide according to the following equation:[3]

$$Ni(OH)_2 + xK^+ + (1 + x)OH^- \rightleftharpoons NiOOH \cdot xKOH \cdot (H_2O) + e$$

The fraction of potassium that is bonded into the nickel oxyhydroxide lattice is represented by x. The value is small (much less than 1.0) and varies according to manufacturing process.

The cadmium hydroxide in the negative electrode is reduced to metallic cadmium during charge:

$$Cd(OH)_2 + 2e \rightleftharpoons Cd + 2OH^-$$

The overall charge-discharge reaction is thus

$$2Ni(OH)_2 + 2xKOH + Cd(OH)_2 \underset{\text{discharge}}{\overset{\text{charge}}{\rightleftharpoons}} 2NiOOH \cdot xKOH \cdot (H_2O) + Cd$$

According to the above equation, one might think that the change in the electrolyte concentration might offer a means of state-of-charge determination by measuring the specific gravity of the electrolyte. Unfortunately, the complication of potassium in the active material, accumulation of carbonates, along with the large volume of electrolyte makes this type of measurement unreliable and impractical.

The positive electrode does not accept charge, and convert nickel hydroxide to nickel oxyhydroxide, at the thermodynamically reversible potential.[3] In fact, with a low enough charge rate, gassing according to the equation below occurs:

$$4OH^- \rightarrow 2H_2O + O_2 + 4e$$

If the rate is increased appreciably, this will result in an oxygen overvoltage sufficiently high to allow the preferred conversion of the nickel hydroxide to the nickel oxyhydroxide instead of oxygen gassing. However, when about 80% conversion of nickel hydroxide to nickel oxyhydroxide is achieved, the competing oxygen generating reaction occurs gradually and remains until 100% state-of-charge is achieved, then the only reaction occurring is oxygen evolution.

The negative electrode accepts charge until it is essentially 100% charged, at which time the favored reaction is hydrogen gassing as shown in the following equation

$$2H_2O + 2e \rightarrow H_2 + 2OH^-$$

The hydrogen gassing reaction, with a $C/10$ charge rate, occurs at a cell voltage close to 1600 millivolts as shown in Fig. 27.1.

The hydrogen overvoltage on the cadmium electrode is quite high, about 110 mV at the $C/10$ rate. Consequently there is a sharp rise in voltage as the negative electrode goes into overcharge. This rise in voltage is used in various charging schemes to control or terminate charging.

During overcharge, all the current is used to electrolyze water to hydrogen and oxygen as shown in the overall reaction

$$2H_2O \rightarrow 2H_2 + O_2$$

This overcharge reaction consumes water and thereby decreases the level of electrolyte in the cell. The water loss can be limited by controlling the amount of overcharge so as to maximize the interval between needed water replenishments.

Cells are constructed with 50% excess capacity in the negative electrodes and thus are positive limited.

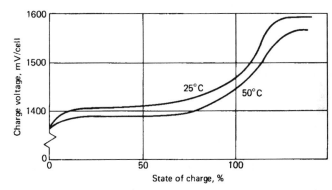

FIGURE 27.1 Constant-current charge voltage of vented sintered-plate nickel-cadmium cell, $C/10$ charge rate.

27.3 CONSTRUCTION

Vented cells are designed so that both electrodes reach full charge at about the same time. The positive electrode, as noted, will begin to evolve oxygen before it is fully charged. If this gas is allowed to reach the negative electrode due to failure of the gas barrier, it will recombine and generate heat. This will not only prevent the negative from reaching a full state of charge, but it will also result in reduced voltage due to depolarization of the cadmium electrode. To maintain the fullest capability, adequate precautions must be taken to prevent oxygen recombination at the negative plate. This is accomplished by providing a gas barrier between the positive and negative plates and by flooding the plates with excess electrolyte.

Figure 27.2 shows details of a typical vented sintered-plate nickel-cadmium cell.

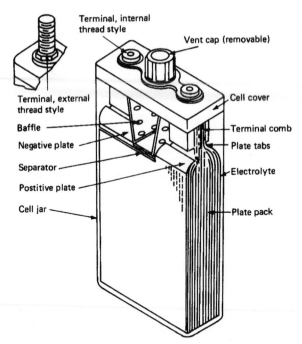

FIGURE 27.2   Cross section of vented sintered-plate nickel-cadmium cell.

27.3.1   Plates and Processes

A variety of plate formulations are used in vented, sintered nickel-cadmium cells produced by different manufacturers. The plates differ according to the nature of the substrate, method of sintering, impregnation process, formation and termination techniques. The predominate plate fabrication process used for vented sintered plate over the years has been described by Fleischer.[4] There are several reviews on electrode fabrication processes that have been used in flooded vented cells.[3,5,6]

Substrate.   The substrate serves as a mechanical support for the sintered structure and as a current collector for the electrochemical reactions which occur throughout the porous sintered portion of the plate. It also provides mechanical strength and continuity during the manufacturing processes. Two types of substrate are typically used: (1) perforated nickel-plated steel or pure nickel strip in continuous lengths, and (2) woven screens of nickel or nickel-clad steel wire. A common perforated type may be 0.1 mm thick with 2-mm holes and a void area of about 40%. A typical screen may use 0.18-mm-diameter wire with 1.0-mm openings.

Plaque.   The sintered structure before impregnation is generally referred to as "plaque." It usually has a porosity of 80 to 85% and ranges in thickness from 0.40 to 1.0 mm. Two generic sintering processes are used: (1) the slurry coating process and (2) the dry-powder process. Both processes employ special low-density battery grades of carbonyl nickel powder.

In the slurry coating process, the nickel powder is suspended in a viscous, aqueous solution containing a low percentage of a thixotropic agent. The nickel-plated strip with the desired perforated pattern is pulled through the suspension. The thickness is controlled by passing it through doctor blades, while wiping the edges free of slurry. The continuous strip is then dried before sintering in a reducing atmosphere at about 1000°C.

The dry-powder processes generally employ wire screen precut to the so-called master plaque dimension. The screens are placed in molds with loose powder on each side. They are then typically sintered in a belt furnace in a reducing atmosphere at 800 to 1000°C.

Impregnation. A review of various impregnation processes, used to load the porous sintered structure of the positive with nickel hydroxide and of the negative with cadmium hydroxide, has been given by Pickett.[6] The plaque is impregnated with a concentrated solution of nitrate that is then converted to hydroxide by chemical precipitation[4] or electrochemical precipitation.[7,8,9] The most widely used process for vented cells involves a chemical precipitation, and with minor variations, follows, in principle, the process described in 1948.[4] The plaque is impregnated with a concentrated solution of the nitrate, briefly rinsed, and the nitrate salts are precipitated as hydroxide with caustic. Following the addition of caustic the plaque is made cathodic. This is called polarization. The polarization cycle usually consists of a high current charge (C rate or higher) for approximately one hour. The plaque is then rinsed and this sequence of steps is repeated a number of times so as to fill about 40 to 60% of the sintered pore volume (or until a targeted weight gain is achieved).

Plate Formation. Following impregnation, the plates are mechanically brushed and electrochemically cleaned and formed by charging and discharging the electrode. In the master-plaque process they are formed against inert counterelectrodes (typically stainless steel or nickel) and can be performed in a loose pack or tight pack configuration. Formation is essential for properly converting hydroxide into the pores of the sinter structure, as well as the reduction of nitrates in the plates. Typical formation cycles for chemical plates consist of a high current cycling. This regime or time may vary for plaque type and capacity. In the case of the continuous-strip process, the formation is done on a machine similar in appearance to a continuous-strip-electroplating machine. Plates blanked from the continuous strip have a clean, wiped area at the top that serves as attachment points for nickel or nickel-plated steel current-collector tabs. In the case of the master-plaque process, a coined or densified area is provided for attachment of these collector tabs.

27.3.2 Separator

The separator system is a thin, multi-layered combination. It consists of a cloth that electrically separates the positive and negative plates and an ion-permeable plastic membrane that serves as the gas barrier.

Electrical and mechanical separation of the plates is typically provided by either woven or felted nylon material. This material is relatively porous in order to provide a good microporous polypropylene ionic conduction path through the electrolyte.

The microporous polypropylene membrane, typically Celgard® (Celgard 3400, manufactured by Celgard LLC, Charlotte, NC, 28273),[10] is utilized as the gas barrier while at the same time it offers minimum ionic resistance. This thin gas barrier, which becomes relatively soft when wetted, is frequently placed between two layers of the cloth separator and receives significant mechanical support from them. Substantial improvements have been made in recent years to the toughness of the plastic membrane gas barrier.

27.3.3 Plate-Pack Cell Assembly

Plate packs are assembled by alternately stacking positive and negative plates with the separator-gas barrier system interleaved between them. The cell terminals are bolted, riveted or welded to the current-collector plate tabs. In the case of cells with many plates, the tabs from the outermost plates may need to be bent quite significantly inward to reach the cell terminals. Spacers at the terminals are sometimes used in these situations to keep the angle of the tabs at a minimum.

27.3.4 Electrolyte

Potassium hydroxide electrolyte is used in a concentration of approximately 31% at full charge (specific gravity 1.30). Performance of the cell, particularly at low temperatures, is significantly dependent on this concentration (see Sec. 27.4.2).

Electrolyte purity can also have significant effects on cell performance. The level of potassium carbonate in the cell relates directly to the cell's performance. Increasing carbonate concentration changes the characteristics of the electrolyte, reducing the high-rate charge and discharge capability of the cell. Fresh electrolyte contains very low levels of carbonate. However, organic components in the cell are slowly oxidized in the presence of the electrolyte and oxygen, forming small amounts of carbonate. The carbonates accumulate as the cell ages and eventually reduce cell performance. Carbonate levels at the time of cell activation are on the order of 80 to 90 grams per liter due to reaction of the electrolyte with residues from the impregnation process. High quality cells are designed with components that do not degrade in KOH. In addition, at least one manufacturer flushes new cells repeatedly with fresh electrolyte, lowering final carbonate levels to 6 to 8 grams per liter.

27.3.5 Cell Container

The plate pack is placed into the cell container with the cell terminals extending through the cover. The cell container is usually made of a low-moisture-absorbent nylon and consists of the cell jar and matching cover that are permanently joined together at assembly by solvent sealing, thermal fusion, or ultrasonic bonding. The container is designed to provide a sealed enclosure for the cell, thus preventing electrolyte leakage or contamination, as well as providing physical support for the cell components. The terminal seal is generally provided by means of O-rings with Belleville washers and retaining clips.

27.3.6 Vent Cap and Check Valve

The vent cap serves as a removable cap to provide the access required for replenishment of water to the electrolyte and also to function as a check valve to release gases generated when water is consumed during overcharge. The check valve prevents atmospheric contamination of the electrolyte. It consists of a nylon body with a hollow center post, through which a cross-hole is drilled and around which an elastomeric sleeve is placed. This functions as a Bunsen valve to allow gas to escape from the cell but not to enter. Typical sleeves used for this application have developed significantly over time, and ethylene-propylene rubber seems to have the best characteristics for vented cells. Neoprene vent sleeves were previously used but the neoprene is attacked by potassium hydroxide and can soften, swell, and split. It also frequently erodes at the interface between the vent cap and sleeve until the neoprene no longer seals. Before erosion occurs, a sleeve surface can soften due to electrolyte at the interface between the sleeve and vent cap, dry during a subsequent storage, and literally glue itself to the vent. When this occurs, the pressure will build up in the cell during charge until the sleeve breaks free or ruptures or the cell explodes.[10]

27.4 *PERFORMANCE CHARACTERISTICS*

27.4.1 Discharge Properties

The discharge curves for a typical vented sintered-plate nickel-cadmium battery at various constant-discharge loads are shown in Fig. 27.3. The discharge curves for a typical battery at various temperatures are shown in Fig. 27.4. The curves for this battery are characterized by a flat voltage profile, even at relatively high discharge rates and low temperatures. Voltages at various constant-current discharge loads and states of discharge are given in Fig. 27.5.

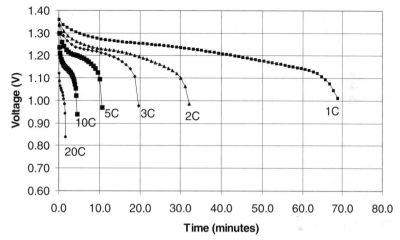

FIGURE 27.3   Typical discharge curves at various *C* rates, 25°C.

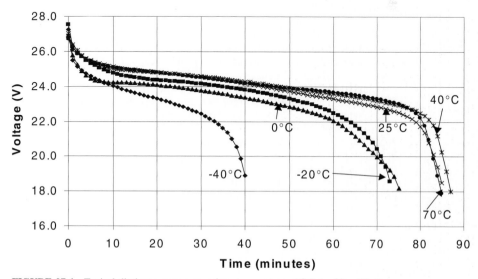

FIGURE 27.4   Typical discharge curves at various temperatures 1C-Rate, 20 cell battery.

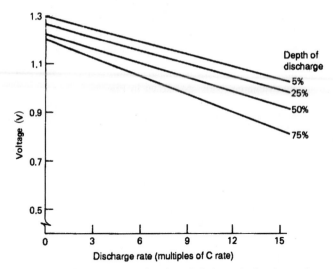

FIGURE 27.5 Voltage as a function of discharge load and at various states of charge at 25°C.

The battery, because of its low internal resistance, is capable of delivering pulse currents as high as the 20 to 40C rate. For this reason it can be used successfully for very high power applications, such as engine starting (see Sec. 27.4.3).

27.4.2 Factors Affecting Capacity

The total capacity that the fully charged sintered-plate vented battery is capable of delivering is dependent on both discharge rate and temperature, although the sintered-plate battery is less sensitive to these variables than most other battery systems. The relationships of capacity to discharge load and temperature are shown in Figs. 27.6 and 27.7, respectively.

Low-temperature performance is enhanced by the use of eutectic 31% KOH (1.30 specific gravity) electrolyte which freezes at −66°C. Higher or lower concentrations will freeze at higher temperatures; for example, 26% KOH freezes at −42°C. As shown in Fig. 27.7, more than 60% of the 25°C capacity is available at −35°C, with the temperature having an increasingly significant effect as it is reduced toward −50°C. At high discharge rates, heat that is generated may cause the battery to warm up, giving improved performance on immediate subsequent discharges than would be expected under the ambient conditions.

Vented sintered-plate batteries can also be discharged at elevated temperatures. Strict control is required, however, when charging at high temperature. As with most chemically based devices, exposure to high temperatures for extended periods of time will detract from the life of the battery (see Sec. 27.7.3).

The combined effects of both increased discharge rate and low temperature may be approximated by multiplying the two derating factors.

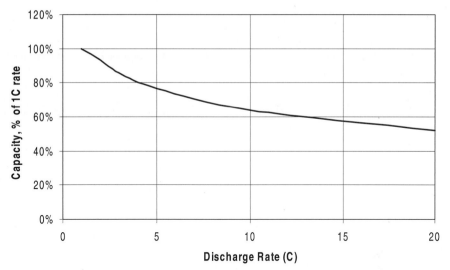

FIGURE 27.6   Capacity derating as a function of discharge rate at 25°C.

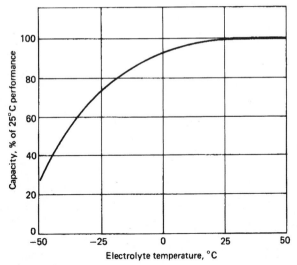

FIGURE 27.7   Capacity derating as a function of discharge temperature at 1C rate discharge.

27.4.3   Variable-load Engine-Start Power

The most common and demanding use of the vented sintered-plate nickel-cadmium battery is as the power source for starting turbine engines on board aircraft. The discharge in this application occurs at relatively high rates for periods of 15 to 45 s. Typically the load resistance when the start is initiated, particularly in low-temperature in a marginal-start situations, is of the same order of magnitude as the effective internal resistance of the battery, R_e. The apparent load resistance increases as the engine rotor gradually comes up to speed.

This results in a typical discharge current which slowly decreases from some high initial value while the battery voltage recovers from an initial drop of perhaps 50% or more, back toward 1.2-V per cell, the effective zero load voltage. A representative graph of the battery-starter voltage and current, expressed as a function of time is shown in Fig. 27.8.

A common and useful measure of battery performance is the maximum power current. This property is generally defined as the load current at which the battery voltage would be 0.6N V, or one-half of the effective open-circuit voltage (1.2 V/cell) and where N is the number of cells in the battery. The instantaneous maximum power current decreases with decreasing state of charge due to rising internal resistance. Its value versus state of charge tends to behave exponentially, as shown in Fig. 27.9. An approximation of I_{mp} may also be measured by performing a "constant-potential" discharge at 0.6N V for 15 to 120 seconds. A typical discharge is shown in Fig. 27.10.

The maximum power delivery P_{mp} and the effective internal resistance R_e are related to the value of I_{mp} as follows:

$$P_{mp} = 0.6N \ I_{mp}$$

and

$$R_e = \frac{0.6N}{I_{mp}}$$

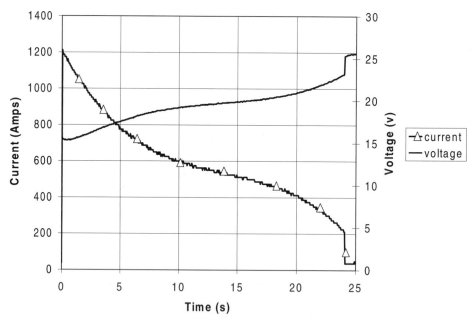

FIGURE 27.8   Battery voltage and current as a function of time for a typical turbine engine start (20 cell battery).

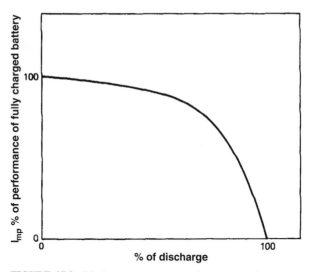

FIGURE 27.9   Maximum power current derating as a function of state of charge at 25°C.

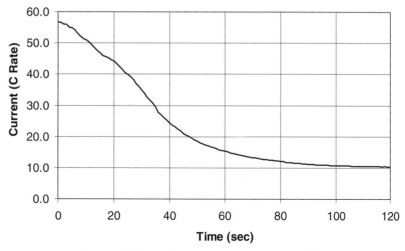

FIGURE 27.10   Representative 0.6V constant potential discharge at 25°C.

27.4.4   Factors Affecting Maximum Power Current

The value of I_{mp}, which a battery is capable of delivering, is maximum at full charge and at 25°C electrolyte temperature. Derating effects due to state-of-charge and electrolyte temperature factors are shown in Figs. 27.9 and 27.11, respectively. It will be noted that both relationships are nonlinear in that the effects on maximum power delivery, per unit of change, increase with decreasing state of charge and with decreasing temperature. As with capacity, the approximate effect of combined low electrolyte temperature and decreased state of charge

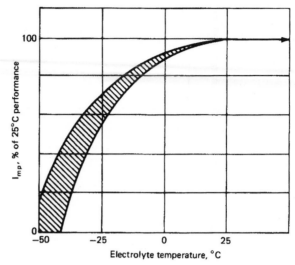

FIGURE 27.11 Maximum power current derating as a function of battery temperature (fully charged).

may be determined by multiplying the individual factors. It should be noted, however, that high-rate discharge at low temperature may increase the battery temperature. This self heating must be accounted for when determining the combined derating factors for a subsequent discharge. A negligible effect on I_{mp} occurs with increases in electrolyte temperature above 25°C.

27.4.5 Energy/Power Density

Typical average values for the energy and power densities of the vented sintered-plate nickel-cadmium battery at 25°C are shown in Table 27.2.

TABLE 27.2 Energy and Power Characteristics of Vented Sintered-Plate Nickel-Cadmium Battery (Single Cell Basis)

Capacity specific (single cell, C rate)	25–31 Ah/kg
Capacity density	48–80 Ah/L
Specific energy (C rate)	30–37 Wh/kg
Energy density	58–96 Wh/L
Power specific (at maximum power)	330–460 W/kg
Power density	730–1250 W/L

27.4.6 Service Life

The service life (discharge time) of the vented sintered-plate nickel-cadmium cell, normalized to unit weight (kilogram) and size (liter) at various discharge rates at 25°C, is approximated in Figs. 27.12 and 27.13.

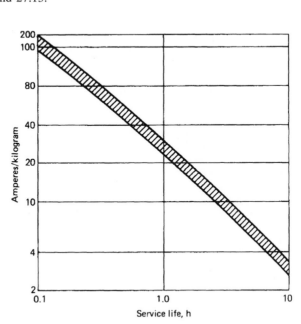

FIGURE 27.12 Service life of typical vented sintered-plate nickel-cadmium battery (gravimetric) at 25°C.

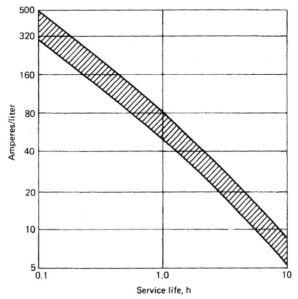

FIGURE 27.13 Service life of typical vented sintered-plate nickel-cadmium battery (volumetric) at 25°C.

27.4.7 Charge Retention

Charge retention or capacity retention refers to the amount of dischargeable capacity remaining in a battery following prolonged storage under open-circuit conditions. Two mechanisms are responsible for the loss of charge, namely, self-discharge and electrical leakage between cells.

Self-discharge rates are an intrinsic property of cells. Typically, experimental results for the capacity retained as a function of open-circuit storage time best fit a semilogarithmic relationship such as that shown in Fig. 27.14. The self-discharge rate of a cell is affected by impurity levels and the electrochemical stability of the electrodes.

The effect of temperature is shown in Fig. 27.15, where the exponential time constant (tc), the time to retention of 36.8% of initial capacity, is plotted against temperature. Storage temperature is the most important factor affecting the self-discharge rate.

The second mechanism, the loss of charge due to electrical leakage, is influenced by the history of the battery's use and maintenance. Charge retention usually improves with cycling of the battery, and this will be true unless this cycling history has been abusive. The maintenance factor influencing charge retention is primarily battery cleanliness. Battery charge can leak from the terminals of one cell to the terminals of other cells across the cell tops if they are wet with potassium hydroxide. Loss of charge from this cause is relatively unpredictable, but it can usually be prevented by good housekeeping practices. Although surface leakage may affect only a portion of the cells in the battery, it is important since the capacity of the battery is limited to that of the lowest-capacity cell. Additionally it unbalances the cells in the timing of the onset of overcharge voltage response (see Fig. 27.1.).

It should be noted that loss of charge through these mechanisms is not permanent since the battery capability can be completely restored through comprehensive maintenance practices and recharging.

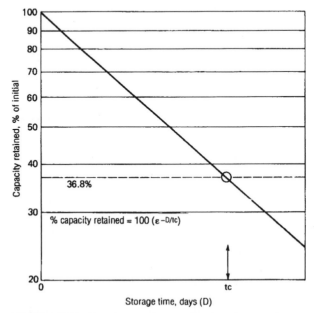

FIGURE 27.14 Capacity retention as a function of storage time.

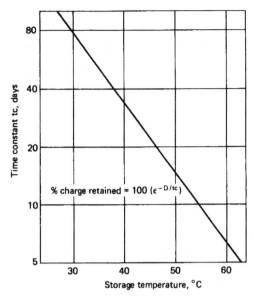

FIGURE 27.15 Charge retention time constant as a function of storage temperature.

27.4.8 Storage

The sintered-plate cell can be stored in any state of charge and over a very broad temperature range ($-60°C$ to $60°C$) for an unlimited period. The battery should be clean and dry before placing it in storage. Intercell hardware may have a light coating of petroleum jelly to prevent corrosion. It should be fully discharged and shorted prior to storage periods greater than 30 days. Fully discharged batteries that have been stored longer than 30 days should be charged by a "slow charge" method. The "slow charge" method typically consists of incremental charge rates to voltage cut-off (i.e. C rate to 1.57 volts, $C/2$ rate to 1.6 volts), with a final charge ($C/10$ or lower) for 2 hours. It is the best practice to store the cells shorted and upright with proper electrolyte level at a temperature between 0 and 30°C. The preferable storage method is to allow the battery to discharge through a resistor until the battery voltage is close to zero. Because a vented Ni-Cd still has considerable power available even at very low states of charge, failure to completely discharge the battery prior to applying a shorting device can create a hazardous situation.

27.4.9 Life

The life of the battery is strongly influenced by factors such as the design, the care with which it is maintained and reconditioned as well as the way it is used; hence it is difficult to predict battery life. Best life performance is obtained with operation at normal temperatures, temperature-controlled charging, and minimum reconditioning. Some design features that improve the life expectancy of a battery are: the use of modern separator materials and gas barriers, the elimination of materials that degrade in KOH (e.g. O-rings), the reduction of electrolyte impurity levels in manufacturing (by electrolyte flushing and replacement), and the use of pure nickel components versus nickel-plated steel.

27.5 *CHARGING CHARACTERISTICS*

The functional design of the vented cell battery differs from that of the sealed cell battery primarily by the inclusion of a gas barrier between the positive and negative electrodes. This gas barrier has one principal function, which is to prevent, as discussed in Sec. 27.3, the cross-plate migration and recombination of generated gases within the cell. Preventing this recombination allows both positive and negative plates to return to full charge. This results in an overvoltage during onset of overcharge, which is used as the feedback signal to control the charging device. Because the gas is driven out of the cell, however, the vented cell consumes water, which must be replenished.

Charging of the vented sintered nickel-cadmium battery following its discharge in cyclic use, has four significant objectives. These may be stated as follows:

1. Restore the charge used during discharge as quickly as possible.
2. Maintain the "fully charged" capacity as high as possible during the use intervals between removals for maintenance.
3. Minimize the amount of water usage during overcharge.
4. Minimize the damaging effects of overcharge.

Fulfillment of the first objective is the principal reason for the design and use of vented cells, since the gas barrier provides the "voltage signal" which may be utilized in several different ways to terminate the fast recharge. The charge may thus be accomplished at the desired high rate, without compromising the battery, by continuing that rate in overcharge. Objective 2 must inherently be balanced against object 3 and 4 in the design and control of the charging method. Generally, a continued good capacity between reconditionings is enhanced by providing more overcharge, while more overcharge inherently utilizes more water and, if sufficiently high, may result in damage to the battery. A compromise must therefore be struck. Usually about 101 to 105% of the Ampere-hours removed on discharge are replaced on the subsequent charge.

Charging techniques which are used in "on-board" systems utilize the "signal" provided by the overvoltage of the vented cell in overcharge. This overvoltage signal is shown in Fig. 27.1. This significant voltage rise is present at all charge rates, and its sharpness actually improves as the cell is cycled in high-rate discharge and recharge. The corollary to this curve, which shows voltage response at constant current, is the current response at constant voltage, which may be expected to be somewhat the reverse of Fig. 27.1 and is illustrated in Fig. 27.16.

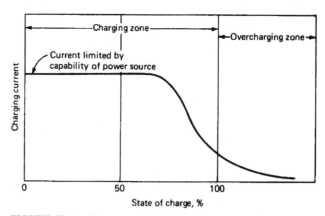

FIGURE 27.16   Constant-potential charge current.

27.5.1   Constant-Potential Recharging

Constant-potential (CP) charging is the oldest of the methods still in use and is typically utilized in general aviation aircraft. Similar to an automobile battery charging system, CP charging utilizes a regulated voltage output from the aircraft DC generator, which is mechanically coupled to the engine. The voltage is typically regulated at 1.40 to 1.50 V per cell. Figure 27.16 illustrates the form of the charge current as a function of the state of charge of the vented cell battery during CP recharging. Although the initial current could be quite high if limited only by the voltage response of the battery, it most frequently is limited by the capability of the source as shown. As the battery approaches full charge, however, the "back EMF" of the battery, illustrated as the charge voltage for the constant-current case in Fig. 27.1, reduces the current to that required by the battery to provide an overvoltage equal to the regulated voltage of the charging source. CP charging requires very careful consideration of the selection of charge voltage and its proper maintenance in order to achieve the balance between objectives 2 and 3. This is particularly difficult to achieve when the battery temperature experiences significant variation, since overvoltage is dependent on battery temperature. This balance may be made essentially independent of battery temperature effects by means of temperature compensation of the CP voltage, as discussed in Sec. 27.5.4.

27.5.2   Constant-Current, Voltage-Controlled Recharging

A number of commercially available chargers based in general on constant-current charging with voltage cutoff control are utilized in modern aircraft. One of the simplest and most effective of these chargers applies in approximately C-rate constant-current charge to the battery and then terminates it when the battery voltage reaches a predetermined voltage cutoff (VCO) value such as 1.50 V per cell. The control also reinitiates the constant-charge current whenever the open-circuit battery voltage falls to a predetermined lower level, such as 1.36 V per cell. The net result is that the charger will recharge the capacity used during an engine start, typically 10% of the battery's total, in approximately 6 min. and then cut the charger off due to the sharp rise in voltage as the cells go into overcharge, as illustrated in Fig. 27.1. Shortly thereafter, as the voltage falls below the turn-on voltage, the charge is reinitiated for a short period of time until the battery voltage again reaches the cutoff voltage. This simple on-off action continues at decreasing frequency and decreasing lengths of on-time, thereby maintaining the battery in a float condition at a completely full state-of-charge.

The battery voltage reduction, due to a discharge, inherently initiates the recharging of the battery without additional controls, thus automatically providing the recharge signal function regardless of the discharge rate or the reason. Adjustment of the cutoff and turn-on voltages as a function of battery temperature, which is described in Sec. 27.5.4, matches this mode of charging to the temperature characteristics of the vented sintered nickel-cadmium battery, thereby maintaining the desired balance of objectives. Both cutoff and turn-on voltages are compensated at the same rate, thereby maintaining a constant differential between turn-on and cutoff. A diagram of the function of this simple basic charge control scheme is shown in Fig. 27.17.

Several other proprietary chargers, based in part on the simple techniques outlined here, may also be found in commercial use. Many of these chargers also provide auxiliary functions such as upper and lower battery temperature charge discontinuation, detection of malfunctioning cells in the battery by detecting half-battery voltage imbalance, and signaling the user in the event of either of these conditions.

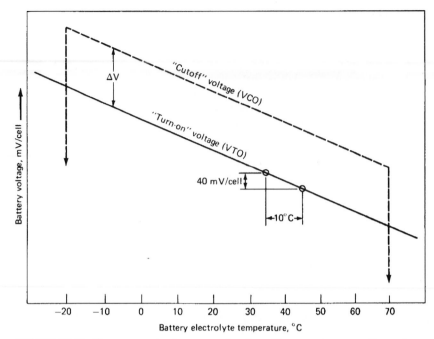

FIGURE 27.17   Charger control voltages as a function of battery temperature, *C*-rate charge (nominal values).

27.5.3   Other Charging Methods

The charging methods outlined in the preceding are those used in order to achieve fast recharging of a battery which has been discharged in normal use. Periodic maintenance of the vented sintered-plated nickel-cadmium battery, however, requires a full and complete discharge of each cell followed by a thorough recharge well into overcharge. This places both positive and negative plates in each cell of the battery in full and complete overcharge. The battery may then be returned to service with all plates of all cells in the same full charged condition, thus enabling the battery to work from the "top down."

The simplest maintenance charge method, requiring the least complex equipment to ensure this fully balanced, fully overcharged condition, is the low-rate charge. This technique utilizes a constant-current, approximately $C/10$ charge-overcharge current without voltage feedback control. At this low rate the charge may be continued safely into overcharge without compromising the physical integrity of the components of the cell. This charge current should be maintained until at least twice the rated capacity of the battery has been replaced. Since this will inherently result in water usage, water level replenishment is best performed on a fully charged battery just prior to placing the battery back into service at the conclusion of this maintenance charging routine.

Batteries in standby service can be maintained in a fully charged condition by a float or trickle charge similar to pocket plate batteries. The float voltage for vented sintered-plate batteries is 1.36 to 1.38 V per cell.

27.5.4 Temperature Compensation of Charge Voltage

In both the constant-potential and the constant-current VCO charging methods, it has been pointed out that the selection of voltage is a compromise between the minimization of water usage and the maintenance of a high state of charge. The inherent change of overcharge voltage as a function of battery temperature increases the difficulty of this compromise significantly. This voltage effect is shown as the Tafel curves in Fig. 27.18. The relationship between the Tafel curves at various temperatures indicates a temperature coefficient of -4 mV/°C at constant-current conditions. In other words, overcharge voltage, at constant current, decreases by approximately 4 mV per cell for each 1°C rise in cell temperature.

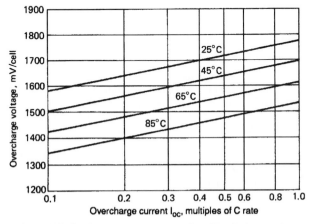

FIGURE 27.18 Overcharge voltage as a function of current and temperature (Tafel curves).

As shown by the slope of these Tafel curves, overcharge voltage is also a linear function of the logarithm of overcharge current. That slope for vented sintered-plate nickel-cadmium batteries is approximately 200 mV per cell per decade of change in overcharge current. Thus with constant-potential charging without temperature compensation, the overcharge current, and therefore both water usage and the overcharge damaging effects, will increase approximately 60% for each 10°C increase in electrolyte temperature.

A convenient technique for avoiding the detrimental effects of an increasing electrolyte temperature is to compensate the "constant"-potential voltage, or the constant-current cutoff/turn-on voltages, for this change in battery temperature at the rate of -4 mV/°C per cell. This may be accurately accomplished through the use of thermistors or other temperature-sensitive electric devices installed in the battery case to signal cell temperature. It is important that cell temperature and not ambient temperature be sensed by this device. There may be significant differences between the two. This function for the constant-current charging system is also shown in Fig. 27.17. The selection and use of the proper value of temperature compensation permits the battery charging system to function as though the battery were maintained at a constant temperature. Care must be exercised in the design and manufacture of these devices, since they must operate in an environment of potassium hydroxide which is electrically conductive and also has a propensity to wet and creep on most surfaces. High-grade potting processes must therefore be used to insulate and protect all auxiliary electronic components and wiring placed inside the battery case.

27.6 MAINTENANCE PROCEDURES

27.6.1 Electrical Reconditioning

The periodic maintenance of the vented sintered-plate nickel-cadmium battery has six specific objectives:

1. Assess the timeliness of the preselected maintenance period schedule.
2. Restore the electrical performance, both capacity and power.
3. Detect and isolate cell failures and facilitate their replacement.
4. Physically clean the battery.
5. Replenish the water in the electrolyte.
6. Maintain the charging-system voltage calibration.

A relatively simple electrical procedure fulfills the first objective, namely a single discharge, initially at a relatively high rate to simulate engine start and second at a relatively low rate, approximating that of the 1-h value. This split-rate discharge of the battery as removed from the aircraft serves as a measure of its performance readiness while it was in the aircraft. The 15-s high-rate portion, at approximately the I_{mp} rate of discharge, while measuring the voltage with the capacity removed, determines the relative engine start power capability. The capacity removed during the low-rate portion of the discharge at approximately the C rate, when added to the Ampere-hour capacity removed during the high rate test, determines the status of the available emergency energy capacity. The battery, prior to performing this discharge should be in the same "fully" charged condition that would typically be encountered when, it is in the normal installation. Comparison of this power delivery, and the total available capacity as removed from the aircraft, with the requirements of the application will allow the user to determine whether the maintenance schedule interval may be increased or whether it needs to be decreased.

The restoration of the electrical characteristics of the battery, known as electrical reconditioning or deep cycle maintenance, objective 2, may be accomplished in two additional simple steps. The first is a thorough and complete discharge of each cell in the battery in order to discharge all the active material. The second step is the complete recharge of each plate of each cell into full gassing overcharge. The first step consists of a C rate discharge to approximately 0.7 volts. Once all cells have reached voltage, then resistors short cells to 0.010 volts per cell or 16 hours, whichever occurs first. The second step is accomplished by charging the battery with constant current at a value of one-tenth its Ampere-hour rating ($C/10$) for at least 20 h. Since the capacity for some of the cells in the battery may be as much as 30 to 40% greater than the rated value, the total charge of 2.0C Ah is sufficient to ensure that both plates of all cells reach the full overcharge required. Adjust electrolyte levels accordingly. The battery should have deep cycle maintenance performed every 1000 flight hours or 500 starts, whichever occurs first or when any abnormal operation of the battery is observed.

There are other procedures used, and recommended by the manufacturers of specialized proprietary equipment, to recondition cells in a shorter period of time. Periodic maintenance in a qualified service center is necessary for optimum performance of the battery. The efficacy of each should be verified and the added expense and complexity entailed in the use of these proprietary reconditioning devices justified in specific applications. Care must always be exercised in their use, however, to avoid sustained high-rate overcharge, which may damage the gas barrier material.

Evaluation of cell-to-battery case leakage current, part of objective 3, when the battery is first received for maintenance will determine the electrical need for cell cleanup as well as the presence of cracked or leaking cell cases. This procedure may be conveniently carried out by the simple expedient of completing a circuit from each cell terminal to the battery case with a fused ammeter. A significant amount of leakage current through the ammeter to the case from anywhere in the cell electric circuit indicates the presence of a conductive path, through potassium hydroxide, on the external surfaces of the cell cases. Such a conductive path may result from spewing of the electrolyte during overcharge, which may indicate either overfilling or the existence of a cracked or leaky cell case. Isolation of the exact cause can be accomplished by determining the leakage "nodal" point in the cell string by repeating the measurements after physical cleanup of the battery.

Detection of the failure of the gas barrier, a very important part of objective 3, may be reliably and conveniently accomplished by extending the $C/10$ charge to 24 h. This overcharge will indicate accurately the failure of the gas barrier by either or both of two principal measurements near the end of that overcharge. First, the overcharge gassing rate is extremely sensitive to gas barrier condition and gas recombination. When measured with a simple ball flowmeter, the 24-h gas rate will be less than 80% of normal if the barrier has failed significantly in the cell. The normal value is 11 ml/min for each ampere of the $C/10$ overcharge rate. The second indicator of gas barrier failure will be a 24-h overcharge voltage of less than 1.5 V if the cell is being charged in a 23°C ambient. Some downward adjustment of this voltage criterion may be made at the rate of 4 m V/°C if the battery is being charged in a higher ambient temperature.

27.6.2 Mechanical Maintenance

The replenishment of water in the electrolyte to return the electrolyte to the level recommended, objective 5, is the most important routine mechanical procedure employed during battery reconditioning. It is best accomplished near the end of the 24 hours of the $C/10$ rate charge by replenishing with deionized water until the electrolyte reaches the recommended level for a battery in overcharge. A record of the amount of water usage in each cell should be maintained and compared with the battery manufacturer's statement of reserve electrolyte level in each cell. If the total water usage between maintenance fillings, after deducting the amount used during the maintenance procedure, exceeds the reserve available in that cell design, the maintenance interval must be shortened to prevent plate dry-out during use and resultant cell failure. Note that the 24-h $C/10$ reconditioning procedure will itself use approximately 0.4 ml of water for each Ampere-hour of rated capacity during the 24-h reconditioning period. For example, 12 mL of water would be used during the reconditioning period for a 30-Ah rated cell, and this must be subtracted from the amount added to determine the amount actually used in service. It should also be noted that a cell with a damaged gas barrier may use less water.

That point in the maintenance procedure, following the thorough short-circuiting of each cell, may be utilized to perform physical maintenance. Cells may only be replaced while in the discharged state. Cleanup generally consists of a thorough rinsing with clear water followed by warm-air drying of the battery. This will dissolve and remove any accumulation of potassium hydroxide and carbonates from the outside of the cell jars. Vent caps should also be washed in warm de-ionized water, warm water forced through the vent, and then dried with warm air. This is safely accomplished only with the cells in the completely discharged state. Replacement of any cells not found defective until the conclusion of the $C/10$ overcharge requires discharging the cells a second time.

Other typical hardware problems include, loose terminal nuts—indicated by burns or arcing on intercell links; terminal seal failure—various heavy deposits around cell terminal, remove all hardware and the O-ring; vent failure—various heavy deposits on or around the vent valve, the valve may have been installed improperly or the vent sleeve or O-ring has failed; also inspect the vent sleeve to insure it is not torn or broken. Power delivery of the battery may be enhanced by removing intercell link hardware, buffing contact surfaces, then replacing and retorquing all connectors.

27.6.3 System Inspection Criteria

Reinstallation of the reconditioned and fully charged battery into the aircraft system presents the opportunity for performing the system voltage calibration check in fulfillment of objective 6. The only measurement required for this on a CP charging system is to record the value of the battery voltage, after reactivating the system and following stabilization of the voltage. This stable value is the regulated float voltage to which the battery will be subjected during extended overcharge in use. This voltage measurement should be made at the battery with the engine running at a sufficiently high speed to produce a representative and stable value.

The battery voltage measurement, on constant-current VCO systems, should be made just as the system reaches cutoff voltage. The regulated voltage, on either of the two systems, should then be adjusted to the manufacturer's recommended value if necessary. These adjustments must consider the effects of any automatic temperature compensation of voltage present in the system.

27.7   RELIABILITY

27.7.1 Failure Modes

The sintered plate construction is very robust and capable of operating in both high and low temperature extremes. The cell is capable of withstanding substantial abuse and still performing as intended. It can be discharged into reversal and given a substantial amount of overcharge, as long as the temperature is controlled. The gas barrier that prevents recombination of the oxygen on the cadmium electrode aids control of the temperature during charge. In the past, cellophane was mainly used as the gas barrier. It had a tendency to hydrolyze in the electrolyte and eventually decompose to carbonate and derivatives of the cellophane structure. In recent times, the cellophane has been replaced with Celgard® 3400 or other similar materials.[10] These materials are polyethylene and polypropylene based, and do not degrade in the 30% KOH electrolyte. Should the cell's gas barrier fail, continuing in operation for enough cycles will result in the battery losing capacity and maximum power capability. Continued temperature increase in the cell can result in fusing, or melting, of the nylon separator and result in an internal short circuit of the cell.

Although the cellophane replacement used for the gas barrier alleviates the above failure mechanism, it introduces another problem, if the cell is not manufactured and maintained properly. Without proper additives in the electrolyte, which find their way into the cadmium electrode during cycling, the cell can lose capacity in the negative electrode. The role of supplying an oxidized cellophane expander for the cadmium electrode needs to be replaced by cellulose derivatives and other additives to maintain the negative capacity.[11]

Several other failure modes which account for a small portion of cell and battery failures include the following: (1) Internally short-circuited cells can result from cut-through of the electrical separator by burrs and other plate irregularities, aggravated by cell interplate pressures and vibration. (2) Cracked and leaking cell cases may result from abusive handling of the cells during cell replacement procedures and maintenance, or from defective manufacturing or sealing procedures. (3) Burned terminal contacts may result from faulty cleaning and buffing procedures during maintenance, insufficient link assembly torquing, terminal screw failure, or conductive articles being dropped on the internal connectors of a charged battery.

27.7.2 Memory Effect

In addition to these permanent failures, there is a reversible effect which may result in a gradual reduction of both power and capacity with cycling. This effect, sometimes referred to as "memory effect," "fading," or "voltage depression," results from charging following repetitive shallow discharges where some portion of the active materials in the cell is not used or discharged, such as in a typical engine-start use. This effect is noticed when the previously undischarged material is eventually discharged. The terminal voltage during the latter part of that full discharge may be lower by approximately 120 mV (hence, "voltage depression"). The total capacity is not reduced, however, if the discharge is continued to the lower voltages, as, for example, to the "knee" of the curve.

This effect is completely reversible by a maintenance cycle consisting of a thorough discharge followed by a full and complete charge-overcharge as described in Sec. 27.6.1.

27.7.3 Factors Influencing Gas Barrier Failure

Gas barrier failure is generally acknowledged to be caused or aggravated by excessive over-charge current, excessive overcharge temperatures, and discharge at high rates with low electrolyte levels. Gas barrier failure may not be detectable during reconditioning with other than the low-rate constant-current procedures. Barrier failures may actually occur during poorly structured maintenance and then manifest themselves at a later time after reinstallation in the aircraft. This possibility emphasizes the importance of an accurate assessment of the condition of the gas barrier at the end of the reconditioning period just prior to reinstallation. This assessment is accurately made by the measurement of overcharge gas flow following the extended $C/10$ overcharge, as described in Sec. 27.6.1.

One indication of the significant importance of the two factors of (1) temperature com-pensation of charger voltage and (2) effective maintenance practices is the existence of large-scale field data which document real-time failure differences of up to 100:1. These life differences exist between well-maintained batteries in temperature-compensated systems on the one hand and identical battery designs in uncompensated CP systems with frequent and poorly managed maintenance procedures on the other. Recent improvements in gas barrier materials have significantly reduced these failures.

27.7.4 Thermal Runaway

The loss of the gas barrier in one or more cells of a vented nickel-cadmium battery can lead to thermal runaway. Loss of this function allows oxygen, generated in overcharge, to reach the negative plate and recombine on it. This generates heat. The temperature increase which follows causes the internal cell voltage to decrease. Charge current then increases exponen-tially to increase cell voltage to match the charger voltage.

Thermal runaway occurs with the use of a voltage-regulated (CP) charge source on a battery containing cells with a failed gas barrier. Thermal runaway begins when the failed cells approach overcharge following recharge. The (over)charge current may reach a mini-mum and then gradually increase. Voltage inequities may exist at this point unless all cells are experiencing similar recombination (gas barrier damage). Oxygen recombination heats, and begins to increase the temperature of the failed cell or cells and thus their neighboring cells unless the battery is effectively air cooled. The resulting temperature increase, however, proceeds slowly due to the large thermal mass involved. It may take 2 to 4 hours of (near) consecutive overcharging for a cell to reach boiling temperature.

If the boiling phase continues long enough, or is repeated, and the failed cell becomes dry, large inequities in cell voltage will appear. The voltage across the cell that has boiled dry will increase, thereby decreasing the charge current and the voltage across the cells that are still wet with electrolyte. The next event probably will be internal short-circuiting of the dried-out cell due to very high temperatures and voltage at the last remaining damp spots with consequent burning of the electrical separator insulation. The (over)charge current then increases sharply due to cell loss, and the process repeats itself with the next cell to go dry.

Because of extensive heating and boil away times, thermal runaway may go undetected for many flight hours following the onset of gas barrier failure if the use of the system is not consecutive or continuous. This can confuse the perceived connection between the cause of the gas barrier damage and the resultant thermal runaway.

27.7.5 Potential Hazards

Potential hazards which may be present during the use and maintenance of the vented sintered nickel-cadmium battery fall into five general categories, as described in this section.

Gas Fire and/or Explosion.   Since all functional vented batteries generate a stoichiometric mixture of hydrogen and oxygen gases during overcharge and expel them normally from the cell into the battery container, a potential for explosion of these gases is always present. Two conditions are necessary for such an explosion, however, and both are recognized and accounted for in the design of a typical system. The first condition is the accumulation of a sufficient quantity of this gas mixture. This condition is minimized in all system designs by the incorporation of adequate battery case ventilation. Some designs rely on supplying a modest quantity of air to purging tubes on the battery from overboard vents in the aircraft. Others incorporate natural convection of the gases from a louvered battery case into a ventilated compartment. Air-cooled designs inherently accomplish the required ventilation due to the large volume of air used.

Unusual circumstances, however, may defeat any of these gas-purging techniques. It should also be remembered that batteries generate a significant amount of explosive gas during the maintenance procedure, and therefore maintenance should always be performed in a well-ventilated shop.

The second condition necessary for explosion of the gases is the presence of a source of ignition. Although normally there are no ignition sources inside the battery case, several abnormal possibilities do exist. One is the internal short-circuiting of a relatively dry cell in overcharge, resulting in an explosion inside the cell with a subsequent ejection of flames into the battery case. A second and more likely source of ignition may exist at an improperly maintained cell terminal due to the high temperatures generated during high-rate discharge. A third source of ignition may occur at the site of stray leakage currents.

Although the coincidence of both a sufficient amount of gas accumulation and a source of ignition is quite rare, it can and has happened. Because of this possibility, many batteries are also designed to be physically capable of managing a hydrogen or oxygen explosion and containing the effects entirely within the battery case. Typically these batteries will also be electrically functional for at least one C-rate discharge following such an explosion.

Arcing and Burning.   This potential hazard concerns excessive leakage currents through electrolyte paths outside the cells. Such currents can occur either between cells which are physically adjacent but with a wide voltage separation in the cell circuit or, more probably, from a cell to a grounded metallic case. Arcing is more likely to occur, however, in the circuit of an inappropriately protected auxiliary device located inside the battery case in the environment containing potassium hydroxide. Some examples of these devices are battery heaters, thermal detectors, and voltage sensors. The proper design of these auxiliary appliances must recognize the conductivity of KOH and the ability of that electrolyte to creep along wires and even into mechanically "sealed" insulation. Devices of this type should be tested with high dielectric voltages while submerged in a water-detergent mix before they are installed in the battery case environment.

The result of a sustained leakage current through relatively localized KOH conducting paths may be the ignition of the explosive environment by arcing, as discussed previously. The result might also be the carbonization of adjacent insulating materials and the subsequent burning of those materials within the battery case.

Electrical Power.   One of the essential functional capabilities required of the vented nickel-cadmium battery is the ability to deliver high-power rates for engine starting. This very capability, however, presents a potential risk in the form of hot spots on improperly torqued cell terminals during high-rate discharge. It is also a potential hazard to the unwary maintenance technician operating with metallic tools or other objects, such as jewelry, in a careless manner in the vicinity of charged batteries. Since the short-circuit current of these cells (batteries) may exceed 1000 to 4000 A, it should be obvious that the exposed conductors of a charged battery should be treated with respectful caution. Very severe burns may occur if, for example, a ring should accidentally make contact between two adjacent cell terminals. Although one of the most obvious, this hazard is one of the most frequently encountered. Insulated cell hardware does provide a partial solution: however, caution and respect for the available power must always be exercised.

Corrosive KOH.   Because of the corrosive nature of the KOH used as an electrolyte, all material employed in the construction of the battery and its accessory appliances must be KOH immune. Materials such as nylon, polypropylene, nickel-plated steel, steel, and stainless steel are therefore used. The potential hazard of KOH corrosiveness, however, is primarily encountered during maintenance. The use of safety glasses and safety face shields should be mandatory while performing maintenance on these batteries. A very small amount of KOH in the eye, for example, without prompt, continued, and adequate flushing followed by medical treatment can result in the loss of eyesight. KOH is also corrosive to the skin, and the affected area should be thoroughly washed and rinsed with water, thereby minimizing the detrimental effect.

Electric Shock.   Most vented sintered nickel-cadmium batteries are arrayed in groups of 10 to 30 cells, presenting maximum voltages of 15 to 45 V. There are applications, however; in which batteries are connected electrically in series strings of 90 to 200 cells or more. It should be apparent that the voltages presented by this number of cells in series may be lethal to anyone exposed to them. Personnel should also be cautioned, because of the high probability of electrolyte being present between the cell circuit and the conductive external surface of the battery, to exercise care by disconnecting series-connected batteries prior to personal exposure. Significant shock currents may be carried by relatively small amounts of KOH.

27.8   CELL AND BATTERY DESIGNS

27.8.1   Typical Vented Sintered-Plate Nickel-Cadmium Cells

A listing of several typical vented sintered-plate nickel-cadmium cells is given in Table 27.3. The 14-, 22-, and 36-Ah sizes are those typically employed in aircraft batteries. Other cells are available in sizes up to about 350 Ah. The larger cells are generally constructed in steel containers rather than the plastic containers now used for the aircraft-size cells.

TABLE 27.3   Typical Vented Sintered Plate Nickel-Cadmium Cell Properties

Rated capacity (A-h)	Cell designation	Height (cm)	Width (cm)	Thickness (cm)	Weight (grams)
1.5	2B02-0	10.16	2.92	1.70	86
2	2A02-0	8.74	3.81	1.83	95
5.5	5A06-0	10.31	5.51	2.39	236
5.5	5C06-1	10.36	5.51	2.39	272
7	0707-0	18.85	6.65	1.29	299
6	0906-1	11.60	5.89	2.69	354
13	1313-1	11.93	7.95	3.02	486
10	1410-1	14.48	5.89	2.69	445
12	1412-0	14.38	5.86	2.69	422
20	2020-1	17.42	7.95	3.53	1067
28	2028-1	17.27	7.95	3.53	1149
23	2223-1	20.57	8.08	2.72	903
36	3030-1	17.42	7.95	5.08	1562
40	4040-1	23.31	7.95	3.53	1453
42	4240-1	23.31	7.95	3.53	1453
100	6060-1	24.48	12.7	3.83	2860

Source: Courtesy of Eagle-Picker Technologies, LLC. Power Systems Dept.

27.8.2   Typical Battery Designs

A typical arrangement of vented sintered nickel-cadmium cells into a battery configuration is the conventional aircraft battery. An example of this use is shown in Fig. 27.19 and in detail in Fig. 27.20. This arrangement generally consists of a completely enclosing battery case and cover made of either stainless steel or steel with a KOH-resistant finish of epoxy or paint. The cover is typically secured with over-center-type latches. The battery case is provided with gas-purging vents or with freely convective gas-diffusion openings for dilution. The cells are encased in nylon-molded cell cases with terminals extending through a nylon cover sealed to the case. The cells are electrically connected in series with nickel-plated copper links from cell terminal to cell terminal and from the first and last cell to the battery termination and disconnect device. This battery termination extends through the battery case wall and is present on the outside surface of the battery case as a recessed double-male, polarized, high-current receptacle. Functional requirements of aircraft batteries are specified in SAE standard AS 8033A.

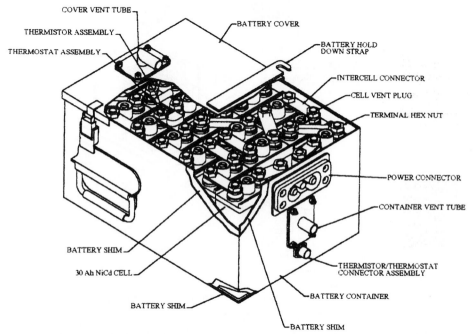

FIGURE 27.19   Vented sintered-plate nickel-cadmium battery. (*Courtesy of Eagle-Picker Technologies, LLC. Power Systems Dept.*)

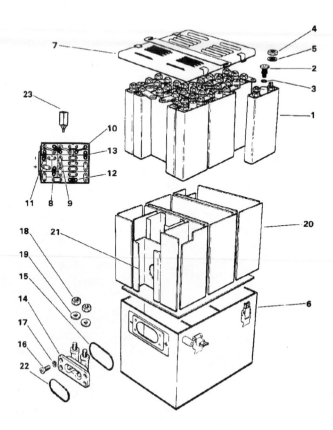

REF NO.	DESCRIPTION	QTY. REQ.	REF NO.	DESCRIPTION	QTY. REQ.
1	Cell Ass'y	20	13	Link, Terminal - Flat	7
2	Valve, Vent	1	14	Receptacle, Battery Connector	1
3	"O" - Ring	1	15	Gasket, Connector Receptacle	1
4	Nut, Terminal	2	16	Screw, Connector Receptacle Mtg.	4
5	Washer, Terminal	2	17	Washer, Connector Receptacle Mtg.	4
6	Case, Battery	1	18	Nut, Terminal Adapter	2
7	Cover Ass'y, Battery Case	1	19	Washer, Terminal Adapter	2
8	Link, Terminal - Flat	2	20	Liner - Spacer Kit	1
9	Link, Terminal - Flat	1	21	Bracket, Spacer	1
10	Link, Terminal - Curved	3	22	Spring, Shorting	1
11	Link, Terminal - Flat	1	23	Wrench, Vent	1
12	Link, Terminal - Flat	7			

FIGURE 27.20   Typical assembly of vented sintered-plate nickel-cadmium aircraft battery.

27.8.3 Air Cooling / Heating

The major battery manufacturers produce some battery designs with provision for forced-air cooling. These designs generally take the form of plenum spaces below and above the cells, with cooling passages between the cells connecting the two plenums. The construction provides a means for the external connection of a high-volume, low-pressure air source. Supplying 23°C air will not only effectively cool an overheating battery, it will also rapidly warm a cold battery. The heat transfer coefficient from the battery core is improved up to 10 times by this technique. The thermal time constant for a battery with this feature may be as short as 10 to 20% of that of a standard non-air-cooled battery.

27.8.4 Temperature Sensors

Sensors may be provided inside the battery case which sense either typical or average cell temperatures. They are equipped with a provision for external electrical connections. These devices may be of the on-off type, such as thermostats, or they may be the continuous type, such as thermistor assemblies. Continuous types have the capability of providing continuous modulation of, for example, the regulated charging voltage of a CP charging source or of the cutoff/turn-on voltages of a constant-current VCO charging system.

27.8.5 Battery Cases

Although the corrosive effect of KOH on bare steel is minimal and cosmetic in effect, additional KOH resistance in the case material is desirable. In addition to stainless steel and steel with a protective finish, some special applications use KOH-resistant plastic materials. It must be emphasized, however, that battery cases may require withstanding significant rough treatment in shock and vibration without losing their KOH-containment capability.

27.8.6 Battery Electrical Termination

Aircraft batteries are normally terminated on the front case surface by a connector of the type shown in Fig. 27.19. Special applications, however, may utilize direct cable connection to the first and last cell terminals, as well as various other special configurations capable of handling the high-current rates available in the event of a short circuit of the external battery circuit.

27.8.7 Battery Heaters

Alternative to the airflow heating discussed in Sec. 27.8.3, heater blankets are sometimes employed on the inside or outside of the battery case. These blankets may be energized with any available electric energy source. Primarily this source will be either the DC bus of the same voltage as the battery or an aircraft AC bus of higher potential. Heaters may also be energized from an auxiliary ground supply. Heater blankets have the inherent poor thermal time constant of nonair-cooled batteries.

27.8.8 Extensions of Vented Sintered-Plate Nickel-Cadmium Designs

To avoid costly maintenance procedures associated with vented nickel-cadmium designs and to improve general battery reliability, the U.S. Air Force has incorporated lessons learned from vented cell anomalies into a Maintenance Free Nickel-Cadmium Aircraft Battery. This battery uses the same geometric cell shape as the vented nickel-cadmium design, but allows gasses to recombine inside the cell. The cell design has a number of features found in aerospace cell technology. It is not strictly a sealed design since the cell will vent at a high pressure then reseal itself well above ambient pressure. To avoid excessive overcharge and associated thermal runaway, the cell is charged and controlled by its own integrated charger and associated electronics. This battery design is now supplied to several military aircraft programs by Eagle-Picher Technologies, LLC, Power Systems Department, in Colorado Springs, Colorado.[12]

REFERENCES

1. J. McBreen, *The Nickel Oxide Electrode,* Modern Aspects of Electrochemistry, No. 21, Ralph E. White, J. O'M. Bockris and B. E. Conway, Eds., p. 29, Plenum Press, 1990, New York.

2. D. F. Pickett and J. T. Maloy, *J. Electrochem. Soc.,* **12:**1026 (1978).

3. S. U. Falk and A. J. Salkind, *Alkaline Storage Batteries,* Wiley, New York, 1969.

4. A. Fleischer, *J. Electrochem. Soc.* **94:**289 (1948).

5. G. Halpert, *J. Power Sources* **12:**117 (1984).

6. D. F. Pickett, in *Proceedings of the Symposium on Porous Electrodes, Theory and Practice,* (eds.) H. C. Maru, T. Katan, and M. G. Klein, The Electrochemical Society, Pennington, NJ, 1982, p. 12.

7. M. B. Pell and R. W. Blossom, U.S. Patent 3,507,699 (1970).

8. R. L. Beauchamp, U.S. Patent 3,573,101 (1971); U.S. Patent 3,653,967 (1972).

9. D. F. Pickett, U.S. Patent 3,827,911 (1974); U.S. Patent 3,873,368 (1975).

10. Mil-B-81757, Performance Specification, Batteries and Cells, Storage, Nickel Cadmium, Aircraft General Specification, Crane Division, NSWC, July 1, 1984.

11. J. J. Lander, personal communication.

12. T. M. Kulin, 33rd *Intersociety Engineering Conference on Energy Conversion* IECEC-98-145, Colorado Springs, CO, August 2-6, 1998.

CHAPTER 28
PORTABLE SEALED NICKEL-CADMIUM BATTERIES

Joseph A. Carcone

28.1 GENERAL CHARACTERISTICS

Sealed nickel-cadmium batteries incorporate specific battery design features to prevent a buildup of pressure in the battery caused by gassing during overcharge. As a result, batteries can be sealed and require no servicing or maintenance other than recharging. These unique characteristics for a secondary battery have created wide acceptance for use in a variety of applications, ranging from lightweight portable power (photography, toys, housewares) to high-rate, high-capacity power (electronic devices such as phones, computers, camcorders, power tools) and standby power (emergency lighting, alarm, memory backup). Some nickel-cadmium batteries are now incorporating smart battery control circuitry to give state-of-charge indication and to control overcharge and overdischarge (see Chap. 5).

The major advantages and disadvantages of the sealed nickel-cadmium battery are summarized in Table 28.1. The important characteristics are described in this section.

TABLE 28.1 Major Advantages and Disadvantages of Seated Nickel-Cadmium Batteries

Advantages	Disadvantages
Batteries are sealed; no maintenance required	Voltage depression or memory effect in certain applications†
Long cycle life	Higher cost than sealed lead-acid battery
Good low-temperature and high-rate performance capability	Poor charge retention
Long shelf life in any state of charge	Sealed lead-acid battery better at high temperature and float service*
Rapid recharge capability	Environmental concern with the use of cadmium
	Lower capacity than other competitive batteries

*High-temperature nickel-cadmium batteries are available (see Sec. 28.6.3).
†See Sec. 28.4.11.

Maintenance-free operation: The batteries are sealed, contain no free electrolyte, and require no servicing or maintenance other than recharging.

High-rate charging: Sealed nickel-cadmium batteries are capable of recharge at high rates within 1 h under controlled conditions. Many batteries can be charged in 3 to 5 h without special controls, and all can be recharged within 14 h.

High-rate discharge: Low internal resistance and constant-discharge voltage make the nickel-cadmium battery especially suited for high-rate discharge or pulse-current applications.

Wide temperature range: Sealed nickel-cadmium batteries can operate over the range from about -40 to $50°C$ and are particularly noted for their low-temperature performance. Premium performance batteries extend this range to $70°C$.

Long service life: Over 500 cycles of discharge or up to 5 to 7 years of standby power are common to sealed nickel-cadmium batteries.

28.2 CHEMISTRY

The active materials of the sealed nickel-cadmium battery are the same as for other types of nickel-cadmium batteries, namely, in the charged state, cadmium for the negative electrode, nickel oxyhydroxide for the positive, and a solution of potassium hydroxide for the electrolyte In the discharged state, nickel hydroxide is the active material of the positive electrode, and cadmium hydroxide that of the negative.

During charge, nickel hydroxide, $Ni(OH)_2$, is converted to a higher-valence oxide,

$$Ni(OH)_2 + OH^- \rightarrow NiOOH + H_2O + e$$

At the negative electrode, cadmium hydroxide, $Cd(OH)_2$, is reduced to cadmium,

$$Cd(OH)_2 + 2e \rightarrow Cd + 2OH$$

The overall discharge/charge reaction is

$$Cd + 2NiOOH + 2H_2O \underset{\text{charge}}{\overset{\text{discharge}}{\rightleftharpoons}} Cd(OH)_2 + 2Ni(OH)_2$$

During operation, the active materials undergo changes in their oxidation states but little change in their physical states. Similarly, there is little if any change in the electrolyte concentration. The active materials of both electrodes, in both charged and discharged states, are relatively insoluble in the alkaline electrolyte, remain as solids, and do not dissolve while undergoing changes in their oxidation states. Because of these and other properties, nickel-cadmium batteries are characterized by long life in both cyclic and standby operation and by a relatively flat voltage profile over a wide range of discharge currents.

The operation of the sealed battery is based on the use of a negative electrode having a higher effective capacity than the positive. During charge, the positive plate reaches full charge before the negative and begins to evolve oxygen. The oxygen migrates to the negative electrode, where it reacts with and oxidizes or discharges the cadmium to produce cadmium hydroxide,

$$Cd + \tfrac{1}{2}O_2 + H_2O \rightarrow Cd(OH)_2$$

A separator permeable to oxygen is used so that oxygen can pass through the separator to the negative electrode. Also, a limited amount of electrolyte is used (starved electrolyte system) as this facilitates the transfer of oxygen. This process is illustrated in Fig. 28.1.

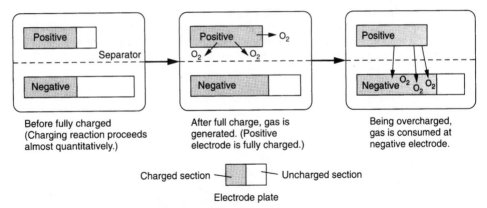

Before fully charged
(Charging reaction proceeds
almost quantitatively.)

After full charge, gas is
generated. (Positive
electrode is fully charged.)

Being overcharged,
gas is consumed at
negative electrode.

Charged section — Uncharged section

Electrode plate

FIGURE 28.1 Oxygen recombination process. (*Courtesy of Sanyo Energy Corp.*)

At a steady state, the recombination reaction rate during overcharge must be no lower than the rate of oxygen generation to prevent buildup of pressure. The internal pressure is sensitive to charge current, the reactivity of the negative electrode, the electrolyte level, and the temperature. Solid cadmium, gaseous oxygen, and liquid water must coexist in mutual contact for the recombination reaction to occur. If, for example, the electrolyte level is too high (the electrodes are in a flooded state), the oxygen is prevented from contacting the electrode, and the reaction rate at a given temperature and pressure is substantially lowered.

A safety venting mechanism is used in the battery design to prevent rupture in case of excessive pressure buildup due to a malfunction, high charge rate, or abuse.

28.3 *CONSTRUCTION*

Sealed nickel-cadmium cells and batteries are available in several constructions. The most common types are the cylindrical shaped batteries (see Table 28.3). Smaller button batteries and rectangular batteries are also manufactured.

28.3.1 Cylindrical Batteries

The cylindrical battery is the most widely used type because the cylindrical design lends itself readily to mass production and because excellent mechanical and electrical characteristics are achieved with this design. Figure 28.2 shows a cross section of the cylindrical battery.

The positive electrode is a highly porous sintered, foam or fibered nickel structure into which the active material is introduced by embedding or impregnation with a molten nickel salt, followed by the precipitation of nickel hydroxide by immersion or electrochemical deposition in an alkaline solution. The negative electrodes are made by several methods: using a sintered-nickel substrate similar to the positive, by pasting or pressing the negative active material onto a substrate, or by a continuous electrodeposition process.

After processing, the positive and negative electrodes are cut to size and wound together in a jelly-roll fashion with a separator material between them. The separator material, usually unwoven nylon or polypropylene, is highly absorbent to the potassium hydroxide electrolyte and permeable to oxygen. The roll is inserted into a rugged nickel-plated steel can, and the electrical connections are made. The negative electrode is welded or press connected to the can and the positive to the top cover. The very small amount of the electrolyte, enough for efficient operation, is absorbed by the separator. There is no free liquid electrolyte. The cover assembly incorporates a fail-safe vent mechanism to prevent rupture in case of excessive pressure buildup, which could result from extreme overcharge or discharge rates.

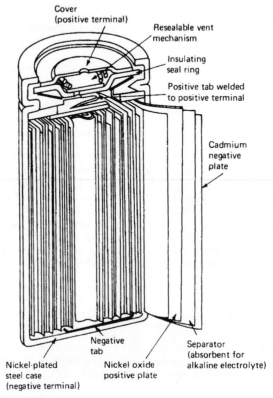

FIGURE 28.2 Construction of sealed nickel-cadmium cylindrical battery.

28.3.2 Button Batteries

Nickel-cadmium button cells and batteries usually have electrodes made from "pressed" plates. The active materials are compressed in molds into disks or plates, and the electrodes are assembled in a sandwich configuration, as shown in Fig. 28.3.

In some cases the electrodes are backed with expanded metal or screen to enhance electrical conductivity and mechanical strength. The button battery does not have a resealable fail-safe device, but its construction allows the battery to expand, either breaking the electrical continuity or opening the seal to relieve excess pressure caused by an abnormal circumstance. Button batteries are very suitable for low-current, low-overcharge-rate applications.

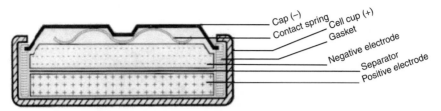

FIGURE 28.3 Nickel-cadmium battery, button configuration.

28.3.3 Slim Rectangular Batteries

The slim or flat rectangular batteries are designed to meet the needs of lightweight and compact equipments. The rectangular shape permits more efficient battery assembly, eliminating the voids that occur with the assembly or cylindrical batteries. The volumetric energy density of batteries can be increased by a factor of about 20%.

Figure 28.4 shows the structure of the slim rectangular battery. The plates are manufactured as described in Sec. 28.3.1, but the finished electrodes are cut to predetermined dimensions and placed in the metal casing. These are then hermetically sealed in place to the cover plate. All sides of the casing and cover-plate assembly are laser-welded together to prevent electrolyte leakage. A resealable safety venting system is built into these batteries similar to the ones used in the cylindrical designs.

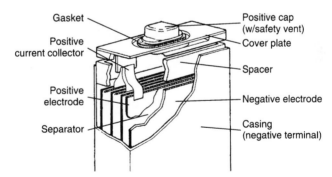

FIGURE 28.4 Construction of sealed slim rectangular battery. (*Courtesy of Sanyo Energy Corp.*)

28.3.4 Rectangular Batteries

The rectangular battery is housed in a nickel-plated steel can using a construction similar to the vented battery, but incorporating the features needed for sealed-battery operation. This construction is particularly suited for high discharge rates because of the large electrode area. Figure 28.5 is an illustration of a sealed rectangular battery, in this case using fiber-structured electrodes (see Sec. 26.6).

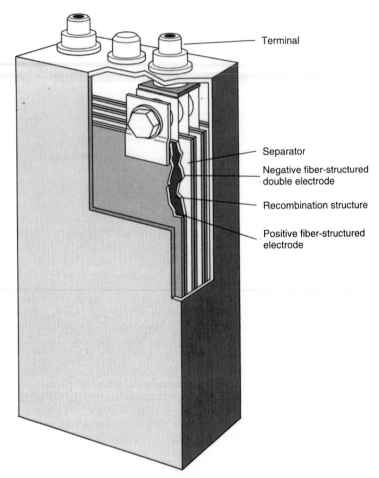

FIGURE 28.5   Sealed rectangular nickel-cadmium battery using fiber-structured electrodes. (*Courtesy of Hoppecke Batteries.*)

28.4   *PERFORMANCE CHARACTERISTICS*

28.4.1   General Characteristics

A typical charge-discharge cycle for the sealed nickel-cadmium cylindrical battery, at 20°C, is shown in Fig. 28.6. The voltage increases slowly but steadily at the $C/10$ charge rate to a steady-state condition, decays slightly during the 1-h rest, and is relatively flat during the 1-h discharge to 1.0 V. The voltage recovers rapidly over the next hour, while at rest, to near 1.2 V.

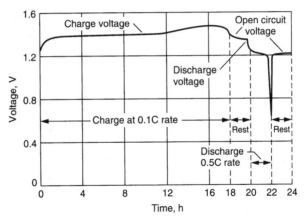

FIGURE 28.6   Voltage profile of nickel-cadmium battery in a typical charge/discharge cycle.

28.4.2   Discharge Characteristics

Typical discharge curves for the cylindrical battery at 20°C at various discharge loads are shown in Fig. 28.7. The flat voltage profile, after the initial voltage drop, is characteristic.

The capacity that can be obtained from a battery is dependent on the rate of discharge, the voltage at which discharge is terminated, the discharge temperature, and the previous history of the battery. Figure 28.8 shows the percentage of rated capacity delivered during discharges at various rates and temperatures. The midpoint voltage during discharge decreases as the discharge rate increases (see Fig. 28.11). If the battery were allowed to discharge to a lower cutoff voltage, a greater percentage of the $C/5$ rate capacity will be obtained. However, batteries or batteries should not be discharged to too low a cutoff voltage as the battery may be damaged (see Sec. 28.4.6).

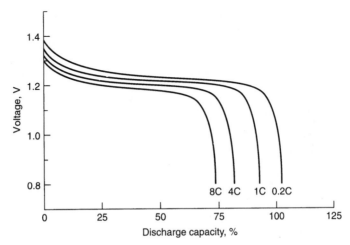

FIGURE 28.7   Constant-current discharge curves for sealed nickel-cadmium batteries at 20°C, charge 0.1C, 16 h. (*Courtesy of Sanyo Energy Corp.*)

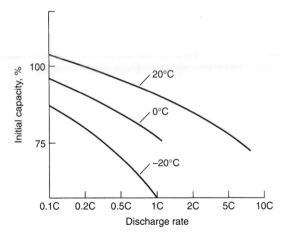

FIGURE 28.8   Percent of $C/5$-rate capacity vs. discharge rate to 1.0-V cutoff for typical sealed nickel-cadmium battery.

28.4.3   Effect of Temperature

The sealed nickel-cadmium battery is capable of good performance over a wide temperature range. Best operation is between -20 and $+30°C$, although usable performance can be obtained beyond this range. The low-temperature performance, particularly at high rates, is generally better than that of the lead-acid battery but usually inferior to that of the vented sintered-plate battery. The reduction in performance at low temperatures is due to an increase in the internal resistance. At high temperatures, the loss can be due to a depressed operating voltage or to self-discharge.

Figure 28.9 shows some typical discharge curves of the sealed battery at various temperatures at the $0.2C$ and $8C$ rates; Fig. 28.10 shows typical discharge curves at $-20°C$. A flat discharge profile is still characteristic, but at a lower operating voltage than at room temperature. Figure 28.11 shows the effect of temperature on the midpoint voltage. Ambient temperatures significantly above or below 20 to $25°C$ have a depressing effect on the average discharge voltage.

The effect of temperature and discharge rate on the capacity of a sealed nickel-cadmium battery is shown in Fig. 28.12. These data are typical of standard batteries. Manufacturers should be contacted to obtain performance characteristics of specific batteries.

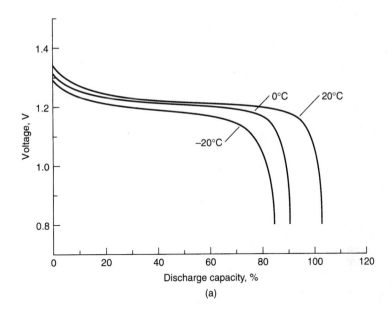

(a)

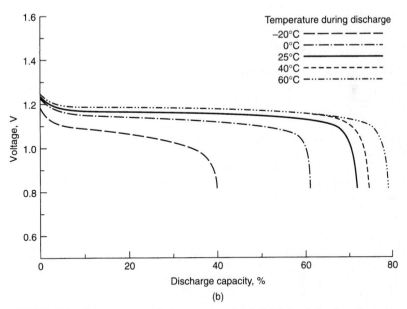

(b)

FIGURE 28.9   Constant-current discharge curves of sealed nickel-cadmium batteries at various temperatures. (*a*) 0.2*C* discharge rate. (*b*) 8*C* discharge rate.

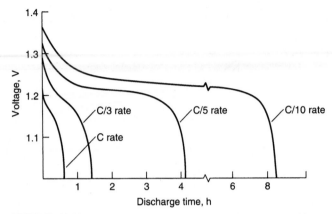

FIGURE 28.10 Constant-current discharge curves of sealed nickel-cadmium batteries at −20°C.

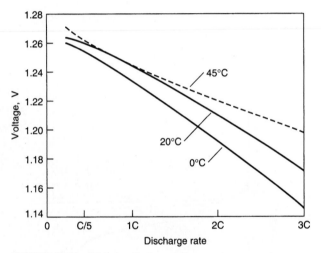

FIGURE 28.11 Midpoint discharge voltage vs. rate at various temperatures for sealed nickel-cadmium batteries, 1-V cutoff.

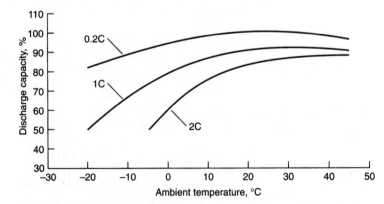

FIGURE 28.12 Percentage of rated capacity vs. temperature at different discharge rates for typical sealed nickel-cadmium batteries, 1.0-V cutoff.

28.4.4 Internal Impedance

The internal impedance of a battery is dependent on several factors, including ohmic resistance (due to conductivity, the structure of the current collector, the electrode plates, separator, electrolyte, or other features of the battery design), resistance due to activation and concentration polarization, and capacitive reactance. In most cases the effects of capacitive reactance can be ignored. Polarization effects are dependent in a complicated way on current, temperature, and time; they decrease with increasing temperature (see Chap. 2). The effect may be negligible for pulses of short duration, that is, less than a few milliseconds.

The nickel-cadmium battery is noted for its low internal resistance due to the use of thin and large-surface-area plates with good electrical conductivity, a thin separator with good electrolyte retention, and an electrolyte having a high ionic conductivity. During discharge, the activation and concentration polarization effects are negligible, at least at low and moderate rates, and the internal resistance of the battery, and the discharge voltage, remains relatively constant from the state of full charge to the point where almost 90% of the capacity has been discharged. At that point the resistance increases due to the conversion of active materials in the electrode plates, which tends to lower electrical conductivity. Figure 28.13 shows the change in internal resistance with the depth of discharge for two batteries of different size and capacity. Figure 28.14 illustrates the effect of temperature. The internal resistance increases as the temperature drops because the conductivity of the electrolyte and other components is lower at the lower temperatures.

With use over time, a nickel-cadmium battery gradually loses capacity, resulting in a gradual increase in internal resistance. This is caused by gradual deterioration of the separator and electrodes and by loss of liquid through the seals, which changes the electrolyte concentration and level. The net effect is an increase in internal impedance.

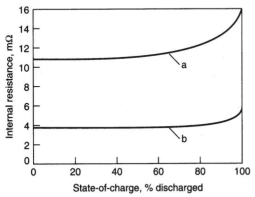

FIGURE 28.13 Resistance vs. state of charge at 20°C, discharged at $0.2C$ rate, for sealed nickel-cadmium batteries. a—AA-size battery, b—sub-C-size battery. (Typical for sintered-plate electrode type batteries.)

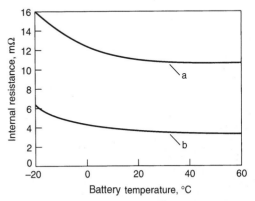

FIGURE 28.14 Resistance vs. temperature for fully charged sealed nickel-cadmium batteries. a—AA-size battery, b—sub-C-size battery. (Typical for sintered-plate electrode type batteries.)

28.4.5 Service Life

The service life of a sealed nickel-cadmium battery, normalized to unit weight (kilograms) and size (liters), at various discharge rates and temperatures is summarized in Fig. 28.15. The curves are based on a capacity, at the $C/5$ discharge rate at 20°C, of 30 Ah/kg and 85 Ah/L, reflecting the performance of standard-type sealed cylindrical batteries. Manufacturers should be contacted for performance characteristics of specific batteries.

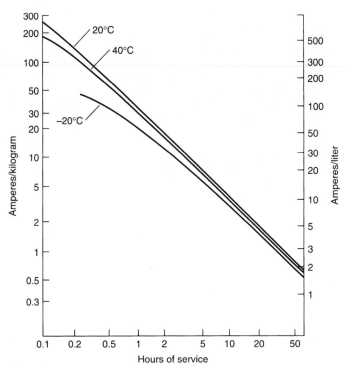

FIGURE 28.15 Service life of sealed nickel-cadmium battery at various constant-current discharge rates and temperatures; end voltage at 1.0 V.

28.4.6 Reversal of Voltage Polarity

When three or more batteries are series-connected, the lowest-capacity battery can be driven into voltage reversal by the others. The larger the number of batteries in series, the greater the possibility of this occurring. During reversal, hydrogen may evolve from the positive electrode and oxygen from the negative. Figure 28.16 shows the complete discharge curve of a battery, including polarity reversal. Section 1 is the normal period of discharge with active materials remaining on both electrodes. Section 2 represents the period when the discharge has been extended and all of the active material on the positive electrode has been discharged and hydrogen gas is generated at this electrode. Active material still remains on the negative electrode and its normal discharge reaction continues. The battery voltage varies with the discharge current, but stays at about -0.2 to -0.4 V. In Section 3 the negative active material has been discharged and oxygen gas is generated at this electrode.

Continued discharging during polarity reversal will lead to high battery pressure and opening of the safety vent. This then results in a loss of gas and electrolyte and breakdown of the capacity balance of the positive and negative electrodes.

Some battery designs provide a limited amount of built-in protection against deep reversal by adding a small amount of cadmium hydroxide to the positive electrode. The term used for the material added to the positive electrode for reversal protection is "antipolar mass" (APM). When the positive electrode is completely discharged, the cadmium hydroxide in that electrode is converted to cadmium, which, combining with the oxygen generated from the negative electrode, depolarizes the positive, preventing hydrogen generation for a time. This reaction occurs at about -0.2 V. This reaction can sustain for only a limited time, after which hydrogen is evolved from the positive electrode. Because hydrogen combines with the battery materials to only a limited extent, repetitive battery reversal will gradually increase a battery's internal pressure, ultimately causing the battery to vent.

Discharging to the point of reversal should be avoided. In order to prevent voltage reversal in any of the cells of a multicell battery, particularly with a series string of more than four cells, the battery should not be discharged to a voltage below 0.8 V per cell. In applications of multicell batteries where it is likely that the battery will be frequently discharged below 1.0 V per cell, a voltage-limiting device is recommended to avoid cell reversal.

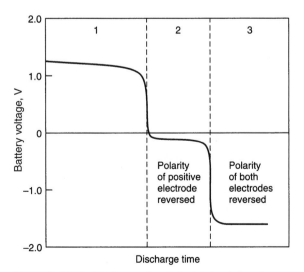

FIGURE 28.16 Discharge of sealed nickel-cadmium battery showing polarity reversal.

28.4.7 Types of Discharge

As discussed in Sec. 3.2.3, a battery may be discharged under different modes (such as constant resistance, constant current, or constant power), depending on the characteristics of the equipment load. The type of discharge mode selected has a significant impact on the service life delivered by a battery in a specified application. The voltage profiles of a nickel-cadmium battery discharged under the three different modes are plotted in Fig. 28.17. The data are based on a discharge of a 650-mAh battery so that, at the end of the discharge (1.0 V per cell), the power output is the same for all modes of discharge. In this example the power output is 130 mW. To discharge at 130 mW at 1.0 V, the constant-current discharge is 130 mA ($C/5$ rate) and the constant-resistance discharge is 7.7 Ω. As shown, the longest service life is obtained under the constant-power mode as the average current is the lowest under this mode of discharge.

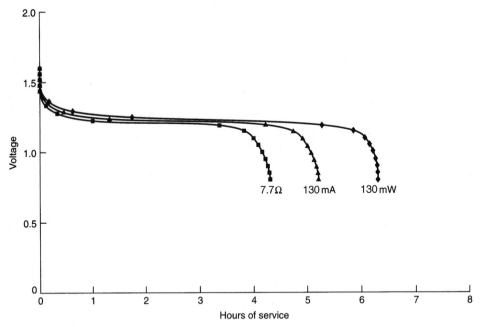

FIGURE 28.17 Discharge curves of AA-size (650-mAh) nickel-cadmium battery—constant power (♦) vs. constant current (▲) vs. constant resistance (■).

28.4.8   Constant-Power Discharge

The discharge characteristics of the nickel-cadmium battery under the constant-power mode, at several different power levels, are shown in Fig. 28.18. These are similar to the data presented in Fig. 28.7 for constant-current discharges, except that the performance is presented in hours of service instead of percent discharge capacity. The power levels are shown based on the E-rate. The E-rate is calculated in a manner similar to calculating the C-rate, but based on the rated watthour capacity. For example, for the $E/5$ power level, the power for a battery rated at 780 mWh is 156 mW.

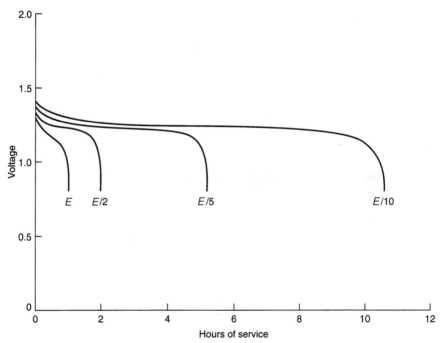

FIGURE 28.18   Constant-power discharge curves at various E rates for sealed nickel-cadmium batteries at 20°C.

28.4.9   Shelf Life (Capacity or Charge Retention)

Nickel-cadmium batteries lose capacity during storage. The rate of this self-discharge is a function of storage temperature and battery design. Figure 28.19 can serve as a guide for the shelf life (capacity or charge retention) at several temperatures for typical standard type nickel-cadmium sealed batteries. Specifically designed batteries, as discussed in Sec. 28.6, may have considerably different charge retention characteristics. For example, the button batteries designed for memory-backup application have significantly better charge retention characteristics than the standard lower-resistance higher-discharge-rate cylindrical batteries.

Sealed nickel-cadmium batteries can be stored in a charged or a discharged condition. Except for extended storage at high temperatures, they can be restored to full capacity after storage by recharging (two or three charge-discharge cycles). Figure 28.20 illustrates the capacity recovery after prolonged storage at several temperatures. The recovery time may be longer after high-temperature storage.

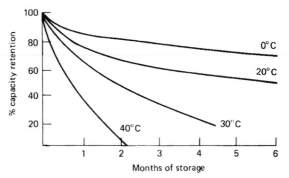

FIGURE 28.19   Capacity retention (shelf life) of standard type sealed nickel-cadmium batteries.

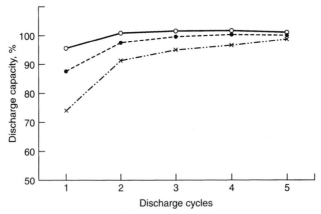

FIGURE 28.20   Capacity recovery of standard type sealed nickel-cadmium batteries, discharge after 2-year storage at 0.2C rate. Storage temperatures: at 20°C (○); 35°C (●); and 45°C (×).

28.4.10 Cycle Life

The cycle life is usually measured to the point when the battery will not deliver more than a given percentage (usually 60 to 80%) of its rated capacity. Sealed nickel-cadmium batteries have long cycle lives. Under controlled conditions over 500 cycles can be expected on a full discharge, as illustrated in Fig. 28.21. On shallow discharges considerably higher cycle life can be obtained, as shown in Fig. 28.22. Cycle life is also very dependent on the many conditions to which the battery has been exposed, including charge, overcharge, and discharge rates, frequency of cycling, the temperatures to which the battery has been exposed, and battery age, as well as battery design and battery components. Specially designed batteries, such as those using alkali-resistant materials, are also manufactured, which have longer life, particularly at the higher temperatures (see Sec. 28.6).

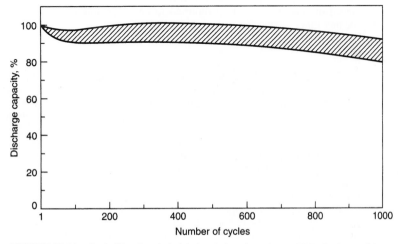

FIGURE 28.21   Cycle life of sealed nickel-cadmium batteries at 20°C. Cycle conditions: charge—$0.1C \times 11$ h: discharge—$0.7C \times 1$ h. Capacity-measuring conditions: charge—$0.1C \times 16$ h; discharge—$0.2C$, end voltage—1 V.

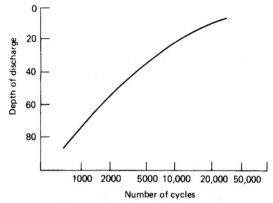

FIGURE 28.22   Cycle life of sealed nickel-cadmium batteries at shallow discharge.

28.4.11 Life Expectancy and Failure Mechanisms

The useful life of a nickel-cadmium battery can be measured either in terms of the number of cycles before failure or in units of time. It is virtually impossible to know all the detailed information necessary to make any kind of accurate prediction of battery life in a given application. The best that can be provided is an estimate based on laboratory test data and field experience or extrapolation of accelerated test data.

Basically, failure of a battery occurs when it ceases to operate the device, for whatever reason, at the prescribed performance level, despite the possibility that the battery may still be useful in another application with less demanding requirements.

Failure of a nickel-cadmium battery can be classified into two general categories: reversible and irreversible failure. When a battery fails to meet the specified performance requirements but can, by appropriate reconditioning, be brought back to an acceptable condition, it is considered to have suffered a reversible failure. Permanent or irreversible failure occurs when the battery cannot be returned to an acceptable performance level by reconditioning or any other means.

Reversible Failures

Voltage Depression (Memory Effect). A sealed nickel-cadmium battery may suffer a reversible loss of capacity when it is cycled repetitively on shallow discharges (discharge terminated before its full capacity is delivered) and recharged. For example, as shown in Fig. 28.23, if a battery is cycled repetitively, but only partially discharged and then recharged, the voltage and delivered capacity will gradually decrease with cycling (curves 2 representing repetitive cycling). If the battery is then fully discharged (curve 3), the discharge voltage is depressed compared to the original full discharge (curve 1). The discharge profile may show two steps and the battery may not deliver the full capacity to the original cutoff voltage. This phenomenon is known as "voltage depression." It also is referred to as "memory effect," as the battery appears to "remember" the lower capacity of the shallow discharge. Operation at higher temperatures accelerates this type of loss.

The battery can be restored to full capacity with a few full discharge-charge reconditioning cycles. The discharge characteristics on the reconditioning cycles are illustrated in Fig. 28.23 (curves 4 and 5).

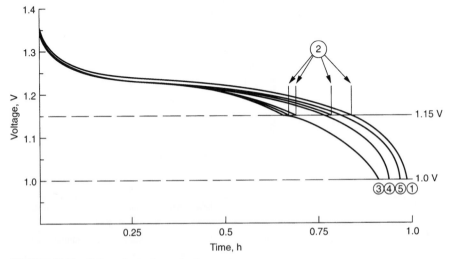

FIGURE 28.23 Voltage depression and subsequent recovery.

The voltage drop occurs because only a portion of the active materials is discharged and recharged during shallow or partial discharging. The active materials that have not been cycled, particularly the cadmium electrode, may undergo a change in physical characteristics and an increase in resistance. The effect has also been ascribed to structural changes in the nickel electrode.[1] Subsequent cycling restores the active materials to their original state.

The extent of voltage depression depends on the depth of discharge and can be avoided or minimized by the selection of an appropriate end voltage. Too high an end voltage, such as 1.16 V per cell, terminates the discharge prematurely. (A high end voltage should be used only if an extended cycle life is desired and the lower capacity can be tolerated, as in some satellite applications.) A small voltage depression may be observed if the discharge is terminated between 1.16 and 1.10 V per cell. The extent of the depression is dependent on the depth of discharge, which is also rate-dependent. Discharging to an end voltage below 1.1 V per cell should not result in a subsequent voltage depression. Discharging to too low an end voltage, however, should be avoided, as discussed in Sec. 28.4.6.

This phenomenon varies with the design and formulation of the electrode and may not be evident with all sealed nickel-cadmium batteries. Modern nickel-cadmium batteries use electrode structures and formation processes that reduce the susceptibility to voltage depression, and most users may never experience low performance due to memory effect. However, the use of the term "memory effect" persists, since it is often used to explain low battery capacity that is attributable to other problems, such as ineffective charging, overcharge, battery aging, or exposure to high temperatures.

Overcharging. A similar reversible failure can occur with long-term overcharging, particularly at elevated temperatures. Figure 28.24 shows the voltage "step" near the end of discharge that can be induced by long-term overcharging. The capacity is still available, but at a lower voltage than when it was freshly cycled. Again, this is a reversible failure; a few charge and discharge cycles will restore normal voltage and expected capacity.

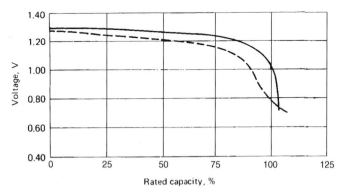

FIGURE 28.24 Typical discharge voltage profile of sealed nickel-cadmium batteries after long-term overcharge (dotted line) vs. 16-h charge, both at C /10 rate.

Irreversible Failures.   Permanent failure in nickel-cadmium batteries results from essentially two causes: short-circuiting and loss of electrolyte. An internal short circuit may be of relatively high resistance and will be evidenced by an abnormally low on-charge voltage and by a drop of voltage as the battery's energy is dissipated through the internal short circuit. A short circuit may also be of such a low resistance that virtually all the charge current travels through it or the battery electrodes are totally shorted internally.

Any loss of electrolyte will cause degradation in capacity. Charging at high rates, repeated voltage reversal, and direct short-circuiting are ways that can cause loss of electrolyte through the pressure relief device. Electrolyte can also be lost over a long period of time through the battery seals, and capacity is lost in proportion to the reduction in electrolyte volume. Capacity degradation caused by electrolyte losses is more pronounced at high discharge rates.

High temperature degrades battery performance and life. A nickel-cadmium battery gives optimum performance and life at temperatures between 18 and 30°C. Higher temperatures reduce life by promoting separator deterioration and increasing the probability of short-circuiting. Higher temperatures also cause more rapid evaporation of moisture through the seals. These effects are all long-term; but the higher the temperature, the more rapid the deterioration. Table 28.2 lists the recommended temperature limits for sealed nickel-cadmium batteries.

TABLE 28.2   Operating and Storage Limits for Sealed Nickel-Cadmium Batteries

Type	Temperature, °C	
	Storage	Operating
Button	−40 to 50	−20 to 50
Standard cylindrical	−40 to 50	−40 to 70
Premium cylindrical	−40 to 70	−40 to 70

28.5 CHARGING CHARACTERISTICS

28.5.1 General Characteristics

Sealed nickel-cadmium batteries are usually charged by means of the constant-current method. The $0.1C$ rate can be used and the battery is charged for 12 to 16 h (140%). At this rate the battery can withstand overcharging without harm, although most sealed nickel-cadmium batteries can be safely charged at the $C/100$ to $C/3$ rate. At higher charge rates care must be taken not to overcharge the battery excessively or develop high battery temperatures.

The voltage profile of a sealed nickel-cadmium battery during charge at the $C/10$ and $C/3$ rates is shown in Fig. 28.25. A sharp rise in voltage to a peak near the end of the charge is evident.

The voltage profile of a sealed nickel-cadmium batteries is different from the one for a vented one, as illustrated in Fig. 28.26 The end-of-charge voltage for the sealed battery is lower. The negative plate does not reach as high a state of charge as it does in the vented construction because of the oxygen recombination reaction.

Constant-potential charging is not recommended for sealed nickel-cadmium batteries as it can lead to thermal runaway. It can, however, be used if precautions are taken to limit the current toward the end of charge.

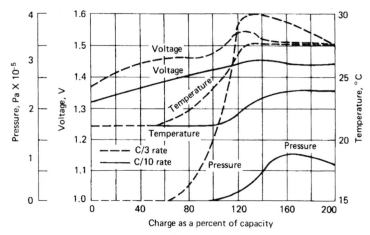

FIGURE 28.25 Typical pressure, temperature, and voltage relationships of sealed nickel-cadmium battery during charge.

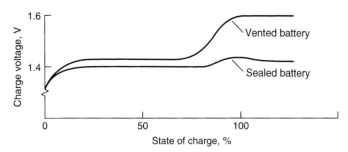

FIGURE 28.26 Charge vs. voltage for sealed and vented nickel-cadmium batteries at 25°C; 0.1*C* charge rate.

28.5.2 The Charge Process

The charge process is summarized in Fig. 28.27. Figure 28.27*a* plots the charge efficiency,

$$\text{Charge efficiency} = \frac{\text{discharge energy (on subsequent discharge)}}{\text{charge input energy}}$$

against total input energy. At the start of the discharge (area 1), the charge energy is consumed by the conversion of the active materials into a chargeable form, and the charge efficiency is low. In area 2 charging is most efficient and almost all of the input energy is used to convert the discharged active material into the charged state. As the battery approaches the charge state (area 3), most of the energy goes to the generation of oxygen, and the charge efficiency is low.

Figure 28.27*b* presents the relationship of charge efficiency to charge rate. It shows that the charge efficiency, as well as the output capacity, is lower at a lower charge rate.

The charge efficiency also depends on the ambient temperature during charge. This relationship is shown in Fig. 28.27*c*. There is a decrease in capacity in the high-temperature range due to a fall in potential for oxygen gas generation at the positive electrode.

The principle of the sealed battery is based on the ability of the negative electrode to recombine this oxygen gas and prevent the buildup of internal gas pressure. The capacity for this recombination is limited. Hence the maximum charge rate that can be tolerated is the one that keeps the rate of oxygen generation below the gas recombination rate so that the internal gas pressure does not build up excessively.

Overcharging, at rates beyond the ability of oxygen recombination or heat dissipation, can result in failure. "Fast" charging methods can be used successfully, but a means must be provided for monitoring and terminating the charge before excessive overcharging occurs. Temperature rise, voltage, or pressure can be monitored and used effectively as a cutoff.

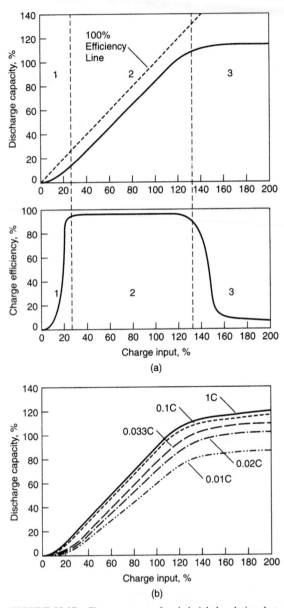

FIGURE 28.27   Charge process of sealed nickel-cadmium batteries. (*a*) Charge efficiency at 20°C. Charge—0.1C × 16 h; discharge—0.2°C; end voltage—1 V. (*b*) Charge efficiency vs. charge rate at 20°C.

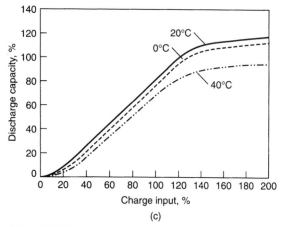

FIGURE 28.27   (c) Charge efficiency vs. ambient temperature during charge at 0.1C rate (*Continued*).

28.5.3 Voltage, Temperature, and Pressure Relationships

Figure 28.25 also shows the relationship of voltage, temperature, and pressure of a typical sealed battery during charging at the $C/10$ and $C/3$ rates. The voltage increases gradually during the charge until the battery is about 75 to 80% charged. The voltage then rises more sharply due to the generation of oxygen at the positive electrode. The temperature remains relatively constant during the early part of the charge as the charge reaction is endothermic. It then rises as the battery reaches the overcharge state due to the heat generated by the oxygen recombination reaction. Similarly the internal pressure remains low until the battery goes into the overcharge condition, when most of the current is used to produce oxygen and the pressure rises. Finally, as the battery reaches full charge, the voltage drops because of a decrease in the battery's internal resistance due to the increase in the battery temperature. This drop in voltage can be used effectively in a control circuit to terminate the charge.

As shown, when the battery is overcharged at acceptable rates, the pressure and temperature tend to stabilize. These steady-state conditions are governed by such factors as ambient temperature, overcharge rate, heat transfer characteristics of the cell and battery, cell design and components such as the separator, recombination capability of the negative electrode, and the resistance of the cell and battery. Charging at higher rates, such as the $C/3$ rate compared to the $C/10$ rate, results in higher temperatures and internal pressures. At higher charge rates, temperature and pressure will rise more significantly, particularly if the oxygen recombination rate is exceeded. Because of the possibility of venting and other deleterious effects on battery performance due to these high temperatures and pressures, it is necessary to terminate the charge before these conditions are reached.

28.5.4 Voltage Characteristics during Charge

The voltage profile of a sealed nickel-cadmium battery during charge at various charge rates at 20°C is shown in Fig. 28.28. The charge voltage also depends on temperature, as shown in Fig. 28.29. The voltage and voltage peak decrease with a rise in temperature. Charging at temperatures between 0 and 30°C is best for sealed batteries. At lower temperatures the voltage increases, recombination of oxygen is slower, and the internal gas pressure tends to increase. Charging rates must be reduced. Above 40°C the charging efficiency is low, and higher temperatures cause battery deterioration.

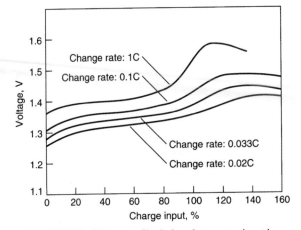

FIGURE 28.28   Voltage profile during charge at various charge rates at 20°C.

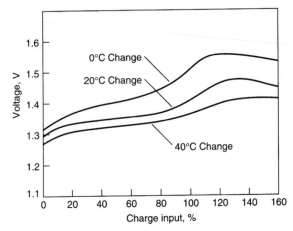

FIGURE 28.29   Voltage profile during charge at 0.1*C* rate at various temperatures.

28.5.5   Charge Methods

There are a number of different methods for charging sealed nickel-cadmium batteries. The standard method is a quasi-constant current charge at a relatively low rate. "Fast" charge methods are becoming more popular in order to reduce the time required for charging. A control circuit is needed when charging at these high rates to cut off the charge or reduce the charge current at the completion of charge. Figure 28.30 shows the voltage and current profiles during charge for each of the charge control methods. (Also see Sec. 5.5.)

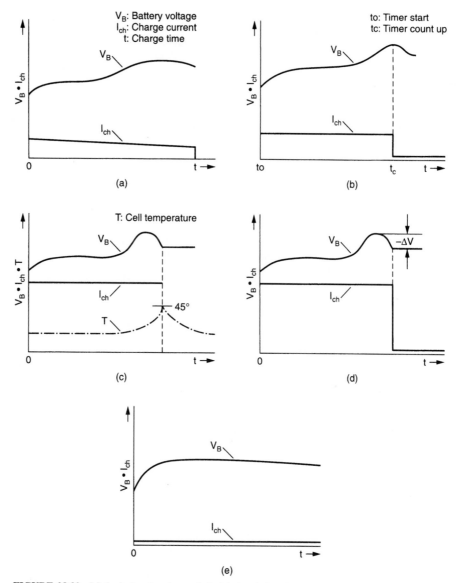

FIGURE 28.30 Methods for charging sealed nickel-cadmium batteries and charge control. (*a*) Semi-iconstant current (V_B—battery voltage; I_{ch}—charge current; t—time). (*b*) Timer control (t_0—timer start; t_c—timer count up). (*c*) Temperature detection (T—battery temperature). (*d*) $-\Delta V$ detection. (*e*) Trickle charge. (*Courtesy of Panasonic Industrial Co., Division of Matsushita Electric Corp. of America.*)

Standard Method. (Fig. 28.30*a*). This is the simplest method, which uses a relatively in-expensive circuit, controlling the charge current by inserting a resistance between the DC power supply and the battery. The battery is charged at a constant current at a low ($C/10$) rate so that the generation of oxygen is below the recombination rate. This rate also limits the temperature rise. Excessive overcharge should be avoided. The battery should be charged to about 140 to 150% charge input.

Timer Control. (Fig. 28.30*b*). For moderate charge rates a timer can be used to cut off the charge or reduce the charge current to the trickle charge level. This is a relatively inexpensive control device and suitable for applications where the battery is usually fully discharged before charging. It is not suitable for applications where the battery is frequently charged without prior deep discharging as this could result in excessive overcharge. A thermal cutoff control should be used when charging at rates higher than the *C*/5 rate or without deep discharging to prevent the battery from reaching high temperatures.

Temperature Detection. (Fig. 28.30*c*). This control system uses a sensor to detect the temperature rise of the battery and terminate the charge. A thermostat or thermistor is used as the detection device, and the detecting temperature is usually set at 45°C. It is important that the sensor be located so that it can accurately determine the battery's temperature. Charging in high ambient temperatures can result in an insufficient charge while low ambient temperatures may result in overcharge. The cycle life with this method may be shorter than with the $-\Delta V$ method or peak voltage methods as the battery could be subjected to more overcharge.

Negative Delta V ($-\Delta V$). (Fig. 28.30*d*). This is one of the preferred charge control systems for sealed nickel-cadmium batteries. The drop in voltage of the battery is detected after the battery voltage has reached its peak during charge. The signal can be used to terminate the charge or reduce the charge current to a trickle charge. The method provides a complete charge regardless of ambient temperature or residual capacity from the previous charge. A value of 10 to 20 mV per cell is usually used for the control. The method is not suitable for charging below the $0.5C$ as the $-\Delta V$ value is too low to be detectable.

Trickle and Float Charge. (Fig. 28.30*e*). Trickle charge systems are used in two different situations (1) in a standby power application where the battery is on continual charge to maintain it in a state of full charge (compensating for self-discharge) until it is connected to the load when the prime power fails, and (2) as a supplementary charge after the termination of rapid charging. Charging is at the 0.02 to $0.05C$ rate, depending on the frequency and depth of discharge. A periodic discharge every six months followed by a charge is advisable to ensure optimum performance.

28.6 SPECIAL-PURPOSE BATTERIES

Special-purpose sealed nickel-cadmium batteries are manufactured with specifically designed characteristics, overcoming some of the limitations of standard batteries, to meet the requirements for certain applications. Manufacturers' recommendations should be followed because of the specific performance characteristics of these batteries.

28.6.1 High-Capacity Batteries

These batteries incorporate design features, such as nickel foam substrate positive plates, pasted negative electrodes, thin-walled battery containers, and increased amounts of active material. These changes result in a 20 to 40% increase in capacity. These batteries are also designed with improved oxygen recombination capability and are capable of being charged at the $0.2C$ rate or less without control. They are capable of fast 1-h charging using $-\Delta V$ charge control. Figure 28.31 compares the discharge characteristics of the high-capacity battery with those of a standard battery. Figure 28.32 shows the relationship of battery capacity and discharge current for the two designs.

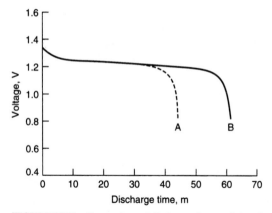

FIGURE 28.31   Comparison of discharge characteristics of sub-C size standard battery (*A*) vs. high-capacity battery (*B*) on discharge at 20°C. Discharge at *C* rate, charge at 0.1*C* rate for 16 hours.

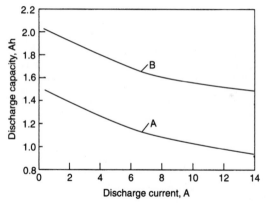

FIGURE 28.32   Comparison of performance of standard battery (*A*) vs. high-capacity battery (*B*) (sub-C size), at 20°C.

28.6.2   Fast-Charge Batteries

These batteries have electrode structures and electrolyte distribution designs to enhance oxygen recombination. They can be charged at the fast 1-h rate with charge control (such as temperature-sensing and $-\Delta V$ techniques) and at the $C/3$ rate without charge control because of their ability to withstand this level of overcharge. They are also capable of performance at high discharge rates, though this is achieved at the expense of a slightly reduced battery capacity. The batteries used in these batteries have improved internal heat conductivity, which results in a faster increase in surface temperature. This feature can be used advantageously in a temperature-sensing fast-charge system. Figure 28.33 shows the charge characteristics of a fast-charge battery compared to a standard one. The internal gas pressure of the standard battery increases quickly during charging whereas that of a fast-charge battery stabilizes.

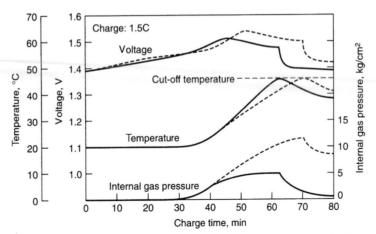

FIGURE 28.33   Comparison of charge characteristics of fast-charge (solid line) battery vs. standard battery (broken line).

28.6.3   High-Temperature Batteries

These batteries are designed to operate at high temperatures without the service life deterioration and charging inefficiencies experienced with conventional designs. Figure 28.34 compares the performance of the high-temperature battery with the standard battery as a function of ambient temperature during charge. This type of battery is capable of charge-discharge cycling at temperatures as high as 35 to 45°C and is particularly designed for trickle charging ($C/20$ to $C/50$ rate) at these high temperatures. The charge voltage of these batteries is slightly higher than that of the standard battery due to the designed-in control of the oxygen-generating potential.

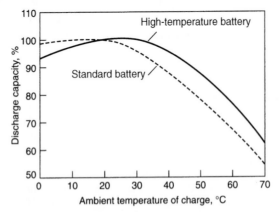

FIGURE 28.34   Comparison of performance of high-temperature battery vs. standard battery. charge—$C/30$ rate; discharge—$1C$ rate at 20°C.

28.6.4 Heat-Resistant Batteries

These batteries are designed for fast charging at high temperatures. For example, charging at the $0.3C$ rate is possible even at temperatures as high as 45 to 70°C. Their performance characteristics are similar to those of the standard battery. However, they have a superior service life when used at high temperatures because of the use of specially selected materials with minimum deterioration at high temperatures. Figure 28.35 compares the service life for standard and heat-resistant batteries throughout the temperature range.

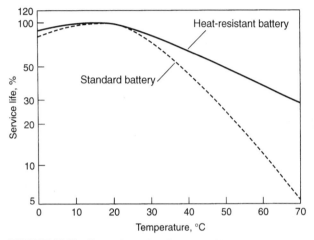

FIGURE 28.35   Comparison of performance of heat-resistant battery vs. standard battery.

28.6.5 Memory-Backup Batteries

These batteries are used to provide battery backup for volatile semiconductor memory devices. The key requirements for this type of battery are long life (up to 10 years in certain applications), low self-discharge, and good performance at low discharge rates. Figure 28.36 shows the storage characteristics of the memory-backup battery. (This can be compared to the characteristics of the standard battery shown in Fig. 28.19.) The low-rate discharge characteristics of the battery are plotted in Fig. 28.37. As the backup battery is designed for low-rate use, its internal resistance is higher than that of the standard battery and its high-rate discharge characteristics are not as good.

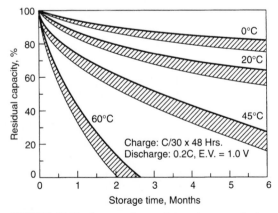

FIGURE 28.36   Storage characteristics of memory-backup batteries.

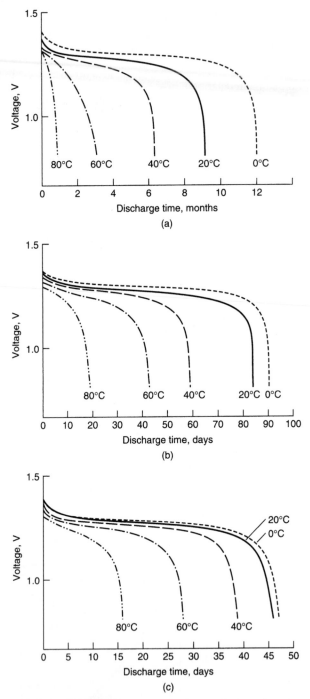

FIGURE 28.37 Performance of memory-backup batteries. Charge—$C/30$ for 48 h at 20°C. Discharge rate: (*a*) $C/10,000$; (*b*) $C/2000$; (*c*) $C/1000$.

28.6.6 Slim Rectangular Batteries

The constructional features of the slim rectangular battery are described in Sec. 28.3.3. The advantage of the rectangular battery is that it permits more efficient battery design, eliminating the voids that occur with the assembly of cylindrical batteries. The volumetric energy density of these batteries can be about 20% higher than a battery using a cylindrical design.

Most of the performance characteristics are similar to those of the standard cylindrical battery, except that it also incorporates some of the features of the high-capacity battery. Gas recombination has been improved to permit charging at the $0.2C$ rate or less and 1-h charging with charge control, preferably with $-\Delta V$ sensing. This is illustrated in Fig. 28.38. The voltage profile on discharge is flat, as with the cylindrical battery, as shown in Fig. 28.39. However, because the resistance of the rectangular battery is higher, performance at rates greater than $4C$ are not as good as with the cylindrical battery. Storage characteristics and cycle life are similar to those of the cylindrical battery.

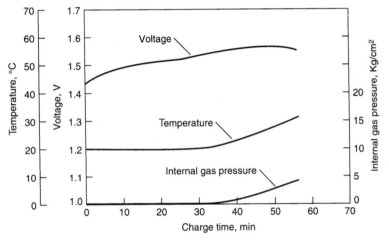

FIGURE 28.38 Charge characteristics of slim rectangular batteries at 20°C. Charge—$1.5C$ rate; $-\Delta V = 10$ mV.

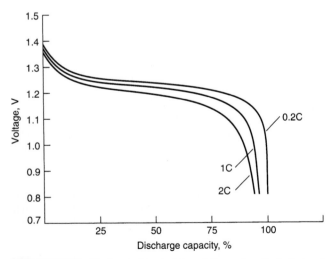

FIGURE 28.39 Discharge characteristics of slim rectangular batteries at 20°C. Charge—$0.1C$ rate for 16 h.

28.7 BATTERY TYPES AND SIZES

Table 28.3 lists some of the types of sealed nickel-cadmium single-cell batteries that are manufactured and some of their physical and electrical specifications. Multicell batteries are also manufactured, using these cells, in a variety of output voltages and configurations.

Figure 28.40 is a guide to determining the approximate battery size required for a given performance requirement or application. These data are based on the performance of a standard battery at 20 to 25°C. Allowance must be factored into the estimate to determine the performance under other discharge conditions.

Manufacturers' data should be consulted for specific details on dimensions, ratings, and performance characteristics as they may be different from those shown.

TABLE 28.3 Specifications of Typical Sealed Nickel-Cadmiuim Single-cell Batteries

Battery size	Capacity at 0.2C rate, mAh	Dimensions, max., mm		
		Diameter	Height	Weight, g
Cylindrical batteries				
Standard batteries: Charging—standard, 0.1C rate, 14–16 h; quick, 0.3C rate, 4–5 h				
N	170	12.0	29.3	9
AAAA	120	8.0	42.5	6
⅓ AAA	55	10.5	15.8	4
AAA	270	10.5	44.4	11
½ AA	300	14.5	30.3	14
AA	650	14.3	50.2	23
A	550	17.0	28.5	19
SC	1450	22.9	43.0	45
SC	1550	22.9	43.0	47
D	4400	33.0	59.5	160
D	4800	33.0	61.1	145
F	7700	33.2	91.0	230
M	12000	43.1	91.0	400
High-capacity batteries: Charging—standard, 0.1C rate, 14–16 h; quick, 0.3C rate, 4–5 h				
AA	880	14.3	50.3	23
AA	1150	14.3	50.3	24
A	650	17.0	28.5	18
A	1200	17.0	43.0	28
A	1550	17.0	43.0	31
SC	1900	22.9	43.0	47
SC	2400	22.9	50.0	58
D	5400	33.2	59.5	150
M	25500	43.1	146.1	700

TABLE 28.3 Specifications of Typical Sealed Nickel-Cadmiuim Single-cell Batteries (*Continued*)

Battery size	Capacity at 0.2C rate, mAh	Dimensions, max., mm		
		Diameter	Height	Weight, g

Cylindrical batteries

Fast-charge batteries: Charging—standard, $0.1C$ rate, 14–16 h; quick, $0.3C$ rate, 4–5 h; fast, $1.5C$ rate, 1 h

Battery size	Capacity	Diameter	Height	Weight, g
A	550	17.0	28.5	19
⅘ SC	1250	22.9	34.0	43
SC	1400	22.9	43.0	52
SC	1850	22.9	43.0	54
SC	2000	22.9	42.9	56
C	3200	26.0	50.0	84
D	4300	33.0	59.5	160

High-temperature batteries: Charging—standard, $0.1C$ rate, 14–16 h

Battery size	Capacity	Diameter	Height	Weight, g
AA	650	14.3	48.9	23
SC	1650	22.9	43.0	49
C	3100	26.0	50.5	78
D	4500	33.2	59.5	145
F	7700	33.2	91.0	230
M	12000	43.1	91.0	400

Heat-resistant batteries: Charging—standard, $0.1C$ rate, 14–16 h; quick, $0.3C$ rate, 4–5 h

Battery size	Capacity	Diameter	Height	Weight, g
⅔ AA	300	14.5	30.3	14
AA	650	14.3	50.2	23
⅘ SC	1350	22.9	43.0	52
SC	1800	22.9	42.9	56
C	2200	26.0	50.0	80

Slim rectangular batteries

Capacity at 0.2 C rate, mAh	Dimensions, max., mm		
	Height	Width	Thickness	Weight, g

Slim rectangular batteries: Charging—standard, $0.1C$ rate, 14–16 h; quick, $0.3C$ rate, 4–5 h; fast, $1.5C$ rate, 1 h

Capacity at 0.2 C rate, mAh	Height	Width	Thickness	Weight, g
450	48.0	17.2	6.3	17
650	48.0	17.2	8.5	22
650	67.0	17.2	6.3	24
900	67.0	17.2	8.5	30
1200	67.0	17.2	10.7	38

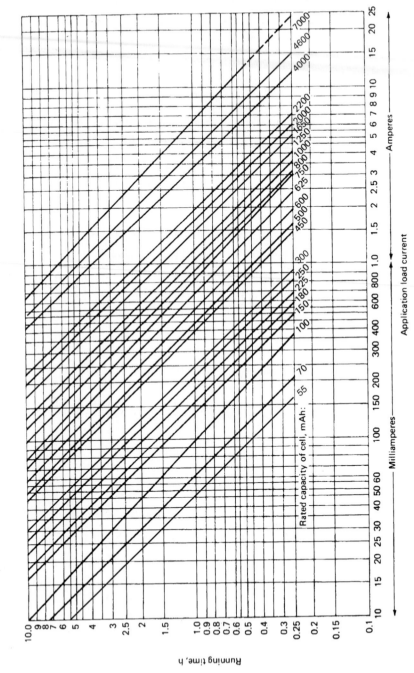

FIGURE 28.40 Selector guide for sealed nickel-cadmium cylindrical batteries. Guide can be used to determine approximate required battery size, given the load and desired run (service) time. Data based on fully charged battery and 20°C operating temperature.

REFERENCE

1. Y. Sato, K. Ito, T. Arakawa and K. Kobaya Kawa "Possible Causes of the Memory Effect Observed in Nickel-Cadmium Secondary Batteries. *J. Electrochemical Society,* **143:**L225 (October 1996).

BIBLIOGRAPHY

Cadnica Sealed Type Nickel-Cadmium Batteries, Sanyo Electric Co., Osaka, Japan.

Ford, Floyd E.: *Handbook for Handling and Storage of Nickel-Cadmium Batteries: Lessons Learned,* NASA Ref. Publ. 1326, Feb. 1994.

Nickel-Cadmium Batteries, Charge System Guide, Panasonic Industrial Co., Secaucus, N.J.

Nickel-Cadmium Batteries, Technical Handbook, Panasonic Industrial Co., Secaucus, N.J.

Sealed NiCad Handbook, SAFT Corp. of America, Valdosta, Ga.

Sealed Nickel-Cadmium Accumulators, Sales Program and Technical Handbook, Varta Batterie AG, Hanover, Germany.

CHAPTER 29
PORTABLE SEALED NICKEL-METAL HYDRIDE BATTERIES

David Linden and Doug Magnusen

29.1 GENERAL CHARACTERISTICS

The rechargeable sealed nickel-metal hydride battery is a relatively new technology with characteristics similar to those of the sealed nickel-cadmium battery. The principal difference is that the nickel-metal hydride battery uses hydrogen, absorbed in a metal alloy, for the active negative material in place of the cadmium used in the nickel-cadmium battery.

The metal hydride electrode has a higher energy density than the cadmium electrode. Therefore the amount of the negative electrode used in the nickel-metal hydride cell can be less than that used in the nickel-cadmium cell. This allows for a larger volume for the positive electrode, which results in a higher capacity or longer service life for the metal hydride battery. Furthermore, as the nickel-metal hydride battery is free of cadmium, it is considered more environmentally friendly than the nickel-cadmium battery and may reduce the problems associated with the disposal of rechargeable nickel batteries.

Most of the operating characteristics of the sealed nickel-metal hydride battery on discharge are similar to those of the nickel-cadmium battery. The sealed nickel-metal hydride battery, however, does not have the very high rate capability of the nickel-cadmium battery. In addition, the behavior of the two systems on charge, particularly on fast charge, is different. The nickel-metal hydride battery is less tolerant of overcharge and requires control of the cutoff of the charge, which may not always be required for nickel-cadmium batteries.

During the past five years, the specific energy and energy density of the nickel-metal hydride battery has been increased by over 35% as a result of improvements in both the positive and negative electrodes. Concurrently, improvements were made in its high-rate performance and cycle life. Because of its higher energy density and other comparable performance characteristics, the nickel-metal hydride battery is replacing the nickel-cadmium battery in computers, cellular phones and other consumer electronic applications with the possible exceptions of high-drain power tools and applications where low battery cost is the major consideration. However, the nickel-metal hydride battery now is being replaced, in turn, by the lithium-ion battery which has an even higher specific energy and energy density. The metal hydride battery in larger sizes is also being considered for use in applications such as electric vehicles, where its higher specific energy and good cycle life approach critical performance requirements.

The advantages and limitations of the sealed nickel-metal hydride battery are summarized in Table 29.1. The main advantage of the nickel-metal hydride battery compared to the nickel-cadmium battery is its higher specific energy and energy density.

TABLE 29.1   Major Advantages and Disadvantages of Sealed Nickel-Metal Hydride Batteries

Advantages	Disadvantages
Higher capacity than nickel-cadmium batteries	High-rate performance not as good as with nickel-cadmium batteries
Sealed construction, no maintenance required	Poor charge retention
Cadmium-free, minimal environmental problems	Moderate memory effect
Rapid recharge capability	Higher cost negative electrodes
Long cycle life	
Long shelf life in any state of charge	

29.2 CHEMISTRY

The active metal of the positive electrode of the nickel-metal hydride battery, in the charged state, is nickel oxyhydroxide. This is the same as the positive electrode in the nickel-cadmium battery.

The negative active material, in the charged state, is hydrogen in the form of a metal hydride. This metal alloy is capable of undergoing a reversible hydrogen absorbing-desorbing reaction as the battery is charged and discharged.

An aqueous solution of potassium hydroxide is the major component of the electrolyte. A minimum amount of electrolyte is used in this sealed cell design, with most of the liquid absorbed by the separator and the electrodes. This "starved-electrolyte" design, similar to the one in sealed nickel-cadmium batteries, facilitates the diffusion of oxygen to the negative electrode at the end of the charge for the oxygen-recombination reaction. This is essentially a dry-cell construction, and the cell is capable of operating in any position.

During discharge, the nickel oxyhydroxide is reduced to nickel hydroxide.

$$NiOOH + H_2O + e \rightarrow Ni(OH)_2 + OH^- \qquad E'' = 0.52 \text{ V}$$

and the metal hydride MH is oxidized to the metal alloy M.

$$MH - OH^- \rightarrow M + H_2O + e \qquad E'' = 0.83 \text{ V}$$

The overall reaction on discharge is

$$MH + NiOOH \rightarrow M - Ni(OH)_2 \qquad E'' = 1.35 \text{ V}$$

The process is reversed during charge.

The sealed nickel-metal hydride cell uses an oxygen-recombination mechanism to prevent the buildup of pressure that may result from the generation of gases toward the end of the charge and overcharge. It is based on the use of a negative electrode (the metal hydride electrode) that has a higher effective capacity than the positive or nickel oxyhydroxide electrode. This is shown schematically in Fig. 29.1. During charge the positive electrode reaches full charge before the negative and begins to evolve oxygen.

$$2OH^- \rightarrow H_2O - \tfrac{1}{2}O_2 + 2e$$

The oxygen gas diffuses through the separator to the negative electrode, the diffusion facilitated by the starved-electrolyte design and the selection of an appropriate separator system.

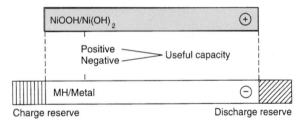

FIGURE 29.1 Schematic representation of electrodes of sealed nickel-metal hydride cell, divided into useful capacity, charge reserve, and discharge reserve. (*Courtesy of Duracell, Inc.*)

At the negative electrode the oxygen reacts with and oxidizes or discharges the hydrogen electrode to produce water, and the pressure does not build up,

$$4MH + O_2 \rightarrow 4M + 2H_2O$$

Furthermore, the negative electrode will not become fully charged, which prevents the generation of hydrogen.

The charge current, however, must be controlled at the end of the charge and during overcharge to limit the generation of oxygen to below the rate of recombination to prevent the buildup of gases and pressure.

The nickel-metal hydride cell also is designed with a discharge reserve in the negative electrode to minimize gassing and degradation of the cell in the event of overdischarge (see Sec. 29.4.6). Overall, as shown in Fig. 29.1, the negative electrode has excess capacity compared to the positive to handle both overcharge and overdischarge. The useful capacity of the battery is thus determined by the positive electrode.

A key component of the sealed nickel-metal hydride cell is the hydrogen storage metal alloy. The composition of the alloy is formulated to obtain a material that is stable over a large number of charge-discharge cycles. Other important properties of the alloy include:

1. Good hydrogen storage to achieve a high-energy density and battery capacity
2. Thermodynamic properties suitable for reversible absorption/desorption
3. Low hydrogen equilibrium pressure
4. High electrochemical reactivity
5. Favorable kinetic properties for high-rate performance
6. High oxidation resistance
7. Stability, with repeated charge/discharge cycles, in alkaline electrolyte

Two types of metallic alloys are generally used. These are the rare-earth (Misch metal) alloys based on lanthanum nickel ($LaNi_5$), known as the AB_5 class of alloys and alloys consisting of titanium and zirconium, known as the AB_2 class of alloys. In both cases, some of the base metals are replaced by other metals to improve performance characteristics.

In the case of the AB_5 class of alloys, substitutions have improved the alloy as follows:[1,2]

1. Ce, Nd, Pr, Gd and Y as a mixed or Misch metal (a naturally occurring mixture of rare-earth metals) are low cost substitutes for La.
2. Ni and Co are major constituents and suppress corrosion resulting in longer cycle life.
3. Al, Ti, Zr and Si are minor constituents and increase corrosion resistance resulting in longer cycle life.

In the case of the AB_2 class of alloys, substitutions have improved the alloy as follows:[3]

1. V, Ti, and Zr improve hydrogen storage.
2. Ni and Cr are major constituents which suppress corrosion and provide longer cycle life.
3. Al, Ti, Zr and Si are minor constituents which also suppress corrosion and provide longer cycle life.

The AB_2 alloy has a higher capacity per unit weight and volume over a moderate operating range of temperature and discharge rate than the AB_5 alloy as shown in Table 29.2. However, over a very broad and demanding operating range of temperature and discharge rate, the lower capacity AB_5 alloys are favored. AB_5 alloys have advantages and better performance[4] under a number of conditions, including high discharge rates, high rate charge acceptance, low and high discharge temperatures and superior high temperature stability.[4] To compensate for some of these advantages of the batteries made with AB_5 alloy, batteries made with the AB_2 alloy have been constructed with larger conductors, more surface area and electrolyte, leaving less room for the active material. As a result, the overall energy density of batteries made with AB_2 alloy may actually be the same or even lower than those made with AB_5 alloy. Therefore, the use of the AB_5 alloy dominates in portable sealed nickel-metal hydride batteries because of their overall advantages. (See discussion of metal hydride alloys in Chap. 30.)

TABLE 29.2 Comparison of Metal Hydride Alloys—Electrochemical Equivalents

Alloy	Gravimetric Ah/kg	Volumetric Ah/L
AB_5	270–290	2200–2400
AB_2	360–400	2500–2800

29.3 CONSTRUCTION

Sealed nickel-metal hydride cells and batteries are constructed in cylindrical, button, and prismatic configurations, similar to those used for the sealed nickel-cadmium battery.

The electrodes are designed with highly porous structures having a large surface area to provide a low internal resistance and a capability for high-rate performance. The positive electrode in the cylindrical nickel-metal hydride cell is a highly porous sintered, or felt nickel substrate into which the nickel compounds are impregnated or pasted and converted into the active material by electrodeposition. Felts and foams have generally replaced sintered plaque electrodes. Expanded metals and perforated sheets are cheaper, but they have poor high rate capability. Sintered structures are much more expensive. The negative electrode, similarly, is a highly porous structure using a perforated nickel foil or grid onto which the plastic-bonded active hydrogen storage alloy is coated. The electrodes are separated with a synthetic nonwoven material, which serves as an insulator between the two electrodes and as a medium for absorbing the electrolyte.

29.3.1 Cylindrical Configuration

The assembly of the cylindrical unit is shown in Fig. 29.2a. The electrodes are spirally wound and the assembly is inserted into a cylindrical nickel-plated steel can. The electrolyte is added and contained within the pores of the electrodes and separator.

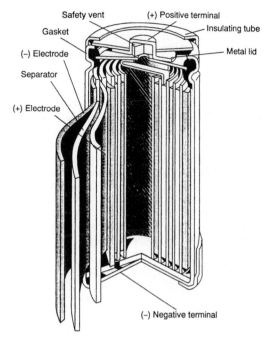

FIGURE 29.2a Construction of a sealed cylindrical nickel-metal hydride battery. (*Courtesy of Duracell, Inc.*)

The cell is sealed by crimping the top assembly to the can. The top assembly consists of a lid, which includes a resealable safety vent, a terminal cap, and a plastic gasket. The can serves as the negative terminal and the lid as the positive terminal, both insulated from each other by the gasket. The vent provides additional safety by releasing any excessive pressure that may build up if the battery is subjected to abuse.

29.3.2 Button Configuration

The button configuration is illustrated in Fig. 29.2b. It is similar in construction to the nickel-cadmium button cell, except that the cadmium is replaced by the hydrogen storage alloy.

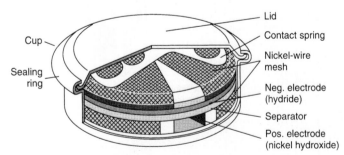

FIGURE 29.2b Construction of a sealed-nickel metal hydride button cell. (*Courtesy of Varta Batteries AG.*)

29.3.3 Prismatic Configuration

The thin prismatic batteries are designed to meet the needs of compact equipment. The rectangular shape permits more efficient battery assembly, eliminating the voids that occur with the assembly of cylindrical cells. The volumetric energy density of the battery can be increased by a factor of about 20%. The prismatic cells also offer more flexibility in the design of batteries, as the battery footprint is not controlled by the diameter of the cylindrical cell.

Figure 29.2c shows the structure of the prismatic battery. The electrodes are manufactured in a similar manner as the electrodes for the cylindrical cell, except that the finished electrodes are flat and rectangular in shape. The flat electrodes are then assembled, with the positive and negative electrodes interspaced by separator sheets, and welded to the cover plate. The assembly is then placed in the nickel-plated steel can and the electrolyte is added. The cell is sealed by crimping the top assembly to the can. The top assembly is a lid which incorporates a resealable safety vent, a terminal cap, and a plastic gasket, similar to the one used on the cylindrical cell. An insulating heat-shrink tube is placed over the metal can (jacket). The bottom of the metal can serves as the negative terminal and the top lid as the positive terminal. The gasket insulates the terminals from each other.

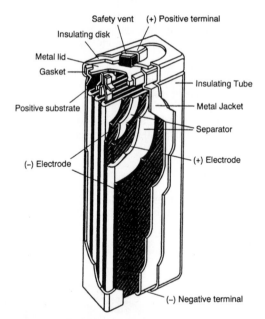

FIGURE 29.2c Construction of a sealed prismatic nickel-metal hydride battery.

29.3.4 9-Volt Multicell Battery

The construction of the 9-volt multicell battery is shown in Fig. 29.2*d*.

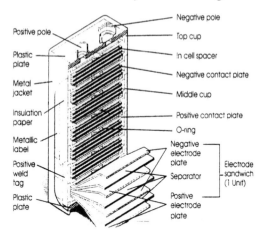

FIGURE 29.2*d*   Construction of a sealed 9-volt nickel-metal hydride battery.

29.3.5 Larger Prismatic Batteries

Larger prismatic cells with flat plates are being developed for hybrid and all electric vehicles and other applications requiring larger batteries in sizes up to about 250 Ah. These cells use the lightweight substrates, such as foams and fibers, for the electrodes in conjunction with mechanical loading of the active materials, lightweight containers, and so on, to achieve the higher specific energy required for these applications. These larger nickel-metal hydride cells and batteries are covered in Chap. 30.

29.4 DISCHARGE CHARACTERISTICS

29.4.1 General Characteristics

The discharge characteristics of the sealed nickel-metal hydride batteries are very similar to those of the sealed nickel-cadmium battery. Several comparisons are illustrated in Chap. 22. The open-circuit voltage of the batteries of both systems ranges from 1.25 to 1.35 V, the nominal voltage is 1.2 V, and the typical end voltage is 1.0 V.

(Note: All batteries were recharged between 20°C and 25°C under the conditions shown, unless specified otherwise.)

29.4.2 Discharge Characteristics

Cylindrical Batteries.   Typical discharge curves for the cylindrical sealed nickel-metal hydride battery under various constant-current loads and temperatures are shown in Fig. 29.3. (The data are based on the rated performance at 20°C at the 0.2*C* discharge rate to 1.0 V.) A flat discharge profile is characteristic. The discharge voltage, as expected, is dependent on the discharge current and discharge temperature. Typically, the higher the current and the lower the temperature, the lower the operating voltage. This is due to the higher *IR* drop with increasing current and the increasing resistance at the lower temperatures. However, because of the relatively low resistance of the nickel-metal hydride battery (as well as the nickel-cadmium battery), this drop in voltage is less than experienced with other types of portable primary and rechargeable batteries.

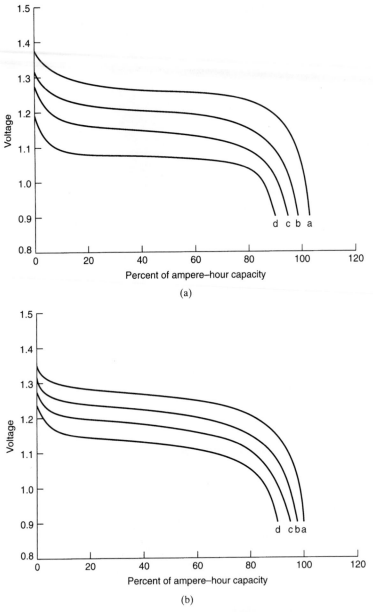

FIGURE 29.3 Discharge performance of sealed cylindrical nickel-metal hydride batteries at (*a*) 20°C; (*b*) 45°C. Curves *a*—0.2*C* rate; curves *b*—1*C* rate; curves *c*—2*C* rate; curves *d*—3*C* rate.

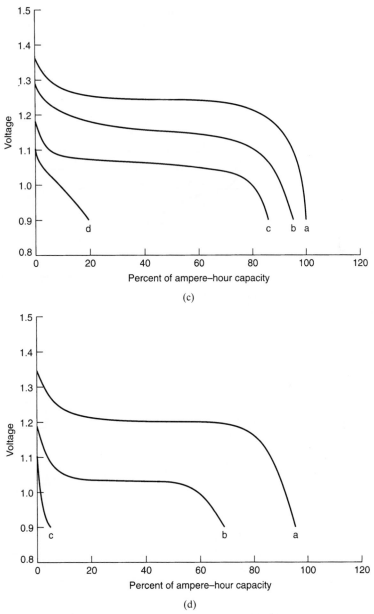

FIGURE 29.3 (*c*) 0°C; (*d*) −20°C. Curves *a*—0.2*C* rate; curves *b*—1*C* rate; curves *c*—2*C* rate; curves *d*—3*C* rate (*Continued*).

Button Batteries.   Typical discharge curves for button-type sealed nickel-metal hydride batteries at room and other temperatures are shown in Figs. 29.4*a* and 29.4*b*.

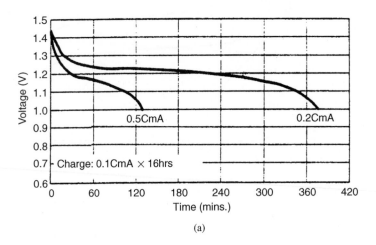

(a)

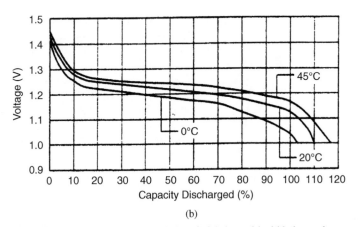

(b)

FIGURE 29.4   Discharge characteristics of nickel-metal hydride button batteries (*a*) Discharge at 20°C. (*b*) Discharge at 0.2 C rate. (*Courtesy of GP Batteries, Inc.*)

Prismatic Batteries.   Typical discharge curves for the prismatic sealed nickel-metal hydride batteries at room and other temperatures are shown in Figs. 29.5a and 29.5b.

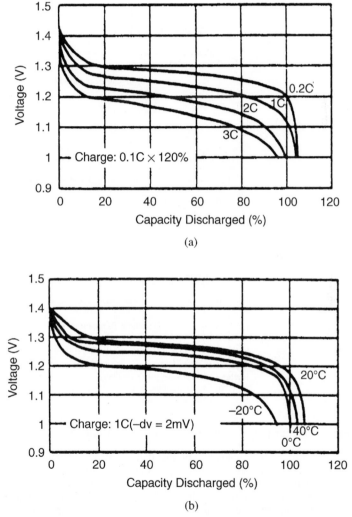

FIGURE 29.5   Discharge characteristics of nickel-metal hydride prismatic batteries (a) Discharge at 20°C. (b) Discharge at 0.2 C rate. (*Courtesy of GP Batteries, Inc.*)

9-Volt Battery.   Typical discharge curves for the 9-volt sealed nickel-metal hydride battery at room and other temperatures are shown in Figs. 29.6*a* and 29.6*b*.

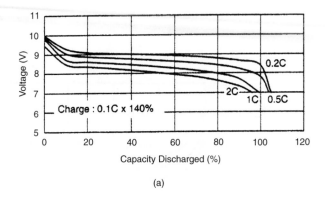

(a)

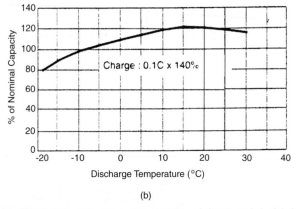

(b)

FIGURE 29.6   Discharge characteristics of 9-volt sealed nickel-metal hydride battery (*a*) Discharge at 20°C. (*b*) Discharge at 0.2 C rate to 7 volts. (*Courtesy of GP Batteries, Inc.*)

29.4.3   Effect of Discharge Rate and Temperature on Capacity

The ampere-hour capacity of the battery also is dependent on the discharge current and temperature, as can be observed in Fig. 29.3 through Fig. 29.6. It should also be noted that the delivered capacity is dependent on the cutoff or end voltage. The capacity can be increased by continuing the discharge to lower end voltages, particularly at the higher current drains and lower temperatures, where the voltage drops off more rapidly than at the lighter drains. However, the battery should not be discharged to too low a cutoff voltage as the cells may be damaged (see Sec. 29.4.6). The cutoff or end voltage for the nickel-metal hydride battery is typically 1.0 V per cell.

The relationship of the ampere-hour capacity (expressed as a percentage of the capacity at 20°C on a discharge at the 0.2*C* rate) of the sealed nickel-metal hydride battery and the discharge temperature and current (expressed in *C* rate) is summarized in Fig. 29.7.

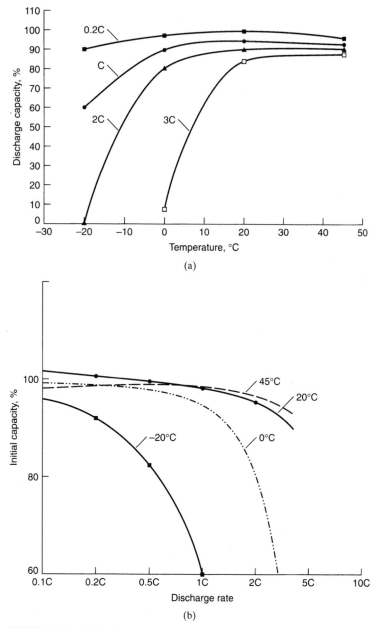

FIGURE 29.7   (*a*) Discharge capacity vs. ambient temperature for sealed cylindrical nickel-metal hydride batteries at various discharge rates; end voltage 1.0 V/cell. (*b*) Discharge capacity (% of 0.2*C* rate) vs. discharge rate (*C*-rate) for sealed cylindrical nickel-metal hydride batteries at various temperatures: end voltage 1.0 V/cell.

Typically, for the nickel-metal hydride battery the best performance is obtained between 0 and 40°C. The performance characteristics of the battery on discharge are affected moderately at higher temperatures, but decrease more significantly at the lower discharge temperatures, mainly due to the increase in internal resistance. Similarly, the effects of a change in temperature are more pronounced at the higher discharge rates. The capacity of the battery decreases more noticeably as the current increases beyond the 3 to $4C$ rate, particularly at the lower temperatures.

A drop in the capacity can be observed at the lower discharge rates and higher temperatures. This loss in capacity is due to the effect of self-discharge, which can become evident under these discharge conditions.

29.4.4 Service Life (Hours of Service)

Figure 29.7 can be used to approximate the capacity and the service life of standard cylindrical nickel-metal hydride batteries at various discharge rates and temperatures if the rated capacity (at 20°C and the $0.2C$ discharge rate) is known. The percentage of the rated capacity delivered under other conditions can be determined directly from this figure. The approximation is valid if the cells are of a similar construction and behave similarly to the standard cell on which the data are based. The specific data for a given cell type obtained from the manufacturer should be used to estimate the performance more precisely.

Another form for presenting these data is shown in Fig. 29.8. The service life, in hours, of a cylindrical nickel-metal hydride battery is plotted against the discharge current, normalized to unit weight (Ah/kg) and size (Ah/L). These data are based on a rated capacity

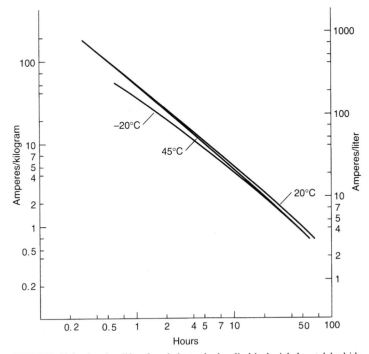

FIGURE 29.8 Service life of sealed standard cylindrical nickel-metal hydride batteries at various discharge rates and temperatures. Based on an energy density under rated conditions at 20°C of 60 Ah/kg and 200 Ah/L: end voltage 1.0 V/cell.

at the $0.2C$ rate of 60 Ah/kg and 200 Ah/L, reflecting the performance of a standard type battery. As discussed in Sec. 3.2.6, this figure provides a convenient nomograph to determine the approximate performance, in service hours, of a battery or to estimate the size of a battery that will deliver the desired performance under specified discharge conditions—again with the caveat that the battery has similar construction and characteristics to the standard battery on which the data are based and an energy density close to the one specified.

29.4.5 Internal Resistance

The nickel-metal hydride battery has a low internal resistance because of the use of thin plates with large surface areas and low resistance and an electrolyte having a high conductivity. Figure 29.9 shows the change in internal resistance with the depth of discharge. The resistance remains relatively constant during most of the discharge. Toward the end of the discharge, the resistance increases due to conversion of the active materials. The internal resistance also increases as the temperature drops because the resistance of the electrolyte and other components is higher at the lower temperatures. The resistance of the nickel-metal hydride battery increases with use and cycling. This is illustrated in Fig. 29.24*a* which shows the drop in midpoint voltage as the battery is cycled.

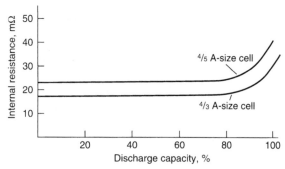

FIGURE 29.9 Internal resistance vs. discharge capacity of sealed cylindrical nickel-metal hydride cells.

29.4.6 Polarity Reversal during Overdischarge

When a multicell series-connected battery is discharged, the lowest-capacity battery will reach the point of full discharge before the others. If the discharge is continued, this lower-capacity cell can be driven into an overdischarged condition through 0 V and its polarity (voltage) reversed. This is illustrated in Fig. 29.10.

Phase 1 of the figure is the normal phase of the discharge with active material remaining on both the positive and the negative electrodes.

During phase 2 the active material on the positive electrode has been discharged and generation of hydrogen gas starts. Some of this gas may be absorbed by the hydrogen storage metal alloy in the negative electrode and the remainder builds up in the cell. Active material, however, still remains on the negative electrode and the discharge continues. The cell voltage is dependent on the discharge current, but remains within -0.2 to about -0.4 V.

In phase 3 the active materials on both electrodes have been depleted and oxygen is produced at the negative electrode. Prolonged overdischarge leads to gassing, higher internal cell pressure, opening of the safety vent, and deterioration of the cell.

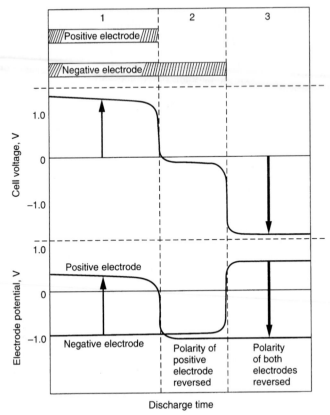

FIGURE 29.10 Polarity reversal during discharge of sealed cylindrical nickel-metal hydride cells.

The larger the number of cells in series in a multicell battery, the greater the possibility that this polarity reversal will occur. To minimize the effect, whenever three or more cells are connected in series, the cells selected should have capacities within the range of ±5%. The process of selecting cells of similar capacity is called "matching." Further, a cutoff voltage of 1.0 V per cell or higher should be used for discharge rates up to the 1*C* rate to prevent the possibility for any cell to go into reversal. Higher cutoff voltages should be used for batteries containing more than 10 cells in series and for discharge rates exceeding 1*C*.

29.4.7 Type of Discharge

As discussed in Secs. 3.2.3 and 3.2.4, a battery may be discharged under different modes (such as constant resistance, constant current, or constant power), depending on the characteristics of the equipment load. The type of discharge mode selected has a significant impact on the service life delivered by a battery in a specified application. The discharge profiles of a nickel-metal hydride battery under the three different modes are plotted in Fig. 29.11. Figure 29.11(*a*) shows the voltage profile, Fig. 29.11(*b*) the current profile, and Fig. 29.11(*c*) the power profile during discharge of the battery. As an example, the data are based on the discharge of a 1000-mAh battery such that, at the end of the discharge (1.0 V per

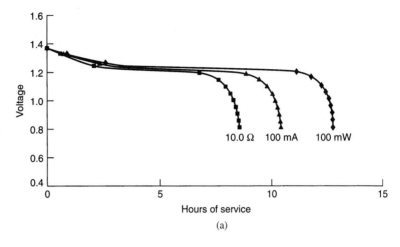

(a)

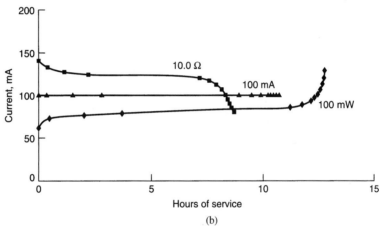

(b)

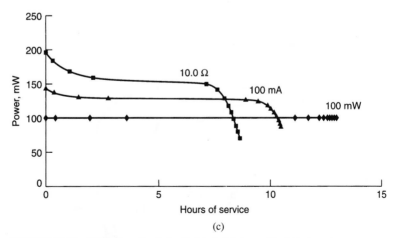

(c)

FIGURE 29.11 Discharge of sealed cylindrical nickel-metal hydride batteries—constant power vs. constant current vs. constant resistance: (*a*) Voltage profile. (*b*) Current profile. (*c*) Power profile. ■—constant resistance: ◆—constant power; ▲—constant current. Based on a battery rated at 1000 mAh.

cell), the power output is the same for all modes of discharge. In this example the power output at the 1.0-V cutoff is 100 mW. To discharge at 100 mW at 1.0 V, the constant-current discharge is 100 mA ($C/10$ rate) and the constant-resistance discharge is 10 Ω. As shown, the longest service life is obtained under the constant-power mode as the average current is the lowest under this mode of discharge.

29.4.8 Constant-Power Discharge Characteristics

The discharge characteristics of the nickel-metal hydride battery under the constant-power mode, at several different power levels, are shown in Fig. 29.12. These are similar to the data presented in Fig. 29.3a for constant-current discharges, except that the performance is

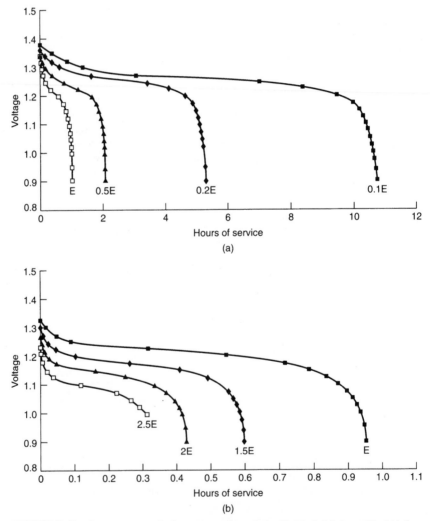

FIGURE 29.12   Constant-power discharge curves for sealed cylindrical nickel-metal hydride batteries at 20°C. (a) 0.1E to 1E discharge rate. (b) 1E to 2.5E discharge rate.

presented in hours of service instead of percent ampere-hour capacity. The power levels are shown on the basis of E-rate. The E-rate is calculated in a manner similar to calculating the C-rate but based on the rated watthour capacity rather than ampere-hour capacity. For example, for the $E/2$ power level, the power for a battery rated at 1200 mWh (rated at 1000 mAh at the $C/5$ rate at 1.2 V) is 600 mW.

29.4.9 Voltage Depression (Memory Effect)

A reversible drop in voltage and loss of capacity may occur when a sealed nickel-metal hydride battery is partially discharged and recharged repetitively without the benefit of a full discharge. This is illustrated in Fig. 29.13. After an initial full discharge (cycle 1) and charge, the battery is partially discharged (in this example to 1.15 V) and recharged for a number of cycles. During this cycling the discharge voltage and the capacity drop gradually (cycles 2 to 18). On a subsequent full discharge (cycle 19) the discharge voltage is depressed compared to the original full discharge (cycle 1). The discharge profile may show two steps, and the cell does not deliver the full capacity to the original cutoff voltage. This phenomenon is known as voltage depression. At times it is referred to as "memory effect," as the battery appears to "remember" the lower capacity. The battery can be restored to full capacity with a few full discharge-charge cycles, as illustrated in Fig. 29.13 (cycles 20 and 21).

The voltage drop occurs because only a portion of the active materials is discharged and recharged during shallow or partial discharging. The active materials that have not been cycled change in physical characteristics and increase in resistance. The active materials are restored to their original state by the subsequent full discharge-charge cycling.

The extent of voltage depression and capacity loss depends on the depth of discharge and can be avoided or minimized by discharging the battery to an appropriate end voltage. The effect is most apparent when the discharge is terminated at the higher end voltages, such as

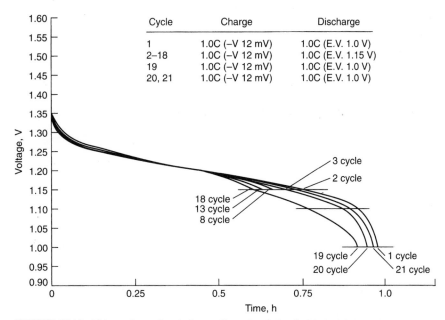

FIGURE 29.13 Voltage depression during cycling of sealed cylindrical nickel-metal hydride battery, 20°C. (*Courtesy of Duracell, Inc.*)

1.2 V per cell. A smaller loss occurs if the discharge is cut off between 1.15 and 1.10 V per cell. Discharging to an end voltage below 1.1 V per cell should not result in a significant voltage depression or capacity loss on the subsequent discharges. Discharging to too low an end voltage, however, should be avoided, as discussed in Sec. 29.4.6.

The effect is also dependent on the discharge rate. To a given end voltage, the depth of discharge will be less on discharges at the higher rates. This will increase the capacity loss as less of the active material is cycled.

While the memory effect may result in reduced battery performance, the actual voltage depression and capacity loss are only a small fraction of the battery's capacity. Most users may never experience low performance due to this behavior of the sealed nickel-metal hydride cell. Often memory effect is used incorrectly to explain a low battery capacity that should be attributed to other problems, such as inadequate charging, overcharge, or exposure to high temperatures.

29.4.10 Self-Discharge and Charge Retention

The state of charge and capacity of the nickel-metal hydride battery decreases during storage due to self-discharge. This is caused by the reaction of the residual hydrogen in the cell (the atmosphere inside the cell is hydrogen) with the positive electrode as well as the slow (but reversible during subsequent charging) decomposition of both electrodes.

The rate of self-discharge is dependent on the storage temperature and time: the higher the temperature, the greater the rate of self-discharge. This is illustrated in Fig. 29.14, which shows the charge retention for sealed cylindrical nickel-metal hydride batteries following storage at different temperatures for varying periods of time. The comparison is for a battery discharged at the rated discharge load (approximately the $0.2C$ rate) at 20°C. It is to be noted also that the charge or capacity retention also is dependent on cell size, cell design, discharge and charge conditions and other such factors. The more stringent the discharge condition such as a higher rate or lower temperature, for example, the lower the capacity retention. The charge retention characteristics for the other types of sealed nickel-metal hydride batteries are similar to those shown for the cylindrical batteries.

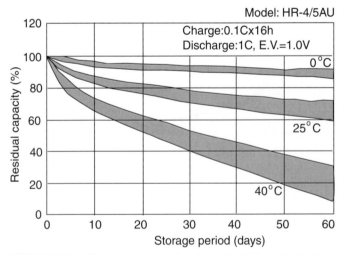

FIGURE 29.14   Charge retention characteristics of sealed cylindrical nickel-metal hydride batteries at various temperatures (*Courtesy of Sanyo Electrical Co. Ltd.*)

In general long-term storage of the nickel-metal hydride cell, in either a charged or a discharged condition, has no permanent effect on its capacity. The capacity losses due to self-discharge are reversible and batteries can recover to full capacity by recharging. This is shown in Fig. 29.15. Full capacity can be restored within two to three charge-discharge cycles.

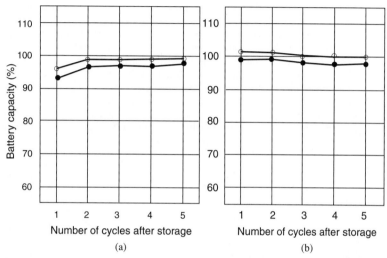

FIGURE 29.15 Capacity recovery after storage. (*a*) Storage for one month in charged state. (*b*) Storage for one month in discharged state. Charge $0.1C \times 16$ h; discharge $1C$; end voltage 1.0 V; 20°C. ○—storage at 25°C ●—storage at 40°C. (*Courtesy of Sanyo Electric Co, Ltd.*)

Long-term storage at high temperatures, similar to operation at elevated temperatures, may deteriorate seals and separators and could cause permanent damage, such as loss of capacity, cycle life, and overall battery lifetime. The recommended temperature range for long-term storage of nickel-metal hydride cells is 20 to 30°C.

29.5 CHARGING SEALED NICKEL-METAL HYDRIDE BATTERIES

29.5.1 General Principles

Recharging is the process of replacing energy that has been discharged from the battery. The subsequent performance of the battery, as well as its overall life, are dependent on effective charging. The main criteria for effective charging are to:

- Recharge the battery to its full capacity
- Limit the extent of overcharge
- Avoid high temperatures and excessive temperature fluctuations

The recharging characteristics of the nickel-metal hydride battery are generally similar to those of the sealed nickel-cadmium battery. There are some distinct differences, however, particularly on the requirements for charge control, as the metal hydride battery is more sensitive to overcharge. Caution should be exercised before using the same battery charger interchangeably for both types of batteries.

The most common charging method for the sealed nickel-metal hydride battery is a constant-current charge, but with the current limited to avoid an excessive rise of temperature or exceeding the rate of the oxygen-recombination reaction.

The voltage and temperature profiles of the nickel-metal hydride and the nickel-cadmium batteries during charge at a moderate constant-current charge rate are compared in Fig. 29.16. The voltage of both battery systems rises as the battery accepts the charge. During the first phase of the charge, the temperature of the nickel-cadmium battery remains relatively constant because its charge reaction is endothermic. The temperature of the nickel-metal hydride battery, on the other hand, rises gradually because its charge reaction is exothermic. As the batteries approach 75 to 80% recharge, the voltage rises more sharply due to the generation of oxygen at the positive electrode, and the temperature, in both cases, rises due to the exothermic oxygen-recombination reaction. This increase in cell temperature causes the voltage to drop as the battery reaches full charge and goes into overcharge.

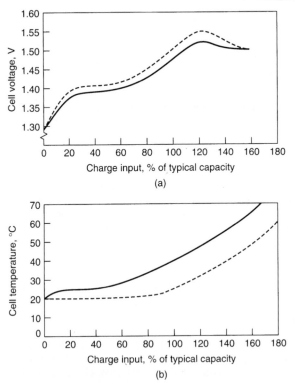

FIGURE 29.16 Comparison of typical charge characteristics of nickel-cadmium and nickel-metal hydride batteries: (*a*) Voltage characteristics. (*b*) Temperature characteristics. Solid line— Ni-MH: broken line—NiCd. (*Courtesy of Duracell, Inc.*)

The voltage profile of the nickel-metal hydride battery does not show as prominent a peak as that of the nickel-cadmium battery. Second, while the temperature of both batteries increases as the batteries reach full charge due to the oxygen-recombination reaction, during overcharge the temperature of the nickel-cadmium battery levels off at reasonable charge rates while the temperature of the nickel-metal hydride battery continues to rise. Both the voltage drop after peaking $(-\Delta V)$ and the temperature rise can be used as methods to terminate the charge. However, while similar charge techniques can be used for both types of batteries, the conditions to terminate the charge may differ because of the different behavior of the two battery systems during charge.

The voltage of the sealed nickel-metal hydride battery during charge depends on a number of conditions, including charge current and temperature. Figure 29.17 shows the voltage profile of the nickel-metal hydride battery at different charge rates and temperatures. The voltage rises with an increase in charge current due to a higher *IR* and overpotential during the electrode reaction. The voltage decreases with increasing temperature as the internal resistance and the overpotential during electrode reaction decreases. The voltage peak is not as evident at the low charge rates and at the higher temperatures.

The increase in battery temperature during the charge at various charge rates is shown in Fig. 29.18. The internal cell pressure increases similarly. This rise in temperature and pressure at the higher charge rates emphasizes the need for proper charge control and effective charge termination when "fast charging" to avoid venting and other deleterious effects.

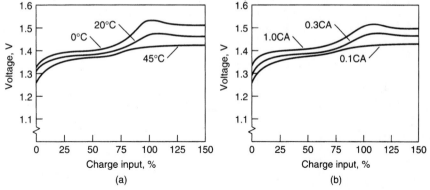

FIGURE 29.17   Cell voltage vs. charge input for sealed cylindrical nickel-metal hydride batteries. (*a*) At various temperatures (charge rate 0.3*C*). (*b*) At various charge rates at 20°C.

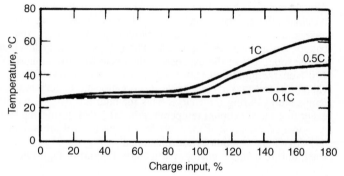

FIGURE 29.18   Battery temperature during charge of nickel-metal hydride cylindrical batteries.

The charge efficiency is also dependent on temperature. It decreases at the higher temperatures due to the increasing evolution of oxygen at the positive electrode. At the lower temperatures, charge efficiency is high due to decreasing oxygen evolution. However, at the lower temperatures, oxygen recombination is slowed down and a rise in internal cell pressure may occur depending on the charge rate. Figure 29.19 shows the available discharge capacity following charging at various temperatures and several charge rates. As shown, the battery capacity is reduced after high-temperature charging. The extent of this effect is also dependent on the conditions of the discharge following the charge, as well as on other charge conditions.

Proper recharging is critical not only to obtain maximum capacity on subsequent discharges, but also to avoid high temperatures, overcharge, and other conditions which could adversely affect battery life.

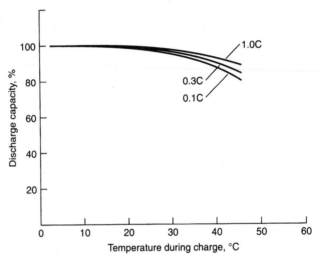

FIGURE 29.19   Charge efficiency vs. charge temperature at various charge rates for sealed cylindrical nickel-metal hydride batteries. Discharge at $0.2C$ rate to 1.0 V.

29.5.2   Techniques for Charge Control

The characteristics of the nickel-metal hydride battery define the need for charge control to terminate the charge to prevent the battery from being overcharged or exposed to high temperatures. The advantage of employing proper charge control to maximize the life of the battery is illustrated in Fig. 29.20. The highest capacity levels are achieved with the 150% charge input, but at the expense of the cycle life. the longest cycle life is attained with the 120% charge input, but with lower capacity due to insufficient charge input. Thermal cutoff charge control may reduce cycle life because the battery is usually allowed to reach higher temperatures during the charge. This method, however, is useful as a backup control in the event that the maximum temperature is exceeded during the charge.

Some of the popular methods for charge control are summarized hereafter. The characteristics of these methods are illustrated in Fig. 29.21. In many cases several methods are used, during a single charge, particularly to control high-rate charging.

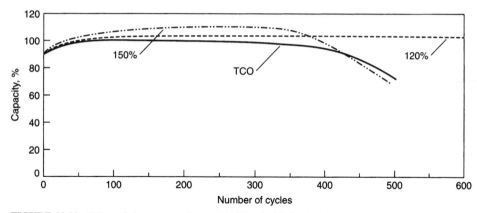

FIGURE 29.20 Effect of charge control on cycle life of sealed cylindrical nickel-metal hydride batteries. 1C charge rate: discharge at 1C to 1.0 V. TCO—charge termination at 40°C; 120%—charge termination at 120% charge input; 150%—charge termination at 150% charge input. (*Courtesy of Duracell, Inc.*)

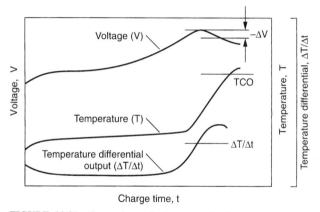

FIGURE 29.21 Comparison of charge termination methods: TCO, $\Delta T/\Delta t$, and $-\Delta V$.

Timed Charge. Under this charge control method the charge is terminated after the battery has been charged for a predetermined amount of time. This method should only be used for charging at low rates to avoid excessive overcharge because the state of charge of the battery, prior to charging, cannot always be determined. This procedure is also used as a "topping" charge to other charge termination methods to ensure a complete recharge.

Voltage Drop $(-\Delta V)$***.*** With this technique, widely used with sealed nickel-cadmium batteries, the voltage during charge is monitored and the charge is terminated when the voltage begins to decrease. This approach can be used with the nickel-metal hydride battery, but as noted in Sec. 29.5.1, the peak with the metal hydride cell is not as prominent and may be absent in charge currents below the 0.3C rate, particularly at elevated temperatures. The voltage signal must be sensitive enough to terminate the charge when the voltage drops, but not so sensitive that it will terminate the charge prematurely due to noise or other normal voltage fluctuations. A 10-mV per cell drop is generally used for the nickel-metal hydride battery.

Voltage Plateau (0ΔV). As the sealed nickel-metal hydride battery does not always show an adequate voltage drop, an alternate method is to terminate the charge when the voltage peaks and the slope is zero rather than waiting for the voltage to drop. The risk of overcharge is reduced as compared to the $-\Delta V$ method. A topping charge can follow to ensure a full recharge.

Temperature Cutoff (T). Another technique for charge control is to monitor the temperature rise of the battery and terminate the charge when the battery has reached a temperature which indicates the beginning of overcharge. It is difficult to determine this point precisely as it is influenced by ambient temperature, cell and battery design, charge rate, and other factors. For example, a cold battery may be overcharged before reaching the cutoff temperature while a warm battery may be undercharged. Usually this method is used in conjunction with other charge control techniques and mainly to terminate the charge in the event the battery reaches excessive temperatures before the other charge controls activate.

Delta Temperature Cutoff (ΔT). This technique measures the battery temperature rise above the ambient temperature during charging and terminates the charge when this rise exceeds a predetermined value. In this way, the influence of ambient temperature is minimized. The cutoff value is dependent on several factors, including cell size, configuration and number of cells in the battery, and the heat capacity of the battery. Therefore, the cutoff value must be determined for each type of battery.

Rate of Temperature Increase (ΔT/Δt). In this method, the change in temperature with time is monitored and the charge is terminated when a predetermined incremental temperature rise is reached. The influence of ambient temperature is virtually eliminated. A $\Delta T / \Delta t$ cutoff is a preferred charge control method for nickel-metal hydride batteries because it provides long cycle life.

Note: Details on the design of batteries using protective devices and a description of the thermal protective devices that can be used for charge control are discussed in Chap. 5.

29.5.3 Charging Methods

Sealed nickel-metal hydride batteries can be charged by several methods, ranging from low rate to fast charging, provided the change is controlled to prevent damaging effects.

Low-Rate Charge. A convenient method to fully charge sealed nickel-metal hydride batteries is to charge at a constant current at about the $0.1C$ rate with time-limited charge termination. At this current level the generation of gas will not exceed the oxygen-recombination rate. The charge should be terminated after 150% capacity input (approximately 15 h for a fully discharged battery). Excessive overcharge should be avoided as this can be injurious to the battery. The temperature range for this charge method is 5 to 45°C, with best performance being obtained between 15 and 30°C.

Quick Charge (4–5 h). Nickel-metal hydride batteries can be recharged efficiently and safely at higher rates. Charge control is required in order to terminate the charge when the rate of oxygen recombination is exceeded or the battery temperature rises excessively. The fully discharged battery can be charged at the $0.3C$ rate for a charge time equivalent to a 150% charge input (approximately 4.5 to 5 h). In addition to the timer control, a thermal cutoff device should be used as a backup control to terminate the charge at about 55 to 60°C to avoid exposing the battery to excessively high temperatures. This charging method may be used in an ambient temperature range of 10 to 45°C.

As a further precaution, the decrease in voltage $-\Delta V$ should also be sensed to ensure that the charge is terminated early enough to minimize overcharge. This is particularly advisable if the battery being charged was not fully discharged. A "topping" charge at the $0.1C$ rate, as described, may then be used to assure 100% recharge.

Charging sealed nickel-metal hydride batteries at rates between 0.1 and 0.3C generally is not recommended. At these rates the voltage and temperature charge profiles may not exhibit characteristics suitable for voltage-based cutoff control and the batteries may otherwise be exposed to harmful overcharge.

***Fast Charge* (1 h).** Another method of charging nickel-metal hydride batteries in an even shorter time is to charge at a constant current at the 0.5 to 1C rates. At these high charge rates it is essential that the charge be terminated early during overcharge. Timer control is inadequate as the time needed for charge cannot be predicted. A partially charged battery could easily be overcharged while a fully discharged one could be undercharged, depending on how the timer control is set.

With fast charging, the decrease in voltage $-\Delta V$ and the increase in temperature ΔT can be used to terminate the charge. For better results, termination of fast charge can be controlled by sensing the rate of temperature increase $\Delta T/\Delta t$ with a thermal cutoff (TCO) backup.

Figure 29.22 shows the advantage of using a $\Delta T/\Delta t$ method compared to $-\Delta V$ in terminating a fast charge. The $\Delta T/\Delta t$ method can sense the start of the overcharge earlier than the $-\Delta V$ method. The battery is exposed to less overcharge and overheating, resulting in less loss of cycle life. A temperature increase of 1°C/min should be used for $\Delta T/\Delta t$; a temperature of 60°C is recommended for the TCO.

In the case of multicell batteries of three cells or more, $-\Delta V$ termination with TCO backup may be adequate. The $-\Delta V$ value usually is 10–15 mV per cell and 60°C for the TCO.

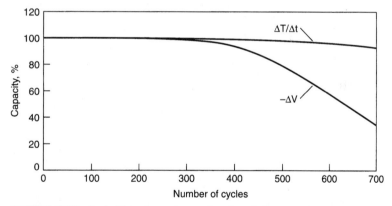

FIGURE 29.22 Cycle life and capacity as a function of charge termination method for sealed cylindrical nickel-metal hydride batteries. Charge 1C; rest 0.5 h; discharge, 1–0 V; rest 2 h. (*Courtesy of Duracell, Inc.*)

Trickle Charge. A number of applications require the use of batteries which are maintained in a fully charged condition. This is done by trickle charging at a rate that will replace the capacity loss due to self-discharge. A trickle charge at a current of between the 0.03 and 0.05C rates is recommended. The preferred temperature range for trickle charging is between 10 and 35°C. Trickle charge may be used following any of the previously discussed charging methods.

Three-Step Charge Procedure. A three-step procedure provides a means of rapidly charging a sealed nickel-metal hydride battery to full charge without excessive overcharge or exposure to high temperatures.

1. The first step is a charge at the $1C$ rate terminated by using the $\Delta T/\Delta t$ method or the $-\Delta V$ method.

2. This is followed by a $0.1C$ topping charge, terminated by a timer after $\frac{1}{2} - 1$ h of charge.

3. The third step is a maintenance charge of indefinite duration at a current of between the 0.05 and $002C$ rates. The battery should also be protected with a thermal cutoff device to terminate the charge so that the temperature does not exceed 60°C.

29.5.4 Smart Batteries

Battery technology has evolved to a point now where batteries are often termed "smart," indicating that a certain level of electronics has been installed within the battery. These controls can provide information on the battery's condition and "state of charge," can control charging and provide data on discharge which can improve service life. Nickel-metal hydride batteries benefit from this because of the difficulty in reaching a full charge under a wide range of operating conditions. Inefficiencies during charge at the positive electrode make it difficult to estimate when to stop charging. The tendency is to charge longer than needed, and the result is higher end-of-charge temperatures. The "smart" electronics "act" by measuring the amount of current being drawn, the temperature, the charge rate, etc. Using a calibration of the chemistry and design peculiar to the manufacturer, the amount of charge accepted by the battery can be estimated to about 1% and the charge can be terminated at the proper point.[5]

The digital electronics can also estimate discharge time using similar calibration tables based on the discharge curves and other relevant data. This information is then displayed as lighted LEDs or communicated over a serial bus to electronics in the laptop computer, for example, for display or control of the device. A typical set of calibration curves is shown in Fig. 29.23. It shows the values used to calculate the charge lost (self-discharge) as a

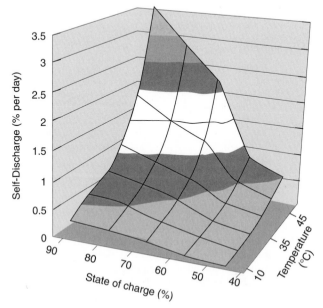

FIGURE 29.23 Characterization of rate of self-discharge for sealed nickel-metal hydride batteries. Self discharge rate vs state of charge and temperature (from reference 6).

function of standtime, temperature and state-of-charge. This information and other relevant data on the battery's characteristics are used in an algorithm to display available capacity, remaining hours of service or the amount of recharge required to achieve a full charge[5] (see Chap. 5 for more details on smart batteries).

29.6 CYCLE AND BATTERY LIFE

29.6.1 Cycle Life

The cycle life of nickel-metal hydride batteries, as all rechargeable batteries, depends on the many conditions to which the battery has been exposed, such as:

- Temperature during charge and discharge
- Depth of discharge
- Charge and discharge current
- Method of charge control
- Exposure to overcharge and overdischarge
- Storage conditions and length of storage

Typically under a standard charge-discharge cycle at the $0.2C$ rate and at normal ambient temperature ($20°C$), about 500 cycles can be achieved with the battery delivering at least 80% of its rated capacity. This is illustrated in Figs. 29.20 and 29.22. The gradual reduction in capacity results from an increase in the cell's internal resistance due to minor irreversible changes in the electrode structures and loss of electrolyte or cell dry-out. The increase in internal resistance is illustrated in Fig. 29.24, which shows the gradual decrease in midpoint voltage during discharge (Fig. 29.24a) and the gradual increase in voltage during charge with cycling (Fig. 29.24b).

For optimum battery and maximum cycle life, the nickel-metal hydride battery should be operated near room temperature. Operation at extreme temperatures during charge or discharge will adversely affect its performance, as shown graphically in Fig. 29.25. Operation at high temperatures, particularly in the overcharge condition, can cause the cell to vent, releasing gas and possibly electrolyte through the safety vent. High temperatures will also hasten the deterioration of the separator and other materials in the cell. At low temperatures the oxygen-recombination reaction slows down, the cell is more sensitive to overcharge, and gas pressure will build up more rapidly.

The charge rate and the amount of charge input during overcharge also are important factors affecting the cycle life. If the battery is charged at a rate that exceeds the oxygen-recombination rate, oxygen that is generated during overcharge will not be reacted and gas pressure and temperature will build up with deleterious effects on battery and cycle life. The more effective the method to terminate the charge promptly when deleterious overcharge begins, the less is the effect on cycle life. This was also illustrated in Figs. 29.20 and 29.22.

Finally cycle life is also affected by the depth of discharge. About 500 cycles can be obtained, depending on the charge termination method, with the battery being fully discharged on each cycle (100% depth of discharge). Considerably higher cycle life can be obtained if the battery is cycled on shallower charges and discharges. Figure 29.26 shows the increase in cycle life on a shallow charge-discharge. In this example the battery is discharged at the $0.25C$ rate to about a 60% depth of discharge. The cycle life is increased to about 1000 cycles. Shallower discharges will further increase the cycle life.

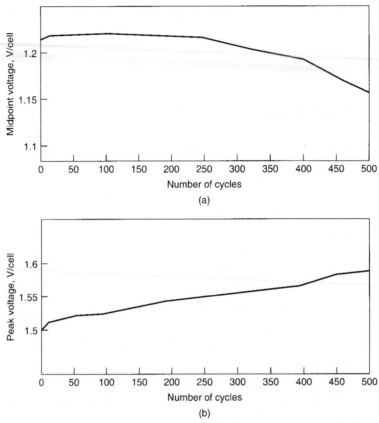

FIGURE 29.24 Effect of cycling on voltage characteristics of sealed cylindrical nickel-metal hydride batteries. (*a*) Decrease in midpoint voltage with cycling. (*b*) Increase in maximum voltage during charge with cycling. (*Courtesy of Duracell, Inc.*)

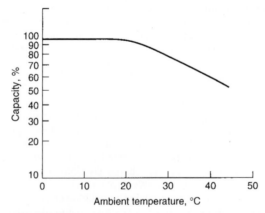

FIGURE 29.25 Effect of ambient temperature on cycle life of sealed cylindrical nickel-metal hydride batteries. (*Courtesy of Duracell, Inc.*)

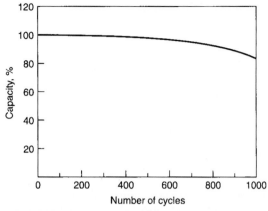

FIGURE 29.26 Effect of depth of discharge on cycle life at 20°C for sealed cylindrical nickel-metal hydride batteries. Cycle conditions: charge 0.25C, 3.2 h; discharge 0.25C, 2.4 h. Capacity measurement: charge 0.3C, 5 h, every 50 cycles, discharge 0.3C, 5 h. (*Courtesy of Duracell, Inc.*)

29.6.2 Battery Life

The same factors that affect cycle life affect overall battery life. Operation or storage at extreme temperatures, overcharging, cell venting, and abuse will reduce battery life. For optimum life, operation and storage should be as close to normal temperatures (20°C) as possible. Recommended and permissible temperature limits are shown in Table 29.3.

TABLE 29.3 Recommended Temperature Limits for Use of Sealed Nickel-Metal Hydride Batteries

	Recommended	Permissible
Standard change	15 to 30°C	0 to 45°C
Quick charge	10 to 30°C	10 to 45°C
Fast charge	10 to 30°C	10 to 45°C
Trickle charge	10 to 30°C	10 to 35°C
Discharge	0 to 40°C	−20 to 60°C
Storage, charged	10 to 30°C	10 to 35°C
Storage, discharged	10 to 30°C	10 to 35°C

29.7 PROPER USE AND HANDLING

Sealed nickel-metal hydride cells and batteries can give long-term maintenance-free, safe, and reliable service if they are used in accordance with recommended procedures and are not abused. They are sealed and can be used in any operating position. Other than charging, the only maintenance that should be required is to keep them clean and dry during operation and storage.

Most important, sealed nickel-metal hydride batteries, as most battery systems, should not be exposed to extreme temperatures for any extended period. They can be stored, in a charged or discharged condition, at moderate temperatures without detrimental effects.

Typically nickel-metal hydride batteries are shipped in a discharged state. However, as some residual capacity may remain, caution should be exercised to avoid short-circuiting the cell or battery.

After storage, or periods during which the battery has not been used, the battery should be charged before being used. Overcharging or overheating the battery should always be avoided.

It is possible that cells may vent if the cell or battery is overcharged or otherwise abused. The nickel-metal hydride cell releases hydrogen gas during venting, which could form potentially explosive mixtures with air. (Note that this differs from the nickel-cadmium cell, which emits oxygen when it vents.) Caution should be exercised to prevent these gases from collecting in the battery or equipment or from exposure to a source of ignition. Airtight compartments should be avoided.

29.8 APPLICATIONS

The characteristics of the nickel-metal hydride battery present opportunities for use over a wide range of applications. In the portable sealed configuration, it is being used in increasing quantities in consumer electronic devices such as cellular phones, transceivers, computers, camcorders, and other portable applications. In larger sizes, it is also one of the candidate battery systems for use in electric vehicles, another important market for rechargeable batteries in the next decade. (See Chaps. 30 and 37.)

A variation of the nickel-metal hydride cell is the silver-metal hydride cell where the positive nickel oxyhydroxide electrode is replaced by a silver oxide electrode. The silver-metal hydride cell has the advantage of a higher-energy density and better high-rate performance, and may be useful in critical applications where its higher cost and shorter cycle life is acceptable.

29.9 BATTERY TYPES AND MANUFACTURERS

Table 29.4(a–d) lists some of the portable sealed nickel-metal hydride cells and batteries that are manufactured and their physical and electrical specifications. Multicell batteries are also manufactured using these cells or batteries in a variety of output voltages and configurations. Manufacturers' data should be consulted for specific details on dimensions, ratings, and performance characteristics as they may differ from those shown.[6]

TABLE 29.4a Specifications of Sealed Cylindrical Nickel-Metal Hydride Single-Cell Batteries, Nominal Voltage: 1.2 V

Type	Capacity (0.2C discharge) (mAh)		Nominal dimension (mm)		Weight (g)
	Minimum	Typical	Diameter ϕ	Height (H)	
Standard series					
AAAA	300	310	8.4	40.6	6.5
7/5 AAAA	500	520		66	10.5
1/4 AAA	70	80	10	10.3	2.4
1/3 AAA	120	130		13.7	3.3
1/2 AAA	210	230		22	5.3
2/3 AAA	280	290		29	7.3
4/5 AAA	400	450	10.3	35.9	8.5
AAA	550	600	10.5	43.5	12.5
	600	630			
9/5 AAAL	1100	1130		80	23
1/3 AA	250	270	14.5	16.6	7
2/3 AA	300	320		28.7	10
	600	660			13
4/5 AA	1200	1250		42.5	22
AA	600	660	14.4	48.2	15
	1300	1400			25
1/2 A	600	650	17	22	14
2/3 A	1000	1090		28.7	22
4/5 A	1810	1870		43	32
A	2300	2350		50.2	40
4/3A, 7/5A	3500	3580	17.2	66.7	52
1/2 N	180	200	11.5	16.7	5.3
N	360	380		29.3	10
18210	800	860	18.3	21.5	17
Sub-C	2200	2420	23	42.6	55
D	4500	4950	33	61.5	140
High capacity series					
2/3 AAA	350	380	10	29	7.1
AAA	650	670	10.5	44.7	13
AAAL	700	720		48	14.5
7/5 AAAL	1000	1050		66.5	19.l5
2/3 AA	750	770	14.5	28.7	15
AA	1500	1540	14.4	48.2	26
AA	1600	1650	14.5	49.2	26
7/5 AA	1900	1960		65	40

TABLE 29.4a Specifications of Sealed Cylindrical Nickel-Metal Hydride Single-Cell Batteries, Nominal Voltage: 1.2 V (*Continued*)

Type	Capacity (0.2C discharge) (mAh)		Nominal dimension (mm)		Weight (g)
	Minimum	Typical	Diameter ϕ	Height (H)	
High capacity series (*Continued*)					
4/5 A	2100	2170	17	43	32
A	2500	2550		50.2	39
7/5 A	3700	3700	17.2	66.7	52
4/5 A, 7/5 A	3700	3800	17.5	67.5	53.5
N	550	580	11.5	29.1	9.7
18650	4100	4200	18.3	65.5	58
Sub-C	3000	3150	23	42.6	61
C	3500	3850	25.8	50	78.5
D	7000	7700	33	61.5	170
High temperature series					
AA	1250	1270	14.4	48.2	25.5
4/5 AF	1600	1660	17	43	31
AF	2100	2200		50.2	37.5
18650	4000	4100	18.3	65.5	62
Sub-C	2200	2310	23	42.6	53

TABLE 29.4b Specifications of Sealed Prismatic Nickel-Metal Hydride Single-cell Batteries

Nominal voltage (V)	Capacity (0.2C discharge) (mAh)		Nominal dimension (mm)			Weight (g)
	Minimum	Typical	Thickness (T)	Width (W)	Height (H)	
1.2	400	440	6.1	17.0	31.3	10.3
	500	550	6.1	17.0	35.3	12.0
	700	750	6.1	17.0	48.0	17
	800	850	6.1	17.0	48.0	17
	1000	1100	8.3	17.0	48.0	22.0
	1100	1200	6.1	17.0	67.0	23.0

TABLE 29.4c Specifications of "9-volt" Nickel-Metal Hydride Batteries

Nominal voltage (V)	Capacity (0.2C discharge) (mAh)		Nominal dimension (mm)	Weight (g)
	Minimum	Typical		
7.2	150	160	17.5(T) × 26.5(W) × 48.5(H)	39
8.4				42

TABLE 29.4d Specifications of Button Type Nickel-Metal Hydride Batteries

Nominal capacity at 5-hr rate (mAh)		11	15	25	40	80	150	250	320
Nominal dimensions	Diameter (mm)	9.40		11.52		15.42		24.98	
	Height (mm)	2.30	2.40	3.00	5.45	16.25	7.50	6.33	8.68
	Weight (g)	0.58	0.61	1.25	1.70	3.50	4.60	11.50	14.50

Source for Tables 29.4 (a to d): GP Batteries (USA) Inc.

REFERENCES

1. T. Sakai, "Hydrogen Storage Alloys for Nickel-metal Hydride Battery." *Zeitschift feur Physikalische Chemie, Bd.* 183, S. 333–346 1994.

2. L. Y. Zhang, "Rare Earth-Based Metal Hydrides and NiMH Rechargeable Batteries," *Rare Earths: Science Technology & Applications* 3rd Symp. Orlando, Fla., Feb. 1997; Minerals, Metals & Materials Soc.

3. D. M. Kim, "A Study of the Development of a High Capacity and High Performance ZR-Ti-MN-V-Ni Hydrogen Storage Alloy for NiMH Batteries," 279 *J. Alloys and Compounds,* Oct. 1998, 209–214.

4. John K. Erbacher, "Nickel-Metal Hydride Technology Evaluation," Final Report Wright Laboratories WL-TR—96-2069, Fairborn, OH, 1996.

5. D. Friel, "How Smart Should a Battery Be?, *Battery Power Products & Technology,* March 1999 and commercial literature and data books from various battery manufacturers.

6. J. Norman Allen, "The Payoff in Smart Battery Electronics" *Battery Power Products & Technology,* July 2000.

CHAPTER 30
PROPULSION AND INDUSTRIAL NICKEL-METAL HYDRIDE BATTERIES

Michael Fetcenko

30.1 INTRODUCTION

Nickel-metal hydride (NiMH) batteries are now in high volume commercial production for small portable battery applications. The key driving forces for the rapid growth of NiMH are environmental and energy advantages over nickel-cadmium and the explosive growth of portable electronic devices such as communication equipment and laptop computers. Because of its technical advantages over other rechargeable systems, this system is currently being developed for large batteries, including industrial and electric vehicle applications.

The first introduction of electric vehicles began at the turn of the century, but was short-lived after losing the competition to the Internal Combustion Engine (ICE). In the mid-1970s, electric vehicles emerged again due to what turned out to be temporary gasoline shortages. In the early 1990s, electric vehicles reemerged due to environmental air quality concerns in urban areas. In particular, the California Air Resources Board (CARB) gave electric vehicles tremendous attention by mandating that 2% of all vehicles sold in California must be zero emission vehicles by 1998 and 10% of all vehicles by 2003. While the original CARB mandate has changed, low and zero emission vehicles are still a high profile development focus for vehicle manufacturers.

The California mandate led to the formation of the United States Advanced Battery Consortium (USABC), an unprecedented cooperation between the major U.S. automakers, the U.S. Department of Energy, and the electric utility companies. Besides providing a central funding source for a multitude of competing technology development organizations, USABC provided the battery performance targets shown in Table 30.1 along with a comparison to current NiMH status.

More recently, there has been a significant shift by the automotive manufacturers from electric vehicles toward hybrid electric vehicles (HEVs) due to several factors including relaxation of the CARB mandate from zero to low emissions, customer reaction to initial low range lead-acid electric vehicles, and the lack of geographically widespread EV charging and fast charging stations.

TABLE 30.1 Primary USABC Midterm Performance Goals for the EV Battery and Actual Performance of the Commercial and Advanced NiMH Battery

Property	USABC	NiMH Commercial	NiMH Prototype
Specific energy (Wh kg^{-1})	80 (100 desired)	63–75	85–90
Energy density (Wh per liter)	135	220	250
Power density (W per liter)	250	850	1000
Specific power (W kg^{-1}) (80% DOD in 30 s)	150 (>200 desired)	220	240
Cycle life (cycles) (80% DOD)	600	600–1200	600–1200
Life (years)	5	10	10
Environmental operating temperature	$-30°$ to 65°C	$-30°$ to 65°C	$-30°$ to 65°C
Recharge time	<6 hours	15 min (60%) <1 hour (100%)	15 min (60%) <1 hour (100%)
Self discharge	<15% in 48 hours	<10% in 48 hours	<10% in 48 hours
Ultimate projected price (dollars per kWh) (10,000 units at 40 kWh)	<$150	$220–400	$150

30.2 GENERAL CHARACTERISTICS

Nickel-metal hydride batteries have become the dominant advanced battery technology for EV and HEV applications by having the best overall performance to meet the requirements set by USABC. In addition to the essential performance targets of energy, power, cycle life and operating temperature, the following features of NiMH[1,2] have established this technology in this market:

- Flexible cell sizes from 0.3 to 250 Ah
- Safe operation at high voltage (320+ V)
- Excellent volumetric energy and power, flexible vehicle packaging
- Readily applicable to series and series/parallel strings
- Choice of cylindrical or prismatic cells
- Safety in charge and discharge, including tolerance to abusive overcharge and overdischarge
- Maintenance free
- Excellent thermal properties
- Capability to utilize regenerative braking energy
- Simple and inexpensive charging and electronic control circuits
- Environmentally acceptable and recyclable materials

Examples of EV NiMH battery packs for the General Motors EV1 and S-10 pickup truck are shown in Fig. 30.1. In addition, nickel metal-hybride batteries are being used in industrial applications, replacing nickel-cadmium batteries and are being considered as a replacement for lead-acid batteries in SLI applications.

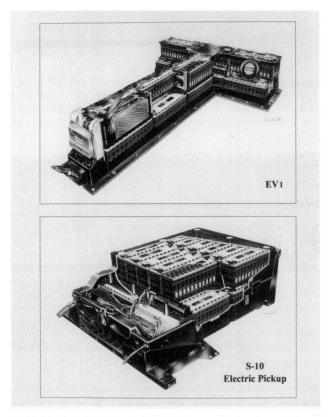

FIGURE 30.1 GM Ovonic NiMH battery packs for General Motors
EV1 (30 kWh) and Chevrolet S-10 pickup truck (30 kWh).

30.3 *CHEMISTRY*

The chemistry and chemical reactions for large propulsion NiMH batteries are the same as
in small portable applications, as discussed in Chap. 29.

The active material in the positive electrode is nickel hydroxide. Usually, the nickel
hydroxide has a conductive network composed of cobalt oxides and a current collector which
commonly is a nickel foam skeleton, but may alternately be a nickel fiber matrix or may be
produced by sintering filamentary nickel fibers.[3]

The active material in the negative electrode is an alloy capable of storing hydrogen, and
usually of the AB_5 (LaCePrNdNiCoMnAl) or AB_2 (VTiZrNiCrCoMnAlSn) type, where the
"AB_x" designation refers to the ratio of the A type elements (LaCePrNd or TiZr) to that of
the B type elements (VNiCrCoMnAlSn). In either case, the materials have complex microstructures that allow the hydrogen storage alloys to operate in the aggressive environment
within the battery where most of the metals are thermodynamically more stable as oxides.
The metal hydride materials are conductive, and are applied to a current collector usually
made of perforated or expanded nickel, or nickel foam substrate. Copper current collectors
have recently emerged for extremely high rate discharge applications.[4]

The electrolyte is also similar to small portable NiMH batteries, predominantly 6 Molar KOH in H_2O with a LiOH additive. Practically all NiMH batteries are sealed and use the "starved electrolyte" design, allowing for fast gas transport through the separator that is vital for gas recombination. The separator itself is usually a highly porous permanently wettable polypropylene, or it may be a polypropylene/polyethylene composite made by either wet laying of the fibers or by the "melt blown" process. In either case, the separator is usually made wettable by a grafting process that bonds a material such as acrylic acid to the base fibers by an energetic process such as radiation grafting. There are many variations within the separators used by manufacturers and the processing details are proprietary.

30.4   CONSTRUCTION

30.4.1   Cylindrical Versus Prismatic Configuration

NiMH batteries are versatile in that both cylindrical and prismatic constructions can be utilized. Each type of construction has advantages and disadvantages, and a particular end use can determine which configuration is used. For NiMH applications below about 10 Ah, cylindrical construction dominates due to low cost and high speed of manufacture. Above 20 Ah, cylindrical construction is extremely difficult and the prismatic configuration dominates. In the 10 to 20 Ah cell size range, manufacturers are offering both cylindrical and prismatic designs although prismatic designs are more common.

Cylindrical cells for industrial and propulsion applications are similar to high volume production consumer batteries in that the well-known "jelly roll" construction is used. However, most small portable cylindrical cells require only low to moderate discharge rate capability and electrode terminal connections are usually quite simple. Conversely, since industrial and propulsion NiMH applications require high to ultra-high discharge rate capability and low internal resistance, multiple tab current collection is used. This type of construction is termed *edge welding* and requires each electrode to have a current collecting strip on one side of the electrode. The current collecting strip for the positive and negative electrodes are on opposite sides of the jellyroll. After coil winding of the jellyroll, the edge current collector is welded in multiple locations to each electrode. Other aspects of cell assembly are virtually identical to small consumer batteries. The net result of the enhanced current collection is a reduction in cell specific energy and an increase in specific power due to decreased cell AC impedance (usually around 8 to 12 mΩ for small portable batteries and around 1 to 2 mΩ for industrial cylindrical cells). For industrial applications such as HEV, motorcycle and electric bicycle applications, the most popular cylindrical cell sizes are standard C and D sizes, although a multitude of height changes within those diameters are also used. Some work on larger size cells such as the F size has also been reported.[5]

Prismatic construction is conventional in that electrode stacks of alternating positive and negative electrodes with intermediate separators are used (Fig. 30.2). The main design alternatives involve the thickness and number of each electrode and the aspect ratio (relative proportions of cell height to width to thickness). Key design variations include the ratio of active materials to inactive components such as the cell can and terminal, and current collectors. In all cases, the cell designer has the objective of emphasizing one or more properties of performance such as energy vs. power, while maintaining a minimum threshold of other performance factors such as cycle life. One example is that for EV NiMH prismatic cells,[6] a specific power of about 200 W/kg is acceptable for most vehicles and consequently, relatively thick positive and negative electrodes can be used to increase the ratio of active material to inactive cell components, allowing specific energy in the 63 to 80 Wh/kg range. Alternately, HEV prismatic NiMH batteries typically must deliver greater than 500 W/kg specific power and therefore electrode thickness must be lower than that for EV use. NiMH batteries for HEV use have a typical specific energy ranging from about 42 to 68 Wh/kg.

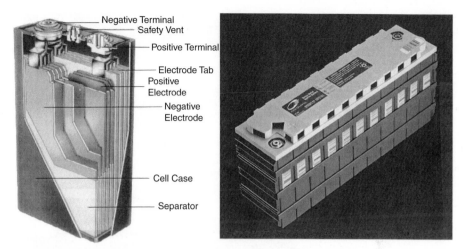

Negative Terminal
Safety Vent
Positive Terminal
Electrode Tab
Positive Electrode
Negative Electrode
Cell Case
Separator

FIGURE 30.2 GM Ovonic 90 Ah Prismatic Cell and 13 V module.

30.4.2 Metal Versus Plastic Cell Cases

NiMH cylindrical cells exclusively use metal cell cases. One important reason is that the can itself is electrically connected to the metal hydride negative electrode and serves as the negative terminal. Another important reason is that many applications require fast charge where gas recombination rates permit considerable internal pressure where only a metal container would suffice without significant volumetric penalties. Finally, the metal crimp seal to the cover plate assembly with a polysulfone seal ring is inexpensive, readily manufacturable, and reliable.

Both metal and plastic cell cases are common for the prismatic NiMH batteries used in automotive applications such as electric vehicles. Unlike cylindrical cells where the can itself is the negative terminal, prismatic design cells have both a positive and negative terminal at the top cover plate. Key criteria for selecting metal cell cases include excellent thermal conductivity, inexpensive prototyping costs for changing cell dimensions, and small volumetric penalties.

Advantages to plastic cell cases are cost and electrical isolation in the 320 V and higher battery packs common to today's electric vehicles, where even high resistance leakage currents to other vehicle components must be considered. Further design considerations for plastic cell cases include development costs for permanent molds, gas permeation, thermal conductivity, and sufficient plastic thickness for gas pressure containment without can wall bulging. NiMH battery modules using plastic cell cases are shown in Fig. 30.3.

FIGURE 30.3   Plastic cell case NiMH module designs of SAFT (93 Ah, 64 Wh/kg) and Matsushita (95 Ah, 63 Wh/kg).

30.4.3 Sintered Versus Pasted Nickel Hydroxide Positive

Positive electrodes for use in NiMH batteries for industrial and transportation application, whether cylindrical or prismatic, can be of the sintered or pasted type also common to NiCd batteries.

Sintered electrodes have the best rate and power capability,[7] but sacrifice capacity on a weight and volume basis and are more expensive to manufacture. Sintered electrodes involve an expensive and complicated sequence of manufacturing steps, and consequently only companies with an existing capital investment for this process would manufacture sintered electrodes. Sintered positives begin with the pasting of filamentary nickel onto a substrate such as perforated foil, where the nickel fibers are then "sintered" under a high temperature annealing furnace in a nitrogen/hydrogen atmosphere and binders from the pasting process are burned away to leave a conductive skeleton of nickel having a typical average pore size of about 30 microns. Nickel hydroxide is then precipitated into the pores of the sinter skeleton using either a chemical or electrochemical impregnation process. The impregnated electrode is then formed or preactivated in an electrochemical charge/discharge process.

Important design variables in the manufacture of sintered nickel hydroxide electrodes include:

- Strength and pore diameter of the filamentary nickel skeleton
- Chemical composition of the nickel hydroxide active material
- Active material loading
- Level of harmful impurities (e.g. nitrate, carbonate)

The more common pasted nickel hydroxide positive electrode is typically produced by mechanically pasting high density spherical nickel hydroxide into the pores of a foam metal substrate (Fig. 30.4). The foam metal substrate is typically produced by coating polyurethane foam with a layer of nickel either by electrodeposition or by chemical vapor deposition, followed by a heat treatment process to remove the base polyurethane. The typical pore size of the foam metal skeleton current collector is on the order of 200 to 400 microns. The foam[8] is then physically loaded with nickel hydroxide having an average diameter of about 10 microns in a paste containing conductive cobalt oxides which form a conductive network to compensate for the large distance from the nickel hydroxide to the metal current collector and for the fact that the nickel hydroxide itself is relatively low in conductivity. A cross-sectional comparison of sintered and pasted positive electrodes is presented in Fig. 30.5 under backscattered electron imaging where the bright areas indicate metallic nickel current collection.

One aspect of sintered versus pasted nickel hydroxide technology is that sintered electrodes require the battery manufacturer to make a significant capital investment in facilities and equipment, and to have a great deal of expertise in processing. Pasted electrodes conversely place a great deal of emphasis on the expertise of the suppliers for both the nickel foam substrate and for the high density spherical nickel hydroxide. Recent development of the pasted electrode has brought about exceptional improvements in power and high rate discharge capability to a level comparable to that of the sintered electrode, making the choice of which electrode configuration one of ease of manufacture and cost; these are areas where the pasted electrode is superior and this is the main explanation for the predominance of this process.

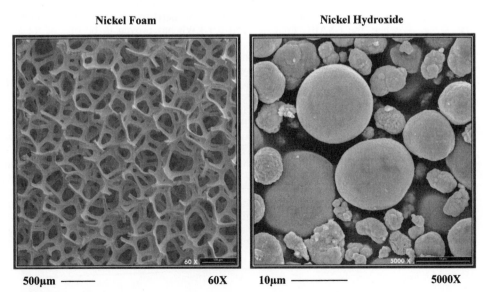

Nickel Foam   **Nickel Hydroxide**

500μm ———   60X   10μm ————   5000X

FIGURE 30.4   SEM micrographs of positive electrode nickel foam substrate and high density spherical nickel hydroxide.

Sintered   **Pasted**

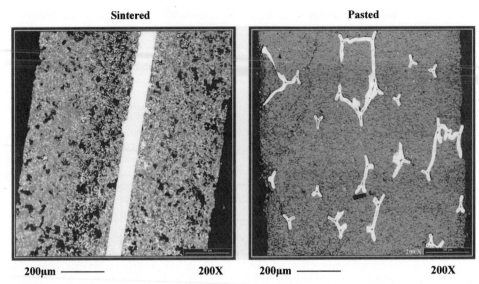

200μm ————   200X   200μm ————   200X

FIGURE 30.5   SEM micrographs under BEI imaging where bright areas indicate nickel metal current collection illustrating difference in active material distance to current collector for pasted and sintered positive electrodes.

30.4.4   Energy Versus Power Design Tradeoffs

As with most other battery technologies, NiMH batteries can be designed to emphasize energy, power, or some combination of the two. The application itself may make the choice apparent, but perhaps not for obvious reasons. In electric vehicle applications, a certain threshold power such as 200 W/kg is required for adequate vehicle performance, and once the power requirement is met competing designs usually provide a specific energy in the range of 62 to 80 Wh/kg. The motivation for the higher energy is of course longer vehicle range, but also cost. The typical figure of merit for electric vehicle battery cost is $/kWh, where the USABC goal of $150 per kWh has proven to be one of the most challenging development targets. NiMH cost is mainly controlled by raw material amounts, costs and utilization rather than processing labor, assembly, packaging, etc. Consequently, an important development activity for many NiMH manufacturers is to get a higher utilization of less expensive active materials on a mAh/g level basis and therefore reduce manufacturing costs.

For HEV applications, specific energy is of far less importance since the battery has a vastly different function than in pure electric vehicles. For HEVs, the main purpose of the battery is to accept and utilize the energy from regenerative braking and assist in acceleration. The battery is commonly exposed to very high current pulses during both charge and discharge, but energy extraction is usually limited to relatively low depths of discharge. The crucial requirement for the NiMH HEV battery is therefore specific power, and the USABC/PNGV (Partnership for a New Generation of Vehicles) goal for battery developers is greater than 1000 W/kg. NiMH batteries available today are in the 500 to 800 W/kg specific power range, with development reports approaching 1000 W/kg. Besides performance required for HEV operation, a crucial figure of merit for NiMH battery developers is power cost in $/kW, and the PNGV goal is below $12 per kW. In this case, specific energy for a typical HEV battery may be in the 32 to 56 Wh/kg range combined with specific power in the 500 to 800 W/kg range.

30.4.5 Metal Hydride Alloy

NiMH batteries are an unusual battery technology in that the metal hydride active material is an engineered alloy made up of many different elements and that the MH alloy formulas vary to a significant degree.[9,10] As stated previously, the active material in the negative electrode is either of the disordered AB$_5$ (LaCePrNdNiCoMnAl) or disordered AB$_2$ (VTiZrNiCrCoMnAlSn) type.[11,12,13,14] AB$_5$ type alloys are more common, despite significantly lower hydrogen storage capacity as compared to AB$_2$ (320 vs. 440 mAh/g). The advantages of the AB$_5$ alloys include low raw material cost, ease of activation and formation, flexibility in electrode processing and high discharge rate capability. On the other hand, there has been significant ongoing development to improve the properties of AB$_2$ materials to take advantage of their inherently higher energy, which is especially important to reduce cost.

Electrochemical utilization of metal hydride materials as anodes in NiMH batteries requires meeting a demanding list of performance attributes including hydrogen storage capacity, suitable metal-to-hydrogen bond strength, acceptable catalytic activity and discharge kinetics, and acceptable oxidation/corrosion resistance to allow for long cycle life. Multi-element, multiphase, disordered alloys of the LaNi$_5$ and VTiZrNiCr type are attractive development candidates for atomic engineering due to a broad range of elemental addition and substitution, availability of alternate crystallographic phases which form the matrix for chemical modification, and a tolerance for non-stoichiometric formulas. Through the introduction of modifier elements, ease of activation and formation have been achieved. Special processing steps, such as alloy melting and size reduction, suitable for these metallurgically challenging materials have also been developed.

The metal hydride active material and construction also have similar special design options. The active materials may be adjusted to influence one or more of the factor of capacity, power, and/or cycle life. For the AB$_5$ system, typical formulas include:

- La$_{5.7}$Ce$_{8.0}$Pr$_{0.8}$Nd$_{2.3}$Ni$_{59.2}$Co$_{12.2}$Mn$_{6.8}$Al$_{5.2}$ (atomic percent, at.%)
- La$_{10.5}$Ce$_{4.3}$Pr$_{0.5}$Nd$_{1.3}$Ni$_{60.1}$Co$_{12.7}$Mn$_{5.9}$Al$_{4.7}$

While the capacity of various AB$_5$ alloys is usually around 290 to 320 mAh/g, other overall performance attributes can be greatly influenced. It is common for the ratio of La/Ce to be reversed to emphasize cycle life and power. The total amount of Co, Mn and Al significantly affect ease of activation and formation but increased cobalt has cost implications. After production of the AB$_5$ alloy ingot, it is common to further refine the microstructure of the material by a post annealing treatment of about 1000°C for 10 hours. The annealing treatment can have a significant effect on capacity, discharge rate and cycle life by adjusting crystallite size and grain boundaries, as well as eliminating unwanted phases precipitated during ingot melting and casting. Further, special processing methods such as melt spinning and other rapid solidification techniques can also promote cycle life, although these processing methods are more expensive and may involve a tradeoff with discharge rate capability.[15]

Commercial AB$_5$ alloys have a predominantly CaCu$_5$ crystallographic structure. However, within that structure, there are a range of lattice constants brought about by compositional disorder[16] within the material which are important for catalysis, storage capacity and stability to the alkaline environment and embrittlement. These materials also precipitate a nickel-cobalt phase which is important for high rate discharge.[17,18]

AB$_2$ alloys also have formula and processing choices. Popular AB$_2$ alloy formulas (at.%) include:

- V$_{18}$Ti$_{15}$Zr$_{18}$Ni$_{29}$Cr$_5$Co$_7$Mn$_8$
- V$_5$Ti$_9$Zr$_{26.7}$Ni$_{38}$Cr$_5$Mn$_{16}$Sn$_{0.3}$
- V$_5$Ti$_9$Zr$_{26.2}$Ni$_{38}$Cr$_{3.5}$Co$_{1.5}$Mn$_{15.6}$Al$_{0.4}$Sn$_{0.8}$

Alloy capacity may range from 385 to 450 mAh/g. Higher vanadium content alloys may suffer from higher rates of self discharge due to the solubility of vanadium oxide and its consequent ability to form a special type of redox shuttle. The concentration of Co, Mn, Al and Sn are important for ease of activation and formation and for long cycle life. The ratio of hexagonal to cubic phase is important to emphasize capacity or power.

For both AB_5 and AB_2 metal hydride alloys, the metal/electrolyte surface oxide interface is a crucial factor in discharge rate capability and cycle life stability. Original $LaNi_5$ and $TiNi$[19] alloys extensively studied in the 1970s and 1980s for NiMH battery applications were never commercialized due to poor discharge rate and cycle life capability.[20,21] Lack of catalytic activity at the surface oxide limits discharge and lack of sufficient oxidation/corrosion resistance is a critical obstacle to long cycle life. The complicated chemical formulas and microstructures of present disordered AB_5 and AB_2 alloys extend to the surface oxide. In the oxide, important factors include thickness, microporosity and catalytic activity. Of crucial importance to discharge rate was the discovery that ultrafine metallic nickel particles having a size on the order of 50 to 70 Angstroms or less dispersed within the oxide are excellent catalysts for the reaction of hydrogen and hydroxyl ions.[22]

The other critical design factor within the surface oxide is to achieve a balance between surface oxide passivation and corrosion. Porosity with the oxide is important to allow ionic access to the metallic catalysts and therefore promote high rate discharge while passivation of the oxide is problematic for high rate discharge and cycle life, unrestrained corrosion is equally destructive. Oxidation and corrosion of the anode metals consume electrolyte, change the state of charge balance and create corrosion products which are sometimes soluble and capable of poisoning the positive electrode. Establishing a balance between passivation and corrosion for stability is a primary function of compositional and structural disorder.

The construction of the metal hydride electrode is quite versatile, in part because the metal hydride active material powder itself is conductive. Therefore, the choice of substrate is partially current collector and partially as a carrier for the application of powder. While metal hydride electrodes have used substrates of foam, fiber mat, and wire mesh all commercially, two substrates dominate: expanded metal and perforated sheet. The choice of substrate is mostly influenced by cost, method of connecting a current collecting tab, and statistical likelihood of a short circuit. For all of these factors, perforated sheet has emerged as the leading substrate. Most manufacturers find that a reduced incidence of sharp edges and metallic burrs occur with perforated sheet and therefore a parts per million level of short circuit defects in cell assembly can be achieved. It is common for the perforated hole patterns to leave solid material at the edge of the substrate to facilitate welding and edge welding for high power cell designs.

Substrate material has been dominated by nickel and nickel-plated steel. For cost reasons, nickel-plated steel is preferred. Further, the metal hydride electrode is unlikely to undergo the low voltage condition due to over discharge because hydrogen gas evolved at the nickel hydroxide electrode during reversal can recombine at the metal hydride electrode to maintain state of charge, except under highly abusive conditions. Consequently, corrosion of iron that poisons the nickel hydroxide is of less concern in a NiMH battery. Taking advantage of the metal hydride electrode voltage protection against corrosion has led to the emergence of a copper substrate for both cost reduction and higher conductivity, vital for high power applications such as HEVs and power tools.

30.4.6 Nickel Hydroxide

Nickel hydroxide for use in NiMH batteries is fundamentally the same as that used in nickel cadmium and nickel-iron batteries, and from a simple viewpoint, the basic compound is the same as that used by Thomas Edison 100 years ago. Today's high performance nickel hydroxide is much more complicated and continues to improve in capacity and utilization, power and discharge rate capability, cycle life, high temperature charging efficiency, simplified manufacturing processes and cost.

As mentioned previously, one type of nickel hydroxide is by far the most common—a high density spherical type for use in pasted electrodes which became commercial around 1990.[23,24] High density spherical nickel hydroxide is made in a precipitation process where metal salts such as nickel sulfate are reacted with caustic such as NaOH in the presence of ammonia. The nickel source may have additives such as cobalt and zinc to enhance performance. The important physical parameters within this type of nickel hydroxide are:

- *Chemical formula.* Common compositions are $Ni_{94}Co_3Zn_3$ (wt%) where cobalt is coprecipitated for the purpose of conductivity enhancement and both cobalt and zinc have the function of oxygen overvoltage adjustment and microstructure refinement.
- *Tap density.* Usually around 2.2 g/cc, tap density is a measure of the dry nickel hydroxide powder packing efficiency and influences the amount of active material which can be loaded into the pores of the nickel foam current collector.
- *Particle size,* usually having an average particle size of about 10 microns.
- *Surface area.* Measured by the BET method, surface area refers not to the geometric area of each nickel hydroxide sphere, but rather to the total surface area of each particle which contribute to the charge/discharge reactions and can thus affect utilization and high rate discharge capability. A typical BET surface area for high-density spherical nickel hydroxide is about 10 m^2/g.
- *Crystallinity.* Each nickel hydroxide sphere has an extremely high surface area corresponding to the nickel hydroxide crystallites themselves. Crystallinity is measured by X-ray diffraction, where the full width at half maximum (FWHM) of a reflection such as the <101> plane may yield a typical crystallite size of about 110 angstroms.

A variety of other factors contribute to performance, including impurities from processing such as residual sulfates, nitrates, sodium sulfate, etc.

The nickel hydroxide active material and electrode formula can be formulated for specific applications. For operation at 65°C, some manufacturers may add $Ca(OH)_2$, CaF_2, or Y_2O_3 to the paste formula to inhibit premature oxygen evolution on charge.[25] Other paste formula modifications may be an adjustment in the type and quantity of conductive network additives such as cobalt metal and cobalt monoxide.[26,27,28] When long-term storage over several years is required, it is common to increase the levels of cobalt additives to combat breakdown of the conductive network and isolation of the active material. For ultra-high power discharge, it is also possible to add metallic nickel fibers to the paste formula to enhance conductivity. In the case of calcium additives, the additive may cause a loss in power and/or cycle life. Cobalt metal and cobalt monoxide are relatively expensive materials, and increased use has cost implications. Addition of metallic nickel fibers lowers the amount of active material and consequently capacity and specific energy are reduced.

The nickel hydroxide active material itself is most commonly a NiCoZn triprecipitate. However, the amounts of cobalt and zinc which are usually about 1 to 5% each can be adjusted for conductivity and oxygen overvoltage with some tradeoffs in terms of active material capacity and cost. Other more complicated multi-element precipitates such as NiCoZnCaMg offer higher capacity, cycle life and high temperature performance, but cannot be manufactured by conventional precipitation processes. Another common active material choice is "cobalt-coated nickel hydroxide." Usually, the paste additives are finely divided cobalt metal and cobalt monoxide that will dissolve and reprecipitate on the nickel hydroxide active material surface. However, the coating may not be uniform and complete in coverage. Consequently, although more expensive and lower in capacity, the nickel hydroxide itself may be coated with cobalt hydroxide by the active material manufacturer. It has been reported the cobalt-coated materials offer increased utilization and high rate discharge performance.[29]

In addition to the nickel hydroxide active material and paste formula, the NiMH battery manufacturer may modify the conductive skeleton current collector, which is most commonly nickel foam. The pore size may be decreased from about 400 microns to 200 microns for better conductivity. The density of the foam may also be adjusted to promote conductivity and power versus capacity and utilization.

30.4.7 Electrolyte

The electrolyte in NiMH batteries of all types is normally a about 30% potassium hydroxide in water, providing high conductivity over a wide temperature range. It is most common for the KOH/H_2O electrolyte to have a lithium hydroxide additive at a concentration of about 17 grams per liter to promote improved charging efficiency at the nickel hydroxide electrode by suppressing oxygen evolution, the competing reaction to charge acceptance.

An important feature of the electrolyte is related to fill fraction. Essentially all NiMH batteries are of the sealed, starved electrolyte design. As with NiCd batteries, the electrodes are nearly saturated with electrolyte, while the separator is only partially saturated to allow for rapid gas transport and recombination.

Special electrolytes are also used in NiMH batteries to enhance high temperature operation. Instead of binary KOH/LiOH electrolytes, it is also possible to substitute a portion of the KOH with NaOH. The ternary KOH/NaOH/LiOH electrolyte is used at a high concentration of about 6 Molar, but the contribution of NaOH promotes high temperature operation charging efficiency although it is typical for this electrolyte to decrease cycle life by increased corrosion of the metal hydride active materials.

30.4.8 Separator

The primary function of the separator is to prevent electrical contact between the positive and negative electrodes, while holding the electrolyte necessary for ionic transport. Original separators for NiMH batteries were standard NiCd and NiH_2 separator materials. However, NiMH batteries were more sensitive to self discharge, especially when conventional nylon separators were used.[30] The presence of oxygen and hydrogen gas cause the polyamide materials to decompose, producing corrosion products which poison the nickel hydroxide, promoting premature oxygen evolution and also forming compounds capable of a redox shuttle between the two electrodes which further increases the rate of self discharge.

Consequently, there was a need for a more stable NiMH separator material to reduce self discharge while still retaining electrolyte crucial for maintaining cycle life. NiMH batteries now have widespread use of what is termed "permanently wettable polypropylene." In fact, the polypropylene is a composite of polypropylene and polyethylene fibers where the base composite is hardly wettable to the KOH electrolyte without surface treatment.

Surface treatments involve the grafting of a chemical such as acrylic acid to the base fibers to impart wettability and accomplished using a variety of techniques such as UV or cobalt radiation.[31] Another method of imparting wettability to the polypropylene is a sulfonation treatment where the base fiber material is exposed to fuming sulfuric acid. The separator surface is designed to be made hydrophilic to the electrolyte.

The separator has a crucial role relative to cycle life. In the starved electrolyte design, it is a common design principle to saturate the electrodes with electrolyte at the assembly stage. The separator is designed to have a high electrolyte fill fraction in order to hold as much electrolyte as possible but not be overfilled so as to inhibit gas recombination. To the battery manufacturer, this has the implication that during the first few charge/discharge cycles ("formation") when the electrodes have not yet absorbed all of the intended electrolyte, charging must be initiated carefully to avoid venting.

The electrolyte design concept relates to capillary theory that the electrolyte will migrate to the smallest pores. In the NiMH battery, this translates to the nickel hydroxide positive electrode having the smallest pores, followed by the metal hydride negative electrode and finally the separator. At cell assembly, it is common for the separator to be about 90% filled, and then reduced to about 70% during the cell formation process of the first few charge/ discharge cycles as both the positive and negative electrode expand and contract opening interior regions for electrolyte absorption. This process continues to some degree over many hundreds of charge/ discharge cycles, where NiMH cell failure is commonly when the separator fill fraction has been reduced to about 10 to 15% of original. Consequently, separators which can absorb larger quantities of electrolyte at cell assembly, have small pores able to retain electrolyte, and maintain surface wettability are highly desirable. Inspection of the separator in failed batteries shows even these types of separators undergo some degradation and loss of electrolyte absorption ability, although certainly not to the same degree as earlier separators.

30.4.9 Monoblock Construction

NiMH technology is especially well suited for monoblock construction due to its high tolerance to overcharge and overdischarge. A monoblock design reduces cost by having a common pressure vessel construction and far fewer cell parts; a single vent assembly and shared hardware can be used in multicell modules. Further attributes of monoblock construction include reduced volume since interior cell walls can be shared, and flexible choices of liquid or air cooling. A water cooled 12 V, 12 Ah plastic monoblock NiMH battery is shown in Fig. 30.6, and has the following characteristics:

Number of cells	10
Nominal voltage	12 V
Rated capacity	12 Ah
Weight	3.2 kg
Volume	2.2 L
Rated energy	150 Wh
Specific energy	48 Wh/kg
Energy density	71 Wh/L
Peak power	1.76 kW
Specific power	550 W/kg
Power density	800 W/L

Issues of monoblock construction include selection of the plastic casing material to avoid gas permeation, the need for individual cells to have well matched capacity and impedance to avoid cell-to-cell imbalance. Further, monoblock construction must recognize that electrodes within each cell expand and contract during charge and discharge and that the ensuing swelling of the electrode stack must be compensated for in the mechanical design and loading of the monoblock.

FIGURE 30.6   GM Ovonic 12 Ah, 12 V NiMH watercooled plastic monoblock.

30.5   EV BATTERY PACKS

30.5.1   USABC Performance Targets

The USABC technical performance requirements are listed in Table 30.1, with the performance of NiMH batteries against these requirements.

In 1991, there were only two industrial size rechargeable batteries that could be used for electric vehicles; lead-acid and nickel-cadmium. Lead acid batteries simply do not have sufficient specific energy and cycle life to be viable for EV use. Industrial NiCd, while somewhat higher in specific energy and in cycle life, was still not high enough in energy and faced the environmental issues of cadmium use, operation in a flooded electrolyte configuration requiring maintenance, and memory effect.

Essentially, an advanced battery technology for EV application must satisfy all minimum performance requirements and then compete on energy and cost. NiMH EV batteries demonstrate:

- *High energy:* Commercial electric vehicles achieve over 200 mile range when using 70 Wh/kg specific energy NiMH. Prototype vehicles using NiMH batteries have demonstrated a 350 mile range.
- *High power:* Commercial electric vehicles using 220 W/kg specific power NiMH batteries have demonstrated acceleration comparable to ICE powered vehicles. Advanced EVs using NiMH routinely set race and demonstration range and speed records.
- *Flexible packaging:* EV NiMH batteries operate well in 320 V AC propulsion or 180 V DC propulsion systems and have a wide variety of cell capacities (Fig. 30.7). The volumetric performance of NiMH is high, offering vehicle designers a small size battery package.
- Long life (600 to 1200 cycles to 80% DOD)
- Wide operating temperature range ($-30°C$ to $+65°C$)
- Fast charge and simple normal charging
- Maintenance free operation

FIGURE 30.7   GM Ovonic "Family of Batteries" NiMH modules of varying capacity and dimensions (12-100 Ah).

The main development activity for NiMH EV batteries has been to reduce cost. Because only prototype production volumes exist to date, initial cost of manufacture for NiMH has been high. Significant investment has been made to decrease cost, with highlights including:

- *Pilot manufacturing capability.* Increased production rates allows reduced overhead and labor, and the incorporation of automated and semi-automated manufacturing.
- *Increased specific energy.* By raising specific energy from 56 to 80 Wh/kg through increased utilization of active materials, the battery cost in $/kWh has decreased significantly (>$1000/kWh to <$350/kWh, even at relatively low production volumes).
- *Low cost nickel hydroxide.* From 1995, where high density spherical nickel hydroxide prices were around $30 per kg, new low cost suppliers and manufacturing processes have allowed nickel hydroxide prices to drop below $10 per kg.
- *Low cost metal hydride materials.* MH development has included improved efficiency melt/ cast processes, scrap recycle, elimination of costly sintering, versatile use of inexpensive substrate materials and significantly reduced battery formation.

30.6 HEV BATTERY PACKS

30.6.1 PNGV Performance Targets

The Partnership for the Next Generation Vehicles (PNGV) was established to enable auto-makers and the government to meet an equivalent fuel economy of 80 miles per gallon of gasoline. For many car manufacturers, development efforts focused on hybrid electric vehicles of a variety of concepts.[32] HEVs provide many of the advantages of electric vehicles without the need to recharge the battery which is carried out on-board the vehicle by a combustion engine. There are basically two major types of HEVs based on how the inter-action between the internal combustion engine (ICE) and the electric drive are incorporated into the system. In a series design, the ICE is connected to a generator to provide electricity for the battery pack and the electric drive motor. In a parallel design, the ICE and the electric motor drive the wheels. In both cases, the electric motor can also be used to recharge the battery. There are a variety of options within each battery approach such as range extender, power assist, or dual mode. It is common for an HEV to be described as a 70:30 or 90:10 vehicle where the ratio is the amount of power supplied by the ICE compared to that of the electric drive and battery.

The various specific operating modes for each type of vehicle make the size of the battery extremely wide ranging, anywhere from 0.9 to 5 kWh. For hybrid vehicles, the smallest cell size used is 6.5 Ah NiMH D cells and the largest HEV NiMH cell size is about 60 Ah. In the Honda Insight HEV, the NiMH battery is 900 Wh and consists of 6.5 Ah D cells at 144 V. In contrast, the Toyota Prius HEV is 1.8 kWh and consists of 6.5 Ah D cells at 288 V. In the General Motors Precept HEV, designed as a 50:50 ICE to electric drive, a 4.2 kWh NiMH battery consisting of twenty-eight 12 V modules having a capacity of 12 Ah is employed.

HEV battery requirements include:

- Power density > 1000 W/L
- Specific power > 500 W/kg up to >1000 W/kg
- Specific energy > 50 Wh/kg
- Regenerative braking charge > 500 W/kg
- HEV cycle efficiency > 85%
- Long life
- Low cost

The crucial targets for HEV as compared to EV goals is a much greater emphasis on power, both on charge and on discharge, and less emphasis on energy. Cycle life is measured at low depth of discharge (~ 2-5% DOD) compared to 80% DOD for an EV. Even cost is calculated on available power in $/kW. Car makers essentially want to use the smallest energy battery capable of supplying the required power. A figure of merit target from the HEV car makers is a power to energy ratio of 40. Present NiMH HEV batteries have attained P/E ratios of about 25. Performance specifications for several NiMH HEV designs are presented in Table 30.2.

TABLE 30.2 Performance Specifications for Ovonic HEV NiMH Battery Modules

	13-HEV-60	7-HEV-28	12-HEV-20
Nominal voltage (V)	13.2	7.2	12
Nominal capacity (Ah)	60	28	20
Weight (kg)	12.2	4.3	5.2
Volume (L)	5.1	2.0	2.3
Specific energy (Wh/kg)	68	50	48
Energy density (Wh/L)	160	102	110
Specific power (W/kg)	600	550	550
Power density (W/L)	1400	1200	1300

30.6.2 Charge Depletion

Charge depletion concept hybrid vehicles are designed to have a certain vehicle range during the complete electric vehicle mode. The motivation for this requirement is that in heavily congested urban areas where air quality is a vital concern, the vehicle be able to operate with zero emission. Therefore, in the charge depletion mode the specific energy of the battery remains important and a specific energy above 60 Wh/kg may be required. As an example, a Toyota Prius HEV using 1.8 kWh NiMH D cells has a pure electric vehicle range of less than 6 miles. However, when replaced with a 6.0 kWh NiMH battery pack, vehicle range on electric drive only may be extended to over 25 miles.

30.6.3 Charge Sustaining

The more common HEV concept is termed charge sustaining, where the vehicle is designed to be operating under the various modes giving the greatest overall fuel economy. In this case, a primary function of the battery is to utilize the otherwise wasted energy of braking and another major function is to assist the vehicle during acceleration. This concept essentially utilizes the knowledge that ICE engines are very efficient on the highway at constant speed but very inefficient under the starting and stopping conditions of the city traffic. It is a common strategy for the engine to be shut off while the vehicle is stopped at a traffic light, and for the battery to supply the energy needed for the initial vehicle acceleration while the ICE is restarted.

For NiMH HEV batteries in the charge sustaining mode, the crucial performance target is a specific power of 1000 W/kg. For these applications, typical NiMH HEV batteries have specific energy in the 45 Wh/kg range and have thus attained a specific power of about 600 to 850 W/kg.

30.7   *FUEL CELL STARTUP AND POWER ASSIST*

Fuel cells are a significant development focus for many companies worldwide, offering the potential for equivalent fuel economy over 100 miles per gallon. The most commonly discussed fuel cell is the Proton Exchange Membrane (PEM) type (see chapters 42 and 43).

The relevance to NiMH batteries is that fuel cell vehicles will still require a substantial (~1 to 10 kWh) battery for several reasons:

- *Startup.* The PEM fuel cell and reformer may operate at a temperature of about 100°C and the battery must have sufficient energy for the vehicle to operate until the system comes up to temperature.
- *Acceleration.* A critical PEM fuel cell obstacle will be cost due to the use of expensive catalysts and membranes. The fuel cell will be sized for an average vehicle power, not for peak power conditions where the battery will dominate. Essentially, the fuel cell in a fuel cell hybrid electric vehicle is used to charge the battery and operate the vehicle, while the battery is used for power load-leveling.
- *Efficiency.* The fuel cell can be operated under close to steady state conditions in an optimum region of fuel utilization.
- *Regenerative braking.* As with ICE powered hybrid vehicles, the fuel cell has no mechanism to absorb the energy of braking without the battery.

30.8   *OTHER APPLICATIONS*

30.8.1   36 to 42 V SLI Batteries

While EV, HEV and Fuel Cell electric vehicles receive tremendous attention, the largest worldwide rechargeable battery application is still the common lead acid Starting Lighting, Ignition (SLI) battery.

As modern ICE vehicles add more electrical devices such as electric power steering, braking, and accessories, carmakers will be using a nominal electrical operating voltage of 36 to 42 Volts compared to the normal 12 V system. Advantages including improved engine performance and efficiency through electrically actuated valves are offset by added battery cost and redesign of vehicle electrical systems. One option would be to use several lead-acid batteries to meet this voltage, but size and weight considerations make this approach impractical. Figure 30.8 illustrates a relative size comparison of lead-acid and NiMH for 36 to 42 V SLI applications. NiMH batteries are considered the leading advanced battery for high voltage SLI applications due to the following advantages compared to lead-acid:

- Higher gravimetric and volumetric energy
- Longer cycle life, especially under high depth of discharge

Commercialization issues will include improving low temperature power at −30°C and initial cost at low to moderate production volumes.

36-V Lead-Acid
Starter Battery

36-V Ovonic NiMH
Starter Battery

FIGURE 30.8 Relative comparison of 36 V lead acid and 36 V NiMH batteries for SLI application.

30.8.2 Aircraft Applications

Prismatic nickel-metal hydride cells with 43 Ah capacity are under development for use in military aircraft.[33] The objective of this program is to demonstrate an environmentally safe retrofit for the U.S. Air Force F-16 Pre-Block 40 Main Aircraft battery and to pursue a specific energy goal of 75 Wh/kg for the More Electric Aircraft Generation II Program. Target performance parameters are listed in Table 30.3. Four-cell test sets have been evaluated for capacity at several charge and discharge rates, for self-discharge and for ambient and environmental cycling behavior. The ambient cycle life tests showed stable performance over 200 cycles and environmental cycling behavior over 80 complete cycles of 10 minutes at 20 amps at specified temperature. Charging efficiency was determined by charging at -20, 0, $+23$ and $+50°C$ at $C/2$, C and $2C$ rates followed by a five-hour rest at $+23°C$ and discharge at $C/2$ to 0.9 V/cell. The results are shown in Fig. 30.9 for the $2C$ rate for four different cell designs. Charge acceptance is good in the range of -20 to $+23°C$ but drops significantly at $+50°C$. This effect is undoubtedly due to the fact that oxygen evolution occurs at lower voltages at higher temperatures. Discharge characteristics were also determined vs. temperature. Cells were charged at $+23°C$ at the $C/2$ rate followed by 3 hours at $C/20$ followed by a five-hour rest and then discharged at -40, -30, -20, 0, $+23$, $+50$ and $+70°C$ at $C/2$, C and $2C$ rates to 0.9 V/cell. Single cell test on four different designs are shown in Fig. 30.10 at the $C/2$ rate and in Fig. 30.11 at the $2C$ rate. The $C/2$ discharge is well behaved between -30 and $+70°C$ with a maximum at $+23°C$ but dropped below 5 Ah at $-40°C$. At the $2C$ rate, the discharge capacity of one configuration dropped to a low value at $-20°C$, while the other design interations provided 40 Ah at that temperature. Self-discharge was also evaluated from -10 to $+50°C$. One design configuration met the requirement of less than 25% self-discharge per week at all temperatures up to $+50°C$. Based upon this test program, one design iteration has been selected for optimization and five baseline batteries projected to provide 43 Ah capacity at 28 V and a specific energy of 64 Wh/kg will be evaluated for full performance characterization.

TABLE 30.3   Envvironmental Nickel-Metal Hydride Aircraft Battery Parameters

Nominal Voltage (V)	28
Capacity (Ah)	50
End of Life Capacity (Ah)	18 (Maintained at 14.5)
Current (A)	48 (Maximum)
Operating Temperature Range (°C)	-40 to $+71$
Battery Energy at $C/2$ (Wh)	1,325
Specific Energy Density (Wh/kg)	75
Volumetric Energy Density (Wh/L)	177
Total Battery Weight (kg {lb})	18.18 {40}
Mean Time Between Failures (MTBF)	>6000 h
Maintenance Interval	Maintenance Free for Three Years
Self-Discharge (<25% over 7 days)	All Operating Temperatures

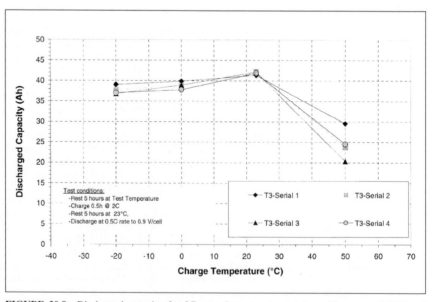

FIGURE 30.9   Discharged capacity for 2C rate charge vs. temperature. (*Courtesy of GRC and SAFT.*)

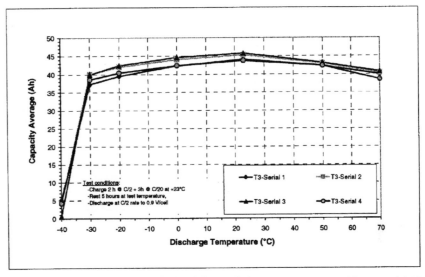

FIGURE 30.10 Discharged capacity for C/2 rate discharge vs. temperature. (*Courtesy of GRC and SAFT.*)

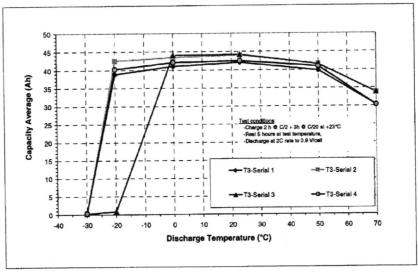

FIGURE 30.11 Discharged capacity for 2C rate discharge vs. temperature. (*Courtesy of GRC and SAFT.*)

30.9 DISCHARGE PERFORMANCE

30.9.1 Specific Energy

NiMH specific energy can vary anywhere from 42 to 100 Wh/kg depending on the particular application requirements. For laptop computers where run time is paramount, NiMH battery makers need not have high power capability or even the need for ultra-long cycle life. On the other hand, for extremely high power charge and discharge, extra current collection, high N/P (negative/positive) ratios, and other cell design and construction decisions can cumulatively affect specific energy. For the most common small portable NiMH batteries, specific energy is usually about 75 to 90 Wh/kg, for EV batteries usually about 65 to 75 Wh/kg, and for HEV batteries and other high power applications about 45 to 60 Wh/kg.

While gravimetric energy usually receives the attention for advanced battery technologies, in many cases volumetric energy density in Watt-hours per liter is actually more important. NiMH has exceptionally good energy density, achieving up to 320 Wh/liter.

30.9.2 Specific Power

NiMH power capability is an advantage relative to other advanced battery chemistries. NiMH is now replacing NiCd in many applications due to higher energy and environmental concerns.

Voltage profiles up to 10-C discharge rate for high power cylindrical NiMH are presented in Fig. 30.12, attaining a specific power of 835 W/kg (Fig. 30.13). NiMH HEV module power is shown in Fig. 30.14 in excess of 600 W/kg for both charge and discharge.

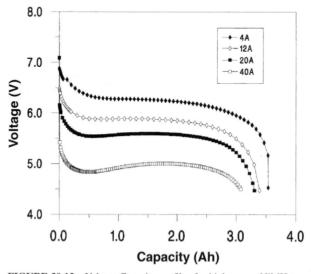

FIGURE 30.12   Voltage-Capacity profiles for high power NiMH cylindrical 3.5 Ah C-size batteries at different rates on continuous discharge.

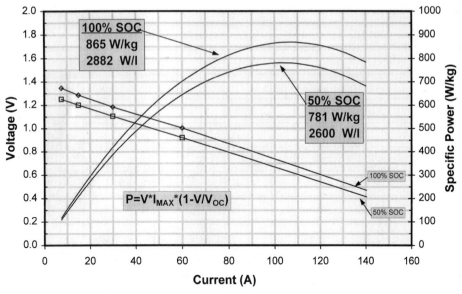

FIGURE 30.13 Specific power for ultra-high power NiMH cylindrical 3.5 Ah C-size batteries.

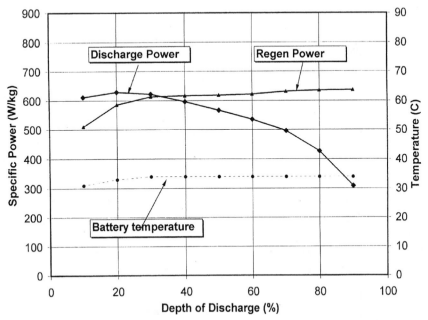

FIGURE 30.14 Specific power of 12 V, 20 Ah NiMH HEV battery module as a function of depth of discharge.

30.9.3 Effect of Temperature

NiMH, due to its KOH/H$_2$O electrolyte, provides performance as function of temperature similar to nickel-cadmium with some exceptions. Capacity dependence as a function of temperature is strongly influenced by the active material selection of the metal hydride and nickel hydroxide, and typical commercial NiMH performance is presented in Fig. 30.15. At low temperature, NiMH has reduced discharge capacity and power compared to NiCd because the metal hydride electrode is polarized due to the generation of water on discharge. On the other hand, whereas NiCd has some difficulty with charging at cold temperatures, and particularly fast charge at cold temperature, NiMH is much less sensitive to this effect.

High temperature charge efficiency is a matter of significant importance for propulsion applications. Electric and hybrid electric vehicles will be used in warm weather climates, where vehicle range in the summer is of critical importance to the end user. First generation NiMH batteries for EV application would lose almost 50% of room temperature capacity when charged at 60°C. The problem with high temperature charge acceptance is the oxygen evolution characteristics of the nickel hydroxide positive electrode. Normally, at room temperature, the nickel hydroxide electrode has almost complete charge acceptance until about 80% charge input when the competing reaction of oxygen evolution begins. At full charge, continued charging causes 100% oxygen evolution at the nickel hydroxide electrode, and the oxygen migration to the metal hydride or cadmium electrode provides the well known oxygen recombination overcharge mechanism.

At elevated temperatures, the issue is therefore that oxygen evolution occurs at much lower states of charge, and total charge acceptance is reduced. To combat this premature oxygen evolution mechanism, manufacturers have introduced oxygen evolution suppressants

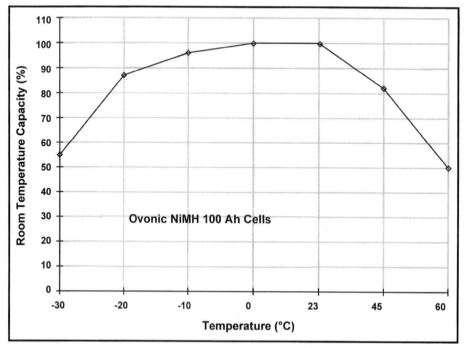

FIGURE 30.15 Temperature performance of 100 Ah NiMH EV prismatic batteries using commercial nickel hydroxide.

such as $Ca(OH)_2$, CaF_2, or Y_2O_3. Another method is to modify the formula of the nickel hydroxide itself. The most common NiMH positive active material is NiCoZn. To improve high temperature performance, multielement formulas such as NiCoZnCaMg materials have been developed. As shown in Table 30.4, introduction of these additives can reduce capacity loss during 65°C charging from about 50% to below 10%. NiMH manufacturers must carefully select the oxygen suppressant type, amount and location to avoid deleterious effects such as power loss and cycle life reduction due to the nonconductive nature of many of these materials.

Another aspect of performance is the effect of storage at temperature on life (discussed later). Extended storage over temperatures of about 45°C can reduce life due to degradation of the separator, oxidation and corrosion of the metal hydride alloy, and disruption of the cobalt conductive network in the positive electrode. Each of these mechanisms is highly dependent on the manufacturer's choices of the materials.

TABLE 30.4 High Temperature Charging Efficiency Performance in NiMH C-Cell Battery

Charging Efficiency for C-cell batteries at 65°C	
Commercial Nickel Hydroxide ($Ni_{94}Co_3Zn_3$)	52%
Commercial Controls + $Ca(OH)_2$ External Additive	83%
Advanced NiCoZnCaMg Formula	92%

30.9.4 Charge Retention

Charge retention is also highly affected by temperature depending on the manufacturers' choice of the specific metal hydride alloy formula, separator material, and the quality of the nickel hydroxide active material. Because these same selection criteria also apply to small portable NiMH batteries, large industrial NiMH batteries have very similar charge retention performance. There are two main self discharge mechanisms; the oxygen instability of the nickel hydroxide electrode and shuttle mechanisms where chemical species may oxidize at the positive, diffuse to the negative electrode and be reduced, with the redox mechanism repeated.[34]

Charge retention is an area where NiMH manufacturers compete and the end user must compare all performance properties for a given design. For example, advanced separator materials have the ability to reduce self-discharge from about 30% loss per month at room temperature to about 15%. However, these same materials may reduce cycle life from 15 to 50%. The chemical mechanisms by which the separator reduces self discharge are complicated, involving chemical grafting agents that can bind and thereby inactivate chemical species which may promote self discharge. However, these separator treatments may also have deleterious effects on the amount of electrolyte absorption and the ability of the separator to retain the electrolyte during cycling.

The metal hydride alloy has similar design tradeoffs. Higher capacity AB_2 alloys also have traditionally had higher rates of self-discharge compared to the lower capacity AB_5 alloys discussed previously. The mechanisms for the MH alloy effect on self-discharge are twofold. First, corrosion products from the hydride alloy may migrate to the nickel hydroxide positive electrode and promote oxygen evolution during storage. Second, other corrosion products such as vanadium with multivalent oxides may form redox shuttle mechanisms similar to that for nitrate ions. The quality of the nickel hydroxide affects self-discharge due to amounts of residual impurities such as nitrates and carbonates, which can influence self-discharge mechanisms. Achievement of ultra-low levels of impurities may incur increased processing costs.

30.9.5 Cycle Life

Cycle life for industrial NiMH batteries has both similarities and differences to those of small portable NiMH batteries. Cycle life for small portable NiMH batteries can vary from manufacturer to manufacturer, but usually falls in the range of 500 to 1000 cycles (100% DOD under 2 hour charge/discharge). Design and chemistry factors affecting cycle life that are common to both large and small NiMH batteries include:

1. Metal hydride electrode
   - Alloy formula (oxidation/corrosion properties)
   - Alloy processing effect on microstructure (particle disintegration)
   - Electrode construction (swelling in x-y-z direction, stability of conduction pathways)
2. Nickel hydroxide electrode
   - Active material formula (swelling and poisoning resistance)
   - Conductive network stability (amount and type of cobalt oxides)
   - Substrate (pore size, strength and resistance to fracture)
3. Cell design
   - N/P ratio (amount of excess negative electrode capacity to influence cell pressure, MH corrosion, disintegration)
   - MH discharge reserve (overdischarge protection)
   - Separator (stability to corrosion, electrolyte absorption and retention, thickness and resistance to short circuit)
   - Electrolyte (composition, amount, and fill fraction)
   - Vent pressure (weight loss, charge imbalance)
   - Electrode stack design (compression, electrode thickness, aspect ratio of height to width)

Factors affecting cycle life which are different from small portable applications for large industrial NiMH applications include:

- Significantly higher typical battery voltages (42 to 320 V vs. 12 V) increases risk of abusive overcharge and overdischarge due to capacity or state-of-charge mismatch.
- Overall higher energy (0.1 kWh vs. 33 kWh) increases heat generation and criticality of thermal management, which can be further influenced by battery pack enclosure heat transfer limitations.
- Typically higher operating temperature. Air cooled and water cooled vehicle batteries are usually operating at a temperature of 35°C or higher whereas operating temperatures for small portable batteries may experience transient high temperatures, but on average are at or near room temperature.
- Series/parallel strings. In small portable batteries, there are a large number of cell sizes available, ranging from 100 mAh button cells to 7 to 12 Ah D and F cells. There are many fewer cell sizes for HEV and EV NiMH batteries. Consequently, vehicle applications may be required to use series/parallel combinations of NiMH cells to meet the required pack voltage and energy demands, and thereby increase the risk of pack imbalances.
- End of life definition for small portable batteries is usually based on capacity loss. In contrast, NiMH for EV and HEV applications find end of life is due to power limitations.
- Qualification and operation testing. The emphasis on power greatly influences testing and methodology. In small portable battery cycle life testing, it is most common to use 1 or 2 hour constant current charge and discharge, usually to 100% DOD each cycle. For EV cycle life testing, discharge is usually a variable current/time profile to simulate driving conditions—the so-called DST driving profile. The significance of pulsed discharge cycle life testing is that the high current pulse dominates the test. On the other hand, most EV

cycle life tests are done at 80% DOD and typical NiMH module cycle life is from 600 to 1200 cycles. HEV testing emphasizes power capability even more, and de-emphasizes depth of discharge further still. Typical HEV mode cycle life testing is under a high current pulse profile with a 2 to 10% state of charge variation. Typical NiMH cycle life under these conditions is over 90,000 cycles, which corresponds to nearly 160,000 kilometers in a vehicle.

During early development, failure modes for EV and HEV batteries can include short circuiting due to mechanical penetration through the separator. The frequency of such events is usually small if sound cell and electrode design is employed and if manufacturing quality control is effective. Another failure mode may be abusive overcharge where venting results in insufficient electrolyte within the separator. Abusive overcharge may result from charge imbalances caused by thermal differences from one part of the large battery to another. The problem may be compounded by the sophistication of the charger, where voltage and temperature sensing is not necessarily done on an individual module or cell basis. Another form of abuse is overdischarge, where a cell or module within a high energy EV battery is discharged below the minimum recommended cell voltage of 0.9 V. Overdischarge is usually caused by state-of-charge imbalance within a high voltage string brought on by thermal gradients within a battery. Another source of abusive overcharge and overdischarge is the "weak cell or weak module" concept. This involves the statistical predictability within a large number of cells as to the decay rate of capacity, power and resistance as a function of cycle number.

A common feature of the above-cited EV/HEV failure modes is the importance of maintaining state-of-charge balance within a battery pack that may contain several hundred individual cells. The method used to maintain equalized state-of-charge within an industrial NiMH battery is a complete charge, in effect to routinely bring all the cells to the same state-of-charge. This method of using overcharge to equalize state-of-charge is ineffective if the cell temperature within a battery pack is extreme or if cell-to-cell temperature gradients are too large. Consequently, one of the biggest factors in replicating the excellent cycle life of small portable NiMH batteries in EV and HEV applications is proper thermal management, which is discussed in Sec. 30.11.

If premature failure due to short circuit and abusive overcharge/overdischarge is prevented, the principal failure mode in large industrial EV and HEV NiMH batteries is increasing internal resistance with cycling. To the end EV user, the observation is that acceleration capability will diminish on long-term use, or that vehicle range will gradually decrease. To the end HEV user, battery failure due to increasing internal resistance and resultant power loss will be observed as inability of the battery to assist acceleration and inability of the battery to utilize regenerative braking energy due to excessive heating caused by the high currents used.

The NiMH battery primary failure mode of increasing internal resistance and power loss during cycling is caused by the same failure mechanism as observed in small portable NiMH batteries: separator dryout as a result of electrolyte redistribution due to swelling of the metal hydride and nickel hydroxide electrodes; consumption of electrolyte due to oxidation of the separator metal hydride active material and positive electrode materials, and loss of electrolyte through venting.[35] These mechanisms may be exaggerated for large NiMH batteries due to their prismatic construction. Cylindrical cells have one positive electrode, one negative electrode and one separator. NiMH prismatic EV batteries may have 20 positives, 21 negatives and a corresponding number of separator sheets. The cylindrical design is more effective for pressure containment than a rectangular container, both for gas pressure and the force applied by the can on the electrode stack itself. Therefore, another critical factor for large NiMH batteries is management of the compressive forces within a module. Typically, restraining bands are used to secure a 10 or 11 cell module which has an endplate to equalize lateral forces on the case end. Failure to adequately equalize the compression within each cell in a module and within the internal cell stack itself will lead to premature failure due to unequal electrolyte distribution within a cell.

30.9.6 Shelf Life

Shelf life, also termed calendar life, for large NiMH batteries is from 5 to 10 years based on a variety of factors including temperature, charge equalization, electrolyte compensation, and gas permeation. Over perhaps 6 months to a year of open-circuit storage, NiMH batteries may completely self-discharge. Further open circuit stand may cause the cell voltage to gradually decline to 0 to 0.4 V, which can cause a breakdown of the cobalt conductive network in the positive electrode and/or increased surface oxidation of the metal hydride active material.[36] The length of time under this low voltage condition, the temperature of low voltage storage, and cell design influence the ease and degree of recovery of the battery. For example, a few cycles of low rate charge and discharge may be needed to recover cell capacity and power. If the degree of low voltage degradation is severe, the battery may not be recoverable.

Design factors which the NiMH manufacturer must consider for good shelf life are the oxidation and corrosion resistance of the metal hydride alloy, the amount of precharge on the metal hydride electrode, the nickel hydroxide active material formula, and the quality of the cobalt conductive network in the positive electrode.

Since large industrial NiMH batteries represent a significant financial investment, most users leave the battery on trickle charge if the battery will not be used for extended periods. Alternately, the battery may receive a periodic top-off charge designed to compensate for normal self-discharge capacity losses.

30.9.7 Coulombic/Energy Efficiency

A strength of NiMH technology is high efficiency due to low internal resistance. As shown in Fig. 30.16, over 90% energy efficiency was observed for 60 Ah prismatic NiMH EV batteries at 100 A; over 75% efficiency at 300 A.

Energy efficiency is largely determined by the linear resistance components in the cell, the electronic and ionic resistances, which can be lowered by further engineering improvements. Coulombic efficiency was determined to be 99% at 50% state of charge under an aggressive simulated HEV driving cycle which is a typical operating point for charge-sustaining HEVs. Under the EPA combined city-highway FTP driving schedule, energy efficiency is about 93 to 95%.

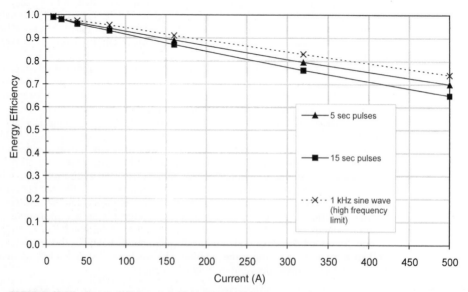

FIGURE 30.16 Energy Efficiency of 60 Ah NiMH HEV battery module as a function of discharge rate.

30.10 CHARGE METHODS

30.10.1 Normal Charge

NiMH batteries are extremely flexible in accepting diverse charge methods. The principal design factor is to prevent excessive overcharge, especially at high rates, in order to avoid heat buildup and electrolyte venting losses. Several methods of sensing overcharge are common, including absolute temperature, ΔT, dT/dt, $-\Delta V$, and pressure rise (see Chap. 29). In all cases, the oxygen evolution/recombination mechanism which creates heat is the basis for each sensing method.

30.10.2 Regenerative Braking

Regenerative braking is a vehicle characteristic which helps NiMH dominate EV and HEV applications. Vehicle range is an area of competition for EV and HEV manufacturers, and a method of extending vehicle range from 5 to 20% is to utilize the energy lost during conventional braking to charge the battery. Regenerative braking energy is available to the battery at extremely high power (~500 W/kg), and many rechargeable battery technologies simply cannot efficiently utilize the energy supplied at such a high power. NiMH batteries are able to accept regenerative braking energy over a wide state of charge range, and over a wide temperature range.

Figure 30.17 shows a plot of voltage-current measurements on the Ovonic NiMH 13-HEV-60 (13 V, 60 Ah) module during an aggressive, simulated HEV driving cycle. In contrast to results of a typical high power lead-acid battery also shown, the V-I slope is relatively constant and of lower slope. The lead-acid battery showed a significantly higher resistance for charge. For the Ovonic 13-HEV-60 module, the effective resistance on charge and discharge were the same, about 6 mΩ. This provides a substantially higher capability for acceptance of regenerative braking energy.

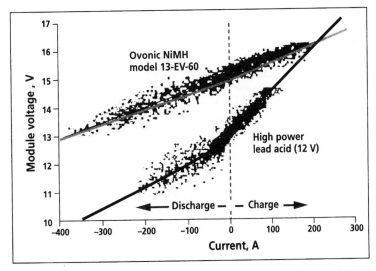

FIGURE 30.17 Regenerative braking charge acceptance comparison of 13 V, 60 Ah, NiMH HEV battery module and high-power lead acid battery.

30.10.3   Charge Algorithms

Charge algorithms refer to the programming used to charge the NiMH EV and HEV battery. The concept is that the battery can accept charge input at extremely high rates until the battery is about 80% fully charged, at which point the charge current must be reduced. In overcharge, the charge current should not exceed the 10-hour charge rate, and generally, the total charge input should be below about 110 to 120%.

One method used to charge NiMH EV batteries reliably utilizes constant current steps to a temperature compensated voltage limit. Practically, this means that charging is done at a high current until a predetermined voltage is reached which is an indication of a certain state of charge. At this voltage, the current is reduced or stepped down, until another predetermined voltage is reached. An important aspect of the charge algorithm is that the prescribed voltage set points are variable, based on temperature and current.

30.10.4   Fast Charge

While the infrastructure to fast charge EV batteries is not fully in place, a strength of NiMH batteries is their ability to accept fast charge if the required power is available. As an example, a typical EV cell capacity may be ~100 Ah. Therefore, if 15 minute charging is desired, currents of ~400 Amps (at 360+ volts) must be available.

NiMH batteries can accept from 60 to 80% charge within 15 minutes, at which point current must be reduced. Temperature rise due to internal resistance heating is small, on the order of 15°C for a 33 kWh battery, whereas heating due to oxygen recombination is very large. Again, the crucial aspect of fast charge is proper sensing of the onset of overcharge.

30.11   THERMAL MANAGEMENT

30.11.1   Cell, Module and Pack Design

A typical 4.0 Ah C size NiMH portable cell has an AC impedance on the order of 8 to 15 mΩ, and a $\Delta V/\Delta I$ resistance of 15 to 30 mΩ. In contrast, a 100 Ah NiMH EV cell has an AC impedance on the order of 0.4 mΩ, and a $\Delta V/\Delta I$ resistance of 0.9 mΩ. Despite the low resistance, heating is still a concern due to the extremely high current pulses resulting from regenerative braking and during vehicle acceleration. Even the I^2R heating effects at these high currents are small in comparison to heating due to overcharge. Consequently, an essential aspect of NiMH EV and HEV battery design involves thermal management.

Proper thermal management begins at the metal hydride negative electrode, with recognition that overcharge heat is generated at the surface of the hydride electrode where oxygen is recombined. Heat must migrate from the negative electrode to the cell case and therefore good thermal conductivity within the electrode and electrode stack bundle is very important. Cells are usually packaged into 12 V modules, with the important concept being the cells are bound together in a back-to-back arrangement. An important design feature is that the cells at each end will have a much higher amount of exposed surface available for convective cooling, and raises the concern of thermal imbalance within a module and resultant state of charge imbalance which will lead to premature failure. Therefore, proper module design must include endplate and heat sink considerations within a module.

Modules are packaged in a variety of configurations. Within the various packaging arrangements, important design considerations are distance between modules, air or water flow channels and, battery tray heat sink characteristics to equalize cooling from module-to-module.

30.11.2 Water Cooling Versus Air Cooling

A key first step in NiMH EV and HEV thermal management is a decision on whether to use water cooling or air cooling, with no clear cut consensus yet available on which approach is preferable. Present electric vehicles using air cooled batteries include the General Motors' EV1 and S-10, the Toyota RAV4 and the Ford Ranger. Present electric vehicles using water cooled batteries include the Honda EV Plus and the Daimler-Chrysler EPIC minivan. Merits of each approach include:

	Air cooling	Water cooling
Advantages	• Light weight • Simple	• More effective heat transfer • Average fluid temperature more consistent • Integrated with vehicle cooling
Disadvantages	• Complicated distribution of air within pack • Incoming air must be free of dirt and water from road • Variable ambient temperature	• Increased weight • Elaborate module design • Higher average fluid temperature

The NiMH battery manufacturer prefers metal cases if air-cooling is to be used for cooling, and prefers plastic cell cases if water cooling is to be used.

30.12 ELECTRICAL ISOLATION

Isolation of the EV and HEV battery from the vehicle is essential due to the high voltages involved. Isolation begins at the cell level, where plastic cell cases are preferred from an isolation perspective, and metal cased cells must have some kind of insulative coating which must be stable and free of pinholes.

The battery module (usually 12 V) must also be isolated from the battery tray; this is usually accomplished with plastic isolation mounts. It is important that the cell and module interconnecting straps also be isolated from the battery tray.

In air cooled battery packs, pack-to-vehicle resistance is strongly influenced by road contaminants such as dirt and salt, and pack case design must take this into consideration. For water cooled designs, the issue is the resistance of the plastic casing materials to the coolant (e.g. ethylene glycol). There is a strong dependence of the plastic resistance to temperature where a pack isolation resistance of 500 MΩ at 20°C may be reduced to 5 MΩ at 65°C. A figure of merit for HEV and EV battery pack isolation is a resistance of 1 to 10 MΩ.

30.13   DEVELOPMENT TARGETS

Despite the huge commercial production of portable NiMH batteries, the technology is still in its infancy.[37,38] Since the initial introduction of portable NiMH in 1987, great strides have been made in terms of specific energy, specific power, cycle life, charge retention, and cell sizes. Beginning with only small low-power cylindrical cells, NiMH is now available in high-power cylindrical cells and large high-power prismatic designs. Initially only metal case cells were available but NiMH is now produced in both metal and plastic prismatic cells and even in plastic monoblock configuration.

Intensive development of NiMH battery technology is underway in many places world-wide. Widespread activity has been reported on magnesium-based metal hydride alloys for capacity and cost advantages, evaluation of bipolar NiMH designs, satellite NiMH battery development, and a myriad of technical goals too lengthy to list here. However, the most intensive NiMH development at this time involves efforts to raise specific power to new levels and to reduce cost.

30.13.1   Cost Reduction

Cost reduction is at the forefront of NiMH development. High volume portable battery production has seen NiMH cost equal to or less than the $/Wh cost of NiCd, previously thought to be unobtainable. Key achievements in NiMH which have allowed this significant cost reduction include:

- Nickel hydroxide cost reduction from $30 per kg to under $10 per kg
- Nickel foam substrate cost reduction of ~50%
- Increased cell capacity by ~50% with virtually no additional cost
- Replacement of pure nickel wire mesh metal hydride electrode substrates with copper and nickel-plated steel on expanded metal and perforated sheet
- Elimination of metal hydride electrode sintering
- Improved NiMH battery activation and formation

Despite these dramatic achievements, EV and HEV applications demand even further cost reduction. Initial NiMH EV battery cost was about $1000 per kWh at a prototype basis, and has steadily progressed to the $250 to 400 per kWh range (at projected production volumes of about 7,000 to 20,000 vehicles per year). At the $250 per kWh level, and with a 30 kWh battery as an example, the vehicle battery cost is $7500. Most vehicle manufacturers agree that the cost of an internal combustion engine vehicle is about the same as a comparable electric version, not including the battery cost. While the operating costs of an electric car are less than half that of an ICE vehicle (comparing the cost of electricity vs. gasoline), the upfront costs are prohibitive. This paradigm has resulted in the emphasis on hybrid electric vehicles where the battery may be only 10 to 25% of the EV battery size, and the tremendous development push for reducing NiMH cost to $150 per kWh, a target set by USABC.

NiMH cost reduction begins with the recognition that the technology cost is primarily dictated by the materials cost. Important development activities to meet this goal include:

- Raising NiMH EV specific energy range from 63 to 75 Wh/kg to 95 Wh/kg
- Replacement of expensive inactive components such as nickel foam substrate and cobalt additives
- Use of inexpensive plastic cell cases, and
- Monoblock construction to reduce the number of parts

Effort to raise specific energy involves development of metal hydride alloys with higher hydrogen storage capacity (from 320 to 385 mAh/g active material to 450 mAh/g) and higher utilization nickel hydroxide (from 240 mAh/g active material to 280 to 300 mAh/g).[39,40] Each of these higher utilization active materials involves innovative materials research involving modified alloy formulas and advanced processing techniques.

Reduction of expensive inactive cell components is focused on the positive electrode, including the nickel foam substrate and cobalt metal and cobalt monoxide used to form the conductive network. Several approaches are being studied: replacement of the cobalt compounds by metallic nickel fibers; use of reduced quantity and less expensive cobalt compounds. An innovative approach being studied is the use of inherently more conductive nickel hydroxide accomplished by multielement modification, and by heterogeneous nickel hydroxide powder particles where filamentary metallic fibers have been embedded into the base nickel hydroxide to make contact with the overall conductive network. To reduce the cost of the foam substrate, development activities include less expensive nickel coating processes and elimination of the foam entirely by enhancing the conductivity of the nickel hydroxide and the conductive network.

Cell construction cost reduction involves development of novel plastic monoblock housings; reduction of piece parts through monoblock construction, low cost plastic materials where possible, shared terminals and vents and standardized sizes.

30.13.2 Ultra-high Power Designs

For many years, it was commonly believed that NiMH would never be able to replace NiCd in portable battery applications where extremely high rates of discharge were required. In particular, power tools such as cordless drills require almost continuous discharge capability at the 10 C rate. Today's NiMH cylindrical cells have even exceeded the power of NiCd. Achievement of such high rate discharge capability has been accomplished through application of low resistance current collectors as in NiCd technology and improved surface catalytic activity of metal hydride materials. High power 3.5 to 7.0 Ah NiMH cylindrical cells for power tools and HEV application have an AC impedance of about 1.7 mΩ and a $\Delta V/\Delta I$ internal resistance of about 4 mΩ. NiMH products in a variety of sizes (most commonly Cs, C, and D) are being employed in portable power tools, scooters and electric bicycle applications.

While NiMH for HEV application already demonstrates specific power in excess of 500 W/kg, there is a worldwide development effort to achieve 1000 W/kg. NiMH cylindrical and prismatic prototype batteries have been announced with room temperature peak power in excess of 700 W/kg and power over 800 W/kg at 35°C. In some cases, power has been increased by the sacrifice of energy to about 44 Wh/kg while in other designs inherently higher power materials and construction have allowed extremely high power without energy tradeoffs. The concept is to reduce battery cost further by recognizing that for power assist HEVs, the energy of the battery is not crucial but rather the size of the battery is set to provide a predetermined power. Battery cost is primarily determined by the amount of the battery materials, and higher power utilization materials reduces the amount of material used.

A crucial factor in high power NiMH batteries has been the development of highly catalytic metal hydride active materials. In particular, the interface between the metal hydride itself and the electrolyte has been identified as essential for low voltage loss under pulse discharge. The surface oxide thickness and microporosity influence the reaction of H$^+$ and OH$^-$, and a critical observation has been made that within the high power surface oxide metallic catalysts of enriched nickel having sizes less than 70 Å are especially important for reducing activation polarization.[41,42]

30.13.3 Other Design Options

In another approach to larger NiMH batteries, a quasi-bipolar design is being developed for a variety of applications.[43] In a true bipolar design, each electrode is made with a positive and negative surface mounted back-to-back on a conductive substrate which serves as an intercell conductor. A separator is placed between each electrode and the bipolar plates are sealed with an insulating material to form the cell stack. The advantage of the bipolar design is a shorter, more uniform current path which reduces the intercell resistance and improves the power capability of the battery. In the quasi-bipolar battery design under development, individual cells are fabricated using one positive and one negative electrode, both on a metallic conductive substrate with a separator between them. An insulating perimeter seal is employed around the edge of each cell to contain the electrolyte and gasses which form during operation. Each cell is fabricated as a discrete element designed for the particular application. The cells are stacked with positive to negative faces in contact to construct multi-cell batteries of the required voltage. The battery stack is held in compression by end plates and the battery housing. The end plates also contain positive and negative terminals for the battery. Conductive heat transfer plates can also be placed within the cell stack to improve thermal management. Electrodes are formed by rolling plastic-bonded mixes of active materials on the conductive substrates. No screens or grids are employed in the electrodes. Figure 30.18 shows this stacked cell concept.

Batteries are being developed for a variety of applications including Low Earth Orbit (LEO) Satellite Batteries, Military Communications Batteries, Aircraft and Heart-Assist Pump Batteries. Table 30.5 shows the requirement for the LEO Satellite battery shown in Fig. 30.19. Charge-discharge cycles for a single-cell battery at different points in the cycle life are shown in Fig. 30.20 at 40% depth of discharge. Five thousand cycles have been demonstrated in test cells at 40% DOD while over ten thousand cycles have been obtained at 25% DOD. Batteries are also being developed for the other applications described above.

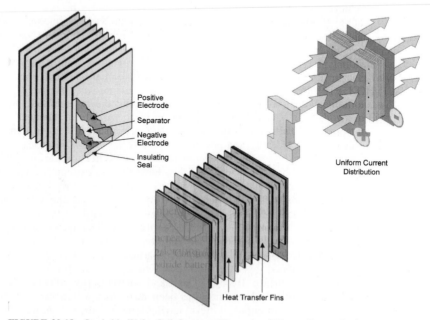

Positive Electrode
Separator
Negative Electrode
Insulating Seal

Uniform Current Distribution

Heat Transfer Fins

FIGURE 30.18   Stackable Wafer Cell Concept. (*Courtesy of Electro-Energy Inc.*)

TABLE 30.5 Satellite Battery Requirements

Voltage	28 V
Current	22.7 A Charge, 35.7 A Discharge
Power	1 kiloWatt
Battery Capacity	52 Ahr (20.8 Ahr at 40% DoD)
Energy Density Goal	100 Wh/kg
Cycle Life Goal	30,000
Operating Temperature	$-10°C$ to $+25°C$
Overcharge	5%
Electrical Insulation Resistance	>1 megOhm @ 100 VDC
Cooling	Conductive baseplate
Maximum Temperature Change	25°C above cold plate

FIGURE 30.19   28 V, 52 Ah NiMH LEO Satellite Battery. (*Courtesy Electro-Energy Inc.*)

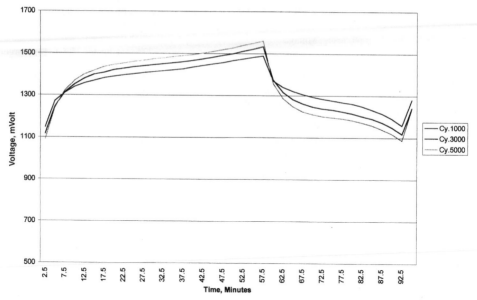

FIGURE 30.20 Cycle life of a single sealed 1.6 Ah single-cell battery operated at 40% DOD, 55 Minutes Charge at 0.72 Amp. and 35 minute discharge at 1.1 Amp. (*Courtesy Electro-Energy Inc.*)

REFERENCES

1. S. R. Ovshinsky, S. K. Dhar, M. A. Fetcenko, K. Young, B. Reichman, C. Fierro, J. Koch, F. Martin, W. Mays, B. Sommers, T. Ouchi, A. Zallen and R. Young, *17th International Seminar & Exhibit on Primary and Secondary Batteries,* Ft. Lauderdale, Florida, March 6–9, 2000.

2. R. C. Stempel, S. R. Ovshinsky, P. R. Gifford, D. A. Corrigan, *IEEE Spectrum,* Volume 35, Number 11, November, 1998.

3. G. Halpert, *Proceedings of the Symposium on Nickel Hydroxide Electrodes,* Electrochemical Society, Hollywood, FL, October 1989 (Electrochemical Society, Pennington, NJ, 1990), pp. 3–17.

4. B. Reichman, S. Venkatesan, M. A. Fetcenko, S. R. Ovshinsky, *United States Patent 5,851,698* (1998).

5. F. J. Kruger, *15th International Seminar on Primary and Secondary Batteries,* Ft. Lauderdale, Florida, (March 1998).

6. D. A. Corrigan, S. Venkatesan, P. Gifford, A. Holland, M. A. Fetcenko, S. K. Dhar, and S. R. Ovshinsky, *Proceedings of the 14th International Electric Vehicle Symposium,* Orlando, Florida, 1997.

7. V. Puglisi, *17th International Seminar & Exhibit on Primary and Secondary Batteries,* Ft. Lauderdale, Florida, March 6–9, 2000.

8. V. Ettel, J. Ambrose, K. Cushnie, J. A. E. Bell, V. Paserin, P. J. Kalal, *United States Patent 5,700,363* (1997).

9. S. R. Ovshinsky, *Materials Research Society,* MRS Fall Meeting, Boston, MA (November 1998).

10. K. Sapru, B. Reichman, A. Reger, S. R. Ovshinsky, *United States Patent 4,623,597* (1986).

11. S. R. Ovshinsky, M. Fetcenko, and J. Ross, *Science,* 260, 176 (1993).

12. S. R. Ovshinsky in *"Disordered Materials: Science and Technology,"* D. Adler, B. Schwartz, and M. Silver, Eds., Institute for Amorphous Studies Series, Plenum Publishing Corporation, New York, 1991.

13. J. R. van Beek, H. C. Donkersloot, J. J. G. Willems, *Proceedings of the 14th International Power Sources Symposium,* 1984.

14. R. Kirchheim, F. Sommer, G. Schluckebier, *Acta metall.* Vol. 30, pp. 1059-1068 (1982).

15. R. Mishima, H. Miyamura, T. Sakai, N. Kuriyama, H. Ishikawa, I. Uehara, *Journal of Alloys and Compounds,* 192 (1993).

16. T. Weizhong, S. Guangfei, *Journal of Alloy and Compounds,* Volume 203, pp. 195–198 (1994).

17. P. H. L. Notten, P. Hokkeling, *Journal of the Electrochemical Society,* Volume 138, No. 7, July 1991.

18. P. H. L. Notten, J. L. C. Daams, R. E. F. Einerhand, *Ber. Bunsenges. Phys. Chem.* (96) No. 5, (1992).

19. K. Beccu, *United States Patent 3,669,745* (1972).

20. M. H. J. van Rijswick, *Proceedings of the International Symposium on Hydrides for Energy Storage* (Pergamon, Oxford, 1978), p. 261.

21. M. A. Gutjahr, H. Buchner, K. D. Beccu, H. Saufferer, *Power Sources 4,* D. H. Collins, Ed. (Oriel, Newcastle upon Tyne, United Kingdom, 1973), p. 79.

22. M. A. Fetcenko, S. R. Ovshinsky, B. Chao, B. Reichman, *United States Patent 5,536,591* (1996).

23. M. Oshitani, H. Yufu, K. Takashima, S. Tsuji, Y. Matsumaru, *Journal of the Electrochemical Society,* Volume 136, No. 6, June 1989.

24. M. Oshitani, H. Yufu, *United States Patent 4,844,999* (1989).

25. K. Ohta, H. Matsuda, M. Ikoma, N. Morishita, Y. Toyoguchi, *United States Patent 5,571,636* (1996).

26. I. Matsumoto, H. Ogawa, T. Iwaki, M. Ikeyama, *16th International Power Sources Symposium,* 1988.

27. S. Takagi, T. Minohara, *Society of Automotive Engineers,* 2000-01-1060, March 2000.

28. K. Watanabe, M. Koseki, N. Kumagai, *Journal of Power Sources,* 58 (1996), pp. 23–28.

29. I. Kanagawa, *15th International Seminar on Primary and Secondary Batteries,* Ft. Lauderdale, Florida, (March 1998).

30. M. A. Fetcenko, S. Venkatesan, S. Ovshinsky, in *Proceedings of the Symposium on Hydrogen Storage Materials, Batteries, and Electrochemistry* (Electrochemical Society, Pennington, NJ, 1992), p. 141.

31. J. Cook, *"Separator—Hidden Talent,"* Electric & Hybrid Vehicle Technology, 1999.

32. R. Elder, R. Moy, M. Mohammed, *16th International Seminar on Primary and Secondary Batteries,* Ft. Lauderdale, Florida, March 1999.

33. J. R. Fetner, B. A. MacRae, and C. Jordy, *Proceedings of the 39th Power Sources Conf.,* pp. 270–273 (2000).

34. P. Leblanc, P. Blanchard, S. Senyarich, *Journal of the Electrochemical Society,* Volume 145, No. 3, March 1998.

35. M. A. Fetcenko, S. Venkatesan, K. C. Hong, B. Reichman, in *Proceedings of the 16th International Power Sources Symposium* (International Power Sources Committee, Surrey, United Kingdom, 1988). p. 411.

36. D. Singh, T. Wu, M. Wendling, P. Bendale, J. Ware, D. Ritter, L. Zhang, *Materials Research Society Proceedings,* Volume 496, pp. 25–36 (1998).

37. T. Doan, *"Nickel Metal Hydride for Power Tools,"* *16th International Seminar on Primary and Secondary Batteries,* Ft. Lauderdale, Florida, March 1999.

38. K. Ishiwa, T. Ito, K. Miyamoto, K. Takano, S. Suzuki, "Evolution and Extension of NiMH Technology," *16th International Seminar on Primary and Secondary Batteries,* Ft. Lauderdale, Florida, March 1999.

39. D. A. Corrigan, S. K. Knight, *Journal of the Electrochemical Society,* Volume 143, No. 5, May 1996.

40. S. R. Ovshinsky, D. A. Corrigan, S. Venkatesan, R. Young, C. Fierro, and M. Fetcenko, *United States Patent 5,348,822,* April 14, 1994.

41. B. Reichman, W. Mays, M. A. Fetcenko, S. R. Ovshinsky, *Electrochemical Society Proceedings,* Volume 97–16, October 1999.

42. K. Young, M. A. Fetcenko, B. Reichman, W. Mays, S. R. Ovshinsky, *Proceedings of the 197th Electrochemical Society Meeting,* May 2000.

43. M. Klein, M. Eskra, R. Pivelich, and P. Ralston, *Proceedings of the 39th Power Sources Conf.,* pp. 274–277 (2000).

CHAPTER 31
NICKEL-ZINC BATTERIES

Dwaine Coates and Allen Charkey

31.1 GENERAL CHARACTERISTICS

The nickel-zinc (zinc/nickel-oxide) battery is an alkaline rechargeable system. It is a combination of the nickel electrode, as used in other batteries such as nickel-cadmium, nickel-iron and nickel-metal hydride, and the zinc electrode, which is similar to that used in the silver-zinc battery system. Currently the nickel-zinc system is capable of delivering about 50 to 60 Watt-hours per kilogram (Wh/kg) and 80 to 120 Watt-hours per liter (Wh/L) depending on the specific design. Nickel-zinc batteries are capable of delivering more than 500 cycles at 100% depth-of-discharge (DOD) and up to several thousand cycles at low DOD. The advantages of the nickel-zinc battery include good specific energy, good cycle capability, abundant low-cost materials and an environmentally acceptable chemistry. The nickel-zinc battery is appropriate for a number of commercial applications including electric bicycles, electric scooters, electric lawn and garden equipment and deep cycle marine applications.

The generalized advantages and disadvantages of nickel-zinc batteries are shown in Table 31.1.

TABLE 31.1 Advantages and Disadvantages of Nickel-Zinc Batteries

Advantages	Disadvantages
Fast recharge capability	Relatively low volumetric energy density
Sealed maintenance-free design	Higher cost than lead-acid
Good specific energy	
Good cycle life	
Low environmental impact	
Abundant raw materials	
Relatively low cost per Watt-hour	

31.1.1 Background

The goal of the initial development of the nickel-zinc battery was to combine the long cycle life associated with nickel-cadmium batteries with the high specific energy of the zinc electrode. Historically, the nickel-zinc battery dates back at least to 1901 in a Russian patent by Michaelowski. Further work was performed by Drumm in Ireland in the 1930s. There was a serious effort in the 1960s to develop nickel-zinc as a longer-life replacement for the silver-zinc battery in military applications. Considerable effort was again focused on nickel-zinc in the 1970s in response to the increased interest in electric vehicles that resulted from an energy crisis and increasing gasoline prices. System development was hampered for many years by the limited cycle life associated with the zinc electrode. This cycle life limitation was primarily due to the solubility of zinc in the alkaline electrolyte. This has been overcome by recent developments in stabilizing the zinc electrode in alkaline electrolyte and thereby reducing its solubility and increasing the cycle life of the battery. The improved cycle life now available has again increased interest in the nickel-zinc battery for a wide variety of consumer applications. There is currently an extensive patent and scientific literature base available concerning the nickel-zinc battery system.[1–3]

31.2 CHEMISTRY

The nickel-zinc battery system uses the nickel/nickel oxide electrode (also known as the nickel-hydroxide/nickel oxyhydroxide electrode) as the positive and the zinc/zinc oxide electrode as the negative. When the battery is discharged, nickel(III) oxyhydroxide is reduced to nickel(II) hydroxide and metallic zinc (0) is oxidized to zinc(II) oxide/hydroxide. The electrochemistry of zinc in alkaline solution is actually quite complex so the reactions presented here are for the purpose of illustration only. The theoretical open-circuit voltage of this electrochemical couple is 1.73 V. When the battery is overcharged, oxygen is produced at the nickel electrode and hydrogen is produced at the zinc electrode. These gases may then recombine to form water. In addition, oxygen produced at the nickel electrode during overcharge may recombine with metallic zinc directly at the zinc electrode. If the battery is overdischarged, hydrogen is formed at the nickel electrode and oxygen may be produced at the zinc electrode. In practical batteries, these reactions are influenced by the balance of active materials present and the active material utilization of the two electrodes. The simplified, representative electrochemical reactions are as follows:

		$E°$
Discharge		
Positive electrode	$2\ NiOOH + 2\ H_2O + 2\ e^- \rightarrow 2\ Ni(OH)_2 + 2\ OH^-$	0.49 V
Negative electrode	$Zn + 2\ OH^- \rightarrow Zn(OH)_2 + 2\ e^-$	1.24 V
Overall reaction	$2\ NiOOH + 2\ H_2O + Zn \rightarrow 2\ Ni(OH)_2 + Zn(OH)_2$	1.73 V

Charge
Overall Reaction $\quad 2\ Ni(OH)_2 + Zn(OH)_2 \rightarrow 2\ NiOOH + 2\ H_2O + Zn$

Overcharge
Positive electrode $\quad 2\ OH^- \rightarrow \frac{1}{2}\ O_2 + H_2O + 2\ e^-$
Negative electrode $\quad 2\ H_2O + 2\ e^- \rightarrow H_2 + 2\ OH^-$
$\qquad\qquad\qquad\ \ Zn + \frac{1}{2}\ O_2 \rightarrow ZnO \quad$ (*Recombination of oxygen from the positive electrode*)
Overall reaction $\quad\ \ H_2O \rightarrow H_2 + \frac{1}{2}\ O_2$

Nickel hydroxide active material has a theoretical specific energy of 289 milliAmpere-hours per gram (mAh/g) and zinc oxide has a theoretical specific energy of 659 mAh/g. Overall, the nickel-zinc battery has a theoretical specific energy of 334 Wh/kg, which makes it a very attractive system for a number of applications. Practical batteries operate at a loaded discharge voltage of 1.55 to 1.65 V and may deliver up to about 70 Wh/kg, depending on the specific design. This is only about 20% of the theoretical specific energy, which indicates that there is room for system improvement.

31.3 CELL COMPONENTS

Nickel-zinc batteries can be constructed in a variety of configurations such as prismatic or cylindrical cell designs. The specific cell or battery design will dictate the cell component configurations used. However, all cell designs have basic similarities in terms of the active materials used, the use of cell containers and covers, the use of metallic conductors and in the basic manufacturing techniques and application of these materials.

31.3.1 Nickel Electrode

The electrochemically active material of the nickel electrode is nickel hydroxide. This material is an amorphous colloid and is only semiconductive at best. It must be supported and contained by a structural component which provides mechanical support, conductivity and current collection for the electrode. Standard types of nickel electrodes can be used in the nickel-zinc system. They can be classified by the type of electrode substrate used and by the method of preparation. These electrodes consist of two basic types, sintered and nonsintered. Each type has different advantages and disadvantages and may be selected based on the application. Other types of nickel electrodes, such as pocket plate, are generally not in common use.

A particular situation with the nickel-zinc battery, which is not present in other nickel batteries, is the effect of the zinc electrode on the nickel electrode.[4] The observed effect is that there is essentially an irreversible initial capacity loss due to decreased capacity in the nickel electrode. There are several possible causes for this effect but the most likely seems to be pore plugging of the nickel electrode by zinc which is deposited from the electrolyte. The effect appears in the first few cycles of the battery and then does not seem to further influence the performance or cycle life of the battery. This must be taken into account in the design of the battery. The effect is apparently independent of the calcium hydroxide level in the calcium containing zinc electrode but is influenced by the electrolyte formulation and concentration.

A similar study showed that the effect of the zinc electrode on the nickel electrode was primarily mechanical.[5] Previous speculation of the effect of nickel-hydroxyzincates or nickel zinc double hydroxides on capacity degradation in the nickel electrode were not found to be significant. However, significant increases in the resistivity of cycled nickel electrodes were observed which corresponded with mechanical damage observed by optical microscopy.

Further work showed that the performance of the nickel electrode is degraded in the nickel-zinc battery because of zinc occlusion in the pores of the nickel electrode and the reduced electrolyte concentration typically used.[6] Deformations in the electric field were also investigated as a possible mechanism for the observed degradation in performance.

Pocket Plate Electrode. This is the same type of electrode used in pocket plate nickel-cadmium and nickel-iron batteries.[2] Electrodes are prepared by loading nickel hydroxide hydrate active material and a conductive additive (graphite and/or nickel flake) into tubular flat pockets which are then assembled into electrodes. Little interest currently exists in using this type of electrode in nickel-zinc cells since modern cells attempt to utilize lightweight electrode construction.

Sintered Nickel Electrodes.   Until recently, the sintered electrode was the most common type of nickel electrode used in commercial nickel-cadmium batteries. It has been widely replaced in cylindrical cells by the pasted type electrode, discussed below. The electrode substrate is prepared by sintering high surface area carbonyl nickel powder into a porous structure. Sintering involves heating the nickel powder to just below its melting point so that the individual particles are fused together. This is done under a reducing atmosphere to avoid oxidation of the metallic nickel. This electrode structure is typically 80% to 84% porous (known as the nickel plaque electrode) before the introduction of the nickel hydroxide active material. The electrode substrate also typically contains a current collector support such as nickel wire mesh, perforated nickel or nickel-plated steel. The nickel hydroxide active material can be introduced into this porous structure in a number of ways. The most common is chemical impregnation. The porous substrate is alternately dipped in a bath of nickel nitrate and a separate bath of potassium hydroxide, with the resultant chemical precipitation of nickel hydroxide into the structure. This method is cumbersome in that as many as 8 to 10 cycles may be required to achieve adequate active material loading. A refinement of this process involved dipping the substrate into the nickel nitrate bath under a vacuum. This was typically done for high-performance aerospace cells but still required multiple cycles. Sintered electrodes are capable of high energy density and long cycle life but are relatively expensive to produce because of the high nickel content and the expensive sintering process.

A further refinement involved the electrochemical deposition of nickel hydroxide into the porous electrode substrate directly from the nitrate bath. This is accomplished by cathodically polarizing the nickel electrode substrate such that a local change in pH inside the substrate pores causes the deposition of the nickel-hydroxide active material. This improvement allows the impregnation process to be completed in a single step. An important advantage of this process is that the electrodes are impregnated with active material from the inside out, resulting in a much more uniform cross-sectional loading. The resultant electrodes display better performance and active material utilization than other impregnation methods. This process is typically used only in aerospace nickel-cadmium and nickel-hydrogen cells and has not generally been adopted in commercial batteries. Other types of impregnation are possible, such as the thermal decomposition of nickel salts, but these have also not found extensive commercial application.

The primary disadvantage of sintered nickel electrodes is that they contain a relatively low ratio of active materials to inactive materials compared to some types of non-sintered electrodes. Sintered electrodes typically contain as much as 60% inactive material, in the form of the porous electrode substrate and/or current collector. Practical sintered nickel electrodes yield about 100 to 120 mAh/g, depending on the method of preparation and the active material utilization. This is only about 40% of the theoretical specific energy of the active material. Also, the sintering process itself is relatively expensive and must be done at high temperatures under a reducing atmosphere.

Non-sintered Nickel Electrodes.   Two types of nonsintered nickel electrodes have been developed for commercial applications. The most common type is the pasted nickel electrode. In this method, the nickel-hydroxide active material formulation is mechanically introduced into a porous nickel substrate. The substrate can be nonwoven nickel fiber or a variety of other materials, but nickel foam is the most commonly used. Nickel foam is produced by depositing metallic nickel on a porous plastic foam and then burning away the polymer in an oven, leaving only the nickel metal. Electrode substrates can be produced in this manner with a porosity as high as 95%. Many small form-factor cylindrical nickel-cadmium and nickel-metal hydride cells use a pasted nickel foam electrode. This type of electrode is generally not used in prismatic flooded cells due to the tendency for the electrode to extrude the active material out of the structure during cycling. The pasted type of electrode suffers from relatively poor active material utilization, a tendency towards significant swelling during cycling and consequently relatively poor cycle life. The advantage is that it is relatively cheap and efficient to produce in large quantities.

A second type of nonsintered nickel electrode which has been developed for commercial applications is the plastic-bonded electrode. An example of this process is shown in Fig. 31.1. This electrode has been developed primarily for the nickel-zinc battery and optimized for lightweight and low cost. In this process, the active materials are blended together with a polytetrafluoroethylene (PTFE) binder and then roll-bonded into a porous three-dimensional structure by fibrillating the binder. PTFE has the unique property of forming nanometer scale fibers through a combination of heat, compression, shear and mechanically working the material. By properly controlling the process, the resultant orientation of the nanofibers and thus the resultant electrode structure can be controlled. The fibrillated PTFE forms an elastic three-dimensional network which locks the active materials into the structure. This unique electrode structure is shown in Figure 31.2. This electrode structure overcomes the major disadvantage of the pasted foam electrode, which is shedding the active material from the electrode structure during cycling and a resultant reduction in cycle life. The active material is essentially locked into place by the nano-fibrillated structure. The composite electrode active material matrix is supported in the electrode by a very thin (typically less than 0.05 millimeter) metallic substrate. This provides mechanical support and current collection and also provides a point of electrical contact.

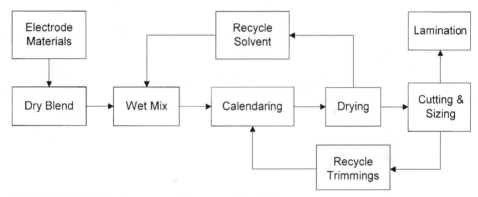

FIGURE 31.1   Roll-bonding process. (*Courtesy of Evercel Corp.*)

The plastic-bonded nickel positive uses a graphite composite structure which replaces the metallic nickel, used in standard sintered and foam electrodes, with high surface area carbon. This results in an electrode with a much lower metallic content, with a savings in cost and weight. Graphite corrosion in the highly oxidizing cell environment is controlled by treating the carbon with a cobalt spinel.[7] The increased impedance of the graphite composite electrode is partially offset by the increased surface area of carbon as compared to nickel powder or foam. The conductivity of the carbon is also enhanced by the cobalt treatment. The increased surface area provides greater interfacial contact between the active material and the carbon support and between the active material composite structure and the electrolyte. The electrode is able to support current densities suitable for high-rate battery applications. Since graphite has a much lower density than nickel metal, the plastic bonded electrode is much lighter in weight than standard nickel electrodes for an equivalent energy storage capacity. Plastic-bonded graphite composite nickel electrodes yield about 140 to 150 mAh/g, depending on the formulation, use of additives, the method of preparation and the active material utilization. By the same token, since graphite has a lower density it also occupies a larger volume for the same weight as metallic nickel. This results in a somewhat lower energy density for the electrode (milliAmpere-hours per cubic centimeter).

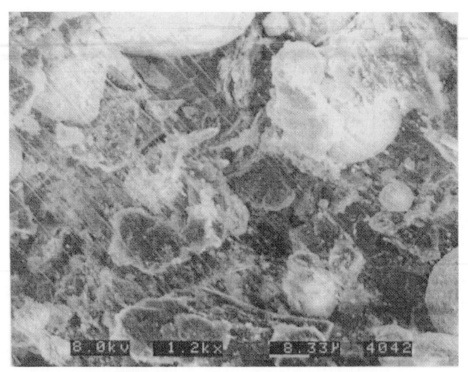

FIGURE 31.2 Scanning electron microphotograph of roll-bonded nickel electrode material. (*Courtesy of Evercel Corp.*)

31.3.2 Zinc Electrode

Traditionally, zinc electrodes have been prepared by a variety of means including pressed powders, electrodeposition and pasting methods. A thorough review of the state of zinc electrode technology has been previously published.[2] Zinc electrodes can be prepared in either the charged or discharged state depending on whether the starting material is metallic zinc (charged) or zinc oxide (discharged). Typically charged zinc electrodes are used only in primary batteries while zinc electrodes for secondary batteries are usually manufactured in the discharged state. Like the nickel electrode, the zinc electrode contains a mechanical support/current collector. This is usually copper or plated copper and can be perforated foil, expanded metal, wire mesh or other similar materials.

Zinc oxide electrodes, which are manufactured in the discharged state, have a relatively high initial impedance due to the semiconductive nature of the oxide. During cell formation, a significant quantity of the oxide is reduced to metallic zinc. This greatly improves electrode performance because of the increased conductivity imparted by zinc metal. Zinc electrodes may take as many as 50 cycles to become fully formed, but performance is adequate after normal battery formation of 3 or 4 cycles. The zinc electrode will continue to improve in performance over the initial usage cycle as the metallic zinc content of the electrode is gradually increased. Alternatively, some metallic zinc can be added to the electrode during manufacture to offset the initial high impedance.

The electrochemistry of the zinc electrode in alkaline solution is actually quite complex. Zinc commonly has a coordination number of 4, such as in zinc oxide, or of 6, as in some crystals. The electronic configuration of zinc makes the +2 oxidation state the most stable, although references are occasionally made to monovalent zinc. Zinc has a tendency towards dissolution in the electrolyte and the subsequent formation of a variety of solvated complex ions. The species involved in zinc electrochemistry are typically oxides and hydroxides, and the products of their interactions with the aqueous alkaline electrolyte.

Zinc is partially soluble in the alkaline electrolyte. Herein lies the basic problem with the zinc electrode, which is manifested as electrode shape change and the formation of dendrites. Redistribution of zinc on the electrode is commonly referred to as "shape change." During discharge, zinc is oxidized from metallic zinc to zinc (II) and dissolves in the electrolyte in the form of zincate anions (e.g. $Zn(OH)_4^{2-}$). During charging, zinc ions in the electrolyte are reduced back to metallic zinc and are replated back onto the electrode. The zinc does not necessarily replate onto the electrode in the same place from which it came. Therefore the "shape" of the electrode changes. Typically, the zinc tends to migrate to the lower one-third of the electrode which becomes much more dense. This densification reduces the electrochemical activity of the zinc and may lead to cell failure through electrical shorting. An associated phenomenon is zinc dendrite formation. If the surface of the zinc electrode is not uniform, zinc may preferentially replate onto the "high" spots. The zinc may gradually build up until the separator is penetrated and a short-circuit occurs. Dendrites will form under diffusion controlled deposition conditions.

Zinc electrodes can also be manufactured by the plastic-bonded method, similar to that of the nickel electrode described above. The zinc oxide dry powder, PTFE binder and other additives are blended with an organic solvent and then the mixture is passed through a calendaring process similar to that in Fig. 31.1. The PTFE fibrillates into a nano-structured three-dimensional fiber matrix. This electrode structure for the calcium zincate electrode is shown in Fig. 31.3 for a freshly prepared electrode. The active materials are locked into the electrode structural matrix which helps to reduce the tendency towards shape change and dendritic growth.

FIGURE 31.3 Scanning electron microphotograph of fresh roll-bonded zinc electrode. (*Courtesy of Evercel Corp.*)

Zinc Electrode Additives.   In addition to its partial solubility in the alkaline electrolyte, metallic zinc has a tendency to oxidize (corrode), evolving hydrogen in the process. This is undesirable in a sealed battery for a number of reasons. Many additives have been investigated to reduce hydrogen evolution and to improve zinc electrode cycle life performance.[8] The traditional mercury amalgamation zinc electrode process has been largely abandoned because of the environmental problems associated with mercury. A wide variety of substitutes has been investigated. These include various mixed oxides and hydroxides of lead, bismuth, indium, thallium, tin and antimony in addition to fluorides, carbonates, borates and others. Currently, lead provides the best observed performance in sealed batteries. Even though lead is present in a relatively small amount in the nickel-zinc battery, it is still an environmentally objectionable material and efforts are underway to reduce or completely eliminate lead from the battery.

The average zinc electrode active material utilization was found to be about 60% of the theoretical capacity.[9] The reason for this relatively low electrochemical utilization has not been reported in the literature but may be related to the solubility and complex chemistry of zinc in alkaline solution. The utilization obtained will be a function of the electrode design and composition and will be affected by the use of any additives either to the electrode or the electrolyte. The utilization of the zinc electrode may also vary for a specific cell design and must be taken into account in the electrochemical design and active material balance of the cell.

Calcium-Zincate Electrode.   The use of calcium as an additive to the zinc electrode dates to at least 1962.[10] Studies have been done on the formation of calcium zincate during the corrosion of zinc in cement. The calcium zincate complex was found to improve the strength of concrete. Calcium zincate materials are also used in a variety of industrial process and materials such as dielectrics, catalysis and glass. A thorough discussion of calcium zincate electrode theory and operation has been published.[4] This work found that zinc electrode shape change was dramatically reduced by the addition of calcium to the zinc electrode. It was further determined that the optimum concentration of calcium is about 25% by weight, which is near the stoichiometric ratio depending on the assumed structure of the zincate complex. It was also found that the rate of decomposition of the zincate complex is relatively fast but that the rate of formation can be slow, and perhaps even rate limiting in some cases. The rate of formation of calcium zincate is favored by lower concentrations of potassium hydroxide. Therefore the use of lower electrolyte concentrations in the nickel-zinc cell has a positive effect on calcium zincate formation which helps to offset the reduced conductivity and ionic strength of the weaker electrolyte.

A study[11] showed that reduced zinc oxide solubility improved the cycle life of $Zn/NiOOH$ batteries. There have been several studies of the rate of formation and decomposition of calcium zincate in solutions of $Ca(OH)_2$, ZnO and KOH.[11–13] The rate equations developed in these studies evaluate the effectiveness of adding $Ca(OH)_2$ with calcium zincate formation and decomposition reactions. At a discharge rate of $C/3$, about 26% of the zincate is still in the solution. At a charge rate of $C/6$, the liberation of zincate from calcium zincate is complete. Because the calcium zincate formation is a slower process, the discharge rate could effect how well the addition of $Ca(OH)_2$ controls shape change in a battery. Since the calcium zincate decomposition is a faster process, the addition of $Ca(OH)_2$ will reduce the growth of dendrites.

A zinc electrode with reduced solubility has been developed and patented.[12] This technology has made nickel-zinc batteries practical from a cycle-life point of view. The "shape change" that occurs in the standard zinc electrode during charge/discharge cycling is due to the formation of the intermediate zincate ion. The zinc active material passes through this soluble phase during oxidation and reduction in alkaline solution. The calcium-zincate electrode design will reduce the formation of the soluble zincate intermediate reaction product by proceeding directly from the metallic oxidation state to an insoluble phase. Calcium hydroxide is used as an additive to the electrode to promote the formation of calcium zincate which is thermodynamically stable and remains substantially insoluble in the electrolyte. Zinc dissolution is mitigated by formation of a passivating layer at the electrode/electrolyte interface. A potential concern is electronic conduction through this passivating, insoluble film on the zinc electrode. This can be addressed by doping the electrode with electronically conductive metallic oxides, such as PbO or Bi_2O_3.

The relative lack of shape change which occurs after cycling in the calcium zincate electrode is shown in Fig. 31.4. Only a slight redistribution of zinc from around the edges of the electrode is observed. There is a slight densification of zinc which occurs over the bottom one-fourth of the electrode, but it is relatively minor compared to the massive changes seen in zinc electrodes which do not contain calcium hydroxide. The detailed structure of a calcium zincate electrode which has been cycled 500 times at 100% depth-of-discharge is shown in the scanning electron microphotograph of Fig. 31.5.

The formation of calcium zincate is given by the reaction:

$$Ca(OH)_2 + 2\ ZnO + 4\ H_2O \rightleftharpoons Ca[Zn(OH)_3]_2 \cdot 2\ H_2O$$

The stoichiometry of calcium zincate has also been reported in the literature as $Ca(OH)_2 \cdot 2\ Zn(OH)_2 \cdot 2\ H_2O$.[13] This reference also contains a detailed study of the kinetics of calcium zincate formation in potassium hydroxide solutions. This study concluded that the rate of formation of calcium zincate was not a function of the dissolution and transport of zinc oxide or of the transport of water, but is a function of the calcium hydroxide concentration.

FIGURE 31.4 Roll-bonded calcium zincate electrode after 500 cycles. (*Courtesy of Evercel Corp.*)

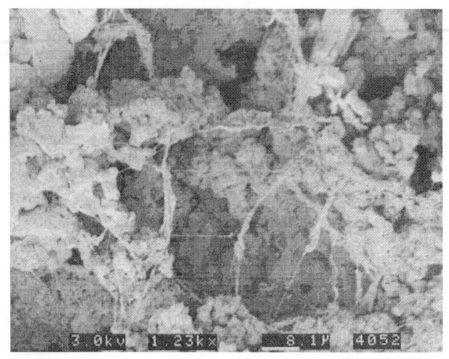

FIGURE 31.5 Scanning electron microphotograph of roll-bonded zinc electrode after 500 cycles. (*Courtesy of Evercel Corp.*)

Calcium zincate is much less soluble in alkaline solution than zinc oxide. This forms the basis for the improvement in cycle life obtained with this technology.

Cells containing calcium zincate electrodes can be manufactured in at least two different ways. Calcium hydroxide can be added to the zinc oxide electrode mixture. In this case, the calcium zincate is formed *in situ* as the electrode is cycled in the cell during the electrochemical formation process. Another method is to form calcium zincate in a separate step and then use this material in the electrode fabrication process.[14] Calcium zincate can be prepared, purified and identified by its X-ray diffraction pattern, shown in Fig. 31.6. This process produces electrodes with a more uniform distribution of calcium zincate and overcomes the problem of zinc dissolution that occurs during the first few cycles before the calcium-zincate has fully formed in the *in situ* method. This serves to increase the overall cycle life and performance of the battery. In either case, an important advantage of the plastic-bonded calcium zincate structure is the reduction of the tendency to form zinc dendrites. The zinc active materials are embedded in the three-dimensional structure of the PTFE nanofibers. This increases the stability of the electrode. Although it is possible for some zinc dissolution and migration to occur in the cell, dendritic deposits which penetrate the separator are rarely seen.

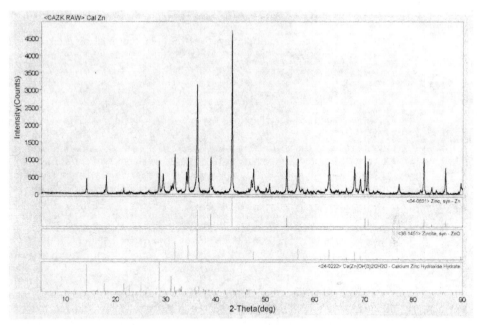

FIGURE 31.6 X-ray diffraction pattern showing calcium zincate in zinc electrode after cycling. (*Courtesy of Evercel Corp.*)

31.3.3 Separator

Nickel-zinc batteries typically use a multicomponent separator system. This is due to the requirements for providing both an electrolyte reservoir and for retarding zinc migration. No single material has been identified as yet that will adequately provide both functions. Even if a reduced solubility zinc electrode is used in the cell design, such as the calcium-zincate electrode described above, zinc still has a finite solubility in the alkaline electrolyte. A zinc migration barrier must be provided in order to prevent shorting of the zinc to the nickel electrode. The tendency towards dendrite formation is significantly reduced through the use of calcium zincate electrode technology. This lessens the requirements imposed on the separator material, particularly that of mechanical strength and penetration resistance.

Traditional nickel-zinc and silver-zinc cell designs used multiple layers of cellulose-based membrane materials. Many different materials were tried, but the most common material used was cellophane which is a cellulose film which acts as a membrane and is capable of resisting zinc penetration. The material has a high molecular weight and varying degrees of crystallinity, which gives it a wide range of distributions of order (pore size). The cycle life of cell designs using this material is severely limited due to hydrolysis of the cellophane in alkaline solution. Various methods have been tried to stabilize cellulose materials, such as chemical treatment and radiation grafting to other polymers, but none have as yet proved economically feasible. Considerable work has been done in the past to develop a suitable separator for nickel- and silver-zinc batteries.[15] An excellent discussion of separator development is contained in a comprehensive review.[2]

The most successful zinc migration barrier material yet developed for the nickel-zinc battery is Celgard®. This is a microporous polypropylene film which has a typical thickness of 0.025 mm. Polypropylene is inherently hydrophobic so the material is typically treated with a wetting agent for aqueous applications. One disadvantage of microporous materials, as compared to membrane materials, is the lack of tortuosity. There is not a tortuous path that the zincate ions must migrate through in order to form an electrical short to the nickel electrode. As a result, multiple layers of the material must be used. Since the holes in the layers are typically not "lined up," some artificial tortuosity is introduced.

31.3.4 Electrolyte

Traditional nickel-zinc cells utilize 31% to 35% concentration aqueous potassium-hydroxide. Typically 1% lithium hydroxide is also added to the electrolyte. Higher electrolyte concentrations reduced the hydrolysis rate of the cellulose-based zinc migration barrier materials. However, the solubility of zinc also increases as electrolyte concentration is increased. This increases the likelihood of zinc penetration through the separator. A variety of electrolyte formulations and additives have been tried with varying degrees of success.[16–17] These included potassium fluoride, potassium carbonate and a variety of other additives. The goal was to reduce the solubility of the zinc to improve cycle life but this goal can also be achieved through zinc electrode additives, as discussed above.

More recently it has been shown that electrolyte formulations in the potassium hydroxide concentration range of 20% to 25% can increase the cycle life obtained from the battery. This is particularly true when used in conjunction with the reduced solubility calcium zincate electrode. The solubility of zinc is substantially reduced by decreasing the concentration of the electrolyte below 25%. This reduces zinc dissolution and therefore zinc migration, stabilizing the zinc electrode and increasing the cycle life of the battery.

Newly manufactured cells are typically vacuum filled with the electrolyte. Commercially manufactured equipment is available for this purpose. In this method, a high vacuum is applied to the cell and the electrolyte is then drawn in by the vacuum. The purpose of the vacuum is to remove air from the microporous cell components, such as the electrodes and separators. This facilitates wetting of these components to make the cell immediately susceptible to charging. The cell can also be electrolyte activated by simply filling the cell and allowing it to soak. However, this takes much longer for the cell components to adequately wet which reduces throughput in manufacturing.

31.4 CONSTRUCTION

Various types of cell and battery design and construction can be used in the nickel-zinc battery system. Cells have been built in both prismatic and cylindrical designs and both vented and sealed designs. However, most current commercial applications require the use of a sealed, maintenance-free design. A typical sealed prismatic battery is shown in Fig. 31.7. This type of construction can be used for a wide range of cell sizes and is particularly suited to larger capacity batteries (e.g. greater than 10 Ampere-hours).

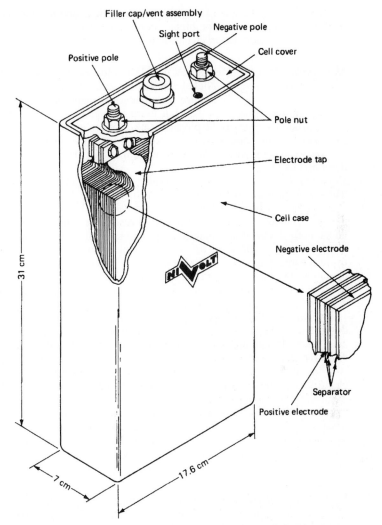

FIGURE 31.7   Sealed nickel-zinc cell. (*Courtesy of Evercel Corp.*)

31.4.1 Sealed Cell Design

Nickel-zinc prismatic cell designs have been developed ranging from 2 Ah to over 100 Ah.[18] The technology is readily scaleable to prismatic cells, cylindrical cells, or monobloc batteries. In a typical cell design, the cell is stacked with alternating nickel and zinc electrodes, with the separator material between. Usually a zinc "half-plate" terminates the electrode stack on either end. Prismatic cell and battery cases and covers are typically plastic. Cylindrical cell and battery cases and covers are typically nickel-plated steel. Metallic conductors (typically nickel or copper) are used for the electrode tabs and the cell and battery terminals. Sealed nickel-zinc cell and battery designs incorporate a resealable safety vent which prevents excessive pressure from building up inside the cell container. This resealable vent provides for an operating pressure in the cell which aids in forcing gas recombination. Plastic cased batteries typically vent at a few atmospheres while cylindrical metal-cased batteries may vent at up to 20 atmospheres or higher. Some nickel-zinc cell designs also use an auxiliary catalytic gas recombination electrode to facilitate gas management.

Mechanical Design.   Mechanical cell design follows the traditional patterns established by other battery chemistries. Nickel-zinc is best suited for larger form-factor, deep-cycle applications and these mostly require rectangular prismatic designs. Some work has been done with cylindrical designs as well but these have not yet been commercialized. Mechanical design primarily involves the physical configuration of the cell along with non-active material issues such as the case and cover, structural integrity, internal electrical connections, other cell parts and components and the external interface. The two major types of cells are briefly described below.

Prismatic Cell Designs.   Nickel-zinc prismatic cell and battery cases and covers are typically molded from commercial grade resins such as Noryl®, depending on the application. Polysulfone has been used in ruggedized designs, but is more expensive than some other plastics. The cover usually incorporates a resealable safety valve. This valve remains closed under normal operating conditions, providing sealed operation, yet provides a necessary safety margin in the event of a catastrophic event such as fire or severe overcharge or overdischarge abuse. The safety relief valve provides a fail-safe leak-before-burst mode of operation. Prismatic cells are very easy to manufacture and automated equipment exists for this purpose. For larger capacity cells (for example greater than 5 to 10 Ah) prismatic cells are more efficient for most applications.

Cylindrical Cell Designs.   Cylindrical nickel-zinc cells have been built in the past.[19] This design incorporated several interesting features such as the addition of bismuth oxide to the zinc electrode mixture and an approach for controlling the internal pressure of the cell. However, cells were never commercialized due to the relatively short cycle life obtained with traditional zinc electrode technology. With the advent of newer, reduced solubility zinc electrode designs, the cycle life of cylindrical cells could be extended in the same manner that has been demonstrated in prismatic cells. Evercel Corporation has developed a sub-C cylindrical cell in its laboratory. Additional work was performed on AA and C size cylindrical cells.[20] Prototype cells were constructed which delivered up to 1.5 Ah depending on load and cut-off voltage. The cells were designed using the RAM battery concept in which the zinc anode was made from a gelled mixture of zinc and potassium-hydroxide (see Sec. 36.3).

Electrochemical Design.   The electrochemical design of the cell consists primarily of balancing the active materials present in the electrodes. This previously has been discussed for each of the two electrodes separately, the nickel positive electrode and the zinc negative electrode. When combined in the cell, the two active materials must be present in some ratio with respect to each other. As with most other alkaline nickel batteries, the nickel-zinc system is typically positive (nickel electrode) limited. This means that the cell contains more zinc active material, on an Ampere-hour basis, than nickel active material. This must take into account the active materials present in the cell in addition to the active material utilization of each.

Excess zinc is included in the cell for a number of reasons. Since zinc has some finite solubility in the electrolyte, additional zinc must be manufactured into the cell in order to compensate for the quantity of zinc that dissolves when electrolyte is added to the cell. Having excess uncharged zinc oxide also minimizes hydrogen evolution during charging because the zinc electrode normally doesn't achieve a full state of charge. In addition, excess metallic zinc serves to react with oxygen produced at the nickel electrode during overcharge as discussed below. Adequate electrochemical design allows sealed operation of the cell by managing the gases produced during charging and discharging.

An interesting effect in sealed cells has been noted by Alekseeva.[21] Detailed studies were performed on the active material balance in vented and sealed nickel-zinc cells as a function of cycling. It was discovered that zinc metal tends to accumulate as the cell is cycled, such that the cell may actually become zinc active material limited at some point in its life. This may be influenced by a number of factors including current distribution, zinc redistribution and variable utilization effects in the zinc electrode as well as other factors. Typically the zinc electrode charges more efficiently than the nickel electrode, which also tends towards the accumulation of zinc metal in the cell. Also, current utilization in the zinc electrode becomes more efficient as the metallic zinc content of the electrode increases. A similar effect has been observed at Evercel Corporation. The zinc electrode typically takes longer to be fully formed than does the nickel electrode and zinc metal tends to accumulate with cycling. This is offset in the sealed cell design by direct oxygen recombination which may consume some of the excess zinc metal in forming zinc oxide. The stoichiometric ratio of zinc to nickel is typically 3-to-1 in long cycle life applications at 100% DOD. This ratio may be as low as 2.5-to-1 for other applications as long as the KOH concentration is kept below 23%.

Gas & Electrolyte Management.   Vented nickel-zinc cell designs have an advantage over sealed designs because gas and electrolyte management issues are minimized due to the flooded electrolyte and vented nature of the battery. Unfortunately vented designs also have a number of significant disadvantages. These include the requirement that the electrolyte be periodically replenished and the hazards associated with venting gases such as fire and explosion or damaging the equipment in which the battery is installed by the entrainment of corrosive electrolyte in the gases vented. Therefore most applications demand a sealed, maintenance-free battery.

A delicate balance must be achieved in cell and battery design in order to maintain sealed operation while providing optimal cycle life. Insufficient electrolyte leads to premature dry-out of the electrodes and separators which results in failure of the battery. Excess electrolyte can introduce gas management issues by reducing the rate of oxygen recombination with the zinc electrode surface. This can result in the venting of gases through the safety vent which also leads to premature dry-out. Most sealed cell designs are either starved or semi-starved with electrolyte. Since cell dry-out is a common long-term failure mode, it is desirable to put as much electrolyte as possible into the battery without adversely affecting the gas recombination characteristics. It is important to have as much excess electrolyte as possible at the beginning of life. This allows a design margin to accommodate the swelling normally observed when the battery electrodes and other components are first wetted and subsequently cycled.

Gas recombination can be managed by several means and a split-negative electrode stack design has been patented.[22] A hydrophobic gas diffusion membrane is used to separate the back-to-back zinc half-plates in the split-negative design. This allows oxygen gas to diffuse into the cell stack and increases the surface area of zinc electrode available to recombine oxygen. A similar stack arrangement has been used in the nickel-hydrogen battery for the same purpose.

An auxiliary catalytic gas recombination electrode can also be used to facilitate gas management. This concept has been patented.[23] The auxiliary gas recombination electrode typically consists of a catalyst, such as platinum or palladium, supported on a high surface area carbon. The reaction of hydrogen, produced at the zinc electrode during charge, and oxygen, produced at the nickel electrode, is exothermic. A heat sink is normally provided to dissipate the heat produced. The recombination of hydrogen and oxygen on the catalyst is a chemical reaction, not an electrochemical one.

31.4.2 Battery Design and Packaging

Single cells can be assembled into a multicell battery by a variety of conventional means or in a monobloc type of battery construction. Single cells provide a 1.65 VDC building block from which any desired battery voltage can be achieved. The monobloc is typically more cost effective and is usually built as a 12 VDC module containing seven or eight cells, depending on the application. Multiple monoblocs can be grouped for systems requiring higher voltages.

Electrical Design.   Battery electrical design is dictated by the system interface require-ments. Battery design can be simple, as in the case of several cells strapped together into a module, or quite complex such as in a hybrid or electric vehicle. The fundamental require-ment for the electrical design of the battery is the output voltage required. This determines the number of cells in the battery and all other design aspects flow from this requirement. Battery electrical design also includes safety and protective devices and components which can be designed to protect the battery, personnel and systems in which the battery is installed. These can include protective components such as thermal cut-off (TCO), fuses, protective diodes and other safety devices. The electrical interface to the battery is also important and should include safety and operational features such as polarized and/or keyed battery con-nectors that impede the improper use of the battery. Also, it should be ensured that the battery polarity is properly labeled and marked on the battery.

Nickel/zinc cells and batteries should not be charged or discharged in parallel configu-rations. The battery should be sized to provide the Ampere-hour energy storage requirement of the application rather than operating smaller batteries in an electrically parallel configu-ration. Operation in parallel presents current sharing and efficiency imbalances which may adversely affect Ni/Zn battery performance and cycle life. Multi-voltage "taps" should not be used on the battery for similar cell balance and safety issues.

Monobloc Design.   Multi-cell batteries can be configured either as single cells connected in series or can be constructed as a monobloc similar to the standard lead-acid automotive-type battery. In monobloc batteries, the entire multi-cell battery case is molded as a single component as opposed to each cell being in an individually molded case. The cells can be interconnected either through the cell wall or the terminals can protrude through the top of each cell and be connected with standard intercell connectors. Normally if cell terminals are exposed on the top of the battery, a protective cover is fitted to prevent electrical hazards from the exposed terminals. In the monobloc design, each cell is individually sealed from the other cells in order to prevent electrolyte bridging. Relief valves are included for safety to prevent excessive pressure from building inside the battery case.

Thermal Design.   As with most batteries, nickel-zinc performance and cycle life are strongly dependent on the thermal environment in which the battery is operated. System level design should minimize the temperature differential (ΔT) experienced by the cells of the battery. Heat is reversibly consumed and generated during charge and discharge, respec-tively, as a result of endothermic and exothermic chemical reactions. However, heat is also generated irreversibly during both charge and discharge as a result of I^2R losses. Thus, the net thermal result of discharge is heat evolution, but the net thermal result of charge is variable. The bulk of the charging process is slightly endothermic but this is compensated for by I^2R heating, resulting in a net rise in battery temperature. Near the end of charge, gas evolution becomes significant and this can generate significant quantities of heat. This is one reason that overcharge should be avoided. The heat produced near the end of charge may be carried over into discharge causing a higher than normal temperature increase during discharge. Whenever possible, heat should be removed from the battery via convection. Severe overcharge or over-discharge can result in runaway exothermic reactions, and should be avoided.

31.5 *PERFORMANCE CHARACTERISTICS*

31.5.1 General Discharge Characteristics

The general characteristics of the nickel-zinc system are presented in Table 31.2. Nickel-zinc batteries are capable of delivering up to about 50 to 60 Wh/kg and 80 to 120 Wh/L depending on the specific design characteristics. The batteries have good high rate and high power discharge capability and very good charge retention characteristics. As with all batteries, cycle life is strongly dependent on the application, environmental conditions, the depth-of-discharge and the charge/discharge regimen experienced by the battery during use. In controlled laboratory testing, nickel-zinc batteries yield about 500 cycles when operated at 100% depth-of-discharge and more than 10,000 cycles at depths as low as 10% DOD. Nickel-zinc's specific energy of 60 Wh/kg is intermediate between nickel-cadmium and nickel-metal hydride.

TABLE 31.2 Characteristics of Nickel-Zinc Batteries

Parameter	Nickel-zinc
Cathode electrochemistry	$Ni(OH)_2/NiOOH$
Anode electrochemistry	ZnO/Zn
Theoretical specific energy (Watt-hours per kilogram)	334
Electrolyte (% potassium hydroxide)	20 to 25
Nominal cell voltage (Volts)	1.65
Operating temperature range (°C)	−20 to 50
Specific energy (Watt-hours per kilogram)	50–60
Energy density (Watt-hours per liter)	80–120
Specific power (Watts per kilogram)	280
Power density (Watts per liter)	420
Charge retention (percent loss per month @ 25°C)	<20
Cycle life (cycles @ 100% DOD)	~500

Rate Capability.   Figure 31.8 shows a series of discharges on a fresh standard design 30 Ah nickel-zinc battery. The cell was discharged at nine different rates ranging from C/20 up to 6C. The capacity on the initial low-rate discharge is greater than 35 Ah. (This battery is a prismatic design, incorporating a lightweight plastic cell case and resealable pressure vent.) The battery was discharged at ambient temperature (about 23°C) with no active cooling. The battery performed extremely well up to the C rate and only started to drop significantly in loaded mid-point voltage above the 2C rate. Data such as this can be used to estimate the load voltage of a battery at a given current. System level designers should consult the manufacturer's data for the specific design that will be used.

Figure 31.9 shows the typical mid-point discharge voltage under load, as a function of discharge rate and at three different temperatures. This data allows an estimate of the battery discharge voltage at different rates and temperatures. Below the 2C rate, the battery voltage is nearly independent of temperature over the range of 0°C to 40°C. As the discharge current is increased, temperature has an increased effect on battery voltage, primarily due to the increasing impedance (and polarization) of the battery at colder temperatures. At high rates (6C) and very cold temperatures (0°C), the battery voltage is 1.32 V per cell. At 40°C and at the 6C rate the average battery voltage is about 1.50 V. This difference is entirely due to the effect of operating temperature. The data also show that the effect of temperature is not linear. For example at the 6C rate, decreasing the temperature from 40°C to 25°C produces

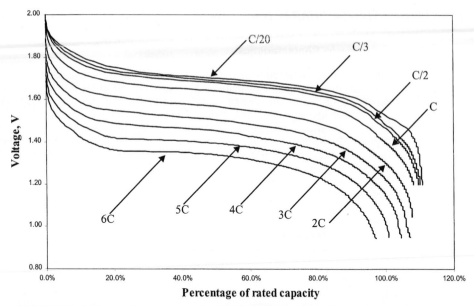

FIGURE 31.8   Multi-rate discharge curves for 1.65 V nickel-zinc battery. Battery discharged at room temperature. C/20, C/3, C/2 and C rate: battery was discharged to 1.2 V cutoff; 2C rate: battery was discharged to 1.05 V; 3C, 4C, 5C and 6C rate: battery was discharged to 0.95 V. (*Courtesy of Evercel Corp.*)

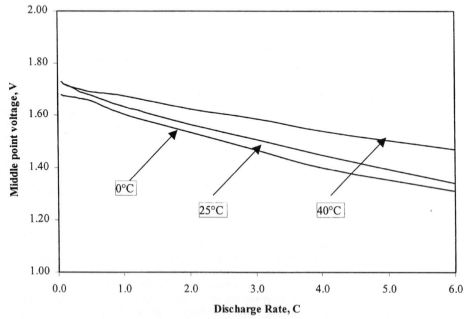

FIGURE 31.9   Typical nickel-zinc battery MPDV as a function of discharge rate and temperature. (*Courtesy of Evercel Corp.*)

a relatively large decrease in battery voltage (approximately 160 mV) while decreasing from 25°C to 0°C only reduces the battery voltage 40 mV. This nonlinearity is due to a variety of effects including the nonlinearity of electrolyte impedance, electrode kinetics and polarization as a function of temperature.

Figure 31.10 indicates the capacity delivered by the nickel-zinc battery as a function of discharge rate and temperature. As the discharge current is increased, the battery output voltage decreases due to impedance and polarization losses. The data show that the capacity is nearly independent of discharge rate, up to the 6C rate, but that the capacity is somewhat dependent on temperature. The discharge capacity drops about 12% when the temperature decreases from 40°C to 0°C.

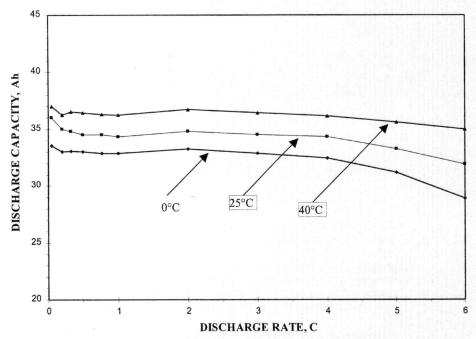

FIGURE 31.10 Discharge capacity of a nominal 30 Ah nickel-zinc battery as a function of discharge rate and temperature. (*Courtesy of Evercel Corp.*)

Figure 31.11 shows the specific energy (Watt-hours per kilogram) obtained from a typical nickel-zinc battery which is a direct function of the battery discharge capacity. Therefore, the specific energy curve versus discharge rate behaves similar to the discharge capacity versus discharge rate data shown above. Specific energy is highly dependent on specific battery design and can vary within the nickel-zinc chemistry depending on the design used. Battery design can be optimized for energy storage capacity, rate capability, power density, cycle life or other specific performance factors. These data show a standard type of battery which has not been specifically designed for energy, rate or power.

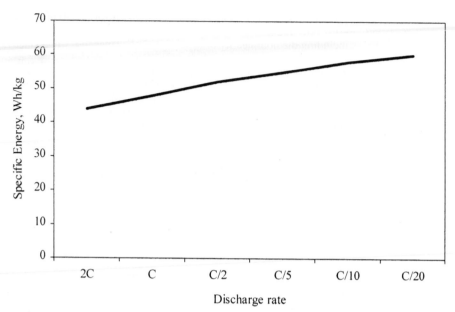

FIGURE 31.11   Nickel-zinc specific energy as a function of discharge rate. Batteries were discharged at room temperature. 2C: battery was discharged to 1.05 V cutoff; others: battery was discharged to 1.2 V. (*Courtesy of Evercel Corp.*)

Temperature Dependence of Performance.   Temperature has a strong influence on battery performance. As a general rule of thumb, nickel-based batteries achieve optimal performance when charged cold and discharged warm. Unfortunately, in many applications, temperature is ambient and uncontrolled. Figure 31.12 shows the effect of temperature on the discharge capacity of the nickel-zinc battery at four different rates. At lower rates, the discharge capacity is a linear function of temperature. At the 6C rate, the relationship starts to become nonlinear, primarily due to the increased conductivity of the electrolyte above 30°C.

Cell Balance on Discharge.   Cell balance becomes an important issue in the discharge performance of multicell batteries. The capacity of the battery will be reduced by weaker (lower capacity) cells. Newly manufactured cells typically fall within a narrow range of capacity (for example, plus or minus two or three percent over a manufacturing lot). As a battery is cycled, the individual cells may diverge in performance. In practical terms this means that the cells in the battery exist at different states-of-charge. This affects battery performance because the discharge capacity of the battery is dominated by the weaker cells, which tend to depress the overall battery voltage. If the cells in a battery become extremely unbalanced, battery performance and cycle life may be adversely affected. This is strongly influenced by system level design, particularly thermal design. Temperature differentials in the battery will cause cells to diverge in performance and thereby reduce the overall performance of the battery. It is therefore recommended that systems be designed such that a minimum temperature difference (ΔT) exists among the cells in the battery.

Reconditioning.   Many alkaline rechargeable nickel-based batteries, such as nickel-cadmium, nickel-hydrogen and nickel-metal hydride, are capable of being reconditioned. Typically this means that the battery is taken to a very low state-of-charge and then recharged at a moderate rate. Sometimes the reconditioning effect is only seen after several such cycles.

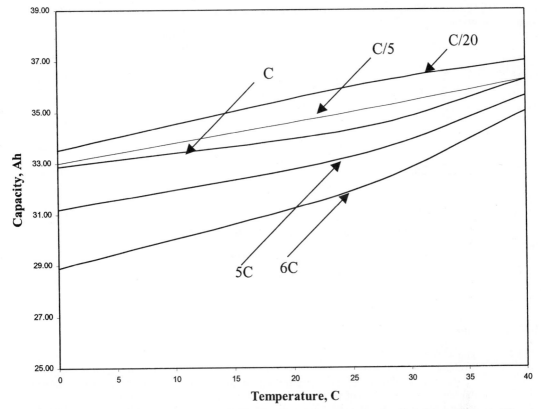

FIGURE 31.12 Discharge capacity of a nominal 30 Ah nickel-zinc battery as a function of temperature at different discharge rate. C/20, C/5, C rate: battery was discharged to 1.2 V, 5C, 6C: battery was discharged to 0.95 V. (*Courtesy of Evercel Corp.*)

Reconditioning is not currently recommended for nickel-zinc batteries. Should more data become available in the future, reconditioning may become an option for the nickel-zinc system. In the meantime, the nickel-zinc battery should not be treated as a generic nickel-based rechargeable battery with respect to reconditioning.

31.5.2 Charge Retention

Fully charged nickel-zinc batteries lose only about 20% of their original capacity within a month on open-circuit stand at 25°C. The self-discharge rate of nickel-zinc, like that of most batteries, increases with temperature.

Typical charge retention data for the nickel-zinc battery is shown in Fig. 31.13. As with most batteries, the voltage decays *exponentially* the first few days and then levels out to a very slow rate of decline. Capacity decay during open-circuit stand amounts to less than 1% per day after the initial higher loss rate that occurs during the first 4 or 5 days. If the battery is allowed to stand for very long periods of time, the battery voltage may reach zero volts. If this happens the battery usually requires 3 or 4 cycles to fully recover capacity when the battery is brought into service again.

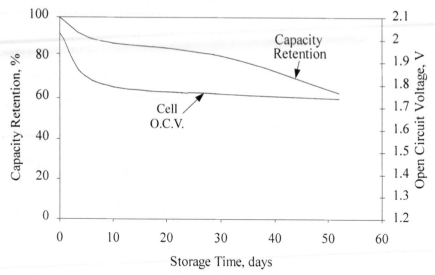

FIGURE 31.13   Self-discharge of a nickel-zinc battery at room temperature. (*Courtesy of Evercel Corp.*)

31.5.3   Cycle Life

The nickel-zinc battery is capable of delivering more than 500 cycles at 100% depth-of-discharge. Cycle life is largely a function of the application, including factors such as duty cycle, depth-of-discharge, the charging regime used, the cumulative amount of overcharge, the level of abuse the battery receives and the thermal and mechanical environment. As the battery is cycled, the capacity gradually declines due to physical changes and degradation processes in the battery. This gradual decline in capacity is both normal and predictable for a given battery design. Fig. 31.14 shows the discharge capacity for a 12 VDC nickel-zinc battery (30 Ah nominal capacity) as a function of cycling. The battery under test is a standard design built by Evercel Corporation using a plastic-bonded graphite composite nickel electrode and a plastic-bonded calcium zincate zinc electrode. The battery achieved more than 600 cycles at 80% DOD while retaining in excess of 80% of its designed capacity. During each cycle, the battery was discharged to 80% of its rated capacity at the C/5 rate (6 Amperes) and was charged using a two-step CC/CV method, described in Sec. 29.6. The discharge capacity of the battery was checked at 25 cycle intervals, which adds approximately 24 additional cycles performed at 100% DOD. The data shows a gradual and predictable decline in performance which is important from a system level. This allows the system designer to account for battery aging in the overall product specification and design. At 100% DOD, the battery achieved 550 cycles before the capacity dropped to 80% of its rated value. The rate of decline of capacity tends to increase near the end-of-life which provides an early indication that failure is imminent.

   Figure 31.15 shows voltage performance as a function of cycle life for a single-cell nickel-zinc battery. The chart plots the mid-point discharge voltage (MPDV) as a function of the number of cycles. The MPDV is defined as the loaded cell voltage at the mid-point of discharge as calculated based on the capacity removed from the cell (i.e. half of the actual capacity of the cell). The battery was cycled at 100% depth-of-discharge, both charging and discharging at the C/2 rate (15 Amperes). This represents an accelerated test in which slightly more than 3 cycles per day can be accumulated, whereas in most applications only a single cycle per day is performed. The loaded discharge voltage of the battery gradually

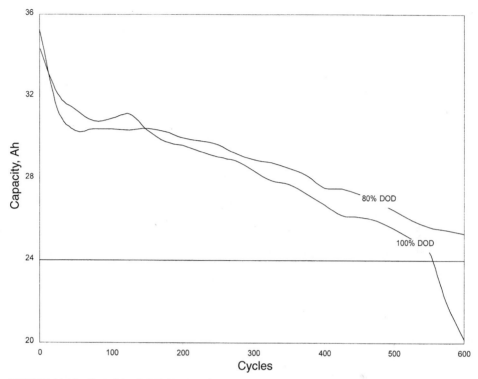

FIGURE 31.14 Cycle life of 12 V, 30 Ah nickel-zinc battery. 80% DOD testing performed by JBI. (*Courtesy of JBI and Evercel Corp.*)

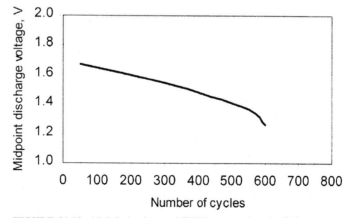

FIGURE 31.15 Nickel-zinc battery MPDV, measured at the C/2 rate, as a function of cycle number. (*Courtesy of Evercel Corp.*)

decreases with cycling. The effect is similar to that observed with the gradual decline in capacity. The voltage declines primarily due to the gradually increasing impedance of the battery as the electrodes slowly degrade as it gradually dries-out. The decay in voltage is linear and predictable until a point is reached at which the battery can no longer support the imposed discharge current. When this happens the battery has essentially "failed" from the standpoint that the required load voltage can no longer be supplied. If the discharge current is reduced, it will continue to function and will still deliver greater than 80% of its rated capacity. The battery shown in Fig. 31.15 delivered nearly 600 cycles under these test conditions.

In each set of data, the batteries were cycled at 100% depth-of discharge (DOD).

Depth of Discharge.   Cycle life is a direct function of depth-of-discharge, as well as being a function of many other factors discussed elsewhere in this chapter. Generally, cycling the battery at deeper depths-of-discharge results in reduced cycle life. The general relationship between cycle life and depth-of-discharge for the nickel-zinc battery is shown in Fig. 31.16. This empirical data can be used as a guide in system level design to achieve the required cycle life by limiting the depth-of-discharge of the battery. Battery life is reduced when cycled at deeper depths-of-discharge because of the higher stress levels induced in the electrodes. Mechanical expansion and contraction, zinc electrode solubility issues and electrochemical issues are all involved in the process.

Temperature Dependence.   Temperature is also an important factor in relation to cycle life. Temperature affects all aspects of battery performance. In general, nickel-based alkaline batteries perform best at moderate temperatures in the range of 10°C to 30°C. Outside of this temperature range, performance and cycle life may be less than optimum. If the system design and the application environment are able to maintain the battery within the optimum range, improved battery performance and cycle life will result.

Failure Mechanisms.   Failure mechanisms for earlier nickel-zinc batteries include zinc migration, shape change, dendritic shorting and hydrolysis of the cellulose-based separator. These have been substantially eliminated in modern nickel-zinc battery technology. Dendritic shorting and shape change have been virtually eliminated through the use of reduced solu-

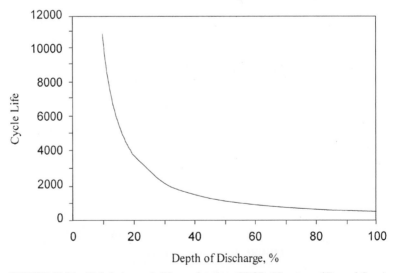

FIGURE 31.16   Nickel-zinc cycle life as a function of DOD. (*Courtesy of Evercel Corp.*)

bility calcium zincate electrode technology. Zinc migration has been substantially reduced. Separator systems have also improved substantially. Stable polymeric zinc migration barrier materials are used in place of cellulose-based separators. There remain two primary failure mechanisms in sealed nickel-zinc batteries, failure of the zinc electrode and cell dry-out.

Even with the use of reduced solubility calcium-zincate technology, the zinc electrode still has some finite solubility in the alkaline electrolyte. Zinc can form complex zincate anions in the electrolyte and diffuse throughout the battery. Some of this zincate is deposited within the pores of the nickel electrode. This may adversely affect the performance of the nickel electrode and thus the performance of the battery. It is possible that the gradual decrease in capacity observed is partially for this reason.

In plastic bonded electrodes, the fibrillated Teflon® structure minimizes mechanical fatigue in the electrode and provides dimensionally stable long-term performance by allowing the electrode active materials to expand and contract during charge and discharge. Conventional sintered or pasted electrodes provide no mechanism for this expansion and contraction. In addition, a fibrillated Teflon® structure prevents zinc migration and shape change by locking the active material in place within a stable three-dimensional structure. This effect also reduces the extrusion of active material from the nickel electrode into the separator.

31.5.4 Memory Effect

Nickel-zinc batteries may exhibit only a very mild memory effect that is associated with the nickel electrode. Nickel-cadmium batteries commonly exhibit what is termed "memory effect" or "fading." This is a reversible phenomenon usually caused by repetitive cycling at less than full depth-of-discharge. The observed effect is a depression in the discharge voltage (as much as 120 millivolts) when the battery is discharged below the depth at which it was previously cycled. Nickel-zinc batteries are only slightly affected by a similar phenomenon.

31.6 CHARGING CHARACTERISTICS

Proper charging of the nickel-zinc battery is a critical factor in achieving maximum performance and cycle life. The goal of recharging any battery is to input the correct quantity of charge to deliver the optimum discharge capacity. Charging beyond this point is not productive and in many cases may cause degradation in battery performance and cycle life. This is particularly true in the nickel-zinc battery because the zinc electrode is susceptible to increased zinc dissolution/migration during extreme overcharge. The critical factor in charging system design is how to detect when the battery has achieved a full state-of-charge. Several methods can be used for charge termination including temperature compensated voltage, the rate of change of voltage with respect to time, increased battery temperature or a variety of other commonly used techniques.

Development of the commercial nickel-zinc battery necessitated the development of charging methods and algorithms in order to supply charging systems for commercial applications. Extensive testing has been performed to fully characterize the nickel-zinc system as a function of both charge rate and temperature. Several charging algorithms and methods of charge termination have been evaluated. The defining characteristic for a charging system is cost, as the cost of the charger must be proportional to the cost of the battery and the system in which the battery is used.

It should be stressed that the manufacturer's recommendations should be strictly adhered to in charging any battery. Excessive overcharge, current which is too high or too low, or the use of an inappropriate charging algorithm may result in reduced performance, reduced cycle life and potential safety hazards. Use only a charger specifically designed for the nickel-zinc battery.

31.6.1 Charging Regimes

The most common charging method is a two-stage constant current/constant voltage (CC/CV) regime, with the CC phase cutting off at a temperature-compensated termination voltage. The CV phase can be terminated when the current falls below a predetermined value or after a specified time. Other charging schemes can also be used with the nickel-zinc system, but it is important to follow the manufacturer's recommendations. The primary charging criterion is the required time for recharging the battery. This affects battery and system design and also determines the cost of the charger.

31.6.2 Fast Charging

For applications where frequent, repeated use is anticipated, a high-rate fast charging system can be used, with more sophisticated charging algorithms and charge termination methods. Fast charge systems are capable of achieving a fully charged battery in as little as 2.5 h. This method works particularly well for smaller systems such as bicycles and small battery applications where extremely high currents are not required. At higher charging currents, charge termination becomes much more critical as Ampere-hours accumulate much faster. Higher currents also may induce higher temperatures in the battery which may reduce its overall charge efficiency. A balanced approach to charging at high rates is required in order to achieve maximum performance and cycle life.

In a typical high-rate charging cycle, the battery is charged at the $C/2$ rate up to nearly 95% state-of-charge. A five-minute rest is included at the end of the higher rate $C/2$ portion of the charge to allow depolarization of the battery and voltage relaxation. When the current is reapplied, the battery is clamped at a constant voltage and the current is allowed to taper, topping off the battery to full capacity. The electrochemistry of the battery is such that large quantities of charge can be accepted at a lower state-of-charge. As the state-of-charge increases, parasitic gas evolution reactions become significant and charge efficiency decreases. This two-step method greatly reduces oxygen evolution at the nickel electrode during overcharge and improves overall charge efficiency. Fig. 31.17 shows a standard two-step charging method used for nickel-zinc batteries. The voltage rise near the end of charge signals that the battery is approaching full charge. The voltage rise occurs because the electrochemistry of the cell is transitioning from the normal charge reactions to oxygen evolution, which occurs at a different characteristic voltage. This phenomenon provides an easily implemented method of charge termination. Once this point is reached, the charge current is reduced to compensate for the reduced charge efficiency of the battery and the lower current thereby reduces the gas evolution rate. Typically only about 10% of the rated capacity of the battery is input during the second step. A back-up charge termination time limit should also be provided to prevent thermal runaway.

Charge termination voltage is a function of both temperature and charge rate. At higher charge rates, the end-of-charge voltage increases as a function of cell impedance and polarization. The charge termination voltage also increases at colder temperatures, primarily due to the decreased conductivity of the electrolyte. Because of this strong temperature dependence, the battery charger must be temperature compensated, requiring the use of a thermistor to detect battery temperature. This slightly increases the cost of the charging system but provides much more efficient charging of the battery.

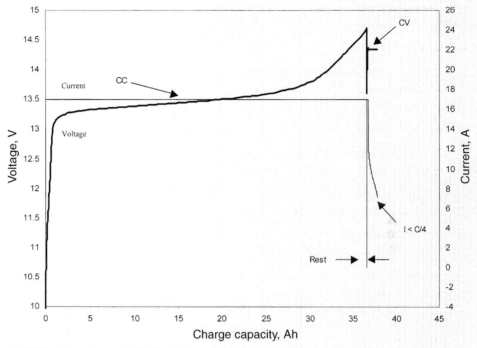

FIGURE 31.17   Seven-cell, 30 Ah nickel-zinc battery charge profile at room temperature. C/1.75 charge to 14.7 V then 14.35 V charge to I < C/4. (*Courtesy of Evercel Corp.*)

31.6.3   Slow Charging

A slow charge method can be used for applications where rapid recharging is not required or where the cost of the charger must be reduced. This might include commercial applications where only one battery discharge is required per day. Slow charging in these types of applications is typically done overnight. Charge termination is less critical when charging at low rate. After charge termination, the battery can be left on a low rate trickle charge, however this is typically not necessary as the self-discharge rate for the battery is low. Trickle charging is generally not recommended for the nickel-zinc system.

31.6.4   Charge Termination

Charge termination is the critical issue for all types of charging methods, either fast or slow charging. It is necessary for the charger to detect when the battery has achieved a full state-of-charge and either stop charging or reduce the charging current. There are a number of charge termination methods which are used in nickel-based alkaline rechargeable batteries such as nickel-cadmium and nickel-metal hydride. These include voltage, $-dV/dt$ (the negative derivative of voltage with respect to time), temperature rise and a variety of other methods based on the current/voltage profile of the battery as it nears a full state-of-charge or other properties of the battery. Similar techniques can be used with the nickel-zinc system.

As seen in Fig. 31.17, the reproducible rise in voltage at the end of charge can be used as a charge termination method. This voltage is a function of several factors including current, temperature and the specific battery design. Fig. 31.18 shows the relationship between the end-of-charge (EOC) termination voltage and charging current at three different temperatures which bracket many normal operating conditions. The charge termination voltage is nearly a linear function of charging current at each temperature. It is apparent that the temperature has a very large effect over the range of 0°C to 40°C. Figure 31.19 shows the

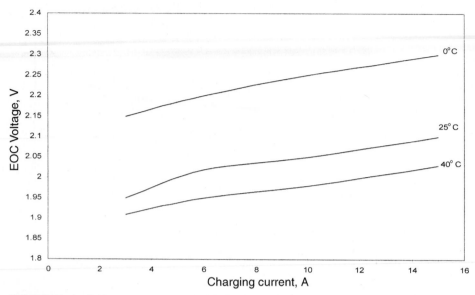

FIGURE 31.18 EOC voltage as a function of charging current and temperature for a 30 Ah single-cell nickel-zinc battery. (*Courtesy of Evercel Corp.*)

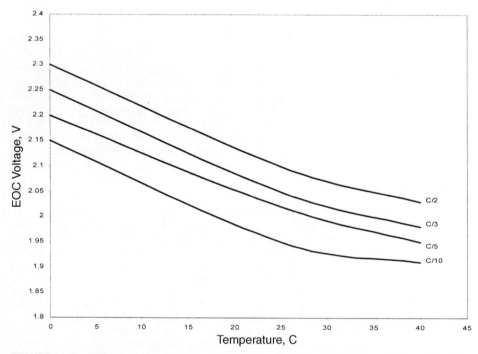

FIGURE 31.19 EOC voltage as a function of temperature and charge rate. (*Courtesy of Evercel Corp.*)

relationship between charge termination voltage and temperature graphically as a function of charging current. The data show that the charge termination voltage is linear as a function of temperature (at rates between C/10 and C/2) up to about 25°C. Above this temperature, the slope of the end-of-charge voltage versus temperature relationship declines significantly. At temperatures above 25°C, the EOC termination voltage varies very little with increasing temperature.

31.6.5 Overcharging

Overcharging is detrimental to most batteries, including nickel-zinc. Overcharge is based on the nominal (or actual) capacity of the battery being defined as 100% state-of-charge. Charge input above the 100% state-of-charge level is defined as overcharge. Overcharge simply means that more Ampere-hours of charge were put into the battery than was removed from the battery on the previous discharge. It is typically expressed as a percentage of the nominal (or actual) capacity. Batteries are more tolerant to overcharge at lower charge rates. But even low current can damage a battery if enough total Ampere-hours are input. As a battery approaches full state-of-charge, parasitic gas evolution reactions occur which generate heat and can cause a battery to vent. Excessive overcharge may also degrade the electrodes and reduce battery performance and cycle life. Therefore, it is highly recommended to prevent severe overcharging of the battery.

31.6.6 Cell Balance During Charging

Cell balance also becomes an issue in multicell batteries from the standpoint of charging. As a battery is cycled, the individual cells may diverge in performance. Many of the concepts from the discussion of the effect of cell balance in discharge performance also apply here. Cell balance affects charging because cells may receive differing amounts of overcharge depending on the state-of-charge of the individual cell. Cell balance may also affect the efficiency of various charge termination methods, making charging the battery less uniform and reliable. If the cells in a battery become extremely unbalanced, battery performance and cycle life may become adversely affected. Thermal imbalance can cause electrical unbalance in the battery. It is therefore recommended that systems be designed such that a minimum temperature difference (ΔT) exists among the cells in the battery.

31.7 APPLICATIONS

Nickel-zinc provides the lowest-cost option for a long-cycle-life alkaline-rechargeable system. The nickel-zinc system is suited for mobile applications such as electric bicycles, electric scooters and electric and hybrid vehicles or other deep cycle applications. Nickel-zinc may also replace other nickel based batteries with a less expensive system.

31.7.1   Temperature and Environmental Considerations

All batteries are limited in the operational and non-operational environments to which they can be exposed. Alkaline batteries are limited in the low temperature extreme primarily by the freezing point of the electrolyte. This can vary from about $-25°C$ for 20% KOH to below $-60°C$ for 31% potassium hydroxide. Twenty-five percent potassium hydroxide freezes at about $-38°C$. Nickel-zinc battery electrolyte is typically in the range of 20% to 25% potassium hydroxide, while most nickel-based alkaline rechargeable batteries use 31% potassium-hydroxide or even higher concentrations in some cases. Dissolved zinc anions in the electrolyte will also effectively lower the normal freezing point of the pure electrolyte. The lower concentration of electrolyte is used in order to reduce zinc solubility and extend cycle life. It is preferable to maintain the battery above the freezing point of the electrolyte or permanent damage to the battery may occur. In some cases, nickel-zinc batteries can be optimized for extreme cold weather applications by the use of electrolyte additives which enhance conductivity at cold temperature and depress the freezing point.

Temperature Extremes.   The operating temperature range for most nickel-zinc batteries is typically specified as $-10°C$ to $+50°C$. Batteries which have been optimized for colder temperatures can operate down to $-30°C$ or even slightly colder. At warmer temperatures, the problem becomes one of charge efficiency. The charge efficiency of the battery drops sharply at temperatures above $+40°C$, resulting in reduced battery capacity. Battery performance is affected by operating temperature as discussed in the performance section of this chapter. The non-operating temperature range for the nickel-zinc battery is typically specified as $-30°C$ to $+50°$. Long-term storage above $+50°C$ is not recommended or reduced battery performance may result.

31.7.2   Electric Bikes and Scooters

Bicycles and scooters are the primary method of personal transportation in many countries, including China, Taiwan and many other parts of Asia. Scooters are also very popular in parts of Europe, such as Italy, and the Mediterranean countries. When a bicycle is used for transportation and for carrying personal goods, human power becomes very limiting in terms of range and convenience, particularly when traveling uphill. For this reason electric power-assist bicycles are being marketed in many countries. A typical electric bicycle is shown in Fig. 31.20.

Similarly, scooters are widely used as a primary source of transportation in many countries. Highly polluting small gasoline-powered scooters are endemic in Asian cities and clean electric scooters are beginning to replace noisy gas-powered scooters. Under strict test conditions at the Energy and Resources Lab of Taiwan Industrial Technology Research Institute, the nickel-zinc battery for electric scooters demonstrated a cycle life for 14,168 km of travel. As a result, scooter manufacturers who use this battery would receive a government incentive of US $750 per scooter. Moreover, the Taiwanese government has stipulated that at least 2% of every custom scooter manufacturer's output must be electrically powered. The first commercial production scooter to use the new deep cycle nickel-zinc technology is the EVT, shown in Fig. 31.21. This scooter is manufactured in China and is currently also being distributed in the United States.

Most standard operating voltages for electric bicycle and scooter batteries are 24 V and 48 V, respectively. Capacity ranges from 10 Ah up to 40 Ah. An electric bike equipped with a nickel-zinc battery has a minimum range of 15 km before recharging, not including additional range supplied by human pedaling. Motors powered by nickel-zinc batteries allow electric bikes to achieve a maximum speed up to 20 miles per hour. An electric scooter equipped with a nickel-zinc battery has a range of 80 km before recharging and can achieve a maximum speed up to 50 miles per hour.

FIGURE 31.20 Nickel-zinc powered ZAP electric bicycle. (*Courtesy of Evercel Corp.*)

FIGURE 31.21 Nickel-zinc powered EVT scooter. (*Courtesy of Evercel Corp.*)

31.7.3 Deep-Cycle Applications

Deep-cycle applications for nickel-zinc batteries include trolling motors, electric bicycles and scooters, wheelchairs, golf carts, electric lawnmowers, electric vehicles and similar uses. In general, nickel-zinc batteries have good deep-cycle capability. Cycle life and performance issues at high depths-of-discharge are discussed elsewhere in this chapter. In general, advanced nickel-zinc batteries are capable of exceeding 500 cycles at 100% depth-of-discharge when operated within the specified temperature limits and charged according to the manufacturer's specifications. It is important to follow the manufacturer's recommendations for charging when operating the battery at high depth-of-discharge because cell imbalance becomes an even more critical issue. Cells tend to diverge in performance more rapidly as the depth-of-discharge is increased. Lawn and garden applications are becoming increasingly important as the disadvantages of the noise and pollution of small gasoline-powered engines becomes more apparent in both cities and suburban neighborhoods. An example of a nickel-zinc powered electric lawnmower is shown in Fig. 31.22.

Deep cycle marine applications, such as trolling motors, are an ideal use for nickel-zinc batteries. Nickel-zinc batteries have the required discharge rate capability, high cycle life at deep depths-of-discharge, lightweight rugged construction and are capable of being stored over the winter months without any maintenance charging. Nickel-zinc batteries are capable of storage for several months in a fully discharged state.

FIGURE 31.22   Nickel-zinc powered electric lawnmower. (*Courtesy of Evercel Corp.*)

31.7.4 Hybrid and Electric Vehicles

As the automotive industry moves forward with the development of electric and hybrid vehicles and other alternatives to the conventional gasoline-powered automobile, existing battery technologies are being stretched to their limits in terms of weight, size, cell balance issues, environmental concerns and cost reduction. Nickel-zinc may be applicable to hybrid and electric vehicle applications because of its light weight, high power capability and deep cycle ability. Nickel-zinc is also lower in cost than other nickel-based rechargeable batteries. Several demonstration electric vehicles have been powered with prototype nickel-zinc batteries. A full-size 11.2 kWh nickel-zinc electric vehicle battery was built by Evercel Corporation. This battery was tested in a PIVCO Citi Bee (Ford purchased PIVCO in 1998 and the vehicle is now sold under the brand name TH!NK). The vehicle achieved ranges of 78.8 to 131.5 km at average cruising speeds of 30 to 50 km/h. This vehicle, along with the nickel-zinc battery which powers it, is shown in Fig. 31.23.

An EV made by Trapos was tested at 40 km/h with both nickel-zinc and lead-acid batteries. A 205 kg (12 kWh) nickel-zinc battery provided a range of 172 km, whereas a 280 kg (7.0 kWh) lead-acid provided a range of only 69 km. The nickel-zinc battery provided a specific energy of 58 Watt-hours per kilogram while the lead-acid battery only provided 25 Wh/kg. The nickel-zinc battery weighed 25% less than the lead-acid battery but provided 70% more energy storage capacity with more than double the range of the lead-acid battery. Reduced battery weight is an important advantage for large batteries such as those used in electric and hybrid vehicles. Eagle-Picher manufactured an 18 kWh, 200 Ampere-hour monobloc nickel-zinc electric vehicle battery. The battery was tested in a Solectria® converted Geo Metro® vehicle. Another program at Eagle-Picher, sponsored by the Air Force

FIGURE 31.23 Nickel-zinc powered PIVCO Citi Bee. (*Courtesy of Evercel Corp.*)

Wright Aeronautical Laboratories, developed a nickel-zinc battery for a remotely piloted vehicle (RPV).[24]

31.7.5 Standby / Float Charge

Nickel-zinc batteries may be used in float charge and standby applications such as emergency lighting, emergency power back-up systems and uninterruptible power supplies (UPSs). If float charging is performed, the float charge rate should balance and offset the self-discharge rate so that the battery receives no net overcharge. This float charge rate may vary slightly depending on the battery design and the environmental conditions of the application, but it is typically in the range of $C/40$ to $C/70$. The battery manufacturer's recommendations should be followed to obtain maximum performance and life.

31.7.6 Military Applications

Nickel-zinc batteries may be useful for military application requiring deep cycle capability. Nickel-zinc batteries may also be employed in torpedoes, swimmer delivery vehicles, and other submersibles. Nickel-zinc has long cycle life and low cost and meets the deep cycle requirements of torpedoes and submersibles.

31.7.7 Special Applications

Microscopic nickel-zinc batteries are being developed for use in micro-electro-mechanical systems (MEMS), remote autonomous sensors, and other microelectronics.[25] Photolithography and patterned electrodeposition are used to fabricate 1000 cells at a time on a silicon chip. Polypropylene laminate is glued to a silicon wafer and etched with oxygen plasma into cell cavities and cell walls. NiOOH is deposited into the cell cavities, while Zn is deposited onto ledges surrounding the NiOOH. The cells are then filled with aqueous 25% KOH and sealed with polypropylene. The specific capacity achieved by these cells is 2 to 4 C/cm^2 and is limited by the positive electrode. Each cell occupies only 10^{-3} to 10^{-2} cm^2. These batteries can sustain a current density of 100 mA/cm^2 for 2 seconds (MEMS require only 10 to 15 ms of high power discharge). Higher current densities can last for a few milliseconds. The low internal surface area of today's nickel-oxyhydroxide microelectrodes results in high polarization at high current densities, rendering thin film nickel-zinc cells incapable of sustaining current densities greater than 1 A/cm^2. An SEM photomicrograph of a microscopic nickel-zinc battery is shown in Fig. 31.24.

FIGURE 31.24 SEM-micrograph of a Ni/Zn cell. *(Courtesy of Bipolar Technologies Corp.)*

31.8 HANDLING AND STORAGE

The nickel-zinc battery does not impose any special handling or storage requirements beyond those of any other commercially available battery. However, it must be realized that any battery does impose some safety and handling issues such as electrical hazards or corrosive electrolyte spillage. A material safety data sheet (MSDS) and operating/handling instructions should be obtained from the battery manufacturer.

31.8.1 Cell Venting Hazards

Nickel-zinc, like many other commercially available batteries, incorporates a safety vent to prevent an overpressure condition in the battery which might result in the rupture of the battery case. During normal operation, the battery operates in a sealed manner. The vent is present as a safety feature only. The nickel-zinc battery typically uses a resealable valve so that venting does not necessarily lead to immediate battery failure. This low-pressure valve also serves to regulate internal pressure, which assists in gas recombination within the battery. Since it is possible for the valve not to open, some potential hazard does exist. The primary concern is the fire and explosion hazard due to potentially flammable gasses being vented. It is also possible that some small quantity of corrosive electrolyte may be entrained within the gas. These potential hazards to personnel and equipment must be recognized and taken into consideration during operation of the battery. Venting is most likely during severe over-charge or overdischarge conditions, but might possibly occur during normal battery operation as well.

31.8.2 Overcharge and Overdischarge Protection

The battery should be protected from excessive overcharge and overdischarge. Extreme abuse may reduce battery performance and cycle life or cause a potential safety hazard. The system in which the battery is installed should provide protection for the battery under adverse conditions. Voltage exceeding the normal charging voltage should not be applied to the battery under any circumstances.

31.8.3 Fire and Explosion Hazard

Batteries may produce potentially explosive gases, particularly near the end of charge or when over-discharged. Any battery which has a safety relief vent has the potential for venting hydrogen or other explosive gases. The nickel-zinc battery presents no special hazard in this respect compared to other batteries which are capable of venting.

31.8.4 Storage

Nickel-zinc batteries are similar to other nickel-based alkaline rechargeable batteries. In general, nickel-zinc batteries are tolerant of storage at low states-of-charge. Because of self-discharge, most storage occurs at low state-of-charge. It is generally advisable to store batteries in a controlled manner when possible or at least within the non-operating environmental conditions specified by the manufacturer. It is preferred to store the batteries in a cool and dry place. Long-term storage at temperature extremes may adversely affect battery cycle life. Long-term storage above +50°C is not recommended or reduced battery performance may result. Smoking should not be permitted in the storage area. If batteries are being placed into storage prior to use, they should be withdrawn from storage on a first in/first out basis (FIFO). Nickel-zinc, like most other battery chemistries, should be stored on open-circuit. If the battery is stored with a load across the terminals, cycle life may be reduced. The non-operating (storage) temperature range for the nickel-zinc battery is typically specified as −20 to 50°C. The battery manufacturer's recommendations should be followed.

Short-term Storage. There are no issues regarding short-term storage of the nickel-zinc battery for periods of less than 6 months. If the battery has been unused for some period of time, a few "wake-up" cycles may be required before the battery recovers full capacity. This effect is typically not noticeable in most applications. It is possible for the battery to exhibit a lower than normal charge voltage which would preclude normal end-of-charge voltage termination. The charger should have a back-up charge termination mode to counter this possibility, including a time limit charge termination as a minimum.

Long-term Storage. Long-term storage is typically considered to be a period exceeding 6 months in which the battery is stored at essentially zero state-of-charge (due to self-discharge). Nickel-zinc batteries can be stored for up to 3 years. Long-term storage should be done in a controlled environment to minimize adverse effects on the battery of temperature extremes. If stored for periods of more than 3 years, some degradation in performance may be observed. A more aggressive charge may be required to fully recover capacity after long-term storage greater than 6 months. If possible a "pre-conditioning" charge can be performed which terminates at a slightly higher voltage than the normal charge termination voltage. This serves to overcome passivating oxide layers that may build up on the electrodes during long periods of non-use. Pre-conditioning allows 60% capacity recovery on the first use after storage and 90% recovery on the second use. If a preconditioning charge is not possible, the battery may require several cycles to achieve full capacity using the normal charging method. As mentioned above, long-term storage above +50°C is not recommended or reduced battery performance may result.

Acknowledgment

The authors would like to acknowledge the significant contributions of Jonathan O'Neill, Frank Cao and Kendra Pruneau, Evercel Corporation, in researching the scientific literature, compiling background information and assisting in the preparation and editing of this chapter.

REFERENCES

1. J. McBreen, *Journal of Power Sources,* **51** (1994), pp. 37–44.

2. F. R. McLarnon and E. J. Cairns, *J. Electrochem. Society.* Vol. 138, No. 2, February 1991.

3. A. Salkind, F. McLarnon, V. Bagotsky, Electrochemical Society, Pennington, N.J. (1996).

4. R. Jain, T. C. Adler, F. R. McLarnon and E. J. Cairns, *J. Applied Electrochem.* **22** (1992), pp. 1039–1048.

5. R. Plivelich, F. McLarnon, E. Cairns, *Journal of Applied Electrochemistry,* Vol. 25, No. 5, pp. 433–440 (1995).

6. D. Ohms, et. al. DE-99:0G3041, NDN-108-0667-6311-7, Deutsche Automobilgesellschaft GmbH (1998).

7. U.S. Patent 4,546,058 (1985) Charkey and Januszkiewicz.

8. K. Bass, P. J. Mitchell and G. D. Wilcox, *Journal of Power Sources,* **35** (1991), pp. 333–351.

9. M. E. Alekseeva, *Russian Journal of Applied Chemistry,* Vol. 71, No. 6, 1998, pp. 969–974.

10. V. V. Romanoff, *J. Appl. Chem* (*USSR*) **35** (1962), 1246.

11. Y-M. Wang and G. Wainwright, *J. Electrochem. Soc.* Vol. 133, No. 9, (1986), pp. 1869–1872.

12. U.S. Patent 5,460,899.

13. R. A. Sharma, *J. Electrochem. Soc.* **135,** No. 8, pp. 1875–1882.

14. U.S. Patent 5,863,676 (1999), Charkey and Coates.

15. U.S. Patents 5,547,779 (1996) and U.S. Patent 5,320,916 (1994). (Charkey et al.)

16. T. C. Adler, F. R. McLarnon and E. J. Cairns, *J. Electrochem. Soc.* Vol. 140, No. 2, February 1993 pp. 289–294; *Ind. Eng. Chem. Res.* 1998, **37,** 3237–3241.

17. E. G. Gagnon *J. Electrochem. Soc.* Vol. 138, No. 11, November 1991.

18. Nickel-Zinc Batteries for Commercial Applications, Dwaine Coates and Allen Charkey, 34th Intersociety Energy Conversion Engineering Conference (IECEC), Vancouver, BC, Canada, August 1999.

19. U.S. Patent 4,552,821 (1985).

20. W. Taucher-Mautner, K. Kordesch, Proceedings of the Symposium on Batteries for Portable Applications and Electric Vehicles, The Electrochemical Society, Pennington, N.J., pp. 710–716 (1997).

21. M. E. Alekseeva, *Russian Journal of Applied Chemistry,* Vol. 71, No. 6, 1998, pp. 969–974.

22. U.S. Patent 5,658,694. (Charkey)

23. U.S. Patent 4,810,598. (Levy and Charkey)

24. D. Dappert, AFWAL-TR-82-2003, Defense Technical Information Center (1982).

25. D. M. Ryan and R. M. LaFollette, Proceedings of the 32nd Intersociety Energy Conversion Engineering Conference, Vol. 1, pp. 77–82, American Institute of Chemical Engineers, New York (1997).

CHAPTER 32
NICKEL-HYDROGEN BATTERIES

James D. Dunlop, Jack N. Brill, and Rex Erisman

32.1 GENERAL CHARACTERISTICS

A sealed nickel-hydrogen (Ni-H_2) secondary battery is a hybrid combining battery and fuel-cell technologies.[1] The nickel oxide positive electrode comes from the nickel-cadmium cell, and the hydrogen negative electrode from the hydrogen-oxygen fuel cell. Major advantages and disadvantages are listed in Table 32.1.

Salient features of this hybrid Ni-H_2 battery are a long cycle life that exceeds any other maintenance-free secondary battery system; high specific energy (gravimetric energy density) compared to other aqueous batteries; high power density (pulse or peak power capability); and a tolerance to overcharge and reversal. It is these features that make the Ni-H_2 battery system the energy storage subsystem currently employed in many aerospace applications, such as geosynchronous earth-orbit (GEO) commercial communications satellites, and low earth-orbit (LEO) satellites, such as the Hubble space telescope.

Application of the Ni-H_2 battery has mainly been directed toward the aerospace field. Recently, however, programs have been started for terrestrial applications, such as long-life stand-alone photovoltaic systems.

TABLE 32.1 Major Advantages and Disadvantages of the Nickel-Hydrogen Battery

Advantages	Disadvantages
High specific energy (60 Wh/kg)	High initial cost
Long cycle life, 40,000 cycles at 40% DOD for LEO applications	Self-discharge proportional to H_2 pressure
Long lifetime in orbit, over 15 years for GEO applications	Low volumetric energy density:
Cell can tolerate overcharge and reversal	50–90 Wh/L (IPV cell)
H_2 pressure gives an indication of state of charge	20–40 Wh/L (battery)

32.2 CHEMISTRY

The electrochemical reactions of the Ni-H$_2$ cell for normal operation, overcharge, and reversal are

Normal operation:

Nickel electrode

$$NiOOH + H_2O + e \underset{charge}{\overset{discharge}{\rightleftharpoons}} Ni(OH)_2 + OH^-$$

Hydrogen electrode

$$\tfrac{1}{2}H_2 + OH^- \underset{charge}{\overset{discharge}{\rightleftharpoons}} H_2O + e$$

Net reaction

$$\tfrac{1}{2}H_2 + NiOOH \underset{charge}{\overset{discharge}{\rightleftharpoons}} Ni(OH)_2$$

Overcharge:

Nickel electrode

$$2OH^- \rightarrow 2e + \tfrac{1}{2}O_2 + H_2O$$

Hydrogen electrode

$$\tfrac{1}{2}O_2 + H_2O + 2e \rightarrow 2OH^-$$

Reversal:

Nickel electrode

$$H_2O + e \rightarrow OH^- + \tfrac{1}{2}H_2$$

Hydrogen electrode

$$\tfrac{1}{2}H_2 + OH^- \rightarrow H_2O + e$$

32.2.1 Normal Operation

Electrochemically, the half-cell reactions at the positive nickel oxide electrode are similar to those occurring in the nickel-cadmium system. At the negative electrode, hydrogen gas is oxidized to water during discharge and is reformed, during charge, from the water by electrolysis. The net reaction shows hydrogen reduction of nickel oxyhydroxide to nickelous hydroxide on discharge with no net change in KOH concentration or in the amount of water within the cell.

32.2.2 Overcharge

During overcharge, oxygen is generated at the positive electrode. An equivalent amount of oxygen is recombined electrochemically at the catalytic platinum electrode. Again, there is no change in KOH concentration or the amount of water in the cell with continuous overcharge. The oxygen recombination rate at the negative platinum electrode is very rapid, sustaining very high rates of continuous overcharge, provided that there is adequate heat transfer away from the cell to avoid thermal runaway. This is one of the operational advantages of the Ni-H$_2$ cell.

32.2.3 Reversal

During cell reversal, hydrogen is generated at the positive electrode and consumed at the negative electrode at the same rate. Therefore the cell can be operated continuously in the cell reversal mode without pressure buildup or net change in electrolyte concentration. This is a unique feature of the system.

32.2.4 Self-Discharge

The electrode stack is surrounded by hydrogen under pressure. A salient feature is that the hydrogen reacts electrochemically but not chemically to reduce the nickel oxyhydroxide. Actually, the nickel oxyhydroxide is reduced chemically, but at such an extremely low rate that performance for aerospace applications is not affected.

32.3 CELL AND ELECTRODE-STACK COMPONENTS

Ni-H$_2$ cell stacks are assembled in three distinct configurations. These include the COMSAT back-to-back, the Air Force recirculating and the hybrid, Mantech back-to-back designs. This section describes the electrode-stack components used for the fabrication of aerospace Ni-H$_2$ cells in these configurations. Figure 32.1 shows the truncated disk electrode-stack components used in the COMSAT design. Figure 32.2a and b shows the circular components for the Air Force recirculating and hybrid, Mantech back-to-back designs.

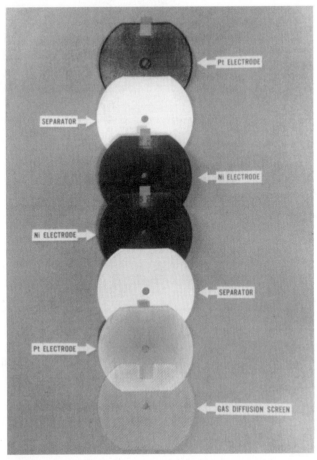

FIGURE 32.1   COMSAT bus-bar-configuration electrode-stack components.

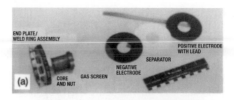

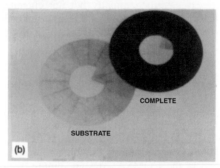

FIGURE 32.2 Air Force pineapple-slice config-uration: (*a*) Stack components. (*b*) Negative elec-trode. (*c*) Pressure-vessel cylinder and dome.

32.3.1 Positive Electrodes (Sintered)

The sintered positive electrode consists of a sintered porous nickel plaque that is impregnated with nickel hydroxide active material. The porous sintered plaque serves to retain the active nickel hydroxide material within its pores and to conduct the electric current to and from the active material. Essential features of the sintered plaque are high porosity, large surface area, and electrical conductivity in combination with good mechanical strength.[2]

Active material is impregnated into the sintered plaque by an electrochemical impreg-nation process. There are two electrochemical impregnation processes used—aqueous and alcoholic. The aqueous impregnation process (Bell Laboratories process)[3] uses an aqueous-based nickel nitrate solution for the impregnation bath. The alcoholic impregnation process (Air Force process)[4] uses an alcohol-based nickel nitrate solution for the impregnation bath. Both processes provide the following advantages:

1. *Loading of active material:* Electrochemical impregnation gives very uniform loading of the active material within the pores of the nickel sinter.

2. *Loading level:* The loading level of active material can be accurately controlled by the electrochemical impregnation process. Typical loading values are

1.67 ± 0.1 g/cm³ void volume for GEO applications
1.65 ± 0.1 g/cm³ void volume for LEO applications

32.3.2 Hydrogen Electrode

The hydrogen electrodes consist of a Teflon-bonded platinum black catalyst supported on a photo-etched nickel substrate with Teflon bonding. The sintered Teflon-bonded platinum electrodes were originally developed at Tyco Laboratories for the fuel-cell industry.[5] For Ni-H₂ cells, a hydrophobic Teflon backing was added to these platinum electrodes to stop water or electrolyte loss from the back side of the negative platinum electrode during charge and overcharge while readily allowing diffusion of hydrogen and oxygen gas. The use of Gortex® as the microporous Teflon membrane resulted from a development contract with HAC (Hughes Aircraft Corporation).[6] The platinum content is normally specified as 7.0 ± 1.0 mg/cm². The physical properties of this hydrogen electrode provide the right interface for the electrochemical reactions to occur without flooding or drying out the electrode at the separator interface.

32.3.3 Separator Materials

Two types of separator materials are being used for aerospace Ni-H₂ cells, (1) asbestos (fuel-cell-grade asbestos paper) and (2) Zircar (untreated knit ZYK-15 Zircar cloth).

Fuel-cell-grade asbestos is a nonwoven fabric with a thickness of 10 to 15 mil. The asbestos fibers are made into a long roll of nonwoven cloth by a paper-making process. As an added precaution, the asbestos can be reconstituted in a blender and then reformed into a cloth to avoid any nonuniformity in the original structure that would allow oxygen to bubble through. The fuel-cell-grade asbestos has a high bubble pressure for oxygen gas; a pressure difference of more than 1.7×10^5 Pa is required across the separator cloth (250 μm thick) to force oxygen bubbles through the material. During overcharge, oxygen gas is forced off the backside of the positive electrode. The oxygen cannot channel or bubble through the separator to cause rapid recombination at the negative electrode.

Zicar fibrous ceramic separators are available in textile product forms (Zircar Products, Inc.). These textiles are composed of zirconia fibers stabilized with yttria. These materials offer the extreme temperature and chemical resistance of the ceramic zirconia. They are constructed of essentially continuous individual filaments fabricated in flexible textile forms. Even with the fibrous structure, the inherently brittle nature of the ceramic material zirconia makes these separators fragile and susceptible to breaking. They must be handled with care. Untreated knit ZYK-15 Zircar cloth material is the tensile form used for Ni-H₂ cells.[7] Either one of two 250 to 380 μm-thick layers of this separator material can be used. The second ZYK-15 layer is normally used as a backup to prevent oxygen channeling in the event of assembly damage to the first layer. The knit Zircar cloth has a very low oxygen bubble-through pressure, and during charge and overcharge, oxygen gas readily permeates through the separator to recombine at the hydrogen platinum electrode to form water.

Both the asbestos and the Zircar separators serve the following functions:

1. They act as separators between positive and negative electrodes.

2. They serve as reservoirs for KOH electrolyte and remain stable in the electrolyte, allowing long-term storage and cycling.

3. They serve as media for charge and discharge current through the separator via ionic conduction of hydroxyl ions in the electrolyte.

32.3.4 Gas Screen

A polypropylene gas diffusion screen is placed behind the hydrogen electrode to allow hydrogen gas and oxygen gas to diffuse to the back side of the negative electrode with the Teflon backing.

32.4 Ni-H₂ CELL CONSTRUCTION

Sealed Ni-H$_2$ cells contain hydrogen gas under pressure within a cylindrical pressure vessel (see Fig. 32.2c). They are referred to as individual-pressure-vessel (IPV) cells because each individual cell is contained within its own pressure vessel. IPV cells are assembled using either single or dual electrode stack configurations inside the pressure vessel. An extension of IPV is the two cell (2.5 V) CPV design made by connecting the dual electrode stacks in series within a single pressure vessel. IPV designs include cells having diameters of 6.35, 8.89, 11.43 and 13.97 centimeters.

Descriptions of the various cell designs follow. These designs represent the first generation of Ni-H$_2$ cell technology, which was developed in the 1970s and utilized in the 1980s along with the baseline designs currently in use.

32.4.1 COMSAT Ni-H₂ Cell

Components of the COMSAT NTS-2 Ni-H$_2$ cell are shown in Fig. 32.3 with the electrode-stack assembly and weld ring positioned in front of the pressure shells.[8,9] These cells were built by Eagle-Picher Industries (EPI) under an INTELSAT/COMSAT licensing agreement. The U.S. Navy's Navigation Technology Satellite-2 (NTS-2), launched on June 23, 1977, was the first flight demonstration of the Ni-H$_2$ battery.

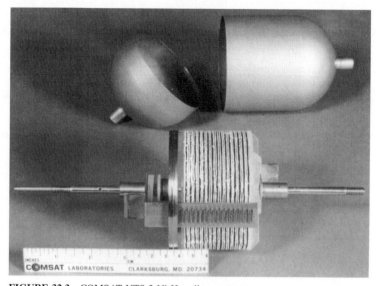

FIGURE 32.3   COMSAT NTS-2 Ni-H$_2$ cell components.

Electrode Stack. Figure 32.1 shows the basic arrangement of the electrode-stack components for the COMSAT back-to-back design. Two positive nickel oxide electrodes are positioned back-to-back. A separator is placed on each side of the positive electrodes. The negative platinum electrodes are placed with the platinum black surface next to the separator material. A plastic diffusion screen is placed on the back side of each negative electrode to facilitate gas diffusion to the back side of this electrode. These components constitute one module of the electrode stack. This arrangement is repeated until the number of modules is reached to provide the required capacity. A complete stack can be seen in Fig. 32.3. The bus bars for the positive and negative electrodes are located along the outside of the electrode stack.

During charge and overcharge, oxygen gas that evolves at the nickel electrodes is forced out between the back-to-back positive electrodes. The oxygen diffuses into the gas space between the electrode stack and the pressure vessel wall into the region of the gas diffusion screens on the back of the negative electrode, and through the porous backing of the negative electrode, where it combines with hydrogen to form water. The partial pressure of oxygen is dependent on this diffusion process. The limiting step is the oxygen diffusion in the gas-phase pores of the Teflon-bonded electrode, not in the Teflon backing.[6] The fraction of oxygen gas should be less than 0.5% in the surrounding hydrogen gas when the cell is continuously overcharged at a C/2 rate.

Pressure Vessel. The pressure vessel (dome and cylinder), terminal bosses and weld rings are all fabricated from Inconel alloy 718. The weld ring is manufactured by one of two methods. The first is manufactured using an investment casting process and is then machined to final dimensions. The second is manufactured by machining to final dimensions from an extruded or wrought Inconel metal. The outside diameter of the weld ring is machined as a T section to position the pressure shells on the weld ring and provide a backup support for an electron beam girth weld. The Inconel 718 pressure vessel shells are manufactured to a near uniform thickness using either a hydroforming or drawn process and then cut to length. The thickness is determined by the operating pressure and cycle requirements for the particular use. The pressure shells are "age hardened" using a standard heat treatment process. The terminal bosses for the compression seals are machined from Inconel 718 material and electron beam welded into the domes of the pressure vessel shells. Nylon plastic is injection molded into these barrels. The Ziegler compression seal[10] is made by crimping the bosses. Cell designs commonly operate under maximum operating pressures between 4.1×10^6 Pa and 8.3×10^6 Pa. Depending on the particular use, the vessels are designed to provide a safety factor between 2:1 to over 4:1.

32.4.2   Air Force Ni-H$_2$ Cell

Typical components for the Air Force NiH$_2$ cell are shown in Fig. 32.2 including the electrode stack components, the negative electrodes with the chem-milled substrate, and the pressure vessel cylinder and dome with the plasma sprayed zirconium oxide wall wick for the electrolyte. These components are typically assembled in one of two configurations. Separators used in these designs are asbestos and Zircar alone or in combination with each other.

The first is commonly referred to as the recirculating electrode stack design. It is made of a number of modules comprised of a gas screen, hydrogen electrode (negative), separator(s), and nickel electrode. The capacity of the cell is determined by the number of modules used to assemble the stack. A single gas screen and hydrogen electrode is used as the final module to maintain a hydrogen electrode for recombination on both sides of the positive throughout the stack. A wall wick is used on the interior of the pressure vessel to return electrolyte to the stack, hence the name recirculating design.

In this design using asbestos or a combination of asbestos and Zircar separators, the oxygen generated on overcharge diffuses off the back of the nickel electrode. The diffusion path is very short; the oxygen gas simply travels through the gas screen to recombine at the next hydrogen electrode. Oxygen comes off the back side of the positive electrode of one module and recombines to form water at the next module. During overcharge, this transfer of water to the next module occurs throughout the electrode stack. The last module in the stack is simply a negative electrode and separator reservoir. The water formed at this electrode-separator combination goes to the wall wick and recirculated back to balance the electrolyte throughout the stack. With this recirculating design, the oxygen concentration is kept very low (below 0.2% in the surrounding hydrogen gas) during continuous overcharge at the C rate.

The second configuration is the Air Force Mantech back-to-back design. It is made in the same manner as the COMSAT design. A separator is placed on each side of the positive electrodes. The negative platinum electrodes are placed with the platinum black surface next to the separator material. A plastic diffusion screen is placed on the back side of each negative electrode to facilitate gas diffusion to the back side of this electrode. These components constitute one module of the electrode stack. This arrangement is repeated until the number of modules is reached to provide the required capacity. This design also uses the wall wick for return of electrolyte to the stack.

Designs using Zircar alone normally extend the separator to the pressure vessel wall. The Zircar contains large enough pores for oxygen gas to permeate through the pores of this separator to the negative electrode, where it recombines to form water. Oxygen can, of course, also emerge off the back side of the nickel electrode and diffuse through the gas screens as before. However, most of the oxygen permeates through the separator and there is little or no recirculation of water in cells with Zircar separators. For this design, the concentration of oxygen in the surrounding hydrogen gas is negligible.

The electrode-stack components are shaped like a pineapple slice (see Fig. 32.2b) with provisions for the tab in the center hole. These electrode-stack components are assembled onto a polysulfone central core (see Fig. 32.2a). The electrode tabs are brought out through this central core. The positive and negative tabs can be in opposite directions or in the same direction depending on the terminal configuration. This center core serves to align the electrode-stack components, provide a conduit for the positive and negative tabs, and insulate the positive and negative tabs from each other and from the electrode-stack components.

The cell capacity in a particular diameter is limited by the ability to manufacture a pressure vessel of sufficient length. In designing cells of larger capacity without increasing the diameter of the pressure vessel, two approaches are used. The first involves the use of a dual stack design to increase the capacities of cells for the individual cell diameters.[15] This is accomplished by using two stacks assembled on a single core as described above. The two stacks are separated by end plates and a weld ring and are connected electrically in parallel to attain a 1.25 volt cell. The second approach utilizes a three piece pressure vessel assembly.[16] A single electrode-stack is made with weld rings at each end. A cylinder is placed over the stack and joined at each end with the weld rings and two dome assemblies.

Heat transfer is better with the pineapple-slice configurations than with the COMSAT back-to-back configuration. Heat is transferred uniformly from the entire circumference of the pineapple-slice electrodes, whereas sections are removed from the circumference of the COMSAT back-to-back electrodes.

Pressure Vessel.   The pressure vessels used for the Air Force designs are essentially the same as those for the COMSAT designs. Certain designs utilize chemical milling to remove material from lower stress areas for weight reduction. Typically the weld areas are not reduced to compensate for the strength reduction of the age hardened, Inconel 718 material in the heat-affected zone of the weld areas. The chemical milling is done prior to heat treating (age hardening) of the pressure vessel. The operating pressures and design margins are similar to those discussed for the COMSAT designs.

Two terminal designs have been used. One involves the terminal design using the compression seal described for the COMSAT designs. The second utilizes a hydraulic seal design. When this is used, the seal area is hydroformed as an integral part of the dome and cylinder. The terminal seals are hydraulic cold-flow Teflon seals.[14]

Electrolyte Management.   There are three mechanisms for the loss of electrolyte from the electrode stack: (1) by entrainment in the hydrogen and oxygen gases evolved during charge and overcharge, (2) by weeping of the negative electrode, and (3) by electrolyte displacement, that is, the electrolyte being pressed out of the positive electrodes in the cell stack by oxygen gas evolved during charge and overcharge.

Electrolyte loss by both entrainment and weeping of the negative electrode was determined to be negligible for negative electrodes with Gortex backing for both back-to-back and recirculating electrode-stack configurations. The major electrolyte loss mechanism is by displacement. When electrolyte is added to the cell, the void volume of the positive electrode is completely saturated with electrolyte. During activation, approximately 25% of the electrolyte in the positive electrodes is displaced by oxygen gas during charge and overcharge of the cell.[6] It was found that electrolyte loss by displacement occurred during initial cycling (activation) but eventually decreased to zero, leaving enough electrolyte to operate the cell efficiently.[11,12]

Water Loss.   Water loss from the electrode stack can result from evaporation and condensation of water vapor from the cell stack to the pressure-vessel wall when a large enough temperature difference exists between stack and wall (approximately 10°C difference). The plasma-sprayed zirconium oxide wall wick[13] shown in Fig. 32.2c provides a return path for any water loss from the cell stack independent of the mechanism.

32.4.3 Specific Energy and Energy Density

Figures 32.4 and 32.5 depict the specific energy and energy density that could be projected for the different nickel-hydrogen cell designs. The actual values may vary depending on manufacturer.

1. In general, the specific energy increases as the capacity increases.

2. The choice and number of separators affect the weight (quantity of electrolyte) and thus the specific energy of the cell.

3. The energy density is primarily a function of the pressure range or free volume in the cell. Cells operating at higher maximum operating pressures have higher energy densities. This increase is a result of the reduction in weight of the pressure vessel at the higher pressures.

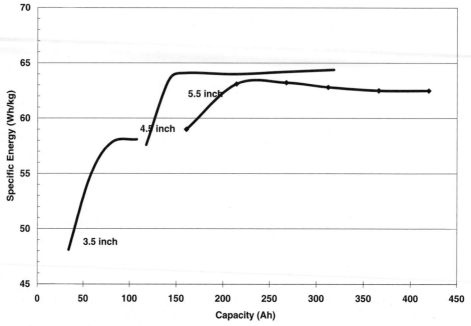

FIGURE 32.4   Specific energy of Ni-H$_2$ cells.

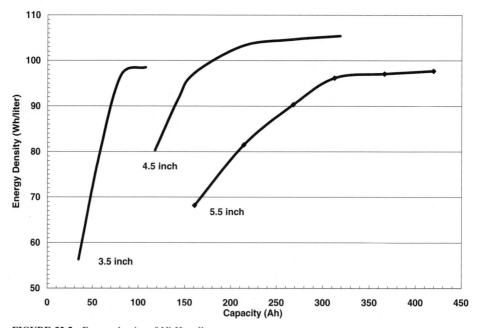

FIGURE 32.5   Energy density of Ni-H$_2$ cells.

32.5 Ni-H$_2$ BATTERY DESIGN

Various Ni-H$_2$ battery designs have evolved through the years. They are tailored to the specific application and interface with the particular satellite. Mechanical and thermal requirements are the primary drivers for the configuration and interface of each battery. The nickel-hydrogen system is sensitive to temperature and performs best between $-10°C$ and $+10°C$. Thus, thermal control of the battery is important to minimize size and weight.

Several features have been integrated into the battery designs which enhance the performance and reliability. These include: pressure monitoring of cells through strain gage or transducers, strain gage voltage amplification circuits, individual cell voltage monitors, temperature monitoring, redundant individual cell heaters, and individual cell diode bypass protection. Bypass diodes on each cell protect the battery against a failure from an open-circuited cell. Protection in the charge direction is provided by three silicon diodes in series, while protection in the discharge direction is provided by one Schottky barrier diode. The diodes are mounted on heat sinks on the thermal sleeves near the base of the cells or on a separate panel attached to the battery base plate.

Several different battery configurations can be seen in Figs. 32.6, 32.7, 32.8, 32.9 and 32.10. One of the earliest Ni-H$_2$ battery designs was flown on the INTELSAT V program. Two 27-cell, 30 Ah Ni-H$_2$ batteries were used to provide the electric energy during launch, transfer orbit and solar eclipses.[17]

The first battery using the dual electrode stack configuration was made by Eagle-Picher Technologies, LLC for the EUTELSAT II Program. The first of two flight sets was delivered in February 1990 and launched in August 1990.[18] This battery is shown in Fig. 32.7.

The photovoltaic power subsystem for the International Space Station will use Ni-H$_2$ batteries for energy storage to support eclipse and contingency operations. These batteries

FIGURE 32.6 DMPS 100-Ah Ni-H$_2$ battery assembly.

FIGURE 32.7 EUTELSAT II 58-Ah Ni-H$_2$ battery.

FIGURE 32.8 International Space Station 81-Ah Ni-H$_2$ 38-cell assembly.

FIGURE 32.9   TRW 81-Ah Ni-H$_2$ battery for a flight program (Photograph provided courtesy of TRW).

FIGURE 32.10   MIDEX 23-Ah CPV Ni-H$_2$ battery assembly.

are designed for LEO operation with a 6.5-year design life expectancy and are configured as orbital replacement units (ORU), permitting replacement of worn-out batteries over the anticipated 30-year station life.[19] The baseline energy storage system design contains 2 batteries of 76 cells of 81-Ah capacity, packaged as 38-cell assemblies (Fig. 32.8), or approximately 184.7 kWh of stored energy.

32.5.1 Design Features

Mechanical Design.   Typically, each battery will have a thermal sleeve around each cell. The cells are mechanically restrained by clamping them in a precision-machined sleeve. These sleeves can be made of either a metal such as aluminum or a composite made in a manner to provide electrical isolation, high thermal conductivity and strength. The sleeve is isolated electrically from the cell by a blanket, such as CHO-THERM which allows thermal transfer, wrapped around the cylindrical portion of the cell between the cell and sleeve. The space between the sleeves, blanket and cell is normally filled with a material such as an RTV 566 to provide better thermal transfer as well as to bond the interfaces mechanically. The sleeves are then either attached mechanically to a base plate which is the interface to the satellite structure or are attached to an interface such as extruded heat pipe assemblies which are a part of the satellite structure. The exposed surfaces of the cells are protected by a coating of Solithane or a combination of paint on the cell pressure vessel and Solithane. The desired battery voltage defines the number of cells used for the assembly.

Thermal Design.   Each battery is designed to operate thermally within a specified set of limits dictated by the mission requirements and the satellite interface. These limits are normally in the range between $-10°C$ and $+15°C$ during the periods of battery operation. During periods of inoperation such as the equinox periods of the geosynchronous missions, the temperature range can be reduced since the thermal output of the battery is less. Heat dissipated from within the cells is conducted radially to the pressure vessel wall, through the insulating blanket/RTV 566 to the thermal sleeves, and down the thermal sleeve to the base plate or mounting interface. Depending on the interface, the heat is transferred through the mounting to second surface mirrors or to thermal heat pipes for dissipation. The charge control and heater assemblies must be operated in conjunction with the passive or active thermal dissipation to regulate operating temperature. The battery surfaces may be anodized to optimize thermal emissivity.

Weight and Energy Density.   Most battery designs are optimized through stress and thermal analysis. Figures 32.11 and 32.12 depict expected specific energies and energy densities for 22-cell, 28-V batteries. These, of course, may vary depending on the manufacturer and the intended use.

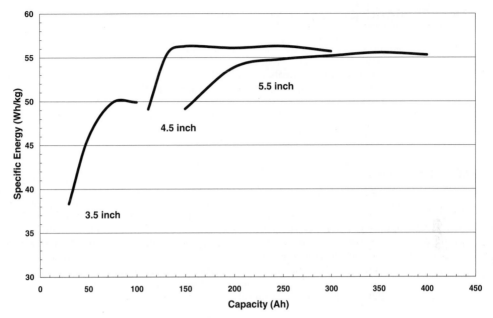

FIGURE 32.11   Specific energy for 28-V Ni-H$_2$ battery.

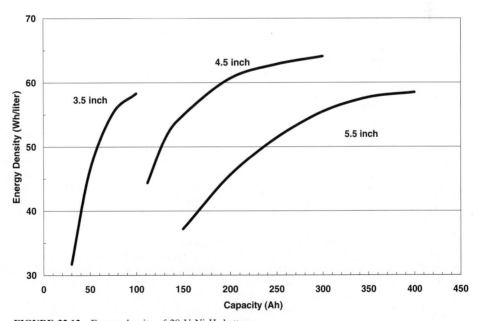

FIGURE 32.12   Energy density of 28-V Ni-H$_2$ battery.

32.6   APPLICATIONS

Aerospace applications of Ni-H$_2$ batteries can be divided into two categories: LEO and GEO applications. These two applications stress different requirements for the batteries. LEO life-time-in-orbit requirements are typically only 3 to 6 years at approximately 6000 cycles per year for a total cycle requirement of 18,000–36,000 cycles. The GEO applications stress lifetime in orbit, namely 15 to 20 years. The batteries are cycled about 100 times per year for a total of 1500 to 2000 cycles. Ni-H$_2$ batteries are also being developed and evaluated for terrestrial applications.[20,21]

32.6.1   GEO Applications

Eclipse Seasons.   Communications satellites are required to operate continuously without interruption, which includes operation during eclipse seasons. These satellites pass through eclipse periods each day during the equinox seasons. An eclipse occurs when the satellite is in the shadow of the earth. The eclipses start with a few minutes duration and build up to a maximum length of 72 min, which occurs midseason (about March 21 or September 23), and then drop off again symmetrically. The total duration of the season is 45 days. The batteries supply power to the spacecraft during eclipse periods and are recharged during the sunlight portion of each eclipse day. During the summer and winter solstice periods (approximately 138 days each) between eclipse seasons, the batteries are kept on trickle charge.

Charge Control.   In the middle of the eclipse season, the battery is typically discharged up to 70% of its beginning-of-life rated capacity. At the end of life, after 15 years of operation, the battery still must meet the same initial load requirements. The preferred charge method for GEO satellites is to recharge the battery at a fixed c/d ratio, returning 105 to 115% of the capacity removed on discharge at a high charge rate, and then switch to a low trickle charge rate for the remainder of the 24-h eclipse day to maintain the batteries at the full (100%) state of charge. For the 135 days between eclipse seasons, the batteries are maintained on trickle charge in the fully charged condition.

Reconditioning.   Batteries are typically reconditioned prior to each eclipse season.

32.6.2   LEO Applications

Battery Requirements.   A 96-min orbit is typically used to characterize LEO satellite applications. The time it takes a satellite to orbit the earth at 555 km is 96 min. The satellite orbits the earth 15 times in one day. The orbital duration remains fixed, but the sunlight and eclipse periods vary with each orbit. For example, a LEO satellite orbiting 555 km above the earth, at an inclination of 28.3°, has an orbital period that is constant at 96 min; but during a given month, such as December 1991, the eclipse durations vary from a maximum of 35.58 min on December 1 to a minimum of 26.97 min on December 30.

Charge Control.   The battery is charged during the sunlight period and discharged during the eclipse period. With this high duty cycle, it is essential to minimize overcharge (heat dissipation) and maximize overall Watthour efficiency for the battery. A charge method is needed to compensate for the variation in the depth of discharge (variation in eclipse duration) and to minimize overcharge. If the battery can be maintained at a low temperature on charge between 0 and 10°C, the Ampere-hour charge efficiency approaches 100%, and the Watthour efficiency approaches 85%.

NASA's Marshall Space Flight Center (MSFC) and Lockheed Missile and Space Company (LMSC) selected a temperature-compensated voltage-limit charging method for charging the 88-Ah Ni-H$_2$ batteries that replaced the Ni-Cd batteries on board the Hubble Space Telescope (HST) satellite.[22,23] Figure 32.13 shows the battery voltage limits used for the HST program as a function of temperature. At the beginning of life, the batteries were charged to the K1-L3 and K2-L3 voltage limits at the high rate, then switched to trickle charge, which is approximately a $C/100$ rate. The K1-L3 setting has a cell voltage limit of 1.513 V at 0°C, or a battery voltage limit of 33.28 V for the 22-cell battery. The battery is not fully charged at this voltage limit but rather charged to about 73 Ah (83% of its rated capacity). The battery is thermally stable with this charging method. The overall battery Watthour efficiency is 80 to 85% during cycling with these level 3 control limits. Charging the battery to a higher voltage limit would decrease the coulombic efficiency, reduce the overall energy efficiency, increase the heat dissipated internally within the battery, and possibly exceed the constraints for thermal control of the battery. The cell pressure serves as an indication of the state of charge of the battery and is very useful in this type of application where the battery is not fully charged.

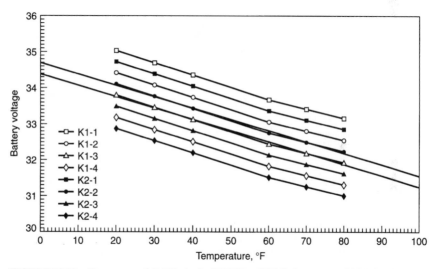

FIGURE 32.13   Charge control V-T limits for HST 22-cell Ni-H$_2$ battery, $y = 34.7412 - 0.0329x$, $R = 1.00$; $y = 34.316 - 0.0318x$, $R = 1.00$.

32.6.3 Terrestrial Applications

The advantages offered by Ni-H$_2$ batteries, including long life, low maintenance, and high reliability, make them very attractive for terrestrial applications such as stand alone photovoltaic systems or stand-by power for emergency or remote site use. The major drawback to the wider use of the Ni-H$_2$ battery is its high initial cost. The following are two examples of terrestrial applications addressing Ni-H$_2$ batteries.

Starting in 1983, Sandia National Laboratories sponsored a cost-sharing program with COMSAT Laboratories and Johnson Controls, Inc., for the design and development of a sealed Ni-H$_2$ battery for deep-discharge terrestrial applications that would be cost-competitive with lead-acid batteries in a system designed for a 20-year life. The main thrust of this program was to reduce the cost of the aerospace technology without comprising the desirable features of the Ni-H$_2$ system. Figure 32.14 shows a 5-cell, 6-V 100-Ah Ni-H$_2$ battery assem-

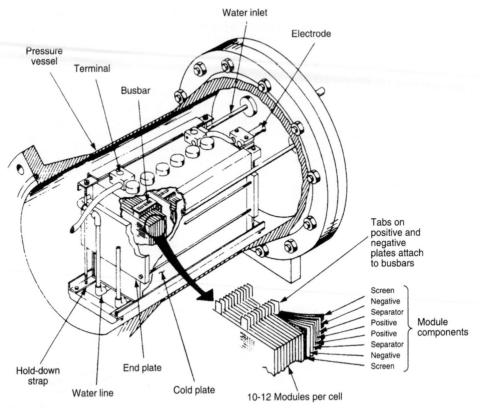

FIGURE 32.14 6-V 100-Ah terrestrial Ni-H$_2$ battery. (*Courtesy of COMSAT Laboratories.*)

bled in a pressure vessel. Assemblies have been tested by Sandia National Laboratories in combination with photovoltaic arrays. Conclusions were expressed that the cells "continue to perform well reinforcing the projection of a 20-year life, matching that of photovoltaic panels."[21]

Eagle-Picher Technologies, LLC has another design for use as stand-by power for a remote site in a terrestrial application. Again the primary effort in the design was to create a more cost-effective, reliable battery for terrestrial use. This design utilizes a combination of the dependent pressure vessel (DPV) and two-cell common pressure vessel (CPV) technologies. Five 2-cell DCPV units are assembled creating a 12-V battery. The nominal capacity of the battery is 44 Ah at 10°C. This battery is shown in Fig. 32.15.

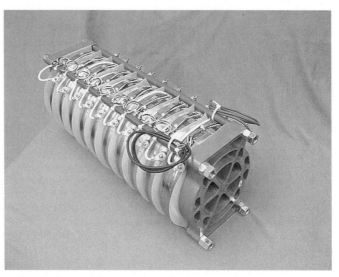

FIGURE 32.15   DPV battery assembly.

32.7 *PERFORMANCE CHARACTERISTICS*

32.7.1 Voltage Performance

Electrochemically impregnated nickel oxide electrodes are used in Ni-H$_2$ cells because of their excellent cyclic performance capabilities.[9,24] The capacity of these electrodes increases as the temperature decreases. The measured capacity of an electrochemically impregnated electrode is about 20% greater at 10°C than at 20°C. The capacity of the NTS-2 35-Ah cell at different temperatures is presented in Fig. 32.16 to show the variation in capacity with temperature. These NTS-2 cells were discharged at the $C/1.67$ rate. Note that the mid-discharge voltage is between 1.2 and 1.25 V.

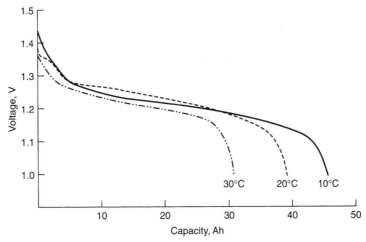

FIGURE 32.16   Capacity of NTS-2 35-Ah cell at different temperatures. Discharge rate, $C/1.67$.

The high-rate discharge of an INTELSAT V cell at 200 A (12-min rate) is shown in Fig. 32.17. The discharge profile is almost flat at 0.6 V; the potential drop of approximately 600 mV is due to the 3-mΩ terminal impedance of the cell (3 mΩ × 200 A = 600 mV). As previously seen in Fig. 32.16, the mid-discharge voltage is between 1.2 and 1.25 V for a cell discharge at the $C/1.67$ rate. The aerospace IPV Ni-H$_2$ cells are not optimized for high-rate discharge but are optimized for maximum specific energy at discharge rates of between $C/2$ and $C/1.5$. At higher rates, the usable energy drops off because of the I^2R losses in the terminals. For example, the INTELSAT V cell is capable of delivering about 50 Wh/kg up to the C rate (1-h rate) of discharge at 0°C. Above the C rate (30-A rate), however, the specific energy starts to drop off, as shown in Fig. 32.18.

A salient feature of the Ni-H$_2$ cell is that the pressure is a direct indication of the state of charge of a cell (Fig. 32.19). On charge, the hydrogen pressure increases linearly until the nickel oxide electrode approaches the fully charged condition. During overcharge, oxygen evolved at the positive electrode recombines at the negative electrode, and the pressure stabilizes. On discharge, the hydrogen pressure decreases linearly until the nickel oxide electrode is fully discharged. If the cell is reversed by overdischarging, hydrogen generated at the positive electrode is consumed at the negative electrode, and again the pressure is constant.

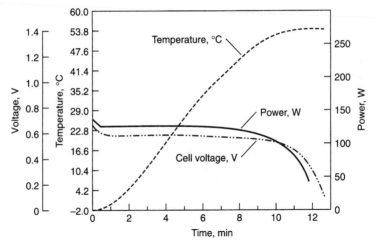

FIGURE 32.17 Discharge of INTELSAT V 30-Ah cell at 200-A rate. 6.7 C-rate discharge.

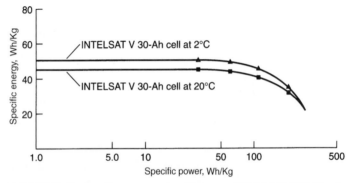

FIGURE 32.18 Specific energy vs. specific power of INTELSAT V Ni-H$_2$ cell.

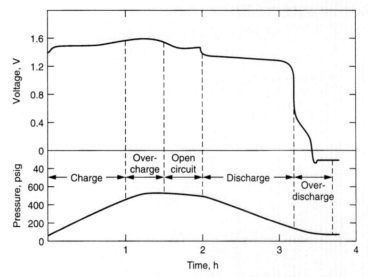

FIGURE 32.19   Pressure and voltage characteristics of NTS-2 Ni-H$_2$ cell at 23°C.

The effects on cell voltage and capacity of discharge at different rates for a 90 Ah Hubble Space Telescope cell can be seen in Fig. 32.20. The data was gathered from cells stabilized at 10°C. As the current increases, the capacity to 1.00 V decreases. Changes in the cell voltage and capacity can be attributed to the impedance of the cell (0.9 mΩ).

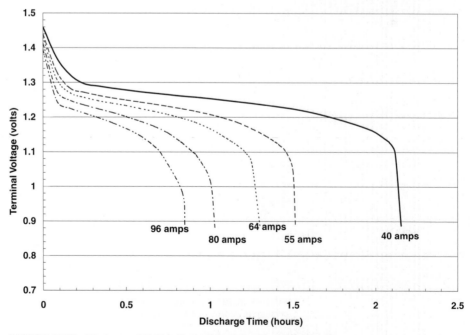

FIGURE 32.20   Discharge of Hubble Space Telescope cell at different rates.

32.7.2 Self-Discharge Characteristics of Ni-H₂ Cells

The self-discharge rate as a function of temperature was determined experimentally for Air Force 50-Ah cells used in the INTELSAT VI program.[25] Figure 32.21 gives the data for the self-discharge of these 50-Ah cells at 10, 20, and 30°C. Figure 32.22 shows the Arrhenius plot for these three temperatures. The slope of the straight-line regression fit to these three data points indicates an activation energy of 13.6 kcal/mol.[25]

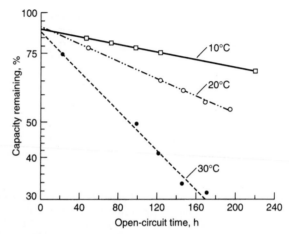

FIGURE 32.21 Self-discharge rates vs. temperature for 50-Ah Ni-H₂ cell. (*Courtesy of COMSAT Laboatories.*)

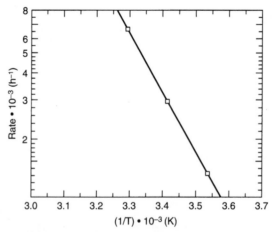

FIGURE 32.22 Arrhenius plot, self-discharge rates vs. temperature (*Courtesy of COMSAT Laboratories.*)

32.7.3   Capacity as a Function of Electrolyte Concentration

The effects of electrolyte concentration on capacity were determined experimentally. Air Force positive electrodes and the Air Force 50-Ah Ni-H$_2$ cells from the INTELSAT VI program were used for this investigation. The Air Force positive electrodes were impregnated with active material by the alcoholic electrochemical impregnation process. The plaque was manufactured using the dry-powder process.

For the Air Force standard 50-Ah Ni-H$_2$ cell, the electrolyte concentration was determined to be 26% KOH in the charged state and 31% KOH in the discharged state.[25] Cells were activated with three different levels of electrolyte concentration: 25, 31, and 38 wt% concentration of KOH. The electrolyte concentration in these cells was determined by analyses in both the charged and the discharged conditions. Table 32.2 presents cell capacity, electrolyte concentration, and average discharge voltage.

TABLE 32.2   Cell Capacity and Voltage vs. Electrolyte Concentration at 10°C

	Electrolyte concentration		
Parameter	38%	31%	25%
Cell capacity, Ah	64	56	43
Number of positive plates	40	40	40
Capacity per plate, Ah	1.60	1.40	1.08
Electrolyte concentration:			
Charged, wt% KOH	32*	26	21*
Discharged, wt% KOH	38	31	25
Average discharge voltage, V	1.247	1.268	1.290

*Estimated.

32.7.4   GEO Performance

The Ni-H$_2$ batteries on the INTELSAT V satellites have completed up to nine years in orbit.

Voltage Performance in Orbit.   In-orbit performance of the INTELSAT V batteries is judged by the minimum end-of-discharge voltage observed during an eclipse season. The minimum battery voltage requirement is 28.6 V, or 1.10 V per cell average with one cell failed short-circuited. The actual minimum battery voltages and the corresponding load currents and depths of discharge for each of the 14 batteries on the F6 through F15 satellites are presented in Table 32.3 for the Fall 1990 eclipse season.[26] Also presented are the minimum cell voltages within each battery and the corresponding average cell voltage per battery. After up to 7 years in orbit, the minimum end-of-discharge battery voltages on the longest eclipse day ranged from 31.2 to 32.4 V for all 14 batteries. The cells within the batteries were well matched at their minimum end-of-discharge voltages during the Fall 1990 eclipse season (maximum deviation is ±20 mV between cells within the same battery). The one exception is cell 22 in battery 1 on the F-6 spacecraft. This cell was 40 mV below the average cell voltage. These battery voltages are well above the minimum voltage requirement.

TABLE 32.3 Battery Loads and Minimum Voltages for Fall 1990 Eclipse Season

	DOD, %		Current, A		Voltage, V		Cell voltage, V			
	Battery 1	Battery 2	Battery 1	Battery 2	Battery 1	Battery 2	Batt. 1 av.	Batt. 1 min.	Batt. 2 av.	Batt. 2 min.
F-6	55.8	53.1	14.2	13.5	32.0	32.4	1.20	1.16	1.20	1.19
F-8	54.0	54.4	13.7	13.8	32.0	32.0	1.20	1.18	1.20	1.18
F-10	56.9	55.7	14.4	14.3	31.8	32.0	1.19	1.18	1.20	1.18
F-11	55.3	60.0	14.1	15.4	32.0	32.0	1.20	1.18	1.20	1.19
F-12	53.5	58.0	13.6	14.8	32.0	31.8	1.20	1.18	1.18	1.18
F-13	67.0	59.0	16.9	15.0	31.2	31.8	1.17	1.15	1.19	1.17
F-15	67.0	62.3	16.9	15.8	31.2	31.8	1.17	1.16	1.18	1.16

Pressure Data. INTELSAT V batteries are reconditioned prior to each eclipse season. The reconditioning capacity and pressure data for INTELSAT V F-6 battery 2 are presented in Table 32.4.[26] The pressure data were measured during the reconditioning discharge. The EOC pressure and the EOD pressure are the pressure at the start and the pressure at the end of the reconditioning discharge, respectively. The pressure constant is ΔP per measured capacity.

The data in Table 32.4 show the following:

1. The strain-gauge bridge circuit provides useful pressure data.

2. No change occurred in the reconditioning EOD pressure with time for these INTELSAT V battery cells. The EOD pressure in Fall 1991 was almost the same as the EOD pressure at the beginning of life in orbit.

3. The significance of these data is that no oxidation or corrosion has occurred within these cells. Any oxidation of the cell components would result in a pressure increase at the end of the reconditioning discharge.

TABLE 32.4 Reconditioning Capacity and Pressure Data for INTELSAT V F-6 Battery 2

Eclipse season	Measured capacity, Ah	Max EOC pressure, lb/in²	Min EOD pressure, lb/in²	ΔP, lb/in²	Pressure constant ΔP/measured capacity, lb/in²/Ah*
F83	38.1	No pressure data in database			
S84	35.4	No pressure data in database			
F84	37.7	516.39	13.87	502.62	13.33
S85	37.6	518.49	17.90	500.59	13.31
F85	37.5	515.14	17.23	497.9	13.27
S86	37.9	519.34	15.32	504.02	13.29
F86	37.6	519.73	22.03	497.70	13.23
S87	38.3	519.34	13.87	505.47	13.19
F87	37.2	519.73	22.03	497.7	13.37
S88	38.3	525.78	16.20	509.58	13.30
F88	37.8	521.86	17.90	503.96	13.33
S89	36.9	526.91	18.67	508.24	13.77
F89	40.2	534.22	−0.57	534.79	13.30
S90	38.6	551.73	19.22	532.51	13.79
F90	36.0	530.87	38.04	492.83	13.68
S91	39.5	546.52	17.23	529.29	13.39
F91	39.0	545.30	17.90	527.40	13.52
					Average 13.37

*1 lb/in² = 6895 Pa.

32.7.5 LEO Performance Data

The HST was launched on April 24, 1990, with six 88-Ah Ni-H$_2$ batteries as the primary energy storage subsystem. This was the first reported nonexperimental mission to use Ni-H$_2$ batteries in an LEO application.[27] The batteries are being charged to a temperature-compensated voltage limit as described in Sec. 32.6.2. The batteries are discharged to 7 to 10% of depth of discharge. As reported at the 1991 IECEC, "To date (April 1991) the performance of the batteries has been flawless."[27]

32.8 ADVANCED DESIGNS

32.8.1 Advanced Designs for IPV Ni-H$_2$ Cells

A number of advanced concepts for IPV Ni-H$_2$ cells are being used to improve cycle life at deep DOD[28] mitigating failure modes commonly found in Ni-H$_2$ cells. They include: (1) the use of alternative methods for recombination (that is, the catalyzed wall wick), (2) the use of serrated-edge separators to facilitate the movement of gas within the stack while maintaining physical contact with the cell wall, (3) the incorporation of Belleville washers to yield an expandable stack capable of accommodating some of the nickel electrode expansion that is known to occur with cycling, and (4) the use of lower KOH concentrations to improve cycle life.

Cells utilizing the catalyzed wall wick are now in use. This concept offers an improved thermal design with recombination taking place on the pressure vessel wall. The heat from recombination is removed immediately through the pressure vessel wall to the cell thermal sleeve. This design also mitigates damage from internal stack popping since the recombination site is outside the cell stack.

The use of serrated edges for the separator is usually employed in designs utilizing asbestos. The irregular edge allows the unrestricted passage of oxygen along the edge of the cell stack while maintaining contact with the wall wick of the pressure vessel for recovery of electrolyte and removal of heat from the stack.

The Belleville washer acts as a spring, allowing further compression to accommodate any plate expansion during cycling.

The effects of KOH concentration on the cycle life of Ni-H$_2$ were investigated.[29] A breakthrough in the LEO cycle life of individual pressure-vessels cells was reported and cell cycle life was improved by greater than a factor of 10 when KOH concentration was reduced from 31% to 26% in the fully discharged state. The lower concentrations, while enhancing the cycle life, result in a slightly higher mid-discharge operating voltage with a lower available capacity to the 1.00 V per cell limit.

32.8.2 Advanced Battery Design Concepts

The common pressure vessel (CPV) Ni-H$_2$ battery and the bipolar Ni-H$_2$ battery are two advanced battery design concepts investigated to improve the gravimetric and volumetric energy densities as compared to the IPV cell and battery Ni-H$_2$ technology.

Common Pressure Vessel.   Conceptually a CPV Ni-H$_2$ battery consists of a number of individual cells connected together in series and contained within one common pressure vessel.[30] For the IPV cell, each individual Ni-H$_2$ cell is contained within its own pressure vessel. Potential advantages for the CPV Ni-H$_2$ batteries include a significant increase in volumetric energy density (a decrease in volume), a decrease in manufacturing cost, a reduction in the complexity associated with the wiring and interconnection of IPV cells, an increase in specific energy, a reduction in the internal impedance of the battery, and improved heat transfer between the electrode stack and the pressure-vessel wall.

Several dual-cell CPV designs have been developed and tested. This design utilizes the dual-stack configuration used for IPV cells. For the CPV cell, the two stacks are connected in series, as shown in Fig. 32.23. This dual cell CPV battery offers a 30% reduction in volume and a 7 to 14% reduction in mass compared to an equivalent battery with IPV cells.[31]

Batteries utilizing these cells have been used in several flight programs including LEO and interplanetary missions. Two batteries used on the Mars Global Surveyor and Mars Polar Lander flight programs can be seen in Figs. 32.24 and 32.25. These are 28 V batteries having capacities of 23 Ah and 16 Ah, respectively.

FIGURE 32.23 EPT CPV design (2.5 volt) (*Courtesy of Power Subsystems Group, Eagle-Picher Technologies, LLC*).

A lightweight CPV Ni-H$_2$ battery was designed and developed jointly by COMSAT and Johnson Controls, Inc.[32] A prototype 10 inch diameter 26 cell, 24 Ah CPV battery was fabricated and tested to demonstrate the feasibility of this lightweight design for LEO applications. This battery used two 13 cell half-stacks connected in series within the single common pressure vessel to provide a nominal 32 V battery. The 10 inch aerospace design used a semicircular cell with a double-tap design to enhance current distribution. The components of the prototype CPV battery are shown in Fig. 32.26 with the fixed heat-fin cavity and lightweight pressure vessel.

Johnson Controls developed a new 5 inch diameter 9.6 Ah CPV battery with loose heat fins (Fig. 32.27). The loose heat fin design was designed to overcome the problems encountered with the insertion of the cells into the heat fin cavity for the 10 inch CPV cell described.[33]

A 5 inch diameter, 28 V, 15 Ah CPV battery with the loose fin design was flown on the Clementine Program. This battery was manufactured by Johnson Controls under contract with the Naval Research Laboratory. This flight was launched and flown successfully in January, 1994.

The advantages of the CPV Ni-H$_2$ battery make it a candidate for use in large multikilo-Watthour LEO energy storage applications, such as the International Space Station or constellation systems such as Iridium®. It also appeals to the other end of the spectrum—the small 100–400 Wh applications that need low-cost lightweight batteries.[34]

FIGURE 32.24 Mars Global Surveyor 23-Ah CPV battery assembly.

FIGURE 32.25 Mars Polar Lander 16-Ah CPV battery assembly.

FIGURE 32.26   COMSAT/JCI CPV Ni-H$_2$ battery (10-in diameter).

(a)

(b)

FIGURE 32.27   JCI CPV Ni-H$_2$ battery (5-in diameter). (*a*) Circular cell and loose heat fin. (*b*) 10-cell stack.

Eagle-Picher Technologies, LLC supplied 10 inch diameter 28 V, 50 and 60 Ah CPV batteries for the Iridium® program. Over 80 satellites using these CPV batteries have been launched to date. A 28 V, 60 Ah CPV battery manufactured for the Iridium® program can be seen in Fig. 32.28. This design offered impedances less than 25 milliohm, a specific energy of 55 Wh/kg, and an energy density of 68 Wh/L.

FIGURE 32.28   IRIDIUM® 60-Ah CPV battery assembly.

32.8.3 Bipolar Ni-H$_2$ Batteries

Studies have shown that the bipolar batteries promise saving in weight and volume as compared to IPV batteries.[35] The research has been directed toward large energy storage requirements for LEO applications such as the International Space Station program. Several bipolar Ni-H$_2$ batteries were designed, fabricated, and tested at NASA Lewis Research Center. The second one, assembled in 1983, was a 10-cell 6.5-Ah bipolar Ni-N$_2$ battery. Useful data were generated from tests of this 10-cell bipolar battery and results should aid the development work needed to improve performance.[35]

REFERENCES

1. J. Dunlop, J. Giner, G. van Ommering, and J. Stockel, "Nickel-Hydrogen Cell," U.S. Patent 3,867,199, 1975.

2. S. U. Falk and A. J. Salkind, *Alkaline Storage Batteries,* Wiley, New York, 1969, sec. 2.5

3. R. L. Beauchamp, "Positive Electrodes for Use in Nickel Cadmium Cells and the Method for Producing Same and Products Utilizing Same," U.S. Patent 3,653,967, Apr. 4, 1972.

4. D. F. Pickett, H. H. Rogers, L. A. Tinker, C. Bleser, J. M. Hill, and J. Meador, "Establishment of Parameters for Production of Long Life Nickel Oxide Electrodes for Nickel-Hydrogen Cells," *Proc. 15th IECEC,* Seattle, Wash., 1980, p. 1918.

5. L. W. Niedrach and H. R. Alford, *J. Electrochem. Soc.* **112:**117–124 (1965).

6. G. Holleck, "Failure Mechanisms in Nickel-Hydrogen Cells," *Proc. 1976 Goddard Space Flight Center Battery Workshop,* pp. 279–315.

7. E. Adler, S. Stadnick, and H. Rogers, "Nickel-Hydrogen Battery Advanced Development Program Status Report," *Proc. 15th IECEC,* Seattle, Wash., 1980, p. 189.

8. G. van Ommering and J. F. Stockel, "Characteristics of Nickel-Hydrogen Flight Cells," *Proc. 27th Power Sources Conf.,* June 1976.

9. J. Dunlop, J. Stockel, and G. van Ommering, "Sealed Metal Oxide-Hydrogen Secondary Cells," *Proc. 9th Int. Symp. on Power Sources,* 1974; in D. H. Collins (ed.), *Power Sources,* Academic, New York, Vol. 5, 1975, pp. 315–329.

10. E. McHenry and P. Hubbauer, "Hermetic Compression Seals for Alkaline Batteries," *J. Electrochem. Soc.* **119:**564–568 (May 1972).

11. H. H. Rogers, S. J. Krause, and E. Levy, Jr., "Design of Long Life Nickel-Hydrogen Cells," *Proc. 28th Power Sources Conf.,* June 1978.

12. G. L. Holleck, M. J. Turchan, and D. DeBiccari, "Improvement and Cycle Testing of Ni/H_2 Cells," *Proc. 28th Power Sources Symp.,* June 1978, pp. 139–141.

13. H. H. Rogers, U.S. Patent 4,177,325, Dec. 4, 1979.

14. S. J. Stadnick, U.S. Patent 4,224,388, Sept. 23, 1980.

15. L. Miller, J. Brill, and G. Dodson, "Multi-Mission Ni-H_2 Battery Cells for the 1990s," *Proc. 24th IECEC,* Washington, D.C., 1989, p. 1387.

16. T. M. Yang, C. W. Koehler, and A. Z. Applewhite, "An 83-Ah Ni-H_2 Battery for Geosynchronous Satellite Applications," *Proc. 24th IECEC,* Washington, D.C., 1989, p. 1375.

17. G. van Ommering, C. W. Koehler, and D. C. Briggs, "Nickel-Hydrogen Batteries for INTELSAT V," *Proc. 15th IECEC,* Seattle, Wash., 1980, p. 1885.

18. P. Duff, "EUTELSAT II Nickel-Hydrogen Storage Battery System Design and Performance," *25th IECEC,* 1990, vol. 6, p. 79.

19. R. J. Hass, A. K. Chawathe, and G. van Ommering, "Space Station Battery System Design and Development," *Proc. 23d IECEC,* 1988, vol. 3, pp. 577–582.

20. D. Bush, "Evaluation of Terrestrial Nickel/Hydrogen Cells and Batteries," SAND88-0435, May 1988.

21. D. Bush, "Terrestrial Nickel/Hydrogen Battery Evaluation," SAND90-0390, July 1990.

22. D. E. Nawrocki, J. D. Armantrout, et al., "The Hubble Space Telescope Nickel-Hydrogen Battery Design," *Proc. 25th IECEC,* Reno, Nev., 1990, vol. 3, pp. 1–6.

23. J. E. Lowery, J. R. Lanier Jr., C. I. Hall, and T. H. Whitt, "Ongoing Nickel-Hydrogen Energy Storage Device Testing at George C. Marshall Space Flight Center," *Proc. 25th IECEC,* Reno, Nev., 1990, pp. 28–32.

24. M. P. Bernhardt and D. W. Mauer, "Results of a Study on Rate of Thickening of Nickel Electrodes," *Proc. 29th Power Sources Conf.,* Electrochemical Society, Pennington, N.J., 1980.

25. J. F. Stockel, "Self-Discharge Performance and Effects of Electrolyte Concentration on Capacity of Nickel-Hydrogen (Ni/H_2) Cells," *Proc. 20th IECEC,* 1986, vol. 1, p. 1.171.

26. J. D. Dunlop, A. Dunnet, and A. Cooper, "Performance of INTELSAT V Ni-H_2 Batteries in Orbit (1983–1991)," *Proc. 27th IECEC,* 1992.

27. J. C. Brewer, T. H. Whitt, and J. R. Lanier, Jr., "Hubble Space Telescope Nickel-Hydrogen Batteries Testing and Flight Performance," *Proc. 26th IECEC,* 1991.

28. J. J. Smithrick, M. A. Manzo, and O. Gonzalez-Sanabria, "Advanced Designs for IPV Nickel-Hydrogen Cells," *Proc. 19th IECEC,* San Francisco, Calif., 1984, p. 631.

29. H. S. Lim and S. A. Verzwyvelt, "KOH Concentration Effects on the Cycle Life of Nickel-Hydrogen Cells," *Proc. 20th IECEC,* Miami Beach, Fla., 1985, p. 1.165.

30. D. Warnock, U.S. Patent 2,975,210, 1976.

31. T. Harvey and L. Miller, private communication on EPI handout.

32. M. Earl, J. Dunlop, R. Beauchamp, J. Sindorf, and K. Jones, "Design and Development of an Aerospace CPV Ni-H$_2$ Battery," *Proc. 24th IECEC,* 1989, vol. 3, pp. 1395–1400.

33. J. Zagrodnik and K. Jones, "Development of Common Pressure Vessel Nickel-Hydrogen Batteries," *Proc. 25th IECEC,* 1990.

34. J. Dunlop and R. Beauchamp, "Making Space Nickel-Hydrogen Batteries Lighter and Less Expensive," AIAA/DARPA Meeting on Lightweight Satellite Systems, Monterey, Calif., Aug. 1987, NTIS N88-13530.

35. R. L. Cataldo, "Life Cycle Test Results of a Bipolar Nickel-Hydrogen Battery," *Proc. 20th IECEC,* 1985, vol. 1, pp. 1346–1351.

BIBLIOGRAPHY

NASA Handbook for Nickel-Hydrogen Batteries, NASA Reference Publ. 1314, September 1993.

CHAPTER 33
SILVER OXIDE BATTERIES

Alexander P. Karpinski, Stephen F. Schiffer, and Peter A. Karpinski

33.1 GENERAL CHARACTERISTICS

The rechargeable silver oxide batteries are noted for their high specific energy and power density. The high cost of the silver electrode, however, has limited their use to applications where high specific energy or power density is a prime requisite, such as lightweight medical and electronic equipment, submarines, torpedoes, and space applications. The characteristics of the silver oxide secondary batteries are summarized in Table 33.1.

The first recorded use of a "silver battery" was by Volta with his now historic silver-zinc pile battery, which he introduced to the world in 1800.[1] This battery dominated the scene in the early nineteenth century, and during the next 100 years many experiments were made with cells containing silver and zinc electrodes. All these cells, however, were of the primary (nonrechargeable) type.

The first person to report a workable secondary silver battery was Jungner in the late 1880s.[2] Although he experimented in the early stages with iron/silver oxide and copper/silver oxide batteries (which reportedly delivered as much as 40 Wh/kg), he settled on the cadmium/silver oxide battery for his experiments with electric car propulsion. The short cycle life and high cost of these batteries, however, made them commercially unattractive. During the next 40 years other scientists experimented with various electrode formulations and separators, but without much practical success. It was the French professor Henri André who provided the key to the practical rechargeable zinc/silver oxide (silver-zinc) battery in 1941.[3,4] He described the use of a semipermeable membrane-cellophane—as a separator which would retard the migration of the soluble silver oxide to the negative plate and also impede the formation of zinc "trees," or dendrites, from the negative to the positive plate, the two major causes of cell short circuits.

In the 1950s, interest was revived in the silver-cadmium battery using the then newly available silver-zinc and nickel-cadmium technologies. This provided improved cycle life over the silver-zinc system. These batteries were first commercialized by Yardney International Corporation. Later, Westinghouse Corporation reported the commercial application of a silver-iron battery (see Chap. 25) in which they sought to "eliminate the zinc plate problems with a trouble-free iron plate, ease the separator materials and life problem and shift the deep discharge capacity stability to that limited by the silver plate."[5] The goal now, as for the past two centuries, is to provide the high energy content and power capability of the silver electrode in an improved-life, lower-cost commercially viable secondary battery.

TABLE 33.1 Advantages and Disadvantages of Silver Oxide Secondary Batteries

Advantages	Disadvantages
Silver-zinc (zinc/silver oxide)	
High energy per unit weight and volume	High cost
High-discharge-rate capability	Relatively low cycle life
Moderate-charge-rate capability	Decreased performance at low temperatures
Good charge retention	Sensitivity to overcharge
Flat discharge voltage curve	
Low maintenance	
Low self-discharge	
Safe	
Silver-cadmium (cadmium/silver oxide)	
High energy per unit weight and volume (approx. 60% of silver-zinc)	High cost
Good charge retention	Decreased performance at low temperatures
Flat discharge voltage curve	
Low maintenance	
Nonmagnetic construction	
Safe	
Silver-iron (iron/silver oxide)	
High energy and power capability	High cost
Good capacity maintenance	Water and gas management requirements
Overcharge capability	Not yet proven in field use

Zinc/silver oxide batteries provide the highest energy per unit weight and volume of any commercially available aqueous secondary batteries. They can operate efficiently at extremely high discharge rates, and they exhibit good charge acceptance at moderate rates and low self-discharge. The disadvantages are low cycle life (ranging from 10 up to 250 deep cycles, depending on design and use), decreased performance at low temperatures, sensitivity to overcharge, and high cost. Rates as high as 20 times the nominal capacity (20C rate) can be obtained from specially designed silver-zinc batteries because of their low internal impedances. These high rates, however, must often be limited in time duration because of a potentially damaging temperature rise within the cells.

Cadmium/silver oxide batteries have been viewed as a compromise between the high energy density but short life of the silver-zinc system and the long cycle life but low energy density of the nickel-cadmium system. Their energy density is roughly 2 to 3 times higher than that of nickel-cadmium, nickel-iron, or lead-acid batteries, with a relatively long cycle life, especially during shallow cycling. Charge retention is excellent. In addition, the ability to fabricate the cells without use of magnetic materials has made them the battery of choice for several scientific satellite programs. The major disadvantage of the silver-cadmium system is cost; the cost per unit energy is even higher than for the silver-zinc battery. In addition their low-temperature discharge characteristics and their high-rate properties are not as good as those of the silver-zinc system.

Iron/silver oxide batteries may provide high energy and power capability with long service life under deep-discharge use. They are capable of withstanding overcharge and over-discharge without damage and can provide good capacity maintenance with cycling. Disadvantages are, once again, cost and also the need for gas and water management in overcharge applications. Their nominal load voltage of 1.1 V is comparable to that of the silver-cadmium system, but lower than the 1.5-V level for silver-zinc. Sufficient data have not been published for these batteries to date to permit complete characterization of their properties.

All three systems also offer the advantages of long dry shelf life and of providing a flat discharge voltage during the major portion of their discharge. This latter characteristic is related to the fact that as the silver oxide is reduced to metallic silver during discharge, the conductivity of the silver electrode increases and serves to counteract polarization effects.

33.2 CHEMISTRY

33.2.1 Cell Reactions

The overall electrochemical cell reactions for the silver-zinc, silver-cadmium, and silver-iron systems, all of which use aqueous solutions of potassium hydroxide (KOH) for electrolyte, can be summarized as follows:

$$AgO + Zn + H_2O \underset{\text{charge}}{\overset{\text{discharge}}{\rightleftharpoons}} Zn(OH)_2 + Ag$$

$$AgO + Cd + H_2O \underset{\text{charge}}{\overset{\text{discharge}}{\rightleftharpoons}} Cd(OH)_2 + Ag$$

$$4AgO + 3Fe + 4H_2O \underset{\text{charge}}{\overset{\text{discharge}}{\rightleftharpoons}} Fe_3O_4 \cdot 4H_2O + 4Ag$$

These are simplified equations since there is still no general agreement on the detailed mechanisms of these reactions or on the exact form of all the reaction products.

33.2.2 Positive-Electrode Reactions

The charge and discharge processes of the silver electrode in alkaline systems are of special interest because they are characterized by two discrete steps which manifest themselves as two plateaus in the charge and discharge curves. The reaction occurring at the silver electrode at the higher (peroxide) voltage plateau is shown as

$$2AgO + H_2O + 2e \underset{\text{charge}}{\overset{\text{discharge}}{\rightleftharpoons}} Ag_2O + 2OH^-$$

and at the lower, monoxide, voltage plateau as

$$Ag_2O + H_2O + 2e \underset{\text{charge}}{\overset{\text{discharge}}{\rightleftharpoons}} 2Ag + 2OH^-$$

As shown, these reactions are reversible.

33.3   CELL CONSTRUCTION AND COMPONENTS

Secondary silver cells have been produced in prismatic, spirally wound cylindrical, and button shape configurations. The most common shape is the prismatic cell. The construction of a typical prismatic cell is shown in Fig. 33.1. This cell contains flat electrodes which are wrapped with multiple layers of separator to provide mechanical separation and inhibit migration of the silver to the zinc plate and the growth of zinc dendrites toward the positive plate. The plate groups are intermeshed, and the pack is placed in a tightly fitting case (Fig. 33.2). Because of the relatively short shelf life of the activated silver cells, they are usually supplied by the manufacturers in the dry charged or dry unformed condition with filling kits and instructions. The cells are filled with electrolyte and activated just prior to use. They may also be supplied in the filled and ready-to-use condition if required by the user.

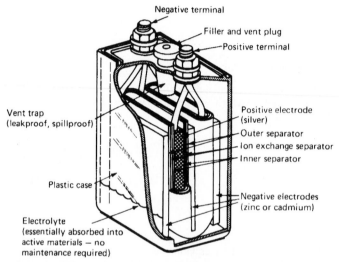

FIGURE 33.1   Cutaway view of typical prismatic zinc/silver oxide or cadmium/silver oxide secondary cell.

FIGURE 33.2   Cell stack being assembled into cell case; model LR-190, 210-Ah silver-zinc battery. (*Courtesy of Yardney Technical Products, Inc.*)

The mechanical strength of these cells is usually excellent. The electrodes are generally strong and are fitted tightly into the containers. The cell containers are made of high-impact plastics. Specific designs of these cells, when properly packaged, have met the high-shock, vibration, and acceleration requirements of missiles and torpedoes with no degradation.

33.3.1 Silver Electrodes

The most common fabrication technique for silver electrodes is by sintering silver powder onto a supporting silver grid. The electrodes are manufactured either in molds (as individual plates or as master plates which are later cut to size) or by continuous rolling techniques. They are then sintered in a furnace at approximately 700°C.

Alternate techniques include dry processing and pressing, as well as slurry pasting with a binder of AgO or Ag_2O onto a grid. If pasted, the plates are often sintered, converting the silver oxide into metallic silver and burning off the organic additives. The grid may be a woven or expanded metal form of silver, or silver-plated copper.

After being cut to size and having wires or tabs hot-forged onto an appropriately coined (compressed) area to carry current to the cell terminals, the electrodes are either electroformed (charged in tanks against inert counterelectrodes) before assembly into cells or assembled into the cells in the metallic state and later charged in the cell.

Grid material, density, and thickness, electrical lead type and size; and final electrode size, thickness, and density are all design variables which depend on the intended application for the cells. The silver powder particle size may be varied, with the finer powders approaching the theoretical silver utilization of 2.0 g/Ah. The use of very fine powder, however, results in an initial voltage dip (typically less than 120 mS) at medium (C) to high discharge rates.

33.3.2 Zinc Electrodes

Zinc electrodes are most widely made by dry pressing, by a slurry or paste method, or by electrodeposition. In the dry-pressing method, a mixture of metallic zinc or zinc oxide, binder, and additives is compressed around a metal grid; this is normally done in a mold. The grid usually has the current-carrying leads prewelded in place. As the unformed powder electrodes have little strength, one component of the separator system, the negative interseparator, is usually assembled around the electrode as part of the fabricating operation. Rolling techniques have also been developed to permit continuous fabrication of dry-powder electrodes.[6]

In the paste or slurry method, a mixture of zinc oxide, binder, and additives is combined with water and applied continuously to a carrier paper or directly to an appropriate metal grid. Again, the negative interseparator is usually integral to provide needed physical strength. After drying, multiple layers of these pasted slabs may be pressed together about a pretabbed grid to form the final electrode. These plates may be assembled unformed into the cell, or they may be electroformed in a tank against inert counterelectrodes.

Electrodeposited negative electrodes are manufactured by plating zinc in tanks onto metallic grids. The plates must then be amalgamated and pressed or rolled to the desired thickness and density, followed by drying.

The zinc electrode is acknowledged as the life-limiting component in both the silver-zinc and the nickel-zinc systems. Accordingly, much work has been done in the area of additives for these electrodes, both to reduce hydrogen evolution and to improve cycle life. The common additive to reduce hydrogen evolution has traditionally been mercury (1 to 4% of the total mix), but this is being replaced, for personnel safety and enviromental reasons, by small amounts or mixes of the oxide of lead, cadmium, indium, thallium, gallium,[7-11] and bismuth.[23] Many other (proprietary) additives have been introduced into the zinc electrode by various manufacturers in attempts to increase life.

Zinc electrodes also suffer capacity loss, which results from "shape change," or the migration of materials from the sides and top to the center and bottom of the electrode.

Several approaches have been taken to improve the stability of the zinc electrode: (1) an excess of zinc is used to compensate for losses during cycling, (2) oversized electrodes are used on the basis that shape change starts on the electrode edges where current densities are higher, (3) binders such as PTFE, potassium titanate, neoprene latex or other polymers are used to hold the active materials together, and (4) electrolyte additives are used.[12-14]

As is the case for the silver electrodes, the grid material, additives, and final electrode size, thickness, and density are all design variables which depend on the final application.

33.3.3 Cadmium Electrodes

Most silver-cadmium cells contain cadmium electrodes that are manufactured by pressed-power or pasting techniques. Although other methods have been used, such as impregnating nickel plaque with cadmium salts, as is done for nickel-cadmium cells, the most common method in silver-cadmium cells is to press or paste a mixture of cadmium oxide or cadmium hydroxide with a binder onto a silver or nickel grid. These processes are similar to those used for the pressed and pasted zinc electrodes.

33.3.4 Iron Electrodes

The iron electrodes used here are generally manufactured by powder-metallurgy techniques (see Chap. 25).

33.3.5 Separators

The separators in the silver cells must meet the following major requirements:

1. Provide a physical barrier between positive and negative electrodes
2. Have minimum resistance to the flow of electrolyte and ions
3. Prevent migration of particles and dissolved silver compounds between positive and negative electrodes
4. Be stable in the electrolyte and cell operating environment

In general, secondary silver-zinc and silver-cadmium cells require up to three different separators, as shown in Fig. 33.1. The inner separator, or positive interseparator, serves both as an electrolyte reservoir and as a barrier to minimize oxidation of the main separator by the highly oxidative silver electrode. This separator is usually made of an inert fiber such as nylon or polypropylene.

The outer separator, or negative interseparator, also serves as an electrolyte reservoir and can also, ideally, stabilize the zinc electrode and retard zinc penetration of the main separator, thus minimizing dendrite growth. Much work has been done to develop improved inorganic positive electrode interseparators utilizing such materials as asbestos and potassium titanate. Improvements in life have been reported as a result of this work.[7-10,11,15] This separator is usually omitted in short and medium life cells.

The main, or ion exchange, separator remains the key to the wet life of the secondary silver cell. It was André's[3] use of cellophane as a main separator that first made the secondary silver cells feasible. The cellulosics (cellophane, treated cellophane, and fibrous sausage casing) are usually employed in multiple layers as the main separators for these cells. Again, much work has been done in recent years to develop improved separators utilizing such materials as radiation-grafted polyethylene,[16] inorganic separators,[10,11,15,17,24] and other synthetic polymer membranes. Improved cell life has been reported through use of these new membranes either alone or in combination with cellulosics. Some of these have yet to be applied extensively to commercial silver cells, however, because of drawbacks usually involving high impedance, availability and cost.

33.3.6 Cell Cases

The cell cases must be chemically resistant to attack by the corrosive concentrated potassium hydroxide electrolyte and to oxidizing effects of the silver electrodes. They must also be strong enough to contain any internal pressure generated in the cells and to maintain structural integrity throughout the anticipated range of environmental conditions that will be experienced by the cells.

The majority of secondary silver cells are assembled in plastic cases. The plastic most commonly used is an acrylonitrile-styrene copolymer (SAN). This material is relatively transparent and can be sealed easily by solvent cement or epoxy. However, its relatively low softening temperature (80°C) precludes it from use in some applications. A wide variety of other plastics have been used for cell cases. Table 33.2 lists some of these materials and gives their characteristics. Metal cases have been used for some sealed cell and button cell applications; however, these present problems in sealing and in electrically isolating the electrodes from the cases and are not used widely.

TABLE 33.2 Properties of Typical Plastic Cell Case and Cover Materials

Type (Trade name)	SAN (SAN 35)	ABS (Cycolac X-37)	ABS (Cycolac GSM)	Nylon (612)	Modified PPO (Noryl SE-1)	Polysulfone* (P-1700)
Specific gravity	1.07	1.06	1.04	1.07	1.08	1.24
Transparency	Yes	No	No	No	No	Yes
Tensile strength, 10^3 lb/in^2	11.5	7.0	6.3	8.8	9.2	10.2
Flexural strength, 10^3 lb/in^2	14.0	12.0	10.7	11.0	15.0	15.4
ZOD impact strength, ft · lb per inch notched	0.45	3.0	7.0	0.85	3.5	1.3
Hardness, Rockwell R	83	109	105	114	119	120
Heat-deflecton temperature at 264 lb/in^2. F	220	230	192	150	255	345
Dielectric strength, V/mil	460	500	427	400	500	425
Thermal conductivity, Btu · in/(h · ft^2 · °F)	0.87	2.38	1.55	1.5	1.5	1.8

*Glass or teflon filled (typically 10%) is also used to increase strength.

33.3.7 Electrolyte and Other Components

The electrolyte used in secondary silver cells is generally an aqueous solution (35 to 45% concentration) of potassium hydroxide (KOH). Lower concentrations of electrolyte provide lower resistivity and thus a higher voltage output under load as well as a lower freezing point. Concentrations below 45% KOH, however, are more corrosive to the cellulosic separators typically used in silver-based batteries and are not used for extended wet-life applications. Table 33.3 depicts the critical parameters of various KOH solutions. Various additives such as zinc oxide, lithium hydroxide, potassium fluoride, potassium borate, tin, and lead have been used to reduce the solubility of the zinc electrode.[14]

Since potassium hydroxide readily combines with carbon dioxide in the air to form potassium carbonate, thus reducing conductivity, cell vents are usually covered with a vent cap or a low-pressure relief valve.

Cell terminals are typically made of steel or brass and are almost always silver- or nickel-plated to improve conductivity and corrosion resistance.

TABLE 33.3 Physical and Electrical Characteristics of KOH Solutions

% KOH	Specific gravity at 15.6°C	Conductivity at 18°C, $\Omega^{-1} \cdot cm^{-1}$	Specific heat at 18°C, cal/ g·°C	Freezing point, °C	Boiling point, °C at 760 mm Hg	Boiling point, °C at 100 mm Hg	Vapor pressure, mm Hg at 20°C	Vapor pressure, mm Hg at 80°C	Viscosity, cP. at 20°C	Viscosity, cP. at 40°C
0	1.0000		0.999	0	100	52	17.5	355	1.00	0.66
5	1.0452	0.170	0.928	−3	101	52.5	17.0	342	1.10	0.74
10	1.0918	0.310	0.861	−8	102	53	16.1	327	1.23	0.83
15	1.1396	0.420	0.801	−14	104	54	15.1	306	1.40	0.95
20	1.1884	0.500	0.768	−23	106	56	13.8	280	1.63	1.10
25	1.2387	0.545	0.742	−36	109	59	11.9	250	1.95	1.31
30	1.2905	0.505	0.723	−58	113	62	10.1	215	2.42	1.61
35	1.3440	0.450	0.707	−48	118	66	8.2	178	3.09	1.99
38	1.3769	0.415	0.699	−40	122	69	7.0	156	3.70	2.35
40	1.3991	0.395	0.694	−36	124	71	6.2	140	4.16	2.59
45	1.4558	0.340	0.678	−31	134	80	4.5	106	5.84	3.49
50	1.5143	0.285	0.660	+6	145	89	2.6	70	8.67	4.85

33.4 Performance Characteristics

33.4.1 Performance and Design Tradeoffs

The secondary silver batteries provide high energy capability combined with minimum weight and volume. The advantages and disadvantages of the various systems have been described earlier in this chapter. The performance of the batteries for specific application will depend on the internal design and history of the cells. It is rare that one can select an "off-the-shelf" battery that will meet all the requirements of a specific application.

Starting with the basic parameters, the cell design will consist of a series of compromises to obtain the most favorable combination of voltage, electrical capacity, and cycle life characteristics within the allowable battery weight and volume.

Assuming, for example, a nominal 1.5 V per silver-zinc battery at low current densities (0.01 to 0.03 A/cm²) and lower voltages at higher currents, the designer selects the number of cells for the application. The problem is increased if high current pulse loads are required and the battery must provide voltage above the minimum allowable at the high rate, while not exceeding the maximum allowable voltage at initial low rates. The size of the cell is then chosen by dividing the allowable volume by the calculated required number of cells.

The voltage, current, electrical capacity, and cycle life requirements must then be reviewed in conjunction with the allowable weight and the environmental conditions that the battery must be able to withstand. Each of these will be a factor in determining the choice of separator material for the cell. The stability and number of layers of separator must be sufficient to provide the desired wet life under these conditions while having a resistance low enough to prevent undue voltage drop at the high current load. Each of these requirements is also a factor in choosing the number of electrodes within the cell. As the number of electrodes (and thus the active electrode area) is increased, the current density during any discharge (Amperes per square centimeter) is decreased, raising the output voltage. It should be noted that a cell design optimized for high discharge rates will, by nature of the design, have a reduced capacity under low discharge rates. This is a result of having many electrodes each having to be wrapped with the required number of layers of separators. Given a fixed internal volume, it follows that in such a high-rate cell, less space is available for active electrode material.

The cell must also be designed to contain enough active electrode material (such as silver and zinc) to supply the required electrical capacity for the desired number of cycles. Theoretically, 2 g of silver and 1.2 g of zinc are required in the cell for each Ampere-hour of electrical capacity desired. Since these values are the theoretical capability of the pure material, and since some of the active materials will go into solution with each charge-discharge cycle, the designer must work with higher values—on the order of 3.5 g of silver and 3.0 g of zinc per nominal Ampere-hour for long cycle-life cells. Other design variables, such as silver powder particle size, will also ultimately affect cell performance.

Because of these considerations, the performance curves shown in the following sections must be viewed as general characteristic of the systems and not necessarily of specific batteries for a specific application.

33.4.2   Discharge Performance for Zinc / Silver Oxide Batteries

The open-circuit voltage of the zinc/silver oxide battery is 1.82 to 1.86 V. The discharge is characterized by two discrete steps, the first corresponding to the divalent oxide and the second to the monovalent oxide, as shown in Fig. 33.3. The flat portion of the curves is referred to as "plateau voltage." This voltage is rate-dependent; at high rates the voltage steps may be obscured.

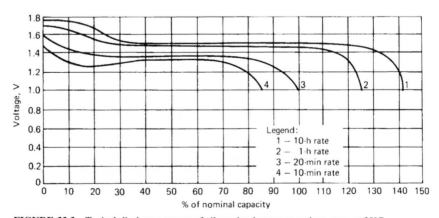

FIGURE 33.3   Typical discharge curves of silver-zinc battery at various rates at 20°C.

The performance of the battery at various discharge rates and temperatures can be seen in Figs. 33.4 to 33.6, which show the effect on plateau voltage and capacity. The high-rate capability of the zinc/silver oxide battery is a complex process which can be characterized as the result of the electrical conductivity of the silver grid and the conductivity of the positive electrode as it is discharged, as well as the thin multiplate design of the cell. The performance of the battery falls off with decreasing temperature, particularly below −20°C. Heating the battery with external heaters or by retaining the internal heat generated during the discharge can improve the performance at low ambient temperatures.

The performance characteristics of the zinc/silver oxide battery are summarized in Figs. 33.7 and 33.8, which can be used to determine the capacity, service life, and voltage under a variety of discharge conditions. These figures present typical performance data. Performance differences can occur for each specific design and even for each battery, depending on cycling history, state of charge, storage time, temperature, and other conditions of use.

Figures 33.3–33.8 are specifically for high-rate (HR) designs. For many applications, tradeoffs can be made to provide longer life at the expense of somewhat lower energy density. Alternative low-rate (LR) designs contain additional layers of separator, meaning, of necessity, fewer electrodes with higher impedance and lower capacity within a given volume. Typically, the LR battery cannot be discharged at higher than the 1-h rate and will provide about 3 to 5% lower average voltage and capacity than its HR counterpart at the 1-h rate. However, the LR batteries do provide substantial wet shelf life and cycle life advantages (see Table 33.4).

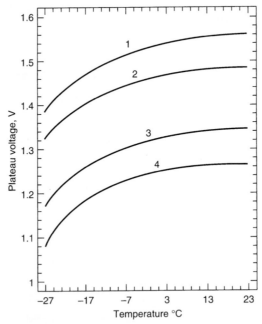

FIGURE 33.4 Typical effect of temperature on plateau voltage for high-rate silver-zinc battery (operated without heaters). Curve 1–10-h rate; curve 2–1-h rate; curve 3–20-min. rate; curve 4–10-min. rate.

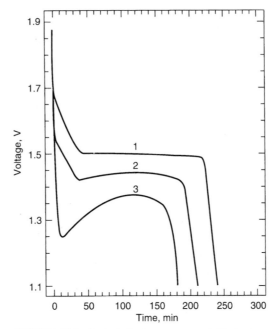

FIGURE 33.5 Typical discharge curves of HR-5 silver zinc battery at $C/3$ (2-A) rate. Curve 1—25°C; curve 2—0°C; curve 3—−18°C.

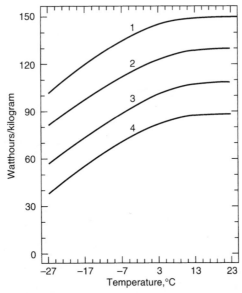

FIGURE 33.6 Typical effect of temperature on energy per unit weight for silver-zinc battery (operated without heaters). Curve 1—10-h rate: curve 2—1-h rate: curve 3—20-min rate: curve 4—10-min rate.

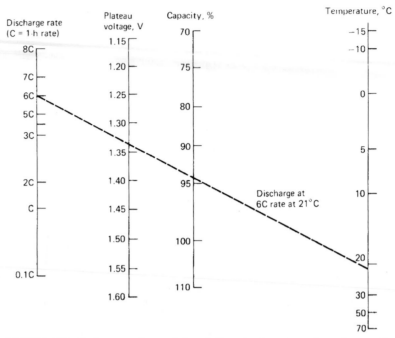

FIGURE 33.7 Performance characteristics of silver-zinc batteries under various conditions. (To find the capacity and the plateau voltage of a silver-zinc battery, draw a straight line between the discharge rate and the ambient temperature at which the battery is stabilized.)

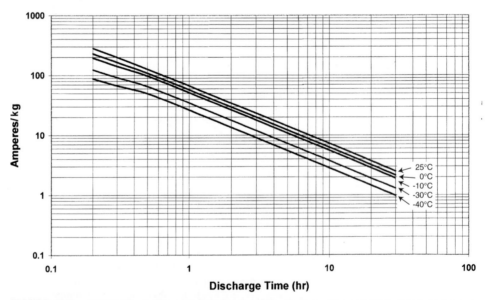

FIGURE 33.8a Service life of silver-zinc batteries at various discharge rates and temperatures. (Amperes/kg vs. discharge time in hours)

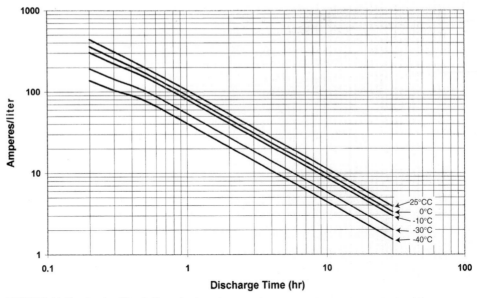

FIGURE 33.8*b*   Service life of silver-zinc batteries at various rates and temperature. (Amperes/liter vs. discharge time in hours)

TABLE 33.4   Nominal Life Characteristics of Secondary Silver Batteries*

	High rate (HR)Ag-Zn	Low rate (LR)Ag-Zn	Ag-Cd	Ag-Fe
Wet shelf life	6–18 months	1–3 years	2–3 years	2–4 years
Cycle life†	10–50 cycles	100–250 cycles	150–1000 cycles	100–300 cycles

*These characteristics are nominal and vary with operating conditions and design of individual models.
†Cycle life characteristics are for deep (80–100% of full capacity) discharge cycles. Cycle life improves considerably with partial discharges.

33.4.3   Discharge Performance for Cadmium/Silver Oxide Batteries

The open-circuit voltage of the cadmium/silver oxide battery is 1.38 to 1.42 V. Typical discharge curves at 20°C are given in Fig. 33.9, showing the two-step flat discharge typical of the silver oxide electrode. The discharge characteristics are similar to those of the zinc/silver oxide battery except for the lower operating voltage; Ampere-hour capacities are about the same.

The capacity and the discharge voltage of the battery are temperature-dependent, again similar to the zinc/silver oxide battery. The effects of temperature and discharge rate on voltage and capacity are shown in Figs. 33.10 and 33.11, respectively. The recommended operational temperature range is −25 to 70°C, with the optimum performance obtained between 10 and 55°C. With heating, the temperature range can be lowered to −60°C.

The performance characteristics of the cadmium/silver oxide battery are summarized in Figs. 33.12 and 33.13*a* and *b*, which can be used to determine the capacity, service life, and voltage levels under a variety of discharge conditions.

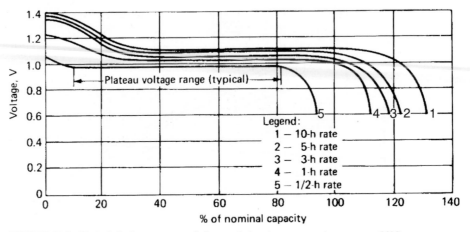

FIGURE 33.9   Typical discharge curves of silver-cadmium battery at various rates at 20°C.

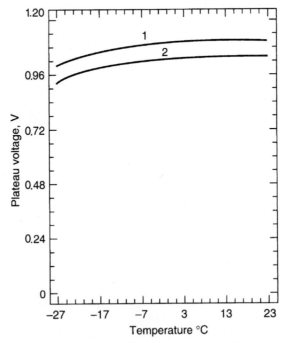

FIGURE 33.10   Typical effect of temperature on plateau voltage for silver-cadmium battery (discharged without heaters). Curve 1—10-h rate; curve 2—1-h rate.

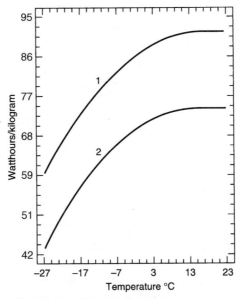

FIGURE 33.11 Typical effect of temperature on energy per unit weight for silver-cadmium battery (discharged without heaters). Curve 1—10-h rate; curve 2–h rate.

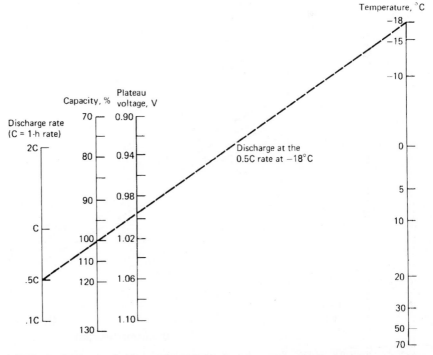

FIGURE 33.12 Performance characteristics of silver-cadmium batteries under various conditions. (To find the capacity and the plateau voltage of a silver-cadmium battery, draw a straight line between the discharge rate and the ambient temperature at which the cell is stabilized.)

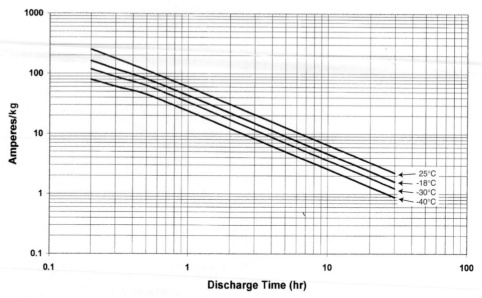

FIGURE 33.13a Service life of silver-cadmium batteries at various discharge rates and temperatures. (Amperes/kg vs. discharge time in hours)

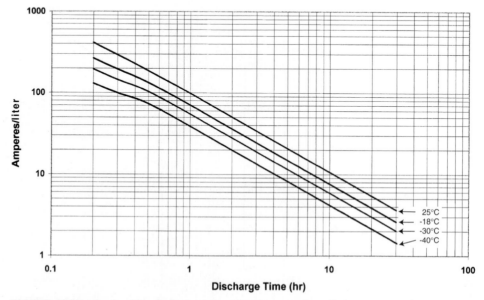

FIGURE 33.13b Service life of silver-cadmium batteries at various discharge rates and temperatures. (Amperes/liter vs. discharge time in hours)

33.4.4 Impedance

The impedance of the silver oxide cells is normally low but can vary considerably with many factors, including the separator system, current density, state of charge, stand time, cell age, temperature, and, importantly, cell size. In a study of the effect of storage time on the impedance of silver-zinc cells, initial values of 5 to 11 mΩ for partially charged cells were reported,[18] with the values rising to as much as 3 Ω following 8 months' storage at 21°C and 9 to 15 Ω following 8 months at 38°C. Cells stored in the fully discharged condition retained their low impedance (ranging from 2 to 10 mΩ) throughout the entire test period. The high-impedance cells returned to normal low values, however, within several seconds of the start of discharge.

The AC impedance of the silver oxide batteries is highly dependent on the frequency of the load, with the impedance rising sharply above 5 kHz. The impedance of a six-cell, 350-Ah silver-zinc battery, discharged to approximately 50% DOD, is shown in Fig. 33.14, and the corresponding phase angle is shown in Fig. 33.15.[19]

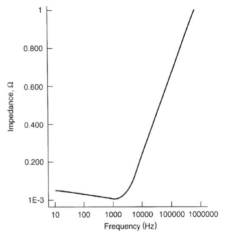

FIGURE 33.14 Impedance magnitude vs. frequency.

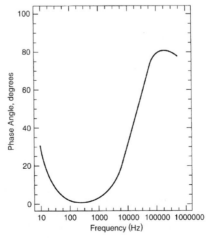

FIGURE 33.15 Impedance phase angle vs. frequency.

33.4.5 Charge Retention

The charge retention of the activated and charged silver oxide cells is better than that of most secondary batteries, with retention of 85% of charge after 3 months' storage at 20°C.

As with other chemical reactions, the rate of loss of charge is dependent on the storage temperature (see Fig. 33.16). Storage at −20 to 0°C is recommended for maximizing charge retention. In the dry and charged condition, properly sealed and stored cells will retain their charge for as long as 5 to 10 years. Here again, low temperature storage is highly recommended.

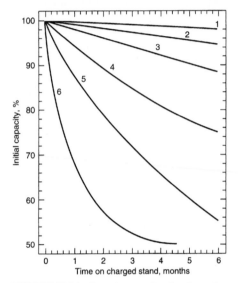

FIGURE 33.16 Capacity retention for silver-zinc and silver-cadmium batteries at various storage temperatures. Curve 1-−10°; curve 2-0°C; curve 3-10°C; curve 4-20°C; curve 5-30°C; curve 6-40°C.

33.4.6 Cycle Life and Wet Life

The separator system and the solubility of the active materials play critical roles in determining the wet and cycle lives of the silver-based cells. The separator must have a low electrolytic resistance for discharges at high rates, yet it must have high resistance to chemical oxidation by the silver species as well as low permeability to colloidal silver, zinc, cadmium, or iron.

Since cadmium and iron are relatively insoluble in concentrated alkaline electrolytes, the life of the silver-cadmium and silver-iron cells is therefore limited by the rate of silver migration through the various layers of the separator system. Failure (internal short circuit) occurs when a metallic bridge is established through the separator between positive and negative electrodes. Multiple layers of separator are used to extend the life capability, however, at the expense of higher internal resistance.

The life of the silver-zinc battery is further hindered by the high solubility of zinc in alkaline electrolytes. The problem manifests itself in two failure mechanism: shape change and growth of dendrites. Shape change is the migration of zinc from the electrode tops and edges where it becomes depleted, to the center and bottom where it densifies, resulting in capacity loss. Dendrites, a sharp, needlelike structure of the metal, are formed during overcharge. They may perforate the separators and cause internal short circuits. The capacity degradation in silver-zinc batteries due to shape change is illustrated in Fig. 33.17. The decline in capacity of silver-cadmium batteries is at a much slower rate, as depicted in Fig. 33.18.

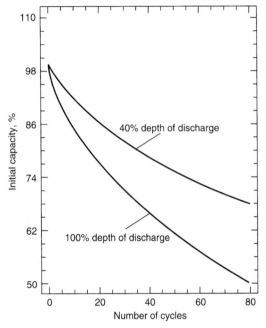

FIGURE 33.17   Typical capacity degradation of low-rate silver-zinc batteries discharged at $C/10$ rate.

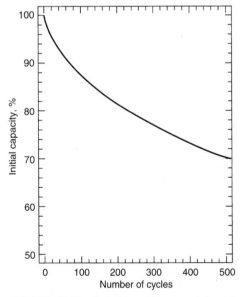

FIGURE 33.18   Typical capacity degradation of sealed silver-cadmium batteries discharged at $C/3$ rate to 100% depth of discharge.

Aside from normal capacity loss due to extended wet or cycle life, dry-charged zinc-silver oxide batteries may exhibit a one-time deviation in capacity performance (typically less than 20% of initial capacity) during the second cycle discharge called "second cycle syndrome." The cause of this deviation is unknown but is generally recognized by the user community.

The nominal life ratings for the silver-zinc, silver-cadmium, and silver-iron batteries are given in Table 33.4. The life of the silver oxide cells will also vary greatly with operating and storage conditions. High rates of discharge to 100% depth of discharge and high-temperature exposure (for more than 30 days) will significantly reduce the wet and cycle lives of the batteries. Cold-temperature storage (at less than $-10°C$), when not in use, on the other hand, will greatly increase the life of the cells. The cycle and wet lives will also increase with a decreasing depth of discharge.

In a study to evaluate the capabilities of silver-cadmium batteries for satellite applications, an extensive test program was run on three-cell, 3-Ah silver-cadmium batteries at various depths of discharge.[20] These results are summarized in Table 33.5, showing the increase in cycle life with decreasing depth of discharge.[20] Another study on 250 Ampere-hour silver-zinc cells, cycled at less than 1% depth of discharge, with 14 full-capacity cycles, resulted in a cycle life of 7280 cycles over a 38-month period.[21]

TABLE 33.5   Cycle Life vs. Depth of Discharge for 3-V, 3-Ah Sealed Cadmium/Silver Oxide Batteries

Depth of discharge, %	Cycle life at first cell failure
65	1800
50	3979
50	>5400 (375 days)
35	>5400 (375 days)
25	>5400 (375 days)

Source: Ref. 20.

33.5 CHARGING CHARACTERISTICS

33.5.1 Efficiency

The *Ampere-hour* efficiency (Ampere-hour output/Ampere-hour input) of the silver-zinc and silver-cadmium systems under normal operating conditions is high—greater than 98% because practically no side reactions occur when charging at normal rates. The *Watthour* efficiency (Watthour output/Watthour input) is about 70% under normal conditions because of the difference between charge and discharge voltages.

33.5.2 Zinc/Silver Oxide Batteries

The manufacturers of these batteries recommend constant-current charging at the 10 to 20-h rate for most applications. However, constant-potential and pulsed charging techniques have also been applied.

A typical charge curve at constant current is shown in Fig. 33.19. The two plateaus reflect the two levels of oxidation of the silver electrode: the first from silver to monovalent silver oxide (Ag_2O), which occurs at a potential of approximately 1.6 V; the second from the

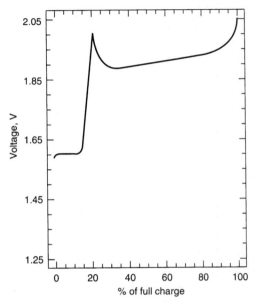

FIGURE 33.19 Typical charge curve of silver-zinc battery at 20°C, 10-h charge rate.

monovalent to the divalent silver oxide (AgO), which occurs at approximately 1.9 V. It should be noted that during this transition from the monovalent to the divalent state of charge, a momentary increase in the charge voltage, of up to 2.00 V, may occur prior to stabilizing at the 1.90 to 1.95-V plateau. To ensure a full charge, the charging system must be designed to ignore this temporary rise in voltage.

Charging is normally terminated when the voltage during charge rises to 2.0 V. Above 2.1 V the cell begins to generate oxygen at the silver electrode and/or hydrogen at the zinc electrode, decomposing water from the electrolyte. Overcharge is also detrimental to cell life in that it promotes the growth of zinc dendrites and subsequent short circuits.

The importance of proper charging to the life of these batteries cannot be overemphasized. Special provisions must be made for those applications which do not permit the use of a constant current with preset voltage cutoff or a similar controlled charging method.

33.5.3 Cadmium/Silver Oxide Batteries

Except for lower voltages on each of the plateaus (1.2 V on the lower level, 1.5 V on the upper level), the charging characteristics of the silver-cadmium battery are similar to those of silver-zinc. A typical charge curve is shown in Fig. 33.20.

As with silver-zinc, silver-cadmium batteries are usually charged at constant current at the 10 to 20-h rates. The recommended cutoff voltage during charge is normally 1.6 V per cell.

The silver-cadmium battery, however, is less sensitive to overcharge than the silver-zinc battery. Other charge methods can be and have been adapted to specific applications.

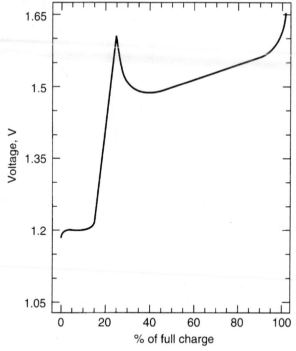

FIGURE **33.20** Typical charge curve of silver-cadmium cell at 20°C, 10-h charge rate.

33.6 CELL TYPES AND SIZES

Tables 33.6 to 33.8 present excerpts from the catalogs of two major silver battery manufacturers—Yardney Technical Products Inc., and Eagle-Picher Industries, Inc. They are intended as a guide only, since the design parameters can be varied to meet specific customer requirements. Many applications for the high-energy silver batteries, in fact, require special designs, which often dictate new cell case and cover designs and tooling. These then become the "available" models for future applications.

TABLE 33.6 Nominal Characteristics of Typical Vented Zinc/Silver Oxide Batteries

Cell type	Capacity, Ah	Cell dimensions, mm including terminals)			Cell weight, g including electrolyte)	Maximum continuous rate, A
		Length	Width	Height		
HR-02	0.2	5.6	16.0	49.3	6.5	2.0
HR-05	1.3	13.7	27.4	39.6	21.3	4.0
HR-1	2.0	13.7	27.4	51.3	31.2	6.0
HR-2	4.5	15.0	43.7	63.5	68.0	20.0
PMV-2(4.5)*	5.0	15.2	43.7	64.3	72.6	100
HR-5	8.5	20.3	52.8	73.7	127.6	60.0
PM-15*	21.8	20.3	58.9	125.5	295	200
HR-21	35.9	20.6	58.4	191.5	439	160
PM-30*	44.0	25.4	77.7	166.4	607	400
HR-40	46.0	25.1	82.6	180.3	646	200
HR-105	121	35.2	96.9	137.4	950	120
HR-140	190	72.4	82.5	183.4	1721	600
PML-170*	221	35.3	97.0	184.4	1520	120
HR-190	238	39.4	152.6	165.4	2217	800
PML-2500*	2750	107.2	107.2	479.0	18,150	600
Low-rate types						
LR-1	2.1	13.7	27.4	51.3	30.1	4.5
LR-4	7.5	15.0	43.7	85.3	99.2	16.0
LR-8	10.0	16.3	29.9	120.1	116.3	16.0
LR-12	16.0	19.1	47.2	100.1	163.0	20.0
LR-40	64.0	25.1	82.6	180.3	638	64.0
LR-70	100	36.1	92.5	155.4	1055	160
LR-90	155	54.9	82.9	179.3	1588	150
LR-190	270	39.1	151.6	162.6	2048	200
LR-350	560	107.4	107.4	222.3	5615	350
LR-360	570	69.9	147.3	162.6	4391	300
LR-660	840	79.2	161.3	177.8	6183	180
Special deep submersible types						
LR-700(DS)†	1060	107	107	486	11,200	900
LR-750(DS)†	1075	142	97	513	12,500	750
LR-850-21	1200	119	114	479	13,200	800
LR-875	1050	160	787	183	7,000	125
LR-1000(DS)†	1072	137	137	513	18,500	1250

*Primary, manually activated.
†Pressure compensated
Source: Yardney Technical Products, Inc.

TABLE 33.7 Nominal Characteristics of Typical Vented Zinc/Silver Oxide Cells

	High rate					Low rate						Physical dimensions, mm		
Cell type	Rated capacity Amp-hr	Nominal capacity Amp-hr rates			Weight, g	Cell type	Rated capacity Amp-hr	Nominal capacity Amp-hr rates			Weight, g	L	W	H
		15 min.	30 min.	60 min.				4 hr	10 hr.	20 hr.				
SZHR 0.8	0.8	0.7	0.7	0.8	22.7	SZLR 0.8	0.8	0.8	0.8	0.8	22.7	10.9	26.9	51.6
1.5	1.5	1.4	1.5	1.5	39.7	1.5	1.5	1.5	1.5	1.5	42.6	12.4	30.7	57.2
2.8	2.8	2.6	2.7	2.8	53.9	3.0	3.0	3.0	3.0	3.0	56.7	14.2	35.1	63.2
5.0	5.0	4.8	5.0	5.0	76.6	5.3	5.3	5.3	5.3	5.3	85.1	16.3	40.1	70.9
6.5	6.5	6.2	6.4	6.5	119.1	7.5	7.5	7.4	7.5	7.5	124.8	14.9	43.7	90.2
10.5	10.5	10.0	10.3	10.5	170.2	11.5	11.5	11.5	11.4	11.5	184.4	20.1	49.5	84.8
15	15	12	14	15	210.0	16.5	16.5	15.5	16.5	16.5	215.6	21.3	41.1	120.7
26	26	20	24	26	312.1	30	30	28.0	30.0	30.0	326.3	25.4	62.7	103.9
48	48	*	40	48	595.9	51	51	48	51	51	624.2	18.5	89.9	167.9
65	65	*	50	65	737.8	70	70	65	70	70	780.3	26.9	83.1	155.4
100	100	*	80	100	1107	115	115	100	110	115	1220	37.3	92.7	150.9
140	140	*	*	140	1944	160	160	*	150	160	2049	74.17	75.7	161.8

*Not applicable at this rate.
Source: Eagle-Picher Technologies.

33.24

TABLE 33.8 Nominal Characteristics of Typical Cadmium-Silver Oxide Batteries

Cell type	Capacity, Ah	Cell dimensions, mm (including terminals)			Cell weight, g (including electrolyte)	Maximum continuous rate, A
		Length	Width	Height		
YS-1	1.5	13.7	27.4	51.3	31.2	5.0
YS-3	4.2	15.2	43.7	72.6	82.2	12.0
YS-5	7.8	19.1	51.1	73.9	130.5	25.0
YS-5 (sealed)	6.8	20.1	52.8	73.9	141.8	15.0
YS-10	14.5	18.8	58.9	122.2	246.7	30.0
YS-16 (sealed)	21.0	20.6	58.4	146.1	348.8	50.0
YS-20	32	43.9	52.1	108.7	450.9	40.0
YS-40	54	25.1	82.6	179.8	745.9	100
YS-100	132	70.6	87.4	122.2	1503	150
YS-150	240	45.2	106.4	272.0	2978	150
YS-300	420	45.2	106.4	444.5	5190	150

Source: Yardney Technical Products, Inc.

33.7 SPECIAL FEATURES AND HANDLING

Silver cells are capable of providing extremely high currents if short-circuited. Accordingly, provisions must be made to insulate all tools used with the batteries and to protect the cells against grounding in their application.

The electrolyte is a caustic solution of potassium hydroxide. Precautions such as the use of gloves and safety goggles are required when handling the electrolyte. In most applications, addition of electrolyte or water is not required. However, the manufacturer's recommendations for periodic maintenance and electrolyte checks should be followed closely.

Proper ventilation of these as well as other vented batteries, although not as much a problem here as with other battery systems, is required to avoid the accumulation of hazardous hydrogen, especially during charge. For larger installations, forced air or fans may also be required to prevent undesirable temperature buildups. When close voltage regulation is required at cold temperatures, thermostatically controlled heaters are often used with the batteries.

Because of the sheer size and power of the batteries used, and because of critical personnel safety requirements (for example, the U.S. Navy's NR-1 submersible has a 240-V, 850-Ah silver-zinc backup battery installed under the deck and venting into the operator's quarters), a whole new body of engineering technology has been developed for the application of these batteries for underwater power. Some of the special features developed include removal of all mercury, provision for fire walls inside cells, and provision for pressure compensation for those batteries that are external to the vessels' pressure hulls.[8] Electronic systems have also been developed to permit continuous scanning of individual cell voltages to maximize battery life. Special applications such as these can be successfully developed only if the battery designers, manufacturers, and users work closely together during the design of the product.

33.8 APPLICATIONS

Because of the high cost of silver, the major applications of these batteries have historically been, and continue to be, governmental. However, because of their high power and energy density, these batteries have found many varied uses where space and weight limitations are critical. In addition, the cost of silver for most applications can be offset by reclaiming the metal after the battery completes its useful life.

One of the original applications for the silver-zinc battery was for use in torpedoes.[22] Much of the original development work was sponsored by the U.S. Navy. Later, development expanded to other underwater applications, including mines, buoys, special test vehicles, swimmer aids, and, at present, deep submergence and rescue vehicles (DSRV), such exploratory underwater vehicles as UUV and NR-1, and various antisubmarine warfare (ASW) applications. The MK 66 torpedo battery is illustrated in Fig. 33.21. The discharge rate for this nominal 245-V battery is 580 A; operating time is 4 min. The deep submergence and rescue vehicle (DSRV) battery illustrated in Fig. 33.22 is a pressure compensated design using mineral oil to fill the cell and battery void during ocean descent. It is a 115 V, 700 Ah rated battery and the pressure compensation allows it to be mounted outside of the pressurized vessel.

FIGURE 33.21   MK 66 torpedo battery. (*Courtesy of Yardney Technical Products, Inc.*)

Silver-zinc batteries have found wide use on numerous space applications, including launch-vehicle guidance and control, telemetry, and destruct power: Apollo lunar spacecraft, lunar and Mars rovers, and lunar drill power; space shuttle payload launch and get-away special batteries: as well as power for the life-support equipment used by the U.S. astronauts during Extra-Vehicular Activities (EVA). Figure 33.23 shows a typical aerospace battery consisting of ten 145-Ah silver-zinc cells housed in a cast magnesium case, equipped with a pressure relief valve, a pressurizing valve, and a battery connector.

Silver-cadmium batteries have been used in a number of space applications requiring nonmagnetic properties. One such battery provided the main power for the *Giotto Halley Comet* intercept spacecraft.

Ground applications include communications equipment, portable television cameras, portable lights, camera drives, medical equipment, vehicle motive power, and similar uses requiring high-energy-density rechargeable batteries.

The user should keep in mind that no one type of battery is suitable for all applications. Optimum performance of a battery in an application can usually be achieved only by meeting the critical needs of the application and subordinating the others. The best approach for battery selection is to work with the battery manufacturers during the early stages of equipment design, rather than asking the battery designers, as is often done, to "design a battery that meets all my requirements and fits into this remaining cavity in my equipment."

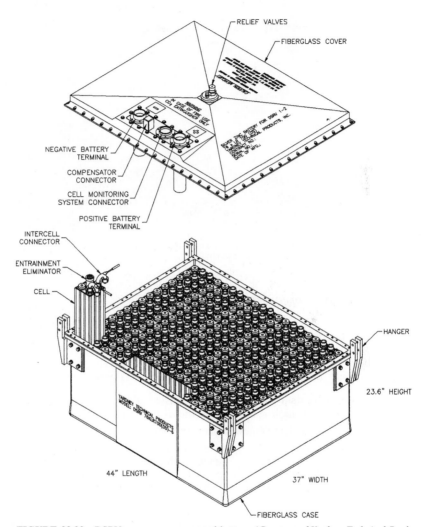

FIGURE 33.22 DSRV pressure compensated battery. (*Courtesy of Yardney Technical Products, Inc*).

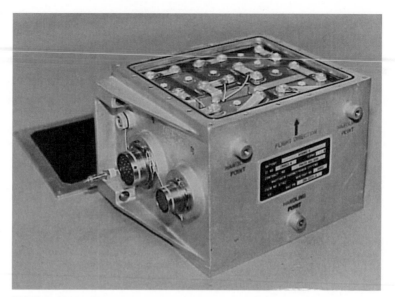

FIGURE 33.23   Silver-zinc aerospace battery, 15 V, 145 Ah. (*Courtesy of Yardney Technical Products, Inc.*)

33.9   *RECENT DEVELOPMENTS*

The vast majority of recent efforts in secondary silver cells has been concentrated in the areas of negative electrode and separator improvements for the zinc/silver oxide system because of its proportionally larger market volume.[7] The major developments are summarized in Table 33.9. Some of the most recent work has been quite successful, including:

- Use of bismuth oxide as an inexpensive additive to the zinc electrodes to improve the capacity retention and the cycle life of the cells.[25]
- The development of new separators, including:
  1. Microporous polypropylene with a proprietary coating,[26] which is very thin and has improved resistance to attack by the electrolyte.
  2. A polyolefin film with inorganic fillers, which has demonstrated to more than double the cycle life of the cells built with it, by inhibiting the "shape change" of the zinc electrodes.

Both separators are currently costly, and require further development work prior to production.

TABLE 33.9 Summary of Recent Developments in Silver Oxide Secondary Batteries and Components

Development area	Advantages	Disadvantages
Zinc electrode		
Increased zinc-to-silver-weight ratio	Delays onset of capacity losses	Gains not proportional to extra material used; reduces energy density
Oversized negatives	Reduce current density at plate edges where shape changes start	Reduce energy density
"Contoured" negatives	Reinforce areas of maximum erosion	Reduce energy density; high cost
PTFE binder	Reduces shape changes and dendrite growth; improves low-temperature performance	High cost; difficult to disperse evenly; may interfere with normal electrode reactions
Potassium titanate fibers	Reduce shape changes and dendritic growth; reduce probability of "hot" short circuits	Somewhat higher cost; hazardous during fabrication; restricted availability
Lead, lead/cadmium, bismuth additives as mercury substitutes	Reduce hazard to health and equipment in cases of "hot" short circuits; improve capacity maintenance	Smaller performance database
Main separator		
Inorganic separators	Resist temperature above 150°C: resist attack by silver oxides and electrolyte	High electrolytic resistance; bulky, difficult to handle: high cost
Cellulosics with metallic groups in molecule	Improved resistance to attack by silver oxides and electrolyte: extend cycle life	Somewhat higher cost
Microporous polypropylene	Proven resistance to attack by the electrolyte	Higher cost
Polyolefin film/inorganic fillers	Major improvement in cycle life	High cost; needs additional development work
Interseparators		
Positive: asbestos	Prevents or reduces magnitude of short circuits: acts as silver stopper	Bulky: reacts with silver oxides; may contaminate cell with iron: hazardous during fabrication
Zirconium oxide	Reduce hazard to health during cell manufacture	Higher cost
Negative: potassium titanate mat	Reduces zinc shape changes; reduces incidence and magnitude of short circuits	Bulky; reduces energy density; somewhat costly; hazardous during fabrication
Silver electrode		
Close control of particle-size distribution	Improves control of cell voltage and capacity	
Hardware		
New plastics for cases and covers (e.g., modified PPO, polysulfone) New adhesives	High-temperature operation: better mechanical properties More efficient at high temperatures: comply with EPA requirements	Higher cost

REFERENCES

1. A. Volta, *Phil. Trans. R. Soc. London* **90**:403 (1800).

2. S. U. Falk, and A. J. Salkind, *Alkaline Storage Batteries,* Wiley, New York, 1967.

3. H. André, *Bull. Soc. Fr. Electrochem.* (6th ser.) **1**:132 (1941).

4. H. André, U.S. Patent 2,317,711 (1943).

5. J. T. Brown, "Iron-Silver Battery—A New High Energy Density Power Source," Westinghouse Corp., Rep. 77-5E6-SILEL-RI, 1977.

6. "Design & Cost Study, Zinc/Nickel Oxide Battery for Electric Vehicle Propulsion, Yardney Electric Corp., Final Rep., Contract 31-109-38-3543, Oct. 1976.

7. R. Serenyi, "Recent Developments in Silver-Zinc Batteries," Yardney Electric Corp., Internal Rep. 2449–79, Oct. 1979.

8. G. W. Work and P. A. Karpinski, "Energy Systems for Underwater Use," *Marine Tech. Expo. Int. Conf,* New Orleans, La., Oct. 1979.

9. A. Himy, "Substitutes for Mercury in Alkaline Zinc Batteries," *Proc. 28th Annual Power Sources Symp.,* 1978, pp. 167–169.

10. R. Serenyi and P. Karpinski, "Final Report on Silver-Zinc Battery Development," Yardney Electric Corp., Contract N00140-76-C-6726, Nov.1978.

11. "Medium Rate Rechargeable Silver-Zinc 850 Ah Cell," Eagle-Picher Industries, Final Rep., USN Conract N00140-76-C-6729, Mar. 1978.

12. R. Einerhand, W. Visscher, J. de Goeij, and E. Barendrecht, "Zinc Electrode Shape Change," *J. Electrochem. Soc.* **138**:7–17 (Jan. 1991).

13. K. Choi, D. Bennion, and J. Newman, "Engineering Analysis of Shape Change in Zinc Secondary Electrodes," *J. Electrochem. Soc.* **123**:1616–1627 (Nov. 1976).

14. K. Bass, P. J. Mitchell, G. D. Wilcox, and J. Smith, "Methods for the Reduction of Shape Change and Dendritic Growth in Zinc-Based Secondary Cells," *J. Power Sources* **35**:333–351 (1991).

15. A. Charkey, "Long Life Zinc-Silver Oxide Cells," *Proc. 26th Ann. Power Sources Symp.,* 1976, pp. 87–89.

16. V. D'Agostino, J. Lee, and G. Orban, "Grafted Membranes," in A. Fleischer and J. Lander (eds.), *Zinc-Silver Oxide Batteries,* Wiley, New York, 1971, chap. 19, pp. 271–281.

17. C. P. Donnel, "Evaluation of Inorganic/Organic Separators," Yardney Electric Corp., Contract NAS3-18530, Oct. 1976.

18. H. A. Frank, W. L.: Long, and A. A. Uchiyama, "Impedance of Silver Oxide-Zinc Cells," J. *Electrochem. Soc.* **123**(1):1–9 (Jan. 1976).

19. J. C. Brewer, R. Doreswamy, and L. G. Jackson, "Life Testing of Secondary Silver-Zinc Cells for the Orbital Maneuvering Vehicle," *Proc. 25th IECEC,* Reno, Nev., Aug. 1990.

20. "Evaluation of Silver-Cadmium Batteries for Satellite Applications," Boeing Co., Test D2-90023, Feb. 1962.

21. A. P. Karpinski and J. A. Patten, "Performance Characteristics of Silver-Zinc Cells for Orbiting Spacecraft," *Proc. 25th IECEC,* Reno, Nev., Aug. 1990.

22. A. Fleischer and J. Lander (eds.), *Zinc-Silver Oxide Batteries.* Wiley, New York, 1971.

23. R. Serenyi, U.S. Patent 5,773, 176.

24. A. P. Karpinski, B. Makovetski, S. J. Russell, J. R. Serenyi and D. C. Williams, "Silver-Zinc: Status of Technology and Applications," *J. of Power Sources* **80**:53–60 (1999).

25. *Proceedings of the 5th Workshop for Battery Exploratory Development,* Burlington, VT, July 1997, 153–157.

26. *Proceedings of the 38th Power Sources Conference,* Cherry Hill, NJ, June 1998, 175–178.

CHAPTER 34
RECHARGEABLE LITHIUM BATTERIES (AMBIENT TEMPERATURE)

Thomas B. Reddy and Sohrab Hossain

34.1 GENERAL CHARACTERISTICS

Rechargeable lithium batteries operating at room temperature offer several advantages compared to conventional aqueous technologies, including

1. Higher energy density (up to 150 Wh/kg, 400 Wh/L)
2. Higher cell voltage (up to about 4 V per cell)
3. Longer charge retention or shelf life (up to 5 to 10 years)

These advantageous characteristics result in part from the high standard potential and low electrochemical equivalent weight of lithium metal.

Ambient-temperature lithium rechargeable batteries, on the other hand, do not have the high-rate capability (because of the lower conductivity of the aprotic organic or inorganic electrolytes that must be used because of the reactivity of lithium in aqueous electrolytes) nor, in some instances, the cycle life of conventional rechargeable batteries. In addition, rechargeable lithium batteries that use lithium metal as the negative electrode present potential safety problems which are more challenging than those with primary lithium batteries. This is due to a three- to fivefold excess of lithium, which is required for these types of cells in order to obtain a reasonable cycle life, and to the reactivity of the high-surface-area lithium that is formed during cycling.

There is another type of rechargeable "lithium" battery, however, which uses a lithiated carbon or other intercalation material for the negative electrode in place of lithium. The absence of metallic lithium in these lithium-ion batteries minimizes these safety concerns. This type of battery is covered in Chap. 35.

The advantages and disadvantages of rechargeable lithium batteries operating at ambient temperature are summarized in Table 34.1. (These may not be applicable in all cases because of the different characteristics of the various rechargeable lithium systems.)

TABLE 34.1 Advantages and Disadvantages of Ambient-Temperature Lithium Rechargeable Batteries

Advantages
High energy density and specific energy
High voltage
Good charge retention, low self-discharge rate

Disadvantages
Low cycle life with metallic lithium systems
Relatively poor high-rate performance (compared to conventional aqueous rechargeable batteries)
Relatively poor low-temperature performance (compared to conventional aqueous rechargeable batteries)
Capacity fading (with some systems)
Potential safety problems with metallic lithium systems

As with primary lithium batteries, a number of different approaches have been taken in the chemistry and design of rechargeable lithium batteries to obtain the desired performance characteristics. These are summarized in Fig. 34.1a for batteries with lithium metal negative electrodes (the anode during discharge) and in Fig. 34.1b for batteries with other materials, such as lithium alloys and lithiated carbon.*

A variety of materials have been investigated for the positive electrode (the cathode during discharge), such as intercalation solid compounds, soluble inorganic cathodes, and polymeric materials. Liquid aprotic organic and inorganic electrolytes are used in many cells. Solid polymer electrolytes are also popular as they may provide a safer design because of their lower reactivity with lithium. These materials are identified in Fig. 34.1.

Rechargeable lithium batteries have been introduced into the market on a limited scale. Coin cells, using lithium-aluminum anodes, are available for special applications mainly for low-power portable applications where they can be conveniently recharged, in some instances by solar cells. A small cylindrical cell, using a lithium anode, was briefly, and perhaps prematurely, introduced in the 1980s for consumer electronics applications, but was withdrawn when safety problems arose. More recently, the lithium-ion cell, which has a safety advantage over other lithium secondary cells as it does not contain lithium in a metallic form, has been marketed as a power source for consumer electronics such as cellular phones and camcorders. This technology has become dominant in the market.

The shipment, use, and disposal of rechargeable lithium batteries are regulated, as for the primary lithium batteries, by international organizations, government agencies, and quasi-government institutions because of the concern for potential safety problems with lithium batteries, particularly if they are physically or electrically abused.[1]

---

*As discussed in Chap. 1, the electrodes in a rechargeable battery are referred to as negative and positive. Sometimes they are referred to as anode and cathode, respectively, the appropriate terminology for the cell during discharge.

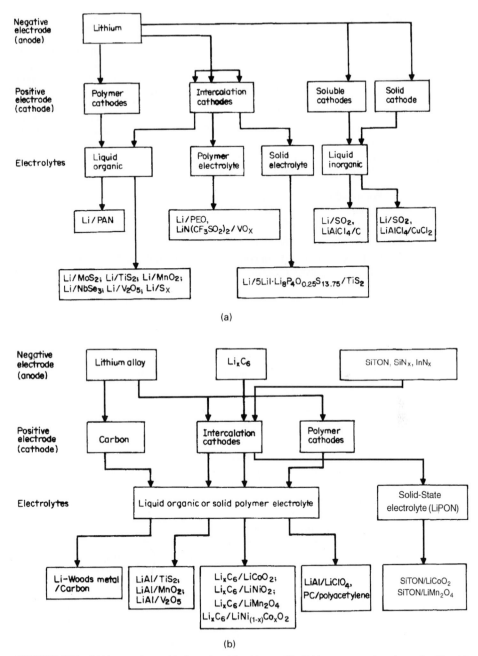

FIGURE 34.1 Lithium rechargeable batteries (*a*) with metallic lithium as negative electrode, (*b*) with lithium alloys or lithiated carbon negative electrode.

34.2 *CHEMISTRY*

The objective of the rechargeable lithium battery development is to produce batteries that have high energy density, high power density, good cycle life and charge retention, and to provide this performance reliably and safely. The selection of cell components and designs is necessarily a compromise to achieve the optimum balance. Many of the characteristics and criteria for selection are similar to those for primary lithium cells covered in Chap. 14. The process, however, is even more complex for rechargeable batteries since the reactions that occur during recharge affect all of the characteristics and the performance on subsequent cycling.

The different types of lithium rechargeable batteries identified in Fig. 34.1 can be classified conveniently into five categories:

1. Solid-cathode cells using intercalation compounds for the positive electrode, a liquid organic electrolyte, and a metallic lithium negative electrode.

2. Solid-cathode cells using intercalation compounds for the positive electrode, a polymer electrolyte, and a metallic lithium negative electrode.

3. Cells using intercalation compounds for both the positive and the negative electrodes and a liquid or polymer electrolyte (lithium-ion cells) (see Chap. 35).

4. Inorganic electrolyte cells, which use the electrolyte solvent or a solid redox couple for the positive and lithium metal for the negative active material.

5. Cells with lithium-alloy anodes, liquid organic or polymer electrolytes, and a variety of cathode materials, including polymers. This technology has been used mainly in small flat or coin cells.

The components and reactions of typical examples of these types of rechargeable lithium batteries are summarized in Sec. 34.3.

34.2.1 Negative Electrodes

Typical negative electrode materials for rechargeable lithium batteries are listed in Table 34.2. Of these, lithium is the lightest and most electropositive, and it has a high specific capacity, 3.86 Ah/g. It is also more easily handled than the other alkali metals.

TABLE 34.2 Negative Electrode Material

Material	Voltage range vs. lithium, V	Theoretical specific capacity, Ah/g	Comments
Li metal	0.0	3.86	Lithium foils readily available
LiAl	0.3	0.8	Generally brittle foils, difficult to handle
$Li_{0.5}C_6$ (coke)	0.0–1.3	0.185	Used for lithium-ion cells
LiC_6 (MCMB (*b*), or graphite)	0.0–0.1	0.372 (*a*)	
$LiWO_2$	0.3–1.4	0.12	Possible use for lithium-ion cells
$LiMoO_2$	0.8–1.4	0.199	
$LiTiS_2$	1.5–2.7	0.266	

(*a*) Based on weight of carbon only.
(*b*) Mesocarbon microbeads.

While metallic lithium has the highest specific capacity, it is more reactive than lithium-aluminum and other alloys. These have been used mainly in small flat or coin cells, as it is difficult to scale up to larger and spirally wound designs because most lithium alloys are brittle and cannot be extruded into thin foils. Another approach is the lithium ion battery which uses a carbon material for the negative electrode. These are attractive from a safety viewpoint as reactive metallic lithium is not present in the cell. A suitable intercalation compound is selected for the positive electrode for these cells so that a high cell voltage is obtained. The cell operates by the lithium ions shuttling back and forth between the electrodes during the discharge-charge cycle. No metallic lithium is plated during the charge and no metallic lithium is present in the cell.

Lithium Metal.   The search for high-energy-density batteries has inevitably led to the use of lithium, as the electrochemical characteristics of this metal are unique. A number of batteries, both primary and rechargeable, using a lithium anode in conjunction with intercalation cathodes, were developed which had attractive energy densities, excellent storage characteristics, and, for rechargeable cells, a reasonable cycle life. Commercial success has eluded all but the primary batteries due to persistent safety problems.

The difficulties associated with the use of metallic lithium stem from its reactivity with the electrolyte and the changes that occur after repetitive charge-discharge cycling. When lithium is electroplated, during recharge, onto a metallic lithium electrode, it forms a mossy and in some cases a dendritic deposit with a larger surface area than the original metal. While the thermal stability of lithium metal foil in many organic electrolytes is good, with minimal exothermic reaction occurring up to temperatures near the melting point of lithium (181°C), after cycling the surface area of the lithium increases significantly with a corresponding increase in the reactivity. This lowers the thermal stability of the system, with the result that cells become increasingly sensitive to abuse as they are cycled.

Another contributing effect is the inability of attaining 100% lithium cycling efficiency. This happens because lithium is not thermodynamically stable in the organic electrolytes and the surface of the lithium is covered with a film of the reaction products between the lithium and the electrolyte. Every time the lithium is stripped and replated during discharge and charge, a new lithium surface is exposed and then passivated with a new film, consuming lithium. Because of the mossy deposit, some lithium becomes electrochemically unreactive on repeated cycling. In order to obtain a reasonable cycle life, a three- to fivefold excess of lithium is required.

The failure to control the surface area of the lithium anode remains a problem, limiting the commercialization of lithium anode cells with liquid organic electrolytes. Given this problem, an alternate solution is to use an electrolyte, such as a solid polymer electrolyte, which is less reactive with lithium. This is covered in Sec. 34.2.3.

Carbon Materials.   In the lithium-ion cell, carbon materials, which can reversibly accept and donate significant amounts of lithium (Li:C = 1:6) without affecting their mechanical and electrical properties, can be used for the anode instead of metallic lithium. Carbon material is used as an anode in lithium-ion cells since the chemical potential of lithiated carbon material is close to that of metallic lithium, as shown in Fig. 34.2. Thus an electrochemical cell made with a lithiated carbon material will have almost the same open-circuit voltage as one made with metallic lithium. In practice, the lithium-ion cell is manufactured in a fully discharged state. Instead of using lithiated carbon material, which is air-sensitive, the anode is made with carbon and lithiation is carried out by subsequent formation of the cell.

The specific energy or capacity of the lithium-ion system depends largely on the type of carbon materials used, the lithium intercalation efficiency, and the irreversible capacity loss associated with the first charge process. Table 34.3 lists the properties of some of these carbon materials. Coke-type carbon, having physical properties such as ash content <0.1%, surface area <10 m^2/g, true density <2.15 g/cm^3, and interlayer spacing >3.45 Å, were used in first-generation lithium-ion system. These types of carbon materials can provide about

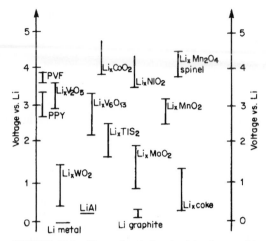

FIGURE 34.2 Electrochemical potential of some Li-intercalation compounds vs. Li metal. PPY = polypyrrole. PVF = polyvinyl furan.

TABLE 34.3 Physical Properties of Coke and Graphite Anodes for Lithium-Ion Cells

Item	Coke		Graphite		
Type	Petroleum coke	Synthetic, KS-15	Synthetic, KS-44	Isotopic, EC-110	Natural
Structure	Disordered		Ordered layer structure		
Physical parameters:					
Interlayer spacing d002, Å	3.46	3.35	3.35	3.34	3.34
Crystalline size L_c, Å	46	900	>1000	>1000	>2000
Surface area, m²/g	6	14	10	10	—
Density, g/cm³	2.14	2.255	2.248	—	—
Ash content, %	0.08	0.05	<0.1	—	—

185 mAh/g capacity (corresponding to LiC_{12}). By controlling the temperature of the heat treatment, carbon material having specific properties such as density and interlayer spacing can be prepared. Doping with nitrogen, boron, or phosphorus can increase the capacity of coke type materials to 350 mAh/g. Graphitic carbons having an interlayer spacing of 3.36 Å can also deliver 350 mAh/g capacity. The capacity delivered by different types of carbon materials, including natural and synthetic graphite, in ethylene carbonate-based electrolytes on the first discharge is shown in Fig. 34.3.[2] This figure shows the advantage of the graphite materials, which have a flatter discharge and a higher capacity than the coke materials. Mesocarbon Microbeads (MCMB), which are also graphitized, have also been employed in lithium ion batteries.

During the first electrochemical intercalation of lithium into the carbon, some lithium is irreversibly consumed forming a solid electrolyte interface (SEI) and cannot be recovered in the following discharge, resulting in a loss of capacity. (See Chap. 35.) Figure 34.4 shows the voltage profile of the first charge and discharge of petroleum coke and graphite electrodes versus a lithium electrode, respectively.[3] This irreversible capacity, which depends on the electrolyte solution and the type of carbon material, is explained on the basis of the reduction

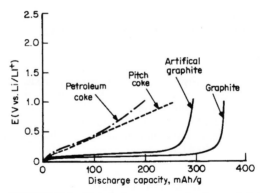

FIGURE 34.3 First discharge curves of carbon materials.

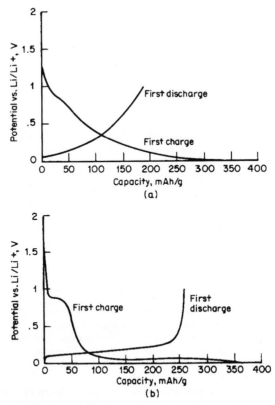

FIGURE 34.4 Representation of irreversible capacity associaed with the first charge/discharge process. (*a*) Coke. (*b*) Artificial graphite.

of the electrolyte solution and the formation of a SEI layer on the Li_xC interface.[4] When the film is sufficiently thick to prevent electron tunneling, the electrolyte reduction is suppressed and the electrode can then be cycled reversibly. The first step is, therefore, critical in order to obtain a uniform passivating film. Very little loss usually occurs after the first intercalation. The capacity on the second and subsequent cycles is about the same, and the lithium intercalation during charge and discharge is nearly 100% reversible.

The theoretical capacity of metallic lithium is much higher than that of lithiated carbon material having a composition of LiC_6. The advantage of the higher capacity of metallic lithium is reduced because a three- to fivefold excess of lithium is required in rechargeable batteries having metallic lithium anodes to achieve a reasonable cycle life. The comparison is shown in Table 34.4.

TABLE 34.4 Comparison of Usable Specific Capacity (Ah/kg) and Capacity Density (Ah/L) for Lithiated Carbon vs. Lithium Metal Anodes

Characteristics	Li metal	LiC_6
Theoretical specific capacity, Ah/kg	3862	372
Theoretical capacity density, Ah/L	2061*	837
Practical specific capacity, Ah/kg		
Fourfold excess of lithium	965	372
95% active material in carbon electrode	965	353
Practical capacity density, Ah/L		
Fourfold excess of lithium	515	837
Porosity of carbon electrode (50%)	515	418

* Density of lithium = 0.534 g/cm³.

Transition Metal Compounds.   Transition metal compounds having layered structures into which lithium ions can be intercalated and deintercalated during charge and discharge and with electrochemical potentials close to those of lithiated carbon materials have also been used as the negative electrode in lithium-ion cells. Figure 34.2 shows the electrochemical potentials of some lithiated transition metal compounds. The electrochemical potentials of Li_xWO_2, Li_xMoO_2, and Li_xTiS_2 are close to that of lithiated carbon and distinctly different from the values for $Li_xMn_2O_4$, Li_xCoO_2, and Li_xNiO_2. WO_2, MoO_2, or TiS_2 can then be used as anodes and $LiMn_2O_4$, $LiCoO_2$, or $LiNiO_2$ as cathodes. Cells of these types have been developed using TiS_2 anodes and $LiCoO_2$ cathodes in an organic electrolyte.[7]

34.2.2 Positive Electrodes

There is a wide choice of materials that can be selected for the positive electrodes of lithium batteries. However, many of these involve reactions which cannot be readily reversed and are limited to primary nonrechargeable batteries. The best cathodes for rechargeable batteries are those where there is little bonding and structural modification of the active materials during the discharge-charge reaction.[8]

Intercalation Compounds.   The insertion or intercalation compounds are among the most useful cathode materials. In these compounds, a guest species such as lithium can be inserted interstitially into the host lattice (during discharge) and subsequently extracted during recharge with little or no structural modification of the host.

The intercalation process involves three principal steps:

1. Diffusion or migration of solvated Li^+ ions
2. Desolvation and injection of Li^+ ions into the vacancy structure
3. Diffusion of Li^+ ions into the host structure

The electrode reactions which occur in a Li/Li_x (HOST) cell, where (HOST) is an intercalation cathode, are

$$yLi \leftrightarrow yLi^+ + ye \qquad \text{at the Li metal anode}$$

$$yLi^+ + ye + Li_x(\text{HOST}) \leftrightarrow Li_{x+y}(\text{HOST}) \qquad \text{at the cathode}$$

leading to an overall cell reaction of

$$yLi + Li_x(\text{HOST}) \leftrightarrow Li_{x+y}(\text{HOST})$$

A number of factors have to be considered in the choice of the intercalation compound, such as reversibility of the intercalation reaction, cell voltage, variation of the voltage with the state of charge, and availability and cost of the compound. Table 34.5 lists the key requirements for the intercalation materials and Table 34.6 presents some of the characteristics of the intercalation and other compounds that have been used in lithium rechargeable batteries. The electrochemical potentials of several lithium intercalation compounds versus those of lithium, metal and the variation of voltage with the amount of intercalation are shown in Fig. 34.2.

Transition metal oxides (MnO_2, $LiCoO_2$, $LiNiO_2$, VO_x), sulfides (MoS_2, TiS_2), and selenides ($NbSe_3$) have been used in rechargeable lithium batteries. The lithiated transition metal oxides (such as $LiCoO_2$, $LiNiO_2$, and $LiMn_2O_4$) are attractive materials used as the cathode in the lithium-ion rechargeable cell. $LiCoO_2$, $LiNiO_2$ and $LiNi_{1-x}Co_xO_2$ have a layered structure, where lithium and transition metal cations occupy alternate layers of octahedral sites in a distorted cubic close-packed oxygen-ion lattice. The layered metal oxide framework provides a two-dimensional interstitial space, which allows for each removal of the lithium ions. The layered structure can be prepared by high-temperature treatment of a mixture of lithium hydroxide or other salts and the metal oxide in air,[9]

$$Co_2O_3 + 2LiOH \xrightarrow{700°C} 2LiCoO_2 + H_2O$$

$$2NiO + 2LiOH + \tfrac{1}{2}O_2 \xrightarrow{700°C} 2LiNiO_2 + H_2O$$

TABLE 34.5 Key Requirements for Positive-Electrode Intercalation Material (Li_xMO_2) Used in Rechargeable Lithium Cells

---

High free energy of reaction with lithium
Wide range of x (amount of intercalation)
Little structural change on reaction
Highly reversible reaction
Rapid diffusion of lithium ion in solid
Good electronic conductivity
No solubility in electrolyte
Readily available or easily synthesized from low-cost reactants

---

TABLE 34.6   Positive-Electrode Materials and Some of Their Characteristics

Material	Average voltage vs. lithium,* V	Lithium/ mole	Practical specific energy, † Wh/kg	Comments
MoS_2	1.7	0.8	230	Naturally occurring
MnO_2	3.0	0.7	650	Inexpensive
TiS_2	2.1	1	550	Costly
$NbSe_3$	1.9	3	450	Costly
$LiCoO_2$	3.7	0.5	500	Good for lithium-ion system
$LiNiO_2$	3.5	0.5	480	Good for lithium-ion system
$LiMn_2O_4$	3.8	0.8	450	Good for lithium-ion system, safe
VO_x	2.3	2.5	300	Good for SPE system
V_2O_5	2.8	1.2	490	Good for SPE system
SO_2	3.1	0.33	220	Good for pulse power applications, safety issues
$CuCl_2$	3.3	1	660	Good for pulse power applications, safety issues
Polyacetylene	3.2	1	340	For polymer electrodes
Polypyrrole	3.2	1	280	For polymer electrodes

* At low rates.
† Based on cathode material only and average voltage and lithium/mole as shown.

Spinel $LiMn_2O_4$ may be obtained by heating a mixture of appropriate amounts of Li_2CO_3 and MnO_2 at 800°C in air.[5,10] The $LiMn_2O_4$ spinel framework possesses a three-dimensional space via face sharing octahedral and tetrahedral structures, which provide conducting pathways for the insertion and extraction of lithium ions.

The removal and insertion of the lithium ion for lithiated transition metal oxides (M) are:

$$LiMO_2 \rightleftharpoons Li_{(1-x)}MO_2 + xLi^+ + xe$$

The reversible value of x for $LiCoO_2$ and $LiNiO_2$ is approximately 0.5, and the value is greater than or equal to 0.85 for lithiated manganese oxide. Thus although the theoretical capacity of $LiCoO_2$ and $LiNiO_2$ (274 mAh/g) is almost twice as high as that of $LiMn_2O_4$, the reversible capacity of the three cathode materials are in the same range. Figure 34.5 compares the reversible discharge capacity of the three cathode materials. It is to be noted that lithium-ion cells made with $LiMn_2O_4$ require a higher charge voltage to achieve full capacity. These batteries are typically charged to 4.2 V vs. 4.1 V for Co and 4.0 V for Ni.

The diffusion coefficients of lithium ion (D_{Li+}) in the three lithiated metal oxides are shown in Table 34.7. The diffusion steps play a significant role in the mechanistic aspects of the intercalation process.

Polymers.   Electronically conductive polymers may also be used as cathode materials in rechargeable lithium batteries. The most popular polymers are polyacetylene, polypyrrole, polyaniline, and polythiophene, which are made conductive by doping with suitable anions. The discharge-charge process is a redox reaction in the polymer. The low specific energy, high cost, and their instability, however, make these polymers less attractive. They have been used in small coin-type batteries with a lithium-aluminum alloy as the anode.

Another potential approach is an all-polymer cell using a cation-doped polymer for the anode, a solid polymer electrolyte, and a polymer for the cathode. These cells can be fabricated in thin sections and in a variety of shapes but, again, most of the development has been in small coin sizes. No progress on these systems has been reported recently.

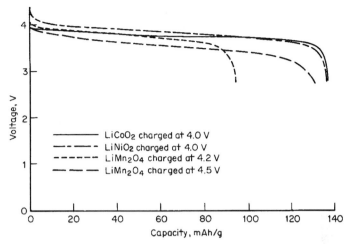

FIGURE 34.5 Reversible discharge capacity of cathode materials for lithium-ion cells (cells fabricated with graphite anode). (*Courtesy Yardney Technical Products, Inc.*)

TABLE 34.7 Diffusion Coefficients of Li-Ion in Lithiated Transition Metal Oxides

Metal oxides	Techniques	D_{Li}^+, cm^2/s	References
LiCoO$_2$	Transient	5×10^{-9}	a
	Transient	5×10^{-9}	b
	Impedance	5×10^{-8}	c
LiNiO$_2$	Impedance	2×10^{-7}	d
LiMn$_2$O$_4$	Transient	10^{-11}	e
	Transient	10^{-9}	f

[a] K. Mizushima, P. C. Jones, P. J. Wieseman, and J. B. Goodenough, *Solid State Ionics* **3/4**:171, 1981.
[b] S. Kikkawa, S. Miyazaki, and M. Koizumi, *J. Power Sources* **14**: 231, 1985.
[c] M. G. S. R. Thomas, P. Bruce, and J. Goodenough, *Solid State Ionics* **18/19**:794, 1986.
[d] P. G. Bruce, A. Lisowska-Oleksiak, M. Y. Saidi, and C. A. Vincent, *Solid State Ionics* **57**:353, 1992.
[e] P. G. Dickens and G. F. Reynolds, *Solid State Ionics* **5**:53, 1981.
[f] D. Guyomard and J. M. Tarascon, *J. Electrochem. Soc.* **139**:937, 1992.

Inorganic Electrolytes.   Another choice of active materials for the cathode is the inorganic electrolytes such as liquid SO$_2$-based LiAlCl$_4$ or LiGaCl$_4$. Reduction and oxidation occur during the discharge-charge process on carbon surfaces. The cell reaction depends on the type of carbon and electrolytes used. On relatively low-surface-area carbon such as acetylene black, a simple SO$_2$/S$_2$O$_4^=$ couple is involved in the charge-discharge process,

$$2SO_2 + 2e \underset{\text{charge}}{\overset{\text{discharge}}{\rightleftharpoons}} S_2O_4^=$$

The reduction of SO$_2$ occurs at about 2.9 V at 1 mA/cm^2.

When high-surface-area carbons such as Ketjen black and SO_2-based $LiAlCl_4$ electrolytes are used,[11] a carbon-electrolyte complex is believed to take part in the discharge-charge process.

$$LiAlCl_4 \cdot 3SO_2 + xC + 3e \underset{\text{charge}}{\overset{\text{discharge}}{\rightleftharpoons}} LiClAl\begin{matrix} \diagup OSO \\ -OSO \cdots \cdots C_x + 3Cl^- \\ \diagdown OSO \end{matrix}$$

Solid Cathodes. Solid-state couples can also be used as cathode materials in SO_2-based inorganic electrolytes. The high reduction potential of SO_2 limits the selection of the positive electrode materials that can be used in SO_2 electrolytes to those compounds that reduce at potential above 2.90 V. Thus $CuCl_2$, which has an open-circuit voltage of 3.43 V and discharges at about 3.35 V at 1 mA/cm², corresponding to the reduction process of

$$Cu^2 + e \underset{\text{charge}}{\overset{\text{discharge}}{\rightleftharpoons}} Cu^+$$

meets these criteria. The second reduction process ($Cu^+ + e \rightarrow Cu$) occurs at 2.5 V. In principle, the discharge of $Li/CuCl_2$ cells should not occur below 2.9 V, since the cathode will be passivated by lithium dithionite (LiS_2O_4) produced from the reduction of SO_2.

$LiCoO_2$ is being investigated in SO_2-based inorganic electrolytes.[12,65,66] This cathode material has a higher voltage and the SO_2-based $LiAlCl_4$ electrolyte is potentially suitable because of its wide electrochemical voltage window.

34.2.3 Electrolytes

The choice of electrolyte for rechargeable lithium batteries is also critical. The electrolyte should have the following characteristics:

1. Good ionic conductivity ($>10^{-3}$ S/cm from -40 to 90°C) to minimize internal resistance.
2. Lithium ion transference number approaching unity (to limit concentration polarization)
3. Wide electrochemical voltage window (0 to 5 V)
4. Thermal stability (up to 70°C)
5. Compatibility with other cell components

Aprotic Organic Electrolytes. Aprotic liquid organic electrolyte solvents, such as dioxolane, propylene carbonate, ethylene carbonate, diethyl carbonate and ethylmethyl carbonate are the most common electrolyte solvents because of their low reactivity with lithium. A list of the electrolyte solvents used in rechargeable lithium batteries with their major characteristics is given in Table 34.8.[8] Choices for the electrolyte solute and their ionic conductivities in various solvents at different temperatures are listed in Table 34.9. These organic liquid electrolytes generally have conductivities that are about two orders of magnitude lower than aqueous electrolytes.

TABLE 34.8 Characteristics of Organic Solvents*

Characteristic	γ-BL	THF	1,2-DME	PC	EC	DMC	DEC	DEE	Dioxolane (DN)
Structural formula	CH₂—CH₂, CH₂, C=O, O	CH₂—CH₂, CH₂, CH₂, O	CH₂—O—CH₃, CH₂—O—CH₃	O=C, O—CH—CH₃, O—CH₂	O=C, O—CH₂, O—CH₂	O=C, O—CH₃, O—CH₃	O=C, O—CH₂—CH₃, O—CH₂—CH₃	CH₂—O—C₂H₅, CH₂—O—C₂H₅	CH₂, O, O, CH₂—CH₂
Boiling point, °C	202–204	65–67	85	240	248	91	126	121	78
Melting point, °C	−43	−109	−58	−49	−39–40	4.6	−43	−74	−95
Density, g/cm³	1.13	0.887	0.866	1.198	1.322	1.071	0.98	0.842	1.060
Viscosity at 25°C, cP	1.75	0.48	0.455	2.5	1.86 (at 40°C)	0.59	0.75	0.65	0.58
Dielectric constant at 20°C	39	7.75	7.20	64.4	89.6 (at 40°C)	3.12	2.82	5.1	6.79
Molecular weight	86.09	72.10	90.12	102.0	88.1	90.08	118.13	118.18	74.1
Typical H₂O content, ppm	<10	<10	<10	<10	<10	<10	<10	<10	<10
Electrolytic conductivity at 20°C, 1M LiAsF₆, mS/cm	10.62	12.87	19.40	5.28	6.97	11.00 (1.9 M)	5.00 (1.5 M)	~10.00†	~11.20†

*γ-BL = γ-butyrolactone; THF = tetrahydrofuran; 1,2-DME = 1,2-dimethoxyethane; PC = propylene carbonate; EC = ethylene carbonate; DMC = dimethyl carbonate; DEC = diethyl carbonate; DEE = diethoxyethane.
†Estimation based on Walden's rule.
Source: From Ref. 8.

34.13

TABLE 34.9 Ionic Conductivity of Some 1 Molar Organic Liquid Electrolytes Used in Secondary Lithium Battery Systems

Salt	Solvents	Solvent, vol %	Conductivities at °C, mS/cm							References
			−40	−20	−0	20	40	60	80	
$LiPF_6$	EC/PC	50/50	0.23	1.36	3.45	6.56	10.34	14.63	19.35	*
	2-MeTHF/EC/PC	75/12.5/12.5	2.43	4.46	6.75	9.24	11.64	14.00	16.22	*
	EC/DMC	33/67	—	1.2	5.0	10.0	25.2	20.0	—	†
	EC/DME	33/67	—	8.0	13.6	18.1	—	31.9	—	‡
	EC/DEC	33/67	—	2.5	4.4	7.0	9.7	12.9	—	‡
$LiAsF_6$	EC/DME	50/50	Freeze	5.27	9.50	14.52	20.64	26.65	32.57	*
	PC/DME	50/50	Freeze	4.43	8.37	13.15	18.46	23.92	28.18	*
	2-MeTHF/EC/PC	75/12.5/12.5	2.54	4.67	6.91	9.90	12.76	15.52	18.18	*
$LiCF_3SO_3$	EC/PC	50/50	0.02	0.55	1.24	2.22	3.45	4.88	6.43	*
	DME/PC	50/50	—	2.61	4.17	5.88	7.46	9.07	10.61	*
	DME/PC	50/50	Freeze	Freeze	5.32	7.41	9.43	11.44	13.20	*
	2-MeTHF/EC/PC	75/12.5/12.5	0.50	0.93	1.34	1.78	2.31	2.81	3.30	*
$LiN(CF_3SO_2)_2$	EC/PC	50/50	0.28	1.21	2.80	5.12	7.69	10.70	13.86	*
	EC/DME	50/50	—	Freeze	7.87	12.08	16.58	21.25	25.97	*
	PC/DME	50/50	—	3.92	7.19	11.23	15.51	19.88	24.30	*
	2-MeTHF/EC/PC	75/12.5/12.5	2.07	3.40	5.12	7.06	8.71	10.41	12.02	*
$LiBF_6$	EC/PC	50/50	0.19	1.11	2.41	4.25	6.27	8.51	10.79	*
	2-MeTHF/EC/PC	75/12.5/12.5	—	0.38	0.92	1.64	2.53	3.43	4.29	
	EC/DMC	33/67	—	1.3	3.5	4.9	6.4	7.8	—	‡
	EC/DEC	33/67	—	1.2	2.0	3.2	4.4	5.5	—	‡
	EC/DME	33/67	—	6.7	9.9	12.7	15.6	18.5	—	‡
$LiClO_4$	EC/DMC	33/67	—	1.0	5.7	8.4	11.0	13.9	—	‡
	EC/DEC	33/67	—	1.8	3.5	5.2	7.3	9.4	—	‡
	EC/DME	33/67	—	8.4	12.3	16.5	20.3	23.9	—	‡

* J. T. Dudley et al., *J. Power Sources*, **35**:59–82, 1991.
† D. Guyomard and J. M. Tarascon, *J. Electrochem. Soc.* **140**:3071–3081, 1993.
‡ S. Sosnowski and S. Hossain, unpublished results, Yardney Technical Products, Inc.

Polymer Electrolytes.   An alternative to the liquid electrolytes is a solid polymer electrolyte (SPE) formed by incorporating lithium salts into polymer matrices and casting into thin films. These can be used as both the electrolyte and separator. These electrolytes have lower ionic conductivities and low lithium-ion transport numbers compared to the liquid electrolytes, but they are less reactive with lithium, which should enhance the safety of the battery. The use of thin polymer films or operation at higher temperatures (60–100°C) compensate in part for the lower conductivity of the polymer film. The solid polymers also offer the advantages of a "nonliquid" battery and the flexibility of designing thin batteries in a variety of configurations.

Initially, high-molecular-weight polymers such as polyethylene oxide (PEO) and lithium salts such as $LiClO_4$ and $LiN(CF_3SO_2)_2$ were used.[13] These PEO-lithium salt electrolytes have good mechanical properties but low conductivities, which are on the order of 10^{-8} S/cm at 20°C. A significant improvement in conductivity to approximately 10^{-5} S/cm has been achieved with the combination of modified comb-shaped PEO structures with lithium salts,[14] but these types of solid polymer electrolytes have poor mechanical properties and their conductivity is still two orders of magnitude lower than that of most organic liquid electrolytes. Further improvement in conductivity was obtained with the addition of liquid plasticizers such as polypropylene carbonate.[15,16] The amount of plasticizer may be as high as 70%, resulting in limited chemical and mechanical stability. Another class of polymer electrolytes called "gelled" electrolytes has been developed by trapping liquid solutions of lithium salts in aprotic organic solvents [for example, $LiClO_4$ in propylene carbonate-ethylene carbonate (PC/EC) solvent] into a solid polymer matrix such as polyacrylonitrile (PAN).[17,18] The "gel" electrolytes are made by inserting liquid electrolyte solutions into polymer cages with an immobilization procedure such as cross-linking, gellification, and casting. Cross-linking may be carried out by ultraviolet, electron-beam, or gamma-ray irradiation. Conductivities as high as 10^{-3} S/cm at 20°C and transference number around 0.6 have been obtained. However, these plasticized and gelled electrolytes are more reactive with lithium than true solid polymers. The conductivites of the various classes of polymer electrolytes and their variations with temperature are presented in Fig. 34.6.

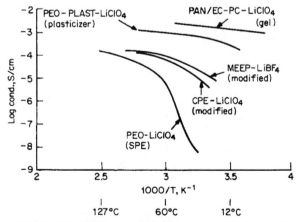

FIGURE 34.6   Variation of conductivity with temperature of different classes of polymer electrolytes. PEO = polyethylene oxide. CPE = cross-linked polyvinylether. MEEP = poly[bis(methoxy ethoxy ethoxide)]. PAN = polyacrylonitrile.

Solid-State Electrolytes.   True all solid-state ionic electrolytes such as lithium phosphorus oxynitride (LiPON) provide adequate conductivity for use in thin film solid-state batteries (see Sec. 35.8.)

Inorganic Electrolytes.   The SO_2-based inorganic electrolytes are another alternative for use in rechargeable lithium batteries. These electrolytes are attractive because they offer the highest ionic conductivity of any electrolyte used in rechargeable lithium batteries. Figure 34.7a shows the ionic conductivity of SO_2-based $LiAlCl_4$ electrolytes at various temperatures.[19]

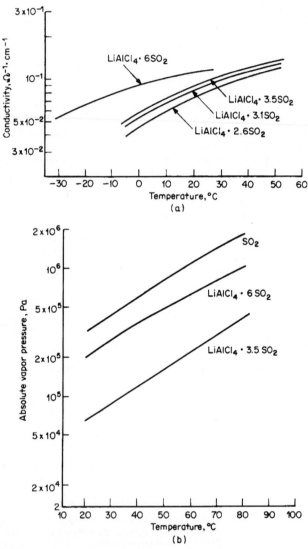

FIGURE 34.7   (a) Conductivity of $LiAlCl_4/SO_2$ electrolytes at various temperatures. (b) Vapor pressure of SO_2 and SO_2-based electrolytes.

The SO_2-based electrolytes, however, present potential safety problems. One disadvantage, the high vapor pressure of SO_2, can be reduced by adding electrolyte salts such as $LiAlCl_4$ and $LiGaCl_4$. These form complexes exothermically with the SO_2 and reduce the vapor pressure. For example, as shown in Fig. 34.7*b*, the vapor pressure of $LiAlCl_4 \cdot 3SO_2$ is below atmospheric pressure at 20°C.[19]

34.3 CHARACTERISTICS OF LITHIUM RECHARGEABLE BATTERIES

A number of different battery systems have been investigated for the development of lithium rechargeable batteries in order to achieve the high specific energy and charge retention that lithium batteries can offer without sacrificing other important characteristics, such as specific power and cycle life, while maintaining safe and reliable operation. These approaches are summarized in Fig. 34.1.

The rechargeable lithium batteries are generally characterized by a high cell voltage, good charge retention, higher specific energy but poorer high-rate performance, and poorer cycle life than conventional aqueous rechargeable batteries.

Rechargeable lithium metal batteries, because of their many potential advantages, have been considered for use in a wide variety of applications. Because of the reactivity of lithium and the possibility of safety problems, emphasis also has been placed on achieving safe operation under normal and abusive conditions.

For these reasons, too, commercialization of rechargeable lithium batteries has been limited. They have been introduced into the market only on a limited scale and in small cell sizes. Coin-type batteries have been commercially available for use in low-power portable applications and as memory backup. Small cylindrical cells, using a lithium metal anode, have been marketed briefly for consumer electronics applications but were withdrawn when safety problems arose. Rechargeable lithium metal batteries (including ambient-temperature as well as high-temperature types), because of their high energy density, have been investigated for applications requiring larger size cells and batteries as, for example, electric vehicles. More recently, the lithium-ion type battery was introduced into the consumer market, again in small cylindrical and prismatic sizes for camcorders, cell phones and other portable electronics. The lithium ion battery became the dominant rechargeable lithium system during the 1990s.

34.3.1 Electrochemical Systems

The different types of ambient-temperature lithium metal rechargeable batteries have been classified into four design categories. These are identified in Table 34.10, which lists representative chemical systems and the key advantages and disadvantages of each class. The components and chemical reactions and the performance characteristics of typical examples of the five types are summarize and compared in Sec. 34.3.2.

Liquid Organic Electrolyte Batteries. These cells use lithium metal for the negative electrode, a liquid organic aprotic solution for the electrolyte, and transition metal compounds (oxides, sulfides, and selenides) for the positive electrode. These transition metal compounds are insertion or intercalation compounds and possess a structure into which lithium ions can be inserted or from which they can be removed during discharge and charge, respectively,

$$x\text{Li} + \text{M}_z\text{B}_y \underset{\text{charge}}{\overset{\text{discharge}}{\rightleftharpoons}} \text{Li}_x\text{M}_z\text{B}_y$$

where M_zB_y is the transition metal compound.

TABLE 34.10 Classification of Rechargeable Lithium Metal Batteries (Ambient Temperature)

1. *Liquid organic electrolyte cells* (solid cathode cells using intercalation compounds for cathode, a liquid organic electrolyte, and a metallic lithium anode):
   High specific energy
   Moderate-rate capability
   Potential safety problems
   Low cycle life
   Low self-discharge rate
       Examples: Li/MoS_2, Li/MnO_2, Li/TiS_2, $Li/NbSe_3$, Li/V_2O_5, $Li/LiCoO_2$, $Li/LiNiO_2$

2. *Polymer electrolyte cells* (cells using polymer electrolyte, intercalation compounds for cathode, and lithium metal for anode):
   High energy density
   Safer design
   Low electrolyte conductivity (poor high-rate capability)
   Poor low-temperature performance
   Low self-discharge rate
       Examples: $Li/PEO-LiClO_4/VO_x$ and others

3. *Inorganic electrolyte cells* (liquid cathode cells using inorganic cathode materials which also function as electrolyte solvent):
   High energy density
   High-rate capability
   Excellent shelf life
   Ability to sustain overcharge
   Safety problems, toxicity
   Capacity fade on cycling
       Examples: Li/SO_2, $Li/CuCl_2$

4. *Lithium alloy cells* (cells with lithium-alloy anodes, liquid organic electrolytes, and a variety of cathodes):
   Coin cell configuration
   Lithium alloy considered safer than metallic lithium
   Low energy density
   Poor cycle life (except on shallow depth of discharge)
       Examples: $LiAl/MnO_2$, $LiAl/V_2O_5$, $LiAl/C$, LiC/V_2O_5, $LiAl/polymer$

The overall discharge process is shown graphically in Fig. 34.8. Lithium ions formed at the negative electrode during discharge migrate through the electrolyte, and are inserted into the crystal structure of the host compound at the positive electrode. The electrons travel in the external circuit to enter into the electronic band structure of the host compound. The process is reversed during charge. The electrode reactions that occur during this process are covered in Sec. 34.2.2.

These batteries have the highest energy density of any of the rechargeable lithium batteries, the highest power density (with the exception of the SO_2-based batteries), and a low self-discharge rate, typically less than 1% per month. Safety, however, is a major concern with this type of lithium battery. Most designs use a three- to five-fold excess of lithium in order to obtain a reasonable cycle life to make up for the fact that the active lithium is lost as it is stripped and replated during cycling. Further, the freshly formed porous high-surface-area lithium that is formed during recharging is highly reactive and susceptible to forming dendrites, which could cause internal short circuits.

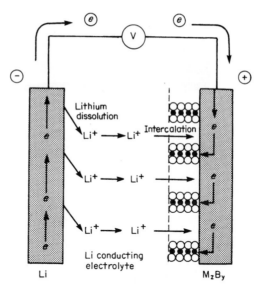

FIGURE 34.8 Scheme of the discharge process of a liquid organic electrolyte Li/M_zB_y cell. $\circ$ = B; $\bullet$ = M. (*From Ref. 8.*)

The liquid electrolyte batteries were among the first lithium rechargeable batteries to be investigated and the first to be introduced commercially. Small coin-type batteries were marketed for use in small electronic devices and for memory backup. These batteries, however, use lithium alloys, which are less reactive than metallic lithium, for the negative electrode. The small cylindrical batteries that were introduced briefly in the 1980s, on a limited basis, were withdrawn as safety problems arose. In the 1990s, cylindrical AA batteries were reintroduced, but have recently also been withdrawn from the market. Applications will probably be limited to coin-type and other small-size batteries until a technological breakthrough occurs.

Polymer Electrolyte Batteries. The rechargeable lithium batteries which use solid polymer electrolytes (SPE) are considered to have a safety advantage over the organic liquid electrolytes because of their lower reactivity with lithium and the absence of a volatile, flammable organic solvent. In their most common form, as shown schematically in Fig. 34.9. these cells use a lithium-ion conducting polymer membrane which acts both as the electrolyte and as the separator, a thin lithium metal foil as the negative or anode material and a transition metal oxide or chalcogenide, such as V_2O_5, TiS_2, or VO_x, blended with carbon and the polymer electrolyte and backed by a metal-foil current collector as the positive electrode. The basic structure can be represented as

$$Li/PEO-LiX/M_xB_y, PEO-LiX, C/M^*$$

where M_xB_y is the intercalation compound, PEO-LiX the polymer electrolyte, and M^* the current collector.

The cell reaction is similar to that in the liquid organic electrolyte cell—intercalation of lithium into the structure of the cathode during discharge and deintercalation of lithium from the charged cathode and deposition on the anode during charge,

$$x\text{Li} + M_zB_y \underset{\text{charge}}{\overset{\text{discharge}}{\rightleftarrows}} Li_xM_zB_y$$

This process is shown schematically in Fig. 34.10.

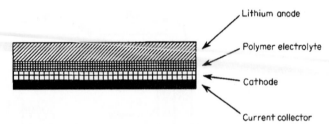

FIGURE 34.9 Schematic cross section of a solid polymer electrolyte (SPE) cell. (*Courtesy Valence Technology, Inc.*)

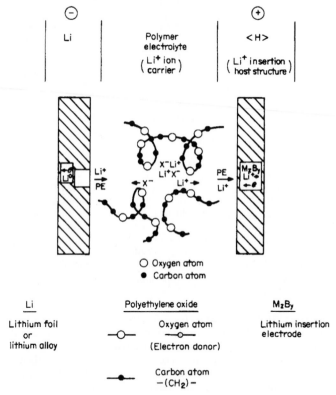

FIGURE 34.10 Schematic of discharge process of a Li/SPE/M_zB_y cell.

The ionic conductivity of most solid polymer electrolytes is significantly lower than that of the liquid electrolytes. Cells must be designed with thin electrodes and cell components to minimize the internal cell resistance. The total thickness of a cell assembly is as low as 200 μm or thinner. An alternative is to operate at higher temperatures where the conductivity is higher. While this may be acceptable for electric-vehicle and stand-by batteries, it will not be acceptable for many portable consumer applications. Newer polymer electrolytes are being developed using plasticizers or gel-type polymers. These methods increase the conductivity of the polymers, but since they contain organic solvents, they will be more reactive with the lithium anode.

One of the major problems of lithium polymer electrolyte systems is the development of high interfacial resistance at the lithium/polymer electrolyte interphase.[20] This resistance grows with time and could be as high as 10 kΩ cm^{-2}.[21] This resistance layer is due to the reactions of lithium with water, other impurities, and the salt anions. Similar to nonaqueous electrolytes, the solid electrolyte interphase (SEI) also exists in the lithium/polymer electrolyte systems.[22] In this case, the solid electrolyte interface (SEI) consists of the inorganic reduction products of the polymer electrolyte and its impurities.

The capacity of a lithium polymer electrolyte battery decreases gradually on cycling. This capacity loss is most likely related to degradation at the electrode interfaces due to repeated plating-stripping cycles at the lithium metal anode and the repeated intercalation-deintercalation cycles at the composite cathode. The cycling performance of lithium polymer electrolyte cells can be improved by reducing the lithium passivation phenomena and by improving the cathode morphology and homogeneity (particle size).

The energy density of the SPE batteries is projected to be close to that of the liquid-electrolyte lithium rechargeable batteries. The thin cell design, however, usually requires a larger percentage of materials of construction and nonreactive components, which tend to lower the energy density. Larger electrode areas are also required for a given cell capacity than with conventional battery design. The thin separator and larger electrode area could adversely affect the cycle life, safety, and reliability of the battery by increasing the chance for internal short circuits, lithium dendrite penetration, and other deleterious effects.

Another potential advantage of the solid polymer battery is that the design lends itself to manufacture by automatic processes and the capability to be easily fabricated in a variety of shapes and forms. Very thin batteries for cell phones, PDAs and similar applications can be manufactured. At the other extreme, large thin plates can be manufactured and assembled in multiplate prismatic or bipolar batteries.

Solid polymer batteries using lithium metal have not yet been introduced in the marketplace, and most of the performance characteristics are based on laboratory prototypes. The technology offers several potential advantages compared to the other rechargeable lithium batteries, including high energy density, low self-discharge rate, improved safety, and thin cell design. The success of this approach will depend on the ability to achieve these advantages, first in small cells and batteries and then to be able to scale up the technology to the larger-size cells and multicell batteries required, for example, for electric vehicles. Lithium ion batteries using polymer electrolytes have been introduced to the market in 2000.

Inorganic Electrolyte Batteries. The rechargeable lithium batteries with SO_2-based inorganic electrolytes are attractive because they can operate at high charge and discharge rates due to the high ionic conductivity of the electrolyte. Other advantages are a high energy density, a low self-discharge rate, the ability to accept limited overcharge and overdischarge, and a wide electrochemical voltage window (5.0 V versus lithium).

Their drawback is the toxicity of the electrolyte and concern with safety problems. High-capacity fade and relatively low capacity at low temperatures are other disadvantages.

Several types of batteries have been investigated using the SO_2-based electrolyte cells (with $LiAlCl_4$ as a solute) and metallic lithium for the negative electrode. These use either carbon[11] or $CuCl_2$[26] for the positive electrode material.

With a high-surface-area carbon black such as Ketjen black, the SO_2-based $LiAlCl_4$ electrolyte forms a complex which is believed to take part in the discharge-charge process as

$$3Li + LiAlCl_4\cdot3SO_2 + xC \underset{\text{charge}}{\overset{\text{discharge}}{\rightleftharpoons}} LiClAl \overset{OSO}{\underset{OSO}{\longleftarrow}} OSO\cdots C_x + 3LiCl$$

During overcharge, more lithium is deposited from the electrolyte solution at the anode, and the decomposition of $AlCl_4^-$ occurs at the cathode with the generation of free chlorine. This free chlorine combines with metallic lithium to form LiCl, which subsequently recombines with $AlCl_3$ and regenerates the electrolyte,

$$Li^+ + e \rightarrow Li$$

$$AlCl_4^- \rightarrow AlCl_3 + \tfrac{1}{2}Cl_2 + e$$

$$Li + \tfrac{1}{2}Cl_2 \rightarrow LiCl$$

$$LiCl + AlCl_3 \rightarrow LiAlCl_4$$

While this scheme provides an overcharge mechanism, the chloride generated will attack the separator and compromise safety.

With $CuCl_2$ as the positive electrode material, the discharge-charge process involves a simple redox reaction. The high reduction potential of SO_2 (2.9 V versus lithium on carbon surfaces) limits the selection of positive electrode materials to only those compounds that reduce at potentials above the reduction potentials of SO_2. Thus $CuCl_2$, which has an open-circuit potential of 3.45 V and discharges at about 3.35 V, corresponding to the reduction process of

$$Cu^{2+} + e \rightarrow Cu^+$$

meets the criteria. The second reduction process,

$$Cu^+ + e \rightarrow Cu$$

occurs at about 2.5 V. In principle, the discharge of this $Li/CuCl_2$ cell in SO_2-based electrolytes should not be permitted beyond 2.9 V, below which the $CuCl_2$ electrode would be passivated by the lithium dithionite produced by the reduction of SO_2.

Although the SO_2-based inorganic electrolyte system has been investigated for some time, the toxicity and the corrosivity of the electrolyte present a potential safety hazard, restricting commercial development. This system is also known to have severe explosive tendencies for the reason cited above.

34.3.2   Summary of Design and Performance Characteristics

The different types of rechargeable lithium metal batteries operating at ambient temperature are classified in four categories (see Table 34.10).

The cell components and the reaction mechanisms of typical cells for each of these categories are presented in Table 34.11, and their characteristics are summarized in Table 34.12. Typical discharge curves are shown in Fig. 34.11. Detailed electrical and performance data are presented in Sec. 34.4.

TABLE 34.11 Chemistry of Rechargeable Lithium Systems

System	Anodes	Cathodes	Electrolytes	Separator	Cell reactions ($\xrightleftharpoons[\text{charge}]{\text{discharge}}$)
1. Liquid organic electrolyte, solid cathode cells:					
Lithium molybdenum disulfide (Li/MoS$_2$)	Li	MoS$_2$	LiAsF$_6$, PC/EC	Polypropylene	$x\text{Li} + \text{MoS}_2 \rightleftharpoons \text{Li}_x\text{MoS}_2$
Lithium manganese dioxide (Li/Li$_{0.3}$MnO$_2$)	Li	Li$_{0.3}$MnO$_2$	LiAsF$_6$, DN, Bu$_3$N,	Polypropylene	$x\text{Li} + \text{MnO}_2 \rightleftharpoons \text{Li}_x\text{MnO}_2$
Lithium titanium diosulfide (Li/TiS$_2$)	Li	TiS$_2$	LiAsF$_6$, 2-MeTHF, THF	Polypropylene	$x\text{Li} + \text{TiS}_2 \rightleftharpoons \text{Li}_x\text{TiS}_2$
Lithium niobium selenide (Li/NbSe$_3$)	Li	NbSe$_3$	LiAsF$_6$, PC/EC	Polypropylene	$x\text{Li} + \text{NbSe}_3 \rightleftharpoons \text{Li}_x\text{NbSe}_3$
Lithium/lithiated cobalt oxide (Li/Li$_x$CoO$_2$)	Li	Li$_x$CoO$_2$	LiAsF$_6$/LiBF$_4$, MF/MA	Polypropylene	$x\text{Li} + \text{Li}_{1-x}\text{CoO}_2 \rightleftharpoons \text{LiCoO}_2$
Lithium/lithiated nickel oxide (Li/Li$_x$NiO$_2$)	Li	Li$_x$NiO$_2$	LiAsF$_6$, MF/MA	Polypropylene	$x\text{Li} + \text{Li}_{1-x}\text{NiO}_2 \rightleftharpoons \text{LiNiO}_2$
2. Polymer electrolyte cells:					
Lithium/vanadium oxide (Li/PEO/VO$_x$)	Li	VO$_x$	SPE	—	$y\text{Li} + \text{VO}_x \rightleftharpoons \text{Li}_y\text{VO}_x$
Lithium/titanium disulfide	Li	TiS$_2$	SPE	—	$x\text{Li} + \text{TiS}_2 \rightleftharpoons \text{Li}_x\text{TiS}_2$
3. Inorganic electrolyte cells:					
Lithium/sulfur dioxide (Li/SO$_2$)	Li	Carbon (Ketjen black)	LiAlCl$_4 \cdot x$SO$_2$	Tefzel	$\text{LiAlCl}_4 \cdot 3\text{SO}_2 + x\text{C} + 3\,\text{Li} \rightleftharpoons$ $\text{LiClAl}\!\left\langle\!\begin{smallmatrix}\text{OSO}\\ \text{OSO} \\ \text{OSO}\end{smallmatrix}\!\right\rangle\!\text{C}_x + 3\text{LiCl}$
Lithium/copper chloride (Li/CuCl$_2$)	Li	CuCl$_2$	LiAlCl$_4 \cdot x$SO$_2$	Tefzel	$\text{Li} + \text{CuCl}_2 \rightleftharpoons \text{LiCl} + \text{CuCl}$
4. Lithium alloy and other coin cells:					
Lithium-aluminum/manganese dioxide (LiAl/MnO$_2$)	LiAl	MnO$_2$	LiAsF$_6$, EC/BC/DME	Polypropylene	$\text{LiAl} + \text{MnO}_2 \rightleftharpoons \text{Li}_x\text{MnO}_2 + \text{Li}_{1-x}\text{Al}$
Lithium-aluminum/vanadium pentoxide (LiAl/V$_2$O$_5$)	LiAl	V$_2$O$_5$	—	Polypropylene	$\text{LiAl} + \text{V}_2\text{O}_5 \rightleftharpoons \text{Li}_x\text{V}_2\text{O}_5 + \text{Li}_{1-x}\text{Al}$
Carbon-lithium cells	Sn$_x$Bi$_y$	Activated carbon	LiClO$_4$, PC	Polypropylene	$\text{Sn}_x\text{Bi}_y + n\text{C} + \text{LiClO}_4 \rightleftharpoons \text{LiSn}_x\text{Bi}_y + \text{C}_n\text{ClO}_4$
Lithium carbon/vanadium pentoxide	Li/C	V$_2$O$_5$	—	Polypropylene	$\text{Li}_y\text{C} + \text{V}_2\text{O}_5 \rightleftharpoons \text{Li}_x\,\text{V}_2\text{O}_5 + \text{Li}_{y-x}\text{C}$
Lithium-aluminum/polymer	LiAl	Polypyrrole (PP) or Polyaniline (PA)	—	Polypropylene	$\text{LiAl} + \text{PP} \rightleftharpoons \text{Li}_x\text{PP} + \text{Li}_{1-x}\text{Al}$

TABLE 34.12 Performance Characteristics of Rechargeable Lithium Metal Systems*

System	Midpoint voltage, V	Size	Weight, g	Capacity, mAh	Specific energy Wh/kg	Energy density Wh/L	Self-discharge†	Cycle life‡
Liquid organic batteries:								
Li/MoS_2	1.75	AA	21	600	50	135	1–2	200
$Li/Li_{0.3}$ MnO_2	3.0	AA	17	800	140	270	1.25	200
Li/TiS_2	2.1	AA	20	900	95	235	1–2	250
$Li/NbSe_3$	1.95	AA	21	1100	100	270	1	250
$Li/LiCoO_2$	3.8	AA	21	500	95	235	See Table 34.14	50
$Li/LiNiO_2$	3.6	D	105	4500	155	325	—	See Fig. 34.20 (b)
Polymer electrolyte batteries:								
$Li/SPE/VO_x$	—	—	—	—	97	110	—	—
Li/SPE/S-based polymer				950	310	350	10–15	200
Inorganic electrolyte batteries:								
Li/SO_2	3.0	AA	20	500	75	200	0.1	>50
$Li/CuCl_2$	3.2	AA	21	500	75	220	0.1	>100
Lithium alloy batteries:								
$LiAl/MnO_2$	2.5	2430	4.0	70	45	120	0.4	See Tables 34.19, 34.20
$LiAl/V_2O_5$	1.8	2320	2.8	30	30	100	0.2	See Table 34.21
LiAl/C	2.4	2320	2.8	1.5	1.6	5.4	—	See Fig. 34.45, Table 34.22
$LiTiO_2/LiMn_2O_4$	1.5	1620	1.3	14	16	52	—	See Table 34.23

* Some data on experimental cells; see text.
† Self-discharge—% capacity loss per month at 20°C (value-dependent on charge/discharge conditions and cycling).
‡ Cycle life—to 80% capacity, 100% depth of discharge.

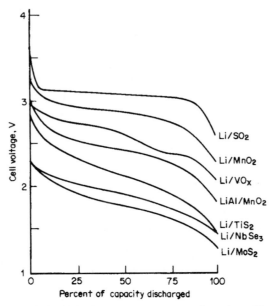

FIGURE 34.11 Typical discharge curves for rechargeable lithium batteries.

34.4 CHARACTERISTICS OF SPECIFIC RECHARGEABLE LITHIUM METAL BATTERIES

34.4.1 Liquid Organic Electrolyte, Solid-Cathode Batteries

These rechargeable batteries were attractive because they can deliver the highest energy densities of all of the ambient-temperature lithium rechargeable batteries. They use a metallic lithium anode, a liquid aprotic organic electrolyte, and one of several different cathode materials. Most of the development of these batteries has focused on the cylindrical spirally wound construction and flat-plate prismatic designs in order to optimize rate capability. The properties of these batteries, in cylindrical AA or comparable sizes, are compared in Table 34.12. A plot comparing the energy density and power density of the different rechargeable batteries and lithium/manganese dioxide primary batteries is given in Fig. 34.12. Although several of these batteries were introduced commercially, they have not been viable in the battery market.

Lithium/Molybdenum Disulfide (Li/MoS₂) Batteries. The Li/MoS_2 system was the first rechargeable lithium battery to be manufactured when it was introduced in the mid-1980s in a cylindrical AA-size.[27] The cell used thin lithium metal anodes (125 μm), with a stoichiometric excess of about three times, and MoS_2 on a thin aluminum foil (150 μm) for the cathode. A spirally wound construction, as illustrated in Fig. 34.13, was used. The electrolyte was $1M$ LiAsF₆ dissolved in a 50:50 mixture of propylene carbonate and ethylene carbonate. Two types of batteries were manufactured. Table 34.12 describes the more advanced of the two.

The discharge characteristics of the battery at various discharge loads temperatures are shown in Fig. 34.14. The sloping voltage profile makes it necessary to discharge the battery over a wide voltage range to obtain full capacity. However, because of this characteristic,

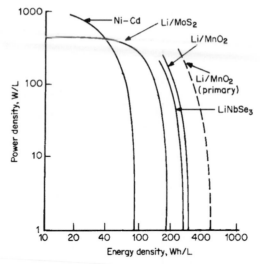

FIGURE 34.12 Comparison of liquid organic electrolytes, solid cathode batteries.

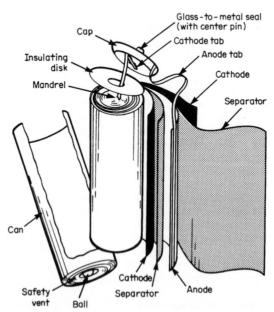

FIGURE 34.13 Construction of the Li/MoS$_2$ cell. (*Courtesy Moli Energy, Ltd.*)

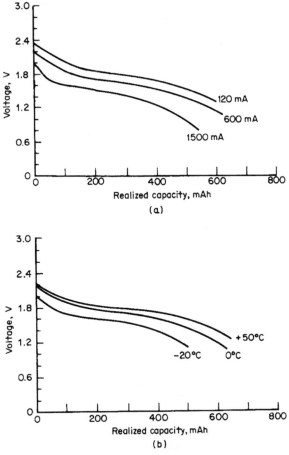

FIGURE 34.14   Discharge characteristics of Li/MoS$_2$ battery. Battery charged at 20°C, 60 mA to 2.4 V. (*a*) Discharge characteristics at 20°C. (*b*) Discharge characteristics at 120 mA. (*From Moli Energy Ltd.*)

the voltage can be used as a state-of-charge indicator. The performance of the battery and its cycle life depend on the manner in which it is discharged as shown in Table 34.13. Discharging the cells to 0 V could result in irreversible damage due to a phase change of the cathode material, and low-voltage cutoff circuitry is recommended. The self-discharge characteristics of the cell as a function of temperature are shown in Fig. 34.15. A constant-current charge at the $C/10$ rate to a voltage limit of 2.2 or 2.3 V is the recommended charging procedure. These batteries were withdrawn from the market after several safety incidents occurred.[28]

Lithium/Manganese Dioxide (Li/Li$_{0.3}$MnO$_2$) Batteries.   Rechargeable spirally wound cylindrical AA-size cells using the Li/MnO$_2$ chemistry have also been developed.[29–35] The same construction was used as with the Li/MoS$_2$ cells. A specially heat-treated Li$_x$MnO$_2$ was used as the positive electrode material with carbon black. Commercial cells used an electrolyte of $1M$ LiAsF$_6$ in dioxolane with added tributylamine as a stabilizer.

TABLE 34.13   Performance Characteristics of Li/MoS$_2$ AA-Size Batteries

User mode	Voltage limits	10th cycle		Cycle life, number of cycles	
		Capacity, Ah	Average energy, Wh	To 80% of 10th-cycle capacity	To 50% of 10th-cycle capacity
High capacity	2.6–1.1	0.82	1.52	100	110
	2.4–1.1	0.70	1.23	180	200
Normal	2.4–1.3	0.60	1.10	400	500
High cycle life	2.1–1.6	0.40	0.74	800	1000
	1.95–1.75	0.15	0.27	1800	3000

Source: Moli Energy Ltd.

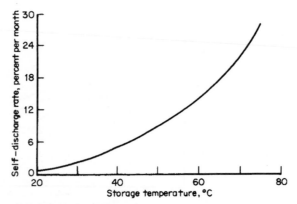

FIGURE 34.15   Self-discharge characteristics of Li/MoS$_2$ AA-size battery. (*From Moli Energy, Ltd.*)

Some of the performance characteristics of this AA-size battery are listed in Table 34.12 and are plotted in Fig. 34.16. The discharge profile is fairly flat with a midpoint voltage, at moderate discharge rates, close to 3 V. The cell can be discharged at current levels of up to about 2A, to a 2.0 V cut-off. The self-discharge rate is low, and is typical of the self-discharge rate of many of the lithium metal/organic electrolyte rechargeable cells. Two hundred cycles were obtained with a $C/3$ discharge and a $C/12.5$ charge rate when cycled to 100% depth of discharge (DOD). Three hundred to 350 cycles can be obtained to 65% of initial capacity at 100% DOD when discharged at 250 mA and changed at 80 mA to 3.40 V.

These AA-size batteries with lithium metal anodes were introduced into the consumer market, with claims of safe performance under abusive electrical conditions due to an intrinsic shut-down mechanism involving polymerization of the electrolyte which protects the cell under short circuit and overcharge (shutdown at 4.0 V).[33,34]

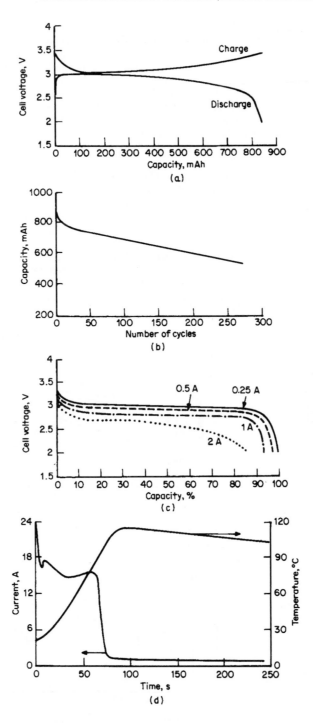

FIGURE 34.16 Performance characteristics of $Li/LiO_{0.3}MnO_2$ rechargeable AA-size battery. (*a*) Charge and discharge curves. $I_c =$ 250 mA; $I_d = 250$ mA (cycle no. 5). (*b*) Capacity vs. cycle number. $I_c = 60$ mA; $I_d = 250$ mA (100% DOD). (*c*) Capacity vs. discharge rate. (*d*) Short-circuit behavior after 50 cycles. (*From Tadiran Electronic Industries, Inc.*).

These AA-size batteries have been withdrawn from the commercial market because it has been found that, at high charging rates, lithium deposition produces small grains which react with the electrolyte in spite of the passivating tendency of the $LiAsF_6$-dioxolane-tributylamine electrolyte. The amount of electrolyte is limited in spiral-wound construction and is spread in a thin layer within the electrode structure. During cycling, the lithium-electrolyte reaction depletes the amount of electrolyte so that only a part of the active material continues to function, leading to a pronounced increase in internal resistance and ultimate failure as a result of the high internal impedance and reduced amount of active material. Another failure mechanism is the high current densities which develop as the area of active material decreases, leading to dendritic short circuits.

Lithium/Titanium Disulfide (Li/TiS₂) Cells. This rechargeable battery system was first developed as a button cell for use in watches with a lithium-aluminum alloy anode. This product never achieved commercial success due to fragmentation of the alloy anode on cycling which limited cycle life. Later the battery was investigated in cylindrical and prismatic designs of up to about 5-Ah capacity.[36–39] Its advantageous characteristics are a relatively flat discharge and high energy density.

Figure 34.17a compares the discharge characteristics of the Li/TiS_2 battery, which used an electrolyte composed of $1.5M$ $LiAsF_6$ in a mixture of 2-methyl tetrahydrofuran and tetrahydrofuran, with that of the Li/MoS_2 cell. Figure 34.17b shows the discharge-charge characteristics of a 2-Ah battery.[36] More than 250 cycles were achieved with a specific energy of about 100 Wh/kg.

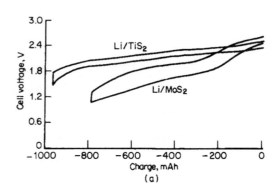

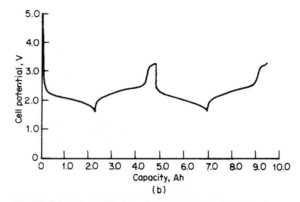

FIGURE 34.17 (a) Comparative performance of AA-size batteries. Discharge rate = 200 mA; 20°C. (b) Successive cycles of a 2-Ah Li/TiS_2 battery. The battery is overcharged at the end of each cycle, with the overcharge plateau time-limited to be at ~0.4 Ah.

Lithium/Niobium Selenide (Li/NbSe$_3$) Batteries.   This cathode material is attractive because it can accept (intercalate) three lithium ions per molecule. The cathode can also be made without any binder and conducting diluent because of its soft, fibrous nature and high electronic conductivity. A AA-size cylindrical battery can deliver about 1.1 Ah with an average voltage of 2.0 V. Figure 34.18 shows the discharge characteristics of a Li/NbSe$_3$ battery at various discharge rates and temperatures.[40] The battery has a mildly canted discharge profile. An important characteristic of this system is its high-rate discharge capability (up to the 2C rate). A typical discharge and charge cycle is shown in Fig. 34.18c. More than 200 cycles were obtained at 75% depth of discharge.

The development of this system was terminated despite its technical promise.

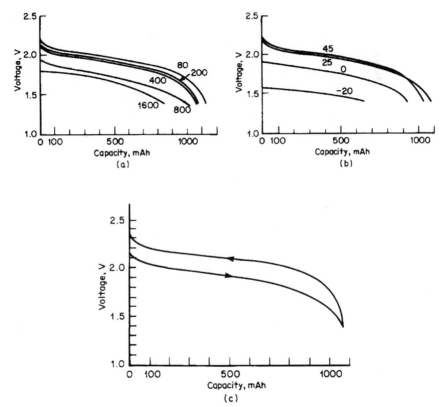

FIGURE 34.18   Discharge characteristics of the Li/NbSe$_3$ AA-size battery. (*a*) Discharge curves at 20°C (mA). (*b*) Discharge curves at different temperatures (°C) at 200 mA. (*c*) Typical discharge and charge curve for the Li/NbSe$_3$ AA-size battery. Discharge: 400 mA; charge: 80 mA.

Lithium/Lithiated Cobalt Dioxide (Lu/Li$_x$CoO$_2$) and Lithium/Lithiated Nickel Dioxide (Li/Li$_x$NiO$_2$) Batteries.   These batteries are characterized by a high cell voltage, high energy density, good low-temperature performance, relatively high charge-discharge rates, and moderate cycle life. The net charge and discharge reactions of the Li/Li$_x$CoO$_2$ cell can be represented by the cell reaction:

$$xLi^+ + L_{1-x}CoO_2 \; \underset{\text{charge}}{\overset{\text{discharge}}{\rightleftharpoons}} \; LiCoO_2$$

The Li/Li$_x$CoO$_2$ battery is manufactured in a fully discharged state and thus requires a charge prior to its usage. When used in combination with oxidation-resistant ester-based solvents (methyl formate or methyl acetate), the battery can reach full charge at 4.2 to 4.3 V depending on the cycle history of the cell.

A Li/Li$_x$CoO$_2$ battery that utilizes an ester-based electrolyte displays a flat discharge at 3.85 to 3.95 V in a cathode compositional range of between $x = 0.5$ and $x = 1.0$, depending on the discharge rate conditions.

Batteries have been built in cylindrical spirally wound configurations in AA (300 to 600 mAh), D (4–6 Ah), and larger sizes. The electrolyte consists of LiAsF$_6$/LiBF$_4$ in methyl formate, methyl acetate, or a mixture of both. Figure 34.19a shows the discharge characteristics of a 5-Ah D-size battery. The effect of various charging procedures on the discharge performance is illustrated in Fig. 34.19b for a 30-Ah battery. Figure 34.19c shows the cycle life characteristics of a D-size battery. The battery also has good storage capability. The effects of storage on both initial capacity after storage and permanent capacity loss are summarized in Table 34.14. Capacity losses of batteries stored at temperatures up to 35°C in the charged state can be fully recovered on the first charge.[41]

The charge and discharge reactions of the Li/Li$_x$NiO$_2$ battery are similar to those of the lithiated cobalt oxide system:

$$xLi^+ + Li_{1-x}NiO_2 \; \underset{\text{charge}}{\overset{\text{discharge}}{\rightleftharpoons}} \; LiNiO_2$$

Batteries have been build in a spirally wound cylindrical D size. Table 34.15 lists the design parameters. This D-cell battery employed an alloy anode of 85% Li-15% Al. Figure 34.20a shows the discharge characteristics of the battery at several discharge rates and Fig. 34.20b shows the cycle life characteristics. The declining capacity of the cell with cycling is attributed to increasing impedance of the cathode.[42]

While there is currently no commercial production, the Li/Li$_x$CoO$_2$ and Li/Ni$_x$CoO$_2$ systems continue to receive attention for military applications because of their high specific energy and energy density.

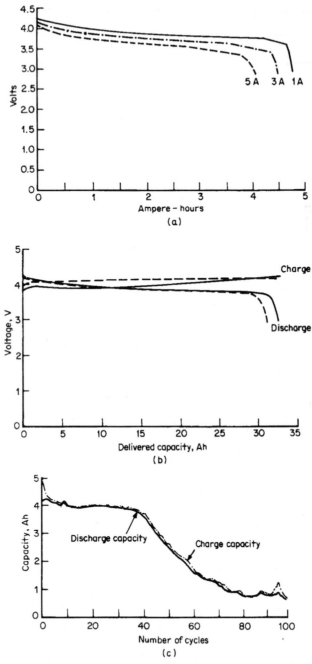

FIGURE 34.19 (*a*) Discharge characteristics of a 5-Ah Li/Li$_x$CoO$_2$ D-size battery at 20°C. (Delivered specific energy at 1.0 A, 70 Wh/lb.) (*b*) Effect of charging rates on discharge performance of 30-Ah Li/Li$_x$CoO$_2$ battery at 20°C. - - -, constant potential at 4.20 V [current limit at 5.0 mA/cm^2 (*C*/2)]. ——, constant current at 1.0 mA/cm^2 (*C*/10). Discharge rate at *C*/6. (*c*) Cycle life characteristics of a Li/Li$_x$CoO$_2$ D-size battery. Charge at 300 mA, 4.3 V. Discharge at 3 A, 3.0 V. (*From Alliant Tech-systems.*).

TABLE 34.14 Effect of Storage at 20°C on Capacity Retention of Charged $Li/Li_x CoO_2$ D-Size Batteries

Charge and prestorage test conditions	Average capacity of first 5 cycles, Ah	Discharge capacity after 15-day (360-h) storage, Ah	Capacity loss during storage period, %	Average capacity of next 5 cycles after storage, Ah	Permanent capacity loss due to storage, %
4.3 V charge (0.5 F/M)	5.1	4.6	9.8	5.2	0
4.1 V charge (0.4 F/M)	4.05	3.65	9.9	4.0	1
4.1 V float charging	4.05	4.2	0	4.0	1

TABLE 34.15 Design Features of $Li/LiNiO_2$ D-Size Batteries

Positive electrode (supported on aluminum grid):
  89.5% $LiNiO_2$, 2% graphite, 5% EC600 carbon,
  0.015 in (0.38 mm)
  Porosity 40%
  Loading 75 mg/cm²

Negative electrode (four copper wires running length of electrode):
  15% Al-85% Li
  0.004 in (0.10 mm)

Electrode capacity ratio:
$$\frac{\text{Li anode}}{\text{LiNiO}_2 \text{ cathode}} = 3$$

Electrolyte:
  1.2M $LiAsF_6$ in PC/EC/DMC
  18 ml

Separator:
  Celgard 2400 (1 ply)
  Polyethylene (2 ply)

General:
  Working area 900 cm²
  Allowance for expansion 5%
  Nickel-plated steel hardware
  Molybdenum pin and TA-23 glass-to-metal seal

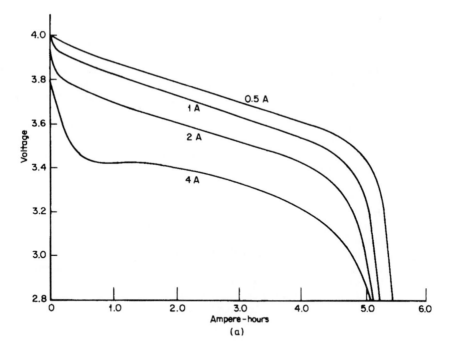

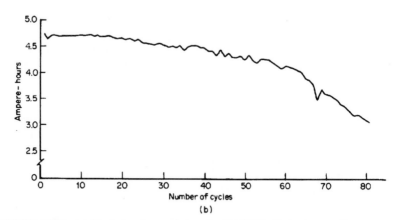

FIGURE 34.20   (*a*) Discharge characteristics of a Li/Li$_x$NiO$_2$ D-size battery. Charge at 0.3 A to 4.1 V. (*b*) Cycle life characteristics of a Li/Li$_x$NiO$_2$ D-size battery. Discharge at 2 A to 2.8 V. Charge at 0.4 A to 4.1 V. Taper to 0.1 mA. (*From Ref. 42.*)

34.4.2 Polymer Electrolyte Cells

The polymer electrolyte lithium batteries contain all solid-state components: lithium as the anode material, a thin polymer film as a solid electrolyte and separator, and a transition metal chalcogenide or oxide, or a sulfur-based polymer as the cathode material. These features offer the potential for improved safety because of the reduced activity of lithium with the solid electrolyte, flexibility in design as the cell can be fabricated in various sizes and shapes, and high energy density.

Figure 34.9 shows a schematic cross section of a solid polymer electrolyte cell. The cathode and electrolyte are coated onto a current collector to form a thin sheet, called the cathode laminate. The lithium metal foil is applied to the cathode laminate to form a layered structure, with the solid polymer separating the lithium from the cathode. These cells are designed in extremely thin components with high surface areas to minimize the internal resistance and compensate for the lower conductivity of the polymer electrolyte. The thickness depends on the specific cell design and required capacity. A thicker laminate delivers a higher capacity per unit area of electrode, but with lower efficiency at the higher current drains.[43]

The early batteries used polyethylene oxide (PEO)-based electrolytes containing a lithium salt, which have an appreciable conductivity at about 100°C but low conductivity at room temperature. Later new polymeric electrolyte materials, such as PEO copolymers, PEO blends, plasticized PEO electrolytes, and gelled electrolytes, with better conductivity were developed. Some of the cathode materials investigated for these batteries are TiS_2, VO_x, V_2O_5, Li_xCoO_2 and sulfur-based polymers.

The solid polymer electrolyte (SPE) battery has been considered for a wide range of applications, from small portable electronics up to and including electric vehicles.

34.4.3 Lithium Batteries Using PEO-based Electrolytes

In the late 1970s, polymer electrolyte materials were proposed for use in solid-state battery designs.[44] A considerable development effort has resulted in a number of review articles[45–47] describing the status of such batteries in some detail. The unique aspect of these batteries is that the electrolyte is a solid flexible film comprised of a polymer matrix and an ionic salt complexed into the matrix. Thin-film solid-polymer electrolyte batteries offer the possibility of an intrinsically safe battery design in combination with good high-rate capability.

Polyethylene oxide was the first material utilized as the matrix material. Initially, it was necessary to operate the cells at elevated temperatures (100°C) to obtain adequate conductivities (10^{-3} S/cm), but a variety of polymeric electrolytes that may be useful at normal ambient temperatures were subsequently developed. This polymer electrolyte battery is based on thin-film components that incorporate large area electrolyte and electrode layers. A general cell can be schematized as:

$$Li_x\langle A\rangle/Li^+ \text{ ion-conductor}/Li_z\langle B\rangle$$

where $Li_x\langle A\rangle$ is a metallic lithium anode or a low-voltage lithium-intercalation anode (negative electrode), the Li^+ ion conductor is a lithium salt-polymer complex, and $Li_z\langle B\rangle$ is a high-voltage lithium-intercalation cathode (positive electrode). Usually, the polymer component of the electrolyte layer is also the binder for the electrode layers. In addition, the polymer component may be gelled by the addition of low molecular weight (liquid) solvents to enhance its ionic conductivity. Such a battery system shows two major advantages with respect to standard batteries that use inert porous separators:

1. The electrode and electrolyte layers are laminated (usually by heating and pressing the layers), thus allowing various battery shapes without loss of contact.
2. Even if a low molecular weight (liquid) plasticizer is added to obtain high conductivity at ambient and sub-ambient temperatures, there is no free liquid present in the battery thus preventing any leakage problems.

Examples of the battery systems under development or available on the marketplace are indicated in Table 34.16. They can be conveniently classified into three categories:

1. Positive and negative intercalation electrodes (lithium-ion) with a gelled polymer electrolyte (PEO-based).
2. Positive and negative intercalation electrodes (lithium-ion) with a porous polymer separator layer filled with a liquid electrolyte (PVDF-based).
3. Positive intercalation electrode-lithium anode (lithium-metal) with a dry polymer electrolyte layer.

The first and the second categories, the latter of which (PVDF-based systems) does not properly belong to the class of polymer electrolyte batteries, are covered in Chap. 35. In this chapter, the attention is focused on the lithium metal, dry polymer electrolyte systems.

The search for high energy density batteries, especially for electric vehicle applications, has led to the consideration of the use of lithium metal anodes. No other anode can attain the same electrochemical performance in terms of specific energy as metallic lithium. However, safety issues as well as poor cycle life performance have prevented the commercial success of lithium-metal batteries with the exception of primary batteries. At the present time, there are three major programs aimed at the development of lithium-anode, dry-polymer electrolyte batteries for electric vehicles (Table 34.17). Within these R&D programs, substantial successes have been obtained in terms of cyclability of the lithium-polymer electrolyte interface[51] as well as the whole battery system.[50] However, Argotech's battery NPS-24V80 (Table 34.18) appears to be the only system close to commercialization. This battery, designed for telecommunication applications (especially for outside plants), is a fall-out of the EV battery module developed within the USABC program.

Figure 34.21 illustrates the internal design of the lithium polymer cell. The cell is made by laminating five thin layers including an insulating layer, a lithium foil anode, a solid polymeric electrolyte, a transition metal oxide cathode (VO_x) and a current collector. The unit cell is made by winding the laminated film into a jelly roll (as in Fig. 34.21) or a flat roll (preferred for multiple cell battery assembly) structure. These cells are then assembled in series and in parallel arrangements to form modules of different sizes and shapes for numerous applications. As an example, Fig. 34.22 illustrates the design of a lithium-polymer battery module for electric vehicle applications. Unit laminated cells are packaged into a stack of flat cells to create a module. The cells can be connected in parallel and/or series arrays within a single container to build the desired module capacity and voltage. The packaging also provides the mechanical, electrical, and thermal controls required for operation. Each module is a fully functional battery system, including intelligent control and monitoring electronics that interface with the battery pack controller.

TABLE 34.16 Status of Lithium and Lithium-Ion Polymer Battery Development ([a] from Ref. 48; [b] from Ref. 49)

Manufacturer	Country	Cathode	Anode	Electrolyte	Voltage V	Energy density Wh/L	Specific energy Wh/kg	Application
Matsushita Battery[a]	Japan	$LiCoO_2$	Graphite	PVDF gel	3.7	250/125		Cellular
Sony[a]	Japan	$LiCoO_2$	Graphite	PVDF gel	3.7	245/125		Cellular, PC
Japan Storage Battery[a]	Japan	$LiCoO_2$	Graphite	PVDF gel	3.6	210/125		Cellular, MiniDisc
Hitachi Maxell[a]	Japan	$LiCoO_2$	Graphite	PEO gel	3.6	130/90		Cellular, PC
Sanyo[a]	Japan	$LiCoO_2$	Graphite	PEO gel?	3.6	200/120		Cellular
Toshiba[a]	Japan	$LiCoO_2$	Graphite	PVDF gel	3.6	245/115		Cellular, PC
Yuasa Corp.[a]	Japan	$LiCoO_2$	Coke	PEO gel	3.6	165/95		Cellular, MiniDisc
Hirion/Mitsubishi Chem.[a]	Japan	$LiCoO_2$	Graphite	PEO gel	3.7	280/130		Cellular, PC
Ultralife[a]	US	$LiMn_2O_4$	Graphite	PVDF gel	3.7	185/105		Cellular, PC
Valence[a]	US	$LiMn_2O_4$	Graphite	PVDF gel	3.7	220/110		PC
Thomas & Betts (HET)[a]	US	$LiCoO_2$	Graphite	PVDF gel	3.7	220/120		Cellular
Lithium Technology[a]	US	$LiCoO_2$	Graphite	PVDF gel	3.6	240/125		PC
ElectroFuel[a]	Canada	$LiCoO_2$	Graphite	PVDF gel	3.6	435/175		PC
Argotech[b] (HydroQuebec/3M)	Canada	VO_x	Li-metal	Dry PEO	24 (2.6)	110/97		Telecommunications facilities
Shubila[a]	Malaysia	$LiCoO_2$	Graphite	PVDF gel	3.6	215/120		Cellular

TABLE 34.17 Major Lithium Metal Anode, Dry Polymer Electrolyte (PEO-based), Battery Technology Development Projects (from Ref. 50)

Sponsor	Project leaders	Cathode	Application	Status
USABC (USA)	IREQ-3M-ANL	VO_x	EV	2.4 kWh prototypes
Bolloré Tech. EDF (France)	CEREM	VO_x, Li_yMnO_x	EV	Prototypes
MICA-MURST (Italy)	ENEA	VO_x, Li_yMnO_x	EV	Scaling-up to 1 kWh prototypes

TABLE 34.18 Specifications of Argo-Tech's Batteries from Ref. 49

	NPS-24V80	EV module
Nominal voltage	24 V/Module 2.65 V/Cell	20 V
Rated capacity	80 Ah (@ C/8)	119 Ah (@ C/3)
Rated energy	1.9 kWh (@ C/8)	2.4 kWh @ C/3)
Maximum discharge current	20 A	365 A
Float voltage	27.9 V/Module 3.1 V/Cell	N/A
Energy density	110 WhL^{-1}	220 WhL^{-1}
Specific energy	97 Whkg^{-1}	155 Whkg^{-1}
Length	394 mm	Application specific
Width	176 mm	Application specific
Height	251 mm	Application specific
Volume	17.4 dm^3	$ca.$ 11 dm^3
Weight	19.9 kg	15.7 kg
Ambient operating temperature	−40 to 65°C	N/A
Storage temperature	−40 to 65°C	N/A
Float life at 40°C	Over 10 years	N/A
Cycle life at 60°C (80% DOD)	200 cycles	600 cycles

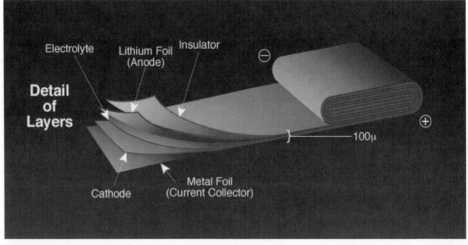

FIGURE 34.21 Laminated lithium-polymer electrolyte cell assembly. (*From the Proceedings of EVS 16, reprinted by permission of the publisher, EVAAP.*)

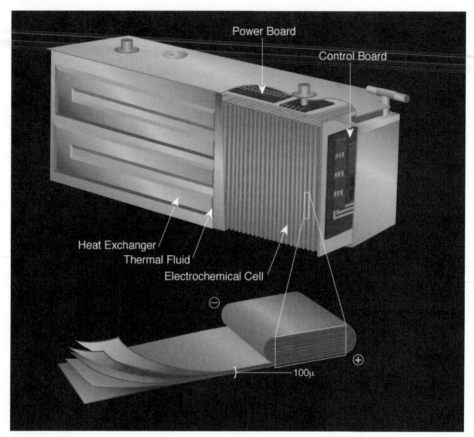

FIGURE 34.22 Design of a Lithium Polymer Battery (LPB) module for EV applications. The module includes an enclosure which provides mechanical support and thermal insulation along with control hardware and vehicle interfaces. (*From the Proceedings of EVS 16, reprinted by permission of the publisher, EVAAP.*)

The specification given for the stationary application NPS-24V80 battery are given in Table 34.18 and compared with the parent EV-module. A typical discharge curve is depicted in Figure 34.23. The battery is able to deliver 80 Ah capacity at a 10 A discharge rate in a 60°C temperature environment. The excellent performance at high ambient temperature is one of the major advantages of this system. The unit cells operate at 40 to 60°C, and do not contain any liquid: The battery is designed to have no degradation in non-ventilated and warm environments (for example: telecommunication centers). It contains no liquid, thus it is hermetically sealed, and dry-out during operation and liquid leakage are impossible, and therefore no maintenance is required. In addition, the battery shows very good safety characteristics. Overcharging tests showed no gas generation.[53] Once more, no liquid leakage is possible even when the module is accidentally crushed. Finally and most important, water immersion tests on entire crushed cells showed only a very slow reaction.

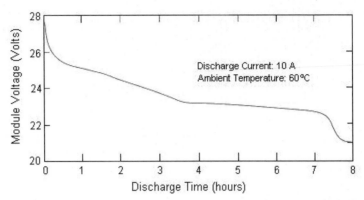

FIGURE 34.23 Typical discharge curve of the Argo-Tech's battery NPS-4V80. (*Reprinted by permission of 3M Company, St. Paul, Minnesota.*)

Polymer Electrolyte Battery Using V_6O_{13}. The design of another configuration of an SPE cell which was being developed for portable applications and for scaling up to larger batteries is illustrated in Fig. 34.24. A folded "bicell," which is formed by folding the cathode laminate and inserting the lithium foil anode, is the basic assembly unit. The bicells are stacked, the terminal leads are attached, and the unit is hermetically sealed into the final package. The size and number of stacks used in each cell assembly depend on the desired capacity and cell and battery dimensions.[43]

The heart of the process was the fabrication of the polymer laminate for the cathode. The cathode formulation, including polymer precursor, vanadium oxide, and carbon, was mixed and coated onto the carbon current collector. A second layer, the electrolyte, was coated on the cathode to provide the ionic conductivity and separate the cathode from the lithium anode. The coated laminate was then passed through an electron beam generator to cross-link the polymer precursors and produce the solid polymer material.

Figure 34.25 shows the discharge-charge curve for this lithium polymer battery, using a lithium foil anode and a vanadium oxide cathode (V_6O_{13}). The capacity of the battery is about 50 mAh. The discharge curves at different rates at 20°C are shown in Fig. 34.26. The sloping discharge profile is characteristic of the intercalation cathode. The voltage plateaus relate to the intercalation of the lithium ions into the vanadium oxide. The midpoint voltage decreases with increasing current, but there is relatively little change in Ampere-hour capacity if the cell is discharged to 1.8 V at rates slower than the 1-h rate. The self-discharge rate of these cells was very low, on the order of 0.5% per month at 20°C. The battery was expected to deliver about 200 cycles, with the capacity dropping slightly with cycling until near the end of life. During cycling, the impedance of the cell increases gradually, accounting for this loss of capacity. Work on this system has been discontinued in favor of lithium-ion technology using PVDF-based electrolytes (see Table 34.16).

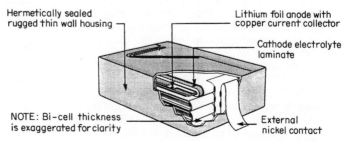

FIGURE 34.24 Schematic of a SPE battery design configuration. (*From Valence Technology, Inc.*)

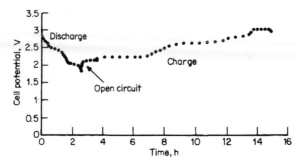

FIGURE 34.25   Laboratory prototype SPE cell (50-mAh size). Typical discharge/charge curve. (*From Valence Technology, Inc.*)

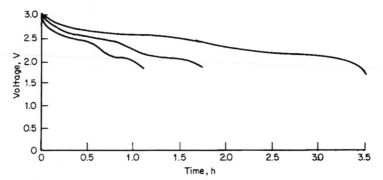

FIGURE 34.26   Laboratory prototype SPE cell, typical discharge curves at 20°C. (*From Valence Technology, Inc.*)

34.4.4   Lithium Batteries Using Sulfur-based Polymeric Positive Electrode Materials

Polymeric sulfur-based positive electrodes are under development for use in rechargeable batteries with lithium metal negatives.[55,56] Both liquid dioxolane-based and proprietary polymer electrolytes have been employed with this system.[55] Examples of the positive electrode polymers used in this system are shown in Fig. 34.27. These positive electrode materials are claimed to have specific capacities of 700 to 1200 mAh/g. Initial work was carried out with AA-size batteries using liquid electrolytes which operated at 2.1 V, provided 1 Ah capacity, a specific energy of 215 Wh/kg and an energy density of 260 Wh/L. This system is claimed to provide an intrinsic overcharge and over-discharge mechanism involving soluble sulfur species which shuttle between the electrodes under these conditions. See Fig. 34.28. The AA-size battery demonstrated its ability to operate on the standardized GSM cell-phone test as shown in Fig. 34.29 Approximately 200 cycles were obtained under C/3 charge-discharge cycling conditions with the AA-size battery. The cylindrical design has been superceded by a tin-foil laminate structure, which has evolved from 66 micron to 47 micron thickness for use in flattened wound cell designs. See Fig. 34.30. A 1 Ah unit is projected to have an electrode surface area of over 1000 cm², allowing operation at low current density. Prismatic designs (34 × 48 × 3.5 mm) were stated to give 950 mAh at 2.1 V, providing 310 Wh/kg and 350 Wh/L. Self-discharge is high, at 10 to 15% per month, probably due to the presence of soluble sulfur species. This design concept provides high power capability for this system, as shown in the Ragone plot of Fig. 34.31, which demonstrates that this system is capable of discharge rates of up to 8C, delivering 800 W/kg at 100 Wh/kg, which makes it potentially useful in hybrid electric vehicle applications. This technology continues under development by several companies.

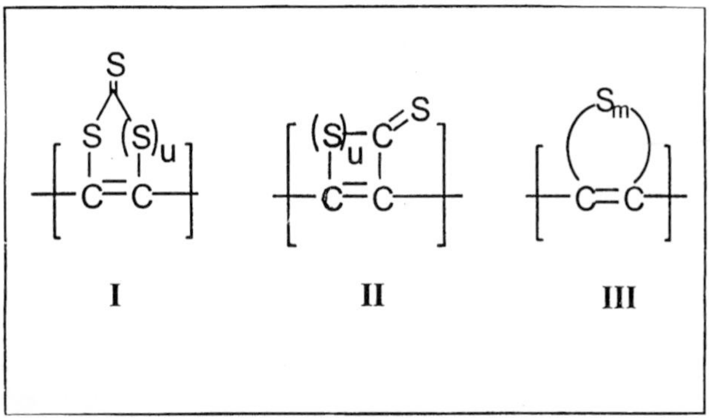

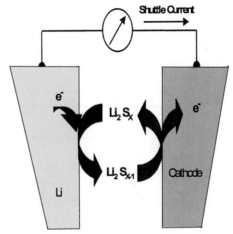

FIGURE 34.27   Sulfur-based electrode materials consist of congurated carbon backbones connected to contiguous sulfur rings or chains. Cathode discharge process is a redox reaction in which the sulfur rings open, providing 700 to 1200 mAh/g. The sulfur rings reform on recharge. (*From Ref. 55.*)

FIGURE 34.28   Redox shuttle protection mechanism provided by sulfur-based polymeric materials. (*From Ref. 56.*)

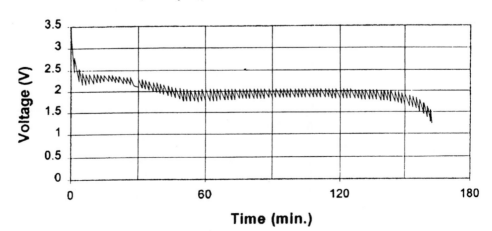

FIGURE 34.29   GSM cell-phone test of AA lithium/sulfur-based cathode AA-size battery. Test consists of 1.5 A pulse for 0.6 millisec followed by 0.1 A for 4.4 millisec. (*From Ref. 55.*)

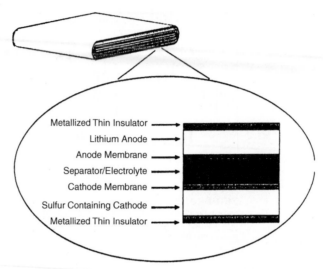

FIGURE 34.30 Schematic illustration of thin, flat rechargeable lithium battery with sulfur-based polymer cathode material. (*From Ref. 56.*)

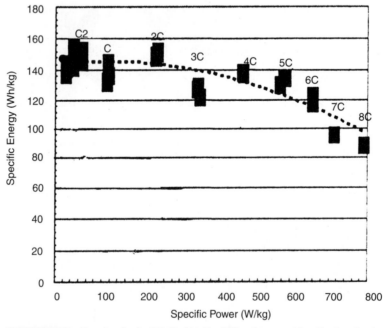

FIGURE 34.31 Ragone plot for 0.8 Ah thin-film lithium battery with sulfur-based cathode material. (*From Ref. 56.*)

The solid polymer electrolyte technology provides flexibility in battery design as the cell can be fabricated in a variety of shapes, configurations, and sizes. The estimated specific energy and the energy density are very attractive. Major interest is in using this battery system for applications requiring high voltages and capacity such as electric and hybrid electric vehicles and for stand-by power applications such as large uninterruptible power supplies.

34.4.5   Inorganic Electrolyte Batteries

Since the introduction of lithium/SO_2 primary batteries, significant research efforts have been directed to the development of rechargeable lithium/SO_2 batteries. Among the different electrolyte salts investigated, $LiAlCl_4$, $LiGaCl_4$, and $Li_2B_{10}Cl_{10}$ are noteworthy because of their good ionic conductivity and lithium cycling efficiency. High-surface-area carbon such as Ketjen black[11,19,57-61] or $CuCl_2$[26,60,62,63] are used as cathode materials. Microporous Tefzel separator was found to be compatible in these electrolytes. This material is no longer available commercially, however.

Inorganic electrolyte batteries offer several advantages, including high-rate capability, excellent shelf life, and the ability to accept limited overcharge through a shuttle mechanism.

The major disadvantages of these batteries are safety issues, capacity fade on cycling, significantly low capacity at low temperature, and high toxicity of electrolytes.

Lithium/Carbon Batteries.   These batteries are made in an hermetically sealed stainless-steel container with a metallic lithium anode, a Ketjen black carbon cathode, and a $LiAlCl_4 \cdot 6SO_2$ electrolyte. The cathode composition is 96% Ketjen black and 4% Teflon.[58] Teflon-rich carbon-coated nickel exmet is used as the cathode substrate. These batteries are made cathode-limited with the anode capacity at least twice the cathode capacity. Batteries are vacuum filled with the electrolyte. The open-circuit voltage of the system is about 3.3 V. The average discharge voltage is 3.1 V at 1 mA/cm². It is believed that high-surface-area carbon forms a complex with the electrolyte, and this complex takes part in the cell reaction.

Typical discharge-charge curves of lithium/carbon batteries of 3 Ah capacity using a $LiAlCl_4 \cdot 6SO_2$ electrolyte are shown in Fig. 34.32a for selected cycles. The change in voltage with time for the cell and the cathode versus the lithium reference are almost identical, indicating little anode polarization at the 1 mA/cm² discharge and charge rate. The gradual loss of capacity and the increased charge plateau on cycling are related to cathode polarization. Scanning electron microscopy and X-ray diffraction analyses of discharged cathodes confirm that the polarization is caused by irreversible deposition of a nonconductive discharge product (LiCl).

The lithium/carbon system in SO_2-based $LiAlCl_4$ electrolyte has excellent shelf life (<1% capacity loss per month) and insignificant voltage delay, even after 3 years of storage at room temperature.[59] The cell delivers significantly lower capacity at low temperature. Figure 34.32c shows the discharge characteristics of a lithium/carbon cell at −20°C. The system is particularly suitable for high-rate pulse power applications. Figure 34.32b shows the pulse discharge behavior of a lithium/carbon battery in $LiAlCl_4 \cdot 6SO_2$ at 20 mA/cm².[61]

Lithium/Copper Chloride Battery.   In this system, the copper chloride cathode is directly reduced and oxidized and no lithium intercalation-deintercalation occurs during the discharge-charge process. The use of $CuCl_2$ as the active cathode material provides a higher discharge voltage (3.3 V) for the reduction of Cu^{2+} to Cu^+. Figure 34.33 shows the discharge-charge behavior of a prismatic D-size $Li/CuCl_2$ battery at 1 mA/cm². If the discharge is limited to a single-electron process, high cycle life (>200) with good capacity ($0.9e/CuCl_2$) can be achieved. This system shows excellent coulombic efficiency (see Fig. 34.33b).

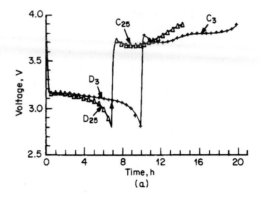

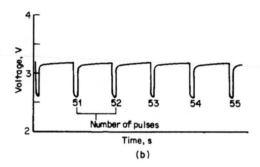

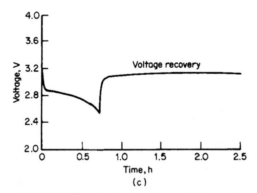

FIGURE 34.32 (*a*) Discharge/charge behavior of a Li/LiAlCl$_4$·6SO$_2$ C-size battery at 1 mA/cm^2. Voltage limits; 2.8–4.0 V. Line—battery, symbols (△)—cathode vs. ref. (*b*) High-rate pulse discharge of a Li/LiAlCl$_4$·6SO$_2$ C-size battery. Pulse rate at 20 mA/cm^2 for 20 s. Rest period for 180 s. (*c*) Discharge characteristics and voltage recovery (without load) of a Li/LiAlCl$_4$·6SO$_2$ C-size battery at −20°C. Discharge at 3.2 mA/cm^2. (*From Yardney Technical Products, Inc.*)

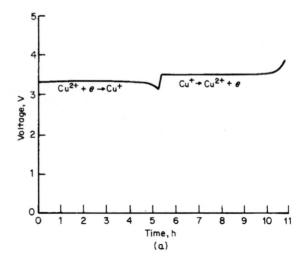

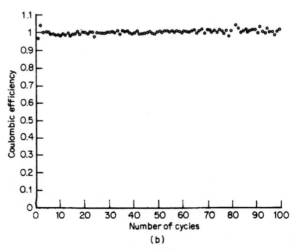

FIGURE 34.33 (*a*) Discharge/charge behavior of a Li/CuCl$_2$ prismatic D-size battery in LiAlCl$_4 \cdot$ 6SO$_2$ electrolyte at 1 mA/cm^2. (*b*) Coulombic efficiency of a Li/CuCl$_2$ battery at 1 mA/cm^2. (*From Yardney Technical Products, Inc.*)

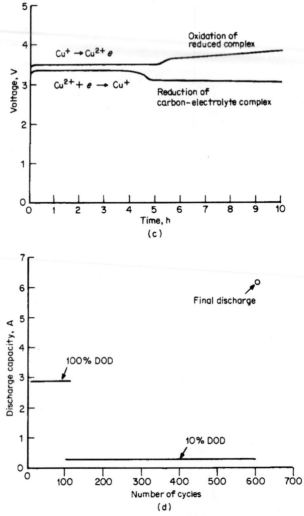

FIGURE 34.33 (*c*) Discharge behavior of a Li/CuCl$_2$ rechargeable battery in LiAlCl$_4$ · 6SO$_4$ electrolyte at 1 mA/cm^2. (*d*) Cycling behavior of a Li/CuCl$_2$ prismatic D-size battery at 1 mA/cm^2. (*From Yardney Technical Products, Inc.*) (*Continued*).

When an optimal combination of Ketjen black carbon and CuCl$_2$ is used as the cathode, limited overdischarge-overcharge protection can be achieved. These batteries can also deliver extra capacity when needed. The discharge-charge as well as the overdischarge-overcharge characteristics of such a battery are shown in Fig. 34.33*c*. Possible cell reactions are as follows:

Discharge:

$$\text{Cathode:} \qquad \text{Cu}^{2+} + e \rightarrow \text{Cu}^+$$

$$\text{Anode:} \qquad \text{Li} \rightarrow \text{Li}^+ + e$$

Overdischarge:

Cathode $LiAlCl_4 \cdot 3SO_2 + xC + 3e \longrightarrow LiClAl{\underset{OSO}{\overset{OSO}{-OSO}}}C_x + 3Cl^-$

Anode $Li \longrightarrow Li^+ + e$

Charge:

Cathode $Cu^+ \rightarrow Cu^{2+} + e$

Anode $Li^+ + e \rightarrow Li$

Overcharge:

Cathode $LiClAl{\underset{OSO}{\overset{OSO}{-OSO}}}C_x + 3Cl^- \longrightarrow LiAlCl_4 \cdot 3SO_2 + xC + 3e$

Anode $Li^+ + e \longrightarrow Li$

The composition of the cathode is typically 67% $CuCl_2$, 25% Ketjen black, and 8% Teflon. The cycling behavior of a Li/$CuCl_2$ Ketjen black prismatic D-size cell in $LiAlCl_4 \cdot 6SO_2$ electrolyte is shown in Fig. 34.33*d* for the first 100 cycles at 100% depth of discharge (the cell was cycled with time limits rather than voltage limits), then 500 cycles at 10% depth of discharge, and finally the total discharge and overdischarge capacity of the cell (see Fig. 34.34). The extra capacity is due to the reduction of the complex formed by the SO_2-based $LiAlCl_4$ with the high-surface-area carbon.

The Li/$CuCl_2$ Ketjen black system in $LiAlCl_4 \cdot 3SO_2$ electrolyte has been investigated in a bipolar configuration for high-rate pulse power applications. Figure 34.35 shows a portion of the cycling behavior of a four-cell-stack bipolar battery. The battery delivered more than 1000 pulse cycles.[63]

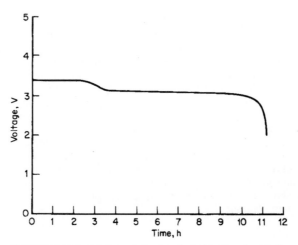

FIGURE 34.34   Discharge behavior of a Li/$CuCl_2$ prismatic D-size battery after 100 cycles at 100% DOD and 500 cycles at 10% DOD. Discharge rate at 1 mA/cm². (*From Yardney Technical Products, Inc.*)

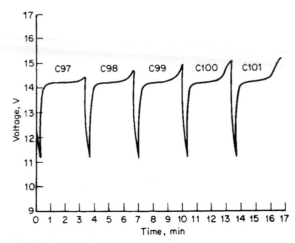

FIGURE 34.35 Discharge/charge behavior of a bipolar Li/ CuCl$_2$ battery (four-cell stack) at 50-mA/cm^2 (total 24 A) discharge for 20 s and 5.56-mA/cm^2 charge for 180 s. Voltage limits: 11.0–16.0 V. (*From Yardney Technical Products, Inc.*)

Laboratory-scale work has continued to improve the safety and cycle-life of the Li/CuCl$_2$ system.[64] Design modifications and electrolyte additives are claimed to reduce the tendency of this battery to form dendrites on overcharge, leading to internal shorts and ventings. Three hundred charge-discharge deep-discharge cycles are claimed with these changes. The electrolyte modification apparently involves the addition of chlorine, which leads to corrosion of 300-series stainless steel case-positive cells during charging, and this ultimately limits cycle life. Proprietary shut-down separators are also stated to be under development to improve the safety of this battery system.

Although the SO$_2$-based inorganic electrolyte systems have been investigated for some time and may be of interest for special application because of their advantageous performance characteristics, the toxicity and corrosivity of the electrolyte present a potential safety hazard limiting commercial applications. Safety problems have also been persistent with these systems and have limited their development beyond the laboratory level.

A lithium-metal LiAlCl$_4$-SO$_2$/LiCoO$_2$ battery is currently under development.[12,65,66] This system is built in the discharged state and is charged by plating lithium metal on a nickel metal substitute. In this way, no excess lithium is present. The lithium electrode has been found to have a SEI layer on lithium of lithium dithionite (Li$_2$S$_2$O$_4$) through which the electrochemical reactions occur.[65] Early versions of this battery used a microporous Tefzel® separator, but this material is no longer commercially available. This system also possesses an overcharge mechanism in which chlorine is evolved at the positive electrode above 4.5 V and reacts with excess lithium on the negative electrode. Lithium is plated in filamentary form during charge using this electrolyte, which contains a SO$_2$/LiAlCl$_4$ ratio of 1.5 to 1.8. Cycling efficiencies of 98 to 99% for the lithium plating and stripping reactions have been found in this electrolyte.[66] Prototype 7 Ah batteries are stated to provide a specific energy of about 200 Wh/kg and reach specific power levels of 1300 W/kg.[66] Development of this system continues.

34.4.6 Rechargeable Lithium Alloy and Other Coin and Microcells

A number of coin-type rechargeable lithium batteries have been developed for portable applications as a power source for small electronic devices, for memory backup, or as a storage device for solar or other types of auxiliary power sources. These batteries typically have a higher voltage of about 3 V (thus reducing the number of cells required for a given voltage) and a significantly higher energy density, and are smaller and lighter than nickel-cadmium and other conventional rechargeable batteries. The self-discharge rate of these cells is usually less than 5% per year compared to about 30% per month for the nickel-cadmium cells. This characteristic eliminates the need for recharging even after long storage periods. The cycle life of most of these cells on deep discharge is relatively poor, but they give a long cycle life and show good storage properties when shallow depths of discharge are used.

These lithium coin cells generally use a metal oxide intercalation compound for the positive active material and a lithium alloy, such as lithium aluminum, which is less reactive than metallic lithium, for the negative electrode. An organic solution is used for the electrolyte.[67] The energy density and specific energy of the different rechargeable lithium coin-type batteries are summarized in Fig. 34.36.

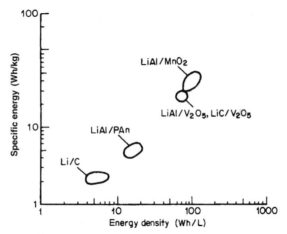

FIGURE 34.36 Energy density and specific energy of rechargeable lithium coin-type batteries.

Lithium-Aluminum/Manganese Dioxide Batteries.   Figure 34.37 shows the construction of a LiAl/MnO$_2$ cell. There is little difference in the design of this cell and that normally found in primary lithium coin cells. A lithium-aluminum alloy, with special additives to enhance the cycling performance and prevent cracking and blistering, is used for the anode. The cathode consists of a mixture of MnO$_2$ and LiOH, which undergoes a heat treatment to yield a high-porosity composite manganese oxide (MnO$_2$ + Li$_2$MnO$_3$), as follows:

$$y\text{MnO}_2 + z\text{LiOH} \rightarrow \left(y - \frac{z}{2}\right)\text{MnO}_2 + \left(\frac{z}{2}\right)\text{Li}_2\text{MnO}_3 + \left(\frac{z}{2}\right)\text{H}_2\text{O}$$

The characteristics of the composite allow the lithium ion to be removed easily from the positive-electrode material during the charge cycle. The electrolyte uses a three-ingredient solvent, consisting of ethylene carbonate, butylene carbonate, and DME. The separator is polypropylene.

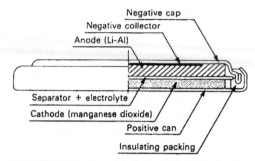

FIGURE 34.37   Cut-away view of a LiAl/MnO$_2$ coin-cell. (*From Sanyo Energy Corp.*)

The physical and electrical characteristics of coin-type batteries are listed in Tables 34.19 and 34.20. The discharge characteristics at several discharge currents and temperatures are shown in Fig. 34.38 for a typical lithium alloy coin-type battery. The cell voltage is close to 2.5 V because of the 0.3-V penalty associated with the use of the lithium-aluminum alloy.

TABLE 34.19   Rechargeable Lithium Aluminum/Manganese Dioxide Coin-Type Batteries

Model	ML 2430	ML 2016
Nominal voltage, V	3	3
Rated capacity, mAh	70	20
Diameter, mm	24.4	20
Height, mm	3.1	1.6
Weight, g	4	1.7
Operating temperature, °C	−20 to 60	−20 to 60
Self-discharge, %	<5 per year at 20°C	
Cycle life	2000 cycles at 3 mAh	2000 cycles at 1 mAh
	500 cycles at 12 mAh	500 cycles at 4 mAh
	200 cycles at 24 mAh	200 cycles a 6 mAh

Source: Sanyo Energy Corp.

TABLE 34.20   Characteristics of Lithium-Aluminum/Manganese Dioxide Coin-Type Rechargeable Batteries

Model no.	Nominal voltage (V)	Nominal capacity (mAh)	Dimensions (mm) External diameter	Height	Basic battery weight (g)	Recommended drain (mA)
ML612S	3	2.3	6.8	1.2	0.15	0.01
ML616	3	2.0	6.8	1.6	0.20	0.01
ML616S	3	2.9	6.8	1.6	0.19	0.01
ML621	3	3.0	6.8	2.1	0.25	0.01
ML621S	3	4.5	6.8	2.1	0.23	0.01
ML2020	3	45.0	20.0	2.0	2.2	0.1

*Nominal capacity shown is based on standard drain and cut off voltage down to 2.0 V at 20°C (68°F)
Source: Panasonic Division of Matsushita Electric Corp. of America.

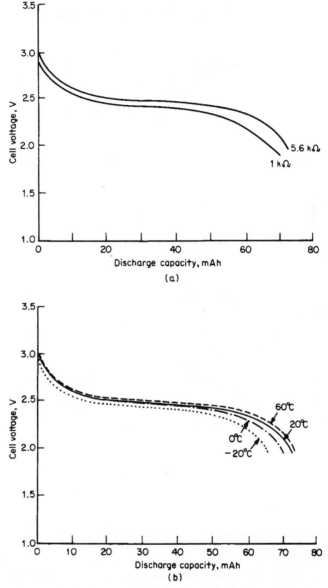

FIGURE 34.38 Discharge characteristics of LiAl/MnO$_2$ coin-type battery (ML2430 size): (*a*) discharge at 20°C, (*b*) at various temperatures, discharge load: 5.6 kΩ. (*From Sanyo Energy Corp.*)

The basic charging design requires a current source with a voltage limit of 3.25 V. Figure 34.39 shows the charging characteristics of a constant-voltage method where the charge current tapers as the battery voltage reaches the limit set by the charger. A protective resistor is incorporated within the charger circuit to limit and control the charge rate during the initial and long-term float condition. Figure 34.40 shows the dependence of the cycle life on the depth of discharge and the high cycle life obtained with a shallow depth of discharge.[68,69]

These batteries are employed for memory back-up in applications such as mobile telephones, memory cards, pagers and communications devices.

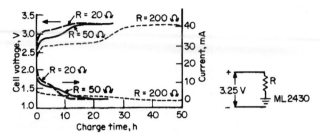

FIGURE 34.39 Constant voltage charging of LiAl/MnO$_2$ coin-type battery (ML2430 size), with a charging voltage of 3.25 V, at 20°C. (*From Sanyo Energy Corp.*)

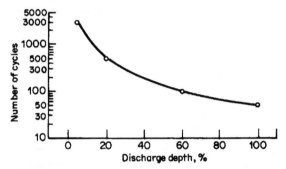

FIGURE 34.40 Cycle life vs. depth of discharge for a LiAl/ MnO$_2$ coin-type battery at 20°C. (*From Sanyo Energy Corp.*)

Lithium-Aluminum/Vanadium Pentoxide Batteries. The LiAl/V$_2$O$_5$ cell is another rechargeable lithium coin-type battery with a low self-discharge rate (2% per year at 20°C) which can be used for low-drain loads. Over 1000 cycles can be obtained at 10% depth of discharge. The construction is similar to the one illustrated in Fig. 34.37, except that V$_2$O$_5$ is used for the active positive material.

The specifications for these batteries are listed in Table 34.21. The discharge characteristics at several discharge rates and temperatures, illustrating the double-plateau discharge profile of these cells, are shown in Fig. 34.41. The ability of this cell to withstand overcharge and overdischarge is shown in Fig. 34.42. The relationship of capacity and depth of discharge is illustrated in Fig. 34.43

These batteries are employed as memory back-up power sources for applications such as facsimile machines, memory cards, personal computers, telephones, tuners and video cameras.

TABLE 34.21   Rechargeable Lithium-Aluminum/Vanadium Pentoxide Coin-Type Batteries

Model no.	Nominal voltage (V)	Nominal capacity (mAh)	Dimensions (mm) External diameter	Height	Basic battery weight (g)	Recommended drain (mA)
VL621	3.0	1.5	6.8	2.1	0.3	0.01
VL1216	3.0	5.0	12.5	1.6	0.7	0.03
VL1220	3.0	7	12.5	2.0	0.8	0.03
VL2020	3.0	20	20.0	2.0	2.2	0.07
VL2320	3.0	30	23.0	2.0	2.8	0.10
VL2330	3.0	50	23.0	3.0	3.7	0.10
VL3032	3.0	100	30.0	3.2	6.3	0.20
Charge/discharge Cycle	About 1000 times with a 10% discharge depth					
Charge	Constant voltage charging					
Operating temperature	$-20°C(-4°F)$ to $60°C(140°F)$					

Source: Panasonic Division of Matsushita Electric Corp. of America.

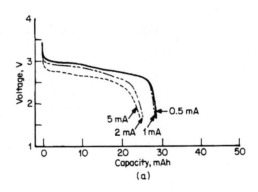

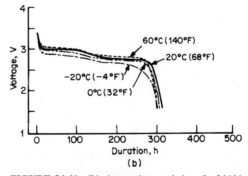

FIGURE 34.41   Discharge characteristics of a LiAl/V_2O_5 coin-type battery (VL2020 size): (*a*) discharges at 20°C, (*b*) discharges at 30 kΩ at various temperters. (*From Panasonic Division of Matsushita Electric Corp. of America.*)

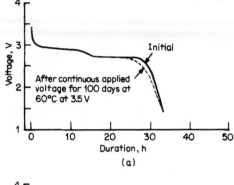

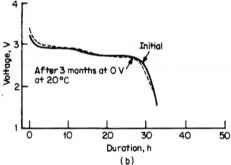

FIGURE 34.42 Overcharge and overdischarge characteristics of LiAl/V$_2$O$_5$ coin-type battery (VL2020 size): (*a*) overcharge characteristics, (*b*) overdischarge characteristics (charge at 200 Ω for 48 h, discharge at 3 kΩ). (*From Panasonic Division of Matsushita Electric Corp. of America.*)

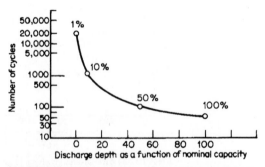

FIGURE 34.43 Cycle life vs. depth of discharge (%) for a LiAl/V$_2$O$_5$ coin-type battery (VL2020 size) at 20°C. (*From Pansonic Division of Matsushita Electric Corp. of America.*)

Carbon-Lithium Rechargeable Batteries.   The carbon-lithium batteries use a lithium alloy for the negative active material, a nonaqueous organic electrolyte, such as propylene carbonate, and activated carbon for the positive electrode. The battery is built in a discharged state. The mode of operation and the reactions during charge and discharge are delineated as follows:

Cell as constructed:

Sn-Bi alloy/$LiClO_4 \cdot PC$/C (activated charcoal powder)

$$\text{Negative} \qquad Sb_xBi_y + Li^+ + e \underset{\text{discharge}}{\overset{\text{charge}}{\rightleftharpoons}} LiSn_xBi_y$$

$$\text{Positive} \qquad nC + LiClO_4 \underset{\text{discharge}}{\overset{\text{charge}}{\rightleftharpoons}} C_nClO_4 + Li^+ + e$$

The construction is also similar to the one shown in Figure 34.37, except that activated carbon is used for the active positive material.

The carbon-lithium system has a low energy density but long cycle life and a low self-discharge rate. It can be charged at a current as low as several microamperes. A key application has been for small electronic equipment such as a remote controller requiring a low-power, long-life power source. The carbon-lithium battery can be combined with a solar cell, which will charge the carbon-lithium battery and maintain it in a charged condition. The charging characteristics of the carbon-lithium battery at various light intensities are shown in Fig. 34.44.

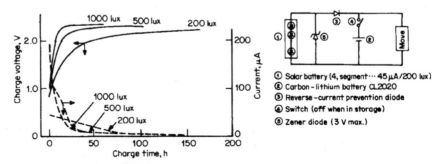

FIGURE 34.44   Charge characteristics of a carbon-lithium coin-type battery (CL2020 size) with solar batteries. (*From Panasonic Division of Matsushita Electric Corp. of America.*)

The discharge characteristics of the carbon-lithium battery at various rates and temperatures are shown in Fig. 34.45. The discharge profile is sloping and the delivered capacity is very dependent on the end voltage of the discharge. The cycle life, as shown in Fig. 34.46, likewise is dependent on the extent of the discharge. More than 10,000 cycles can be obtained on a very shallow discharge, for example, on a discharge to 0.1 mAh for a CL2020 cell. The specifications are listed in Table 34.22. These products are no longer commercially available.

Lithium-manganese-titanium rechargeable coin-type batteries which operate at 1.5 V are available commercially. These employ a lithiated manganese oxide positive material, an organic electrolyte and a lower voltage lithium titanium oxide intercalation compound as the negative electrode material. These are lithium ion systems in which the lithium ions shuttle between the oxides as the battery is charged and discharged. Characteristics of these batteries are listed in Table 34.23. These batteries are used as the main power supply for products such as rechargeable watches and as memory backup power supplies for pagers and timers.

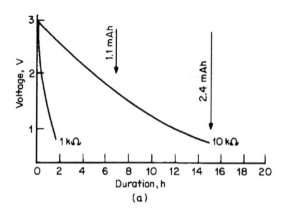

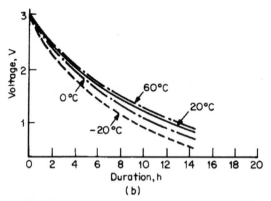

FIGURE 34.45 Discharge characteristics of carbon-lithium rechargeable coin-type battery (CL2020 size): (*a*) discharge at 20°C, (*b*) discharge at 10 kΩ. (*From Panasonic Division of Matsushita Electric Corp. of America.*)

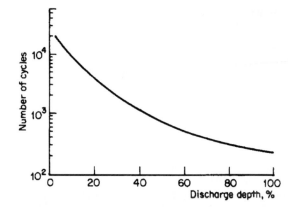

FIGURE 34.46 Cycle life vs. depth of discharge for a carbon-lithium coin-type battery (CL2020 size). (*From Panasonic Division of Matsushita Electric Corp. of America.*)

TABLE 34.22 Rechargeable Carbon-Lithium Coin-Type Batteries

Model	CL 1220	CL 2020	CL 2320	CL 2330
Nominal voltage, V	3	3	3	3
Rated capacity (to 2 V), mAh	0.3	1.0	1.5	2.5
Diameter, mm	12.5	20	23	23
Height, mm	2.0	2.0	2.0	3.0
Weight, g	0.8	1.9	2.8	3.7
Operating temperature, °C		−20 to + 60		
Cycle life		See Fig. 36.65		
Recommended drain, μA	1–30	1–100	1–200	1–200
Charging method		Voltage control		

Source: Panasonic Division of Matsushita Electric Corp. of America.

TABLE 34.23 Rechargeable Lithium-Manganese-Titanium Coin-Type Batteries

Model no.	Nominal voltage (V)	Nominal capacity (mAh)	Dimensions (mm) External diameter	Height	Basic battery weight (g)	Recommended drain (mA)
MT516	1.5	0.9	5.8	1.6	0.15	0.1
MT616	1.5	1.05	6.8	1.6	0.2	0.1
MT621	1.5	1.5	6.8	2.1	0.3	0.1
MT920	1.5	4.0	9.5	2.0	0.5	0.2
MT1620	1.5	14.0	16.0	2.0	1.3	0.5

*Nominal capacity shown is based on standard drain and cut off voltage down to 2.0 V at 20°C (68°F).
Source: Panasonic Division of Matsushita Electric Corp. of America.

ACKNOWLEDGMENT

Section 34.4.3 dealing with rechargeable lithium batteries using PEO-based electrolytes was contributed by Stefano Passerini. Jeffrey R. Dahn and Sid A. Megahed provided material for this chapter in the second edition of this Handbook.

REFERENCES

1. International Civil Aviation Organization (Montreal), "Technical Instructions for Safe Transport of Dangerous Goods by Air," U.S. Depart. of Transportation Code of Federal Regulations, CFR49.

2. M. Fujimoto, K. Ueno, T. Mohma, M. Takahashi, K. Nishio, and T. Saito, Extended Abstracts, in *Electrochem. Soc. Spring Meet.,* Honolulu, May 16–21, 1993, p. 108.

3. S. Hossain, Yardney Technical Products, Inc., unpublished data.

4. R. Fong, U. Von Sacken, and J. R. Dahn, *J. Electrochem. Soc.* **137:**2009 (1990).

5. D. Guyomond and J. M. Tarascon, *J. Electrochem. Soc.* **139:**937 (1992).

6. T. Uchida, Y. Morikawa, H. Ikuta, and M. Wakihara, Extended Abstracts, *Electrochem. Soc. Spring Meet.,* Honolulu, May 16–21, 1993, p. 109.

7. F. Croce, B. Scrosati, and M. Salomon, in *Proc. 36th Power Sources Conf.,* Cherry Hill, N.J., June 6–9, 1994, p. 57; E. J. Plichta and W. Behl, Extended Abstracts, in *Electrochem. Soc. Fall Meet.,* Toronto, October 1992.

8. B. Scrosati and S. Megahed, Electrochemical Society Short Course, New Orleans, Oct. 10, 1993.

9. A. Lecerf, M. Broussely, and J. P. Gabano, U.S. Patent 4,980,080 (1989).

10. M. M. Thackeray et al., *J. Electrochem. Soc.* **139:**363 (1992).

11. A. N. Day, H. C. Kuo, P. Pilliero, and M. Kalianidis, *J. Electrochem. Soc.* **135:**2115 (1988).

12. G. Hambitzer, J. Dreher, J. Dunger, and U. Schriever, in *Proc. 35th Int. Power Sources Symp.,* Cherry Hill, N.J., June 22–25, 1992.

13. M. B. Armand, J. M. Chubagno, and M. Duclot, in P. Vashista, J. M. Mundy, G. K. Sherroy (eds.), *Fast Ion Transport in Solid,* North-Holland, Amsterdam, 1979; M. B. Armand, *Solid State Ionics* **9810:**745 (1979).

14. M. B. Armand, in J. R. MacCallum and C. A. Vincent (eds.), *Polymer Electrolyte Reviews—1,* Elsevier Applied Science, New York, 1987.

15. K. M. Abraham and M. Alamgir, *J. Electrochem. Soc.* **136:**1657 (1990).

16. R. Koksbang, M. Gauthier, A. Belanger, in K. M. Abraham and M. Salomon (eds.), *Proc. Symp. Primary and Secondary Lithium Batteries,* The Electrochemical Society, Pennington, N.J., 1991.

17. K. M. Abraham, in B. Scrosati (ed.), *Applications of Electroactive Polymers,* Chapman and Hall, London, 1993.

18. D. H. Shen, G. Nagasubramanian, C. K. Huang, S. Surampudi, and G. Halpert, in *Proc. 36th Power Sources Conf.,* Cherry Hill, N.J., 1994.

19. D. L. Foster, H. C. Kuo, C. R. Schlaijker, and A. N. Dey, *J. Electrochem. Soc.* **135:**2682 (1988).

20. F. M. Gray, *Solid Polymer Electrolytes Fundamental and Technological Applications,* VCH, Cambridge, 1991.

21. F. Croce and B. Scrosati, *J. Power Sources* **43/44:**9 (1993).

22. E. Peled, D. Golodnitsky, C. Menachem, and G. Ardel, in *Proc. 7th Int. Meet. Lithium Batteries,* Boston, May 1994.

23. J. R. Dahn, U. Von Sacken, M. W. Juzkow, and H. Al-Janaby, *J. Electrochem. Soc.* **138:**2207 (1991).

24. D. Guyomard and J. Tarascon, *J. Electrochem. Soc.* **138:**2864 (1991).

25. D. Guyomard and J. Tarascon, *J. Electrochem. Soc.* **140:**3071 (1993).

26. W. L. Bowden and A. N. Dey, U.S. Patent 4,515,875 (1985).

27. D. Fouchard, in *Proc. 33rd Power Sources Symp.,* The Electrochemical Society, Pennington, N.J., 1988; J. A. R. Stilb, *J. Power Sources* **26:**233 (1989).

28. For example, see "Cellular Phone Recall May Cause Setback for Moli," *Toronto Globe and Mail,* August 15, 1989.

29. K. Brandt, in *Proc. 4th Int. Seminar Lithium Battery Technology and Applications,* Deerfield Beach, Fla., Mar. 6–8, 1989.

30. T. Nohma, T. Saito, N. Furukawa, and H. Ikeda, *J. Power Sources* **26:**389 (1989).

31. T. Nohma and N. Furukawa, in K. M. Abraham and M. Saloman (eds.), *Proc. Symp. Primary and Secondary Lithium Batteries,* The Electrochemical Society, Pennington, N.J., 1991.

32. J. Yamusa, *Proc. 5th Int. Seminar on Lithium Battery Technology and Applications,* Deerfield Beach, Fla., 1991.

33. P. Dan, E. Mengeritski, Y. Geronov, and D. Aurbach, in *proc. 7th Int. Meet. on Lithium Batteries,* Boston, May 15–20, 1994.

34. Y. Geronov, E. Mangeritski, I. Yakupov, and P. Dan, in *Proc. 36th Power Sources Conf.,* Cherry Hill, N.J., June 6–9, 1994.

35. D. Aurbach, E. Zinigrad, H. Teller and P. Dan, *J. Electrochem. Soc.,* **147:**1274 (2000).

36. K. M. Abraham, T. N. Nguen, R. J. Hurd, G. L. Holleck, and A. C. Makrides, in *Proc. 3rd Int. Rechargeable Battery Seminar,* Deerfield Beach, Fla., Mar. 5–7, 1990.

37. D. Zuckerbrod, R. T. Gionvanni, and K. R. Grossman, in *Proc. 34th Power Sources Symp.,* Cherry Hill, N.J., June 25–28, 1990.

38. S. Surampudi, D. H. Shen, C. K. Huang, F. Deligiannis, and G. Halpert, in *Proc. 34th Power Sources Symp.,* Cherry Hill, N.J., June 25–28, 1990.

39. C. J. Post and E. S. Takeuchi, in *Proc. 35th Power Sources Symp.,* Cherry Hill, N.J., June 25–28, 1992.

40. J. Broadhead, in *Proc. 3rd Ann. Battery Conf. on Applications and Advances,* California State University, Long Beach, Jan. 12–14, 1988; S. Basu and F. A. Trumbore, *J. Electrochem. Soc.* **139:**3379 (1992); F. A. Trumbore, *J. Power Sources* **26,** 65–75 (1989).

41. K. W. Beard, W. A. DePalma, and J. P. Buckley, in *Proc. 35th Int. Power Sources Symp.,* Cherry Hill, N.J., June 22–25, 1992; H. W. Lin and D. L. Chua, in *Proc. 3rd Lithium Battery Exploratory Development Workshop,* Lake Placid, N.Y., June 23 and 24, 1993.

42. R. J. Staniewicz, A. Romero, and A. Gambrell, in *Proc. 3rd Lithium Battery Exploratory Development Workshop,* Lake Placid, N.Y., June 23 and 24, 1993.

43. R. J. Brodd, J. T. Lundquist, J. L. Morris, and D. R. Shackle, in *Proc. 9th Annual Battery Conf.,* Long Beach, Calif., January 1994.

44. M. B. Armand, J. M. Chabagno, and M. Duclot, "Extended Abstracts," *2nd Int. Meeting on Solid Electrolytes,* St. Andrews, Scotland, Sept. 1978.

45. F. M. Gray in "Polymer Electrolytes" published by The Royal Society of Chemistry (ISBN 0-85404-557-0), Cambridge, U.K. (1997).

46. S. Passerini and B. B. Owens, in *Exploratory Research on Advanced Batteries and Supercapacitors for Electric/Hybrid Vehicles,* Outlook Document 1998; Electric Vehicle: Technologies and Programs, Annex V of The International Energy Agency, 1998.

47. M. Broussely, P. Biensan and B. Simon, *Electrochimica Acta,* **45:**322 (1999).

48. T. Osaka, *Interface,* **8:**3, 9 (1999).

49. C. A. Donnelly, "Lithium Polymer Batteries Development for Large Battery Applications," *2nd Int. Meeting on Application of Conducting Polymers,* Minneapolis, USA, July 1999.

50. B. B. Owens and S. Passerini, "International Development Trends of Energy Storage Technology for EV/HEV," *4th Symp. of Advanced Technology of Energy Storage for EV,* Tokyo, Japan, Nov. 1999.

51. P. P. Prosini, S. Passerini, R. Vellone and W. H. Smyrl, *J. of Power Sources,* **75:**73–83 (1998).

52. C. St-Pierre, R. Rouillard, A. Belanger, B. Kapfer, M. Simoneau, Y. Choquette, L. Gastonguay, R. Heiti, C. Behun, "Lithium Polymer Battery for Electric Vehicle and Hybrid Electric Vehicle Applications" *EVS 16,* China, Oct. 1999.

53. Argo-Tech (HydroQuebec & 3M) NPS-24V80 Battery Preliminary Specification Brochure.

54. J. L. Morris and D. R. Shackle, in *Proc. 7th Int. Meet. on Lithium Batteries,* Boston, May 15–20, 1994.

55. J. Broadhead and T. Skotheim, *Proc. 15th Int. Seminar on Primary and Secondary Batteries,* March 2–5, 1998.

56. P. Blonsky and T. A. Skotheim, *Proc. 39th Power Sources Symp.,* pp. 132–135, 1999.

57. R. J. Mammone and M. Binder, in *Electrochem. Soc. Fall Meet.,* San Diego, Oct. 1986.

58. S. Hossain, P. Harris, R. McDonald, and F. Goebel, in *Proc. 34th Power Sources Symp.,* Cherry Hill, N.J., June 1990.

59. S. Hossain, G. Kozlowski, and F. Goebel, Extended Abstracts, in *Electrochem. Soc. Fall Meet.,* Toronto, Oct. 1992.

60. R. McDonald, P. Harris, S. Hossain, and F. Goebel, in *Proc. 35th Power Sources Symp.,* Cherry Hill, N.J., June 1992.

61. C. R. Schlaijker, in V. R. Koch, B. B. Owens and W. H. Smyrl (eds.). *Rechargeable Lithium Batteries,* The Electrochemical Society, Pennington, N.J., 1990.

62. S. Hossain, P. Harris, F. Goebel, and R. McDonald, Extended Abstracts, in *Electrochem. Soc. Fall Meet.,* Seattle, Oct. 14–19, 1990.

63. S. Hossain, G. Kozlowski, and F. Goebel, *NASA Aerospace Workshop,* Huntsville, Ala., Nov. 15–19, 1992.

64. J. P. Campbell, J. R. Cormier, H. A. Hobbs, L. M. Toomey, and F. W. Dampier, *Proc. 6th Workshop for Battery Exploratory Development,* pp. 139–144, June 21–24, 1999.

65. B. Hefer, G. Hambitzer and C. Lutz, *Proc. 37th Power Sources Conf.,* pp. 203–207, 1996.

66. G. Hambitzer, V. Doge, I. Stassen, K. Pinkwart and C. Ripp. *Proc. 39th Power Sources Conf.,* pp. 200–202, 2000.

67. Z. Takehara, *J. Power Sources* **26:**257–266 (1989).

68. J. Carcone, *PowerTechnics Mag.* pp. 18–21, June 1990.

69. J. Carcone, in *Proc. 3rd Int. Rechargeable Battery Seminar,* Deerfield Beach, Fla., Mar. 1990.

CHAPTER 35
LITHIUM-ION BATTERIES

Grant M. Ehrlich

35.1 GENERAL CHARACTERISTICS

Lithium-ion (Li-ion) batteries are comprised of cells that employ lithium intercalation compounds as the positive and negative materials. As a battery is cycled, lithium ions (Li^+) exchange between the positive and negative electrodes. They are also referred to as rocking-chair batteries as the lithium ions "rock" back and forth between the positive and negative electrodes as the cell is charged and discharged. The positive electrode material is typically a metal oxide with a layered structure, such as lithium cobalt oxide ($LiCoO_2$), or a material with a tunneled structure, such as lithium manganese oxide ($LiMn_2O_4$), on a current collector of aluminum foil. The negative electrode material is typically a graphitic carbon, also a layered material, on a copper current collector. In the charge/discharge process, lithium ions are inserted or extracted from interstitial space between atomic layers within the active materials.

The first batteries to be marketed, and the majority of those currently available, utilize $LiCoO_2$ as the positive electrode material. Lithium cobalt oxide offers good electrical performance, is easily prepared, has good safety properties, and is relatively insensitive to process variation and moisture. More recently lower cost or higher performance materials, such as $LiMn_2O_4$ or lithium nickel cobalt oxide ($LiNi_{1-x}Co_xO_2$), have been introduced, permitting development of cells and batteries with improved performance. The batteries that were first commercialized employed cells with coke negative electrode materials. As improved graphites became available, the industry shifted to graphitic carbons as negative electrode materials as they offer higher specific capacity with improved cycle life and rate capability.

The Li-ion battery market has grown in a decade from an R&D interest to sales of over 400 million units in 1999. Market value at the OEM level was estimated to be $1.86 billion in 2000.[1] By 2005, the market is expected to grow to over 1.1 billion units with value of over $4 billion (¥455 billion),[2] while the average unit price is expected to fall 46% from 1999 to 2005. Market interest in this cost-effective, high performance, and safe technology has driven spectacular growth, as illustrated in Fig. 35.1. This technology has rapidly become the standard power source in a broad array of markets, and battery performance continues to improve as Li-ion batteries are applied to an increasingly diverse range of applications. To meet market demand, an array of designs has been developed, including spiral wound cylindrical, wound prismatic and flat plate prismatic designs in small (0.1 Ah) to large (160 Ah) sizes. Applications now addressed with Li-ion batteries include consumer electronics, such as cell phones, laptop computers, and personal data assistants, as well as military electronics, including radios, mine detectors and thermal weapons sights. Anticipated applications include aircraft, space craft, satellites, and electric or hybrid electric vehicles.

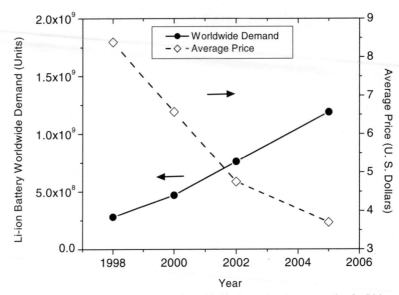

FIGURE 35.1 Current and anticipated worldwide demand and average price for Li-ion batteries. (*From Ref. 1.*)

The major advantages and disadvantages of Li-ion batteries, relative to other types of batteries, are summarized in Table 35.1. The high specific energy (~150 Wh/kg) and energy density (~400 Wh/L) of commercial products makes them attractive for weight or volume sensitive applications. Li-ion batteries offer a low self-discharge rate (2% to 8% per month), long cycle life (greater than 1000 cycles) and a broad temperature range of operation (charge at −20°C to 60°C, discharge at −40°C to 65°C), enabling their use in a wide variety of applications. A wide array of sizes and shapes is now available from a variety of manufacturers. Single cells typically operate in the range of 2.5 to 4.2 V, approximately three times that of NiCd or NiMH cells, and thus require fewer cells for a battery of a given voltage. Li-ion batteries can offer high rate capability. Discharge at 5C continuous, or 25C pulse, has been demonstrated. The combination of these qualities within a cost effective, hermetic package has enabled diverse application of the technology.

TABLE 35.1 Advantages and Disadvantages of Li-ion Batteries

Advantages	Disadvantages
Sealed cells; no maintenance required	Moderate initial cost
Long cycle life	Degrades at high temperature
Broad temperature range of operation	Need for protective circuitry
Long shelf life	Capacity loss or thermal runaway when over-
Low self-discharge rate	charged.
Rapid charge capability	Venting and possible thermal runaway when
High rate and high power discharge capability	crushed
High coulombic and energy efficiency	Cylindrical designs typically offer lower power
High specific energy and energy density	density than NiCd or NiMH
No memory effect	

A disadvantage of Li-ion batteries is that they degrade when discharged below 2 V and may vent when overcharged as they do not have a chemical mechanism to manage overcharge, unlike aqueous cell chemistries. Li-ion batteries typically employ management circuitry and mechanical disconnect devices to provide protection from over-discharge, overcharge or over temperature conditions. Another disadvantage of Li-ion products is that they permanently lose capacity at elevated temperatures (65°C), albeit at a lower rate than most NiCd or NiMH products.

35.1.1 Designation and Markings

The International Electrotechnical Commission (IEC) is developing standards for the designation, marking, electrical testing, and safety testing of Li-ion cells and batteries. A proposed designation and marking system for Li-ion cells utilizes five figures in the case of cylindrical cells and six figures in the case of prismatic cells. The first letter designates the type of negative electrode employed, I for an intercalation electrode. The second letter designates the type of positive electrode employed, such as C for a cobalt type, N for Nickel type, M for Manganese type or V for a Vanadium type. The third letter will designate the shape of the cell, R for round or P for prismatic. In the case of round cells, the next two figures will designate the diameter in millimeters and then the next three figures the height of the cell, in tenths of millimeters. For prismatic cells, the next three figures will designate the thickness in tenths of millimeters and width and height of the cell, in millimeters. In the event that any dimension exceeds 100 mm, a solidus ($/$) is to be placed between the figures. If a dimension is less than 1 mm, the dimension is preceded by a t (t7 = 0.7 mm). For example the common round cells that use the $C/LiCoO_2$ cell chemistry are designated ICR18650 as they are 18 mm in diameter and 65.0 mm in height.

The proposed designation standard for Li-ion batteries will involve a number that will designate the number of series connected cells, followed by three letters that will designate the negative electrode system (I for intercalation), the positive electrode system (C for cobalt, N for nickel, M for manganese), and the battery's shape (R for round, P for prismatic). The letters will be followed by two figures in the case of round batteries, to designate the diameter and height of the battery, in millimeters and tenths of millimeters or three figures in the case of prismatic batteries, to designate the thickness, width and height of the battery, in millimeters. If a dimension exceeds 100 mm, a solidus is used between the figures. If the battery comprises two or more parallel cells, the number of parallel connected cells is added after the figures that designate the dimensions. For example, a battery designated 1ICP206870-2 would be a prismatic Li-ion battery with two $C/LiCoO_2$ cells connected in parallel, with thickness 20 mm, width 68 mm and height 70 mm.

The latent IEC standards should be obtained for detailed information on nomenclature, performance and safety guidelines.

ANSI standard C18.2M "Standard for Portable Rechargeable Cells and Batteries" includes standards for portable Li-ion batteries.

35.2 *CHEMISTRY*

The electrochemically active electrode materials in Li-ion batteries are a lithium metal oxide for the positive electrode and lithiated carbon for the negative electrode. These materials are adhered to a metal foil current collector with a binder, typically polyvinylidene fluoride (PVDF) or the copolymer polyvinylidene fluoride–hexafluoropropylene (PVDF-HFP), and a conductive diluent, typically a high-surface-area carbon black or graphite. The positive and negative electrodes are electrically isolated by a microporous polyethylene or polypropylene separator film in products that employ a liquid electrolyte, a layer of gel-polymer electrolyte in gel-polymer batteries, or a layer of solid electrolyte in solid-state batteries.

Since the commercialization of Li-ion batteries by Sony in 1990, a broad array of variants has been introduced. One type, gel-polymer Li-ion batteries, utilizes the same active materials as products that employ liquid electrolytes, but in a different construction that enables the fabrication of cells with a thin form factor. Gel-polymer batteries, also referred to as polymer Li-ion batteries in the marketplace, are products where the microporous separator film used in conventional batteries is substituted by a layer of PVDF-HFP, or other polymer, impregnated with liquid electrolyte and the solid current collector foil is typically substituted with an open expanded metal current collector grid. In gel-polymer cells, the positive, separator, and negative layers are bound by the polymer, typically PVDF-HFP, and can be laminated together to form a monolithic device. Despite these differences, the active cell chemistry may remain identical to that in cylindrical or prismatic Li-ion batteries.

35.2.1 Intercalation Processes

The active materials in Li-ion cells operate by reversibly incorporating lithium in an intercalation process, a topotactic reaction where lithium ions are reversibly removed or inserted into a host without a significant structural change to the host. The positive material in a Li-ion cell is a metal oxide, with either a layered or tunneled structure. The graphitic carbon negative materials have a layered structure similar to graphite. Thus the metal oxide, graphite, and other materials act as hosts, incorporating lithium ions, guests, reversibly to form "sandwich" like structures.[3]

Intercalation materials, originally discovered by the Chinese 2700 years ago,[4] have been the subject of contemporary chemical research for only the last half-century. Today, intercalation compounds form the basis for a host of technologies ranging from superconductors to catalysis. Intercalation materials in common use include graphite,[5,6] layered silicates such as talc $(Mg_3(OH)_2(Si_4O_{10}))$,[7] clays, and layered transition metal dichalcogenides, such as TiS_2.[8] The intercalation of a variety of electron donors, including lithium, and electron acceptors, such as halogens, into graphite has been studied.[6,7] The field of graphite intercalation compounds is especially rich,[9] both in the diversity of the chemistry and the depth of study. Of particular interest to the field of Li-ion batteries is the work on alkali metal intercalation of graphite and related carbons, in particular Li_xC_6 $(0 < x < 1)$.[10] When a Li-ion cell is charged, the positive material is oxidized and the negative material is reduced. In this process, lithium ions are de-intercalated from the positive material and intercalated into the negative material, as illustrated in Fig. 35.2. In this scheme, $LiMO_2$ represents the metal oxide positive material, such as $LiCoO_2$, and C the carbonaceous negative material, such as graphite. The reverse happens on discharge. As metallic lithium is not present in the cell, Li-ion batteries are chemically less reactive, safer, and offer longer cycle life than possible with rechargeable lithium batteries that employ lithium metal as the negative electrode material. The charge-discharge process in a Li-ion cell is further illustrated graphically in Fig. 35.3. In the figure, the layered active materials are shown on metallic current collectors.

Positive: $LiMO_2 \underset{\text{discharge}}{\overset{\text{charge}}{\rightleftharpoons}} Li_{1-x}MO_2 + xLi^+ + xe^-$

Negative: $C + xLi^+ + xe^- \underset{\text{discharge}}{\overset{\text{charge}}{\rightleftharpoons}} Li_xC$

Overall: $LiMO_2 + C \underset{\text{discharge}}{\overset{\text{charge}}{\rightleftharpoons}} Li_xC + Li_{1-x}MO_2$

FIGURE 35.2   Electrode and cell reactions in a Li-ion cell.

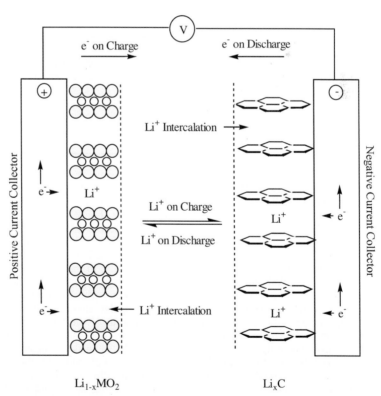

FIGURE 35.3   Schematic of the electrochemical process in a Li-ion cell.

35.2.2 Positive Electrode Materials

Positive electrode materials in commercially available Li-ion batteries utilize a lithiated metal oxide as the active material. The first Li-ion products marketed by Sony used $LiCoO_2$. Goodenough and Mizushima developed this material, as described in a series of patents.[11] Recently, cells have been developed that utilize less costly materials, such as $LiMn_2O_4$ (spinel), or materials with higher coulombic capacity, such as $LiNi_{1-x}Co_xO_2$. Commercial interest in $LiNiO_2$ has waned as its instability, driven by the energetic formation of NiO and oxygen, has been shown to contribute to safety issues.[12]

Viable electrode materials must satisfy a number of requirements, as summarized in Table 35.2. These factors guide the selection and development of positive electrode materials. To enable high capacity, materials must incorporate a large amount of lithium. Further, the materials must reversibly exchange that lithium with little structural change to permit long cycle life, high coulombic efficiency, and high energy efficiency. To achieve high cell voltage and high energy density, the lithium exchange reaction must occur at a high potential relative to lithium. When a cell is charged or discharged, an electron is removed or returned to the positive material. So that this process can occur at a high rate, the electronic conductivity and Li^+ mobility in the material must be high. Also, the material must be compatible with the other materials in the cell; in particular it must not be soluble in the electrolyte. Finally, the material must be of acceptable cost. To minimize cost, preparation from inexpensive materials in a low-cost process is preferred.

TABLE 35.2 Requirements for Li-ion Positive Electrode Materials

High free energy of reaction with lithium
Can incorporate large quantities of lithium
Reversibly incorporates lithium without structural change
High lithium ion diffusivity
Good electronic conductivity
Insoluble in the electrolyte
Prepared from inexpensive reagents
Low cost synthesis

Characteristics of Positive Electrode Materials.   A variety of positive electrode materials has been developed and many of these are commercially available. All commercially available materials have one of two structure types. $LiCoO_2$, $LiNiO_2$, and related materials such as $LiNi_{1-x}Co_xO_2$, have layered structures whereas $LiMn_2O_4$ or "spinel" materials have a three-dimensional "framework" structure. The term *spinel* formally refers to the mineral ($MgAl_2O_4$), although the term is used for materials with equivalent structure. The ideal layered structure of MnO_2 is shown in Fig. 35.4. In the case of $LiCoO_2$ or $LiNi_{1-x}Co_xO_2$, the cobalt or nickel atoms would reside within oxygen octahedra, and the lithium atoms would reside in the space between the oxygen layers. The $LiMn_2O_4$ (spinel) materials, however, have a three-dimensional framework or tunneled structure based on λ-MnO_2, as illustrated in Fig. 35.5. In spinel, lithium fills one-eighth of the tetrahedral sites within the λ-MnO_2 structure as Mn-centered oxygen octahedra fill one-half of the octahedral sites.

Commercially, the structural difference between these materials has had unusually high impact because the original patent by Goodenough and Mizuchima,[11] granted to the United Kingdom Atomic Energy Agency (AEA) in 1981, referred to "mixed oxides having the layered structure of α-$NaCrO_2$," thus cells using spinel did not fall within the scope of Goodenough and Mizuchima's patents.

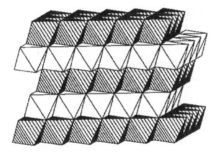

FIGURE 35.4 The idealized structure of layered-MnO_2. (*Courtesy of CISR.*)

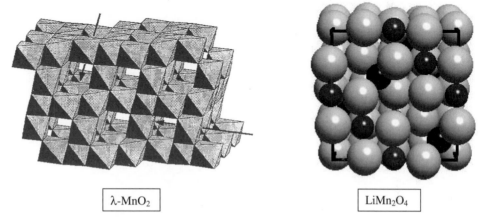

| λ-MnO_2 | $LiMn_2O_4$ |

FIGURE 35.5 The idealized structure of λ-MnO_2 and $LiMn_2O_4$ spinel. The model on the left shows the manganese centered oxygen octahedra of λ-MnO_2. In the model on the right of $LiMn_2O_4$, oxygen is grey and lithium is black. (*Courtesy of CISR and Michael Tucker.*)

The voltage and capacity characteristics of common positive electrode materials are summarized in Table 35.3. The most commonly used positive electrode material, $LiCoO_2$ (Fig. 35.6), offers good capacity, 155 mAh/g, and high voltage, 3.9 V vs. Li. The $LiNi_{1-x}Co_xO_2$ materials offer higher capacity, up to 220 mAh/g, albeit at ~0.2 V lower than $LiCoO_2$ or $LiMn_2O_4$.

Spinel, $LiMn_2O_4$, is also of commercial interest, particularly for applications that are cost sensitive or require exceptional stability upon abuse. It has lower capacity, 120 mAh/g, slightly higher voltage, 4.0 V vs. Li, but has higher capacity loss on storage, especially at elevated temperature, relative to cells that use $LiCoO_2$ or $LiNi_{1-x}Co_xO_2$, described below.

Properties of $LiNiO_2$-Based Positive Electrode Materials. Despite its high capacity and low cost, $LiNiO_2$ is not widely used commercially because of the energy evolved upon decomposition, the relatively low temperature at which self-heating ensues, and the difficulty of preparing the material consistently in quantity.[13] The synthesis, electrochemistry, and structural properties of $LiNiO_2$ have been described,[14] as have the morphological changes that can occur when the material is charged to 4.2 V vs. lithium.[15]

TABLE 35.3 Characterisitcs of Positive Electrode Materials*

Material	Specific Capacity (mAh/g)	Midpoint V vs. Li (at 0.05C)	Advantages or disadvantages
$LiCoO_2$	155	3.88	Most common commercially, Co is expensive
$LiNi_{0.7}Co_{0.3}O_2$	190	3.70	Intermediate cost
$LiNi_{0.8}Co_{0.2}O_2$	205	3.73	Intermediate cost
$LiNi_{0.9}Co_{0.1}O_2$	220	3.76	Highest specific capacity
$LiNiO_2$	200	3.55	Most exothermic decomposition[17]
$LiMn_2O_4$	120	4.00	Mn is inexpensive, low toxicity, least exothermic decomposition

*Experimental values.

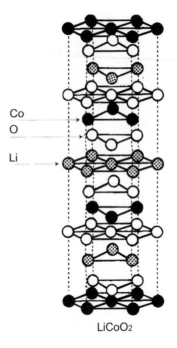

Co
O
Li

$LiCoO_2$

FIGURE 35.6 The idealized layered structure of $LiCoO_2$, Li is speckled, O white and Co black.

The energy evolved on decomposition is illustrated in Fig. 35.7, which shows Differential Scanning Calorimetry (DSC) data from electrodes that were charged at 4.5 V for 40 hours prior to heating in a DSC. As shown, $LiNiO_2$ evolves the most energy and has a low onset temperature. The decomposition of $LiNiO_2$ has been shown to follow two paths, both leading to NiO;[12] low temperature (LT)-$LiNiO_2$ has been shown to be less stable than high temperature (HT)-$LiNiO_2$, and the stability is dependent on the particle size. To mitigate the instability of $LiNiO_2$, cobalt substituted materials are now commonplace. As shown in Fig. 35.7, $LiCoO_2$ has a higher onset temperature and evolves less energy than $LiNiO_2$. As expected, the intermediate phase $LiNi_{0.8}Co_{0.2}O_2$ has an onset temperature and energy evolution between that of $LiNiO_2$ and $LiCoO_2$.

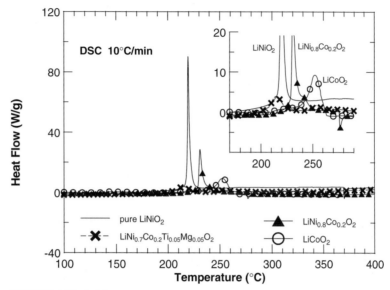

FIGURE 35.7 DSC data showing the onset temperature and energy released from charged cathode materials upon heating. Electrode float charged to 4.5 V for 40 hours. (*Courtesy of FMC.*)

To reduce the energy evolved during positive material decomposition, new materials have been developed that offer either less evolved energy or higher onset temperatures. Examples include Al or Ti and Mg doped lithium nickel cobalt oxides. DSC data for $LiNi_{0.7}Co_{0.2}Ti_{0.005}Mg_{0.005}O_2$ is included in Fig. 35.7. As shown, it evolves much less energy than even $LiCoO_2$, however its reversible capacity is 25 mAh/g less than $LiNi_{0.8}Co_{0.2}O_2$. Others have developed magnesium oxide coated $LiNi_{1-x}Co_xO_2$ type materials that offer electrochemical performance comparable to un-modified materials, but with a higher self-heating temperature (219°C versus 211°C) and a reduced exotherm when heated in a DSC.[16]

Electrical Properties of Positive Electrode Materials. The voltage characteristics of $LiMn_2O_4$, $LiCoO_2$, and $LiNi_{1-x}Co_xO_2$, when cycled versus a Li counter electrode are illustrated in Fig. 35.8 (Charge) and Fig. 35.9 (Discharge). As indicated, $LiMn_2O_4$ offers the highest voltage (4.0 V) but the lowest capacity (~120 mAh/g), $LiNi_{1-x}Co_xO_2$ the lowest average voltage (~3.75), but the highest capacity (~205 mAh/g), $LiCoO_2$ is intermediate (3.88 V, ~155 mAh/g). Spinel is unique in that it exhibits two distinct voltage plateaus on charge and discharge, more evident in the differential capacity plot shown in Fig. 35.10. Also shown in Fig. 35.10 is the differential capacity for $LiNi_{0.8}Co_{0.2}O_2$. As shown, $LiNi_{0.8}Co_{0.2}O_2$ reversibly incorporates lithium over the range of 3.7 V to 4.25 V on charge (deintercalation), and 3.4 V to 4.25 V on discharge (intercalation). Shown in Fig. 35.11 is the differential capacity for $LiCoO_2$ when cycled versus Li. It intercalates the majority of its capacity at 3.93 V with small peaks at 4.07 V and 4.20 V that correspond to small peaks observed in the case of the $LiNi_{0.8}Co_{0.2}O_2$ material.

A comparison of three different $LiNi_{1-x}Co_xO_2$ materials, $LiNi_{0.7}Co_{0.3}O_2$, $LiNi_{0.8}Co_{0.2}O_2$, and $LiNi_{0.9}Co_{0.1}O_2$, is illustrated in Fig. 35.12, which shows first cycle charge curves, and Fig. 35.13, which shows first cycle discharge curves. On charge, these $LiNi_{1-x}Co_xO_2$ materials offer similar voltage, except that as the cobalt content is reduced, higher capacity results, up to 222 mAh/g. This trend in capacity is also observed on discharge. The synthesis, properties and electrochemical performance of $LiNi_{1-x}Co_xO_2$ ($x = 0, 0.1, 0.2, 0.3$) materials, including charge, discharge, and irreversible capacities, and their thermal stability have been reported.[13]

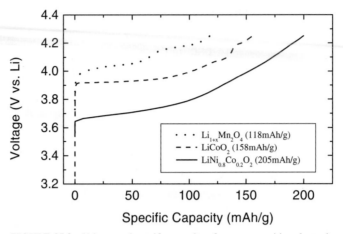

FIGURE 35.8   Voltage and specific capacity of common positive electrode materials on the first charge at 25°C. All materials were charged at approximately C/20, the LiCoO$_2$ at 7.2 mA/g, the LiNi$_{1-x}$Co$_x$O$_2$ materials at 0.16 mA/cm^2, and the LiMn$_2$O$_4$ at 0.09 mA/cm^2. (*Courtesy of FMC and Yardney Technical Products, Inc.*)

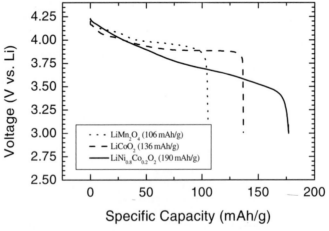

FIGURE 35.9   Voltage and specific capacity of common positive electrode materials on the first discharge at the C/20 rate. The LiNi$_{1-x}$Co$_x$O$_2$ materials were discharged at 0.16 mA/cm^2, the LiCoO$_2$ at 7.5 mA/g, and the LiMn$_2$O$_4$ at 0.09 mA/cm^2. (*Courtesy of FMC and Yardney Technical Products, Inc.*)

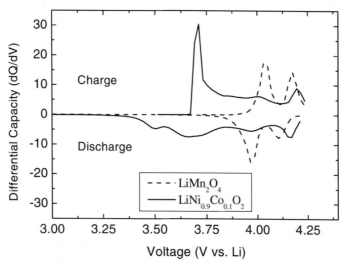

FIGURE 35.10 Differential capacity on the first cycle for $LiMn_2O_4$ and $LiNi_{0.9}Co_{0.1}O_2$ when cycled versus Li at the C/15 rate. (*Courtesy of Yardney Technical Products, Inc.*)

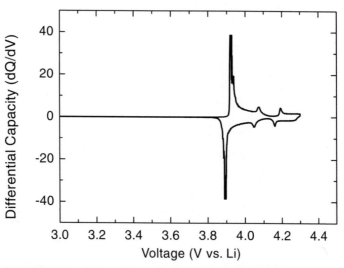

FIGURE 35.11 Differential capacity on the first cycle for $LiCoO_2$ when cycled versus Li at the C/20 rate. (*Courtesy of FMC and Yardney Technical Products, Inc.*)

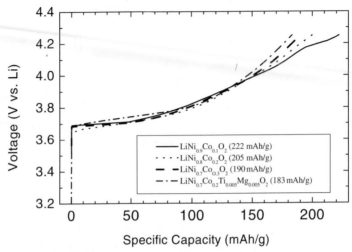

FIGURE 35.12 Charge curves for various $LiNi_{1-x}Co_xO_2$ materials charged at the C/20 rate. (*Courtesy of Yardney Technical Products, Inc. and FMC.*)

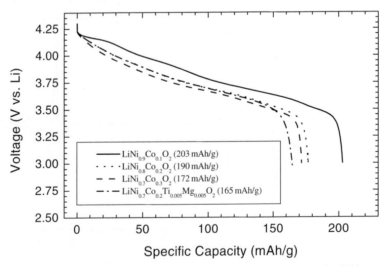

FIGURE 35.13 Discharge curves for various $LiNi_{1-x}Co_xO_2$ materials at the C/20 rate except $LiNi_{0.7}Co_{0.2}Ti_{0.005}Mg_{0.005}O_2$ which was discharged at the C/7 rate. (*Courtesy of Yardney Technical Products, Inc. and FMC.*)

The $LiNi_{0.7}Co_{0.2}Ti_{0.005}Mg_{0.005}O_2$ material, engineered to offer improved safety through reduced exotherm on decomposition, offers slightly lower capacity, 183 mAh/g on the first deintercalation, and 165 mAh/g reversibly, but with voltage similar to the other $LiNi_{1-x}Co_xO_2$ materials described.

Synthesis of Lithiated Metal Oxides.   The synthesis of lithiated metal oxides, including $LiCoO_2$ and $LiMn_2O_4$, has be achieved through a wide variety of routes, although those practiced commercially use inexpensive starting materials, such as lithium carbonate, lithium hydroxide and the metal oxide. The physical and electrochemical properties of the materials may be controlled by the choice of starting materials and the preparation conditions.[18]

The easiest to prepare is $LiCoO_2$. Its thermodynamic stability results in the desired phase after treatment of a range of lithium and cobalt reagents, including carbonates,[19] oxides,[20] hydroxides, nitrates,[21] and organic acid complexes,[22] in the appropriate ratio at high temperature, 600°C to 1100°C, in air. $LiCoO_2$ can be prepared in bulk at lower temperatures, 400°C, from either cobalt nitrate[23] or the acetates.[24] Preparation of $LiCoO_2$ at 300°C has been achieved from hydroxide mixtures.[25] $LiCoO_2$ has also been prepared via nonaqueous routes from nitrates,[26] under hydrothemal conditions,[27] and as thin films by laser ablation[28] or spray pyrolysis.[29] An overview of the preparation and properties of lithium cobalt oxides is available.[30]

Lithium nickel cobalt oxides can be prepared using routes similar to those used in the preparation of $LiCoO_2$, although the properties of the material are more sensitive to the preparative method. Preparations of lithium nickel cobalt oxides are designed to achieve molecular mixing of the cobalt and nickel materials prior to their reaction. Lithium cobalt nickel oxides have been prepared from lithium, nickel and cobalt hydroxide co-precipitates from nitrate solutions, treated between 400°C and 800°C after removal of excess water.[31] Another preparation includes treatment of Li_2CO_3, $CoCO_3$ and $Ni(NO_3)_2 \cdot 6H_2O$ at 400°C.[32]

The electrochemical properties of $LiMn_2O_4$ materials are sensitive to the preparative method, motivating the development of preparations that yield single phase material with controlled Li, Mn, and O stoichiometry with the desired structure. The effect of preparation conditions on the properties of spinels prepared from Li_2CO_3 and MnO_2, such as electrolytic MnO_2 (EMD), at 600°C to 900°C,[33] has been the subject of numerous investigations.[34] Materials have also been prepared from LiOH and γ-MnO_2 (CMD),[35] or Mn_2O_3,[36] towards spinels with improved capacity retention.

35.2.3   Capacity Fade in C/LiMn₂O₄ Cells

Manganese-based materials continue to receive industrial and commercial interest because of their low cost, benign environmental qualities,[37] good electrochemical properties, and excellent safety properties, despite the higher capacity and improved high temperature stability possible when cobalt- or nickel-based materials are used. The complex chemistry of the lithium manganese oxides has been the focus of many academic[38] and industrial studies. The most significant differences in performance between typical manganese- and cobalt-based cells are their lower capacity, their higher rate of capacity fade when cycled or stored in the charged or discharged state, and the ability of spinel cells to sustain abuse, attributed to the stability of λ-MnO_2 to deoxygenation, relative to NiO_2 or CoO_2.

Capacity fade in spinel cells is the result of multiple processes,[39] including those related solely to the spinel material and others involving the interaction of spinel with the electrolyte and negative electrode materials. Most fade mechanisms can be attributed to three factors:[40]

1. Dissolution of Mn^{2+} into the electrolyte after disproportionation of $Li_xMn_2O_4$:

$$2Mn^{3+}_{(solid)} \rightarrow Mn^{4+}_{(solid)} + Mn^{2+}_{(solution)}$$

Dissolution of spinel into the electrolyte,[41,42] is promoted by acid-induced delithiation, resulting in disproportionation and formation of λ-MnO$_2$. Li-ion electrolytes that employ LiPF$_6$ are acidic, as the salt reacts with adventitious water to form HF. This reaction has a secondary effect of reducing electrolyte conductivity:

$$4H_2O + LiPF_6 \rightarrow LiF + 5HF + H_3PO_4$$

However, studies of the material losses from spinel electrodes, and the resulting Mn^{2+} concentration in the electrolyte, found dissolution can account for only a fraction of the capacity fade observed in spinel cells.[42] For example, when spinel cells were cycled at 50°C, analysis of the electrolyte for manganese found spinel dissolution could account for only 34% of the capacity fade observed.[43]

2. Instability of the electrolyte

Decomposition of the electrolyte includes reaction of the solvent at electrode surfaces[42,44,45] to yield a passivation layer (Solid-Electrolyte Interphase or SEI), resulting in an increase in electrode resistance, cell polarization, and apparent capacity loss.[43]

The presence of lithium alkoxides within the SEI formed on the surface of graphite negative electrodes, as well as LiOH, Li$_2$CO$_3$, and lithium alkyl carbonates, has been established.[45,46] One proposed spinel degredation mechanism[47] starts with the reaction of lithium alkoxides at the graphite negative with adventitious water to yield alcohol:

$$LiOMe + H_2O \rightarrow MeOH + LiOH$$

The alcohol can then react at the positive electrode where it is oxidized to H$_2$O and CO$_2$, the spinel electrode serving as a source of oxygen:

$$MeOH + 1/\delta\ Mn_2O_4 \rightarrow CO_2 + 2H_2O + 1/\delta\ Mn_2O_{4-3\delta}$$

This water and CO$_2$ may then return to the negative electrode and react with Li$_x$C$_6$ to yield immobilized Li in the form of LiOH or Li$_2$CO$_3$:

$$2H_2O + 1/\delta\ LiC_6 \rightarrow 2LiOH + 1/\delta\ Li_{1-2\delta}C_6 + H_2$$

$$H_2CO_3 + 1/\delta\ LiC_6 \rightarrow Li_2CO_3 + 1/\delta\ Li_{1-2\delta}C_6 + H_2$$

Any H$_2$ generated may be oxidized at the positive electrode to form more water:

$$H_2 + 1/\delta\ Mn_2O_4 \rightarrow H_2O + 1/\delta\ Mn_2O_{4-\delta}$$

This mechanism does not require positive electrode weight loss, Jahn-Teller distortion,[48] or lattice contraction. Consistent with this mechanism, analysis of C/LiMn$_2$O$_4$ cells cycled at 45°C found the majority (75%) of the immobilized lithium within the graphite negative electrode, the remainder (25%) was found within the positive electrode, implicating involvement of the negative electrode in spinel cell capacity fade.

3. Jahn-Teller distortion in discharged cells (Li$_1$Mn$_2$O$_4$)

Jahn-Teller distortion[49] occurs in LiMn$_2$O$_4$ at 7°C (280 K).[50] This phase transition results in a transformation from the cubic space group Fd3m to the tetragonal group I4$_1$/amd. The structural distortion results from interaction of the Jahn-Teller active species Mn^{3+} (t_{2g}^3-e_g^1), in contrast, Mn^{4+} (t_{2g}^3-e_g^0) and Mn^{2+} (t_{2g}^3-e_g^2) are not Jahn-Teller active. Because of the low temperature of this transition, and that modified spinels have in general lower transition temperatures, this mechanism may not be as relevant as others for currently used spinel materials. Related mechanisms can cause strain and structural failure,[51] resulting in electrically disconnected particles.

Capacity fade in spinel cells is prominent when cells are stored in either the charged or discharged state.[43] An unmodified spinel will lose over 20% of its capacity when stored for three weeks at 50°C. Coated spinels, such as those protected by an inorganic coating of Li_2CO_3 or $LiCoO_2$, offer improved stability, loss of 1% to 3% per week at 50°C is typical. The coatings inhibit electrolyte decomposition and acid formation, thus are most effective in mitigating capacity loss in the charged state.

Modified spinels that contain excess lithium, and preferably an admetal ($Li_{1+x}M_yMn_{2-x-y}O_4$, M$=Al^{3+}$, Cr^{3+}, Ga^{3+}), offer improved storage stability in the discharged state as manganese disproportionation is inhibited when the $Mn^{3+}:Mn^{4+}$ ratio is reduced. Materials with admetal content offer fade rates at 55°C of 0.05%/cycle (0.05 mAh/g per cycle),[52] whereas current coated materials offer irreversible capacity loss on storage of less than 1% per week at 55°C in the discharged state, or 20% capacity loss after 500 cycles at 55°C.[53]

Alternatively, improved electrolytes, in particular those with additives to reduce water and HF impurity levels, such as hexamethyldisilazane,[54] permit spinel cells to offer improved performance. To illustrate the ability of the improved spinel materials to cycle at elevated temperature (55°C), Fig. 35.14 shows the specific capacity of a spinel material when cycled versus lithium at 23°C and 55°C. As shown, at 23°C the material demonstrated a fade rate of 0.04%/cycle and at 55°C, 0.15%/cycle. While these fade rates are higher than those demonstrated for $LiCoO_2$, they are viable for applications that value the low cost and benign safety properties of the manganese oxides.

Since reaction with the electrolyte or dissolution processes occur at the particle surface, materials with lower surface area and specially coated surfaces have been developed, such as $LiCoO_2$ or Li_2CO_3 coated $LiMn_2O_4$. To illustrate the ability of these materials to sustain cycling at 55°C, Fig. 35.15 shows the specific capacity of un-coated and $LiCoO_2$ coated $LiMn_2O_4$ materials cycled at C/2 at 23°C or 55°C. As illustrated, at 55°C the un-coated $LiMn_2O_4$ material initially provides 126 mAh/g but the fade rate increased after 10 cycles to 1.3%/cycle. In contrast, the $LiCoO_2$ coated material demonstrated lower capacity fade at 55°C, 0.6%/cycle.

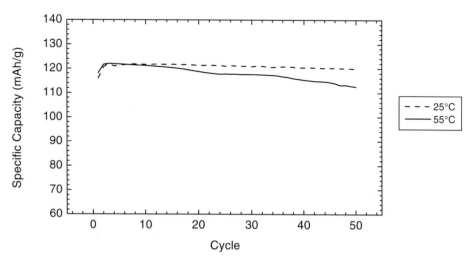

FIGURE 35.14 The specific capacity of a manganese spinel cycled at 55°C and 23°C at the C/2 rate between 4.3 and 3.5 V. (*Courtesy of Carus Chemical.*)

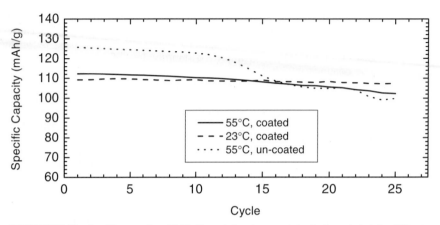

FIGURE 35.15   Specific capacity of LiCoO$_2$ coated and un-coated spinels cycled at the C/2 rate between 4.2 and 3.5 V at 23°C or 55°C. (*Courtesy of Carus Chemical.*)

35.2.4 Negative Electrode Materials

Historical Overview.   Since the early 1970s, intercalation compounds have been considered as electrode materials for secondary lithium batteries. However, secondary lithium battery development effort throughout the 1970s and early 1980s focused on the use of lithium metal as the negative electrode because of the high specific capacity of the metal. Cells with impressive performance were developed and some were commercialized, however safety issues with lithium metal batteries[55] caused the industry to concentrate on using lithium intercalation into carbon at the negative electrode instead of lithium metal.[56] The safety issues with lithium metal have been attributed to the changing morphology of lithium as a cell is cycled. As described in Chapter 34, the safety properties of negative electrodes may be correlated to their surface area, thus while the properties of lithium metal negative electrodes change with use, carbon electrodes offer stable morphology resulting in consistent safety properties over their useful life.[56] By utilizing low surface area carbons, electrodes with acceptable self-heating rates may be fabricated.

The first Li-ion batteries marketed by Sony utilized petroleum coke at the negative electrode. Coke-based materials offer good capacity, 180 mAh/g, and are stable in the presence of propylene carbonate (PC)-based electrolytes, in contrast to graphitic materials. The disorder in coke materials is thought to pin the layers inhibiting reaction or exfoliation in the presence of propylene carbonate.[56] In the mid-1990s most Li-ion cells utilized electrodes employing graphitic spheres, in particular a Mesocarbon Microbead (MCMB) carbon. MCMB carbon offers higher specific capacity, 300 mAh/g, and low surface area, thus providing low irreversible capacity and good safety properties. Recently, a wider variety of carbon types has been used in negative electrodes. Some cells utilize natural graphite, available at very low cost, while others utilize hard carbons that offer capacities higher than possible with graphitic materials.

Types of Carbon.   Many types of carbon materials are industrially available and the structure of the carbon greatly influences its electrochemical properties, including lithium intercalation capacity and potential. The basic building block for carbon materials is a planar sheet of carbon atoms arranged in a hexagonal array, as shown in Fig. 35.16. These sheets are stacked in a registered fashion in graphite. In Bernal graphite, the most common type, ABABAB stacking occurs, resulting in hexagonal or 2H graphite. In a less common polymorph, ABCABC stacking occurs, termed rhombohedral or 3R graphite.[3]

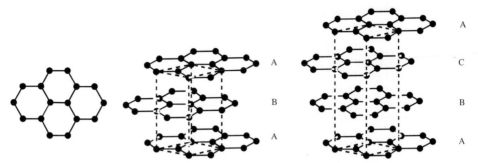

FIGURE 35.16 The hexagonal structure of a carbon layer and the structures of hexagonal (2H) and rhombohedral (3R) graphite.

Most real materials contain disorder, including the 2H and 3R stacking orders as well as random stacking, thus a more precise way to identify a graphite is to specify the relative fractions of 2H, 3R and random stacking. Forms of carbon have been developed with a range of stacking disorders and different morphologies. Stacking disorders include those where the graphitic planes are parallel but shifted or rotated, termed *turbostratic disorder,*[57] or those in which the planes are not parallel, termed *unorganized carbon.*[56] Particle morphologies range from the flat plates of natural graphites, to carbon fibers, to spheres.

Carbon materials can be considered as different aggregations of a basic structural unit (BSU) consisting of two or three parallel planes with a diameter of 2 nm.[58] The BSU's may be oriented randomly, resulting in carbon black, or oriented to a plane, axis or point, resulting in a planar graphite, a whisker or a spherule.

The types of carbon may alternatively be organized based on the type of precursor material, as illustrated in Fig. 35.17 as the precursor material, and the processing parameters determine the nature of carbon produced. Materials that can be graphitized by treatment at high temperature (2000°C to 3000°C) are termed *soft carbons.* Upon graphitization, the turbostratic disorder is removed and strain in the material relieved.[59] Hard carbons, such as those prepared from phenolic resin, cannot be readily graphitized, even when treated at 3000°C. Coke type materials are prepared at ~1000°C, typically from aromatic petroleum precursors.

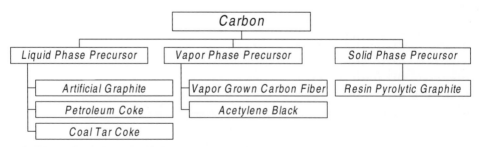

FIGURE 35.17 Carbons classified by the precursor phase.

Staging and Electrochemical Intercalation into Carbon. When lithium is intercalated into graphite, the ABAB structure transforms to an AAAA structure and distinct voltage plateaus are observed. This is illustrated in Fig. 35.18, which shows the voltage of a Li/graphite cell over one cycle at low rate for a highly ordered graphite. Voltage plateaus are observed upon lithium intercalation as distinct phases (stages) are formed.[60] A classical model of lithium staging is illustrated in Fig. 35.19. As shown, lithium forms "islands" within graphite instead of distributing homogeneously. The most lithium rich stage, LiC_6, is termed stage 1 and is formed at the lowest voltage, as indicated in Fig. 35.18. As lithium is removed from the graphite, higher stages are formed, as indicated in the figure.

In graphites used in Li-ion cells, less distinct stages are observed and a flat discharge profile results. In contrast, when petroleum coke or another disordered material is used, a continuous, sloping voltage profile is observed. This is illustrated in Fig. 35.20, which shows the first intercalation (charge) and deintercalation (discharge) for coke and artificial graphite. As shown, the coke material does not exhibit distinct stages and has a higher average voltage, 0.3 V versus lithium.

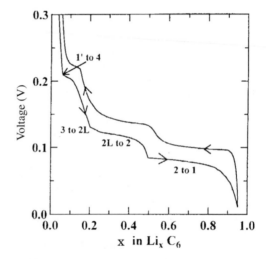

FIGURE 35.18 The voltage of a Li/graphite cell illustrating Li staging upon intercalation of graphite. (*Reproduced with permission from Physical Review B. From Ref. 59.*)

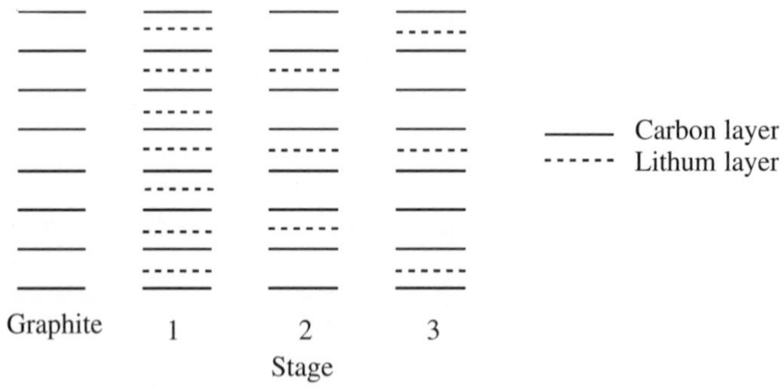

FIGURE 35.19 Schematic diagram of lithium staging in graphite.

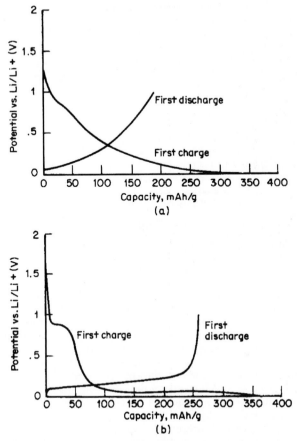

FIGURE 35.20 Potential of the carbon negative electrode in a Li-ion cell on the first cycle illustrating the irreversible capacity associated with (a) coke or (b) artificial graphite materials.

On the first cycle, passivation layers are formed on the surface of the electrodes. These layers have been shown to result from the reaction of the electrolyte with the electrode surface. The passivation layers contain lithium that is no longer electrochemically active, thus their formation results in irreversible capacity, an undesirable property of all current materials that occurs largely on the first cycle. The capacity difference between the charge and discharge curves in Fig. 35.20 results from irreversible capacity.

To emphasize the impact of the negative electrode material on cell voltage, Fig. 35.21 shows the discharge voltage of commercial 18650-type $C/LiCoO_2$ Li-ion cells with different electrode materials. As shown, cells with graphite negatives have flatter discharge curves than cells with coke negative electrodes. Since most commercial products now on the market have a flat discharge curve and high average voltage, they apparently use a graphitic negative electrode material.

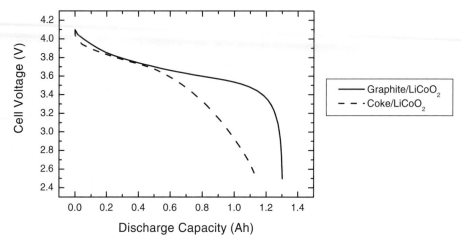

FIGURE 35.21   Effect of the carbon type on the discharge profile of Li-ion cells. (*Courtesy of the University of South Carolina.*)

Properties of Carbons.   Performance and physical characteristics of various carbons are included in Table 35.4. An ideal material would offer high specific capacity without irreversible capacity. The carbon used in the cells commercialized by Sony in 1990 was a petroleum coke. Cokes are compatible with a wide variety of electrolyte solvents, including PC, but have lower capacity than graphitic materials. MCMB carbons offer good capacity, ~300 mAh/g, and low irreversible capacity, ~20 mAh/g. Lower cost graphites offer higher capacity, ~350 mAh/g, but higher irreversible capacity, ~50 mAh/g, and have higher fade rates than MCMB carbons, thus not necessarily higher energy density.[61] This is illustrated in Fig. 35.22, which compares the reversible and irreversible capacity and energy density of two MCMB materials and an artificial graphite. In this case, the graphite offers higher capacity but also higher irreversible capacity than the MCMB's, thus intermediate energy density. In general, irreversible capacity may be correlated to the surface area of a material, thus the interest in low surface area, spherical materials. The MCMB 25-28 has lower specific surface area than MCMB 10-28, thus less irreversible capacity. In practice, a particle size less than ~30 μm is required for rate capability to the C rate. MCMB carbon can have a variety of structures, depending on how the graphite planes are oriented within the sphere. The performance of MCMB is related to its structure. The laboratory preparation and properties of a variety of MCMB carbons have been reported.[62]

TABLE 35.4   Properties and Performance of Various Carbons (Experimental Values). (*From Ref. 67.*)

Carbon	Type	Specific capacity (mAh/g)	Irreversible capacity (mAh/g)	Particle size $D_{50}(\mu m)$	BET surface area (m²/g)
KS6	Synthetic graphite	316	60	6	22
KS15	Synthetic graphite	350	190	15	14
KS44	Synthetic graphite	345	45	44	10
MCMB 25-28	Graphite sphere	305	19	26	0.86
MCMB 10-28	Graphite sphere	290	30	10	2.64
Sterling 2700	Graphitized Carbon Black	200	152	.075	30
XP30	Petroleum coke	220	55	45	N/A
Repsol LQNC	Needle coke	234	104	45	6.7
Grasker	Carbon fiber	363	35	23	11
Sugar carbon	Hard carbon	575	215	N/A	40

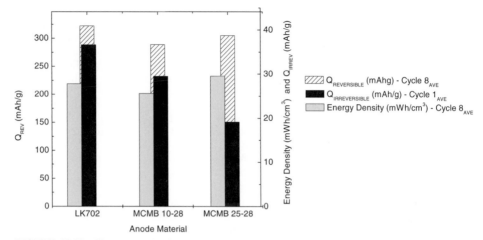

FIGURE 35.22 The energy density, reversible and irreversible capacity of carbons commonly used as negative electrode materials.

The theoretical specific capacity of carbon (LiC_6) is 372 mAh/g. Hard carbon materials offer higher capacity, over 1000 mAh/g, but have not achieved broad acceptance because they have greater irreversible capacity and higher voltage, ~1 V vs. lithium,[63] than graphitic materials. Hard carbons are highly disorganized. Mechanisms have been proposed to explain incorporation of Li above the theoretical capacity of graphite. That proposed by Sato suggests lithium occupies nearest neighbor sites between pairs of graphene sheets.[64] A mechanism proposed by Dahn et al. suggests additional lithium may bind hydrogen containing regions of carbon,[65] as supported by theoretical studies which illustrate the importance of hydrogen-terminated edge regions.[66]

35.2.5 Electrolytes

Four types of electrolytes have been used in Li-ion batteries: liquid electrolytes, gel electrolytes, polymer electrolytes and ceramic electrolytes. Liquid electrolytes are solutions of a lithium salt in organic solvents, typically carbonates. As described in a monograph,[68] a polymer electrolyte is a liquid- and solvent-free material, where an ionically conducting phase is formed by dissolving a salt in a high molecular weight polymer, whereas a gel electrolyte is an ionically conductive material wherein a salt and a solvent are dissolved or mixed with a high molecular weight polymer. Gel electrolytes developed for Li-ion batteries are typically films of PVDF-HFP, $LiPF_6$ or $LiBF_4$ salt, and carbonate solvent. Fumed silica may be added to the PVDF-HFP film for additional structural integrity. Potential advantages of polymer electrolytes include improved safety properties resulting from their low volatility and high viscosity, as they do not contain a volatile, flammable solvent component. A possible advantage of gel electrolytes is that the liquid phase is absorbed within the polymer, thus less likely to leak from a battery, however, in a typical Li-ion battery employing a liquid electrolyte, the electrolyte is almost completely absorbed into the electrode and separator materials. In the marketplace and the literature, gel electrolytes are often termed gel-polymer electrolytes, and cells that employ gel (or gel-polymer) electrolytes are termed gel-polymer or simply polymer cells. Ceramic electrolytes refer to inorganic, solid-state materials that are ionically conductive.

Most Li-ion electrolytes in current use utilize $LiPF_6$ as the salt as its solutions offer high ionic conductivity, $>10^{-3}$ S/cm, high lithium ion transference number (~0.35), and acceptable safety properties. As reviewed below, many other salts have attracted industrial interest, notably $LiBF_4$. Electrolytes in current use are formulated with carbonate solvents. Carbonates are aprotic, polar, and have a high dielectric, thus can solvate lithium salts to high concentration (>1 M). They also provide compatibility with cell electrode materials over a broad range of potential. While the industry focused initially on propylene carbonate (PC)-based solutions, current formulations utilize other carbonates, notably ethylene carbonate (EC), dimethyl carbonate (DMC), ethyl methyl carbonate (EMC) and diethyl carbonate (DEC), as PC causes degradation in graphite electrodes as it co-intercalates with lithium, resulting in exfoliation. The choice of solvents for a Li-ion electrolyte is also influenced by any low temperature requirements of the application. Low temperature electrolytes utilize low viscosity solutions with low freezing points.

Salts. Salts commonly used in Li-ion cells are listed in Table 35.5. Most cells currently marketed use $LiPF_6$ as its solutions have high conductivity and good safety properties. However, the salt is costly, hygroscopic, and $LiPF_6$ yields hydrofluoric acid (HF) upon reaction with water, thus must be handled in a dry environment. Organic salts have also been developed. They are more stable to water, thus easier to handle. In particular, BETI (lithium bisperfluoroethanesulfonimide) has received significant attention as its solutions offer high conductivity, it is stable to water, it can be easily dried, and it does not cause aluminum corrosion—an issue with other organic salts such as lithium triflate.

TABLE 35.5 Salts Used in Electrolytes for Li-ion Cells

Common name	Formula	Mol. wt. (g/mol)	Typical impurities	Comments
Lithium hexafluorophosphate	$LiPF_6$	151.9	H_2O (15ppm) HF (100ppm)	Most commonly used
Lithium tetrafluoroborate	$LiBF_4$	93.74	H_2O (15ppm) HF (75ppm)	Less hygroscopic than $LiPF_6$
Lithium perchlorate	$LiClO_4$	106.39	H_2O (15ppm) HF (75ppm)	When dry, less stable than alternatives
Lithium hexafluoroarsenate	$LiAsF_6$	195.85	H_2O (75ppm) HF (15ppm)	Contains arsenic
Lithium triflate	$LiSO_3CF_3$	156.01	H_2O (100ppm)	Al corrosion above 2.8 V, stable to water
Lithium bisperfluoroethane-sulfonimide (BETI)	$LiN(SO_2C_2F_5)_2$	387	N/A	No Al corrosion below 4.4 V, stable to water

Solvents. A wide variety of solvents, including carbonates, ethers and acetates, has been evaluated for non-aqueous electrolytes. The industry has now focused on the carbonates as they offer excellent stability, good safety properties and compatibility with electrode materials. Neat carbonate solvents typically have intrinsic solution conductivity less than 10^{-7} S/cm, dielectric constant >3, and solvate lithium salts to high concentration. Table 35.6 presents the properties of some commonly used solvents. The properties of these and other materials are included in a recent review.[69]

TABLE 35.6 Characteristics of Organic Solvents

Characteristic	EC	PC	DMC	EMC	DEC	1,2-DME	AN	THF	γ-BL
Structure									
BP (°C)	248	242	90	109	126	84	81	66	206
MP (°C)	39	−48	4	−55	−43	−58	−46	−108	−43
Density (g/ml)	1.41	1.21	1.07	1.0	0.97	0.87	0.78	0.89	1.13
Viscosity (cP)	1.86 (40°C)	2.5	0.59	0.65	0.75	0.455	0.34	0.48	1.75
Dielectric constant	89.6 (40°C)	64.4	3.12	2.9	2.82	7.2	38.8	7.75	39
Donor number	16.4	15	8.7[70]	6.5[70]	8[70]	—	14	—	—
Mol. wt.	88.1	102.1	90.1	104.1	118.1	90.1	41.0	72.1	86.1

*EC = ethylene carbonate, PC = propylene carbonate, DMC = dimethyl carbonate, EMC = ethyl methyl carbonate, DEC = diethyl carbonate, DME = dimethylether, AN=acetonitrile, THF=tetrahydrofuran, γBL = γ-butyrolactone.
(*From Refs. 69, 70, 71 and 72.*)

Conductivity of Electrolytes.   Electrolyte formulations in current Li-ion cells typically utilize two to four solvents. Formulations with multiple solvents can provide better cell performance, higher conductivity, and a broader temperature range than possible with a single
solvent electrolyte. For example, ethylene carbonate (EC) is associated with low irreversible
capacity and low capacity fade when used in conjunction with graphitic negative electrodes.
Ethylene carbonate is found in many commercial electrolyte formulations, but is a solid at
room temperature. Multiple solvent formulations often include EC, thereby incorporating its
desirable properties, while using other solvents to lower the freezing point and viscosity of
the mixture.

The conductivity of 1 M $LiPF_6$ solutions in solvents commonly used in Li-ion electrolytes
is given in Table 35.7 and plotted in Fig. 35.23 for temperatures between $-40°C$ and $80°C$.
In general these solutions offer high conductivity, $10^{-2} S/cm$, and a few solvents, such as
PC and EMC, offer good conductivity at low temperature and a high boiling point. MA and
MF offer high conductivity, but cell performance is poor if either is used at levels greater
than 25% by weight.

TABLE 35.7   Conductivity, in mS/cm, of 1M $LiPF_6$ Solutions in Various Solvents.
(*Courtesy of Merck KGaA, Darmstadt, Germany.*)

Solvent	$-40.0°C$	$-20.0°C$	$0.0°C$	$20.0°C$	$40.0°C$	$60.0°C$	$80.0°C$
DEC	—	1.4	2.1	2.9	3.6	4.3	4.9
EMC	1.1	2.2	3.2	4.3	5.2	6.2	7.1
PC	0.2	1.1	2.8	5.2	8.4	12.2	16.3
DMC	—	1.4	4.7	6.5	7.9	9.1	10.0
EC	—	—	—	6.9	10.6	15.5	20.6
MA	8.3	12.0	14.9	17.1	18.7	20.0	—
MF	15.8	20.8	25.0	28.3	—	—	—

*DEC = diethyl carbonate, EMC = ethyl methyl carbonate, PC = propylene carbonate, DMC =
dimethyl carbonate, EC=ethylene carbonate, MA=methyl acetate, MF = methyl formate.

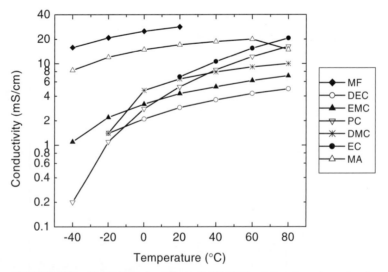

FIGURE 35.23   Conductivity, in mS/cm, of 1M $LiPF_6$ solutions in various solvents.
(*Courtesy of Merck KGaA, Darmstadt, Germany.*)

The conductivity of binary 1:1 mixtures of EC with common Li-ion electrolyte solvents over a range of salt concentrations and temperatures is given in Table 35.8. As indicated, for many solvent pairs, conductivity is highest with 1 M LiPF$_6$, and these formulations are liquid from −40°C to 80°C. The conductivity of 1 M LiPF$_6$ binary solutions with EC are plotted in Fig. 35.24. As shown, the EC:MA solution offers the highest conductivity although such levels of MA are associated with high capacity fade.[74] Other mixtures, including EC: DEC, EC:DMC, and EC:EMC offer good conductivity and low capacity fade. In particular, EC:EMC mixtures offer 0.9 mS/cm conductivity at −40°C, and low capacity fade.

The conductivity of PC:DME 1:1 solutions of LiPF$_6$, and a variety of organic salts, is illustrated in Fig. 35.25, which shows the conductivity of the solutions from 0.25 M to 2 M. As shown, LiPF$_6$ offers the highest conductivity, 13 mS/cm at 1.2 M, although solutions with the organic salts offer comparable conductivity, up to 11 mS/cm.

TABLE 35.8 Conductivity, in mS/cm, of LiPF$_6$ Solutions in Binary Mixtures, 1:1 by Weight. C = partially crystallized, S = saturated. (*Courtesy of Merck KGaA, Darmstadt, Germany.*)

Solvents	Concentration	−40°C	−20°C	0°C	20°C	40°C	60°C	80°C
EC:DEC	0.25 M	—	—	1.7 (C)	4.2	5.8	7.3	8.8
	0.50 M	—	2.5 (C)	3.0	6.4	8.7	11.1	13.6
	1.00 M	0.7	2.2	4.2	7.0	10.3	13.9	17.5
	1.25 M	0.4	1.7	3.6	6.4	9.7	13.5	17.4
	1.50 M	—	—	—	5.6	—	—	—
	1.75 M	—	—	—	4.8 (S)	—	—	—
EC:DMC	0.25 M	—	—	4.2	5.8	7.8	9.7	11.5
	0.50 M	—	—	6.5	9.3	12.8	16.0	19.1
	0.75 M	—	3.8	6.9	10.3	14.0	17.9	21.6
	1.00 M	—	3.7	7.0		15.0	19.5	24.0
	1.25 M	0.7	2.7	5.6	9.3	13.7	18.4	23.3
	1.50 M	—	2.2	5.4	9.3	14.1	19.2	24.7
	1.75 M	—	—	—	7.5	—	—	—
	2.00 M	—	—	—	6.7	—	—	—
	2.25 M	—	—	—	0.9 (S)	—	—	—
EC:EMC	0.25 M	—	—	3.7	5.3	7.2	9.1	10.9
	0.50 M	—	3.0	5.1	7.5	10.2	12.8	15.4
	1.00 M	0.9	2.7	5.3	8.5	12.2	16.3	20.3
	1.25 M	0.6	2.3	4.7	8.0	12.0	16.2	20.6
	3.50 M	—	—	—	0.9 (S)	—	—	—
EC:MA	0.25 M	2.4 (C)	4.6	6.3	8.3	10.4	12.4	—
	0.50 M	3.1 (C)	6.7	9.8	13.1	16.0	19.3	—
	1.00 M	3.8	7.8	12.2	17.1	22.3	27.3	—
	1.25 M	—	7.1	11.8	17.2	22.7	28.4	—
	3.0 M	—	0.5	2.1	5.2	—	15.4	21.8
	3.5 M	—	—	—	3.4 (S)	—	—	—
EC:MPC	1.00 M	C	1.5	3.6	6.3	9.5	12.9	16.8

*DEC = diethyl carbonate, EMC = ethyl methyl carbonate, DMC = dimethyl carbonate, EC = ethylene carbonate, MA = methyl acetate.

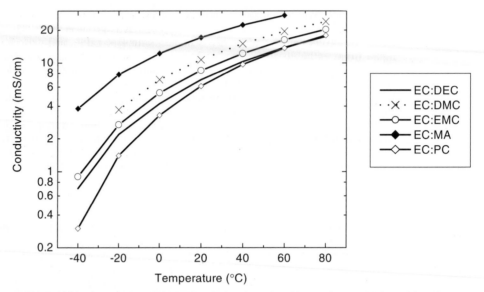

FIGURE 35.24   Conductivity of 1M LiPF$_6$ solutions in various binary mixtures, 1:1 by weight. (*Courtesy of Merck KGaA, Darmstadt, Germany.*)

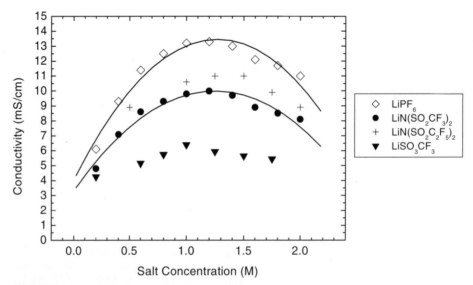

FIGURE 35.25   Conductivity of electrolytes at 20°C prepared with PC:DME (1:1, V/V) and various salts. (*Courtesy of 3M, St. Paul, MN.*)

The conductivity of selected ternary 1 M LiPF$_6$ solutions is given in Table 35.9 and plotted in Fig. 35.26. These mixtures contain 33% EC, as is typical in current Li-ion electrolytes, but still offer high conductivity and a broad temperature range, illustrating the utility of multiple component mixtures. For example, four of these mixtures offer at least 1 mS/cm at −40°C, of which three are liquid at 80°C. Quaternary mixtures have also been developed to provide electrolytes with better low temperature performance. The conductivity of 1.0 M LiPF$_6$ solutions in various quaternary solvent mixtures is shown in Fig. 35.27. As illustrated, these solutions offer over 1 mS/cm at −40°C and up to 0.6 mS/cm at −60°C.

A survey of the conductivity and solvent properties of electrolytes utilizing either LiAsF$_6$, LiPF$_6$, LiSO$_3$CF$_3$, or LiN(SO$_2$CF$_3$)$_2$ and a wide variety of solvents, including EC, DME, ethyldiglyme, triglyme, tetraglyme, sulfolane, Freon, and methylene chloride has been published.[73]

TABLE 35.9 Conductivity, in mS/cm, of M LiPF$_6$ Solutions in Various Ternary Solvent Mixtures. (*Courtesy of Merck KGaA, Darmstadt, Germany.*)

Solvent	Wt. ratio	−40°C	−20°C	0°C	20°C	40°C	60°C	80°C
EC:PC:DMC	20:20:60	—	—	6.9	10.6	14.5	18.4	22.2
EC:PC:EA	15:25:60	3	6.2	9.8	13.7	17.8	21.6	25.1
EC:PC:EMC	15:25:60	1	2.8	5.3	8.1	11.5	14.6	17.8
EC:PC:MA	15:25:60	4.1	8.1	12.9	17.8	22.8	27.6	boils
EC:PC:MPC	15:25:60	0.5	1.4	3.3	5.6	8.2	10.9	13.9
EC:DMC:EMC	15:25:60	1.4	3.2	5.3	7.6	10	12.1	14.1
EC:DMC:MPC	15:25:60	0.7	1.8	3.4	5.3	7.2	9	10.9

*DEC = diethyl carbonate, EMC = ethyl methyl carbonate, PC = propylene carbonate, DMC = dimethyl carbonate, EC = ethylene carbonate, MA = methyl acetate, MPC = methyl propyl carbonate, EA = ethyl acetate.

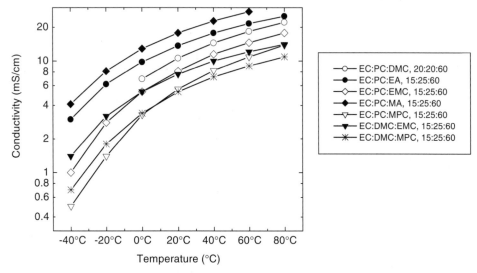

FIGURE 35.26 Conductivity of 1 M LiPF$_6$ solutions in ternary solvent mixtures. (*Courtesy of Merck KGaA, Darmstadt, Germany.*)

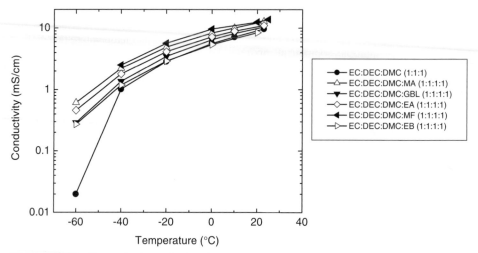

FIGURE 35.27   The conductivity of 1.0 M LiPF$_6$ solutions in quaternary solvent mixtures. (*Courtesy of JPL.*)

Electrolyte Formulation, Irreversible Capacity and the SEI.   Various electrolytes have been used in Li-ion batteries. The solvents used must be stable at both the anodic and cathodic potentials found in Li-ion cells, 0 V to 4.2 V vs. lithium. No practical solvents are thermodynamically stable with lithium or Li$_x$C$_6$ near 0 V vs. Li, but many solvents undergo a limited reaction to form a passivation film on the electrode surface. This film spatially separates the solvent from the electrode, yet is ionically conductive, and thus allows passage of lithium ions. The passivation film, termed the solid electrolyte interphase (SEI), therefore imparts extrinsic stability to the system allowing the fabrication of cells that are stable for years without significant degradation.[74]

When the SEI is formed, lithium is incorporated into the passivation film. This process is irreversible and is thus observed as a loss of capacity, primarily on a cell's first cycle. The amount of irreversible capacity is dependent on the electrolyte formulation and the electrode materials, in particular the type of carbon used in the negative electrode. As the reaction occurs at the particle surface, materials with low specific surface area typically offer lower irreversible capacity.

Cells with electrolyte formulations that contain alkyl carbonates, in particular EC, have been shown to offer low capacity fade, low irreversible capacity and high capacity.[75] In EC containing electrolytes, the passivation film formed on the surface of Li-ion electrodes is formed with a minimum amount of lithium. This SEI has been shown to consist primarily of Li$_2$(OCO$_2$(CH$_2$)$_2$OCO$_2$)$_2$,[76] and related reaction products, including Li$_2$CO$_3$ and LiOCH$_3$,[77] of the electrolyte solvent with either lithium or a lithiated species such as Li$_x$C$_6$. While solvents other than EC, typically esters or alkyl carbonates such as EMC or MPC,[78] also form stable passivation films, most solvents do not. If an ester or alkyl carbonate is not used, graphite can be cycled in a solvent that does not form a stable passivation film if an additive, such as a crown ether[79] or CO$_2$,[76,80] is added to the electrolyte.

35.2.6 Separator Materials

Li-ion cells use thin (10 to 30 μm), microporous films to electrically isolate the positive and negative electrodes. To date, all commercially available liquid electrolyte cells use microporous polyolefin materials as they provide excellent mechanical properties, chemical stability and acceptable cost. Nonwoven materials have also been developed but have not been widely accepted, in part due to the difficulty in fabricating thin materials with uniform, high strength.[81]

Requirements for Li-ion separators include:

- High machine direction strength to permit automated winding
- Does not yield or shrink in width
- Resistant to puncture by electrode materials
- Effective pore size less than 1 μm
- Easily wetted by electrolyte
- Compatible and stable in contact with electrolyte and electrode materials

Microporous polyolefin materials in current use are made of polyethylene, polypropylene or laminates of polyethylene and polypropylene. Also available are surfactant coated materials, designed to offer improved wetting by the electrolyte. These materials are fabricated by either a dry, extrusion type process or a wet, solvent based process.[82] The properties of commercial materials, including pore dimensions, porosity, and permeability, have been reported.[83] Commercial materials offer pore size of 0.03 μm to 0.1 μm, and 30 to 50% porosity, as illustrated by the SEM micrograph of a commercial material in Fig. 35.28.

FIGURE 35.28 SEM micrograph of Celgard 3501 separator. (*Courtesy of Yardney Technical Products, Inc.*)

The low melting point of polyethylene (PE) materials enables their use as a thermal fuse. As the temperature approaches the melting point of the polymer, 135°C for polyethylene and 165°C for polypropylene (PPE), porosity is lost.[84] Tri-layer materials (PPE/PE/PPE) have been developed where a polypropylene layer is designed to maintain the integrity of the film, while the low melting point of polyethylene layers is intended to shutdown the cell if an over-temperature condition is reached.

35.2.7 Additives

To further improve battery performance, electrolyte additives have been developed. Some, such as BF_3,[85] and related complexes, are designed to passivate the surface of electrode materials thereby reducing their propensity to degrade. Others, such as hexamethyldisilazane (HMDS), have been used to reduce interfacial resistance,[86] and react with and immobilize water and HF, thus improving cell performance:[87]

$$(CH_3)_3SiN(H)Si(CH_3)_3 + H_2O \rightarrow (CH_3)_3SiOSi(CH_3)_3 + NH_3$$

$$NH_3 + HF \rightarrow NH_4F$$

To illustrate the utility of one such additive, Fig. 35.29 shows the specific capacity of a Li/LiMn$_2$O$_4$ battery cycled at 55°C using an electrolyte with and without HMDS additive. As shown, the cell with the HMDS additive provided lower capacity fade in this case, as the additive removed the water and HF impurities that contribute to the degradation of the electrode materials.

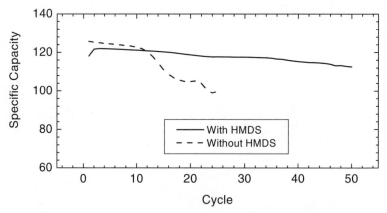

FIGURE 35.29   The specific capacity of C/LiMn$_2$O$_4$ battery when cycled at 55°C in an electrolyte with and without HMDS additive. Cycled at C/2 rate, 4.2 V to 3.5 V. (*Courtesy of Carus Chemical.*)

35.3 *CONSTRUCTION OF CYLINDRICAL AND PRISMATIC Li-ION CELLS AND BATTERIES*

Cylindrical and prismatic Li-ion batteries have been developed. Wound designs are typical in small cells (<4 Ah); however in large cell designs, prismatic configurations with flat plate construction are more common. For prismatic Li-ion batteries two cell design types are practiced, flat mandrel wound pseudo-prismatic designs and flat-plate true prismatic designs.

Since Li-ion cells are fabricated in the discharged state, they must be charged before use.

35.3.1 Construction of Wound Li-ion Cells

The construction of a cylindrical wound Li-ion cell is illustrated in Fig. 35.30. The fabrication of wound prismatic cells is similar to cylindrical versions except that a flat mandrel is used instead of a cylindrical mandrel. A schematic diagram of a wound prismatic cell is shown in Fig. 35.31. The construction consists of a positive and negative electrode separated by a 16 μm to 25 μm thick microporous polyethylene or polypropylene separator. Positive electrodes consist of 10 μm to 25 μm Al foil coated with the active material to a total thickness typically ~180 μm. Negative electrodes are typically 10 μm to 20 μm Cu foil coated with a carbonaceous active material to a total thickness ~200 μm. The thin coatings and separator are required because of the low conductivity of non-aqueous electrolytes, ~10 mS/cm,[88] and slow Li$^+$ diffusion in the positive and negative electrode materials, about ~10^{-10} m^{2}s^{-1}. Typically a single tab at the end of the wind is used to connect the current collectors to their respective terminals. The case, commonly used as the negative terminal, is typically Ni-plated steel. When used as the positive terminal, the case is typically aluminum. Most commercially available cells utilize a header that incorporates disconnect devices, activated by pressure or temperature, such as a PTC device, and a safety vent. One design is illustrated in Fig. 35.32. These devices can limit cell performance at high rates, i.e. in a typical 18650 cell, 12 A discharge results in disconnect after 20 seconds once the temperature of the disconnect device reaches 70°C due to resistive heating. The header-can seal is typically formed through a crimp.

In a typical 18650 cell, the positive current collector is coated on both sides with 12 g of LiCoO$_2$ resulting in an electrode thickness of ~7.0 mil. Typically 6.5 g of carbon is used on the negative electrode and the positive to negative ratio is such that the carbon negative is utilized at a maximum of 270 mAh/g, 10% less than the capacity typical for MCMB carbon and 100 mAh/g less than the theoretical capacity of carbon, (372 mAh/g).[89] The mass distribution for the components in two 18650 products is illustrated in Fig. 35.33.

For specialized applications, such as for satellites, larger cylindrical cells have been developed. The "25 Ah" cells developed by Blue Star Advanced Technology, are depicted in Fig. 35.34. These products utilize a LiCoO$_2$ based positive and a graphite negative. As shown, the header incorporates two glass-to-metal seal terminals, a rupture disk and a fill port. The mass of the major components of one cell is described in Table 35.10. As indicated, the cell container accounts for 14% of the mass while the electrodes and the electrolyte account for 81% of the cell mass. This cell delivers 121 Wh/kg and 280 Wh/L, slightly less than the lower aspect "25 Ah Design II" cell which offers 125 Wh/Kg and 265 Wh/L.

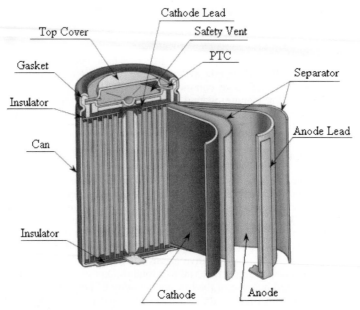

FIGURE 35.30 Cross-sectional view of a cylindrical Li-ion cell. (*Courtesy of the University of South Carolina. Reproduced with permission from the Journal of Power Sources.*)

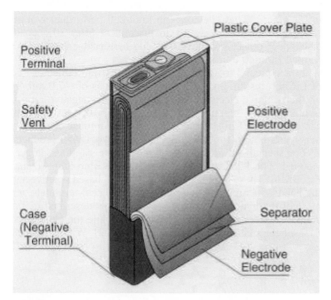

FIGURE 35.31 Schematic drawing of a wound prismatic cell. (*Courtesy of Japan Storage Battery Co., Ltd.*)

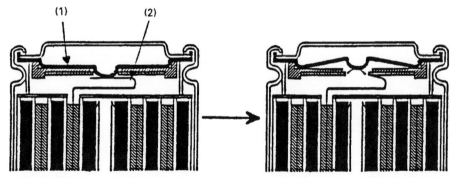

FIGURE 35.32 Detail of the construction of a cell header with a breaker and vent mechanism for an abnormal rise of internal pressure, (1) Aluminum burst disk, (2) Aluminum lead. (*Courtesy of Sony Corp.*)

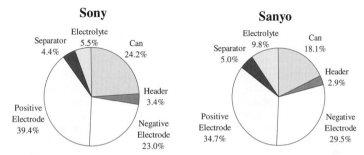

FIGURE 35.33 Mass distribution in 18650 Li-ion products. The batteries weighed 39.4 g and 37.7 g respectively. (*Courtesy of the University of South Carolina.*)

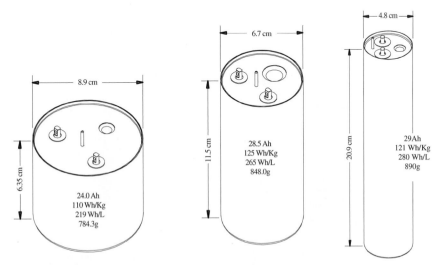

FIGURE 35.34 25 Ah cylindrical Li-ion cells. (*Courtesy Blue Star Advanced Technologies. Reproduced with permission from SAE paper 01-1390, 1999.*)

TABLE 35.10 Mass Analysis of a 29 Ah Cylindrical Cell. (*Courtesy of Blue Star Advanced Technologies. Reproduced with permission from SAE paper 01-2640, 1999.*)

Component	Mass (g)	% of total
Case	108.5	12.2
Cap assembly	15.5	1.8
Electrolyte	217.9	24.5
Positive electrode	339.2	38.1
Negative electrode	165.0	18.5
Miscellaneous	43.4	4.9
Total	889.5	100

35.3.2 Construction of Flate-plate Prismatic Li-ion Batteries

The construction of flat-plate prismatic cells is illustrated in Fig. 35.35. As in a wound cell, a microporous polyethylene or polypropylene separator separates the positive and negative electrodes. Typically each plate in the cell has a tab, the tabs are bundled and welded to their respective terminals or to the cell case. Cell cases of either nickel-plated steel or 304L stainless steel have been used. As shown, the cover typically incorporates one or two terminals, a fill port and a rupture disk. The terminal may be a glass-to-metal seal, for low cost applications compression type seals have been used, or the terminal may incorporate devices similar to those found in the header of cylindrical products to provide pressure, temperature and over current interrupt in one component. The case to cover seal is typically formed either by TIG or laser welding.

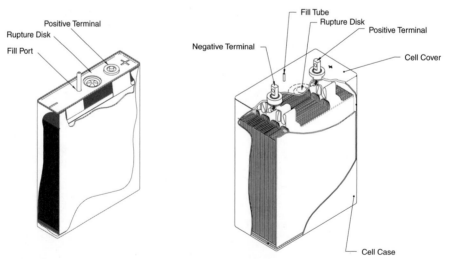

FIGURE 35.35 Schematic view showing the header and electrodes of 7 Ah (case negative) and 40 Ah (case neutral) flat plate prismatic Li-ion cells. (*Courtesy of Yardney Technical Products, Inc.*)

35.4   Li-ION BATTERY PERFORMANCE

The general performance characteristics of Li-ion batteries are outlined in Table 35.11. As indicated in the table, Li-ion batteries have a high voltage, typically operating in the range of 2.5 to 4.2 V, approximately three times that of NiCd or NiMH. As such, fewer cells are required for a battery of a given voltage. Li-ion batteries offer high specific energy and energy density, batteries with specific energy over 150 Wh/Kg and energy density over 400 Wh/L are commercially available. Multiple-tabbed Li-ion batteries also offer high rate capability, up to 5C continuous or 25C pulse, thus high power density, and low self-discharge rate, years of calendar life, no memory effect, and a broad temperature range of operation. Li-ion batteries can be charged from −20°C to 60°C and discharged from -40°C to 65°C. The combination of these qualities within a cost effective, hermetic package has enabled the diverse applicability of the technology.

TABLE 35.11   General Performance Characteristics of Li-ion Batteries

Characteristic	Performance range
Operational cell voltage	4.2 to 2.5 V
Specific energy	100 to 158 Wh/kg
Energy density	245 to 430 Wh/L
Continuous rate capability	Typical: 1C
	High rate: 5C
Pulse rate capability	Up to 25C
Cycle life at 100% DOD	Typically 3000
Cycle life at 20 to 40% DOD	Over 20000
Calendar life	Over 5 years
Self discharge rate	2 to 10%/month
Operable temperature range	−40°C to 65°C
Memory effect	None
Power density	2000 to 3000 W/L
Specific power	700 to 1300 W/Kg

35.4.1   Characteristics and Sizes of Lithium Ion Batteries

As illustrated in Table 35.12 and Table 35.13, Li-ion batteries are available in a wide range of sizes from 0.6 to 160 Ah in both cylindrical and prismatic designs with a range of aspect ratios. Li-ion battery performance has steadily improved, in the period 1996 to 1999, the specific energy of 18650-type cells increased 8% per year while energy density increased 14% per year, on average.

TABLE 35.12   Electrical and Physical Characteristics of Typical Cylindrical $C/LiCoO_2$ or $C/LiNi_{1-x}Co_xO_2$ Li-ion Batteries

Type	14500	14650	17500	17670	18500	18650	26650	33600
Height, mm	50.0	65	50	67	50	65	65	60
Diameter, mm	14	14	17	17	18	18	26	33
Volume, ml.	7.7	10	11.3	15.2	12.7	16.5	34.5	51.4
Mass, g.	19	26	25	35	31	42	93	125
Capacity (Ah)	.65	.90	0.83	1.25	1.1	1.8	3.2	5.0
Specific energy (Wh/kg)	126	128	123	132	131	155	354	150.4
Energy density (Wh/L)	312	333	273	306	320	410	131	366

TABLE 35.13 Electrical and Physical Characteristics of Typical Prismatic C/LiCoO$_2$ or C/LiNi$_{1-x}$Co$_x$O$_2$ Li-ion Batteries

Cell type	61/19/48	46/30/48	55/30/48	81/31/48	65/35/67	103/34/50	160/61/78	280/95/151	460/89/128	500/130/208
Shape	Prismatic (thickness (0.1 mm)/width (mm)/height (mm))									
Height, mm	48	48	48	48	67.3	50	78	151	128	208
Width, mm	19.5	30	30	30.5	35.1	34.1	61	95	89	130
Thickness, mm	6.1	4.6	5.5	8.1	6.5	10.3	16	28	46	50
Volume, ml.	5.70	6.6	7.9	11.9	15.3	18.4	76	136	465	1352
Mass, g.	11.5	14	16	24	33	38	185	870	1108	3650
Capacity (Ah)	0.42	.52	0.6	0.9	.96	1.5	7.0	35	40	160
Specific energy (Wh/kg)	135	137	138	139	104	146	145	145	156	160
Energy density (Wh/L)	272	291	281	280	226	301	345	344	372	430

The most significant challenges to the broader application of Li-ion technologies are related to either stability at high temperature or safety. While batteries may be exposed to temperatures as high as 70°C for short periods, the rate of degradation of current Li-ion batteries is significant above 60°C. Li-ion batteries are generally safe, although venting can occur if they are overcharged or crushed. To vent from overcharge, batteries must typically be charged to greater than 200% of their rated capacity. Protective devices are employed to prevent ventings under abusive conditions.

In current Li-ion batteries, the overcharge, overdischarge and over temperature issues have been largely addressed by the incorporation of management circuits into batteries to provide protection from over-charge, over-discharge or over-temperature. In addition, the circuits may also perform fuel gauge functions and can record battery history. The controller in the management circuit typically monitors the voltage of each cell or string of cells in a battery. In addition, circuits typically include a thermistor or other thermostat for restorable over-temperature control and a thermal fuse for non-restorable over-temperature control. Additional information on battery management circuits is found in Sec. 5.6.

35.4.2 Performance of Commerical Cylindrical Batteries

Discharge Rate Capability.   The rate capability and capacity of Li-ion batteries are dependent on their design, and varies considerably between manufacturers. Discharge curves for 18650-type $C/LiCoO_2$ batteries discharged at rates from 0.33 A to 3.6 A at 21°C are shown in Fig. 35.36. At low rates (C/2), the battery provided 1.67 Ah, and at 1.65 A (the 1 C rate) 1.60 Ah while at the C rate the average voltage was 3.5 V.

Batteries utilizing the $C/LiMn_2O_4$ cell chemistry have also been commercialized. The spinel cathode materials, as described in the section on cathode materials, are regarded as providing lower cost and more benign safety properties relative to $LiCoO_2$, although they possess lower specific capacity. Shown in Fig. 35.37 is the rate capability and capacity of 18650-type $C/LiMn_2O_4$ batteries. The products provide 1.44 Ah at the C/5 rate (0.28 A) and 1.34 Ah at the C rate (1.4 A), comparable to, but less than the capacity of $C/LiCoO_2$ batteries. The discharge curves have a flat voltage profile with average voltage at the C rate of 3.65 V, 0.15 V higher than $C/LiCoO_2$ cells at the C rate.

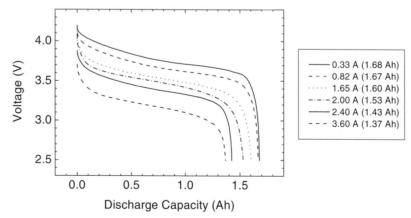

FIGURE 35.36   Discharge capability of 18650 type $C/LiCoO_2$ batteries at constant current at 21°C to 25°C. (*Courtesy of the University of South Carolina and NEC Moli Energy.*)

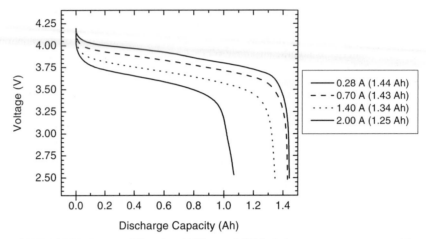

FIGURE 35.37 Rate capability of an 18650-type $C/LiMn_2O_4$ battery at constant current. The battery was charged in a CCCV (constant current-constant voltage) regime at 1.4 A to 4.2 V for 2.5 hours, then discharged at 21°C. (*Courtesy of NEC Moli Energy.*)

The performance of Li-ion batteries under constant power discharge is illustrated in Fig. 35.38 for 18650-type $C/LiCoO_2$ batteries, and in Fig. 35.39 for an 18650-type $C/LiMn_2O_4$ battery. At 1 W the $C/LiCoO_2$ battery provided 6.3 Wh, or 145 Wh/kg and 380 Wh/L. At 1 W the $C/LiMn_2O_4$ battery provided 5.5 Wh, or 130 Wh/kg and 332 Wh/L.

The rate capability of several 18650-type batteries is compared in Fig. 35.40 which illustrates the higher capacity of newer products and the 3C rate capability of the A&T LSR18650 product. The average discharge voltage is lowered significantly at the higher rates.

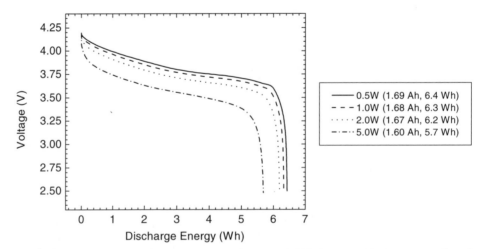

FIGURE 35.38 Constant power discharge of an 18650-type $C/LiCoO_2$ battery. The battery was charged in a CCCV regime at 1.65 A to 4.2 V for 2.5 hours at 21°C. (*Courtesy of NEC Moli Energy.*)

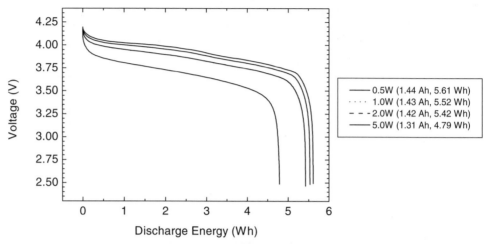

FIGURE 35.39 Constant power discharge of a $C/LiMn_2O_4$ 18650 battery. The battery was charged in a CCCV regime at 1.4 A to 4.2 V for 2.5 hours at 21°C. (*Courtesy of NEC Moli Energy.*)

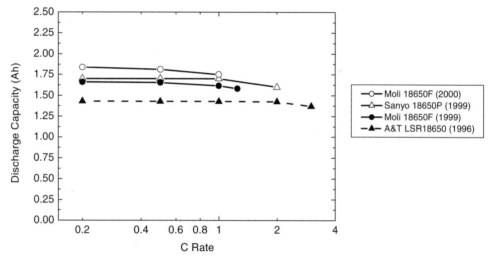

FIGURE 35.40 Comparison of the rate capability at constant current for $C/LiCoO_2$ 18650 batteries.

High and Low Temperature Discharge.   The voltage and capacity on discharge at the C/5 rate for 18650-type $C/LiMn_2O_4$ and $C/LiCoO_2$ batteries at high and low temperatures are illustrated in Fig. 35.41, for a $C/LiMn_2O_4$ battery, and in Fig. 35.42 for a $C/LiCoO_2$ product. Comparing the $C/LiMn_2O_4$ and $C/LiCoO_2$ batteries, the spinel product provides nominally 0.25 V higher voltage, although the $C/LiCoO_2$ product provides typically 20% more coulombic capacity at any given temperature than the $C/LiMn_2O_4$ unit. Between $-10°C$ and 60°C, the $C/LiMn_2O_4$ 18650 battery delivered 1.4 Ah with an average voltage of 3.9 V above 20°C, and 3.7 V at $-10°C$. At $-20°C$, the $C/LiMn_2O_4$ product provided 1.3 Ah at 3.5 V, on average.

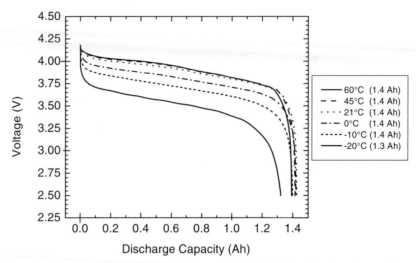

FIGURE 35.41 Discharge of an 18650-type $C/LiMn_2O_4$ battery at various temperatures when charged in a CCCV (constant current-constant voltage) regime at 1.4 A to 4.2 V for 2.5 hours and discharged at 0.28 A (C/5). The average voltage at 21°C was 3.9 V, at −20°C, 3.5 V. (*Courtesy of NEC Moli Energy.*)

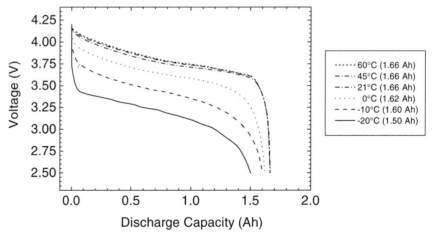

FIGURE 35.42 Discharge at 0.2C (0.33 A) of an 18650-type $C/LiCoO_2$ battery at various temperatures when charged in a CCCV regime at 1.65 A to 4.2 V for 2.5 hours and discharged at 0.33 A (C/5 rate). The average voltage at 21°C was 3.6 V, at −20°C, 3.2 V. (*Courtesy of NEC Moli Energy.*)

As illustrated by Fig. 35.42, $C/LiCoO_2$ batteries provided similar capacities over the range of temperatures evaluated. At or above −10°C, the battery provided 1.6 Ah and 1.5 Ah at −20°C. The average voltage at 21°C was 3.6 V, and at −20°C, 3.2 V.

At high rates, self-heating effects are observed upon discharge at low temperature. This is illustrated in Fig. 35.43 which shows discharge curves at 1.5 A (0.9C rate) for an 18650-type $C/LiCoO_2$ battery at temperatures from 60°C to −20°C. At 21°C or above, the discharge curves appear similar to those in Fig. 35.42 for the C/5 rate as the battery delivered similar capacity (1.64 Ah) with similar average voltage (3.6 V). At −10°C and −20°C, the voltage recovered in mid-discharge as self-heating ensued. At −10°C, 1.2 Ah was delivered at 3.0 V on average. At −20°C, 1.0 Ah was delivered at 2.8 V, on average.

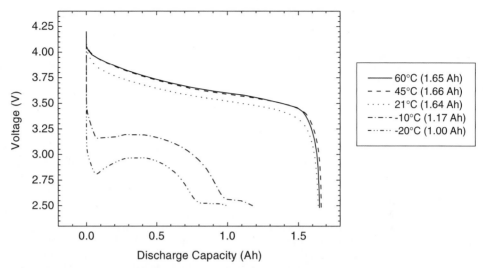

FIGURE 35.43 Approximate C-rate discharge of an 18650-type $C/LiCoO_2$ battery at various temperatures when charged in a CCCV regime at 1.65 A to 4.2 V for 2.5 hours, then discharged at 1.5 A. The average voltage at 21°C was 3.6 V, and at −20°C, 2.9 V. (*Courtesy of NEC Moli Energy.*)

The capacity and energy of 18650-type $C/LiMn_2O_4$ and $C/LiCoO_2$ batteries when discharged at 0.2C rate are summarized in Fig. 35.44. Illustrated in the figure is the higher capacity and energy available from the $C/LiCoO_2$ chemistry, although because of the higher voltage of the $C/LiMn_2O_4$ product, the energy difference is less than the capacity difference at any given temperature. As shown, the batteries offer consistent capacity and energy between 0°C and 60°C, and 90% of the 21°C capacity at −20°C. The average cell voltage is depressed significantly at the lower temperatures.

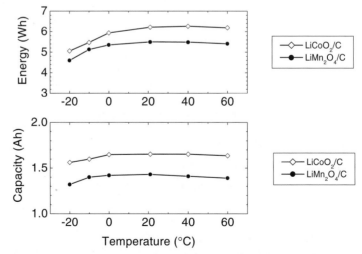

FIGURE 35.44 Discharge capacity and energy of 18650-type $C/LiMn_2O_4$ batteries at 0.2 C (0.28 A), and $C/LiCoO_2$ batteries at 0.2 C rate (0.33 A), at temperatures from 60°C to −20°C to 2.5 V. (*Courtesy of NEC Moli Energy.*)

Cycle Life of Commercial Products.   A desirable characteristic of Li-ion batteries is their long cycle life and ability to cycle continuously at ambient, low, or elevated temperature. This is illustrated in Fig. 35.45 for a $C/LiCoO_2$ battery which shows the discharge capacity when cycled at either 21°C or 45°C at either the 1C, 0.5C or 0.2C rate. As shown, the fade rate in all cases was comparable and relatively linear. At 21°C, the batteries had a fade rate of 0.024%/cycle, while the battery at 45°C had a fade rate of 0.028%/cycle. Thus after 500 cycles, the battery cycled at the 1C rate at 21°C delivered 1.42 Ah, 91% of the initial value, and the one cycled at 45°C at the 1C rate delivered 1.35 Ah on the 500th cycle, or 87% of the initial value.

The capacity of 18650-type $C/LiMn_2O_4$ batteries cycled at 21°C or 45°C at the 1C, 0.5C or 0.2C rates is illustrated in Fig. 35.46. At 21°C at the 1C rate, after 75 cycles, the fade rate was linear at 0.42%/cycle, and at the 0.5C rate, 0.32%/cycle. After 500 cycles at the 1C rate, the battery delivered 1.01 Ah, or 70% of the initial capacity. For comparison, the cell cycled at the 0.5C rate at 21°C delivered 1.16 Ah after 500 cycles, or 80% of its initial capacity. Higher fade rates are observed at 45°C.

As batteries are cycled their impedance increases, resulting in reduced voltage and lower delivered energy. Illustrated in Fig. 35.47 are discharge curves at various cycles for a $C/LiCoO_2$ battery cycled at the 1C rate at 21°C. As shown, the battery voltage decreased by typically 0.3 mV/cycle. For comparison, Fig. 35.48 shows discharge curves at various cycles for an 18650-type $C/LiMn_2O_4$ battery cycled at the 1C rate at 21°C. As the cycling proceeded, the average voltage decreased, by 0.3 mV/cycle, comparable to that observed with the $C/LiCoO_2$ chemistry.

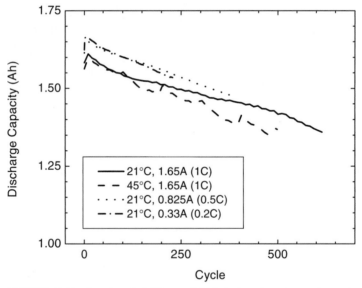

FIGURE 35.45   Capacity of 18650-type $C/LiCoO_2$ batteries when cycled at either 21°C or 45°C at rates from 0.2C (0.33 A) to 1.0C (1.65 A) between 4.2 V and 2.5 V. (*Courtesy of NEC Moli Energy.*)

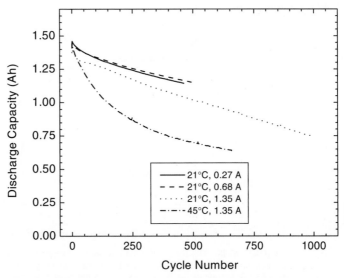

FIGURE 35.46 Capacity fade of 18650-type $C/LiMn_2O_4$ batteries when cycled at various rates at either 21°C or 45°C between 4.2 V and 2.5 V. (*Courtesy of NEC Moli Energy.*)

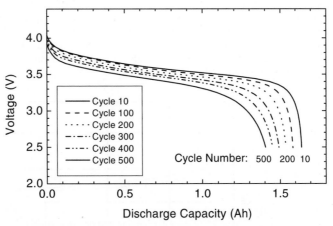

FIGURE 35.47 Discharge curves for an 18650-type $C/LiCoO_2$ battery when cycled at 1.65 A (1C) charge and discharge illustrating the increase in resistance upon cycling. (*Courtesy of NEC Moli Energy.*)

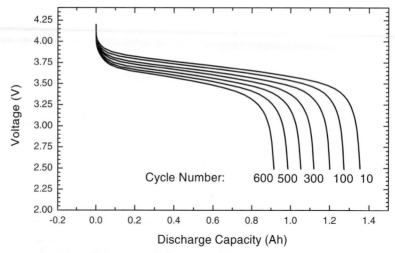

FIGURE 35.48 Discharge curves from an 18650-type $C/LiMn_2O_4$ battery at various cycle numbers illustrating the increase in resistance upon cycling. The battery was charged in a CCCV regime at 1.4 A to 4.2 V for 2.5 hours, and discharged at 1.35 A. (*Courtesy of NEC Moli Energy.*)

Storage Performance.   Two criteria that can be used to characterize the storage performance of batteries are their self-discharge rate and capacity recovery after storage. The capacity of a battery may be determined by discharge before storage. After storage in the charged state, discharge enables determination of self-discharge, subsequent charge and discharge enables determination of capacity recovery, also termed reversible capacity. These effects are illustrated by the discharge curves in Fig. 35.49. As shown, the voltage is minimally reduced (0.1 V) by extended storage (6 months) at 20°C and while self-discharge of 10% occurred over six months, after the battery was charged, 97% of the original capacity was delivered. For a 17500-type $C/LiCoO_2$ Li-ion battery, the self-discharge performance (capacity retention) is illustrated in Fig. 35.50 and capacity recovery further illustrated in Fig. 35.51. Self-discharge for 18650-type $C/LiCoO_2$ batteries over a range of temperatures is illustrated in Fig. 35.52. As shown, at low and ambient temperatures, low self-discharge was observed, 11.5% over six months at 25°C, and 25% at 40°C. However, over 6 months at 60°C, 80% self-discharge occurred.

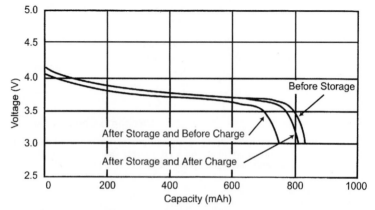

FIGURE 35.49 Discharge of a 17500-type cylindrical $C/LiCoO_2$ battery before and after six months storage at 20°C in a fully discharged state. Charge conditions: Constant voltage/constant current, 4.2 V, 550 mA (max.), 2 hours, 20°C. Discharge conditions: Constant current 156 mA to 3.0 V at 20°C. (*Courtesy of Panasonic.*)

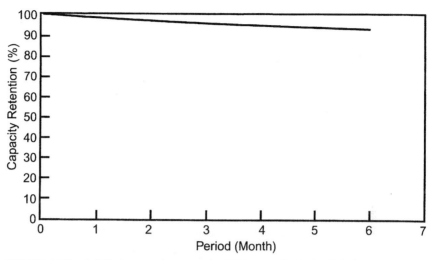

FIGURE 35.50 Self-discharge performance of a 17500-type cylindrical C/LiCoO$_2$ battery at 20°C when discharged after storage in the charged state for periods of up to six months. Charge conditions: Constant voltage/constant current, 4.2 V, 550 mA (max.), 2 hours, 20°C. Discharge conditions: Constant current 156 mA to 3.0 V at 20°C. (*Courtesy of Panasonic.*)

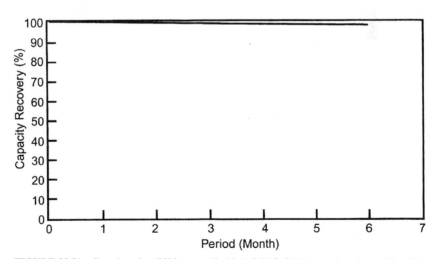

FIGURE 35.51 Capacity of a 17500-type cylindrical C/LiCoO$_2$ battery when charged then discharged after storage at 20°C in the fully charged state for periods of up to six months. Charge conditions: Constant voltage/constant current, 4.2 V, 550 mA (max.), 2 hours, 20°C. Discharge conditions: Constant current 156 mA to 3.0 V at 20°C. (*Courtesy of Panasonic.*)

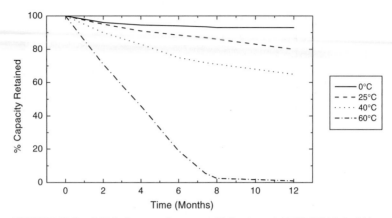

FIGURE 35.52   Self-discharge performance of fully charged 18650 $C/LiCoO_2$ Li-ion batteries over a range of temperatures. (*Courtesy of Sanyo.*)

35.4.3   Large Cylindrical Batteries

To power space probes, such as those intended for use by NASA to explore Mars, 25 Ah cylindrical $C/LiCoO_2$ Li-ion batteries have been developed. Given the mission's low temperature requirements, these batteries have been designed to charge and discharge at low temperature. An electrolyte of 1 M $LiPF_6$ in EC:EMC is used. The capacity of a 25 Ah cylindrical cell when cycled at 23°C at 5 A, on charge and discharge, is shown in Fig. 35.53. The cell delivered 26.8 Ah initially and 24.1 Ah on the 500th cycle, thus provides a fade rate of 0.02%/cycle. Figure 35.54 shows the capacity of another 25 Ah cell when cycled at the C/5 rate (charge and discharge) at 0°C and −20°C. This 890 g cell, with a nominal capacity of 29 Ah at 20°C (121 Wh/kg and 280 Wh/L), delivered comparable capacity at 0°C and over 25 Ah at −20°C.

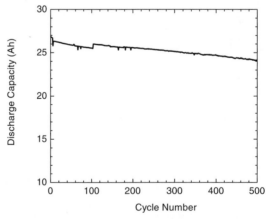

FIGURE 35.53   Capacity of a 25 Ah cylindrical Li-ion battery when cycled at the C/5 rate between 4.1 V with taper charge to C/50 and 3.0 V for 500 cycles at 23°C. (*Courtesy of Blue Star Advanced Technologies. Reproduced with permission from SAE paper 01-1390, 1999.*)

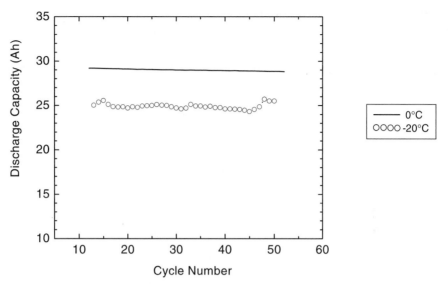

FIGURE 35.54 Capacity of a 25 Ah cylindrical Li-ion cells when cycled at 0°C and −20°C at the C/5 rate between 4.1 and 3.0 V. (*Courtesy of Blue Star Advanced Technologies. Reproduced with permission from SAE paper 01-1390, 1999.*)

40 Ah Cylindrical Batteries for Low Earth Orbit Satellites. A unique battery application is in Low Earth Orbit (LEO) satellites. This application is more completely described in the chapter on Nickel Hydrogen batteries (Chap. 32). Satellites in low earth orbit (550 km) orbit the earth approximately every 100 minutes, or approximately 15 times per day. When in sunlight, the batteries are charged, while during eclipse they are discharged. Over the course of the year, depending on the specific orbit, the eclipse duration varies, typically from 26 to 36 minutes. Given this regime in which the discharge rate is nominally twice the charge rate, LEO batteries are typically required to provide 30,000 to 50,000 cycles enabling a 5- to 10-year operational life. LEO applications are currently addressed almost entirely using nickel-hydrogen batteries as they offer high cycle life, despite their low specific energy (64 Wh/kg) and low energy density (105 Wh/L). Li-ion cells are being developed for LEO applications as their use will enable the development of dramatically lighter, smaller and less expensive satellite batteries, if the required cycle life can be obtained. Because of the high energy efficiency of Li-ion cells (95 to 98%) on discharge and charge relative to Ni-H_2 cells (80 to 85%), further reduction in spacecraft weight and cost is anticipated as improved energy efficiency permits a reduction of other systems, including solar arrays, the electrical buss and thermal controls.

Described here is the performance of 40 Ah cylindrical Li-ion cells tested over 1.33 years under LEO requirements. The cells provided over 40 Ah (41 Ah to 43 Ah) or 120 Wh/kg when charged to 4.0 V and discharged at the C/3 rate to 2.7 V. The case negative cells utilized a stainless steel case and glass to metal seals. The cell chemistry was based on a $LiNiO_2$ positive electrode, coated onto Al with a PVDF binder, and graphite negative coated onto copper. The separator was Celanese 2300, a microporous tri-layer laminated PPE/PE/PPE film.[90] The C/$LiNiO_2$ chemistry is not used in commercial cells due to safety issues.

Three accelerated LEO regimes were used to shorten the typical 100 minute orbit time. One set utilized a 24 minute orbit and cycled at 20% DOD (8 Ah), a second set utilized a 44 minute orbit and cycled at 20% DOD (8 Ah), the third set cycled within a 47 minute orbit and cycled at 30% DOD (12 Ah). The test parameters are summarized in Table 35.14. In this regime, cells cycling at 20% DOD occurred between 60% and 40% state of charge (SOC), and those at 30% DOD cycled between 60% and 30% SOC. The accelerated regime was used for 48 of every 50 cycles, the remaining two cycles were performed at normal LEO rates to determine the residual capacity. In this testing, the cells were operated at 25°C with an end of charge (EOC) voltage of 3.75 V.

The residual energy of the cells (the discharge energy to 100% DOD at C/3 rate), with the recommended end of charge voltage of 3.75 V, is shown in Fig. 35.55. The data presented in this figure are the average of three cells. As shown, the energy fade rate for the longest regime (20% DOD, 44 minute) is one-third less than that of the faster regimes. In the worst case, the cells cycled at 30% DOD had an average fade rate of 0.00092%/cycle.

Using this data, the cycle life of the batteries in a LEO application has been estimated. Here, the cycle life was defined as that when residual energy upon discharge from 3.75 V to 2.7 V was 42 Wh for cells cycled at 30% DOD or 28 Wh for cells cycled to 20% DOD. The results of a calculation based on the energy fade data, considered to be the most conservative, are presented in Table 35.15. As shown, cycle lives between 55,000 and 137,000 cycles are predicted, suggesting lithium ion technology is capable of meeting LEO satellite requirements. These projections are based upon cycling these batteries in the middle of the useable voltage range.

TABLE 35.14   Parameters for the Accelerated LEO Testing

	Discharge rate	Discharge time	Charge rate	Charge time
20% DOD	60 A	8 min.	35 A	16 min.
20% DOD	60 A	8 min.	15 A	36 min.
30% DOD	60 A	12 min.	25 A	35 min.

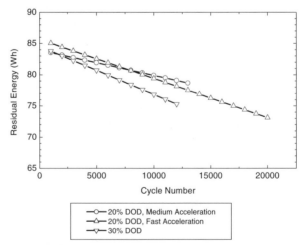

FIGURE 35.55   The residual energy of a 40 Ah Li-ion cell cycled in an accelerated LEO regime. (*Courtesy of SAFT.*)

TABLE 35.15 Summary of LEO Test Results and Predicted Cycle Life. (*Courtesy of SAFT.*)

Test regime	Cycles completed	Predicted cycle life
20% DOD, 24 minute	21,000	91,833
20% OD, 44 minute	12,500	137,663
30% DOD, 47 minute	12,000	55,071

35.4.4 Prismatic Li-ion Batteries

Prismatic Li-ion batteries also offer a high level of performance, although design considerations related to the geometry result in differences in performance. For example, current cylindrical 18650 cells manufactured by Sanyo provide 1.7 Ah and weigh 42 g, thus offer 150 Wh/kg and 380 Wh/L. A comparably sized prismatic battery, such as the UF103450P product by Sanyo, provides 1.5 Ah and weighs 38 g, thus offers similar specific energy (146 Wh/kg) but lower energy density (301 Wh/L). Some of this difference may result from unused space remaining in the cell case, such as that resulting from the insertion of a wound electrode stack into a prismatic can, while other differences occur because in cylindrical designs the cell case may be loaded with hoop stress without deformation, whereas if pressure is applied to the broad face of a prismatic design, bulging will result.

These factors change when comparing flat-plate to wound prismatic cell designs. As described below, many of the larger batteries utilize a flat-plate design whereas small cells almost universally utilize a wound design consistent with commercial cost requirements.

When $LiNi_xCo_{1-x}O_2$-type positive electrode materials are used, the charge and discharge curves are sloping and a higher battery capacity can be achieved by charging to higher voltage, although this shortens battery life. The increased capacity is illustrated by the change in the specific energy or energy density of INCP 61/16/78 batteries charged to either 4.2 or 4.1 V as illustrated in Table 35.16 below. After charge to 4.2 V, the cells delivered 145 Wh/kg and 345 Wh/L, whereas after charge to 4.1 V, the batteries delivered 131 Wh/kg and 325 Wh/L. Similarly sized batteries fabricated using lithium manganese oxide spinel positive electrode material delivered 84 Wh/kg and 168 Wh/L after charge to 4.2 V.

TABLE 35.16 Specific Energy and Energy Density of Prismatic Li-ion Cells. (*Courtesy of Yardney Technical Products, Inc.*)

Cell	Voltage limits	Specific energy	Energy density
INCP 61/16/78 (7 Ah)	4.2 V–2.5 V	145 Wh/kg	345 Wh/L
INCP 61/16/78 (6 Ah)	4.1 V–2.5 V	131 Wh/kg	325 Wh/L
IMP 61/16/78 (4 Ah)	4.2 V–2.5 V	84 Wh/kg	168 Wh/L

Discharge Rate Capability of Wound Prismatic Batteries.   The rate capability of a 1.4 Ah wound prismatic $C/LiCoO_2$ battery at 20°C is illustrated by the discharge curves shown in Fig. 35.56. At the 0.2C rate (270 mA), the battery provided 1.4 Ah, at the C rate (1.35 A), 1.36 Ah, and at the 2C rate (2.7 A), 0.8 Ah. The average voltage at the C/5 rate was 3.76 V, 3.5 V at the C rate, and 3.35 V at the 2C rate.

Discharge of Wound Prismatic Batteries at High and Low Temperature.   The discharge capability of wound prismatic batteries is indicated by the discharge curves shown in Fig. 35.57 for a 1.4 Ah wound prismatic design. When discharged at 0.945 A (0.7C), at 45°C and 60°C, the battery delivered 1.4 Ah; at 20°C, 1.34 Ah; at 0°C, 1.16 Ah; and at −10°C, 0.85 Ah. The average voltage at 20°C was 3.56 V; at 0°C, 3.43 V and at −10°C, 3.2 V.

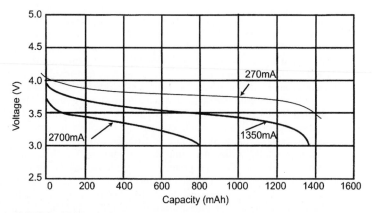

FIGURE 35.56   Performance of a 1.4 Ah wound prismatic CGP345010 $C/LiCoO_2$ battery at various discharge rates to 3.0 V at 20°C, after being charged at 945 µA to 4.2 V followed by a taper charge for 2 hours total at 20°C.

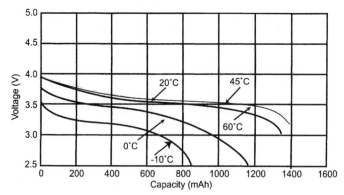

FIGURE 35.57   High and low temperature discharge of a 1.4 Ah CGP345010 $C/LiCoO_2$ wound prismatic battery at 1350 mA. Charged at 945 mA to 4.2 V followed by taper charge for 2 hours total at 20°C. (*Courtesy of Panasonic.*)

Storage Performance of Wound Prismatic Cells.   The storage stability of wound prismatic C/LiCoO$_2$ batteries is indicated by the discharge curves shown in Fig. 35.58. As shown, after charge to 4.1 V then storage for 4 months at 20°C, the 0.63 mAh battery delivered 0.57 mAh, with a self discharge rate of 2.4%/month. After being recharged, the battery delivered 0.63 mAh, demonstrating that no irreversible capacity loss occurred on storage. The self-discharge performance of the battery is shown in Fig. 35.59. As shown, the self-discharge rate is typically 2.4%/month and after 6 months storage at 20°C, 90% of the battery's capacity was delivered.

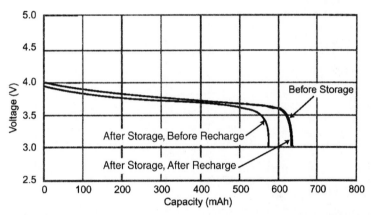

FIGURE 35.58   Discharge at 120 mA to 3.0 V of a wound prismatic 30486 C/LiCoO$_2$ battery before and after 4 month storage at 20°C. The battery was charged at 420 mA to 4.1 V on a CCCV regime followed by a taper charge for 2 hours total at 20°C. (*Courtesy of Panasonic.*)

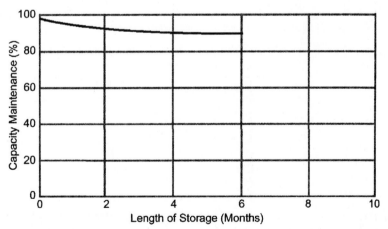

FIGURE 35.59   Self-discharge performance of a 30486 C/LiCoO$_2$ battery at 20°C. Charging conditions: Constant voltage/constant current, 4.1 V, 420 mA (max.) 2 hours, 20°C. Discharge conditions: Constant current 120 mA to 3.0 V at 20°C. (*Courtesy of Panasonic.*)

Discharge of Flat-Plate Prismatic Spinel Batteries.   Low temperature capability has been demonstrated in prismatic batteries that utilize a manganese-based spinel positive electrode material. Shown in Fig. 35.60 are discharge curves at 0.8 A (C/5) for a 4 Ah spinel cell at 25°C to −40°C. At −20°C, the capacity delivered was 96% of that delivered at 25°C (4.2 Ah). At −40°C, 3.3 Ah was delivered, or 75% of the value at 25°C. As is characteristic of the spinel chemistry, the voltage was higher than designs that employ $LiCoO_2$ or $LiNi_xCo_{1-x}O_2$. At 25°C, the average cell voltage was 3.85 V; at −30°C, 3.43 V; and at −40°C, 3.16 V.

Prismatic cells that utilize the spinel chemistry offer good cycle life at 25°C as illustrated in Fig. 35.61. Although spinel designs have fade rates typically twice that of cells that use cobalt-based chemistry, as illustrated, hundreds of cycles can be delivered without significant fade. However, spinel cells typically have more capacity fade at high temperature than those that use cobalt-based chemistry.

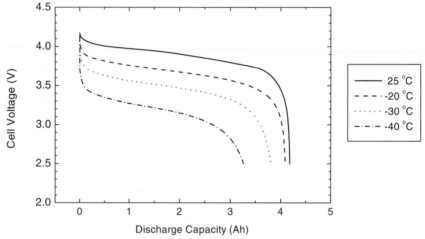

FIGURE 35.60   Discharge curves for a 4 Ah prismatic spinel ($C/LiMn_2O_4$) battery when discharged at the 0.8 A rate at several temperatures.

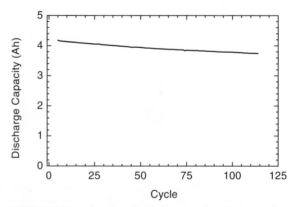

FIGURE 35.61   Capacity of a 4 Ah prismatic spinel cell when cycled at 0.8 A between 4.2 and 3.0 V at 25°C. (*Courtesy of Yardney Technical Products, Inc.*)

Pulse Discharge Performance of 6 Ah C/LiNi$_{1-x}$Co$_x$O$_2$ Flat-Plate Prismatic Cells. Many applications, including power tools, radios, thermal imaging devices, engine starters, and medical implants require high-rate pulse discharge capability, although the duty cycles vary greatly. To indicate the pulse discharge capability of a flat-plate prismatic C/LiNi$_{1-x}$Co$_x$O$_2$ Li-ion battery, Fig. 35.62 shows the cell voltage during 0.2 second pulses at 50 A (8.3C) to 125 A (20.8C) on a 6 Ah cell. The positive electrode in these cells had a geometric area of 2106 cm^2. As shown, the minimum voltage at 50 A (24 mA/cm^2) was 2.81 V, at 75 A (36 mA/cm^2) 2.45 V, at 100 A (47 mA/cm^2) 2.11 V and at 125 A (59 mA/cm^2), 1.76 V. Thus at 125 A the cell delivered 1170 W/kg and 2933 W/L. A plot of the minimum voltage versus the discharge rate is shown in Fig. 35.63. As shown, for short, high-rate pulses, the minimum voltage is linearly related to the discharge rate. Figure 35.64 shows the voltage during a high-rate pulse discharge using a duty cycle of 25 A, 0.1 second pulses with 7.5 A continuous between pulses. In this regime, the cell delivered 5.28 Ah and the voltage drop during the pulse was typically less than 0.3 V, again illustrating the ability of this technology to provide high power.

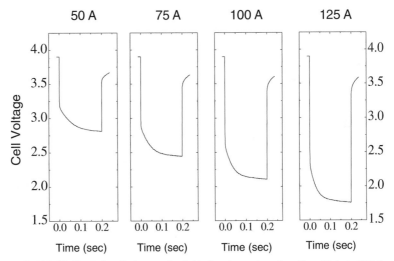

FIGURE 35.62   Pulse discharge of a 6 Ah flat-plate prismatic cell at 50 A to 125 A for 0.2 seconds at 25°C. Prior to the test the cell was aged for 36 weeks at room temperature then charged at 5 A. (*Courtesy of Yardney Technical Products, Inc.*)

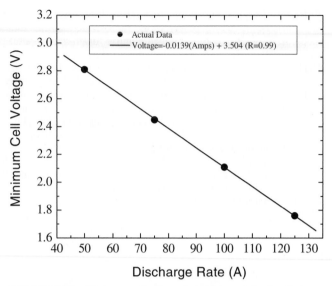

FIGURE 35.63 Minimum cell voltage during 0.2 second pulse discharge of 6 Ah prismatic cell at various rates at 25°C. (*Courtesy of Yardney Technical Products, Inc.*)

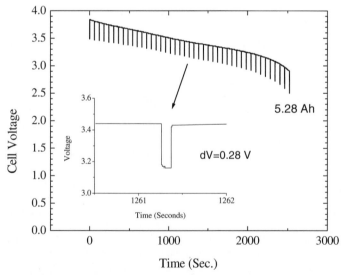

FIGURE 35.64 Pulse discharge of a 6 Ah prismatic cell at 25°C using a duty cycle with 25 A, 0.1 second pulses and 7.5 A between cycles. (*Courtesy of Yardney Technical Products, Inc.*)

Low-Temperature Discharge of Flat-Plate Prismatic 6 Ah Cells.   Flat-plate prismatic cells can provide improved low temperature performance if they employ a ternary electrolyte specially formulated to broaden their temperature capability. Discharge curves at 1 A at temperatures from 25°C to −40°C for a 6 Ah cell charged to 4.1 V are shown in Fig. 35.65. The capacity, energy and average voltage provided by the cell at temperatures from 25°C to −40°C are shown in Table 35.17. At −20°C, the product delivered 4.90 Ah and 16.8 Wh, or 82% of the capacity and 79% of the energy delivered at 25°C, and at −40°C, the cell delivered 4 Ah and 11.6 Wh.

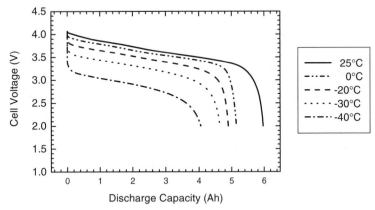

FIGURE 35.65   Voltage of a 6 Ah prismatic cell charged to 4.1 V at 25°C, then discharged at 1 A at several temperatures. (*Courtesy of Yardney Technical Products, Inc.*)

TABLE 35.17   Capacity and Energy of a 6 Ah Cell at Low Temperatures

Temperature (°C)	Capacity (Ah)	Energy (Wh)	V_{avg} (V)
25	5.97	21.35	3.57
0	5.15	18.34	3.56
−20	4.90	16.78	3.42
−30	4.64	14.92	3.22
−40	4.06	11.57	2.85

Discharge Rate Capability of 7 Ah Flat-Plate Prismatic Cells.   Flat-plate prismatic cells can provide higher rate capability than common wound products, as illustrated below for 7 Ah cells. This cell employs a different electrode configuration from the 6 Ah unit in the same case. The high rate capability is attributable to the cell design as flat-plate prismatic designs have many current collector tabs, versus the single or small number of tabs typical in wound units, minimizing ohmic polarization. These cells utilized the $LiNi_{1-x}Co_xO_2$-type positive electrode materials and have a sloping charge curve, characteristic of Li-ion cells that use this material. As a result, cells can be charged to higher voltage, yielding higher capacity, although this typically adversely affects cell life. To illustrate the effect of charge voltage, data for cells charged to either 4.1 V or 4.2 V are compared.

Constant-current discharge curves for cells charged to either 4.1 V or 4.2 V at 1 A at 25°C are shown in Fig. 35.66. At 0.36 A, the unit charged to 4.1 V delivered 6.8 Ah and 24.3 Wh, 10.5% less capacity and 12.2% less energy than the unit charged to 4.2 V. Both delivered high-rate capability; here discharge at rates up to 24 A (3.4C) are shown. For the cell charged to 4.2 V, the average voltage during the 24 A discharge was 3.24 V, 0.4 V less than that at 0.36 A (3.64 V).

The performance characteristics of this 7 Ah prismatic cell as a function of discharge rate after charging to 4.2 V are summarized in Table 35.18.

At constant power, more significant differences in cell performance are observed between units charged to 4.1 V or 4.2 V. Constant power discharge curves for INCP 611678 cells are shown in Fig. 35.67; the energy delivered by these products is tabulated in Table 35.19 and Table 35.20. As illustrated, at rates up to the 1 E rate (22.5 W), the batteries delivered energy comparable to that delivered at the lowest rate evaluated. For example, at 22.5 W, the unit charged to 4.2 V delivered 90% of the energy delivered at 1.35 W, and at 22.5 W the unit charged to 4.1 V delivered 86% of the energy delivered at 1.35 W. However, the unit charged to 4.2 V offered better performance at high rates. At 90 W (4 E), the cell charged to 4.2 V delivered 21.9 Wh, or 78% of the value delivered at 1.35 W, whereas the cell charged to 4.1 V delivered 12.3 Wh at 90 W, or 50% of the value delivered at 1.35 W. This performance improvement for prismatic batteries charged to a higher voltage is achieved, however, at the expense of reduced cycle life. It must be emphasized that charging to higher voltages must be controlled to avoid lithium plating on the negative electrode, particularly at lower temperatures.

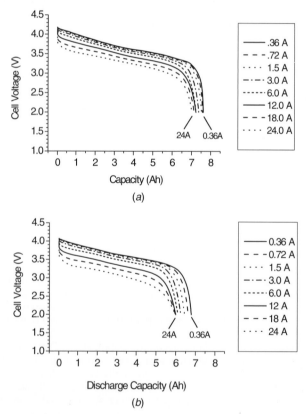

FIGURE 35.66   Voltage vs. capacity for a 7 Ah INCP 61/16/78 cell when discharged at constant current at 25°C after charging to (a) 4.1 V or (b) 4.2 V at 1 A. (*Courtesy of Yardney Technical Products, Inc.*)

TABLE 35.18 The Capacity and Energy Delivered by a 7 Ah Cell Upon Constant Current Discharge When Charged to 4.2 V at 1 A. (*Courtesy of Yardney Technical Products, Inc.*)

Discharge rate (A)	Capacity (Ah)	Energy (Wh)	Energy density (Wh/L)	Specific energy (Wh/kg)	Average voltage (V)
0.36	7.60	27.63	324	147	3.64
3.00	7.38	26.36	309	140	3.57
6.00	7.23	25.55	300	136	3.53
12.00	7.23	24.79	291	132	3.43
24.00	7.02	22.76	267	121	3.24

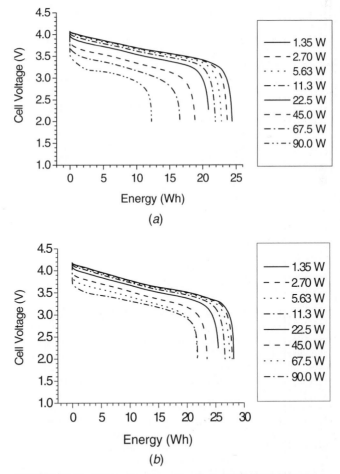

FIGURE 35.67 Voltage vs. energy for a 7 Ah INCP 60/16/78 cell upon discharge at constant power at 25°C after charging to (*a*) 4.1 V or (*b*) 4.2 V at 1 A. (*Courtesy of Yardney Technical Products, Inc.*)

TABLE 35.19 Energy Delivered by a 7 Ah INCP 160/61/78 Cell on Constant Power Discharge after Charging to 4.1 V at 1 A at 25°C. (*Courtesy of Yardney Technical Products, Inc.*)

Power (W)	Energy (Wh)	Energy density (Wh/L)	Specific energy (Wh/kg)
1.35	24.39	286	130
11.3	21.90	256	116
22.5	20.92	245	111
45.0	18.82	221	100
90.0	12.25	144	65

TABLE 35.20 Energy Delivered by a 7 Ah INCP 160/61/78 Cell on Constant Power Discharge after Charging to 4.2 V at 1 A at 25°C. (*Courtesy of Yardney Technical Products, Inc.*)

Power (W)	Energy (Wh)	Energy density (Wh/L)	Specific energy (Wh/kg)
1.35	28.13	329	149
11.3	26.62	312	141
22.5	25.45	298	135
45.0	23.53	276	125
90.0	21.90	257	116

Cycle Life of Flat-Plate Prismatic Cells. Prismatic Li-ion cells offer high cycle life and can be cycled continuously at high rates. The capacity of INCP 60/61/78 cells cycled at the 1 A or 6 A rate for charge and discharge, with charge limited at either 4.2 V or 4.1 V and discharge limited to 3.0 V or 2.5 V is illustrated in Fig. 35.68. The average fade rate over the initial 300 cycles is listed in Table 35.21. As shown, the cells cycled at 1 A demonstrated better capacity retention than cells cycled at 6 A, and the cells cycled to 4.1 V demonstrated better capacity retention than those cycled to 4.2 V. However, in all cases the fade rate is low. For example, the cell cycled at 6 A (1C rate) between 4.2 V and 3.0 V delivered 5.2 Ah on the 300th cycle, or 83% of its initial capacity, and the cell cycled at 1 A between 4.1 V and 3.0 V delivered 5.8 Ah on the 300th cycle, or 90% of the initial value. To illustrate the cycle life of these cells over longer periods, Fig. 35.69 shows data from this test to 3000 cycles. As shown, after 300 cycles the fade rate of a cell cycled at 6 A was linear at 0.017%/cycle. As the cell was cycled, the average voltage decreased as resistance increased, as illustrated by the discharge curves in Fig. 35.70.

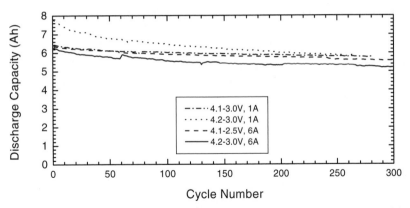

FIGURE 35.68 Capacity of INCP 160/61/78 cells when cycled at a constant current of 1 A and 6 A, for charge and discharge, between 4.2 or 4.1 V and 3.0 V or 2.5 V at 25°C. (*Courtesy of Yardney Technical Products, Inc.*)

TABLE 35.21 Voltage Limits and Fade Rate of INCP 160/61/78 Cells Cycled at 1 A or 6 A at 25°C. (*Courtesy of Yardney Technical Products, Inc.*)

Charge limit	Discharge limit	Rate	Fade rate cycles 1–300	Fade rate cycles >300
4.1 V	3.0 V	1.0 A	0.035%/cycle	—
4.2 V	3.0 V	1.0 A	0.092%/cycle	—
4.1 V	2.5 V	6.0 A	0.037%/cycle	0.017%/cycle
4.2 V	3.0 V	6.0 A	0.056%/cycle	0.016%/cycle

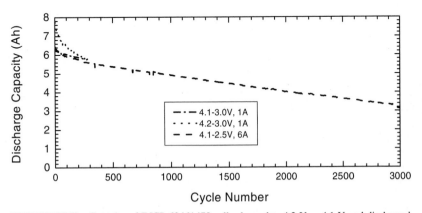

FIGURE 35.69 Capacity of INCP 60/61/78 cells charged to 4.2 V or 4.1 V and discharged to 3.0 V or 2.5 V at 25°C at 1 A or 6 A. The cells cycled at 1 A completed 300 cycles, whereas the cell cycled at 6 A completed 3000 cycles. (*Courtesy of Yardney Technical Products, Inc.*)

Charged Stand Performance of 6 Ah Flat-Plate Prismatic Cell.   The ability of a 6 Ah INCP 160/61/78 cell to provide discharge capability after prolonged charged stand is indicated by the data shown in Fig. 35.71. Prior to the test, the cell delivered 6.1 Ah. After a four month stand at full charge at 25°C, the cell provided 5.3 Ah, thus a self-discharge rate of 3.5% per month. After recharge, the cell provided 5.8 Ah, or 95% of the capacity provided four months earlier. When a similar test was performed on a cell stored at 50% DOD, 98% capacity recovery was observed after four months storage at 25°C.

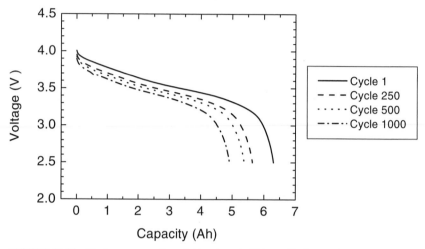

FIGURE 35.70   Discharge curves for a 6 Ah prismatic Li-ion cell when cycled at 6 A between 4.1 V and 2.5 V at 25°C. (*Courtesy of Yardney Technical Products, Inc.*)

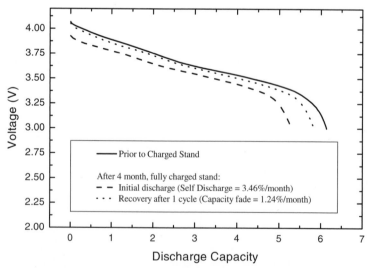

FIGURE 35.71   Discharge of a 6 Ah INCP 160/61/78 cell before and after 4-month stand fully charged, and discharge after charge, all at 25°C. (*Courtesy of Yardney Technical Products, Inc.*)

Constant Current Discharge Rate Capability of 35 Ah C/LiNi$_{1-x}$Co$_x$O$_2$ Flat-Plate Prismatic Cells. The high rate, continuous discharge capability of a 35 Ah C/LiNi$_{1-x}$Co$_x$O$_2$ flat-plate prismatic cells at 25°C is indicated by the discharge curves shown in Fig. 35.72 and the plot of discharge capacity versus rate in Fig. 35.73. These cells weigh 870 g and have outside dimensions of 9.45 × 14.0 × 2.74 cm. At 50 A, they provide an average voltage of 3.5 V, at 125 A, 3.3 V, at 175 A, 3.2 V and at 250 A, 3.0 V. As illustrated in Fig. 35.73, on discharge at up to 50 A the cells offered over 32 Ah and at 125 A, 29 Ah. At 25 A, they provided 134 Wh/kg and 323 Wh/L, and at 50 A 130 Wh/kg and 313 Wh/L. At 250 A, the voltage plateaued at 3.35 V yielding a specific power and power density of 960 W/kg and 2300 W/L.

Larger flat-plate prismatic cells of similar design offer comparable relative performance. See Fig. 35.76, which describes the capacity of a 35 Ah flat-plate prismatic battery at 0.1C to C rates at 40°C, 20°C and −20°C. At the C rate at 20°C, the capacity was 12% less than that at the C/5 rate.

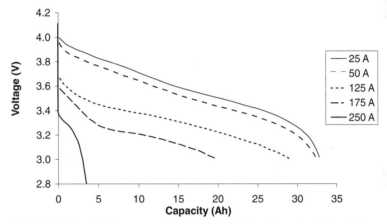

FIGURE 35.72 Discharge curves for a 35 Ah flat-plate prismatic cell at rates from 25 A to 250 A at 25°C. (*Courtesy of Yardney Technical Products, Inc.*)

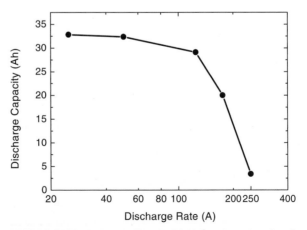

FIGURE 35.73 Rate capability of a 35 Ah flat-plate prismatic cell at 25°C. (*Courtesy of Yardney Technical Products, Inc.*)

Pulse Discharge Performance of 35 Ah Prismatic Cells. Flat-plate prismatic cells can deliver high-rate pulse discharge at temperatures as low as −40°C. The voltage of a 35 Ah prismatic cell during 10 second pulse discharge at 25 A, 50 A, 75 A and 100 A at 20°C and −20°C is shown in Fig. 35.74. At 20°C, the voltage drop from the open circuit voltage is 0.2 V at 50 A and 0.36 V at 100 A. At −20°C, the voltage drop from the open circuit voltage is 0.84 V at 50 A and 1.24 V at 100 A. The ability to deliver the 10 pulse at temperatures from 30°C to −40°C is illustrated in Fig. 35.75. As shown, the minimum cell voltage at 100 A at −40°C was 1.6 V.

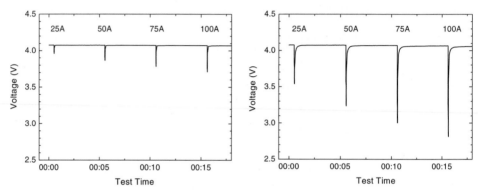

FIGURE 35.74 Cell voltage during 10 second pulse discharge at 25 A, 50 A, 75 A and 100 A at 20°C and −20°C for a 35 Ah prismatic design.

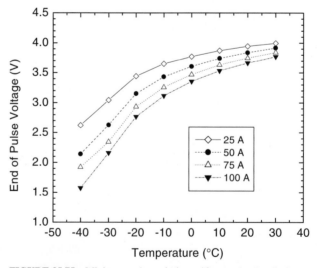

FIGURE 35.75 Minimum voltage during a 10 second pulse discharge of a 35 Ah prismatic cell at 25 A, 50 A, 75 A and 100 A.

Continuous Low Temperature Discharge of Flat-Plate Prismatic 35 Ah Cells. Figure 35.76 shows the discharge capacity of a 35 Ah prismatic cell at 40°C, 20°C and −20°C. As shown, the cell provided rate capability to 1C at −20°C, delivering 80% of the capacity provided at the C/5 rate. From the 0.1C to the 1C rate, the capacity at −20°C was typically 30% less than that at 20°C.

Prismatic 35 Ah cells that employ electrolytes specifically formulated for low temperatures have demonstrated charge and discharge capability at low temperature. Figure 35.77 shows the discharge of a 35 Ah cell at 25°C or −20°C after charge at 25°C. When discharged at −20°C, the battery delivered 24.5 Ah, 72% of that provided at 25°C (34.1 Ah). After charge at −20°C, the cell provided 22 Ah when discharged at −20°C, 91% of that provided after charge at 25°C, as shown in Fig. 35.78. After charge at −30°C, discharge at −30°C provided 17.7 Ah, as shown in Fig. 35.79, 81% of that delivered after charge at 25°C and discharge at −30°C (21.9 Ah).

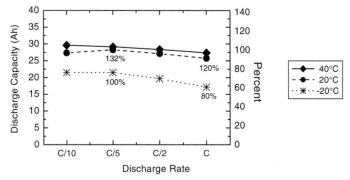

FIGURE 35.76 Rate capability of a 35 Ah flat-plate prismatic cell at 40°C, 20°C, and −20°C. (*Courtesy of Yardney Technical Products, Inc.*)

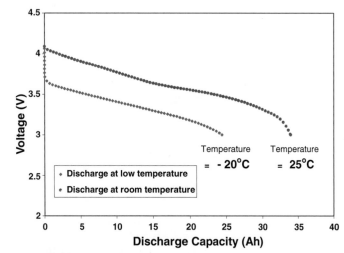

FIGURE 35.77 Discharge of a Yardney 35 Ah prismatic Li-ion battery at 5 A at 25°C and −20°C after charge at 5 A to 4.1 V at 25°C. (*Courtesy of Jet Propulsion Laboratory (JPL).*)

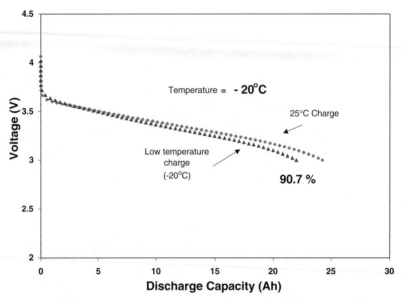

FIGURE 35.78   Discharge of a Yardney 35 Ah prismatic Li-ion cell at 5 A at −20°C after charge at 5 A to 4.1 V at either 25°C or −20°C. (*Courtesy of JPL.*)

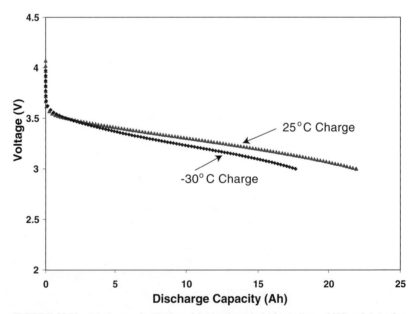

FIGURE 35.79   Discharge of a Yardney 35 Ah prismatic Li-ion cell at −30°C at 2.5 A after charge at 2.5 A at 25°C or −30°C. (*Courtesy of JPL.*)

The 35 Ah cells also provide high rate capability at low temperatures, as indicated by discharge curves at 40°C, shown in Fig. 35.80, those at −20°C, shown in Fig. 35.81, and at −30°C, shown in Fig. 35.82. When charged and discharged at 40°C, the discharge curves for rates from 2.5 A to 12.5 A are very similar, the delivered capacity ranged from 35.8 Ah to 34.4 Ah. The discharge voltages were also comparable. At −20°C, the delivered capacity ranged from 25.2 Ah to 21.2 Ah for these rates. The average voltage at the 2.5 A rate, 3.5 V, was 0.22 V higher than that at 12.5 A, 3.27 V. After charge at −30°C, discharge at −30°C at 2.5 A yielded 17.9 Ah, and 9.8 Ah at 12.5 A. The average voltage at −30°C and 2.5 A was 3.29 V, 0.06 V higher than at the 12.5 A rate at −30°C (3.23 V). The capacity data is summarized in Fig. 35.83, which compares the discharge capacity on charge and discharge at 40°C, −20°C and −30°C, at rates from 2.5 A to 12.5 A.

Storage Stability of 20 Ah Flat-Plate Prismatic Batteries. The ability of flat-plate prismatic batteries to retain capacity on storage is indicated by the data presented in Table 35.22 which lists capacity retention after storage at 0°C, 40°C, and 50°C at either 50% or 100% state of charge for 8 or 16 weeks. These cells were on the $C/LiCo_xNi_{1-x}O_2$ chemistry and used a ternary carbonate electrolyte. At or below 40°C, the batteries retained over 97% of their capacity. At 50°C, storage at 100% DOD resulted in 81% capacity retention, while storage at a 50% state of charge resulted in 91% capacity retention.

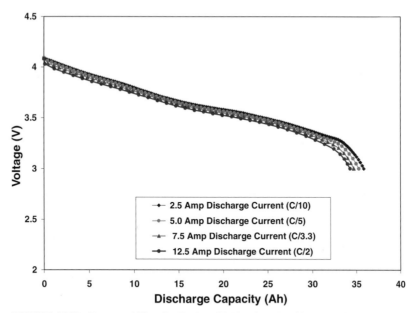

FIGURE 35.80 Rate capability of a Yardney 35 Ah prismatic Li-ion cell when charged at 2.5 A to 4.1 V at 40°C and discharged at 40°C. (*Courtesy of JPL.*)

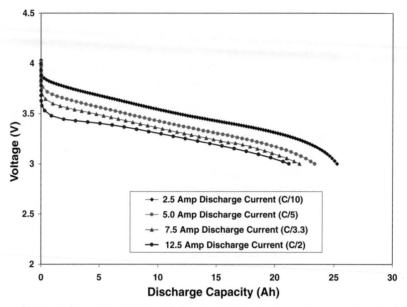

FIGURE 35.81 Rate capability of 35 Ah prismatic Li-ion cells when charged at 2.5 A to 4.1 V at −20°C and discharged at −20°C. (*Courtesy of JPL.*)

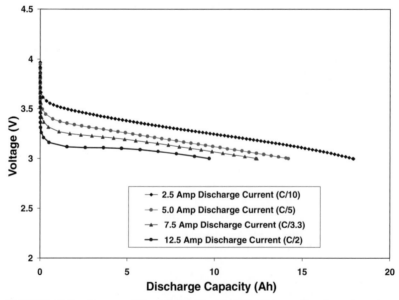

FIGURE 35.82 Rate capability of 35 Ah prismatic Li-ion cells when charged at 2.5 A to 4.1 V at −30°C and discharged at −30°C. (*Courtesy of JPL.*)

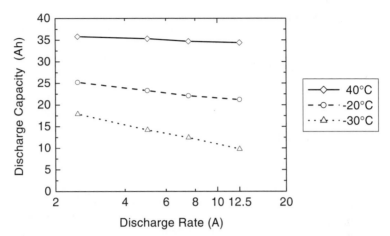

FIGURE 35.83 Summary of the rate capability of Yardney 35 Ah prismatic Li-ion cells when charged and discharged at 40°C, −20°C and −30°C. (*Courtesy of Yardney Technical Products, Inc.*)

TABLE 35.22 Storage Stability of 20 Ah Prismatic Li-ion Batteries. (*Courtesy of JPL.*)

Storage temperature	State of charge	Storage time	Reversible capacity retention
0°C (32°F)	50%	8 weeks	98%
0°C (32°F)	100%	8 weeks	97%
40°C (104°F)	50%	8 weeks	99%
40°C (104°F)	100%	8 weeks	98%
50°C (122°F)	100%	16 weeks	81%
50°C (122°F)	50%	16 weeks	91%

35.5 CHARGE CHARACTERISTICS OF Li-ION BATTERIES

Li-ion cells are fabricated in the discharged state and thus must be charged before use. Li-ion cells are typically charged using either a constant current (CC) or constant current, constant voltage with a taper charge (CCCV) regime. This charge regime is accomplished through the use of battery management circuitry. At low rates (C/5), the CC regime approximates the CCCV regime as the cell is fully charged when the upper voltage limit is attained.

Li-ion cells are typically charged to either 4.1 V or 4.2 V. While $LiCoO_2$ is structurally stable at both potentials,[91] the cutoff voltage affects cell performance when $LiNi_{1-x}Co_xO_2$ is used.[92] When charged to 4.2 V, cells with $LiNi_{1-x}Co_xO_2$ positive electrode material can offer higher capacity, but with reduced cycle life and storage stability, relative to cells charged to 4.1 V. The capacity of $C/LiNi_{1-x}Co_xO_2$ cells charged to 4.1 V or 4.2 V is illustrated in Fig. 35.66 for constant current discharge; delivered energy on constant power discharge is illustrated in Fig. 35.67. At low rates (C/10), cells charged to 4.2 V delivered about 14% more capacity than those charged to 4.1 V, whereas at high rates (2C), typically 18% more capacity was delivered when the higher charge voltage was used. Although the cells charged to

4.2 V delivered more capacity, their capacity fade was higher. The capacity of the cells when cycled at constant current is illustrated in Fig. 35.68 and Fig. 35.69. As shown, even though the capacity fade rate of cells charged to 4.2 V was higher, up to 300 cycles they delivered greater capacity. Only when cycled for more than 500 cycles did cells charged to 4.1 V outperform cells charged to 4.2 V. Also illustrated by this data is the performance of Li-ion cells on high rate (6 A, C rate) or low rate (1 A, C/6 rate) charge. As shown, cells charged at the lower rate did offer higher capacity, although this difference was reduced after 300 cycles. Further, the capacity fade rates of the cells was comparable.

The voltage, percent charged and current profile for the CCCV charging of a $C/LiMn_2O_4$ 18650 cell at 1.4 A with a voltage limit of 4.2 V, and a $C/LiCoO_2$ cell at 1.65 A and 4.2 V, is shown in Fig. 35.84. Initially, as the cells are charged at constant current, the state of charge increases linearly while the cell voltage asymptotically approaches 4.2 V. In the constant voltage portion of the charge, the current decays as the cell approaches full charge. Li-ion cells have high coulombic efficiency, typically 99.9%, and high energy efficiency, typically 95 to 98%. Because of the absence of parasitic processes, such as the gas forming reactions common in aqueous systems, more complex charge methodologies, such as pulse charging, offer little additional benefit.

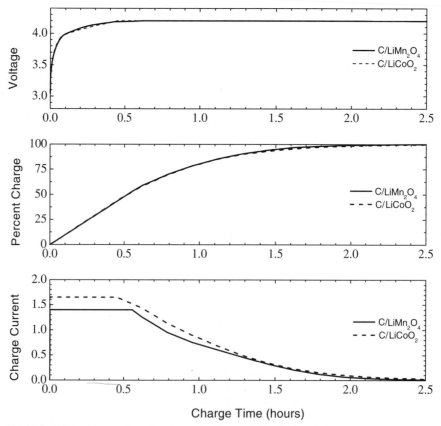

FIGURE 35.84   Voltage, state of charge and current for $C/LiMn_2O_4$ and $C/LiCoO_2$ 18650 cells in a 2.5h CCCV charge profile with 1.4 A ($C/LiMn_2O_4$) or 1.65 A ($C/LiCoO_2$) maximum current and and 4.2 V maximum voltage. (*Courtesy of NEC Moli Energy.*)

Because Li-ion cells are degraded irreversibly by overcharge or overdischarge, and may vent if overcharged, Li-ion batteries typically employ battery management circuitry to ensure safe operation and prevent overcharge. This circuitry can also provide other features such as a state-of-charge gauge and temperature monitoring. Many manufacturers recommend charging at low rates (less than 0.1C) if the cell voltage is below 2.5 V.

Many applications require batteries to retain capacity on continuous float (constant voltage) charge. In this scenario, the cells are charged whenever their open circuit voltage is less than a predetermined threshold value. The ability of wound $C/LiCoO_2$ cells to retain capacity on float charge is indicated in Fig. 35.85, which shows the capacity retained by cells float charged for up to 21 months.[93] Capacity retention was temperature dependent and followed an Arrhenius relationship. The capacity loss rate increased by a factor of ~1.3 for each 10°C rise in temperature.

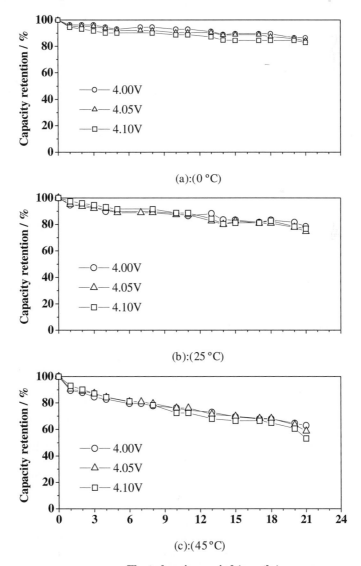

(a):(0 °C)

(b):(25 °C)

(c):(45 °C)

Float-charging period (months)

FIGURE 35.85   Capacity retention of wound ICP22846 $C/LiCoO_2$ Li-ion cells float charged at 0°C, 25°C or 45°C at 4.0 V, 4.05 V or 4.1 V. (*Courtesy of Japan Storage Battery Co., Ltd.*)

The ability of flat plate prismatic designs to retain capacity on "float" is indicated by the data shown in Fig. 35.86. After 12 weeks, discharge of the 5 Ah cells yielded 4.3 Ah, indicating 18% self-discharge had occurred. Subsequent cycling demonstrated that they retained over 98% of their reversible capacity.

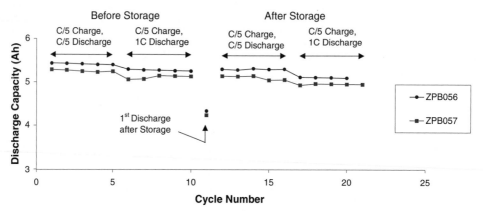

FIGURE 35.86   Float charge test of two different 5 Ah prismatic cells over 12 weeks at 25°C. (*Courtesy of Yardney Technical Products, Inc.*)

35.6 *SAFETY TESTING OF CYLINDRICAL C/LiCoO$_2$ BATTERIES*

Cylindrical 18650-type C/LiCoO$_2$ batteries have been exposed to a variety of safety test regimes including UN, UL, IATA and military safety tests. These organizations, as well as the IEC, are currently developing standards for Li-ion cells and batteries that will address electrical testing, safety evaluation and transportation tests for Li-ion cells and batteries. The safety evaluation tests include electrical, mechanical and environmental tests. The results are summarized in Table 35.23.

Table 35.24 gives the results of typical safety tests performed on commercial 18650-type C/LiCoO$_2$ Li-ion batteries. On short circuit, performed by discharging through a minimum length of 16 AWG wire, the voltage dropped to zero as the temperature rose to a maximum of 69°C after 130 minutes, after which the temperature returned to room temperature. Similar results are reported for nail penetration. When the battery was exposed to heat, up to 150°C, leakage resulted. On crush, the cell voltage dropped to 0.8 V within 30 seconds as the temperature rose to 116°C over 5 minutes after which the temperature returned to ambient as the voltage decayed to 0 V. Performance on impact was similar to crush, a maximum temperature of 117°C was reached after 4 minutes after which the temperature decayed to ambient as the voltage decayed to 0 V. In both the crush and impact tests, leakage resulted. On forced discharge at 1.6 A, the battery voltage reached a minimum of −2 V after 8 minutes then recovered to −0.6 V, the maximum temperature on forced discharge was 58°C. In the overcharge test reported here, the cell was overcharged to a maximum of 12 V at 4.8 A (3C). In this test, the cell reached its maximum temperature of 56°C two minutes after it was fully charged. After full charge was achieved, the voltage quickly reached the prescribed maximum of 12 V while the charging current went to zero for the remaining duration of the test (30 minutes). Note that in none of these tests did the cell spontaneously disassemble, and pressure was released through the vent mechanism when required.

TABLE 35.23 Summary of Safety Tests Included for Li-ion Batteries

Type	Test	Conditions
Electrical	Continuous charge	Charge at 20°C per manufacturers recommendation, hold at end of charge voltage for 28 days.
	Short circuit	<50 mΩ to 0.1 V at 20°C and 50°C.
	Forced discharge	Discharge at 0.2C, for 12.5 h.
	Over charge	Charge at manufacturer recommend rate to 2.5 times rated capacity
	High rate charge	Charge at three times manufacturer's recommended rate.
Mechanical	Shock	3 axis, minimum 75 g, peak 125 g to 175 g.
	Vibration	0.8 mm amplitude, 10 Hz to 55 Hz
	Crush	Between flat surfaces to 13 kN
	Free fall	1 m to hard wood floor, 6 times
Environmental	High temp. storage	Store at 75°C for 48 h
	Thermal shock	75°C for 48 h, 20°C for 30 h
	Altitude	11.6 kPa for 6 h
	Exposure	Heat to 130°C, hold at 130°C for 60 min.

TABLE 35.24 Typical Safety Tests Performed on Commercial 18650 Li-ion Batteries. (*Courtesy of Sanyo.*)

Test	Method	Result
Short circuit	16 AWG wire, 20°C	No event, Tmax = 69°C
Short circuit	16 AWG wire, 60°C	No event
Heat	Heat at 5°C/min to 150°C, soak at 150°C for 10 min.	Leakage
Crush	Flat surface to 2500 psig (17.2 Mpa).	Leakage, Tmax = 116°C
Impact	20 lbs from 2 ft on a 5/16 in bar	Leakage, Tmax = 117°C
Humidity	65°C, >95% humidity	No event
Vibration	0.03 in amplitude, 10 to 55 Hz	No event
Drop	6 ft, for ten times onto concrete surface	No event
Forced discharge	1.6 A, −12 V minimum	No event, Tmax = 58°C
Over charge	12 V max, 4.8 A max	No event, Tmax = 56°C
Fire exposure	Expose to fire	Fire, leakage, smoke

35.7 POLYMER Li-ION BATTERIES

Polymer Li-ion batteries provide the performance characteristics of Li-ion batteries, including their high specific energy and high energy density, in a thin, high aspect-ratio form factor. The technology addresses applications, in particular portable communications and computing devices which require a thin, large footprint rechargeable battery. While polymer Li-ion cells utilize the same active materials as cylindrical or prismatic Li-ion cells, in polymer Li-ion cells, flat, bonded electrodes are used to enable the fabrication of thin cells packaged within a barrier film, in contrast to the steel or aluminum cell case used in other Li-ion technologies. This construction is shown schematically in Fig. 35.87, which illustrates the various layers of the construction. A commercial product is shown in Fig. 35.88.

An attractive feature of $C/LiMn_2O_4$ polymer Li-ion cells is their ability to sustain abuse. Safety and abuse tests passed by $C/LiMn_2O_4$ polymer Li-ion cells are listed in Table 35.25. In addition, $C/LiMn_2O_4$ polymer Li-ion cells can sustain nail penetration in the fully charged state or the overcharged state without explosion or fire.

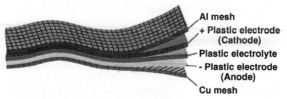

FIGURE 35.87   Schematic diagram showing the construction of a polymer Li-ion cell. (*Courtesy of Telecordia.*)

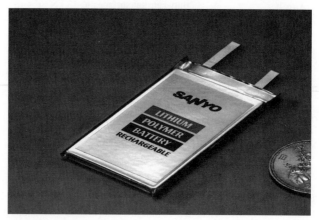

FIGURE 35.88   A 0.57 Ah polymer Li-ion battery. (*Courtesy of Sanyo.*)

TABLE 35.25   Safety and Abuse Tests Passed by C/LiMn$_2$O$_4$ Polymer Li-ion Batteries

Test	Description
Flat crush	13 kN between flat plates
External short, 23°C	<50 mOhm, 23°C
External short, 60°C	<50 mOhm, 60°C
Forced discharge	Discharge to −12 V, 250% of rated capacity
Shock	75 G initial, 125 to 175G peak
Vibration	10-55-10 Hz, 90 min, 1.6 mm
Drop	Ten times, 6 ft onto concrete
Impact	20 lbs from 2 ft onto a 8 mm insulated bar
Thermal cycling	−40°C to 70°C, ten times
Altitude	Six hours at 11.6 kPa
Heating	5°C/min to 150°C, soak
Abnormal charge	Three times recommended rate to 250% of rated capacity
Humidity	Cycle 30°C to 65°C, 85% to 95% relative humidity

The polymer Li-ion cells described here may be more accurately described as employing a gel electrolyte, as the electrolyte contains a monomeric, volatile liquid component absorbed into a polymeric host, in contrast to technologies which do not employ a volatile, liquid component, such as solid polymer electrolyte batteries.[94] Because of the poor conductivity of currently available solid polymer electrolytes (solid polymer lithium batteries developed to date operate at 40°C to 80°C to accommodate the low conductivity of the electrolyte) (see Sec. 34.4.2), current "polymer" Li-ion batteries incorporate less viscous, liquid components to improve the conductivity of the electrolyte, enabling their use at ambient temperatures.

The electrochemistry of polymer Li-ion cells covers a wide range of active materials and electrolyte compositions, comparable to those used in liquid-electrolyte Li-ion cells. Active materials include lithiated manganese oxides, such as $Li_{1.05}Mn_{1.95}O_4$, $LiCoO_2$, $LiNiO_2$ and its Co-doped derivatives, as positive (cathode) active materials, and graphitic and non-graphitic carbons capable of reversibly intercalating lithium as the negative (anode) active materials. Liquid electrolytes used in polymer Li-ion cells are comparable to those used in cylindrical or prismatic Li-ion cells, typically ca. 1 M solutions of $LiPF_6$ in mixtures of ethylene carbonate (EC), dimethyl carbonate (DMC), and other carbonate esters or other co-solvents.

35.7.1 Electrode and Battery Fabrication

The electrodes for polymer Li-ion cells may be cast from a viscous mixture (a slurry) composed of an active material (e.g., $LiMn_2O_4$, $LiCoO_2$, $LiAl_{0.05}Co_{0.15}Ni_{0.8}O_2$, etc. for the positive electrode, and a microbead mesophase graphite, artificial graphite or milled graphite fiber for the negative electrode); a conductive additive (e.g., Super P carbon); a dissolved polymeric binder (e.g., PVDF-HFP, such as Kynar FLEX® 2801); a medium-to-low-volatility plasticizer, such as dibutyl phthalate or propylene carbonate; and processing aids, including surfactants, antioxidants etc.; and a volatile solvent such as acetone or methyl ethyl ketone (MEK). A plastic separator film may be cast from a homogenized slurry of the same polymeric binder (PVDF-HFP), silanized fumed silica, or alumina, and a plasticizer, in a similar volatile solvent.[95]

In the Telcordia (formerly Bellcore) technology, an ancillary plasticizer such as dibutyl phthalate is incorporated into the resin that temporarily remains in the separator and/or electrode layers after their respective films are cast using a more volatile solvent. The plasticizer facilitates the densification of the electrodes under lower temperature and/or pressure than is required for the full densification of the electrode compositions typical in liquid-electrolyte Li-ion cells. This, in turn, prevents damage to the delicate expanded-metal current collector grids typically used in polymer Li-ion cells. The plasticizer also acts as a porosity modifier and preserver in the electrodes and the separator during processing. Further, the plasticizer facilitates formation of a bond between the electrodes and the separator. Often, identical or similar polymeric binders, typically based on commercially available poly(vinylidene fluoride-co-hexafluoropropylene) (PVDF-HFP) copolymers, are used in both electrodes and the separator, facilitating the bonding of layers during the lamination step. The use of other polymers, copolymers and their mixtures, e.g., PVDF-CTFE (CTFE = chlorotrifluoroethylene), poly(vinyl chloride) (PVC) and polyacrylonitrile (PAN) has also been described.[96]

Electrode tapes are then densified and bonded to their corresponding expanded-metal current collector grids, aluminum in the case of the positive electrode and copper for the negative, using heated double-roll laminators or parallel-platen presses. The current collector grid can be either embossed into the outside surface of the electrode, or preferably, embedded between two layers of electrode tape. Finally, the electrodes are laminated to a plastic or microporous separator interposed between the electrodes.

The electrode-separator bonding can be achieved by lamination (calendering), again between a pair of preheated rollers or platens, followed by removal of the plasticizer. The plasticizer is removed by either supercritical or solvent extraction with an appropriate low-boiling solvent of the plasticizer, but non-solvent of the polymeric binder, such as propane, carbon dioxide, methanol, ether, or hexanes. Alternatively, the plasticizer is removed by simple evaporation, preferably under reduced pressure and at elevated temperature.

The de-plasticized (extracted or otherwise) cells are then packaged into vapor-impermeable, flexible, multilayer polymer-aluminum bags, dried under reduced pressure and elevated temperature to remove any adsorbed water, activated with a measured amount of liquid electrolyte solution, and sealed. The liquid electrolyte is rapidly absorbed into the microporous structure of the electrodes and separator, thus making the spillage of the liquid electrolyte from an open battery highly improbable.

A larger-capacity cell can contain one or more flat plates, each plate composed of one positive electrode and one negative electrode bonded to two opposite sides of an ionically conductive separator. The plates can be stacked as individual plates, Z-folded, or folded in other ways as permitted by the mechanical properties of the individual component layers. A popular option is a "bicell" configuration, where the central plate is a double-thickness electrode, typically a negative electrode, sharing a single current collector grid. Both faces of the central electrode are bonded via two separator layers to two slightly smaller counter-electrodes, typically positive electrodes, each having its separate current collector.

A prismatic battery of larger capacity can be assembled from one or more stacked bicells, whose current collector tabs are welded inside the packaging and thus present only a single metal foil feed-through tab connected to each of the two multi-plate electrodes. To facilitate multi-plate battery assembly and alignment, individual bicells may be Z-folded to prevent electrodes and current collectors from breaking at the fold line; they may also be connected by common current collector(s), separator(s), electrode layer(s), etc.

35.7.2 Energy Density of Polymer Li-ion Batteries

Current implementations of polymer Li-ion batteries technology achieve specific energy and energy density values that are somewhat higher than those reported for liquid-electrolyte Li-ion batteries. Since the electrochemistry as well as the volume and weight fractions of various components in these two technologies are similar, this gain may be attributed to the lighter, thinner packaging materials used in flat, bonded-electrode batteries, and possibly more efficient utilization of space.

The physical dimensions, specific energy and energy density of a selection of $C/LiCoO_2$ and $C/LiMn_2O_4$ polymer Li-ion batteries are listed in Table 35.26. Polymer Li-ion batteries offer a thin form factor, batteries with thickness from 1.2 mm to 8.4 mm are available, although 3.5 mm to 4.0 mm thick batteries are typical; available capacities range from 0.5 Ah to 8 Ah. The practical lower limit to polymer battery thickness is 0.5 mm. The batteries offer specific energy and energy density comparable to that delivered by cylindrical and prismatic Li-ion batteries, $C/LiCoO_2$ batteries deliver 145 to 190 Wh/kg, and 270 to 400 Wh/L, whereas $C/LiMn_2O_4$ batteries deliver 130 to 144 Wh/kg and 235 to 300 Wh/L.

TABLE 35.26 Electrical and Physical Characteristics of C/LiCoO$_2$ or C/LiMn$_2$O$_4$ Polymer Li-ion Batteries

Battery type	ICP* 36/35/62	ICP* 36/34/50	ICP* 48/50/80	ICP* 60/100/100	IMP 30/25/110	IMP 30/34/48	IMP 30/103/103	IMP 30/36/65	IMP 12/70/140	IMP 57/70/140
Shape	C/LiCoO$_2$ (Source: Sony, Sanyo, Telcordia)				C/LiMn$_2$O$_4$ (Source: Valence Technologies)					
Length, mm	62	50	80	100	110	48	103	65	140	140
Width, mm	35	34	50	100	25	34	103	36	70	70
Thickness, mm	3.6	3.6	4.8	6.0	3.0	3.0	3.0	3.0	1.2	5.7
Volume, ml.	7.8	6.1	19.2	60	8.25	4.9	31.8	7.0	11.8	55.9
Mass, g.	14.5	15	48	150	16	9.0	63	14	23	116
Capacity (Ah)	0.57	0.70	2.40	7.7	0.55	0.30	2.29	0.47	0.73	4.39
Specific energy (Wh/kg)	145	170	180	190	130	130	138	127	120	144
Energy density (Wh/L)	270	360	380	400	253	236	273	254	235	300

* Prototypes, not available in quantity.

35.7.3 Performance of C/LiCoO₂ Polymer Li-ion Batteries.

Charge Performance.   Polymer Li-ion batteries, like cylindrical or prismatic Li-ion batteries, can be charged using either a constant current (CC), or constant current-constant voltage (CCCV) regime. The charge current and battery voltage during the charge of a 0.120 Ah C/LiCoO₂ polymer Li-ion battery at the 1.3C, 1C, and 0.7C rates is illustrated in Fig. 35.89. In each case, the batteries charged at constant current until the battery voltage reached the 4.2 V voltage limit, at which time the current was reduced to maintain 4.2 V. At the 1C rate, 40% of the two hour regime was at constant current.

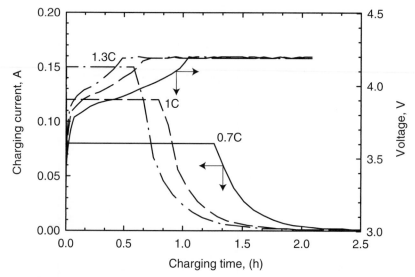

FIGURE 35.89 Charge current and battery voltage during the charge of a 0.120 Ah C/LiCoO₂ polymer Li-ion battery at the 1.3C, 1.0C and 0.7C rates, at 21°C. (*Courtesy of Telcordia.*)

Discharge Rate Capability.   The rate capability of a C/LiCoO₂ polymer Li-ion battery at rates from 3C to 0.2C is illustrated in Fig. 35.90. As shown, the battery provided over 95% of its capacity at the 1C rate, 87% at 2C, and 77% at 3C. At low rates, the battery provided an average voltage of 3.8 V; at the 2C rate, 3.55 V; and at the 3C rate, 3.45 V on average. The battery provided over 80% of its available energy at 2C.

Polymer Li-ion batteries offer similar capability when discharged at constant power, as illustrated in Fig. 35.91, which shows discharge curves for a 0.136 Ah (0.5 Wh) C/LiCoO₂ polymer Li-ion battery discharged at rates from 2 E to 0.25 E. At low rates (0.25 E), the battery provided 0.51 Wh. At the 1 E rate, the battery provided 0.46 Wh, and 0.3 Wh at 2 E, or 58% of the available energy. As shown, the average voltage at low rates was 3.75 V, whereas at the 2 E rate, 3.5 V was obtained.

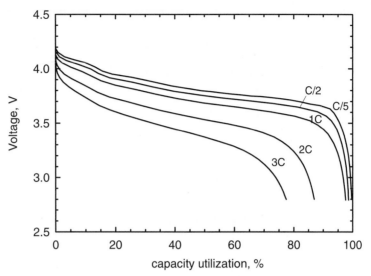

FIGURE 35.90 Discharge rate capability of a 0.12 Ah C/LiCoO$_2$ polymer Li-ion battery at 3C to 0.2C rates, at 21°C. (*Courtesy of Telcordia.*)

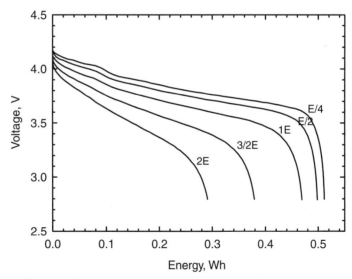

FIGURE 35.91 Discharge rate capability of a 0.136 Ah C/LiCoO$_2$ polymer Li-ion battery at 2 E to 0.25 E, at 21°C. (*Courtesy of Telcordia.*)

Low Temperature Performance.   Discharge curves for a 0.57 Ah Sanyo C/LiCoO$_2$ Li-ion polymer battery are shown at the 1C rate at temperatures from −20°C to +25°C in Fig. 35.92. As shown, the battery provided 58% its rated capacity at −10°C.

The low temperature discharge capability of C/LiCoO$_2$ polymer Li-ion batteries is summarized in Fig. 35.93 for rates from 0.125C to 2C at temperatures from −20°C to 21°C. At 0°C, when discharged at the 1C rate, the batteries provided 72% of their capacity. At −20°C, at low rates (0.125C) the batteries provided 90% of their capacity, whereas at the C/2 rate at −20°C, the battery provided 37% of its capacity. After the low temperature discharge, the battery provided its original capacity when discharged at 21°C indicating the system was not degraded by thermal cycling.

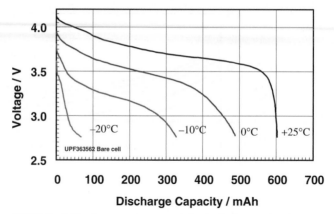

FIGURE 35.92 Battery voltage on 0.57 A discharge of a 0.57 Ah polymer Li-ion battery. The battery was charged in a CCCV regime at 0.57 A to 4.2 V (0.028 A cutoff). (*Courtesy of Sanyo.*)

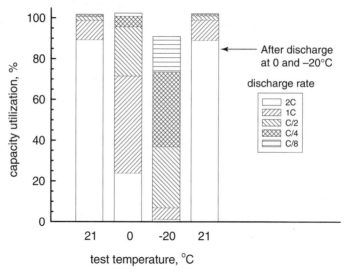

FIGURE 35.93 Low temperature discharge capability of C/LiCoO$_2$ polymer Li-ion batteries. Batteries were charged at 21°C. (*Courtesy of Telcordia.*)

Cycle Life. Polymer C/LiCoO$_2$ batteries offer high cycle life, as illustrated in Fig. 35.94 for a 0.136 Ah battery. This battery was charged at the 1C rate to 4.2 V (CCCV, 90 min.), then discharged at either the 0.2C, C, or 2C rate. After discharge at 2C, the battery was further discharged at 1C. Over 300 cycles, the battery capacity faded nominally 10%. The effect of repeated cycling on the battery voltage and capacity is illustrated in Fig. 35.95, which shows discharge curves at the 0.2C, C or 2C rates at cycles 20, 80, 130, 180, 240 and 290. As shown, little change in average voltage was observed, less than 0.1 V change was observed, although the effect of capacity fade was more significant at lower rates. The decrease in voltage is due to the increase in internal resistance with repeated cycling. Also, the thickness of polymer Li-ion batteries increases 2 to 3% over the first 100 cycles. The effect of these changes is illustrated in Fig. 35.96, which shows the capacity of a 0.57 Ah polymer Li-ion battery when cycled at the 1C rate for 500 cycles, and Fig. 35.97, which shows the internal resistance of the battery and the battery thickness during this regime.

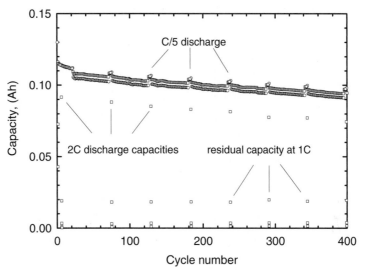

FIGURE 35.94   Capacity of a 0.136 Ah C/LiCoO$_2$ polymer Li-ion battery when charged at the 1C rate to 4.2 V (CCCV, 90 min.), and discharged at the 1C rate to 2.8 V, with periodic discharge at the 0.2C and 2C rates, at 21°C. (*Courtesy of Telcordia.*)

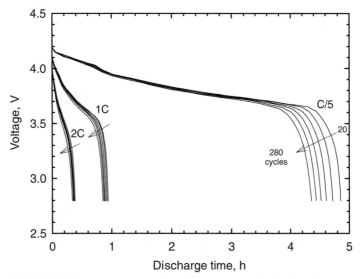

FIGURE 35.95   Battery voltage when discharged at the 0.2C, C or 2C rates at cycle numbers from 20 to 290, at 21°C. (*Courtesy of Telcordia.*)

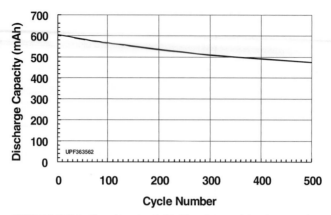

FIGURE 35.96 Capacity of a 0.57 Ah polymer Li-ion battery when charged at 0.57 A in a CCCV regime to 4.2 V (0.028 A cutoff), and discharged at 0.57 A to 2.57 V. (*Courtesy of Sanyo.*)

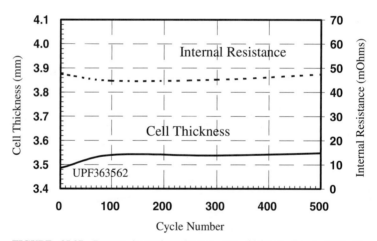

FIGURE 35.97 Battery internal resistance and thickness for a 0.57 Ah polymer Li-ion battery charged at 0.57 A in a CCCV regime to 4.2 V (0.028 A cutoff) and discharged at 0.57 A to 2.75 V. (*Courtesy of Sanyo.*)

Storage Stability of Polymer C/LiCoO$_2$ Batteries.   Polymer C/LiCoO$_2$ Li-ion batteries can sustain storage at elevated temperature, as illustrated in Fig. 35.98, for storage at 60°C, and in Fig. 35.99 for storage at 80°C. In these tests, batteries were cycled at the C/5 rate, stored, then cycled at either the C/2 or C/5 rates. As shown, after storage at 60°C or 80°C, approximately 7% capacity loss was observed.

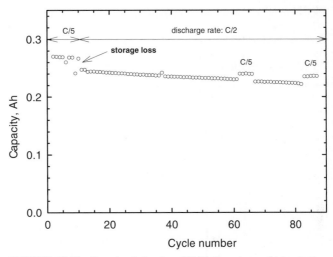

FIGURE 35.98 Capacity fade of a C/LiCoO$_2$ polymer Li-ion battery when cycled at 21°C after storage fully charged at 60°C for seven days. (*Courtesy of Telcordia.*)

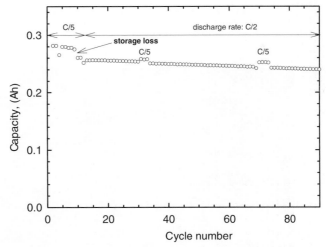

FIGURE 35.99 Capacity fade of a C/LiCoO$_2$ polymer Li-ion battery when cycled at 21°C after storage fully charged at 80°C for seven days. (*Courtesy of Telcordia.*)

35.7.4 Performance of C/LiMn$_2$O$_4$-type Polymer Li-ion Batteries

The electrical performance of polymer Li-ion batteries that employ the C/LiMn$_2$O$_4$ chemistry is comparable to other Li-ion batteries. The voltage of a C/LiMn$_2$O$_4$ battery when discharged at 0.5C then charged in a CCCV regime at the 0.5C rate to 4.2 V is indicated in Fig. 35.100. As shown, the batteries provide slightly higher average voltage than C/LiCoO$_2$-type batteries, 3.8 V, although their specific energy and energy density are lower than comparable batteries that employ a cobalt-based chemistry. The batteries also provide a low self discharge rate, 2.5% per month at 23°C and 6% per month at 40°C. The manganese-based cathode materials these batteries employ are regarded as environmentally more benign than their cobalt or nickel-based analogs, and the batteries are claimed to offer improved safety properties. Figure 35.101 is an illustration of several polymer $C/LiMn_2O_4$ Li-ion batteries.

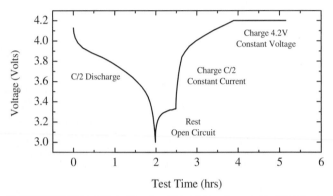

FIGURE 35.100   Voltage as a function of time for a 3 Ah $C/LiMn_2O_4$ polymer Li-ion battery discharged and charged at the 0.5C rate. (*Courtesy of Valence Technology.*)

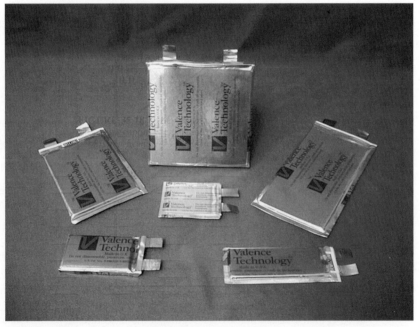

FIGURE 35.101   Polymer $C/LiMn_2O_4$ Li-ion batteries. (*Courtesy of Valence Technology.*)

Discharge Rate Capability. The rate capability of a $C/LiMn_2O_4$ batteries charged in a CCCV regime at the 0.5C rate to 4.2 V is illustrated in Fig. 35.102. As shown, the batteries provide 98% of their capacity at the 0.5C rate, and 85% at the 1C rate. Also, these batteries provide a flat voltage profile with average voltage of 3.8 V at low rates (0.2C); at higher rates (1C) the average voltage is lower, typically 3.6 V.

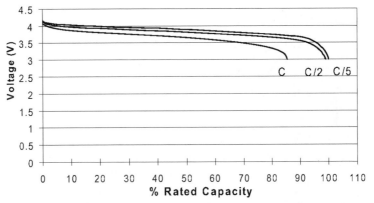

FIGURE 35.102   Rate capability of a $C/LiMn_2O_4$ polymer Li-ion battery charged at the 0.5C rate to 4.2 V in a CCCV regime at 23°C. (*Courtesy of Valence Technology.*)

High and Low Temperature Discharge Capability. The ability of $C/LiMn_2O_4$ polymer Li-ion batteries to provide 0.2C discharge capability at temperatures from 60°C to −20°C is illustrated in Fig. 35.103. As shown, at temperatures at or above 23°C, the battery voltage and capacity was little changed. At 0°C, the battery provided over 95% of its capacity; at −10° the battery provided 86% of its capacity, and at −20°C the battery provided 46% of the capacity delivered at 23°C. At −20°C the average voltage was 3.5 V, 0.3 V lower than provided at 23°C. The ability of a $C/LiMn_2O_4$ polymer Li-ion cell to provide 1C discharge capability at temperatures from 23°C to −20°C is indicated by the data shown in Fig. 35.104. As shown, at 0°C the 0.6 Ah cell provided 40% of its capacity, and at −10°C, 23%.

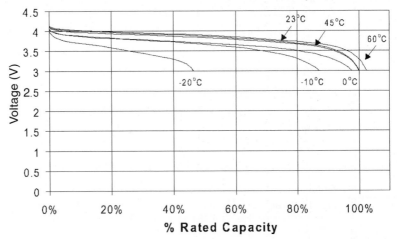

FIGURE 35.103   Voltage of a 0.6 Ah $C/LiMn_2O_4$ polymer Li-ion battery when discharged at the 0.2C rate at temperatures from 60°C to −20°C. The battery was charged in a CCCV regime at 0.5C to 4.2 V at 23°C. (*Courtesy of Valence Technology.*)

Cycle Life. Polymer $C/LiMn_2O_4$ batteries also provide long cycle life, as illustrated in Fig. 35.105 for a battery cycled at the 0.5C rate. As shown, the battery provided 90% of its initial capacity after 300 cycles, and 84% after 600 cycles.

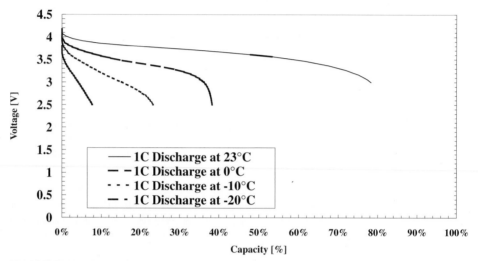

FIGURE 35.104 Voltage of a 0.6 Ah $C/LiMn_2O_4$ polymer Li-ion battery when discharged at the 1C rate at temperatures from 23°C to −20°C. The battery was charged in a CCCV regime at 0.5C to 4.2V at 23°C. (*Courtesy of Valence Technology.*)

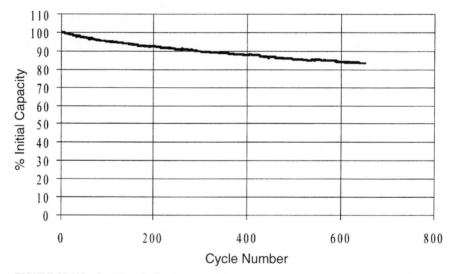

FIGURE 35.105 Capacity of a $C/LiMn_2O_4$ polymer Li-ion battery when charged in a CCCV regime at the 0.5C rate to 4.2 V and discharged to 3.0 V at the 0.5C rate, at 23°C. (*Courtesy of Valence Technology.*)

35.7.5 Conclusions: Polymer Li-ion Batteries

Polymer Li-ion batteries offer specific performance comparable to cylindrical and prismatic Li-ion batteries within a unique, thin form factor. In addition to providing a thin form factor, the batteries offer many properties desirable for commercial applications including good rate capability, low self-discharge, long storage stability and the ability to safely sustain physical or electrical abuse. These products are now being produced in high volume and are being used in increasing quantities in cell phones and PDAs.

35.8 THIN-FILM, SOLID-STATE Li-ION BATTERIES

A specialized type of Li-ion battery developed for semi-conductor and printed circuit board (PCB) applications are thin-film, solid-state devices. These batteries which employ ceramic negative, solid electrolyte and positive electrode materials, can sustain high temperatures (250°C), and can be fabricated by high volume manufacturing techniques on silicon wafers which are viable as on-chip or on-board power sources for microelectronics.[97,98] Batteries of this type can be very small, 0.04 cm × 0.04 cm × 2.0 μm. For microelectronics applications, all components must survive solder re-flow conditions, nominally 250°C in air or nitrogen for 10 minutes. Cells with liquid or polymer electrolytes cannot sustain these conditions because of the volatility or thermal stability of organic components. Further, cells that employ lithium metal also fail as solder re-flow conditions exceed the melting point of lithium (180.5°C).

A schematic drawing of a thin-film, solid-state Li-ion cell is shown in Fig. 35.106. These cells are fabricated by sequential layer deposition of the cell components using rf magnetron sputtering, except for the metallic current collector components which are deposited by DC magnetron sputtering. The deposition conditions for $LiCoO_2$[99] and lithium phosphorous oxynitride (LiPON) electrolyte are reported in the literature.[98] The cells are fabricated on a substrate, typically alumina, quartz, soda-lime glass, or silicon. Positive current collectors of gold or platinum (0.1 μm to 0.3 μm), over a layer of cobalt (0.01 μm to 0.05 μm, to improve adhesion), have been used. Cells using either $LiCoO_2$ or $LiMn_2O_4$ positive electrode materials have been fabricated. The positive electrode layer of laboratory test cells is typically 0.05 μm to 5 μm thick and 0.04 cm^2 to 25 cm^2 in area, depending on the capacity required by the application. The electrolyte layer, LiPON, is typically 0.7 μm to 2 μm thick.

Cells with and without negative electrode materials have been fabricated.[100] In cells with a negative electrode material, $SiSn_{0.87}O_{1.20}N_{1.72}$ (SiTON), SnN_x, InN_x or Zn_3N_2 have been used. To accommodate the difference in volumetric capacity of positive and negative electrode materials, the thickness of the negative electrode is typically 7% of the positive electrode. Negative current collectors of copper, titanium, or titanium nitride (0.1 μm to 0.3 μm) are typical. To enhance the hermeticity of the cell, protective overlayers of LiPON (1 μm) or parylene (6 μm) and titanium or aluminum (0.1 μm) have been used.

Alternatively, the negative electrode materials may be omitted. A schematic diagram of this type of cell prior to and after initial charge is illustrated in Fig. 35.107. In these cells, lithium metal is plated onto the negative current collector when the cell is charged. The lithium plating-stripping process is efficient with this technology. Thus cells with a negative electrode material are typically engineered to oversaturate the negative electrode material, or the negative electrode material is omitted completely.

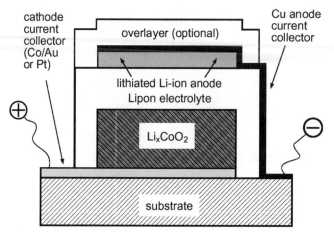

FIGURE 35.106 Schematic drawing of the construction of a thin-film, solid-state Li-ion cell after an initial charge that lithiates the negative electrode material. (*Courtesy of Oak Ridge National Laboratory.*)

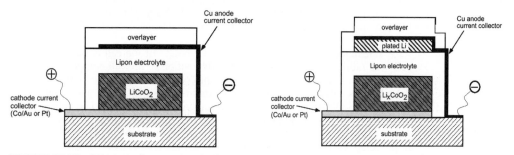

FIGURE 35.107 Schematic diagram of a thin-film "lithium free" solid-state battery prior to and after initial charge. (*Courtesy of Oak Ridge National Laboratory. Reproduced by permission of The Electrochemical Society, Inc. From Ref. 100.*)

35.8.1 Electrical Performance of Thin-film, Solid-state Li-ion Batteries

SiTON/LiCoO$_2$ Batteries. The rate capability of a SiTON/LiCoO$_2$ battery before any thermal treatment is illustrated in Fig. 35.108. As shown, the battery provided a discharge current density up to 5 mA/cm^2, comparable to cylindrical Li-ion batteries, within a voltage range of 4.2 V to 2.7 V. The utilization of the negative electrode material (anode) was high, 600 mAh/g at 2 mA/cm^2.

These batteries are designed for use in microelectronics applications where all components must sustain solder reflow processes, typically a heat treatment at 250°C for 10 minutes. To demonstrate the ability to sustain these conditions, Figure 35.109 shows discharge curves before and after heat treatment at 250°C for 10 minutes or 1 hour. As shown, the capacity of the battery increased by 20% as a result of the 10 minute treatment, indicating that solder re-flow processes improve battery performance. This improvement has been suggested to result from improved ionic conductivity and decreased charge transfer resistance of the LiPON electrolyte as a result of the heat treatment.[98] Heat treatment also improved the rate capability of the battery, as illustrated in Fig. 35.110. As shown, at 5 mA/cm^2, between 4.2 V and 2.7 V, the utilization of the SiTON was 450 mAh/g, 30% higher that observed before heat treatment. In addition, discharge at 10 mA/cm^2 and 20 mA/cm^2 was demonstrated.

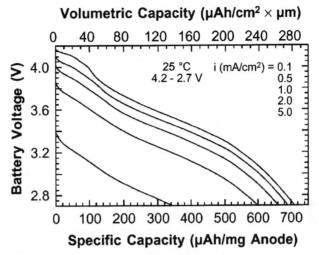

FIGURE 35.108   Rate capability of a SiTON/LiPON/LiCoO$_2$ battery. Specific capacity and volumetric capacity are based on the as-deposited mass and volume of SiTON. (*Courtesy of Oak Ridge National Laboratory. Reproduced from the Journal of Power Sources, Vol. 81, 1999, with permission from Elsevier Science.*)

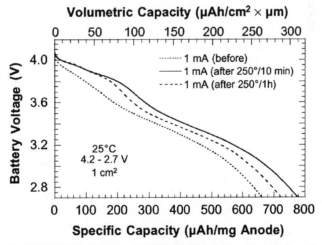

FIGURE 35.109   Voltage of a SiTON/LiCoO$_2$ battery when discharged at 25°C at 1 mA/cm^2 before and after heat treatment at 250°C for 10 minutes or 1 hour. (*Courtesy of Oak Ridge National Laboratory. Reproduced from the Journal of Power Sources, Vol. 81, 1999, with permission from Elsevier Science.*)

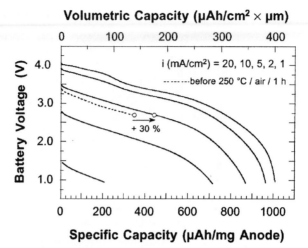

FIGURE 35.110 Rate capability of a SiTON/LiPON/LiCoO$_2$ battery after treatment at 250°C in air for 1 hour. Specific capacity and volumetric capacity are based on the as-deposited mass and volume of SiTON. (*Courtesy of Oak Ridge National Laboratory. Reproduced from the Journal of Power Sources, Vol. 81, 1999, with permission from Elsevier Science.*)

Solid-state SiTON/LiCoO$_2$ batteries offer high cycle life, as illustrated in Fig. 35.111 for a cathode heavy cell cycled initially between 3.93 V and 2.7 V for 3000 cycles, then between 4.1 V and 2.7 V for 10,000 cycles, all at 0.08 mA/cm^2 at 25°C. When the upper voltage limit was set to 3.93 V, no Li-plating occurred, whereas with the upper voltage limit set to 4.1 V, 30% of the capacity occurred while the negative electrode was at 0 V vs. lithium. In the initial 3000 cycles, the capacity fade rate was 0.001% per cycle, whereas with the higher upper voltage limit, the capacity fade rate was 0.002% per cycle while the cell capacity was initially increased over 40%.

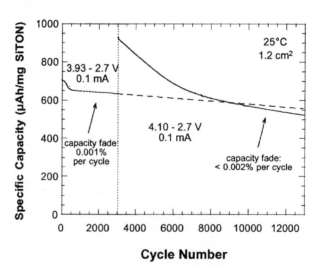

FIGURE 35.111 SiTON specific capacity when cycled in a SiTON/LiCoO$_2$ battery between either 3.93 V and 2.7 V or 4.1 V and 2.7 V, at 0.08 mA/cm^2 at 25°C. (*Courtesy of Oak Ridge National Laboratory. Reproduced from the Journal of Power Sources, Vol. 81, 1999, with permission from Elsevier Science.*)

LiCoO$_2$ Batteries with In-situ Electroplated Negative Electrodes.   Solid-state batteries fabricated without a negative electrode material offer performance comparable to SiTON/ LiCoO$_2$ cells, the simplicity of fewer manufacturing steps and materials, avoid complications or irreversible capacity associated with oxide or oxynitride negative electrode materials, and avoid the limitations of batteries that contain lithium metal. When fabricated, lithium metal is not present, they can be processed at high temperature. When these batteries are charged, lithium metal is plated onto the negative current collector, a highly reversible process in these batteries.

The battery voltage on the initial charge cycle and subsequent discharge at 0.1 mA/cm^2, 1 mA/cm^2, and 5 mA/cm^2 at 25°C is illustrated in Fig. 35.112. As anticipated, the battery has a voltage characteristic for LiCoO$_2$ when cycled versus lithium, and can operate at current densities comparable to cylindrical C/LiCoO$_2$ batteries. The rate capability of batteries with three different thicknesses of LiCoO$_2$ is further illustrated in Fig. 35.113. As shown, the capacity and rate capability of the cells scales linearly with the thickness of the LiCoO$_2$ layer. Discharge at current densities as high as 5 mA/cm^2 was demonstrated.

The capacity of a thin-film LiCoO$_2$ battery with an in-situ electroplated lithium negative electrode cycled at high rate (4C charge, 20C discharge) is illustrated in Fig. 35.114. As shown, the fade rate was 0.02%/cycle, comparable to that typical for cylindrical C/LiCoO$_2$ batteries cycled at the 1C rate.

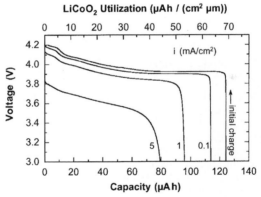

FIGURE 35.112   Battery voltage for a solid-state thin-film LiCoO$_2$ battery with an in-situ electroplated lithium negative electrode. (*Courtesy of Oak Ridge National Laboratory. Reproduced by permission of The Electrochemical Society, Inc. From Ref. 100.*)

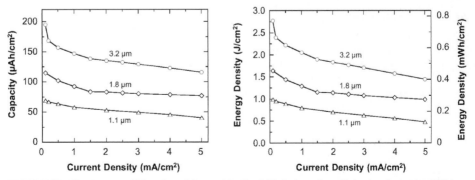

FIGURE 35.113   Rate capability of a solid-state thin-film LiCoO$_2$ batteries with in-situ electroplated lithium negative electrodes. The cells were cycled between 4.2 V and 3.0 V at 25°C. (*Courtesy of Oak Ridge National Laboratory. Reproduced by permission of The Electrochemical Society, Inc. From Ref. 100.*)

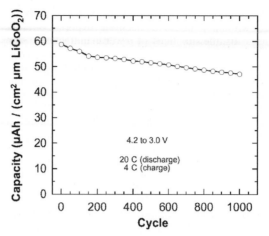

FIGURE 35.114 Capacity of a LiCoO$_2$ battery with an in-situ electroplated lithium negative electrode, normalized to 1 cm^2 and 1 μm LiCoO$_2$. The battery was cycled between 4.2 V and 3.0 V at 25°C, charged at the 4C rate and discharged at the 20C rate. (*Courtesy of Oak Ridge National Laboratory.*)

35.9 CONCLUSIONS AND FUTURE TRENDS

Li-ion batteries have quickly evolved from an R and D interest to a significant and growing fraction of the worldwide battery market. The acceptance of the technology has been driven by its unique ability to offer a high level of performance in many aspects, including energy density, specific energy, rate capability, cycle life, and storage life, in a safe, low cost product. As costs are reduced, the diversity of available designs increases, and performance improves, the range of applications addressed with Li-ion batteries is anticipated to increase.

Further improvements in cell performance will be made possible through both more efficient mechanical designs and improved materials. Li-ion materials are currently a subject of great interest in the R and D community. Improved positive electrode materials that offer higher capacity and improved safety properties are in development, as are new negative electrode materials, such as the tin-based materials, that offer the potential for further improvement in specific energy, energy density, rate capability and longevity.

ACKNOWLEDGEMENT

The author thanks the many individuals and organizations that contributed the data presented in this manuscript, Jan Riemers, Rick Howard and Jordan Gonchar for their comments, and Elizabeth Ehrlich for her patience.

REFERENCES

1. "Studies of the Li-ion Battery Market-Year 1999". Yano Research Institute Ltd. of Japan (2000) and D. MacArthur, G. Blomgren and R. Powers, "Lithium and Lithium Ion Batteries 2000." Power Associates (2000).

2. *Purchasing,* **128:**100, 2000.

3. M. S. Wittingham and M. B. Dines, *Surv. Prog. Chem.* **9:**55, (1980).

4. A. Weiss, *Angew. Chem.* **75:**755–61 (1963).

5. A. Herold, *Bull. Soc. Chim. Fr., 999* (1955).

6. A. Herold, in *Intercalated Materials,* F. Levy (ed), D. Reidel Publishing, Dordrecht, The Netherlands, 1979, p. 323.

7. D. M. Adams, *Inorganic Solids,* Wiley, New York, 1974.

8. A. R. West, *Solid State Chemistry,* Wiley, New York, 1984, pp. 25–29.

9. H. Selig and L. B. Ebert, *Advances in Inorganic Chemistry and Radiochemistry*, **23:**281, 1980.

10. J. R. Dahn, A. K. Sleigh, H. Shi, B. M. Way, W. J. Weydanz, J. N. Reimeres, Q. Zong, and U. von Sacken in "Lithium Batteries—New Materials, Developments and Perspectives," G. Pistoia (ed.) 1994, pp. 1–97.

11. U.S. Patent 4,357,215. U.S. Patent 4,302,518.

12. M. R. Palacin, D. Larcher, A. Audemer, N. Sac-Epee, G. G. Amatucci, and J.-M. Tarascon, *J. Electrochem. Soc.* **144:**4226 (1997).

13. J. Cho, H. Jung, Y Park, G. Kim, and H. S. Lim, *J. Electrochem. Soc.,* **147:**15–20 (2000).

14. T. Ohzuku, A. Ueda, and M. Nagayama, *J. Electrochem. Soc.,* **140:**1862–1870 (1993).

15. K. Dokko, M. Nishizawa, S. Horikoshi, T. Itoh, M. Mohamedi, and I. Uchida, *Electrochemical and Solid State Letters,* **3:**125–127 (2000).

16. H. J. Kweon and D. G. Park, *Electrochemical and Solid State Letters,* **3:**128–130 (2000).

17. J. Cho, H. Jung, Y. Park, G. Kim, and H. S. Lim, *J. Electrochem. Soc.,* **147:**15–20 (2000).

18. R. Koksbang, J. Barker, H. Shi, and M. Y. Saidi, *Solid State Ionics,* **84:**1–21, 1996.

19. K. Mizushima, P. C. Jones, P. J. Wiseman, and J. B. Goodenough, *Mat. Res. Bull.,* **15:**783–789 (1980).

20. W. D. Johnson, R. R. Heikes, and D. Sestrich, *Phys. Chem. Solids,* **7:**1–13 (1958).

21. E. Jeong, M. Won, and Y. Shim, *J. Power Sources,* **70:**70–77 (1998).

22. M. Yoshio, H. Tanaka, K. Tominaga, and H. Noguchi, *J. Power Sources,* **40:**347–353 (1992); E. Zhecheva, R. Stoyanova, M. Gorova, R. Alcantra, J. Moales, and J. L. Tirado, *Chem. Mater.,* **8:** 1429–1440 (1996).

23. B. Garcia, J. Farcy, J. P. Pereira-Ramos, J. Perichon, and N. Baffier, *J. Power Sources,* **54:**373–377 (1995).

24. P. N. Kumta, D. Gallet, A. Waghray, G. E. Blomgren, and M. P. Setter, *J. Power Sources,* **72:**91–98 (1998).

25. Y. Chiang, Y. Jang, H. Wang, B. Huang, D. Sadoway, and P. Ye, *J. Electrochem. Soc.,* **145:**887 (1998).

26. T. J. Boyle, D. Ingersoll, T. M. Alam, C. J. Tafoya, M. A. Rodriguez, K. Vanheusden and D. H. Doughty, *Chem. Mater.,* **10:**2270–2276 (1998).

27. G. G. Amatucci, J. M. Tarascon, D. Larcher, and L. C. Klein, *Solid State Ionics,* **84:**169–180 (1996); D. Larcher, M. R. Palacin, G. G. Amatucci, and J.-M. Tarascon, *J. Electrochem. Soc.,* **144:**408 (1997).

28. M. Antaya, J. R. Dahn, J. S. Preston, E. Rossen, and J. N. Reimers, *J. Electrochem. Soc.,* **140:**575 (1993); M. Antaya, K. Cearns, J. S. Preston, J. N. Reimers, and J. R. Dahn, *J. Appl. Phys.,* **75:**2799 (1994).

29. P. Frajnaud, R. Nagarajan, D. M. Schleich, and D. Vujic, *J. Power Sources,* **54:**362–366 (1995).

30. E. Antolini, *J. Eur. Ceram. Soc.,* **18:**(10):1405–1411 (1998).

31. D. Caurant, N. Baffier, B. Garcia, and J. P. Pereira-Ramos, *Solid State Ionics,* **91:**45–54 (1996).

32. R. J. Gummow and M. M. Thackeray, *J. Electrochem. Soc.,* **140:**3365–3368 (1993).

33. Y. Gao and J. R. Dahn, *J. Electrochem. Soc.,* **143:**100–114 (1996). Y. Gao and J. R. Dahn, *Solid State Ionics,* **84:**33–40 (1996).

34. J. M. Tarascon, W. R. McKinnon, F. Coowar, T. N. Bower, G. Amatucci, and D. Guyomard, *J. Electrochem. Soc.,* **141:**1421–1431 (1994). J. M. Tarascon, F. Coowar, G. Amatuci, F. K. Shokoohi, and D. G. Guyomard, *J. Power Sources,* **54:**103–108 (1995).

35. R. J. Gummow, A. de Kock, and M. M. Thackeray, *Solid State Ionics,* **69:**59–67 (1994). M. H. Rossow, A. de Kock, L. A. de Picciotto, and M. M. Thackeray, *Mat. Res. Bull.,* **25:**173–182 (1990).

36. A. Yamada, K. Miura, K. Hinokuma, and M. Tanaka, *J. Electrochem. Soc.,* **142:**2149–2156 (1995).

37. J. F. Brady (ed.), "The Book of Chemical Lists," *Business & Legal Reports,* Madison, CT, 1990.

38. R. Koksbang, J. Barker, H. Shi and M. Y. Saidi, *Solid State Ionics,* **84:**1–21 (1996).

39. A. Antonini, C. Bellitto, M. Pasquali, and G. Pistoia, *J. Electrochem. Soc.,* **145:**2726 (1998).

40. R. J. Gummow, A. de Kock, and M. M. Thackeray, *Solid State Ionics,* **69:**59–67 (1994); M. M. Thackeray, Y. S. Horn, A. J. Kahaian, K. D. Kepler, E. Skinner, J. T. Vaughey, and S. A. Hackney, *Electrochemical and Solid-State Letters,* **1:**7–9 (1998).

41. J. M. Tarascon, W. R. McKinnon, F. Coowar, T. N. Bowmer, G. Amatucci, and D. Guyomard, *J. Electrochem. Soc.,* **141:**1421 (1994). R. J. Gummow, A. de Kock, and M. M. Thackeray, *Solid State Ionics,* **69:**59 (1994). Y. Xia, Y. Zhou, and M. Yoshio, *J. Electrochem. Soc.,* **144:**2204 (1977). G. G. Amatucci, C. N. Schmutz, A. Blyr, C. Sigala, A. S. Gozdz, D. Larcher, and J. M. Tarascon, *J. Power Sources,* **69:**11 (1997).

42. D. H. Jang, Y. J. Shin, and S. M. Oh, *J. Electrochem. Soc.,* **143:**2204 (1996).

43. Y. Xia, Y. Zhou, and M. Yoshio, *J. Electrochem. Soc.,* **144:**2593 (1997).

44. S. Mori, H. Asahina, H. Suzuki, A. Yonei, and K. Yokoto, *J. Power Sources,* **68:**59–64 (1997). R. Imhof and P. Novak, *J. Electrochem. Soc.,* **146:**1702–06 (1999).

45. D. Aurbach, Y. Ein-Eli, B. Markovsky, A. Zaban, S. Luski, Y. Carmeli, and H. Yamin, *J. Electrochem. Soc.,* **142:**2882 (1995).

46. D. Aurbach, Y. Ein-Eli, O. Chusid, M. Babai, Y. Carmeli, and H. Yamin, *J. Electrochem. Soc.,* **141:** 603 (1994). S. Mori, H. Asahina, H. Suzuki, A. Yonei, and K. Yokoto, *J. Power Sources,* **68:**59–64 (1997).

47. Y. Wang and J. Reimers, "Proposed Mechanism for Cycling Fade in LiMn₂O₄ Li-ion Cells." *9th Int. Meeting on Lithium Batteries,* July 12–17, 1998, Edinburgh.

48. J. Reimers, Y. Wang, M. Gee, M. Zhang, L. Ting, and M. Tao, presented at the 196th Meeting of the Electrochemical Society, Oct. 17–22, 1999, Honolulu, Hawaii.

49. M. M. Thackeray, *J. Electrochem. Soc.,* **142:**2558 (1995). A. Blyr, A. Du Pasquier, G. Amatucci, and J.-M. Tarascon, *Ionics,* **3:**321 (1977). Y. Sun, Y. Jeon, H. J. Lee, *Electrochemical and Solid State Letters,* **3:**7 (2000).

50. A. Yamada and M. Tanaka, *Mat. Res. Bull,* **30:**715–721 (1995).

51. A. Yamada, K. Miura, K. Hinokuma, and M. Tanaka, *J. Electrochem. Soc.,* **142:**2149 (1995). M. M. Thackeray, Y. Shao-Horn, A. J. Kahaian, K. D. Kepler, E. Skinner, J. T. Vaughey, and S. A. Hackney, *Electrochemical and Solid State Letters,* **1:**7 (1998).

52. G. Pistoia, A. Antonini, R. Rosati, and C. Bellitto in "Batteries for Portable Applications and Electric Vehicles," C. F. Holmes and A. R. Langrebe (eds.), *The Electrochemical Society Proc. Series,* Pennington, N.J., 1997, PV 97-18, p. 406.

53. W. F. Howard, Jr., S. W. Sheargold, M. S. Jordan, and S. H. Tyler, *Proc. 17th Int. Seminar on Primary and Secondary Batteries,* Ft. Lauderdale, Fla., March 6–9, 2000.

54. T. Inoue, H. Yamane, and M. Sano, *Proc. NPC Conf.,* Japan, p. 347ff.

55. "Cellular Phone Recall May Cause Setback for Moli," Toronto Globe and Mail, August 15, 1989 (Toronto, Canada); *Adv. Batt. Technology,* **25**(10):4 (1989).

56. J. R. Dahn, A. K. Sleigh, H. Shi, B. M. Way, W. J. Weydanz, J. N. Reimers, Q. Zong, and U. von Sacken, in "Lithium Batteries—New Materials, Developments and Perspectives," G. Pistoia (ed.), pp. 1–97.

57. R. E. Franklin, *Proc. Roy. Soc.* (London), **A209:**196 (1951).

58. M. Inagaki, *Solid State Ionics,* **86–88:**833–839 (1996).

59. T. Zheng, J. N. Reimers, and J. R. Dahn, *Phys. Rev. B,* **51:**734 (1995).

60. H. Selig and L. B. Ebert, *Advances in Inorganic Chemistry and Radiochemistry,* **23:**281 1980.

61. H. Shi, J. Barker, M. Y. Saidi, and R. Koksbang, *J. Electrochem. Soc.,* **143:**3466 (1996).

62. Y. C. Chang, J. J. Sohn, C. H. Ku, Y. G. Wang, Y. Korai, and I. Mochida, *Carbon,* **37:**1285 (1999).

63. Y. Jung. M. Suh, S. Shim, and J. Kwak, *J. Electrochem. Soc.,* **145:**3123 (1998); W. Xing, R. A. Dunlap, and J. R. Dahn, *J. Electrochem. Soc.,* **145:**62 (1998).

64. K. Sato, M. Noguchi, A. Demachi, N. Oki, and M. Endo, *Science,* **264:**556 (1994).

65. J. R. Dahn, T. Zheng, Y. Liu, and J. S. Xue, *Science,* **270:**590 (1995).

66. P. Papanek, M. Radosavljevic, and J. E. Fischer, *Chem. Mater,* **8:**1519–1526 (1996).

67. T. D. Tran and J. H. Feikert, *J. Electrochem. Soc.,* **142:**3297 (1995); G. Chung. S. Jun, K. Lee, and M. Kim, *J. Electrochem. Soc.,* **146:**1664–1671 (1999).

68. F. M. Gray, *Polymer Electrolytes,* The Royal Society of Chemistry, 1997.

69. A. B. McEwen, H. L. Ngo, K. LeCompte, and J. L. Goldman, *J. Electrochem. Soc.,* **146:**1687–1695 (1999); A. B. McEwen, S. F. McDevitt, and V. R. Koch, *J. Electrochem. Soc.,* **144:**L84 (1997).

70. H. Nakamura, H. Komatsu, and M. Yoshio, *J. Power Sources,* **62:**219–222 (1996).

71. B. Scrosati and S. Megahed, Electrochemical Society Short Course, New Orleans, Oct. 10, 1993.

72. D. Linden (ed.), "The Handbook of Batteries," 2d ed. McGraw-Hill, New York, 1995, p. 36.14.

73. J. T. Dudley, D. P. Wilkinson, G. Thomas, R. LaVae, S. Woo, H. Blom, C. Horvath, M. W. Juzkow, B. Denis, P. Juric, P. Aghakian, and J. R. Dahn, *J. Power Sources,* **35:**59–82 (1991).

74. S. T. Mayer, H. C. Yoon, C. Bragg, and J. H. Lee, "Low Temperature Ethylene Carbonate Based Electrolyte for Lithium-ion Batteries," Polystor Corporation, Dublin, CA, 1997.

75. D. Guyomard and J. M. Tarascon, *J. Electrochem. Soc.,* **54:**92 (1995); T. Zheng, Y. Liu, E. W. Fuller, U. von Sacken, and J. R. Dahn, *J. Electrochem. Soc.,* **142:**2581 (1995); D. Aurbach, B. Markovsky, A. Schechter, Y. Ein-Eli, and H. Cohen, *J. Electrochem. Soc., 143:*3809 (1996).

76. D. Aurbach, Y. Ein-Eli, B. Markovsky, A. Zaban, S. Luski, Y. Carmeli, and H. Yamin, *J. Electrochem. Soc.,* **142:**2882 (1995).

77. H. Yoshida, T. Fukunaga, T. Hazama, M. Terasaki, M. Mizutani, and M. Yamachi, *J. Power Sources,* **68:**311–315 (1997).

78. Y. Ein-Eli, S. F. McDevitt, D. Aurbach, B. Markovsky, and A. Schechter, *J. Electrochem. Soc.,* **144:**L180 (1997).

79. Z. X. Shu, R. S. McMillian, and J. J. Murray, *J. Electrochem. Soc.,* **140:**922 (1993).

80. O. Chusid, Y. Ein-Eli, M. Babai, Y. Carmeli, and D. Aurbach, *J. Power Sources,* **43–44:**47 (1993).

81. R. Spotnitz, in "Handbook of Battery Materials," J. O. Besenhard (ed.), VCH Wiley, Amsterdam and New York, 1999.

82. H. S. Bierenbaum, R. B. Isaacson, M. L. Druin, and S. G. Plovan, *Ind. Eng. Chem. Prod. Res. Dev.,* **13:**2 (1974).

83. G. Venugopal, J. Moore, J. Howard, and S. Pendalwar, *J. of Power Sources,* **77:**34–41 (1999).

84. R. P. Quirk and M. A. A. Alsamarraie, in "Polymer Handbook," J. Brandrup and E. H. Immergut (eds.), Wiley, New York, 1989.

85. Y. Wang, M. Zhang, U. Von Sacken, and B. M. Way, European Patent Application 98108301.7 (1999).

86. S. E. Sloop and M. M. Lerner, *J. Electrochem. Soc.,* **143:**1292 (1996).

87. C. G. Pillai, A. K. Padhi, C. Homan, K. Pisarczyk, and M. Osborne, *Proc. 17th Int. Seminar on Primary and Secondary Batteries,* Ft. Lauderdale, Fla., March 6–9, 2000; Japan Patent 07296847 JP, Fuji Elelctrochem Co. Ltd., 1995.

88. M. C. Smart, B. V. Ratnakumar, and S. Surampudi, *J. Electrochem. Soc.,* **146:**486–492 (1999).

89. B. A. Johnson and R. E. White, *J. Power Sources,* **70:**48–54 (1998).

90. Courtesy SAFT.

91. E. Plichta, S. Slane, M. Uchiyama, M. Salomon, D. Chu, W. B. Ebner, and H. W. Lin, *J. Electro-chem. Soc.,* **136:**1865 (1989); E. Plichta, M. Salomon, S. Slane, and M. Uchiyama, *J. Power Sources,* **21:**25 (1987).

92. J. Cho, H. Jung, Y. Park, G. Kim, and H. S. Lim, *J. Electrochem. Soc.,* **147:**15–20 (2000).

93. T. Sasaki, N. Imamura, M. Terasaki, M. Mizutani, and M. Yamachi, "Studies on the Characteristics of Float-Charged Li-ion Battery [sic]", presented at the 196th meeting of the Electrochemical Society, Oct. 17–22, 1999.

94. F. M. Gray, "Polymer Electrolytes," RSC Materials Monograph, J. A. Connor (ed.) 1997.

95. A. Gozdz. of Telcordia Technologies, Red Bank, N.J. Private Communication.

96. A. Gozdz, U. S. Patents 5,571,634 and 5,540,741.

97. B. J. Neudecker and R. A. Zuhr, *Proc. Electrochemical Society: Intercalation Compounds for Battery Materials.* (1999) Submitted for publication.

98. B. J. Neudecker, R. A. Zhur, J. B. Bates, *J. Power Sources,* **81–82:**27–32 (1999).

99. J. B. Bates, N. J. Dudley, B. J. Neudecker, F. X. Hart, H. P. Jun, and S. A. Hackney, *J. Electrochem. Soc.,* **147:**59–70 (2000).

100. B. J. Neudecker, N. J. Dudney, and J. B. Bates, *J. Electrochem. Soc.,* **147:**517–523 (2000).

CHAPTER 36
RECHARGEABLE ZINC/ALKALINE/ MANGANESE DIOXIDE BATTERIES

Karl Kordesch and Josef Daniel-Ivad

36.1 GENERAL CHARACTERISTICS

The rechargeable zinc/alkaline/manganese dioxide battery is an outgrowth of the primary battery. Zinc is used as the negative active material (the anode during discharge), manganese dioxide for the positive active material (the cathode during discharge), and a potassium hydroxide solution for the electrolyte.

The original design of this rechargeable battery closely followed the cylindrical inside-out design of the alkaline primary battery and retained its advantages of long shelf life, good current density and safety.[1] This battery was marketed in the mid-1970's, but only briefly, for 6-volt lanterns and portable TV sets. Its advantages were its lower cost compared to other rechargeable batteries and that it was manufactured in a fully charged state. The problems that limited the commercialization of this design were that the cells were not strictly zinc-limited and would lose their ability to be recharged due to the expansion of the cathode if the discharge continued below the voltage level corresponding to a one-electron discharge of the manganese dioxide ($MnO_{1.5}$). Thus voltage control was needed to limit the discharge to 1.1 to 1.0 V per cell, depending on the load and age of the battery and the capacity was reduced because of this higher end voltage. Further, the capability of catalytic hydrogen-gas recombination was not included in the cell design.

A way to control the cathodic discharge is to limit the capacity of the zinc electrode. This can, however, cause poor rechargeability of the zinc electrode. Further study of this battery system led to the development of reliable techniques for limiting the capacity of the zinc electrode.[2–4] Present-day batteries can be discharged to lower end voltages.

The major advantages and disadvantages of the rechargeable zinc/alkaline/MnO_2 battery are listed in Table 36.1.

TABLE 36.1 Major Advantages and Disadvantages of Rechargeable Zinc/Alkaline-Manganese Dioxide Batteries

Advantages	Disadvantage
Low initial cost (and possible lower operating cost than other rechargeable batteries)	Useful capacity about two-thirds of primary battery but higher than most rechargeable batteries
Manufactured in a fully charged state	Limited cycle life
Good retention of capacity (compared to other rechargeable batteries)	Available energy decreases rapidly with cycling and depth of discharge
Completely sealed and maintenance-free	Higher internal resistance than NiCd and NiMH
No "memory effect" problem	

36.2 CHEMISTRY

The discharge mechanism of electrolytic manganese dioxide, which is essentially γ-MnO_2, has been studied extensively.[1] It is generally assumed that the first electron discharge step proceeds in a homogeneous reaction by the movement of protons and electrons into the lattice, resulting in a gradually decreasing value of x in MnO_x from $x = 2$ to $x = 1.5$. The reaction is a conversion of one solid structure (MnO_2) into another ($MnOOH$), with manganese (formally) in the tri-valent state,[5]

$$MnO_2 + H_2O + e \rightarrow MnOOH + OH$$

Soluble manganese species begin to appear if the discharge is continued, especially when the lower voltage second-electron range is approached. The manganese ions find their way to the zinc anode increasing the corrosion reaction and reducing the shelf life characteristics.

When electrolytic manganese dioxide is recharged, the process is reversed. The number of discharge and charge cycles obtainable depends on the depth of discharge, indicating that irreversible electrode processes are occurring. The relationship between cycle number and depth of discharge was found to be essentially logarithmic. Coulometric studies indicate that the recharge efficiency reaches nearly 100% after a few cycles. The initial losses are probably related to "formation problems" as manganese dioxide is a nonconductor and an interface with the graphite structure must be formed.

The cathode-dominated cycle characteristic of the rechargeable batteries and the logarithmic relationship of loss of capacity with cycling are shown in Fig. 36.1. A zinc capacity of about 2 Ah is usually provided in an AA-size battery to maintain a high first-cycle discharge capacity. This still prevents the MnO_2 from discharging beyond the one-electron capacity. In this special experiment the zinc anode is not limiting because only a predetermined capacity of 1.0 Ah is removed at each cycle until the cutoff voltage of 0.9 V is reached.

Charging of the zinc/alkaline/manganese dioxide battery must be controlled to limit the charge voltage to 1.65 to 1.68 V. Charging to higher voltages produces a hexa-valent manganate and oxygen gas. The soluble manganate disproportionates into tetra-valent MnO_2 and a nonrechargeable divalent manganese species which results in a loss in cycle capacity. The oxygen gas reacts with zinc to produce ZnO.

The use of special catalysts is now being investigated to suppress the manganate formation and overcome the poor reversibility of the MnO_2 discharge. One approach is to use cobalt-spinel type catalysts.[6–10] Another is to replace the electrolytic manganese dioxide (EMD) with a specially prepared bismuth-doped form of manganese dioxide (BMD). This modified cathode permits a deep two electron $MnO_2 \rightarrow Mn(OH)_2$ reversible discharge.

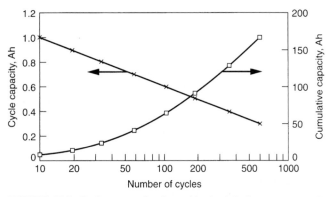

FIGURE 36.1   Performance of rechargeable zinc/alkaline-manganese dioxide battery on cycling at 20°C: recharging after each discharge. (*Courtesy of Battery Technologies, Inc.*)

Bismuth is incorporated as Bi_2O_3 at about 10 weight-percent of the MnO_2. A cathode of BMD formulated with high surface area, oxidation-resistant carbon can give several hundred charge-discharge cycles with a depth of discharge equivalent to over 80% of the theoretical two-electron capacity of the MnO_2. The bismuth appears to act as a redox catalyst, extending the heterogeneous discharge regime as well as blocking the formation of the non-rechargeable manganese compounds. This produces a markedly flatter voltage discharge profile relative to EMD and, in addition, a higher capacity as illustrated in Fig. 36.2. The presence of bismuth is essential in order for the charging to proceed through the reaction sequence:

1. $Bi \rightarrow Bi^{3+} + 3e-$ (electrochemical)

2. $Bi^{3+} + 3Mn^{2+} \rightarrow 3\ Mn^{3+} + Bi$ (chemical)

3. $3Mn^{3+} \rightarrow 1.5\ Mn^{2+} + 1,5\ Mn^{4+}$ (chemical)

Mn^{2+} in the last disproportionation step recycles to step 2 until charging is complete. Soluble Mn^{3+} and Bi^{3-} species must be retained at the cathode. Special chemically functionalized separators are being developed for the BMD cathode.[11]

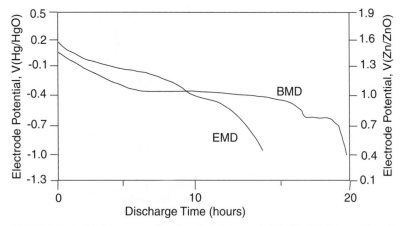

FIGURE 36.2   Discharge curve for electrolytic Manganese Dioxide (EMD) and Bismuth modified Manganese Dioxide (BMD) in KOH. (*Courtesy of Rechargeable Battery Corp.*)

36.3 CONSTRUCTION

The construction of the cylindrical rechargeable alkaline cell is shown in Fig. 36.3. The construction is similar to the primary cell, using an inside-out design. The positive electrode consists of three or four cathodic rings, which are formed under high pressure to a slightly oversized diameter and then inserted into the steel can. The cathodic mix formulation uses electrolytic MnO_2 and 10% graphite. The negative electrode, consisting of a powder zinc mass containing gelled KOH, is in the center. A nail, located in the center of the gel, serves as the negative current collector. The cell is crimped-sealed and contains a vent mechanism.

The following features distinguish the rechargeable cell. The cathode contains additives, such as $BaSO_4$, which increase the cathode capacity and improve cycling. The cathode also contains a catalyst, such as silver on acetylene black or carbon, to recombine any hydrogen that may form. The limiting zinc powder anode contains KOH and a gelling agent. The amount of zinc determines the depth of discharge and thereby the capacity of the cell. Mercury is not added to the anode and special zinc alloys and/or organic inhibitors are used to reduce the zinc corrosion. However, for rechargeable cells, this is not sufficient as the zinc deposited after the first charge is much finer than the original granulated zinc powder. The silver catalysts added to the cathode recombine the hydrogen that results from the zinc corrosion. Zinc oxide is dissolved into the KOH to assure that on charge (or overcharge) only oxygen, but not hydrogen, can be formed by electrolysis. That is, ZnO reduction occurs instead of the generation of hydrogen. The separator, which is multilayered, contains regenerated cellulose with a high oxidation resistance to caustics. It also prevents internal short circuiting due to zinc dendrite formation.[12]

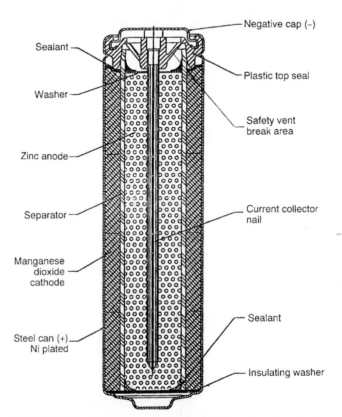

FIGURE 36.3 AA-size rechargeable zinc/alkaline-manganese dioxide battery. (*Courtesy of Battery Technologies, Inc.*)

36.4 PERFORMANCE

36.4.1 First Cycle Discharge

The rechargeable batteries are manufactured and shipped in a charged state, as are the primary cells. Because of their good shelf life, they can retain most of this capacity (depending on storage prior to use) and do not need to be recharged before first use. The discharge characteristics of the rechargeable zinc/alkaline/manganese dioxide batteries are similar to those of the primary batteries. However, due to the zinc-limited design of the rechargeable cell, its terminal voltage on a medium- or high-load discharge drops rapidly once 0.8 V is reached. On a low-load regime, a slope to about 0.6 to 0.7 V is sometimes noticeable before the voltage drops practically to zero. Figure 36.4 shows the discharge curves of fresh AA-size rechargeable batteries on the first cycle of discharge at several constant-current discharge loads. Figure 36.5 shows similar discharge curves for discharges at constant-resistance loads.

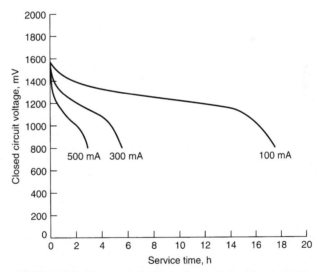

FIGURE 36.4 First-cycle discharge characteristics of rechargeable zinc/alkaline/manganese dioxide AA-size batteries discharged continuously at different constant-current loads at 22°C. (*Courtesy of Battery Technologies, Inc.*)

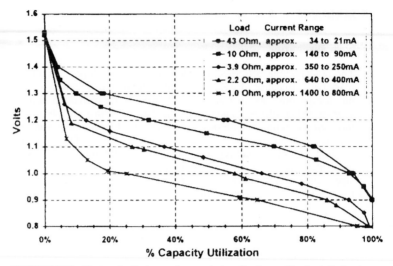

FIGURE 36.5   First-cycle discharge characteristics of rechargeable zinc/alkaline/manganese dioxide AA-size batteries discharged continuously at different constant-resistance loads at 22°C: (*Courtesy of Battery Technologies, Inc.*) (From Ref. 12a.)

36.4.2   Cycling

The rechargeable battery has its highest capacity on the first cycle, and that value at 20°C is about 70 to 80% of the capacity of the primary cell. On subsequent charge-discharge cycles, if the cells are completely discharged before being recharged, 20 discharge cycles can be obtained until the capacity drops to about 50% of the initial capacity. The shape of the discharge curve changes slightly during cycling, but the voltage level drops with cycling, as shown in Fig. 36.6. Additional deep discharge cycles can be obtained, but at reduced capacity, if the cycling is continued.

The number of useful cycles and the cycle capacity increases when the batteries are only partially discharged and recharged after use. Figure 36.7*a* shows the increased cycle life obtained on a intermittent discharge of a AA-size battery on a 10-ohm load, applied daily for 4h (about a 25% depth of discharge). The cells are charged overnight on a constant voltage charger set at about 1.65 V. Although the same drop in voltage level with cycling as shown in Fig. 36.6 is present, more than 200 cycles can be obtained with the terminal voltage about 0.9 volts.

The cycle life, when the battery is discharged to other depth of discharge, is shown in Fig. 36.7*b*. This figure shows the results of repeated discharge of the rechargeable AA-size battery to approximately 12%, 18% and 25% depth-of-discharge, followed by recharge. The cycle life increases with reducing depth-of-discharge and the voltage drop increases with lowering the depth-of-discharge.

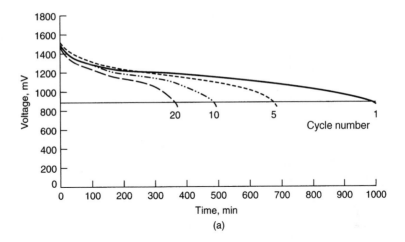

(a)

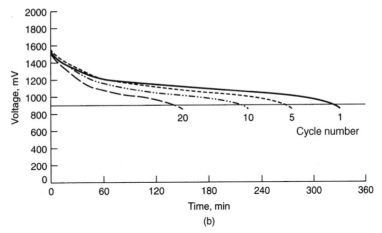

(b)

FIGURE 36.6 Continuous-discharge characteristics of rechargeable zinc/alkaline/manganese dioxide AA-size batteries after cycling at 20°C. (*a*) 10-Ω cycling. (*b*) 4-Ω cycling. (*Courtesy of Battery Technologies, Inc.*)

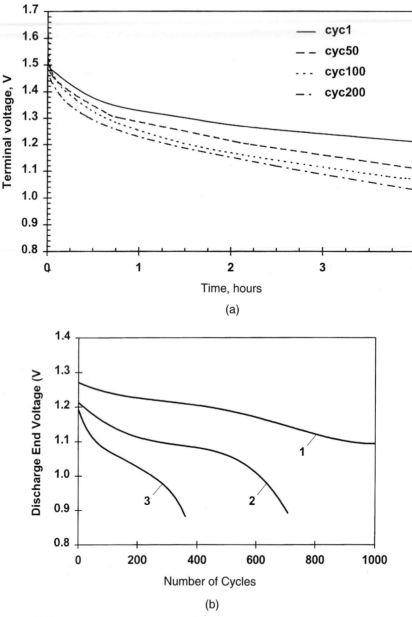

FIGURE 36.7 (*a*) Discharge characteristics of rechargeable zinc/alkaline/manganese dioxide AA-size batteries after cycling at 20°C. Cells discharged 4h per day at 10 Ω; 1.65 V constant voltage charging overnight. (*b*) The cycle life of AA-size batteries on a 10 Ω load. Curve 1 discharged for 200 mAh, then recharged: curve 2 discharged for 300 mA then recharged: curve 3 Discharged for 400 mAh, then recharged. (*Courtesy of Battery Technologies, Inc.*)

36.4.3 Performance of Different Sizes of Batteries

The rechargeable alkaline-manganese dioxide batteries are available in other sizes. Figures 36.8 and 36.9 show the performance of C- and D-size batteries and Fig. 36.10 that of AAA-size.

Note that the AA-size and AAA-size batteries, with their thinner positive electrodes give relatively better performance than the larger diameter C-size and D-size batteries. The efficiency of utilization of the manganese dioxide and the deep cycling performance of the rechargeable battery are related to the thickness of this electrode. This is further illustrated in Fig. 36.11, which shows the utilization of the manganese dioxide and the ampere-hour capacity delivered from the batteries as a function of load current. The thinner batteries deliver a higher percentage of their capacity than the larger diameter batteries when discharged at the higher discharge currents.[13–15]

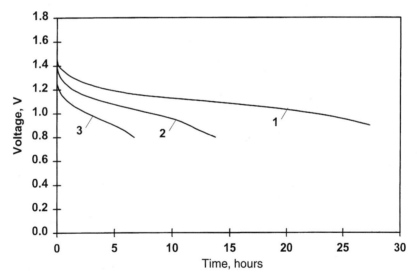

FIGURE 36.8 First cycle discharge characteristics of rechargeable zinc/alkaline/manganese dioxide C-size batteries discharged continuously at different constant resistance loads at 20°C: Curve 1 − 6.8 Ω. 160 mA (approx.): curve 2 − 3.9 Ω. 270 mA (approx.): curve 3 − 2.2 Ω. 450 mA (approx.). (*Courtesy of Battery Technologies Inc.*)

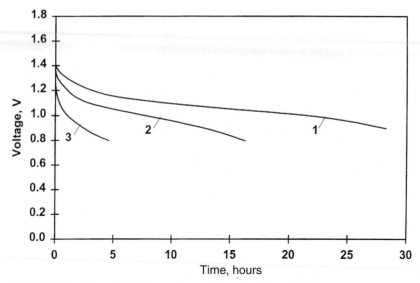

FIGURE 36.9 First cycle discharge characteristics of rechargeable zinc/alkaline/manganese dioxide D-size batteries discharged continuously at different constant resistance loads at 20°C. Curve 1 − 3.9 Ω. 280 mA (approx.): curve 2 − 2.2 Ω. 460 mA (approx.): curve 3 − 1.0 Ω. 1A (approx.). (*Courtesy of Battery Technologies Inc.*)

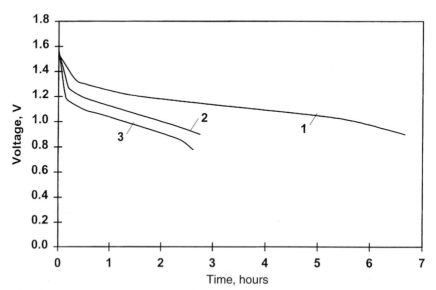

FIGURE 36.10 First cycle discharge characteristics of rechargeable zinc/alkaline/manganese dioxide AAA-size cells discharged continuously at different constant resistance loads at 20°C. Curve 1 − 10 Ω. 110 mA (approx.): curve 2 − 5.1 Ω. 190 mA (approx.): curve 3 − 3.9 Ω. 260 mA (approx.). (*Courtesy of Battery Technologies Inc.*)

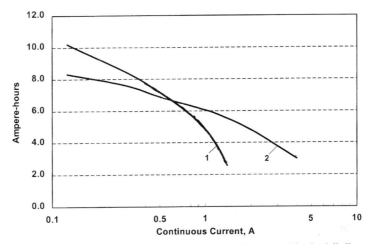

FIGURE 36.11   Comparison of 20°C performance of rechargeable zinc/alkaline /manganese dioxide D-size battery (curve 1) to output of four AA-cells connected in parallel (in D-size can) to 0.9 V end-voltage (curve 2). (*Courtesy of Battery Technologies Inc.*)

36.4.4   Bundle (Parallel) Multicell Batteries

Figure 36.11 also illustrates the advantage of using a multicell battery design, with smaller cells in parallel, than a larger single cell battery. The battery with four AA-size cells in a container having the same dimensions as a D-size battery weighs about 90 grams, compared to a single D-size battery which weighs about 125 grams, but will outperform it at the higher discharge currents. The beneficial effect of using several small diameter cylindrical cells in parallel in a multicell battery instead of a single larger diameter battery also results in a gain in total internal electrode interface, thereby decreasing the current density at a given load and improving performance.[15]

36.4.5   Effect of Temperature

The performance of the rechargeable zinc/manganese dioxide batteries (AA and D-size) at various temperatures is shown in Fig. 36.12. The relative performance of the AA-size battery at low temperatures again is superior to that of the D-size cell because of its thinner cathode and proportionately larger interface area. The performance of the AA-size cell at 45°C and 65°C is shown in Fig. 36.13. It should also be noted that the capacity and high current drain capability are higher at the higher temperatures due to better diffusion and higher MnO_2 utilization.[16,17]

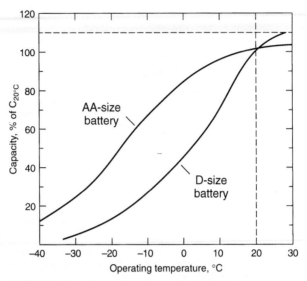

FIGURE 36.12 Comparison of performance of rechargeable zinc/alkaline/manganese dioxide D and AA-size batteries at various temperatures. Discharge at $C/20$ rate, to 0.9-V end voltage. (*Courtesy of Battery Technologies, Inc.*)

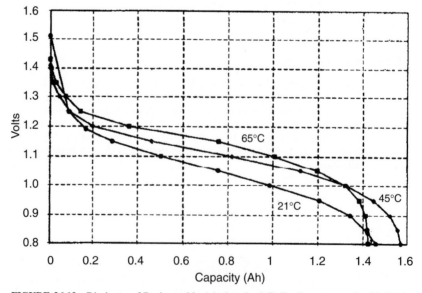

FIGURE 36.13 Discharge of Rechargeable AA-size zinc/alkaline/manganese dioxide batteries at different temperatures at a 3.9 ohm load. (*From Ref. 16*)

36.4.6 Shelf Life

The shelf life of fresh, unused (charged) rechargeable alkaline-manganese dioxide batteries is about the same as that of the primary batteries (20 to 25% loss after 3 to 4 years when stored at room temperature). The data for capacity losses after high-temperature storage are shown in Fig. 36.14.

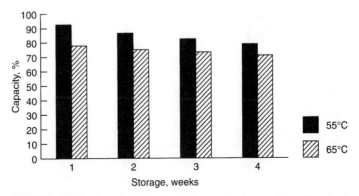

FIGURE 36.14 Capacity retention of mercury-free rechargeable zinc/alkaline/manganese dioxide AA-size batteries at 20°C. (*Courtesy of Battery Technologies, Inc.*)

The shelf life of cycled cells depends on whether they are stored in a charged or a discharged condition. Batteries stored in a charged state after cycling show about the same losses as a fresh uncycled battery. Storage of batteries in a discharged condition, particularly at elevated temperatures (65°C) may be detrimental to the anode performance on subsequent cycles. However, under normal usage, batteries can be recharged close to the capacity level of the previous cycle.

36.5 CHARGE METHODS

In the charging process for the zinc/alkaline/manganese dioxide cell, the discharged positive active material, manganese oxyhydroxide (MnOOH), is oxidized to manganese dioxide (MnO_2) and the zinc oxide (ZnO) in the negative is reduced to metallic zinc. Manganese dioxide can be further oxidized to higher oxides (Mn^{+6} compounds) which are soluble, resulting in loss of rechargeability. Therefore proper recharging is important to obtain optimum life. Charging over 1.72 V per cell for days or over 1.68 V per cell for weeks can damage the battery. Batteries should not be charged after 105% of the ampere-hours removed have been replaced. Batteries can be float-charged for extended periods at 1.65 V per cell.[17]

36.5.1 Constant-Potential Charging

Constant-potential charging is the preferred method. This is equivalent to a taper current charge method. The voltage on charge should not exceed 1.65 to 1.68 V. If charging is continued at higher voltages, current will continue to flow and some damage to the cell can be expected due to increased anode corrosion by the soluble Mn^{+6} species. If the end voltage is set to below 1.65 V, the charging takes longer and the cell may not be fully charged overnight, but the cycle life of the battery is improved. Figure 36.15 shows the voltage and current profiles during the charge. A constant potential charger using a voltage regulator such as a LM317 device is shown in Fig. 36.16.

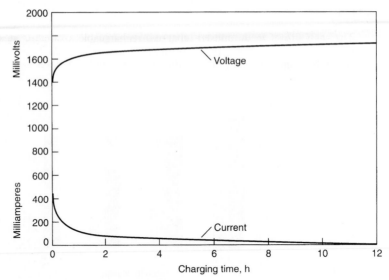

FIGURE 36.15   Constant-potential charging of AA-size rechargeable zinc/alkaline/manganese battery at 20°C. (*Courtesy of Battery Technologies, Inc.*)

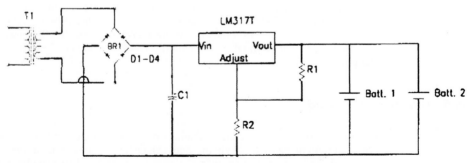

FIGURE 36.16   Principle circuit diagram for a constant potential charger utilizing a LM317T voltage regulator. (*Courtesy of Battery Technologies, Inc.*)

36.5.2   Constant-Current Charging

Uncontrolled constant-current charging over an extended period of time leads to electrolysis of the electrolyte, which causes a buildup of internal gas pressure and results in the rupturing of the safety vent, allowing the release of the gases. Constant-current charging is feasible if the charge voltage is limited to 1.65 V per cell (resistance-free) and a shutoff control is incorporated in the charge circuit.

36.5.3 Pulse Charging

Pulse charging can be used to permit quick charging of the rechargeable zinc-alkaline batteries. The pulse-charging method utilizes halfwave rectified 60-Hz alternating current. During pulse pause the circuit measures the cell voltage. Because no charge current flows through the battery during the pause, the true electrochemical voltage, without any ohmic resistance, is measured. This voltage is often called the "resistance-free" voltage. The pulse-charger circuit regulates the time period when the charge is on by comparing the actual resistance-free cell voltage to the preset cutoff voltage. As long as the resistance-free voltage is lower than the cutoff voltage, charge current passes into the battery. If the resistance-free voltage is equal to or higher than the charge cutoff voltage, the charge current is cut off. The charge voltage can be much higher than the specified charge cutoff voltage as long as the resistance-free voltage does not exceed the charge cutoff voltage. This causes the initial charge current to be much higher, making fast charging possible. Pulse charging has also been shown to increase the cycle life of the battery due to improved replating of zinc.[18] Figure 36.17 shows the current profile for pulse charging of one to four AA-size cells in parallel.

It was found that the total charging time can be lowered if the electrodes are given time to equalize their internal charge polarization gradients by "recovery perods" in the order of minutes. After the recovery period, the charge current, which had dropped to low values, will start and continue at the higher values until the polarization gradients reach the previous high level. Overall, the charging can be carried out with a higher average current over a shorter time period. These "charge and recovery periods" in the time range of minutes act different from the usual pulse charge methods which interrupt for only fractions of a second. Diffusion and interface-concentration gradients in the porous manganese dioxide graphite electrodes (with a slow proton transport) change gradually.

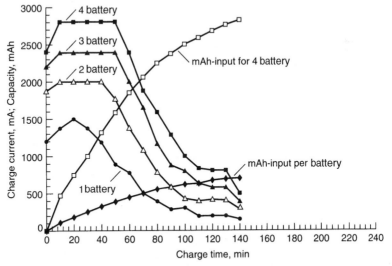

FIGURE 36.17   Pulse charging of AA-size rechargeable zinc/alkaline/manganese dioxide batteries. One to four batteries are charged in parallel to 1.68-V end voltage at 20°C. [Batteries discharged at 100% DOD at 1 Ω.] The charge current flow is determined by a regulating circuit that reads the battery voltage between current pulses. (*Courtesy of Battery Technologies, Inc.*)

36.5.4 Overflow Charging

The term "overflowing charging" describes a process for charge control using electronic devices which become conductive at a given voltage and then divert the charging current from the battery as it becomes fully charged. LED's, Zener diodes and/or other diodes can be used to provide this overcharge protection. For example, a red LED will start to conduct at 1.6 V at 1 mA, increase to 70 mA at 1.65 V and to 100 mA at 1.68 V. This technique can be used for charging multicell batteries connected in parallel or series configurations.

Figure 36.18 illustrates a circuit for charging six cells in a series connection equipped with LED overcharge protection. In addition, diodes are connected across each cell to protect against deep voltage reversal on overdischarge. In a constant current charge limited to 100 mA, as a cell nears full charge, the LED starts to conduct at 1.62 V, diverting several milliamperes from charging the cell. At 1.65 V, about 70 mA goes into the cell and 70 mA are diverted through the LED. At 1.68 V, 100 mA "overflow" through the LED and practically nothing into the cell which is now fully charged.

Another example, a charging circuit for a 3-cell AAA-size zinc-manganese dioxide battery protected by a 5.1 V Zener diode, is illustrated in Fig. 36.19. The switch is required to prevent self-discharge of the battery pack through the Zener diode in off-load storage.

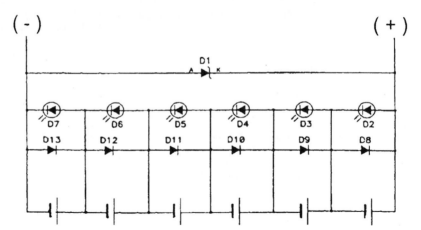

FIGURE 36.18 Charger circuit for a 6-cell battery with LED overflow and diode reversal protection against deep voltage reversal on overdischarge. (*From Ref. 19*)

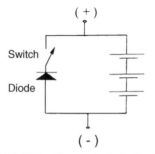

FIGURE 36.19 Charger circuit for 3-cell battery using a 5.1 volt 0.2 A Zener diode, 174733A. (*From Ref. 19*)

Other chargers are available using a combination of diode and Darlington transistors with overflow protection up to 500 mA or even higher using larger diodes. Some of these protect each cell independently or, in some cases, switch off the entire string when the first cell reaches the designated charge voltage—or similarly, on discharge, when the first cell drops under the designated discharge voltage. This is a useful method if the cells are uniform, but may not offer the same protection as the cells will become unbalanced with cycling if the circuitry does not provide cell equalization.[19–21]

36.6 TYPES OF CELLS AND BATTERIES

The characteristics of commercially available rechargeable zinc/alkaline-manganese dioxide batteries are listed in Table 36.2.

TABLE 36.2 Typical Rechargeable Zinc/Alkaline/Manganese Dioxide Batteries

| Cell type | Dimensions, mm | | Weight, G | Rated capacity, Ah* (initial discharge) |
	Height	Diameter		
AAA	44	10	11	0.90 at 75 Ω
AA	50	14	22	1.8 at 10 Ω
C	50	26	63	5.0 at 10 Ω
D	60	34	128	10.0 at 10 Ω
Bundle-C	50	26	50	3.0 at 2.2 Ω
Bundle-D	60	34	100	6.0 at 1.0 Ω

*Based on discharge, through specified resistance, to 0.8 V per cell.
Source: Battery Technologies, Inc.

REFERENCES

1. K. Kordesch (ed.), *Batteries,* vol.1, *Manganese Dioxide,* Dekker, New York, 1974.

2. K. Kordesch, J. Gsellmann, R. Chemelli, M. Peri, and K. Tomantschger, *Electrochim. Acta* **26**:1495–1504 (1981).

3. K. Kordesch et al., "Rechargeable Alkaline Zinc Manganese Dioxide Batteries," *33d Int. Power Sources Symp.,* Cherry Hill, N.J., June 13–16, 1988.

4. Environmental Battery Systems, Richmond Hill, Ont., L4B 1C3, Canada.

5. A. Kozawa, "Electrochemistry of Manganese Oxide," in K. Kordesch (ed.), *Batteries,* vol. 1, Dekker, New York, 1974, chap. 3.

6. K. Kordesch and J. Gsellmann, German Patent DE 3337568 (1989).

7. M. A. Dzieciuch, N. Gupta, and H. S. Wroblowa, J. *Electrochem. Soc.* **135**:2415 (1988), also U.S. Patents 4,451,543 (1984), 4,520,005 (1985).

8. D. Y. Qu, B. E. Conway, and L. Bai, *Proc. Fall Meet. of the Electrochemical Soc.,* Toronto, Ont., Canada, Oct. 1992, abstract 8; Y. H. Zhou and W. Adams, *ibid.,* abstract 9.

9. B. E. Conway et al., "Role of Dissolution of Mn(III) Species in Discharge and Recharge of Chemically Modified MnO_2 Battery Cathode Materials," J. *Electrochem. Soc.* **140** (1993).

10. E. Kahraman, L. Binder, and K. Kordesch, *J. Power Sources* **36**:45–56 (1991).

11. B. Coffey, Rechargeable Battery Corp., College Station, Tx.

12. K. Kordesch et al., "Rechargeable Alkaline Zinc-Manganese Dioxide Batteries" and T. Messing et al., "Improved Components for Rechargeable Alkaline Manganese–Zinc Batteries," *36th Power Sources Conference,* Palisades Institute for Research Services, Inc., New York, 1994.

12a. J. Daniel-Ivad, K. Kordesch, and E. Daniel-Ivad, "An Update on Rechargeable Alkaline Manganese RAM® Batteries," *Proc. 39th Power Sources, Conf.,* Cherry Hill, N.J., 2000, pp. 330–333.

13. K. Kordesch et al., *Proc. 26th JECEC,* Boston, Mass., 1991, vol. 3, pp. 463–468.

14. K. Kordesch, L. Binder, W. Taucher, J. Daniel-Ivad, and Ch. Faistauer, *18th Int. Power Sources Symp.,* Stratford-on-Avon, England, Apr. 19–21, 1993.

15. K. Kordesch, J. Daniel-Ivad, and Ch. Faistauer, "High Power Rechargeable Alkaline Manganese Dioxide-Zinc Batteries," *Proc. 182 Meet. of the Electrochemical Soc.,* Toronto, Ont., Canada, Oct. 11–16, 1992, abstract 10.

16. J. Daniel-Ivad, K. Kordesch, and E. Daniel-Ivad, "Performance Improvements of Low-cost RAM® Batteries," *Proc. 38th Power Sources Conf.,* Cherry Hill, NJ, pp. 155–158, (1998).

17. J. Daniel-Ivad, K. Kordesch and E. Daniel-Ivad, "High-rate Performance Improvements of Rechargeable Alkaline (RAM®) Batteries," *Proc. Vol. 98-15 Aqueous Batteries of the 194th Electrochem. Soc. Meeting,* Boston Mass., Nov. 1–6, 1998.

18. K. V. Kordesch, "Charging Methods for Batteries Using the Resistance-Free Voltage as Endpoint Indication," *J. Electrochem. Soc.* **119:**1053–1055 (1972).

19. K. Kordesch et al., "Charging Systems for Rechargeable Alkaline Manganese Dioxide Batteries," *Proc. 38th Power Source Conf.,* Cherry Hill, NJ, June 8–11, 1998, pp. 291–294.

20. J. Daniel-Ivad, Karl Kordesch, "In-application Use of Rechargeable Alkaline Manganese Dioxide/Zinc (RAM®) Batteries," *Portable By Design Conference,* Santa Clara, CA March 24–27, 1997.

21. J. Daniel-Ivad, Karl Kordesch, and David Zhang, "Protection Circuit for Rechargeable Battery Cells," Patent Appl. WO9749158 (1996).

P · A · R · T · 5

ADVANCED BATTERIES FOR ELECTRIC VEHICLES AND EMERGING APPLICATIONS

CHAPTER 37
ADVANCED BATTERIES FOR ELECTRIC VEHICLES AND EMERGING APPLICATIONS— INTRODUCTION

Philip C. Symons and Paul C. Butler[a]

37.1 PERFORMANCE REQUIREMENTS FOR ADVANCED RECHARGEABLE BATTERIES

The types and number of applications requiring improved or advanced rechargeable batteries are constantly expanding. The new and evolving applications include electric and electric hybrid vehicles, electric utility energy storage, portable electronics, and storage of electric energy produced by renewable energy resources such as solar or wind generators. In addition, the performance, life and cost requirements for the batteries used in many of these new and existing applications are becoming increasingly more rigorous. Commercially available batteries may not be able to meet these performance requirements. Thus, a need exists for both conventional battery technology with improved performance and advanced battery technologies with characteristics such as high energy and power densities, long life, low cost, little or no maintenance, and a high degree of safety.

Battery performance requirements are application dependent. For example, electric vehicle batteries need: (1) high specific energy and energy density to provide adequate vehicle driving range; (2) high power density to provide acceleration; (3) long cycle life with little maintenance; and (4) low cost. On the other hand, batteries for hybrid electric vehicles require: (1) very high specific power and power density to provide acceleration; (2) capability of accepting high power repetitive charges from regenerative braking; (3) very long cycle life with no maintenance under shallow cycling conditions; and (4) moderate cost. Batteries for electric-utility applications must have: (1) low first cost; (2) high reliability when operated in megawatt-hour-size systems at 2000 V or more; and (3) high volumetric energy and power densities. Portable electronic devices require low-cost and readily available, lightweight batteries that have both high specific energy and power and high energy and power densities. Safe operation and minimal environmental impact during manufacturing, use and disposal are mandatory for all applications.

---
[a]The authors acknowledge the support of the U.S. Department of Energy, Energy Storage Systems Program, in the preparation of this chapter.

37.1.1 Batteries for Electric and Hybrid Electric Vehicles

The major advantages of the use of electric vehicles (EVs) and hybrid electric vehicles (HEVs) are reduced dependence on fossil fuels and environmental benefits. For electric vehicles, energy from electric utilities or renewable sources would be used for battery charging. These facilities can be operated more efficiently and with better control of effluents than automotive engines. Hybrid vehicles are expected to require less fuel per mile of travel than current vehicles. This not only results in lower petroleum consumption, but also in lower emissions of undesirable pollutants.

Deteriorating air quality in a number of regions of the U.S. in the mid- to late-1980s led to an increasing number of federal and state regulations designed to effect reductions of emissions from automobiles. The most important of the regulations from the perspective of the developers of EV batteries was the "EV Mandate" promulgated by the California Air Resources Board (CARB). In 1990, CARB issued a regulation requiring, among other things, that 2% of the passenger cars and light trucks offered for sale in 1998 would have to be zero-emission vehicles (ZEVs). Many other states are planning to follow these guidelines. As the only practical means of achieving this requirement was through the introduction of battery-powered EVs, the U.S. auto companies, GM, Ford, and Chrysler, formed the United States Advanced Battery Consortium (USABC),[b] to expedite the development of EV batteries. In 1996, and again in 2000, the date for the first level (2%) of EV offerings and the other provisions of the EV Mandate were delayed by three to four years, in part because it took longer than expected to develop EV batteries with the characteristics defined by the USABC. However, the delays were also necessitated by the poor sales of the EVs that were offered by both domestic and foreign auto makers. In fact, the most recent EV regulation from CARB appears to make the offering of EVs to be voluntary, rather than mandatory, apparently because HEVs are now regarded as a more viable competitor to gasoline-fueled internal combustion engine vehicles than all-battery EVs. In the year 2000, several automobile manufacturers began working nationally on HEVs.

During the 1990s, several battery development programs were conducted by the USABC. These programs were directed toward developing mid-term and long-term battery options for EVs. The batteries for the mid-term were originally intended to achieve commercialization of electric vehicles competitive with existing internal combustion vehicles by 1998. The long-term battery program was directed toward developing advanced batteries projected for commercialization starting in 2002. Both of these objectives were later relaxed due to continuing technical challenges, difficulties in meeting cost goals, and the changing political climate. The USABC criteria for performance of electric-vehicle batteries are shown in Table 37.1.[1]

The severity of the performance requirements for EV batteries is typified by the Dynamic Stress Test (DST) to which batteries developed with USABC funding were subjected. One cycle for the DST is shown in Fig. 37.1.[2] The DST simulates the pulsed power charge (negative percentage, required for regenerative braking) and discharge (positive percentage, for acceleration and cruising) environment of electric vehicle applications and is based on the Federal Urban Driving Schedule (FUDS) automotive test regime. The power levels are based on the maximum rated discharge power capability of the cell or battery under test. The vehicle range on a single discharge can be projected from the number of repetitions a battery can complete on the DST before reaching the discharge cut-off criteria. This test provides more accurate cell or battery performance and life data than constant-current testing because it more closely approximates the application requirements.

---

[b]The USABC is a partnership between General Motors Corporation, Ford Motor Company, and Daimler-Chrysler Corporation with participation by the Electric Power Research Institute and several utilities. It is funded jointly by the industrial companies and the U.S. Department of Energy.

TABLE 37.1 USABC Criteria for Performance of Electric Vehicle Batteries

	Mid-term	Long-term
Specific energy, Wh/kg (C/3 discharge rate)	80 (100 desired)	200
Energy density, Wh/L (C/3 discharge rate)	137	300
Specific power, W/kg (80% DOD/30 s)	150 (200 desired)	400
Power density, W/L	250	600
Life, years	5	10
Cycle life, cycles (80% DOD)	600	1000
Ultimate price, $/kWh	<$150	<$100
Operating environment	−30–65°C	−40–85°C
Recharge time, h	<6	3–6
Continuous discharge in 1 h, % (no failure)	75 (of rated energy capacity)	75
Power and capacity degradation, % of rated specifications	20	20
Efficiency, % C/3 discharge, 6-h charge	75	80
Self-discharge	<15%/48 h	<15%/month
Maintenance	No maintenance (service by qualified personnel only)	No maintenance (as mid-term)
Thermal loss at 3.2 W/kWh (for high-temperature batteries)	15% of capacity per 48-h period	15% of capacity per 48-h period
Abuse resistance	Tolerant (minimized by on-board controls)	Tolerant (minimized by on-board controls)
Specified by contractor: packaging constraints; environmental impact; safety; recyclability; reliability; overcharge/overcharge tolerance		

Source: Ref. 1.

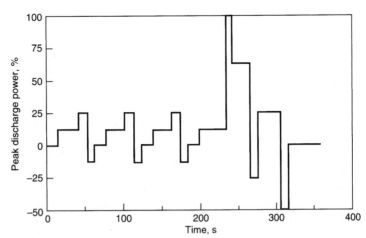

FIGURE 37.1 Typical cycle of dynamic stress test for electric-vehicle batteries. (*From Ref. 2.*)

A multi-year program to develop HEV batteries was initiated by a government-industry cost-shared program in 1993. The HEV battery program is conducted by the Partnership for Next Generation Vehicles (PNGV). The technical targets that were released by the PNGV for HEV batteries in 1999 are shown in Table 37.2.[3] The requirements for HEVs are even more stringent than indicated by the DST for EVs. The severity of these targets, particularly with regard to power capability, is more readily appreciated when it is realized that the power-assisted HEV targets translate to a specific power requirement of almost 750 W/kg. As described in Table 37.2, two HEV operating modes are being considered: "power assist" and "dual mode." The power assist mode involves partial load leveling between the two power systems and includes recovery of braking energy. In this operating scenario, the battery power demands are very high in order to contribute to the acceleration demands of the vehicle. The dual mode option involves extensive load leveling by the two power systems and a second mode to operate the vehicle on battery power only. In this mode, the battery power demands are lower and the energy requirements are more significant in order to provide an appreciable range for the vehicle when powered by the battery only.

TABLE 37.2 PNGV Technical Targets* for Power-Assisted (Targets Shown in Parentheses) and for Dual-Mode Hybrid Electric Vehicle Batteries. Targets are Shown for a 400 V-Battery System

Characteristics	Units	Calendar year		
		2000	2004	2006
18-second power/energy ratio	W/Wh	(83) 27	(83) 27	(83) 27
Specific energy	Wh/kg	(8) 23	(8) 23	(10) 24
Energy density	Wh/L	(9) 38	(9) 38	(12) 42
Cycle life**	Thousand of cycles	(200) 120	(200) 120	(200) 120
Calendar life	Years	(5) 5	(10) 10	(10) 10
Cost***	$/kWh	(1670) 555	(1000) 333	(800) 265

*From Ref. 3.
**For cycles corresponding to the minimum excursion of state-of-charge during an urban driving cycle.
***Based on cost per available energy.

37.1.2 Electric-Utility Applications

The use of battery energy storage in utility applications allows the efficient use of inexpensive base-load energy to provide benefits from peak shaving and many other applications. This reduces utility costs and permits compliance with environmental regulations. Analyses have determined that battery energy storage can benefit all sectors of modern utilities: generation, transmission, distribution, and end use.[4] The use of battery systems for generation load leveling alone cannot justify the cost of the system. However, when a single battery system is used for multiple, compatible applications, such as frequency regulation and spinning reserve, the system economics are often predicted to be favorable.

The energy and power requirements of batteries for typical electric-utility applications are shown in Table 37.3. The concept of load leveling is illustrated in Fig. 37.2a, and a simplified test regime simulating a frequency regulation and spinning reserve application is illustrated in Fig. 37.2b. The frequency regulation and spinning reserve test profile simulates these two utility applications on a sub-scale battery in order to predict performance, life, and thermal effects. Charge (positive power) and discharge (negative power) vary according to a specified regime and provide a realistic environment for batteries used in these applications.

TABLE 37.3 Utility Energy Storage Applications and Corresponding Requirements

	Energy capacity, MWh	Average discharge time, h	Maximum discharge rate, MW
Load leveling	>40	4–8	>10
Spinning reserve	<30	0.5–1	<60
Frequency regulation	<5	0.25–0.75	<20
Power quality	<1	0.05–0.25	<20
Substation applications, transformer deferral, feeder or customer peak shaving, etc.	<10	1–3	<10
Renewables	<1	4–6	<0.25

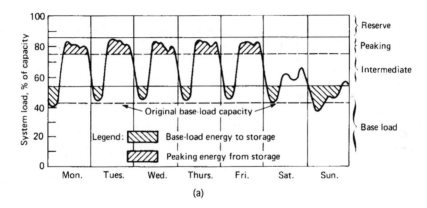

(a)

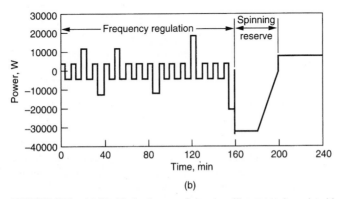

(b)

FIGURE 37.2 (*a*) Weekly load curve of electric-utility generation mix with energy storage. (*b*) Test regime typical of frequency regulation and spinning reserve application for electric utilities. (*Courtesy of Sandia National Laboratories. See Ref. 4.*)

Commercially available lead-acid batteries can satisfy the requirements for certain utility energy storage applications and are being used in several demonstration projects worldwide. The use of advanced batteries offers still greater potential for reduced cost and could enable market opportunities to be enhanced. These opportunities result from the predicted advantages of advanced batteries for lower cost, smaller system footprint, no maintenance, and high reliability even when used with highly variable duty cycles.

37.1.3   Renewable Applications

Battery storage provides significant benefits in solar, wind, and other renewable generation systems where the energy source is intermittent. The battery is charged when the source is generating energy. This energy can then be discharged when the source is not available. Operating characteristics vary widely depending on application. For photovoltaic systems, typical applications include village power, telemetry, telecommunications, remote homes, and lighting. Operating characteristics for photovoltaic systems are shown in Table 37.4.[5] Detailed requirements are being developed, and considerations such as high energy efficiency, low self-discharge, low cost, long cycle and calendar lives, and no maintenance are important.

TABLE 37.4   Operating Characteristics for Photovoltaic Systems

Characteristic	Value	Comments
System:		
Storage capacity	0.05–1000 kWh	
Voltage	6–250 V DC	
Battery:		
Capacity	30–2000+ Ah	
Charge rate	C/15–C/500	Charge regulation mechanisms: on-off, constant-voltage, pulse-width-modulated, multi-step
Discharge rate	C/5–C/300	27% of systems discharge battery at C/50
		46% of systems discharge battery at C/100
		15% of systems discharge battery at C/200
Average daily DOD (depth of discharge)	1–30%	Dependent on economics and battery chemistry
Temperature range	−40°–60°C	Geographically dependent
Average life	4 years	For <350-Ah cells
	7–10 years	For >350-Ah cells
Average cost	$67/kWh	For flooded/vented lead-acid
	$97/kWh	For valve-regulated lead-acid

Source: Data from Ref. 4.

37.1.4   Portable Electronics

The demand for batteries used in portable electronics, such as communication, photographic and electronic equipment, computers, and many other consumer, industrial, and military devices, has been increasing dramatically since the mid-1980s and is now a substantial market for advanced rechargeable batteries. Progress in the miniaturization of electronics resulted in a demand for batteries that are smaller, lighter, and offer longer service. Also important are power output, storage life, reliability, safety, and cost. Currently available conventional primary and secondary batteries did not meet all of these needs, and new battery systems with advanced performance characteristics are required. Valve-regulated lead-acid

(VRLA) batteries (primarily in Europe) and nickel-cadmium batteries (more popular in North America) are still used extensively for applications such as power tools and electric tooth-brushes. During the 1990s, nickel-metal hydride batteries became the system of choice for applications requiring higher performance (cell phones, laptop computers), but by the early 2000s, sales of lithium-ion batteries became comparable with those for nickel-metal hydride batteries. Portable fuel cells, larger versions of which are being developed for advanced HEV and stationary (distributed electricity generation) applications, may become a factor for powering portable electronic equipment in the future (see Chaps. 42 and 43).

37.2 CHARACTERISTICS AND DEVELOPMENT OF RECHARGEABLE BATTERIES FOR EMERGING APPLICATIONS

A number of battery chemistries and technologies are being explored and developed in order to meet the requirements described in the previous section. These activities can be categorized as follows:

1. Near-term activities to improve the performance of existing conventional technologies for use within the next few years.
2. Mid-term activities to complete the development of those advanced battery technologies that are not commercialized but, with necessary progress, can be introduced to the market within 5–10 years.
3. Long-term activities to develop new electrochemical technologies, such as refuelable batteries and fuel cells, which offer the potential of higher energy and power, but which require significant development before commercialization.

The performance of some of the candidate battery systems for those emerging applications is compared in Figure 37.3(*a*) in a plot of specific energy vs. specific power. Figure 37.3(*b*) plots the specific characteristics of the nickel-metal hydride battery for both the EV and HEV applications. The nickel-metal hydride battery is currently the battery of choice by many of the developers of hybrid electric cars and light trucks (see Chapter 30).

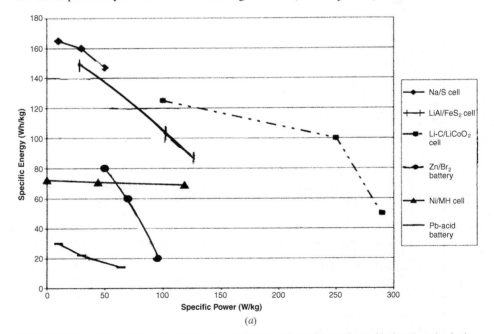

(*a*)

FIGURE 37.3 (*a*) Plot of specific energy vs. specific power for various rechargeable battery technologies. (*Courtesy of Sandia National Laboratories.*)

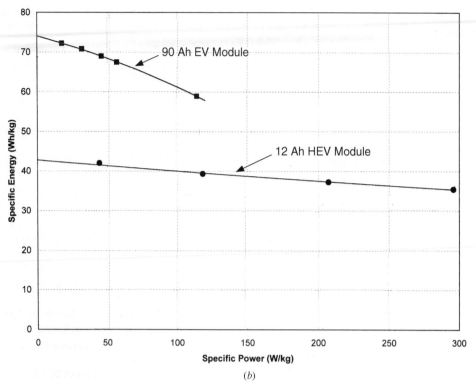

FIGURE 37.3   (*b*) Plot of specific energy vs. specific power for nickel-metal hydride EV and HEV modules. (*Courtesy of Ovonic Battery Co.*)

In the United States, the development of batteries for EVs and HEVs has been mostly carried out under the auspices of the USABC since the early 1990s (see Sec. 37.1.1). In addition to the USABC, the Advanced Lead-Acid Battery Consortium (ALABC) was formed by the International Lead Zinc Research Organization (ILZRO) and the lead-acid battery industry to develop that technology for EV applications.

Significant effort has gone into the development of many advanced batteries for these applications. In recent years, decisions were made to focus on lead-acid, nickel-metal hydride, and lithium-ion. These technologies have become the most likely to be used in either EVs or HEVs due to a combination of performance capability, safety, life, and cost. Earlier development of high temperature and flowing electrolyte technologies for EVs or HEVs has been mostly redirected or discontinued due to these decisions.

In a 2001 survey[6] of alternative propulsion vehicles, of 68 vehicle models, about 2/3 are hybrid or all electric. Only 26 of the 68 are currently available; the rest are being planned. Of those planned, about 35% are EVs and the rest are either gasoline/hybrid or diesel/hybrid. Three battery types were identified as being used in these vehicles. About half use lead-acid batteries, about 40% use nickel-metal hydride, and the rest use lithium-ion. Of the vehicles available at this time, over 60% use lead-acid, 30% use nickel-metal hydride, and the rest use lithium-ion batteries. Some of the vehicles in the planning stage may use fuel cells as part of the power system.

The Massachusetts Institute of Technology (MIT) in 2000[7] evaluated the possible advances in vehicle technologies by the year 2020 with respect to alternative propulsion systems, and characterized their potential for efficiency improvements, carbon emissions reductions, and cost changes. While the uncertainty in the estimates is significant (as high as plus or minus 30%), the hybrid electric system is predicted to have about a 33% lower life-cycle energy use and about 20% lower life-cycle carbon emissions compared to a pure electric. The predicted cost per km driven of the pure electric is about 15% higher than the hybrid. These results agree in principle with cost and lifetime experiences with battery-powered electric vehicles. Further, due to the inherently limited range of pure electrics and uncertainty regarding possible battery breakthroughs in the foreseeable future, the emphasis in alternative propulsion technologies has changed to focus on HEV concepts.

In stationary applications, there has been significant support for developing batteries for electric utility energy storage from the U.S. DOE through Sandia National Laboratories since the 1980s, and from EPRI (formerly the Electric Power Research Institute) in the 1980s and early 1990s. In 1991, the DOE/Sandia and EPRI cooperatively worked with the utility industry to form the Utility Battery Group that promoted the exchange of information and data on technologies for these applications. Now named the Electricity Storage Association, this group includes electricity providers, technology developers, and international participants carrying out the objectives for a wide range of energy storage technologies.

The DOE has continued to provide research and development support for batteries, and recently, other energy storage technologies, for utility energy storage applications.[8] In the mid-1990s, the DOE program broadened its scope and became the Energy Storage Systems Program. Working through Sandia, the Program has collaborated with industry to develop battery technologies, power electronics, and controls, and is now evaluating flywheels and superconducting magnetic energy storage concepts. Battery technologies such as lead-acid, zinc/bromine, and sodium/sulfur have been intensively developed and installed in complete systems for operation in utility and off-grid systems. Applications of interest include power quality, peak shaving, back-up power, and a number of other utility-related uses. In partnership with industry, systems ranging in capacity from hundreds of kW/kWh to tens of MW /MWh have been successfully built, tested, and characterized and some are now being commercialized by industry. The DOE Program continues to work closely with industry (ILZRO) and the Electricity Storage Association to develop and test promising technologies and systems for many increasingly important utility energy storage uses.

In Japan, the development of advanced secondary battery systems for electric-utility applications was carried out from 1981 to 1991 as a part of the "Moonlight Project."[9] Development on four systems proceeded through 60-kW class modules, and 1-MW pilot plants were built for two systems: sodium/sulfur and zinc/bromine. Testing was satisfactorily concluded in March 1992 with 76% energy efficiency (211 cycles) for sodium/sulfur and 66% energy efficiency (158 cycles) for zinc/bromine batteries. Areas for further research were identified, and improvements in reliability, maintainability, compactness, and cost reduction were expected to yield systems that would be practical for utility applications. Following completion of the Moonlight Project, work in Japan focused on sodium/sulfur batteries with funding from the Tokyo Electric Power Company and, to a much smaller extent, on redox batteries with funding from the Kansai Electric Power Company. There were, as far as is known, no other national efforts on batteries for the advanced applications, although there were privately-funded efforts on redox batteries for utility applications in the United Kingdom and Australia during the 1990s.

Several test facilities are in existence in the U.S. for the evaluation of improved and advanced battery systems. Batteries of all types are tested at Argonne National Laboratory, Idaho National Engineering Laboratory, Lawrence Berkeley National Laboratory, and Sandia National Laboratories. Certain tests for satellite and military applications are conducted at the Naval Surface Warfare Center in Crane, Ind. There are also specialized testing facilities established by companies in the private sector and test facilities in other countries.

The major battery technologies that have been considered from time to time for electric-vehicle, utility energy storage, and renewable energy storage applications are listed in Table 37.5, together with the chapter numbers in the Handbook in which each is discussed. The companies that are most active in the development of improved and/or advanced batteries for these applications are listed in Table 37.6. U.S. National Laboratories and similar organizations that are involved in advanced battery R&D are also shown in Table 37.6.

TABLE 37.5 Index of Rechargeable Battery Systems and Refuelable Technologies Chapters in this Handbook

	Chapter or section
Conventional battery systems	
Lead-acid	23 and 24
Nickel-iron*	25
Nickel-hydrogen*	32
Nickel-cadmium	26 and 27
Nickel-zinc	31
Nickel-metal hydride	30
Zinc/silver oxide	33
Aqueous batteries	
Metal/air	38
Iron/air*	25
Zinc/air	38
Flow Batteries	
Zinc/chlorine*	Sec. 37.4.1
Zinc/bromine	39
Iron/chromium redox*	Sec. 37.4.1
Vanadium-redox	Sec. 37.4.1
Polysulfide/bromine redox (Regenesys)	Sec. 37.4.1
High-temperature batteries	
Lithium/sulfur*	41
Lithium-aluminum/iron sulfide	41
Lithium-aluminum/iron disulfide*	41
Sodium/sulfur	40
Sodium/metal chloride*	40
Lithium ambient-temperature batteries	
Liquid electrolyte	34
Lithium-ion	35
Lithium-polymer	34 and 35
Refuelable systems	
Fuel cells**	41 and 42
Zinc/air batteries	37 and 38
Aluminum/air batteries	37 and 38
Lithium/air batteries*	38

*No significant work underway on this system.
**Portable fuel cells are discussed in Chap. 43. Fuel cells for EVs and large-scale power generation are beyond the scope of this Handbook (see Bibliography, Appendix F).

TABLE 37.6 Organizations with Major Development Projects on Advanced Rechargeable Batteries for EVs/HEVs and/or Electric Utility Storage

Private sector			
Organization, Country	Major funders of work	Advanced batteries	Application of interest
Avestor, Canada	Avestor (Hydro-Quebec), DOE	Lithium-polymer	EV, HEV, utility storage
Innogy, United Kingdom	Innogy (formerly National Power)	Regenesys: Polysulfide/bromine redox (flow)	Utility storage
NGK, Japan	Tokyo Electric Power Co., NGK	Sodium/sulfur	Utility storage
Ovonics	USABC, DOE	Nickel-metal hydride	EV, HEV
Powercell, USA, Austria	Powercell, DOE	Zinc/bromine	Utility storage
SAFT, France, USA	SAFT, DOE	Nickel-metal hydride	EV, HEV
		Lithium-ion	EV, HEV, utility storage, telecomm
Sony Energetic, Japan	Sony	Lithium ion	EV
Sumitomo Electric, Japan	Sumitomo Elec., Kansai Electric Power Co.	Vanadium-redox (flow) battery	Utility storage
ZBB, USA, Australia	DOE, ZBB	Zinc/bromine	Utility storage

Universities & U.S. DOE National Laboratories	
Organization	Main programmatic interest
Argonne National Laboratory	Batteries for EV, HEV Lithium/iron sulfide battery R&D
Case Western Reserve University	Basic battery and fuel cell research for EVs and HEVs
Lawrence Berkeley Laboratory	Basic battery research for all applications
Lawrence Livermore National Laboratory	Basic battery research for advanced applications
Sandia National Laboratories	Batteries and other advanced technologies for utility storage systems and HEV
Texas A&M University	Basic battery and fuel cell research for EVs and HEVs

Comparative background data for rechargeable battery technologies are listed in Table 37.7.[10] More information on each technology is contained in Table 37.8 with data for technologies for current and emerging applications, and Table 37.9, which describes other technologies of interest.

TABLE 37.7 Comparative Background Data for Rechargeable Battery Technologies[e]

Technology	Open-circuit voltage, V	Approx. closed-circuit voltage,[a] V	Theoretical specific capacity,[b] Ah/kg	Theoretical specific energy,[b] Wh/kg	Operating temperature, °C	Recharge time, h	Self-discharge, % per month @ 20°C
Lead-acid	2.1	1.98	120	252	−20–50	8–24	3
Nickel-cadmium	1.35	1.20	181	244	−40–60	1–16	10
Nickel-iron	1.4	1.20	224	314	−10–60	5	25
Nickel-hydrogen	1.5	1.20	289	434	−10–30	1–24	60
Nickel-metal hydride	1.35	1.20	178	240	−30–65	1–2	30
Nickel-zinc	1.73	1.60	215	372	−20–50	8	15
Zinc/silver oxide	1.85	1.55	283	524	−20–60	8–18	5
Zinc/bromine	1.83	1.60	238	429	10–50	–	12–15[c]
Regenesys (polysulfide/bromine)	1.5	1.2	27	41	10–50	8–12	5–10
Vanadium-redox	1.4	1.25	21	29	10–50	6–10	5–10
Zinc/air	1.6	1.1	825[d]	1320[e]	0–45	–	–
Aluminum/air	2.73	1.4	2980[d]	8135[e]	10–60	–	–
Iron/air	1.3	1.0	960[d]	1250[c]	−20–45	–	15
Sodium/sulfur	2.08	2.0	375	755	300–350	5–6	–
Sodium/nickel chloride	2.58	2.47	305	787	250–350	3–6	–
Lithium-aluminum/iron monosulfide	1.33	1.30	345	459	375–500	5–8	–
Lithium-aluminum/iron disulfide	1.73	1.73	285	490	375–450	5–8	–
Li-C/LiCoO$_2$	3–4	3–4	100	360	−20–60	–	–
Li-C/LiNi$_{1-x}$Co$_x$O$_2$	3–4	3–4	–	–	−20–45	2.5	<3.5
Li-C/LiMn$_2$O$_4$-polymer elect.	3–4	3–4	105	400	−20–60	3	<2.5

[a] At C/5 rate.
[b] Calculated values based on the electrochemical cell reactions and the mass of active material.
[c] Finite self-discharge. This value applies if electrolyte is not circulating. Self-discharge is limited to that due to the amount of bromine in the cell stacks.
[d] Based on metal negative electrode only.
[e] See Ref. 10.

TABLE 37.8 Comparative Data for Rechargeable Battery Technologies for Current & Emerging Applications

Technology	Cycle life,[a] cycles	Configuration	Specific energy,[b] Wh/kg	Energy density,[b] Wh/L	Specific power,[c] W/kg	Applications	Advantages/disadvantages
Lead-acid	800	Cell	35	80	200	Electric/hybrid vehicles, utility energy storage, consumer	Commercially available, no maintenance/low specific energy
Nickel-cadmium	1000	Cell	35	80	260	Electric/hybrid vehicles, aerospace, consumer	Commercially available/low energy, high cost
Nickel-metal hydride	900	Cell	65	220	850	Electric/hybrid vehicles, aerospace, consumer	High specific power/high cost
Nickel-iron	1000	Cell	30	60	100	Industrial	Commercially available/high maintenance, significant H_2 evolution
Nickel-hydrogen	2000	Cell	55	60	100	Aerospace, military	Long life/very high cost, high self-discharge
Zinc/silver oxide	40–50	Cell	90	180	500	Aerospace, military, consumer	High specific energy and power/high cost, very short life
Zinc/bromine	1250	Battery	65	60	90	Utility energy storage	Low cost/low specific energy density
Zinc/air	Mech. Rech.	Battery	150	160	95	Industrial	Mod. specific energy/short life, low sp. power
Regenesys (polysulfide/bromine)	2000	Battery	20	20	–	Utility energy storage	Very large scale
Vanadium-redox	3000	Battery	10	10	–	Utility energy storage	Very large scale
Sodium/sulfur	1500 / 1000	Cell / Battery	170 / 115	345 / 170	250 / 240	Utility energy storage	High specific energy and energy density/high temperature
Li-C/LiCoO$_2$	600	Cell	155	410	–	Consumer, electric/hybrid vehicles, utility storage	High specific energy/uncertain cost
Li-C/LiNi$_{1-x}$Co$_x$O$_2$	400	Cell	150	400	–	Consumer, electric/hybrid vehicles	High specific energy
Li-C/LiMn$_2$O$_4$ (polymer electrolyte)	600	Cell	140	300	–	Consumer, electric/hybrid vehicles	High specific energy/development needed
Li/MnO$_2$ (liquid electrolyte)	300	Cell	120	265	–	Consumer	High specific energy safety concern

[a] At approximately the $C/5$ rate to 80% of rated capacity.
[b] At approximately the $C/5$ rate.
[c] Short-duration pulse, fully charged to half-charged except sodium/sulfur, which is 50–80% charged. The values listed do not reflect the maximum that is achievable if batteries are purposely designed for high specific power.

TABLE 37.9 Comparative Data for Other Rechargeable Battery Technologies of Interest

Technology	Cycle life,[a] cycles	Configuration	Specific energy,[b] Wh/kg	Energy density,[b] Wh/L	Specific power,[c] W/kg	Advantages/disadvantages
Nickel-zinc	200	Battery	60	120	300	High specific energy/high cost, short life
Aluminum/air	Mechanically rechargeable	Battery	200–250	150–200	–	High specific energy/low specific power, not electrically rechargeable
Iron/air	300	Battery	65	100	–	Good life, low cost/low voltage per cell, low coulombic efficiency on charge
Sodium/nickel chloride	2500	Cell	115	190	260	High specific energy/high temperature
	1000	Battery	95	150	170	
Lithium/iron monosulfide	1000	Cell	130	220	240	High energy density/low specific power, high temperature
Lithium/iron disulfide	1000	Cell	180	350	400	High specific energy and power/high temperature

[a] At C/5 rate to 80% of rated capacity.
[b] At C/5 rate.
[c] Short-duration pulse, fully charged to half-charged, except lithium/iron monosulfide, and lithium/iron disulfide, which are 50–80% charged. The values listed do not reflect the maximum that is achievable if batteries are purposely designed for maximum specific power.

37.3 NEAR-TERM RECHARGEABLE BATTERIES

The major candidates for the electric vehicle (EV), hybrid electric vehicle (HEV) and electric utility applications in the near term are those rechargeable battery technologies that are now available commercially. Many of these have been improved over the last decade to meet the needs of the emerging applications. Further improvement may, in most cases, be necessary to effect economic viability.

37.3.1 Lead-Acid Batteries

Of the currently commercialized battery chemistries, the lead-acid battery is the most widely used and economical and has an established manufacturing base. It is being used in both mobile and stationary applications. Its main disadvantage is low specific energy. Lead-acid batteries with improved performance are being developed for EVs and HEVs. High surface area electrodes with thin active material layers are being investigated using materials and designs such as lightweight fiber-glass reinforced lead wire grids, thin metal foils, bipolar plates, forced-flow electrolyte systems and unique cell assemblies. Methods for fast charging lead-acid batteries are being developed as the ability to rapidly recharge batteries in as little as an hour is considered an important factor for the acceptance of electric vehicles.

37.3.2 Nickel-based Batteries

Nickel-cadmium batteries are being considered for use in large HEVs such as city buses[11] and for electric utility storage applications.[12] They offer good power density, maintenance-free operation over a wide temperature range, long cycle life, and a relatively acceptable self-discharge rate. Another advantage is their capability for rapid recharge. The specific energy of the nickel-cadmium battery is higher than that of the lead-acid battery but, as with most nickel batteries, their initial cost is much higher. Their longer cycle life may offset some of this cost on a life cycle basis. New electrode developments such as plastic-bonded and nickel foam electrodes promise to improve performance and reduce costs. The use of cadmium presents environmental challenges that will have to be resolved.[13]

Some of these limitations are overcome by the nickel-metal hydride battery. This battery has characteristics similar to the nickel-cadmium battery, but it is cadmium-free and has a higher specific energy, about twice that of the nickel-cadmium battery. It is comparable to the nickel-cadmium battery in power density, although there is a more pronounced voltage drop at very high rates. It also requires more careful charge control to prevent overcharge and overheating. Costs should be similar to those of other nickel batteries.

Nickel-metal hydride batteries are currently the energy storage system of choice for many of the developers of hybrid electric cars and light trucks. Several commercial HEVs, including the Toyota PRIUS and the Honda Insight, are using nickel-metal hydride batteries as their energy storage system.

Nickel-hydrogen batteries have been used primarily in satellite applications. They are highly reliable, have a long cycle life, and are able to tolerate deep discharges. They have a high initial cost due to expensive catalysts used in the hydrogen electrode and the requirement that the cell container be a high pressure vessel. Their low volumetric energy density and high self-discharge rate as well as the need to store hydrogen in the interior of the cell are barriers to the wider deployment of this system. Attempts have been made to employ this system in load-levelling applications, but they were unsuccessful.

37.4 ADVANCED RECHARGEABLE BATTERIES-
GENERAL CHARACTERISTICS

Advanced rechargeable batteries can be classified into three main types: advanced aqueous electrolyte systems, or as they are more-commonly known, flow batteries; high-temperature systems; and ambient-temperature lithium batteries.

37.4.1 Flow Batteries

These advanced aqueous-electrolyte battery systems have the advantage of operating close to ambient temperature. Nevertheless, complex system design and circulation of electrolyte are needed to meet performance objectives. Work on developing flow batteries started with the invention of the zinc/chlorine hydrate battery in 1968.[14] This system was the subject of development for EV and electric utility storage applications[15] during from the early-1970s to the late-1980s in the United States, and from 1980 to 1992 in Japan,[16] but has now been abandoned in favor of other flow battery chemistries that appear more attractive. Three main types of flow batteries continue to be developed: zinc/bromine, vanadium-redox, and Regenesys.

Zinc/Bromine Batteries.   The *zinc/bromine* battery technology is currently being developed primarily for stationary energy storage applications (see Chap. 39). The system offers good specific energy and design flexibility, and battery stacks can be made from low-cost and readily available materials using conventional manufacturing processes. Bromine is stored remotely as a second-phase polybromide complex that is circulated during discharge. Remote storage limits self-discharge during standby periods. An added safety benefit of the complexed polybromide is greatly reduced bromine vapor pressure compared to that of pure bromine.

Redox Batteries.   Another type of aqueous flowing electrolyte system is the redox flow technology. There are several systems of this type, only one of which, the *vanadium redox battery* or VRB as it is known, has any significant development continuing as of 2001. Work on this category of flow battery started with a development program at NASA[17] on a system using $FeCl_3$, as the oxidizing agent (positive) and $CrCl_2$, as the reducing agent (negative). The aim of this work was to develop the redox flow batteries for stationary energy storage applications. The term "redox" is obtained from a contraction of the words "reduction" and "oxidation." Although reduction and oxidation occur in all battery systems, the term "redox battery" is used for those electrochemical systems where the oxidation and reduction involves only ionic species in solution and the reactions take place on inert electrodes. This means that the active materials must be mostly stored externally from the cells of the battery. Although redox systems are capable of long life, their energy density is low because of the limited solubility of the active materials typically involved.[18]

In Japan, development of *iron/chromium redox flow battery* technology was included as part of the Moonlight Project[9] in the 1980s. The goal of this work was electric utility energy storage. Improvements made in the course of the Moonlight Project included new electrode materials and a reduction in the requirement for pumping power.[19] A 60 kW battery was tested[20] and 1-MW system was designed,[21] but the redox flow technology was not chosen for development to a 1-MW pilot plant stage.[22]

Other redox systems were also proposed in the past, such as the *zinc/alkaline sodium ferricyanide* [$Na_3Fe(CN)_6 \cdot H_2O$] couple, and initial development work was performed.[23] However, none of these efforts proved successful, mainly because of difficulties resulting from the efficacy and resistance of the ionic exchange membranes, until the development of the vanadium redox battery, by the University of New South Wales, Australia, in the late 1980s. Almost concurrently with this, development work started on VRBs at Sumitomo Electric Industries (SEI) of Osaka, Japan.[24] Starting in the mid-1990s, VRB development has also been conducted at Mitsubishi Chemical's Kashima-Kita facility, although at a lower level of effort than at SEI.

The electrolytes in the positive and negative electrode compartments of VRBs are different valence states of vanadium sulfate. Both solutions are 2 M in concentration and contain sulfuric acid as a supporting electrolyte. The electrode reactions occur in solution, with the reaction at the negative electrode in discharge being:

$$V^{2+} \rightarrow V^{3+} + e$$

and at the positive electrode:

$$V^{5+} + e \rightarrow V^{4+}$$

Both reactions are reversible on the carbon felt electrodes that are used. An ion-selective membrane is used to separate the electrolytes in the positive and negative compartments of the cells. Cross-mixing of the reactants would result in a permanent loss in energy storage capacity for the system because of the resulting dilution of the active materials. Migration of other ions (mainly H^+) to maintain electroneutrality, however, must be permitted. Thus, ion-selective membranes are required.

A schematic of a VRB system is shown in Fig. 37.4.[24] The construction of the cell stacks is bipolar. The electrolyte solutions are stored remotely in tanks and are pumped through the cells.

Several multi-kW systems have been built and tested by SEI and Mitsubishi Chemical. A photograph of an SEI 100 kW-8hr VRB system is shown in Fig. 37.5. Two electrolyte tanks are installed in a sub-basement below the level of the battery stacks and the AC-DC-AC converter.

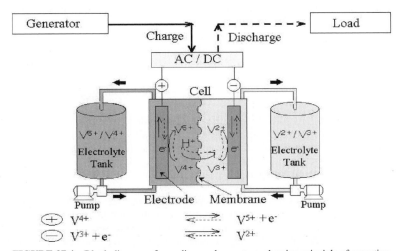

FIGURE 37.4 Block diagram of vanadium-redox system, showing principle of operation.

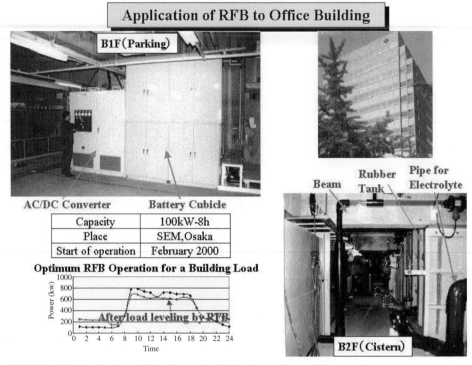

FIGURE 37.5   Application of Vanadium redox battery in an office building.

Regenesys System.   The third type of flow battery that it is being actively developed is the *polysulfide-bromine,* or *Regenesys* system of Innogy (formerly National Power) in the United Kingdom. Innogy has been involved in the development of this redox-like system, in which both reactants and products of the electrode reactions remain in solution, since the early 1990s. Regenesys is similar to a redox system but both the positive and negative reactions involve neutral species. The discharge reaction at the positive electrode is:

$$NaBr_3 + 2Na^+ + 2e \rightarrow 3NaBr$$

and that at the negative is:

$$2Na_2S_2 \rightarrow Na_2S_4 + 2Na^+ + 2e$$

Sodium ions pass through the cation exchange membranes in the cells to provide electrolytic current flow and to maintain electroneutrality. The sulfur that would otherwise be produced in discharge dissolves in excess sodium sulfide that is present to form sodium polysulfide. The bromine produced at the positives on charge dissolves in excess sodium bromide to form sodium tribromide. A block diagram of a Regenesys energy storage plant is shown in Fig. 37.6.

   Innogy built many multi-kW batteries in their development program in the 1990s, with this part of the effort culminating in construction of 100 kW cell stacks (modules) with electrodes of up to one square meter in area[25] (see Fig. 37.7). Innogy has announced that by the end of 2002 they should have completed construction and acceptance testing of a 15 MW, 120 MWh Regenesys energy storage plant at the Little Barford power station in the United Kingdom.

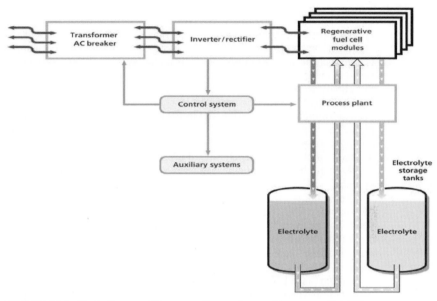

FIGURE 37.6   Block diagram of Regenesys Energy Storage Plant. (*From Ref. 25.*)

FIGURE 37.7   Regenesys modules of varying sizes. (*From Ref. 25.*)

37.4.2   High-Temperature Systems

High-temperature systems operate in the range of 160 to 500°C and have high-energy density and high specific power compared to most conventional ambient-temperature systems. The negative electrode material is an alkali metal, such as lithium or sodium, which has a high voltage and electrochemical equivalence. Aqueous electrolytes cannot be used because of the chemical reactivity of water with alkali metals. Molten salt or solid electrolytes that require high temperatures are used instead. Benefits are high ionic conductivity, which is needed for high power density, and insensitivity to ambient temperature conditions. However, high operating temperatures also increase the corrosiveness of the active materials and cell components and thereby shorten the life of the battery. Also, thermal insulation is needed to maintain operating temperatures during standby periods.

The main high-temperature battery systems are the sodium/beta and lithium/iron sulfide systems:

The sodium/beta battery system includes designs based on either the sodium/sulfur or the sodium/metal chloride chemistries (see Chapter 40). The sodium/sulfur technology has been in development for over 30 years and multi-kW batteries are now being produced on a pilot plant scale for stationary energy storage applications.[26] At least two 8 MW/40 MWh sodium/sulfur batteries have been put into service for utility load leveling by TEPCO in Japan.

Sodium/nickel chloride is a relatively new variation of the sodium/beta technology and was being developed mainly for electric-vehicle applications. There has not been nearly the effort on this chemistry as on the sodium/sulfur battery.

Sodium/sulfur and sodium/metal chloride technologies are similar in that sodium is the negative electrode material and beta-alumina ceramic is the electrolyte. The solid electrolyte serves as the separator and produces 100% coulombic efficiency. Applications are needed in which the battery is operated regularly. Sodium/nickel chloride cells have a higher open-circuit voltage, can operate at lower temperatures, and contain a less corrosive positive electrode than sodium/sulfur cells. Nevertheless, sodium/nickel chloride cells are projected to be more expensive and have lower power density than sodium/sulfur cells.

The lithium/iron sulfide rechargeable battery system is another high-temperature system and must be operated above 400°C so that the salt mixture (LiCl-KCl) used as an electrolyte remains molten (see Chapter 41). The negative electrode is lithium, which is alloyed with aluminum or silicon, and the positive electrode can be either iron monosulfide or iron disulfide. No development is being performed on these technologies at this time because room temperature battery systems are showing comparable performance.

37.4.3   Ambient-Temperature Lithium Batteries

Rechargeable lithium batteries, which operate at or near ambient temperature, have been and continue to be developed because of their advantageous energy density and charge retention compared to conventional aqueous batteries. The lithium-ion version of this chemistry has been commercialized for consumer electronics and other portable equipment in small button and prismatic cylindrical sizes. The attractive characteristics of rechargeable lithium batteries make them promising candidates for aerospace, electric vehicles, and other applications requiring high-energy batteries. High energy and power densities have been achieved with rechargeable lithium batteries, despite the lower conductivity of the organic and polymer electrolytes that are used to ensure compatibility with the other components of the lithium cell. Scaling up to the sizes and power levels, and achieving the cycle life required for electric vehicles and maintaining the high degree of safety needed for all batteries, remains a challenge.

A number of different approaches are being taken in the design of rechargeable lithium batteries. The rechargeable lithium cell that can deliver the highest energy density uses metallic lithium for the negative electrode, a solid inorganic intercalation material for the positive electrode, and an organic liquid electrolyte. Manganese dioxide appears to be the best material for the positive electrode based on performance, cost, and toxicity. Poor cycle life and safety, however, are concerns with this type of battery because the porous, high-surface-area lithium that is plated during recharge is highly reactive and susceptible to forming dendrites which could cause internal short-circuiting and they are no longer marketed commercially.

Another approach is the use of a solid polymer electrolyte. These electrolytes are considered to have a safety advantage over the liquid electrolyte because of their lower reactivity with lithium and the absence of a volatile and sometimes flammable electrolyte. These electrolytes, however, have a lower conductivity that must be compensated for by using thinner electrodes and separators and by having larger electrode areas.

The approach that has been commercialized successfully for portable-sized batteries is the "lithium-ion" battery. This battery uses a lithiated carbon material in place of metallic lithium. A lithiated transition metal intercalation compound is used for the positive active material, and the electrolyte is either a liquid aprotic organic solution or a gel polymer electrolyte. Lithium ions move back and forth between the positive and negative electrodes during charge and discharge. As metallic lithium is not present in the cell, lithium-ion batteries are less chemically reactive and are safer and have a longer cycle life than other options.[27] These systems require battery management circuitry to prevent overcharge, overdischarge and to provide cell balancing and other safety features. Larger lithium-ion batteries, in sizes up to 100 Ah, are under development.

The ambient temperature lithium battery technologies, especially the lithium-ion, are among the most promising for EVs, HEVs, electric utility energy storage, and other such applications. The scaling, safety, life issues, and cost remain a significant challenge to their use in these emerging applications.

37.5 REFUELABLE BATTERIES AND FUEL CELLS— AN ALTERNATIVE TO ADVANCED RECHARGEABLE BATTERIES

Another category of aqueous battery systems is the metal-air battery. These batteries are noted for their high specific energy as they utilize ambient air as the positive active material, and light metals, most commonly aluminum or zinc, as the negative active material. Except for the iron/air battery, on which earlier development work for EV applications has now been abandoned, metal-air batteries have either limited capability for recharge, as for zinc/air, or they cannot be electrically recharged at all, as in the case of the aluminum/air system.

The zinc/air system is commercially available as a primary battery. For EV and other applications, there are efforts underway to develop a "mechanically" rechargeable battery where the discharged electrode is physically removed and replaced with a fresh one. Recycling or recharging of the reaction product is done remotely from the battery. There also was a significant effort in the 1980s and 1990s to develop an aluminum/air battery with mechanical recharging,[28] but this work is now continuing at a reduced level.

Fuel cells can in a sense be regarded as refuelable batteries, and are being considered for use in portable electronic equipment (see Chaps. 42 and 43).

REFERENCES

1. Proc. Annual Automotive Technology Development Contractors' Coordination Meeting, Society of Automotive Engineers, Dearborn, Mich., Nov. 1992, pp. 371–375.

2. USABC Electric Vehicle Battery Test Procedures Manual Revision 2, U.S. Advanced Battery Consortium, DOE/ID-10479, Rev. 2, Jan. 1996.

3. R. A. Sutula, et al., "Recent Accomplishments of the Electric and Hybrid Vehicle Energy Storage R&D Programs at the U.S. Department of Energy: A Status Report," *17th Int. Electric Vehicle Symp.*, EVAA, Oct. 2000.

4. P. C. Butler, "Battery Energy Storage for Utility Applications: Phase I—Opportunities Analysis," Sandia National Laboratories, SAND94-2605, Oct., 1994.

5. R. L. Hammond, S. R. Harrington, and M. Thomas, "Photovoltaic Industry Battery Survey," Photovoltaic Design Assistance Center, Sandia National Laboratories, Albuquerque, N. Mex., Apr. 1993.

6. USCAR Mileposts, *www.uscar.org,* Winter, 2001, pp. 4–5.

7. M. A. Weiss, et al., "On the Road in 2020," Energy Laboratory Report # MIT EL 00-003, Oct., 2000.

8. J. D. Boyes, "Energy Storage Systems Program Report for FY99," SAND2000-1317, June 2000.

9. S. Furuta, "NEDO's Research and Development on Battery Energy Storage System," Utility Battery Group Meeting, Valley Forge, Pa., Nov. 1992.

10. See specific battery chapters for detailed data.

11. R. D. King, R. A. Koegl, L. Salasoo, K.B. Haefner, and A. Hamilton, "Heavy Duty (225 kW) Hybrid-Electric Propulsion System for Low-Emission Transit Buses—Performance, Emissions, and Fuel Economy Tests," *Proc. 14th Int. Electric Vehicle Symp.*, Orlando, Fla., published by EVAA, Dec., 1997.

12. S. Sostrom, "Update on the Golden Valley BESS," *Proc. of Conf. on Electric Energy Storage Applications and Technologies,* Orlando, Fla, Sept. 2000.

13. "Changing Perceptions of Ni-Cad Batteries," *Batteries International,* Jan. 1994.

14. U.S. Patent 3,713,888, "Process for Electrical Energy Using Solid Halogen Hydrate," P. C. Symons, 1973.

15a. Energy Development Associates, "Development of the Zinc Chloride Battery for Utility Applications," Electric Power Research Institute, EPRI AP-5018, Jan. 1987.

15b. C. C. Whittlesey, B. S. Singh, and T. H. Hacha, "The FLEXPOWER™ Zinc-Chloride Battery: 1986 Update," *Proc. 21st IECEC,* San Diego, Calif., 1986, pp. 978–985.

16a. T. Horie, H. Ogino, K. Fujiwara, Y. Watakabe, T. Hiramatsu, and S. Kondo, "Development of a 10 kW (80 kWh) Zinc-Chloride Battery for Electric Power Storage Using Solvent Absorption Chlorine Storage System (Solvent Method)," *Proc. 21st IECEC,* San Diego, Calif., 1986, vol. 2, pp. 986–991.

16b. Y. Misawa, A. Suzuki, A. Shimizu, H. Sato, K. Ashizawa, T. Sumii, and M. Kondo, "Demonstration Test of a 60kW-Class Zinc/Chloride Battery as a Power Storage System," *Proc. 24th IECEC,* Washington, D.C., 1989, vol. 3, pp. 1325–1329.

16c. H. Horie, K. Fujiwara, Y. Watakabe, T. Yabumoto, K. Ashizawa, T. Hiramatsu, and S. Kondo, "Development of a Zinc/Chloride Battery for Electric Energy Storage Applications," *Proc. 22nd IECEC,* Philadelphia, Pa., 1987, vol. 2, pp. 1051–1055.

17. L. H. Thaller, "Recent Advances in Redox Flow Cell Storage Systems," DOE/NASA/1002-79/4, NASA TM 79186, Aug. 1979. N. Hagedorn, "NASA Redox Storage System Development Project," U.S. Dept. of Energy, DOE/NASA/12726-24, Oct. 1984.

18. M. Bartolozzi, "Development of Redox Flow Batteries. A Historical Bibliography," *J. Power Sources* **27:**219–234 (1989).

19. Z. Kamio, T. Hiramatsu, and S. Kondo, "Research and Development of 10-kW Redox Flow Battery," *Proc. 22nd IECEC,* Philadelphia, Pa., 1987, vol. 2, pp. 1056–1059.

20. T. Tanaka, T. Sakamoto, N. Mori, T. Shigematsu, and F. Sonoda, "Development of a 60-kW Class Redox Flow Battery System." *Proc. 3d Int. Conf. of Batteries for Utility Energy Storage,* Kobe, Japan, 1991, pp. 411–423.

21. H. Izawa, T. Hiramatsu, and S. Kondo, "Research and Development of 10 kW Class Redox Flow Battery," *Proc. 21st IECEC,* San Diego, Calif., 1986, vol. 2, pp. 1018–1021.

22. T. Hirabayashi, S. Furuta, and H. Satoh, "Status of the 'Moonlight Project' on Advanced Battery Energy Storage System," *Proc. 26th IECEC,* Boston, Mass., 1991, vol. 6, pp. 88–93.

23. R. P. Hollandsworth, "Zinc-Redox Battery, A Technology Update," The Electrochemical Society, Fall Meeting, Oct., 1987.

24. N. Tokuda, et al., "Vanadium Redox Flow Battery for Use in Office Buildings," *Proc. of Conf. on Electric Energy Storage Applications and Technologies,* Orlando, Fla., Sept. 2000.

25. I. Whyte, S. Male, and S. Bartley, "A Utility Scale Energy Storage Project at Didcot Power Station," *Proc. 6th Int. Conf. on Batteries for Utility Energy Storage,* Gelsenkirchen, Germany, 1999.

26. E. Kodama, et al., "Advanced NaS Battery for Load Leveling," *Proc. 6th Int. Conference on Batteries for Utility Energy Storage,* Gelsenkirchen, Germany, 1999.

27a. N. Doddapaneni, "Technology Assessment of Ambient Temperature Rechargeable Lithium Batteries of Electric Vehicle Applications," Sandia National Laboratories, Rep. SAND91-0938, July 1991.

27b. *Proc. 7th Int. Meeting on Lithium Batteries,* Boston, May, 1994.

28. E. J. Rudd and S. Lott, "The Development of Aluminum-Air Batteries for Application in Electric Vehicles Final Report," Sandia National Laboratories, SAND91-7066, Dec., 1990.

CHAPTER 38
METAL/AIR BATTERIES

Robert P. Hamlen and Terrill B. Atwater

38.1 GENERAL CHARACTERISTICS

The electrochemical coupling of a reactive anode to an air electrode provides a battery with an inexhaustible cathode reactant and, in some cases, very high specific energy and energy density. The capacity limit of such systems is determined by the Ampere-hour capacity of the anode and the technique for handling and storage of the reaction product. As a result of this performance potential, a significant effort has gone into metal/air battery development.[1,2] The major advantages and disadvantages of the metal/air battery system are summarized in Table 38.1.

Primary, reserve, electrically rechargeable, and mechanically rechargeable metal/air battery configurations have been explored and developed. In the mechanically rechargeable designs (that is, replacing the discharged metal electrode) the battery essentially functions as a primary battery and can use relatively simple "unifunctional" air electrodes which need to operate only in a discharge mode. Conventional electrical recharging of metal/air batteries requires either a third electrode (to sustain oxygen evolution on charge) or a "bifunctional" electrode (a single electrode capable of both oxygen reduction and evolution).

Table 38.2 lists the metals that have been considered for use in metal/air batteries with several of their electrical characteristics. Of the potential metal/air battery candidates, zinc has received the most attention because it is the most electropositive metal which is relatively stable in aqueous and alkaline electrolytes without significant corrosion, provided the appropriate inhibitors are used. It has been used for many years in commercial primary zinc/air batteries. Initially the products were large batteries using alkaline electrolytes for such applications as railroad signaling, remote communications, and ocean navigational units requiring long-term, low-rate discharge. As thin electrodes were developed, the technology was applied to small (button-type), high-capacity primary cells which are used in hearing aids, pagers, and similar applications (see Chap. 13).

Zinc is also attractive for electrically rechargeable metal/air systems because of its relative stability in alkaline electrolytes and also because it is the most active metal that can be electrodeposited from an aqueous electrolyte. The development of a practical rechargeable zinc/air battery with an extended cycle life would provide a promising high-capacity power source for many portable applications (computers, communications equipment) as well as, in larger sizes, for electric vehicles. Problems of dendrite formation, nonuniform zinc dissolution and deposition, limited solubility of the reaction product, and unsatisfactory air electrode performance have slowed progress toward the development of a commercial rechargeable battery. However, there is a continued search for a practical system because of the potential of the zinc/air battery.

Other metals have also been investigated as electrode materials for metal/air batteries. Calcium, magnesium, lithium, and aluminum have attractive energy densities. Lithium/air,[3,4] calcium/air, and magnesium/air batteries[5,6] have been studied, but cost and problems such as anode polarization or instability, parasitic corrosion, nonuniform dissolution, safety, and practical handling have so far inhibited the development of commercial products. The voltage and the specific energy of the iron/air battery are relatively low, and its cost is high compared to the other metal/air batteries. Development on this battery, therefore, has concentrated on an electrically rechargeable system (see Chap. 25) as the iron electrode is long-lived and more adapted to recharging.

Aluminum is attractive for use because of its geological abundance (third most abundant element in the earth's crust), its potentially low cost, and its relative ease of handling.[7–9] However, the aluminum/air battery has too high a charging potential to be electrically recharged in an aqueous system (water is preferentially electrolyzed). Therefore the effort has been directed to reserve and mechanically rechargeable designs.

Table 38.3 summarizes the work on the various types and designs of metal/air batteries.

TABLE 38.1   Major Advantages and Disadvantages of Metal/Air Batteries

Advantages	Disadvantages
High energy density	Dependent on environmental conditions:
Flat discharge voltage	Drying-out limits shelf life once opened to air
Long shelf life (dry storage)	Flooding limits power output
No ecological problems	Limited power output
Low cost (on metal use basis)	Limited operating temperature range
Capacity independent of load and temperature	H_2 from anode corrosion
when within operating range	Carbonation of alkali electrolyte

TABLE 38.2   Characteristics of Metal/Air Cells

Metal anode	Electrochemical equivalent of metal, Ah/g	Theoretical cell voltage,* V	Valence change	Theoretical specific energy (of metal), kWh/kg	Practical operating voltage, V
Li	3.86	3.4	1	13.0	2.4
Ca	1.34	3.4	2	4.6	2.0
Mg	2.20	3.1	2	6.8	1.2–1.4
Al	2.98	2.7	3	8.1	1.1–1.4
Zn	0.82	1.6	2	1.3	1.0–1.2
Fe	0.96	1.3	2	1.2	1.0

*Cell voltage with oxygen cathode.

TABLE 38.3 Metal/Air Batteries

Primary zinc/air cells
- Button cells
  200–400 Wh/kg
  650–1000 Wh/L
- Prismatic cells
  270–375 Wh/kg
  380–460 Wh/L
- Hybrid MnO_2 cells
  350–400 Wh/kg
  420–600 Wh/L
- Industrial cells
  200–300 Wh/kg
  225–330 Wh/L
- Cylindrical (development)
  250–350 Wh/kg

Secondary zinc/air cells
- Mechanically rechargeable
  - Anode replacement
  - Zinc powder (packed bed)
    100–225 Wh/kg
- Electrically rechargeable
  - Bifunctional air electrode
    130–180 Wh/kg
    130–160 Wh/L
  - Metal foam negative
    100 Wh/kg

Primary aluminum/air cells
- Saline electrolyte
  ~600 Wh/kg (dry)
  ~400 Wh/L
- Al/O_2 for undersea use
  640 Wh/kg (dry)
- Alkaline electrolyte
  200–250 Wh/kg
  150–200 Wh/L
- Al/O_2 for underwater provision
  265 Wh/kg
  265 Wh/L

Primary magnesium/air cells
- Mg/O_2 cell for undersea use
  2700 Wh/kg (dry)

Primary and secondary lithium/air cells
- Lithium/air
  N/A
- Lithium/water for undersea use
  2200 Wh/kg of Li
- Lithium/O_2 with polymer electrolyte (secondary)

Secondary iron/air cells
- Electrically rechargeable
  60–75 Wh/kg
  100 Wh/L

38.3

38.2 CHEMISTRY

38.2.1 General

The metal/air batteries being developed use neutral or alkaline electrolytes. The oxygen-reduction half-cell reaction during discharge may be written

$$O_2 + 2H_2O + 4e \rightleftharpoons 4OH^- \qquad E^0 = +0.401 \text{ V}$$

The theoretical cell voltages, the equivalent weights of the metals, and the theoretical specific energies obtained when this oxygen electrode is coupled with various metal anodes are given in Table 38.2. Polarization effects at both electrodes degrade these voltages to those shown in the table at practical operating discharge rates. Note that the theoretical specific energy of metal/air batteries is based only on the negative electrode (anode or fuel electrode during discharge) as this is the only reactant that has to be carried in the battery. The other reactant, oxygen, is introduced into the battery from ambient air during discharge.

The discharge reaction at the negative or metal electrode (anode during discharge) is dependent on the specific metal used, the electrolyte, and other factors in the cell chemistry. The discharge reaction at the negative electrode can be generalized as

$$M \rightarrow M^{n+} + ne$$

The generalized overall discharge reaction may be written

$$4M + nO_2 + 2nH_2O \rightarrow 4M(OH)_n$$

where M is the metal and the value of n depends on the valence change for the oxidation of the metal, as listed in Table 38.2.

Most metals are thermodynamically unstable in an aqueous electrolyte and react with the electrolyte to corrode or oxidize the metal and generate hydrogen as follows:

$$M + nH_2O \rightarrow M(OH)_n + \frac{n}{2} H_2$$

This parasitic corrosion reaction, or self-discharge, degrades the coulombic efficiency of the anode and must be controlled to minimize this loss of capacity.

Other factors which can affect the performance of the metal/air battery are the following:

Polarization.   The voltage of a metal/air battery drops off more sharply with increasing current than that of other types of batteries because of diffusion and other limitations in the oxygen or air cathode. This means that these air systems are more suited for low- to moderate-power applications than to high-power ones.

Electrolyte Carbonation.   As the cell is open to air, carbon dioxide can be absorbed. This can result in the crystallization of carbonate in the porous air electrode, which may impede air access and cause mechanical damage and a decreasing electrode performance. Potassium carbonate is also less conductive than the KOH electrolyte normally employed in metal/air batteries.

Water Transpiration.   Again, as the cell is open to air, water vapor can be transferred if a vapor partial pressure difference exists between the electrolyte and the surrounding environment. Excessive water loss increases the concentration of the electrolyte and leads to drying out and premature failure. Gain of water can lead to dilution of the electrolyte. This gain can cause flooding of the air electrode pores and electrode polarization due to the inability of the air to reach the reaction sites.

Efficiency.   The oxygen electrode at moderate temperatures displays a significant irreversibility during both charge and discharge. As a result there is generally about a 0.2 V difference between the actual charging voltage and the reversible potential, with the same situation on discharge. For example, a zinc/air battery generally discharges at a voltage of about 1.2 V, while the charging voltage is about 1.6 V or higher. This results in a loss of overall energy efficiency even before any other factors are considered.

Charging.   Oxidation of catalysts and electrode supports during charging can be a problem for those systems which are recharged electrically, such as zinc/air and iron/air. Approaches to solving this problem generally involve either the use of oxidation-resistant substrates and catalysts, the use of a third electrode for charging, or charging the negative (metal) electrode material external to the cell.

38.2.2 Air Electrode

Successful operation of metal/air batteries depends on an effective air electrode. As a result of the interest in gaseous fuel cells and metal/air batteries over the past 30 years, a significant effort has been aimed at improved high-rate, thin air electrodes, including the development of better catalysts, longer-lived physical structures, and lower-cost fabrication methods for such gas diffusion electrodes.

An alternative approach is to use a low-cost air cathode with more modest performance, but this requires a greater cathode area in each cell. Figure 38.1 shows a type of electrode which is produced by a continuous process using low-cost materials.[10-12] This electrode is composed of two active layers bonded to each side of a current-collecting screen, with a microporous Teflon layer bonded to the air side of the electrode. The active layers are fabricated by passing a nonwoven web of carbon fibers (see Fig. 38.1b) through a slurry containing the catalyst, a dispersing agent, and a binder in a continuous process, with a drying and compacting step built into the process. The active layers, the screen, and the Teflon layer are then bonded in the continuous process. These electrodes are used in the aluminum air reserve standby batteries (see Sec. 38.4.2).

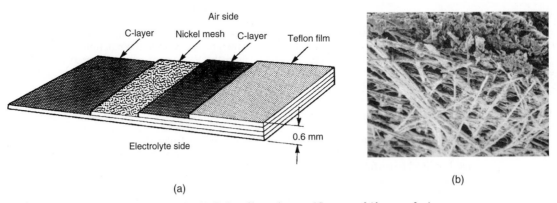

FIGURE 38.1   (*a*) Laminated air cathode. (*b*) Carbon fiber substrate. (*Courtesy of Alupower, Inc.*)

38.3   ZINC/AIR BATTERIES

38.3.1   General

Zinc/air batteries are commercially available in primary button type batteries (see Chap. 13), and in the late 1990s 5 to 30 Ahr prismatic batteries as well as larger primary industrial-type batteries. Electrically rechargeable batteries are being considered for both portable and electric-vehicle applications, but the control of the recharging (replating) of zinc and the development of an efficient high-rate bifunctional air electrode remain a challenge. In some designs, a third oxygen-evolving electrode is used for recharging, or recharging is done external to the cell to avoid the need for the bifunctional air electrode. Another approach to avoid the difficulties with electrical recharging is the mechanically rechargeable battery, where the spent zinc electrode and/or the discharged products are removed and physically replaced. Table 38.3 contains a summary of the different types of zinc/air batteries.

The overall cell reaction for a zinc/air battery on discharge in an alkaline electrolyte may be represented as

$$Zn + \tfrac{1}{2}O_2 + H_2O + 2(OH)^- \rightarrow Zn(OH)_4^{-2} \qquad E^0 = 1.62 \text{ V}$$

The initial discharge reaction at the zinc electrode can be simplified to

$$Zn + 4OH^- \rightleftharpoons Zn(OH)_4^{-2} + 2e$$

This reaction occurs as a result of the solubility of the zincate anion in the electrolyte and proceeds until the zincate level reaches the saturation point. There is no well-defined solubility limit, since the degree of supersaturation is time-dependent. After partial discharge, the solubility exceeds the equilibrium solubility level, with subsequent precipitation of zinc oxide, as follows:

$$Zn(OH)_4^{-2} \rightarrow ZnO + H_2O + 2(OH)^-$$

The overall cell reaction then becomes

$$Zn + \tfrac{1}{2}O_2 \rightleftharpoons ZnO$$

This transient solubility is one of the main reasons for the difficulty in making a successful rechargeable zinc/air battery. The location of the precipitation of the reaction product cannot be controlled, so that on a subsequent recharge the amount of zinc deposited on different parts of the electrode area of the cell can vary.

38.3.2   Portable Primary Zinc/Air Batteries

Primary zinc/air button-type batteries are described in Chap. 13. This configuration is an effective way to package the zinc/air system in small sizes, but scaling up to larger sizes tends to lead to performance and leakage problems, but these can be overcome with prismatic cell designs. Figure 38.2 shows the basic schematic of a prismatic zinc/air cell. A typical prismatic cell uses a metal or plastic tray, which holds the zinc anode/electrolyte blend while the separator and cathode are bonded onto the rim of the tray. The anode/electrolyte blend is similar to the anode blend used in zinc/alkaline primary cells, containing zinc powder in a gelled aqueous potassium hydroxide electrolyte. The cathode is a thin gas diffusion electrode comprising two layers, an active layer and a barrier layer. The active layer of the cathode, which interfaces with the electrolyte, uses a high surface area carbon and a metal oxide catalyst bonded together with Teflon. The high surface area carbon is required for oxygen reduction and the metal oxide catalyst (MnO_2) for peroxide decomposition. The barrier layer, which interfaces with air, consists of carbon bonded together with Teflon. A high concentration of Teflon prevents electrolyte from weeping from the cell. Prismatic zinc/air cells have been designed with moderately high rate and high capacity. The thickness of the cell determines the anode capacity of the cell and the cross-sectional surface area determines the maximum rate capability.[13,14]

In addition to prismatic cell designs, cylindrical zinc/air cells (see Fig. 38.3), have been designed.[15–17]

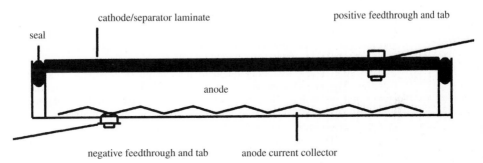

FIGURE 38.2 Design of a Prismatic Primary Zinc-Air Cell (*Courtesy of Electric Fuel Corp.*)

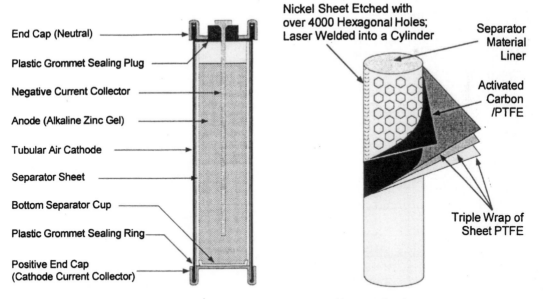

FIGURE 38.3 Design of a Cylindrical Primary Zinc-Air Cell (*Courtesy of Rayovac Corp.*)

The high specific energy, low cost and safety of the zinc/air primary battery make it an attractive choice for many portable electronics applications. It is particularly advantageous for applications where the battery energy is consumed within a range of one to fourteen days, since the high specific energy and energy density on the zinc/air system can be realized and the impact of environmental interactions (dryout, flooding and carbonation) is low. Typical cell discharge curves at 25°C are shown in Fig. 38.4. The cell voltage is relatively flat throughout most of the discharge, with little capacity remaining beyond 0.9 volts per cell. Figures 38.5 and 38.6 show the specific energy of prismatic zinc/air as a function of drain rate. Figure 38.5 shows the specific energy for 5 Ah zinc/air batteries over typical current ranges for portable tape players and analog cellular phones. Figure 38.6 shows the specific energy for 30 Ah zinc/air batteries over typical current ranges for portable stereo systems and camcorders. A summary of discharge characteristics of representative state-of-the-art prismatic zinc/air cells is given in Table 38.4.

Two approaches are being taken to the design of prismatic zinc/air cells for portable batteries. The first is a metal case prismatic cell. This design is essentially an adaptation of button cell technology. In this design a cathode subassembly, contained in a nickel-plated steel can, is crimp sealed onto an anode subassembly, contained in a copper lined nickel plated stainless steel can. A molded plastic insulator seal separates the anode and cathode assemblies. This design has performed well for smaller sizes (5 Ah or less). Figure 38.7 shows a battery designed for cellular telephone applications while the characteristics are listed in Table 38.5.

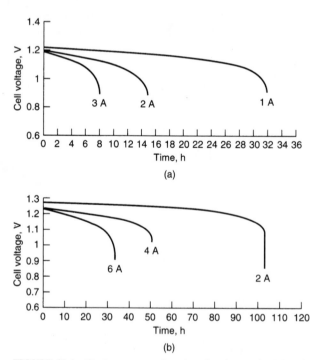

FIGURE 38.4 Discharge curves for prismatic primary zinc/air cells at 25°C. (*a*) High-rate cell. (*b*) Large-capacity cell. (*Courtesy of Matsi, Inc.*)[18]

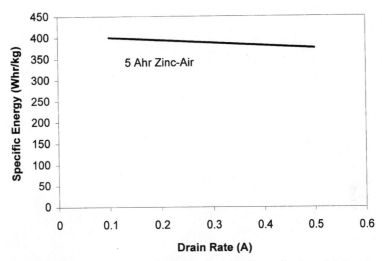

FIGURE 38.5 Specific energy for 5 Ahr zinc/air cell as a function of drain rate. (*Courtesy Electric Fuel Corp.*)

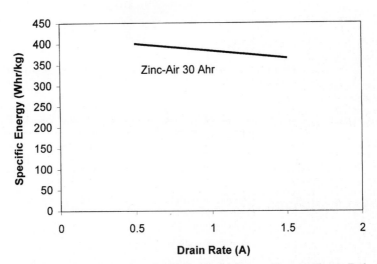

FIGURE 38.6 Specific energy for 30 Ahr zinc/air battery. (*Courtesy Electric Fuel Corp.*)

TABLE 38.4   Specifications of Prismatic Zinc/Air Cells

Variable	Cellular phone cell	Field charger cell
Facial Dimension, cm (length × width)	4.6 × 2.7	7.6 × 7.6
Height, cm	0.43	0.6
Weight, g	15	87
Capacity, Ah	3.6	30
Energy Density[1], Wh/L	800	1000
Specific Energy[1], Wh/kg	300	400

[1] At nominal voltage.

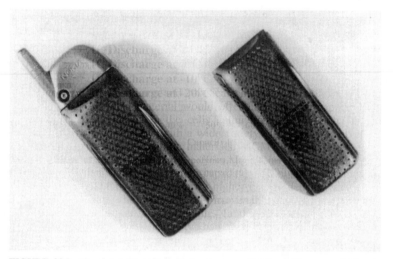

FIGURE 38.7   Zinc/air battery for cellular phone applications. (*Courtesy of Electric Fuel Corp.*)

TABLE 38.5   Characteristics of Prismatic Zinc/Air Batteries

Variable	Cellular phone battery (Nokia)	Field charger battery
Number of Cells	4	24
Voltage, V (nominal)	4.8	28
Capacity, Ah	3.6	30
Dimensions, cm		
Length	10.4	31
Width	4.5	18.5
Height	1.5	6
Weight, g	78	2400
Volume, cm³	70	3500
Energy Density[1], Wh/L	250	240
Specific Energy[1], Wh/kg	220	350

[1] At nominal voltage.

The second design uses plastic for the case of the prismatic zinc/air cell. This design employs adhesive technology to bond the cell anode and cathode subassemblies. The plastic cell design is preferred for large capacity cell sizes (>5 Ah) due to technological limitations imposed on the metal cell design. In particular, leak tight crimp seals become a challenge as cell dimensions increase due to the need for close dimensional tolerances. The key challenges for the plastic cell include the development of the proper designs and materials for the cathode and cell seals and for the current feed-throughs. The latter is required for the plastic cell but not the metal cell, in which the cans serve as terminals for electrical contact. Figures 38.8 and 38.9 shows cell and battery prototype under development for remote applications. The characteristics of this field charger battery are also listed in Table 38.5.

Prismatic cells are designed so they can be stacked as multicell batteries for use in various portable electronic equipment. Stacking of the cells requires a provision, such as a spacer, to permit air access to the cathode and a fan to provide forced flow of air. The thickness of the spacer is dependent on the dimensions of the cell and the required current density. If the spacer is too thin, the cell can become oxygen starved, while if too thick, it increases the battery weight and volume unnecessarily. An alternative approach to dealing with oxygen diffusion is by providing a positive pressure of air by designing a fan and air channels into the battery design.

Cylindrical zinc/air cells (Fig. 38.3) have been designed primarily in the "AA" cell size. These cells allow for the direct replacement of zinc alkaline manganese dioxide cells. The zinc/air technology uses a very thin cathode allowing for the bulk of the cell to contain the anode/electrolyte mixture. The relatively high surface area of "AA" cells allows for high power discharge rates. Batteries constructed from arrays of these cells do not provide for forced flow of air, but it has been shown that thermal gradients within the battery pack do provide convective flow.

Figure 38.10 shows a typical discharge curve for two 12 Volt zinc/air battery configurations: 1) twelve 30 Ampere-hour prismatic zinc/air cells in series and 2) forty-eight "AA" zinc-air cells consisting of four parallel strings of 12 cells in series. Figure 38.11 shows the discharge characteristics for three single-cell zinc/air batteries designed for portable electronic equipment.

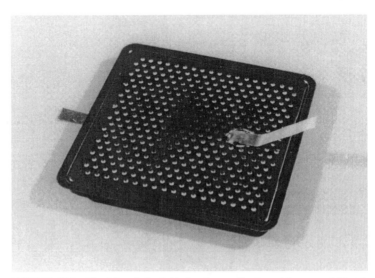

FIGURE 38.8   Zinc/air cell for field charging applications. (*Courtesy of Electric Fuel Corp.*)

FIGURE 38.9 Zinc/air battery for field charging applications. (*Courtesy of Electric Fuel Corp.*)

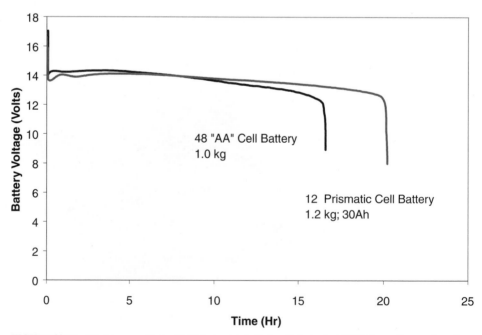

FIGURE 38.10 Discharge profile for 12-Volt zinc/air batteries discharged at 18 Watts continuous. Data from US Army tests.

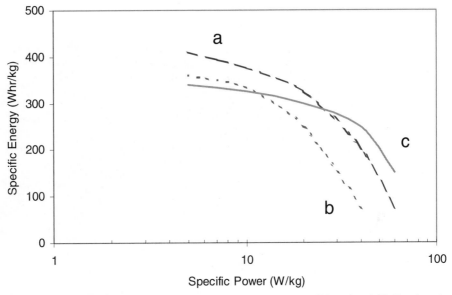

FIGURE 38.11   Discharge characteristics for primary zinc/air single-cell batteries. a) 35 Ahr prismatic cell, b) 5 Ahr prismatic cell and c) 5 Ahr "AA" cell. Data from US Army tests.

38.3.3   Industrial Primary Zinc/Air Batteries

Large primary zinc/air batteries have been used for many years to provide low-rate, long-life power for applications such as railroad signaling, seismic telemetry, navigational buoys, and remote communications. They are available in either water-activated (containing dry potassium hydroxide) or preactivated versions.[19] Preactivated versions are also available with a gelled electrolyte to minimize the possibility of leakage. Until recently, the zinc contained a few percent of mercury to minimize self-discharge after activation. The newer batteries use additives and alloys to eliminate the mercury and minimize hydrogen generation and corrosion.

Preactivated and Water-Activated Types.   A typical preactivated industrial-type zinc/air cell, the Edison Carbonaire cell, is manufactured in a 1100-Ah size and is available in two- and three-cell configurations, as illustrated in Fig. 38.12. The cell case and cover are molded from a tinted transparent acrylic plastic. The construction features are shown in Fig. 38.13 identifying the wax-impregnated carbon cathode block, the solid zinc anodes, and the lime-filled reservoir. These cells normally have a bed of lime to absorb carbon dioxide and to remove soluble zinc compounds from solution and precipitate them as calcium zincate. They are made with transparent cases so that the electrolyte level and the state of charge can be monitored visually. The state of charge can be monitored by observing the condition of the zinc plates and the condition of the lime bed. The bed turns darker as it is converted to zincate.

You can *watch* the activating water reach its proper level — and stop filling. No more overfilling or underfilling. No guessing, no gauges, no dip-sticks.

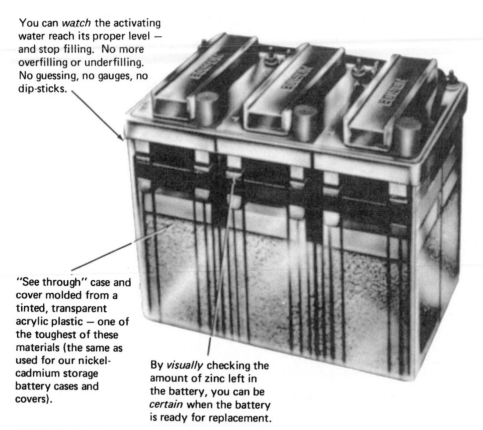

"See through" case and cover molded from a tinted, transparent acrylic plastic — one of the toughest of these materials (the same as used for our nickel-cadmium storage battery cases and covers).

By *visually* checking the amount of zinc left in the battery, you can be *certain* when the battery is ready for replacement.

FIGURE 38.12   Edison Carbonaire zinc/air battery. (*Courtesy of SAFT America, Inc.*)

Water-activated cells and batteries are supplied sealed. The caustic (potassium hydroxide) electrolyte and the lime flake are present in the dry form. The cell is activated by removing the seals and adding the appropriate amount of water to dissolve the potassium hydroxide. Periodic inspection and addition of water are the only required maintenance.

These cells are manufactured in a 1100-Ah size and are available in multicell batteries, with the cells connected in series or parallel. The maximum continuous discharge rate for this battery at 25°C is 0.75 A. A preactivated 1100-Ah three-cell battery weighs about 2 kg, giving an energy density of about 180 Wh/kg. The physical and electrical characteristics of these batteries are listed in Tables 38.6 and 38.7.

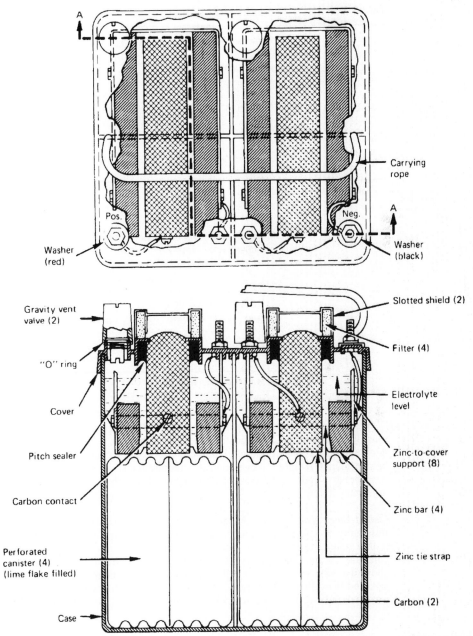

FIGURE 38.13 Top and side views of type ST-2 Carbonaire zinc/air battery. (*Courtesy of SAFT America, Inc.*)

TABLE 38.6   Edison Carbonaire Zinc/Air Batteries

Type	Dimensions, cm			Weight (filled), kg	Connection	Nominal voltage, V	Nominal capacity, Ah
	Length	Width	Height				
Two cells:							
ST-22-1100	21.9	20.0	28.9	14	Series	2.5	1100
ST-22-2200					Parallel	1.25	2200
Three cells:							
ST-33-1100	32.4	20.0	28.9	21	Series	3.75	1100
ST-33-3300					Parallel	1.25	3300

Source: SAFT America, Inc.

TABLE 38.7   Maximum Discharge Rates (Amperes) to 1.0 V per Cell
for Edison Carbonaire ST Type Zinc/Air Batteries

	Duty cycle			
	10% on (up to 0.5 s)	20% on (up to 2.0 s)	50% on (up to 1 s)	100% on (continuous)
20°C	3.5	2.8	2.3	1.25
−5°C	2.4	1.9	1.6	0.75

Source: SAFT America, Inc.

Voltage versus time curves plotted in Fig. 38.14 depict the performance obtained at various discharge rates. Capacities obtained are quite consistent over the range of 0.15 to 1.25 A continuous discharge, although at the higher rates, voltage variation with temperature must be considered. If tight voltage control is required, a series-parallel assembly of batteries may be needed to reduce the current drain per cell and thus minimize the voltage fluctuations with temperature.

Gelled-Electrolyte Types.   An alternative version uses a gelled electrolyte to eliminate the possibility of leakage during operation. The zinc electrode is composed of zinc powder mixed with a gelling agent and the electrolyte, and the reaction product is zinc oxide rather than calcium zincate. The battery is filled with electrolyte during manufacture. Figure 38.15 showed a cross section of the cell. The Gelaire battery is manufactured in a nominal 1200-Ah size and is available in multicell batteries with the cells connected in series or parallel. The physical and electrical characteristics of these batteries are listed in Table 38.8.

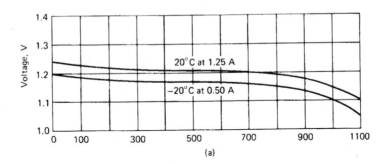

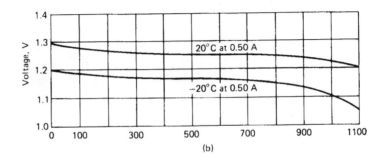

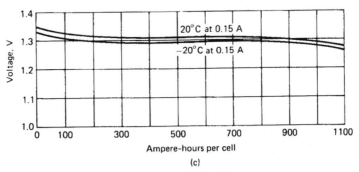

FIGURE 38.14 Typical discharge characteristics of Carbonaire 1100-Ah zinc/air batteries. (*a*) Maximum continuous rates. (*b*) Moderate continuous rates. (*c*) Low continuous rates. (*Courtesy of SAFT America, Inc.*)

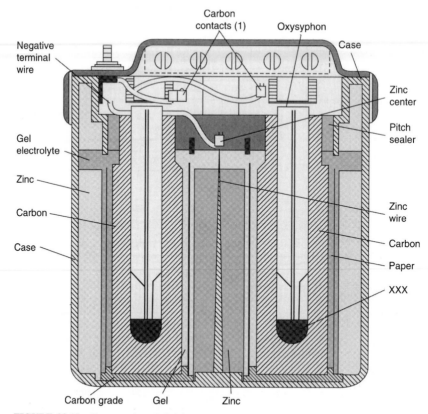

FIGURE 38.15 Cross section of Gelaire cell. (*Courtesy of SAFT America, Inc.*)

TABLE 38.8 Physical and Electrical Characteristics of Gelaire Battery

	NT 1000X (single cell)	2NT 1000X (two-cell block)*	3NT 1000X (large-cell block)*
Dimensions, mm:			
Length	108	216	324
Width	200	200	200
Height	213	213	213
Weight, kg	5.4	10.9	16.4
Nominal voltage, per cell, V	1.3	1.3	1.3
Nominal capacity per cell, Ah	1200	1200	1200

*Units can be connected either in series or in parallel.
Source: SAFT America, Inc.

Figure 38.16 shows the discharge characteristics of the 1200-Ah cell at several discharge loads. As a result of the use of a porous zinc electrode, and the direct formation of zinc oxide as the reaction product, the specific energy of this type of battery is higher than for the water-activated types. At low rates of discharge it is 285 Wh/kg.

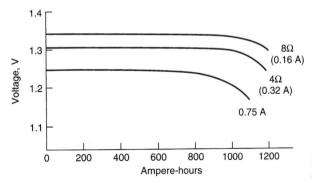

FIGURE 38.16 Discharge performance of Gelaire battery (NT1000X). (*Courtesy of SAFT America, Inc.*)

38.3.4 Primary Hybrid-Air/Manganese Dioxide Batteries

Another approach to primary zinc/air batteries is to use a hybrid cathode which contains a significant amount of manganese dioxide.[20] During low-rate operation, the battery functions as a zinc/air system. At high rates, as the oxygen may be depleted, the discharge function at the cathode is taken over by the manganese dioxide. This means that such a battery should essentially have the capacity of a zinc/air battery when discharged at low rates, but it should have the pulse current capability of a manganese dioxide battery. After the high-current pulse, the manganese dioxide is partially regenerated by air oxidation so that the pulse current capability is restored.

Figure 38.17 is a side view of a flat-pack cell. Figure 38.18 compares the performance of a 6-V four-cell hybrid "lantern" battery with similar alkaline and zinc-carbon batteries at an intermittent low-rate discharge. The specific energy of this battery is about 350 Wh/kg. Single and multicell batteries are available in capacities of 40 to 4800 Ah.

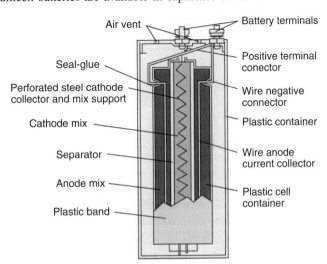

FIGURE 38.17 Side view of hybrid zinc/air-manganese dioxide cell. (*Courtesy of Celair Corp.*)

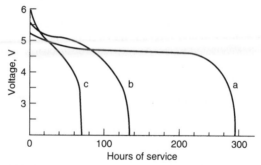

FIGURE 38.18 Comparison of performance of hybrid "lantern" battery with zinc-carbon and alkaline-manganese dioxide battery. Curve *a*—air-alkaline; curve *b*—alkaline-MnO$_2$; curve *c*—heavy-duty zinc-carbon. (*Courtesy of Celair Corp.*)

38.3.5 Electrically Rechargeable Zinc/Air Batteries

Electrically rechargeable zinc/air batteries use a bifunctional oxygen electrode so that both the charge process and the discharge process take place within the battery structure.

The basic reactions of an electrically rechargeable zinc/air cell using a bifunctional oxygen electrode are shown in Fig. 38.19. Advances in electrically rechargeable zinc/air cells have concentrated on the bifunctional air electrode.[21-23] Electrodes based on La, Sr, Mn and Ni perovskites have demonstrated good cycle life. Figure 38.20 shows the gains in cycle life for the bifunctional air cathode achieved going from Phase I to Phase II of a recent research and development program.

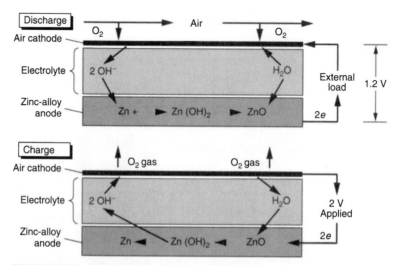

FIGURE 38.19 Basic operation of electrically rechargeable zinc/air cell. (*Courtesy of AER Energy Resources, Inc.*)

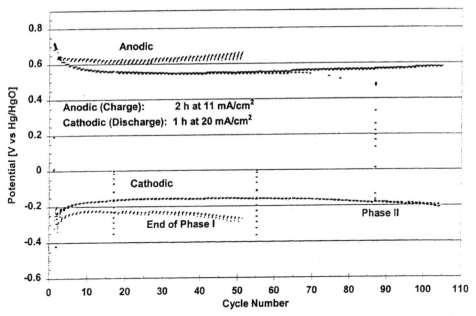

FIGURE 38.20 Advances in bifunctional air electrodes. LSNC Perovskite plus Shawinigan Black Carbon. Area = 25 cm². 8 M KOH at room temperature. (*Courtesy of Alupower, Inc.*)

Portable Electrically Rechargeable Batteries. An electrically rechargeable zinc/air cell having a bifunctional oxygen electrode, designed for use in computers and other electronic communication equipment, is shown in Fig. 38.21. The cell is a prismatic or thin rectangular design. A high-porosity zinc structure, which maintains its integrity and morphology during cycling, is used. The air electrode is a corrosion-resistant carbon structure, containing a large number of small pores, and a catalyst. The structure is permeable to oxygen, hydrophobic, and supported by a low-impedance current collector. The flat zinc negative and the air electrode plates face each other, separated by a high-porosity separator with low electrochemical resistance and the ability to absorb and retain the potassium hydroxide electrolyte. The cell case is injection-molded polypropylene with openings to permit the inflow of oxygen during discharge and release of the oxygen generated during charge.

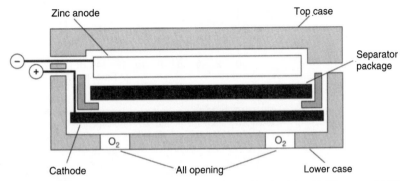

FIGURE 38.21 Cross section of electrically rechargeable zinc/air cell. (*Courtesy of AER Energy Resources, Inc.*)

A critical factor in the design of the cell and battery is the means of controlling the flow of air into and out of the cell, which must be matched to the needs of the application. Excessive amounts of air could result in drying out the cell; too little air (oxygen-starved) will result in a drop-off of performance. The stoichiometric quantity of air required is 18.1 cm^3/min per Ampere of continuous current. An air manager is used to control the flow of air by opening air access to the cathode during discharge and sealing the battery from the air when it is not in use to minimize self-discharge. A fan, powered by the battery, also is used to assist the airflow.

A discharge and charge profile of a 20-Ah zinc/air battery is shown in Fig. 38.22. The battery is typically discharged at the $C/20$ rate or lower to about 0.9 V. The voltage profile is flat with a sharp drop at the end of the discharge. Deep discharge to 0 V may be detrimental. Discharging at rates higher than the $C/10$ rate is not feasible because of the battery's relatively high internal resistance. This power limitation dictates the minimum size and weight, and the battery cannot be designed efficiently to operate for less than 8 to 10 h. When discharged at the acceptable loads, the battery can deliver about 150 Wh/kg and 160 Wh/L.

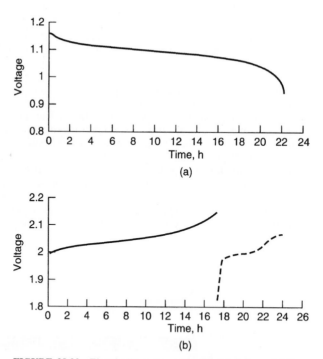

FIGURE 38.22 Electrically rechargeable zinc/air battery. (*a*) Representative discharge profile, 1-A discharge. (*b*) Representative charge profile, 1.25-A charge followed by 0.5-A charge. (*Courtesy of AER Energy Resources, Inc.*)

The battery is recharged by constant-current methods, using a two-step process, as shown in Fig. 38.22b. A moderate rate is used initially until the battery is about 85% charged. This is followed by a low rate to complete the charge. Charging a fully discharged battery takes about 24 h. Both charge rate and overcharging must be controlled. Overcharging will result in the generation of hydrogen at the negative electrode. It will damage the cell and shorten life due to the corrosion of the air cathode. Energy efficiency is about 50% due to the large difference between discharge and charge voltages. The overall life of the battery is independent of the number of cycles. About 400 h of operation has been demonstrated, but further development of this battery has been terminated because of its limited cycle life.

A sketch of a battery design for a notebook computer, fitting into the base of the computer case, is shown in Fig. 38.23. The battery contains five cells and is rated at 5 V and 20 Ah. Table 38.9 provides the physical and electrical characteristics of this battery.[24]

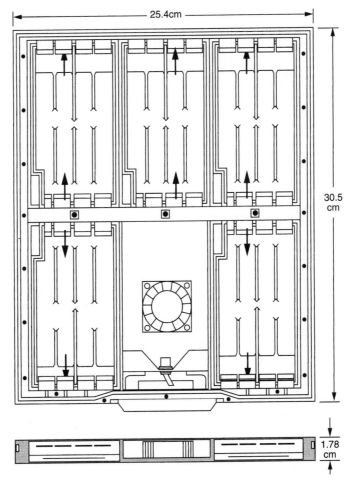

FIGURE 38.23 Prototype electrically rechargeable zinc/air battery for notebook computer. (*Courtesy of AER Energy Resources, Inc.*)

TABLE 38.9 Physical and Electrical Characteristics of Electrically Rechargeable Zinc/Air Battery

Open-circuit voltage	1.45 V
Nominal operating voltage	1.2–0.9 V (design using 1 V)
Cutoff voltage	0.9 V
Capacity	20 Ah (1 A)
Capacity retention	Capacity loss less than 2% per month when stored sealed at room temperature
Max current:	
Continuous	2 A
Pulse	3A
Energy density	130 Wh/kg
Specific energy	160 Wh/L
Cycle life	400 h
Charge characteristics	2.0 V/cell at 750 mA
Overcharge sensitivity	Overcharge degrades battery life
Charge termination	dV/dT and maximum voltage
Cathode air rate	100 cm³/min per cell at 1-A rate
Weight	155 g
Dimensions:	
Length	13.5 cm
Width	7.6 cm
Height	1.22 cm
Temperature:	
Operating	5–35°C
Storage	−20 to 55°C
Relative humidity:	
Operating	20–80%
Storage	5–95%

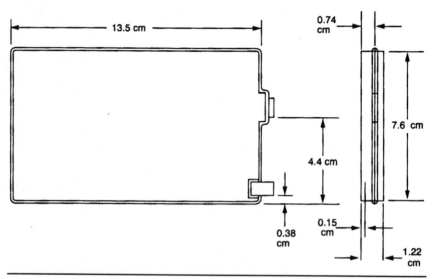

Source: AER Energy Resources, Inc.

Electrically Rechargeable Systems for Electric Vehicles.   A similar rechargeable zinc/air cell, operating at room temperature, was being developed for use in electric vehicles. The cell uses a planar bipolar configuration. The negative electrode consists of zinc particles in a paste form, similar to the electrode used in alkaline-manganese dioxide primary cells. The bifunctional air electrode consists of a membrane of carbon and plastic with appropriate catalysts. The electrolyte is potassium hydroxide with gelling agents and fibrous absorbing materials. A typical cell is rated at 100 Ah with an average operating voltage of 1.2 V.

Specific energies up to 180 Wh/kg at the 5 to 10-h discharge rates and a battery life of about 1500 h have been achieved. Technical limitations are limited power density and a relatively short separator life. The air must be managed to remove carbon dioxide, and to provide humidity and thermal management. Table 38.10 provides some of the characteristics of this battery which is no longer under development.[25,26]

TABLE 38.10   Characteristics of Zinc/Air Traction Battery

Physical characteristics:	
Cell size	$33 \times 35 \times 0.75$ cm
Cell weight	1.0 kg (typical)
Cell voltage:	
Open-circuit	1.5 V
Average	1.2 V
High load	1.0 V
Charging	1.9 V
Configurations:	
General-purpose	120 Wh/kg, 120 W/kg peak power
High energy	180 Wh/kg at 10 W/kg
High power	200 W/kg peak at 100 Wh/kg

Source: Dreisbach Electromotive, Inc. (DEMI).

38.3.6   Mechanically Rechargeable Zinc/Air Batteries

Mechanically rechargeable or refuelable batteries are designed with a means to remove and replace the discharged anodes or discharge products. The discharged anode or discharge products can be recharged or reclaimed external to the cell. This avoids the need for a bifunctional air cathode and the shape change problems resulting from the charge/discharge cycling of an in situ zinc electrode.

Mechanically Refueled Systems—Anode Replacement.   Mechanically replaceable zinc/air batteries were seriously considered for powering portable military electronic equipment in the late 1960s because of their high specific energy and ease of recharging. This battery contained a number of bicells connected in series to provide the desired voltage. Each bicell, as illustrated in Fig. 38.24, consisted of two air cathodes connected in parallel and supported by a plastic frame which together formed an envelope for the zinc anode. The anode, which was a highly porous zinc structure enclosed in an absorbent separator, was inserted between the cathodes. The electrolyte, KOH, was contained in a dry form in the zinc anode and only water was needed to activate the cell. "Recharging" was accomplished by removing the spent anode, washing the cell, and replacing the anode with a fresh one. These batteries were never deployed because of their short activated life, poor intermittent operation, and the development of new high-performance primary lithium batteries which were superior in rate capability and ease of handling in the field.[27,28]

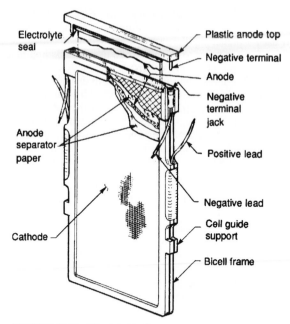

FIGURE 38.24 Zinc/air bicell.

A design similar to the portable mechanically rechargeable zinc/air battery has been considered for electric vehicle applications. The battery would be refueled "robotically" at a fleet servicing location or at a public service station by removing and replacing the spent anode cassettes. The discharged fuel would be electrochemically regenerated, using a modified zinc electrowinning process, at central facilities that serve regional distribution networks.[29]

This zinc/air battery consisted of modular cell stacks, each containing a series of individual bicells. Each bicell consists of an anode cassette containing a zinc-based electrolyte slurry, contained between air cathodes, and a separator system. The slurry is maintained in a static bed without circulation. In addition, the battery contains subsystems for air provision and heat management and is adapted for fast mechanical replacement of the cassettes.

The technology has been evaluated in a full-size 264 V, 110 kWh battery weighing 650 kg in a van that was converted to electric drive. The battery delivered 230 Wh/kg and 230 Wh/L with a power density of 100 W/kg.

Another approach to powering electric vehicles with mechanically rechargeable zinc/air batteries is a hybrid configuration where the zinc/air battery is combined with a rechargeable battery, such as a high-power lead-acid battery.[30] With this approach the performance of each battery can be optimized, using the high specific energy zinc/air battery as the energy source with a high specific power rechargeable battery to handle the peak power requirements. The power battery can also be sized to handle the anticipated peak load and duty cycle. In operation, during periods of light load, the zinc/air battery handles the load and recharges the rechargeable battery through a voltage regulator. The load is shared by both batteries during peak load conditions. When fully discharged, the zinc/air battery is recharged by removing and replacing the zinc oxide discharge product which can be regenerated externally and efficiently in designated facilities. The advantage of this hybrid design is illustrated in Fig. 38.25, a Ragone plot comparing the performance of the hybrid with the performance of the individual batteries. In this example, the hybrid lead-acid battery is one specifically designed for high rate performance (see also Sec. 6.4.4).

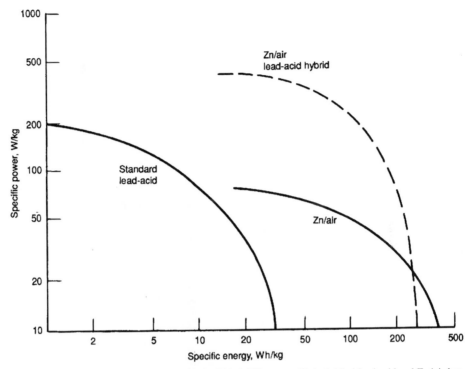

FIGURE 38.25 Comparison of Zn/air lead-acid hybrid battery with individual lead-acid and Zn/air batteries. Lead-acid battery uses special high-rate design.

Mechanically Refueled System—Zinc Powder Replacement.[31-33] Figure 38.26 is a sketch of an 80-cm² laboratory cell using a packed bed of zinc powder, which can be replaced when depleted. Natural convection is utilized for electrolyte circulation. During operation, electrolyte flows downward through the zinc bed and upward around the back of the current collector, which is either graphite or copper. Figure 38.27 shows the voltage profile on constant-current discharge for each of these current collectors.

The cell was designed so that the zinc bed and electrolyte could be pumped out at the end of discharge and replaced with a fresh charge of zinc and electrolyte to simulate operation in an electric vehicle. The cell was discharged at 2 A for 4 h. Most of the electrolyte and the residual particles were then sucked out of the anode side of the cell through a tube passing through a hole in the top of the cell and connected to a water jet aspirator. Without rinsing, fresh particles and electrolyte were placed in the cell through the hole and a second discharge was carried out. Following this, about 90% of the particles were removed less carefully, and the cell was refilled and discharged for a third time. The data in Fig. 38.28 shows that the three discharges were essentially the same.

Based on these experiments a conceptual design was made for a 55 kW (peak power) electric-vehicle battery. Projected specific energy of the battery was 110 Wh/kg at 97 W/kg under a modified Simplified Federal Urban Driving Schedule (SFUDS). These values were increased to 228 Wh/kg at 100 W/kg when the battery was designed for optimum capacity, and to 100 Wh/kg at 150 W/kg when designed for optimum power output, based on the results of discharge experiments at 45°C.

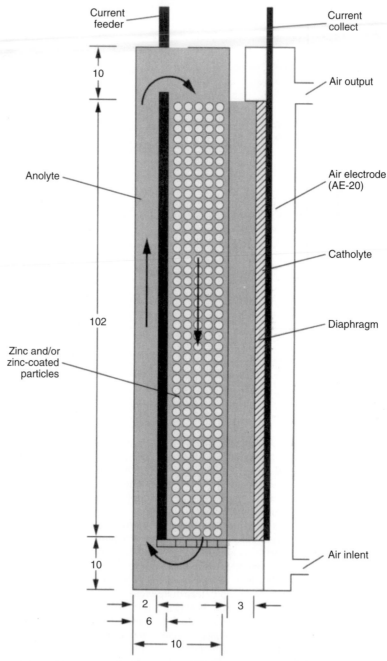

FIGURE 38.26 Schematic of mechanically refueled 80-cm² laboratory zinc/air cell. (*Courtesy of Lawrence Berkeley Laboratory.*)

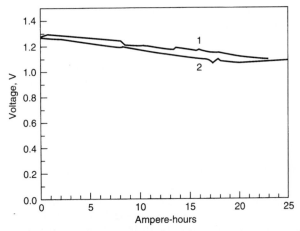

FIGURE 38.27 Constant-current discharges of mechanically refueled zinc/air battery, graphite vs. copper feeders. Anolyte/catholyte—45% KOH; anode—30-mesh zinc; cathode—AE-20 air electrode; $I = 2$ A; $A = 78$ cm². Curve 1—1.5-mm copper current feeder; curve 2—4.0-mm graphite current feeder. (*Courtesy of Lawrence Berkeley Laboratory.*)

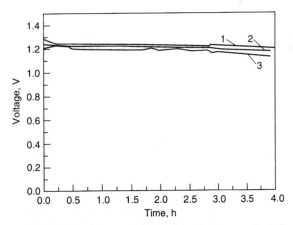

FIGURE 38.28 Voltage vs. time during subsequent mechanical recharging for mechanically refueled zinc/air battery. Anolyte/catholyte—45% KOH; anode—20-mesh zinc particles; cathode—AE-20 air electrode; $I = 2$ A; $A = 76$ cm². Curve 1—first run; curve 2—100% of anolyte/particles suctioned out, cell refilled with fresh ones, no rinsing; curve 3–90% of anolyte/particles suctioned out, cell refilled with fresh ones, no rinsing. (*Courtesy of Lawrence Berkeley Laboratory.*)

Efficient regeneration of the zinc particles is required to provide for a practical, efficient system. It is projected that for the practical application of this system, the spent electrolyte and residual particles would be removed at local service centers and the vehicle would be quickly refueled by the addition of regenerated zinc powder and electrolyte. The system under development[33] involves stopping the discharge of the battery described when the voltage falls below a practical value rather than when the voltage becomes zero. Under these conditions, no precipitation has occurred and the electrolyte is clear. The processing of products removed from the cell is then simply one of redeposition of zinc onto the particles.

38.4 ALUMINUM/AIR BATTERIES

Aluminum has long attracted attention as a potential battery anode because of its high theoretical Ampere-hour capacity, voltage, and specific energy. While these values are reduced in a practical battery because of the inability to operate aluminum and the air electrodes at their thermodynamic potentials and because water is consumed in the discharge reaction, the practical energy density still exceeds that of most battery systems. The inherent hydrogen generation of the aluminum anode in aqueous electrolytes is such that the batteries are designed as reserve systems with the electrolyte added just before use, or as "mechanically" rechargeable batteries with the aluminum anode replaced after each discharge. Electrically rechargeable aluminum/air batteries are not feasible using aqueous electrolytes.

The discharge reactions for the aluminum/air cell are

Anode $Al \rightarrow Al^{+3} + 3e$

Cathode $O_2 + 2H_2O + 4e \rightarrow 4OH^-$

Overall $4Al + 3O_2 + 6H_2O \rightarrow 4Al(OH)_3$

The parasitic hydrogen-generating reaction is

$$Al + 3H_2O \rightarrow Al(OH)_3 + \tfrac{3}{2}H_2$$

Aluminum can be discharged in neutral (saline) solutions as well as in caustic solutions. The neutral electrolytes are attractive because of the relatively low open-circuit corrosion rates and the reduced hazards of these solutions compared with concentrated caustic. Saline systems were under development for relatively low power applications, such as ocean buoys and portable battery applications, with specific energies of a "dry" battery as high as 800 Wh/kg. Seawater batteries for underwater vehicle propulsion and other applications, using oxygen present in the ocean, rather than air, or operating as corrosion cells, also are of interest because of the potentially high energy output.

Alkaline systems have an advantage over saline systems because the alkaline electrolyte has a higher conductivity and a higher solubility for the reaction product, aluminum hydroxide. Thus the alkaline aluminum/air battery is a candidate for high-power applications such as standby batteries, propulsion power for unmanned underwater vehicles, and has been proposed for electric vehicle propulsion. The specific energy can be as high as 400 Wh/kg. Aluminum/air batteries (as well as zinc/air batteries), because of their high energy densities, also can be used as power sources for recharging lower-energy rechargeable batteries in remote areas where line power is not available.

38.4.1 Aluminum/Air Cells in Neutral Electrolytes

Aluminum/air cells using neutral electrolytes have been developed for portable equipment, stationary power sources, and marine applications. Aluminum alloys are now available for saline cells with low polarization voltages, which can operate with coulombic efficiencies in the range of 50 to 80%. Alloying elements are required to enhance the disruption of the anodic surface film when current is drawn. Interestingly, in neutral electrolytes the corrosion reaction, resulting in the direct evolution of hydrogen, occurs at a rate linearly proportional to the current density, starting from near zero at zero current.[34]

Cathodes, such as those described earlier, are satisfactory. However, there are some extra limitations which apply in a saline solution. Nickel is not a suitable substrate where extensive periods on open circuit are involved. Under these conditions the potential of the active material in contact with the screen is high enough to oxidize the screen. One way to minimize this problem is to continue to draw, during no-load periods, a very low current, which is sufficient to keep the cathode potential from rising to its open-circuit value.

A suitable neutral electrolyte is a 12 wt % solution of sodium chloride, which is near the maximum conductivity. Current densities are limited to 30 to 50 mA/cm^2 as a result of the limitation imposed by the conductivity of the electrolyte. Such batteries may also be operated in seawater, with obvious limitations in current capability as a result of the lower conductivity of seawater.

Electrolyte management is required because of the behavior of the reaction product, aluminum hydroxide. It has a transient high solubility in the electrolyte and tends to become gellike when it first precipitates. In an unstirred system the electrolyte starts to become "unpourable" when the total charge produced exceeds 0.1 Ah/cm^3. Up to this point the electrolyte and the reaction product can be poured out of a cell and more saline solution added to continue the discharge until all of the aluminum is consumed. If the discharge is continued without draining the electrolyte, it will proceed satisfactorily until the total discharge reaches approximately 0.2 Ah/cm^3. At this point the cell contents are nearly solid.

Approaches to minimizing the amount of electrolyte required have been studied.[35] In one approach the electrolyte was stirred in a reciprocating manner, which minimized gel formation and produced a finely divided product which was dispersed in the electrolyte. A total electrolyte capacity of 0.42 Ah/cm^3 was achieved using reciprocated 20% potassium chloride electrolyte. A similar result was achieved by injecting a pulsed air stream at the bottom of each cell. This has the additional advantage that it sweeps the hydrogen out of each cell in a concentration below the flammability limit. An electrolyte utilization of 0.2 Ah/cm^3 was achieved in a system from which the electrolyte could be easily drained.

Portable Aluminum/Air Batteries. A number of batteries using saline electrolytes have been designed. In general, they are built as reserve batteries and activated by adding the electrolyte to the battery.

A saltwater battery, illustrated in Fig. 38.29 was designed for field recharging of nickel-cadmium and lead-acid storage batteries. Figure 38.30 shows the charge and discharge characteristics of a 2-Ah 24-V sealed nickel-cadmium battery being charged within 4 h. The aluminum/air battery can recharge this size nickel-cadmium battery about seven times before the aluminum is depleted. The specific energy of a dry battery, with enough metal for the anode and salt for the electrolyte to provide for a complete discharge, is about 600 Wh/kg.

Ocean Power Supplies. Batteries based on the use of oxygen dissolved in seawater have an advantage over others as all reactants, except for the anode material, are supplied by the seawater. In these batteries a cathode, which is open to the ocean, is spaced around an anode so that the reaction products can fall out into the ocean.[37] Relatively large surface areas are required as there is not much oxygen in seawater. In addition, because of the conductivity of the ocean, there can be no series arrangement of cells. Higher voltages are obtained by the use of a DC-to-DC converter.

FIGURE 38.29   Aluminum/air field recharger. 600 Wh, 6 V. (*Courtesy of Alupower, Inc.*)

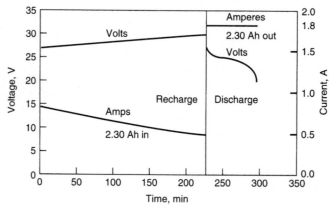

FIGURE 38.30   Charge and discharge of a nickel-cadmium battery, aluminum/air field recharger. (*Courtesy of Alupower, Inc.*)

Many instruments and devices used in the ocean have to operate over long periods of time, and aluminum is a candidate for the anode for missions requiring months or years of service.

Figure 38.31 shows a flat-plate aluminum/dissolved oxygen battery.[38] The battery, about 1.5 m high, has a dry specific energy of 500 Wh/kg and can operate at power densities of up to 1 W/m^2. This battery can be installed beneath a buoy, as shown in the illustration, and used with a DC-to-DC converter to charge a lead-acid battery.

FIGURE 38.31   Aluminum dissolved oxygen flat-plate battery (attached to Woods Hole Oceanographic Institution buoy.) (*Courtesy of Alupower, Inc.*)

38.4.2 Aluminum/Air Cells in Alkaline Electrolytes

The concept of operating aluminum/air batteries at high energy and power densities was described in the early 1970s, but successful commercialization was impeded because of technological limitations, including the high open-circuit corrosion rate of aluminum alloys in alkaline electrolyte, the nonavailability of thin, large-dimension air cathodes, and the difficulty of handling and removing the cell reaction products (precipitated aluminum hydroxide) to prevent cell clogging.

Significant advances have been made in reducing the corrosion of aluminum alloys in caustic electrolytes.[39,40] An aluminum anode, containing magnesium and tin, has approximately two orders of magnitude reduction in corrosion current at open circuit and operates at greater than 98% coulombic efficiency over a wide range of current densities. Even at open circuit, the alloy can remain in a caustic electrolyte with a relatively low rate of self-discharge. Prior to this development, the amount of hydrogen and heat evolved during open-circuit stand was usually so high as to require that the electrolyte be drained from the alloy during this period to prevent it from boiling.

Techniques to handle the aluminum so that it can be continuously fed as chips or pellets into the electrochemical system have also been developed.[41] One design uses aluminum particles having diameters of 1 to 5 mm.[42] The electrode is a pocket whose walls are composed of cadmium-plated expanded steel screen. The electrode is fed by a system which keeps the cell maintained with an aluminum particulate at an optimum level. Figure 38.32 shows the performance of a cell operating at 50°C using 8*N* KOH-containing stannate. The battery, with an electrode area of 360 cm², was able to deliver a current of 56 A at 1.35 V for 110 h, with the automatic addition of aluminum every 20 min.

Management of the electrolyte to remove the reaction product from it is required as the conductivity of the electrolyte decreases with increasing aluminate concentration. As shown in Fig. 38.33, the voltage decreases if the aluminate is not removed. Several techniques have been developed to remove the reaction products and are discussed later in this section.

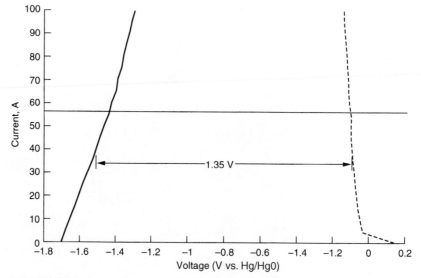

FIGURE 38.32   Polarization curves of particulate-feed aluminum/air battery. - - -, positive electrode; ——, negative electrode. (*Courtesy of Sorapec.*)

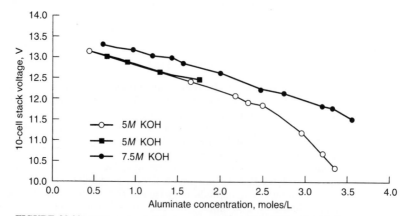

FIGURE 38.33   Voltage vs. aluminate concentration. (*Courtesy of Eltech Systems.*)

Applications of Alkaline Aluminum/Air Batteries.   The alkaline aluminum/air batteries being developed cover a wide range of applications from emergency power supplies to field-portable batteries for remote power applications and underwater vehicles. Most of these are designed as reserve batteries, which are activated before use, or "mechanically" recharged by replacing the exhausted aluminum anodes.

Reserve Power Units, Standby Battery.[43,44]   This is a reserve battery, used with conventional lead-acid batteries, to provide a standby power supply with extended service life. The aluminum/air battery is about one-tenth the weight and one-seventh the volume of a lead-acid battery containing the same energy. The basic elements of the power supply design are shown in Fig. 38.34. The aluminum/air battery consists of an upper cell stack, a lower electrolyte reservoir (which is isolated from the stack during periods of nonuse), and auxiliary systems to pump and cool the electrolyte and circulate air through the battery. The electrolyte is an $8M$ solution of potassium hydroxide containing a stannate additive. During operation the electrolyte becomes increasingly saturated and then supersaturated with potassium aluminate. Eventually the conductivity of the electrolyte decreases to the point where the battery is unable to sustain the load. At that point it has reached the end of its capacity based on total electrolyte volume (VLD). The electrolyte can be changed at this point, and the discharge continued to the point where the aluminum in the anode is exhausted (ALD). Figure 38.35 shows the performance for a nominal 1200-W battery discharged in the two modes. Operation in the electrolyte-capacity limited mode will yield a total discharge time of 36 h, while operation to anode exhaustion, incorporating one electrolyte change, will result in a total discharge time of 48 h. The overall energy density and specific energy are greater than 150 Wh/L and 250 Wh/kg.

The control system for this power unit is arranged so that in the event of a power outage, the lead-acid batteries provide the backup for the first 1 to 3 h. As the voltage of the lead-acid battery begins to fall, the aluminum/air battery is activated by a controller which initiates the pumping of electrolyte from a reservoir through the aluminum/air cell stack. Once the aluminum/air battery reaches full power (about 15 min from activation), it provides full power to the load and recharges the depleted lead-acid batteries. The electrical characteristics of the unit are shown in Figure 38.36. The aluminum/air battery has limited restart capability but can be refurbished by replacing the cell stack and the electrolyte.

Battlefield Power Unit.   This power source, called the Special Operations Forces Aluminum Air (SOFAL) battery,[45] was developed as a reserve system to support specialized military communications equipment in covert field operations. The SOFAL weighs approximately 7.3 kg after activation and powers 12 and 24 VDC equipment with pulse currents up to 10 Amps, continuous drains up to 4 Amps and has a design capacity of 120 Ah. To minimize weight, this battery is carried to the field dry and can be activated with any source of water.

The SOFAL unit consists of sixteen series-connected cells (Fig. 38.37*a*) with intercell connections provided by a printed circuit board. The cell stack, which weighs 3.5 kg dry, is shown in Fig. 38.37*b*. Activation of the system is accomplished with 2.5 L of water through a manifold system to each cell where it dissolves a cast block of stannated potassium hydroxide giving a 30% (w/w) KOH solution. After activation, each cell has an open circuit voltage of 1.7 v or 27.2 v for the cell stack. Electrochemistry of the battery is similar to that described earlier. Dissolution of the KOH and corrosion of the aluminum provide heat to operate the system even at low temperature. The unit has been designed to access 1.6 L/min of air, which provides for low-power operation. If ambient air flow is insufficient, a small fan, activated by the battery, will provide the required airflow and will dissipate excess heat during high power use. The SOFAL unit provides up to two weeks of service after activation.

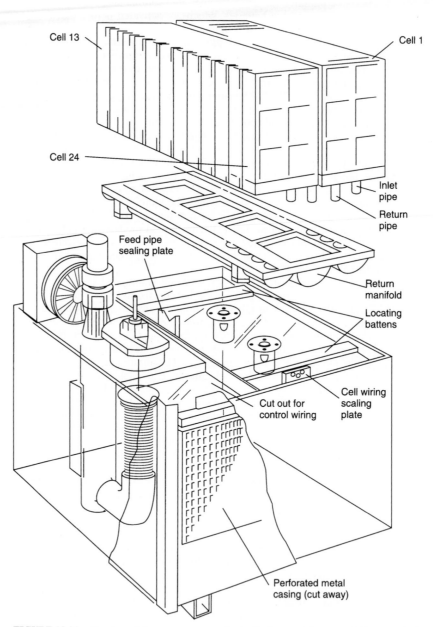

FIGURE 38.34   Aluminum/air reserve power unit, standby battery. (*Courtesy of Alupower, Inc.*)

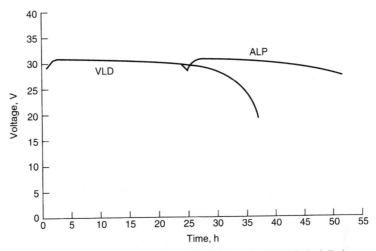

FIGURE 38.35 Comparison of volume (VLD) vs. anode (ALD) limited discharges of aluminum/air reserve power unit. (*Courtesy of Alupower, Inc.*)

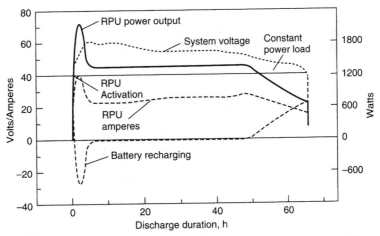

FIGURE 38.36 Discharge profile of 6-kW aluminum/air reserve power unit. (Direct connection to power system.) (*Courtesy of Alupower, Inc.*)

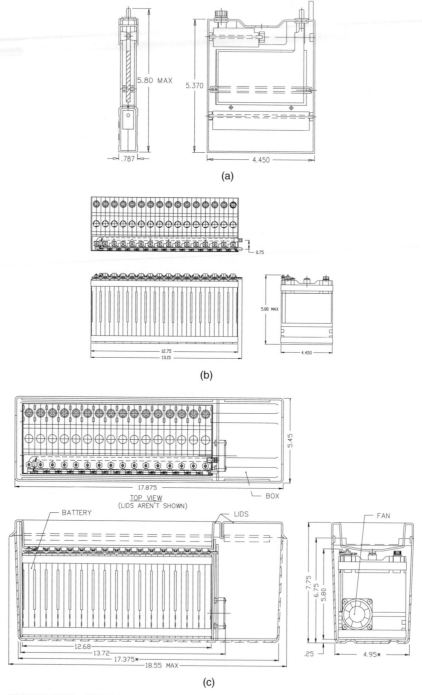

FIGURE 38.37   SOFAL battery. (*a*) Cell design configuration, (*b*) Battery block design configuration, (*c*) Full scale design layout.

The Electronics Module Package (EMP) shown in Fig. 38.37c contains the electronic components, provides the mounting for an internal rechargeable battery, and houses the fan. The electronics package contains the power management circuitry to keep the internal secondary battery fully charged, provides both 24 V and a regulated 12 V output and can be used to power electronics directly or recharge external batteries. Figure 38.38a shows the discharge profile of the SOFAL battery on a 2 Amp continuous load on 24 Volt operation. Figure 38.38b shows the discharge curves for two cells from the SOFAL battery. Cell No. 1 was activated with 8 molar KOH, while cell No. 2 employed KOH pellets in the cell and was activated with water. Both cells were discharged at 0.5 Amps, and provided 135 Amphours capacity, but cell No. 1 operated at slightly higher voltage, particularly at the end of discharge.

Following discharge, the cell stack can be replaced to provide a new battlefield power unit.

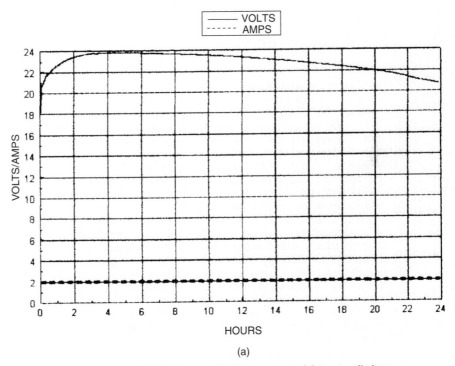

(a)

FIGURE 38.38 Sixteen cell SOFAL battery. (a) Constant current 2.0 Ampere discharge.

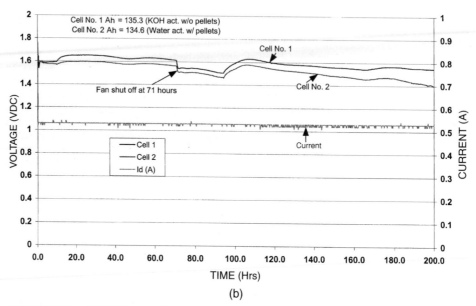

FIGURE 38.38   (*b*) SOFAL test cells on 0.5 Amp resistive discharge (*Continued*).

Underwater Propulsion.   Another area of application for alkaline aluminum batteries is in self-contained extended-duration power supplies for underwater vehicles such as un-manned vehicles for submarine and mine surveillance, long-range torpedoes, swimmer delivery vehicles, and submarine auxiliary power.[46,47] In these applications, the oxygen can be carried in pressurized or cryogenic containers, or it can be obtained from the decomposition of hydrogen peroxide or from oxygen candles. The aluminum/oxygen system can produce almost twice as much energy per kilogram of oxygen as a hydrogen/oxygen fuel cell, as the operating voltage of 1.2 to 1.4 V is almost twice that of the fuel cell. One type of an aluminum/oxygen battery for the propulsion of underwater vehicles is shown in Fig. 38.39, and its characteristics are listed in Table 38.11.

This battery uses a "self-managing" electrolyte system, where the required electrolyte circulation and precipitation take place within the cell chamber, without the requirement for external pumps. This has the advantage of allowing the design of battery systems without electrolyte circulation pumps. There are no shunt currents between cells as each cell is independent, and there are no electrolyte paths between cells. In addition, the cells can be conformal with the system which they are designed to power. Figure 38.40 shows the design of a system using 19-in diameter (48.25-cm) cells. Each cell is about 0.5 in (1.25 cm) thick. Thermal and concentration gradients and the resulting convection currents within the cell precipitate the reaction product to the bottom. With this type of system, it is possible to utilize about 0.8 Ah/cm^3 of electrolyte. Figure 38.41 gives a discharge curve for a cell which is exactly one-half the cell shown in Fig. 38.40. The cell was divided down the center to provide for redundancy. The discharge was carried out at a constant power level of 18 W and a current density of about 50 mA/cm^2. The figure shows that the voltage remains relatively constant, between 1.4 and 1.5 V, for most of the run.

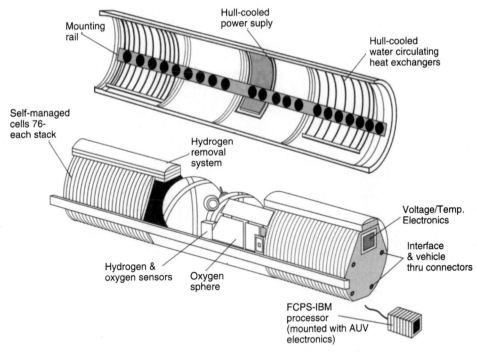

FIGURE 38.39 Aluminum/oxygen power system. (*Courtesy of Alupower, Inc.*)

TABLE 38.11 Characteristics of Aluminum/Oxygen Battery

Performance:	
Power	2.5 kW
Capacity	100 kWh
Voltage	120 V nominal
Endurance	40 h at full power
Fuel	25-kg aluminum anodes
Oxidant	22 kg oxygen at 4000 lb/in^2
Buoyancy	Neutral, including aluminum hull section
Time to refuel	3 h
Dimensions:	
Mass	360 kg
Battery diameter	470 mm
Hull diameter	533 mm
System length	2235 mm
Performance:	
Energy density	265 Wh/L
Specific energy	265 Wh/kg

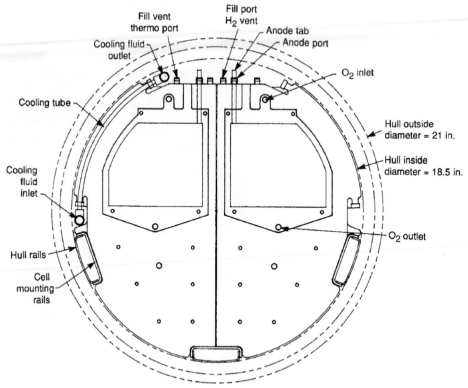

FIGURE 38.40   Self-managing cell system. (*Courtesy of Alupower/Alcan.*)

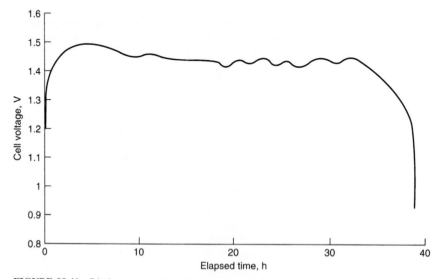

FIGURE 38.41   Discharge curve for self-managing cell. (*Courtesy of Alupower/Alcan.*)

To maximize the capacity, the amounts of aluminum and electrolyte are matched so that at the end of discharge the aluminum is consumed and the electrolyte is completely filled with reaction product. Under this condition, the module is either discarded or rebuilt after use. Alternatively, a higher concentration of electrolyte can be used and discharge stopped before the onset of precipitation. In this mode of operation, the amount of aluminum incorporated into the cell can be sufficient for several runs with only the electrolyte being replaced between runs.

Another requirement of the underwater power system is the hydrogen-removal system, which is needed to safely remove the hydrogen that is generated by corrosion of the anode. Catalytic recombination is especially attractive for applications where space and energy efficiency are needed.[48] The unit shown in Fig. 38.39 uses a hydrogen-removal system, but the amount of hydrogen generated is not excessive since a low-corrosion aluminum alloy is used.

Another approach to removing the reaction product is a filter/precipitator system,[48] as shown in Fig. 38.42. The aluminate concentration is controlled by pumping the electrolyte out of the cell stack and through the filter/precipitator. The filter promotes the growth of the aluminum trioxide and regeneration of KOH as follows:

$$KAl(OH)_4 \rightarrow KOH + Al(OH)_3$$

The crystal cake gradually increases in thickness with a subsequent increase in the pressure drop across the filter. When the pressure drop reaches a predetermined level, the cake is pulsed off the filter by backflushing (flow reversal) and collected in the bottom of the precipitate tank.

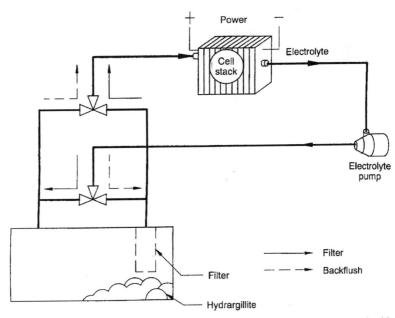

FIGURE 38.42 Conceptual design of filter/precipitator system when integrated with aluminum/oxygen battery. (*Courtesy of Eltech Systems.*)

38.5 *MAGNESIUM/AIR BATTERIES*

The discharge reaction mechanisms of the magnesium/air battery are

Anode

$$Mg \rightarrow Mg^{+2} + 2e$$

Cathode

$$O_2 + 2H_2O + 4e \rightarrow 4(OH)^-$$

Overall

$$Mg + \tfrac{1}{2}O_2 + H_2O \rightarrow Mg(OH)_2$$

The theoretical voltage of this reaction is 3.1 V, but in practice, the open-circuit voltage is about 1.6 V.

Magnesium anodes tend to react directly with the electrolyte with the formation of magnesium hydroxide and the generation of hydrogen,

$$Mg + 2H_2O \rightarrow Mg(OH)_2 + H_2$$

This reaction stops in alkaline electrolytes because of the formation of an insoluble film of magnesium hydroxide on the electrode surface which prevents further reaction. Acid tends to dissolve the film. An important consequence of the film on magnesium electrodes (see also Chap. 9) is that there is a delayed response to an increase in the load because of the need to disrupt the film to create new bare surfaces for reaction. "Pure" magnesium anodes usually do not give good cell performance, and several magnesium alloys have been developed for use as anodes tailored to provide the desired characteristics.

Magnesium/air batteries have not been successfully commercialized and an effort has been directed to undersea applications using the dissolved oxygen in seawater as the reactant. The battery uses a magnesium alloy anode and a catalytic membrane cathode, and is activated by the seawater. The main advantage of this system is that, with the exception of the magnesium, all of the reactants are supplied by the seawater. The battery can have a specific energy of about 700 Wh/kg.

The concentration of oxygen in seawater is only 0.3 mol/m³, corresponding to 28 Ah per ton of seawater. Therefore the cathode must have an open structure to ensure that there is sufficient contact with the seawater. Further, as seawater is highly conductive, it is not feasible to use more than one cell. A DC-to-DC converter is used to increase the low cell voltage to the required voltage range.

Figure 38.43 illustrates a cell design for an undersea mission designed to deliver 3 to 4 W for one year or longer at a total weight of 32 kg. In this design, the oxygen-reduction cathode is positioned on the circumference of a cylinder with a total cathode area of 3 m². The anode is a 19-kg cylinder of magnesium, located internal to the air cathode. The weight of the cathode is about 1.8 kg, and the remainder of the weight is for support structure and other necessary hardware. The single-cell battery has a long shelf life in its dry, unactivated state. It is immediately active on deployment when it is immersed in the seawater electrolyte. Figure 38.44 shows the discharge of a test battery with periodic voltage spikes when the load is increased.[49]

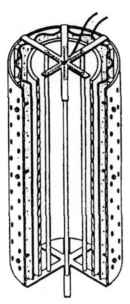

FIGURE 38.43 Schematic of concentric cylinder configuration of seawater cell. The output cylinder is comprised of porous fiberglass coated with an anti-foulant. The corrugated structure is the air cathode, which is exterior to the magnesium anode. The entire structure is open to the seawater electrolyte. (*Courtesy of Westinghouse Corp.*)

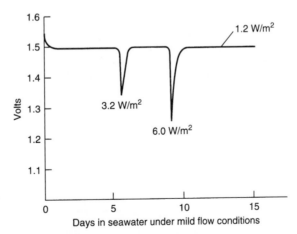

FIGURE 38.44 Discharge profile of seawater power source at 80 μA, 20°C. Current spikes represent charging of a small silver-iron battery. Neutral pH was maintained by periodic addition of hydrochloric acid. (*Courtesy of Westinghouse Corp.*)

38.6 *LITHIUM/AIR BATTERIES*

The lithium/air battery is attractive because lithium has the highest theoretical voltage and electrochemical equivalence (3860 Ah per kilogram of lithium) of any metal anode considered for a practical battery system. The cell discharge reaction is

$$2Li + H_2O + \tfrac{1}{2}O_2 \rightarrow 2LiOH \qquad E^0 = 3.35 \text{ V}$$

Lithium metal, atmospheric oxygen, and water are consumed during the discharge, and excess LiOH is generated. The cell can operate at high coulombic efficiencies because of the formation of a protective film on the metal that retards rapid corrosion after formation. On open-circuit and low-drain discharge, the self-discharge of the lithium metal is rapid, due to the parasitic corrosion reaction

$$Li + H_2O \rightarrow LiOH + \tfrac{1}{2}H_2$$

This reaction degrades the anode coulombic efficiency and must be controlled if the full potential of the lithium anode is to be realized. This self-discharge also necessitates the removal of the electrolyte during stand.

The theoretical open-circuit voltage of the lithium/air cell is 3.35 V, but in practice, this value is not achieved because the lithium anode and the air cathode exhibit mixed potentials. Figure 38.45 shows that the actual open-circuit voltage is near 2.85 V. It is also evident that the major voltage loss is at the air cathode and that the cell cannot be discharged efficiently at voltages much above 2.2 V unless a substantial reduction in cathode polarization is achieved.

The principal advantage of the lithium/air battery is its higher cell voltage, which translates into higher power and specific energy. However, in view of their availability, cost and safety advantages, the development of metal/air batteries has concentrated primarily on zinc and aluminum.

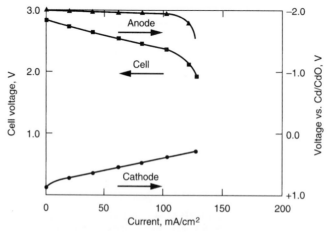

FIGURE 38.45 Typical individual electrode and cell polarization curve of lithium/air cell.

38.6.1 Lithium-Water Batteries

The high energy density of lithium is attractive as a reserve battery for undersea applications using its reaction with water.[50] In general the combination of lithium with water may be considered hazardous because of the high heat of reaction. However, in the presence of hydroxyl (OH^-) ion at concentrations greater than $1.5M$, a protective film is formed which exists in a dynamic steady state. The film is pseudo-insulating, which permits the cathode to be pressed against it without causing a short circuit, thus reducing IR losses and concentration polarization and achieving high current outputs.

The major difficulty in the use of lithium in aqueous solution is the fact that it will not discharge efficiently at current densities less than about 0.2 to 0.4 kA/m^2 and the battery cannot be placed on open circuit with electrolyte present within it. Also, under certain conditions the lithium will passivate. This feature can be used to advantage, however, to temporarily terminate reaction for standby with an electrolyte-filled battery.

The aqueous lithium systems are in principle quite simple, but in practice the electrolyte management subsystem and internal cell features involve sophisticated design and a level of complexity characteristic of fuel cells, requiring a reservoir, pump, heat exchanger, and controller, and using a flowing electrolyte. The electrodes are held in close juxtaposition, and frequently they are pressed together. A film on the lithium prevents short-circuiting, but permits high flux rates. The rate of discharge is inversely proportional to the concentration of electrolyte, and the output of the cell can be uniquely controlled by adjusting the molarity M of the LiOH produced at the anode. The voltage of the batteries is influenced more by the characteristics of the cathode than by the lithium anode.

The discharge reactions of the lithium-water battery are

Anode $$Li \rightarrow Li^+ + e$$

Cathode $$H_2O + e \rightarrow OH^+ + \tfrac{1}{2}H_2$$

Overall $$Li + H_2O \rightarrow LiOH + \tfrac{1}{2}H_2$$

Parasitic corrosion $$Li + H_2O \rightarrow LiOH + \tfrac{1}{2}H_2$$

Precipitation of LiOH occurs as the monohydrate crystal,

$$LiOH + H_2O \rightarrow LiOH \cdot H_2O$$

The overall electrochemical reaction has a thermodynamic potential of 2.21 V and a theoretical specific energy of 8530 Wh/kg based on lithium, which is the only reactant that has to be supplied with the battery. Dissolved oxygen is not necessary to depolarize the cathode as lithium possesses a high enough voltage to reduce water to hydrogen.

The parasitic corrosion reaction is highly undesirable as it produces no electric energy but consumes lithium. This highly exothermic reaction (-53.3 kcal/g·mol of lithium) can accelerate corrosion detrimentally. Efficient minimum-weight batteries require that this parasitic reaction be minimized.

The overall battery concept is shown in Fig. 38.46.[51] The neoprene bellows enable pressure equalization between the inside of the battery and the surrounding ocean. The bellows can also expand or contract to take up any changes in the cell volume with time. During operation, water is pumped into the battery to satisfy the water requirements. Due to the nature of the water-pumping concept, some hydroxide will be slowly lost to the surrounding seawater. However, hydroxide is continually being generated by the reaction and the rate of hydroxyl-ion loss will be slower than its rate of generation.

The discharge characteristics of a 2-month test of a prototype battery are shown in Fig. 38.47. The operating voltage on a discharge of 2.0 mA/cm^2 (equivalent to a 2-W discharge on a full-size battery) was between 1.4 and 1.43 V.

A typical lithium-water battery uses a 30-cm diameter, 30-cm thick solid cylindrical anode with a weight of approximately 11.5 kg. This battery is designed to deliver 2 W at 1.4 V for about a year with a specific energy of 1800 to 2400 Wh/kg.

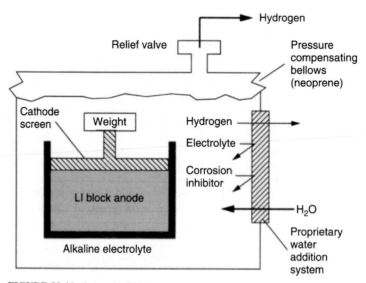

FIGURE 38.46   Low-rate lithium-water battery concept. (*From Shuster.*[51])

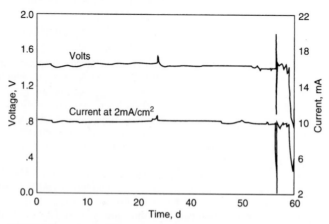

FIGURE 38.47   Typical performance of 28-cm diameter low-rate lithium-water test battery.

38.6.2 Lithium/Oxygen Battery with Polymer Electrolyte

A novel, rechargeable Li/O$_2$ battery has recently been described.[52] It comprised a lithium-ion conductive polymer electrolyte membrane between a thin Li metal foil anode and a thin carbon composite on which oxygen, the electro-active cathode material, is reduced during discharge. Fig. 38.48 shows the cell structure which is encapsulated in a metallized plastic envelope with pores on the cathode surface which are covered by tape prior to activation. The cathode is composed of 20 wt % acetylene black (or graphite powder) and 80 wt % polymer electrolyte catalyzed with cobalt phthalocyanine in some cases, and pressed on a Ni or Al screen current collector. The electrolyte was composed of 12% polyacrylonitrile (PAN), 40% ethylene carbonate, 40% propylene carbonate and 8% LiPF$_6$, all by weight, formed into a film 75–100 microns thick. The lithium electrode was 50 microns thick. Fig. 38.49 shows an intermittent discharge curve for the Li/PAN-based electrolyte/O$_2$ battery at a current density of 0.1 mA/cm using an acetylene black cathode in an atmosphere of flowing oxygen. The capacity was found to be proportional to the carbon weight. This battery exhibited an open circuit voltage (OCV) of 2.85 V prior to discharge. The OCV remained steady during intermittent discharge, indicating a two-phase equilibrium at the electrode surface. Raman spectra of the reaction product absorbed on the electrode surface showed it was Li$_2$O$_2$ and that the following reaction is occurring during discharge:

$$2Li + O_2 \rightarrow Li_2O_2 \qquad E = 3.10V$$

It was also determined that the absorbed lithium peroxide on a catalyzed electrode could be re-oxidized to oxygen. Fig. 38.50 shows the first discharge and recharge, followed by the second discharge of one battery. Although this system is of considerable technological interest, its active life was limited by the diffusion of oxygen through the PAN electrolyte, where it reacted chemically with lithium. Further development has not occurred.

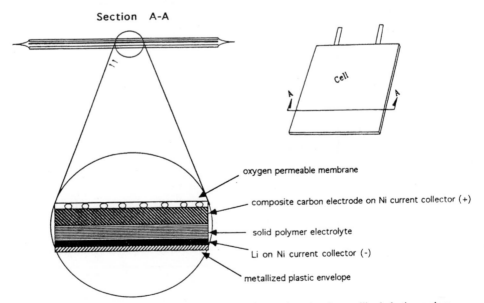

Section A-A

oxygen permeable membrane

composite carbon electrode on Ni current collector (+)

solid polymer electrolyte

Li on Ni current collector (-)

metallized plastic envelope

Cell

FIGURE 38.48 Lithium/oxygen battery with solid polymer electrolyte in metallized plastic envelope.

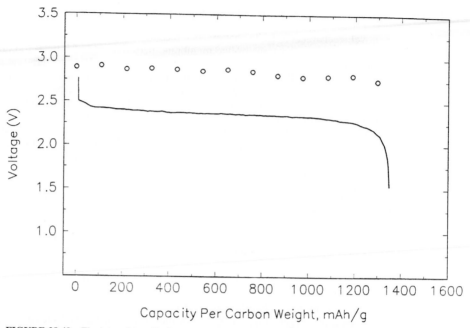

FIGURE 38.49 The intermittent discharge curve and the open-circuit voltages of a Li/PAN-based polymer electrolyte/oxygen battery at a current density of 0.1 mA/cm² at room temperature in an atmosphere of oxygen. The cathode contained Chevron acetylene black carbon. The cell was discharged in 1.5 h increments with an open-circuit stand of about 15 min between discharges. Open circles: OCV; solid line: load voltage.

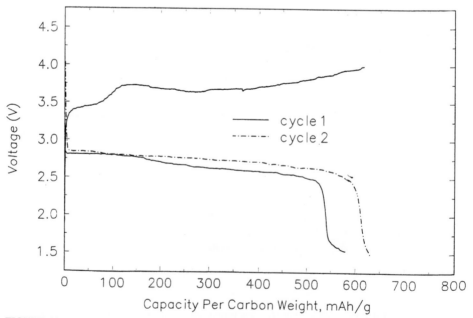

FIGURE 38.50 Cycling data for a Li/PAN-based polymer electrolyte/oxygen battery at room temperature in an atmosphere of oxygen. The cathode contained 20 w/o catalyzed Chevron carbon black and 80 w/o polymer electrolyte. The battery was discharged at 0.1 mA/cm² and charged at 0.05 mA/cm².

REFERENCES

1. D. A. J. Rand, "Battery Systems for Electric Vehicles: State of Art Review," *J. Power Sources* **4:** 101 (1979).

2. K. F. Blurton and A. F. Sammells, "Metal/Air Batteries: Their Status and Potential—A Review," *J. Power Sources* **4:**263 (1979).

3. H. F. Bauman and G. B. Adams, "Lithium-Water-Air Battery for Automotive Propulsion," Lockheed Palo Alto Research Laboratory, Final Rep., COO/1262-1, Oct. 1977.

4. W. P. Moyer and E. L. Littauer, "Development of a Lithium-Water-Air Primary Battery," *Proc. IECEC,* Seattle, Wash., Aug. 1980.

5. W. N. Carson and C. E. Kent, "The Magnesium-Air Cell," in D. H. Collins (ed.), *Power Sources,* 1966.

6. R. P. Hamlen, E. C. Jerabek, J. C. Ruzzo, and E. G. Siwek, "Anodes for Refuelable Magnesium-Air Batteries," *J. Electrochem. Soc.* **116:**1588 (1969).

7. J. F. Cooper, "Estimates of the Cost and Energy Consumption of Aluminum-Air Electric Vehicles," ECS Fall Meeting, Hollywood, Fl., Oct. 1980, Lawrence Livermore, UCRL-84445, June 1980; update UCRL-94445 rev. 1, Aug. 1981.

8. R. P. Hamlen, G. M. Scamans, W. B. O'Callaghan, J. H. Stannard, and N. P. Fitzpatrick. "Progress in Metal-Air Battery Systems," International Conference on New Materials for Automotive Applications, Oct. 10–11, 1990.

9. A. S. Homa and E. J. Rudd, "The Development of Aluminum-Air Batteries for Electric Vehicles," *Proc. 24th IECEC,* vol. 3, 1989, pp. 1331–1334.

10. W. H. Hoge, "Air Cathodes and Materials Therefore," U.S. Patent 4,885,217, 1989.

11. W. H. Hoge, "Electrochemical Cathode and Materials Therefore," U.S. Patent 4,906,535, 1990.

12. W. H. Hoge, R. P. Hamlen, J. H. Stannard, N. P. Fitzpatrick, and W. B. O'Callaghan, "Progress in Metal-Air Systems," Electrochem. Soc., Seattle, Wash., Oct. 14–19, 1990.

13. T. Atwater, R. Putt, D. Bouland, and B. Bragg, "High-Energy Density Primary Zinc/Air Battery Characterization," *Proc. 36th Power Sources Conf.,* Cherry Hill, NJ, 1994.

14. R. Putt, N. Naimer, B. Koretz, and T. Atwater, "Advanced Zinc-Air Primary Batteries," *Proc. 6th Workshop for Battery Exploratory Development,* Williamsburg, VA, 1999.

15. J. Passanitti, "Development of a High Rate Primary Zinc-Air Cylindrical Cell," *Proc. 5th Workshop for Battery Exploratory Development,* Burlington, VT, 1997.

16. J. Passanitti, "Development of a High Rate Primary Zinc-Air Cylindrical Cell," *Proc. 38th Power Sources Conf.,* Cherry Hill, NJ, 1998.

17. J. Passanitti and T. Haberski, "Development of a High Rate Primary Zinc-Air Battery," *Proc. 6th Workshop for Battery Exploratory Development,* Williamsburg, VA, 1999.

18. R. A. Putt and G. W. Merry, "Zinc-Air Primary Batteries," *Proc. 35th Power Sources Symp.,* IEEE, 1992.

19. Sales literature, SAFT, Greenville, N.C.

20. Celair Corp., Lawrenceville, Ga.

21. A. Karpinski, "Advanced Development Program for a Lightweight Rechargeable "AA" Zinc-Air Battery," *Proc. 5th Workshop for Battery Exploratory Development,* Burlington VT, 1997.

22. A. Karpinski, B. Makovetski, and W. Halliop, "Progress on the Development of a Lightweight Rechargeable Zinc-Air Battery," *Proc. 6th Workshop for Battery Exploratory Development,* Williamsburg, VA, 1999.

23. A. Karpinski, and W. Halliop, "Development of Electrically Rechargeable Zinc/Air Batteries," *Proc. 38th Power Sources Conf.,* Cherry Hill, NJ, 1998.

24. AER Energy Resources, Inc., Atlanta, Ga.

25. L. G. Danczyk, R. L. Scheffler, and R. S. Hobbs, "A High Performance Zinc-Air Powered Electric Vehicle," SAE Future Transportation Technology Conference and Exposition, Portland, Ore., Aug. 5–7, 1991, paper 911633.

26. M. C. Cheiky, L. G. Danczyk, and M. C. Wehrey, "Second Generation Zinc-Air Powered Electric Minivans, SAE International Congress and Exposition, Detroit, Mich., Feb. 24–28, 1992, paper 920448.

27. S. M. Chodosh et al., "Metal-Air Primary Batteries, Replaceable Zinc Anode Radio Battery," *Proc. 21st Annual Power Sources Conf.,* Electrochemical Society, Pennington, N.J., 1967.

28. D. Linden and H. R. Knapp, "Metal-Air Primary Batteries, Metal-Air Standard Family," *Proc. 21st Annual Power Sources Conf.,* Electrochemical Society, Pennington, N.J., 1967.

29. Electric Fuel, Ltd. Jerusalem, Israel.

30. R. A. Putt, "Zinc-Air Batteries for Electric Vehicles," *Zinc/Air Battery Workshop,* Albuquerque, NM, Dec. 1993.

31. H. B. Sierra Alcazar, P. D. Nguyen, G. E. Mason, and A. A. Pinoli, "The Secondary Slurry-Zinc/Air Battery," LBL Rep. 27466, July 1989.

32. G. Savaskan, T. Huh, and J. W. Evans, "Further Studies of a Zinc-Air Cell Intended for Electric Vehicle Applications, Part I: Discharge," *J. Appl. Electrochem.* (Aug. 1991).

33. T. Huh, G. Savaskan, and J. W. Evans, "Further Studies of a Zinc-Air Cell Intended for Electric Vehicle Applications, Part II: Regeneration of Zinc Particles and Electrolyte by Fluidized Bed Electrodeposition," *J. Appl. Electrochem.* (Aug. 1991).

34. A. R. Despic, "The Use of Aluminum in Energy Conversion and Storage," *First European East-West Workshop on Energy Conversion and Storage,* Sintra, Portugal, Mar. 1990.

35. N. P. Fitzpatrick and D. S. Strong, "An Aluminum-Air Battery Hybrid System," *Elec. Vehicle Develop.,* **8:**79–81 (July 1989).

36. T. Dougerty, A. Karpinski, J. Stannard, W. Halliop, V. Alminauskas, and J. Billingsley, "Aluminum-Air Battery for Communications Equipment," *Proc. 37th Power Sources Conf.,* Cherry Hill, NJ, 1996.

37. C. L. Opitz, "Salt Water Galvanic Cell With Steel Wool Cathode." U.S. Patent 3,401,063, 1968.

38. D. S. Hosom, R. A. Weller, A. A. Hinton, and B. M. L. Rao, "Seawater Battery for Long Lived Upper Ocean Systems," *IEEE Ocean Proc.,* vol. 3 (Oct. 1–3, 1991).

39. J. A. Hunter, G. M. Scamans, and J. Sykes, "Anode Development for High Energy Density Aluminium Batteries," *Power Sources,* vol. 13 (Bournemouth, England, Apr. 1991).

40. R. P. Hamlen, W. H. Hoge, J. A. Hunter, and W. B. O'Callaghan, "Applications of Aluminum-Air Batteries," *IEEE Aerospace Electron. Mag.* **6:**11–14 (Oct. 1991).

41. S. Zaromb, C. N. Cochran, and R. M. Mazgaj, "Aluminum-Consuming Fluidized Bed Anodes," *J. Electrochem. Soc.* **137:**1851–1856 (June 1990).

42. G. Bronoel, A. Millott, R. Rouget, and N. Tassin, "Aluminum Battery with Automatic Feeding of Aluminium," *Power Sources,* vol. 13, Bournemouth, England, Apr. 1991; also French Patents 88.15703, 1988; 90.07031, 1990; 90.14797, 1990.

43. W. B. O'Callaghan, N. Fitzpatrick, and K. Peters, "The Aluminum-Air Reserve Battery—A Power Supply for Prolonged Emergencies," *Proc. 11th Int. Telecommunications Energy Conf.,* Florence, Italy, Oct. 15–18, 1989.

44. J. A. O'Conner, "A New Dual Reserve Power System for Small Telephone Exchanges," *Proc. 11th Int. Telecommunications Energy Conf.,* Florence, Italy, Oct. 15–18, 1989.

45. A. P. Karpinski, J. Billingsley, J. H. Stannard, and W. Halliop, *Proc. 33rd IECEC,* 1998.

46. K. Collins et al., "An Aluminum-Oxygen Fuel Cell Power System for Underwater Vehicles," Applied Remote Technology, San Diego, 1992.

47. D. W. Gibbons and E. J. Rudd, "The Development of Aluminum/Air Batteries for Propulsion Applications," *Proc. 28th IECEC,* 1993.

48. D. W. Gibbons and K. J. Gregg, "Closed Cycle Aluminum/Oxygen Fuel Cell with Increased Mission Duration," *Proc. 35th Power Sources Symp.,* IEEE, 1992.

49. J. S. Lauer, J. F. Jackovitz, and E. S. Buzzelli, "Seawater Activated Power Source for Long Term Missions," *Proc. 35th Power Sources Symp.,* IEEE, 1992.

50. E. L. Littauer and K. C. Tsai, "Anodic Behavior of Lithium in Aqueous Electrolytes, ii. Mechanical Passivation," *J. Electrochem. Soc.* **123:**964 (1976); "Corrosion of Lithium in Aqueous Electrolytes," *ibid.* **124:**850 (1977); "Anodic Behavior of Lithium in Aqueous Electrolytes, iii. Influence of Flow Velocity, Contact Pressure and Concentration," *ibid.* **125:**845 (1978).

51. N. Shuster, "Lithium-Water Power Source for Low Power Long Duration Undersea Applications," *Proc. 35th Power Sources Symp.,* IEEE, 1992.

52. K. M. Abraham and Z. Jiang, J. Electrochem. Soc. *143,* 1 (1996).

CHAPTER 39
ZINC/BROMINE BATTERIES

Paul C. Butler, Phillip A. Eidler, Patrick G. Grimes, Sandra E. Klassen, and Ronald C. Miles

39.1 GENERAL CHARACTERISTICS

The zinc/bromine battery is an attractive technology for both utility-energy storage and electric-vehicle applications. The major advantages and disadvantages of this battery technology are listed in Table 39.1. The concept of a battery based on the zinc/bromine couple was patented over 100 years ago,[1] but development to a commercial battery was blocked by two inherent properties: (1) the tendency of zinc to form dendrites upon deposition and (2) the high solubility of bromine in the aqueous zinc bromide electrolyte. Dendritic zinc deposits could easily short-circuit the cell, and the high solubility of bromine allows diffusion and direct reaction with the zinc electrode, resulting in self-discharge of the cell.

Development programs at Exxon and Gould in the mid-1970s to early 1980s resulted in designs which overcame these problems, however, and allowed development to proceed.[2] The Gould technology was developed further by the Energy Research Corporation but a high level of activity was not maintained.[3–5] In the mid-1980s Exxon licensed its zinc/bromine technology to Johnson Controls, Inc., JCI (Americas), Studiengesellschaft für Energiespeicher und Antriebssysteme, SEA (Europe), Toyota Motor Corporation and Meidensha Corporation (Japan), and Sherwood Industries (Australia). Johnson Controls sold their interest in zinc/bromine technology in 1994 to ZBB Energy Corporation, which is located in the United States and Australia. Powercell Corporation was formed in 1993, including SEA (now Powercell GmbH), and is located in the United States, Austria, and Malaysia. The technology discussed in this chapter is based on the original Exxon design.

TABLE 39.1   Major Advantages and Disadvantages of Zinc/Bromine Battery Technology

Advantages	Disadvantages
Circulating electrolyte allows for ease of thermal management and uniformity of reactant supply to each cell	Auxiliary systems are required for circulation and temperature control
Good specific energy	System design must ensure safety as for all batteries
Good energy efficiency	Initially high self-discharge rate when shut down while being charged
Made of low-cost and readily available materials	Improvements to moderate power capability may be needed
Low-environmental-impact recyclable/reusable components made using conventional manufacturing processes	
Flexibility in total system design	
Ambient-temperature operation	
Adequate power density for most applications	
Capable of rapid charge	
100% depth of discharge does not damage battery but improves it	
Near-term availability	

39.2 DESCRIPTION OF THE ELECTROCHEMICAL SYSTEM

The electrochemical reactions which store and release energy take place in a system whose principal components include bipolar electrodes, separators, aqueous electrolyte, and electrolyte storage reservoirs. Figure 39.1 shows a schematic of a three-cell zinc/bromine battery system that illustrates these components (plus other features which are discussed in Sec. 39.3). The electrolyte is an aqueous solution of zinc bromide, which is circulated with pumps past both electrode surfaces. The electrode surfaces are in turn separated by a microporous

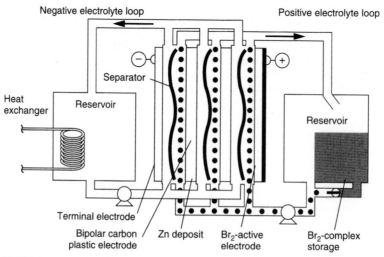

FIGURE 39.1   Schematic of three-cell zinc/bromine module. (*Courtesy of Exxon Research and Engineering Co. and Sandia National Laboratories.*)

plastic film. Thus two electrolyte flow streams are present—one on the positive side and one on the negative side. The directions of the flow streams may differ depending on the designs of different companies.

The electrochemical reactions can be simply represented as follows:

		(25°C)	(50°C)
		$E^0 = 0.763$ V	0.760 V

Negative electrode $Zn^{2+} + 2e \underset{\text{discharge}}{\overset{\text{charge}}{\rightleftharpoons}} Zn^0$

Positive electrode $2Br^- \underset{\text{discharge}}{\overset{\text{charge}}{\rightleftharpoons}} Br_2(aq) + 2e$ $E^0 = 1.087$ V   1.056 V

Net cell reaction $\underset{\text{discharged state}}{ZnBr_2(aq)} \underset{\text{discharge}}{\overset{\text{charge}}{\rightleftharpoons}} Zn^0 + Br(aq)$ $E^0_{cell} = 1.85$ V

During charge, zinc is deposited at the negative electrode, and bromine is produced at the positive electrode. During discharge, zinc and bromide ions are formed at the respective electrodes. The microporous separator between the electrode surfaces impedes diffusion of bromine to the zinc deposit, which reduces direct chemical reaction and the associated self-discharge of the cell.

The chemical species present in the electrolyte are actually more complex than that described. In solution, elemental bromine exists in equilibrium with bromide ions to form polybromide ions, Br_n^-, where $n = 3, 5, 7$.[6] Aqueous zinc bromide is ionized, and zinc ions exist as various complex ions and ion pairs. The electrolyte also contains complexing agents which associate with polybromide ions to form a low-solubility second liquid phase. The complex reduces the amount of bromine contained in the aqueous phase 10 to 100-fold, which, in addition to the separator, also reduces the amount of bromine available in the cell for the self-discharge reaction. The complex also provides a way to store bromine at a site remote from the zinc deposits and is discussed further in the next section. Salts with organic cations such as N-methyl-N-ethylmorpholinium bromide (MEMBr) are commonly used as the complexing agents. One researcher has proposed a mixture of four quaternary ammonium salts for use in zinc/bromine batteries. The proposed electrolyte has favorable properties with regard to aqueous bromine concentration, resistivity, and bromine diffusion and does not form solid complexes at low temperatures (5°C and above).[7] Complexes with quaternary ammonium ions are reversible and also have an added safety benefit due to a much reduced bromine vapor pressure (see Sec. 35.6).

The electrodes are bipolar and are typically composed of carbon plastic. The presence of bromine precludes the use of metal electrodes—even titanium can corrode in this environment.[8] A high-surface-area carbon layer is added to the positive side of the electrode to increase the area for reaction. On charge, circulation of the electrolyte removes the complexed polybromide as it is formed, and on discharge complexed polybromide is delivered to thee electrode surface. Circulation of the electrolyte also reduces the tendency for zinc dendrites to form and simplifies thermal management of the battery. Thermal management will be needed in many applications of present and advanced batteries.

The optimum operating pH range is set by the occurrence of undesirable mossy zinc plating and bromate formation above pH = 3, and by an increased zinc corrosion rate at lower pH. Hydrogen evolution due to the reaction of zinc with water has sometimes been observed during operation of zinc/bromine batteries. The hydrogen overpotential on zinc is large, however, and the reaction is slow in the absence of metals with low hydrogen overpotential, such as platinum.[9] During the development program it was found that the amount of hydrogen generated was small in the absence of impurities and had a minimal effect on the capacity of the battery.[10] Because of the circulating electrolytes, it would be easy to add water or acid to the system to compensate for any hydrogen formed, but this has not been necessary.

In a system where the cells are connected electrically in series and hydraulically in parallel, an alternate pathway for the current exists through the common electrolyte channels and manifolds during charge, discharge, and at open circuit. These currents are called shunt currents and cause uneven distribution of zinc between end cells and middle cells. This uneven distribution causes a loss of available capacity because the stack will reach the voltage cutoff upon discharge sooner than if the zinc were evenly distributed. Also shunt currents can lead to uneven plating on individual electrodes, especially the terminal electrodes. This uneven plating can in turn lead to zinc deposits that divert or even block the electrolyte flow.

Shunt currents can be minimized by designing the cells to make the conductive path through the electrolyte as resistive as possible. This is done by making the feed channels to each cell long and narrow to increase the electrical resistance. This, however, also increases the hydraulic resistance and thus the pump energy. Good battery design balances these factors. Higher electrolyte resistance reduces shunt currents but also reduces battery power. Since the cell stack voltage is the driving force behind the currents, the number of cells in series can be set low enough that the magnitude of the shunt currents is minimal. In a specific utility battery design with 60 cells or less per cell stack, the capacity lost to shunt currents can be held to 1% or less of the total input energy. When these approaches are not sufficient to control the shunt currents, protection electrodes can be used to generate a potential gradient in the common electrolyte equal to and in the same direction as that expected from the shunt current.[10] Several modeling approaches to calculate the currents for various applications have been proposed.[11–14]

39.3 CONSTRUCTION

In general terms the battery is made up of cell stacks and the electrolyte along with the associated equipment for containment and circulation. The primary construction materials are low-cost readily available thermoplastics. Conventional plastic manufacturing processes such as extrusion and injection molding are used to make most of the battery components. Because terminal electrodes must also collect the current from over the surface and deliver it to a terminal connection, the lateral conductivity must be higher than in bipolar electrodes, where the current only passes perpendicularly through the electrodes. A copper or silver screen is molded into the end block to serve as a current collector. Plastic screens are placed as spacers between the electrodes and separators. The components are assembled into a battery stack either by compression using bolts and gaskets, by using adhesives, or by thermal or vibration welding.[15,16] Assembly of a leak free stack using vibration welding has been demonstrated by manufacturing cells that can withstand three times the normal operating pressure before bursting.[17] Figure 39.2 is a schematic showing the components and assembly of a cell stack.

Various materials have been used for the separator. Ideally a material is needed which allows the transport of zinc and bromide ions, but does not allow the transport of aqueous bromine, polybromide ions, or complex phase. Ion-selective membranes are more efficient at blocking transport than nonselective membranes; thus higher columbic efficiencies can be obtained with ion-selective membranes. These membranes, however, are more expensive, less durable, and more difficult to handle than microporous membranes.[10] In addition use of ion-selective membranes can produce problems with the balance of water between the positive and negative electrolyte flow loops. Thus battery developers have generally used nonselective microporous materials for the separator.[3,4,10,15]

As shown in Fig. 39.1, two electrolyte circulation circuits are needed for the battery and include pumps, reservoirs, and tubing. The positive electrolyte side has an additional pro-

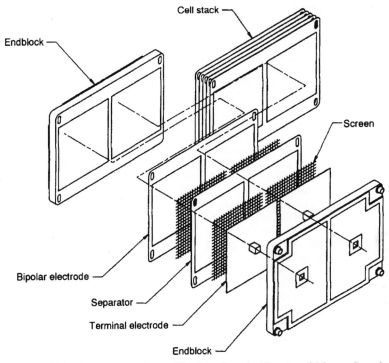

FIGURE 39.2   Components and assembly of a cell stack. (*Courtesy of Johnson Controls Battery Group, Inc.*)

vision to store polybromide complex, which settles by gravity into a lower part of the reservoir. In Fig. 39.1 complexed polybromide is being delivered to the electrode surfaces during discharge. During charge, the bulk of the polybromide complex is not recirculated. The polybromide which is formed at the positive electrode associates with the aqueous-phase complexing agent and is collected in the storage area of the reservoir. This limits the potential self-discharge of the battery to only that complex which is in the cell stack at the termination of the charge process. The bromine may be dissolved in the aqueous phase, absorbed on the electrode surface, or complexed as polybromide.

A heat exchanger, located in the negative electrolyte reservoir, as shown in Fig. 39.1, provides for the thermal management of the battery. In general plastic heat exchangers can be used, and even though titanium corrodes when used as electrode material, titanium has been used successfully as the tubing material for the heat exchanger.

Ultimately the battery parts will be reclaimed or sent for disposal. The most significant parts of the battery in this respect are the cell stacks and electrolyte. The battery stacks are nearly all plastic and can be recycled by conventional processes and new processes that are being developed by the plastics industry. The electrolyte is not consumed in the battery. It will be removed and reused in other batteries.

39.4 PERFORMANCE

Zinc/bromine batteries are typically charged and discharged using rates of 15 to 30 mA/cm². A charge-discharge profile for a 50-cell stack is shown in Fig. 39.3. The amount of charge is set based on the zinc loading that is defined at 100% state of charge. This amount is always less than the total zinc ion dissolved in the electrolyte. Thus rate of charge, time duration of charge, and charge efficiency are used to determine the end of charge. The voltage rises at the end of charge, and severe overcharge will electrolyze water. The discharge is usually terminated at about 1 V per cell since the voltage is falling rapidly at this point.

The capacity of a battery is directly related to the amount of zinc that can be deposited on the negative electrodes, and zinc loadings can range from 60 to 150 mAh/cm². One hundred percent state of charge is defined as a specific zinc loading and can vary depending on the battery. Considerable effort has been expended to ensure good-quality dense zinc plating. It is important to control the pH to avoid undesirable mossy zinc deposits. Circulation of the electrolyte reduces the occurrence of dendritic deposits. Studies have shown that current density, zinc bromide concentration, electrolyte additives, and operating temperature also affect the quality of the zinc deposit.[4,15] With these studies and improvements, the problems are being managed or have been eliminated.

Zinc deposited onto a clean carbon plastic surface is smoother than when deposited on top of zinc; but zinc can be completely removed by total discharge to renew the surface. This is, in effect, a 100% depth of discharge and does not damage the battery but improves it. In practical applications the battery should complete many cycles before a strip cycle is run. A plot of the cycling efficiencies of a 15-kWh battery is shown in Fig. 39.4. The periodic nature of cycles 60 to 200 is a result of multiple tests in which five cycles were followed by a strip cycle and also occasionally a baseline cycle. The first new cycle is only slightly lower in efficiency because the base coat of zinc is being replated.

The amount of $ZnBr_2$ electrolyte that can be reacted at the electrodes is called the utilization and varies depending on the application. For utility applications battery efficiency is a primary concern, and percent utilization is about 50 to 70% to maximize efficiency. For electric-vehicle applications battery size and weight are more important, and the percent utilization can be as high as 80 to 90%. High utilization results in solutions of lower conductivity at the end of charge, which lowers voltaic and energy efficiencies. Attempts to charge to very high utilization result in electrolysis of water as a competing reaction, and high utilization cycles are also opposed because some of the reactant material is isolated in the opposing electrolyte chamber.[18]

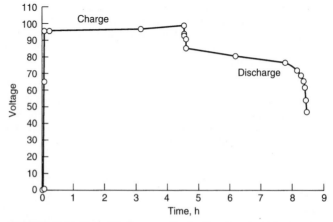

FIGURE 39.3 Charge-discharge profile for 50-cell stack. 80% electrolyte utilization; 30°C; 90-mAh/cm² zinc loading; 20-mA/cm² or *C*/4.5 charge rate; 20-mA/cm² or *C*/4 discharge rate. (*Courtesy of Sandia National Laboratories.*)

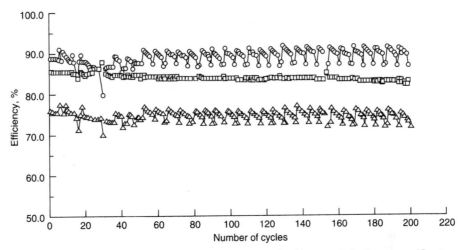

FIGURE 39.4   Cycle efficiencies for 15-kWh battery. ○—coulombic; □—voltaic, △—energy. (*Courtesy of Sandia National Laboratories.*)

TABLE 39.2   The Effect of Discharge Rate on Temperature and Energy Output for a 60-cell Battery Stack

Discharge current (A)	Discharge time (hours)	Maximum temperature (°C)	Energy output (kWh)
35.5	5.61	30.6	19.83
42.8	4.67	31.6	19.71
53.3	3.75	33.2	19.43
71.2	2.82	35.0	19.15
104.9	1.87	39.5	17.86
209.9	0.82	50.9	13.54

Source: From Clark, Eidler, and Lex.[19]

Battery performance varies with discharge rate. As the discharge rate becomes higher, the energy efficiency decreases and the temperature of the battery increases. Table 39.2 shows the effect of rate on temperature and energy output for a 60 cell stack.[19] Figure 39.5 shows the charge and discharge voltage profiles for a 50 A and 100 A discharge of a 60 cell stack.[20] The average voltage and energy efficiency for the 50 A discharge were 98 V and 77%, respectively. The values for the 100 A discharge are 92 V and 72%, respectively. Calculations are based on a discharge of 60 V (1 V per cell), and the battery had been charged at a constant current of 50 A for 4.5 hours prior to both cycles. The average charge voltage was 112 V.

In a battery system a portion of the energy will be diverted to auxiliary systems such as thermal management, pumps, valves, controls, and shunt current protection as required. The energy needed for auxiliaries depends on a number of factors, including the efficiency of pumps and motors, pump run time, and system design. Little publicly available data exist

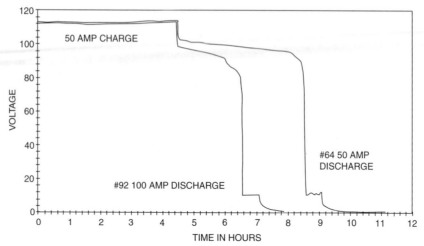

FIGURE 39.5 Charge and discharge profiles for a 50 A and 100 A discharge of a 60 cell stack. (*Courtesy of ZBB Energy Corporation.*)

on total energy requirements of auxiliary systems, although the energy devoted to auxiliaries is projected to be less than a few percent of the total battery energy. ZBB Energy Corporation, for example, reports that the pump power requirements for a 60 cell stack are just over 1% of the battery power during discharge at a 100 A rate.[21] In another study, ZBB compared the performance of the battery when the pumps and controls were powered by the stacks versus AC.[22] The battery output energy and energy efficiency were 49.5 kWh and 64.5% when powered by the stacks and 53.3 kWh and 69.3% when powered by AC. It was also shown that energy efficiency could be raised from 61.3% to 69.8% by using larger pumps and by making modifications to the plumbing, controls, and manifold system. It should be noted that other systems have auxiliaries and balancing inefficiencies as well.

Energy will also be lost during stand time. This was measured in one zinc/bromine battery system to be about 1%/h (watt-hour capacity lost) over an 8-h period.[16] During the test, electrolyte, which did not contain the complexed bromine phase, was circulated periodically to remove heat. The self-discharge reaction ceases once bromine in the stacks has been depleted.

Zinc/bromine batteries normally operate between 20 and 50°C. Typically the operating temperature has little effect on energy efficiency, as shown in Fig. 39.6. At low temperature the electrolyte resistivity increases, resulting in lower voltaic efficiency. This is offset by slowed bromine transport, which results in higher coulombic efficiency. At high temperature the resistance decreases and the bromine transport increases, again partially compensating for each other. Temperature control is accomplished with a heat exchanger and the circulating electrolyte. The optimum temperature will vary depending on the individual battery design and electrolyte used.

For applications which require high-power discharges, such as electric vehicles, the conductivity of the electrolyte can be enhanced by using additives such as KCl or NH_4Cl. In this way internal ohmic energy losses are decreased. A test using NH_4Cl supporting electrolyte showed more peak power capability than unsupported electrolyte over a range of depths of discharge, as shown in Fig. 39.7. A Ragone plot of data from sustained power tests is shown in Fig. 39.8. Batteries with supporting electrolyte, however, do have disadvantages. Overall efficiencies are about 2% lower than with unsupported electrolyte. Also plating

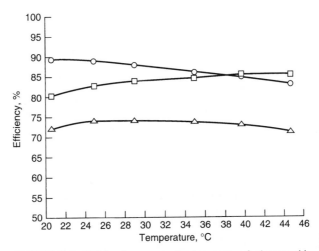

FIGURE 39.6 Efficiencies vs. operating temperature for battery with load-leveling electrolyte. ○—coulombic; □—voltaic; △—energy. (*Courtesy of Johnson Controls Battery Group, Inc.*)

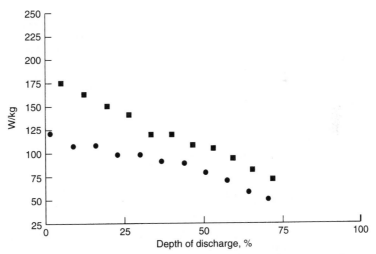

FIGURE 39.7 Zinc/bromine battery peak power for NH_4Cl supporting (■) and unsupported (●) electrolyte. Peak power—maximum power that can be achieved for 20 s. 80% electrolyte utilization; 30°C; 90-mAh/cm² zinc loading. (*Courtesy of Johnson Controls Battery Group, Inc.*)

additives are needed to counteract the tendency of supported electrolytes to produce rougher zinc deposits.[15] Since the supporting salts increase the weight and cost of the battery without increasing the energy content, they would not be added unless the extra power was necessary. Multicycle and long-term testing is needed to determine the specifications for optimum operation.

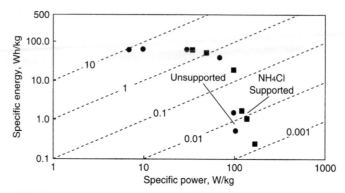

FIGURE 39.8   Zinc/bromine sustained power discharge. 80% electrolyte utilization; 30°C; 90-mAh/cm² zinc loading. Diagonal lines denote hours. (*Courtesy of Johnson Controls Battery Group, Inc.*)

The largest factor influencing the lifetime of zinc/bromine batteries is most likely the long-term compatibility of the components with bromine. Improvements have been made in dealing with degraded seals, corrosion of the terminal current collectors, and warpage of the electrodes (which can interfere with the flow of the electrolytes), and in many cases problems have been solved. Studies have been done in a variety of plastic compositions.[3,4,15,23,24] Additives which may affect battery performance can be leached out of polyvinylchloride. Fluorinated polyolefins are generally chemically stable, but they are expensive, and carbon-loaded materials are not dimensionally stable in the presence of bromine. High-density polyethylene with glass fibers appears to be a good choice of materials for battery components from the viewpoint of chemical and dimensional stability and has displaced polypropylene. With control of warpage, a battery lifetime of more than 2000 cycles is possible. Studies have also been done on the stability of the quaternary ammonium salts used as bromine complexing agents, and decomposition was not found.[10,25] Another study showed that a carbon-plastic bipolar electrode could tolerate 3000 cycles of zinc deposition and removal without degradation.[3] Battery companies claim 10 to 20 year service lives, which are, of course, influenced by the application and number of cycles required per year.[20,26]

39.5   TRADEOFF CONSIDERATIONS

The zinc/bromine battery, as do all battery systems, offers a tradeoff between high-rate discharges and lower-rate discharges; i.e., power and energy. Other additional design tradeoffs can be made. The two most important are increasing the ratio of electrolyte volume to electrode area to favor energy storage over power, and adding a conductivity salt to the electrolyte to favor power over energy. There is no clear evidence that operating a zinc/bromine battery at high power necessarily reduces life. If, however, the battery is allowed to operate at higher temperatures, plastic material degradation has been observed which does lead to reduced life. Therefore thermal management is a key issue that has to be addressed in high-power applications. To be assured of adequate cooling in a high-power application, larger cooling systems may be needed, which would also increase the battery weight.

The specific energy of the battery may decrease depending on the degree of safety required. Containment can be enhanced by incorporating multiple barriers to minimize electrolyte loss through a breach. Impact and leak sensors with shut-down controls can be incorporated to further ensure that electrolyte circulation ceases in the event of an accident. All of these additions, however, add weight which contributes to lower specific energy.

39.6 SAFETY AND HAZARDS

Very little free bromine exists in the battery. Bromine is present as polybromide ions dissolved in the aqueous portion of the electrolyte or bound with complexing agents in a second phase. Any remaining bromine is dissolved in the aqueous electrolyte. Liquid or gaseous bromine is hazardous; it injures through physical contact, especially inhalation.[27] In the complexed condition, however, the chemical reactivity and the evaporation rate are greatly reduced from those of pure bromine. For example, at 20°C the vapor pressure of bromine over the complex with MEMBr is more than 20 times lower than for elemental bromine.[28] If spilled, the charged electrolyte will slowly release bromine from the complex, which in turn will form vapor downwind from the spill site. Bromine has a strong odor, and it is readily detected at low levels. Spilled electrolyte can be treated using methods recommended by qualified industrial hygienists or methods listed in material safety data sheets. The EPA DOT reportable quantity for zinc bromide spills is 1000 lb.[29]

Runaway chemical reactions are unlikely because the polybromide complexes are stored away from the zinc. Even if the zinc electroplate were somehow flooded with polybromide complex, the reaction rate of the complex would be relatively slow because of the low zinc surface are available for reaction.

39.7 APPLICATIONS AND SYSTEM DESIGNS

A great deal of flexibility is available when designing zinc/bromine battery systems. Batteries can be custom built for a particular application, where multiple modules share a single set of electrolyte reservoirs or where each module contains a complete system of cell stacks, reservoirs, and controls. Modules can be stacked to conserve the footprint in energy storage applications, and reservoirs can be made to match the space available in electric vehicles.

39.7.1 Electric-Vehicle Applications

The Studiengesellschaft für Energiespeicher und Antriebssysteme (SEA, now Powercell GmbH) in Mürzzuschlag, Austria, has been developing zinc/bromine batteries for electric vehicles since 1983 and has produced batteries with capacities ranging between 5 and 45 kWh.[30] SEA replaced the original Exxon molding of electrolyte channels into the flow frames with an external tubing manifold system. This allowed a flexible tubing connection of the stack to the reservoirs and ease of assembly and disassembly. The stacks are in horizontal layers, allowing low-profile construction. Programmable microprocessor controllers have been developed to allow thorough, safe, and reliable operation of the systems under various loads. Systems designed by SEA demonstrate a range of characteristics, as listed in Table 39.3.

SEA has installed a 45-kWh 216-V battery in a Volkswagen bus which the Austrian Postal Service has been using to deliver packages in the mountains around Mürzzuschlag (Fig. 39.9). The battery weighs about 700 kg, and the maximum speed achieved by the bus is 100 km/h. The maximum range at 50 km/h is 220 km.

Hotzenblitz, a German company, has designed an electric vehicle to be powered specifically by a zinc/bromine battery (Fig. 39.10). The battery is 15 kWh 114 V and is located in a compartment under the passenger area, as shown in Fig. 39.11. It is sealed from the passenger area and is located between side impact barriers for safety. Specifications and performance data are given in Table 39.4.

TABLE 39.3 System Properties for SEA Designed Zinc/Bromine Batteries

Cell voltage, theoretical	1.82 V
Cell voltage, nominal	1.5 V
Coulombic efficiency	88–95%
Voltaic efficiency	80–86%
Energy efficiency	68–73%
Gravimetric energy density	65–75 Wh/kg
Volumetric energy density	60–70 Wh/L
Specific power	90–110 W/kg
Operating temperature	Ambient

Source: From Tomazic.[23]

FIGURE 39.9 Volkswagon bus powered by SEA zinc/bromine battery. (*Courtesy of SEA.*)

TABLE 39.4 Specifications and Performance Data for Hotzenblitz EL SPORT

Vehicle	Battery
Length, 2700 mm	Type, zinc-bromine battery
Width, 1480 mm	Charging time, approx. 5–6 h
Height, 1500 mm	Onboard charger, 3 kW
Gross vehicle weight; approx. 650 kg	Battery weight, approx. 240 kg
Payload capacity, 300 kg	Speed max. >100 km/h
Range, 150–180 km	Gradability, approx. 25%
Motor, 12 kW/144 V	

Source: From Tomazic.[30]

FIGURE 39.10 SEA 15-kWh 114-V zinc/bromine battery used to power Hotzenblitz EL SPORT. (*Courtesy of SEA.*)

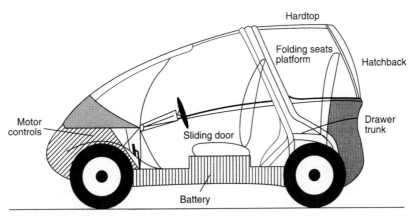

FIGURE 39.11 Schematic of Hotzenblitz EL SPORT showing location of zinc/bromine battery. (*From Tomazic.*[23])

In the United States, a team at the University of California, Davis, installed a Powercell 30 kWh Zinc-Flow® battery in a Geo Prizm with a Dolphin drive system for the Endura project.[31] The battery consisted of two stack towers, each containing two 54 cell bipolar stacks connected in series, two electrolyte tanks and associated pumping systems, and control and cooling systems. Selected battery specifications are given in Table 39.5. At 391 V, this is the highest voltage zinc/bromine battery used in an electric vehicle, and the team claims the Endura is one of the few electric vehicles that is capable of travelling in the high speed lane of the Los Angeles freeway system. The Endura travelled 175 miles on a single charge, and the vehicle was capable of accelerating from zero to 100 kilometers (62 miles) per hour in 15 seconds, and of reaching a velocity of 125 kilometers (78 miles) per hour.[32] The gross vehicle weight was 1595 kg with a front to rear weight distribution of approximately 50/50.

TABLE 39.5   Selected Specifications for the Powercell Endura Electric Vehicle Battery

Open circuit voltage @ 100% SOC	391 V
Maximum power @ 50% SOC	~40 kW
Charge method	Constant current
Nominal charge rate	30 amps DC
End of charge voltage	432 V
Nominal charge power	12 kW
Nominal charge time	5 hours (0 to 100% SOC)
Battery specific energy @ C/3 rate	65.8 Wh/kg

Source: From Swan, et al.[31]

After an extensive safety review,[33] the Endura was entered into several competitions with successful results. The team placed third in the 1994 Arizona Public Service Electric 500, first in the 1994 World Clean Air Rally, second in the 1994 American Tour de Sol, and first in the 1995 Arizona Public Service Electric 500.[31,33]

Powercell GmbH asserts that the zinc/bromine technology will be one of the most affordable for electric vehicles.[32] In direct comparison using the same test vehicle fitted with lead-acid batteries versus Zinc-Flow® batteries, Powercell has been able to show that the test vehicle will have a 2 to 3 fold greater range when fitted with a Zinc-Flow® battery due to this technology's reduced weight and higher energy density.

Toyota Motor Corporation has also been developing zinc/bromine batteries for electric vehicles.[35,39] A concept urban transportation vehicle, called the EV-30, has been designed for use with Toyota's zinc/bromine battery and has been displayed at motor shows in Japan. This two-seater vehicle would transport people in buildings, shopping centers, small communities, and to and from train stations—a "horizontal elevator" concept. The front-wheel-drive system uses an AC induction motor built by Toyota Motor Corporation. The battery system is modular zinc/bromine at 106 V and 7 kWh.

39.7.2   Energy Storage Applications

The use of zinc/bromine batteries in energy storage applications is also being demonstrated. A study by Sandia National Laboratories rated the zinc/bromine battery as excellent for these four utility applications: storage of energy generated by renewable sources, transmission facility deferral, distribution facility referral, and demand peak reduction.[36]

ZBB Energy Corporation has designed and manufactured a 50 kWh battery module that serves as a building block for larger systems. Each module is made up of three 60 cell stacks connected in parallel, an anolyte reservoir, a catholyte reservoir, and an electrolyte circulation system as shown in Fig. 39.12.[20] These modules offer flexibility in building larger batteries because they can be placed in a variety of series and parallel arrangements. A 400 kWh battery has been designed for utility demonstrations and consists of two strings connected in parallel with each string composed of four 50 kWh modules in series. Specifications for the battery and module can be found in Table 39.6.[20,37] Operation of the battery is flexible because the power conversion system (PCS) is connected directly to each of the strings which allows either simultaneous or independent operation.

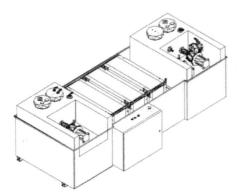

FIGURE 39.12   Schematic and photograph of ZBB Energy Corporation's 50 kWh battery module. (*Courtesy of ZBB Energy Corporation.*)

TABLE 39.6   ZBB Energy Corporation Battery and Module Specifications

	Battery system	Module
DC interface	504 volts DC maximum	126 volts DC maximum
Open circuit voltage	432 volts	109 volts
Baseline charge	300 amps for 4.5 hours	150 amps
Capacity	400 kWh—2 hour discharge	50 kWh—2 hour discharge
Dimensions	Approx. 2.44 meters × 2.44 meters × 5.18 meters	Approx. 2.44 meters × 0.91 meters × 0.91 meters
Weight	Approx. 18200 kg	Approx. 1360 kg

Source: From Lex and Jonshagen[20] and Lex.[37]

Future costs for this battery, excluding the PCS, are projected to be $400/kWh or lower at modest product levels. The battery is expected to operate for 10 years if cycled five days per week (2500 cycles) before the stacks and pumps need to be replaced. Replacement costs are anticipated to be about 20% of the initial cost.

A 400 kWh zinc/bromine battery will be installed at the United Energy Ltd., Nunawading Electrical Distribution Substation in Box Hill, Victoria, Australia, to shave peaks in the load curve.[37] A load profile from the Nunawading Substation is shown in Fig. 39.13. Discharge of the battery during times of peak demand will reduce the transformer load so that equipment ratings aren't exceeded. The battery can be recharged at night when demand is lower. Operation in this manner allows costly system upgrades to be deferred.

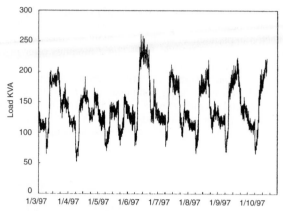

FIGURE 39.13   Load profile from the Nunawading Substation, Australia, January 3–10, 1997. (*Courtesy of ZBB Energy Corporation.*)

ZBB Energy Corporation also plans a 400 kWh demonstration in the United States and is working with the Department of Energy and Sandia National Laboratories to evaluate potential sites.[37] A transportable battery system is an advantage to a utility company because it can be moved to locations as needs arise. The goal of this demonstration is to use the battery for peak reduction in the summer at one site and load levelling and power quality applications during the fall and winter at another site.

Powercell Corporation is also in the process of demonstrating and commercializing zinc/bromine battery technology for utility applications.[26,38] The company introduced Power-Block® in 1998, which combines the Zinc-Flow® battery technology with solid state power electronics for quality power output, and a global monitoring system to provide real-time display of performance as well as data storage. PowerBlock® can be applied to a variety of electric power issues. PowerBlock® can be used to correct voltage disturbances seen with intermittent and incessant power quality issues, and can be used to provide protection against momentary and extended outages. PowerBlock® can also be applied to peak shaving and other energy service management applications.

A photograph of PowerBlock® is shown in Fig. 39.14, and selected specifications are shown in Table 39.7.[26] PowerBlocks® can serve as modules to build larger systems that can provide 1 MW of power for 2 hours, for example.[39] These energy storage systems are self-contained and transportable to allow flexible application. PowerBlock® connections are 480 V, 3-phase (delta) and 112.5 kVA isolation transformers are placed between PowerBlock® and the load and between PowerBlock® and the grid or other generating source.[38] PowerBlock® is designed for a 20-year service life, and 1250 cycles have been demonstrated in a single battery.[26] Replacement of the electrode stacks and pump motors will be approximately 15% of the system cost.

A PowerBlock® module has been installed near the Denver International Airport to store energy produced by two microturbines that are powered by natural gas from a well on the property as shown in Fig. 39.15.[40] Installation was completed in three days.

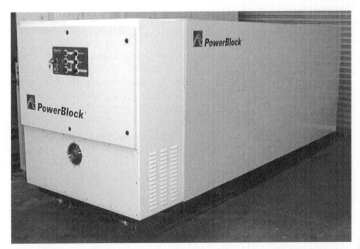

FIGURE 39.14   Photograph of PowerBlock®. (*Source: From Reference 22.*)

TABLE 39.7   Selected PowerBlock® Specifications

Continuous power rating	100 kW
Short term (10s) rating	150 kW
Energy capacity	100 kWh @ 25 kW
Recharge time	5 hours
Energy storage efficiency	>70%
Ambient conditions	0 to 35°C
Dimensions	3.41 meters × 1.12 meters × 1.31 meters
Mass	2670 kg

   Source: From Winter.[26]

FIGURE 39.15   Aerial view of the PowerBlock® module with two microturbines near the Denver International Airport. (*Source: From Reference 22.*)

In Japan a long-term project to develop zinc/bromine battery technology for electric-utility applications has been part of the Moonlight Project under the sponsorship of the Ministry of International Trade and Industry (MITI).[41,42] During the 1980s research and development resulted first in 1-kW batteries, then 10-kW batteries, and finally 60-kW battery modules. The modules were used as components of a larger system, and in 1990 a 1-MW 4-MWh battery was installed at the Imajuku substation of the Kyushu Electric Power Company in Fukuoka City by the New Energy and Industrial Technology Develop Organization, the Kyushu Electric Power Company, and the Meidensha Corporation (Fig. 39.16). The battery room of the Imajuku energy storage test plant is shown in Fig. 39.17. The system is composed of 24 25-kW submodules connected in series and is presently the largest zinc/bromine battery in the world. Design specifications are given in Table 39.8.[43]

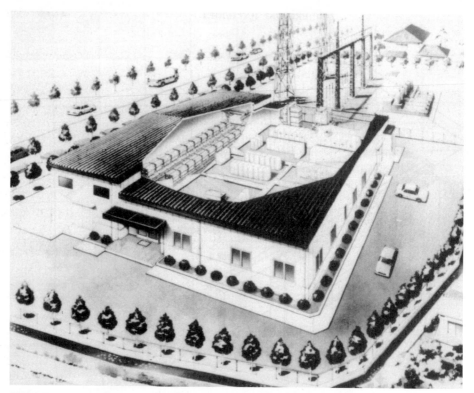

FIGURE 39.16 Artist's conception of Imajuku energy storage test plant. (*Courtesy of NEDO, Kyushu Electric Power Co., and Meidensha Corp.*)

A typical charging voltage (DC) nominally is about 1400 V with a current of 520 A, while the discharge starts at 1186 V at an average current of about 900 A.[42,44] Discharge is terminated when the voltage drops to 720 V. Discharge can be carried out for 8 h at 500 kW or for 4 h at 1000 kW. The 1100-V DC battery output is supplied to a self-commutated inverter of 1000 kVA, and the output transformer is a self-cooled 1200-kVA type. The AC output is fed to the Kyushu utility grid in times of peak demand. In periods of low demand the batteries are recharged from the grid. The battery completed over 1300 cycles with an overall energy efficiency of 65.9%.[45]

MITI has also sponsored a project with Meidensha Corporation and Aisin Seiki Company, Ltd., to evaluate a 30 kWh zinc/bromine battery module for storage of electrical energy produced by a solar stirling generator.[46] The intent was to use the battery for lighting at night and to charge the battery using electric power generated during the day. Results of field tests at Miyako Island were unsatisfactory due to bad weather; however, tolerance of this technology to the uneven charge currents generated by solar energy was demonstrated.

FIGURE 39.17   Battery room of Imajuku energy storage test plant. (*Courtesy of NEDO, Kyushu Electric Power Co., and Meidensha Corp.*)

TABLE 39.8   Design Specifications for Imajuku Energy Storage Test Plant

Power	1 MW AC
Capacity	4 MWh AC (1000 V AC, 4 h)
Cell electrode area	1600 cm²
Current density	13 mA/cm² (nominal)
Stack	30 cells bipolar
Submodule	25 kW (30 cells in series, 24 stacks in parallel)
Dimensions (height × width × length)	3.1 × 1.67 × 1.6 m
Weight	6380 kg
Module	50 kW (25-kW submodule, 2 series)
Pilot plant system	50 kW, 12 series
Total weight	153 tons

Source: From Fujii et al.[43]

39.8 DEVELOPMENTS AND PROJECTIONS

In the United States, the most likely near-term market for zinc/bromine batteries is electric-utility applications. Sales of millions of kilowatthours of capacity per year may be possible at an estimated cost of $150/kWh or less. Application of zinc/bromine technology to electric vehicles in the United States is a longer-term market possibility, and the USABC (United States Advanced Battery Consortium) has presently chosen not to fund development of zinc/bromine technology. In Europe, however, interest in electric-vehicle applications for zinc/bromine technology is stronger and represents the near-term market.

Research and development issues are being addressed and will result in further improvements to the performance of the zinc/bromine battery system. For example, materials for separators which would allow improved transport selectivity while maintaining low resistivity would result in higher efficiencies. Development of a high-surface-area carbon layer with enhanced stability will lead to improved electrodes and longer lifetimes. Rigid plastics for frames and electrodes will minimize warpage and allow even flow of electrolyte. With any materials development, tradeoffs between performance, cost, and manufacturability need to be considered. Development of more efficient auxiliary systems is needed. System design is also important and includes factors such as footprint, structural integrity, number of cells per stack, number of stacks, and number of stacks per electrolyte reservoir.

The challenges for the zinc/bromine battery technology are the following:

* Scale-up to battery capacities of interest to large customers
* Identify separator material and source
* Engineer the system to eliminate leaks, maintain flow uniformity, and maximize reliability and life
* Develop robust power electronics and controllers
* Reduce costs of manufacturing and installation

REFERENCES

1. C. S. Bradley, U.S. Patent 312,802, 1885.

2. R. A. Putt and A. Attia, "Development of Zinc Bromide Batteries for Stationery Energy Storage," Gould, Inc., for Electric Power Research Institute, Project 635-2, EM-2497, July 1982.

3. L. Richards, W. Vanschalwijk, G. Albert, M. Tarjanyi, A. Leo, and S. Lott, "Zinc-Bromine Battery Development," final report, Sandia Contract 48-8838. Energy Research Corporation, Sandia National Laboratories, SAND90-7016, May 1990.

4. A. Leo, "Zinc Bromide Battery Development," Energy Research Corporation for Electric Power Research Institute, Project 635-3, EM-4425, Jan. 1986.

5. P. C. Butler, D. W. Miller, C. E. Robinson, and A. Leo, "Final Battery Evaluation Report: Energy Research Corporation Zinc/Bromine Battery," Sandia National Laboratories, SAND84-0799, Mar. 1984.

6. D. J. Eustace, "Bromine Complexation in Zinc-Bromine Circulating Batteries," *J. Electrochem. Soc.* **528** (Mar. 1980).

7. K. Cedzynska, "Properties of Modified Electrolyte for Zinc-Bromine Cells," *Electrochimica Acta,* **40**(8): 971–976 (1995).

8. R. Bellows, H. Einstein, P. Grimes, E. Kantner, P. Malachesky, K. Newaby, H. Tsien, and A. Young, "Development of a Circulating Zinc-Bromine Battery Phase II," final rep. Exxon Research and Engineering Company, Sandia National Laboratories, SAND83-7108, Oct. 1983.

9. M. Pourbaix, *Atlas d'Equilibres Electrochimiques,* Gauthier-Villars, Paris, France, 1963, p. 409.

10. R. Bellows, H. Einstein, P. Grimes, E. Kantner, P. Malachesky, K. Newby, and H. Tsien, "Development of a Circulating Zinc-Bromine Battery Phase I, final rep. Exxon Research and Engineering Company, Sandia National Laboratories, SAND82-7022, Jan. 1983.

11. E. A. Kaminski and R. F. Savinell, "A Technique for Calculating Shunt Leakage and Cell Currents in Bipolar Stacks Having Divided or Undivided Cells," *J. Electrochem. Soc.* **130:**1103 (1983).

12. H. S. Burney and R. E. White, "Predicting Shunt Currents in Stacks of Bipolar Plate Cells with Conducting Manifolds," *J. Electrochem. Soc.* **135:**1609 (1988).

13. K. Kanari et al., "Numerical Analysis on Shunt Current in Flow Batteries," *Proc. 25th IECEC,* Reno, Nev., 1990, vol. 3, p. 326.

14. C. Comminellis, E. Platter, and P. Bolomey, "Estimation of Current Bypass in a Bipolar Electrode Stack from Current-Potential Curves," *J. Appl. Electrochem.* **21:**415–418 (1991).

15. J. Bolsted, P. Eidler, R. Miles, R. Petersen, K. Yaccarino, and S. Lott, "Proof-of-Concept Zinc/Bromine Electric Vehicle Battery," Johnson Controls, Inc., Advanced Battery Engineering, Sandia National Laboratories, SAND91-7029, Apr. 1991.

16. N. J. Magnani, P. C. Butler, A. A. Akhil, J. W. Braithwaite, N. H. Clark, and J. M. Freese, "Utility Battery Exploratory Technology Development Program Report for FY91," Sandia National Laboratories, SAND91-2694, Dec. 1991.

17. N. Clark, P. Eidler, and P. Lex, "Development of Zinc/Bromine Batteries for Load-Leveling Applications: Phase 2 final report," Sandia National Laboratories, SAND99-2691, Oct. 1999, p. 3–1.

18. H. F. Gibbard, "Physical Chemistry of the Zinc-Bromine Battery II. Transference Numbers of Aqueous Zinc Solutions, "*Proceedings of the Symposium on Battery Design and Optimization,* edited by S. Grosse, Electrochem. Soc. Proc. vol. 79–1, p. 212.

19. N. Clark, P. Eidler, and P. Lex, "Development of Zinc/Bromine Batteries for Load-Leveling Applications: Phase 2 final report," Sandia National Laboratories, SAND99-2691, Oct. 1999, p. 5–7.

20. P. Lex and B. Jonshagen, "The Zinc/Bromine Battery System for Utility and Remote Area Applications," ZBB Energy Corporation, *Proc. of The Electrical Energy Storage Systems Applications & Technologies (EESAT) Conference,* Chester, United Kingdom, June 16–18, 1998. Also found in *Power Engineering Journal,* June, 1999, pp. 142–148.

21. B. Jonshagen, "The Zinc Bromine Battery," ZBB Technologies, Inc., *Proc. of the Australian and New Zealand Solar Energy Society's 34th Annual Conf.,* Darwin, Australia, Oct., 1996, pp. 248–257.

22. Proc. of the DOE Energy Storage Systems Program, Annual Peer Review, Arlington, Virginia, Sept. 8–9, 1999.

23. C. Arnold, Jr., "Durability of Polymeric Materials Used in Zinc/Bromine Batteries," *Proc. 26th IECEC,* Boston, Mass., 1991, vol. 3, p. 440.

24. C. Arnold, Jr., "Durability of Carbon-Plastic Electrodes for Zinc/Bromine Storage Batteries," Sandia National Laboratories, SAND92-1611, Oct. 1992.

25. P. M. Hoobin, K. J. Cathro, and J. O. Niere, "Stability of Zinc/Bromine Battery Electrolytes," *J. Appl. Elecrochem.* **19:**943 (1989).

26. R. Winter, "Commercializing Zinc-Flow® Technology," Powercell Corporation, *Proc. of the 6th International Conference—Batteries for Utility Energy* Storage, Energiepark Herne, Germany, Sept. 21–23, 1999.

27. N. I. Sax, *Dangerous Properties of Industrial Materials,* 6th ed., Van Nostrand Reinhold, New York, 1984.

28. R. G. Zalosh and S. N. Bajpai, "Comparative Hazard Investigation for a Zinc-Bromine Load-Leveling Battery," Factory Mutual Research Corp. for Electric Power Research Institute, Project RP1198-4, Oct. 1980.

29. "40CFR302 EPA Designated Reportable Quantities," May 13, 1991.

30. G. S. Tomazic, "The Zinc-Bromine-Battery Development by S.E.A.," *Proc. 11th Int. Electric Vehicle Symp.,* Florence, Italy, Sept., 1992.

31. D. H. Swan, B. Dickinson, M. Arikara, and M. Prabhu, "Construction and Performance of a High Voltage Zinc Bromine Battery in an Electric Vehicle," Institute of Transportation Studies, University

of California at Davis, Proceedings of the 10th Annual Battery Conference on Applications and Advances, Long Beach, California, Jan. 10–13, 1995, pp. 135–140.

32. G. Tomazic, "Zinc-Bromine Systems for EV-Batteries Advances and Future Outlook," Powercell GmbH, *Proc. of the Symposium on Exploratory Research and Development of Batteries for Electric and Hybrid Vehicles,* The Electrochemical Society, San Antonio, Texas, Oct. 6–11, 1996, pp. 212–221.

33. a. N. Marincic, "Safety Aspects of Zinc/Bromine Batteries Powering Electric Vehicles," Powercell Corporation, Powercell Tech Paper, TD-94005a
    b. N. Marincic, "Safety Aspects of Zinc/Bromine Batteries Powering Electric Vehicles," Powercell Corporation, *Proc. of the International Symposium on Automotive Technology and Automation,* Aachen, Germany, Oct. 31–Nov. 4, 1994, pp. 175–182.

34. Toyota Motor Corp., leaflet describing the EV-30 and zinc/bromine battery.

35. Japan Electric Vehicle Association, brochure describing electric vehicles manufactured by Japanese automobile companies.

36. P. C. Butler, "Battery Energy Storage for Utility Applications: Phase 1—Opportunities Analysis," Sandia National Laboratories, Oct. 1994, SAND94-2605, p. 21.

37. P. J. Lex, "Utility Sized Zinc/Bromine Battery Demonstrations," ZZB Energy Corporation, *Proc. of the 6th International Conference—Batteries for Utility Energy Storage,* Energiepark Herne, Germany, Sept. 21–23, 1999.

38. http://www.powercell.com.

39. *PowerBlock™ Mobile Energy Storage for the Future,* Powercell brochure.

40. *Company Overview,* a presentation by Powercell Corporation, Dec., 1999.

41. S. Furuta, T. Hirabayashi, K. Satoh, and H. Satoh, "Status of the 'Moonlight Project' on Advanced Battery Electric Energy Storage System," *Proc. 3d Int. Conf. on Batteries for Utility Energy Storage,* Kobe, Japan, 1991, pp. 49–63.

42. Y. Yamamoto, S. Kagata, T. Taneba, K. Satoh, S. Furuta, and T. Hirabayashi, "Outline of Zinc-Bromide Battery Energy Storage Pilot Plant," *Proc. 3d Int. Conf. on Batteries for Utility Energy Storage,* Kobe, Japan, 1991, pp. 107–125.

43. T. Fujii, M. Igarashi, K. Fushimi, T. Hashimoto, A. Hirota, H. Itoh, K. Jin-nai, T. Hashimoto, I. Kouzuma, Y. Sera, and T. Nakayama, "Zinc/Bromine Battery Development for Electric Power Storage," *Proc. 25th IECEC,* Reno, Nev., 1990, vol. 6, pp. 136–142.

44. Y. Ando, M. Igarashi, K. Fushimi, T. Hashimoto, A. Hirota, H. Itoh, K. Jinnai, T. Fujii, T. Nabetani, N. Watanabe, Y. Yamamoto, S. Kagata, T. Iyota, S. Furuta, and T. Hirabayashi, "4 MWh Zinc/Bromine Battery Development for Electric Power Storage," *Proc. 26th IECEC.* Boston, Mass., 1991.

45. S. Furata, "New Type Battery Power Storage System," *Enerugi,* **26:**3, 1993, English abstract. Paper is published in Japanese.

46. H. Itoh, K. Fushimi, Y. Kataoka, and H. Hashiguchi, "Zinc/Bromine Battery for Solar Stirling Power Generation System," Meidensha Corporation, *Proc. of the 29th Intersociety Energy Conversion Engineering Conf.,* Monterey, California, Aug. 7–11, 1994, pp. 795–800.

CHAPTER 40
SODIUM-BETA BATTERIES

Jeffrey W. Braithwaite and William L. Auxer

40.1 GENERAL CHARACTERISTICS

Rechargeable high-temperature battery technologies that utilize metallic sodium offer attractive solutions for many relatively large-scale energy-storage applications. Candidate uses include several that are associated with electric-power generation and distribution (e.g., utility load leveling, power quality and peak shaving) and some that involve powering motive devices (e.g., electric cars, buses and trucks and hybrid buses and trucks) and space power (e.g., aerospace satellites). The uses related to electric power are collectively referred to as stationary applications to differentiate them from the motive applications.

A number of sodium-based battery options have been proposed over the years, but the two variants that have been developed the furthest are referred to as sodium-beta batteries. This designation is used because of two common and important features: liquid sodium is the active material in the negative electrode and the ceramic beta″-alumina (β''-Al_2O_3) functions as the electrolyte. The sodium/sulfur technology was introduced in the mid-1970s (see forward in Ref. 1). Its advancement has been pursued in a variety of designs since that time. The attractive properties and the primary limitations of the sodium/sulfur system are summarized in Table 40.1. A decade later, the development of the sodium/metal-chloride system was launched.[2] This technology offered potentially easier solutions to some of the development issues that sodium/sulfur was experiencing at the time. A feature comparison between the sodium/nickel-chloride and the sodium/sulfur technologies is presented in Table 40.2.

From the time of their discovery through the mid-1990s, these two technologies were among the leading candidates believed to be capable of satisfying the needs of a number of emerging battery energy-storage applications that had very promising markets. Because of the prime importance of the intended applications to the development process, some of the important aspects of this subject are addressed in Sec. 40.5. The one application, however, that generated the most interest centered on powering the electric vehicle (EV). Given the large potential size of the market in combination with the inherent environmental advantages, many industrial companies and government organizations heavily invested in the technology development process. Significant advancements were made and because of acceptable performance, durability, safety, and manufacturability, at least four automated pilot-production facilities were built and were functioning by the mid-1990s. However, during this same time period, the realization was made that the public (especially in the U.S.) would not purchase a pure battery-powered electric vehicle in high volumes. Its shorter range, lower power, and especially higher cost compared with conventional internal-combustion powered vehicles were believed to ultimately be unacceptable. Existing environmentally driven government

TABLE 40.1 Advantages and Limitations of Sodium/Sulfur Battery Technology

Characteristic	Comments
Advantages	
Potential low cost relative to other advanced batteries	Inexpensive raw materials, sealed, no-maintenance configuration
High cycle life	Liquid electrodes
High energy and good power density	Low-density active materials, high cell voltage
Flexible operation	Cells functional over wide range of conditions (rate, depth of discharge, temperature)
High energy efficiency	80+% due to 100% coulombic efficiency and reasonable resistance
Insensitivity to ambient conditions	Sealed high-temperature systems
State-of-charge identification	High resistance at top of charge and straightforward current integration due to 100% coulombic operation
Limitations	
Thermal management	Effective enclosure required to maintain energy efficiency and provide adequate stand time
Safety	Reaction with molten active materials must be controlled
Durable seals	Cell hermeticity required in a corrosive environment
Freeze-thaw durability	Due to the use of a ceramic electrolyte with limited fracture toughness that can be subjected to high levels of thermally driven mechanical stress

TABLE 40.2 Characteristic Comparison between Sodium/Nickel-Chloride and Sodium/Sulfur Battery Technologies

Higher cell open-circuit voltage—2.59 V open circuit (2.076 V for sodium/sulfur).

Wider operating temperature range—sodium/nickel-chloride can function at temperatures from as low as 220'dgC to 450°C, whereas sodium/sulfur is limited to a range from 290°C to approximately 390°C. However, the range over which practical power levels and long service life has been established is between 270°C to 350°C versus 310°C to 350°C for sodium/sulfur.

Safer products of reaction—the exothermic heats of reaction are lower and the vapor pressure of the reactants less than atmospheric up to a temperature level of 900°C.

Less metallic component corrosion—the chemistry of the positive electrode is non-aggressive compared to molten Na_2S_x.

Assembly in the fully discharged state without the handling of metallic sodium—a discharged positive electrode can be used.

Reliable failure mode—if the electrolyte fails, sodium will react with the secondary electrolyte to short-circuit the cell.

No freeze/thaw limitation—the thermally induced mechanical stress on the electrolyte is lower due to: (a) the positive electrode is located inside of the electrolyte; (b) a smaller difference exists between the solidification temperature of the positive electrode and ambient and (c) less mismatch in thermal expansion between the secondary and primary electrolyte.

Easier reclamation—primarily because of the value of the nickel in spent batteries (< 2 kg/kWh), reclamation is an economic necessity.[3] Due to the cell configuration, recycling is a straightforward process. Recovery of the nickel pays for the recycling.

The relatively low power density observed in early 1990s cells that incorporated a tubular electrolyte configuration has been overcome. The lower power was caused by higher cell resistance, especially near the end of discharge. The improved cell design using a cruciform shaped electrolyte and incorporate a doping addition in the positive electrode material.[4]

Slightly lower energy density.

mandates did not require the introduction of the EV market until the turn of the century (e.g., the mandate in California to sell zero-emission vehicles). Economic analyses performed by the funding organizations showed that without any income for at least another five years, they could expect to end up with a net negative cash flow. At this point, further support for the two most-advanced developers of the sodium/sulfur technology for motive applications was terminated and the company overseeing the development of the one existing sodium/metal-chloride program changed. Although some observers feel that the corporate decisions regarding the sodium/sulfur technology were ultimately the result of technical shortcomings (e.g., reliability, power, safety, need to keep the batteries tethered to a power supply during off-use periods), the time delay to market coupled with the lack of a definable and significant-sized market provided the actual motivation for the termination of the two major sodium/sulfur EV development efforts. However, as is described later, development of the sodium/metal-chloride technology along with several sodium/sulfur efforts in Japan for stationary applications is still underway. For reference, the major sodium-beta battery developers that existed during the 1990s are listed in Table 40.3 along with their primary application and program status.

The objectives of this chapter are to describe the two technologies, show why the application-based interest has existed now for over a quarter century, provide insight into the cell and battery design process, and finally, offer guidance about how to best manage and utilize their advantages. This objective requires the incorporation of legacy information from several of the now terminated development programs. Because of the existence of several common features and more available published information, the general emphasis of this chapter is on the sodium/sulfur technology.

TABLE 40.3 Principle Sodium-Beta Battery Developers

Company	Abbreviation	Country	Primary application	Status
Sodium/Sulfur				
NGK Insulator, Ltd— Tokyo Electric Power Co.	NGK	Japan	Stationary	Active
Yuasa Corp.	YU	Japan	Stationary	Active
Hitachi Ltd.	HIT	Japan	Stationary	Active
Silent Power Ltd	SPL	U.K. U.S.	Motive Stationary	Terminated
Asea Brown Boveri	ABB	Germany	Motive	Terminated
Ford Aerospace	FACC	U.S.	Motive and Stationary	Terminated
Eagle-Picher Technologies	EPT	U.S.	Aerospace	Dormant
Sodium/Nickel-Chloride				
MES-DEA SA	MES	Switzerland	Motive	Active

40.2 DESCRIPTION OF THE ELECTROCHEMICAL SYSTEMS

The basic cell structure and associated electrochemistry of the two sodium-beta technologies are depicted in Fig. 40.1. As introduced in the previous section, both sodium-beta cells use the solid sodium ion-conducting β''-Al_2O_3 electrolyte. Cells must be operated at a sufficiently high temperature (270 to 350°C) to keep all (Na/S) or portions (Na/MeCl$_2$) of the active electrode materials in a molten state and to ensure adequate ionic conductivity through the β''-Al_2O_3 electrolyte.

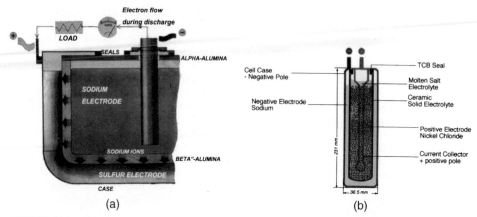

FIGURE 40.1   Diagrams showing the basic functionality of the two types of sodium-beta cells: (*a*) sodium/sulfur, and (*b*) sodium/nickel-chloride *(Diagram (b) is courtesy of MES-DEA SA)*

40.2.1   Sodium/Sulfur

During discharge, the sodium (negative electrode) is oxidized at the sodium/β''-Al$_2$O$_3$ interface, forming Na$^+$ ions. These ions migrate through the electrolyte and combine with the sulfur that is being reduced in the positive electrode compartment to form sodium pentasulfide (Na$_2$S$_5$). The sodium pentasulfide is immiscible with the remaining sulfur, thus forming a two-phase liquid mixture. After all of the free sulfur phase is consumed, the Na$_2$S$_5$ is progressively converted into single-phase sodium polysulfides with progressively higher sulfur content (Na$_2$S$_{5-x}$). During charge, these chemical reactions are reversed. The two half-cell and full-cell reactions are as follows:

Negative electrode:
$$2\mathrm{Na} \underset{\text{charge}}{\overset{\text{discharge}}{\rightleftharpoons}} 2\mathrm{Na}^+ + 2e^-$$

Positive electrode:
$$x\mathrm{S} + 2e^- \underset{\text{charge}}{\overset{\text{discharge}}{\rightleftharpoons}} \mathrm{S}_x^{-2}$$

Overall cell: $2\mathrm{Na} + x\mathrm{S} \underset{\text{charge}}{\overset{\text{discharge}}{\rightleftharpoons}} \mathrm{Na}_2\mathrm{S}_x$ $(x = 5 - 3)$ $E_{\mathrm{ocv}} = 2.076 - 1.78$ V

Although the actual electrical characteristics of sodium/sulfur cells are design dependent, the general voltage behavior follows that predicted by thermodynamics. A typical cell response is shown in Fig. 40.2. This figure is a plot of the equilibrium potential or open-circuit voltage and the working voltages (charge and discharge) as a function of the depth of discharge. The open-circuit voltage is constant (at 2.076 V) during the 60 to 75% of the discharge when the two-phase mixture of sulfur and Na$_2$S$_5$ is present. The voltage then linearly decreases in the single phase Na$_2$S$_x$ region to the selected end-of-discharge point. End of discharge is normally defined at open-circuit voltages of 1.78 to 1.9 V. The approximate sodium polysulfide composition corresponding to 1.9 V per cell is Na$_2$S$_4$; for 1.78 V per cell, it is Na$_2$S$_3$. Many developers choose to limit the discharge to less than 100% of theoretical (such as to 1.9 V) for two reasons: (1) the corrosivity of Na$_2$S$_x$ increases as x decreases and (2) to prevent local cell overdischarge due to possible non-uniformities within the battery (e.g., temperature or depth-of-discharge). If the discharge is continued past Na$_2$S$_3$, another two-phase mixture again forms, except that now the second phase is solid Na$_2$S$_2$. The for-

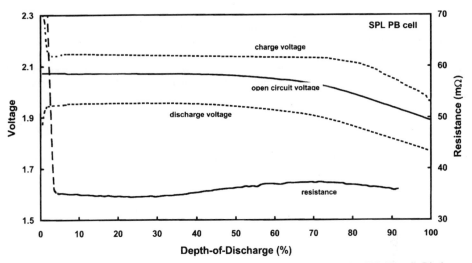

FIGURE 40.2   Cell voltage and resistance as a function of depth of discharge for SPL PB cell. Discharge rate C/3; charge rate C/5.

mation of Na_2S_2 is undesirable in the cell because high internal resistance, very poor rechargeability, and structural damage to the electrolyte can result. Several other important characteristics of the sodium/sulfur electrochemical couple are evident from this figure. At high states of charge, the working voltage during charge increases dramatically due to the insulating nature of pure sulfur (shown also by the higher cell resistance). This same factor also causes a slight decrease in cell voltage at the start of discharge. At the C/3 discharge rate, the average cell working voltage is approximately 1.9 V. The theoretical specific energy of the electrochemical couple is 755 Wh/kg (to 1.76 V open circuit). Although not all of the sodium is recovered during the initial charge, cells subsequently deliver 85 to 90% of their theoretical ampere-hour capacity. Finally, the existence of wholly molten reactants and products eliminates the classical morphology-based electrode aging mechanisms, thus yielding an intrinsically high cycle life.

40.2.2 Sodium/Metal-Chloride

The prime electrochemical difference between the two sodium-beta technologies is the sodium/metal-chloride positive electrode. This component contains a molten secondary electrolyte ($NaAlCl_4$) and an insoluble and electrochemically active metal-chloride phase (Fig. 40.1b). The secondary electrolyte is needed to conduct sodium ions from the primary β''-Al_2O_3 electrolyte to the solid metal-chloride electrode. Cells using positive electrodes with two transition metal-chlorides, nickel and iron, have been developed. These specific metals were selected based on their insolubility in the molten $NaAlCl_4$ secondary electrolyte.[2,3] During discharge, the solid metal-chloride is converted to the parent metal and sodium chloride crystals. The overall cell reactions for these two chemistries are as follows:

Nickel-based cell: $NiCl_2 + 2Na \underset{\text{charge}}{\overset{\text{discharge}}{\rightleftharpoons}} Ni + 2NaCl \qquad E_{ocv} = 2.58 \text{ V}$

Iron-based cell: $FeCl_2 + 2Na \underset{\text{charge}}{\overset{\text{discharge}}{\rightleftharpoons}} Fe + 2NaCl \qquad E_{ocb} = 2.35 \text{ V}$

The voltage behavior as a function of rate and depth-of-discharge for an early 1990s vintage sodium/nickel-chloride cell is shown in Fig. 40.3. This figure also includes the thermodynamic potentials for the overall cell reaction and two additional cell reactions that only become active during overcharge and overdischarge, respectively. Past the end of normal discharge, the working voltage quickly drops when all of the nickel chloride is consumed. At this point, the reduction of the $NaAlCl_4$ to aluminum begins to occur according to the following reaction:

$$3Na + NaAlCl_4 \leftrightarrows 4NaCl + Al \qquad E_{ocv} = 1.58V$$

If continued to the point of complete sodium depletion, electrolyte fracture will occur. But because of the presence of the metallic aluminum, the cell will remain electrically conductive. This characteristic permits batteries to be configured with long series strings (for more information refer to section 40.4.1 on "Electrical Networking"). The quick decrease in cell voltage functions as a reliable indicator for the end of discharge and is used to provide overdischarge protection.

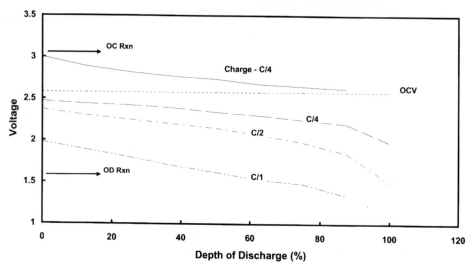

FIGURE 40.3 Cell voltage as a function of depth of discharge for a 1990 vintage sodium/nickel-chloride cell. Lower curves are for three discharge rates and the two arrows indicate the voltage at which the intrinsic overcharge (OC) and overdischarge (OD) reactions occur.

If the cell is overcharged, excess nickel chloride will be produced by the decomposition of the secondary electrolyte according to this reaction:

$$Ni + 2NaAlCl_4 \leftrightarrows 2Na + 2AlCl_3 + NiCl_2 \qquad E_{ocv} = 3.05 \ V$$

Although degradation of the positive electrode will occur during excessive overcharge, this reaction will prevent voltage-induced fracturing of the β''-Al_2O_3 electrolyte. In practice, cells and batteries can be safely overcharged by more than 50%.

40.3 *CELL DESIGN AND PERFORMANCE CHARACTERISTICS*

The goal of the information contained in this section is to first provide a better understanding of the technical factors that must be addressed in the design of sodium-beta battery cells and then to document the performance characteristics of many of the most advanced cells. Importantly, the two sodium-beta battery technologies possess many common features. As such, most sodium/sulfur cell and battery technical considerations are applicable to the sodium/metal-chloride system and the basic approach taken with the design of the sodium /nickel-chloride cells has paralleled that used for the sodium/sulfur cells. Thus, cell design considerations are presented only for the more generic sodium/sulfur technology. A similar approach is used in Sec. 40.4 which addresses battery-level designs and performance. Many of the subjects contained in the two sodium/sulfur design-consideration sections are described in much more detail in Ref. 1.

40.3.1 Sodium/Sulfur Cell Design Considerations

Most of the cell design criteria are imposed directly from battery requirements (such as energy, power, voltage, service life, weight, and size). These requirements can be termed "battery down." Those key criteria originating at the cell level, termed "cell up," include cost and manufacturability, safety, and durability. The durability factor is defined as the ability to accommodate mechanical stresses resulting from occasional freeze-thaw cycles or those imposed by vibration and shock. Usually the battery-down and cell-up requirements conflict, therefore forcing compromises to be made. Those parameters that influence the cell design process are shown graphically in Fig. 40.4, and the major factors are discussed in more detail in this section.

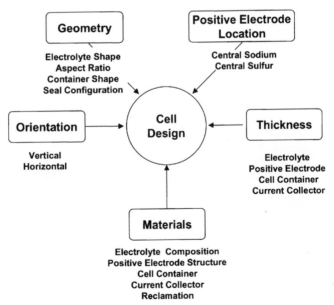

FIGURE 40.4 Diagram of principal variables to consider during design of sodium/sulfur cells.

Major Components.   A simplified schematic diagram of a standard monopolar cell design is shown in Fig. 40.5*a*. Referring to this figure, every cell contains the following components:

Electrolyte: A solid β''-Al_2O_3 ceramic functions as the electrolyte by conducting sodium ions (ionic charge transfer) between the positive and negative electrodes. At operating temperature, the conductivity of the β''-Al_2O_3 is approximately equivalent to that for the electrolytes used in most aqueous batteries. In addition, this component is impermeable to the molten reactants and has negligible electronic conductivity. Thus, the electrolyte also functions as an excellent separator for the molten electrodes, preventing any direct self-discharge and permitting 100% coulombic efficiency. In all modern cell designs, the electrolyte is formed in the shape of a closed-end tube.

Negative (sodium) electrode: This component contains the sodium metal that is electrochemically oxidized during discharge and an inert conductive metal component for transferring current to a terminal (current collector). In Fig. 40.5*a* the current collector is a combination of the metal rod and the metal container. The sodium is placed on the inside of the electrolyte tube. This cell configuration is preferred by all developers and is termed central sodium. The internal metal container restricts the flow of sodium to the electrolyte, thus effectively limiting the quantity of sodium that can be exposed to sulfur in the event of seal or electrolyte failure to the volume in the very small container-electrolyte gap.

Positive (sulfur) electrode: The constituents of this elecrode include the sulfur that is reduced during discharge and a current collector. In Fig. 40.5*a*, the current collector consists of the external metal container and a layer of compressed carbon or graphite felt within the space between the electrolyte and the container (not shown). The carbon fibers are needed to conduct electrons through the insulating sulfur.

External metal container: This component facilitates packaging and safe handling and, as noted, functions as a current collector. In a central sodium cell, the external container must resist corrosive attack by molten sodium polysulfides (discharge reaction product). The best corrosion-resistant materials identified to date use aluminum or chromium or one of their alloys. Usually chromium-containing layers are coated on an inexpensive substrate.

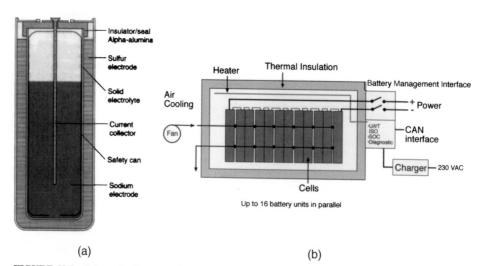

(a)   (b)

FIGURE 40.5   Schematic diagrams of reference sodium-beta cell and battery configurations: (*a*) monopolar sodium/sulfur cell (SPL XPB cell), and (*b*) a sodium/nickel-chloride battery (*Diagram (b) is courtesy of MES-DEA SA*).

Seals: To prevent exposure of cell reactants to the external atmosphere, hermetic seals are required. Their physical design and materials of construction are important because of the high operating temperature and the presence of internal corrosive liquids. Two types of seals are needed: one to join the β''-Al_2O_3 electrolyte to an insulating α-Al_2O_3 component; and the other to join the metal current collectors to the same α-Al_2O_3. The two ceramic components are normally sealed with a glass that is stable in sodium, sulfur, and sodium polysulfides and has a similar coefficient of thermal expansion to the ceramic components. The metal-to-ceramic seals typically use some form of mechanical or thermocompression bonding.

Physical Configuration

Electrolyte Shape. The cell design with a single-electrolyte tube is preferred from the perspective of the electrolyte primarily because of manufacturing and service-life considerations. Attempts to manufacture flat-plate electrolyte shapes with satisfactory quality have been unsuccessful. Significant problems have also been encountered with the seals around the perimeter of the plates because of thermal expansion mismatch and poor chemical durability. Finally, the capability of flat-plate cells to accommodate freeze-thaw cycles and vibration has not been validated. As is discussed in Sec. 40.3.3, sodium/nickel-chloride cells now use a single electrolyte tube with a cruciform/fluted cross-section shape to improve cell power. Multitube approaches have also been proposed as a technique to achieving high power levels, and experimental assemblies have been constructed. However, the problems that have precluded serious consideration of the multitube configuration include difficulty in fabricating the complex alpha-alumina headers, sharing of positive-electrode reactants, overfilling of some tubes during recharge, and an inability to prevent self-discharge of the cell after an electrolyte failure.

Aspect Ratio. Several factors influence the practical limits of the aspect ratio (length to diameter) of a cell, including electrolyte manufacturability, cell power-to-energy ratio, and cell durability. From the perspective of cost and energy density, cells with a large capacity are advantageous. However, the following considerations relative to each of these factors force a compromise solution to be sought.

- Longer electrolyte lengths are more difficult to manufacture with acceptable minimum thickness, diameter tolerances, and manufacturing yield. In the next subsection, NGK Insulators Ltd. has recently had remarkable success reliably fabricating very long electrolyte tubes.

- If length and thickness are maintained, as the electrolyte (and cell) diameter decreases, power density increases. However, a practical limit exists because electrolyte manufacturing yields, gravimetric energy density, and cell cost per unit energy deteriorate. Cell length does not, in general, affect energy density or power significantly, but longer cells cost less per unit of energy. Most cell lengths are limited by an application-related space constraint, by manufacturing yield considerations, or by the inability to keep the resistance of the current collectors low.

- In standard tubular cell designs, increasing the aspect ratio may have important negative implications on the ability of the cell to survive thermal cycling and vibration or shock. These detrimental effects occur because of higher mechanical stresses on the seals and the electrolyte-header interface as the length of the electrolyte increases. Also, non-uniform conditions associated with longer cells can produce reduced safety performance and greater thermal gradients.

Positive/Negative Electrode Location. In the tubular cell design, either sodium or sulfur can be contained within the electrolyte, and the corresponding configurations are referred to as central sodium and central sulfur, respectively. Each configuration has advantages and disadvantages. For example, in the normally preferred central-sodium approach, the cost of a corrosion-resistant cell container is a significant portion of the total material costs. A central sulfur design would reduce the required quantity of corrosion-resistant material substantially and thus provide potential savings. The major disadvantages of the central sulfur design are its questionable ability to survive freeze-thaw cycling and its lower power (due to a limited area for conduction). Central sodium and central sulfur cells also possess different thermal characteristics because of the relative thermal conductivities of sodium and sulfur. In the central sulfur design, the sodium between the electrolyte and the container provides good heat conduction, whereas in the central sodium design the sulfur acts as an insulating blanket.

Cell Orientation. Cells can theoretically be designed for operation in a horizontal or a vertical orientation, although success has only been attained with vertical cells. The advantage of horizontal orientation is greatly improved battery-packaging flexibility. For example, the limited height of electric-vehicle batteries eliminates the potential to use long vertical cells. In addition, with a horizontal operation the electrical path length between cells would be minimized, reducing battery resistance and weight. Unfortunately, reactants segregate in horizontal cells due to gravity, producing reduced freeze-thaw durability and increased sensitivity to vibration and shock. This, coupled with poor sodium distribution, eliminated horizontal cells from serious consideration.

Component Thickness

Electrolyte. The electrolyte contributes a significant portion to the total cell resistance, a factor directly related to its wall thickness and area. A thinner electrolyte tube reduces cell resistance, although a penalty on manufacturing yields and mechanical durability is usually incurred.

Electrodes. The structure and thickness of the positive electrode can have a substantial effect on cell resistance, polarization (effect of charge and discharge rate), and recharge characteristics. Increasing thickness degrades all three of these performance parameters. The relationship between positive electrode thickness and performance requires accurate modeling to ensure that the correct projections of performance are made when changing cell dimensions. The thickness of the negative electrode compartment is not an electrical factor because it is filled with highly conductive sodium metal.

Container. The cost and weight of the cell container are related to its wall thickness. Cost is influenced mainly by the fabrication method and the cell aspect ratio. Typically, containers with an aspect ratio of less than 4 can be pressed from relatively thin material.

Construction Materials

Positive Electrode. Two different electrode structures have been developed for constructing sulfur electrodes that have good recharge characteristics. The purpose of both methods is to reduce the amount of elemental sulfur that forms on the electrolyte surface during recharge. One method uses physical components that are preferentially wet by sodium polysulfides and the other alters the electric-potential distribution in the sulfur electrode by using graded-porosity carbon or graphite felt. In general, the first method produces cells with better recharge characteristics, but because the cells have a more complicated structure, this method can be more costly. The primary advantages of the graded felt approach are lower cost and better lower-temperature operation.

Container. A major materials challenge associated with central sodium cells is the identification of a suitable sulfur container. The solution to this problem is difficult because the container must be very corrosion-resistant, have a surface with reasonable electrical conductivity, possess good mechanical properties, and yet be lightweight and inexpensive. Corrosion is detrimental not only because of its direct impact on the cell lifetime, but also because corrosion products can cause the electrical performance of the cell to continuously degrade. A container constructed with a single material has proven impractical because the two known corrosion-resistant, electrically conductive metals (molybdenum and chromium) are generally too expensive or difficult to fabricate. In addition, molybdenum sometimes produces irreversible cell polarization problems, a phenomenon that still is not fully understood. Aluminum has excellent durability in the presence of corrosive sodium polysulfides, but forms a nonconducting sulfide product layer. Although a few nickel-based superalloys and stainless steels have fair resistance to molten sodium polysulfides, their corrosion rate is too high for the long life desired for these cells (more than five years). Thus with unacceptable single-component containers, manufacturers have been forced to select and use composite materials. These are usually an inexpensive substrate (such as aluminum, carbon steel, or stainless steel) that has been coated, plated, or sheathed with at least one corrosion-resistant material. The key to success of these composites is to ensure that the corrosion-resistant layer is defect free, thus preventing undermining, substrate attack, and spalling.

40.3.2 Sodium/Sulfur Cell Technology

General.   The majority of the work on the sodium/sulfur technology has been directed to electric-vehicle and stationary energy-storage applications. Cells intended for aerospace and defense-related applications (e.g., satellites, submarines, and tanks) have primarily been based on electric-vehicle designs. As noted in Sec. 40.1, the only active development of this technology is currently being performed in Japan for stationary applications and a photograph of three of NGK's cells is shown in Fig. 40.6a. The electrical performance characteristics

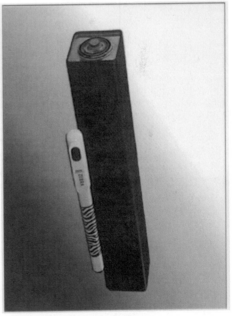

(a)   (b)

FIGURE 40.6 Modern sodium-beta battery cells: (*a*) 3 NGK sodium/sulfur cells (left to right—T4.1, T4.2, T5.1), and (*b*) an MES-DEA sodium/nickel-chloride cell (ML3). For reference, the dimensions of the largest NGK cell are 91 mm in diameter × 515 mm long while the MES ML3 cell is 36 mm square × 232 mm long. (*Photographs courtesy of Tokyo Electric Power Company and NGK Insulators, Ltd. (a) and MES-DEA SA (b)*).

of this technology are demonstrated in the graphs presented in Figs. 40.7 to 40.11. These figures are based on data collected at Sandia National Laboratories for two of the advanced electric-vehicle cell designs developed by ABB and SPL respectively. The relative insensitivity of the cell voltage-current behavior and the peak power to the depth of discharge ("stiffness") is demonstrated in Figs. 40.7 and 40.8. A Ragone plot (relation of specific power to specific energy) is shown in Fig. 40.9. When considering the Ragone plot, it is important to remember that these data are only for cells (does not include weight burdens associated with battery hardware). Typically, cells constitute 50 to 60% of the battery weight. Long-term cycling stability is exhibited by the capacity and resistance response given in Fig. 40.10. Finally, Fig. 40.11 shows that capacity is, in general, not greatly affected by practical discharge or charge rates. However, the capacity of many cells does decline at charge rates greater than C/5.

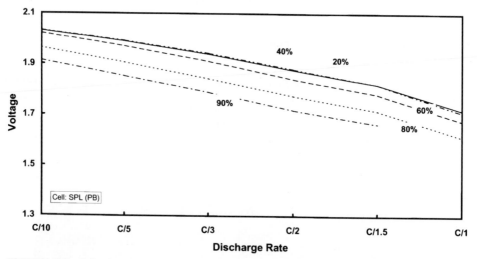

FIGURE 40.7 Cell voltage as a function of discharge rate for an SPL PB sodium/sulfur cell. The parameter is depth of discharge.

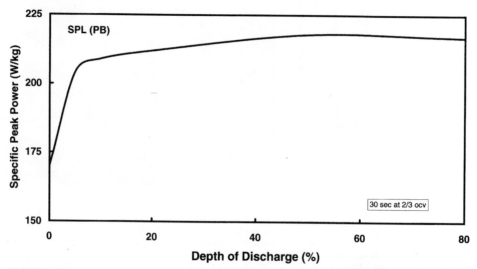

FIGURE 40.8 Specific peak power as a function of depth of discharge for an SPL PB sodium/sulfur cell.

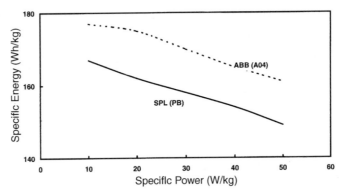

FIGURE 40.9   Relationship of specific power to specific energy (Ragone plot) for two sodium/sulfur cells.

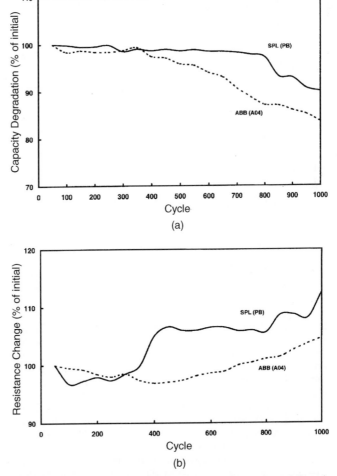

FIGURE 40.10   The effect of cycle life on (*a*) cell capacity and (*b*) end-of-discharge resistance (discharge rate: C/3, charge rate: C/5).

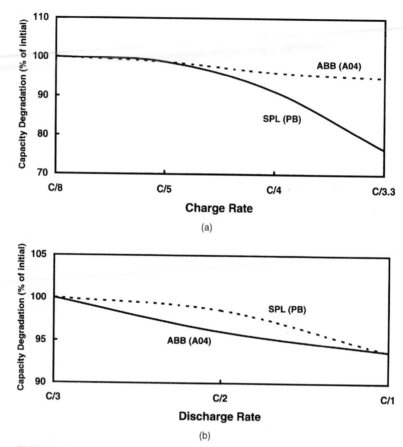

FIGURE 40.11   The effect of (*a*) charge and (*b*) discharge rate on cell capacity retention.

Sodium/Sulfur Cell Development

Stationary Energy-Storage Cells.   The physical and performance specifications for the leading cells designed for use in stationary applications are listed in Table 40.4. As will be further documented in Sec. 40.4, the dominant organization that is presently developing and commercializing the sodium/sulfur technology is a Japanese collaboration between NGK Insulator, Ltd and Tokyo Electric Power Company (TEPCO) that started in 1984.[5] Their goal was to develop cells with sufficient energy capacity for use in utility-based load-leveling and peak-shaving applications (e.g., TEPCO's) that require up to an 8-hr discharge period. The critical technology for such cells involved the manufacture of large diameter beta-alumina tubes of very high quality and precise dimensions. The capability to manufacture 40-mm diameter tubes for 160-Ah cells (designated "T4.1") was confirmed and was soon extended to 45-mm diameter and slightly longer units for use in 248-Ah cells (designated "T4.2"). By 1998, NGK was in the process of developing yet larger cells. This effort has culminated in a 632-Ah cell (designated "T5") based on 58-mm diameter beta-alumina tubes.

The other active Japanese developers include Yuasa Corporation and Hitachi Ltd. These organizations originally targeted large-scale utility load leveling as their prime intended use. Yuasa's initial effort was part of the national Moonlight Project that resulted in the design and fabrication of a large number of 300-Ah cells.[6-8] Hitachi has been developing several

TABLE 40.4 Specifications for Sodium/Sulfur Cells Designed for Stationary-Energy-Storage Applications

Manufacturer	NGK	NGK	NGK	YU	HIT	SPL
Cell designation	T4.1	T4.2	T5	—	—	XPB
Capacity, Ah	160	248	632	176	280	30
Diameter, mm	62	68	91	64	75	44
Length, mm	375	390	515	430	400	114
Weight, g	2000	2400	5400	2700	4000	345
Energy density, Wh/L	285	340	370	240	300	345
Specific energy, Wh/kg	160	202	226	120	133	170
Power density, W/L	36	43	46	60	—	360

sodium/sulfur cell designs since 1983. Lately, they have targeted other applications, including renewable energy storage.[9,10] Before the discontinuation of their programs in the mid-1980s, two U.S. companies, Ford Aerospace and Communications Corporation (FACC) and General Electric, had also been developing large central-sodium cells for utility load-leveling applications. Silent Power Ltd., a company operating in the U.K. and the U.S, developed a cell specifically for utility applications that had a nominal capacity of 30 Ah.[11] This cell (the XPB) was essentially an extended version of their electric-vehicle cell.

Motive-Power (Electric-Vehicle) Cells. Although all development of sodium/sulfur for these applications has ceased, the two major participants up through the mid-1990s were Asea Brown Boveri (ABB) and Silent Power Limited (SPL). The performance of their main EV cells was discussed in the previous section on Sodium/Sulfur Cell Technology. A compilation of the physical and performance specifications of their cells and Eagle-Picher's aerospace cell is included in Table 40.5. During the 1980s, ABB's A04 cell technology cell configuration was the building block of their battery program. Innovations included the use of thermocompression bonding for the electrode seals and aluminum as the primary cell container. Then, ABB made some subtle, but important, changes to the A04 cell and created the A08. Discharge of this cell was terminated at 1.87 V open circuit rather than the A04's

TABLE 40.5 Specifications for Sodium/Beta Cells Designed for Motive-Power Applications

Manufacturer	ABB	SPL	EPI	MES
Cell designation	A04	PB	—	ML3
Prime application[a]	EV	EV	Aero	EV
Chemistry	Na/S	Na/S	Na/S	Na/NiCl$_2$
Capacity, Ah	38	10.5	55	32
Electrolyte shape[b]	cyl	cyl	cyl	cruc
Diameter, mm	35	44	36	36.5
Length, mm	220	45	240	232
Weight, g	410	120	590	715
Resistance, Ω	6	32	5	6-20
Specific energy, Wh/kg	176	178	150	116
Specific peak power, W/kg[c]	390	250	—	260

[a] *EV: electric vehicle; Aero: aerospace*
[b] *cyl: cylindrical, cru: cruciform/fluted*
[c] *at two-thirds open-circuit voltage and 80% DOD*

1.76 V. This higher open-circuit cutoff voltage significantly decreased the corrosivity of the sodium polysulfide active material, permitting the use of lower-cost container coatings.[12] During this same time period, SPL adopted a major change in cell design philosophy to address problems with freeze-thaw durability and battery reliability (networking considerations). A relatively small, nominally 10-Ah cell resulted that was designated the PB.

40.3.3 Sodium / Nickel-Chloride Cell Technology

A schematic diagram of a sodium/nickel-chloride cell was shown previously in Fig. 40.1b and a photograph of a modern cell in Fig. 40.6b. In this standard configuration, the sodium is located on the outside of the β''-Al$_2$O$_3$ electrolyte (outside sodium). An inside sodium configuration would require the use of an expensive nickel container. Another advantage of the outside-sodium cell configuration is that an external cell case with a square cross-section can be used. This cell geometry permits maximum volumetric packing efficiency within the battery enclosure to be attained. A third advantage of this configuration is the cell behavior during thermal "freeze/thaw" cycling. Here, detrimental tensile stresses on the electrolyte do not develop, thus effectively eliminating this potential failure mode. The positive electrode itself is then contained within the electrolyte. In a fully charged cell, this electrode is a porous nickel matrix that is partially chlorinated to nickel dichloride. The remaining nickel backbone serves as part of the positive-electrode current collector. About 30% of the nickel is used in the cell reactions. The matrix is impregnated with the NaAlCl$_4$ molten salt. The sodium compartment is less complex than that of the sodium/sulfur cell because safety features are not needed. The approach to primary containment for both the outer container and the electrode seals is similar to that with sodium/sulfur. As discharge proceeds, the reactions in the positive electrode occur from the outside of the solid nickel structure and proceed inqard thorugh an ever-increasing thickness of reduced nickel. This shrinking-core process resuls in an increasing electrical resistance as the cell discharges because the effective area of the reacting nickel chloride is constantly being reduced. The chemistry of the discharge process provides another significant advantage relative to sodium/sulfur: cells can be safely assembled with the discharge products (nickel metal and salt), and then charged.

Also in contrast to the sodium/sulfur technology, the development of the sodium/metal chloride system has been pursued by a successive progression of single, integrated organizations for one primary application—electric vehicles. Currently, the prime developer is the Swiss company, MES-DEA SA. They purchased the technology from AEG Anglo Battery Holdings (AABH), an organization formed by the German company AEG in cooperation with the original developers (Zebra Power Systems and Beta R&D Ltd.). This technology is often referred to as ZEBRA because of its origins. The acronym ZEBRA stands for *Zero Emission Battery Research Activities*. To date, development has focused almost exclusively on the higher voltage nickel variant but using iron as a doping addition to the positive electrode. As such, the pure iron-based system will not be discussed further in this chapter.

The cell designs that have been developed to date have capacities ranging from 20 to 200 Ah. Specifications for the actual cell design that incorporates the advancements made during the past five years are listed in Table 40.5. These advancements resulted in an improved pulse-power capability (especially near the end of discharge) and energy content.[4] At 80% depth-of-discharge, the power of a modern cell (ML3) is 2.5 times that of a 1992 vintage cell (SL09). This enhanced level of power performance can be determined by comparing the cell data previously presented in Fig. 40.3 with modern performance data shown in Fig. 40.12 for a full-sized battery. The most important modifications involved: (1) the use of a fluted or cruciform-shaped electrolyte that minimized the thickness of the positive electrode and increased the area of the β''-Al$_2$O$_3$ electrode; and (2) the introduction of iron as a dopant to the nickel positive electrode. Optimization of the design and improvements in the chemistry of the positive electrode also resulted in a 20 to 40% increase in energy content. The demonstrated reliability of the early 1990s ZEBRA cells was outstanding with cell failures virtually non-existent.

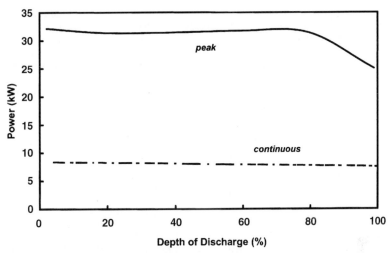

FIGURE 40.12 Power capability (peak and continuous) as a function of depth of discharge for a modern sodium/nickel-chloride battery (Z5C). (*Courtesy of MES-DEA SA*).

40.4 *BATTERY DESIGN AND PERFORMANCE CHARACTERISTICS*

This section on battery-level information is organized the same as Sec. 40.3. That is, battery-level design considerations specific to the sodium/sulfur technology are presented first. Then, brief descriptions of modern battery configurations and performance for both sodium-beta technologies are provided. For reference, a schematic diagram of an integrated sodium/nickel-chloride battery system was shown previously in Fig. 40.5*b*.

40.4.1 Sodium/Sulfur Battery Design Considerations

Battery-level components include mechanical supports for the cells, a thermal management system (incorporating the thermal enclosure) to ensure that each cell is maintained at a relatively high temperature (e.g., for Na/S from 300°C to 350°C), electrical interconnects (cell-cell, cell-module, module-battery), possibly cell-failure devices, and safety-related hardware (such as thermal fuses). As discussed in the next paragraph, batteries are assembled by connecting cells into series and series-parallel arrays to produce the required battery voltage, energy, and power. Electrical heaters are installed within the enclosures to initially warm the cells and then to offset heat loss during periods while the battery is at temperature, but idle. Normally, extra heat is not required during regular discharging and charging due to ohmic heating and chemical reaction effects.

Electrical Networking.   During the lifetime of a battery, individual cells will fail, resulting in a degradation of electrical performance. The rate of degradation is primarily a function of the failure characteristics of the cells and the electrical networking configuration of the battery. Some compensation can be provided with additional capacity at the beginning of life. However, with some battery configurations, especially those with higher voltages (such as 200+ V), the impact of small numbers of cell failures can be significant. The effects of cell failure must therefore be mitigated or controlled in the battery.

Cells must be connected in series to produce the required battery voltage. If the cell capacity is less than the required battery capacity, then further connection of strings in parallel is necessary. Basic options available for configuring cells to produce similar battery output are shown in Fig. 40.13. Each configuration can satisfy the same nominal requirements of voltage and capacity, but each also has benefits and disadvantages that depend on the reliability of the cell population, the electrical behavior of the cell during and after failure, and the cell capacity. In addition, the battery configuration can influence safety.

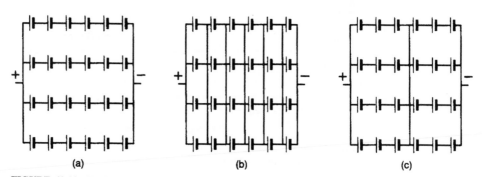

FIGURE 40.13 Basic options for networking sodium-beta cells: (*a*) long-series strings, (*b*) total parallel connection, and (*c*) series-parallel array.

Long Series Strings. A cell must fail with a low resistance and be capable of carrying normal operating currents if the long series string option (Fig. 40.13*a*) is used. Otherwise a single cell failure will force the entire string to become inoperative. If this condition cannot be guaranteed, a short-circuiting device is needed in parallel with each cell, whose function is triggered by cell failure and which then effectively allows current to bypass the cell. Some success has been attained with the development of a proprietary device. An important aspect of the failure devices is that their reliability be high. Otherwise premature cell short-circuiting is possible. Also, strings with large numbers of cells are not recommended because the first cell to recharge can experience a high overvoltage, which is proportional to the number of cells in a string. An even higher overvoltage condition can be produced if a string contains numerous failed cells.

Total Parallel Connection. If the total parallel option (Fig. 40.13*b*) is used, then the resistance of each failed cell must be high to prevent discharge of the remaining cells in parallel. Because this condition hardly ever exists in practice, a device must be placed in series with each cell, which causes the cell to go to an open-circuit condition on cell failure. Such a device must have a very low electrical resistance prior to triggering. To date no success has been achieved with the development of such a device.

Series-Parallel Combination. The series-parallel combination (Fig. 40.13*c*), where parallel connections are provided at frequent intervals along the series string, provides a compromise between the first two options and can eliminate the need for separate and very reliable cell-failure devices. For example, SPL used a four- or five-cell string subunit with parallel interconnection at the ends of the string. A cell failure results in recharge of the remaining cells in the string by the other parallel strings. One of these cells, in turn, becomes highly resistant at the top of charge, preventing further current flow in the string. In effect, a surviving cell acts as an open-circuit device. The major disadvantage of this configuration is that the failure of one cell removes the remaining cells from the string along with its contribution to the battery energy. Furthermore, if large-capacity cells are used, the number of parallel paths in a battery is small, forcing the battery to become vulnerable to cell failure.

Industrial Practice. ABB and SPL used similar approaches for networking the cells in the battery. For example, the ABB B11 batteries (1985 vintage) had six strings in parallel that were cross-connected about every five cells. This approach was basically similar to the approach that SPL used. Because the ABB cell capacity was nominally 40 Ah compared to 10 Ah for the SPL PB cell, the ABB battery had four times fewer strings in parallel. ABB employed a cell failure device (short circuit) in conjunction with this networking approach to minimize the effect of individual cell failures.

Battery/Cell Reliability Requirements. Battery developers must ensure that their designs will result in a product that has adequate service life. Again, the rate of battery degradation is mainly a function of the failure characteristics of the cells and the electrical networking configuration of the battery. These characteristics are often quantitatively lumped together into a term called *reliability*. The sodium/sulfur developers commonly used the two-parameter Weibull statistic to describe the reliability of individual cells. This intrinsic information (generated during cell testing) in combination with a selected cell networking configuration and the number of required series and parallel strings was, in turn, used to estimate battery life.

Battery life increases as the number of cell strings in parallel increases. In this regard the stationary battery applications are much more accommodating because a large number of cells (and strings) is normally required. Electric-vehicle batteries are more limited, especially with the current trend to higher-voltage systems (>300 V). The Weibull cell reliability statistics required to achieve a four-year battery life for two battery designs and two interconnection strategies are presented in Fig. 40.14. In the underlying calculations, assumptions included a cycle rate of 250 cycles per year, a 200-V electric-vehicle battery, and incorporation of a cell bypass/short-circuiting device with each cell in the long-series string networking approach. The Weibull characteristic life is related to the average cell life and the shape factor is related to the rate of failure. Higher shape factor values indicate increasing failure rates with time and lower incidence of early failures. The increased reliability of

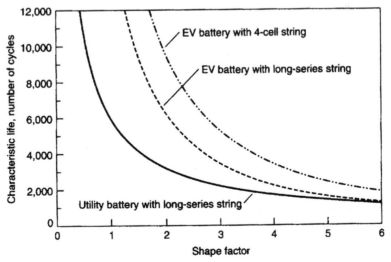

FIGURE 40.14 Cell reliability requirements (Weibull statistical parameters) to achieve a 4-year battery life. Acceptable region is to the right and above appropriate curve.

using a large number of cells is shown by comparing the stationary (utility) battery to the electric-vehicle battery with long series strings. As shown, the utility-battery design allows cells with a significantly lower reliability to be used (unless the shape factor is large (e.g., >5)). The impact of cell interconnect strategies is demonstrated by the two electric-vehicle battery requirements. The four-cell string approach requires a more reliable cell than does the long-series string configuration. However, in the long string approach, a very reliable cell failure device is needed.

Thermal Management.   All batteries used in large-scale energy storage (utility-based, electric vehicle), including those based on lithium or nickel, will probably need some form of thermal management. This requirement is obvious for the high-temperature sodium-beta technologies. Because of space and cost constraints, the electric-vehicle applications impose particularly difficult demands on the hardware used to maintain and control the temperature within the battery. This hardware is often referred to as the thermal management system (TMS). The TMS must minimize heat loss from the battery under normal operating conditions and idle periods in order to maintain the high efficiency of the technology, yet permit sustained high-power discharge periods without reaching unacceptably high temperature levels or creating undesirable temperature differentials within the battery. To satisfy these technical requirements, the TMS in a sodium/sulfur battery usually includes the following components:

- A thermally insulated battery enclosure
- An active or passive cooling system
- A method of distributing heat within the battery enclosure
- Heaters to warm the battery to operating temperature and to maintain it at operating temperature during long idle periods, if necessary

The amount and type of thermal insulation used in the thermal enclosure (e.g., conventional, evacuated, variable conductance) is dependent on the intended application. Relevant application requirements that must be considered include the physical size of the battery, the power-to-energy ratio, the duty cycle, and the duration of any "idle" periods. For example, utility-energy-storage applications are not as constrained with respect to weight and volume as those associated with electric vehicles. Therefore, utility batteries can consider using conventional insulation materials.

Batteries developed for electric vehicles employed evacuated insulation to minimize thickness and weight. Both ABB and SPL used a double-walled, evacuated thermal enclosures with either a fiber board or microporous insulation. Chemical gettering agents were placed within the enclosure to maintain the needed levels of vacuum. This type of system was the only identified design that adequately minimized heat loss while providing the necessary load-bearing capability.

The need for a cooling system is determined by the quantity of heat generated during sustained high-rate discharge, the thermal capacitance of the battery, and the upper temperature limit of the battery. The techniques that have been used and/or contemplated include direct and indirect heat exchange with air, indirect liquid-based heat exchange, heat pipes, thermal shunts, latent-heat-storage, evaporative cooling, and variable conductance insulation systems. ABB utilized an active cooling scheme in its family of EV batteries. The cells were mounted on a flat-plate liquid heat exchanger that transferred excess heat to an external oil/air or oil/water heat sink. If the electrical resistance of the cells and their interconnections are sufficiently low, the temperature rise incurred by the battery could be accommodated by the thermal capacitance of the battery. This was the desired approach for SPL although active cooling was employed in their later designs.

Battery Safety. The approach taken to ensure that the cell and battery designs would be safe under both normal and accident conditions focused on the prevention of electrical short-circuiting and on minimizing exposure to and interaction with any reactive materials. Relative to accidents involving mobile applications, a firm requirement has been mandated by the automobile industry and various governing organizations: the presence of the battery cannot increase its hazard (contribute to the severity or consequence). The specific safety factors that have been addressed in various sodium/sulfur cell and battery designs included the following:

- Selecting proper construction materials (such as low reactivity, high melting point)
- Limiting the availability and flow of sodium to an electrolyte or seal failure site, thus reducing the potential for large thermal excursions ($>100°C$) in cells, which can cause damaging cell breaches
- Using components that minimize the effect of cell failure on adjacent cells (e.g., porous or sand filler between cells in stationary batteries)
- Including thermal and electrical fuses to eliminate the potential for catastrophic short-circuiting
- Providing protection against the environmental hazards associated with each application; in the case of the sodium/sulfur technology, the thermal enclosure is effective against many of these factors
- Including functional redundancy to ensure that improbable or overlooked phenomena do not result in an unwanted consequence

Reclamation. Ultimately all batteries, including sodium/sulfur, will reach their end of life and must be reclaimed or disposed of in some manner. In addition to sodium and sulfur, sodium polysulfide is classified as a hazardous material because of its corrosivity. Chromium or chromium compounds are still being used as coatings for containment corrosion protection. Therefore sodium/sulfur batteries used in all terrestrial applications have to be returned to a processing center for proper recycling, reclamation, or disposal.

40.4.2 Sodium/Sulfur Battery Technology

Stationary Energy-Storage Batteries. As discussed in Sec. 40.5, many stationary applications represent a very promising use for the sodium/sulfur technology primarily because of its small relative footprint (high energy density), excellent electrical efficiency if routinely used, ease of thermal management, lack of required maintenance, and cycling flexibility. All of the developers have now adopted a similar design approach for their battery systems that involves the use of self-contained modules each with 10 to 50 kW of power and 50 to 400 kWh of energy. That is, independent battery-level modules are manufactured that consist of various series-parallel configurations of cells within a thermal enclosure. An analogy to these modules is an integrated electric-vehicle battery. The battery itself is then constructed by connecting these modules again in a series-parallel arrangement (often in a common structure) to give the desired voltage, energy, and power. The resultant battery is combined with a power-conversion system (PCS) to form an integrated facility that can be connected to an electrical system (either utility or customer side).[11] The specifications for the principal stationary modules presently being used are given in Table 40.6. Three of the modules described in this table are associated with NGK Insulator Ltd. The modular concept and its implementation is pictorially shown in Fig. 40.15 which contains one of the NGK modules along an integrated system that contains these same 50-kW modules.

TABLE 40.6 Specifications for Japanese Stationary Sodium/Sulfur Battery Modules

Manufacturer	NGK	NGK	NGK	Yuasa	Hitachi
Battery designation	12.5 kW	25 kW	50 kW	25 kW	12.5 kW
Prime application*	LL; PS; PQ	LL; PS; PQ	LL; PS; PQ	LL	renew
Cell type	4.1	4.2	5	—	—
Number of cells	336	480	384	320	216
Capacity, Ah	2280	2272	3624	1408	290
Energy, kWh	105	211	421	100	100
Cell connection	(6Sx14P) ×4S	(6S × 10P) ×8S	(8S × 6P) ×8S	(4S × 8P) ×10S	72S × 3P
Voltage, V	48	96	128	80	144
Width	996	1435	2200	710	1800
Depth	1853	1940	1762	1371	1800
Height	510	520	640	1117	800
Weight, kg	1400	2000	3620	1700	1900
Specific density, Wh/kg	75	105	116	59	53
Energy density, Wh/L	112	145	170	92	40

*LL: utility-based load leveling; PS: peak shaving; PQ: power quality; renew: renewable energy hybrid

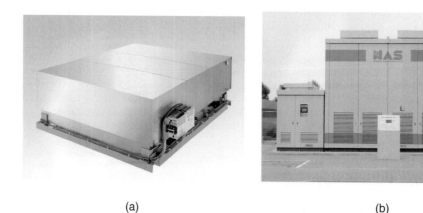

(a)   (b)

FIGURE 40.15 NGK stationary-energy-storage batteries: (*a*) the 50 kW modular battery component (1760 mm wide × 640 mm high × 2200 mm deep) and (*b*) an integrated 500 kW/4 MWh demonstration battery system that uses 10 of these modular batteries (*Courtesy of Tokyo Electric Power Company and NGK Insulators, Ltd.*)

NGK is nearing the commercialization stage. They have constructed a highly automated pilot production facility and have built and are currently operating 18 significant-sized integrated battery systems (see Table 40.7).[5,13] In order to be cost effective, their prime customer, TEPCO, places demanding requirements on these systems. For example, their 50kW, 400kWh module (Fig. 40.15*a*) is designed for use in an integrated system that must achieve an AC-to-AC energy efficiency of more than 70% (including losses to maintain operating temperatures) after operation for 15 years and 2250 full charge-discharge cycles. These systems are intended to be routinely used for 8-hour daily peak shaving applications at utility and industrial sites (i.e., the system stores energy during periods of excess supply and discharges it during periods of peak power demand). TEPCO is presently operating the largest

TABLE 40.7 List of NGK Battery Demonstrations On-line as of May 2000. All batteries have 8 hours of discharge capacity at given power rating (*Courtesy of NGK Insulators, Ltd.*).

Utility	Test site	Power, kW	Start date	Cell designation
Tokyo Electric Power Company	Kawasaki Electric Energy Storage Test Facilities	500	8/95	T4.1 (160Ah)
	Kawasaki Electric Energy Storage Test Facilities	250	12/95	T4.1 (160Ah)
	TEPCO New Energies Park	50	12/95	T4.2 (248Ah)
	Kawasaki Electric Energy Storage Test Facilities	200	6/96	T4.2 (248Ah)
	Tsunashima Substation	6000	3/97	T4.2 (248Ah)
	Kinugawa Power Station	200	1/98	T4.2 (248Ah)
	Ohito Substation	6000	3/98	T5 (632Ah)
	TEPCO R&D Centers	200	9/99	T5 (632Ah)
Chubu Electric Power Company	Electric Power Research & Development Center	100	10/95	T4.1 (160Ah)
	Odaka Substation	1000	2/00	T5 (632Ah)
Touhoku Electric Power Company	Research & Development Center	100	6/96	T4.1 (160Ah)
Hokuriku Electric Power Company	Engineering Research and Development Center	100	2/98	T4.2 (248Ah)
Kandenko	Tsukuba Technology Research & Development Institute	50	5/98	T4.2 (248Ah)
Chugoku Electric Power Company	Technical Research Center	50	7/98	T4.2 (248Ah)
Okinawa Electric Power Company	Miyako Photovoltaic Power Generation System	200	9/98	T4.2 (248Ah)
Kansai Electric Power Company	Tatsumi Substation	200	11/98	T4.2 (248Ah)
Kyushu	Imajuku Substation	100	3/00	T4.2 (248Ah)
NGK Insulators, Ltd	Headquarters Office	500	6/98	T5 (632Ah)

Na/S system ever constructed, a 6MW / 48MWh system shown in Fig. 40.16. This battery system consists of three 2000kW "slide-along" units, each comprised of 40 50kW, 400kWh modules. This field demonstration is confirming performance of the battery system under automatic control. With respect to durability, four 12.5 kW, 100 kWh modules comprised of "T4.1" cells has been in operation for 7 years and continue to exhibit DC-to-DC energy efficiencies of more than 87%, including losses to maintain cell operating temperatures. The "T4.1" cells themselves operate at DC-to-DC efficiencies of more than 92% after these seven years of operation. NGK has performed extensive safety engineering of their modules and has not experienced any safety-related issues.[14]

FIGURE 40.16   A photograph a 6 MW/ 48 MWh NGK sodium/sulfur battery system that is operating in a load-leveling mode at Ohito, Japan. This battery is one of the two largest sodium/sulfur batteries ever built and has been operating since early 1998 (*Courtesy of Tokyo Electric Power Company and NGK Insulators, Ltd.*).

Yuasa built and operated the first large sodium/sulfur battery, an integrated 1-MW 8-MWh system, in the early 1990s that contained 26,880 cells.[6–8] As noted earlier, this effort was part of the Japanese national research program called the Moonlight Project. Battery performance was demonstrated in a utility load-leveling application. The basis for this system was a 50-kW 400-kWh module and the battery itself consists of two of 10-module strings connected in parallel.[7,8] Similar to the other Japanese developers, Yuasa is now advancing the modular concept using smaller (~25 kW) modules.[9] Hitachi is now pursuing renewable applications for their Na/S batteries. These are hybrid power applications and need batteries to level power fluctuations from energy sources like wind and solar. Their program has reached the proof-of-concept stage.[10]

Motive Batteries.   Prior to the discontinuance of their programs in the mid-1990s, the development of the ABB and SPL electric-vehicle battery technologies had reached a relatively advanced state. Their batteries were satisfying the primary performance-based vehicle requirements at the time, i.e. those specified by the U.S. Advanced Battery Consortium (US-ABC). The specifications and proven performance for two of their designs are presented in Table 40.8 and a picture of two of the integrated ABB batteries is shown in Fig. 40.17*a*. Because of their non-commercialized state, neither organization was able to demonstrate that the USABC cost target ($150/kWh) could be attained. By the end of their effort, the in-vehicle test experience of ABB was quite comprehensive. Their batteries had been fitted into a variety of vehicles, including the 300 series BMW, the purpose-designed BMW E-1 electric car, the Chrysler minivan T115, the Daimler Benz 190, two VW vehicles, and a number of Ford Ecostar demonstration vehicles. SPL had more limited in-vehicle testing centered on a number of 60-kWh batteries that were installed in Bedford CF or Griffon electric vans. Their most advanced design (HP battery) was tested extensively in a VW CitySTROmer vehicle.

TABLE 40.8 Specifications for Sodium-Beta Electric-Vehicle Batteries

Manufacturer	ABB	SPL	MES
Battery designation	B17	HP	Z5C
Chemistry	Na/S	Na/S	Na/NiCl$_2$
Cell type	A08	PB	ML3
Energy, kWh	19.2	27.7	17.8
Number of cells	240	1408	216
Cell connection	48S × 5P	(4-5S × 16P) × 20S	216S × 1p and 108S × 2p
Voltage, V	96	176	557 and 278
Dimensions, mm			
Length	730	728	755
Width	540	695	533
Height	315	365	300
Weight, kg	175	250	189
Specific density, Wh/kg	118	117	94
Energy density, Wh/L	155	171	147
Specific power, W/kg[a]	215	240	170

[a] *Defined at two-thirds open-circuit voltage and 80% depth-of-discharge.*

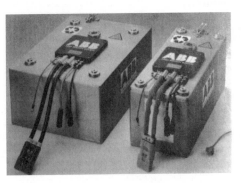

(a)   (b)

FIGURE 40.17 Electric-vehicle batteries: (*a*) ABB sodium sulfur – B16 (left) and B17 (right) and (*b*) Zebra Z5 sodium/nickel-chloride (*Courtesy of Asea Brown Boveri (a) and MES-DEA SA (b)*).

Importantly, safety-related performance at the cell and battery level was addressed by both developers during the design process. The approach to safety that was pursued was described earlier in this and the previous section. Although the active constituents of the sodium/sulfur systems are very reactive, the primary concerns involved electrical short-circuiting and potential thermal excursions. Both organizations conducted extensive safety testing on their final products that encompassed mechanical shock (crash), deformation, external short-circuiting, and fire exposure.[15] Other sets of destructive evaluations were subsequently performed on the newer battery designs. A typical null response from a safety test performed on an SPL HP battery is shown in Fig. 40.18*a*. The satisfactory results from this type of testing allowed ABB and SPL to put sodium/sulfur battery powered electric vehicles on public streets in Europe and the U.S.

Relative to other motive applications, the U.S. Air Force designed and conducted a flight experiment aboard a space shuttle to determine the effect of a micro-G environment on the performance of a small, 4-cell sodium/sulfur battery. This successful test used cells manufactured by Eagle-Picher Technologies (Table 40.5).[16]

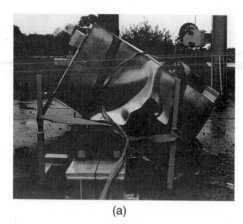
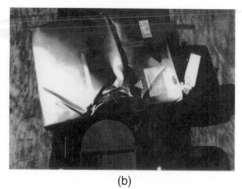

(a)   (b)

FIGURE 40.18 Two sodium-beta batteries following safety/abuse testing that was performed while the batteries were charged and at operating temperature: (*a*) an SPL HP sodium/sulfur battery dropped onto a steel pole, and (*b*) a Zebra battery after a steel beam was pushed through its mid-section (*Photograph* (*b*) *is courtesy of MES-DEA SA*).

40.4.3 Sodium/Nickel-Chloride Battery Technology

Some of the important specific aspects contained within a ZEBRA battery were shown previously in Fig. 40.5*b*. The design of the thermal management system is similar to that which was used with the sodium/sulfur technology. The double walled vacuum insulated box uses a very efficient insulation material with less than 0.006 W/mK thermal conductivity. This material is heat resistant to above 1000°C so that it provides containment even at worst case fault conditions. The thermal management uses air cooling via battery internal cooling plates that separate the air from the cells. For heating, there is an AC heater that is externally powered and a DC heater that is internally powered by the battery. Although both sodium/sulfur and sodium/nickel-chloride batteries are high-temperature systems utilizing cells of comparable voltage and capacity, several battery-level design considerations are different. A few of the noteworthy ones are as follows:

- *Electrical Networking.* During the lifetime of a battery, individual cells may fail. Such an occurrence will result only in the reduction of the open circuit voltage by 2.58 V per cell failure because the failure mode of ZEBRA cells is an internal short (due to the reaction of the secondary liquid electrolyte with sodium forming a solid aluminum shunt). Because of this characteristic, long series chains with 216 cells and 557 V can be built. Intercell connections or voltage taps are not necessary.

- *Battery/Cell Reliability.* Module tests with cells have demonstrated that the rated service life of 2500 electrical cycles can be expected. At the battery level, 1300 electrical cycles using an ECN European vehicle test regime have been made and the calendar life is expected to be greater than 10 years. This expectation is supported by a test battery that, in June 2000, had been in operation for nine years. The end-of-life criteria are when actual capacity or power reaches 80% of rated or when 5% cell failures occur. The battery controller is designed to detect cell failures and automatically adapt operation parameters. The ZEBRA battery is regarded as failure tolerant and, similar to sodium/sulfur, is designed for no maintenance throughout its lifetime.

- *Battery Recycling.* The recycling options for ZEBRA batteries have been comprehensively studied.[17] The result is that very simple process has been developed: the batteries are disassembled and then the cells are cut and put into a standard pyrometallurgical stainless-steel production process. The steel and nickel goes into the metal melt and the salt, aluminum chloride, and ceramic partition into the slag.

As noted earlier, the ZEBRA sodium/nickel-chloride battery technology has been developed almost exclusively for electric-vehicle applications. The first batteries constructed by the ZEBRA development team used both the iron chloride and the nickel chloride Beta/55 cells.[18] A significant problem with these batteries was their very low power capability. Batteries containing an improved "Slimline" cell design were built and extensively tested in actual vehicles during the first part of the 1990's.[19-21] Ten's of thousands of kilometers were logged with these systems in a variety of vehicle types, including the Mercedes 190. Of great significance at the time was the demonstrated service life of these batteries. The development team solved an early problem (observed in the late 1980s) with the durability of the cell seal. Cycle life for full-voltage batteries then exceeded 1000 cycles with zero or only a few cell failures. In addition at that time, three batteries powered vehicles for over a year without any degradation in cell or battery performance.[21]

The latest ZEBRA batteries contain the cells that have incorporated changes to improve dramatically the battery power and energy (refer to Sec. 4.3.3).[4,22,23] The physical and performance specifications for a modern battery design are listed in Table 40.8 and a corresponding picture is shown in Fig. 40.17*b*. The ZEBRA batteries now meet or exceed all of the mid-term EV requirements of the three major U.S. automakers (specified through the USABC).

Relative to safety, ZEBRA batteries must satisfy the requirement noted earlier in the section on "Battery Safety" that the presence of the battery cannot contribute to the hazard of an accident. Some of the specific safety tests that must be passed include the following:[24]

- Crash simulation by dropping a fully charged and operational battery at a 50 km/h impact speed onto a pole (Fig. 40.18*b*)
- 50% overcharge
- Immersion in water without contamination of the water
- Exposure of the entire battery to an external fuel fire for 30 minutes without material release or any combustion
- External short using a metal spike pressed into the battery internal components

Other testing has been used to investigate the response of the batteries to freeze/thaw cycling, overheating, and vibration, and shock. The ZEBRA battery has passed all of these tests and, as reported, "no hazardous effects have been observed and in many cases the battery continued to function."[21] The primary reason for this high level of performance is the intrinsic tolerance of the cells to unintentional abuse. The factors that certainly contribute to this tolerance include: (1) in the event of an electrolyte failure, the sodium reacts with the $NaAlCl_4$ secondary electrolyte to form solid products (aluminum and NaCl) around the site of the failure that restrict any further reaction; (2) sodium does not react with the nickel or the sodium chloride in the positive electrode; (3) the reaction products are relatively non-corrosive to the other metallic components; and (4) all reactants and products have low vapor pressures even at elevated temperatures. For example, at 800°C the vapor pressure for sodium and $NaAlCl_4$ is less than 1 atm. The current health and safety issues in the United States of sodium/metal-chloride batteries have been investigated by NREL.[25]

Finally, some limited attention has been given to applications other than electric vehicles. A number of years ago, development of sodium/nickel-chloride cells for aerospace applications was undertaken and, more recently, the use of this technology for powering submarines was evaluated.[26,27] The aerospace cells are essentially electric-vehicle cells with an optimized positive electrode and wicks for the sodium, and the secondary electrolyte that ensure operation in micro-g space environments.

40.5   APPLICATIONS

As noted in the first section of this chapter, the sodium-beta battery technologies have been developed for use in relatively large-scale energy storage applications (i.e. those requiring 10's to 1000's kWh). The prime attributes that make these technologies attractive candidates for such uses include the high energy density, lack of required maintenance, performance independent of ambient temperature, 100% coulombic efficiency, potential for low cost (relative to other advanced batteries), and operating flexibility compared to existing conventional rechargeable systems. The relevant applications can be grouped into the following two broad categories:

- *Stationary Energy Storage:* A number of storage applications involving electric power generation, distribution, and consumption are emerging that require a battery that remains stationary during use. Examples include some located at generation utilities (e.g., load-leveling, spinning reserve, area regulation), at renewable generation facilities (e.g., solar, wind), at distribution facilities (e.g., line stability, voltage regulation), or at a customer site (e.g., demand peak reduction, power quality). An assessment of the opportunities for battery-based energy storage in these types of applications in the United States was performed in 1995 by Sandia National Laboratories.[28] For reference, this report also includes detailed descriptions of the requirements for each application.

- *Motive Power:* The primary motive-power application is for a true zero-emission electric vehicle. For this potentially high-volume market to develop, the vehicle range probably must exceed 180 km and the vehicle initial cost cannot greatly exceed that for conventionally powered vehicles. As noted earlier, the inability to satisfy this latter factor probably led to the termination of several of the sodium/sulfur development efforts. For a given physical envelope, sodium-beta batteries can provide a significantly greater energy (range) at a reduced weight while meeting the vehicle power requirements compared with conventional battery technologies. Another attractive motive application is in forklifts. Here, the use of more energetic sodium-beta batteries could result in dramatically fewer battery change-outs with associated reduced workforce and capital investment costs. A final significant application class is for aerospace power. The lower weight for a given energy and power requirement coupled with potential cost savings (compared to existing nickel/hydrogen systems) and long life enable new types of satellite applications to become feasible. Currently, interest for this application has switched primarily to lithium-ion battery technology.

From a technical perspective, the main charactristic that limits the optimal use of the sodium-beta technologies relative to a number of other candidate applications (e.g., consumer electronics) involves thermal management. Thermal considerations are important because of their potentially significant impact on the "battery overhead"—size, weight, and overall battery electrical efficiency. Thermal losses are similar, from a system perspective, to self discharge. Thermal-related design issues were discussed in greater detail in the secton above on "Thermal Management." In general, an effective thermal management system must have; (1) a very low-conductivity insulated enclosure to maintain the battery at its operating temperature; (2) heaters to provide make-up heat during long standby periods; and (3) in certain high-power applications, cooling to reject the generated heat. Those applications that are volume or envelope constrained are most impacted. In general, consideration of thermal issues leads to the conclusion that the best theoretical use of sodium-beta technologies from the perspective of energy-efficiency or service life for those applications that minimize need of thermal management in which:

1. Operation (charge-discharge cycling) is relatively frequent. Examples are daily peak shaving or load-leveling.

2. The energy requirement is large (greater than about 5 kWh).

3. The battery is infrequently cooled to ambient temperature, allowing the reactants to freeze (applicable only to sodium/sulfur).

Of prime importance, however, is that these conclusions are primarily technical not economic. The ultimate commercial criterion for any of these applicatons is operating economics. That is, many applications that require large thermal management penalties may indeed be economic from the perspective of life-cycle costs. For example, if the cost of energy to maintain temperature is low and the thermal management system is efficient, the other attributes of sodium-beta may make it a cost-effective choice for some infrequent use standby and/or uninterruptible power-supply applications. Other examples of the cost-effective use of thermal management systems are: (1) the energy requirement limit of several kWh has been shown not to apply for some aerospace applications because size is not a primary design criterion and the high operating temperature is a system benefit because of the ease of heat rejection; and (2) specialized sodium-beta cells can be designed for high-rate discharge applications such as defense pulse power by utilizing ultra-thin sulfur electrodes.

A detailed assessment using the information contained in Ref. 28 was performed for the projected U.S. stationary energy-storage market. This study concluded that the sodium-beta technologies would be a very attractive candidate for use in renewable energy systems, customer-side system peak reduction (shaving), and to defer the need for distribution facilities.[11] Another important finding of this study was that to be economically attractive to U.S. customers, the battery system would probably need the capability to simultaneously satisfy two or more individual applications. In fact, recent changes in the marketplace have created higher value opportunities for protection from short-duration power losses than has traditionally been recognized for UPS markets. Specifically, the broad dependence of the commercial and industrial businesses on computer-based systems has made much of the world economy vulnerable to power disturbances that may last only a few seconds, but can disrupt expensive automated data handling and manufacturing processes. Accordingly, NGK has recently begun investigating the "pulse power" capability of their sodium/sulfur technology. As shown in Fig. 40.19, testing has shown that their T5 cell can deliver 500% of its nominal power rating for periods lasting up to 30 seconds each hour during an 8-hour discharge cycle.[5] These results show promise for satisfying both power-quality and peak-shaving functions with the same installation and thus for being able to penetrate high value power quality markets. Confirmatory battery-level testing is in progress in Japan and related market research is underway in the U.S.

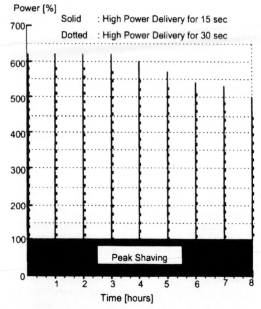

FIGURE 40.19 Capability of NGK sodium/sulfur cells to satisfy requirements of a dual-use application potentially important to emerging U.S. markets: peak shaving and power quality (*Courtesy of Tokyo Electric Power Company and NGK Insulators, Ltd.*).

REFERENCES

1. J. Sudworth and R. Tilley, *The Sodium/Sulfur Battery,* Chapman and Hall, London, 1985.

2. J. Coetzer, "A New High-Energy-Density Battery System," *J. Power Sources,* **18:**377–380, 1986.

3. J. Prakash, L. Redy, P. Nelson, and D. Vissers, "High Temperature Sodium Nickel Chloride Battery for Electric Vehicles," *Electrochemical Society Proceedings,* **96:**14, (1996).

4. R. Galloway and S. Haslam, "The ZEBRA Electric Vehicle Battery: Power and Energy Improvements," *J. Power Sources,* **80:**164–170, (1999).

5. T. Oshima and H. Abe, "Development of Compact Sodium Sulfur Batteries," *Proc. of the 6th Int. Conf. on Batteries for Utility Energy Storage,* Wissenschaftspark Gelsenkirchen Energiepark Herne, Germany, Sept. 1999.

6. A. Kita, "An Overview of Research and Development of a Sodium Sulfur Battery," *Proc. DOE/ EPRI Beta Battery Workshop VI,* pp. 3-23–3-27, May 1985.

7. K. Takashima et al., "The Sodium Sulfur Battery for a 1 MW/8 MWh Load Leveling System," *Proc. Int. Conf. on Batteries for Utility Energy Storage,* Mar. 1991, pp. 333–349.

8. E. Nomura et al., "Final Report on the Development and Operation of a 1 MW/8 MWh Na/S Battery Energy Storage Plant," *Proc. 27th IECEC Conf.* **3:**3.63–3.69, 1992.

9. R. Okuyama and E. Nomura, "Relationship Between the Total Energy Efficiency of a Sodium-Sulfur Battery System and the Heat Dissipation of the Battery Case," *J. Power Sources,* **77:**164–169, (1999).

10. A. Araki and H. Suzuki, "Leveling of Power Fluctuations of Wind Power Generation Using Sodium Sulfur Battery," *Proc. 6th Int. Conf. on Batteries for Utility Energy Storage,* Wissenschaftspark Gelsenkirchen Energiepark Herne, Germany, Sept. 1999.

11. A. Koenig, "Sodium/Sulfur Battery Engineering for Stationary Energy Storage, Final Report," Sandia National Laboratories *SAND Rep. 96-1062,* Albuquerque, NM, April 1996.

12. T. Hartkopf and H. Birnbreir, "The New Generation of ABB's High Energy Batteries," *11th Int. Vehicle Symp. (EVS-11)*, paper 14.04, Florence, Italy, Sept. 1992.

13. E. Kodama and K. Nakahata, "Advanced NaS Battery System for Load Leveling," *Proc. 6th Int. Conf. on Batteries for Utility Energy Storage,* Wissenschaftspark Gelsenkirchen Energiepark Herne, Germany, Sept. 1999.

14. K. Tanaka et al., "Safety Test Results of Compact Sodium-Sulfur Batteries," *Proc. 6th Int. Conf. on Batteries for Utility Energy Storage,* Wissenschaftspark Gelsenkirchen Energiepark Herne, Germany, Sept. 1999.

15. W. Dorrscheidt et al., "Safety of Beta Batteries," *Proc. DOE/EPRI Beta Battery Workshop VII*, pp. 46-1–46-7, June 1990.

16. J. Garner et al., "Sodium Sulfur Battery Cell Space Flight Experiment," *IECEC Paper No. AP-339,* pp. 131–136, ASME 1995.

17. H. Böhm, "Recycling of ZEBRA Batteries," EVS-14, Orlando, Fla., 1997.

18. A. Tilley and R. Bull, "The Design and Performance of Various Types of Sodium/Metal Chloride Batteries," *Proc. 22nd IECEC Conf.,* **2:**1078–1084 (1987).

19. D. Sahm and J. Sudworth, "Track and Road Testing of a Zebra Battery Powered Car," *Proc. DOE /EPRI Beta Battery Workshop VIII*, EPRI Rep. GS-7163, 31-1–31-19, 1991.

20. R. Bull and W. Bugden, "Reliability Testing of Zebra Cells," *Proceedings of the DOE/EPRI Beta Battery Workshop VIII*, EPRI Rep. GS-7163, 33-1–33-22, 1991.

21. C.-H. Dustmann and J. L. Sudworth, "Zebra Powers Electric Vehicles," *11th Int. Vehicle Symp. (EVS-11)*, Paper 14.05, Florence, Italy, Sept. 1992.

22. A. van Zyl, "Review of the Zebra Battery System Development," *Solid State Ionics,* **86–88:**883–889, 1996.

23. C.-H. Dustmann, "ZEBRA Battery Meets USABC Goals," *J. Power Sources,* **72:**26–31 (1998).

24. J. Gaub and A. V. Zyl, "Mercedes-Benz Electric Vehicles with ZEBRA Batteries," EVS-14, Orlando, Fla. 1997.

25. D. Trickett, "Current Status of Health and Safety Issues of Sodium/Metal Chloride (ZEBRA) Batteries," NREL/TP-460-25553, National Renewable Energy Laboratory, Golden, Col., Nov. 1998.

26. R. Bones, et al., "Development of Na/NiCl$_2$ (Zebra) Cells for Space Applications," *Extended Abstracts,* Electrochemical Society Meeting, Toronto, Ont., Canada, 1992, Abs. 69, p. 107.

27. E. Kluiters et al., "Testing of a Sodium/Nickel Chloride (Zebra) Battery for Electric Propulsion of Ships and Vehicles," *J. Power Sources,* **80:**261–264, 1999.

28. P. Butler, "Battery Energy Storage for Utility Applications: Phase I—Opportunities Analysis," Sandia National Laboratories *SAND Rep. 94-2605,* Albuquerque, NM, Nov. 1995.

CHAPTER 41
LITHIUM/IRON SULFIDE BATTERIES*

Gary L. Henriksen and Andrew N. Jansen

41.1 GENERAL CHARACTERISTICS

Investigations of electrochemical cells having alkali-metal electrodes and molten-salt electrolytes began in 1961 at Argonne National Laboratory (ANL) in an effort to develop thermally regenerative galvanic cells for the direct conversion of heat to electricity.[1] This work led to the invention of the lithium/sulfur cell in 1968, in which elemental lithium and sulfur were used as the active materials in the electrodes with moltent LiCl-KCl electrolyte.[2] The melting point of the LiCl-KCl eutectic electrolyte is 352°C, which requires that the cell operating temperature be maintained above 400°C. Lithium and sulfur are attractive materials for a high-performance battery because they possess low equivalent weights, develop a high voltage (approximately 2.3 V), and are capable of high current densities (>1 A/cm²). Attempts to develop the lithium/sulfur cell were abandoned in 1973 because of problems in containing the active materials of both electrodes.[3] These problems were overcome by substituting alloys of lithium for elemental lithium in the negative electrode and metal sulfides for elemental sulfur in the positive electrode.[4]

Lithium alloy/metal sulfide batteries employ a molten-salt electrolyte and solid porous electrodes. Depending on electrolyte composition, they operate over a temperature range of 375 to 500°C. Operation at these temperatures with molten-salt electrolytes achieves high power densities, due to the high electrolyte conductivities and fast electrode kinetics. A shift from prismatic battery designs to bipolar designs enhances the power characteristics further by reducing the battery impedance.

Alloys of lithium with aluminum or silicon are typically used as reversible solid negative electrodes. These alloys reduce the lithium activity—below that of lithium metal—to a controllable level, yielding high coulombic efficiencies. Development of a two-component lithium alloy ($\alpha + \beta$ Li-Al and $Li_5Al_5Fe_2$) negative electrode material provides an in situ overcharge tolerance capability by increasing the lithium activity in the negative electrode at the

---

*This chapter has been authored by a contractor of the U.S. Government under contract No. W-31-109-ENG-38. Accordingly, the U.S. Government retains a nonexclusive, royalty-free license to publish or reproduce the published form of this contribution, or allow others to do so, for U.S. Government purposes.

end of charge. This initiates a high-rate self-discharge reaction that allows continued passage of charging current—at or below the self-discharge rate—without further charge acceptance.[5] Development of this in situ overcharge tolerance capability has rendered the bipolar design a viable option for lithium/iron sulfide batteries.

Several metal sulfides—iron, nickel, cobalt, and so on—can be used as positive electrodes. These metal sulfides relieve the vapor pressure and corrosion issues associated with the use of sulfur. Cost considerations lead to the selection of FeS or FeS_2 for commercial applications, while other metal sulfides remain viable options for specialty battery applications where cost is less important. Combining the dense FeS_2 positive electrode, operating only on its upper voltage plateau, with the low-melting LiCl-LiBr-KBr electrolyte has achieved a stable, reversible, and high-performance lithium/iron disulfide cell technology.

Refinements in the composition of this low-melting electrolyte and development of stable chalcogenide ceramic/sealants have led to the development of sealed electrolyte-starved bipolar Li/FeS and Li/FeS_2 cells and stacks.[6] These cells and stacks exhibit high power and energy densities. As a result of the advances that have been made in high-performance bipolar cell and stack technology, it has replaced the prismatic design. In addition to its higher performance, the bipolar design is likely to be more cost-effective than the prismatic design, because it significantly reduces the quantity of nonactive materials used in the battery stack and thermal enclosure. The major advantages and disadvantages of bipolar lithium/iron sulfide batteries are summarized in Table 41.1. In addition to their high power and energy densities, they are tolerant to most types of abuse encountered in electric-vehicle applications. Due to their low internal impedances and favorable electrode kinetics, both ampere-hour and watt-hour capacities of these bipolar batteries are relatively independent of load. Also, they possess characteristics that render them inherently safe and reliable,[7] whereas many other technologies have to engineer safety and reliability into their systems. Because it is a high-temperature battery, it is housed within a thermal enclosure and, therefore, is independent of environmental conditions. Its major disadvantages are those associated with high-temperature batteries—mainly the need for a thermal management system to maintain its temperature within acceptable limits and the need to consume stored energy to keep the battery hot during extended stand periods in locations where a source of external electric power is not available.

TABLE 41.1 Major Advantages and Disadvantages of Bipolar Lithium/Iron Sulfide Cells and Batteries

Advantages	Disadvantages
Combines high power and energy densities	Requires thermal management system to maintain operating temperature within acceptable window
Tolerant to overcharge, overdischarge, and freeze-thaw abuse	Consumption of stored energy may be required to maintain temperature, during extended stand periods
Capacity independent of load	
Inherently safe and reliable	
Independent of environmental conditions	

The main interest in high temperature batteries such as lithium/iron sulfide, sodium/sulfur, and sodium/nickel chloride is for electric vehicle applications due to their high specific power and energy possibilities. The replacement of the liquid lithium electrode with a solid LiAl alloy alleviated many of the safety concerns that plagued the other two systems, which are based on a liquid sodium electrode. In 1991, the United States Advanced Battery Consortium (USABC) selected the bipolar molten-salt $LiAl/FeS_2$ battery to be developed as a long-term battery technology for electric vehicles. Although significant advances were made in this technology[8] at the end of 1995, it was decided to discontinue R and D efforts in this technology in favor of the rapidly developing lithium-ion and lithium-polymer battery technologies. These newer technologies were also considered to be long-term developments by the USABC with power and energy capabilities similar to that of the LiAl/FeS$_2$ battery but which operate at much lower temperatures. Should interest revive in stationary energy storage (SES) batteries, lithium/iron sulfide batteries remain viable candidates for this application. The primary version of the lithium/iron sulfide battery continues to find wide usage as thermal batteries (see Chap. 21).

Development of the rechargeable lithium/iron sulfide battery is still continuing but no longer with molten salt electrolytes. The most notable work[9,10] involves the replacement of the molten salt electrolyte and magnesia separator with a composite polymer electrolyte that operates at temperatures of 90 to 135°C. There are many advantages to operating this system at this reduced temperature range. This lower temperature permits the use of lithium foil instead of cold-pressed pellets of lithium-aluminum alloy. The positive electrode also does not need to be a cold pressed pellet; it can be cast from slurry as is common for lithium-polymer and lithium-ion technologies. Steel and polymer seals can be used for the cell hardware instead of the molybdenum and ceramic seals required for high-temperature molten-salt batteries. This new direction for the lithium/iron sulfide battery looks promising and will hopefully result in a commercial battery. Although there are many similarities between the two technologies, the majority of the information presented in this chapter is based on the molten salt battery technology that was under development for nearly three decades.

41.2 DESCRIPTION OF ELECTROCHEMICAL SYSTEM

Secondary Li/Fe$_x$ cells that were developed for electric-vehicle applications at ANL, incorporate cold-pressed FeS or dense FeS$_2$ positive-electrode pellets, two-component Li-Al/Li$_5$Al$_5$Fe$_2$ negative-electrode pellets, and LiCl-rich LiCl-LiBr-KBr/MgO electrolyte/separator pellets. Electrolyte of the same composition is incorporated in the positive- and negative-electrode pellets. The LiCl-rich electrolyte (34 mol % LiCl–32.5 mol % LiBr–33.5 mol % KBr) possesses a low melting point, broad liquidus region, and high conductivity.[5] The high conductivity yields low area-specific impedance (ASI) from low-burdened electrolyte-starved cells. Li/FeS$_x$ cells employing this electrolyte can operate in the 400 to 425°C temperature range, compared with operation in the 450 to 475°C range for cells employing LiCl-KCl electrolyte. Use of this electrolyte with a dense FeS$_2$ positive electrode, which operates only on the upper voltage plateau (U.P.), has extended the cycle life for the Li/FeS$_2$ technology to more than 1000 cycles in flooded cells.[11] Comparable cycle life was demonstrated previously for the Li/FeS chemistry operating with LiCl-KCl electrolyte.[12]

The overall electrochemical reactions for the Li/FeS and U.P. Li/FeS$_2$ cells are

$$2\text{Li-Al} + \text{FeS} \underset{\text{charge}}{\overset{\text{discharge}}{\rightleftharpoons}} \text{Li}_2\text{S} + \text{Fe} + 2\text{Al}$$

$$2\text{Li-Al} + \text{FeS}_2 \underset{\text{charge}}{\overset{\text{discharge}}{\rightleftharpoons}} \text{Li}_2\text{FeS}_2 + 2\text{Al}$$

The theoretical specific energies for these two reactions are approximately 460 and 490 Wh/kg, respectively. The corresponding voltages for these reactions are approximately 1.33 and 1.73 V, respectively. The reactions are more complex than shown and involve the formation of intermediate compounds.[9,10,13–17]

41.3   *CONSTRUCTION*

The original hardware development programs used prismatic cell designs (Fig. 41.1) similar to those employed in most automotive SLI lead-acid batteries. Several flat-plate positive and negative electrodes are positioned vertically and separated by porous separator sheets to form a multiplate cell with the desired combination of ampere-hour capacity (for energy) and electrode surface area (for power). Historically BN (Boron Nitride) cloth or felt has been used as the separator in flooded-electrolyte cells, while MgO pressed-powder plaques have been used in starved-electrolyte cells. Perforated metal sheets are positioned between separators and electrodes to restrain the movement of particulate matter from the electrodes into the separator. Photoetching is used to perforate the metal sheets. Steel is used as the current collector for FeS positive electrodes, while molybdenum is the most commonly used current collector for FeS_2 positive electrodes. In these prismatic cells the negative electrodes are typically grounded to the cell case, while the positive electrodes are connected to a feedthrough terminal that is electrically insulated from the cell case. In many designs "picture-frame" structures are used to locate and hold together electrode components as a unit.

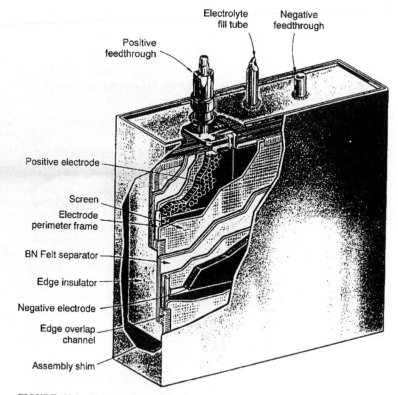

FIGURE 41.1 Cutaway view of flooded prismatic Li-Al/FeS cell showing arrangement of electrodes, separators, and internal current-collector system. (*Courtesy of Argonne National Laboratory.*)

Lithium/iron sulfide prismatic cells can be assembled in a charged, uncharged, or partially charged state. When assembled in the charged state, the negative electrodes are cold-pressed using a mixture of the two-component alloy, $(\alpha + \beta)$ Li-Al and Al_5Fe_2, and electrolyte powders. The positive electrodes are assembled in a similar manner, using a mixture of FeS_x and electrolyte powders. When assembled in the uncharged state, the FeS positive-electrode plaque is pressed using an appropriate mixture of Li_2S, Fe, and electrolyte powders, while the FeS_2 positive-electrode plaque is formed using a mixture of $LiFeS_2$ and electrolyte powders. The negative-electrode plaque is pressed using an appropriate mixture of α-Al, Al_5Fe_2, and electrolyte powders. Partially charged cells can be assembled using appropriate mixtures of the charged and uncharged starting materials.

As a result of several technical advances in the late 1980s and early 1990s (Table 41.2) the design of choice today is the bipolar stack (Fig. 41.2).[6,15] Cell stacks are formed by placing positive-electrode, electrolyte/separator, and negative-electrode pressed-powder plaques into a prefired seal-ring assembly and performing a final weld on the metal bipolar plate (current collector/cell wall) of one cell to a metal ring buried in the seal-ring assembly of an adjacent cell. In a manner similar to the prismatic design, particle retainer screens can be positioned between the electrode and the separator/electrolyte plaques. In an effort to minimize weight and volume, bipolar cells and stacks employ MgO separators in an electrolyte-starved configuration. Also, cells employing overcharge-tolerant two-component lithium alloy negative electrodes require the use of separators made of MgO, or other electrically insulating materials that are stable to the higher lithium activity associated with the overcharge alloy. As described for the prismatic cells, bipolar cells can be assembled in the charged, uncharged, or partially charged state.

The key to designing and constructing practical bipolar stacks is the development of a suitable seal material for making metal-to-ceramic peripheral seals. This was accomplished in 1990 with the development of chemically stable chalcogenide sealants that bond tenaciously to both metals and ceramics.[16] These sealants can be specially formulated to accommodate differences in thermal expansion between metals and ceramics by forming graded seals. They exhibit more than 95% coverage and wetting angles approaching 0° for both steel and molybdenum. Compared to commercially available high-temperature bonding agents, the standard chalcogenide sealant materials exhibit bond strengths which are an order of magnitude greater.[17,18]

TABLE 41.2   Major Technical Advances in Development of Bipolar Li-Al/FeS_x Cells

Time frame, year	Major technical advance	Practical implication
1986	Low-temperature electrolyte and upper-plateau dense FeS_2 cathode	Achieves > 1000cycles
1988	Electrochemical overcharge tolerance	Makes bipolar design variable
1989	Lithium-rich electrolyte in starved cell	Enhances performance
1990	Chalcogenide seal material	Makes bipolar design practical

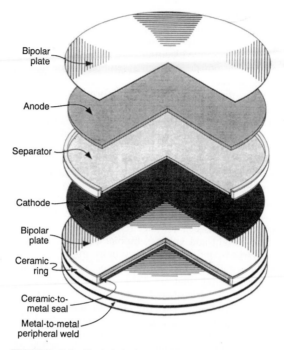

Bipolar plate

Anode

Separator

Cathode

Bipolar plate

Ceramic ring

Ceramic-to-metal seal

Metal-to-metal peripheral weld

FIGURE 41.2 Exploded view of four-cell bipolar Li-Al/ FeS$_x$ stack under development for electric-vehicle applications shown electrodes, separator, and bipolar plate current collector. (*Courtesy of Argonne National Laboratory.*)

41.4 *PERFORMANCE CHARACTERISTICS*

41.4.1 Voltage

The open-circuit voltage of the Li-Al/LiCl-LiBr-KBr/FeS cell is 1.33 V at 425°C, while that of the Li-Al/LiCl-LiBr-KBr/FeS$_2$ (U.P.) cell is 1.73 V.[6,15] The operating voltages of these cells are in the ranges of 0.9 to 1.3 V for the Li-Al/FeS cell and 1.2 to 1.7 V for the Li-Al/FeS$_2$ U.P. cell. Charge voltage cutoffs are approximately 1.6 and 2.0 V, respectively. Charging at these voltages can be conducted for extended durations due to the overcharge tolerance capability provided by the two-component lithium alloy negative electrode.[17]

41.4.2 Discharge Characteristics

A typical set of cell voltage versus delivered ampere-hour capacity discharge curves for 13-cm-diameter bipolar U.P. Li-Al/FeS$_2$ and Li-Al/FeS cells is given in Fig. 41.3.[17] The step in both discharge curves, occurring at 5 to 7 Ah into the discharge, is associated with the transition from the overcharge-tolerant negative-electrode alloy phase to the normal $\alpha + \beta$ Li-Al alloy negative-electrode phase. Both types of cells exhibit good voltage stability throughout the discharge.

The effect of discharge rate on delivered capacity for both types of cells is illustrated in Figs. 41.4 and 41.5 for two 3-cm-diameter cells. The reduction in capacity associated with increasing the discharge rate is less for these technologies than it is for most other battery technologies.

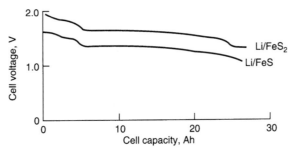

FIGURE 41.3 Voltage vs. delivered capacity plots for 13-cm-diameter bipolar Li-Al/FeS$_x$ cells at 425°C. (*From Kaun et al.*)[17]

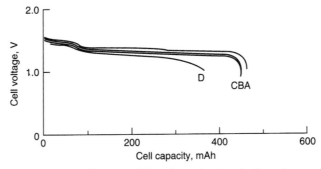

FIGURE 41.4 Voltage vs. delivered capacity plots for 3-cm-diameter bipolar Li-Al/FeS cell for different discharge rates at 425°C. Curve *A*—25 mA/cm^2; curve *B*—70 mA/cm^2; curve *C*—100 mA/cm^2; curve *D*—200 mA/cm^2. (*From Kaun et al.*)[17]

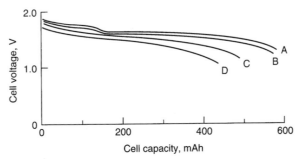

FIGURE 41.5 Voltage vs. delivered capacity plots for a 3-cm-diameter bipolar Li-Al/FeS$_2$ cell for different discharge rates at 425°C. Curve *A*—50 mA/cm^2; curve *B*—100 mA/cm^2, curve *C*—150 mA/cm^2; curve *D*—200 mA/cm^2. (*From Kaun et al.*)[17]

A very attractive characteristic of bipolar Li-Al/FeS$_x$ cells is their low impedance, which leads to high power capabilities. Figures 41.6 and 41.7 provide cell impedance data as a function of the depth of discharge for bipolar Li-Al/FeS and Li-Al/FeS$_2$ cells, respectively.[17] The area-specific impedance (ASI) of Li-Al/FeS cells remains low through 70% depth of discharge and then increases through the remainder of the discharge. The ASI of Li-Al/FeS$_2$ cells remains low throughout the discharge, indicating its ability to deliver high power even at 90% depth of discharge. A comparison of the 25-ms data to the 1-s data reveals that the bulk of the impedance is due to an electronic component rather than an ionic component. This is a direct result of the high conductivity of the electrolyte,[5] 1.7 S/cm at 425°C, which is nearly a thousand times greater than that of the lithium-ion organic electrolyte.

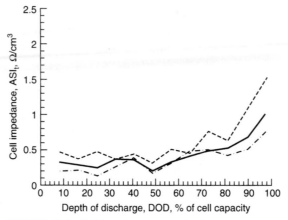

FIGURE 41.6   Cell impedance vs. depth of discharge for 13-cm-diameter bipolar Li-Al/FeS cell at 425°C. Relaxation time t: - - -, 15 s; ——, 1 s; — · —, 25 ms. (*From Kaun et al.*)[17]

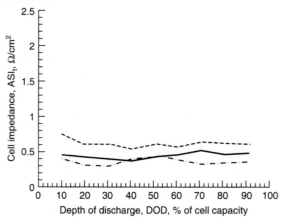

FIGURE 41.7   Cell impedance vs. depth of discharge for 13-cm-diameter bipolar Li-AL/FeS$_2$ cell at 425°C. Relaxation time t: - - -, 15 s; ——, 1 s; — · —, 25 ms. (*From Kaun et al.*)[17]

41.4.3   Effect of Temperature

One of the advantages of switching from the LiCl-KCl eutectic electrolyte to the LiCl-rich LiCl-LiBr-KBr electrolyte is to lower the acceptable operating temperature to 400°C. The LiCl-LiBr-KBr electrolyte has a melting point of about 320°C and a broad liquidus region to allow large variations in the Li$^+$/K$^+$ ratio at 400°C. The LiCl-KCl electrolyte has a melting point of 354°C and requires a cell operating temperature of about 450°C. Lower operating temperatures help extend the calendar and cycle life of lithium/iron sulfide cells. Ambient conditions have little influence on battery performance because a thermal management system is required for this technology.

41.4.4 Self-Discharge

The self-discharge rate of lithium/iron sulfide cells is controlled by the lithium activity of the Li-Al anode and the rate of transport of dissolved lithium to the cathode. Typical self-discharge rates for starved-electrolyte cells with 2-mm-thick MgO separators at 425°C are in the range of 0.1 to 0.2 mA/cm². As described in Sec. 41.4.8, this self-discharge undergoes a stepwise 20-fold increase as the cell enters into the overcharge-tolerant state. This allows fully charged cells to endure extended trickle charge at 2 to 5 mA/cm² without adding capacity to the cell.

41.4.5 Power and Energy Characteristics

The starved-electrolyte bipolar configuration leads to high-performance low-burdened Li-Al/FeS and U.P. Li-Al/FeS$_2$ cells. The specific energy and specific power characteristics of 13-cm-diameter sealed bipolar cells are given in Table 41.3. These performance levels are higher than those achieved in smaller (3-cm-diameter) bipolar cells, indicating an ability to scale up this technology without sacrificing performance. No additional battery hardware weight was included in these performance values.

TABLE 41.3 Specific Energy and Power of 13-cm-Diameter Bipolar Cells

Cell technology	Specific energy, Wh/kg at W/kg	Specific power at 80% DOD, W/kg
Li-Al/FeS	130 at 25	240
Li-Al/FeS$_2$	180 at 30	400

41.4.6 Cycle Life

The end of life, as typically defined for electric-vehicle batteries, is a 20% loss of capacity on a standard simulated vehicle driving profile. A commonly used profile is the simplified federal urban driving schedule (SFUDS). Bipolar 13-cm diameter Li-Al/FeS$_2$ cells obtained over 300 cycles on a modified version of the SFUDS, denoted the Dynamic Stress Test (DST). Both the FeS and U.P. FeS$_2$ chemistries have demonstrated the ability to achieve more than 1000 cycles in flooded-electrolyte prismatic cells when discharged at constant current.[11,12] As is true for other high-temperature batteries, cycle life is not likely to be strongly influenced by cycle type.

41.4.7 Efficiency

The coulombic efficiency of Li-Al/FeS$_x$ cells is controlled by the lithium activity of the negative electrode and the rate at which dissolved lithium can diffuse across the separator to the positive electrode. Typically this rate is only 0.1 to 0.2 mA/cm² at 425°C. This low self-discharge rate leads to high coulombic efficiency. Similarly, the low impedance of bipolar cells (0.5 to 0.7 $\Omega \cdot$ cm²) leads to high voltaic efficiency. Overall the major source of inefficiency is the heat loss associated with high-temperature operation. Development of a highly efficient thermal enclosure is necessary for all high-temperature batteries.

41.4.8 Charging

Development of overcharge-tolerant cells in 1987 to 1988 has significantly minimized previous charging concerns associated with cell balancing in series-connected strings of cells, and concerns with regard to preventing the formation of liquid lithium on the negative electrode during overcharge. A "smart" charger, capable of monitoring individual cells of a battery and electrically bypassing fully charged cells, was developed and demonstrated on a 36-V prismatic Li-Al/FeS battery.[19] Although technically viable, this approach adds complexity and cost to the battery and the charger. This type of charger is no longer needed because present cells employ the two-component lithium alloy negative electrodes. When fully charged, present cells transition to this overcharge-tolerant state, where the lithium activity of the negative electrode is increased to a level that induces a 20-fold increase in the self-discharge rate.[5,17] This allows cells to be charged at 2 to 5 mA/cm² without accepting any additional ampere-hour capacity. Also these cells can be charged at higher rates for short durations, accepting additional ampere-hour capacity without having liquid lithium form on the negative electrode. Figure 41.8 shows a voltage versus time trace for a four-cell stack over three discharge-charge cycles. The second charge half-cycle incorporates an extended (3-h) trickle charge.

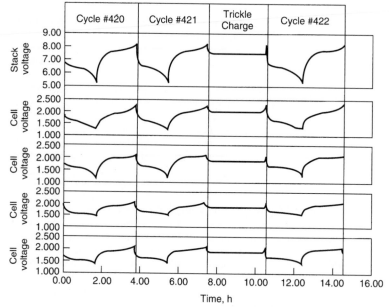

FIGURE 41.8   Stack and cell voltage vs. time plots for four-cell 13-cm-diameter bipolar Li-Al/FeS$_2$ stack over three cycles, with 3-h trickle charge at end of second charge half-cycle. (*From Kaun et al.*[17])

41.4.9 Influence of Additives

Additives to the positive electrode can be classified as either inert or active.[8] Inert additives investigated that were found to be beneficial include graphite fibers and magnesium oxide powder. Magnesium oxide powder in the positive electrode matrix had the effect of acting like an electrolyte sponge. It was observed that in cells without the MgO additive, the positive electrode became electrolyte starved upon repeated cycling. Graphite fibers provided an electronically conductive matrix within the positive electrode but did not appear to retain as much electrolyte as the MgO additive.

Active additives CoS_2 and chalcopyrite ($CuFeS_2$) reduced the cell area-specific-impedance (ASI) significantly while providing nearly the same capacity as the FeS_2 removed in its place. Chalcopyrite is favored over CoS_2 based on its abundance in nature and, hence, low cost. ASI as a function of depth of discharge (DOD) is presented in Figs 41.9 to 41.11 for cells with positive electrodes that consist of FeS_2 only, 12% CoS_2 additive, and 18% $CuFeS_2$ additive, respectively. Concern existed over the deposition of Cu in the separator from the chalcopyrite. To test the impact of this, a 3-cm diameter bipolar cell with 25 mol % $CuFeS_2$ in the positive electrode was cycled for over 1000 cycles at a $C/1$ charge and discharge current rate. The coulombic efficiency remained greater than 99% throughout the life of the cell. A post-test analysis showed small particles of metallic Cu in the separator, but were too dispersed to create a short circuit.[8]

An optimum concentration of CoS_2 additive in the FeS_2 positive electrode was determined to be approximately 12 mol %. The optimum concentration of $CuFeS_2$ additive was found to be around 12 to 18 mol %. The effect of CoS_2 on the ASI is shown in Figure 41.12 over the range of 0 to 100% with a salt concentration of 33-wt. % in the cold pressed positive pellet. Two other data points with different salt concentrations were added to this Figure, one at 28-wt. % salt in 12:88 molar ratio of CoS_2: FeS2, and the other at 43-wt. % salt in 100-mol % CoS_2. The more than doubling of the ASI at 12-mol % CoS_2 additive resulting from the reduction of the salt concentration by only 5% overwhelms the relatively small improvement from the CoS_2 additive. Similar comments can be made for the 43-wt. % data point at 100-mol % CoS_2.

The addition of CoS_2 required an increase in the concentration of salt in the cold pressed pellet to prevent the electrode from being too electrolyte starved. This was due to the much finer particle size and morphology of the synthetic CoS_2 relative to the ground-up naturally occurring pyrite. These points stress the importance of establishing an optimum salt concentration when operating near starved-electrolyte conditions. Too much electrolyte and the electrode becomes too fluid (difficult to contain) and the energy density is reduced needlessly. Not enough electrolyte in the positive electrode results in excessive ASI and poor utilization of the active materials. Salt concentrations between 28 to 32-wt. % in the positive electrode were found to be the most practical.

Improvements were made to the LiCl-rich LiCl:LiBr:KBr electrolyte by addition of small amounts of LiI. The addition of LiF also provided an increase in rate capability of the electrolyte, but not as much as the LiI.[5] The combination of increasing the electrolyte concentration from 28 to 32-wt.% and replacing the CoS_2 with $CuFeS_2$ had a marked improvement in power capability for this cell technology, as can be seen from an inspection of Fig. 41.13, which is a plot of the specific energy that results from constant power level discharges. LiI was added to each of these full-size 13-cm bipolar cells. The specific energy and power levels shown in this figure are based on the total chemistry weight only; no hardware was added to the weight basis. It was estimated that the hardware would constitute approximately 30-wt. % of the total battery weight. Thus, 160 Wh/kg on a chemistry weight basis would correspond to approximately 112 Wh/kg on the total battery weight basis.*

---

*"Total Chemistry Weight" is defined as the total weight of all chemicals in the cell components (negative, separator, and positive), which includes active and inactive materials. For this system, it is the sum of the LiAl, MgO, Al_5Fe_2, LiCl, LiBr, KBr, LiF, LiI, FeS_2, CoS_2, and $CuFeS_2$ throughout the whole cell. The weight of any hardware or the weight of the thermal management system are not included mostly because these were not fully developed but were expected to be approximately 30 percent of the total battery weight. Hence, 160 Wh/kg on a chemistry weight basis would correspond to approximately 112 Wh/kg on a total battery weight basis as noted in Section 41.4.9. This 30 percent estimate is based on a large multicell battery, such as in an all electric vehicle. The hardware and thermal management weight would be a much higher percentage of the battery weight for a battery with only a few cells.

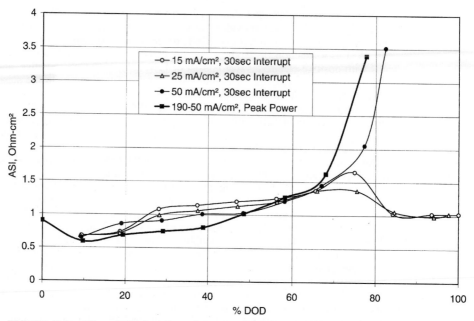

FIGURE 41.9   ASI vs. DOD for a 3-cm diameter cell with FeS_2 positive electrode. ASI determined from 30-second current interrupts at various discharge current rates and also from USABC peak power test. (*From Henriksen, et al.*[8])

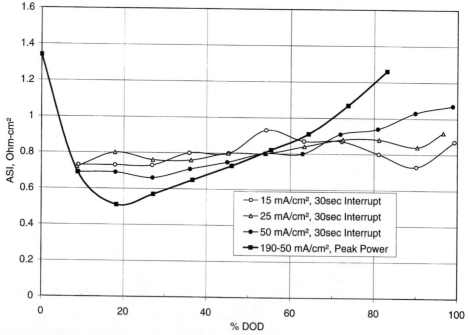

FIGURE 41.10   ASI vs. DOD for a 3-cm diameter cell with 12-mol % CoS_2 additive to the FeS_2 positive electrode. ASI determined from 30-second current interrupts at various discharge current rates and also from USABC peak power test. (*From Henriksen, et al.*[8])

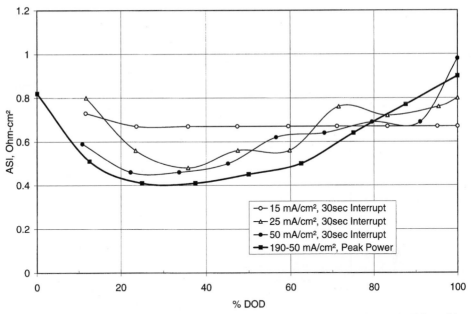

FIGURE 41.11   ASI vs. DOD for a 3-cm diameter cell with 18-mol % $CuFeS_2$ additive to the FeS_2 positive electrode. ASI determined from 30-second current interrupts at various discharge current rates and also from USABC peak power test. (*From Henriksen, et al.[8]*)

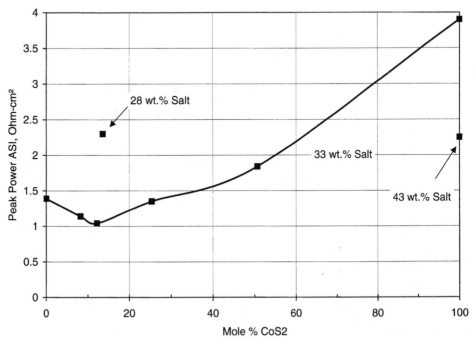

FIGURE 41.12   Peak power ASI evaluated at 80% DOD as a function of mole % CoS_2 additive. (*From Henriksen, et al.[8]*)

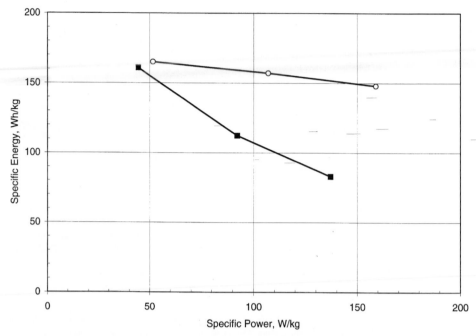

FIGURE 41.13   Specific energy vs constant power discharge level for 13-cm diameter bipolar LiAl/FeS$_2$ with additive CoS$_2$ and 28-wt.% salt in positive electrode —■—, and with additive CuFeS$_2$ and 32-wt.% salt in positive electrode —○—. Each cell was incorporated with 5-wt.% of LiI (based on total chemistry weight of cell) (*From Henriksen, et al.*[8]).

41.4.10   Chalcogenide Bipolar Seal

A key element of the bipolar battery design is an electrically insulating peripheral seal that is chemically resistant to attack by the electrode and electrolyte materials. The seal for a molten-salt bipolar battery must also survive high temperatures and thermal cycling. These added conditions necessitate the need of a ceramic-to-metal seal. Two seal designs were developed at ANL;[8] the baseline seal and the advanced low-cost seal depicted in Fig. 41.14. The baseline seal consists of a Mo-ring/ceramic-ring/Mo-ring construction with an alumina locator ring that is used to position the ring components while the "green" seal is fired in a high temperature furnace. Molybdenum bipolar plates are welded to the outside perimeter of the baseline seal to contain the electrode pellets. The advanced low-cost design replaces

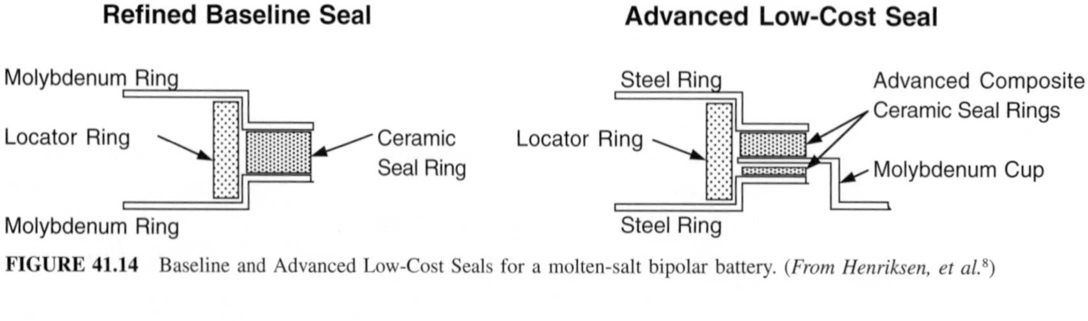

Refined Baseline Seal   **Advanced Low-Cost Seal**

Molybdenum Ring   Steel Ring   Advanced Composite
                                                                 Ceramic Seal Rings
Locator Ring   Ceramic   Locator Ring
                            Seal Ring   Molybdenum Cup
Molybdenum Ring   Steel Ring

FIGURE 41.14   Baseline and Advanced Low-Cost Seals for a molten-salt bipolar battery. (*From Henriksen, et al.*[8])

the expensive Mo rings with steel rings and the Mo bipolar plate with a Mo cup to hold the positive electrode and an optional steel bipolar plate to contain the negative electrode. The replacement of the Mo rings with steel rings had the additional benefit of welding a steel-steel peripheral seam instead of a Mo-Mo peripheral seam. Edge welding 125 micron thick Mo rings together proved to be difficult; the resulting weld was often brittle and rarely pinhole free.

The ceramic rings used to seal the metal rings together are composed of a mixture of metal sulfides and ceramic filler. By adjusting the type and ratio of metal sulfides to ceramic filler, the coefficient of thermal expansion of the ceramic seal is tailored to match that of the metal rings—a necessity if the peripheral seal is to survive thermal cycling without cracking. Suitable metal sulfides used include $CaAl_2S_4$, $Li_2CaAl_2S_5$, $YAlS_3$, $Ca_2Al_2SiS_7$, $LiAlS_2$, and CaY_2S_4. Ceramic fillers include MgO, CaO, Al_2O_3, AlN, and BN.[18] Combinations of these materials were found to be relatively stable to lithium, mechanically strong, and good electrical insulators. AlN-based ceramic seals were the most appropriate for the baseline seal design, while a blend of ceramic fillers was required for the Mo-steel ceramic "glue" used in the advanced low-cost seal design.

41.5 APPLICATIONS AND BATTERY DESIGNS

The predominant interest in bipolar molten-salt lithium/iron sulfide batteries has been for use in electric vehicles and to a lesser extent, pulse-power applications. This technology can be readily adapted to various applications that cover a wide range of power-to-energy ratios. An electric vehicle requires a power-to-energy ratio of approximately 2:1, while a pulse-power source requires a much higher ratio. A hybrid electric vehicle requires an intermediate ratio of at least 10:1. These demands could be met by merely varying the thickness of the electrodes. As is true for other high-temperature batteries, this technology is limited in terms of its ability to downsize to very small battery sizes. However, this technology may be capable of downsizing to a greater extent than other high-temperature batteries because of its shape and compactness.

41.5.1 Pulse-Power Battery Designs

Westinghouse Electric Corporation investigated the use of a bipolar lithium-alloy/metal disulfide battery to be used by the U.S. Army for pulse-power applications.[20] Tape casting methods were used to fabricate the electrodes and separators, rather than the cold-pressed powder plaque methods described in Sec. 41.3, which required presses with >500 tons of force capability. A high melting-point (445°C) high-conductivity LiF-LiCl-LiBr electrolyte was used to achieve maximum power. This effort resulted in the demonstration of a 40-V, 5-kW, 67-Wh module that consisted of 20 cells each with an area of 150 cm². Northrop Grumman Corporation, which acquired Westinghouse's battery operation, redirected this effort to the investigation of rechargeable fused salt batteries for undersea vehicles for the U.S. Office of Naval Research.[21] Molten salt batteries are well suited for this application because of their inherently long storage life and high power capability. CoS_2 was used in the positive electrode instead of FeS_2 because of its superior thermal and chemical stability in the high melting point electrolyte. A sintered aluminum-nitride separator was developed with Advanced Refractory Technologies, Inc. as an alternative to the MgO powder and boron-nitride felt separators. Sintered separators measuring 15-in diameter were produced with a thickness of 0.05 in and a porosity of 36%. However, this new type of separator had limited success during cell operation.

SAFT R&D Center also directed efforts into the development of rechargeable high-power LiAl/FeS$_2$ batteries for military use.[22] The SAFT pulse-power battery technology is more similar to the electric-vehicle battery technology described in Sec. 41.3. Cold-pressed powder plaques were used for the electrodes and the separators. Like Westinghouse and Northrop Grumman, SAFT used the high melting-point (445°C) high-conductivity LiF-LiCl-LiBr electrolyte. Unfortunately, due to the appeal of the relatively new ambient-temperature rechargeable batteries, all three organizations have discontinued further efforts in developing rechargeable high-temperature batteries.

41.5.2 Electric-Vehicle Battery Designs

The design and construction of bipolar lithium/iron sulfide cells and stacks are described in Sec. 41.3 and illustrated in Fig. 41.2. Cells and stacks employing electrodes and electrolyte/separator pellets with 125-cm^2 area have been built and tested. Four-cell stacks of this size have been fabricated in the manner described in Sec. 41.3. The chalcogenide-based ceramic-to-metal peripheral seals have been successfully implemented on these stacks, the diameter of which appears suitable in size for many electric-vehicle applications. The optimal cell size, stack size, and module size will be dictated by the specific requirements and constraints established by the vehicle and its power electronics and control system. Fully integrated bipolar battery modules that incorporate their own thermal management systems have not been built and tested for either the electric-vehicle or pulse-power applications. However, several bipolar battery design and analysis studies have been conducted based on the engineering experience gained in the design and fabrication of prismatic lithium/iron sulfide electric-vehicle battery modules.

The U.S. Advanced Battery Consortium (USABC) established primary and secondary criteria for mid-term and long-term electric-vehicle batteries. These criteria, listed in Tables 41.4 and 41.5, do not incorporate any specifications in terms of battery size, capacity, or voltage. Three types of vehicles are selected here for use in discussing battery designs: a light-duty electric van, a high-performance passenger car, and a hybrid vehicle. These three vehicles were selected because they span the range of power-to-energy ratios being considered for on-road transportation vehicles that utilize battery energy storage. General battery requirements for these vehicles are provided in Table 41.6. The power-to-energy ratio for the light-duty van battery is approximately 1:1, while that for the hybrid vehicle is approximately 6:1. This hybrid vehicle is of the type where the battery possesses the full power capability to accelerate the vehicle and provide an appreciable zero-emission (battery-only) range. If employed as part of a hybrid-vehicle propulsion system in a high-performance high-efficiency passenger car, similar to the General Motors Impact, a battery meeting the hybrid-vehicle requirements could provide a greater than 190 km zero-emission range. If a car similar to the Impact were used as the high-performance all-electric passenger car, a battery meeting the high-performance electric-vehicle requirements could provide a 290 km zero-emissions range. The battery requirements for a light-duty van are those established by the U.S. Department of Energy for high-performance advanced batteries in the IDSEP van.[23] While the hybrid-vehicle requirements listed in Table 41.6 were appropriate during the development of the LiAl/FeS$_2$ battery, the current strategy of the U.S. auto industry under the Partnership for a New Generation of Vehicles (PNGV) is to develop dual-mode hybrid-electric vehicles that have much reduced zero emission range. The PNGV battery requirement of such a vehicle is an available energy of 1.5 kWh in a 65-kg battery pack.

TABLE 41.4 U.S. Advanced Battery Consortium Primary Criteria for Mid-Term and Long-Term Advanced Battery Technologies

Primary criteria	Mid term	Long term
Power density, W/L	250	600
Specific power, W/kg (80% DOD/30 s)	150 (200 desired)	400
Energy density, Wh/L ($C/3$ discharge rate)	135	300
Specific energy, Wh/kg ($C/3$ discharge rate)	80 (100 desired)	200
Life, years	5	10
Cycle life, cycles (80% DOD)	600	1000
Power and capacity degradation, % of rated spec.	20	20
Ultimate price, $/kWh (10,000 units at 40 kWh)	<150	<100
Operating environment, °C	−30 to 65	−40 to 85
Recharge time, h	<6	3–6
Continuous discharge in 1 h, % of rated energy capacity (no failure)	75	75

TABLE 41.5 U.S. Advanced Battery Consortium Secondary Criteria for Mid-Term and Long-Term Advanced Battery Technologies

Secondary criteria	Mid term	Long term
Efficiency, % ($C/3$ discharge, 6-h charge)	75	80
Self-discharge, %	<15 in 48 h	<15 per month
Maintenance	No maintenance; service by qualified personnel only	No maintenance; service by qualified personnel only
Thermal loss (for high-temperature batteries)	3.2 W/kWh 15% of capacity in 48-h period	3.2 W/kWh 15% of capacity in 48-h period
Abuse resistance	Tolerant; minimized by on-board controls	Tolerant; minimized by on-board controls

TABLE 41.6 General Battery Requirements for Three Types of Electric and Hybrid Vehicles

Battery requirements	Van, 160 km, delivery*	Passenger vehicles 290 km, commuter	Passenger vehicles Hybrid vehicle‡
Maximum weight, kg	440	395	200
Maximum volume, L	402	165	90
Minimum power, kW	47	90	90
Minimum energy, kWh	44	20†	15

  *DOE battery goals for high-performance advanced batteries in IDSEP van.[23]

  †Represents 50% increase for GM Impact over high-performance lead-acid batteries.

  ‡Dual Mode Hybrid Vehicle with 190 km zero emission range.

A spreadsheet model which accounts for all the components required to design and build a fully operational lithium/iron sulfide battery was used to conduct design studies on batteries for these and other applications.[24] The model uses the battery requirements for the vehicles and performance data from laboratory cells to develop battery designs that fall within the battery weight and volume envelopes specified for these vehicles. Different assumptions are used regarding the design, type, and thickness of containment hardware, etc., to develop designs with different developmental risk factors—the more conservative designs possessing lower risk. Results from this model are provided in Table 41.7 and Fig. 41.15. The dots represent projected performance levels for individual battery designs developed using the model. All the dots for each type of lithium/iron sulfide battery are enclosed in a shaded envelope to illustrate the effects of changing the power density on the delivered energy density of the battery. The energy available from the bipolar Li-Al/FeS$_2$ battery is affected the least by changes in the power. Figure 41.15 provides also the battery requirements for these three vehicles as well as the USABC mid-term and long-term performance objectives.

TABLE 41.7   Projected Capabilities of Lithium/Iron Sulfide Batteries for Electric-Vehicle Applications

Typical vehicle or mission	Typical battery requirements	Li/FeS$_x$ projected performance		
		Prismatic Li/FeS	Bipolar Li/FeS	Bipolar Li/FeS$_2$
Power:				
W/kg	107	110	140	180
W/L	117	165	224	330
Energy:				
Wh/kg	100	100	125	165
Wh/L	109	150	200	300
Commuter electric vehicle				
Power:				
W/kg	228	—	290	300
W/L	545	—	550	550
Energy:				
Wh/kg	51	—	>80	>155
Wg/L	121	—	150	270
Hybrid vehicle				
Power:				
W/kg	450	—	—	580
W/L	1000	—	—	1000
Energy:				
Wh/kg	75	—	—	130
Wh/L	167	—	—	225

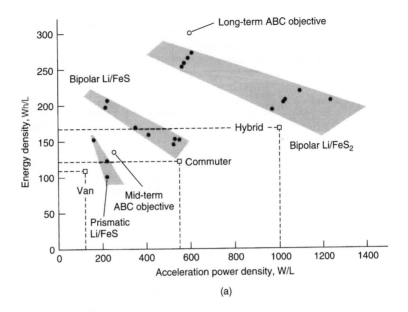

(a)

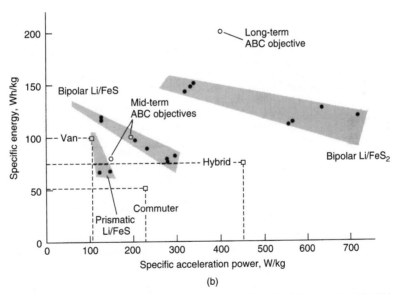

(b)

FIGURE 41.15 Energy vs. acceleration power projections for fully operational Li-Al/FeS$_x$ battery designs based on laboratory cell data. (*a*) Volumetric. (*b*) Gravimetric. (*From Henriksen et al.*)[15]

The prismatic Li-Al/FeS batteries, previously under development for electric-van applications, are projected to meet the light-duty van requirements and those of slightly more demanding electric-vehicle applications. Bipolar Li-Al/FeS batteries are projected to meet or surpass the performance requirements of most electric-vehicle applications, including the very demanding requirements of the high-performance passenger car. Bipolar Li-Al/FeS$_2$ batteries are projected to surpass the requirements of all three vehicle batteries, including the extremely demanding requirements of the hybrid-vehicle battery. The modeling results indicate that a bipolar Li-Al/FeS$_2$ battery, designed to fit the space available in the Impact, could deliver 40 kWh of energy, which corresponds to an estimated 560 km zero-emissions range. Also, at the power-to-energy ratio of 2, this battery technology is projected to approach the long-term performance objective of the USABC. Examples of battery design summary information for electric and hybrid vehicles are provided in Table 41.8. The information contained in this table indicates the high degree of packaging flexibility associated with this battery technology. For example, the three hybrid-vehicle battery designs vary rather dramatically in shape, while retaining high volumetric and gravimetric power and energy densities.

TABLE 41.8 Selected Summary Information from Computer-Aided Battery Design Model Used in Projecting Battery Performances

	High-Performance electric vehicle			Hybrid vehicle		
Battery designation	EV-1	EV-2	EV-3	Hybrid-1	Hybrid-2	Hybrid-3
Power, Kw	90.0	90.0	90.0	90.0	90.0	90.0
Energy store, kWh	40.0	40.0	40.0	18.0	15.0	15.0
Number of parallel strings	2	3	3	2	2	3
Shape of battery cross section	Oval	Triangle	Triangle	Oval	Rect.	Rect.
Stacks per cross section	2	3	3	2	4	6
Total number of stacks	6	6	6	6	4	6
Total number of cells	378	570	570	378	380	570
Cell parameters:						
Cell diameter, cm	16.1	13.4	13.4	16.1	16.3	13.4
Cell thickness, cm	1.126	1.108	1.025	0.587	0.505	0.424
Separator thickness, mm	1.2	1.2	0.5	1.2	1.2	0.5
Welding ring type	Channel	Channel	Channel	Flat	Flat	Flat
Battery performance summary						
Battery dimensions:						
Height, cm	36.8	29.5	29.5	36.8	37.1	31.4
Width, cm	20.4	31.4	31.4	20.4	37.1	45.0
Length, cm	232.0	226.8	211.1	130.1	31.4	53.6
Battery volume, L	155.9	158.6	147.6	87.4	79.6	72.9
Battery weight, kg	282.4	281.5	265.2	161.3	137.0	125.2
Specific power:						
Per unit volume, W/L	577.3	567.5	609.7	1030.0	1130.4	1234.2
Per unit weight, W/kg	318.7	319.7	339.4	558.0	657.0	718.7
Specific energy:						
Per unit volume, Wh/L	256.6	252.2	271.0	206.0	188.4	205.7
Per unit weight, Wh/kg	141.6	142.1	150.8	111.6	109.5	119.8

Source: From Nelson et al.[24]

An artist's rendering of a full-size 50-kWh van battery, using prismatic Li-Al/FeS cells, is shown in Fig. 41.16. These multiplate prismatic cells contain the full ampere-hour capacity required for the van application and are all series-connected electrically to provide the desired voltage. As illustrated, metal intercell connectors are used to connect adjacent cells electrically, one to the other. A double-walled steel vacuum jacket, filled with compressed multifoil insulation, provides thermal insulation for the battery. A forced-air heat exchanger is located inside the thermal enclosure in direct contact with the cells. This heat exchanger is used to prevent overheating during high-power operation. Resistance heaters (not shown) are provided inside the thermal enclosure for heating the battery to operating temperature and providing supplemental heat during extended stand periods. All leads from inside the battery are brought through a thermally insulated end cap or cover.

For bipolar lithium/iron sulfide electric-vehicle batteries, the packaging flexibility is significantly enhanced. An artist's concept of a self-contained thermally insulated 114-V 5-kWh bipolar battery module is provided in Fig. 41.17. The compactness and the cylindrical shape of bipolar stacks allow consideration of small self-contained modules that utilize lightweight loosely wrapped multifoil insulating jackets, of the type shown.[23] The walls of these small cylindrical insulating jackets are sufficiently strong to negate the need for compressed multifoil insulation layers between the walls to provide structural support. These lightweight jackets exhibit excellent insulating properties when under vacuum. Also, a simple concept is shown for altering the pressure between the walls of the jacket by controlling the temperature of a chemical getter material.[25] By heating the getter, the gas pressure between the jacket walls can be raised to permit a higher rate of heat rejection through the jacket for the purpose of cooling the battery without a forced-air heat exchanger. Use of these small self-contained thermally insulated modules does not significantly increase the weight and volume burden associated with the thermal enclosure over the type described for prismatic batteries, yet they appear to offer significant advantages from the standpoint of packaging the battery in the vehicle. Although detailed reliability and safety studies have not yet been performed on bipolar lithium/iron sulfide batteries, they possess favorable inherent characteristics in both areas. In the reliability area, cells routinely develop low-resistance internal short circuits when they reach the end of life. This characteristic permits bipolar stacks—employing strings of series-connected cells—to remain operational following the loss of one or more cells. The loss of energy for the stack is essentially limited to the energy content of the failed cell or cells. Therefore complex cell interconnect schemes, or other methods of making battery reliability and life less sensitive to cell failure, are not needed for lithium/iron sulfide batteries.

In the safety area, the active materials (lithium alloy anodes and FeS_x cathodes) are solids at both room temperature and operating temperature. Once the battery is heated to operating temperature, the molten-salt electrolyte provides a protective coating over the active materials. Even in the case of a very severe accident, where the battery enclosure and cell walls are breached, this electrolyte coating provides protection against the direct exposure of ambient air to the active materials. This was demonstrated in tower drop tests, where the cases of prismatic Li-Al/FeS cells—preheated to operating temperature—ruptured upon impact.[19] The contents of these cells were spewn on the ground, but no combustion of active materials or release of toxic fumes occurred. Other factors concerning safety, such as chemical, thermal, and electrical hazards (short circuit, overdischarge, and overcharge) have been examined. While extensive safety tests have not been conducted on battery systems, those that have reaffirm the development experience which indicates that these batteries are inherently quite safe.[7]

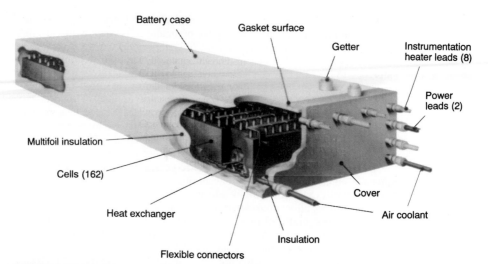

FIGURE 41.16 Artist's concept of full-scale Li-Al/FeS electric van battery with cutaway showing prismatic cells, heat exchanger, and compressed multifoil insulation. (*From Chilenskas et al.;*[19] *Courtesy of Argonne National Laboratory, University of Chicago.*)

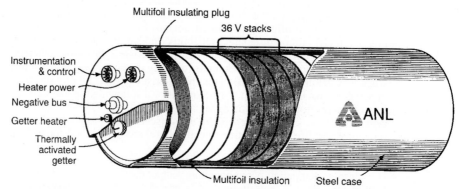

FIGURE 41.17 Artist's concept of thermally insulated electric-vehicle battery module with cutaway showing biploar Li-Al/FeS$_x$ stack, lightweight loosely wrapped multifoil insulation, and thermally activated getter system for controlling heat transfer from module. (*Courtesy of Argonne National Laboratory.*)

REFERENCES

1. H. Shimotake and E. J. Cairns, "A Lithium/Tin Cell with an Immobilized Fused-Salt Electrolyte: Cell Performance and Thermal Regeneration Analysis," *3rd Annual Intersociety Energy Conversion Engineering Conference (IECEC) Record,* Boulder, CO, Aug. 13–17, 1968, p. 76.

2. H. Shimotake, A. K. Fischer, and E. J. Cairns, "Secondary Cells with Lithium Anodes and Paste Electrolyte," *Proc. 4th IECEC,* Washington, D.C., Sept. 22–26, 1969, p. 538.

3. D. R. Vissers, Z. Tomczuk, and R. K. Steunenberg, "A Preliminary Investigation of High Temperature Lithium/Iron Sulfide Secondary Cells," *J. of the Electrochemical Society,* Vol. 121, 1974, p. 665.

4. E. C. Gay, D. R. Vissers, F. J. Martino, and K. E. Anderson, "Performance Characteristics of Solid Lithium-Aluminum Alloy Electrodes," *J. of the Electrochemical Society,* Vol. 123, 1976, p. 1591.

5. T. D. Kaun, A. N. Jansen, G. L. Henriksen, and D. R. Vissers, "Modification of LiCl-LiBr-KBr Electrolyte for LiAl/FeS$_2$ Batteries," *Electrochemical Society Proceedings,* Vol. 96-7, 1996, p. 342.

6. T. D. Kaun, P. A. Nelson, L. Redey, D. R. Vissers, and G. L. Henriksen, "High Temperature Lithium/Sulfide Batteries," *Electrochimica Acta,* Vol. 38, 1993, p. 1269.

7. G. L. Henriksen et al., "Safety Characteristics of Lithium-Alloy/Metal Sulfide Batteries," *Proc. 7th Int. Meet. On Lithium Batteries,* Boston, May 1994.

8. G. Henriksen, T. Kaun, A. Jansen, J. Prakash, and D. Vissers, "Advanced Cell Technology for High Performance LiAl/FeS$_2$ Secondary Batteries," *Electrochemical Society Proceedings,* Vol. 98-11, 1998, p. 302.

9. D. Golodnitsky and E. Peled, "Pyrite as Cathode Insertion Material in Rechargeable Lithium/Composite Polymer Electrolyte Batteries," *Electrochimica Acta,* Vol. 45, 1999, p. 335.

10. E. Straus, D. Golodnitsky, and E. Peled, "Study of Phase Changes During 500 Full Cycles of Li/Composite Polymer Electrolyte/FeS$_2$ Battery," *Electrochimica Acta,* Vol. 45, 2000, p. 1519.

11. T. D. Kaun et al., "Lithium/Disulfide Cells Capable of Long Cycle Life," *Proc. Materials and Processes Symp.,* K. M. Adams and B. B. Owens (eds.), Vol. 89-4, Electrochemical Society, Pennington, N.J., 1989, p. 383.

12. A. A. Chilenskas et al., "Status of the Li-Al/FeS Battery Manufacturing Technology," Argonne National Laboratory, Argonne, Ill, Final Rep., ANL-84-23, Aug. 1984.

13. P. A. Nelson et al., "Development of Lithium/Metal Sulfide Batteries at Argonne National Laboratory: Summary Report for 1978," Argonne National Laboratory, Argonne, Ill., Rep. ANL-79-64, July 1979.

14. D. L. Barney et al., "High Performance Batteries for Electric Vehicle Propulsion and Stationary Energy Storage," Argonne National Laboratory, Argonne, Ill., Rep. ANL-79-94, Mar. 1980.

15. G. L. Henriksen et al., "Lithium/Metal Sulfide Technology Status," *Proc. Annual Automotive Technology Development Contractor's Coordination Meeting 1991,* Society of Automotive Engineers, June 1992, p. 95.

16. T. D. Kaun et al., "Development of a Sealed Bipolar Li-Alloy/FeS$_2$ Battery for Electric Vehicles," *Proc. 25th IECEC,* Reno, Nev., Aug. 12–17, 1990, Vol. 3, p. 335.

17. T. D. Kaun et al., "Development of Prototype Sealed Bipolar Lithium/Sulfide Cells," *Proc. 26th IECEC,* Boston, Mass., Aug. 4–9, 1991, p. 417.

18. T. D. Kaun, M. C. Hash, and D. R. Simon, "Sulfide Ceramics in Molten-Salt Electrolyte Batteries," *Role of Ceramics in Advanced Electrochemical Systems,* P. N. Kumta, G. S. Rohrer, and U. Balachandran, eds., *Ceramic Transactions,* Vol. 65, 1996, p. 293.

19. A. A. Chilenskas et al., "Test Results for 36-V Li/FeS Battery," Argonne National Laboratory, Argonne, Ill., Rep. ANL-90/1, 1990.

20. N. Papadakis et al., "Performance of Rechargeable Lithium-Alloy/Metal Disulfide Pulse Power Bipolar Batteries," *Proc. IEEE 35th Int. Power Sources Symp.,* Cherry Hill, N.J., June 22–25, 1992, p. 285.

21. P. Keller, P. Smith, C. Winchester, N. Papadakis, G. Barlow, and N. Shuster, "Rechargeable Fused Salt Batteries for Undersea Vehicles," *Proc. 38th Power Sources Conf.,* Cherry Hill, N.J., June 8–11, 1998, p. 248.

22. J. D. Briscoe et al., "Rechargeable Pulse Power Lithium-Alloy/Iron Disulfide Batteries," *Proc. IEEE 35th Int. Power Sources Symp.,* Cherry Hill, N.J., June 22–25, 1992, p. 294.

23. "Mission Directed Goals for Electric Vehicle Battery Research and Development," U.S. Department of Energy, Rep. DOE/CE-0148, rev. 1, Nov. 1987.

24. P. A. Nelson et al., "Modeling of Lithium/Sulfide Batteries for Electric Vehicles and Hybrid Vehicles," *Proc. 26th IECEC,* Boston, Mass., Aug. 4–9, 1991, Vol. 3, p. 423.

25. P. A. Nelson et al., "Variable Pressure Insulating Jackets for High-Temperature Batteries," *Proc. 27th IECEC,* San Diego, Calif., Aug. 3–7, 1992, Vol. 3, p. 3.57.

P · A · R · T · 6

PORTABLE FUEL CELLS

CHAPTER 42
PORTABLE FUEL CELLS—INTRODUCTION

David Linden and Thomas B. Reddy

42.1 GENERAL CHARACTERISTICS

A fuel cell is a galvanic device that continuously converts the chemical energy of a fuel (and oxidant) to electrical energy. Like batteries, fuel cells convert this energy electrochemically and are not subject to the Carnot cycle limitation of thermal engines, thus offering the potential for highly efficient conversion. The essential difference between a fuel cell and a battery is the manner for supplying the source of energy. In a fuel cell, the fuel and the oxidant are supplied continuously from an external source when power is desired. The fuel cell can produce electrical energy as long as the active materials are fed to the electrodes. In a battery, the fuel and oxidant (except for metal/air batteries) are an integral part of the device. The battery will cease to produce electrical energy when the limiting reactant is consumed. The battery must then be replaced or recharged.

The electrode materials of the fuel cell are inert in that they are not consumed during the cell reaction, but have catalytic properties which enhance the electroreduction or electrooxidation of the reactants (the active materials).

The anode active materials used in fuel cells are generally gaseous or liquids fuels (compared with the metal anodes generally used in most batteries), such as hydrogen, methanol, hydrocarbons, natural gas, which are fed into the anode side of the fuel cell. As these materials are like the conventional fuels used in heat engines, the term "fuel cell" has become popular to describe these devices. Oxygen, most often air, is the predominant oxidant and is fed into the cathode

Fuel cell technology can be classified into two categories:

1. Direct systems where fuels, such as hydrogen, methanol and hydrazine, can react directly in the fuel cell (see Sec. 43.4).

2. Indirect systems in which the fuel, such as natural gas or other fossil fuels, is first converted by reforming to a hydrogen-rich gas which is then fed into the fuel cell.

Fuel cell systems can take a number of configurations depending on the combinations of fuel and oxidant, the type of electrolyte, temperature of operation, application, etc. Table 42.1 is a summary of various types of fuel cells distinguished by the electrolyte and operating temperature. The PEM fuel cell is currently the predominant one for portable and small fuel cells as it is the only one operating near ambient conditions.*

---

*Large fuel cells are beyond the scope of this Handbook. See Appendix F, Bibliography for references.

TABLE 42.1 Types of Fuel Cells

---

1. **Solid Oxide (SOFC):** These cells use a solid oxygen-ion-conducting metal oxide electrolyte. They operate at about 1000°C, with an efficiency of up to 60%. They are slow to start up, but once running, provide high grade waste heat which can be used to heat buildings. They may find application in industrial and large-scale applications.

2. **Molten Carbonate** (MCFC): These cells use a mixed alkali-carbonate molten salt electrolyte and operate at about 600°C. They are being developed for continuously operating facilities, and can use coal-based or marine diesel fuels.

3. **Phosphoric** Acid (PAFC): This is the most commonly used type of fuel cell for stationary commercial sites such as hospitals, hotels, and office buildings. The electrolyte is concentrated phosphoric acid. The fuel cell operates at about 200°C. It is highly efficient and can generate energy at up to 85% (40% as electricity and another 45% if the heat given off is also used).

4. **Alkaline (AFC):** These are used by NASA on the manned space missions, and operate well at about 200°C. They use alkaline potassium hydroxide as the electrolyte and can generate electricity up to 70% efficiency. A disadvantage of this system is that it is restricted to fuels and oxidants which do not contain carbon dioxide.

5. **Proton** Exchange Membrane (PEM): These cells use a perfluorinated ionomer polymer membrane electrolyte which passes protons from the anode to the cathode. They operate at a relatively low temperature (70 to 85°C), and are especially notable for their rapid start-up time. These are being developed for use in transportation applications and for portable and small fuel cells.

6. **Direct Methanol** (DMFC): These fuel cells directly convert liquid methanol (methyl alcohol) in an aqueous solution which is oxidized at the anode. Like PEMs, these also use a membrane electrolyte, and operate at similar temperatures. This fuel cell is still in the development stage.

7. **Regenerative (RFC):** These are closed-loop generators. A powered electrolyzer separates water into hydrogen and oxygen, which are then used by the fuel cell to produce electricity and exhaust (water). That water can then be recycled into the powered electrolyzer for another cycle.

Source: Connecticut Academy of Science and Engineering Reports, Vol. 15, No. 1, 2000.

A practical fuel cell power plant consists of three basic subsystems:

1. A power section, which consists of one or more fuel cell stacks—each stack containing a number of individual fuel cells, usually connected in series to produce a stack output ranging from a few to several hundred volts (direct current). This section converts the fuel and the oxidant into DC power.

2. A fuel subsystem that manages the fuel supply to the power section. This subsystem can range from simple flow controls to a complex fuel-processing facility. This subsystem processes fuel to the type required for use in the fuel cell (power section).

3. A power conditioner that converts the output from the power section to the type of power and quality required by the application. This subsystem could range from a simple voltage control to a sophisticated device that converts the DC power to an AC power output.

In addition, a fuel cell power plant, depending on size, type, and sophistication, may require an oxidant subsystem, thermal and fluid management subsystems, and ether ancilliary subsystems.

Fuel cells have been of interest for over 150 years as a potentially more efficient and less polluting means for converting hydrogen and carbonaceous or fossil fuels to electricity compared to conventional heat engines. A significant application of the fuel cell has been the use of the hydrogen/oxygen fuel cell by NASA, using cryogenic fuels, in space vehicles for over 40 years including the current fleet of space shuttles. Use of the air-breathing fuel cell in terrestrial applications, such as for utility power and electric vehicles, has been ongoing for some time but has been developing slowly. Recent advances have now revitalized interest for these and other new applications.

During this past decade, interest in small air-breathing fuel cells has arisen for dispersed or on-site electric generators, remote devices and other such applications in the subkilowatt power range, replacing engine-generators, power sources and larger-sized batteries. At lower power levels, from below 1 to 50 watts, historically the domain of batteries, fuel cell technology is seen as an approach to achieve higher specific energy than those delivered by batteries. Progress has been made with fuel cell systems in the sizes above 50 watts, especially for extended long-term service (see Chap. 43). However, the development of yet smaller portable fuel cells (which can be "recharged," for example, by replacing a small container of fuel), competitive in size and performance with batteries remains a challenge (see Sec. 42.3).

42.2 OPERATION OF THE FUEL CELL

42.2.1 Reaction of Mechanisms

A simple fuel cell is illustrated in Fig. 42.1a. Two catalyzed electrodes are immersed in an electrolyte (acid in this illustration) and separated by a gas barrier. The fuel, in this case hydrogen, is bubbled across the surface of one electrode while the oxidant, in this case oxygen from ambient air, is bubbled across the other electrode. When the electrodes are electrically connected through an external load, the following events occur:

1. The hydrogen dissociates on the catalytic surface of the fuel electrode, forming hydrogen ions and electrons.

2. The hydrogen ions migrate through the electrolyte (and a gas barrier) to the catalytic surface of the oxygen electrode.

3. Simultaneously, the electrons move through the external circuit, doing useful work, to the same catalytic surface.

4. The oxygen, hydrogen ions, and electrons combine on the oxygen electrode's catalytic surface to form water.

The reaction mechanisms of this fuel cell, in acid and alkaline electrolytes, are shown in Table 42.2. The major differences, electrochemically, are that the ionic conductor in the acid electrolyte is the hydrogen ion (or, more correctly, the hydronium ion, H_3O^+) and the OH^- or hydroxyl ion in the alkaline electrolyte. The only by-product of a hydrogen/oxygen fuel cell is water; in the acid electrolyte water is produced at the cathode and, in the alkaline electrolyte fuel cell, it is produced at the anode.

The net reaction is that of hydrogen and oxygen producing water and electrical energy. As in the case of batteries, the reaction of one electrochemical equivalent of fuel will theoretically produce 26.8 Ah of DC electricity at a voltage that is a function of the free energy of fuel-oxidant reactions. At ambient conditions, this potential is ideally 1.23 V DC for a hydrogen/oxygen fuel cell.

Figure 42.1(b) is a schematic of the Proton Exchange Membrane Fuel Cell (PEMFC), presently the best candidate for use in small portable fuel cells. Passing through a gas diffuser, hydrogen reacts on a catalyst (small circles) at the anode, sending protons, and electrons to the cathode. The protons migrate through the membrane and the electrons through the external circuit. The protons react with the oxygen, supplied at the cathode, to form water. Products and unused reactants exit through the gas vents.

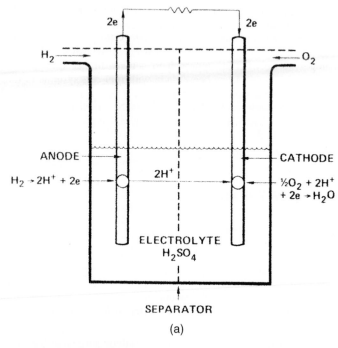

(a)

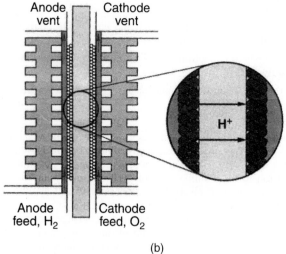

(b)

FIGURE 42.1   Operation of the fuel cell. (*a*) Reactions in an acid electrolyte. (*b*) Based on a proton exchange membrane. (*Source: Chemical and Engineering Areas,* American Chemical Society, Washington D.C., June 14, 1999.)

TABLE 42.2 Reaction Mechanisms of the H_2/O_2 Fuel Cell

	Acid electrolyte	Alkaline electrolyte
Anode	$H_2 \rightarrow 2H^+ + 2e$	$H_2 + 2OH^- \rightarrow 2H_2O + 2e$
Cathode	$1/2 O_2 + 2H^+ + 2e \rightarrow H_2O$	$1/2 O_2 + 2e + H_2O \rightarrow 2OH^-$
Overall	$H_2 + 1/2 O_2 \rightarrow H_2O$	$H_2 + 1/2 O_2 \rightarrow H_2O$

42.2.2 Major Components of the Fuel Cell

The important components of the individual fuel cell are:

1. The *anode* (fuel electrode) must provide a common interface for the fuel and electrolyte, catalyze the fuel oxidation reaction, and conduct electrons from the reaction site to the external circuit (or to a current collector that, in turn, conducts the electrons to the external circuit).

2. The *cathode* (oxygen electrode) must provide a common interface for the oxygen and the electrolyte, catalyze the oxygen reduction reaction, and conduct electrons from the external circuit to the oxygen electrode reaction site.

3. The *electrolyte* must transport one of the ionic species involved in the fuel and oxygen electrode reactions while preventing the conduction of electrons (electron conduction in the electrolyte causes a short circuit). In addition, in practical cells, the role of gas separation is usually provided by the electrolyte system. This is often accomplished by retaining the electrolyte in the pores of a matrix. The capillary forces of the electrolyte within the pores allow the matrix to separate the gases, even under some pressure differential. Currently, the technology in use for portable ambient temperature fuel cells is the electrolyte membrane Nafion®.

42.2.3 General Characteristics

The performance of a fuel cell is represented by the current density vs. voltage (or "polarization") curve (Fig. 42.2). Whereas ideally a single H_2/O_2 fuel cell could produce 1.23 V DC at ambient conditions, in practice, fuel cells produce useful voltage outputs that are somewhat less than the ideal and which decrease with increasing discharge rate (current density). The losses or reductions in voltage from the ideal are referred to as "polarization," as illustrated in Fig. 42.2 (also see Chap. 2).

These losses include:

1. Activation polarization represents energy losses that are associated with the electrode reactions. Most chemical reactions involve an energy barrier that must be overcome for the reactions to proceed. For electrochemical reactions, the activation energy lost in overcoming this barrier takes the form

$$\eta_{act} = a + b \ln i$$

where η_{act} = activation polarization, mV
$\quad a, b,$ = constants
$\quad\quad i$ = current density, mA/cm^2

Activation polarization is associated with each electrode independently and

$$\eta_{act(cell)} = \eta_{act(anode)} + \eta_{act(cathode)}$$

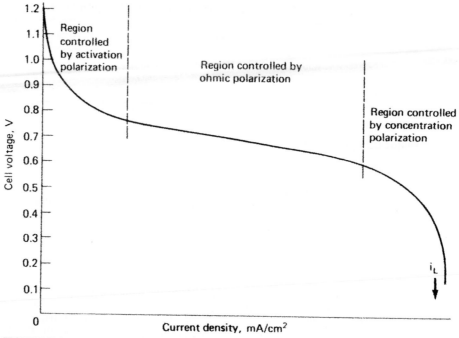

FIGURE 42.2   Fuel cell polarization curve.

2. Ohmic polarization represents the summation of all the ohmic losses within the cell, including electronic impedances through electrodes, contacts, and current collectors and ionic impedance through the electrolyte. These losses follow Ohm's law

$$\eta_{ohm} = iR$$

where η_{ohm} = ohmic polarization, mV
i = current density, mA/cm^2
R = total cell impedance, $\Omega \cdot$cm^2

3. Concentration polarization represents the energy losses associated with mass transport effects. For instance, the performance of an electrode reaction may be inhibited by the inability for reactants to diffuse to or products to diffuse away from the reaction site. In fact, at some current, the limiting current density i_L, a situation will be reached wherein the current will be completely limited by the diffusion processes (see Fig. 42.2). Concentration polarization can be represented by

$$\eta_{conc} = \frac{RT}{nF} \ln \left(1 - \frac{i}{i_L} \right)$$

where η_{conc} = concentration polarization, mV
R = gas constant
T = temperature, $^\circ$K
n = number of electrons
F = Faraday's constant
i = current density, mA/cm^2
i_L = limiting current density, mA/cm^2

Concentration polarization occurs independently at either electrode. Thus, for the total cell

$$\eta_{conc(cell)} = \eta_{conc(anode)} + \eta_{conc(cathode)}$$

The net result of these polarizations is that fuel cells generally operate between 0.5 and 0.9 V DC. Fuel cell performance can be increased by increasing cell temperature and reactant partial pressure. However, for small or portable fuel cells, operation at ambient conditions is usually a requirement, particularly when the fuel cell is to be used as a replacement for batteries.

42.3   SUB-KILOWATT FUEL CELLS

The fuel cell has many attractive features that have increased interest in its use in small and/or portable power sources below a kilowatt in power output. At the same time, because of the unique requirements of portable devices, there are limitations in fuel cell technology that will present challenges for its deployment, particularly in the sizes below 20 watts as replacements for batteries.

These include:

1. *Hydrogen and Energy-rich Fuels.* The use of hydrogen and other energy-rich fuels that have a higher energy density than the active materials normally used in batteries:

Table 42.3 lists the theoretical specific energy and energy density of several of these materials which are significantly higher than those of batteries and which are practical for use in portable fuel cells. Of these, hydrogen stands out not only because of its high specific energy, but because it can be directly converted to electrical energy in a fuel cell operating at ambient temperatures. Natural gas, propane, gasoline and other fossil fuels cannot be considered as they cannot be converted directly, except at very high temperatures. Incorporating a fuel processing unit would not be feasible for a small portable device for battery replacement. Methanol is the only liquid fuel that, at this time, shows promise for direct conversion at reasonable temperatures (see Sec. 43.4.4).

The necessity for containing and supplying hydrogen to the fuel cell, in a practical and safe method, substantially reduces its practical specific energy. A number of methods are being used, including compressed gas cylinders, storage in hydrides, and chemical methods for generating hydrogen, each of which requires a specific method for generating and controlling the generation of hydrogen (see Sec. 43.4.1). Table 42.3 also lists the theoretical values of the various methods for supplying hydrogen and, as applicable, the status of current technology. While these values, albeit much lower than that of hydrogen gas, are still higher than those of most battery systems, a comparison to the specific energy of battery systems, which is often done, is not a correct one. It compares only the fuel supply of the fuel cell (omitting the fuel cell stack and other fuel cell components) to a complete battery system. A more reasonable method of comparison is discussed below and illustrated in Fig. 42.3.

TABLE 42.3 Characteristics of Fuels for Use In Portable Fuel Cells

	Theoretical[1]		Current State-of-Art[2]	
	Wh/kg	Wh/L	Wh/kg	Wh/L
Hydrogen				
Hydrogen (gas)	32705	—		
Hydrogen (liquid) cryogenic	32705	2310		
Pressurized H_2 containers				
34.5×10^6 Pa	1023		400	250
Metal Hydride			160	420
$MH(2\%H_2)$	655			
$MH(7\%H_2)$	2290	3400		
Chemical Hydrides			400[3]	500[3]
$LiH + H_2O$	2540	—		
$NaBH_4 + 2H_2O$	3590	—		
30% $NaBH_4$ solution	2375	2080		
Carbon-based H_2 Storage				
Carbon nanofibres (6.5% H_2)	2130			
Methanol (MeOH)				
100% MeOH	6088	4810		
MeOH, H_2O solution (equal molar)	~3900	~3350		

[1] Based on 1.23 V for H_2/O_2 fuel cell
[2] Based on actual watt-hour output of a fuel cell running on the specified H_2 source
[3] Includes container/packaging and required water

2. *Electrochemical Conversion.* The fuel cell converts chemical energy to electrical energy electrochemically at high conversion efficiencies (in the order of 30 to 60% depending on the output voltage) in small as well as large units and even when it is operating at partial power. However, while conversion efficiency may not be affected, scaling down to the lower power levels may not result in a proportional decrease in the weight and size of the power unit or the auxiliary devices.

3. *Operating Temperature.* Portable fuel cells, from a practical point of view, should operate at ambient temperatures. Based on the current technology, as summarized in Table 42.1, this limits the choice to the Proton Exchange Membrane Fuel Cell (PEMFC). This also limits the choice of fuels to those that can be directly converted to electrical energy at that temperature. It precludes the use of those fuels that have to be reformed to a hydrogen-rich fuel because of the high operating temperatures of the reformer. But operation at the relatively low temperatures results in a lower conversion efficiency and the use of the polymer limits operation at sub-freezing temperatures. Water embedded in the polymer will tend to freeze and, while the polymer can withstand this condition, start-up and operation are impeded and an external energy source, such as external heat or a battery, may be required.

4. *Modular Features.* The modular features of the fuel cell, with separate units for power conversion and fuel storage, facilitate designing the fuel cell system to meet application requirements and equipment footprints. The power conversion unit (the fuel cell stack) can be sized to meet the power requirement and the fuel container can be sized to contain sufficient fuel to meet the service time requirement.

Figure 42.3 compares the performance of several primary and secondary batteries with that of a fuel cell, showing the total weight of each system designed, in this example, to deliver 50 watts, for different times of operational service. The secondary battery systems deliver their rated capacity even at the highest discharge rates shown in the figure; hence, their performance is characterized by a straight sloping line. The slope is equivalent, as shown, to their specific energy and the weight of the battery is reduced almost proportional to the reduction in service time. The primary battery systems, which generally do not perform as well at high discharge rates, show the straight sloping line over the longer periods of service. At the high discharge rates or at the short hours of operational life, the curve levels off indicating little reduction in weight with decreasing service life. The curves for the fuel cell look similar. At the low operational time, the weight reflects the "dead" weight of the system, i.e. the fuel cell stack and other auxiliaries required to produce the required power, the weight of the fuel being inconsequential. At the longer service times, the weight of the power unit becomes insignificant and the system weight increases by the specific energy of the fuel.

This figure graphically illustrates the respective advantages of batteries and fuel cells. The battery shows its advantage on the relatively short term life applications as the fuel cell is penalized by the weight of the power unit. On longer term applications, the fuel cell benefits as the replacement fuel has a higher specific energy than most of the battery systems.

A similar relationship exists if the comparison is made on a volumetric basis.

This figure points out the direction that fuel cell development must take to compete successfully with battery systems in the low power range. The weight and size of the power unit must be reduced substantially as the emphasis in the design of portable equipment is towards lower size and weight even at the sacrifice of service time. Unless, this is done, the advantage of the fuel cell, the lighter weight fuel replacement, will not be significant.

An interesting consideration is a possible tradeoff in the design of the fuel cell component and the fuel source. In the case of the direct methanol fuel cell (DMFC), for example, water is required for the reaction of methanol. The discharge product of the fuel cell, water, can be used if the water management or recovery is incorporated in the fuel cell; a one-time cost of increased size, weight and complexity of the fuel cell. Or water can be added to the fuel source at the expense of a recurring lower specific energy of the diluted fuel source.

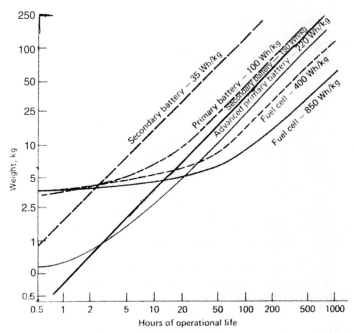

FIGURE 42.3 Comparison of electrochemical systems—weight vs. service life (based on 50 watt W output and stated specific energy.)

5. *Air-Breathing Systems.* Most terrestrial fuel cells are air-breathing and an oxidant does not have to be stored and carried with the fuel cell, thus keeping the size and weight of the system to a minimum. Depending on the power level, the air-flow may be insufficient, necessitating the addition of fans or other means of forced convection for the electrochemical reaction, cooling and water balance.

6. *Environmentally Friendly.* Fuel cells are *environmentally friendly* and, while the large sizes can be complex, much like a chemical plant, in the small and portable sizes they can be quiet and relatively simple in design. While these characteristics are superior to the engine-generators and other heat engines they may replace, for these characteristics they offer no advantage over batteries. Further, the need to provide a method for attaching the fuel supply and a mechanism to supply the fuel makes it more complex as these components are not required for the battery which is self-contained.

7. *Cost.* Cost will be a major factor for the acceptance of fuel cell as a replacement for batteries. The cost of the fuel cell is determined by its two components: the fuel cell and auxiliaries, and the fuel source. At this time, the cost of the fuel cell is high compared to batteries, not only because it has not attained commercial production status, but also because the polymers, catalysts and its other components are expensive. A potential advantage of the fuel cell again focuses on the fuel supply. If the cost of fuel replacement can be reduced so that it is lower than that of battery replacement, fuel cell deployment may be a cost effective approach for extended long term periods of operation.

42.4 INNOVATIVE DESIGNS FOR LOW WATTAGE FUEL CELLS

The development of fuel cells that may be competitive with batteries in the low power range for use in cellular phones, laptop computers and other similar equipments will require innovative designs incorporating thinner components and smaller light-weight auxiliaries. Scale-down of current fuel cell technology will not be adequate to meet the size and weight requirements of these portable devices.

One new concept[1,2] as shown in Fig. 42.4, is a micro fuel cell using diluted methanol, which will be supplied in replaceable ampoules, as a fuel. It is planned to use manufacturing techniques employed in the electronics industry to mass-produce the fuel cell. In this design, a thin film of plastic is bombarded with nuclear particles and then chemically etched to form fine pores (the cells) into which a polymeric electrolyte is added. Chipmaking techniques, including vacuum deposition, are used to layer and etch on the plastic structure, a preferentially permeable barrier (to limit methanol leakage to the cathode), two electrode plates, a catalyst material and a conductive grid to connect the individual cells. Start-up is still sluggish and power levels are low, but a portable charger has been developed to charge a cell phone battery, the advantage being that fuel cell has a higher specific energy than rechargeable batteries.

In another approach,[3] a proton exchange membrane fuel cell (PEMFC) has been fabricated on a silicon chip, again using methods of the microelectronics industry. Porous gas diffusion electrodes, electrical interconnects and gas manifolds were created using lithography and etching processes and solid polymer electrolytes were deposited using thin film deposition techniques. Several novel fuel cell architectures were investigated, including a planar design in which the anodes and cathodes are side-by-side in the same plane. Combined with a hydride as a source for hydrogen, a prototype power source demonstrated an energy density higher than batteries. These small silicon PEMFCs offer advantages including flexible form factors and manufacturability.

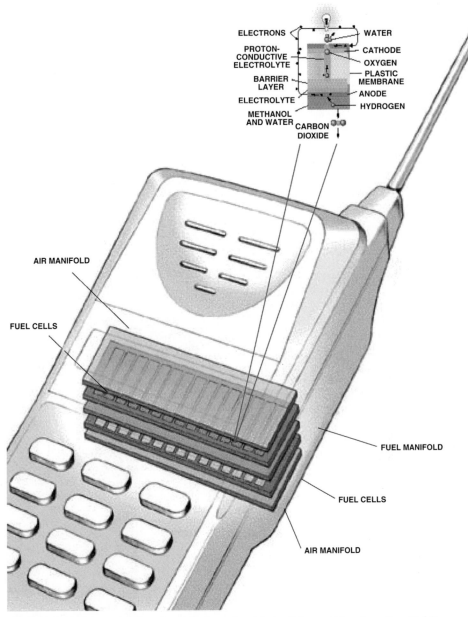

ELECTRONS

WATER

PROTON-
CONDUCTIVE
ELECTROLYTE

CATHODE

OXYGEN

BARRIER
LAYER

PLASTIC
MEMBRANE

ELECTROLYTE

ANODE

HYDROGEN

METHANOL
AND WATER

CARBON
DIOXIDE

AIR MANIFOLD

FUEL CELLS

FUEL MANIFOLD

FUEL CELLS

AIR MANIFOLD

FIGURE 42.4   Concept drawing of a miniature methanol fuel cell for a cellular phone (from Ref 1).

Figure 42.5 illustrates another novel thin-film fuel cell that generates power from a gas mixture that contains both hydrogen and oxygen, as opposed to conventional designs which require separate gas supplies.[1,4] The design relies on a gas permeable electrolyte that is extremely thin (less than one micron). This membrane is sandwiched between two thin layers of platinum (the two electrodes), one that is exposed to the gas mixture and one that sits on a substrate. Hydrogen oxidizes faster at the exposed platinum electrode than at the other electrode, because the former is in direct contact with the gas mixture. Thus, at the exposed electrode, the anode, there is a supply of hydrogen ions that then rapidly diffuse through the membrane to the "slower" electrode (the cathode) on the substrate. There, the hydrogen ions and the electrons from the external circuit combine with oxygen that has diffused through the porous membrane, producing water with a voltage of about 1 V. The mechanism relies on the closeness of the two electrodes, otherwise the hydrogen ions are unable to reach the cathode from the anode. Figure 42.6 is a polarization curve showing the performance of a test cell, with a surface area of 6.5 cm^2, which produced 123 mW. Based on this and similar test samples, now using a palladium catalyst, a 500 mW stack with a volume of 15 cm^3 could be made with approximately a 30% energy conversion efficiency, with the advantage that it could be safely run on 1 to 3% hydrogen in air.

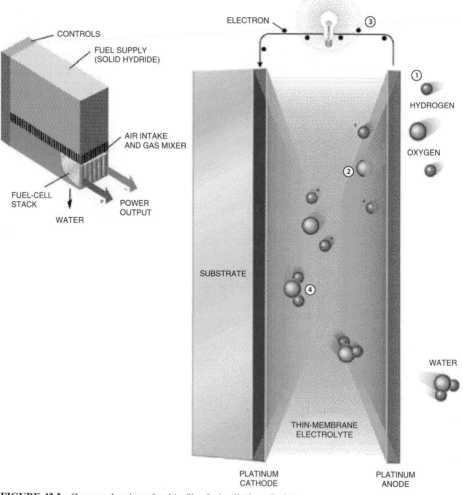

FIGURE 42.5   Concept drawing of a thin-film fuel cell (from Ref 1).

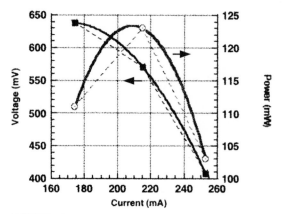

FIGURE 42.6 Polarization curve of the thin film fuel cell with Pt electrodes (from Ref 4).

These various miniature fuel cells illustrate some of the work that is ongoing, but which is still in the very early development stages. Although initial prototypes have been built, none has yet been definitely demonstrated as a viable, compact device that would be competitive in performance, cost and convenience with a battery. The objective remains a challenging, but potentially fruitful goal.

REFERENCES

1. Christopher F. Dyer, *Replacing the Battery Portable Electronics,* Scientific American, July 1999, p. 88.
2. Manhattan Scientifics, New York, N.Y.
3. Sandia National Laboratories.
4. Joseph Bostaph *et al.*, "Thin Film Fuel Cells for Low Power, Portable Applications," *Proc. of the 39th Power Sources Conf.,* p. 152, June 2000.

BIBLIOGRAPHY

See Appendix F.

CHAPTER 43

SMALL FUEL CELLS (LESS THAN 1000 WATTS)

Arthur Kaufman and H. Frank Gibbard

43.1 GENERAL

Small fuel cells that provide power at ratings below 1000 watts, are being introduced into and groomed for application areas such as portable power sources, mobile power sources, remote or unattended power sources, and propulsion power for small, off-the-road vehicles. These existing and candidate applications may include power systems that are fuel-cell-only, fuel-cell/battery hybrids, and fuel-cell/solar/battery hybrids, depending on the nature of the system's requirements. Representative examples of such applications are illustrated in Table 43.1.

The interest in small fuel cells stems from their potential to replace engine-generators with more efficient and environmentally friendly conversion systems and, as replacements or supplements to batteries, because of their potentially higher specific energy.

The energy-storage and power-generating elements of a fuel cell system are separate entities—the fuel storage and the fuel cell stack plus its auxiliaries, respectively. In a battery, on the other hand, the energy-storage and power-generating elements are the same. Hence, the fuel cell system can be designed to relate optimally to its operating mode—the fuel cell stack to satisfy the power requirements, and the fuel storage to satisfy the energy require-

TABLE 43.1 Small Fuel Cell Applications

Application	Power range	Status	Reference
Backup power supply	250 W, 500 W	Limited production	9, 10, 11
Military field battery charger	50 W, 100 W	Limited production	2, 12
General purpose portable power supply	10–100 W	Limited production	1, 2, 8, 10
Briefcase power supply	35 W	Prototype	13
Professional video camera power supply	35 W	Prototype	13
Backup for solar photovoltaic power in telecommunications	20–100 W	Prototype	9, 10
Backup for solar photovoltaic traffic sign	50–100 W	Commercial	8, 9

ments. This can be particularly advantageous in applications where the energy requirement is great and the power requirement is minimal, that is, in applications of long duration (see Sec. 43.4.1). In such applications the fuel cell stack, with its auxiliaries, becomes relatively insignificant within the overall system; and the system's energy density and specific energy approach that of the fuel storage subsystem alone (see Fig. 42.3). The mission duration beyond which fuel cells would tend to be favored over batteries, by providing a smaller and/ or lighter system, depends on the specific application requirements.

Certain applications are well suited to a fuel-cell/battery hybrid system by nature of their duty-cycle. Those that have high peak-to-average load ratios and relatively short-duration peaks are generally attractive candidates. Such a system allows the fuel cell to be rated near its average power while a relatively small battery provides excess power on demand and is recharged by the fuel cell during normal-load operation. Hybrid systems thus exploit the strengths of both batteries and fuel cells—the wide dynamic power range of the former, and the high energy content per unit weight or volume of the latter.

Solar/battery power systems can also be combined advantageously with fuel cells in various applications. Use of the fuel cell can often allow solar power to be exploited without the need for excessive battery size and weight associated with prolonged or unpredictable lack of availability of solar energy.

Small fuel cells are also expected to become an alternative to small engine-generator sets in some applications. In this sector, the fuel cell is unlikely to offer any advantage from an energy density or initial-cost point of view. However, in applications where system life, reliability, efficiency (fuel consumption), noise, and emissions are important, the fuel cell system could become competitive.

43.2   APPLICABLE FUEL CELL TECHNOLOGIES

Various types of fuel cell systems are either in use or under development, and these are generally distinguished on the basis of their electrolyte. These systems exhibit viable operation in different temperature regimes:

1. *Phosphoric acid fuel cells (PAFC)*—use highly concentrated aqueous phosphoric acid electrolyte and generally operate in the 160 to 200°C range

2. *Molten carbonate fuel cells (MCFC)*—use mixed alkali-carbonate molten salt electrolyte, operate typically at about 600°C

3. *Solid oxide fuel cells (SOFC)*—use solid oxygen-ion-conducting metal oxide electrolyte at about 1000°C, although some development activity is focusing on somewhat lower temperatures

4. *Alkaline fuel cells (AFC)*—typically use liquid solutions of potassium hydroxide electrolyte at temperatures ranging from ambient to about 80°C

5. *Proton-exchange membrane fuel cells (PEMFC)*—use solid-polymer proton-conducting membrane electrolyte at temperatures generally ranging from ambient to 90°C. Today's technology primarily uses the trifluoromethanesulfonic-acid-based electrolyte membrane, such as DuPont's Nafion®.

Small fuel cells can be exploited most effectively if they can stand by and operate at ambient temperatures (and can therefore start rapidly), can operate on ambient air, can respond rapidly to load changes, have a non-migrating (solid) electrolyte, and have a reasonably high power density and specific-power. The fuel cell type that best suits these criteria is clearly the PEMFC, despite a drawback related to the fact that liquid water embodied in the solid polymer tends to freeze, and thereby impedes proton conduction, when its temperature drops below the freezing point. The PEMFC can stand by under freezing conditions, however, and can generally operate under these conditions as well, taking advantage of self-generated heat; external means, such as power from a battery or an electric grid, are sometimes required to execute a sub-freezing start-up or to prevent freezing.

Proton-exchange membrane fuel cells are indeed the type that has received the predominant share of development and implementation in the small fuel cell arena.

As the electrolyte, these PEM electrochemical cells use ion-exchange membranes in the acidic form; i.e., with protons as the exchangeable ions. The electrolyte membrane supports cell current generation via proton transport from the anode at one membrane surface to the cathode at the opposite surface, as illustrated in Fig. 43.1. Hydrogen ions are produced at the anode-electrolyte interface from hydrogen gas that diffuses through the anode structure. Electrons generated in this process migrate through the electronically conductive phase of the anode to an adjacent current collector. Hydrogen ions, reaching the cathode-electrolyte interface, react with electrons returning from the external load and with oxygen gas that diffuses through the cathode structure, producing water.

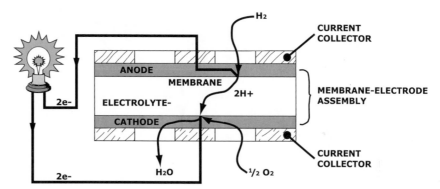

FIGURE 43.1   Schematic representation of an individual proton-exchange membrane fuel cell.

Since cell voltages generated in this process are typically in the 0.6 to 0.8 V range, multiple cells are generally stacked in a series-bipolar array to obtain practical voltage levels; this is illustrated in Fig. 43.2. The electrochemically active cell elements, shown schematically in Figs. 43.1 and 43.2, are commonly referred to as membrane-electrode assemblies, or MEAs. The MEAs are interleaved with bipolar plates that have multiple functions: conduction of electronic current from cell to cell, dispersion of the active hydrogen and oxygen gases through flow channels on the two opposing surfaces, prevention of mixing of these gases, and in many cases providing a means for removing the heat that is generated during the operation of the fuel cell. The voltage or current of the stack shown in Fig. 43.2 can be raised by increasing the number of cells or their active area, respectively.

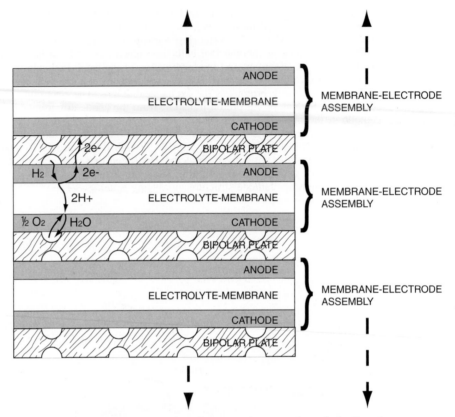

FIGURE 43.2 Schematic representation of proton-exchange membrane fuel cell stack.

43.3 SYSTEM REQUIREMENTS

The requirements for small fuel cell systems vary from application to application. The more important requirements for a given rated-power output include size, weight, cost, energy storage, voltage output, ambient-temperature operating range, and start-up characteristics. The approach followed for small fuel cell system design stresses simplicity of operation and maintenance as well as compactness.

43.3.1 Fuel Supply

To date small fuel cell systems have, for the most part, utilized some form of hydrogen fuel storage. For simplicity, fuel conservation, and safety, hydrogen is provided to the anodes at a modest pressure (typically at a gauge pressure from 10 to 50 kPa) and dead-ended, allowing the anodes to consume the amount of hydrogen required to sustain the electrochemical reaction at any given rate. A momentary "purge" of the exit port is implemented to allow accumulated impurities and water to be discharged from the anode compartment at selected time intervals. (See Sec. 43.4 on Fuel Processing and Storage.)

43.3.2 Air Supply

Small fuel cell stacks can operate on either diffused or forced reactant air. Diffused-air stacks are generally limited in their applicability because of air supply rate issues and their impact on geometry; the requisite openness of the air compartments in such devices also tends to render them vulnerable to atmospheric conditions. Hence, diffused-air (static) fuel cells are usually practical only in certain particularly low-power applications, up to perhaps 25 Watts.[1]

Forced-air fuel cell stacks are practical over the entire power range of small fuel cells. The reactant air is delivered to the fuel cell stack at whatever pressure is necessary to overcome the pressure drop through the stack and associated plumbing; this is typically in the range of 1 to 20 kPa (a small fraction of atmospheric pressure), depending on stack design characteristics. The air-moving devices are usually small air pumps, such as rotary-vane or diaphragm types. The stack's utilization rate of the oxygen in the reactant air will vary in accordance with operating conditions, but a typical rate is about 50 percent. The exit air is generally discharged to the atmosphere.

43.3.3 Water Management

A key design issue for small PEM fuel cells (and other PEM fuel cells as well) is the management of water with respect to the fuel cell stack. The trifluoromethanesulfonic-acid-based electrolyte membrane requires a certain level of water content in order to conduct protons efficiently, with water molecules effectively serving as carriers for the protons in their migration across the membrane. Accordingly, the system design must provide for a reasonably high relative humidity in the reactant passages that are in communication with the membrane.

The moisture requirements for the reactant streams place significant limitations on operating regimes. Ambient (non-humidified) reactant air is highly preferred in small fuel cells in order to achieve the simplicity and compactness that are generally sought in small power sources. The use of ambient air, however, requires design measures to prevent the membrane from drying out. The flow rate of air must be limited to reduce the drying effect, and the cell design must be tailored to take advantage of the product water. The threat of drying clearly becomes far more acute as the ambient temperature increases and as the current density of the stack is increased, thereby increasing the stack's temperature in relation to ambient.

The water management burden is not limited to preventing membrane dryout. The need to operate at relatively high oxygen utilization rates (relatively low air flow rates) increases the tendency to form water droplets within the cell from the formation of product water at the cathode. This can lead to accumulation of water on the surface of or within the electrode substrate or in the air distribution channels of the cathode flow-field. Such events could result in serious performance losses from the ensuing restriction of air access to impacted regions of the electrocatalyst. Consequently, the cell design approach must also serve to prevent such accumulation of water droplets.

43.3.4 Thermal Management

The thermal management requirements for small PEM fuel cells (and other PEM fuel cells as well) are intimately associated with water management. As discussed above, the level of hydration of the electrolyte membrane must be maintained in order to prevent dryout and thus loss of proton conduction in the membrane. The temperatures within the stack must accordingly be restricted. Cooling of the individual cells of the stack is carried out to assure that temperatures are moderate and rather uniform throughout the stack.

Stack cooling in larger PEM fuel cells is generally carried out using a liquid coolant, often water. This methodology is very effective because of the relatively high thermal conductivity and high volumetric heat capacity of liquids. However, as discussed above, small fuel applications predominantly require simple and compact systems, and a circulating liquid loop with external heat rejection is not compatible with these requirements. The preferred approach for small fuel cell stack cooling is based on the use of ambient air delivered across the external surface of each cell via a fan. The effectiveness of this method is typically enhanced by expanding the cells' external surface-area (as by providing finned extensions of the bipolar plates; see Fig. 43.3 and Sec. 43.5 on Stack Design).

FIGURE 43.3 Air-cooled fuel cell stack.

43.3.5 Operational Control

Small fuel cell systems, especially those operating on hydrogen fuel, are characteristically simpler than larger systems, as discussed earlier. Nevertheless, certain control elements must be imposed to foster stable operation. Representative methodology for these is as follows:

1. The reactant-air flow rate must be controlled as a function of load to assure that neither excessive water build-up (low flow rate) nor cell dry-out (high flow rate) is encountered. This requires measurement of stack current and corresponding speed adjustment in the air-moving device.

2. Stack temperature must be controlled to prevent operation in a dry-out condition (too hot). This requires that the stack cooling fan (or fans) be either turned on or ramped to a higher speed in response to a high-temperature signal from the stack's temperature sensor.

3. Since hydrogen-fueled systems run dead-ended, a timer (or coulometer) is utilized to impose a brief open-close cycle to a solenoid valve in the exit line to purge the anode compartment of accumulated impurities and water on a regular basis.

4. Other control means are provided on an application-specific basis as appropriate.

43.4 FUEL PROCESSING AND STORAGE TECHNOLOGIES

The advantages of small fuel cells often depend on the method used for the supply of fuel. Currently small fuel cell systems predominantly use some form of hydrogen storage. It is anticipated, however, that in the future, small systems will also utilize processed forms of common fuels (e.g., methanol, natural gas, and liquefied petroleum gas) as well as unprocessed methanol.

43.4.1 Hydrogen Storage

The hydrogen storage means that are presently being implemented in small fuel cell systems include compressed hydrogen gas canisters and metal hydride canisters. The challenge is to develop lightweight and small "containers" with high specific energy to make the overall system more attractive.

Compressed Hydrogen Gas Storage.   The simplest form of fuel storage and utilization for fuel cell systems is compressed hydrogen. Such storage is impractical for larger fuel cells, except in cases where operation is limited to brief periods in a back-up mode, because of the higher cost and inferior logistics of transport for hydrogen in comparison with common fuels. With respect to small fuel cells, however, compressed hydrogen is sometimes an acceptable storage option, especially at very low power levels and in back-up service.

Compressed hydrogen is generally selected for missions in which compactness is not particularly critical and in which storage pressures of 10 to 30 MPa are not objectionable. Since the weight of the active material is negligible, compressed hydrogen storage provisions can be exploited where light weight is a priority. This would require the use of lightweight canisters and pressure regulators, and the associated cost factors must be taken into account.

Metal Hydrides.   The storage of hydrogen in the form of hydrides of metal-alloy powders is often an attractive and convenient energy storage mode for small fuel cell systems. This is attributable to their simplicity of operation and compactness; the benefits are most realizable in particularly small systems. The energy densities of these materials can range up to 500 to 1000 Watt-hours (elec.) per liter, substantially higher than that of compressed hydrogen. This reflects a hydrogen loading approaching 2 percent by weight, which is characteristic of the maximum obtainable in alloys (typically AB_2 type, where, for example, A is Zr or a mixture of Zr and Ti, and B is a mixture of transition metals) that have useful hydrogen pressures at ambient temperatures. (Magnesium-based alloys have been formulated to obtain hydrogen loadings on the order of 5 percent by weight; but these require discharge temperatures in the neighborhood of 300°C, which necessitates combustion means with a percentage of the hydrogen being sacrificed to generate heat.)

The equilibrium hydrogen pressure of the hydride at the desired operating temperature can be appropriately tailored by adjusting the composition of the alloy mix. Pressure is generally selected in the range of 400 to 2000 kPa, a small fraction of that used in the case of compressed hydrogen. Nevertheless, the containment for the alloy usually constitutes a significant fraction of the overall weight of the energy-storage provisions; and depending on the size and discharge rates required, additional means might be needed for internal hydrogen manifolding and for internal heat transfer. (The latter relates to the endothermic reaction associated with the discharge of hydrogen from the metal hydride lattice. For most small systems, the heat necessary to sustain the reaction is simply supplied through the walls of the canister from ambient air; but, even here, the size and discharge rate might be high enough to require internal heat-conduction enhancement to prevent excessive temperature gradients along with unacceptable diminution of hydrogen pressure.) The AB_2 type metal hydrides (and AB_5 type, e.g., $LaNi_5$ as well) are quite dense (powder bulk densities on the order of 3 g/cm^3), and their specific-energies are not particularly high (perhaps 200 to 300 Watt-hours per kilogram). As indicated above, the containment and other design provisions will significantly reduce the specific energy for the energy-storage system as a whole.

The cost of the active metal hydride materials is usually quite acceptable in the context of small fuel cell systems ($5 to 25 per kilogram). Clearly, however, as in the case of compressed hydrogen, provisions must be made for recharging or reserve storage when the fuel cell is used in continual operation. The charging pressure is quite low, but the associated exothermic reaction limits the charging rate. Depending on size, canister design characteristics, and degree of charging completeness, this process typically requires 5 to 30 minutes using a conventional compressed hydrogen supply. Future methodology for very small systems (e.g., consumer electronic applications) could employ pressurized water electrolysis for canister recharging.

Chemical Hydrides.   Primary hydride systems of various types are being developed for use in small fuel cell systems. These are irreversible (throw-away) chemical systems that generate hydrogen on demand. The active reactant is generally an alkali or alkaline earth metal hydride (sometimes in complex form). This typically is caused to react with water to form hydrogen and a metal oxide (or mixed metal oxide). Analogous chemical hydride systems are also being explored.

The potential advantages of chemical hydride systems include high specific energy, since the hydrogen yield by weight can be a far higher percentage of the reactant weight in comparison with reversible metal hydrides. The hydrogen-generating reaction, theoretical hydrogen yield and the theoretical value for the specific energy of the fuel are shown in Table 43.2 for representative chemical hydrides. (Note that the mass of water required to generate the hydrogen is included in the calculations of percent hydrogen and specific energy for each chemical hydride.) This advantage can be enhanced if product water from the fuel cell can be recovered for use in the chemical hydride reaction, in which case the system could be refueled simply via a stored reserve of reactant powder or granules. On the other hand, energy densities (Wh/L) for these systems are not as attractive because of the relatively low densities of the reactants.

TABLE 43.2 Representative Chemical Hydride Reactions and Theoretical Hydrogen Yields and Specific Energy

Reaction	Theoretical hydrogen yield	Theoretical specific energy
$LiH + H_2O \rightarrow LiOH + H_2$	7.8 percent	2540 Wh/kg
$CaH_2 + 2H_2O \rightarrow Ca(OH)_2 + 2H_2$	5.2 percent	1700 Wh/kg
$NaBH_4 + 2H_2O \rightarrow NaBO_2 + 4H_2$	10.9 percent	3590 Wh/kg

Theoretical, including weight of hydride and water, in H_2/air fuel cell based on 1.23 V per cell.

The challenges associated with chemical hydrides include the requirement that the respective reactants be brought into contact so that the rate of reaction just meets the fuel cell's hydrogen needs. Also, since the reaction products are disposed of, as opposed to being regenerated, the cost of the replenishing reactant chemicals must be taken into account in evaluating the economics of system operation.

Carbon-Based Hydrogen Storage.   Hydrogen-storage methodologies based on carbon materials are currently the subject of considerable technology development and could provide opportunities for attractive energy-storage systems in the future. Some investigations have focused on the sorption of hydrogen on activated carbons.[2] This phenomenon is greatly enhanced by carrying out the sorption (and storage) at cold temperatures (typically below −100°C). Under these conditions, the loading of hydrogen can be rather high (substantially higher in weight percent than that of metal hydrides). Consequently, the specific energies for such storage systems are potentially attractive, although, because of their relatively low densities, the energy per unit volume parameters are not especially high. More important for small fuel cells is the relative impracticality of cryogenic storage at this scale.

In recent years much attention has been devoted to hydrogen sorption on nanoscale carbon or graphitic structures.[3] Hydrogen loadings at room temperature have been reported ranging from a few percent by weight to far higher levels. Results appear to be irreproducible and highly dependent on preparation methodologies. It is too early to predict the ultimate promise of such approaches, but research is proceeding actively toward optimization of loading as well as adsorption-desorption characteristics as a function of pressure and temperature. These activities could play a significant role in determining future energy-storage concepts for small fuel cells (and perhaps larger fuel cells as well).

43.4.2   Fuel Processing

The application of small fuel cells could clearly be greatly enhanced if compact systems using conventional fuels are implemented. In most cases, this requires a fuel processor to convert the fuel into a hydrogen-rich gas that would be delivered to the fuel cell. Much of the challenge in such an approach relates to attaining a sufficiently compact and low-cost fuel processor. In the case of methanol in particular, there is also the potential for systems that use direct fuel feed to the fuel cell anodes.

A variety of common fuels and chemicals can be considered as candidate fuels for small fuel cells. These include the following.

Ammonia.   Ammonia is commonly used in industry and agriculture in the form of a liquid stored at its own, modest vapor pressure. Liquid ammonia offers high specific energy and energy density based on a relatively simple thermocatalytic hydrogen-generating dissociation reaction

$$2NH_3 \rightleftharpoons 3H_2 + N_2$$

Thus, ammonia can yield hydrogen at about 17 percent of its own weight. This corresponds to about 3 kWh (elec.) per kilogram and almost 2 kWh (elec.) per liter, assuming a practical cell operating voltage of 0.7 V. However, a fraction of the hydrogen formed must be consumed in generating heat to sustain the endothermic dissociation reaction. LPG is generally preferred as a fuel over ammonia because of its greater distribution infrastructure, even higher energy density, and lower cost per unit of energy content, and because of ammonia's reputation as a toxic chemical. Nevertheless, ammonia could play a role in selected small fuel cell applications as a result of its far simpler hydrogen-generating process.

Methanol.   Methanol is a widely available chemical that is relatively easy to handle and store as a liquid at atmospheric pressure. It can also be converted to a hydrogen-rich gas by a process that is simplest among those for carbon-containing fuels. The endothermic reaction

$$CH_3OH + H_2O \rightleftharpoons CO_2 + 3H_2$$

is carried out at modest temperature (about 250°C) since methane (favored by thermodynamic equilibrium) cannot be formed when conventional methanol steam-reforming catalysts are used. Accordingly, little downstream processing is needed because very little carbon monoxide is formed at these temperatures. Here again, whereas LPG is favored from the points of view of energy content and cost, methanol is an attractive fuel for small fuel cells because its processing burden is greatly eased.

Ethanol.   Ethanol is similar to methanol in its handling and storage. Its reaction with steam can be expressed as

$$C_2H_5OH + H_2O \rightleftharpoons 2CO + 4H_2$$

Unlike methanol, it is considered non-toxic. Its availability and cost as a fuel are irregular; more important, ethanol cannot be processed at the low temperatures characteristic of methanol steam reforming. Therefore, its potential attractiveness as a fuel for compact fuel cell systems is substantially diminished.

Liquefied Petroleum Gases (LPG).   As indicated above, LPG (principally propane in the U.S., but sometimes principally butane, as in Japan) is often the preferred fuel for dispersed fuel cell systems. It has substantially higher specific energy density than both ammonia and methanol along with lower price per unit energy. Indeed, in the absence of pipeline natural gas, it is the fuel of choice for stationary fuel cells. However, small fuel cells generally have a different set of criteria. In a relative sense, fuel cost is perhaps less important; and compactness, simplicity, and hardware cost more important (although this cannot be over-generalized; e.g., missions with very long durations could be an exception).

Natural Gas.   Small fuel cell systems that are stationary and have ready access to a natural gas pipeline will predominantly take advantage of the natural gas availability (just as in the case of a larger stationary fuel cell). The principal constituent of natural gas is generally methane, CH_4. The cost per unit energy for natural gas is the most attractive, and compactness is presumably not a major issue. Since its storage characteristics are not attractive and its processing system is no more favorable than that of LPG, these are likely to be the only circumstances under which natural gas would be utilized in small fuel cells.

Aviation Fuel or Diesel Fuel.   Aviation and diesel fuels are preferred for military applications since they are the most available and safest (very low vapor pressure). Their availability is also a key factor in certain under-developed regions. On the other hand, these are the most difficult to process (leaving out heavier fuels, like heating oil). This is attributable to the difficulty of breaking down their larger molecules and to their relatively high sulfur content. Therefore, such fuels would generally be avoided for small fuel cell systems.

43.4.3   Methodologies For Fuel Processing

The generation of hydrogen from carbon-containing fuels (with the exception of methanol) requires a high-temperature (usually catalytic) process. The fuel is reacted with steam catalytically (steam reforming, or SR), with sub-stoichiometric oxygen from air (partial oxidation, homogeneous or catalytic, POX), or with steam and oxygen catalytically (autothermal, ATR). SR is an endothermic reaction carried out typically at 650°C, or higher; POX is an exothermic process, usually carried out at higher temperatures (perhaps 1000°C), while the ATR process is almost thermally neutral and typically operates at or somewhat above SR temperatures. Representative reactions for these processes using methane as the fuel can be expressed as follows:

$$SR: CH_4 + H_2O \rightleftharpoons CO + 3H_2$$

$$POX: CH_4 + 0.5O_2 \rightleftharpoons CO + 2H_2$$

$$ATR: CH_4 + 0.25\ O_2 + 0.5H_2O \rightleftharpoons CO + 2.5H_2$$

In simplistic terms, the SR process generally provides the highest hydrogen yield and consequently the highest efficiency. Because of the endothermic reaction, its thermal management tends to be the most complex and bulky. Conversely, POX is typically the least efficient but has the simplest configurations. The ATR process tends to be intermediate to the other two in both respects.

The selection of a preferred fuel processor type depends heavily on the application requirements. For example, a conventional stationary fuel cell system operating continuously on natural gas might be best suited to the SR processor to minimize fuel cost, while a small, mobile system requiring rapid start-up might be best served via a POX, or ATR, system.

Process Gas Upgrading.   The high-temperature processes described above yield a reformate gas that is high in carbon monoxide (CO) content (usually greater than 10 percent). In order to maximize the hydrogen yield (and, in the case of low-temperature fuel cells like PEM, minimize the fuel cell anode-catalyst inhibiting effects of CO), the reformate gas is then passed through a catalytic water-gas shift-converter at lower temperature (perhaps in two stages) where the following reaction takes place

$$CO + H_2O \rightleftharpoons CO_2 + H_2$$

Here again, as in the case of methanol steam reforming, the formation of methane under these conditions is prevented via the specificity of the shift-converter catalyst. For PEM fuel cells, further reduction in CO concentration is necessary (from about 0.5 percent down to less than 100 ppm). This is often carried out by way of catalytic preferential oxidation of CO in the presence of hydrogen with the addition of air at a flow rate that is a small multiple of the stoichiometric rate required for complete oxidation of the CO.

It is evident that steam or water vapor is an essential player in the fuel processor, whether it be in the reforming reaction itself or in the subsequent shift-conversion stage. The source of this water must come from storage, make-up, or condensation and recovery from the fuel cell system. In any event, the design and logistics for water management must be provided for the system, and the selected mode must best reflect the requirements for the specific application.

The complexity of the overall fuel processing system for carbonaceous fuels indicates the challenge associated with adapting conventional fuels for use in small fuel cell systems. Considerable effort is required in designing and optimizing the system to achieve the requisite miniaturization and low cost in these applications.

43.4.4 Direct-Methanol Fuel Cell (DMFC) Technology

It is clear that a system that can utilize a liquid fuel directly at the fuel cell anodes would be particularly appealing in small fuel cell applications. As mentioned earlier, at this time methanol is the only carbonaceous species that can both serve as a practical fuel and provide reasonable electrochemical performance at the fuel cell anode. Substantial research and development efforts have been directed at direct-methanol fuel cells using PEM electrolyte for more than a decade, in the context of small, field-deployable systems.

While the direct-methanol approach is inherently appealing for small fuel cells, this technology is challenging and complex. Specifically, (a) the electrochemical activity at the anode is poor, producing low cell voltages and requiring relatively high precious metal loadings to attain desirable current densities; (b) methanol is soluble in conventional PEM electrolytes, and its "cross-over" to the cathode results in wasted fuel consumption along with inhibition of normal oxygen transport and electrochemical reduction at this electrode; and (c) maintenance of water balance within the cells requires that water vapor discharged at the cathode flowfield exit be condensed and returned to the circulating methanol-water anode feed solution which, in order to retard methanol migration across the electrolyte membrane to the cathode, is maintained in a very dilute state, typically about 3 percent by weight.

Despite the issues cited above, the DMFC is considered to have much potential for implementation in small fuel cell systems. Research and development work that specifically focuses on these challenges is proceeding[4] and meaningful advances are anticipated in the period ahead. Also, for applications requiring extremely low power, approaches involving the feed of methanol, or methanol and water, without recirculation can be considered since heat removal (generally via a heat-exchanger in the recirculation loop) could be accomplished via natural convection.[4,5,6,7] In any event, when DMFCs are ready for commercial service, they will offer attractive incentives, especially for extended-duration missions, because of their simplicity of storage and relatively high energy densities and specific energies (approximately 2000 Wh/kg at a practical cell voltage of 0.4 V).

43.5   FUEL CELL STACK TECHNOLOGY

The technology of the PEM fuel cell stack must be in keeping with the small fuel cell system design approach described in Sec. 43.3.

43.5.1   Design

The requirements of the small fuel cell system translate into a fuel cell stack that is compact, externally air-cooled and utilizes unconditioned ambient air as a reactant. For hydrogen-fueled systems (which are the majority to date) the anode compartments of the stack are virtually dead-ended. The need for compactness calls for internal reactant manifolding and bolting as well as thin end-plates. A representative stack for a small fuel cell system is shown in Fig. 43.3. The cell components, of course, should also be as thin as possible.

43.5.2   Electrolyte

The most suitable electrolyte for small fuel cell systems is the proton-exchange membrane (PEM). For the most part, these are based on trifluoromethanesulfonic acid in a tetrafluoroethylene-based polymer backbone. The material's equivalent weight is typically 1100, although versions with an equivalent weight near 1000 are also in use. Membranes with lower equivalent weight tend to have superior proton-exchange and water-transport properties but they also have greater vulnerability to dry-out conditions. The thickness of membranes in use in fuel cells of all types today is roughly in the 20 to 200 μm range. Thinner membranes, of course, exhibit higher proton-conductance as well as rate of water transport, while posing a somewhat higher risk of failure, either in processing or in operation. Membranes used in small fuel cells are generally on the thinner side, under 100 μm and sometimes far thinner.

43.5.3 Electrodes

Electrodes used in PEM fuel cells typically employ an electrocatalyst layer with a porous, carbonaceous, electronically-conductive substrate that has been rendered hydrophobic. The catalyst layer usually comprises platinum, or a platinum-containing alloy, on a carbonaceous support (typically carbon black), dispersed ionomeric material similar in constituency to the electrolyte-membrane, and dispersed hydrophobic polymer such as polytetrafluoroethylene. The substrate, which serves as a reactant-gas diffusion layer, may be a carbon paper or a woven or non-woven cloth. The catalyst layer may be deposited directly onto the electrolyte-membrane or onto the substrate and later placed in contact with the membrane.

Anodes are usually very similar to, if not identical to those that serve as cathodes. Anodes that operate on reformed-hydrocarbon fuels, which contain some carbon monoxide, generally utilize a platinum-alloy catalyst to enhance co-tolerance. The catalyst-layer structure is sometimes altered between anodes and cathodes to adjust their respective hydrophobicity and reactant-diffusion properties. The thickness of the catalyst layer typically ranges from 10 to 20 μm, that of the substrate from 0.1 To 0.5 mm (uncompressed).

43.5.4 Bipolar Plates

The typical stack construction for PEM (and other) fuel cells is a series-connection of cells with bipolar plates interposed between adjacent cells (or membrane-electrode assemblies). The bipolar plate must provide for electronic conduction from one cell to the next; isolation of fuel on one side from oxidant (air) on the other side; and distribution of fuel and air reactants to the respective adjacent anode and cathode. In an edge-cooled stack of the type described above, the bipolar plate also contributes to heat rejection by conducting heat laterally to its finned edges, where forced-air convection is employed.

Graphite-based bipolar plates are often preferred for small fuel cell stacks because of their relatively high thermal conductivity as well as their electrochemical stability. These plates are generally comprised of graphite and a polymeric resin that have been compression-molded from powders into a nominally pore-free structure. High resin content promotes impermeability, but graphite content in the neighborhood of 80 percent by weight is usually necessary in order to achieve acceptable electronic conductivity. Their thickness requirements must take into account the depth of fuel and air reactant channels as well as structural soundness and avoidance of reactant permeability. Nevertheless, thicknesses down to about 1 mm can be implemented in small stacks.

43.5.5 Seals

Sealing means must be provided at the edges of the cells to prevent reactants from escaping to the atmosphere from porous elements of the electrodes. Also, since reactant manifolding is usually internal to the stack in small fuel cells (for the sake of compactness), sealing around manifold holes is required to prevent the mixing of reactants between manifolds and electrodes; the considerations are the same for internal tie-bolt holes.

The sealing methodology generally employed involves a polymeric frame element (gasket) on each side of the electrolyte-membrane, separating the respective, truncated electrodes from the outside. These elements contain appropriate holes to allow for reactant manifolding and tie-rods in these regions.

43.6 HARDWARE AND PERFORMANCE

The general requirements for small fuel cells include simplicity and compactness, especially for hydrogen-based systems. Accordingly, small fuel cells typically use reactant air, ambient pressure and temperature with no external humidification, and are cooled via ambient air. Operating current densities are, therefore, usually modest; e.g. in the range of 100 to 250 mA/cm². Since fuel storage capacity is often a limiting factor, the design point tends to provide relatively high cell voltage, generally higher than 0.7 V and sometimes as high as 0.8 V, depending on system requirements.

43.6.1 Electrical Characteristics

A representative voltage-current curve for a fuel cell stack in a nominal 50-Watt system is shown in Fig. 43.4. This stack has an active area of 16.5 cm² per cell with 32 cells and a cell pitch of six per centimeter (i.e. a stacking density of six cells per centimeter). The stack uses hydrogen as a fuel (dead-ended) the stoichiometric ratio for the air is 2.0 to 2.5.

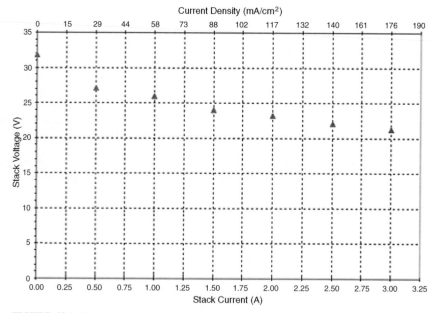

FIGURE 43.4   Voltage vs. current density for 32-cell stack.

Figure 43.5 illustrates a typical performance curve for a stack in a nominal 250-Watt system. This stack, shown in Fig. 43.6, has an active area of 77 cm² per cell with 40 cells and a cell pitch of more than four per centimeter. Hydrogen is the fuel and the air is at atmospheric pressure.

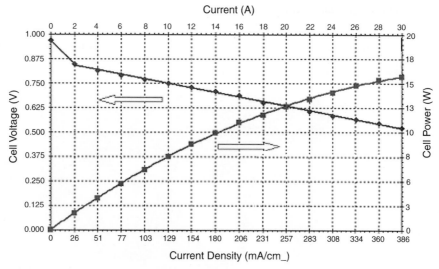

FIGURE 43.5 Performance of 40-cell fuel cell stack.

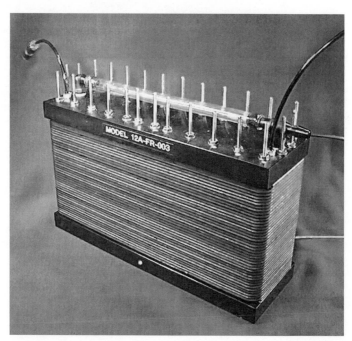

FIGURE 43.6 40-cell fuel cell stack. Dimensions (cm): 15.5(H), 8.6(W), 23.2(L). Weight: 4.5 kg.

43.6.2 50-Watt Fuel Cell, Compressed Hydrogen Cylinder

Figure 43.7 is an illustration of a lightweight 50-Watt system developed for military field use incorporating two compressed hydrogen cylinders for the fuel source. The characteristics of this fuel cell system are summarized in Table 43.3.

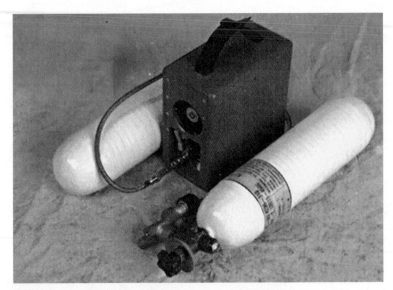

FIGURE 43.7   50-Watt fuel cell system with compressed hydrogen gas. (*Courtesy of Ball Aerospace and Technologies Corp.*)

TABLE 43.3   Characteristics of 50-Watt Fuel Cell

Size (cm)—fuel cell only	10.9(W), 19.6(H), 20.3(L)
Weight (kg)—fuel cell only	2.95
Weight (kg)—fuel source	
(1) High pressure H_2 tanks:	2.3 kg/kWh
(2) Chemical hydride H_2 source:	2.5 kg/kWh
Total system weight (kg) with high pressure H_2 tanks (for 1 kWh of operation)	5.22
Power (Watts) at 12 V	50
Peak power (Watts), at >10 V	65

Source: Ball Aerospace and Technologies Corp.

43.6.3 20-Watt Fuel Cell, Metal Hydride Canister

An example of a small fuel cell system fueled via a metal hydride canister is illustrated in Fig. 43.8. This unit was assembled to fit within the dimensions of military Battery BA-5590 (see Sec. 6.4.2). The fuel cell is rated at approximately 20 Watts continuous and 40 Watts peak. The canister delivers hydrogen for about 110 Watt-hours of operation at a specific energy of about 160 Wh/kg. The specific energy of the overall system is about 75 Wh/kg. On a volume basis, the energy density of the fuel cell system is 110 Wh/L and the replacement canisters deliver about 420 Wh/L.

Although the weight of the fuel cell system is high compared to Battery BA-5590, which has a specific energy of 160 Wh/kg, the fuel cell could be advantageous when used over an extended period. This is illustrated in Fig. 42.3. The specific energy of the replacement canisters is close to that of the battery and its cost, when commercialized, should be significantly lower than that of the battery. Improved devices for hydrogen storage could make the fuel cell system more attractive.

FIGURE 43.8 Metal hydride fueled system using BA-5590 type case. (The figure is of a system design having larger dimensions than the BA-5590.)

43.6.4 Commercial Fuel Cells

Manufacturing and testing of prototype small PEM forced-air fuel cells for commercial application, mainly using hydrogen as a fuel, began in the mid-1990s. Some examples of current developments include the following.

50-Watt to 500-Watt Fuel Cell Systems. Fuel cell systems with higher power ratings, up to 500 Watts, are illustrated in Figs. 43.9 and 43.10. These units have either AC or DC outputs and are manufactured in both portable and rack-mounted configurations. The weight of the 250-Watt unit is 10 kg. The 500-Watt AC unit (120 VAC) is specifically suited for back-up service in the home and operates on hydrogen fuel stored in metal hydride canisters.

The 50-Watt and 100-Watt portable fuel cell systems, illustrated in Fig. 43.11, are designed for military field use. The figure shows the units operating and recharging batteries of military radios in the field.

FIGURE 43.9   250-Watt fuel cell power system. Dimensions (cm): 24.7(H), 15.8(W), 40.7(L). Weight: 10.2 kg. (*Courtesy H-Power Corp.*)

FIGURE 43.10   500-Watt fuel cell power system. Dimensions (cm): 29.2(H), 20.3(W), 40.6(L). Weight: 17.3 kg.

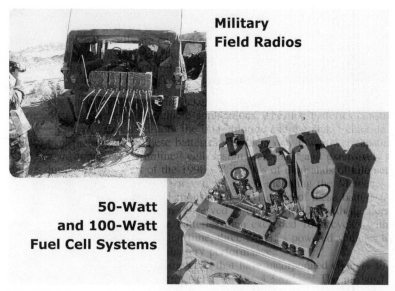

Military Field Radios

50-Watt and 100-Watt Fuel Cell Systems

FIGURE 43.11   50-Watt and 100-Watt systems charging batteries of military field radios.

Remote or Unattended Application.   Another fuel cell application for remote telecommunications is the use of fuel cells to back-up the solar/battery array at weather stations and microwave-repeater stations. Fuel cells in the 20 to 100 Watt range are used with hydrogen supplied in the form of compressed gas cylinders.

An example of an unattended application is illustrated in Fig. 43.12, a power source for variable-message highway signs. This system consists of a 50 Watt hydrogen-air fuel cell stack and its auxiliary components. A second system is provided for redundancy. The two systems operate together under normal circumstances, sharing the load, but one unit could provide full power in the case of failure of the other. These systems are designed to start when the output of the main solar-battery power system is inadequate to maintain the batteries at an acceptable state of charge. Thus, when solar unavailability is sustained long enough to allow the voltage of the batteries of the hybrid system to drop below a cutoff point, the fuel cell takes over and supplies power to the load and charges the batteries to the extent possible.

The hydrogen fuel is supplied in the form of four small aluminum industrial cylinders which are placed in a compartment positioned to replace a portion of the battery bank. The fuel supply could sustain operation without any solar power for 12 days of continuous operation. In actual practice, it tends to last for about six weeks during the winter, when solar availability is most diminished. It is usually not needed during other seasons of the year.

Prognosis.   The introduction of small subkiloWatt fuel cells into commercial and industrial applications has demonstrated the operational viability of the PEM technology and the potential advantage that could be gained in energy density as a result of the use of hydrogen and other high energy-rich fuels. In some cases, e.g., replacement of lead-acid batteries used for wheelchair propulsion, prototype fuel cell power sources have shown clear superiority in operating time for a given weight and volume. The actual depth of penetration into the market traditionally served by batteries will be strongly influenced by a number of factors, including cost reduction, fuel logistics, demonstration of reliability, and identification of applications where fuel cells provide geater performance at acceptable costs.

FIGURE 43.12 Twin 50-watt fuel cell systems. (*Courtesy H-Power Corp.*) Dimensions (cm): 33.0(H), 40.6(W), 45.7(L). Weight: 13.6 kg.

REFERENCES

1. Dougherty, M., D. Haberman, N. Stetson, S. Ibrahim, O. Lokken, D. Dunn, M. Cherniack, and C. Salter, *Proc. of Conference on Portable Fuel Cells,* Lucerne, Switzerland, pp. 69–78, June 21–24, 1999.

2. Hynek, S., W. Fuller, and J. Bentley, *Int. J. of Hydrogen Energy,* **22:**6, pp. 601–610, 1997.

3. Kim, M. S., C. Park, A. Chambers, R. T. K. Baker, and N. M. Rodriguez, *Proc. of Workshop on Hydrogen Storage and Generation for Medium-Power and Energy Applications,* Orlando, Fla., pp. 343–361, April 8–10, 1997.

4. Gottesfeld, S., X. Ren, P. Zelaney, H. Dinh, J. Davey, and F. Guyon, *Proc. of 2000 Fuel Cell Seminary,* Portland, OR, pp. 799–802, October 30–November 2, 2000.

5. Koschany, P., *Proc. of Conf. on Portable Fuel Cells,* Lucerne, Switzerland, pp. 61–65, June 21–24, 1999.

6. Hockaday, R., M. DeJohn, C. Navas, P. Turner, H. Vaz, and L. Vazul, *Proc. of 2000 Fuel Cell Seminar,* Portland, OR, pp. 791–794, October 30–November 2, 2000.

7. Narayanan, S. R., T. I. Valdez, and F. Clara, *Proc. of 2000 Fuel Cell Seminar,* Portland, OR, pp. 795–798, October 30–November 2, 2000.

8. Gibbard, H. F., *Proc. of Conf. on Portable Fuel Cells,* Lucerne, Switzerland, pp. 107–112, June 21–24, 1999.

9. Kaufman, A., P. Terry, and T. Lomma, *Proc. of 2000 Fuel Cell Seminar,* Portland, OR, pp. 811–812, October 30–November 2, 2000.

10. Dubois, R., *Proc. of Fuel Cell 2000 Conf.,* Lucerne, Switzerland, L. Blomen (ed.), pp. 71–74, July 10–14, 2000.

11. Gibbard, H. F., *Proc. of the Energy 2000 Conf.,* Pittsburgh, PA, August 21–23, 2000.

12. Chu, D., *Proc. of Conf. on Small Fuel Cells and Battery Technologies for Use in Portable Applications,* New Orleans, LA, April 26–28, 2000.

13. Kaufman, A., *Proc. of Conf. on Small Fuel Cells and the Latest Battery Technologies for Portable Applications,* New Orleans, LA, April 29–30, 1999.

APPENDICES

APPENDIX A
DEFINITIONS

Adsorption   The taking up or retention of one material or medium by another by chemical or molecular action.

Accumulator   See SECONDARY BATTERY.

Activated Stand Life   The period of time, at a specified temperature, that a battery can be stored in the charged condition before its capacity falls below a specified level.

Activation   The process of making a reserve battery functional, either by introducing an electrolyte, by immersing the battery into an electrolyte, or by other means (see Part 3).

Activation Polarization   Polarization resulting from the charge-transfer step of the electrode reaction (see POLARIZATION).

Active Cell or Battery   A cell or battery containing all components and in a charged state ready for discharge (as distinct from a RESERVE CELL or BATTERY).

Active Material   The material in the electrodes of a cell or battery that takes part in the electrochemical reactions of charge or discharge.

Aging   Permanent loss of capacity due either to repeated use or to passage of time.

Ambient Temperature   The average temperature of the surroundings.

Ampere-Hour Capacity   The quantity of electricity measured in Ampere-hours (Ah) which may be delivered by a cell or battery under specified conditions.

Ampere-Hour Efficiency   The ratio of the output of a secondary cell or battery, measured in Ampere-hours, to the input required to restore the initial state of charge, under specified conditions (also coulumbic efficiency).

Anion   Ion in the electrolyte carrying a negative charge.

Anode   The electrode in an electrochemical cell where oxidation takes place. During discharge, the negative electrode of the cell is the anode. During charge, the situation reverses and the positive electrode of the cell is the anode.

Anolyte   The portion of the electrolyte in a galvanic cell adjacent to the anode; if a diaphragm is present, the electrolyte on the anode side of the diaphragm.

Aprotic Solvent   A nonaqueous solvent that does not contain any reactive protons although it may contain hydrogen atoms in the molecule.

Available Capacity   The total capacity (Amp-hours) that will be obtained from a cell or battery at defined discharge rates and other specified discharge or operating conditions.

Battery   One or more electrochemical cells electrically connected in an appropriate series/parallel arrangement to provide the required operating voltage and current levels including, if any, monitors, controls and other ancillary components (fuses, diodes), case, terminals and markings (see p. 1.3).

Bipolar Plate   An electrode construction where positive and negative active materials are on opposite sides of an electronically conductive plate.

Bobbin   A cylindrical electrode (usually the positive) pressed from a mixture of the active material, a conductive material, such as carbon black, the electrolyte and/or binder with a centrally located conductive rod or other means for a current collector.

Boost Charge   Charging of batteries in storage to maintain their capacity and counter the effects of self-discharge.

Boundary Layer   The volume of electrolyte solution immediately adjacent to the electrode surface in which concentration changes occur due to the effects of the electrode process.

C Rate   The discharge or charge current, in Amperes, expressed as a multiple of the rated capacity in ampere-hours.

$$I = M \times C_n$$

where I = current, A
  C = numerical value of rated capacity of a battery in ampere-hours (Ah)
  n = Time in hours for which rated capacity is specified
  M = multiple or fraction (of C)

For example, the $0.05C$ or $C/20$ discharge current for a battery rated at 5 Ah at the $0.2C$ or $C/5$ rate is 250 mA.

$$I = M \times C_{0.2} = (0.05)(5) = 0.250 \text{ Amperes}$$

Conversely, a battery rated at 300 mAh at the $0.5C$ or $C/2$ rate, discharged at 30 mA, is discharged at the $0.1C$ or $C/10$ rate, which is calculated as follows:

$$M = \frac{I}{C_{0.5}} = \frac{0.030}{0.300} = 0.1 \text{ or } C/10$$

Capacitance Current   The fraction of the cell current consumed in charging the electrical double layer.

Capacity   The total number of Ampere-hours (Ah) that can be withdrawn from a fully charged cell or battery under specified conditions of discharge. (See also AVAILABLE CAPACITY, RATED CAPACITY.)

Capacity Fade   Gradual loss of capacity of a secondary battery with cycling.

Capacity Retention   The fraction of the full capacity available from a battery under specified conditions of discharge after it has been stored for a period of time.

Cathode   The electrode in an electrochemical cell where reduction takes place. During discharge, the positive electrode of the cell is the cathode. During charge, the situation reverses, and the negative electrode of the cell is the cathode.

Catholyte   The portion of an electrolyte in a galvanic cell adjacent to a cathode; if a diaphragm is present, the electrolyte on the cathode side of the diaphragm.

Cation   Ion in the electrolyte carrying a positive charge.

Cell   The basic electrochemical unit providing a source of electrical energy by direct conversion of chemical energy. The cell consists of an assembly of electrodes, separators, electrolyte, container and terminals (see p 1.3.)

Charge   The conversion of electrical energy, provided in the form of a current from an external source, into chemical energy within a cell or battery.

Charge Acceptance   Ability of a battery to accept charge. May be affected by temperature, charge rate, and state of charge.

Charge Control   Techniques for effectively terminating the charging of a rechargeable battery.

Charge Efficiency   See EFFICIENCY.

Charge Rate   The current applied to a secondary cell or battery to restore its capacity. This rate is commonly expressed as a multiple of the rated capacity of the cell or battery. For example, the $C/10$ charge rate of a 500-Ah cell or battery (rated at the 0.2 rate) is expressed as

$$\frac{C_{0.2}}{10} = \frac{500 \text{ Ah}}{10} = 50 \text{ A}$$

Charge Retention   See CAPACITY RETENTION.

Closed-Circuit Voltage (CCV)   The potential or voltage of a cell or battery when it is discharging, normally under a specified load.

Concentration Polarization   Polarization caused by the depletion of ions in the electrolyte at the surface of the electrode. (See also POLARIZATION.)

Conditioning   Cycle charging and discharging of a battery to ensure that it is fully formed and fully charged. Sometimes indicated when a battery is first placed in service or returned to service after prolonged storage.

Constant Current Charge   A method of charging the battery using a current having little variation.

Constant Voltage Charge   A method of charging the battery by applying a fixed voltage, and allowing variations in the current. Also called constant potential charge.

Continuous Test   A test in which a battery is discharged to a prescribed end-point voltage without interruption.

Coulometer   Electrochemical or electronic device, capable of integrating current-time, used for charge control and for measurement of charge inputs and discharge outputs. Results usually reported in Ampere-hours.

Counter Electromotive Force   A voltage of an electrochemical cell opposite to the applied external voltage. Also referred to as back EMF.

Couple   Combination of anode and cathode materials that engage in electrochemical reactions that will produce current at a voltage defined by the reactions.

Creepage   The movement of electrolyte onto surfaces of electrodes or other components of a cell with which it is not normally in contact.

Current Collector   An inert member of high electrical conductivity used to conduct current from or to an electrode during discharge or charge.

Current Density   The current per unit active area of the surface of an electrode.

Cutoff Voltage   The battery voltage at which the discharge is terminated. Also called end voltage.

Cycle   The discharge and subsequent or preceding charge of a secondary battery such that it is restored to its original conditions.

Cycle Life   The number of cycles under specified conditions which are available from a secondary battery before it fails to meet specified criteria as to performance.

Cycle Service   A duty cycle characterized by frequent and usually deep discharge-charge sequences, such as motive power applications.

Deep Discharge   Withdrawal of at least 80% of the rated capacity of a battery.

Density   The ratio of a mass of material to its own volume at a specified temperature.

Depolarization   A reduction in the polarization of an electrode.

Depolarizer   A substance or means used to prevent an increase polarization. The term "depolarizer" is often used to describe the positive electrode or cathode of a primary cell.

Depth of Discharge (DOD)   The ratio of the quantity of electricity (usually in Ampere-hours) removed from a cell or battery on discharge to its rated capacity.

Desorption   The opposite of absorption, whereby the material retained by a medium is released.

Diaphragm   A porous or permeable material for separating the positive and negative electrode compartments of an electrochemical cell and preventing admixture of catholyte and anolyte.

Diffusion   The movement of species under the influence of a concentration gradient.

Discharge   The conversion of the chemical energy of a cell or battery into electrical energy and withdrawal of the electrical energy into a load.

Discharge Rate   The rate, usually expressed in Amperes, at which electrical current is taken from the cell or battery.

Double Layer   The region in the vicinity of an electrode-electrolyte interface where the concentration of mobile ionic species has been changed to values differing from the bulk equilibrium value by the potential difference across the interface.

Double-Layer Capacitance   The capacitance of the electrical double layer at an electrode-electrolyte interface.

Dry Cell   A cell with immobilized electrolyte. The term "dry cell" is often used to describe the Leclanché cell.

Dry Charged Battery   A battery in which the electrodes are in a charged state, ready to be activated by the addition of the electrolyte.

Duplex Electrode or Plate   See BIPOLAR PLATE.

Duty Cycle   The operating regime of a cell or battery including factors such as charge and discharge rates, depth of discharge, cycle length, and length of time in the standby mode.

E Rate   The discharge or charge power, in Watts, expressed as a multiple of the rated capacity in Watthours.

$$P = M = E_n$$

where P = power, W
E = numerical value of the rated energy of a battery in Watt-hours (Wh)
n = time in hours, at which the battery was rated
M = multiple or fraction (of E)

For example, the $0.05E$ or $E/20$ discharge power for a battery rated at 5 h at the $0.2E$ or $E/5$ rate is 250 mW.

$$P = M \times E_{0.2} = (0.05)(5) = 0.250 \text{ Watts}$$

Conversely, a battery rated at 300 mWh at the $0.5E$ or $E/2$ rate, discharged at 30 mW, is discharged at the $0.1E$ or $E/10$ rate, which is calculated as follows:

$$M = \frac{I}{E_{0.5}} = \frac{0.030}{0.300} = 0.1$$

Efficiency   The ratio of the output of a secondary cell or battery on discharge to the input required to restore it to the initial state of charge under specified conditions. (see also AMPERE-HOUR EFFICIENCY, ENERGY EFFICIENCY, VOLTAGE EFFICIENCY, and WATTHOUR EFFICIENCY.)

Energy Efficiency See WATT HOUR EFFICIENCY.

Electrical Double Layer See DOUBLE LAYER.

Electrocapillarity The surface tension between liquid mercury and an electrolyte solution is modified by the potential difference across the interface. The effect is termed "electrocapillarity."

Electrochemical Cell A cell in which the electrochemical reactions are caused by supplying electrical energy or which supplies electrical energy as a result of electrochemical reactions: if the first case only is applicable, the cell is an electrolysis cell; if the second case only, the cell is a galvanic cell.

Electrochemical Couple See COUPLE.

Electrochemical Equivalent Weight of one equivalent of a substance being electrolyzed which is its gram atomic weight or its gram molecular weight divided by the number of electrons in the electrode reaction (see Faraday).

Electrochemical Series A classification of the elements according to the values of the standard potentials of specified electrochemical reactions.

Electrode The site, area, or location at which electrochemical processes take place.

Electrode Potential The voltage developed by a single plate either positive or negative against a standard reference electrode typically the standard hydrogen electron. The algebraic difference in voltage of any two electrodes equals the cell voltage.

Electroformation A term applied to the conversion of the material in both the positive and negative plates to their respective active materials. Also referred to as formation.

Electrolyte The medium which provides the ion transport mechanism between the positive and negative electrodes of a cell.

Electromotive Force (EMF) The standard potential of a specified electrochemical action.

Electromotive Series See ELECTROCHEMICAL SERIES.

Electron The elemental particle of an atom having a negative charge.

Element The negative and positive electrodes together with the separators of a single cell. It is used almost exclusively in describing lead-acid cells and batteries.

End Voltage The prescribed voltage at which the discharge (or charge, if end-of-charge voltage) of a battery may be considered complete (also cutoff voltage).

Energy Density The ratio of the energy available from a battery to its volume (Wh/L). See SPECIFIC ENERGY.

Equalization The process of restoring all cells in a battery to an equal state of charge.

Equilibrium Electrode Potential The difference in potential between an electrode and an electrolyte when they are in equilibrium for the electrode reaction which determines the electrode potential.

Equivalent Circuit An electrical circuit that models the fundamental properties of a device (e.g., a cell) or a circuit.

Exchange Current Under open circuit conditions, the forward and backward current of an electrochemical process are equal and opposite. This equilibrium current is defined as the exchange current.

Faraday One gram equivalent weight of matter is chemically altered at each electrode of a cell for each 96,494 international coulombs, or one Faraday, of electricity passed through the electrolyte.

Fast Charge   A rate of charging which returns full capacity to a rechargeable battery, usually within an hour.

Fauré Plate   See PASTED PLATE.

Flash Current   See SHORT-CIRCUIT CURRENT.

Flat-Plate Cell   A cell fabricated with rectangular flat-plate electrodes (also called a Prismatic Cell).

Float Charge   A method of maintaining a battery in a charged condition by continuous, long-term constant-voltage charging, at a level sufficient to balance self-discharge.

Flooded Cell   A cell design which incorporates an excess amount of electrolyte.

Forced Discharge   Discharging a cell or battery below zero Volts into voltage reversal.

Formation   Electrochemical processing of a battery plate or electrode which transforms the active materials into their usable form.

Fuel Cell   A galvanic cell in which the active materials are continuously supplied from a source external to the cell and the reaction products continuously removed converting chemical energy to electrical energy.

Galvanic Cell   An electrochemical cell that converts chemical energy into electrical energy by electrochemical action.

Gas Recombination   Method of suppressing hydrogen generation during charging by recombining oxygen gas on the negative electrode as the cell approaches full charge.

Gassing   The evolution of gas from one or more of the electrodes in a cell. Gassing commonly results from local action (self-discharge) or from the electrolysis of the electrolyte during charging.

Grid   A framework for a plate or electrode which supports or retains the active materials and acts as a current collector.

Group   An assembly of positive or negative plates which fit into a cell.

Half-Cell   An electrode (either the anode or cathode) immersed in a suitable electrolyte.

Hourly Rate   A discharge rate, in Amperes, of a battery which will deliver the specified hours of service to a given end voltage.

Hydrogen Electrode   An electrode of platinized platinum saturated by a stream of pure hydrogen, immersed in an electrolyte of known acidity (pH).

Hydrogen Overvoltage   The activation overvoltage for hydrogen discharge on an electrode.

Initial (Closed-Circuit) Voltage   The on-load voltage at the beginning of a discharge.

Inner Helmholtz Plane   The plane of closest approach of ions in solution. It corresponds to the plane which contains the adsorbed ions and the innermost layer of water molecules.

Intermittent Test   A test during which a battery is subjected to alternate periods of discharge and rest according to a specified discharge regime.

Internal Impedance   The opposition or resistance of a cell or battery to an alternating current of a particular frequency.

Internal Resistance   The opposition or resistance to the flow of an electric current within a cell or battery; the sum of the ionic and electronic resistances of the cell components.

Ion   A particle in solution which carries a negative or positive charge.

"IR"   A voltage which is the product of the electrical resistance (R) of a cell or battery and the current (I). The value is the product of the resistance in Ohms and the current in Amperes.

Life   For rechargeable batteries, the duration of satisfactory performance, measured in years (float life) or in the number of charge/discharge cycles (cycle life).

Load   A term used to indicate the current drain on a battery.

Local Action   Chemical reactions within a cell that convert the active materials to a discharged state without supplying energy through the battery terminals (self-discharge).

Luggin Capillary   The bridge from an external reference electrode to a cell solution often has a capillary tip. The capillary which is often situated close to the working electrode to minimize the IR drop, is called a Luggin capillary.

Maintenance-Free Battery   A secondary battery which does not require periodic "topping up" to maintain electrolyte volume.

Maximum-power Discharge Current, I_{mp}   Discharge rate at which maximum power is transferred to the external load. This is the discharge rate when the discharge voltage is approximately equal to one-half of the open circuit voltage if the discharge is purely ohmic.

Mean Diffusion Current   In polarography, the periodic detachment of mercury drops from the dropping mercury electrode impart an oscillation to the measured current. The average value of this current is termed the mean diffusion current.

Mechanical Recharging   Restoring the capacity of a cell by replacing a spent or discharged electrode with a fresh one.

Memory Effect   A phenomenon in which a cell, operated in successive cycles to the same, but less than a full, depth of discharge experiences a depression of its discharge voltage and temporarily loses the rest of its capacity at normal voltage levels. See Secs. 28.4.11 and 29.4.9.

Midpoint Voltage   The voltage of a battery midway in the discharge between the fully charged state and the end voltage.

Migration   The movement of a charged species under the influence of a potential gradient.

Motive Power Battery   See TRACTION BATTERY.

Negative Electrode   The electrode acting as an anode when a cell or battery is discharging.

Negative-Limited   The operating characteristics (performance) of the cell is limited by the negative-electrode.

Nominal Voltage   The characteristic operating voltage or rated voltage of or battery (as distinct from MIDPOINT VOLTAGE, WORKING VOLTAGE, etc.)

Off-Load Voltage   See OPEN-CIRCUIT VOLTAGE.

Ohmic Overvoltage   Overvoltage caused by the ohmic drop in an electrolyte.

On-Load Voltage   The difference in voltage between the terminals of a cell or battery when it is discharging.

Open-Circuit Voltage (OCV)   The difference in voltage between the terminals of a cell or voltage when the circuit is open (no-load condition).

Outer Helmholtz Plane   The plane of closest approach of those ions which do not contact-absorb but approach the electrode with a sheath of solvated water molecules surrounding them.

Overcharge   The forcing of current through a battery after all the active material has been converted to the charged state. In other words, charging continued after 100% state of charge is achieved.

Overdischarge   Discharge past the point where the full capacity of the battery has been obtained.

Overvoltage   The potential difference between the equilibrium potential of an electrode and that of the electrode under an imposed polarization current.

Oxygen Recombination   The process by which oxygen generated at the positive plate during charge is reacted at the negative plate.

Paper-Lined Cell   Construction of a cell where a layer of paper, wetted with electrolyte, acts as the separator.

Parallel   Term used to describe the interconnection of cells or batteries in which all of the like terminals are connected together. Parallel connections increase the capacity of the resultant battery as follows:

$$C_p = n \times C_n$$

where C_p = the resultant capacity
       n = the number of cells or batteries connected in parallel
       C_u = capacity of the unconnected cell or battery.

Passivation   The phenomenon by which an electrode, although in conditions of thermodynamic instability, remains unattacked because of its surface condition.

Paste   Mixtures of various compounds that are applied to positive and negative grids of lead batteries. These pastes are then converted to positive and negative active materials. (See also FORMATION.)

Paste-Lined Cell   Leclanché cell constructed so that a layer of gelled paste acts as the separator.

Pasted Plate   A plate, manufactured by coating a grid or support strip with active materials in paste form.

Planté Plate   A plate for a lead-acid battery in which the active materials are formed directly from a lead substrate by electrochemical processing.

Plate   A structure containing active materials held firmly to a grid or conductor.

Pocket Plate   A plate for a secondary battery in which active materials are held in perforated metal pockets on a support strip.

Polarity   Denoting either positive or negative potential.

Polarization   The change of the potential of a cell or electrode from its equilibrium value caused by the passage of an electric current.

   Activation Polarization   That part of electrode or battery polarization arising from the charge-transfer step of the electrode reaction.

   Concentration Polarization   That part of electrode or battery polarization arising from concentration gradients of battery reactants and products caused by the passage of current.

   Ohmic Polarization   That part of electrode or battery polarization arising from current flow through ohmic resistances within an electrode or battery.

Positive Electrode   The electrode acting as a cathode when a cell or battery is discharging.

Positive-Limited   The operating characteristics (performance) of the cell is limited by the positive electrode.

Power Density   the ratio of the power available from a battery to its volume (W/L). See SPECIFIC POWER.

Primary Cell or Battery   A cell or battery which is not intended to be recharged and is discarded when the cell or battery has delivered all its electrical energy.

Prismatic Cell   See FLAT-PLATE CELL.

Rate Constant   At equilibrium, the forward and backward Faradic currents of an electrode process are equal and referred to as the exchange current. This exchange current can be defined in terms of a rate constant called the standard heterogeneous rate constant for the electrode process.

Rated Capacity   The number of Ampere-hours a battery can deliver under specific conditions (rate of discharge, end voltage, temperature); usually the manufacturer's rating.

Recharge   See CHARGE.

Rechargeable Battery   See SECONDARY BATTERY.

Recombination   A term used in a sealed cell construction for the process whereby internal pressure is relieved by reaction of oxygen with the negative active material.

Recovery   See RECUPERATION.

Recuperation   The lowering of the polarization of a cell during rest periods.

Redox Cell   A secondary cell in which two soluble ionic reactants, separated by a membrane, form the active materials.

Reference Electrode   A specially chosen electrode which has a reproducible potential against which other electrode potentials may be measured. (See HYDROGEN ELECTRODE).

Reserve Cell or Battery   A cell or battery which may be stored in an inactive state and made ready for use by adding electrolyte, another cell component, or, in the case of a thermal battery, melting a solidified electrolyte.

Reversal   The changing of the normal polarity of a cell or battery.

Secondary Battery   A galvanic battery which, after discharge, may be restored to the charged state by the passage of an electric current through the cell in the opposite direction to that of discharge.

Self-Discharge   The loss of useful capacity of a cell or battery due to internal chemical action (local action).

Semi-Permeable Membrane   A film that will pass selected ions.

Separator   An ion permeable, electronically nonconductive, spacer or material which prevents electronic contact between electrodes of opposite polarity in the same cell.

Series   The interconnection of cells or batteries in such a manner that the positive terminal of the first is connected to the negative terminal of the second, and so on. Series connections increase the voltage of the resultant battery as follows:

$$V_s = n \times V_u$$

where V_s = the resultant voltage
n = the number of cells or batteries connected in series
V_u = voltage of the unconnected cell or battery

Service Life   The period of useful life of a primary battery before a predetermined endpoint voltage is reached.

Shallow Discharge   A discharge on a secondary battery equalling only a small part of its total capacity.

Shape Change   Change in shape of an electrode due to migration of active material during charge/discharge cycling.

Shedding   The loss of active material from a plate during cycling.

Shelf Life   The duration of storage under specified conditions at the end of which a cell or battery still retains the ability to give a specified performance.

Short-Circuit Current   The initial value of the current obtained from a battery in a circuit of negligible resistance.

Sintered Electrode   An electrode construction in which active materials are deposited in the interstices of a porous metal matrix made by sintering metal powder.

SLI Battery   A battery designed to start internal combustion engines and to power the electrical systems in automobiles when the engine is not running (starting, lighting, ignition).

Specific Energy   The ratio of the energy output of a cell or battery to its weight (Wh/kg). (See also ENERGY DENSITY.)

Specific Gravity   The specific gravity of a solution is the ratio of the weight of the solution to the weight of an equal volume of water at a specified temperature.

Specific Power   The ratio of the power delivered by a cell or battery to its weight (W/kg). (See also POWER DENSITY.)

Spirally Wound Cell   A cylindrical cell which uses an electrode structure made by winding the electrodes and separators into a cylindrical "jelly-roll" construction. (See Fig. 28.2 and Fig. 14.8).

Standard Electrode Potential   The equilibrium value of an electrode potential when all the constituents taking part in the electrode reaction are in the standard state.

Standby Battery   A battery designed for emergency use in the event of a main power failure.

Starved Electrolyte Cell   A cell containing little or no free fluid electrolyte. This enables gases to reach electrode surfaces during charging and facilitates gas recombination.

State-of-Charge (SOC)   The available capacity in a battery expressed as a percentage of rated capacity.

Stationary Battery   A secondary battery designed for use in a fixed location.

Storage Battery   See SECONDARY BATTERY.

Storage Life   See SHELF LIFE.

Sulfation   Process occurring in lead batteries that have been stored and allowed to self-discharge for extended periods of time. Large crystals of lead sulfate grow that interfere with the function of the active materials.

Taper Charge   A charge regime delivering moderately high rate charging current when the battery is at a low state of charge and tapering the charging current to lower rates as the battery is charged.

Thermal Runaway   A condition whereby a battery on charge or discharge will overheat and destroy itself through internal heat generation caused by high overcharge or overdischarging current or other abusive condition.

Traction Battery   A secondary battery designed for the propulsion of electric vehicles or electrically operated mobile equipment operating in a deep-cycle regime.

Transfer Coefficient   The transfer coefficient determines what fraction of the electrical energy of a system resulting from the displacement of the potential from the equilibrium value that affects the rate of electrochemical transformation. (See Chap. 2.)

Transition Time   The time of an electrode process from the initiation of the process at constant current to the moment an abrupt change in potential occurs signifying that a new electrode process is controlling the electrode potential.

Transport Number   The fraction of the total cell current carried by the cation of an electrolyte solution is called the "cation transport number." Similarly, the fraction of the total current carried by the anion is referred to as the "anion transport number." Also called Transference Number.

Trickle Charge   A charge at a low rate, balancing losses through a local action and/or periodic discharge, to maintain a battery in a fully charged condition.

Tubular Plate   A battery plate in which an assembly of perforated metal or polymer tubes holds the active materials.

Unactivated Shelf Life   The period of time, under specified conditions of temperature and environment, that an unactivated or reserve battery can stand before deteriorating below a specified capacity.

Vent   A normally sealed mechanism which allows for the controlled escape of gases from within a cell.

Vented Cell   A cell design incorporating a vent mechanism to relieve excessive pressure and expel gases that are generated during the operation or abuse of the cell.

Voltage Delay   Time delay for a battery to deliver the required operating voltage after it is placed under load.

Voltage Depression   An abnormal low voltage, below the expected value, during the discharge of a battery.

Voltage Efficiency   The ratio of average voltage during discharge to average voltage during recharge under specified conditions of charge and discharge.

Watthour Capacity   The quantity of electrical energy measured in Watthours which may be delivered by a cell or battery under specified conditions.

Watthour Efficiency   The ratio of the Watthours delivered on discharge of a battery to the Watthours needed to restore it to its original state under specified conditions of charge and discharge. Also called Energy Efficiency.

Wet Shelf Life   The period of time that a battery can stand in the charged or activated condition before deteriorating below a specified capacity.

Working Voltage   The typical voltage or range of voltages of a battery during discharge.

APPENDIX B
STANDARD REDUCTION POTENTIALS

TABLE B.1 Standard Reduction Potentials of Electrode Reactions at 25°C

Electrode reaction	E^0, V
$Li^+ + e \rightleftharpoons Li$	-3.01
$Rb^+ + e \rightleftharpoons Rb$	-2.98
$Cs^+ + e \rightleftharpoons Cs$	-2.92
$K^+ + e \rightleftharpoons K$	-2.92
$Ba^{2+} + 2e \rightleftharpoons Ba$	-2.92
$Sr^2 + 2e \rightleftharpoons Sr$	-2.89
$Ca^{2+} + 2e \rightleftharpoons Ca$	-2.84
$Li + 6C + e \rightleftharpoons LiC_6$	-2.8
$Na^+ + e \rightleftharpoons Na$	-2.71
$Mg(OH)_2 + 2e \rightleftharpoons Mg + 2OH^-$	-2.67
$Mg^{2+} + 2e \rightleftharpoons Mg$	-2.38
$Al(OH)_3 + 3e \rightleftharpoons Al + 3OH^-$	-2.34
$Ti^{2+} + 2e \rightleftharpoons Ti$	-1.75
$Be^{2+} + 2e \rightleftharpoons Be$	-1.70
$Al^3 + 3e \rightleftharpoons Al$	-1.66
$Zn(OH)_2 + 2e \rightleftharpoons Zn + 2OH^-$	-1.25
$Mn^{2+} + 2e \rightleftharpoons Mn$	-1.05
$Fe(OH)_2 + 2e \rightleftharpoons Fe + 2OH^-$	-0.88
$2H_2O + 2e \rightleftharpoons H_2 + 2OH^-$	-0.83
$H^+ + M + e \rightleftharpoons MH$	-0.83
$Cd(OH)_2 + 2e \rightleftharpoons Cd + 2OH^-$	-0.81
$Zn^{2+} + 2e \rightleftharpoons Zn$	-0.76
$Ni(OH)_2 + 2e \rightleftharpoons Ni + 2OH^-$	-0.72
$Ga^{3+} + 3e \rightleftharpoons Ga$	-0.52
$S + 2e \rightleftharpoons S^{2-}$	-0.48
$Fe^{2+} + 2e \rightleftharpoons Fe$	-0.44
$Cd^{2+} + 2e \rightleftharpoons Cd$	-0.40

TABLE B.1   Standard Reduction Potentials of Electrode
Reactions at 25°C (*Continued*)

Electrode reaction	E^0, V
$PbSO_4 + 2e \rightleftharpoons Pb + SO_4^{2-}$	−0.36
$In^{3+} + 3e \rightleftharpoons In$	−0.34
$Tl^+ + e \rightleftharpoons Tl$	−0.34
$Co^{2+} + 2e \rightleftharpoons Co$	−0.27
$Ni^{2+} + 2e \rightleftharpoons Ni$	−0.23
$Sn^{2+} + 2e \rightleftharpoons Sn$	−0.14
$Pb^{2+} + 2e \rightleftharpoons Pb$	−0.13
$O_2 + H_2O + 2e \rightleftharpoons HO_2^- + OH^-$	−0.08
$D^+ + e \rightleftharpoons \frac{1}{2}D_2$	−0.003
$H^+ + e \rightleftharpoons \frac{1}{2}H_2$	0.000
$HgO + H_2O + 2e \rightleftharpoons Hg + 2OH^-$	0.10
$CuCl + e \rightleftharpoons Cu + Cl^-$	0.14
$AgCl + e \rightleftharpoons Ag + Cl^-$	0.22
$\gamma\text{-}MnO_2 + H_2O + e \rightleftharpoons \alpha\text{-}MnOOH + OH^-$	0.30
$Cu^{2+} + 2e \rightleftharpoons Cu$	0.34
$Ag_2O + H_2O + 2e \rightleftharpoons 2Ag + 2OH^-$	0.35
$\gamma\text{-}MnO_2 + H_2O + e \rightleftharpoons \lambda\text{-}MnOOH + OH^-$	0.36
$\frac{1}{2}O_2 + H_2O + 2e \rightleftharpoons 2OH^-$	0.40
$NiOOH + H_2O + e \rightleftharpoons Ni(OH)_2 + OH^-$	0.45
$Cu^+ + e \rightleftharpoons Cu$	0.52
$I_2 + 2e \rightleftharpoons 2I^-$	0.54
$2AgO + H_2O + 2e \rightleftharpoons Ag_2O + 2OH^-$	0.57
$LiCoO_2 + 0.5e \rightarrow Li_{0.5}CoO_2 + 0.5Li^+$	~0.70
$Hg^{2+} + 2e \rightleftharpoons 2Hg$	0.80
$Ag^+ + e \rightleftharpoons Ag$	0.80
$O_2 + 4H^+(10^{-7}\ M) + 4e \rightleftharpoons 2H_2O$	0.82
$Pd^{2+} + 2e \rightleftharpoons Pd$	0.83
$Ir^{3+} + 3e \rightleftharpoons Ir$	1.00
$Br_2 + 2e \rightleftharpoons 2Br^-$	1.08
$O_2 + 4H^+ + 4e \rightleftharpoons 2H_2O$	1.23
$MnO_2 + 4H^+ + 2e \rightleftharpoons Mn^{2+} + 2H_2O$	1.23
$Cl_2 + 2e \rightleftharpoons 2Cl^-$	1.36
$PbO_2 + 4H^+ + 2e \rightleftharpoons Pb^{2+} + 2H_2O$	1.46
$PbO_2 + SO_4^{2-} + 4H^- + 2e \rightleftharpoons PbSO_4 + 2H_2O$	1.69
$F_2 + 2e \rightleftharpoons 2F^-$	2.87

APPENDIX C

ELECTROCHEMICAL EQUIVALENTS OF BATTERY MATERIALS

TABLE C.1 Electrochemical Equivalents of Battery Materials

Material	Symbol	Atomic no.	Atomic wt., g	Density, g/cm³	Valence change	Electrochemical equivalents		
						Ah/g	g/Ah	Ah/cm³
Elements								
Aluminium	Al	13	26.98	2.699	3	2.98	0.335	8.05
Antimony	Sb	51	121.75	6.62	3	0.66	1.514	4.37
Arsenic	As	33	74.92	5.73	3	1.79	0.559	10.26
Barium	Ba	56	137.34	3.78	2	0.39	2.56	1.47
Beryllium	Be	4	9.01		2	5.94	0.168	
Bismuth	Bi	83	208.98	9.80	3	0.385	2.59	3.77
Boron	B	5	10.81	2.54	3	7.43	0.135	18.87
Bromine	Br	35	79.90		1	0.335	2.98	
Cadmium	Cd	48	112.40	8.65	2	0.477	2.10	4.15
Cesium	Cs	55	132.91	1.87	3	0.574	1.74	1.07
Calcium	Ca	20	40.08	1.54	2	1.34	0.748	2.06
Carbon (graphite)	C	6	12.01	2.25	4	8.93	0.112	20.09
Chlorine	Cl	17	35.45		1	0.756	1.32	
Chromium	Cr	24	52.00	6.92	3	1.55	0.647	10.72
Cobalt	Co	27	58.93	8.71	2	0.910	1.10	7.93
Copper	Cu	29	63.55	8.89	2	0.843	1.19	7.49
					1	0.422	2.37	3.75
Fluorine	F	9	19.00		1	1.41	0.709	
Gold	Au	79	197.00	19.3	1	0.136	7.36	2.62
Hydrogen	H	1	1.008		1	26.59	0.0376	
Indium	In	49	114.82	7.28	3	0.701	1.43	5.10
Iodine	I	53	126.90	4.94	1	0.211	4.73	1.04
Iron	Fe	26	55.85	7.85	2	0.96	1.04	7.54
					3	1.44	0.694	11.30
Lead	Pb	82	207.2	11.34	2	0.259	3.87	2.94
Lithium	Li	3	6.94	0.534	1	3.86	0.259	2.06
Magnesium	Mg	12	24.31	1.74	2	2.20	0.454	3.83
Manganese	Mn	25	54.94	7.42	2	0.976	1.02	7.24
Mercury	Hg	80	200.59	13.60	2	0.267	3.74	3.63
Molybdenum	Mo	42	95.94	10.2	6	1.67	0.597	17.03
Nickel	Ni	28	58.71	8.6	2	0.913	1.09	7.85
Nitrogen	N	7	14.01		3	5.74	0.174	
Oxygen	O	8	16.00		2	3.35	0.298	
Platinum	Pt	78	195.09	21.37	4	0.549	1.82	11.73
Potassium	K	19	39.10	0.87	1	0.685	1.46	0.59
Silver	Ag	17	107.87	10.5	1	0.248	4.02	2.60
Sodium	Na	11	22.99	0.971	1	1.17	0.858	1.14
Sulfur	S	16	32.06	2.0	2	1.67	0.598	3.34
Tin	Sn	50	118.69	7.30	4	0.903	1.11	6.59
Vanadium	V	23	50.95	5.96	5	2.63	0.380	15.67
Zinc	Zn	30	65.38	7.1	2	0.820	1.22	5.82
Zirconium	Zr	40	91.22	6.44	4	1.18	0.851	7.60

TABLE C.1 Electrochemical Equivalents of Battery Materials (*Continued*)

Material	Symbol	Molecular wt., g	Density, g/cm³	Valence change	Electrochemical equivalents Ah/g	g/Ah	Ah/cm³
			Compounds				
Bismuth trioxide	Bi_2O_3	466	8.5	6	0.35	2.86	2.97
Calcium chromate	$CaCrO_4$	156.1		2	0.34	2.91	
Carbon monofluoride	$(CF)_n$	$(31)_n$	2.7	1	0.86	1.16	2.32
Copper chloride	$CuCl$	99	3.5	1	0.27	3.69	0.95
Copper chloride	$CuCl_2$	134.5	3.1	2	0.40	2.50	1.22
Copper fluoride	CuF_2	101.6	2.9	2	0.53	1.87	1.52
Copper oxide	CuO	79.6	6.4	2	0.67	1.49	4.26
Copper sulfate	$CuSO_4$	159.6	3.6	2			
Copper sulfide	CuS	95.6	4.6	2	0.56	1.79	2.57
Iron monosulfide	FeS	87.9	4.84	2	0.61	1.64	2.95
Iron disulfide	FeS_2	119.9	4.87	4	0.89	1.12	4.35
Lead bismuthate	$Bi_2Pb_2O_5$	912	9.0	10	0.29	3.41	2.64
Lead chloride	$PbCl_2$	278.1	5.8	2	0.19	5.18	1.12
Lead dioxide	PbO_2	239.2	9.3	2	0.22	4.45	2.11
Lead iodide	PbI_2	461	6.2	2	0.12	8.60	0.72
Lead oxide	Pb_3O_4	685	9.1	8	0.31	3.22	2.85
Lead sulfide	PbS	239.3	7.5	2	0.22	4.46	1.68
Lithiated carbon	LiC_6	79.0		1	0.372*	2.69*	
Lithium cobalt oxide	$LiCoO_2$	98		0.5	0.137	7.29	
Manganese dioxide	MnO_2	86.9	5.0	1	0.31	3.22	1.54
Mercuric oxide	HgO	216.6	11.1	2	0.247	4.05	2.74
Molybdenum trioxide	MoO_3	143	4.5	1	0.19	5.26	0.84
Nickel oxide	$NiOOH$	91.7	7.4	1	0.29	3.42	2.16
Nickel sulfide	Ni_3S_2	240		4	0.47	2.12	
Silver chloride	$AgCl$	143.3	5.56	1	0.19	5.26	1.04
Silver chromate	Ag_2CrO_4	331.8	5.6	2	0.16	6.25	0.90
Silver oxide (monovalent)	Ag_2O	231.8	7.1	2	0.23	4.33	1.64
Silver oxide (divalent)	AgO	123.9	7.4	2	0.43	2.31	3.20
Sulfur dioxide	SO_2	64	1.37	1	0.419	2.39	
Sulfuryl chloride	SO_2Cl_2	135	1.66	2	0.397	2.52	
Thionyl chloride	$SOCl_2$	119	1.63	2	0.450	2.22	
Vanadium pentoxide	V_2O_5	181.9	3.6	1	0.15	6.66	0.53

*Based on weight of carbon only.

APPENDIX D
STANDARD SYMBOLS AND CONSTANTS

TABLE D.1 SI Base Units

Quantity	Unit	Symbol
Length	meter	m
Mass	kilogram	kg
Time	second	s
Electric current	ampere	A
Thermodynamic temperature*	kelvin	K
Amount of substance	mole	mol
Luminous intensity	candela	cd

*Celsius temperature is, in general, expressed in degrees Celsius (symbol °C).

Source: From D. G. Fink and W. Beaty (eds.), *Standard Handbook for Engineers,* 12th ed., McGraw-Hill, N.Y., 1987; reproduced from IEEE Standard 268-1982, by permission.

TABLE D.2 SI Prefixes Expressing Decimal Factors

Factor	Prefix	Symbol	Factor	Prefix	Symbol
10^{18}	exa	E	10^{-1}	deci	d
10^{15}	peta	P	10^{-2}	centi	c
10^{12}	tera	T	10^{-3}	milli	m
10^{9}	giga	G	10^{-6}	micro	μ
10^{6}	mega	M	10^{-9}	nano	n
10^{3}	kilo	k	10^{-12}	pico	p
10^{2}	hecto	h	10^{-15}	femto	f
10^{1}	deka	da	10^{-18}	atto	a

Source: From D. G. Fink and W. Beaty (eds.), *Standard Handbook for Engineers,* 12th ed., McGraw-Hill, N.Y., 1987; adapted from IEEE Standard 268-1982, by permission.

TABLE D.3 Greek Alphabet

Greek letter		Greek name	English equivalent	Greek letter		Greek name	English equivalent
A	α	Alpha	a	N	ν	Nu	n
B	β	Beta	b	Ξ	ξ	Xi	x
Γ	γ	Gamma	g	O	o	Omicron	ŏ
Δ	δ	Delta	d	Π	π	Pi	p
E	ε	Epsilon	ĕ	P	ρ	Rho	r
Z	ζ	Zeta	z	Σ	σ	Sigma	s
H	η	Eta	ē	T	τ	Tau	t
Θ	θ	Theta	th	Υ	υ	Upsilon	u
I	ι	Iota	i	Φ	ϕ	Phi	ph
K	κ	Kappa	k	X	χ	Chi	ch
Λ	λ	Lambda	l	Ψ	ψ	Psi	ps
M	μ	Mu	m	Ω	ω	Omega	ō

TABLE D.4 Standard Symbols for Units

Unit	Symbol	Notes
Ampere	A	SI unit of electric current.
Ampere-hour	Ah	
Angstrom	Å	$1\ \text{Å} = 10^{-10}$ m.
Atmosphere, standard	atm	$1\ \text{atm} = 101{,}325$ N/m^2 or Pa.
Atmosphere, technical	at	$\text{at} = \text{kg}_f/\text{cm}^2$.
Atomic mass unit (unified)	u	The (unified) atomic mass unit is defined as one-twelfth of the mass of an atom of the ^{12}C nuclide. Use of the old atomic mass unit (amu), defined by reference to oxygen, is deprecated.
Atto	a	SI prefix for 10^{-18}.
Bar	bar	$1\ \text{bar} = 100{,}000$ N/m^2.
Barn	b	$1\ \text{b} = 10^{-28}$ m^2.
Barrel	bbl	$1\ \text{bbl} = 9702\ \text{in}^3 = 0.15899$ m^3. This is the standard barrel used for petroleum, etc. A different standard barrel is used for fruits, vegetables, and dry commodities.
British thermal unit	Btu	
Calorie (International Table calorie)	cal$_{IT}$	$1\ \text{cal}_{IT} = 4.1868$ J. The 9th Conférence Générale des Poids et Mesures adopted the joule as the unit of heat. Use of the joule is preferred.
Calorie (thermochemical calorie)	cal	$1\ \text{cal} = 4.1840$ J (see note for International Table calorie).
Centi	c	SI prefix for 10^{-2}.
Centimeter	cm	
Coulomb	C	SI unit of electric charge.
Cubic centimeter	cm^3	
Cycle	c	
Cycle per second	Hz, c/s	See hertz. The name "hertz" is internationally accepted for this unit; the symbol Hz is preferred to c/s.
Day	d	
Deci	d	SI prefix for 10^{-1}.
Decibel	dB	

TABLE D.4   Standard Symbols for Units (*Continued*)

Unit	Symbol	Notes
Degree (temperature):		
Degree Celsius	°C	Note that there is no space between the symbol ° and the letter. The use of the word *centigrade* for the Celsius temperature scale was abandoned by the Conférence Générale des Poids et Mesures in 1948.
Degree Fahrenheit	°F	
Degree Kelvin		See kelvin.
Degree Rankine	°R	
Deka	da	SI prefix for 10.
Dyne	dyn	
Electron	e	This symbol is used in this Handbook to denote an electron. More conventionally shown as e^-
Electronvolt	eV	
Erg	erg	
Farad	F	SI unit of capacitance.
Femto	f	SI prefix for 10^{-15}.
Gauss	G	The gauss is the electromagnetic CGS unit of magnetic flux density. Use of SI unit, the tesla, is preferred.
Giga	G	SI prefix for 10^9.
Gilbert	Gb	The gilbert is the electromagnetic CGS unit of magnetomotive force. Use of the SI unit, the ampere (or ampere turn), is preferred.
Gram	g	
Gram per cubic centimeter	g/cm^3	
Hecto	h	SI prefix for 10^2.
Henry	H	SI unit of inductance.
Hertz	Hz	SI unit of frequency.
Hour	h	
Joule	J	SI unit of energy.
Joule per kelvin	J/K	SI unit of heat capacity and entropy.
Kelvin	K	In 1967 the CGPM gave the name "kelvin" to the SI unit of temperature which had formerly been called "degree Kelvin" and assigned it the symbol K (without the symbol °).
Kilo	k	SI prefix for 10^3.
Kilogram	kg	SI unit of mass.
Kilogram-force	kg_f	In some countries the name *kilopond* (kp) has been adopted for this unit.
Kilohm	kΩ	
Kilometer	km	
Kilometer per hour	km/h	
Kilovolt	kV	
Kilowatt	kW	
Kilowatthour	kWh	
Liter	L	$1 \text{ L} = 10^{-3} \text{ m}^3$.
Liter per second	L/s	
Lumen	lm	SI unit of luminous flux.
		SI unit of illuminance.
Maxwell	Mx	The maxwell is the electromagnetic cgs unit of magnetic flux. Use of the SI unit, the weber, is preferred.
Mega	M	SI prefix for 10^6.
Megohm	MΩ	
Meter	m	SI unit of length.
Mho	mho	CGPM has adopted the name "siemens" (S) for this unit.
Micro	μ	SI prefix for 10^{-6}.
Microampere	μA	
Microgram	μg	

TABLE D.4 Standard Symbols for Units (*Continued*)

Unit	Symbol	Notes
Micrometer	μm	
Micron	μm	See micrometer. The name "micron" was abrogated by the Conférence Générale des Poids et Mesures, 1967.
Microsecond	μs	
Microwatt	μW	
Milli	m	SI prefix for 10^{-3}.
Milliampere	mA	
Milligram	mg	
Milliliter	ml	
Millimeter	mm	
Conventional millimeter of mercury	mmHg	1 mmHg = 133.322 N/m².
Millimicron	nm	Use of the name "millimicron" for the nanometer is deprecated.
Millisecond	ms	
Millivolt	mV	
Milliwatt	mW	
Minute (time)	min	Time may also be designated by means of superscripts as in the following example: $9^h46^m20^s$.
Mole	mol	SI unit of amount of substance.
Nano	n	SI prefix for 10^{-9}.
Nanoampere	nA	
Nanometer	nm	
Nanosecond	ns	
Newton	N	SI unit of force.
Newton meter	N · m	
Newton per square meter	N/m²	SI unit of pressure or stress; see pascal.
Newton second per square meter	N · s/m²	SI unit of dynamic viscosity.
Oersted	Oe	The oersted is the electromagnetic cgs unit of magnetic field strength. Use of the SI unit, the ampere per meter, is preferred.
Ohm	Ω	SI unit of resistance.
Pascal	Pa	Pa = N/m². SI unit of pressure or stress. This name accepted by the 14th Conférence Générale des Poids et Mesures.
Pico	P	SI prefix for 10^{-12}.
Picowatt	pW	
Revolution per second	r/s	
Second (time)	s	SI unit of time.
Siemens	S	S = Ω^{-1}. SI unit of conductance. This name and symbol were adopted by the 14th Conférence Générale des Poids et Mesures. The name "mho" is also used for this unit in the United States.
Square meter	m²	
Tera	T	SI prefix for 10^{12}.
Tesla	T	SI unit of magnetic flux density
Tonne	t	1 t = 1000 kg. (in USA: ton, metric)
(Unified) atomic mass unit	u	The (unified) atomic mass unit is defined as one-twelfth of the mass of an atom of the ^{12}C nuclide. Use of the old atomic mass unit (amu), defined by reference to oxygen, is deprecated.
Volt	V	SI unit of voltage.
Volt per meter	V/m	SI unit of electric field strength.
Voltampere	VA	IEC name and symbol for the SI unit of apparent power.
Watt	W	SI unit of power.
Watt per meter kelvin	W/(m · K)	SI unit of thermal conductivity.
Watthour	Wh	

Source: From D. G. Fink and W. Beaty (eds.), *Standard Handbook for Engineers,* 12th ed., McGraw-Hill, N.Y., 1987; adapted from ANSI/IEEE Standard 260-1982.

APPENDIX E
CONVERSION FACTORS

TABLE E.1 Length Conversion Factors*

A. Length units decimally related to one meter

	Meters (m)	Kilometers (km)	Decimeters (dm)	Centimeters (cm)	Millimeters (mm)	Micrometers (μm)	Nanometers (nm)	Angströms (Å)
1 meter =	1	0.001	10	100	1000	1,000,000	10^9	10^{10}
1 kilometer =	1000	1	10,000	100,000	1,000,000	10^9	10^{12}	10^{13}
1 decimeter =	0.1	0.0001	1	10	100	100,000	10^8	10^9
1 centimeter =	0.01	0.00001	0.1	1	10	10,000	10^7	10^8
1 millimeter =	0.001	10^{-6}	0.01	0.1	1	1000	1,000,000	10^7
1 micrometer (micron) =	10^{-6}	10^{-9}	0.00001	0.0001	0.001	1	1000	10,000
1 nanometer =	10^{-9}	10^{-12}	10^{-8}	10^{-7}	10^{-6}	0.001	1	10
1 angström =	10^{-10}	10^{-13}	10^{-9}	10^{-8}	10^{-7}	0.0001	0.1	1

B. Nonmetric length units less than one meter

	Meters (m)	Yards (yd)	Feet (ft)	Inches (in)	Mils (mil)	Microinches (μin)
1 meter =	1	1.09361330	3.28083939	39.3700787	3.93700787×10^4	3.93700787×10^7
1 yard =	0.9144	1	3	36	36,000	3.6×10^7
1 foot =	0.3048	$1/3 = 0.3333$	1	12	12,000	1.2×10^7
1 inch =	0.0254	$1/36 = 0.0277$	$1/12 = 0.0833$	1	1000	1,000,000
1 mil =	2.54×10^{-5}	$2.7\overline{77} \times 10^{-5}$	$8.33\overline{3} \times 10^{-5}$	0.001	1	1000
1 microinch =	2.54×10^{-8}	$2.7\overline{77} \times 10^{-8}$	$8.33\overline{3} \times 10^{-8}$	10^{-6}	0.001	1

C. Nonmetric length units greater than one meter (with equivalents in feet)

	Meters (m)	Rods (rd)	Statute miles (mi)	Nautical miles (nmi)	Astronomical units (AU)	Parsecs (pe)	Feet (ft)
1 meter =	1	0.19883878	$6.21371192 \times 10^{-4}$	$5.39956904 \times 10^{-4}$	$6.68449198 \times 10^{-12}$	$3.24073317 \times 10^{-17}$	3.28083989
1 rod =	**5.0292**	**1**	**0.003125**	$2.71555076 \times 10^{-3}$	$3.36176471 \times 10^{-11}$	$1.62982953 \times 10^{-16}$	**16.5**
1 statute mile =	**1609.344**	**320**	**1**	0.86897624	$1.07576471 \times 10^{-8}$	$5.21545450 \times 10^{-14}$	**5280**
1 nautical mile =	**1852**	368.249423	1.15077945	**1**	$1.23796791 \times 10^{-8}$	$6.00183780 \times 10^{-14}$	6076.11548
1 astronomical unit† =	$\mathbf{1.496 \times 10^{11}}$	$2.97462817 \times 10^{10}$	92,957,130.3	80,777,537.8	**1**	$4.84813682 \times 10^{-6}$	$4.90813648 \times 10^{11}$
1 parsec =	$3.08572150 \times 10^{16}$	$6.13561102 \times 10^{15}$	$1.91737844 \times 10^{13}$	$1.66615632 \times 10^{13}$	206,264.806	**1**	$1.01237582 \times 10^{17}$
1 foot =	**0.3048**	0.06060$\underline{6}$	$1.89\underline{3939} \times 10^{-4}$	$1.64578833 \times 10^{-4}$	$2.03743316 \times 10^{-12}$	$9.87775472 \times 10^{-18}$	**1**

D. Other length units

1 cable = **720** feet = **219.456** meters
1 cable (U.K.) = **608** feet = **185.3184** meters
1 chain (engineers') = **100** feet = **30.48** meters
1 chain (surveyors') = **66** feet = **20.1168** meters
1 fathom = **6** feet = **1.8288** meters
1 fermi = **1** femtometer = 10^{-15} meter
1 foot (U.S. Survey) = 0.3048006 meter
1 furlong = **660** feet = **201.168** meters

1 hand = **4** inches = **0.1016** meter
1 league (international nautical) = 3 nautical miles = **5556** meters
1 league (statute) = 3 statute miles = **4828.032** meters
1 league (U.K. nautical) = 3 nautical miles = 5559.552 meters
1 light-year = 9.4608952×10^{15} meters (= distance traveled by light in vacuum in one sideral year)
1 link (engineers') = **1** foot = **0.3048** meter
1 link (surveyors') = **7.92** inches = **0.201168** meter
1 micron = 1 micrometer = 10^{-6} meter

1 millimicron = **1** nanometer = 10^{-9} meter
1 myriameter = **10,000** meters
1 nautical mile (U.K.) = **1853.184** meters
1 pale = 1 rod = **5.0292** meters
1 perch (linear) = 1 rod = **5.0292** meters
1 pica = **1/6** inch (approx.) = 4.217518×10^{-3} meter
1 point = **1/72** inch (approx.) = 3.514598×10^{-4} meter
1 span = **9** inches = **0.2286** meter

*Exact conversions are shown in boldface type. Repeating decimals are underlined. The SI unit of length is the meter.
†As defined by the International Astronomical Union, 1964.
Source: D. G. Fink and W. Beaty (eds.). *Standard Handbook for Electrical Engineers*, 12th ed., McGraw-Hill, N.Y., 1987.

TABLE E.2 Area Conversion Factors*

A. Area units decimally related to one square meter

	Square meters (m)²	Square kilometers (km)²	Hectares (square hectometers) (hn)	Square centimeters (cm)²	Square millimeters (mm)²	Square micrometers (μm)²	Barns (b)
1 square meter =	**1**	10^{-6}	**0.001**	**10,000**	**1,000,000**	10^{12}	10^{28}
1 square kilometer =	**1,000,000**	**1**	**100**	10^{10}	10^{12}	10^{18}	10^{34}
1 hectare =	**10,000**	**0.01**	**1**	10^{8}	10^{10}	10^{16}	10^{32}
1 square centimeter =	**0.001**	10^{-10}	10^{-8}	**1**	**100**	10^{8}	10^{24}
1 square millimeter =	10^{-4}	10^{-12}	10^{-10}	**0.01**	**1**	10^{6}	10^{22}
1 square micrometer =	10^{-12}	10^{-18}	10^{-16}	10^{-8}	10^{-6}	**1**	10^{16}
1 barn =	10^{-28}	10^{-34}	10^{-32}	10^{-24}	10^{-22}	10^{-16}	**1**

B. Nonmetric area units (with square meter equivalents)

	Square meters (m)²	Square statute miles (mi)²	Acres (acrea)	Square rods (rd)²	Square yards (rd)²	Square feet (ft)²	Square inches (in)²	Circular mils (cmil)
1 square meter =	**1**	$3.86102159 \times 10^{-7}$	$2.47105382 \times 10^{-4}$	$3.95368610 \times 10^{-2}$	1.19599005	10.7639104	1550.00310	1.97342524×10^{9}
1 square statute mile =	$2,589,988.1$	**1**	**640**	$102,400$	$3,097,600$	$27,878,400$	4.01448960×10^{9}	$5.11140691 \times 10^{15}$
1 acre =	4046.85641	**1/640 = 0.0015625**	**1**	**160**	**4840**	**43,560**	**6,272,640**	$7.98657330 \times 10^{12}$
1 square rod =	25.2928526	9.765625×10^{-6}	**1/160 = 0.00625**	**1**	**30.25**	**272.25**	**39,204**	$4.99160831 \times 10^{10}$
1 square yard =	**0.83612736**	$3.22830579 \times 10^{-7}$	$2.06611570 \times 10^{-4}$	$3.30578512 \times 10^{-2}$	**1**	**9**	**1296**	1.65011845×10^{9}
1 square foot =	**0.09290304**	$3.58700643 \times 10^{-8}$	$2.29568411 \times 10^{-5}$	$3.67309458 \times 10^{-3}$	**1/9 = 0.111111**	**1**	**144**	1.83346495×10^{8}
1 square inch =	6.4616×10^{-4}	$2.49097669 \times 10^{-10}$	$1.59422508 \times 10^{-7}$	$2.55076013 \times 10^{-5}$	$7.71604938 \times 10^{-4}$	**1/144 = 0.00694444**	**1**	1.27323955×10^{6}
1 circular mil =	$5.06707479 \times 10^{-10}$	$1.95640851 \times 10^{-16}$	$1.25210145 \times 10^{-13}$	$2.00336232 \times 10^{-11}$	$6.06017101 \times 10^{-10}$	$5.45415391 \times 10^{-9}$	$7.85398163 \times 10^{-7}$	**1**

Exact conversions are:
1 acre = **4046.8564224** square meters
1 square mile = **2,589,988.110336** square meters

C. Other area units

1 are = **100** square meters
1 centiare (centare) = **1** square meter
1 perch (area) = **1** square rod = **30.25** square yards = 25.2928526 square meters
1 rood = **40** square rods = 1011.71411 square meters
1 section = **1** square statute mile = 2,589,988.1 square meters
1 township = **36** square statute miles = 93,239,572 square meters

*Exact conversions are shown in boldface type. Repeating decimals are underlined. The SI unit of area is the square meter.

Source: D. G. Fink and W. Beaty (eds.), *Standard Handbook for Electrical Engineers*, 12th ed., McGraw-Hill, N.Y., 1987.

TABLE E.3 Force Conversion Factors*

	Newtons (N)	Kips (kip)	Slugs-force (slug)	Kilograms-force (kg)	Avoirdupois pounds-free (lb$_f$ avdp)	Avoirdupois ounces-force (oz$_f$ advp)	Poundals (pdl)	Dynes (dyn)
1 newton =	**1**	$2.24808943 \times 10^{-4}$	$6.98727524 \times 10^{-3}$	0.10197162	0.22480894	3.59694309	7.2330142	**100,000**
1 kip =	4448.22162	**1**	31.080949	453.592370	**1000**	**16,000**	32,174.05	444,822,162
1 slug-force =	143.117305	0.03217405	**1**	14.593903	32.17405	514.78480	1035.1695	14,311,730
1 kilogram-force =	**9.806650**	$2.20462262 \times 10^{-3}$	6.8521763×10^{-2}	**1**	2.20462262	35.2739619	70.9316384	**980.665**
1 avdp pound-force =	4.44822162	**0.001**	$3.10809488 \times 10^{-2}$	0.45359237	**1**	**16**	32.17405	444,822.162
1 avdp ounce-force =	0.27801385	**1/16,000 = 0.0000625**	$1.94255930 \times 10^{-3}$	2.834952×10^{-2}	**1/16 = 0.0625**	**1**	2.01087803	27,801.385
1 poundal =	0.13825495	3.1080949×10^{-5}	9.6602539×10^{-4}	0.14098081	0.03108095	0.49729518	**1**	13,825.495
1 dyne =	**0.00001**	$2.24808943 \times 10^{-9}$	$6.98727524 \times 10^{-8}$	$1.01971621 \times 10^{-6}$	$2.24808943 \times 10^{-6}$	$3.59694310 \times 10^{-5}$	7.2330142×10^{-5}	**1**

The exact conversion is: 1 avdp pound-force = **4.448221652605** newtons

*Exact conversions are shown in boldface type. The SI unit of force is the newton (N).

Source: D. G. Fink and W. Beaty (eds.), *Standard Handbook for Electrical Engineers*, 12th ed., McGraw-Hill, N.Y., 1987.

TABLE E.4 Volume and Capacity Conversion Factors

A. Volume units decimally related to one cubic meter

	Cubic meters (steres) (m)³	Cubic decimeters (dm)³	Cubic centimeters (cm)³	Liters (L)	Centiliters (cL)	Milliliters (mL)	Microliters (μL)
1 cubic meter =	1	1000	1,000,000	1000	100,000	1,000,000	10^9
1 cubic decimeter =	0.001	1	1000	1	100	1000	1,000,000
1 cubic centimeter =	0.000001	0.001	1	0.001	0.1	1	1000
1 liter =	0.001	1	1000	1	100	1000	1,000,000
1 centiliter =	0.00001	0.01	10	0.01	1	10	10,000
1 milliliter =	0.000001	0.001	1	0.001	0.1	1	1000
1 microliter =	10^{-9}	0.000001	0.001	0.000001	0.0001	0.001	1

B. Nonmetric volume units (with cubic meter and liter equivalents)

	Cubic meters (steres) (m)³	Liters (L)	Cubic inches (in)³	Cubic feet (ft)³	Cubic yards (yd)³	Barrels (U.S.) (bbl)	Acre-feet (acre-ft)	Cubic miles (mi)³
1 cubic meter =	1	1000	6.10237441×10^4	35.314666	1.30795062	6.28981097	$8.10713194 \times 10^{-4}$	$2.39912759 \times 10^{-10}$
1 liter =	0.001	1	61.10237441	0.03531466	$1.30795062 \times 10^{-3}$	$6.28981097 \times 10^{-3}$	$8.10713193 \times 10^{-7}$	$2.39912759 \times 10^{-13}$
1 cubic inch =	1.6387064×10^{-5}	1.6387064×10^{-2}	1	$1/1728 =$ $5.78703703 \times 10^{-4}$	$1/46656 =$ $2.14334705 \times 10^{-5}$	$1.03071532 \times 10^{-4}$	$1.32852090 \times 10^{-8}$	$3.93146573 \times 10^{-15}$
1 cubic foot =	$2.83168466 \times 10^{-2}$	28.3168466	1728	1	$1/27 = 0.037037$	0.17810761	$1/43560 =$ $2.29568411 \times 10^{-12}$	$6.79357278 \times 10^{-12}$
1 cubic yard =	0.76455486	764.554858	46,656	27	1	4.80890538	$6.19834711 \times 10^{-4}$	$1.83426465 \times 10^{-10}$
1 barrel (U.S.) =	0.15898729	158.987294	9702	5.61458333	0.20794753	1	$1.28893098 \times 10^{-4}$	$3.81430805 \times 10^{-11}$
1 acre-foot =	1233.48184	1.23348184×10^6	7.52716800×10^7	43,560	1613.33333	7758.36734	1	$2.95928030 \times 10^{-7}$
1 cubic mile =	4.16818183×10^9	$4.16818183 \times 10^{12}$	$2.54358061 \times 10^{14}$	$1.47197952 \times 10^{11}$	5.451776×10^9	26.2170749×10^9	3,379,200	1

Exact conversion: 1 cubic foot = **28.316846592 liters**

C. U.S. liquid capacity measures (with liter equivalents)

	Liters (L)	Gallons (U.S. gal)	Quarts (U.S. qt)	Pints (U.S. pt)	Gills (U.S. gi)	Fluid ounces (U.S. floz)	Fluidrams (U.S. fldr)	Minims (U.S. minim)
1 liter =	1	0.26417205	1.056688	2.113376	8.453506	33.814023	270.51218	16,230.73
1 gallon, U.S. =	3.7854118	1	4	8	32	128	1024	61,440
1 quart, U.S. =	0.9463529	$1/4 = 0.25$	1	2	8	32	256	15,360
1 pint, U.S. =	0.4731765	$1/8 = 0.125$	$1/2 = 0.5$	1	4	16	128	7680
1 gill, U.S. =	0.1182941	$1/32 = 0.03125$	$1/8 = 0.125$	$1/4 = 0.25$	1	4	32	1920
1 fluid ounce, U.S. =	2.957353×10^{-2}	$1/128 = 0.0078125$	$1/32 = 0.03125$	$1/16 = 0.0625$	$1/4 = 0.25$	1	8	480
1 fluidram, U.S. =	3.696912×10^{-3}	$1/1024 = 9.765625 \times 10^{-4}$	$1/256 = 3.90625 \times 10^{-3}$	$1/128 = 0.0078125$	$1/32 = 0.03125$	$1/8 = 0.125$	1	60
1 mimim, U.S. =	6.161152×10^{-5}	$1/61440 =$ $1.62760416 \times 10^{-5}$	$1/15360 =$ $6.51041666 \times 10^{-5}$	$1/7680 =$ $1.30208333 \times 10^{-4}$	$1/19200 =$ 5.208333×10^{-4}	$1/480 =$ 2.083333×10^{-3}	$1/60 =$ 0.0166666	1

Exact conversion: 1 liquid quart, U.S. = **0.946352946 liters**

D. British Imperial liquid capacity measures (with liter equivalents)

	Liters (L)	Gallons (U.K.)	Quarts (U.K. qt)	Pints (U.K. pt)	Gills (U.K. gi)	Fluid ounces (U.K. floz)	Fluidrams (U.K. fldr)	Minims (U.K. minim)
1 liter =	1	0.2199692	0.8798766	1.759753	7.039018	35.19506	281.5605	16,893.63
1 gallon, U.K. =	4.546092	1	4	8	32	160	**1280**	**76,800**
1 quart, U.K. =	1.136523	**1/4 = 0.25**	1	2	8	40	**320**	**19,200**
1 pint, U.K. =	0.5682615	**1/8 = 0.125**	**1/2 = 0.5**	1	4	20	**160**	**9600**
1 gill, U.K. =	0.1420654	**1/32 = 0.03125**	**1/8 = 0.125**	**1/4 = 0.25**	1	5	**40**	**2400**
1 fluid ounce, U.K. =	2.841307×10^{-2}	**1/160 = 0.00625**	**1/40 = 0.025**	**1/20 = 0.05**	**1/5 = 0.2**	1	**8**	**480**
1 fluidram, U.K. =	3.551634×10^{-3}	**1/1280 = 7.8125 $\times 10^{-4}$**	**1/320 = 0.003125**	**1/160 = 0.00625**	**1/40 = 0.025**	**1/8 = 0.125**	1	**60**
1 minima, U.K. =	5.919391×10^{-5}	**1/76800 = 1.302098333 $\times 10^{-5}$**	**1/1920 = 5.2083333 $\times 10^{-5}$**	**1/9600 = 1.04166666 $\times 10^{-4}$**	**1/2400 = 4.16666666 $\times 10^{-4}$**	**1/480 = 2.08333333 $\times 10^{-3}$**	**1/60 = 0.01666666**	1

E. U.S. and British dry capacity measures (with liter equivalents)

U.S. dry measures

	Liters (L)	Bushels (U.S. bu)	Pecks (U.S. peck)	Quarts (U.S. qt)	Pints (U.S. pt)
1 liter =	1	0.02837759	0.11351037	0.90808299	1.81816598
1 bushel, U.S. =	35.239070	1	4	32	64
1 peck, U.S. =	8.8097675	**1/4 = 0.25**	1	8	16
1 quart, U.S. =	1.1012209	**1/32 = 0.03125**	**1/8 = 0.125**	1	2
1 pint, U.S. =	0.5506105	**1/64 = 0.015625**	**1/16 = 0.0625**	**1/2 = 0.5**	1
1 bushel, U.K. =	36.36873	1.032057	4.128228	33.02582	66.95165
1 peck, U.K. =	9.092182	0.2580143	1.032057	8.256456	16.51291
1 quart, U.K. =	1.136523	0.03225178	0.1290071	1.032057	2.0641142
1 pint, U.K. =	0.5682614	0.01612589	0.0645036	0.5160184	1.032057

Exact conversion: 1 dry pint, U.S. = **33.6003125** cubic inches

British dry measures

	Bushels (U.K. peck)	Pecks (U.S. peck)	Quarts (U.K. qt)	Pints (U.K. pt)
1 liter =	0.0274961	0.1099846	0.8798766	1.7597534
1 bushel, U.S. =	0.9689387	3.8757549	31.00604	62.01208
1 peck, U.S. =	0.2422347	0.9689387	7.751509	15.50302
1 quart, U.S. =	0.03027934	0.1211173	0.9689387	1.937878
1 pint, U.S. =	0.01513967	0.06055867	0.4844693	0.9689387
1 bushel, U.K. =	1	4	32	64
1 peck, U.K. =	**1/4 = 0.25**	1	8	16
1 quart, U.K. =	**1/32 = 0.03125**	**1/8 = 0.125**	1	2
1 pint, U.K. =	**1/64 = 0.015625**	**1/16 = 0.0625**	**1/2 = 0.5**	1

F. Other volume and capacity units

1 barrel, U.S. (used for petroleum, etc.) = **42** gallons = 0.158987296 cubic meter
1 barrel ("old barrel") = **31.5** gallons = 0.119240 cubic meter
1 board foot = **144** cubic inches = 2.359737×10^{-3} cubic meter
1 cord = **128** cubic feet = 3.624556 cubic meters
1 cord foot = **16** cubic feet = 0.4530695 cubic meter
1 cup = **8** fluid ounces, U.S. = 2.365882×10^{-4} cubic meter

1 gallon (Canadian liquid) = 4.546090×10^{-3} cubic meter
1 perch (volume) = 24.75 cubic feet = 0.700842 cubic meter
1 stere = **1** cubic meter
1 tablespoon = **0.5** fluid ounce, U.S. = 1.478677×10^{-5} cubic meter
1 teaspoon = **1/6** fluid ounce, U.S. = 4.928922×10^{-6} cubic meter
1 ton (register ton) = **100** cubic feet = 2.83168466 cubic meters

*Exact conversions are shown in boldface type. Repeating decimals are underlined. The SI unit of volume is the cubic meter.

Source: D. G. Fink and W. Beaty (eds.), *Standard Handbook for Electrical Engineers*, 12th ed., McGraw-Hill, N.Y., 1987.

TABLE E.5 Mass Conversion Factors*

A. Mass units decimally

	Kilograms (kg)	Tonnes (metric tons) (t)	Grams (g)
1 kilogram =	**1**	0.01	1000
1 tonne =	1000	1	1,000,000
1 gram =	0.001	0.000001	1
1 decigram =	0.0001	10^{-7}	0.1
1 centigram =	0.00001	10^{-8}	0.01
1 milligram =	0.000001	10^{-9}	0.001
1 microgram =	10^{-9}	10^{-12}	0.000001

B. Nonmetric mass units less than one

	Grams (g)	Avoirdupois ounces-mass (oz_m, avdp)	Troy ounces-mass (oz_m, troy)
1 gram =	**1**	0.035273962	0.032150747
1 avdp ounce-mass =	38.3495231	**1**	0.91145833
1 troy ounce-mass =	31.1031768	1.09714286	**1**
1 avdp dram =	1.77184520	**1/16 = 0.0625**	0.056966.15
1 apothecary dram =	3.88793458	0.137142857	**1/8 = 0.125**
1 pennyweight =	1.55517383	0.054863162	**1/20 = 0.05**
1 grain =	**0.06479891**	**1/437.5 =** $2.28571429 \times 10^{-3}$	**1/480 =** 0.00208333
1 scruple =	1.29597820	$4.57142858 \times 10^{-2}$	**1/24 = 0.041\overline{6666}**

C. Nonmetric mass units of one pound-mass

	Kilograms (kg)	Long tons (long ton)	Short tons (short ton)
1 kilogram =	**1**	9.842065×10^{-4}	$1.10231131 \times 10^{-3}$
1 long ton =	1016.0469	**1**	**1.12**
1 short ton =	**907.18474**	**200/224 =** 0.89285714	**1**
1 long hundredweight =	50.802.3454	**0.05**	**0.056**
1 short hundredweight =	**45.359237**	**10/224 =** 0.04464286	**0.05**
1 slug =	14.593903	0.01436341	0.01608702
1 avdp pound-mass =	**0.45359237**	**1/2240 =** $4.46428571 \times 10^{-4}$	**0.0005**
1 troy pound-mass =	0.37324172	$3.67346937 \times 10^{-4}$	$4.11428570 \times 10^{-4}$

Exact conversions: 1 long ton = **1016.0469088** kilograms
 1 troy pound-mass = **0.3732417216** kilogram

D. Other

1 assay ton = 29.166667 grams
1 carat (metric) = **200** milligrams
1 carat (troy weight) = **31/6** grains = 205.19655 milligrams
1 mynagram = **10** kilograms
1 quintal = **100** kilograms
1 stone = **14** pounds, advp = **6.35029328** kilograms

*Exact conversions are shown in boldface type. Repeating decimals are <u>underlined</u>. The SI unit of mass is the kilogram.
 Source: D. G. Fink and W. Beaty (eds.), *Standard Handbook for Electrical Engineers,* 12th ed., McGraw-Hill, N.Y., 1987.

related to kilogram

Decigrams (dg)	Centigrams (cg)	Milligrams (mg)	Micrograms (μg)
10,000	100,000	1,000,000	10^9
10^7	10^8	10^9	10^{12}
10	100	1000	1,000,000
1	10	100	100,000
0.1	1	10	10,000
0.01	0.1	1	1000
0.00001	0.0001	0.001	1

pound-mass (with gram equivalents)

Avoirdupois drams (dr avdp)	Apothecary drams (dr apoth)	Pennyweights (dwt)	Grains (grain)	Scruples (scruple)
0.56438339	0.25720597	0.64301493	15.4323584	0.77161792
16	7.29166666	18.2271667	437.5	21.875
17.5542857	**8**	**20**	480	24
1	0.45572917	1.13932292	27.34375	1.3671875
2.19428570	**1**	**2.5**	60	3
0.87771428	**1/2.5 = 0.4**	**1**	24	1.2
3.65714285 × 10⁻²	**1/60 =** 0.010666666	**1/24 =** 0.04166666	1	0.05
0.73142857	**1/3** = 0.33333333	**5/6** = 0.83333333	20	1

and greater (with kilogram equivalents)

Long hundredweights (long cwt)	Short hundredweights (short cwt)	Slugs (slug)	Avoirdupois pounds-mass (lb$_m$, avdp)	Troy pounds-mass (lb$_m$, troy)
1.96841131 × 10⁻²	2.20462262 × 10⁻²	0.06852177	2.20462262	2.67922889
20	**22.4**	69.621329	**2240**	2722.22222
400/224 = 17.8571429	**20**	62.161901	**2000**	2430.55555
1	**1.12**	3.4810664	**112**	136.111111
100//112 = 0.89285714	**1**	3.1080950	**100**	121.527777
0.2872683	0.3217405	**1**	32.17405	39.100406
1/112 = 8.92857143 × 10⁻³	**0.01**	3.1080950 × 10⁻²	**1**	1.215277777
7.34693879 × 10⁻³	8.22857145 × 10⁻³	0.02557518	0.82285714	**1**

mass units

TABLE E.6 Pressure/Stress Conversion Factors

A. Pressure units decimally related to one pascal

	Pascals (Pa)	Bars (bar)	Decibars (dbar)	Millibars (mbar)	Dynes per square centimeter (dyn/cm²)
1 pascal =	1	0.00001	0.0001	0.01	10
1 bar =	100,000	1	10	1000	1,000,000
1 decibar =	10,000	0.1	1	100	100,000
1 millibar =	100	0.001	0.01	1	1000
1 dyne per second centimeter =	0.1	0.000001	0.00001	0.001	1

B. Pressure units decimally related to one kilogram-force per square meter (with pascal equivalents)

	Kilograms-force per square meter (kg/m²)	Kilograms-force per square centimeter (kg/cm²)	Kilograms-force per square millimeter (kg/mm²)	Grams-force per square centimeter (g/cm²)	Pascals
1 kilogram-force per square meter =	1	0.001	0.000001	0.1	9.80665
1 kilogram-force per square centimeter =	10,000	1	0.01	1000	98,066.5
1 kilogram-force per square millimeter =	1,000,000	100	1	100,000	9,806,650
1 gram-force per square centimeter =	10	0.001	0.00001	1	98.0665
1 pascal =	0.10197162	1.0197162×10^{-5}	1.0197162×10^{-7}	1.0197162×10^{-2}	1

NOTE: 1 atmosphere (technical) = 1 kilogram-force per square centimeter = 98,066.5 pascals.

C. Pressure units expressed as heights of liquid (with pascal equivalents)

	Millimeters of mercury at 0°C (mmHg, 0°C)	Centimeters of mercury at 60°C (cmHg, 60°C)	Inches of mercury at 32°F (inHg, 32°F)	Inches of mercury at 60°F (inHg, 60°F)	Centimeters of water at 4°C (cmH$_2$O, 4°C)	Inches of water at 60°F (inH$_2$O, 60°F)	Feet of water at 39.2°F (ft H$_2$O, 39.2°F)	Pascals (Pa)
1 millimeter of mercury, 0°C =	1	0.100282	0.0393701	0.0394813	1.359548	0.5357756	0.0446046	133.3224
1 centimeter of mercury, 60°C =	9.971830	1	0.3925919	0.3937008	13.55718	5.342664	0.4447895	1329.468
1 inch of mercury, 32°F =	**25.4**	2.547175	1	1.0028248	34.53252	13.60870	1.132957	3386.389
1 inch of mercury, 60°C =	25.32845	**2.54**	0.9971831	1	35.43525	13.57037	1.129765	3376.85
1 centimeter of water, 4°C =	0.735539	0.073762	0.028958	0.0290400	1	0.3940838	0.0328084	98.0638
1 inch of water, 60°F =	1.866453	0.187173	0.073482	0.0736900	2.537531	1	0.0832524	248.840
1 foot of water, 39.2°F =	22.4192	2.248254	0.882646	0.885139	30.47998	12.01167	1	2988.98
1 pascal =	7.500615×10^{-3}	7.521806×10^{-4}	2.952998×10^{-4}	2.96134×10^{-4}	1.01974×10^{-2}	4.01865×10^{-3}	3.34562×10^{-4}	1

NOTE: 1 torr = 1 millimeter of mercury at 0°C = 133.3224 pascals.

D. Nonmetric pressure units (with pascal equivalents)

	Atmospheres (atm)	Avoirdupois pounds-force per square inch (psi)	Avoirdupois pounds-force per square foot (lb/ft², avdp)	Poundals per square foot (pd/ft²)	Pascals (Pa)
1 atmosphere =	1	14.69595	2116.217	68,087.24	**101,325**
1 avdp pound-force per square inch =	6.80460×10^{-2}	1	**144**	4633.063	6894.757
1 avdp pound-force per square foot =	4.725414×10^{-4}	1/144 = 0.006944	1	32.17405	47.88026
1 poundal per square foot =	1.468704×10^{-3}	2.158399×10^{-4}	0.0310809	1	1.488165
1 pascal =	9.869233×10^{-6}	1.450377×10^{-4}	0.0208854	0.6719689	1

NOTE: 1 normal atmosphere = 760 torr = **101,325** pascals.

*Exact conversions are shown in boldface type. Repeating decimals are underlined. The SI unit of pressure or stress is the Pascal (Pa).

Source: D. G. Fink and W. Beaty (eds.), *Standard Handbook for Electrical Engineers,* 12th ed., McGraw-Hill, N.Y. 1987.

TABLE E.7 Energy/Work Conversion Factors

A. Energy/work units decimally related to one joule

	Joules (J)	Megajoules (MJ)	Kilojoules (kJ)	Millijoules (mJ)	Microjoules (μJ)	Ergs (ergs)
1 joule =	1	0.000001	0.001	1000	1,000,000	10^7
1 megajoule =	1,000,000	1	1000	10^9	10^{12}	10^{13}
1 kilojoule =	1000	0.001	1	1,000,000	10^9	10^{10}
1 millijoule =	0.001	10^{-9}	10^{-6}	1	1000	10,000
1 microjoule =	0.000001	10^{-12}	10^{-9}	0.001	1	10
1 erg =	10^{-7}	10^{-13}	10^{-10}	0.0001	0.1	1

NOTE: 1 watt-second = 1 joule.

B. Energy/work units less than ten joules (with joule equivalents)

	Joules (J)	Foot-poundals (ft · pdl)	Foot-pounds-force (ft · lb$_F$)	Calories (International Table) (cal, IT)	Calories (thermochemical) (cal, thermo)	Electronvolts (eV)
1 joule =	1	23.73036	0.7375621	0.2388459	0.2390057	6.24146×10^{18}
1 foot-poundal =	4.2104011×10^{-2}	1	3.108095×10^{-2}	1.006499×10^{-2}	1.007173×10^{-2}	2.63016×10^{17}
1 foot-pound-force =	1.355818	32.17405	1	0.3238316	0.3240483	8.46228×10^{18}
1 calorie (Int. Tab.) =	4.1868	99.35427	3.088025	1	1.000669	2.61317×10^{19}
1 calorie (thermo) =	4.184	99.28783	3.085960	0.9993312	1	2.61143×10^{19}
1 electronvolt =	1.60219×10^{-19}	3.80205×10^{-18}	1.18171×10^{-19}	3.82677×10^{-20}	3.82933×10^{-20}	1

C. Energy/work units greater than ten joules (with joule equivalents)

	Joules (J)	British thermal units, International Table (Btu, IT)	British thermal units, thermochemical (Btu, thermo)	Kilowatthours (kWh)	Horsepower-hours, electrical (hp·h, elec)	Kilocalories, International Table (kcal, IT)	Kilocalories, thermochemical (kcal, thermo)
1 joule =	1	9.478170×10^{-4}	9.4845165×10^{-4}	$1/3.6 \times 10^{6} = 2.777 \times 10^{-7}$	3.723562×10^{-7}	2.388459×10^{-4}	2.3900574×10^{-4}
1 British thermal unit, Int. Tab. =	1055.056	1	1.000669	2.9307111×10^{-4}	3.928567×10^{-4}	0.2519958	0.2521644
1 British thermal unit (thermo) =	1054.35	0.999331	1	2.928745×10^{-4}	03.925938×10^{-4}	0.2518272	0.2519957
1 kilowatthour =	**3,600,000**	3412.141	3414.426	**1**	**1/0.746** = 1.3404826	859.8452	860.4207
1 horsepower hour, electrical =	**2,685,600**	2545.457	2547.162	**0.746**	**1**	641.4445	641.8738
1 kilocalorie, Int. Tab. =	**4186.8**	3.968320	3.970977	**0.001163**	1.558981×10^{-3}	1	1.000669
1 kilocalorie, thermochemical =	4184	3.965666	3.968322	0.0011622	1.5579386×10^{-3}	0.999331	1

The exact conversion is 1 British thermal unit, International Table = **1055.05585262** joules.

*Exact conversions are shown in boldface type. Repeating decimals are underlined. The SI unit of energy and work is the joule (J).

Source: D. G. Fink and W. Beaty (eds.), *Standard Handbook for Electrical Engineers*, 12th ed., McGraw-Hill, N.Y., 1987.

TABLE E.8 Power Conversion Factors

A. Power units decimally related to one watt

	Watts (W)	Megawatts (MW)	Kilowatts (kW)	Milliwatts (mW)	Microwatts (μW)	Picowatts (pW)	Ergs per second (ergs/s)
1 watt =	1	0.000001	0.001	1000	1,000,000	10^9	10^7
1 megawatt =	1,000,000	1	1000	10^9	10^{12}	10^{15}	10^{13}
1 kilowatt =	1000	0.001	1	1,000,000	10^9	10^{12}	10^{10}
1 milliwatt =	0.001	10^{-9}	0.000001	1	1000	1,000,000	10,000
1 microwatt =	0.000001	10^{-12}	10^{-9}	0.001	1	1000	10
1 picowatt =	10^{-9}	10^{-15}	10^{-12}	0.000001	0.001	1	0.01
1 erg per second =	10^{-7}	10^{-13}	10^{-10}	0.0001	0.1	100	1

NOTE: 1 watt = 1 joule per second (J/s).

B. Nonmetric power units (with watt equivalents)

	British thermal units (International Table) per hour (Btu/h, IT)	British thermal units (thermochemical) per minute (Btu/min, thermo)	Avoirdupois foot-pounds-force per second (ft·lb/s avdp)	Kilocalories per minute (thermochemical) (kcal/min, thermo)	Kilocalories per second (International Table) (kcals, IT)	Horsepower (electrical) (hp, elec)	Horsepower (mechanical) (hp, mech)	Watts (W)
1 British thermal unit (Int. Tab.) per hour =	1	0.0166778	0.2161581	4.2027405×10^{-3}	6.9998831×10^{-5}	3.9285670×10^{-4}	3.930148×10^{-4}	0.2930711
1 British thermal unit (thermo) per minute =	59.959853	1	12.960810	0.2519957	4.1971195×10^{-3}	0.0235556	0.0235651	17.57250
1 foot-pound-force per second =	4.6262426	0.0771557	1	0.0194429	3.2383157×10^{-4}	1.817504×10^{-3}	1/550 = 1.818181×10^{-3}	1.355818
1 kilocalorie per minute (thermo) =	237.93998	3.9683217	51.432665	1	0.0166555	0.0934763	0.0935139	69.733333
1 kilocalorie per second (Int. Tab.) =	14,285.953	238.25864	3088.0251	60.040153	1	5.6123324	5.6145911	4186.800
1 horsepower (electrical) =	2545.4574	42.452696	550.22134	10.697898	0.1781790	1	1.0004024	746
1 horsepower (mechanical) =	2544.4334	42.435618	550	10.693593	0.1781074	0.9995977	1	745.6999
1 watt =	3.4121413	0.0569071	0.7375621	0.0143403	2.3884590×10^{-4}	1/746 = 1.3404826×10^{-3}	1.3410220×10^{-3}	1

NOTE: The horsepower (mechanical) is defined as a power equal to **550** foot-pounds-force per second.
Other units of horsepower are:
1 horsepower (boiler) = 9809.40 watts
1 horsepower (metric) = 735.499 watts
1 horsepower (water) = 746.043 watts
1 horsepower (U.K.) = 745.70 watts
1 ton (refrigeration) = 3516.8 watts

*Exact conversions are shown in boldface type. Repeating decimals are underlined. The SI unit of power is the watt (W).
Source: D. G. Fink and W. Beaty (eds.), *Standard Handbook for Electrical Engineers*, 12th ed., McGraw-Hill, N.Y., 1987.

TABLE E.9 Temperature Conversions*

Celsius (°C) °C = 5(°F − 32)/9	Fahrenheit (°F) °F = [9(°C)/5] + 32	Absolute (K) K = °C + 273.15
−273.15	−459.67	**0**
−200	−328	73.15
−180	−292	93.15
−160	−256	113.15
−140	−220	133.15
−120	−184	153.15
−100	−148	173.15
−80	−112	193.15
−60	−76	213.15
−40	−40	233.15
−30	−22	243.15
−20	−4	253.15
−17.77	**0**	255.372
−10	14	263.15
−6.66	**20**	266.483
0	32	273.15
5	41	278.15
10	50	283.15
15	59	288.15
20	68	293.15
25	77	298.15
30	86	303.15
35	95	308.15
40	104	313.15
45	113	318.15
50	122	323.15
55	131	328.15
60	140	333.15
65	149	338.15
70	158	343.15
75	167	348.15
80	176	353.15
85	185	358.15
90	194	363.15
95	203	368.15
100	212	373.15
105	221	378.15
110	230	383.15
115	239	388.15
120	248	393.15
140	284	413.15
160	320	433.15
180	356	453.15
200	392	473.15
250	482	523.15
300	572	573.15
350	662	623.15
400	752	673.15
450	842	723.15
500	932	773.15
1000	1832	1273.15
5000	9032	5273.15
10,000	18,032	10,273.15

*Conversions in boldface type are exact. Continuing decimals are underlined. Temperature in kelvins equals temperature in degrees Rankine divided by 1.8 [K = °R/1.8].

Source: D. G. Fink and W. Beaty (eds.), *Standard Handbook for Electrical Engineers,* 12th ed., McGraw-Hill, N.Y., 1987.

APPENDIX F
BIBLIOGRAPHY

BOOKS

Barak, M. (ed.): *Electrochemical Power Sources—Primary & Secondary Batteries,* Peter Peregrinus, Stevenage, UK, 1980.

Berndt, D.: *Maintenance-Free Batteries,* 2d ed. SAE, Warrendale, PA, 1997.

Brockris, J. O'M., and A. K. N. Reddy: *Modern Electrochemistry,* vols. I and II, Plenum Publishing Corp., New York, 1970; Plenum/Rosetta Ed., 1973.

Bode, H.: *Lead-Acid Batteries* (translated from German by R. J. Broad and K. V. Kordesch), Wiley, New York, 1977.

Bro, P., and S. C. Levy: *Quality and Reliability Methods for Primary Batteries,* John Wiley, New York 1990, Electrochemical Society Monograph Series.

Broadhead, J., and B. Scrosati: *Lithium Polymer Batteries,* vol. 96-17, The Electrochemical Society, Pennington, NJ, 1997.

Conway, B. E.: *Theory and Principles of Electrode Processes,* Ronald, New York, 1965.

Dameja, S.: *Electric Vehicle Battery Systems,* Newnes (Butterworth-Heineman) Woburn, MA, 1999.

Falk, S. U., and A. J. Salkind: *Alkaline Batteries,* Wiley, New York, 1969.

Fleisher, A., and J. J. Lander (eds.): *Zinc-Silver Oxide Batteries,* Wiley, New York, 1971.

Gabano, J. P.: *Lithium Batteries,* Academic Press, Ltd., London, 1983.

Heise, G. W., and N. C. Cahoon (eds.): *The Primary Batter,* vols. I and II, Wiley, New York, 1971 and 1976.

Himy, A.: *Silver-Zinc Battery: Phenomena and Design Principles,* Vantage Press, New York, 1986.

———: *Silver-Zinc Battery: Best Practices, Facts and Reflections,* Vantage Press, New York, 1995.

Jackish, H.-D.: *Batterie-Lexikon,* Pflaum Verlag, Munich, 1993.

Kiehne, H. A.: *Battery Technology Handbook,* Dekker, New York, 1989.

Kordesch, K. V. (ed.): *Batteries,* vol. I: *Manganese Dioxide,* vol. II: *Lead-Acid Batteries and Electric Vehicles,* Dekker, New York, 1974 and 1977.

Kordesch, K., and G. Simander: *Fuel Cells and Their Applications,* VCH Publishers, New York, 1996.

Levy, S. C., and P. Bro: *Battery Hazards and Accident Prevention,* Plenum Press, New York, 1994.

Mantell, C. L.: *Batteries and Energy Systems,* 2d ed., McGraw-Hill, New York, 1983.

Osaka, T., and M. Datta: *Energy Storage Systems for Electronics,* Gordon and Breach Science Publishing, London, 1999.

Pistoia, G.: *Lithium Batteries,* Elsevier, New York, 1994.

Rand, D. A. J., R. Woods, and R. M. Dell: *Batteries for Electric Vehicles,* SAE, Warrendale, PA, 1998.

Tuck, Clive D. S.: *Modern Battery Technology,* Ellis Horwood, Ltd, Chichester, West Sussex, England, 1991.

Venkatasetty, H. V.: *Lithium Battery Technology,* Wiley, New York, 1984.

Vinal, G. W.: *Primary Batteries,* Wiley, New York, 1950.

———: *Storage Batteries,* Wiley, New York, 1955.

Vincent, C., and B. Scrosati: *Modern Batteries,* 2d ed., Newnes (Butterworth-Heinemann), Woburn, MA (1997).

PERIODICALS

Energy Storage Systems, U.S. Dept. of Energy, Washington DC.
Interface, The Electrochemical Society, Pennington, NJ.
Journal of the Electrochemical Society, Pennington, NJ.
Journal of Power Sources, Elsevier Sequoia, S. A., Lausanne, Switzerland.
Progress in Batteries and Solar Cells, JEC Press, Cleveland.

PROCEEDINGS OF ANNUAL/BIENNIAL CONFERENCES

Meeting Abstracts of the ECS Battery Division, The Electrochemical Society, Pennington, NJ.
Proceedings of the International Power Sources Symposium, Broadway House, The Broadway, Crowborough, TN6 1HQ. UK.
Proceedings of the Power Sources Conference, Power Sources Branch, U.S. Army CECOM, Ft. Monmooth, NJ. Available from National Technical Information Service, Springfield, VA 22161-0001.
Proceedings of the Annual Battery Conference on Applications and Advances, California State University, Long Beach, CA.
Proceedings of the Annual International Seminary on Primary and Secondary Batteries, Florida Educational Seminars, Boca Raton, FL.
Proceedings of the Battery and Electrochemical Contractors Conference, U.S. Dept. of Energy, Washington, DC.
Proceedings of the NASA Aerospace Battery Workshop, Marshall Space Flight Center, AL.
Proceedings of the Intersociety Energy Conversion Engineering Conference, Also available in CDROM.
Proceedings of the Biennial Workshop for Battery Exploratory Development, Sponsored by the Office of Naval Research and Naval Surface Warfare Center Carderock Div., West Bethesda, MD.

OTHER REFERENCE SOURCES—HANDBOOKS AND BIBLIOGRAPHIES

Advanced Batteries for Electric Vehicles: An Assessment of Performance and Availability of Batteries for Electric Vehicles: A Report of the Battery Technical Advisory Panel, prepared for the California Air Resource Board (CARB), F. R. Kalhammer et al., El Monte, CA, June 22, 2000.
Batteries, Fuel Cells, and Related Electrochemistry: Books, Journals, and Other Information Sources, 1950–1979, U.S. Dept. of Energy Rep. DOE/CS-0156, Washington, DC, 1980.
Electric Batteries, a Bibliography, Energy Research and Development Administration Report no. TID-3361, Washington, 1977.
Cost of Lithium-Ion Batteries for Vehicles, L. Gaines and R. Cuenca, Argonne National Laboratory, Argonne, IL 60439, Report No. ANL/ESD-42 (May 2000).
Fuel Cells, a Bibliography, Energy Research and Development Administration, Report no. TID-3359, Washington, 1977.
Fuel Cell Handbook, 4th ed., J. D. Stauffer, R. Engleman, and M. Klett, U.S. Dept. of Energy-FETL, Morgantown, WV, 1998.
Fuel Cell Handbook, A. J. Appleby and F. R. Foulkes, Van Nostrand Reinhold, New York 1989; republished by Kreiger Publishing Co., Melbourne, FL, 1993.

Handbook for Handling and Storage of Nickel-Cadmium Batteries: Lessons Learned, NASA Ref. Publ. 1326, Feb. 1994.

Lithium and Lithium-Ion Batteries 2000, D. MacArthur, G. Blomgren and R. Powers, Chemac International, Sterling Heights, MI, 2000 (published biennially).

NASA Handbook of Nickel-Hydrogen Batteries, NASA Ref. Pub. 1314, Sept. 1993.

NAVSEA Batteries Document: State of the Art, Research and Development, Projections, Environmental Issues, Safety Issues, Degree of Maturity, Department of the Navy Publ. NAVSEA-AH-300, July 1993.

Navy Primary and Secondary Batteries—Design and Manufacturing Guidelines, Department of the Navy Publ. NAVSO P-3676, Sept. 1991.

Rechargeable Batteries: Application Handbook, Newnes (Butterworth-Heinemann), Woburn, MA, 1997.

U.S. Battery Market, Business Trend Analysts, Comack, New York, 2000.

1999 Review of Electric Vehicles, Batteries and Fuel Cells (A Powers Report), D. MacArthur, Chemac International, Sterling Heights, MI, 1999.

2000 Battery Industry Developments: Review and Analysis, D. MacArthur and G. Blomgren, Chemac International, Sterling Heights, MI, 2001 (published biennially).

An excellent source for technical information and performance data on batteries are the publications and data books issued by the battery manufacturers. The addresses of these manufacturers are listed in Appendix G. Also see the websites of these manufacturers.

Standards

See Chap. 4.

APPENDIX G
BATTERY MANUFACTURERS AND R&D ORGANIZATIONS‡

Prepared by Vaidevutis Alminauskas and Austin Attewell

AUSTRIA

Banner GmbH
Postfach 777
Salzburger Straβe 298
A-4021 Linz
Tel: +43 732 3888-0
Fax: +43 732 3888-74
http://www.banner.co.at

Bären Batterie GmbH
Feistriz I.
A-9181 Rosental
Tel: +43 4228 20 36
Fax: +43 4228 29 15

Exide Batteriewerke GmbH
Franz Schubert Gasse 7
A-2345 Brunn
Tel: +43 2236 35610-0
Fax: +43 2236 35610-45
http://www.sonnenschein.at

Institut fur Chemische Technologie*
Technischen Universitat
A-8010
Graz

AUSTRALIA

Apollo Batteries–Australia
Unit 1, 45 Powers Road
Seven Hills
NSW 2147
Tel: +61 2 9674 6322
Fax: +61 2 9674 6277
http://www.apollobatteries.com.au

CSIRO Energy Technology*
Bayview Avenue
PO Box 312
Clayton South
Victoria 3169
Tel: +61 3 9545 8500
Fax: +61 3 9562 8919
http://www.det.csiro.au

Exide
PO Box 73
Elizabeth
SA 5113
Tel: +61 88 30 74444
Fax: +61 88 30 74490
(formerly GNB)

---

‡See also: Tables 4.1(a)–(d) for a list of organizations preparing battery standards and Table 37.8, "Organizations with Major Development Projects on Rechargeable Batteries for EVs/HEVs and/or Electric Utility Storage."
*Government Laboratory

Supercharge Batteries (Parent co—Ramcar)
36 Roberna Road
Moorabbin,
Victoria 3189
Tel: +61 3 9555 9000
Fax: +61 3 9555 7194

BELGIUM

Duracell Benelux BV
Ikaroslaan 31
B-1930 Zaventen
Tel: +32 725 51 30
Fax: +32 725 08 30

BRAZIL

Acumuladores Ajax Ltda
R Joaquim Marques De Figueiredo, 5 57
Dt Industrial-Bauru-Sp-17034-290
Tel: +55 14 230 1933
http://www.ajax.com.br

Accumulatories Moura S/A
Rua Diario De Pernambuco 195
Belo Jardim
Pernambuco 55
150-000
Tel: +55 81 476 1644
Fax: +55 81 476 1113
http://www.moura.com.br

Saturnia Hawker Sistemas Energia Ltda
Rua Fidêncio Ramos
100-120 Andar-Vila Olímpia
Cep 04551-010
São Paulo
Tel: +55 11 829 0944
Fax: +55 11 820 0305
http://www.saturnia-hawker.com.br

BULGARIA

Central Laboratory for Electrochem. Power
    Sources, Bulg. Acad. Sciences
Acad.G, Bonchev Street
Block 10
Sofia
Tel: +359 2 718 651
Fax: +359 2 731 552

CANADA

Battery Technologies Inc.
30 Pollard Street
Richmond Hill, Ontario L4B 1C3
Tel: (905) 881-5100
Fax: (905) 881-6043
http://www.bti.ca

Eagle-Picher Energy Products Corp.
13136-82A Avenue
Surrey, British Columbia V3W 9Y6
Tel: (604) 543-4350
Fax: (604) 543-8122
http://www.epi-tech.com

Electrofuel
21 Hanna Avenue
Toronto, Ontario M6K 1W9
Tel: +(416) 535-1114
Fax: +(416) 535-2361

E-One Moli Energy (Canada) Ltd.
20000 Stewart Crescent
Maple Ridge, British Columbia V2X 9E7
Tel: (604) 466-6654
Fax: (604) 466-6600
http://www.molienergy.bc.ca

Institut de recherché d'Hydro-Quebec
1800 montée Ste-Julie
Varennes, Quebec J0L 2P0
Tel: (514) 652-8011
http://www.ireq.ca

Sanford Battery
470 Beach Road
Hamilton, Ontario L8H 3K7
Tel: (905) 544-4220
Fax: (905) 544-4225
http://www.sanfordbattery.on.ca

Surrette Battery Company Ltd.
PO Box 2020, 1 Station Road
Springhill, Nova Scotia B0M 1X0
Tel: (902) 597-3767
Fax: (902) 597-8447
http://www.surrette.ca

University of Ottawa
P.O. Box 450
Ottawa, Ontario K1N 6N5
(613) 564-9536

VHB Industrial Batteries Ltd.
72 Devon Road, Unit 22
Brampton, Ontario L6T 5B4
Canada
Tel: (905) 790-0730
Fax: (905) 790-0744
http://www.vhbbatteries.com
(Invenys plc-HAWKER Energy Storage)

CHINA (PRC)

Baoding Jin Storage Battery Co. Ltd.
8 Fuchang Road
Baoding
Hebei Province 071057
Tel: +86 312 3232 008
Fax: +86 312 3236 051
http://www.fengfan.com.cn

BYD Battery Co. Ltd.
Building 6-7, Zienghua No. 6
Scien-Tech Ind. City
Shenzhen
Guandong 518116
Tel: +86 755 4898 888
Fax: +86 755 4890 125
http://www.byd.co.cn

Changzhou Daily-Max Battery Co. Ltd.
39-1 Huayuan Road
Changzhou
Jiangsu 213016
Tel: +86 519 3270 441
Fax: +86 519 3270 425
http://www.daily-max.com

Chendu Jianzhong Lithium Battery Co.
30 Zangwei Road
Shuangliu
Sichuan
Tel: +86 28 5822 465
Fax: +86 28 5822 429

China Storage Battery Industry Association
(see Shenyang Storage Battery Research
    Inst. China)

China Industrial Assoc. of Power Sources
PO Box 296
18 Lingzhuangzi Road
Liqizhuang
Nankai District
Tianjin 300381
Tel: +86 22 2338 0938
Fax: +86 22 2238 0938
http://www.chinabatteryonline.com

China National Battery Industry
    Information Center
No. 1 Yangtianhu Xincun
Changsha
Hunan 410007
Tel: +86 731 5141901

Chongqing Battery Co.
168 Zhongshan 3 Road
Yuzhong District
Chongqing, 400015
Tel: +86 23 6386 7741
Fax: +86 23 6386 6957
http://www.cqmec.com/battery.htm

Chongqing Storage Battery Factory
Kuzhuba
Tuqiao
Chongqing 630054
Tel: +86 23 6285 4905
Fax: +86 23 6285 1155

Fujian Nanping Nanfu Battery Co. Ltd.
109 Industry Road
Nanping
Fujian 353000
Tel: +86 599 8735 162
Fax: +86 599 8735 117
http://www.nanfu.com

Guangdong Jiangmen JJJ Battery Co. Ltd.
Baisha Industrial Development Area
Western Area, Jiangmen
Guangdong 529000
Tel: +86 750 3532 350
Fax: +86 750 3534 305
http://www.asiansources.com/jjjbat.co

Guangzhou Storage Battery Ind. Co. Ltd.
1 Hua Yuan Road
San Yuan Li
Guangzhou
Guangdong 510400
Tel: +86 20 6661 760
Fax: +86 20 6678 113

Hangzu Narada Battery Co. Ltd.
459 Wensan Road
Hangzhou 310013
Tel: +86 571 3388 415
Fax: +86 571 5128 965
http://www.cnnd.com

Harbin Guangyu Group
68 Dianian Street
Xeuru Road
Nangang District
Harbin 150086
Tel: +86 451 668 8168
Fax: +86 451 667 8032

Lixing Power Sources Co.
7 Land, Guangdong Industrial Park
East Lake Hi-tech Development Zone
Wuhan
Tel: +86 27 7414 024
Fax: +86 27 7414 024
http://www.lisun.com/jieshao

National Engineering Development Center
  of High Technology Energy Storage
  Material*
Zhongshan Port
Zhongshan City
Guangdong
Tel: +86 760 5598 010
Fax: +86 760 5598 544

National Sichuan Wuzhou Battery Factory
No. 5, Nanta Road, Mianyang
Sichuan, 621000
Tel: 0816 2222957
Fax: 0816 2223636

Ningbo Baowang Battery Co., Ltd.
No. 6 Qixiang (N) Road
Yuelong Economic Development Zone
Ninghai, Ningbo
Zhejiang 315600
Tel: +86 574 5579 907
Fax: +86 574 5582 146
http://baowang.com

Ningbo Donghai Storage Battery Factory
Gaoqian Town
Qianhu
Ningbo City, 315124
Tel: +86 574 8499 993

Ningbo Osel Battery Factory
254 Wangtong Road
Ningbo 315010
Tel: +86 0574 7150 274
Fax: +86 0574 7151 765
http://www.ningbobattery.com

Qingdao Storage Battery Factory
No. 63, Third Yan'an Road, Qingdao
Shandong, 266022
Tel: 0532 3864644
Fax: 0532 3830712

Shantou Humei Battery Factory
Pumei Industrial Zone
Chenghai County
Guangdong 515800
Tel: +86 754 5722 789
Fax: +86 754 5722 789

Shanghai White Elephant Swan Battery
  Co., Ltd.
(Formerly Shanghai Battery Factory)
110 Qing Yun Road, Shanghai 200081
Tel: +86 21 5662 7600
Fax: +86 21 5663 4385
*http://www.shanghai-window.com/shanghai
  /ele_app/sbf/sbf.html*

Shenyang Storage Battery Factory
28 Baogong Street
Tiexi District
Shenyang
Liaoning Province 110026
Tel: +86 24 2587 5578
Fax: +86 24 2585 0217

Shenyang Storage Battery Research
  Institute*
33 Bei Er Zhong Road
Tiexi District
Shenyang, Liaoning Province, 110026
Tel: +86 24 574 282
Fax: +86 24 550 217

Shenzhen Center Power Tech. Co. Ltd.
Room 502
Bairong Building
1073 Cuizhu Road
Shenzhen 518020
Tel: +86 755 5164 318
Fax: +86 755 5606 044
http://www.vision-batt.com

Tianjin Institute of Power Sources*
P.O. Box 296, Tianjin
Lingzhuangzi Road
Lingzhuangzi, Nankai District
Tianjin, 300381
Tel: +86 22 382851
Fax: +86 22 383783

Weihai Wenlong Battery Co. Ltd.
Puji Town
Wendeng City 264200
Tel: +86 631 8842 068
Fax: +86 631 8842 068
http://www.chinabatterygroup.com

Yuyao Yao Hu Power Supply Factory
Ma Zhu town, Yu Yao City
Zhe Jiang, 315450
Tel: +86 574 2468 688
Fax: +86 574 2468 688

Wuhan Changguang Battery Co. Ltd.
Economic & Technology Develepment
  Zone
Han Qiang Street
Wuhan City 430056
Tel: +86 27 4893 121
Fax: +86 27 4891 320
http://www.cgbbattery.com

Wuzhou Storage Battery Factory
32 Xiji, Longgulu, Wuzhou, Guangxi
Wuzhou City 543000
Tel: +86 774 3823 700
Fax: +86 774 3823 700

Zhejiang Lighthouse Storage Battery
  Group Co.
946 Yun Dong Road
Shaoxing
Zhejiang, 312000
Tel: +86 575 8649 521
Fax: +86 575 8649 528
http://www.cnlighthouse.com

Zibo Storage Battery Factory
PO Box 9
Zhangdian District
Zibo
Shandong 255056
Tel: +86 533 298 0035
Fax: +86 533 298 0136

CZECH REPUBLIC

VHB Prumyslove Baterie, s.r.o.
Chodovska 7
Praha 4 141 00
Phone +420 2 769328-9
Fax +420 2 71760139
http://www.economy.cz/vhb
(Invenys plc-HAWKER Energy Storage)

DENMARK

Alkaline Batteries A/S
Tigervej 1
DK-7700
Thisted
Tel: +45 97 92 31 22
Fax: +45 97 92 18 99

Grinsted Akkumulator Fabrik A/S
Heimdalsvel 19
DK-7200 Grinsted
Tel: +45 75 32 19 55
Fax: +45 75 31 02 19

FINLAND

Oy Hydocell Ltd
Minnkkiatu 1-3
FIN-04430 Järvenpää
Tel: +358 89 271 02 50
Fax: +358 89 291 10 51

FRANCE

ASB Batteries
Allée Sainte Helène
18021 Bourges
Tel: +33 2 48 55 74 00
Fax: +33 2 48 55 74 01

Chloride Batteries Industrielles
Zone de Fret No. 4
Rue Du Te–Bat. 3450A
BP 10119
95701 Rossy CDG Cedex
Tel: +33 1 4816 3951
Fax: +33 1 4816 3959

CEAC Compagnie Européenne
  d'Accumulateurs
5, Allée des Pierres Mayettes
92230 Gennevilliers
Tel: +33 1 41 21 23 00
Fax: +33 1 41 21 24 05

Exide Holding Europe
5, Allée des Pierres Mayettes
Gennevilliers 92230
France
Tel: +33 1 41 21 23 00
Fax: +33 1 41 21 24 05
http://www.exideworld.com

Oldham France
Rue Alexander Fleming
BP 962
62033 Arras Cedex
Tel: +33 3 21 60 25 25
Fax: +33 3 21 73 16 51
http://www.oldham-france.fr

SAFT
156 Avenue de Metz
93230 Romainville
Tel: +33 1 49 15 36 00
Fax: +33 1 49 15 34 10
http://www.saft.alcatel.com

VHP Batteries Industrielle
Parc Evolic–Rue de Petit Albi
Cergy-St. Christophe
BP 8247
96801 Cergy-Pontoise Cedex
Tel: +33 130 7539
Fax: +33 130 3212 15

GERMANY

ABB Asea Brown Boveri AG
PO Box 100164
D-6800 Mannheim
Tel: +49 621 43810

AEG Anglo Batteries GmbH
Soflinger Strasse 100
D-89077 Ulm
Tel: +49 731 933 1652
Fax: +49 731 933 1852
http://www.zebrabat.de

Accumulatorenfabrik Berga GmbH
Industriestraße 34
D-76437 Rastatt
Tel: +49 7222 9 57 01
Fax: +49 7222 95 71 05

Accumulatorenwerke Hoppeche
Carl Zoellner & Sohn GmbH & Co. KG
Bontkirchener Straße 2
D-59929 Brilon
Tel: +49 2963 61 0
Fax: +49 2963 61 449
http://www.hoppecke.com

Akkumulatorenfabrik Moll GmbH + Co.
 KG
Angerstraße 50
D-96231 Staffelstein
Tel: +49 9573 801
Fax: +49 9573 3832

Accumulatorenfabrik Sonnenschein GmbH
Im Thiergarten
D-63654 Büdingen
Tel: +49 6042 81 313
Fax: +49 6042 81 216
http://www.sonnenschein.de

Banner Batterien Deutschland
Kesselboden 3
D-85391 Allershausen
Tel: +49 8166 1091

DETA Akkumulatorenwerk GmbH
Odertal 35
D-37431 Bad Lauterberg
Tel: +49 5524 82-0
*http://www.motorcarweb.com/germany/
 deta*

Diehl & Eagle-Picher GmbH Batterie-
 Systeme
Fischbachstraße 20
D-90552 Röthenbach
Tel: +49 911 9 57 21 00
Fax: +49 911 9 57 30 70
http://www.battery.de

Emmerich GmbH
6000 Frankfurt-Main 50
Tel: +49 69 655 717
Fax: +49 69 652 431

FRIWO Silberkraft für Batterietechnik
 GmbH
Meidericher Straße 6-8
D-47058 Duisburg
Tel: +49 203 30 02 -178
Fax: +49 203 30 02-241
http://www.friwo-batterien.de

Hagen Batterie AG
Coesterweg 45
D-59494 Soest
Tel: +49 2921 70 30
Fax: +49 2921 70 34 23
http://www.hagenbatterie.de

Homeyer GmbH
Reiherdamm 46
D-20457 Hamburg
Tel: +49 40 32 33 650
Fax: +49 40 31 15 35

Litronik Batterietechnologie
Birkwitzer Strasse 79
D-01796 Pirna
Tel: +49 3501 530 50
Fax: +49 3501 530 599

STECO Batterien Vertriebs GmbH
Vertriebsleitung Wiesbaden
Bismarckring 3
Wiesbaden 65183
Tel: +49 611 3 90 71
Fax: +49 611 30 67 18

Sonnenschein Lithium GmbH
Industriestrasse 22
D-63654 Büdingen
Tel: +49 6042 81-522
Fax: +49 6042 81-523
http://www.sonnenschein-lithium.de

Varta Batterie AG
AM Leinenfor 51
D-30419 Hanover
Tel: +49 5 11 79 03-0
Fax: +49 5 11 79 03-7 66
http://www.varta.com

Yuasa Batteries (Europe) GmbH
Tiefenbroicher Weg 28
D-40472 Düsseldorf
Tel: +49 211 35 00 47

ZSW
Helmholtzstrasse 8
D-89081 Ulm
Tel: +49 731 9530 602
Fax: +49 731 9530 666
http://www.zsw.uni-ulm.de

HONG KONG

Gold Peak Industries Ltd.
30 Kwai Wing Road
Kwai Chung
New Territories
Hong Kong
Tel: +852 2427 1133
Fax: +852 2489 1879
http://www.goldpeak.com

GasTon Battery Industrial Ltd.
7/F, Block B, Edwick Ind. Centre
30 Lei Muk Road
Kwai Chung, Hong Kong
Tel: +852 2447 7507
Fax: +852 2617 2465
http://www.gaston.com.hk

Hi-Watt Battery Industry Co. Ltd.
21 Tung Yuen Street
Yau Tong Bay
Kowloon
Tel: +852 2348 0111
Fax: +852 2772 7703
http://www.hi-watt.com.hk

Manton Technology Ltd.
14/F Wah Hung Centre
41 Hung To Road
Kwun Tong
Kowloon
Tel: +852 2763 5713
Fax: +852 2357 4728
http://www.diamec.com

HUNGARY

Perion Battery Factory Co., Ltd.
Perion Akkumulátorgyár Rt.
1138 Budapest
Tel: +36 1270-0811
Fax: +36 1320-2279

INDIA

Amara Raja Batteries Ltd.
Renigunta Cuddapah Road
Karakambadi
Tirupati 517520
Tel: +91 8574 75561
Fax: +91 8574 75360

AMCO Batteries Ltd.
105, Dr. Radhakrishnan Salai
Chennai 600 004
Tel: +91 44 855 5578
Fax: +91 44 855 5512

AMCO Power Systems Ltd.
Hebbal-Bellary Jakkur Road
Byatrayanapura
Bangalore 560 092
Tel: +91 80 856 1790

AUTOBAT Batteries Ltd.
Autobat House
BT Bus Strand
Nadiad 387 001
Tel: +91 268 66167
Fax: +91 268 60717

Exide Industries Ltd.
Exide House
59E, Chowringee Road
Calcutta 700 020
Tel: +91 33 247 8320
Fax: +91 33 247 9819
http://www.exideindustries.com

Hyderabad Battery Ltd. (HBL)
8-2-601 Road No. 10
Banjara Hills
Hyderabad-500 034
Tel: +91 842 248961
Fax: +91 842 248493

Mysore Alkaline Batteries (Pty). Ltd.
F-2, Krishna Towers, 1st floor, Gandhi Ngr
Bangalore 560 009
Tel: +91 80 2261367
Fax: +91 80 3343150

Shetron Limited
A-6 MIDC, Street No. 5, Andheri (E)
Mumbai-400 093
Tel: +91 22 832 6228
Fax: +91 22 837 2145
http://www.shetrongroup.com

Sharana Industries
Corpn. Shopping Complex
77 C.P.Ramaswamy Road
Madras 600 018
Tel: +91 44 4997568
Fax: +91 44 4995469
http://www.sahrana.co.in

IRELAND

ABL
Kore Development Park
Bluebell
Dublin 12
Tel: +353 1 4602142
Fax: +353 1 4602654

ISRAEL

E.C.R. Kiryat Weitzman,
Building 18 PO Box 1803
Rehovot 76117
Tel: +972 8 9407920
Fax: +972 8 9409683
http://www.ecr.co.il

Electric Fuel Ltd.
Western Industrial Park
PO Box 641
Beit Shemesh 99000
Tel: +972 2 9906 622
Fax: +972 2 9906 688
http://www.electric-fuel.com

E. Schnapp & Co. Works Ltd.
22 Shecterman St.
Netania 42379
Tel: +972 9 860 6111
Fax: +972 9 861 9069
http://www.schnapp.co.il

Tadiran Batteries Ltd. (Division of SAFT)
Kiryat Ekron
Rehovot 76100
Tel: +972 8 944 4315
Fax: +972 8 941 3062
http://www.tadiran.com

Vulcan Batteries Ltd.
PO Box 17
Migdal Tefen 24959
Tel: +972 4 987 2205
Fax: +972 4 987 2672

ITALY

Fabbrica Accumulatori Uranio SpA
Viale del Lavoro, 20
37040 Veronella
Verona
Tel: +39 0442 489111
Fax: +39 0442 480374
http://www.uranio.com

Midac SpA
Zona Industriale
37038 Soave
Verona
Tel: +39 045 6132132
Fax: +39 045 6132134
http://www.midacbatteries.com

Acumulatori Ap SpA
Via del Lavoro—Z.I.
36075 Montecchio Maggiore
Vicenza
Tel: +39 0444 698400
Fax: +39 0444 696787

FIAMM. SpA
Viale Europa, 63
36075 Montecchio Maggiore
Vicenza
Tel: +39 0444 709311
Fax: +39 0444 699237
http://www.fiamm.com

Pilazeta SpA
10024, Moncalieri
Turin
Tel: +39 011 641 12 9
Fax: +39 011 640 35 44

Nouva Scaini
090039 Villacidro Z.I
Cagliari
Tel: +39 070 9311225
Fax: +39 070 9311228
http://www.farplus.ch

JAPAN

Furukawa Battery Co. Ltd.
4-1, 2-Chome Hoshikawa
Hodogaya-Ku
Yokohama 240 006
Tel: +81 45 336 5088
Fax: +81 45 333 2534

Government Research Institute*
Osaka, Midorigaoka 1
Ikeda-shi
Osaka-fu 563
Tel: +81 7 2751 8351

Hitachi Maxell Ltd.
2-12-24 Shibuya-ku
Tokyo 150
Tel: +81 3 5467 9327
Fax: +81 3 5467 9328
http://www.maxell.co.jp

Japan Batteries Industry Association
Kikai Shinko-Kaikan
5-8 Shibakoen, 3-Chrom
Minao-ku
Tokyo 105
Tel: +81 3 343 40261
Fax: +81 3 343 42691
http://www.pweb.in.aix.or.ip/-battery

Japan Storage Battery Co. Ltd.
1, Inobanba-cho
Nishinosho, Kisshoin
Minami-ku
Kyoto 601-8520
Tel: +81 75 312 1211
Fax: +81 75 316 3101
http://www.nippondenchi.co.jp

Matsushita Battery Industrial Co., Ltd.
1, Matsushita-cho
Moriguchi
Osaka 570-8511
Tel: +81 6 994 4531
Fax: +81 6 994 7056
http://www.mbi.panasonic.co.jp

Sanyo Electric Co., Ltd.
5 Keihan-Hondori, 2-Chome
Moriguchi City
Osaka, 570-8677
Tel: +81 6 991 1181
Fax: +81 6 991 6566
http://www.sanyo.co.jp

Sony Energytec Inc.
6-7-35 Kitashinagawa
Shinagawa-ku
Tokyo 141
Tel: +81 3 5448 2111
Fax: +81 3 5448 2244
http://www.world.sony.com

Toshiba Battery Co. Ltd.
Koei Bldg. 13-10
Ginza 7-Chome
Chuo-ku, Tokyo 104
Tel: +81 3 546 0758 0
Fax: +81 3 546 0759 1
http://www.tbcl.co.jp

Yuasa Battery Co. Ltd.
6-6, Josai-cho
Takatsuki
Osaka 569
Tel: +81 726 75 5503
Fax: +81 726 75 5508
http://www.yuasa-jpn.co.jp

JORDAN

United Industries Corp.
PO Box 188
Amman 11118
Tel: +962 6 4892294/5
Fax: +962 6 4890259
http://uic.jordanmall.com

LITHUANIA

Akumuliatoriu Centras
Kauno g. 32
Vilnius 2006
Tel: +370 2 263716
Fax: +370 2 263716

MALAYSIA

Federal Batteries Sdn. Bhd.
Lot 12, Jalan Perusahaan Dua
Batu Caves Industrial Area
68100 Batu Caves
Selangor Darul Ehsan
Tel: +60 3 689 8828
Fax: +60 3 689 8277
*http://www.jaring.my/mbp/directory/ads/
    federalbatteries/federal.html*

MEXICO

Accumuladores Ennermex, S.A. de C.V.
Eugenio Garza Sada Sur 3431
Col. Arroyo Seco
Monterrey, N.L. 64740
Tel: +52 83 299 500
Fax: +52 83 358 9179
http://www.enermex.com

NETHERLANDS

Batterij Aandrijf Techniek—BAT
Edisonstraat 18
Nijkerk (GLD) 3861 NE
Netherlands
Tel: +31 33 2450990
Fax: +31 33 2450991

Philips Research Laboratories
Prof Holslaan 4 (WA p 319)
5656 AA
Eindhoven
Tel: +31 40 274 2701
Fax: +31 40 274 3352

Thales Munitronics (Formerly Signaal
    UFSA)
Hoogezidje 14
5626 DC Eindhoven
P.O. Box 6034 5600 HA Eindhoven
Tel: +31 40 2503 603
Fax: +31 40 2503 777

TNO Environmental Sciences*
PO Box 342
7300 AH Apeldoorn
Tel: +31 55 549 3493

VB Autobatterijen BV
Bovendijk 137
3045 PC Rotterdam
Tel: +31 10 285 9272
Fax: +31 10 422 1385

NEW ZEALAND

Pacific Lithium Ltd.
P.O. Box 90725
Auckland Mail Centre
Tel: +64 9 309 5221
Fax: +64 9 307 1749

NORWAY

Exide Sonnak AS
PO Box 418
Okern
0513 Oslo
Tel: +47 22 72 24 30
Fax: +47 22 64 45 55

Kongsberg Simrad
Dyrmyrgata 35
PO Box 483
3601 Kongsberg
Tel: +47 32 86 50 00
Fax: +47 32 73 60 59
http://www.kongsberg-simrad.com

Hoppecke Norge AS
PO Box 166
Oken
0508 Oslop
Tel: +47 22 63 88 79
Fax: +47 22 63 88 83
http://www.hoppecke.no

PAKISTAN

Exide Pakistan Ltd.
40-K, Block 6, PECH Society
Dr. Hussain Mahmood Road
Off Sharea Faisal
Karachi 75400
Tel: +92 21 453 6750
Fax: +92 21 453 8948

POLAND

Centralne Laboratorium Akumulatorow i
    Ogniw*
ul.Forteczna 12
Poznan 61-362
Tel: +48 61 8790517
Fax: +48 61 8793012

Centra SA
ul.Gdynska 31/33
Poznan 60-960
Tel: +48 61 8786100
Fax: +48 61 8786506
http://www.centra.com.pl

Danish Polish Batteries Sp. zo.o.
ul.Zielona 22
Starogard Gdanski 83-200
Tel: +48 58 5623057
Fax: +48 58 5613127

GP Battery Poland Sp. zo.o.
ul.Grazyny 13/15
Warszawa 02-548
Tel: +48 22 8454095
Fax: +48 22 8455869
http://www.goldpeak.com

Philips Matsushita Battery Poland
ul.Swoneczna 42
62-200 Gniezno
Tel: +61 426.00.00
Fax: +61 425.65.32
http://www.lighting.philips.com.pl/html/
  gniezno.html

Zap
ul. Warsawska 47.05-820
Piastow
Tel: +48 22 723 6011
Fax: +48 22 723 6244
http://www.zap.po

RUSSIA

Electrozariad
ul. Ogareva 5
103918 Moscow K-9
Tel: +7 095 291 5785
Fax: +7 095 291 5785

International Assn. of Chemical Power
Sources Manufacturers
1-St.Smolensky Byst 7
123099 Moscow
Tel: +7 095 244 0735
Fax: +7 095 244 0369

Istochnik Batteries Ltd.
ul.Marxistskaya, 20, str. 9, etazh 1
109147 Moscow
Tel: +7 095 912 54 71
Fax: +7 095 911 91 31
http://www.istochnik.ru

Power International Concern ISC
ul.Novoslobodskaya, 62, korp.11, str. 27
103055 Moscow
Tel: +7 095 973 05 23
Fax: +7 095 200 46 74
http://www.battery.ru

Rigel Battery Co.
38 Prof. Popov Str., 197376
St. Petersburg
Tel: +7 812 234 05 56
Fax: +7 812 234 06 38

SAUDIA ARABIA

National Battery Company
P.O. Box 177
Riyadh 11383
Tel: +966 1 2650019
Fax: +966 1 2650057
http://www.nbc-sa.com

SINGAPORE

GP Batteries
50 Gul Crescent
Singapore 629543
Tel: +65 8622 088
Fax: +65 8622 313
http://www.gpbatteries.com

ST Battery Pte. Ltd.
77 Science Park Drive
02-15 Cintech III
Singapore 118256
Tel: +65 870 1665
Fax: +65 870 1665

SLOVENIA

Tovarna akumulatorskih baterij d.d.
Polena 6, SLO-2392 Mezica
Tel: +386 0 602 8700 201
Fax: +386 0 602 8700 205
http://www.tab-rm.si

SOUTH AFRICA

First National Battery Co (Pty) Ltd.
64 Liverpool Road
Industrial Sites, Benoni South
1501 Benoni
Tel: +27 11 741 3600
Fax: +27 11 421 1625
http://www.battery.co.za

Alpha Power Systems (Pty) Ltd.
Unit A, Mount Royal
James Crescent
1685 Midrand
Tel: +27 11 805 5008
Fax: +27 11 805 5054
http://www.powerman.co.za

Beacon Battery Manufacturers
535 HF Verwoed Drive
Gezina
0031 Pretoria
Tel: +27 12 335 6755
Fax: +27 12 335 6755

Willard Batteries
PO Box 1844
6000 Port Elizabeth
Eastern Cape
Tel: +27 41 451 4491
Fax: +27 41 451 2622/1918

SOUTH KOREA

Dae Sung Industrial Battery Co.
207-55 Junpo 1 dong
Pasanjin-gu
Pusan-si,
Tel: +82 51 807 6663
Fax: +82 51 807 6664
http://www.dsvolcano.co.kr

Global & Yuasa Battery Co., Ltd.
08-8 Yeok sam-dong
Kang nam-gu
Seoul
Tel: +82 2 3451 6201
Fax: +82 2 538 4353
http://www.gybc.co.kr

Korea Battery Industry Cooperative
1304-4 Seocho-dong
Seocho-ku
Seoul 137
Tel: +82 2 5532 401
Fax: +82 2 5561 290

Korea Storage Battery Ltd.
40-42 Dae Hwa-dong
Daeduk-gu
Daejeon
Tel: +82 2 579 0451
Fax: +82 2 579 1050
http://www.ksb.co.kr

Kyungwon Battery Co. Ltd.
6-1 San Galgot-ri
Pyungtaek-gun
Kyungki-do 451 860
Tel: +82 339 72 1321
http://www.solite.co.kr

Namil Battery Co. Ltd.
Samho Mulsan Bldg.
Yangjae-dong
137-130 Socho-Gu
Seoul
Tel: +82 2 2279 61 41
Fax: +82 2 589 1890

SPAIN

Cegasa
Artapadura, 11
01013 Vitoria
Tel: +34 945 129510
Fax: +34 945 129514
http://www.cegasa.com

Se. del Accumulador Tudor
Condesa de Venadito 1
28027 Madrid
Tel: +34 91 566 48 00
Fax: +34 91 404 78 50

SWEDEN

AB Tudor
44981 Nol
Tel: +46 303 33 10 00
Fax: +46 303 74 23 20

SWITZERLAND

Eurobat – Assn. of European Accumulator
  Manufacturers. c/o ATAG Ernst &
  Young
Belpstrasse 23
PO Box 5032
CH-3001
Berne
Tel: +41 31 320 61 61
Fax: +41 31 320 68 43
http://www.aey.ch

Leclanché SA
48, Avenue de Grandson
CH-1401 Yverdon-les-Bains
Tel: +41 24 447 22 72
Fax: +41 24 445 24 42
http://www.leclanche.com

Ralston Energy Systems (Europe) SA
P.O. Box 230
1218 Le Grand-Saconnex
Geneva
Tel: +41 22 929 94 38
Fax: +41 22 929 94 11
http://www.energiser.ch

Renata SA
CH-4452 Itingen
Tel: +41 61 975 75 75
Fax: +41 61 975 75 95
http://www.renata.com

TAIWAN (ROC)

Evervigor Battery Corp.
Room 603, 6F-4
13 Wu-Chiuan 1st Road
Hsien Chung City
Taipei
Tel: +886 2 299 9507
Fax: +886 2 299 1980

NEXcell Battery Co., Ltd.
5 Innovation Road
Science-based Industrial Park
Hsinchu
Tel: +886 3 5783800
Fax: +886 3 5786645
http://www.battery.com.tw

TURKEY

Mutlu Aku ve Malzemeleri San.A.S.
Cavusoglu Mah.Cobanyildizi Cad.
No:55 Kartal
81450 Istanbul
Tel: +90 216 389 5980
Fax: +90 216 353 9547

UNITED KINGDOM

AEA Technology Batteries
Culham Science Centre
Abingdon
Oxon OX14 3DB
Tel: +44 1235 463469
Fax: +44 1235 463400
http://www.aeatbat.com

AEA Technology Battery Systems Ltd.
Ruben House
Crompton Way
Crawley
West Sussex RH10 2QR
Tel: +44 1293 446800
Fax: +44 1293 403396
http://www.aeat.co.uk

Beta Research & Development Ltd.
50 Goodsmoor Road
Sinfin
Derby, DE24 9GN
Tel: +44 1332 770500
Fax: +44 1332 771591
http://www.betard.co.uk

British Battery Manufacturers Association
26 Grosvenor Gardens
London
SW1W 0TG
Tel: +44 20 7838 4800
Fax: +44 20 7838 4871
http://www.bbma.co.uk

CMP Batteries Ltd.
Salford Road
Over Hulton
Bolton
Lancashire BL5 1DD
Tel: +44 1204 656726
Fax: +44 1204 62981
http://www.cmpbatt.co.uk

Chloride Industrial Batteries Ltd.
Rake Lane
Clifton Junction
Manchester
Lancashire M27 8LR
Tel: +44 161 7944611
Fax: +44 161 7936606
http://www.hawker.co.uk

DERA Tech. Management Services*
Aquila, Golf Road
Bromley
Kent BR1 2JB
Tel: +44 20 8285 7127
Fax: +44 20 8285 7346
http://www.dera.gov.uk

Duracell UK Ltd.
Gillette House
Great Western Road
Isleworth
Middlesex TW7 5NP
Tel: +44 20 8560 1234
http://www.duracell.com

Electrochemical Power Sources Centre*
DERA Haslar
Gosport
Hampshire PO12 2AG
Tel: +44 1705 335358
Fax: +44 1705 335102
http://www.dera.gov.uk

Energizer UK Ltd.
Ever Ready House
93 Burleigh Gardens
Southgate
London
N14 5AQ
Tel: +44 20 8882 8661
http://www.energizer-eu.com

Exide Batteries Ltd. (Customer Services
    Division)
Pontyfelin Industrial Estate
New Road
Pontypool
Gwent NP4 5DG
Tel: +44 1495 750075
Fax: +44 1495 750222

GP Batteries (UK) Ltd.
Summerfield Avenue
Chelston Business Park
Wellington
Somerset TA21 9JF
Tel: +44 1823 660044
Fax: +44 1823 665595
http://www.gpbatteries.com

Hawker Energy Products Ltd.
Stephenson Street
Newport, Gwent NP19 4XJ
Tel: +44 1633 277673
Fax: +44 1633 281787
http://www.hawker.co.uk

Hawker Eternacell Ltd.
River Drive
South Shields
Tyne and Wear NE33 2TR
Tel: +44 191 4561451
Fax: +44 191 4566383
http://www.hawker.co.uk

Hawker Ltd.
Rake Lane
Clifton Junction
Swinton
Manchester M27 8LR
Tel: +44 161 794 4611
Fax: +44 161 793 6606

ITS Testing and Certification Ltd.
Cleeve Road
Leatherhead
Surrey KT22 7SA
Tel: +44 1372 367000
Fax: +44 1372 367099
http://www.era.co.uk

Moltech Rechargeable Products
Unit 20 Loomer Road
Newcastle, Staffordshire ST5 7LB
Tel: +44 1782 566688
Fax: +44 1782 565910
http://www.moltechpower.com
(*formerly Energizer Rechargeable
    Products*)

MSA
East Shawhead
Coatbridge
ML5 4TD
Tel: +44 1236 424966
Fax: +44 1236 436650

Oldham Crompton Batteries Ltd.
Edward Street
Manchester
Lancashire M34 3AT
Tel: +44 161 3350999
Fax: +44 161 3350020
http://www.hawker.co.uk

Panasonic Industrial Europe (UK) Ltd.
Willoughby Road
Bracknell
Berks.
RG12 8FB
Tel: +44 1344 853259

Philips Electronics Ltd.
420-430 London Road
Croydon
CR9 3QR
Tel: +44 20 8665 6655

Rayovac Vidor (UK) Ltd.
Galleon House
King Street
Maidstone
Kent ME14 1BG
Tel: +44 1622 688331
Fax: +44 1622 68552
http://www.rayovac.com

SAFT Ltd.
97-101 Peregrine Road
Ilford
Essex IG6 3XJ
Tel: +44 20 84981177
Fax: +44 20 8783 0494
http://www.saft.alcatel.com

Sanyo Energy (UK) Ltd.
Unit 1
Maxted Corner
Hemel Hempstead
Hertfordshire
HP2 7RA
Tel: +44 1442 213121

SEC Industrial Battery Co. Ltd.
Thorney Weir House
Iver, Buckinghamshire SL0 9AQ
Tel: +44 1895 431543
Fax: +44 1895 431880
http://www.battery-vrla.com

TDI Batteries (Europe) Ltd.
Claverhouse Industrial Park
Forfar Road
Dundee
Angus DD4 9UB
Tel: +44 1382 500050
Fax: +44 1382 500055
http://www.tdibatteries.com

Ultralife Batteries (UK) Ltd.
18 Nuffield Way
Abingdon
Oxfordshire OX14 1TG
Tel: +44 1235 542600
Fax: +44 1235 535766
http://www.ulbi.com

Varta Ltd.
Cropmead Industrial Estate
Crewkerne
Somerset TA18 7HQ
Tel: +44 1460 73366
Fax: +44 1460 72320
http://www.varta.com

VHB Industrial Batteries Ltd.
Unit 20, Monks Path Business Park
   Highlands Road
Solihull B90 4NZ
Tel: +44 121 7451234
Fax: +44 121 7451645
http://www.vhb.co.uk

Yuasa Automotive Batteries (Europe) Ltd.
Formans Road
Sparkhill
Birmingham B11 3DA
Tel: +44 121 777 3292
Fax: +44 121 777 7787
http://www.yuasa.co,uk

Yuasa Battery Sales UK Ltd.
Hawksworth Ind. Estate
Swindon
Wiltshire SN2 1EG
Tel: +44 1793 645700
Fax: +44 1793 645701
http://www.yuasa.co.uk

USA

ACME Electric Corp. Aerospace Division
528 W. 21st Street
Tempe, AZ 85282
Tel: (480) 894-6864
Fax: (480) 921-0470
http://www.acmeelec.com/aerospace/

ACR Electronics, Inc.
5757 Ravenswood Road
Ft. Lauderdale, FL 33312
Tel: (954) 981-3333
Fax: (954) 983-5087
http://www.acrelectronics.com

AER Energy Resources
4600 Highlands Parkway Suite G
Smyma, GA 30082
Tel: (770) 433-2127
Fax: (770) 433-2286
http://www.aern.com/

Alexander Manufacturing Co.
P.O. Box 1508
Mason City, IA 50401
Tel: (800) 247-1821
Fax: (800) 577-2539
http://www.alexmfg.com

Air Force Research Laboratory*
Air Force Wright Aeronautical Laboratory
Aerospace Power Division
Wright-Paterson AFB, OH 45435
Tel: (937) 255-7770
Fax: (937) 656-7529

Alliant Power Sources Center Company
104 Rock Road
Horsham, Pennsylvania 19044
Tel: (215) 674-3800
Fax: (215) 373-5444
http://www.atk.com

Alupower, Inc.
82 Mechanic St.
Pawcatuck, CT 06379
Tel: (860) 599-1100
Fax: (860) 599-3903

Argonne National Laboratory*
Electrochemical Technology Program
9700 S. Cass Ave
Argonne, IL 60439
Tel: (630) 252-7563
Fax: (630) 252-4176
http://www.anl.gov

Arthur D. Little, Inc.
Acorn Park
Cambridge, MA 02140-2390
Tel: (617) 498-5000
Fax: (617) 498-7200
http://www.adlittle.com

BAE SYSTEMS*
Battery Technology Center
1601 Research Blvd.
Rockville, MD 20850-3173
Tel: (301) 838-6200
Fax: (301) 838-6222
http://www.tracor.com

Battery Council International
401 North Michigan Avenue
Chicago, Illinois 60611
Tel: (312) 644-6610
Fax: (312) 321-6869
http://www.batterycouncil.org

Battery Engineering, Inc. (Division of
   Wilson Greatbatch Ltd.)
100 Energy Drive
Canton, MA 02021
Tel: (781) 575-0800
Fax: (781) 575-1545
http://www.batteryeng.com

Bolder Technologies, Inc.
4403 Table Mountain Drive
Golden, CO 80403
Tel: (303) 215-7200
Fax: (303) 215-2500
http://www.boldertmf.com

Bren-Tronics Inc.
10 Brayton Court
Commack, NY 11725
Tel: (516) 499-5155
Fax: (516) 499-5504
http://www.bren-tronics.com

BST Systems Inc.
78 Plainfield Pike Road
Plainfield, CT 06374
Tel: (860) 564-4078
Fax: (860) 564-1380
http://www.bstsys.com

C & D Technologies
1400 Union Meeting Road
Blue Bell, PA 19422
Tel: (215) 619-2700
Fax: (215) 619-7840
http://www.cdtechno.com

Concorde Battery Corp.
2009 San Bernardino Road
West Covina, CA 91790
Tel: (626) 813-1234
Fax: (626) 813-1235
http://www.concordebattery.com

Covalent Associates
10 State Street
Woburn, MA 01801
Tel: (781) 938-1140
Fax: (781) 938-1364
http://www.covalentassociates.com

Crown Battery Manufacturing Company
1445 Majestic Drive
Fremont, OH 43420
Tel: (419) 334-7181
Fax: (419) 334-7124
http://www.crownbattery.com

Delphi Energy & Engine Management
   Systems
Delphi Automotive Systems
5725 Delphi Drive
Troy, MI 48098-2815
Tel: (248) 813-2000
Fax: (248) 813-2670
http://www.delphiauto.com

DME Corp.
6830 N.W. 16th Terrace
Ft. Lauderdale, FL 33309
Tel: (954) 975-2100
Fax: (954) 979-3313
http://www.dmecorp.com

Douglas Battery Manufacturing Company
500 Battery Drive
P.O. Box 12159
Winston-Salem, NC 27107
Tel: (336) 650-7000
Fax: (336) 650-7072
http://www.douglasbattery.com

Duracell, Inc.
8 Research Drive
Bethel, CT 06801
(international callers see website for local
   dist.)
Tel: (800) 243-9540
Fax: (800) 796-4565
http://www.duracell.com

Eagle-Picher Technologies, LLC
C & Porter Streets
Joplin, MO 64802
Tel: (417) 623-8000
Fax: (417) 623-5319
http://www.eaglepicher.com

East Penn Manufacturing Co., Inc.
Deka Road
Lyon Station, PA 19536
Tel: (610) 682-6361
Fax: (610) 682-4781
http://eastpenn-deka.com

Eastman Kodak Co.
343 State Street
Rochester, NY 14650
Tel: (716) 724-4000
http://www.kodak.com

EIC Laboratories, Inc.
111 Downey Street
Norwood, MA 02062
Tel: (781) 769-9450
Fax: (781) 551-0283
http://www.eiclabs.com

Electric Power Research Institute
3412 Hillview Avenue
Palo Alto, CA 94304
Tel: (650) 855-2121
http://www.epri.com

Electro Energy, Inc.
Shelter Rock Lane
Danbury, CT 06810
Tel: (203) 797-2699
Fax: (203) 797-2697

Electrochem Industries
Wilson Greatbatch Ltd.
10,000 Wehrle Drive
Clarence, NY 14031
Tel: (716) 759-6901
Fax: (716) 759-8579
http://www.greatbatch.com

Electrochemical Society Inc.
10 South Main Street
Pennington, NJ 08534-2896
Tel: (609) 737-1902
Fax: (609) 737-2743
http://www.electrochem.org

Electrosource Inc.
2809 Interstate 35 South
San Marcos, TX 78666
Tel: (512) 753-6500
Fax: (512) 353-3391
http://www.electrosource.com

EEMB USA Representative Office, Bridge
   R&D Corp.
12601 Monarch Street
Garden Grove, CA 92841
Tel: (714) 891-6509
Fax: (714) 890-8590
http://eemb.com

EMF Systems
14670 Highway 9
Boulder Creek CA 95006
Tel: (408) 338-1800
Tel: (408) 338-2955

Evercel Corp.
2 Lee Mac Avenue
Danbury, CT 06810
Tel: (203) 825-3900
Fax: (203) 730-4842
http://www.evercel.com
(Formerly Energy Research Corporation)

Energizer Inc.
Checkerboard Square
St Louis, MI 63164
Tel: (314) 982-2000
or
R&D Center
P.O. Box 45035
Westlake, OH 44145
Tel: (216) 835-7500
http://www.energizer.com

Exide Corp.
645 Penn Street
Reading, PA 19601
Tel: (610) 378-0500
Fax: (610) 378-0748
http://www.exideworld.com
(see Exide Europe, facilities world wide,
GNB, The Tudor Group, CEAC, Emisa)

G. S. Battery U.S.A., Inc.
17253 Chestnut Street
City of Industry, CA 91748-1004
Tel: (626) 964-8348
Fax: (626) 913-6277
http://www.gsbattery.com

Gould Electronics Inc.
34099 Melinz Parkway
Eastlake, OH 44095
Tel: (440) 918-6060
Fax: (440) 918-6030
http://www.gouldelectronics.com

GP Batteries
11235 West Bernardo Court
San Diego, CA 92127
Tel: (619) 674-5620
Fax: (619) 674-7237
http://www.gpbatteries.com

Harding Energy, Inc.
One Energy Centre
Norton Shores, MI 49441
Tel: (231) 798-7033
Fax: (231) 798-7044
http://www.hardingenergy.com

Hawker Energy Products
617 No. Ridgeview Drive
Warrensburg, MO 64093
Tel: (660) 429-6437
Fax: (660) 429-6397
http://www.hepi.com
(Invenys plc—HAWKER Energy Storage)

Hawker Eternacell Inc. (Division of SAFT)
495 Boulevard
Elmwood Park, NJ 07407
Tel: (201) 796-4800
Fax: (201) 796-6243
http://www.eternacell.com
(Formerly Power Conversion, Inc.)

Hawker Powersource
P.O. Box 808
9404 Ooltewah Industrial Dr.
Ooltewah, TN 37363
Tel: (423) 238-5700
Fax: (423) 238-6060
http://www.hawkerpowersource.com
(Invenys plc—HAWKER Energy Storage)

HED Battery Corporation
3355 Woodward Avenue
Santa Clara, CA 95112
Tel: (408) 980-1877
Fax: (408) 980-1804
http://www/hedb.com

Independent Battery Manufacturers Assoc.
100 Larchwood Drive
Largo, FL 33770
Tel: (727) 586-1408
Fax: (727) 586-1400
http://www.thebatteryman.com/index.html

JBRO Batteries, Inc.
1938-A University Lane
Lisle, IL 60532-2150
Tel: (630) 323-3779
Fax: (630) 964-0678
http://www.jbro.com

Jet Propulsion Laboratory*
Electrochemical Technologies Group
California Institute of Technology
4800 Oak Grove Drive
Pasadena, CA 91109-8099
Tel: (818) 354-4321
Fax: (818) 393-6951
http://www.jpl.nasa.gov

Johnson Controls, Inc.
Battery Group
5757 North Green Bay Avenue
P.O. Box 591
Milwaukee, WI 53201-0591
Tel: (414) 228-1200
http://www.johnsoncontrols.com/bg

Koehler Bright Star Inc.
380 Stewart Road
Hanover Industrial Estates
Wilkes-Barre, PA 18706-1459
Tel: (570) 825-1900
Fax: (570) 825-1984
http://www.flashlight.com

Lawrence Berkley Laboratory*
University of California
Berkley, CA 94720-0001
Tel: (510) 486-4636
http://www.lbl.gov/Science-Articles/
  Batteries-and-FuelCells.html

Lithium Power Technologies, Inc.
20955 Morris Avenue
Manvel, Texas 77578-3819
Tel: (281) 489-4889
Fax: (281) 489-6644
http://www.lithiumpower.com

Lithium Technology Corporation
5115 Campus Drive
Plymouth Meeting, PA 19462
Tel: (610) 940-6090
Fax: (610) 940-6091
http://www.lithiumtech.com

Lucent Technologies (Battery Group part of
  Tyco, Inc.)
3000 Skyline Drive
Mesquite, TX 75149
Tel: (972) 284-2000
Fax: (972) 284-2900
http://www.lucent.com/networks/power/
  batteries.html
(formerly Bell Laboratories)

Maha Communications, Inc.
545-C W. Lambert Road
Brea, CA 92821
Tel: (714) 990-4557
Fax: (714) 990-1325
http://www.maha-comm.com

Marathon Power Technologies
PO BOX 8233
8301 Imperial Drive
Waco, TX 76710
Tel: (254) 776-0650
Fax: (254) 776-6558
http://www.mptc.com

Maxcell Corp. of America
22-08 Route 208
Fair Lawn, NJ 07410
Tel: (201) 794-5900
(Division of Hitachi Maxcell)

Medtronic Inc.—Promeon Div.
6700 Shingle Creek Parkway
Brooklyn Center, MN 55430
Tel: (612) 574-4000
http://www.medtronic.com

Mine Safety Appliances Co.
38 Loveton Circle
Sparks, MD 21152
Tel: (410) 472-7700
Fax: (410) 472-7800
http://www.msanet.com

Moltech Corp.
9062 S. Rita Road
Tucson, AZ 85747
Tel: (520) 799-7500
Fax: (520) 799-7501
http://www.moltech.com

Moltech
Power Systems
P.O. Box 147114
Gainsville, FL 32614-7114
or
HWY 441 No.
Alachua, FL 32615
Tel: (904) 462-3911
Fax: (904) 462-6210
http://www.moltechpower.com

Motorola Energy Systems Group
1700 Belle Meade Court
Lawrenceville, GA 30043
Tel: (770) 338-3000
Fax: (770) 338-3555
http://www.motorola.com/ies/ESG

NASA Glenn Research Center*
21000 Brookpark Road
Cleveland, OH 44135
Tel: (216) 433-5246
http://www.lerc.nasa.gov/

Naval Sea Systems Command, Crane
  Division*
Power Systems Department
Code 609
300 HWY 361
Crane, IN 47522
Tel: (812) 854-1593
Fax: (812) 854-1212

New Castle Battery Mfg. Co.
P.O. Box 5040
3601 Wilmington Road
New Castle, PA 16105
Tel: (724) 658-5501
Fax: (724) 658-5559
http://www.turbostart.com

Optima Batteries, Inc.
17500 East 22nd Avenue
Aurora, CO 80011
Tel: (303) 340-7400
Fax: (303) 340-7470
http://www.optimabatteries.com

Ovonic Battery Company
1707 Northwood Drive
Troy, MI 48084
Tel: (248) 280-1900
Fax: (248) 280-1456
http://www.ovonics.com

Panasonic Industrial Co.
Div of Matsushita Electric Corp of America
P.O. Box 1511
Secaucus, NJ 07096-1511
Tel: (877) 726-2228
Fax: (847) 468-5750
http://www.panasonic.com
(see also Matsushita Battery Industrial Co.,
  Ltd., Japan)

Plainview Batteries, Inc.
23 Newton Road
Plainview, NY 11803
Tel: (516) 249-2873
Fax: (516) 249-2876

Polaroid Corporation
868 Winter Street
Waltham, MA 02154-1274
Tel: (781) 386-6454
Fax: (781) 386-1996
http://www.polaroid.com

PolyStor Corporation
230 South Vasco Road
Livermore, CA 94550
Tel: (925) 245-7000
Fax: (925) 245-7001
http://www.polystor.com

Portable Energy Products
940 Disc Drive
Scotts Valley, CA 95066
Tel: (831) 439-5100
Fax: (831) 439-5118
http://www.portable-energy.com

Portable Rechargeable Battery Association
1000 Parkwood Circle, Suite 430
Atlanta, GA 30339
Tel: (770) 612-8826
Fax: (770) 612-8841
http://www.prba.org

Power Battery, Inc
543 E. 42nd Street
Paterson, NJ 07513
Tel: (973) 523-8630
http://www.powbat.com

Power Conversion Inc. (see Hawker
  Eternacell Inc.)

Power-Sonic Corporation
3106 Spring Street
P.O. Box 5242
Redwood City, CA 94063
Tel: (650) 364-5001
Fax: (650) 366-3662
http://www.power-sonic.com

Preferred Power Technologies, Inc.
850 Clark Drive
Mount Olive, NJ 07828
Tel: (973) 347-7888
Fax: (973) 347-7303
http://www.ecocel.com

Rayovac Corp.
601 Rayovac Drive
PO Box 4960
Madison, WI 53744-4960
Tel: (608) 275-3340
http://www.rayovac.com

RBC Technologies
809 University Drive
Suite 100-E
College Station,
TX 77840
Tel: (979) 260-1120
Fax: (979) 260-1322
http://www.rbctx.com

Rutgers University*
College of Engineering
Battery Materials and Engineering
  Laboratory
98 Brett Road
Piscataway, NJ 08855-8058
Tel: (732) 445-6858
Fax: (732) 445-1669

Sandia National Laboratories*
Power Sources Technology Group
MS 0613, PO Box 5800
Albuquerque, NM 87185—0613
Tel: (505) 845-8105
Fax: (505) 844-6972
http://www.sandia.gov/pstg/battery.html

Sanyo Energy Corp.
2055 Sanyo Avenue
San Diego, CA 92154
Tel: (619) 661-6620
Fax: (619) 661-6743
http://www.sanyobatteries.net
(see also Sanyo Electric Co., Ltd., Japan)

SAFT America Incorporated
711 Gill Harbin Industrial Blvd.
Valdosta, GA 31601
Tel: (912) 247-2331
Fax: (912) 245-2827
and
SAFT America Inc.
313 Crescent Street
Valdese, NC 28690
Tel: (828) 874-4111
Fax: (828) 974-2431
(see also SAFT France)
http://www.saftamerica.com

SEC Industrial Battery Co., Inc.
768 N. Bethlehem Pike, Suite 105
Lower Gwynedd, PA 19002
Tel: (215) 654-9334
Fax: (215) 654-9871
http://www.battery-vrla.com

Sparton Electronics
P.O. Box 788
O5612 Johnson Lake Road
Deleon Springs, FL 32130
Tel: (904) 985-4631
Fax: (904) 985-5036
http://www.sparton.com

Storage Battery Systems
N56 W16665 Ridgewood Drive
P.O. Box 160
Menomonee Falls, WI 53052
Tel: (262) 703-5800
Fax: (262) 703-3073
http://www.sbsbattery.com

Superior Battery Manufacturing Company,
  Inc.
P.O. Box 1010
2515 Hwy, KY 910
Russell Springs, KY 42642
Tel: (502) 866-6056
Fax: (502) 866-6066
http://www.superiorbattery.com

Tadiran Electronic Industries
2 Seaview Blvd.
Suite 102
Port Washington, NY 11050
Tel: (516) 621-4980
Fax: (516) 621-4517
http://www.tadiranbat.com

TDI Batteries, Inc.
1120 Windham Parkway
Romeoville, IL 60446
Tel: (630) 679-8200
Fax: (630) 759-0379
http://www.tdibatteries.com

Teledyne Battery Products
840 W Brockton Avenue
Redlands, CA 92375
Tel: (909) 793-3131
Fax: (909) 793-5818
*http://www.tcmlink.com/batteryProducts/
  index.htm*

Toshiba America Electronic Components, Inc.
One Parkway North
Suite 500
Deerfield, IL 60015
Tel: (847) 945-1500
Fax: (847) 945-1044
http://www.toshiba.com
(see also Toshiba Battery Co., Ltd., Japan)

Trojan Battery Company
12380 Clark Street
Santa Fe Springs, CA 90670
Tel: (562) 946-8381
Fax: (562) 946-8381
http://www.trojan-battery.com

U.S. Army CECOM
Research, Development and Engineering Center*
ATTN: AMSRL-RD-C2-AP
Fort Monmouth, New Jersey 07703-5021
Tel: (732) 544-2084
Fax: (732) 427-3665
http://www.monmouth.army.mil/cecom/lrc/lrchq/power.html

U.S. Army Research Laboratory*
Sensors and Electronic Devices Directorate
2800 Powder Mill Road
Adelphi, MD 20783-1197
Tel: (301) 394-0339

U.S. Battery Manufacturing Co.
1675 Sampson Avenue
Corona, CA 91719
Tel: (909) 371-8090
Fax: (909) 371-4671
http://www.usbattery.com

U.S. Microbattery
74 Batterson Park Road
Farmington, CT 06032
Tel: (860) 487-3838
Fax: (860) 429-5911

Ultralife Batteries, Inc.
2000 Technology Parkway
Newark, NY 14513 USA
Tel: (315) 332-7100
Fax: (315) 331-7800
http://www.ultralifebatteries.com

Undersea Sensor Systems
2101 So. 600 East
Columbia City, IN 46725
Tel: (219) 429-5778
Fax: (219) 248-3510
http://www.ultra-electronics.co.uk/Air/ussi/Default.htm
or
http://www.ultra-electronics.co.uk

Underwriters Laboratories, Inc.
333 Pfingsten Road
Northbrook, IL 60062-2096
Tel: (847) 272-8800
Fax: (847) 272-8129
http://www.ul.com

US Nanocorp
20 Washington Avenue, Suite 106
North Haven, CT 06473-2342
Tel: (203) 234-8024
Fax: (203) 234-0383
http://www.usnanocorp.com

Valence Technology
301 Canestoga Way
Henderson, NV 89015
Tel: (702) 558-1000

Wilson Greatbatch Ltd. (see Electrochem Industries)

Yardney Technical Products, Inc.
82 Mechanic St.
Pawcatuck, CT 06379
Tel: (860) 599-1100
Fax: (860) 599-3903
http://www.yardney.com

INDEX